49 181 $ 52,25
9928
D1722968

LAWS OF HADRONIC MATTER

Subnuclear Series volume 11

PART A

editor
A. ZICHICHI

SUBNUCLEAR SERIES
A. Zichichi - Editor

1. 1963 course:	« Strong, Electromagnetic and Weak Interactions » published by Benjamin, Inc., New York.
2. 1964 course:	« Symmetries in Elementary Particle Physics » published by Academic Press, New York and London.
3. 1965 course:	« Recent Developments in Particle Symmetries » published by Academie Press, New York and London.
4. 1966 course:	« Strong and Weak Interactions - Present Problems » published by Academic Press, New York and London.
5. 1967 course:	« Hadrons and their Interactions » published by Academic Press, New York and London.
6. 1968 course:	« Theory and Phenomenology in Particle Physics » published by Academic Press, New York and London.
7. 1969 course:	« Subnuclear Phenomena » published by Academic Press, New York and London.
8. 1970 course:	« Elementary Processes at High Energy » published by Academic Press, New York and London.
9. 1971 course:	« Properties of the Fundamental Interactions » published by Editrice Compositori, Bologna, Italy.
10. 1972 course:	« Highlights in Particle Physics » to be published by Editrice Compositori, Bologna, Italy.
11. 1973 course:	« Laws of Hadronic Matter » published by Academic Press, New York and London.
12. 1974 course:	"Lepton and Hadron Structure" in preparation

LAWS OF HADRONIC MATTER

1973 International School of Subnuclear Physics
a NATO-MPI-MRST Advanced Study Institute
Sponsored by the Regional Sicilian Government
and the Weizmann Institute of Science
Erice, Trapani - Sicily: 8-26 July 1973

editor

A. Zichichi

1975

Distributed by
ACADEMIC PRESS NEW YORK AND LONDON
A Subsidiary of Harcourt Brace Jovanovich Publishers

Copyright © by A. Zichichi
All rights reserved
No part of this book may be reproduced in any form,
by photostat, microfilm, or any other means, without
written permission from the publishers.

Subnuclear Series volume 11

Italian Edition
PERIODICI SCIENTIFICI
Viale Ezio 5 Milano

~~6A UHXC 1265 - 11+2~~

Foreign Edition
Distributors
ACADEMIC PRESS, Inc.
111 Fifth Avenue, New York, New York 10003

ACADEMIC PRESS, Inc. (London) Ltd
24-28 Oval Road, London N.W. 1

Gesamthochschulbibliothek
- Siegen
Aus dem Bestand der
Universitätsbibliothek
Siegen ausgeschieden

41 = UHXS

ISBN 0-12-780588-5

Standort:
Signatur:
Akz.-Nr.: 79/25883
Id.-Nr.: N1971166
031 12 N
Weiteres Ex. Bloch

Printed in Italy by Litografia Leschiera, Cologno Monzese

CONTRIBUTORS

PART A

Theoretical Lectures and Seminars

W. Thirring
R. Jackiw
C. Rebbi
P.H. Frampton
S. Coleman
J.J. Sakurai
L. Caneschi
H. Kleinert
F. Buccella

PART B

Review Lectures and Seminars on Experimental Topics

D. H. Miller
A.H. Rosenfeld
L. Montanet
E.L. Berger
U. Amaldi
N. Byers
S.C.C. Ting
E.A. Paschos
N.F. Ramsey
A.C. Melissinos
A. Berthelot

Closing speech

A. Zichichi

Foreword

During 20 days in July 1973, 120 physicists from 27 countries met in ERICE, to attend the 11th Course of the International School of Subnuclear Physics "Ettore Majorana".

The countries represented at the School were: Australia, Austria, Belgium, Brazil, Canada, China, Denmark, Finland, France, Germany, Greece, Hong Kong, Hungary, India, Israel, Italy, Japan, Korea, Poland, Portugal, Rumania, South Africa, Spain, Sweden, The United Kingdom, The United States of America, and Yugoslavia.

As usual in our courses, the programme was devoted to the presentation and the discussion of the main results obtained in experimental and theoretical physics during the past year. The greatest news in experimental physics came from Ugo Amaldi, who lectured on experiments done at the ISR to measure the total proton-proton cross-section. Contrary to most peoples' expectation this cross-section was found to rise by $\sim$ 10% in the ISR energy range. In the experimental field we had the reviews of mesons and baryons, including a report on the status of the so-called daughter trajectories, together with a phenomenological attempt to understand the ISR results on short- and long-range correlations, on multiplicities, and on high transverse momentum phenomena; experimental results on electromagnetic interactions including muon-proton and muon-nucleus scattering were also presented and discussed, together with the most recent neutrino experiments. On the theoretical side, a series of interesting models -- even if not necessarily connected with Nature -- were presented, and we were also warned of the fact that eventually one should seriously think of new field theories. On the more general theoretical side we had lectures on spontaneous symmetry breakdown and gauge fields (the secret of Sid Coleman) and on some exact results on the structure of matter, including cosmic bodies. Going back to the theoretical models, the new duality of Sakurai and the work of Buccella on constituents versus current quarks, created a lot of interest.

The above is a brief summary of the material presented at the School. I hope the reader will enjoy the book as much as the participants enjoyed the lectures and the discussion sessions. Thanks to the work of the scientific secretaries, these discussions have been reproduced as well as possible.

I would like to thank all those people who helped me in Bologna, Geneva, and Erice, with the organization and the administrative affairs of the School, and with all the detailed work connected with the Course itself. I would like, in particular, to mention Miss M. Zaini, Drs. A. Gabriele and U.M. Wambach, together with Mme. M. Denzler and Miss P. Savalli, whose hard work and collaboration is highly appreciated.

A. Zichichi - March 1974

CONTENTS

PART A

PART B

REVIEW LECTURES AND SEMINARS ON EXPERIMENTAL TOPICS

CLOSING SPEECH

PART A

THEORETICAL LECTURERS AND SEMINARS

EXACT RESULTS ON THE STRUCTURE OF MATTER

W. Thirring

Table of Contents

EXACT RESULTS ON THE STRUCTURE OF MATTER[≠)]

W. Thirring
Institut für Theoretische Physik der Universität Wien, Wien, Austria

INTRODUCTION AND RESULTS

i) Position of the Problem

A system of N non-relativistic particles interacting with Coulomb and Newton potentials is governed by the Hamiltonian

$$H_N = \sum_{i=1}^{N} \frac{p_i^2}{2m_i} + \sum_{\substack{i,j=1 \\ i>j}}^{N} (e_i e_j - \kappa m_i m_j)|x_i - x_j|^{-1} = K + V \quad . \qquad (1.1)$$

Treated quantummechanically it should give a good description ($\sim 1\%$ accuracy) of atoms, molecules, macroscopic and cosmic bodies: The non-relativistic treatment of electricity and gravitation should be a useful approximation in most circumstances. Strong and weak forces are usually hidden in nuclei which in this level of description are treated as elementary particles. We shall simply call matter systems described by (1) and ask what can be rigorously deduced from this Hamiltonian.

In quantum mechanics where x_i and p_j are unbounded operators we first have to define the domain $\mathcal{D}$ on which H_N is self-adjoint. For K this is known to consist of functions with square integrable generalized first and second derivatives. On some occasions we shall consider our system enclosed in a box Ω. Due to the singularity of $1/|x|$ V is also an unbounded operator. However, the usual intuitive argument with the uncertainty relation suggests that in quantum mechanics this singularity is harmless. This argument can be phrased mathematically[1)] in the form $\forall\, b > 0\ \exists\, a > 0$ such that

$$||V\psi|| \leq a||\psi|| + b||K\psi|| \qquad \forall\, \psi \in \mathcal{D} \quad . \qquad (1.2)$$

As a consequence one can deduce readily that K + V is self-adjoint on the same domain and as K bounded from below. Thus the problem is mathematically well defined:

$$e^{iH_N t}$$

is unitary and the spectrum of H_N contained in an interval $[E_o, \infty)$. With respect to the latter Combes and Balslev[2)] proved the following

≠) Supported in part by "Fonds zur Förderung der wissenschaftlichen Forschung in Österreich", Projekt Nr. 1724.

results which have been partly anticipated as physically reasonable:

1) The singular continuous part (e.g. the one concentrated on a set of (Lebesgue) measure zero) is absent.
2) $(H-z)^{-1}$ can be continued through the real axis and may have poles in the second sheet ("resonances") but only for Re z < O.
3) There are no discrete points of the spectrum for Re z > O (but for $N \geq 3$ there are some embedded in the continuum for Re z < O. They correspond to states whose decay is forbidden by selection rules).

I called the poles in the second sheet resonances although the usual scattering theory does not apply: $e^{-iHt} e^{iKt}$ does not[3] (strongly) converge for $t \to \pm\infty$.

In the following I shall concentrate on the most elementary properties of H_N, namely the position of E_o and later some properties of level densities as they are reflected in the thermodynamic functions.

ii) Atoms

For N = 2 we have the Hamiltonian of the hydrogen atom and everybody knows how to solve that. But for more complicated atoms no exact solution is available and one might ask whether one can at least give rigorous bounds within which E_o has to be. For Helium there are indeed excellent bounds available, only about 1‰ apart[4]. For more complicated atoms there are very good upper limits as result of Hartree-Fock calculations. However, I am not aware of any useful lower bounds. Of course, there are always the trivial lower bounds one obtains just neglecting the Coulomb repulsion, but they are about 40% too low.

iii) Macroscopic Bodies ($N \sim 10^{24}$)

The first empirical fact to account for is the saturation character of the chemical forces, e.g. one wants an inequality of the type

$$- A \leq \frac{E_o}{N} \leq - B \quad . \tag{1.3}$$

An upper limit within some percent of the empirical value is easily available by taking atomic (or molecular) trial functions. The lower bounds are notoriously hard to get and only Dyson and Lenard[5] were able to prove (1.3), however with $A \sim 10^{14}$! It may be a matter of dispute whether we can say that we understand a system when we have reached 50% or 10% or 1% accuracy, but a factor 10^{14} clearly means we have not solved the problem.

The difficulty of proving (1.3) is connected with the fact that it is generally not true. As we shall see shortly it is spoiled by gravitation, thus we need $\kappa = O$. Furthermore, the system has to be electrically neutral (if in a box with volume $\sim N$), i.e.

$$\sum_{j=1}^{N} e_j = O$$

(or small in some sense). Even then one finds for bosons[6)]

$$- A N^{5/3} \le E_o \le - B N^{7/5} \quad . \tag{1.4}$$

To get (1.3) one needs the additional assumption that the particles of one sign of charge (say all negative ones, the electrons) obey the exclusion principle. The factor 10^{14} is just due to the complexity of the analysis. If on every page you give away by an estimate a factor 2 or 3, after 40 pages you have lost 10^{14}.

Regarding the thermodynamic behaviour of matter it has been proved by Lebowitz and Lieb[7)] that the relevant quantities exist and have the desired convexity properties. Their analysis is based on the result of Dyson and Lenard and thus holds if

1) $\kappa = 0$

2) $\sum_j e_j = 0$ (or small)

3) All negative (or positive) particles are fermions.

Starting with a finite system in a box Ω one defines a corresponding Hamiltonian $H_N(\Omega)$. It has a purely discrete spectrum and its eigenvalues (in ascending order) are denoted $\varepsilon_i(\Omega)$. The thermodynamic quantities in the canonical and microcanonical ensemble are

$$e^{-\beta F_N(\beta,\Omega)} = \mathrm{Tr}\; e^{-\beta H_N(\Omega)} \tag{1.5}$$

$$E_N(S,\Omega) = e^{-S} \sum_{i=1}^{e^S} \varepsilon_i(\Omega) \quad . \tag{1.6}$$

Lebowitz and Lieb could show the existence of

$$\lim_{N\to\infty} N^{-1} F_N(\beta,\tfrac{N}{\rho}) = f(\beta,\rho) \tag{1.7}$$

$$\lim_{N\to\infty} N^{-1} E_N(Ns,\tfrac{N}{\rho}) = \eta(s,\rho) \quad . \tag{1.8}$$

In fact, lower bounds for the sequences (1.7) and (1.8) can be derived from (1.3) and all the considerable effort goes into showing that these sequences are monotonically decreasing. Furthermore they show that $\eta(s,\rho)$ is convex increasing in s and ρ. This implies that temperature, specific heat, pressure and compressibility are positive. These fundamental properties of matter are by no means automatic and one can easily find examples where one or the other of these quantities become negative.

If we do not neglect gravitation ($\kappa > 0$) the picture changes completely. For fermions (if $\sum e_i = 0$) one can show[8)]

$$- N^{7/3} A \leq E_o \leq - N^{7/3} B \tag{1.9}$$

for N sufficiently large ($N > 10^{54}$) so that gravitation dominates electricity. Thus the N behaviour is the same as for atoms. It comes about in the same way, namely the radius of the system goes as $N^{-1/3}$ and there are $\sim N^2$ contributions to V. For bosons where the zero-point pressure is smaller $R \sim N^{-1}$ and $E \sim N^3$. The fact that the system contracts to a state of high density can be used to compute the thermodynamic quantities exactly[9]. I shall demonstrate in the last section that for fermions $N^{-7/3} E_N(Ns, N^{-1}\Omega)$ and $N^{-7/3} F_N(N^{4/3}\beta, N^{-1}\Omega)$ tend (for $N \to \infty$) to what one gets from solving the (temperature dependent) Thomas-Fermi equation. A computer solution[10] of this non-linear integral equation shows indeed the behaviour of f and η one expects from a star. In particular there is a region where η is not convex in s. This implies a negative heat capacity and reflects the interesting feature of gravitating systems to become hotter when they give off energy.

Since a computer only spits out numbers and does not give explanations I shall try to give some insight into this remarkable happening by a rough argument. Let us guess the free energy of fermions in a box Ω with gravitational interactions[≠]. The well known expression of the thermodynamic quantities for free fermions involve the somewhat awkward Fermi function, but it is sufficient for our rough argument to interpolate between the high temperature limit where the system behaves like a classical ideal gas and the zero temperature limit where the zero point motion dominates. Therefore, we will construct the free energy per particle by adding to the zero point energy the free energy of the ideal Fermi gas and the potential energy corresponding to a uniform density in the volume V. It turns out that this procedure is numerically an acceptable approximation for the exact free energy of free fermions and the approximation of uniform density is not so bad because the system collapses altogether once the density becomes too inhomogeneous. In this way we get for the free energy per particle

$$\frac{F}{N} = \left(\frac{N}{V}\right)^{2/3} - T \ln \frac{V}{N} T^{3/2} - \frac{\alpha_G N}{V^{1/3}} . \tag{1.10}$$

From this we can deduce by differentiation with the help of the standard thermodynamical relations the following values for entropy per particle, energy per particle and pressure:

$$\frac{S}{N} = \ln \frac{V}{N} T^{3/2} + \frac{3}{2} = \frac{3}{2} \ln\left(\frac{V}{N}\right)^{2/3} \frac{2}{3}\left(E + \frac{\alpha_G N}{V^{1/3}} - \left(\frac{N}{V}\right)^{2/3}\right) + \frac{3}{2} \tag{1.11}$$

≠) From now on we shall use natural units ($k = m = \hbar = c = 1$). $\alpha_G = \kappa m^2$.

$$\frac{E}{N} = (\frac{N}{V})^{2/3} + \frac{3}{2} T - \frac{\alpha_G N}{V^{1/3}} \qquad (1.12)$$

$$P = (\frac{N}{V})^{5/3} \frac{2}{3} + \frac{NT}{V} - \frac{\alpha_G N^2}{3V^{4/3}} = \frac{E_K + \frac{1}{2} E_P}{3V/2} = \frac{E - \frac{1}{2} E_P}{3V/2} \quad . \qquad (1.13)$$

The pressure has been rewritten in various ways to show that our approximation conforms with the virial theorem. Note that the pressure has three contributions. The thermal pressure $\sim V^{-1}$, the zero point pressure $\sim V^{-5/3}$, and the negative contribution due to the gravitational interaction $\sim V^{-4/3}$. Because of their different V-dependence the thermal pressure will dominate for large V and for small V the zero point pressure will win, but for sufficiently large N there will be a region in between where the pressure becomes negative. This is certainly physically impossible because the system is not glued to the walls but what will happen is that the system contracts by itself and thereby reaches a state of lower free energy.
Let us first consider the microcanonical case where the energy is the free variable. From Eq.(1.13) we see that for

$$E > - \frac{\alpha_G N}{2V^{1/3}} \qquad (1.14)$$

the pressure is positive whereas for

$$E < - \frac{\alpha_G N}{2V^{1/3}} \qquad (1.15)$$

it becomes formally negative. But this means that the system will contract to a volume V_o such that

$$E = - \frac{\alpha_G N}{2V_o^{1/3}} \quad . \qquad (1.16)$$

This volume has now also to be inserted in the other thermodynamic relations and the relation between energy and temperature

$$\frac{3}{2} T = \frac{E}{N} + \frac{\alpha_G N}{V^{1/3}} - (\frac{N}{V})^{2/3} \qquad (1.17)$$

becomes by substituting V_o for V

$$\frac{3}{2} T = - \frac{E}{N} - \frac{4E^2}{\alpha_G^2 N^{4/3}} \quad . \qquad (1.18)$$

Thus, the temperature is a linear function of energy for high energies and for low energies a parabolic function as indicated in Fig. 1. One sees that actually the region with pressure zero begins with negative specific heat (T increasing with decreasing E), but at low energies the dominating zero point pressure makes the behaviour again normal.

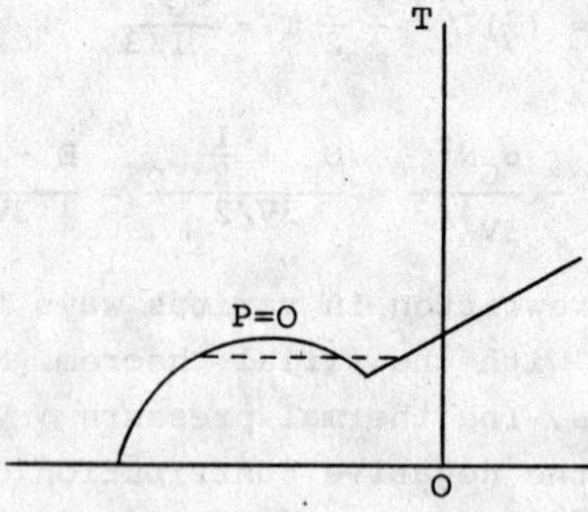

Fig. 1

Let us now consider the situation if we keep the temperature fixed and change the radius. The condition P = O now gives two solutions for the value of the volume. Since the pressure is proportional to the derivative of the free energy with respect to the volume, we see that the free energy is not monotonic in V. This implies that at a certain volume we can gain free energy by collapsing the system and thus the non-monotonic part of the curve will be bridged by a straight line. This corresponds to a phase transition where at constant temperature a finite amount of energy is released. This situation is illustrated in Fig.2 and in Fig.1 we have also drawn the corresponding pointed line which bridges the region of negative specific heat.

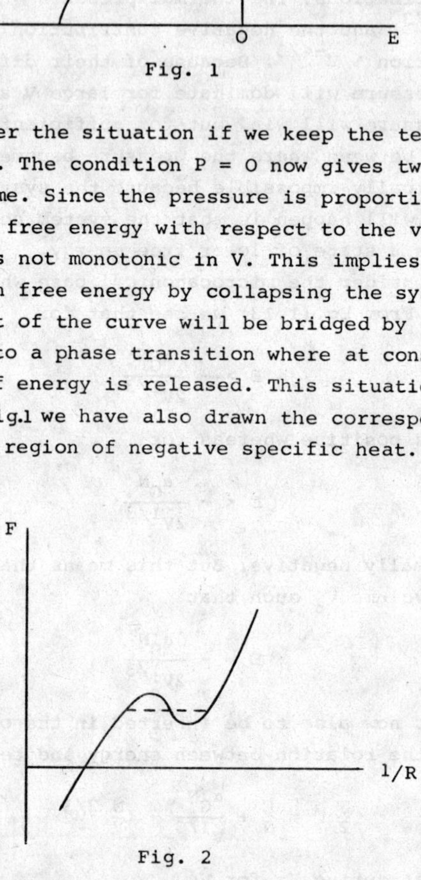

Fig. 2

In the following two sections I shall give some more details on the results for those systems where one can obtain more explicit results and not only abstract existence theorems. Although I will not be able to give all ε's of the proofs I hope to convey the ideas behind the methods used. I shall only quote exact bounds and consider results of uncontrolled approximations as irrelevant.

2. THE METHOD OF MINIMAL CONCAVITY

i) General Inequalities for Coulomb Systems

It is known that E is a concave function of the coupling constant (e.g. $e^2 = \alpha$). More generally the average of the first e^S energy levels (= energy at fixed entropy) possesses this property. This is a simple consequence of the generalized Minimax[11] principle:

$$\frac{1}{n}\sum_{1}^{n} E_n = \frac{1}{n} \inf_{\mathcal{H}_n} \mathrm{Tr}_{\mathcal{H}_n} H \quad , \qquad n \in Z \tag{2.1}$$

where $\mathcal{H}_n$ is an n-dimensional subspace and $\mathrm{Tr}_{\mathcal{H}_n} H$ is the trace of H restricted to this subspace. H and any trace of it is linear in α and the infimum of linear functions is concave. We shall denote the quantity defined in (2.1) by E(α), and concavity tells us for instance α > o

$$\alpha(V)_\alpha \leq E(\alpha) - E(O) \leq \alpha(V)_o \tag{2.2}$$

where $(V)_\alpha = \partial E/\partial\alpha$ is the microcanonical average of the potential evaluated for $H = H_o + \alpha V$. In general α cannot be varied in reality but is a God-given constant. However, for Coulomb systems the α-dependence is linked with the volume dependence for dimensional reasons. Since the only lengths in the game are the Bohr-radius α^{-1} (if $\hbar = m = 1$) and the size of the system L we have

$$E(\alpha) = \alpha^2 f(\alpha L) \tag{2.3}$$

where

$$f(x) \sim \begin{cases} 1/x^2 & \text{for} \quad x \to O \\ \text{const} & \text{for} \quad x \to \infty \end{cases} .$$

The convexity inequality (2.2) can then be expressed in terms of the pressure $P = -\frac{1}{3L^2}\frac{\partial E}{\partial L}$:

$$P \geq \frac{E(\alpha,L) + E(O,L)}{3L^3} . \tag{2.4}$$

Similarly since E(α) is concave

$$\frac{\partial^2 E}{\partial \alpha^2} \leq O \tag{2.5}$$

which gives for the compressibility

$$\kappa^{-1} = -\frac{L}{3}\frac{\partial P}{\partial L} \leq 2P - \frac{2E}{9L^3} \tag{2.6}$$

as mentioned above the quantities in (2.4) and (2.6) are understood at fixed entropy. For free fermions they become in the limit $\alpha \to 0$ (or $L \to 0$) the equalities

$$P = \frac{2E}{3L^3} \quad , \qquad \kappa^{-1} = \frac{10E}{9L^3} \quad . \tag{2.7}$$

For normal densities (2.4) and (2.6) are easily satisfied, which just means that then one is rather far from the region where first order perturbation theory is any good. In the next section we shall therefore try to strengthen (2.5).

It should be remarked that these inequalities hold for any Coulomb system irrespective of its size. They also apply to one diatomic molecule in the Born-Oppenheimer approximation where the extremal length is now the distance between the two atoms. The equilibrium position r_o is defined by $f'(\alpha r_o) = 0$ and there (2.5) becomes

$$\alpha^2 r_o^2 f''(\alpha r_o) \leq -2f(\alpha r_o) \quad . \tag{2.8}$$

Let M be the reduced mass of the molecule. (2.8) then can be rewritten as an upper limit of the vibrational frequency ω:

$$\omega \leq \sqrt{\frac{4|E|}{M r_o^2}} \quad . \tag{2.9}$$

Similarly (2.6) can be expressed as a general upper limit of the velocity of sound in (non-relativistic) Coulomb systems.

ii) Bogoliubov-type Inequalities for Unbounded Operators

There is a general inequality[13)] for $\partial^2 E/\partial\alpha^2$ which is just what we are after:

$$\frac{\partial^2 E}{\partial \alpha^2} \leq - \frac{(\alpha|[D,V]|\alpha)^2}{(\alpha|[D^\dagger[H,D]]|\alpha)} \quad . \tag{2.10}$$

Here $|\alpha)$ is the eigenvector belonging to the eigenvalue $E(\alpha)$ and, surprisingly enough, D is an arbitrary operator. This relation is usually deduced by methods legitimate in a finite dimensional space. In Hilbert space one can easily deduce absurd results from (2.10) if V, H and D are unbounded operators. Paradoxes are well known[14)] which one can obtain with formal manipulations of the canonical commutation relations by ignoring domain questions. Before using (2.10) further we shall prove (2.10) such that our application[12)] is covered.

For arbitrary potentials (2.10) is probably useless since the calculation of the expectation values with the disturbed eigenvectors $|\alpha)$ will be impossible. For the Coulomb potential we may, however, use for D the generator of the dilatation group. There the commutators can be expressed by H and V and correspondingly the expectation values by E and $\partial E/\partial\alpha$. The corresponding differential inequality will be integrated in the next section.

For a rigorous deduction of (2.10) we follow the general strategy that it is better to use the group elements rather than the (unbounded) generators. The group in question is ($\beta \in R$)

$$x \to e^{\beta} x \quad , \qquad\qquad p \to e^{-\beta} p \quad . \tag{2.11}$$

It is now essential whether we work in infinite space and our Hilbert space are the (properly antisymmetrized) functions of $L^2(R^{3N})$ or whether we work in a finite volume $-L \leq x_i \leq L$. In the first case (2.11) is represented by a group of unitary transformations whereas in the second case it exists only for $\beta \leq 0$ and then gives a semi-group of partial isometries. The generator is in both cases a hermitian operator of the form xp + px. However, the domains are different: in the first case it is self-adjoint whereas in the second case it is maximal symmetric. This shows immediately that formal manipulations with the canonical commutation rules are no good because they would not make any distinction between these possibilities. However, we know that they should lead to the virial theorem with external virial for the finite volume and without it in infinite space. Thus one pays for ignoring domain questions by loosing something not of the size ε but of the size of the effect one wants to calculate.

We shall only consider here the case $L = \infty$ which is appropriate for atoms, for instance. We shall derive a generalization of (2.10) for the sum of the first n levels. Consider a Hamiltonian depending linearly on two parameters

$$H(\alpha,\beta) = T(1 + 2\beta) + V_p(1 + \beta) + V_e(1 + \alpha + \beta) \tag{2.12}$$

with

$$V_p = -Z \sum_{i=1}^{N} |x_i|^{-1} \quad , \qquad V_e = \sum_{i>j} |x_i - x_j|^{-1} \quad . \tag{2.13}$$

According to the previous argument the matrix of the second derivatives of

$$E(\alpha,\beta) = \frac{1}{n} \inf_{\mathcal{H}_n} \mathrm{Tr}_{\mathcal{H}_n} H(\alpha,\beta) \tag{2.14}$$

must be non positive≠. Using that $\inf_{\mathcal{H}_n} \mathrm{Tr}_{\mathcal{H}_n}$ is invariant under unitary

≠) The second derivatives may not exist for a finite number of points. In any case they will exist as negative distributions which will be good enough for our purpose.

transformations and that the scale transformation (2.11) produces $T \to e^{-2\beta}T$, $V_{e,p} \to e^{-\beta}V_{e,p}$ we have ($V = V_e + V_p$)

$$E(\alpha,\beta) = \frac{1}{n} \inf_{H_n} \mathrm{Tr}_{H_n} (T(1+2\beta)e^{-2\beta} + V_p(1+\beta)e^{-\beta} + V_e(1+\alpha+\beta)e^{-\beta}) =$$

$$= \frac{1}{n} \inf_{H_n} \mathrm{Tr}_{H_n} (H - \frac{\beta^2}{2}(4T+V) + \alpha(1-\beta)V_e + O_3) \quad . \tag{2.15}$$

O_3 are terms of higher than second power in α, β. By the usual rules of perturbation theory, which are applicable because $V_{p,e}$ are bounded with respect to K, and by the virial theorem we have

$$\left.\frac{\partial^2 E}{\partial \beta^2}\right|_o = - \langle 4T + V \rangle = 2E$$
$$\left.\frac{\partial^2 E}{\partial \beta \partial \alpha}\right|_o = - \langle V_e \rangle = - \frac{\partial E}{\partial \alpha} \tag{2.16}$$

where $\langle\ \rangle$ is the average of the expectation values of the first n levels. (In case of level crossing take n sufficiently large to cover the degenerate levels). Thus we have finally

$$\frac{\partial^2 E}{\partial \alpha^2} \leq (\frac{\partial^2 E}{\partial \alpha \partial \beta})^2 / \frac{\partial^2 E}{\partial \beta^2} = (\frac{\partial E}{\partial \alpha})^2 / 2E \tag{2.17}$$

as could be deduced by using formally $D = xp + px$ in (2.10). We have shifted in our derivation the point $\alpha = 0$ but obviously (2.17) holds $\forall\ \alpha$.

iii) Integration of (2.17)

We put

$$E(\alpha) = E(0)\ e^{\int_o^\alpha d\alpha' \rho(\alpha')} \tag{2.18}$$

since $V_e > 0$ and $E(0) < 0$ we have $\rho(0) = \frac{E'(0)}{E(0)} < 0$. (2.17) becomes

$$\rho' \geq - \rho^2/2 \quad . \tag{2.19}$$

If we take a solution $\bar{\rho}$ of

$$\bar{\rho}' = - \bar{\rho}^2/2 \tag{2.20}$$

with $\bar{\rho}(0) = \rho(0)$ we have

$$\bar{\rho}(\alpha) \leq \rho(\alpha)$$

or

$$E(\alpha) \leq E(0)\ e^{\int_o^\alpha d\alpha' \bar{\rho}(\alpha')} \quad . \tag{2.21}$$

(2.20) can easily be integrated and we obtain a parabolic upper bound

$$E(\alpha) \le E(0)(1 + \frac{\alpha}{2}\frac{E'(0)}{E(0)})^2 \qquad \forall\, \alpha > 0 \ .$$

Similarly one deduces $\forall\, \alpha_0 > 0$

$$E(\alpha) \ge E(0)[1 + \frac{\alpha}{\alpha_0}(\sqrt{\frac{E(\alpha_0)}{E(0)}} - 1)]^2 \qquad \forall\, 0 \le \alpha \le \alpha_0 \ . \tag{2.22}$$

Thus a lower bound for $\alpha = \alpha_0$ can be used to give a lower bound for all α inbetween and this is better than the trivial linear bound

$$E(\alpha) \ge \frac{\alpha}{\alpha_0}(E(\alpha_0) - E(0)) \tag{2.23}$$

which follows from concavity.
As illustration let us just quote the simple example of N = 2, n = 1. The problem can be rescaled such that

$$HZ^{-2} \to H = \frac{p_1^2 + p_2^2}{2} - r_1^{-1} - r_2^{-2} + \frac{1}{Z} r_{12} \ .$$

This means that changing Z with N fixed we can actually vary α, although not continuously. There $E'(0) = 5/8Z$ and for $E(\alpha) < E_1(0) = -5/8$ one has the linear bound $E(\alpha) \ge E(0) + \alpha(0|V_e^{-1}|0)^{-1} = -1 + \alpha\frac{16}{35Z}$. Taking $\alpha_0 = \frac{105}{128}Z$ such that $E(\alpha_0) = E_1(0)$ we have from (2.22) and (2.21)

$$-[1 - \frac{0.2553}{Z}]^2 = -[1 - \frac{128}{105Z}(1 - \sqrt{\tfrac{5}{8}})]^2 \le E \le -[1 - \frac{5}{16Z}]^2 = -[1 - \frac{0.315}{Z}]^2 . \tag{2.24}$$

In Fig. 3 I have given the parabolic and the linear bounds together with some experimental numbers. The concavity aspects are obvious. Of course, using elaborate trial functions these results can be improved[4]. However, the present method also works in more complicated situations[12] where useful trial functions are not available. Furthermore our inequalities also give an upper bound for $(\alpha|V_e|\alpha) = \alpha \frac{\partial E}{\partial \alpha}$, in our example (N = Z = 2)

$$|(\alpha|V_e|\alpha)/(\alpha|V_p|\alpha)| \le \frac{1}{6.4} \ .$$

In the Thomas-Fermi model the latter ratio is 1/7. In any case we see that for atomic systems we talk about differences between the rigorous bounds of the % order whereas for macroscopic systems the lower one is off by 10^{14}.

Fig. 3

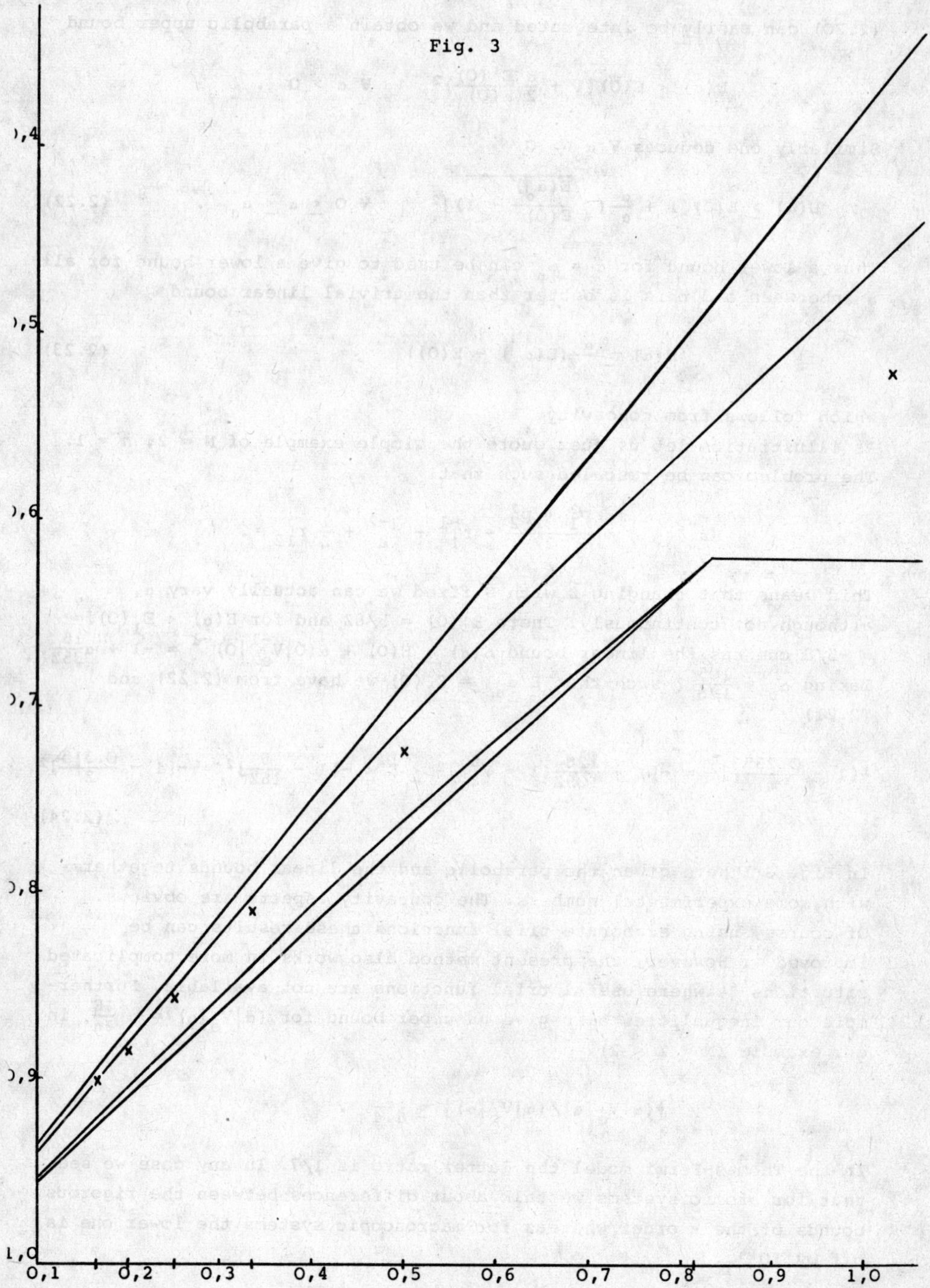
0,4
0,5
0,6
0,7
0,8
0,9
1,0
0,1
0,2
0,3
0,4
0,5
0,6
0,7
0,8
0,9
1,0
C^{4+}
B^{3+}
Be^{2+}
Li^{+}
He
α_0
H^-

3. COSMIC BODIES

i) Finally we shall consider systems dominated by gravitation ((1.1) with $N \sim 10^{57}$). They have some similarity with atoms insofar that their radius shrinks $\sim N^{-1/3}$ and therefore their energy goes like $N^{7/3}$. By this compression atoms are squashed and in the resulting high-density plasma the motion of the particles is mainly determined by the mean gravitational field. Thus one expects the Thomas-Fermi equation to become exact in the limit $N \to \infty$. This can indeed be rigorously demonstrated by a rather lengthy analysis. I can give here only the main points and have to refer you for details to 9). The strategy is the following: You want to calculate the thermodynamic functions and for this you have first to get rid of the singularity in the potential since this will lead to local inhomogenuities. Furthermore for our method a finite self-energy is essential. In fact one shows that approximating the potential by a step function things change little. Little means small compared with the leading term which is $\sim N^{7/3}$. Secondly one separates by infinite walls the boxes within which the potential is constant. Then one has separate free systems where the constant potentials give a contribution proportional to the occupation number of the boxes. The partition functions are now sums over the occupation numbers and one has to demonstrate that the former are almost given by their value for a self-consistently determined set of occupation numbers. These self-consistency equations become, in the limit of infinitely many boxes, the Thomas-Fermi equations. The tactics consists mainly in keeping track of the various ε's and watching the order in which the different limits are to be taken.

ii) Regularizing the Potential

As basis for our further analysis we have to establish the lower bound (1.11). This can be done[8] by the simple device of rewriting ($\hbar = m_i = 1$)

$$H \geq \sum_{i=1}^{N} \frac{p_i^2}{2} - \alpha \sum_{i \neq j} |x_i - x_j|^{-1} , \qquad \alpha = (\kappa + e^2)\frac{1}{2} \tag{3.1}$$

in the form

$$H \geq \sum_{i=1}^{N} \sum_{j \neq i} \left(\frac{p_j^2}{2N} - \alpha |x_i - x_j|^{-1}\right) = \sum_{i=1}^{N} h_i . \tag{3.2}$$

The

$$h_i = \sum_{j \neq i} \left(\frac{p_j^2}{2N} - \alpha |x_i - x_j|^{-1}\right) \tag{3.3}$$

represent N-1 non-interacting particles (with mass N) moving in the Coulomb field of the particle i (which has no kinetic energy). h_i can be bounded from below with the reasoning we have used for atoms.

Filling the Balmer levels $-\frac{1}{2} N \alpha^2 \frac{1}{n^2}$ according to the exclusive principle up to $n_o \sim N^{1/3}$ we find $h_i \geq -\gamma N^{4/3} \alpha^2$ and thus

$$H \geq -\gamma N^{7/3} \alpha^2 \quad . \tag{3.4}$$

The numerical factor γ is of the order 1.
Remark: H is, of course, invariant under permutations of the particles, but h_i is not. Thus the exclusion principle has not been fully taken into account in treating h_i. Nevertheless the bound (3.4) holds a fortiori for all totally antisymmetric functions. With the same method one can show that the addition of a narrow Yukawa potential $-e^{-\mu r}/r$ to $1/r$ changes little. Suppose one uses the latter in (3.1) to (3.3) instead of $1/r$. The argument goes through in the same way except that we now can use $n_o \sim N_b^{1/3}$ where N_b is the number of bound states which for a central potential is known to be bounded by[15])

$$N_b \leq 2\int_0^\infty dr\ 2mr|V(r)|[\sup_r r^2\ 2m|V(r)|]^{1/2} \quad . \tag{3.5}$$

We shall not derive this formula but only remark that for the $\alpha e^{-\mu r}/r$ the corresponding bound $\sim (\frac{2m\alpha}{\mu})^{3/2}$ is about the "classical" value

$$\int \frac{d^3p}{(2\pi)^3}\, d^3x\ \Theta(2m|V(r)| - p^2) \quad .$$

Using the formula we have to remember that m in (3.3) is N and that our system will shrink $\sim N^{-1/3}$. Correspondingly we put $\mu = sN^{1/3}$ (s^{-1} a fixed fraction of the total size of the system) and obtain

$$N_b \leq \gamma_1 N \left(\frac{\alpha}{s}\right)^{3/2}$$

and thus

$$H \geq -\gamma_2 N^{7/3} \alpha^2 \left(\frac{\alpha}{s}\right)^{1/2} \quad , \tag{3.6}$$

again $\gamma_{1,2} = \Theta(1)$. Thus for s sufficiently large the influence of a Yukawa potential becomes arbitrarily small on the $N^{7/3}$-scale. This can be rephrased by saying that even if $\alpha \sim s^{1/5}$ its influence is bounded $\forall$ s:

$$0 \leq \sum_{i>j} \frac{e^{-s|x_i-x_j|N^{-1/3}}}{|x_i-x_j|} \leq s^{-1/5}(K + \gamma_2 N^{7/3}) \tag{3.7}$$

or

$$K - \sum_{i>j} |x_i-x_j|^{-1} \leq K - \sum_{i>j} \frac{1 - e^{-s|x_i-x_j|N^{1/3}}}{|x_i-x_j|} \leq$$

$$K(1 + s^{-1/5}) - \sum_{i>j} |x_i - x_j|^{-1} + s^{-1/5} \gamma_2 N^{7/3} \quad . \tag{3.8}$$

Thus replacing the 1/r by the continuous function $\frac{1-e^{-\mu r}}{r}$ has no larger effect than adding a small constant and changing the mass slightly. Since the thermodynamic functions F (1.5) and E (1.6) are monotonic in H the corresponding inequalities also hold for them.

The next step is approximating

$$\frac{1 - e^{-\mu|x_i - x_j|}}{|x_i - x_j|}$$

by a piecewise constant function. This can be done such that the operators change in norm arbitrarily little so that there is no question that F and E also survive this operation with negligible change.

iii) Putting Up Walls

Next we have to separate the various regions by infinitely high δ-function barriers so that the subsystems communicate only via the potential. These walls are mathematically represented by the condition that the wave function has to vanish at the boundaries. To see in this formulation that the walls are something positive one has to express the Hamiltonian as a quadratic form rather than as an operator. (Although they are the same formal expression they differ in domain). There the domain is restricted by the introduction of walls and thus the eigenvalues increase since they can be expressed as infima taken over the domain. The harder part is to show that the energies do not increase too much. For this one constructs[16] for each state of the system without walls a wave-function for the system with walls such that the expectation value of the Hamiltonian is only slightly increased. This can be done since the number of walls is kept fixed in the limit of $N \to \infty$. Of course, one cannot brutally cut the wave functions at the 3h walls since this would cost too much kinetic energy. To do this in a gentle fashion the wave function is multiplied by a function f which goes smoothly to zero at the walls. Since this would decrease the norm of the state one increases the size of the system with walls to $L' = L + 2b(h-1)$. The region within a distance b from the walls where the wave function goes to zero is used twice in the system with walls by folding it back. This means we define a mapping $L'^3 \to L^3$ which in one dimension looks like Fig. 4 and is given in formulae by

$$(Rx)_r \equiv \begin{pmatrix} x_r - 2(n-1)b \\ 2\zeta_n - 2nb - x_r \\ 2\zeta_n - 2(n-1)b - x_r \end{pmatrix} \quad \text{if} \quad \begin{pmatrix} \zeta_{n-1} + b \le x_r \le \zeta_n - b \\ \zeta_n - b \le x_r \le \zeta_n \\ \zeta_n < x_r \le \zeta_n + b \end{pmatrix} \tag{3.9}$$

where r = 1,2,3 labels the three spatial dimensions, n = 0,1,...,h the walls (including those of the original cube) in a given spatial dimension, and $\zeta_n = n\frac{L}{h} + (2n-1)b$ the location of such a wall.

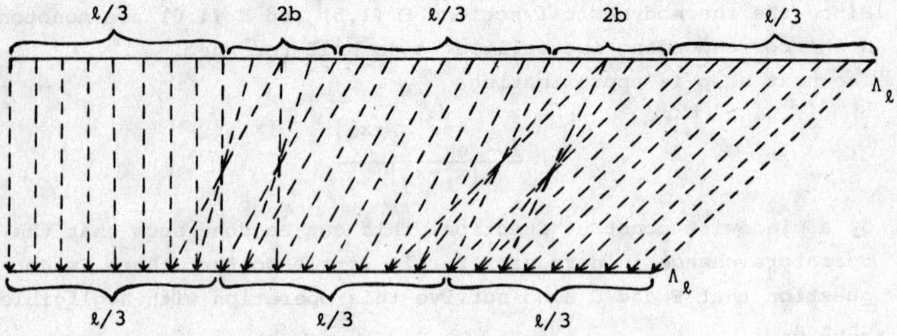

Fig. 4 The mapping R from L' to L for h = 3.

Consider next a function f: $[0,L+2(h-1)b] \to \mathbb{R}$ with the following properties: f is once continuously differentiable, $f^2(x) \leq 1$; for $n = 1,2,\ldots,h-1$: $f(x) \neq 1$ for $x \in [\zeta_n-2b,\zeta_n+2b]$ only; $f(\zeta_n) = 0$; $\xi \to f(\zeta_n+\xi)$ is even for $|\xi| \leq 2b$; $f^2(\zeta_n+\xi) + f^2(\zeta_n+2b-\xi) = 1$ for $0 \leq \xi \leq b$; $(f'(x))^2 \leq 1/b^2$. One can then show by an elementary calculation that I: $H_{N,L^3} \to H_{N,L'^3}$ defined by

$$(I\Phi)(\vec{x}_{11},\ldots,\vec{x}_{2N}) \equiv \prod_{\alpha i} \prod_{r=1}^{3} f(x_{\alpha i,r})\Phi(R\vec{x}_{11},\ldots,R\vec{x}_{2N}) \tag{3.10}$$

is an isometry:

$$(I\Phi,I\psi) = (\Phi,\psi) \quad . \tag{3.11}$$

One can show that the conditions imposed on f ensure that the range of I is in the appropriate domain and that the expectation value of the energy is only changed by a small amount. From the standard Minimax principles for the thermodynamic quantities it follows that they also are shifted only slightly and one can give exact bounds on the change.

iv) The Result

We have now arrived at a situation where particles move in separated cubes within which the potential is constant and depends only on the occupation numbers. Since the kinetic energy is explicitly known the thermodynamic quantities can be written as sums over all possible occupation numbers of well-known expressions. By the usual manipulations of statistical mechanics one then shows that in the limit $N \to \infty$ they are given by the leading term. The latter is obtained if the

occupation numbers satisfy a certain self-consistency relation. This goes over into the Thomas-Fermi equation if one retraces the various steps, i.e. goes to the limit of infinitely many boxes so that the potential becomes continuous again and then takes away the $e^{-\mu r}/r$ so that one is left with the Coulomb potential. I have to refer you to 9), 17) for a demonstration that the result survives all these manipulations. The outcome of all this can be stated as follows:
For given $\beta > 0$, $L > 0$ there are solutions of the Thomas-Fermi equation

$$\rho_\alpha(\vec{x}) = \int \frac{d^3q}{(2\pi)^3} \frac{1}{1 + e^{\beta[\frac{q^2}{2M_\alpha} + W_\alpha(\vec{x}) - \mu_\alpha]}} \tag{3.12}$$

with

$$W_\alpha(\vec{x}) = \sum_\beta \int d^3x' \, \frac{e_\alpha e_\beta - \kappa M_\alpha M_\beta}{|\vec{x}-\vec{x}'|} \rho_\beta(\vec{x}') \quad , \tag{3.13}$$

which are normalized

$$\int_{L^3} d^3x \; \rho_\alpha(\vec{x}) = 1 \quad . \tag{3.14}$$

One then calculates

$$u = \sum_\alpha \int_{L^3} d^3x \; \rho_\alpha(\vec{x}) \, W_\alpha(\vec{x}) \quad , \tag{3.15}$$

$$f = -u + \sum_\alpha \{\mu_\alpha - \frac{1}{\beta} \int_{L^3} d^3x \int \frac{d^3q}{(2\pi)^3} \log(1 + e^{-\beta[\frac{q^2}{2M_\alpha} + W_\alpha(\vec{x}) - \mu_\alpha]})\} \tag{3.16}$$

$$\eta = u + \sum_\alpha \int_{L^3} d^3x \int \frac{d^3q}{(2\pi)^3} \frac{q^2}{2M_\alpha} \frac{1}{1 + e^{\beta[\frac{q^2}{2M_\alpha} + W_\alpha(\vec{x}) - \mu_\alpha]}} \quad , \tag{3.17}$$

and

$$s = \beta(\eta - f) \quad . \tag{3.18}$$

If (3.12) to (3.14) has more than one solution that which minimizes f has to be chosen, and the minimal f coincides with $f(\beta,L)$. From $\beta \to s$, $\beta \to \eta$ the graph $\eta(s,L)$ is constructed by choosing the smaller one of two possible η's. The microcanonical and the canonical thermodynamical functions are Legendre-transforms of each other if the Legendre-transform is possible, i.e. for those values of the entropy s where

$$\frac{\partial^2 \eta(s,L)}{\partial s^2} > 0 \quad .$$

To solve these equations analytically is impossible but I can show you the results of a computer solution for the special case e = 0 and only one species of fermions: For amusement we used for the mass of particles the mass of neutrons and the number N of particles corresponding to a star to illustrate that for this situation these phenomena happen at energies of several MeV per particle and of sizes for the system of several km. In Fig. 5 we plot the entropy as a function of the energy and show that for a radius of 100 km we have indeed a region where the function becomes convex.

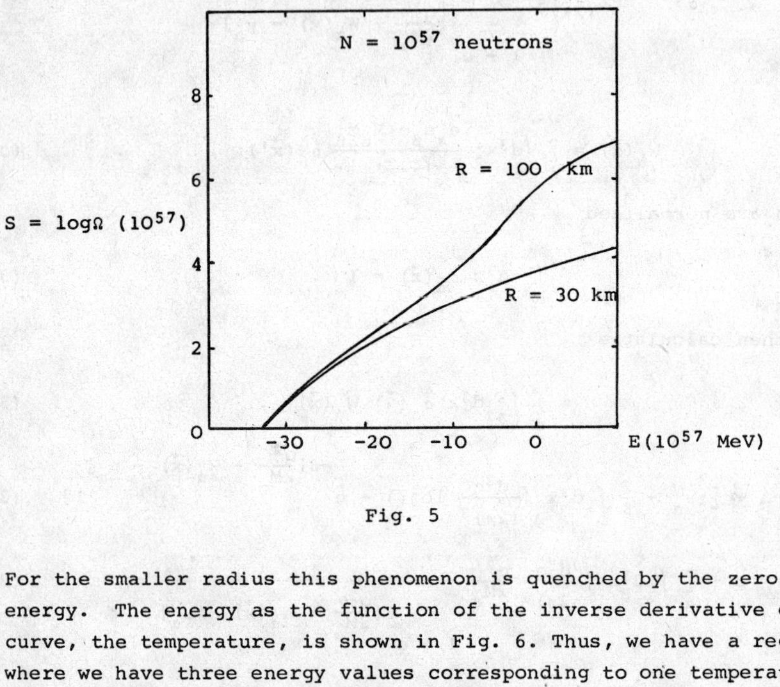

Fig. 5

For the smaller radius this phenomenon is quenched by the zero point energy. The energy as the function of the inverse derivative of this curve, the temperature, is shown in Fig. 6. Thus, we have a region where we have three energy values corresponding to one temperature, and at a certain temperature the system in the canonical ensemble will jump from one branch of the curve to the other. The exact analysis verifies the usual statement that this happens when the free energy of the lower branch becomes lower than the free energy of the upper branch. In the next curve we plot the free energy as function of the temperature which clearly exhibits the point at which the free energy of one branch goes below the one of the other branch. The exact solution also gives the density as function of the radius and in Fig. 7 we plot the density for various temperatures. Note that above the transition point the density is reasonably homogeneous whereas at the phase transition the ratio between the density in the middle to the density of the surface increases by about five orders of magnitude. Below the phase transition a kind of star develops with a reasonably well defined surface and a very thin atmosphere. This is also emphasized by a plot of the measure of the degeneracy of the Fermi gas (Fig. 8).

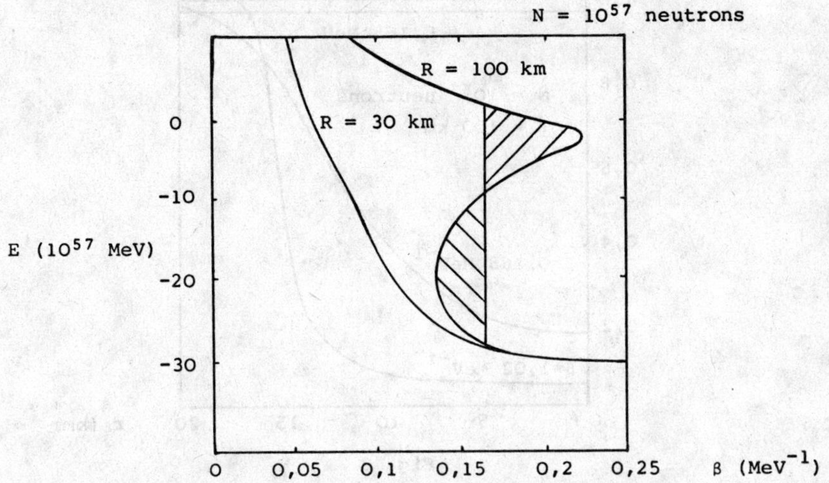

Fig. 6

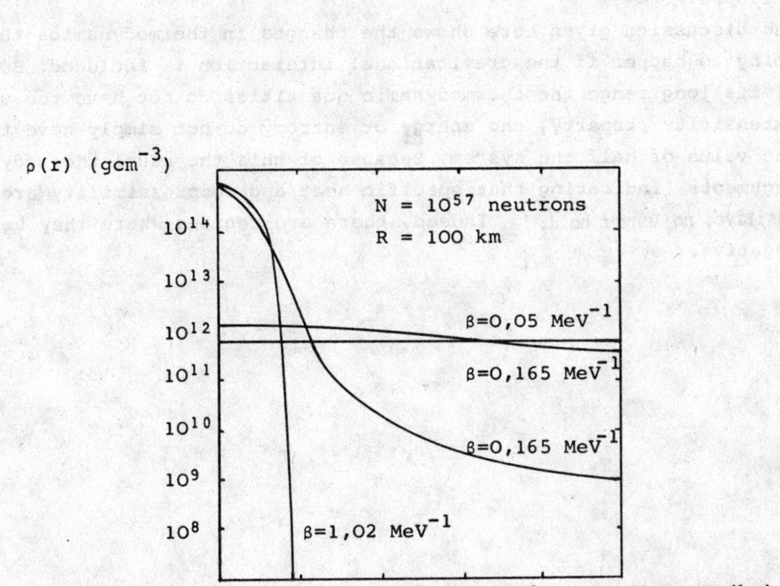

Fig. 7

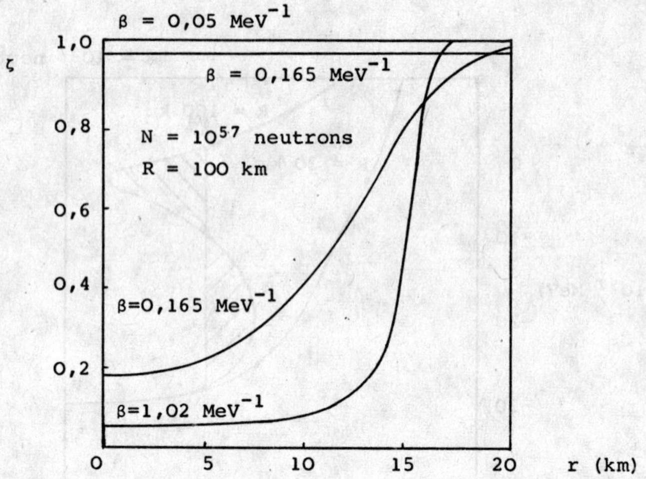

Fig. 8

The atmosphere always remains a Boltzmann whereas the centre becomes degenerate since it is only the zero point pressure which can stop the collapse.

The discussion given here shows the changes in thermodynamics that are going to happen if the gravitational interaction is included. Because of its long range the thermodynamic quantities do not have the usual extensivity property, and energy or entropy do not simply have twice the value of half the system. Because of this the usual thermodynamic arguments, indicating that specific heat and compressibility are positive, no longer hold. Indeed, there are regions where they become negative.

I would like to thank H. Narnhofer for her permission to use our unpublished results. Furthermore I am grateful to P. Hertel for stimulating discussions and to M. Breitenecker and H.-R. Grümm for help with some calculations.

REFERENCES

1) T. Kato, Perturbation Theory for Linear Operators, Berlin, Springer-Verlag 1966

2) E. Balslev, J.M. Combes, Comm. Math. Phys. 22, 280 (1971)

3) J.D. Dollard, J. Math. Phys. 7, 802 (1966)

4) N.W. Bazley, D.W. Fox, Phys. Rev. 124/2, 483 (1961)
P.O. Löwdin, Phys. Rev. 139A, 357 (1965)
T. Kinoshita, Phys. Rev. 105/5, 1490 (1957)
C.L. Pekeris, Phys. Rev. 112, 1649 (1958)

5) F.J. Dyson, A. Lenard, J. Math. Phys. 8, 423 (1967)

6) F.J. Dyson, In Statistical Physics, Phase Transition and Superfluidity, Brandeis University Summer School in Theoretical Physics, Lecture notes (1966)

7) J.L. Lebowitz, E.H. Lieb, Advances in Mathematics 9, 316 (1972)

8) J.M. Lévy-Leblond, J. Math. Phys. 10, 806 (1969)

9) P. Hertel, H. Narnhofer, W. Thirring, Comm. Math. Phys. 28, 159 (1972)

10) P. Hertel, W. Thirring, in Dürr, Quanten und Felder, Braunschweig, Vieweg 1971

11) A. Wehrl, to be published in Acta Phys. Austr.

12) H. Narnhofer, W. Thirring, in preparation;
W. Kohn, Phys. Rev. 71, 635 (1947)

13) S. Okubo, J. Math. Phys. 12, 1123 (1971)

14) M. Reed, B. Simon, Methods of Modern Mathematical Physics, New York and London 1972

15) B. Simon, J. Math. Phys. 10, 1123 (1969)

16) D.W. Robinson, The Thermodynamic Pressure in Statistical Mechanics, Springer lecture notes (1971)

17) P. Hertel, W. Thirring, Comm. Math. Phys. 24, 22 (1971)

D I S C U S S I O N S

CHAIRMAN: Prof. W. Thirring

Scientific Secretaries: J. Nyiri, J. Rosen

DISCUSSION No. 1

- *WILCZEK:*

You stated it as a problem: what cuts off the infinities of QED? Is not QED as it stands (in the form of renormalized perturbation theory) a consistent theory, perhaps even finite? Of course we would like to have a theory which enables us to compute $m_{\pi^+} - m_{\pi^0}$, but it is not so clear that one needs a finite number for Z_3^{-1}. Do you feel there is some difficulty with an infinite Z_3^{-1}?

- *THIRRING:*

The question is, does renormalized perturbation theory converge? That is, can one use it to define a unitary S-matrix etc.? The study of models leads me to the feeling that the infinity of Z_3^{-1} is a serious problem. There is, however, no compelling reason to believe that.

- *WILCZEK:*

As you stated, the construction of the scattering operator for the case of Coulomb interactions requires some care; because of the long-range nature of the force, one requires a different "free" Hamiltonian from the kinetic energy in constructing the Möller operator. Do you have any feeling about what happens in the case of Yang-Mills fields?

- *THIRRING:*

If the gauge particles acquire a mass, there is no difficulty (the force is then no longer of long range).

- *WILCZEK:*

How does the (complex) dilation group enter into the results on spectra proved by Combes and Balslev?

- *THIRRING:*

1) The dilation transformation is $x \to e^{\beta}x$, $p \to e^{-\beta}p$ with β in general complex. Under this transformation $T \to e^{-2\beta}T$, $V \to e^{-\beta}V$. This transformation is of course not unitary for $\mathrm{Im}\,\beta \neq 0$. It may be defined by analytic continuation from the case of real β.

2) We consider first the effect on T. Since V is in a sense small compared to T, it does not change the point where the continuous spectrum begins. The effect of the complex dilation is to rotate the spectrum of T in the complex plane. This permits the analytic continuation of the resolvent.

3) It is probable that the spectrum is two-sheeted; there will be an essentially singularity at 0, although that was not proved in the general case.

4) The same method can be extended to a wider class of potentials than the Coulomb potential; these are the "dilatation analytic" ones.

- *HENDRICK:*

In the zero-mass scattering case you mentioned studying the operator $e^{iHt}\,e^{-iH_0t}$ and its convergence to an operator Ω. A more familiar description of the problem of zero-mass scattering, at least for the case N = 2, is that you lose analyticity along the real axis in the energy plane. Can you indicate how these two descriptions are equivalent?

- *THIRRING:*

Yes. Instead of considering just $e^{iHt}\, e^{-iH_0 t}$, you construct the S-matrix

$$S = \lim_{t\to\infty} e^{-iH_0 t}\, e^{2iHt}\, e^{-iH_0 t}$$

or equivalently

$$S = \lim_{\varepsilon\to 0} \varepsilon \int_0^\infty dt\, e^{-\varepsilon t}\, e^{-iH_0 t}\, e^{2iHt}\, e^{-iH_0 t} .$$

The matrix element between momentum states p and q is

$$\langle q|S|p\rangle = \lim_{\varepsilon\to 0} \left\langle q\middle|\varepsilon \int_0^\infty dt\, e^{-\varepsilon t}\, e^{-it(p^2+q^2)}\, e^{2iHt}\, \middle|p\right\rangle$$

$$= \lim_{\varepsilon\to 0} \left\langle q\middle| \frac{i\varepsilon}{H - \frac{p^2+q^2}{2} + i\varepsilon} \middle|p\right\rangle .$$

Using the expression

$$\frac{1}{H - Z} = \frac{1}{H_0 - Z} - \frac{1}{H_0 - Z} V \frac{1}{H - Z}$$

twice, you come to study

$$\langle q|T_{q^2}|p\rangle\, \delta(p^2 - q^2)$$

with

$$T_Z = V + V \frac{1}{H - Z} V ,$$

from which you can deduce the usual analyticity properties in the energy. However, for the Coulomb potential the situation is more complicated, since $H_0(t)$ contains a term $\sim \ln t$. Therefore, the theory indicated above does not apply and there is a considerable confusion in the literature. Occasionally even the unitarity of the S-matrix is questioned.

- *MENDES:*

Concerning the outstanding problems of QED as discussed in the opening lecture, I think that a problem which is much more important than the inifities (which are well taken care of by renormalized perturbation theory), is the problem of why there are two electrons -- the electron and the muon. Could you comment on where we stand today in this problem?

- *THIRRING:*

I do not think the mathematical problems of QED have anything to do with the electron-muon problem. The structure of QED alone will probably have nothing to say about the existence of two distinct charged leptons with their distinct neutrinos and separately conserved quantum numbers. Perhaps in gauge theories it will be possible to say something about this. Would you like to comment on this question, Coleman?

- *COLEMAN:*

We have been able to construct a gauge field theory model where things like the electron-muon mass ratio become computable.

- *MENDES:*

Is the muon put in by hand in that theory, or do you find in a natural way some sort of spectrum that would predict the existence or non-existence of heavier leptons?

- *COLEMAN:*

We can put as many leptons as we wish in the model; so nothing can be said about a spectrum of leptons.

- *MENDES:*

You have stated that the interpretation of the continuum states as scattering states is not as clear for long-range potentials as it was for short-range potentials. Could you summarize what the situation is, i.e. what are the classes of potentials where the usual interpretation has been checked and those where it fails?

- *THIRRING:*

For potentials of the form $1/r^{\alpha}$ the usual scattering theory of the continuum states and the construction of the S-matrix is all right for $\alpha > 1$ but not for $\alpha \leq 1$.

- *MENDES:*

I did not quite understand the last part of your derivation of the N-dependence of the energy for the case of fermions ($E \sim N^{7/3}$). I agree this is the N-dependence of both the kinetic and the potential energy that minimizes the total energy. However, could not the leading N-dependence be cancelled when the two terms are subtracted?

- *THIRRING:*

No. In the actual proof one shows by the virial theorem that one term is half of the other and the $N^{7/3}$-dependence is preserved for the total energy. What was presented in the lecture was not really a proof.

For the simple case of atoms the proof of such a bound goes as follows. Write the Hamiltonian

$$H = T + \sum_{i>j} \frac{e^2}{|x_i - x_j|} - N\sum_i \frac{e^2}{x_i} .$$

H is greater than $T - N \Sigma_i c^2/x_i$, so a lower bound for this expression is also a lower bound for H. Using now the Balmer formula, this expression is found to be proportional to:

$$-\sum_{n=1}^{n_0} n^2 \frac{N^2}{n^2} = -N^2 n_0 .$$

On the other hand, from

$$\sum_{n=1}^{n_0} n^2 = N$$

it follows that $n_0 \sim N^{1/3}$. Substituting for n_0 in the previous equation, the $N^{7/3}$-dependence follows. Notice that for bosons the result would have been N^3, because without the exclusive principle all electrons would occupy the lowest state.

- *MADARAS:*

I have a philosophical question about the axiomatic field theory that you mentioned. How does one know which axioms should be included in the list of those axioms which must be satisfied?

- *THIRRING:*

I do not really know. The list must be modeled according to what we know from theories such as QED, and should yield some specific results (such as the TCP theorem, spin and statistics, etc.). The present list is restrictive enough so that you can derive these theories, but may be so restrictive that only free fields (in four dimensions) satisfy them.

- *KUPCZYNSKI:*

You mentioned that the divergences of the type of $\alpha/\pi \ln E/m$ are still very small even for the large energies available now. However, for these large energies electrodynamics becomes an open theory due to the possible production of $\mu^+\mu^-$ and hadron-antihadron pairs. So perhaps one should try to build not only a unified theory of weak and electromagnetic interactions, but also of strong interactions.

- *THIRRING:*

I encourage you to do so.

DISCUSSION No. 2

- *NAHM:*

Most of your proofs make use of the dilatational properties of the Coulomb potential. Some of your results certainly have a more general validity, for example, if you add some $V > 0$ you still should have $E_0 \sim N$. Also if you want to compare your results with measured properties of matter, you should be able to estimate the effects of additional terms describing magnetism. Can you say anything about the properties of systems with such more general Hamiltonians?

- *THIRRING:*

The convexity of $E(\alpha)$ is valid generally, but it cannot be used for general potentials. I do not know how to treat the general case. Ferromagnetism is included in the Hamiltonian I used, because it is the consequence of the exchange Coulomb interaction which is included. Spin-orbit forces could probably be treated by similar methods, since they behave simply under dilatations.

- *GENSINI:*

I have been wondering if the convexity property you were talking about this morning had some deeper rooting even outside quantum theory. This kind of property could perhaps be reconciled with the similar relations known in thermodynamics as the Le Chatelier-Braun principle. Also the tendency of these inequalities to work as equalities when $\alpha \to 0$ reminded me of the tendency of classical inequalities to work as equalities in the neighbourhood of the critical point -- a phenomenon which often goes under the name of critical scaling -- and made me wonder whether this could even be a general consequence of general principles, surviving a possible failure of the presently accepted quantum theory.

- *THIRRING:*

Stability of matter against various kinds of perturbations does in fact require many thermodynamical functions to be convex functions of one variable or the other. If the apparent similarity between such relations has some deeper physical meaning, this has been verified only in special cases.

DISCUSSION No. 3

- *HENDRICK:*

For what class of potentials do you think you can derive exact results as you did in your lecture today?

- *THIRRING:*

So far, we have obtained these exact results only for the case of a $1/r$ potential. It is my belief that the same kind of precise results can be obtained for potentials of the type $1/r^n$ for $1 \leq n < 2$. The difficulty with higher values of n is that you lose the first condition, the lower boundedness on the energy.

- *HENDRICK:*

In what sense is the phase transition you mentioned at the end of today's lecture a real phase transition, and could you give it a physical interpretation?

- *THIRRING:*

This picture, which I will redraw, is for the microcanonical ensemble. The phase transition is a real phase transition in the following sense. For a certain portion of the curve, from A to B the specific heat is negative. This means that for energies in this range, as you extract energy from the system it actually becomes hotter rather than cooler. A physical example of such a system is a star. Normally the production of energy within the star will cause heating, but in the region of negative specific heat the production of energy actually causes the star to cool off.

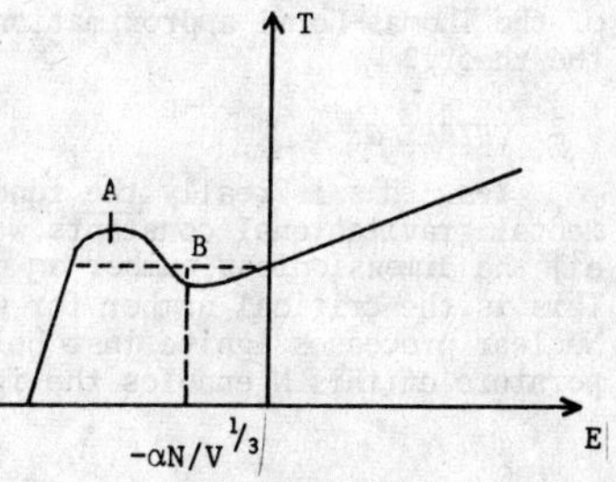

This behaviour is a direct result of the virial theorem which says that the kinetic energy is equal to the absolute value of the total energy, and the equipartition theorem which says the kinetic energy is proportional to the temperature. There are two possible problems with this argument:

a) The virial theorem holds in this form only for infinite systems.

b) The equipartition theorem holds only for classical systems.

However, in an in-between region, the argument remains valid. In the canonical ensemble this region of negative specific heat is jumped over by a phase transition since there the system cannot be in equilibrium with the heat bath. Thus, the specific heat is always positive there and we see a phase transition instead.

- *EYLON:*

We have seen that the energy of a star is not proportional to N (the number of particles). Therefore we are not allowed to divide the star into domains and to assume nearest neighbour interactions. This means that the interaction between widely separated particles is important. Are we allowed to treat a single star separately, ignoring its interactions with other stars?

- *THIRRING:*

Our case is really different from usual thermodynamics, where the energy is additive; the energy of a star is not the sum of energies of its two halves. The reason why a star is made of about 10^{57} particles does not derive directly from the Hamiltonian we began with. If the star were smaller, the maximum temperature reached upon contraction would be lower, and it would not be sufficient to ignite the nuclear reactions. Were the star larger, the radiation pressure we have neglected would become enormous and all sorts of instabilities might occur.

As for the galaxy as a whole, the effect of accumulation of stars in its centre is reminiscent of the above effect (although the galaxy is not composed of fermions), but in this case it is not clear what the relevant relaxation times are. Energy loss of a star by emission of photons is a very slow process, and such a star can be treated as a thermally isolated system. On the other hand, when neutrinos become the effective mechanism of getting energy out, then the phase transition is very quick (but then the relativistic effects are important).

- *RABINOVICI:*

Is there a simple physical reason why the Thomas-Fermi approximation is only good for the systems with attractive forces that you described, and does not work as well for macroscopic systems like a metal?

- *THIRRING:*

The density in ordinary matter is too low and therefore the Thomas-Fermi approximation is not so good. In this connection the results of Simon and Lieb, who investigated the Thomas-Fermi equation (not only for atoms, but also for molecules) are interesting. They established that to the accuracy of $E/z^{7/3}$ it is also exact for molecules and crystals. However, the limit where this becomes too large -- and therefore the density of the electrons becomes very high -- is not reached. In particular, in the Thomas-Fermi equation, there is no molecular binding and this tells you that the chemical forces are not proportional to $z^{7/3}$ but to a much lower power, and this therefore does not show up in the Thomas-Fermi equation.

- *RABINOVICI:*

Does the number of particles you mentioned, 10^{57} that is, necessary for the validity of the Thomas-Fermi approximation, have any connection with the fundamental constants of the theory?

- *THIRRING:*

Yes, this is really <u>the</u> fundamental constant in gravitational theory. From the fundamental gravitational constants we can construct (in analogue to the fine structure constant e^2) the dimensionless number $\alpha_G = K\, m_p^2 = \frac{1}{2}\, 10^{-38}$; through it one derives that $N = (\alpha_G)^{-3/2}$. This is the critical number for which at the maximal temperature things become relativistic. Nuclear processes ignite just before the electrons become relativistic. The maximal temperature at this N enables the ignition of nuclear reactions.

- *NAHM:*

Today you showed that for $K > 0$ one can control the effects of small changes of the potential by means of operator inequalities. You have not done this for $K = 0$. Are there special difficulties for this case?

- *THIRRING:*

I do not think so. However, I have been a little hesitant about spin orbit forces since they are rather singular at the origin. Of course they are not effective for s-electrons, and they are not so dangerous for p-electrons as their wave functions vanish at the centre. However, one has to be careful in this case.

- *GENSINI:*

With regard to the phase transition phenomenon you were speaking about, could you please give us the order of magnitude of the critical parameters for such a phenomenon.

- *THIRRING:*

The order of magnitude where the phase transition sets in is given by a criterion, due to Jeans, namely that the gravitational energy of the system is of the order of the thermal energy.

As far as our results are concerned, we have not investigated critical parameters and critical indices, since what we have is not an analytic expression but just a numerical solution of the Thomas-Fermi equations.

INVARIANT QUANTIZATION, SCALE SYMMETRY AND EUCLIDIAN FIELD THEORY

R. Jackiw

Table of Contents

INVARIANT QUANTIZATION, SCALE SYMMETRY AND EUCLIDIAN FIELD THEORY

R. Jackiw

Laboratory for Nuclear Science and Department of Physics,
Massachusetts Institute of Technology, Cambridge, Massachusetts 02139

Introduction

A method for extracting interesting results from field theoretic models, without explicitly solving the equations of motion which govern these models - an impossible task at the present time - is to study the algebraic relations which follow from canonical commutators. In the past, equal-time canonical commutators were used to derive the very important and successful algebra of currents at equal times.[1] More recently, canonical light-cone commutators led to current algebra on the light cone, which supplements the old equal time algebra and also gives a succinct description of Bjorken scaling in deep inelastic processes.[2]

The fortunate circumstance that canonical commutators can be determined a priori without solving the equations of motion, is a consequence of the quantization procedure. In the conventional approach, time is selected as the direction of evolution of the system, while the quantization conditions - the canonical commutators - are specified a priori on the space-like surface t = constant . In light-cone quantization, the system is allowed to develop along the x^+ direction[3], and commutators are given their a priori values on the surface x^+ = constant . It is natural to inquire whether yet other quantization procedures exist which provide useful information.

In these lectures I shall report on a new method of quantization which was developed by S. Fubini, A. Hanson and me.[4] The surface of quantization is chosen to be the space-like hyperboloid x^2 = positive constant, and propagation of the system proceeds along the normal direction, from one hyperboloid to the next. Of course the physical content of the theory is not changed by the choice of quantization surface, and we shall show that the formula for the S matrix in our approach, coincides with the conventional Feynman-Dyson expression. Nevertheless, as with light-cone quantization, the present technique organizes the theory in a novel fashion and offers new insights into the structure of field theory.

The results which have been obtained up to now are the following. In addition to exhibiting new canonical commutators, we shall give an unexpected interpretation for anomalous dimensions. An operator basis for Euclidean field theory has been discovered which differs from the one used in the past.[5] Especially interesting is the application of our methods to 2 -dimensional field theories where we find that the present ideas provide a bridge between the formalism of conventional field theory and that of the dual resonance model. Explicitly we can derive the Virasoro algebra for a large class of 2-dimensional field theories, and with further hypotheses, the Hagedorn spectrum is observed.

II. Review of Quantization Methods

Let me begin by giving a general discussion of our quantization procedure, comparing and contrasting it with other approaches. In the conventional method one considers the planes x^0 = constant . The evolution operator, that is the operator which takes the physical system from one surface to the next, is the time translation generator P^0 (the Hamiltonian). Also there exist six kinematical operators, that is operators generating motions which leave the surface invariant. These are space translations, generated by P^i and space rotations generated by $M^{ij} = -M^{ji}$ $(i,j = 1,2,3)$. The quantum conditions, that is the commutators, are specified on the selected surface; in the conventional case this is at x^0 = constant, i.e. at equal times.

In light-cone quantization, the selected surfaces are x^+ = constant . The evolution operator is P^- , while the six kinematical operators are P^+, P^i, M^{ij}, M^{i+} $(i,j = 1,2)$. Canonical commutators are specified at equal x^+.

A third possibility, which is the basis of our theory, is a family of hyperboloids x^2= constant > 0 . Clearly the six kinematical operators are the six Lorentz generators $M^{\mu\nu}$ $(\mu,\nu = 0,1,2,3)$, while the evolution operator is the dilatation generator D . In other words if we wish to proceed from one hyperboloid $x^2 = \tau_1{}^2$ to another $x^2 = \tau_2{}^2$, this can be achieved by a dilatation $x^\mu \to \rho x^\mu$. Canonical commutators are given their a priori values on surfaces of equal x^2.

Some attractive and interesting aspects of our quantization are already apparent. First, note that the method is explicitly Lorentz invariant at every stage. Unlike in the previous cases, the quantization surface does not select a preferred direction in space-time; clearly the surface x^2 = constant transforms into itself under a Lorentz transformation. Second, and more striking, is the crucial role that dilatations play in the present discussion: the evolution of the system is governed by the dilatation generator. Currently, dilatation transformations are the focus of much theoretical activity, as a consequence of the famous MIT-SLAC deep inelastic experiments. In the conventional approach to quantum field theory, scale transformations are somewhat peripheral to the main lines of development; thus we are pleased that these interesting transformations are central to the present theory.

An object of great importance in the discussion of the evolution of the system is the propagator function (Green's function). This is given by the vacuum expectation value of the ordered product of two fields, where the ordering is along the direction of propagation: time ordering or x^+ ordering in the familiar examples. For us the ordering will be along the surfaces $x^2 = \tau^2$.

When quantizing a field, it is convenient to expand it on the quantization surface in a complete set of functions associated with three of the six kinematical operators. In the usual equal time case, the three kinematical operators are chosen to be the translation generators P^i $(i = 1,2,3)$, and the expansion functions are eigenfunctions of $\frac{\partial}{\partial x^i}$, i.e. they are $e^{-i\, \underset{\sim}{k} \cdot \underset{\sim}{x}}$. For a free theory, one further expands in eigenfunctions associated with the evolution operator. This is P^0 in the conventional approach and one expands in eigenfunctions of $\frac{\partial}{\partial t} : e^{ik^0 t}$. Thus a free field is expanded in e^{ikx}.

In the light-cone quantization, the three kinematical operators are P^+, P^i $(i = 1,2)$ and the evolution operator is P^- . Hence a free field is again expanded in terms of e^{ikx}.

The present theory is characterised by the kinematical operators $M^{\mu\nu}$; hence we shall expand in eigenfunctions associated with three objects constructed from $M^{\mu\nu}$. The evolution operator is D ; thus the free field will further be expanded in eigenfunctions of $x^\mu \frac{\partial}{\partial x^\mu}$. Consequently our expansion functions, which replace the conventional exponentials, are powers and 4-dimensional harmonics.

All this may appear extremely bizzare; yet in fact our approach is quite common in Schrödinger theory. The Schrödinger wave function, which satisfies

$$-\tfrac{1}{2}\vec{\nabla}^2\psi = (E - V)\psi,$$

can be expanded in momentum eigenfunctions.

$$\psi(r) = \int \frac{d^3k}{(2\pi)^3} \; e^{i\underset{\sim}{k}\cdot\underset{\sim}{x}} \; \phi(\underset{\sim}{k})$$

$$\left(\frac{k^2}{2} - E\right)\phi(\underset{\sim}{k}) = -\int \frac{d^3q}{(2\pi)^3} \; V(\underset{\sim}{k} - \underset{\sim}{q}) \; \phi(\underset{\sim}{q})$$

This would be the analogy to the conventional approach in field theory. Yet practical calculations are not performed in momentum space. Rather one expands the wave function in terms of harmonics.

$$\psi(\underset{\sim}{r}) = \sum_{\ell m} \frac{R_{\ell m}(r)}{r} \; Y_{\ell m}(\Omega)$$

and furthermore, the radial wave function is frequently expanded in powers.

$$R_{\ell m}(r) = \sum_n C_{n\ell m} r^n$$

This is quite similar to the expansion procedure in our approach to field theory. Moreover, the Green's function for the Laplacian is given by

$$\frac{1}{|\underset{\sim}{r} - \underset{\sim}{r}'|} = \theta(r - r') \sum_{\ell m} \frac{1}{r}\left(\frac{r'}{r}\right)^\ell \frac{4\pi}{2\ell+1} Y_{\ell m}(\Omega)\, Y^*_{\ell m}(\Omega')$$

$$+ \; \theta(r' - r) \sum_{\ell m} \frac{1}{r'}\left(\frac{r}{r'}\right)^\ell \frac{4\pi}{2\ell+1} Y_{\ell m}(\Omega)\, Y^*_{\ell m}(\Omega')$$

It is seen that this has a representation as an ordered product, where the ordering is along the radial direction, in complete analogy to the ordering which we shall employ. Thus it is recognized that our method of quantization has already been widely used in Schrödinger theory, and one may view our work as an attempt to use in field theory techniques which proved themselves so successful in Schrödinger theory.

III. Euclidean Field Theory

A technical difficulty presents itself in carrying out quantization on hyperboloids. The problem is that the space-like surfaces $x^2 = \tau^2 > 0$ do not span all of space-time; as τ^2 varies from 0 to ∞, the region outside the light-cone $x^2 < 0$ is not reached. Thus the evolution of the system in that region must be discussed separately, a task made difficult by the fact that the surfaces $x^2 = -\tau^2$ are time-like.[6] In order to overcome this obstacle, we simply continue our field theory to Euclidean space, where of course the surfaces $x^2 = \tau^2$ span all of space.

The continuation in no way excludes application of the general method to physical Minkowski-space theories. Indeed even in the conventional approach, all practical computations are always performed in Euclidean space. Furthermore I shall show explicitly that our formula for the S matrix, coincides with the conventional Feynman-Dyson expression, continued to Euclidean space.

Our definition of Euclidean field theory will now be presented since it differs in all respects from other approaches.[5] We replace x^0 by ix^4, and for every field in Minkowski space $\phi(x_0, \underset{\sim}{x})$, we define a corresponding field in Euclidean space $\phi_E(x_4, \underset{\sim}{x})$.

$$\phi_E(x_4, \underset{\sim}{x}) = \phi(ix_4, \underset{\sim}{x}) \tag{3.1}$$

Thus the free, massless equation in Minkowski space

$$\Box\phi(x) = (\partial_0^2 - \vec{\nabla}^2)\,\phi(x) = 0 \tag{3.2}$$

goes over in Euclidean space to[7]

$$\Box_E \phi_E(x) = (\partial_4^2 + \vec{\nabla}^2)\,\phi_E(x) = 0 \tag{3.3}$$

The Green's function for a free massless and spinless field in Minkowski space is

$$G(x,y) = -\frac{1}{4\pi^2}\,\frac{1}{(x-y)^2 - i\varepsilon} \tag{3.4}$$

In Euclidean space this becomes

$$G_E(x,y) = \frac{1}{4\pi^2}\,\frac{1}{(x-y)^2}. \tag{3.5}$$

It is required that this Green's function be given by the ordered vacuum-expectation value of two Euclidean fields.

$$G_E(x,y) = \langle 0|R\phi_E(x)\phi_E(y)|0\rangle$$
$$\equiv \theta(x^2-y^2)\ \langle 0|\ \phi_E(x)\phi_E(y)|0\rangle$$
$$+\ \theta(y^2-x^2)\ \langle 0|\ \phi_E(y)\phi_E(x)|0\rangle \tag{3.6}$$

Clearly the insistence on ordering makes sense only if the Euclidean fields are non-commuting operators.[7] In the subsequent, I shall suppress the subscript E , since the discussion will be confined to Euclidean fields exclusively. Also vectors will be frequently written as

$$x^\mu = (r,\ \alpha^\mu)$$
$$x^2 = r^2,\ \alpha^\mu = x^\mu/r$$

IV. Quantization of Free Massless Field

The form of canonical field commutators can be deduced from the Green's function, (3.5) and (3.6), which satisfies

$$\Box_x \langle 0|R\ \phi(x)\phi(y)|0\rangle = -\delta^4(x-y) \tag{4.1}$$

However since the field itself satisfies $\Box_x\phi(x) = 0$, the nonvanishing result (4.1) can arise only from the effect of $\Box_x$ on the step functions which define the R ordering. The differentiation can be carried out with the help of the identity

$$\Box = \frac{1}{r^3}\left[\left(\frac{d}{d\log r}\right)^2 - (L^2+1)\right]r \tag{4.2}$$

where L^2 is a purely angular derivative operator defined by

$$L^{\mu\nu} = i(x^\mu\partial^\nu - x^\nu\partial^\mu)$$
$$L^2 = \tfrac{1}{2}\,\ell^{\mu\nu}\ell_{\mu\nu} \tag{4.3}$$

Eq.(4.1) may be regained, provided we set

$$[\chi(r,\alpha_1),\ \chi(r,\alpha_2)] = 0 \tag{4.4a}$$
$$[\dot\chi(r,\alpha_1),\ \chi(r,\alpha_2)] = -\delta^3(\alpha_1-\alpha_2) \tag{4.4b}$$
$$[\dot\chi(r,\alpha_1),\ \dot\chi(r,\alpha_2)] = 0 \tag{4.4c}$$

Here the notation is

$$\chi(r,\alpha) = r\ \phi(x)$$
$$\dot\chi(r,\alpha) = \frac{d}{d\log r}\ \chi(r,\alpha)$$

The δ function is defined on the sphere

$$\int d\alpha_1\, f(\alpha_1)\delta^3(\alpha_1-\alpha_2) = f(\alpha_2) \quad .$$

The commutators (4.4) can also be obtained from the action principle. The action is

$$I = \tfrac{1}{2}\int d^4x\ \partial_\mu\phi\partial^\mu\phi = -\tfrac{1}{2}\int d^4x\ \phi\,\Box\,\phi \tag{4.5a}$$

With the help of (4.2), this may be rewritten

$$I = -\tfrac{1}{2}\int_0^\infty \frac{dr}{r}\int d\alpha\ \chi(r,\alpha)\left[\left(\frac{d}{d\log r}\right)^2 - (L^2+1)\right]\chi(r,\alpha)$$

$$= \tfrac{1}{2}\int_{-\infty}^{\infty} d\log r \int d\alpha\ \left[\left(\frac{d}{d\log r}\chi\right)^2 + \chi(L^2+1)\chi\right] \tag{4.5b}$$

Clearly the canonical coordinate should be identified with χ , while the canonical momentum is $\dot{\chi}$. The results (4.4) now follow.

Next I wish to discuss the properties of the field under Hermitian conjugation. We notice that the right-hand side of Eq.(4.4b) is real. This means that $\chi(r,\alpha)$ can not be an Hermitian operator, which is not surprising when it is recalled that the passage from Minkowski space to Euclidean space introduces factors of i . It is postulated that the field $\chi(r,\alpha)$ satisfies the relation

$$\chi^+(r,\alpha) = \chi(\tfrac{1}{r},\alpha) \tag{4.6}$$

which is consistent with (4.4b). Thus another transformation, which is currently much studied - inversion and its relative the conformal transformation - plays a central role in our theory. The <u>Ansatz</u> (4.6) will have far reaching consequences. As will be seen later, the momentum operator P^μ will be the negative of the Hermitian conjugate of the conformal operator K^μ .

Finally an expansion for the fields is postulated in terms of 4-dimensional harmonics, which are complete and orthonormal.[8]

$$\chi(r,\alpha) = \sum_{\ell nm}\{\, g^{(+)}_{\ell nm}(r)\, Y_{\ell nm}(\alpha) + g^{(-)}_{\ell nm}(r)\, Y^*_{\ell nm}(\alpha)\} \tag{4.7}$$

The dependence of the operators $g^{(\pm)}_{\ell nm}(r)$ on r is deduced from $\Box\phi = 0$. This gives an expansion for $g^{(\pm)}_{\ell nm}(r)$ in terms of powers

$$\chi(r,\alpha) = \sum_{\ell nm}\left\{\frac{a^{(+)}_{\ell nm}\, r^{\ell+1}}{\sqrt{2\ell+2}}\, Y_{\ell nm}(\alpha) + \frac{a^{(-)}_{\ell nm}\, r^{-(\ell+1)}}{\sqrt{2\ell+2}}\, Y^*_{\ell nm}(\alpha)\right\} \tag{4.8}$$

The Hermiticity Ansatz (4.6) implies that

$$a^{(+)}_{\ell nm} = [a^{(-)}_{\ell nm}]^{\dagger} \tag{4.9}$$

Let me now determine the algebraic properties of our operators. We shall postulate the fundamental commutators

$$[\, a^{(\pm)}_{\ell nm} \, , \, a^{(\pm)}_{\ell' n' m'} \,] = 0$$

$$[\, a^{(-)}_{\ell nm} \, , \, a^{(+)}_{\ell' n' m'} \,] = \delta_{\ell\ell'}\delta_{nn'}\delta_{mm'} \tag{4.10}$$

It is now easy to verify that the field χ satisfies the commutation relations (4.4). We introduce a vacuum state $|0\rangle$, annihilated by $a^{(-)}_{\ell nm}$.

$$a^{(-)}_{\ell nm} \; |0\rangle = 0 \tag{4.11a}$$

The Hermiticity conditions (4.6) and (4.9) insure the existence of a dual vacuum state $\langle 0|$ annihilated by $a^{(+)}_{\ell nm}$

$$\langle 0| \, a^{(+)}_{\ell nm} = 0 \tag{4.11b}$$

We are now in a position to check whether the above formalism correctly reproduces the Green's function according to (3.5) and (3.6). From (4.8), (4.10) and (4.11) it follows that

$$\langle 0| \, R \, \phi(x_1)\phi(x_2) \, |0\rangle = \sum_{\ell nm} \frac{1}{2\ell+2} Y^{*}_{\ell nm}(\alpha_1) \, Y_{\ell nm}(\alpha_2) \; \times$$

$$\left\{ \frac{\theta(r_1-r_2)}{r_1^2} \left(\frac{r_2}{r_1}\right)^{\ell} + \frac{\theta(r_2-r_1)}{r_2^2} \left(\frac{r_1}{r_2}\right)^{\ell} \right\} \tag{4.12}$$

With the help of the addition formula for 4-dimensional spherical harmonics,[8] (4.12) is recognised to be $\frac{1}{4\pi^2} \, \frac{1}{(x_1-x_2)^2}$. This shows that we have succeeded in giving a consistent quantization scheme, and have provided a new and intuitively attractive operator basis for Euclidean field theory.

The expansion (4.12) can be used to give a spectral representation for the propagator. By use of the identity

$$\frac{1}{\pi} \int_{\infty}^{\infty} d\gamma \, \frac{z^{i\gamma}}{\gamma^2+(\ell+1)^2} = \frac{z^{\ell+1}\theta(1-z) + z^{-(\ell+1)}\theta(z-1)}{\ell+1}$$

$$z > 0, \; \ell > -1 \tag{4.13}$$

Eq.(4.12) can be cast in the form

$$G(x_1,x_2) = \frac{1}{4\pi^2}\frac{1}{(x_1-x_2)^2} = \frac{1}{2\pi}\int_{-\infty}^{\infty} d\gamma \sum_{\ell nm} \frac{F^*_{\gamma\ell nm}(x_1)\, F_{\gamma\ell nm}(x_2)}{\gamma^2 + (\ell+1)^2}$$

where the functions

$$F_{\gamma\ell nm}(x) = r^{i\gamma-1}\, Y_{\ell nm}(\alpha) \tag{4.14}$$

define the transformation between the coordinate space and the conjugate $(\gamma\ell nm)$ space in which the dimensionality γ and the angular momentum variables (ℓnm) are diagonalized. In this new space, particularly convenient for dilatation invariant problems, the propagator is diagonal. It is

$$D(\gamma,\ell,n,m) = \frac{1/2\pi}{\gamma^2 + (\ell+1)^2} \tag{4.15}$$

This 4 dimensional $(\gamma\ell nm)$ space plays the same role in the present context as 4 dimensional momentum space in the usual quantization schemes.

Although the above has been derived for the free massless case, the introduction of masses poses no special difficulty. The main difference is that the expansion functions are no longer simple powers; rather they are Bessel functions.[9] How to deal with interactions will be explained below.

Conserved Quantities and Generators of Motion

As always, symmetries of the action lead to currents J^μ, which are conserved according to Noether's theorem.

$$\partial_\mu J^\mu = \frac{\partial}{\partial t} J^0 + \vec{\nabla}\cdot\vec{J} = 0 \tag{5.1}$$

In order to obtain the "charge" associated with this symmetry, the current is integrated over the quantization surface.

$$Q = \int d\tau^\mu\, J_\mu(x) \tag{5.2}$$

where $d\tau^\mu$ is the infinitesimal surface element. For us, the surface is given by $x^2 = \tau^2$, and

$$d\tau^\mu = d^4x\, \delta(x^2-\tau^2)\, 2x^\mu \tag{5.3}$$

(The normal to the surface is $2x^\mu$). Hence the charge is

$$Q = \int d^4x\, \delta(x^2-\tau^2)\, 2x_\mu\, J^\mu(x) \tag{5.4a}$$

Of course, as a consequence of current conservation, Q does not depend on τ , which may be set equal to 1. Thus (5.4a) may be cast into a purely angular integral

$$Q = \int d\alpha \; \alpha_\mu J^\mu(1,\alpha) \tag{5.4b}$$

An especially important object in our theory is the dilatation generator

$$D = i\int d\tau^\mu \; x^\nu \theta_{\mu\nu} \tag{5.5a}$$

where $x_\nu \theta^{\mu\nu}(x)$ is the dilatation current, expressed in terms of the new improved energy momentum tensor.[10] For the free massless theory governed by the action (4.5) this current is conserved. Upon expressing $\theta^{\mu\nu}$ in terms of the field ϕ , expanding ϕ in creation and annihilation operators as in (4.8), and performing the integral indicated in (5.5a), we get

$$D = -i \sum_{\ell n m} (\ell+1) \; a^{(+)}_{\ell n m} \, a^{(-)}_{\ell n m} \tag{5.5b}$$

Thus the dilatation generator is diagonal and anti-Hermitian. Frequently it will be convenient to define a Hermitian generator.

$$\Delta = i\, D \tag{5.6}$$

(If the theory is not scale invariant, D will no longer be independent of the surface over which $x_\nu \theta^{\mu\nu}$ is integrated; i.e. it will depend on τ . This in no way changes the fact that $D(\tau)$ describes the evolution of the system. The analogous situation in the ordinary theory occurs when time-dependent interactions are present. The Hamiltonian is then time-dependent, yet it still governs evolution in time.)

The remaining generators of the conformal group can also be constructed

$$M^{\mu\nu} = i\int d\tau_\alpha (x^\mu \theta^{\nu\alpha} - x^\nu \theta^{\mu\alpha}) \tag{5.7a}$$

$$P^\mu = i\int d\tau_\nu \theta^{\mu\nu} \tag{5.7b}$$

$$K^\mu = i\int d\tau_\nu (2x^\mu x_\alpha - g^\mu_\alpha x^2)\theta^{\nu\alpha} \tag{5.7c}$$

Their expression in terms of creation and annihilation operators is quite complicated; I shall not exhibit it here. However, the Hermiticity properties can be easily deduced. We find

$$(M^{\mu\nu})^\dagger = M^{\mu\nu} \tag{5.8a}$$

$$(P^\mu)^\dagger = -K^\mu \tag{5.8b}$$

$$(K^\mu)^\dagger = -P^\mu \tag{5.8c}$$

As promised, conformal transformations are as important in our theory as translations, since the two are related by Hermitian conjugation. The geometrical meaning of (5.8b) and

(5.8c) is clear, once it is remembered that Hermitian conjugation is just coordinate inversion, see (4.6): a translation in the inverted coordinate system is equivalent to a conformal transformation.

VI. States of the System

A "single particle" state is defined by

$$|\ell nm\rangle = a^{(+)}_{\ell nm}\,|0\rangle \tag{6.1}$$

Naturally, this state does not coincide with the physical one particle state defined by Minkowski space quantization. Observe that $|\ell nm\rangle$ is an eigenstate of D with eigenvalue $-i(\ell+1)$. Furthermore, it is $(\ell+1)^2$-fold degenerate: for fixed value of ℓ, there are $(\ell+1)^2$ different values of n and m.[8]

The singularity structure of the dilatation propagator, (4.15), may now be understood. Observe that the propagator has poles at $\gamma = \pm i(\ell+1)$. This is just the dilatation eigenvalue of the single particle state. Thus the propagator in our conjugate $(\gamma\ell nm)$ space has poles at the eigenvalues of the evolution operator. This is in complete analogy with the conventional situation in momentum space, where the propagator has poles at $p^0 = \pm\sqrt{\underset{\sim}{p}^2+m^2}$, i.e. at the eigenvalues of the Hamiltonian.

In the usual theory, interactions modify the position of the pole in the propagator - a phenomenon known as mass renormalization. We expect that in the present theory a similar shift should occur, once interactions are taken into account. Clearly this is the origin of anomalous dimensions in our formalism. Thus the intriguing possibility presents itself of computing anomalous dimensions as the poles of the complete dilatation propagator.

VII. Perturbation Theory and the S Matrix

The above considerations give a complete solution to the problem of quantizing a free theory. Now we must describe how interactions should be included. Suppose the equation of motion for the interacting field Φ is

$$\Box\Phi = \eta \tag{7.1}$$

where η is the source. Free field canonical commutators (4.4) may still be imposed on $\Phi(x)$ and $\frac{d}{d\log r}[r\Phi(x)]$, at equal r.

To solve the equations of motion (7.1) we define the operator $U(r,r_0)$ by the equations

$$\frac{d}{d\log r}U(r,r_0) = -iD^{int}(r)\,U(r,r_0)$$

$$U(r_0,r_0) = 1 \tag{7.2}$$

where $D^{int}(r)$ is the interaction part of the dilatation generator, constructed from the free fields ϕ. In a familiar fashion, one proves that $U(r,r_o)\,\Phi(x)\,U^{-1}(r,r_o)$ is a free field. Since we have solved the free-field problem, all that remains is to integrate (7.2).

Integration of (7.2) gives

$$U(r,r_o) = R \exp -i\int_{r_o}^{r} \frac{dr'}{r'} D^{int}(r') \tag{7.3}$$

where the R ordering is along the spheres. Recalling the definition of D^{int}, we see that

$$\begin{aligned} -i\int_{r_o}^{r} \frac{dr'}{r'} D^{int}(r') &= -i\int_{r_o}^{r} \frac{dr'}{r'} \int d\tau^\mu\, ix^\nu \theta^{int}_{\mu\nu} \\ &= -i\int_{r_o}^{r} \frac{dr'}{r'} \int d^4x\, \delta(x^2-r'^2)2x^\mu(ix^\nu\theta^{int}_{\mu\nu}(x)) \\ &= -\int_{r_o}^{r} \frac{dr'}{r'} \int d^4x\, \delta(x^2-r'^2)2r'^2\, L^{int}(x) \\ &= -\int_{r_o}^{r} d^4x\, L^{int}(x) \end{aligned}$$

$$U(r,r_o) = R \exp -\int_{r_o}^{r} d^4x\, L^{int}(x) \tag{7.4}$$

The integration is over all space bounded by the spheres $r^2>x^2>r_o^2$.

We have regained the Schwinger-Tomonaga result that the evolution of the system from $x^2=r_o^2$ to $x^2=r^2$ is governed by the ordered exponential of the integral of the interaction Lagrangian. Clearly the S matrix is $U(\infty,0)$.

$$\begin{aligned} S &= R \exp -\int d^4x\, L^{int}(x) \\ &= \sum_{n=0}^{\infty} \frac{(-1)^n}{n!} \int d^4x_1 \ldots d^4x_n\, R\, L^{int}(x_1) \ldots L^{int}(x_n) \end{aligned} \tag{7.5}$$

Applying Wick's theorem to this, we regain the familiar Feynman-Dyson perturbation theory. Thus we see that our quantization procedure is entirely successful and consistent with the conventional approach.

The evaluation of the perturbation series may be carried out in position space, or in our conjugate $(\gamma\ell nm)$ space. Calculations in the $(\gamma\ell nm)$ space are especially convenient for conformally invariant interactions. We have not yet researched questions of divergences and renormalization in this new conjugate space.

I. Counting States, Asymptotic Degeneracy

It has already been observed that the single particle state $|\ell nm\rangle$ is an eigenstate of $\Delta = iD$, with eigenvalue $\ell+1$ and degeneracy $(\ell+1)^2$. Consider now the N particle state

$$|N\rangle = a^{(+)}_{\ell_1 n_1 m_1} \dots a^{(+)}_{\ell_N n_N m_N} |0\rangle \tag{8.1}$$

This is an eigenstate of Δ with eigenvalue d_N, and degeneracy $g(N)$ ($d_1 = \ell+1$, $g(1) = (\ell+1)^2$). We wish to compute $g(N)$ for large N in terms of d_N.

The computation is facilitated by introducing a generating function

$$F(\beta) = \mathrm{Tr}\ \exp -\beta\Delta \tag{8.2}$$

From its definition, two expressions for $F(\beta)$ can be given. Clearly from (8.2) it follows that

$$F(\beta) = \sum_N g(N)\ \exp -\beta d_N \tag{8.3}$$

An alternate expression is got for $F(\beta)$ by using the formulas (5.5) and (5.6) for Δ.

$$\begin{aligned} F(\beta) &= \mathrm{Tr}\ \exp -\beta \sum_{\ell nm} (\ell+1)\ a^{(+)}_{\ell nm}\, a^{(-)}_{\ell nm} \\ &= \mathrm{Tr} \prod_{\ell nm} \exp -\beta(\ell+1)\ a^{(+)}_{\ell nm}\, a^{(-)}_{\ell nm} \\ &= \prod_{\ell nm} \mathrm{Tr}\ \exp -\beta(\ell+1)\ a^{(+)}_{\ell nm}\, a^{(-)}_{\ell nm} \end{aligned} \tag{8.4a}$$

Recalling that for each mode $\mathrm{Tr}\ x^{a^+a} = (1-x)^{-1}$, and that each mode is $(\ell+1)^2$-fold degenerate, it follows from (8.4a) that

$$\begin{aligned} F(\beta) &= \prod_\ell \{1 - e^{-\beta(\ell+1)}\}^{-(\ell+1)^2} \\ \log F(\beta) &= -\sum_{\ell=0}^{\infty} (\ell+1)^2 \log [1 - e^{-\beta(\ell+1)}] \\ &= \sum_{n=1}^{\infty} \frac{1}{n} \frac{e^{\beta n/2} + e^{-\beta n/2}}{[e^{\beta n/2} - e^{-\beta n/2}]^3} \end{aligned} \tag{8.4b}$$

Our task is therefore to obtain a relationship between $g(N)$ and d_N for large N, which must hold if (8.3) and (8.4) are simultaneously true. This can be done by an elaborate analytic procedure based on the classic work of Hardy and Ramanujan.[11] However, it is more instructive, to obtain the answer by a thermodynamical analogy.

Observe that $F(\beta)$ defined by (8.2) is a partition function where β plays the role of inverse temperature, $\beta \to 1/T$, and Δ is the analog of the energy operator in the usual theory.

This is as it should be -- we have repeatedly emphasized that the dilatation generator replaces the Hamiltonian in our method. We may use this analogy to develop a thermodynamics based on the partition function $F(\beta)$. This thermodynamics is manifestly covariant. With the thermodynamical technique an expression for $g(N)$ is easily obtained.

We define the "free energy" by

$$A = -\frac{1}{\beta}\log F(\beta) \tag{8.5}$$

A is related to the "entropy" S and to the "internal energy" U by the well known equations

$$S = \beta^2 \frac{\partial}{\partial\beta} A \tag{8.6}$$

$$U = A + \frac{1}{\beta} S = -\frac{\partial}{\partial\beta}\log F(\beta) \tag{8.7}$$

Since we seek $g(N)$ only for large N, we are effectively in the high temperature limit $T\to\infty$ or $\beta\to 0$. In this region the "entropy" should be identified with $\log g(N)$ and the "internal energy" with the dilatation eigenvalue d_N. Furthermore the asymptotic form for $F(\beta)$ as $\beta\to 0$ is easy to obtain from (8.4b)

$$\log F(\beta) \to \frac{\pi^4}{45\beta^3} \tag{8.8}$$

Hence we get from (8.5), (8.6), (8.7) and (8.8)

$$A \to -\frac{\pi^4}{45\beta^4} \tag{8.9}$$

$$S \to \frac{4\pi^4}{45\beta^3} \tag{8.10}$$

$$U \to \frac{\pi^4}{15\beta^4} \tag{8.11}$$

Eliminating the temperature $1/\beta$ between (8.10) and (8.11) gives a relationship between the entropy and the internal energy, valid in the high temperature limit.

$$S = c\,U^{\frac{3}{4}}$$

$$c = \frac{4\pi}{3(15)^{\frac{1}{4}}} \tag{8.12}$$

With our interpretation for S and U, we finally derive

$$g(N) \underset{N\to\infty}{\to} \exp c\, d_N^{\frac{3}{4}} \tag{8.13}$$

The same answer emerges, after much effort, when the computation is performed analytically,[4] which verifies the consistency and correctness of our thermodynamics.

The degeneracy, seen to be of the exponential variety, is reminiscent of the Hagedorn expression. To make contact with the latter formula, let us first recompute the degeneracy in δ dimensions, generalizing the 4-dimensional derivation given above. The answer, simply obtained by the same method, is

$$g_\delta(N) \underset{N\to\infty}{\to} \exp c_\delta \, d_N^{1-1/\delta}$$

$$c_\delta = \frac{\delta}{\delta-1} \left[2(\delta-1) \sum_{n=1}^{\infty} \frac{1}{n^\delta}\right]^{1/\delta} \tag{8.14}$$

In particular for 2-dimensions

$$g_2(N) \underset{N\to\infty}{\to} \exp \frac{2\pi}{\sqrt{3}} d_N^{\frac{1}{2}} \tag{8.15}$$

Let us now postulate, that for reasons that are by no means apparent, there exists a "Law of Nature" relating physical mass to scale- dimensionality in this 2-dimensional space.[12]

$$d_N = \frac{3\kappa}{4\pi^2} m^2 \tag{8.16}$$

With this postulate, we find that the degeneracy of levels is exp κm, which is just the Hagedorn formula.

Since the derivation of the Hagedorn degeneracy is an open challenge to theoreticians, it is useful to elaborate on the possible circumstances that would lead to the new "Law of Nature" (8.16). Imagine that the correct, as yet unknown, physical theory is a field theory defined on some 2-dimensional, Euclidean parameter space. This parameter space is not the physical 4-dimensional space-time, though perhaps it is related to it. On the 2-dimensional space, dilatations can be defined; they are generated by D . In addition, since this hypothetical theory is to describe the 4-dimensional physical world, there must exist a momentum operator, generating physical 4-dimensional translations. Associated with the momentum operator will be the physical mass operator M^2. The postulate (8.16) is then equivalent to the statement that the expression for M^2 in terms of the fundamental fields is proportional to the formula for D .

The observation that 2-dimensional models are in some way relevant to particle theory of the 4-dimensional physical world seems to be a very important though not understood feature of Nature. Historically the relevance of a 2-dimensional transverse space was noted first in the analysis of multi-particle production in the "pionization" region.[13] A consistent, satisfactory description of single particle production in this region is obtained by cutting off large, transverse momenta and using simply the longitudinal, 2-dimensional phase space factor dE/E. More recently, the success of the parton model in interpreting many of the important features of high energy reactions has called attention to 2-dimensional field theories as a means of describing the parton distribution functions.[12] Indeed it was in this context that the relation (8.16) was first suggested. Finally, recent investigations of dual resonance models make essential use of the general formalism of 2-dimensional field theories.[14]

This collection of facts leads me to exhibit further applications of our techniques to 2-dimensional field theories.

IX. Two-Dimensional Field Theories, The Virasoro Algebra

Field theories in 2 dimensions have been first introduced as an interesting theoretical laboratory where explicitly solvable models can be studied. More recently, however, the interest in such models has become much less academic. We apply our formalism to 2-dimensional field theories and demonstrate that our Lorentz-invariant quantization rules in Euclidean space lead directly to the operator structure present in dual models.

It has been shown that the higher moments of the energy-momentum tensor, in a large class of 2-dimensional field theories, obey essentially the Virasoro algebra.[14,15] However, with a conventional approach, some difficulties persist in the interpretation of various integrals. With our formalism, we can obtain a clear, unambiguous derivation of the Virasoro algebra, without the difficulties previously encountered.

Another point of contact between the present approach and dual models is seen in the structure of the R ordered product appearing in our Euclidean perturbation theory, Eq. (7.5). This is identical to that of the operator form for the integrals in the n-point Veneziano formulas.

I shall begin by considering a theory in δ dimensions with a traceless, symmetric energy-momentum tensor (for the moment the metric is immaterial) and I construct the current

$$J^{\mu}_{f}(x) = \theta^{\mu\nu}(x)\ f_{\nu}(x) \tag{9.1}$$

where $f^{\mu}(x)$ is a function of x. Special cases of (9.1) are well known. These are the currents of the conformal group

$$\begin{aligned} &\text{Translation current} : f^{\mu}(x) = a^{\mu} \\ &\text{Lorentz current} : f^{\mu}(x) = \omega^{\mu\nu} x_{\nu}, \omega^{\mu\nu} = -\omega^{\nu\mu} \\ &\text{Dilatation current} : f^{\mu}(x) = a x^{\mu} \\ &\text{Conformal current} : f^{\mu}(x) = 2x^{\mu} a_{\alpha} x^{\alpha} - a^{\mu} x^2 \end{aligned} \tag{9.2}$$

Of course all these currents are conserved, when the energy-momentum tensor is traceless, symmetric and conserved. Are there any other forms for f^{μ}, such that $\partial_{\mu} J^{\mu}_{f}(x) = 0$? To answer this question, we differentiate (9.1)

$$\begin{aligned} \partial_{\mu} J^{\mu}_{f} &= \theta^{\mu\nu} \partial_{\nu} f_{\mu} \\ &= \tfrac{1}{2}\theta^{\mu\nu}\ [\partial_{\mu} f_{\nu} + \partial_{\nu} f_{\mu} - \tfrac{2}{\delta}\, g_{\mu\nu} \partial_{\alpha} f^{\alpha}] \end{aligned}$$

Clearly the current will be conserved when

$$\partial_\mu f_\nu(x) + \partial_\nu f_\mu(x) - \frac{2}{\delta} g_{\mu\nu} \partial_\alpha f^\alpha(x) = 0 \qquad (9.3)$$

Eq.(9.3) for $\delta>2$ is solved only by the forms (9.2).[16] However for $\delta=2$ a wider class of solutions exists. Thus if we wish to have the large symmetry which leads to the conserved currents J^μ_f, with f^μ satisfying (9.3), we are again led to 2-dimensional theories. Since, as will emerge presently, this symmetry is related to the Virasoro algebra,[14] I shall now confine myself to 2-dimensional models.

Examples of field theories in 2 dimensions with a traceless, symmetric and conserved energy-momentum tensor are the following:

I. Free Boson Field

$$L = \tfrac{1}{2}\partial_\mu \phi\, \partial^\mu \phi$$

$$\theta^{\mu\nu} = \partial^\mu \phi\, \partial^\nu \phi - \tfrac{1}{2} g^{\mu\nu} \partial^\alpha \phi\, \partial_\alpha \phi \qquad (9.4)$$

II. Free Fermion Field

$$L = \frac{i}{2} \bar\psi \gamma^\mu \overleftrightarrow{\partial}_\mu \psi$$

$$\theta^{\mu\nu} = \frac{i}{4} \{\bar\psi \gamma^\mu \overleftrightarrow{\partial}^\nu \psi + \bar\psi \gamma^\nu \overleftrightarrow{\partial}^\mu \psi\} \qquad (9.5)$$

III. Gradient Coupling Model

$$L = \tfrac{1}{2}\partial_\mu \phi\, \partial^\mu \phi + \frac{i}{2}\bar\psi\gamma^\mu \partial_\mu \psi - \lambda\bar\psi\gamma^\mu\psi\partial_\mu\phi$$

$$\theta^{\mu\nu} = \partial^\mu\phi\partial^\nu\phi + \frac{i}{4}\{\bar\psi\gamma^\mu\overleftrightarrow{\partial}^\nu\psi + \bar\psi\gamma^\nu\overleftrightarrow{\partial}^\mu\psi\} - \frac{\lambda}{2}\{\bar\psi\gamma^\mu\psi\partial^\nu\phi + \bar\psi\gamma^\nu\psi\partial^\mu\phi\} - \tfrac{1}{2}g^{\mu\nu}\partial_\alpha\phi\partial^\alpha\phi \qquad (9.6)$$

IV. Thirring Model

$$L = \frac{i}{2}\bar\psi\gamma^\mu \partial_\mu\psi - \lambda\bar\psi\gamma^\mu\psi\, \bar\psi\gamma_\mu\psi$$

$$\theta^{\mu\nu} = \frac{i}{4}\{\bar\psi\gamma^\mu\overleftrightarrow{\partial}^\nu\psi + \bar\psi\gamma^\nu\overleftrightarrow{\partial}^\mu\psi\} - \lambda g^{\mu\nu}\bar\psi\gamma^\alpha\psi\, \bar\psi\gamma_\alpha\psi \qquad (9.7)$$

V. The "Covariant String"

$$L = \{\tfrac{1}{2}[\varepsilon_{\mu\nu}\partial^\mu\phi^A\partial^\nu\phi^B]\,[\varepsilon_{\alpha\beta}\partial^\alpha\phi^A\partial^\beta\phi^B]\}^{\frac{1}{2}}$$

$$= \{\det \partial_\mu\phi^A\, \partial_\nu\phi^A\}^{\frac{1}{2}} \qquad (9.8)$$

Here $\varepsilon^{\mu\nu}$ is the totally antisymmetric tensor, $\varepsilon^{01}=1$. The indices A,B describe some additional degree of freedom and are summed over. (In applications to the dual model, A,B refer to physical space-time[17] - one may of course allow models I-IV to possess such further degrees of freedom also.) In this model the energy-momentum tensor vanishes identically.

For all the above theories the currents J_f^μ will be conserved, provided f^μ satisfies (9.3). I shall discuss first the situation in Minkowski space, and shall demonstrate that a consistent algebraic description of this symmetry is not possible. In Euclidean space, when the theory is quantized according to our method, no difficulties arise.

To find the solutions to (9.3) take the ++ component and the -- component of that equation. (The +- component vanishes identically.) In Minkowski space we thus find

$$\partial_+ f_+ = 0, \quad \partial_- f_- = 0 \tag{9.9a}$$

The integrals clearly are (recall that $f_\pm = f^\mp$)

$$f^+ = f^+(x^+)$$
$$f^- = f^-(x^-) \tag{9.9b}$$

The reason for the existence of all these conserved currents is that 2-dimensional theories with a traceless, symmetric and conserved energy-momentum tensor are invariant under the coordinate transformation

$$\delta_f x^\mu = f^\mu(x) \tag{9.10a}$$

where f^μ is of the form (9.9) and the field is transformed as follows

$$\delta_f \phi(x) = f^\mu(x)\partial_\mu\phi(x) + \frac{\partial_\mu f_\alpha(x)}{2}[g^{\alpha\mu}d - \varepsilon^{\alpha\mu}\Sigma]\phi(x) \tag{9.10b}$$

Here d is the scale dimension of the field ϕ and $\varepsilon^{\alpha\mu}\Sigma$ is the spin matrix associated with ϕ. (In 2 dimensions we may set $\Sigma^{\alpha\mu}=\varepsilon^{\alpha\mu}\Sigma$.) The symmetry can be seen directly in models I-IV above. For the string model, example V above, the currents J_f^μ vanish identically since $\theta^{\mu\nu}$ is zero. The model is invariant under the transformation (9.10), even if f^μ does not satisfy (9.9). This is a gauge symmetry which does not lead to any conserved currents.[17] In the following we consider only theories with non-vanishing $\theta^{\mu\nu}$.

The composition law for the transformation (9.10) is

$$\delta_g\delta_f x^\mu = f^\mu(x) + g^\mu(x) + f^\nu(x)\partial_\nu g^\mu(x)$$

$$\delta_f\delta_g x^\mu = g^\mu(x) + f^\mu(x) + g^\nu(x)\partial_\nu f^\mu(x) \tag{9.11a}$$

It is easy to see that if f^μ and g^μ satisfy (9.9), so does

$$h^\mu = g^\nu\partial_\nu f^\mu - f^\nu\partial_\nu g^\mu \tag{9.11b}$$

and the transformations form a group. By choosing

$$f : \begin{array}{l} f^+ = 0 \\ f^- = f(x^-) \end{array} \tag{9.12a}$$

$$f : \begin{array}{l} f^+ = f(x^+) \\ f^- = 0 \end{array} \tag{9.12b}$$

we recognize the fact that one is dealing with a direct product of two groups of transformation, and it suffices to consider only one of the factors. Henceforth, we make the choice (9.12a).

The charges can be constructed as follows.

$$\begin{aligned} Q_f &= \int_{-\infty}^{\infty} dx^1\ \theta^0{}_\mu(x) f^\mu(x) \\ &= \int_{-\infty}^{\infty} dx^1\ \theta^{0+}(x)\ f(x^-) \\ &= \int_{-\infty}^{\infty} \frac{dx^1}{\sqrt{2}}\ [\theta^{++}(x) + \theta^{-+}(x)]\ f(x^-) \end{aligned} \tag{9.13a}$$

$\theta^{+-}(x)$ vanishes since the stress tensor is traceless. Also from the conservation of the stress tensor, $\partial_\mu \theta^{\mu+}(x) = 0 = \partial_+ \theta^{++}(x)$, we learn that θ^{++} depends only on x^-. Hence the charge is

$$Q_f = \int_{-\infty}^{\infty} dx^-\ \theta^{++}(x)\ f(x^-) \tag{9.13b}$$

From the composition law (9.11) one would expect that the charges satisfy the commutation relations

$$i[Q_f, Q_g] = Q_h$$

$$h = f'g - g'f \tag{9.14}$$

In order to verify (9.14), consider

$$\int_{-\infty}^{\infty} dx^- dy^- f(x^-) g(y^-) [\theta^{++}(x), \theta^{++}(y)] = \int_{-\infty}^{\infty} dx^- dy^- f(x^-) g(y^-) [\theta^{++}(x), \theta^{++}(y)]\Big|_{x^+=y^+} \tag{9.15}$$

where the equality is true simply because θ^{++} does not depend on x^+. The form of the local commutator of two energy momentum tensors which will insure the validity of (9.15) is

$$[\theta^{++}(x), \theta^{++}(y)]\big|_{x^+=y^+} = i[\theta^{++}(x) + \theta^{++}(y)]\ \partial_- \delta(x^- - y^-) \tag{9.16}$$

and canonical evaluation in simple models yields this result.

However from general principles one can show that the commutator of the stress tensor with itself cannot be of the form given in (9.16). That the term proportional to the first derivative of the delta function is as indicated insures Poincaré invariance of the theory; indeed (9.16) is the analog of the Dirac-Schwinger commutator for the present problem. In addition, positivity and Lorentz covariance insure that a triple derivative is necessarily present. This is the analog of the usual Schwinger term in current commutators, and the failure of canonical manipulations to expose it is of course familiar.[18] The correct expression for the commutator which replaces (9.16) is

$$[\theta^{++}(x), \theta^{++}(y)]|_{x^+=y^+} = i[\theta^{++}(x) + \theta^{++}(y)]\partial_-\delta(x^- - y^-) - \tfrac{ia}{2}\partial_-^3\delta(x^- - y^-) \tag{9.17}$$

$$a > 0$$

The commutator of the charges therefore is

$$i[Q_f, Q_g] = Q_h - \tfrac{a}{2}\int_{-\infty}^{\infty} dx^- f'''(x^-)g(x^-) \tag{9.18}$$

and for arbitrary f and g the charges do not satisfy the "classical" algebra.

The additional c number poses also the problem of convergence of the integral. If f(x) or g(x) are powers of x, as they are for the Virasoro algebra $\int_{-\infty}^{\infty} dx^- g(x^-)f'''(x^-)$ does not converge. Furthermore, one may question the convergence of the formula (9.13) for the charges, when f is a positive or negative power. Thus we conclude that the symmetry (9.10) and the associated algebra cannot be given an operator basis in a field theory defined on 2-dimensional Minkowski space.

The difficulties which have been encountered in giving a coherent field theoretic basis for the Virasoro algebra in a Minkowski space field theory do not exist if the field theory is quantized by our method, in Euclidean space. I shall now show this. The solution of Eq. (9.3) for f^μ in Euclidean space is

$$\begin{aligned} f_1(x) + i\, f_2(x) &= f(re^{i\theta}) \\ f_1(x) - i\, f_2(x) &= g(re^{-i\theta}) \\ x_1 = r\cos\theta, \quad x_2 &= r\sin\theta. \end{aligned} \tag{9.19}$$

Here f and g are arbitrary functions of the argument. As before we may set g to zero, and confine our attention to a non-vanishing f. The charge is given by an angular integral, (see (5.4))

$$\begin{aligned} Q_f &= i\int_0^{2\pi} d\theta\, x_\mu J_f^\mu(x) \\ &= i\int_0^{2\pi} d\theta\, x_\mu \theta^{\mu\nu}(x)\, f_\nu(x) \\ &= i\int_0^{2\pi} d\theta\, re^{i\theta}\, f(re^{i\theta})\Theta(x) \end{aligned} \tag{9.20}$$

where

$$\Theta(x) \equiv \tfrac{1}{2}[\theta^{11}(x) - i\theta^{12}(x)] \tag{9.21}$$

It is easy to show that Θ is only a function of $x^1 + ix^2 = re^{i\theta}$, as a consequence of conservation tracelessness and symmetry of $\theta^{\mu\nu}$. Hence Q_f is also given by

$$Q_f = \int_{\mathcal{C}} dz\, f(z)\Theta(z) \tag{9.22}$$

and the integration contour is over the circle of radius r. Clearly Q_f is independent of r, if $f(z)\Theta(z)$ is analytic for $0<|z|<\infty$.

To establish the algebraic properties of the charges, in our Euclidean space theory, we compute the $[Q_f, Q_g]$ commutator. Again what is needed is the $[\Theta, \Theta]$ commutator at equal r. The Euclidean space analog of (9.17) is

$$i[\Theta(x), \Theta(x')]\big|_{r=r'} = \frac{1}{r^2}\{e^{-2i\theta}\Theta(x) + e^{-2i\theta'}\Theta(x')\}\partial_\theta\delta(\theta-\theta')$$

$$+ \frac{a}{2r^4}\, e^{-2i(\theta+\theta')}\{\partial_\theta\delta(\theta-\theta') + \partial^3_\theta\delta(\theta-\theta')\}$$

$$a > 0 \tag{9.23}$$

Consequently we find

$$i[Q_f, Q_g] = Q_h - \frac{ia}{2}\int_0^{2\pi} d\theta\, re^{i\theta} g(re^{i\theta})\, f'''(re^{i\theta})$$

$$= Q_h - \frac{a}{2}\int_{\mathcal{C}} dz\, g(z)\, f'''(z) \tag{9.24}$$

When f and g are analytic for $0<|z|<\infty$, the integral is independent of contour. In contrast to the Minkowski space formula (9.18), the present result (9.24) is free from divergences. Also the expression for the charge, (9.20) or (9.22), involves a finite range integral and no question about convergence need be raised. Thus the Euclidean quantization provides a consistent algebraic description of the large symmetry present in many 2-dimensional field theories. The only anomaly is the c number addition proportional to a .

The discrete Virasoro algebra is obtained by setting

$$f(z) = iz^{1-n}$$

$$g(z) = iz^{1-m} \tag{9.25}$$

Then from (9.24) one gets

$$[Q_n, Q_m] = (m-n)Q_{n+m} - \pi a\delta_{n+m,o}\,(n^3-n) \tag{9.26}$$

Notice that the commutator anomaly is absent for n or m = 0, $\pm$ 1. These cases correspond to a constant, linear or quadratic function f; i.e. they give rise to the usual conformal group, Eq. (9.2). Thus the ordinary conformal algebra is realized in a conventional way, both for the Minkowski-space theory and for our Euclidean-space theory. Only when $|n|>1$ does the additional c number become relevant, and in that instance only the Euclidean method gives non-divergent results. The presence of a non-vanishing, c number anomaly shows that Q_n cannot annihilate the vacuum, for all n, and the symmetry is spontaneously broken.

Solutions of invariant equations, that is of equations arising in theories which possess this large symmetry such as the models I-IV given before, provide representations for the group. Explicit examples of spin 0 and spin ½ fields exist, see (9.4) to (9.8). However, I have not been able to construct a spin 1 theory which leads to an invariant equation. The best I can do is the Lagrangian.

$$L = \tfrac{1}{2}\partial_\mu A^\nu \partial^\mu A_\nu - (\partial_\mu A^\mu)^2$$

The action is invariant under the conformal group, i.e. $n=0, \pm 1$ in Eq.(9.25), but the larger group of transformations (9.10) is not a symmetry of this model.

Conclusion

The aspects of this investigation that we find most interesting are the following. The development of a manifestly covariant quantization procedure which can replace the conventional non-covariant method provides a wealth of covariant canonical commutators. It will be interesting to examine current commutators in this context and to see whether previous successes of equal-time and light-like current algebra can be extended.

We have given an operator basis for Euclidean quantum field theory. The quantization procedure follows closely the method of solving partial differential equations of the Laplacian type in Euclidean space, where the initial value data is specified on a sphere, rather than on a plane. Our approach gives special prominence to the dilatation operator. It makes contact with the explicitly conformally covariant formalism of Johnson and Adler.[19] Furthermore it offers the intriguing possibility of studying anomalies of scale invariance as modifications of the dilatation propagator.

A covariant thermodynamics has put us closer to the goal of deriving the Hagedorn spectrum from first principles. What still is missing is an a priori justification of the use of 2-dimensional field theory and of the identification of mass squared with the dilatation eigenvalue. Nevertheless our approach to field theory in Euclidean space offers an attractive alternative to the conventional field theoretic formulations of the dual resonance model.

References

1) For a summary see V. d'Alfaro, S. Fubini, G. Furlan and C. Rossetti, Currents in Hadron Physics, North Holland, Amsterdam (1973).

2) For a summary see R. Jackiw, Springer Tracts in Modern Physics, 62, 1 (1972).

3) Our notational conventions in Minkowski space are $g^{oo}=-g^{ii}=1$; $x^{\pm}=\frac{1}{\sqrt{2}}(x^o \pm x^3)$

In Euclidean space, the metric is positive and no distinction between upper and lower indices need be made.

4) S. Fubini, A. Hanson and R. Jackiw, Phys.Rev. D7, 1732 (1973). In the initial stages of the investigation, we collaborated with E. del Guidice. Light-cone quantization as well as the present method have been discussed by P.A.M. Dirac, Rev.Mod. Phys. 21, 392 (1949).

5) J. Schwinger, Proc.Nat.Acad.Sci. 44, 956 (1958); K. Symanzik, "A Method for Euclidean Quantum Field Theory", in Mathematical Theory of Elementary Particles, R. Goodman and I. Segal, editors, The MIT Press, Cambridge, Mass. (1966).

6) A related problem is that translations are hard to define. It will be seen that our method diagonalizes D. Since D and P^μ satisfy the commutation relation $i[D,P^\mu] = P^\mu$, either D is not Hermitian or P^μ does not exist. In this connection see W.W. MacDowell and R. Roskies, J.Math.Phys. 13, 1585 (1972).

7) It is at this stage that we part company with previous approaches to Euclidean field theory, ref.5. In these investigations the non-interacting Euclidean field does not satisfy the free equation (3.3). Also, for us the fields are non-commuting operators,while previously they were taken to commute.

8) The properties of the harmonics are as follows.

$$Y_{\ell nm}(\alpha) = Y_{\ell nm}(\theta,\phi,\psi)$$
$$= N_{\ell nm}\, e^{im\phi}\, (\sin\theta)^n\, C^{n+1}_{\ell-n}(\cos\Theta)\, (\sin\psi)^m\, C^{m+\frac{1}{2}}_{n-m}(\cos\psi)$$

The $C^\lambda_\nu(z)$ are Gegenbauer polynomials and $N_{\ell nm}$ is a normalization factor defined by

$$(N_{\ell nm})^{-2} = 2\pi\, E_0(\ell,n)\, E_1(n,m)$$

$$E_k(\ell,n) = \frac{\pi 2^{k-2n}\,\Gamma(\ell+n-k+2)}{(2\ell+2-k)(\ell-n)!\,[\Gamma(n+1-k/2)]^2}$$

The range of the indices is

$$\ell = 0, 1, \ldots, \infty$$
$$n = 0, 1, \ldots, \ell$$
$$m = -n, -n+1, \ldots, n$$

The harmonics are complete

$$\sum_{\ell nm} Y^*_{\ell nm}(\alpha)\, Y_{\ell nm}(\alpha') = \delta^3(\alpha-\alpha')$$
$$= \frac{1}{\sin^2\theta\,\sin\psi}\,\delta(\theta-\theta')\delta(\psi-\psi')\delta(\phi-\phi')$$

and orthogonal

$$\int_0^\pi d\theta\,\sin^2\theta \int_0^\pi d\psi\,\sin\psi \int_0^{2\pi} d\phi\, Y^*_{\ell nm}(\alpha)\, Y_{\ell' n' m'}(\alpha') = \delta_{\ell\ell'}\delta_{nn'}\delta_{mm'}$$

They satisfy the addition theorem

$$\sum_{n=0}^{\ell}\sum_{m=-n}^{n} Y^*_{\ell nm}(\alpha)\, Y_{\ell nm}(\alpha') = \frac{\ell+1}{2\pi^2}\, C^1_\ell(\alpha\cdot\alpha')$$

The harmonic is an eigenfunction of the angular derivative defined in (4.3).

$$L^2\, Y_{\ell nm} = \ell(\ell+2)\, Y_{\ell nm}$$

9) A. di Sessa, MIT Preprint, to be published.

10) C.G. Callan,Jr., S. Coleman and R. Jackiw, Annals of Physics (N.Y.) 59, 42 (1970).

11) G.H. Hardy and S. Ramanujan, Proc.London Math.Society (2) 17, 75 (1918).

12) E. del Guidice, P. di Vecchia, S. Fubini and R. Musto, Nuovo Cimento 12A, 813 (1972).

13) D. Amati, S. Fubini and A. Stanghellini, Nuovo Cimento 26, 896 (1962); S. Fubini, Lectures at the 1963 Scottish Universities Summer School, R.G. Moorhouse, editor, Oliver and Boyd, Edinburgh (1966).

14) For a summary see M.A. Virasoro, Proceedings of the 1971 International Conference on Duality and Symmetry in Hadron Theories, The Weizmann Science Press, Jerusalem (1971).

15) S. Ferrara, R. Gatto and A.F. Grillo, Nuovo Cimento 12A, 959 (1972).

16) L. Bianchi, Lezioni di Geometria Differenziale, p.375, Spoetti, Pisa (1902).

17) Y. Nambu, Lectures at the Copenhagen Summer Symposium (1970); O. Hara, Progr. Theoret.Phys. (Kyoto) 46, 1549 (1971); T. Goto, Progr.Theoret.Phys. (Kyoto) 46, 1560 (1971); L.N. Chang and F. Mansouri, Phys.Rev. D5, 2535 (1972); F. Mansouri and Y. Nambu, Phys. Letters 39B, 375 (1972); P. Goddard, J. Goldstone C. Rebbi and C.B. Thorn , Nucl.Phys. B56, 109 (1973).

18) The analogous anomaly for commutators in 4-dimensions was given by D. Boulware and S. Deser, J.Math.Phys. 8, 1468 (1967).

19) K. Johnson, unpublished; S.L. Adler, Phys.Rev. D6, 3445 (1972).

DISCUSSIONS

CHAIRMAN: Prof. R. Jackiw

Scientific Secretaries: S. Jackson, E. Hendrick, S. Elitzur, B. Cahalan

DISCUSSION No. 1

- *DAMMERER:*

1) You defined a Hermitian conjugate operation χ^+. What scalar product did you use to derive this?

2) Your Hermitian conjugate operator involved the inverse of the radius r, therefore you must have a fixed radius r_0 in your calculation. Does this mean that all physical quantities are independent of this radius?

- *JACKIW:*

1) I did not use an explicit functional form for the scalar product. We defined the Hermitian conjugate in this fashion. Suppose you have $\langle 0|0_1(r_1)0_2(r_2)|0\rangle$ where 0_1 and 0_2 are operators expressible in terms of the basic field operator $\chi(r)$. Then one can ask: What is $\langle 0|0_1(r_1)0_2(r_2)|0\rangle^*$ which, from the definition of Hermitian conjugation, is equal to $\langle 0|0_2^\dagger(r_2)0_1^\dagger(r_1)|0\rangle$? We say that if $0_1(r_1) = \chi(r_1)$ and $0_2(r_2) = \chi(r_2)$, this becomes $\langle 0|\chi(1/r_2)\chi(1/r_1)|0\rangle$. The scalar product is the conventional abstract form used in quantum mechanics.

2) You are worried about dimensional balance in the formula. I have a field $\chi(r)$ which is dimensionless. $\chi(r)$ is expanded schematically (without spherical harmonics) as

$$\sum \left[r^n a_n^{(+)} + r^{-n} a_n^{(-)} \right],$$

and you ask r over what? Dimensional balance is maintained by assigning dimension to a_n itself [i.e. $a_n^{(+)}$ has the dimension r^{-n}] in the sense that $i[D, a_n^{(+)}] = n\, a_n^{(+)}$, and the commutation relation between $b_n^{(+)}$ and $b_n^{(-)}$ would be the same as that between $a^{(+)}$ and $a^{(-)}$, so commutation relations are maintained; the scale just does not matter. I have a scale-invariant theory, and it is a feature of scale-invariant theories that the scale is unimportant.

- *CHU:*

I think I can clarify the question. You express the operator $\chi(r)$ as:

$$\chi(r) = \sum \left[r^n a_n^{(+)} + r^{-n} a_n^{(-)} \right].$$

Since $\chi(r)$ must be a dimensionless operator, the operators $a_n^{(+)}$ must have dimensions of length to the minus n^{th} power. How can you define the dimension of the operator

$$\chi\left(\frac{1}{r}\right) = \sum \left[\left(\frac{1}{r}\right)^n a_n^{(+)} + \left(\frac{1}{r}\right)^{-n} a_n^{(-)} \right].$$

- *JACKIW:*

The field $\chi(r)$ has dimensionality defined in the following sense:

$$i[D,\chi(r)] = r\frac{\partial}{\partial r}\chi + 0(\chi) .$$

I discuss what is $i[D,\chi(1/r)]$, given that

$$i[D,\chi(r)] = r\frac{\partial}{\partial r}\chi .$$

Let us take the Hermitian conjugate. We get

$$-i[\chi^+(r),D^+] = r\frac{\partial}{\partial r}\chi^+(r) ,$$

which gives

$$-i\left[\chi\left(\frac{1}{r}\right),-D\right] = r\frac{\partial}{\partial r}\chi\left(\frac{1}{r}\right)$$

since D is anti-Hermitian. Cancelling minus signs, we obtain

$$i\left[D,\chi\left(\frac{1}{r}\right)\right] = -r\frac{\partial}{\partial r}\chi\left(\frac{1}{r}\right) .$$

If we define a new variable f = 1/r, we get

$$i[D,\chi(f)] = f\frac{\partial}{\partial f}\chi(f) .$$

This does work out consistently with the dimensionality rule if we keep track of the peculiar hermiticity property of D itself, i.e. D is an anti-Hermitian operator.

- *REBBI:*

I have a comment about this answer. The role of the radial unit in the new method of quantization is, I believe, no more important than the choice of the origin of time in the standard method of quantization. To see this even better, think of using log r instead of r as a variable. There a change of the radial unit implies only a translation in the independent variable.

- *JACKIW:*

I agree with that. We are discussing a scale-invariant theory, consequently any scale which you wish to use will disappear from any physical process.

- *DRECHSLER:*

Could you comment on the physical interpretation of the relation $(p^\mu)^+ = -K^\mu$, and connected with this, what is the role ordinary space-time translation plays in your description, which as you noted is sacrificed as a manifest invariance at the expense of having the dilatation generator in the game. To what extent is the theory translation-invariant in the end?

- *JACKIW:*

I sacrificed manifest translation invariance at the expense of keeping manifest Lorentz invariance. Regarding translation invariance at the end: this morning I wrote down the formula for the S-matrix:

$$S = R\exp\int d^4x\, H_{int}(x) .$$

This is manifestly translation invariant, except for the ordering. But when I expand this formula in a series and apply Wick's theorem term by term to the series, then the answers I get are identical to the Feynman-Dyson expansion which by inspection is translation-invariant. Regarding the relation $(p^\mu)^+ = -K^\mu$, we can put everything together with this formula:

$$i[D,p^\mu] = p^\mu .$$

Now take the Hermitian conjugate, we get

$$-i[p^{\mu+},D^+] = (p^\mu)^+ .$$

Since D is anti-Hermitian, this is:

$$i[-D,p^{\mu+}] = (p^\mu)^+ ;$$

which leads to

$$i[D,p^{\mu+}] = -(p^\mu)^+ .$$

But I remind you that the commutation relation of K with D is

$$i[D,K^{\mu}] = -K^{\mu} ,$$

so the commutation relation

$$i[D,p^{\mu +}] = -(p^{\mu})^{+}$$

is consistent with

$$(p^{\mu})^{+} = -K^{\mu} .$$

- *NAHM:*

It seems plausible that you get equivalent answers if you quantize on arbitrary space-like surfaces. This is not clear for quantization on a light-like surface, which is particularly suited for treating Bjorken scaling. So your method of quantization is probably not convenient for investigating scaling.

- *JACKIW:*

The surface χ^2 equals a positive constant (x^2 = const. > 0) is a space-like surface. That one gets the same results when quantizing on different space-like surfaces has been proven by Schwinger and Tomonaga. However, in their proof they needed the technical assumption that the surfaces have to be asymptotically flat. The surfaces $\chi^2 = C > 0$ are not, so we had to prove the equivalence explicitly by deriving the formula for the S-matrix and showing that one gets the same results. That one gets the same results for light-like surfaces has been proven by Bjorken, Kogut and Soper, as well as by Rohrlich and collaborators.

- *LOSECCO:*

Does your theory eliminate the anomalies associated with Bjorken scaling?

- *JACKIW:*

One should not call the phenomena observed in perturbative calculations "anomalies". They are not anomalies, rather they are simply the answers that follow from the theory; namely, the deep inelastic cross-section when it is computed in perturbation theory does not scale. Any specific cross-sections I compute, including that for deep inelastic scattering, if calculated in the new theory, will be the same formula as in the conventional theory and in the high-energy limit will behave in the same way. When one says that light-cone quantization leads to Bjorken scaling, one does not mean that if I compute cross-sections in perturbation theory in a theory quantized on the light cone, I get Bjorken scaling One means, if I assume the commutation relations are true in perturbation theory, then I can prove light-cone scaling. The trouble is that if I consider $i[\pi,q]$ where π is the canonical momentum, canonical formalism says that this commutator equals 1,but when it is computed in perturbation theory it does not equal 1.

- *CAHALAN:*

I have two somewhat related questions. First, your quantization surface is $x^2 = \tau^2$ where τ is a positive constant. As Dirac already mentioned in his 1949 article, it is quite natural to consider the limit $\tau \to 0$, i.e. quantization on the light cone. Light signals propagate on this surface, so one generally expects interactions to affect the commutation relations similar to the eikonal phase due to the tangent line in the light-front formulation. What happens in the present case when $\tau \to 0$? Second, what explicit calculations have been done to support the formal equivalence argument?

- *JACKIW:*

Quantizing directly on the light cone would produce an extremely singular theory, and I do not know how it could be done. It is interesting to note that the present formalism interpolates between the light-front and equal-time formalisms, although it does not quantize at the point $x^2 = 0$. The point $x^+ = 0$ can be chosen to coincide with the surface $x^2 = 0$ and the point $t = \infty$ can be chosen to coincide with the surface $x^2 = \infty$. The point $t = \infty$ is the famous LSZ point where all interactions disappear since one is at large times and interactions do not matter much out there. The point $x^2 = 0$ is the point where one hopes that interactions disappear since short-distance amplitudes are canonical. One hopes that this is what Bjorken scaling is telling us. $x^2 = 0$ and $x^2 = \infty$ are the two asymptotic points of no interactions.

As for equivalence, I showed explicitly that one obtains the usual S-matrix with all propagators in Euclidean space. In the usual formalism one always rotates to Euclidean space anyway, and there is presumably a theorem that this rotation is valid for any Feynman diagram. It is also interesting to investigate how formal arguments work out in perturbation theory. Explicit calculations often involve detailed cancellation of anomalies before one obtains agreement with the formal results.

- *KUPCZYNSKI:*

You obtained the S-matrix which enables you to calculate the transition probabilities between some states in the Fock space of your problem. How do you translate the transition probabilities which you have, into the transition probabilities between physical states?

- *JACKIW:*

We do not wish to calculate the transition probability between dilatation eigenstates, and then knowing this transition probability inquire as to how to compute the scattering amplitude for a particular process, say, photon-electron scattering. I would calculate the Euclidean vacuum expectation value of a product of field operators, Fourier transform it, analytically continue into Minkowski space, and then find the physical transition amplitude as the residue of the poles of the analytically continued vacuum expectation value.

- *DERMAN:*

In connection with Kupczynski's question, are the two vacua in the two different quantization schemes the same?

- *JACKIW:*

I do not know. They both have zero energy, both are annihilated by all the generators etc. In any practical calculation -- if I use my vacuum, calculate the result, and continue to Minkowski space -- one gets the same result in the usual equal time quantized theory with its vacuum. But there is not even a unitary transformation between the two vacua -- they are in different Fock spaces because of the analytic continuation.

- *DERMAN:*

The actual question I wanted to ask was the following. You mentioned the advantages to be gained by quantizing on different surfaces, in terms of abstracting information from the field theory which may be true more generally. Could you be a bit more specific about how and what you abstract additionally if all schemes are really equivalent? It seems as though if you get more information from one extra way of quantizing, you should get even more by quantizing on an arbitrary surface.

- *JACKIW:*

For example, if one quantized at equal times, say in the quark model, one can derive:

$$[J_a^0(x),\, J_b^0(y)]_{x^0=y^0} = i\, f_{abc}\, J_c^0 \tag{1}$$

from the equal time commutation relations. One then hopes that these relations can be abstracted and found to be true in nature even when the model itself is not true. If one quantizes on the light cone, one obtains analogously from the commutation relations:

$$[J_a^+(x),\, J_b^-(y)]_{x^+=y^+} = i\, f_{abc}\, J_c^- + \text{bilocal operators} \;. \tag{2}$$

Now formula (1) alone does not imply formula (2). One needs an infinite class of relations of type (1) derived from equal-time commutation relations to deduce formula (2). It is more economical simply to quantize on the light cone in order to abstract formula (2); since, given a complete dynamical model (it does not matter how it is quantized), one chooses a convenient way of summarizing properties one wishes to investigate.

- *CHU:*

You indicated in your lecture, a formal proof of the equivalence of the S-matrix as computed in your method of quantization, and the S-matrix as computed with the Feynman-Dyson formulation. When you do actual calculations in perturbation theory, does this equivalence survive?

- *JACKIW:*

Yes, the explicit proof is really a proof of the equivalence of the perturbation expansions, up to the validity of the analytic continuation to Euclidean space.

- *CHU:*

How do you know that your analytic continuation is valid? When one calculates a Feynman integral by rotating to Euclidean space, it is easy to see that the rotation is valid by looking at the analytic structure of the integral. But your rotation, in the operator language, to Euclidean space does not seem to be so easily checked.

- *JACKIW:*

We can only assume that there exists a theorem, not yet proven, which allows us to make the rotation. I can say that the procedure has been checked in perturbation theory up to sixth order.

- *CHOUDHURY:*

You pointed out that, even in a world where there is no scale invariance, the dilatation operator still makes sense. As I understand, if the single particle states are eigenstates of the dilatation operator, then the mass spectrum must be either zero or continuous? Will you elaborate on this point?

- *JACKIW:*

In Schroedinger theory the Hamiltonian is the generator of time evolution. If there are time-dependent interactions in this theory, then the Hamiltonian H is not a constant of the motion. In spite of this, it is still the operator generating time translation on the basic coordinates of the problem. Similarly, in the present theory the evolution operator is the dilatation operator. It need not be a constant of the motion. If scale invariance does not hold, one has to go back and see how to do perturbation theory with a time-dependent interaction and carry over the technique to do perturbation theory in this scheme for a non-scale-invariant interaction.

- *HENDRICK:*

What is your motivation for choosing the definition of $\chi^+(r) = \chi(1/r)$?

- *JACKIW:*

This choice of $\chi^+(r)$ gives the consistent results. For example, I have $\langle 0|\chi(r_1)\chi(r_2)|0\rangle$ and I can inquire what the complex conjugate of this object is: $\langle 0|\chi(r_1)\chi(r_2)|0\rangle^*$ which by definition equals $\langle 0|\chi^+(r_2)\chi^+(r_1)|0\rangle$. Then I want to know what this, for example, how do I express $\chi^+(r)$ in terms of my basic set of fields such that all complex conjugation properties, hermiticity relations, and commutation relations, work out correctly. The correct choice is $\chi^+(r) = \chi(1/r)$. Again consider:

$$i[D,\chi(r)] = r\frac{\partial}{\partial r}\chi \ .$$

Take the Hermitian conjugate:

$$-i[\chi^+(r),D^+] = r\frac{\partial}{\partial r}\chi\left(\frac{1}{r}\right) ,$$

since D is anti-Hermitian

$$-i\left[\chi\left(\frac{1}{r}\right),-D\right] = r\frac{\partial}{\partial r}\chi\left(\frac{1}{r}\right)$$

$$i\left[D,\chi\left(\frac{1}{r}\right)\right] = -r\frac{\partial}{\partial r}\chi\left(\frac{1}{r}\right) .$$

Defining $f = 1/r$, we obtain:

$$i[D,\chi(f)] = f\frac{\partial}{\partial f}\chi(f) \ .$$

This demonstrates that the choice $\chi^+(r) = \chi(1/r)$ is consistent with the commutation relations being maintained.

- *HENDRICK:*

Is this choice $\chi^+(r)$ unique?

- *JACKIW:*

There may be a class of choices resulting from introducing a scale, which would still have our commutation relation intact. We have not studied the uniqueness of our choices, but I do know that with our prescription all the hermiticity requirements on the field are met.

DISCUSSION No. 2

- *NAHM:*

Dual models tend to have a vector particle of mass zero. Thus it seems odd that you have no theory involving vector particles.

- *JACKIW:*

First let me make it clear that we did not derive dual models, dual amplitudes, or anything of the sort. What we did was derive the Hagedorn spectrum and the Virasoro algebra. We do not have massless vector particles. The closest I can come is a Lagrangian of the form:

$$\mathcal{L} = \frac{1}{2}\,\partial_\mu A_\nu\, \partial^\mu A^\nu - (\partial \cdot A)^2 ,$$

where we are in two dimensions, so

$$A^\mu(x) = \begin{pmatrix} A^0 \\ A^1 \end{pmatrix}, \quad x^\mu = \begin{pmatrix} x^0 \\ x^1 \end{pmatrix} .$$

This Lagrangian is invariant under the transformation

$$\partial x^\mu = f^\mu(x)$$

where

$$f^\mu(x) = a^\mu + \omega^{\mu\nu} x_\nu + bx^\mu + c^\mu x^2 - 2c \cdot xx^\mu ,$$

that is, f^μ has the form of a conventional conformal transformation with terms linear in x, quadratic in x, and independent of x. The more general sets of transformations which leave invariant all the Lagrangians I mentioned in my lecture today, do not leave this Lagrangian invariant. I have not been able to construct a theory having a traceless, symmetric energy-momentum tensor for spin-1 fields. I do not know whether this is impossible for deep reasons, or whether I simply have not been clever enough. It is worth noting that while there are dual models based on Fermion and scalar operators, there are no dual models based on spin-1 operators.

- *NAHM:*

Anomalous dimensions will change the eigenvalues of Δ, the dilatation operator. Can interactions also change the asymptotic behaviour of the spectrum?

- *JACKIW:*

I do not know, that is an interesting question. We obtained a formula for the connection between degeneracy and the dilatation eigenvalues in the case of a non-interacting theory. What happens when you put in interactions has not been checked. One could try to do it in the trivial interaction cases such as the gradient coupling model or the Thirring model. Then one could look to see if, in such models with anomalous dimensions, the spectrum is modified.

- *WILCZEK:*

You showed us several examples of scale-invariant theories in two dimensions with the common features of having a large number of conserved currents and being exactly solvable. Another exactly solvable theory is two-dimensional QED. Do you know if there is a similar set of conserved currents in QED?

- *JACKIW:*

The only formally conserved currents I know of in two-dimensional QED are J^μ and $J^{\mu 5}$, the vector and axial-vector currents. The reason there are none of the conserved currents I discussed earlier is that the energy-momentum tensor for two-dimensional QED is not traceless and symmetric. This is because in two dimensions $\bar{\psi}\gamma^\mu\psi A_\mu$ is not a scale-invariant interaction.

There are recent examples of possibly solvable two-dimensional field theories. Dashen and Frishman have been working on the Thirring model with isospin, which apparently has a traceless, symmetric energy-momentum tensor. Consequently, it would have all the symmetries we have been discussing, and formally, at least, a large number of conserved currents. It would be interesting to see how the Virasoro algebra comes out in that model.

- *WILCZEK:*

Is there any relation between your work and Adler's work in formulating QED on the five-dimensional hypersphere?

- *JACKIW:*

There may very well be. Let us go back to Weinberg's work. He took the ordinary Feynman rules and rearranged them to obtain his infinite momentum rules. Then, many years later, Bjorken, Kogut and Soper, as well as Rohrlich and collaborators, gave an operator derivation of the Weinberg rules. I suspect that what Adler has done can also be derived by the operator technique which I have discussed here. Adler has taken the ordinary Feynman rules and rewritten them introducting a new notation. It would not suprise me in the least if our formalism is an operator basis for the Adler rules, simply because our formalism uses conformal invariance and scale invariance in an essential fashion.

- *FRAMPTON:*

You have been establishing connections between field theory and dual resonance models, in particular, connections with the degeneracy and with the Virasoro algebra. So far, these connections have been demonstrated only for the earliest dual model having the Euler B function. It would be interesting to establish a connection with other dual models such as the Neveu-Schwarz model or even more recent models; for example, by introducing anticommuting Fermion fields. More generally, I would like to have some feeling for whether your field theory can provide inspiration for the construction of new dual resonance models.

- *JACKIW:*

The Thirring model does have anticommuting operators, but I do not know if it has been established that this is analogous to the Neveu-Schwarz model.

- *REBBI:*

I would like to make a comment here. A dual model analogous to the Thirring model has been studied by Gervais and Andric, if I remember correctly. What they do is introduce, besides the fields $x^\mu(\sigma,\tau)$, the additional fields $\psi^\mu(\sigma,\tau)$ which transform like spinors in the internal space. As long as they do not introduce the four-Fermion interaction, they recover the standard model of Neveu and Schwarz. Introducing the four-Fermion interaction causes a shift in the intercept of the trajectories with odd G-parity (not the leading trajectory). The change of intercept is related to the appearance of anomalous dimensions, that is, it results in a change in the eigenvalues of the dilatation operator.

- *HENDRICK:*

If this thermodynamic analogy you draw is correct, you should have a number of useful relations coming from thermodynamics (in addition to the one you derived in your lecture). For example, the first and second laws of thermodynamics, the Maxwell relations, and so on. Are these sensible statements when you translate them back into your field theory description?

- *JACKIW:*

We have not checked all these equations. The only one we have checked is the connection between entropy and internal energy.

- *GOTTLIEB:*

Could there be any connection between the internal variables of Dirac's new positive energy spinor wave equation and the internal variables of your two-dimensional field theory?

- *JACKIW:*

I am not familiar with Dirac's spinor wave equation. However, inasmuch as Dirac's work was the first work along the lines I have been lecturing on, and his work on the general quantization of systems was the first relevant to the quantization of the string, it would not surprise me in the least if his more recent work is also relevant.

- *EYLON:*

Is the commutator $i[D,p^\mu] = p^\mu$ consistent with the fact that you get simultaneous eigenstates of M^2 and D?

- *JACKIW:*

Yes. The M^2, of which I had eigenstates, is the square of the translation operator in the four-dimensional physical space. The p^μ in the commutator is the translation operator in the two-dimensional internal parameter space. There cannot be a connection between M^2 and p^μ, because they affect translation in two completely different spaces.

- *MENDES:*

You mentioned this morning the possibility of using your formalism to develop a kind of current algebra at x^2 different from zero. In which kinematical limit would this algebra be useful?

- *JACKIW:*

I do not know. This is the first thing to look at, but we have not done that yet.

- *MENDES:*

I have not looked into the problem in detail, but I have the feeling that the current commutators at $x^2 \neq 0$ will not be very useful. If one looks at the short distance or the light-cone commutators, one sees that their usefulness comes from the fact that in this limit, the right-hand side is dominated by singularities that are largely interaction-independent.

If one looks at the current commutators of a free-field theory, one sees that the invariant functions on the right-hand side are non-singular at $x^2 \neq 0$, so that lacking a dominant leading singularity, one should expect the x^2 = constant $\neq 0$ commutators to be highly interaction-dependent.

- *JACKIW:*

The form of these commutators will be the same in an interacting theory, because these commutators are canonical. Whether this current algebra will be useful, I do not know because I have not investigated it. It seems to me that at the very least it will reproduce ordinary equal time results, simply because this quantization is very close to the equal time quantization; it is also a quantization on a space-like surface. So, it is likely that the current algebra will be as useful as the equal time current algebra. It may not be as useful as the light-cone algebra, which has additional information coming from the fact that you are quantizing on a light-like surface rather than on a space-like surface.

THE PHYSICAL INTERPRETATION OF DUAL MODELS

C. Rebbi

Table of Contents

THE PHYSICAL INTERPRETATION OF DUAL MODELS

C. Rebbi*)
CERN, Geneva, Switzerland.

1. INTRODUCTION

The theory of dual resonance models developed according to a pattern rather different than the usual: it is a theory where the scattering amplitude was found before we could describe the systems that actually scatter. Indeed, the many-body extension of the original Veneziano amplitude[1)] was produced already in 1968 [2)]. Since then, a long path of investigations brought to a better and better understanding of the physical properties of the intermediate states that appeared as resonant poles of the scattering amplitude. Milestones of these investigations[3)] are the proof of factorization[4)], the introduction of a suitable operator technique to exibit the factorized structure of the amplitude[5)], the discoveries of the gauge equations satisfied by the physical states and the associated algebraic structure[6)]. The introduction of a particular set of physical states[7)], the transverse states, was also a fundamental step, and opened the way to the proof of the no-ghost theorem[8)].

Parallel to these studies, extensions of the original model to other models were made [9-11)]; and it was also pointed out that a relevant feature of the dual model was that it seemed to describe objects with a one dimensional continum of degrees of freedom[12)].

Today, thanks to these and other investigations, we are in a position to reverse the historical path followed by the theory. We can define a dynamical system in space-time, and work out its properties, so as to obtain the spectrum of states of the dual resonance model, give a meaning to the gauge identities, understand better the importance of the transverse states and their role in the no-ghost theorem. We are also able to understand the interaction amplitudes within this physical picture, although some work in this direction must still be done.

We define our dynamical system in the most classic way: by introducing a Lagrangian function and an action functional. This will be our starting point.

The plan is to proceed so as to make the lectures self-contained. No prior knowledge of the theory of dual resonance models should be necessary to understand the properties of the system we are describing. But we shall of course mention the connections of our results with the theory of dual resonance models. For an independent study of the theory of dual resonance models, there exist many lectures and review articles, some of which are mentioned in Ref. (13).

2. THE PHYSICAL SYSTEM

We shall consider a one-dimensional system in space, a sort of string, with no other degrees of freedom than the positions of its points. Kinematically, we describe our system by assigning the spacial coordinates x_i as functions of a parameter σ:

---

*) On leave of absence from Istituto di Fisica Teorica dell' Università di Trieste, Italy.

$$x_i = x_i(\sigma) \qquad\qquad i = 1,2,3 \;.- \tag{2.1}$$

Equation (2.1) give a parametric representation of a curve in space. Any range of variation can be taken for the prameter σ, and in these lectures we shall fix $0 \le \sigma \le \pi$.

As the system evolves in time, the functions $x_i(\sigma)$ will change: the time evolution of the system is described by equations

$$\begin{aligned} & x_i = x_i(\sigma, \tau) \\ \text{with} \qquad & x_0 = \tau \;.- \end{aligned} \tag{2.2}$$

Notice that Eqs. (2.2) give a parametric representation of a surface in space time, the surface spanned by the string during its time evolution. A more general parametrization, of the form

$$x_\mu = x_\mu(\sigma, \tau) \quad , \tag{2.3}$$

can be used to describe the same surface. It is convenient to use this more general parametrization. Indeed, if we describe the evolution of the string with Eqs. (2.2) in a given Lorentz frame, in another frame we shall have

$$x_0' = \Lambda_0^{\;\mu} x_\mu(\sigma, \tau) \; ,$$

so that x_0' will depend on τ and σ, and the specific parametrization (Eqs. 2.2) will be no longer valid.

Vice versa, given the surface spanned by the string in space time, we can describe its time evolution in any Lorentz frame by intersecting the surface with the family of planes

$$n_\mu x^\mu = \tau \;.-$$

To be the surface of evolution in time of a spacial curve, the surface $x_\mu = x_\mu(\sigma\tau)$ must be time-like: at any of its points there must be at least a tangent vector pointing in the time-like direction (or light-like direction allowing for a motion of some points of the string with the speed of light).

We shall use parametrizations where τ is one evolution parameter: we impose then

$$\left(\frac{\partial x}{\partial \tau}\right)^2 \le 0 \quad , \quad \left(\frac{\partial x}{\partial \sigma}\right)^2 > 0 \;.- \tag{2.4}$$

We use a metric $g_{ii} = 1 = -g_{00}$.

We introduce a dynamic by defining an action functional. Given an initial and final configuration of the string:

$$x_\mu^i(\sigma) = x_\mu(\sigma, \tau_0) \quad , \quad x_\mu^f(\sigma) = x_\mu(\sigma, \tau_1) \, , \tag{2.5}$$

an action functional assigns a number, the action, to any possible surface of evolution between $x_\mu^i(\sigma)$ and $x_\mu^f(\sigma)$. The action should be relativistic invariant and an intrinsic property of the surface of evolution, independent from the choice of parametrization.

In reference 14 (see also Ref. 15) it was suggested that the action was chosen proportional to the area of the surface of evolution:

$$S = \frac{1}{2\pi\alpha'} \int_{\tau_0}^{\tau_1} d\tau \int_0^{\pi} d\sigma \sqrt{-\left(\frac{\partial x}{\partial \tau}\right)^2 \left(\frac{\partial x}{\partial \sigma}\right)^2 + \left(\frac{\partial x}{\partial \tau} \cdot \frac{\partial x}{\partial \sigma}\right)^2} \, .- \tag{2.6}$$

The expression

$$\sqrt{-\left(\frac{\partial x}{\partial \tau}\right)^2 \left(\frac{\partial x}{\partial \sigma}\right)^2 + \left(\frac{\partial x}{\partial \tau} \cdot \frac{\partial x}{\partial \sigma}\right)^2} \, d\sigma d\tau = dA \tag{2.7}$$

is the area element between two infinitesimal displacements $\partial x_\mu / \partial \tau \, d\tau$ and $\partial x_\mu / \partial \sigma \, d\sigma$.

A dimensional constant α' with the dimensions of a length square is needed to make S a dimensional (we use units where $\hbar = c = 1$). We shall see that α' measures the slope of the Regge trajectories.

In the following it will be convenient to use an abbreviated notation for derivatives with respect to τ and σ:

$$\frac{\partial f}{\partial \tau} = \dot{f} \quad , \quad \frac{\partial f}{\partial \sigma} = f' \, .- \tag{2.8}$$

Our aim is to study the equations of motion that follow from Hamilton's principle

$$\delta S = 0$$

and to give them a quantum formulation.

We follow the treatment of Ref. (16). Let us define a Lagrangian density

$$\mathcal{L} = - \frac{1}{2\pi\alpha'} \sqrt{-\dot{x}^2 x'^2 + (\dot{x}x')^2} \, .- \tag{2.9}$$

Then $\delta S = 0$ together with Eq. (2.6) implies

$$\int_{\tau_0}^{\tau_1} d\tau \int_0^{\pi} d\sigma \left\{ -\frac{\partial}{\partial \tau} \frac{\partial \mathcal{L}}{\partial \dot{x}_\mu} - \frac{\partial}{\partial \sigma} \frac{\partial \mathcal{L}}{\partial x'_\mu} \right\} \delta x_\mu + \int_{\tau_0}^{\tau_1} d\tau \left\{ \frac{\partial \mathcal{L}}{\partial x'_\mu} \delta x_\mu \right\} \Big|_{\sigma=0}^{\pi} = 0 \, .- \tag{2.10}$$

Eq. (2.10) is obtained in the standard way by an integration by parts; the boundary terms at $\tau = \tau_0$ and $\tau = \tau_1$ can be dropped, because there $\delta x_\mu = 0$; but the boundary term at $\sigma = 0$ and π must be retained.

Eq. (2.10) is satisfied for arbitrary δx if

$$\frac{\partial}{\partial\tau}\frac{\partial\mathcal{L}}{\partial\dot{x}_\mu} + \frac{\partial}{\partial\sigma}\frac{\partial\mathcal{L}}{\partial x'_\mu} = 0 \tag{2.11}$$

and

$$\left.\frac{\partial\mathcal{L}}{\partial x'_\mu}\right|_{\sigma=0}^{\pi} = 0 \ . \tag{2.12}$$

Eqs. (2.11) and (2.12) are the Euler-Lagrange equations for the motion of the string. To understand the meaning of these equations, we note that the Lagrangian density $\mathcal{L}$ is invariant under the Poincaré group, and that there will be therefore locally conserved quantities corresponding to the generators of the group. In particular, by performing an infinitesimal translation, we find that the energy-momentum flowing across an infinitesimal line element ($d\sigma$, $d\tau$) is given by

$$dp^\mu = \frac{\partial\mathcal{L}}{\partial\dot{x}_\mu}\, d\sigma + \frac{\partial\mathcal{L}}{\partial x'_\mu}\, d\tau \ . \tag{2.13}$$

Let us define the two-components of an energy-momentum current flowing on the surface

$$\mathcal{P}_\mu = \frac{\partial\mathcal{L}}{\partial\dot{x}_\mu} = \frac{1}{2\pi\alpha'}\frac{\dot{x}_\mu x'^2 - x'_\mu(\dot{x}x')}{\sqrt{-\dot{x}^2x'^2 + (\dot{x}x')^2}} , \tag{2.14}$$

$$\Pi_\mu = \frac{\partial\mathcal{L}}{\partial x'_\mu} = \frac{1}{2\pi\alpha'}\frac{x'_\mu \dot{x}^2 - \dot{x}_\mu(\dot{x}x')}{\sqrt{-\dot{x}^2x'^2 + (\dot{x}x')^2}} \ . \tag{2.15}$$

The Euler-Lagrange Eqs. (2.11) and (2.12) tell us that the energy-momentum current (P_μ, Π_μ) is conserved on the surface, and that no energy momentum flows out of the ends of the string. It follows that the total momentum of the string

$$P_\mu = \int_c \mathcal{P}_\mu d\sigma + \Pi_\mu d\tau = \int_{\tau\ \text{fixed}} d\sigma\, \mathcal{P}_\mu \tag{2.16}$$

is conserved.

Notice that from Eq. (2.15) one derives immediately

$$2\pi\alpha'\,\Pi^2 + \frac{\dot{x}^2}{2\pi\alpha'} = 0 \ . \tag{2.17}$$

$\Pi_\mu |_0^\pi = 0$ implies then $\dot{x}^2(0) = \dot{x}^2(\pi) = 0$, i.e., the end points of the string move with the speed of light. If our space had an Euclidean metric, the condition (Eq. 2.12) could not be implemented: in an Euclidean space a surface of minimal area cannot be found unless it has a fixed closed boundary. Before we proceed we want to give an example of a solution to the equations of motion; we try to see whether they can be satisfied by a rigidly rotating string.

We make the

$$\begin{aligned} x_0 &= \tau \\ x_1 &= A\left(\sigma - \frac{\pi}{2}\right)\cos\omega\tau \\ x_2 &= A\left(\sigma - \frac{\pi}{2}\right)\sin\omega\tau \end{aligned} \qquad (2.18)$$

and find

$$\mathcal{P}_\mu = \frac{A}{2\pi\alpha'}\frac{1}{\sqrt{1-\omega^2A^2(\sigma-\frac{\pi}{2})^2}}\left(1, -\omega A(\sigma-\tfrac{\pi}{2})\sin\omega\tau, -\omega A(\sigma-\tfrac{\pi}{2})\cos\omega\tau\right) \qquad (2.19)$$

$$\Pi_\mu = -\frac{1}{2\pi\alpha'}\frac{\sqrt{1-\omega^2A^2(\sigma-\frac{\pi}{2})^2}}{A}\left(0, A\cos\omega\tau, A\sin\omega\tau\right). \qquad (2.20)$$

The equation

$$\dot{\mathcal{P}} + \Pi' = 0 \qquad (2.21)$$

is satisfied; the condition

$$\Pi\Big|_{\sigma=0}^{\pi} = 0 \qquad (2.22)$$

requires

$$\frac{\pi}{2}\omega A = 1. \qquad (2.23)$$

We can compute the energy and angular momentum of the rotating string. We use Eqs. (2.16) and (2.23) to evaluate

$$E = M\ (\text{because } \vec{p} = 0) = \frac{A\pi}{4\alpha'}.$$

We compute J through a formula analogous to Eq. (2.16) which gives

$$M_{\mu\nu} = \int_{\tau\ \text{fixed}} (x_\mu \mathcal{P}_\nu - x_\nu \mathcal{P}_\mu)\, d\sigma. \qquad (2.24)$$

We find

$$J = \frac{A^2\pi^2}{16\alpha'};$$

eliminating A we obtain

$$J = \alpha' M^2 . - \tag{2.25}$$

Solving Eqs. (2.11) and (2.12) for a general motion of the string may seem rather hard, but there is a very large group of invariance of the system, which may be used to simplify the equations of motion and which is fundamental in the Hamiltonian treatment of the problem. The action integral, defined in Eq. (2.6), depends only on the geometry of the surface of evolution, and not on the particular choice of the parameters σ and τ we used to describe it. S and the equations of motion must be therefore invariant under the group of reparametrizations of the surface

$$x_\mu(\sigma,\tau) \Rightarrow \tilde{x}_\mu(\sigma,\tau) = x_\mu(\tilde{\sigma},\tilde{\tau}) , \tag{2.26}$$

where

$$\begin{aligned} \tilde{\sigma} &= \tilde{\sigma}(\sigma,\tau) \\ \tilde{\tau} &= \tilde{\tau}(\sigma,\tau) \ .- \end{aligned} \tag{2.27}$$

The infinitesimal generators of these transformations are obtained by

$$\begin{aligned} \sigma &\Rightarrow \tilde{\sigma} = \sigma + \epsilon f(\sigma,\tau) \\ \tau &\Rightarrow \tilde{\tau} = \tau + \epsilon g(\sigma,\tau) \end{aligned} \tag{2.28}$$

where $f(0,\tau) = f(\pi,\tau) = 0$ to preserve the range $0 \le \sigma \le \pi$. The transformations induced on the coordinates $x_\mu(\sigma\tau)$ are

$$x_\mu(\sigma\tau) \Rightarrow \tilde{x}_\mu(\sigma\tau) = x_\mu(\sigma\tau) - \epsilon f(\sigma\tau) x'_\mu(\sigma\tau) - \epsilon g(\sigma\tau)\dot{x}_\mu(\sigma\tau) \ .- \tag{2.29}$$

We shall refer to the group of reparametrizations as gauge group of the string.

A general consequence of the invariance of the action under some group of transformations is that the relation expressing the conjugate momenta in terms of the velocities cannot be inverted, and there exist dependences among the momenta and coordinates, which are algebraic consequences of the form of the Lagrangian. Dirac calls these dependences primary constraints[17]. In our case

$$\mathcal{P}_\mu(\sigma) = \frac{\delta}{\delta \dot{x}^\mu(\sigma)} \int d\sigma \, \mathcal{L} = \frac{\partial \mathcal{L}}{\partial \dot{x}^\mu}(\sigma)$$

is the momentum conjugate to $x_\mu(\sigma)$. It is immediate to check from Eq. (2.14), that the components of $\mathcal{P}_\mu(\sigma)$ are not independent, but that the following constraints hold

$$x'_\mu \mathcal{P}^\mu = 0 \quad , \tag{2.30a}$$

$$2\alpha' (\pi \mathcal{P})^2 + \frac{x'^2}{2\alpha'} = 0 \text{ .-} \tag{2.30b}$$

When there are primary constraints the passage to a Hamiltonian formalism and to the quantization of the system is not straightforward. We are faced with two problems. First of all, the constraints are generally incompatible with canonical Poisson brackets or commutation relations where all the p's and x's are treated as independent. Secondly, when there are constraints of the form $h_j(p,q) = 0$, the effective Hamiltonian that generates the evolution of the system is not necessarily given by

$$H_0 = \Sigma_i p_i \dot{x}_i - L$$

but can contain a linear combination of the constraint functions[16])

$$H(p,q) = H_0(p,q) + \Sigma_j c_j h_j(p,q) \text{ .-} \tag{2.31}$$

There are two ways to proceed, to establish a Hamiltonian formalism. One consists in reducing the number of degrees of freedom of the system by eliminating redundant variables. Then we are left with a smaller set of independent coordinates and momenta, for which alone we assume canonical Poisson brackets and later commutation relations. We shall illustrate this method of quantization in the next lecture.

The other possibility is to deal with all of the coordinates and momenta as if they where independent quantities, disregarding the existence of the constraints when establishing canonical Poisson brackets or commutation relations. The constraints are imposed as initial conditions among the classical variables or on the wave function. The Hamiltonian is then chosen in the form (2.31), and must be such that the constraint equations are preserved during the evolution of the system. (This will automatically happen if H_0 and the constraint Equation form a closed algebra of Poisson brackets or commutators; otherwise it may be necessary to introduce new constraints, but this does not happen in our case).

In this lecture we follow this second method of quantization. We define canonical Poisson brackets

$$\{x^\mu, x^\nu\} = \{\mathcal{P}^\mu, \mathcal{P}^\nu\} = 0 \, , \tag{2.32}$$

$$\{x^\mu(\sigma), \mathcal{P}^\nu(\sigma')\} = g^{\mu\nu} \delta(\sigma - \sigma') \text{ .-} \tag{2.33}$$

Notice the appearance of $g^{\mu\nu}$ in Eq. (2.33), which is necessary to preserve Lorentz covariance. It is instructive, at this stage, to study the algebra of the constraints Eqs. (2.30) under Poisson brackets, and also the Poisson brackets of the constraint functions with the dynamical variables.

To compute the various Poisson brackets of the constraint functions in a compact form it is useful to extend the range of variation of σ from $0 \leq \sigma \leq \pi$ to $-\pi \leq \sigma \leq \pi$ by the following definitions

$$\mathcal{P}_\mu(-\sigma) = \mathcal{P}_\mu(\sigma) \,, \qquad x'_\mu(-\sigma) = -x'_\mu(\sigma) \,. - \tag{2.34}$$

Then the two equations (2.30) are equivalent to the single equation

$$\left(\sqrt{2\alpha'}\,\pi\mathcal{P}(\sigma) + \frac{x'(\sigma)}{\sqrt{2\alpha'}}\right)^2 = 0 \quad , -\pi \leq \sigma \leq \pi \,. - \tag{2.35}$$

We define the functionals

$$L_h = \frac{1}{4\pi}\int_{-\pi}^{\pi} d\sigma\, h(\sigma)\left(\sqrt{2\alpha'}\,\pi\mathcal{P}(\sigma) + \frac{x'(\sigma)}{\sqrt{2\alpha'}}\right)^2 \,. - \tag{2.36}$$

It is clear that the simultaneous validity of Eqs. (2.30a) and (2.30b) is equivalent to $L_h = 0$ for any function h. With the aid of the Poisson brackets [Eqs. (2.32) and (2.33)] it is easy to compute

$$\{L_h , L_k\} = L_{h \otimes k} \,, \tag{2.37}$$

where

$$h \otimes k = h k' - h' k \,, \tag{2.38}$$

and

$$\{L_h , x(\sigma)\} = -\frac{h(\sigma)}{2}\left(2\pi\alpha'\mathcal{P}(\sigma) + x'(\sigma)\right) + (\sigma \leftrightarrow -\sigma) =$$

$$= -\frac{h(\sigma)+h(-\sigma)}{2}\,2\pi\alpha'\mathcal{P}(\sigma) - \frac{h(\sigma)-h(-\sigma)}{2}\,x'(\sigma) \,. - \tag{2.39}$$

There is one point to mention, about these results. In getting to Eqs. (2.37) and (2.39) we must assume that nothing singular happens at $\sigma = -\pi, 0, \pi$. In particular, we must have $x'(0) = x'(\pi) = 0$. We shall see later that this boundary condition does indeed arise from the equations of motion; for the moment, we just wish to warn that a proper dynamical treatment must include proper care of the boundary conditions at the end points of the string.

We see from Eqs. (2.37) and (2.38) that the Poisson bracket of two constraint functionals is again a constraint functional. The constraint functionals form a closed algebra. Equation (2.39) exibits the transformations induced on the coordinates $x_\mu(\sigma)$ by the constraint functionals. We can consider the L_h's as generators of infinitesimal canonical transformations. Their role as such is very important, in fact the Hamiltonian of our system

$$H_0 = \int_0^\pi d\sigma\, \dot{x}(\sigma)\mathcal{P}(\sigma) - \int_0^\pi d\sigma\, \mathcal{L}(\sigma) \tag{2.40}$$

vanishes identically. It follows that the whole evolution of the system must be generated by some superposition of the constraint functions[17]; we will have

$$H = L_R \tag{2.41}$$

for some appropriate h. The choice of h, in Eq. (2.41), can be quite arbitrary: in fact, a definite choice of H implies the choice of a definite gauge for the system. We recall that the gauge invariance manifests itself with the fact that the equations

$$\mathcal{P}_\mu = \mathcal{P}_\mu(x_\mu', \dot{x}_\mu)$$

cannot be inverted to yield a unique

$$\dot{x}_\mu = \dot{x}_\mu(\mathcal{P}_\mu, x_\mu')$$

The velocities can be expressed as functions of the momenta and coordinates only if some further specification is made on the parametrization, i.e., if we fix a gauge. But, when the Hamiltonian is known, the velocities can be expressed in terms of coordinates and momenta, through Hamilton's equations

$$\dot{x}_\mu = \{x_\mu, H\} \;.$$

A choice of H is then equivalent to a choice of a gauge.

Inspection of Eq. (2.39) reveals that the even part of L_h transforms x into $\mathcal{P}$, and generates a dynamical evolution of the system. The odd part corresponds instead to a reparametrization of the points of the string, at fixed τ. Taking h odd we find

$$\delta x_\mu(\sigma) = \epsilon\{L_R, x_\mu(\sigma)\} = -\epsilon h(\sigma)\, x'(\sigma) \;, \tag{2.42}$$

which is the change of parametrization induced by $\sigma \to \tilde{\sigma} = \sigma + \epsilon\, h(\sigma)$ [see Eqs. (2.28) and (2.29)].

We shall fix a gauge for the string by taking h = const = 1. This choice of gauge is very convenient, because it gives a particularly simple form to the equations of motion. With

$$H = \frac{1}{4\pi}\int_{-\pi}^{\pi} d\sigma \left(\sqrt{2\alpha'}\,\pi\,\mathcal{P}(\sigma) + \frac{x'(\sigma)}{\sqrt{2\alpha'}}\right)^2 = \pi\int_{0}^{\pi} d\sigma \left(\alpha'\,\mathcal{P}^2(\sigma) + \frac{x'^2(\sigma)}{4\pi^2\alpha'}\right), \tag{2.43}$$

the equations of motion become

$$\dot{x} = 2\pi\alpha'\mathcal{P}, \qquad \dot{\mathcal{P}} = \frac{1}{2\pi\alpha'}\,x'' . \tag{2.44}$$

We may eliminate $\mathcal{P}$ to obtain

$$\ddot{x} - x'' = 0 . \tag{2.45}$$

The initial values of $\dot{x}$ and x' are not arbitrary, but are subject to

$$\dot{x}x' = 0 , \qquad \dot{x}^2 + x'^2 = 0 , \tag{2.46}$$

which are the constraint equations, expressed in terms of $\dot{x}$ and x'. Notice that, whereas Eqs. (2.30) are always valid, Eqs. (2.46) are valid only in the gauge we have chosen.

Once Eqs. (2.46) are assumed at $\tau = \tau_i$, they continue to be valid at any τ; this follows from Eqs. (2.37) and (2.38), which imply

$$\dot{L}_k = \{L_k, H\} = L_k . \tag{2.47}$$

In our gauge

$$\Pi_\mu = \frac{\partial \mathcal{L}}{\partial x'^\mu} = -\frac{x'_\mu}{2\pi\alpha'} \tag{2.48}$$

[as can be seen from Eqs. (2.16) and (2.46)]. The boundary condition $\Pi_\pi = 0 \,|_{\sigma=0,\pi}$ becomes then $x'_\mu = 0\,|_{\sigma=0,\pi}$, which is consistent with the reflection principle we have assumed. Equations (2.44) and (2.48) show that the harmonic equation satisfied by x_μ in the chosen gauge is equivalent to the equation of conservation of the momentum flow. (Note: there is one geometrical reason why the equations of motion simplify so much in the gauge we have chosen. It is known from the theory of Euclidean surfaces[18] that the equations for a surface of minimal area simplify and become Laplace equations if one requires that the chosen net of coordinates form a locally orthonormal system: the gauge conditions Eq. (2.46) are the equivalent of this orthonormality requirement for a time-like surface.) Having established a Hamiltonian formalism to describe the classical motion of the string, we could quantize the system by the replacement:

$$i\,\{\text{Poisson bracket}\} \to [\text{commutator}] .$$

It is convenient however to expand first the string coordinates and momenta into modes, so as to deal with a discrete set of dynamical variables.

We expand $x(\sigma)$ and $\mathcal{P}(\sigma)$ in $\cos n\sigma$, as required by $x'(0) = x'(\pi) = 0$ and by $\mathcal{P} = 1/(2\pi\alpha')\,\dot{x}$:

$$x_\mu(\sigma) = x_{0,\mu} + \Sigma_n\, x_{n,\mu} \cos n\sigma \ ,$$
$$\mathcal{P}_\mu(\sigma) = \frac{p_\mu}{\pi} + \Sigma_n\, \mathcal{P}_{n,\mu} \cos n\sigma \ . \tag{2.49}$$

Inserting into Eq. (2.43) we obtain

$$H = \alpha' p^2 + \Sigma_n \left(\frac{\alpha' \pi^2}{2} \mathcal{P}_n^2 + \frac{n^2 x_n^2}{8\alpha'} \right) . \tag{2.50}$$

We diagonalize H by introducing

$$a_{n,\mu} = \frac{\pi}{2} \sqrt{\frac{2\alpha'}{n}}\, \mathcal{P}_{n,\mu} - \frac{i}{4} \sqrt{\frac{n}{2\alpha'}}\, x_{n,\mu} \tag{2.51}$$

In terms of the normal coordinates $a_{n,\mu}$ we have

$$H = \alpha' p^2 + \Sigma_n\, n\, a^*_{n,\mu}\, a_n^\mu \tag{2.52}$$

with Poisson brackets

$$\{ a_{m,\mu} \, , \, a^*_{n,\nu} \} = - i\, g_{\mu\nu}\, \delta_{m,n} \ , \tag{2.53}$$

all other vanishing.

This implies a τ evolution

$$a_{n,\mu}(\tau) = e^{-in\tau} a_{n,\mu}(0) . \tag{2.54}$$

It is convenient to introduce the notation

$$\begin{aligned} \alpha_n &= \sqrt{n}\, a_n \\ \alpha_{-n} &= \alpha_n^* \end{aligned} \tag{2.55}$$

Then, we have the following expansions

$$x_\mu(\sigma,\tau) = x_{0,\mu} + 2\alpha' p_\mu \tau + i \sum_{\substack{n=-\infty \\ n\neq 0}}^{\infty} \sqrt{2\alpha'} \frac{\alpha_n}{n} \cos n\sigma\, e^{-in\tau} , \tag{2.56}$$

$$\mathcal{P}_\mu(\sigma,\tau) = \frac{p_\mu}{\pi} + \frac{1}{\pi\sqrt{2\alpha'}} \sum_{\substack{n=-\infty \\ n\neq 0}}^{\infty} \alpha_n \cos n\sigma\, e^{-in\tau} . \tag{2.57}$$

In the theory of dual resonance models one often encounters the adimensional fields [introduced by Fubini and Veneziano[19] and Gervais[20]].

$$Q_\mu(\tau) = \frac{x_{0,\mu}}{\sqrt{2\alpha'}} + \sqrt{2\alpha'}\, p_\mu \tau + i \sum_n \frac{\alpha_n}{n} e^{-in\tau} , \tag{2.58}$$

$$P_\mu(\tau) = \frac{d}{d\tau} Q_\mu(\tau) = \sqrt{2\alpha'}\, p_\mu + \sum_n \alpha_n e^{-in\tau} , \tag{2.59}$$

(frequently Q_μ and P_μ are expressed as functions of $z = e^{i\tau}$).

The meaning of Q and P is clear: if we introduce units where $2\alpha' = 1$ we have

$$Q_\mu(\tau) = x_\mu(0,\tau) \tag{2.60}$$

$$P_\mu(\tau) = \dot{x}_\mu(0,\tau) = \pi \mathcal{P}_\mu(0,\tau) \tag{2.61}$$

We can also write

$$P_\mu(\sigma) = \sqrt{2\alpha'}\, \pi \mathcal{P}_\mu(\sigma) + \frac{x'_\mu(\sigma)}{\sqrt{2\alpha'}} \tag{2.62}$$

where the reflection principle [Eq. (2.34)] has been used to extend the range of σ.

Summarizing, we have the following dynamical description of our system. In the gauge we have chosen the coordinates and momenta of the string are given as a superposition of normal modes. The normal coordinates evolve according to the Hamiltonian

$$H = \alpha' p^2 + \sum_{n>0} n a_n^* a_n = \alpha' p^2 + \frac{1}{2} \sum_{n\neq 0} \alpha_{-n} \alpha_n \ .-$$

Their initial values however are not entirely arbitrary, but are subject to the constraint Eqs. (2.30) [or equivalently Eq. (2.35)].

We transform the constraint equations into a discrete set of conditions by requiring

$$L_n = L_{h=e^{in\tau}} = 0. \tag{2.63}$$

This gives

$$L_n = \sqrt{2\alpha'}\, p\cdot\alpha_n + \frac{1}{2} \sum_{m\neq 0,n} \alpha_{n-m}\alpha_m = 0\ , \quad n\neq 0\ , \tag{2.64}$$

and

$$H = L_0 = 0\ .- \tag{2.64a}$$

The algebra of the L_n's is inferred from Eqs. (2.37) and (2.38) and is

$$\{L_m, L_n\} = -i(m-n) L_{m+n}\ .- \tag{2.65}$$

Notice that Eq. (2.64a) determines the spectrum of the classical system; it gives in fact

$$-\alpha' p^2 = \sum_{n>0} n\,|a_n|^2\ .-$$

It is immediate now to quantize the system. The $a_{n\mu}^*$, $a_{n\mu}$ are replaced by operators $a_{n\mu}$, $a_{n\mu}^+$ with commutation relations

$$[a_{m,\mu}, a^+_{n,\nu}] = g_{\mu\nu}\,\delta_{m,n}\ , \tag{2.66}$$

all other commutators vanishing.

We also consider p_μ and $x_{o\mu}$ as operators with commutation relations

$$[x_{o,\mu}, p_\nu] = i g_{\mu\nu}\ .- \tag{2.67}$$

We introduce a ground state vector $|\phi\rangle$ satisfying

$$a_{m,\mu}|\phi\rangle = p_\mu|\phi\rangle = 0\ .- \tag{2.68}$$

All the quantum excitations of the system are then described by vectors $|\varphi\rangle$ belonging to the Fock space spanned by the $a^+_{m,\mu}$ with basis

$$|\lambda, k_\mu\rangle = a^+_{1,\mu_1} a^+_{1,\mu_2} \cdots a^+_{1,\mu_{\lambda_1}} a^+_{2\nu_1} \cdots a^+_{2\nu_{\lambda_2}} \cdots e^{ikx} |\phi\rangle . - \tag{2.69}$$

The excitations of the system are not arbitrary; the physical states are constrained by

$$\langle\varphi | L_n | \varphi\rangle = 0 \quad , \quad n \neq 0 . - \tag{2.70}$$

This equation replaces the condition $L_n = 0$, which is incompatible with canonical commutators for the a's and a^+'s, and can be separated into a creation and annihilation part:

$$L_n |\varphi\rangle = 0 \quad , \quad n > 0 , \tag{2.71}$$

and

$$\langle\varphi| L_{-n} = \langle\varphi| L^+_n = 0 . -$$

We expect Eq. (2.65) also to have a quantum counterpart. However, the definition of the quantum H is not unambiguous, because of ordering problems.

We define

$$L_0 = : L_0 : = \alpha' p^2 + \sum_{n>0} n a^+_n a_n . - \tag{2.72}$$

We expect Eq. (2.64a) to be valid up to c-numbers. We require then

$$L_0 |\varphi\rangle = (\alpha' p^2 + \sum_{n>0} n a^+_n a_n) |\varphi\rangle = \alpha_0 |\varphi\rangle . - \tag{2.73}$$

In the next lecture we shall see that self consistency of the theory fixes $\alpha_0 = 1$. We conclude with some remarks: the generators of Lorentz transformations are

$$M_{\mu\nu} = i x_\mu p_\nu + i \sum_{n>0} a_{n\mu} a^+_{n\nu} - \mu \leftrightarrow \nu \; ; \tag{2.74}$$

it follows that the spectral condition (2.73) assigns the physical states to Regge trajectories with slope α' and intercept α_0. The conditions (2.71) are the Virasoro gauge conditions on the on-mass-shell states of the dual resonance model[6]. The algebra of the quantum L_n is analogous to the quantum version of Eq. (2.66): however a very important c-number term arises, due to the normal ordering of operators. We have

$$[L_m, L_n] = (m-n) L_{m+n} + \frac{d}{12}(m^3 - m)\delta_{m,-n} . - \tag{2.75}$$

d is the number of different Lorentz indices, i.e., the number of dimensions of space-time. The c-number in Eq. (2.75) may be evaluated by taking the ground state expectation value. One obtains

$$\langle \emptyset | [L_n, L_{-n}] | \emptyset \rangle = \langle \emptyset | L_n, L_{-n} | \emptyset \rangle =$$

$$= \frac{1}{4} \langle \emptyset | (\sum_{\mu=0}^{d-1} \sum_{m=1}^{n-1} \sqrt{m(n-m)}\, a_{m\mu}\, a^{\mu}_{n-m})\, (\text{idem})^{+} | \emptyset \rangle =$$

$$= \frac{d}{12}\, n^2 (n-1) . -$$

3. THE TRANSVERSE GAUGE

The alternative way to establish a Hamiltonian formalism and to quantize the system is based on the elimination of all redundant degrees of freedom. This requires a complete gauge specification: we must choose a parametrization such that no further invariance under reparametrization (and primary constraints) are left. In this complete gauge all the dynamical variables will be expressed in terms of a smaller set of independent ones.

We choose a gauge imposing again the orthonormality conditions [Eqs. (2.46)]

$$\dot{x} x' = 0 ,$$
$$\dot{x}^2 + x'^2 = 0 . - \tag{2.46}$$

Notice that this choice of gauge still leaves the possibility of further reparametrizations. A choice of new parameters $\tilde{\sigma} = \tilde{\sigma}(\sigma\tau)$ and $\tilde{\tau} = \tilde{\tau}(\sigma\tau)$ leaves [Eqs. (2.46)] invariant if

$$\tilde{\sigma}' = \dot{\tilde{\tau}} ,$$
$$\dot{\tilde{\sigma}} = \tilde{\tau}' . - \tag{3.1}$$

Equations (3.1) demands that the new parameters $\tilde{\tau}$ and $\tilde{\sigma}$ satisfy the D'Alembert equation in the old ones

$$\ddot{\tilde{\tau}} - \tilde{\tau}'' = 0 , \tag{3.2a}$$
$$\ddot{\tilde{\sigma}} - \tilde{\sigma}'' = 0 . - \tag{3.2b}$$

Vice versa, given a new parameter $\tilde{\tau}$(or $\tilde{\sigma}$) satisfying Eqs. (3.2a) and (3.2b), we can, using Eqs. (3.1), define a new parametrization preserving the orthonormality condition.

We exploit this further freedom of reparametrization by demanding proportionality between τ and the time coordinate in some frame:

$$t = n_\mu x^\mu , \tag{3.3}$$

where n_μ is a fixed Lorentz vector, with $n^2 \leq 0$.

We impose

$$n_\mu x^\mu (\sigma,\tau) = \text{const} \times \tau \ . - \tag{3.4}$$

This choice of parametrization is compatible with the orthonormality conditions because $x^\mu(\sigma\tau)$ does satisfy the D' Alembert equation, and so we can always choose $n_\mu \ x^\mu(\sigma\tau)$ as new τ parameter, without spoiling Eqs. (2.46).

From Eq. (3.4) we deduce

$$n_\mu \dot{x}^\mu = \text{const} \ .. \tag{3.5}$$

In the orthonormal gauge we have

$$\mathcal{P}_\mu = \frac{1}{2\pi\alpha'} \dot{x}_\mu \ .- \tag{2.44}$$

It follows

$$n_\mu \mathcal{P}^\mu (\sigma\tau) = \frac{\text{const}}{2\pi\alpha'} = \frac{n_\mu p^\mu}{\pi} \ .- \tag{3.6}$$

As a consequence, const = $2\alpha'(p_\mu n^\mu)$ and

$$n_\mu x^\mu = (2\alpha' n_\mu p^\mu)\,\tau , \tag{3.7}$$

where p_μ is the total momentum of the string. Equations (3.6) and (3.7) give a meaning to the choice of parametrization. We take as evolution parameter for the string the time (apart from a constant of proportionality) in a definite Lorentz frame (i.e., we define the τ coordinate lines by intersecting the surface of evolution with planes $n_\mu x^\mu$ = const). Then we label the points of the string in such a way that the density of momentum $n_\mu \mathcal{P}^\mu$ is a constant. Since

$p_\mu^{\,0\to\sigma}$ = momentum flowing between

the points labelled by 0 and σ =

$$= \int_0^\sigma d\sigma \ \mathcal{P}_\mu(\sigma) ,$$

we have

$$(n_\mu p^\mu)^{0\to\sigma} = \int_0^\sigma \frac{n_\mu p^\mu}{\pi} d\sigma = \frac{(n_\mu p^\mu)}{\pi} \sigma ,$$

i.e., we use $(n_\mu p^\mu)^{0\to\sigma}$ as σ-coordinate for the points of the string (apart from a constant of proportionality). Equations (3.6) and (3.7) obviously constitute a complete gauge specification. It can be shown[16] that they imply the orthonormality conditions (2.46). We define the transverse gauge by going to the limit where n_μ becomes light-like. For every Lorentz vector u_μ, we define

$$u_\pm = \frac{u_0 \pm u_1}{\sqrt{2}}$$

and denote by u_i or $u_\perp$ the residual transverse components. Then we choose $n_- = 1$, $n_+ = n_i = 0$, so that Eqs. (3.6) and (3.7) become

$$\mathcal{P}_+(\sigma,\tau) = \frac{p_+}{\pi} \; . \tag{3.8}$$

(3.9)

In the sequel of these lectures we shall use units where $2\alpha' = 1$. Equations (3.8), (3.9) and the constraint Eq. (2.30) give us

$$x_-' \frac{p_+}{\pi} - x_\perp' \mathcal{P}_\perp = 0 \, , \tag{3.10a}$$

$$2p_+(\pi \mathcal{P}_-) - (\pi \mathcal{P}_\perp)^2 - x_\perp'^2 = 0 \; . \tag{3.10b}$$

These equations can be used to express x_-' and $\mathcal{P}_-$ as dependent variables

$$x_-' = \frac{1}{p_+} x_\perp' (\pi \mathcal{P}_\perp) \, , \tag{3.11a}$$

$$\mathcal{P}_- = \frac{1}{2\pi p_+} \left[(\pi \mathcal{P}_\perp)^2 + x_\perp'^2 \right] . \tag{3.11b}$$

$x_\perp(\sigma)$ and $\mathcal{P}_\perp(\sigma)$ are left as independent variables. They still evolve according to

$$\dot{x}_\perp = \pi \mathcal{P}_\perp \, , \qquad \dot{\mathcal{P}}_\perp = \frac{1}{\pi} x_\perp'' \, , \tag{3.12}$$

which are general equations valid in any orthonormal gauge. The τ-evolution of the transverse coordinates and momenta is generated by

$$H_{tr} = \frac{1}{2\pi} \int_0^\pi d\sigma \left[(\pi \mathcal{P}_\perp)^2 + x_\perp'^2 \right] . \tag{3.13}$$

We assume of course canonical Poisson brackets for the independent variables $x_\perp(\sigma)$ and $\mathcal{P}_\perp(\sigma)$. Notice that

$$H_{tr} = p_+ \int_0^\pi \mathcal{P}_-(\sigma)\, d\sigma = p_+ p_- \tag{3.14}$$

i.e., the Hamiltonian is proportional to p_-, with constant of proportionality p_+: it had to be so, since $x_+ = p_+\tau$, and the infinitesimal x_+ displacements are generated by p_-.

The transverse variables do not form a complete set of dynamical variables, because p_+ has to be further specified, as well as some initial baricentric x_- coordinate

$$q_- = \frac{1}{\pi}\int_0^\pi d\sigma \; x_-(\sigma, \tau=0) .-$$

The need to consider q_- as independent variable arises from the fact that the constraint Eq. (3.11a) contains only x'_-.

We are lead then to the following description of our system: independent variables are $x_\perp(\sigma)$, $\mathcal{P}_\perp(\sigma)$, p_+ and q_-, with canonical Poisson brackets:

$$\{ x_i(\sigma), \mathcal{P}_j(\sigma') \} = \delta_{ij}\, \delta(\sigma-\sigma') , \tag{3.15}$$

$$\{ q_-, p_+ \} = -1 .- \tag{3.16}$$

The Hamiltonian is given by

$$H = \frac{1}{2\pi}\int_0^\pi d\sigma \left[(\pi \mathcal{P}_\perp)^2 + x_\perp'^2 \right] .\,.$$

The + coordinates and momenta are given by the gauge specification

$$\mathcal{P}_+(\sigma,\tau) = \text{const} = \frac{p_+}{\pi} ,$$

$$x_+(\sigma,\tau) = p_+\tau ,$$

whereas $\mathcal{P}_-$ and x_- are dependent variables

$$\mathcal{P}_- = \frac{1}{2\pi p_+}\left[(\pi \mathcal{P}_\perp)^2 + x_\perp'^2 \right] \tag{3.11b}$$

$$x_-(\sigma,\tau) = q_- + p_-\tau + \int_0^\pi d\sigma' \left[\sigma' - \theta(\sigma'-\sigma)\right] \frac{(\pi \mathcal{P}_\perp) x_\perp'}{p_+} .- \tag{3.17}$$

In this last equation we have used $\dot{x}_- = \pi \mathcal{P}_-$ to express

$$\frac{1}{\pi}\int_0^\pi d\sigma \; x_-(\sigma,\tau) = q_- + \tau\int_0^\pi d\sigma \; \mathcal{P}_-(\sigma) = q_- + p_-\tau .-$$

The transverse coordinates can be expanded into modes, and the normal coordinates can be quantized. The expansions are quite analogous to the expansions used in the previous section [Eqs. (2.56) and (2.57), with $2\alpha' = 1$]:

$$x_\perp = x_{0\perp} + p_\perp\tau + i\sum_{n>0} \frac{\cos n\sigma}{\sqrt{n}}\left[a_{n\perp} e^{-in\tau} - a^+_{n\perp} e^{in\tau} \right] , \tag{3.18}$$

$$\mathcal{P}_\perp = \frac{p_\perp}{\pi} + \frac{1}{\pi}\sum_{n>0} \sqrt{n}\cos n\sigma \left[a_{n\perp} e^{-in\tau} + a^+_{n\perp} e^{in\tau} \right] .- \tag{3.19}$$

H is given by

$$H = p_+ p_- = \frac{p_\perp^2}{2} + \sum_{n>0} n\, a_{\perp n}^{+} a_{\perp n} - \alpha_0 \ .- \tag{3.20}$$

We must allow for a constant α_0 in Eq. (3.20) because the classical form of H cannot fix ordering ambiguities. Notice that Eq. (3.20) is an equation for the spectrum. It is analogous to Eq. (2.73):

$$\left(\frac{p^2}{2} + \sum_{n>0} n\, a_{\mu n}^{+} a_n^{\mu} \right) |\varphi\rangle = \alpha_0 |\varphi\rangle ,$$

however only transverse excitations are present in Eq. (3.20). We argue that the role of the subsidiary conditions $L_n|\varphi\rangle = 0$ must be of removing time-like and longitudinal excitations from the spectrum of physical states.

The $a_{-,n}$ normal coordinates are dependent variables. We expand x_- and $\mathcal{P}_-$ according to Eqs. (3.18) and (3.19) (with q_- replacing $x_{0\perp}$, $a_{n-} \leftrightarrows a_{n\perp}$, $p_- \leftrightarrows p_\perp$); then we obtain, from Eqs. (3.11b) and (3.17):

$$a_{n,-} = \frac{1}{p_+} L_n^{tr} , \tag{3.21}$$

$$p_- = \frac{1}{p_+} \left(L_0^{tr} - \alpha_0 \right) .- \tag{3.22}$$

The L_n^{tr} are the same bilinear forms as the L_n, defined in Section 2, where only the transverse variables are kept. The only ambiguity in going from the classical Equations to the quantum expressions (3.21) is in the definition of p_-, where a constant α_0 can appear, which is of course the same as in Eq. (3.20).

This procedure of quantization is clearly non covariant. It will be internally self-consistent only if it admits a representation of Lorentz transformations. To understand what this representation may be, let us imagine we perform an infinitesimal Lorentz transformation

$$\Lambda_\mu{}^\nu = \delta_\mu{}^\nu + \epsilon\, m_\mu{}^\nu$$

Denoting by ~ the new values of the variables after the Lorentz transformation, we have

$$\begin{aligned} \tilde{x}_+(\sigma,\tau) &= x_+ + \epsilon\, m_+{}^\nu x_\nu(\sigma,\tau) = \\ &= p_+\tau + \epsilon\, m_+{}^\nu x_\nu(\sigma,\tau) \ .- \end{aligned} \tag{3.23}$$

We see that the gauge condition

$$\tilde{x}_+ = \tilde{p}_+\tau = (p_+ + \epsilon\, m_+{}^\nu p_\nu)\tau$$

is no longer satisfied in the new frame. We introduce a transverse gauge in the new frame by an infinitesimal change of parametrization, defining

$$\tilde{\tau} = \frac{\tilde{x}_+}{\tilde{p}_+} = \tau + \epsilon \frac{\mathcal{M}_+{}^{\nu}}{p_+} x_\nu(\sigma\tau) - \epsilon \frac{\mathcal{M}_+{}^{\nu}}{p_+} p_\nu \tag{3.24}$$

and

$$\tilde{\sigma} = \sigma + \frac{\epsilon}{p_+} \int_0^\sigma d\sigma' \mathcal{M}_+{}^{\nu} x_\nu'(\sigma\tau) - \epsilon \frac{\sigma \mathcal{M}_+{}^{\nu} p_\nu}{p_+} \;. \tag{3.25}$$

In the new frame, the evolution of the system will be described by the functions

$$\tilde{x}_\mu(\sigma(\tilde{\sigma},\tilde{\tau}), \tau(\tilde{\sigma},\tilde{\tau})).$$

This means an overall variation (Lorentz transformation plus change of gauge)

$$\delta x_\mu = \tilde{x}_\mu(\sigma(\tilde{\sigma},\tilde{\tau}), \tau(\tilde{\sigma},\tilde{\tau})) - x_\mu(\tilde{\sigma},\tilde{\tau}) \,.$$

$$= \epsilon\left[\mathcal{M}_\mu{}^{\nu} x_\nu - \frac{\mathcal{M}^+{}_\nu}{p_+}\left\{x'_\mu\left(\int_0^\sigma d\sigma' \dot{x}^\nu - \sigma p^\nu\right) + \dot{x}_\mu\left(x^\nu - \tau p^\nu\right)\right\}\right] \;. \tag{3.26}$$

Equation (3.26) (together with the corresponding equation for $\delta\mathcal{P}_\mu = \frac{\delta\dot{x}}{\pi}$) gives a non linear realization of Lorentz transformations in the space of the transverse variables. It can be checked that the transformation (3.26) is generated (through Poisson brackets) by the angular momentum of the system

$$M = \int d\sigma \, \mathcal{M}_{\mu\nu} \mathcal{P}^\mu(\sigma) \, x^\nu(\sigma) \,,$$

if we substitute for x_+, x_-, $\mathcal{P}_+$, $\mathcal{P}_-$ their expressions in terms of p_+, q_-, $x_\perp$, $\mathcal{P}_\perp$; we have then

$$\delta x_\mu = \{M, x_\mu\} \;.$$

The quantities

$$M_{\mu\nu} = \int d\sigma \left(\mathcal{P}_\mu x_\nu - \mu \leftrightarrow \nu\right) , \tag{3.27}$$

when expressed in terms of p_+, q_-, $x_\perp$, $\mathcal{P}_\perp$, have an algebra of Poisson brackets identical to the algebra of Lorentz generators. We could argue, from all this, that the Lorentz covariance of the transverse quantization will result from the existence of a representation of the Lorentz group in the Hilbert space of transverse excitations, obtained by replacing the quantities $M_{\mu\nu}$ by their quantum counterparts.

We have to be careful, however, because we know that the normal ordering of operators can produce differences between Poisson brackets and commutators. We had an example where we saw that, going to the q-system, the algebra of the L_n constraints acquired a c-number term.

We take over from Eq. (3.27) quantum expressions for the generators of Lorentz transformations, but check explicitly that their algebra closes correctly.

For simplicity, we consider only the generators of the little group of a string with momentum

$$P_\mu = (P_+ , P_- , P_\perp = 0) .-$$

These are

$$M_{ij} = i \sum_{n>0} a^+_{in} a_{jn} - (i \leftrightarrow j) , \tag{3.28}$$

$$M_{Li} = i \sum_{n>0} \left(L^{tr +}_n \frac{a_{in}}{\sqrt{n}} - \frac{a^+_{in}}{\sqrt{n}} L^{tr}_n \right) .- \tag{3.29}$$

Notice that no ordering ambiguity is present in the definition of M_{ij} and M_{Li}.

We expect, from the algebra of Lorentz generators,

$$[M_{Li}, M_{Lj}] = -2i \, p_+ p_- M_{ij} \; .- \tag{3.30}$$

Computing the commutator we find[21)]

$$[M_{Li}, M_{Lj}] = -2i \left(L^{tr}_0 - \frac{d^{tr}}{24} \right) M_{ij} - i \left(\frac{d^{tr}}{24} - 1 \right) \sum_{n>0} n^2 (a^+_{ni} a_{nj} - (i \leftrightarrow j)), \tag{3.31}$$

where d^{tr} is the number of transverse indices. We see that the algebra does not close correctly unless $d^{tr} = 24$; if $d^{tr} = 24$, comparison of $p_+ p_- = (L^{tr}_0 - 1)$ with Eq. (3.22) fixes the c-number α_0 in the definition of p_- to be equal to one.

We see that the transverse quantization is self consistent only in a space of d = 26 dimensions, with intercept $\alpha_0 = 1$ for the Regge trajectories. These two critical numbers come in many different points in the theory of dual resonance models. We can give an explanation of $\alpha_0 = 1$ (but we have no physical understanding of $d^{tr} = 24$).

We want to represent the whole Lorentz group with transverse excitations only. At the first excited level we have $d^{tr} = d - 2$ spin states $a^+_{1,i} |\phi\rangle$; the little group of a particle of finite mass is O(25). At the first level we are short of one spin state to represent O(25), and the particle must have mass zero, which is equivalent to $\alpha_0 = 1$. It is interesting to notice that the states at the second level $a^+_{1i} a^+_{1j} |\phi\rangle$ [(d - 2)(d - 1)/2 states] and

$a^{+}_{2i}\ |\phi\rangle$ (d - 2 states) arrange themselves in a single irreducible representation of O(25) (of "spin 2", with (d - 1)d/2 - 1 states).

Let us now come back to the covariant method of quantization explained in Section 2. We want to define transverse states in the covariant formalism. The physical states satisfy the equations

$$L_n |\varphi\rangle = 0 \,.- \tag{2.71}$$

To define a transverse gauge we impose the additional conditions

$$a_{+,n} |\varphi\rangle = 0 \,.- \tag{3.32}$$

Equation (3.32) implies that only the "zero modes" x_{0+} and p_+ contribute to the expectation value of $x_+(\sigma,\tau)$ and $\mathcal{P}_+(\sigma,\tau)$; Eq. (3.32) is equivalent to requiring

$$\langle\varphi| x_+(\sigma\tau) |\varphi\rangle = \langle\varphi| x_{0+} + p_+\tau |\varphi\rangle , \tag{3.33}$$

$$\langle\varphi| \mathcal{P}_+(\sigma\tau) |\varphi\rangle = \langle\varphi| \frac{p_+}{\pi} |\varphi\rangle , \tag{3.34}$$

which we interpret as the quantum equivalent of Eqs. (3.8) and (3.9). The simultaneous solutions to Eqs. (2.71) and (3.32), which we define as transverse (physical) states, have been discovered by Del Giudice, Di Vecchia and Fubini [7]. Since their formalism is based on the theory of interaction, whereas we study for the moment the free spectrum, we just illustrate their solution, without deriving it. We use the fields [see Eqs. (2.58) and (2.59)]

$$Q_\mu(z) = x_{0\mu} - i p_\mu \log z + i \sum_n \frac{\alpha_n}{n} z^{-n} ,$$

$$P_\mu(z) = i z \frac{d}{dz} Q_\mu(z)$$

and define

$$A_{\perp,n} = \frac{1}{2\pi i} \oint_0 \frac{dz}{z} P_\perp(z)\, e^{i n k_- Q_+(z)} \qquad n \gtrless 0 \,.- \tag{3.35}$$

Since $e^{in\,k_-\,Q_+(z)}$ contains a factor

$$e^{n p_+ k_- \log z} = z^{n p_+ k_-}$$

which is not necessarily single valued, we use the convention that A_n should operate only on states such that $k_- p_+ = 1$. Notice that

$$[\alpha_{+,m}, A_{\perp,n}] = 0, \tag{3.36}$$

because $A_{\perp,n}$ contains no α_- operators; and also that

$$[L_m, A_{\perp,n}] = 0 . \tag{3.37}$$

This follows from

$$[L_m, P(z)] = z\frac{d}{dz} z^m P(z) ,$$

$$[L_m, Q(z)] = z^{m+1}\frac{d}{dz} Q(z) ,$$

which in turn imply

$$[L_m, A_{\perp,n}] = \frac{1}{2\pi i}\oint \frac{dz}{z}\Big\{\Big(z\frac{d}{dz} z^m P_\perp(z)\Big) e^{ink_- Q_+(z)} +$$

$$+ P_\perp(z)\, z^{m+1}\frac{d}{dz} e^{ink_- Q_+(z)}\Big\} = \frac{1}{2\pi i}\oint dz \frac{d}{dz}\Big\{z^m P_\perp(z) e^{ink_- Q_+(z)}\Big\} = 0 .$$

It is easy to prove that

$$A_n |\emptyset\rangle = 0 \qquad n > 0 , \tag{3.38}$$

and that

$$[A_{mi}, A_{nj}] = m\,\delta_{ij}\,\delta_{m,-n} . \tag{3.39}$$

Equations (3.36) to (3.39) imply that a basis for the whole set of transverse states is obtained by acting with the A^+_{ni} operators on ground state vectors carying p_+ momentum equal to $1/k_-$:

$$|\varphi\rangle_{tr} = A^+_{n_1 i_1} A^+_{n_2 i_2} \cdots\cdots \quad e^{i\Pi_\mu x_0^\mu} |\emptyset\rangle, \tag{3.40}$$

with $\Pi_+ = 1/k_-$.

The transverse states are on mass shell if the additional condition $(L_0 - \alpha_0)\,|\varphi\rangle = 0$ is satisfied. This is clearly equivalent to $\Pi_\mu \Pi^\mu = \alpha_0$. The existence of the transverse states is the starting point for the proof of the no-ghost theorem[8], i.e., that the conditions $L_n|\varphi\rangle = 0$ and $(L_0 - \alpha_0)|\varphi\rangle = 0$, with a suitable α_0, eliminate all the states of non positive norm from the spectrum of physical states. From the quantization in the transverse gauge we would expect the transverse states to form a complete set of physical states, in the sense that it should be possible to write every physical state as the sum of a transverse state with a state of zero norm, decoupled from all physical states

$$|\varphi\rangle = |\varphi_{tr}\rangle + |n.n.\rangle \tag{3.41}$$

with

$$\langle n.n.|n.n.\rangle = \langle n.n.|\varphi\rangle = 0 .-$$

Equation (3.41) is indeed true, provided the intercept α_0 is taken equal to one and the number of dimensions d is 26. This was proven by Goddard, Thorn and Brower[8], who first established the no-ghost theorem for the theory of dual resonance models.

It is not surprising that the conditions $\alpha_0 = 1$, $d = d^{tr} + 2 = 26$ are the same as one finds from self consistency requirements for the transverse quantization. We shall not reproduce here the non-ghost theorem, for which we refer to the original articles[8] and to the lectures given last year at Erice by Prof. Fubini[22].

We outline instead a recent work of Brink and Olive[23], from which also the transverse states can be deduced, and has lead to interesting developments. Brink and Olive construct the projection operator over the space of on-mass shell transverse states.

We consider the space t of transverse states spanned by the basis vectors $|\varphi\rangle_{tr}$ defined in Eq. (3.40). All of these vectors have definite $p_+ = 1/k_-$ and $p_\perp = 0$, and satisfy the on mass shell condition $(L_0 - \alpha_0)|\varphi\rangle = 0$. We therefore consider the space h of all states having the same values of p_+, $p_\perp$ and obeying the condition $(L_0 - \alpha_0)|\varphi\rangle$, and look for a projection operator τ from h onto t.

The main idea in the work of Ref. 23 is to compare the excitation number operators

$$R = \Sigma_{n,\mu}\, n\, a^+_{n\mu}\, a^\mu_n \tag{3.42}$$

and

$$R_{tr} = \Sigma_{n,i}\, A^+_{ni}\, A_{ni} \ , \tag{3.43}$$

[remember that the A's operators are normalized to $[A_{ni}, A^+_{ni}] = n$]. We have

$$L_0 = R - \frac{p^2}{2} = (\text{within } h) = R - p_+p_- .- \tag{3.44}$$

R has non negative integer eigenvalues r, which determine the values of p_- and M^2. The on-mass shell condition implies

$$p_- = \frac{M^2}{2p_+} = -\frac{\alpha_0}{p_+} + \frac{R}{p_+} .- \tag{3.45}$$

But for the transverse states we have also

$$p_- = \frac{M^2}{2p_+} = -\frac{\alpha_0}{p_+} + \frac{R_{tr}}{p_+} \tag{3.46}$$

because the ground state has momentum $p_- = -\alpha_0/p_+$, the operator A^+_{ni} increases p_- by

$$n k_- = \frac{n}{p_+}$$

and

$$[R_{tr}, A^+_{n,i}] = n A^+_{ni} \; .-$$

It follows that R and R_{tr} take the same value on t: all over t we have. $R - R_{tr} = 0$; it also follows that

$$[R - R_{tr}, A_{n,i}] = 0 \; .- \tag{3.47}$$

Notice now that $R - R_{tr}$ has non negative integer eigenvalues all over h: indeed, we have

$$[R - R_{tr}, a_{+,n}] = [R, a_{+,n}] = -n a_{+,n} \, , \tag{3.48}$$

$$[R - R_{tr}, L_n] = [R, L_n] = [L_0, L_n] = -n L_n \, . \tag{3.49}$$

Then, if $|\alpha\rangle$ is an eigenvector of $R - R_{tr}$ with eigenvalue α, by acting with $A_{i,n}$, $a_{+,n}$ and L_n on $|\alpha\rangle$ we obtain eigenvectors of $R - R_{tr}$ with eigenvalue $\alpha - m$, where m is an integer ≥ 0. But, by acting repeatedly with $A_{i,n}$, $a_{+,n}$ and L_n we eventually reach the ground state, with $(R - R_{tr})|\phi\rangle = 0$. It follows α = -non negative integer.

We obtain the projection operator τ onto t by a contour integral, in the following way:

$$\tau = \frac{1}{2\pi i} \oint dx \; x^{R - R_{tr} - 1} \tag{3.50}$$

Having derived τ in the form (3.50), the authors of Ref. 23 are able to express the exponent $R - R_{tr}$ in a form that involves the gauge operators L_n. Their result is

$$R_{tr} - R = (D_0 - 1)\left(L_0 - \frac{d^{tr}}{24}\right) + \sum_{n>0} (D_{-n} L_n + L_{-n} D_n) + \\ + \left(1 - \frac{d_{tr}}{24}\right) \oint \frac{dy}{2\pi i y} \frac{1}{k_- P_+(y)} \left[y \frac{d}{dy} \ln(k_- P_+(y)) \right]^2 , \tag{3.51}$$

where

$$D_n = \frac{1}{2\pi i} \oint \frac{dz}{z}\, z^n\, k_- P_+(z)\,. \tag{3.52}$$

We briefly outline their method of derivation. One writes the explicit form for R_{tr}

$$R_{tr} = \sum_{n,i} \frac{1}{(2\pi i)^2} \oint\oint_{|x|>|y|} \frac{dx}{x}\frac{dy}{y}\, P_i(x)\, e^{-in k_- Q_+(x)} P_i(y)\, e^{in k_- Q_+(y)} =$$

$$= \frac{1}{(2\pi i)^2} \oint\oint_{|x|>|y|} \frac{dx}{x}\frac{dy}{y} \left\{ :P_\perp(x) P_\perp(y): + \frac{d^{tr}\, xy}{(x-y)^2} \right\} \sum_n e^{-in k_-(Q_+(x) - Q_+(y))}\,. \tag{3.53}$$

The sum over n can be performed and gives

$$R_{tr} = \frac{1}{(2\pi i)^2} \oint\oint_{|x|>|y|} \frac{dx}{x}\frac{dy}{y} \left\{ :P_\perp(x) P_\perp(y): + \frac{d^{tr}\, xy}{(x-y)^2} \right\} \cdot \tag{3.54}$$

$$\cdot \frac{e^{-ik_- Q_+(x)}}{e^{-ik_- Q_+(y)} - e^{-ik_- Q_+(x)}}\,.$$

The term with: $P_\perp(x)P_\perp(y)$: in integrand of Eq. (3.54) has a simple pole for x = y. Displacing the contours to $|y| > |x|$ one picks up the residence at the pole, and obtains an integral which is exactly the negative of the original one. It follows:

$$\frac{1}{(2\pi i)^2} \oint\oint_{|x|>|y|} \frac{dx}{x}\frac{dy}{y} :P_\perp(x)P_\perp(y): \frac{e^{-ik_- Q_+(x)}}{e^{-ik_- Q_+(y)} - e^{-ik_- Q_+(x)}} =$$

$$= \frac{1}{2}\frac{1}{2\pi i} \oint \frac{dx}{x} \operatorname*{Res}_{x=y} :P_\perp(x)P_\perp(y): \frac{e^{-ik_- Q_+(x)}}{e^{-ik_- Q_+(y)} - e^{-ik_- Q_+(x)}} = \tag{3.55}$$

$$= \frac{1}{2}\frac{1}{2\pi i} \oint \frac{dx}{x} \frac{:P^2(y):}{k_- P_+(y)} - \frac{p^2}{2}\,.$$

(one uses $\Sigma_i P_i P_i = P^2 + 2P_+P_-$). This is the most relevant part of the computation. Having evaluated this term, the dimension dependent term can be computed and one finally arrives to Eq. (3.51). Notice that for $d^{tr} = 24$, Eq. (3.51) reduces to

$$R_{tr} - R = (D_0 - 1)(L_0 - 1) + \sum_{n>0} (D_{-n} L_n + L_{-n} D_n) \quad .- \qquad (3.56)$$

If the on mass shell condition is $(L_0 - 1)|\varphi\rangle = 0$, i.e., if $\alpha_0 = 1$, the matrix elements of $R_{tr} - R$, and also of $x^{R_{tr}-R}$, between on-mass shell physical states vanish.

We have then

$$\langle\varphi_1|\varphi_2\rangle = \langle\varphi_1|\tau|\varphi_2\rangle = {}_{tr}\langle\varphi_1|\varphi_2\rangle_{tr} \quad , \qquad (3.57)$$

if

$$|\varphi_1\rangle \ , \ |\varphi_2\rangle$$

are physical and defining

$$|\varphi\rangle_{tr} = \tau\,|\varphi\rangle \quad .-$$

Equation (3.57) expresses the completeness of the transverse states when $d_{tr} = 24$, $\alpha_0 = 1$. When these two critical conditions are met any physical state can be expressed as a sum of a transverse state with a decoupled state of zero norm.

4. THE INTERACTION AND RELATED PROBLEMS

The quantization of the dynamical system described in Section 3 and 4 (the relativistic string) gives origin to the same spectrum of states, which one finds in the factorization of the conventional dual resonance model.

One would expect that, introducing a suitable interaction among quantized strings, it should be possible to reobtain the dual amplitudes. Although the problem of giving a string interpretation to the dual resonance amplitudes has not yet been completely solved, some progress has been recently made by the authors of Ref. 24, and we shall outline here their results. To explain the results obtained in Ref. 24 and to understand the problems which remain unsolved, we need some basic notions from the theory of dual resonance models*).

---

*) Only a rather general knowledge of the properties of the dual amplitudes is assumed in what follows. The reader completely unfamiliar with the subject can consult the works of Ref. 13.

The dual amplitude $A_N(k_1, \ldots, k_N)$ for the scattering of N ground state particles with momenta $k_1, \ldots, k_N$, has poles in all the planar channels associated with the ordering $k_1, \ldots, k_N$ of the external momenta, and can be fully factorized, to exibit the maximum number (N-3) of compatible resonant poles.

A very useful factorization is the one associated with the multiperipheral configuration, exhibited in Fig. 1:

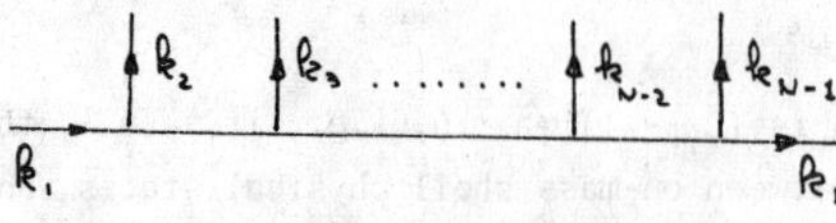

Fig 1)

This factorization is made explicit by the following representation

$$A(k_1 \ldots k_N) = \langle k_N | V(k_{N-1}) D \ldots . V(k_3) D V(k_2) | k_1 \rangle . \quad (4.1)$$

In Eq. (4.1) $|k_1\rangle$ and $|k_N\rangle$ are two vectors representing the initial and final hadrons of the multiperipheral chain, $D = 1/(L_0 - 1)$ is a propagator, and $V(k)$ is a vertex for the emission of a ground state particle with momentum k. Notice that the representation (4.1) remains valid also if the initial and final hadrons are in some excited state [the appropriate vectors $|k_1\rangle$ and $|k_N\rangle$, satisfying the Virasoro gange conditions $L_m|k_1\rangle = L_m|k_N\rangle = 0$ $(m \geq 1)$, must of course be inserted in Eq. (4.1)]

We can then interpret A as the amplitude for the scattering of a system with all the degrees of freedom of the intermediate resonant states, accompanied by the emission of quanta of some external hadronic field.

In Ref. 24 it is shown that, introducing a suitable interaction of the string with an external field, one can obtain, in the quantized version of the theory, A_N as transition amplitude *,**).

Before explaining these results, let us briefly comment on the problems that remain open.

---

*) To be precise, the amplitude obtained in Ref. 24 is not A_N, but the analogous amplitude for the emission of the first excited state of the theory.

**) Results in the same direction have been obtained before in a work of Gervais and Sakita [25], which however make use of different techniques, based on functional integration.

A more complete treatment of the interaction among strings with the method of functional integration can be found in a recent article of S. Mandelstam[36].

The representation (4.1) obscures somehow the duality properties of A_N. From thé theory of dual resonance model we know that A_N can be equally well factorized in any planar tree configuration. In more general factorizations, the vertex W among three excited states (the three-Reggeon vertex) can appear, as illustrated in the example of Fig. 2:

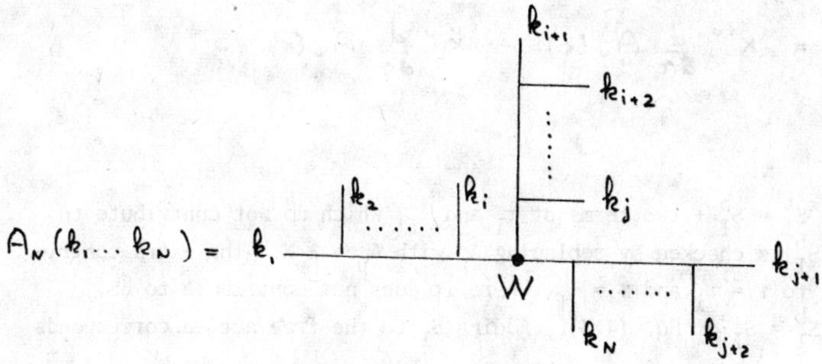

Fig 2)

A problem yet to be solved is to derive the form of W from the theory of the quantum string. The work we are going to describe may be a first step in this direction, because it gives W when one of the legs is in the first excited level.

In the theory of dual resonance models the hadrons emitted in the multiperipheral chain of Fig. 1 can be in any excited state. The vertex V(k) will be of course correspondingly modified. It is convenient to think of the amplitude for the emission of hadrons of the first excited level. These are massless vector particles, having the same properties of the photons. (The vertex is $\varepsilon_\mu V^\mu(k)$, with $\varepsilon_\mu k^\mu = 0$, $k^2 = 0$). This leads us to consider the interaction of the string with an external electromagnetic field. In the case of a plane electromagnetic wave the equations of motion can be explicitly solved[24], and the quantum mechanical transition amplitude is the dual amplitude. We couple the electromagnetic field to the string by adding a term to the free action S [Eq. (2.6)]. We want to add a term S_I invariant with respect of changes of parametrization of the surface of evolution $X_\mu(\sigma,\tau)$, dependent on $\dot{X}_\mu$, X'_μ, $F_{\mu\nu}$, or on A_μ in a gauge invariant way. The possible choices for S_I are the following ones:

$$S_I^a = -\frac{e}{\pi}\int_{\tau_i}^{\tau_f} d\tau \int_0^{\pi} d\sigma \; F_{\mu\nu}(x)\,\dot{x}^\mu x'^\nu \,, \tag{4.2}$$

and

$$S_I^b = -\frac{g_0}{\pi}\int_{\tau_i}^{\tau_f} d\tau\, A_\mu(x(\sigma=0))\,\dot{x}^\mu(\sigma=0) - \frac{g_\pi}{\pi}\int_{\tau_i}^{\tau_f} d\tau\, A_\mu(x(\sigma=\pi))\,\dot{x}^\mu(\sigma=\pi). \tag{4.3}$$

However, S_I^a is equivalent to S_I^b for $g_0 = g_\pi = e$. This is seen by writing

$$F_{\mu\nu}(x)\,\dot{x}^\mu x'^\nu = x'^\nu(\dot{x}^\mu \partial_\mu)\,A_\nu(x) - \dot{x}^\nu(x'^\mu \partial_\mu)\,A_\nu(x) =$$

$$= x'^\nu \frac{d}{d\tau} A_\nu(x) - \dot{x}^\nu \frac{d}{d\sigma} A_\nu(x) \,.-$$

Integrating by part we obtain $S_I^a = S_I^b$ + two terms at τ_i and τ_f, which do not contribute to δS. The gauge invariance of S_I^b is checked by replacing A_μ with $A_\mu + \partial_\mu X$. The extra contribution can be integrated away to $\tau = \tau_i$ and $\tau = \tau_f$, where it does not contribute to δS. We keep therefore the action $S_I^b = S_I$ of Eq. (4.3). Adding S_I to the free action corresponds to adding a term

$$\mathcal{L}_I = \dot{x}_\mu A^\mu \Delta(\sigma) \,, \tag{4.4}$$

with

$$\Delta(\sigma) = -\frac{g_0}{\pi}\delta(\sigma) - \frac{g_\pi}{\pi}\delta(\pi-\sigma) \tag{4.5}$$

to the free Lagrangian density Eq. (2.9). The Euler-Lagrange Eqs. (2.11) for $0 < \sigma < \pi$ are unaffected, but the boundary conditions (2.13):

$$\left.\frac{\partial \mathcal{L}}{\partial x'}\right|_0^\pi = 0$$

are replaced by

$$-\frac{g_\pi}{\pi}\left(\frac{d}{d\tau}A_\mu - \dot{x}^\nu \frac{\partial}{\partial x^\mu} A_\nu\right) = -\frac{g_\pi}{\pi}\dot{x}^\nu F_{\nu\mu}(x) = \frac{\partial \mathcal{L}}{\partial x'^\mu} \qquad \sigma = \pi, \tag{4.6a}$$

$$-\frac{g_0}{\pi}\left(\frac{d}{d\tau}A_\mu - \dot{x}^\nu \frac{\partial}{\partial x^\mu} A_\nu\right) = -\frac{g_0}{\pi}\dot{x}^\nu F_{\nu\mu}(x) = -\frac{\partial \mathcal{L}}{\partial x'^\mu} \qquad \sigma = 0.- \tag{4.6b}$$

The new boundary conditions are non-linear in the variable x_μ, because x_μ appears as argument of the external field $A_\mu(x)$. Still, in the case of a plane wave

$$A_\mu(x) = \epsilon_\mu e^{ik_\mu x^\mu} \,, \tag{4.7}$$

with

$$k^2 = k_\mu \epsilon^\mu = 0 \quad , \tag{4.8}$$

the Eqs. (2.11) and (4.6) can be solved exactly. The clue is that we can separate

$$x_\mu (\sigma,\tau) = x_{0\mu} (\sigma\tau) + x_{I\mu} (\sigma\tau) , \tag{4.9}$$

where x_0^μ satisfies the free equations for the motion of the string in the orthonormal gauge (in particular $x'_\mu \,|_{\sigma=0}^{\pi} = 0$), whereas x_I^μ consists of two terms, proportional to ϵ^μ and k^μ respectively

$$x_I^\mu = \epsilon^\mu a(\sigma,\tau) + k^\mu b(\sigma,\tau) \ .- \tag{4.10}$$

Since in the argument of the plane wave only $k_\mu x^\mu$ appears, and $k_\mu \epsilon^\mu = k^2 = 0$, we see that only $x_0^\mu(\sigma\tau)$ comes in the argument of the field, and it is this fact that allows one to solve the equations of motion in a compact form. It is convenient (not necessary) to consider circularly polarized waves, so that

$$\epsilon^2 = 0 \ .-$$

We still require that the complete solution to the equation of motion

$$x_\mu(\sigma\tau) = x_{0\mu}(\sigma\tau) + x_{I\mu}(\sigma\tau)$$

satisfies the orthonormality conditions

$$\dot{x}^2 + x'^2 = 0$$

and

$$\dot{x}\, x' = 0 \ .- \tag{4.11}$$

Then the equations of motion become

$$\ddot{x}_\mu - x''_\mu = 0 \tag{4.12}$$

and the boundary conditions (setting $\alpha' = \frac{1}{2}$):

$$-x'_\mu = i g_0 (\epsilon_\mu k^\nu - \epsilon^\nu k_\mu) \dot{x}_\nu e^{ikx} \qquad \sigma = 0, \tag{4.13a}$$

$$x'_\mu = i g_\pi (\epsilon_\mu k^\nu - \epsilon^\nu k_\mu) \dot{x}_\nu e^{ikx} \qquad \sigma = \pi \ .- \tag{4.13b}$$

Eqs. (4.11), (4.12) and (4.13) have the solution (which we express in terms of $\dot{x}_\mu$ and x'_μ):

$$\dot{x}_\mu(\sigma,\tau) = \dot{x}_{0\mu}(\sigma\tau) + (\epsilon_\mu k^\nu - \epsilon^\nu k_\mu)\times$$
$$\times\left[R_+\dot{x}_{0\nu}(0,\tau-\sigma)e^{ikx_0(0,\tau-\sigma)} + R_-\dot{x}_{0\nu}(0,\tau+\sigma)e^{ikx_0(0,\tau+\sigma)}\right], \tag{4.14a}$$

$$x'_\mu(\sigma,\tau) = x'_{0\mu}(\sigma\tau) + (\epsilon_\mu k^\nu - \epsilon^\nu k_\mu)\times$$
$$\times\left[-R_+\dot{x}_{0\nu}(0,\tau-\sigma)e^{ikx_0(0,\tau-\sigma)} + R_-\dot{x}_{0\nu}(0,\tau+\sigma)e^{ikx_0(0,\tau+\sigma)}\right], \tag{4.14b}$$

where $x_{0\mu}(\sigma,\tau)$ is any solution to the free equation of motion, p_μ is the free momentum $[p_\mu = 1/\pi \int_0^\pi d\sigma\, \dot{x}_0(\sigma,\tau)]$, and

$$R_\pm = \frac{g_\pi + g_0 e^{\pm i(p\cdot k)\pi}}{2\sin(p\cdot k)\pi}\,. \tag{4.15}$$

The constants $R_\pm$ are fixed by demanding the validity of the boundary conditions (4.13). It is important to remember that (see Eq. 2.56):

$$x_{0\mu}(\sigma\tau) = \text{const} + p_\mu\tau + f_\mu(\sigma+\tau) + g_\mu(\sigma-\tau) \tag{4.16}$$

where both f_μ and g_μ are periodic functions with period 2π.

Equation (4.12) is clearly satisfied; to check the validity of Eqs. (4.11) it is more useful to write them in the equivalent form $(\dot{x} \pm x')^2 = 0$. One notices that [because of Eq. (4.16)]

$$(\dot{x}_0 \pm x'_0)(\sigma,\tau) = \dot{x}_0(0,\tau\pm\sigma)\,;$$

then Eqs. (4.11) give

$$\left(\dot{x}_0^\mu(0,\tau\pm\sigma) + 2(\epsilon^\mu k^\nu - \epsilon^\nu k^\mu)R_\mp\,\dot{x}_{0\nu}(0,\tau\pm\sigma)e^{ikx_0(0,\tau\pm\sigma)}\right)^2 = 0$$

which in turn reduces to

$$\dot{x}_0^2(0,\tau\pm\sigma) = (\dot{x}_0 \pm x'_0)^2 = 0\,,$$

i.e. to the orthonormality conditions for the free coordinates $x_{0\mu}$.

It is amusing to notice that even at the classical level the interaction of the string with the external electromagnetic field gives origin to a resonant phenomenon. The amplitude becomes infinite whenever $p \cdot k = n$ (n integer). The condition $p \cdot k = n$ is equivalent to

$$(p+k)^2 - p^2 = 2n$$

But $2n = n/\alpha'$ is the characteristic spacing between the mass square of the nth excited level and the mass square of the ground state, which is now found as a condition for a resonance even before quantization.

We omit at this point many of the details of the quantization of the interacting system, for which we refer to Ref. 24.

One sets up a canonical formalism, introducing

$$\mathcal{P}_\mu = \frac{\partial \mathcal{L}}{\partial \dot{x}^\mu} = \mathcal{P}_{0\mu} - \frac{g_0}{\pi} A_\mu \delta(\sigma) - \frac{g_\pi}{\pi} A_\mu \delta(\pi - \sigma) \tag{4.17}$$

and postulating canonical Poisson brackets first and then commutation relations between $\mathcal{P}_\mu$ and x_μ.

The normal coordinates are introduced by expanding $x_{0\mu}(\sigma\tau)$ and $\mathcal{P}_{0\mu}(\sigma\tau)$ in a cosine series, as done in Eqs. (2.56) and (2.57). Choosing a Hamiltonian is again equivalent to choosing a gauge. In the orthonormal gauge H is given by

$$H = H_0 - L_I \tag{4.18}$$

i.e., the Hamiltonian of interaction is

$$H_I(\tau) = \frac{g_0}{\pi} \epsilon_\mu \dot{x}_0^\mu(\sigma=0,\tau) e^{ik_\mu x_0^\mu(\sigma=0,\tau)} + \frac{g_\pi}{\pi} \epsilon_\mu \dot{x}_0^\mu(\sigma=\pi,\tau) e^{ik_\mu x_0^\mu(\sigma=\pi,\tau)} \tag{4.19}$$

Notice that the solution to the equations of motion remains valid even if we allow for a dependence of the polarization vector ε_μ and wave vector k_μ on τ. This permits us to evaluate the amplitude for emission of different photons in successive laps of τ. Standard perturbation theory gives us

$$A_{if} \propto \int_{-\infty}^{\infty} d\tau_1 \int_{\tau_1}^{\infty} d\tau_2 \cdots \int_{\tau_{n-1}}^{\infty} d\tau_n \langle f | H_I(\epsilon_n k_n, \tau_n) \ldots H_I(\epsilon_1 k_1, \tau_1) | i \rangle \tag{4.20}$$

This amplitude, after factoring out a δ function due to invariance under τ-translations, can be reduced to the conventional form of the dual amplitudes. Considering for instance $g_\pi = 0$, we obtain

$$A_{if} = \int_0^\infty d\tau_2 \int_{\tau_2}^\infty d\tau_3 \cdots \int_{\tau_{n-1}}^\infty d\tau_n \langle f | H_I(\epsilon_n k_n, \tau_n) \ldots H_I(\epsilon_1 k_1, 0) | i \rangle =$$

$$= \langle f | H_I(\epsilon_n k_n, 0) \frac{1}{L_0 - 1 - i\epsilon} \cdots H_I(\epsilon_1 k_1, 0) | i \rangle, \quad (4.21)$$

which coincides with the dual amplitude for emission of n photons. We have used L_0 to translate the H_I's to $\tau = 0$, $L_0 |f\rangle = |f\rangle$, and performed the τ integrations. The vertex for a photon emission in the dual theory is in fact

$$V(\epsilon_\mu k_\mu) = \epsilon_\mu P^\mu(\tau=0)\, e^{i k_\mu Q^\mu(\tau=0)} \, . \quad (4.22)$$

We have used in this equation the fields $P_\mu(\tau)$ and $Q_\mu(\tau)$, but, using Eqs. (2.58) and (2.59), we see that

$$\frac{g_0}{\pi} V(\epsilon_\mu k_\mu) = H_I(\epsilon_\mu k_\mu ; g_\pi = 0) \, .$$

If we allow for emission with equal strength at $\sigma = 0$ and $\sigma = \pi$, we obtain the amplitudes associated with generalized multiperipheral diagrams, with emissions from the two sides of the line (see Fig. 3):

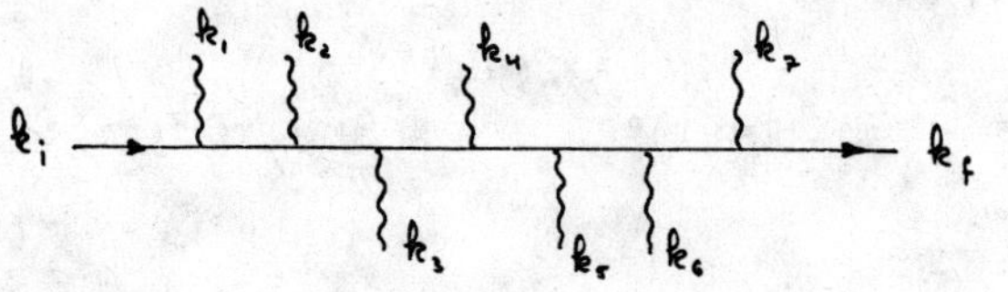

Fig 3)

If we let the τ_i values cover all of the τ range, dropping the restrictions $\tau_i < \tau_{i+1}$ (but of course τ-ordering the product of the interaction Hamiltonians) we obtain the full dual amplitude, sum of all the planar contributions.

It seems appropriate to redraw Fig. 3 in the following way

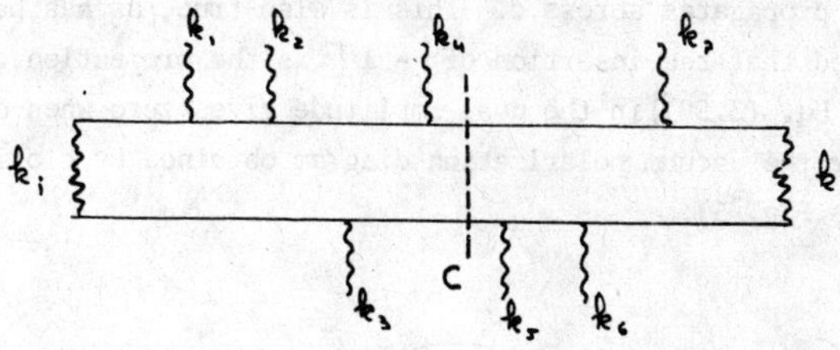

Fig. 4)

In Fig. 4 we represent the evolution of the string through a surface of evolution, which is coupled at the boundaries to the external field. This is said without pretense of rigour*). In the rest of this section we shall use this idea of the surface of evolution to give an intuitive explanation of some features of the dual amplitudes, which it would take too long to explain with more rigour.

The factorization of the amplitude corresponds to a cutting of the surface, as exemplified by line c in Fig. 4.

Notice that we should have only states satisfying the Virasoro gauge conditions propagating through c. This is indeed true.

We have

$$\langle k_f | V D V L_n D V \ldots D V | k_i \rangle =$$

$$= \langle k_f | V D V D L_n^{+} V \ldots D V | k_i \rangle = 0 \quad (n > 0) \tag{4.23}$$

Notice the positioning of the gauge operators. An expression $L_n/(L_0 - 1)$ would of course be meaningless. The L_n gauges $(n > 0)$, which are destruction gauges, must be placed after a propagator. Equation (4.23) is a consequence of the equation

$$(L_n - L_0 + 1) D V \ldots D V | k_i \rangle = 0$$

which is proven in the theory of dual resonance models, and of the fact that suppressing a propagator in a dual amplitude gives zero.

---

*) In fact, the surface of evolution to which we refer should be a sort of quantum mechanical surface, defined by the expectation values of $x_\mu(\sigma\tau)$, an operator for which many matrix elements are ill-defined.

For 26 dimensions of space time we would expect an even stronger result, i.e. that only transverse states propagates across c. This is also true, as has been shown by Brink and Olive[23], who proved that the insertion of $\tau - 1$ [τ is the projection operator on transverse states, given in Eq. (3.50)] in the dual amplitude gives zero when d = 26. It is interesting to consider the vacuum polarization diagram obtained by closing the surface of Fig. 4 into itself (see Fig. 5):

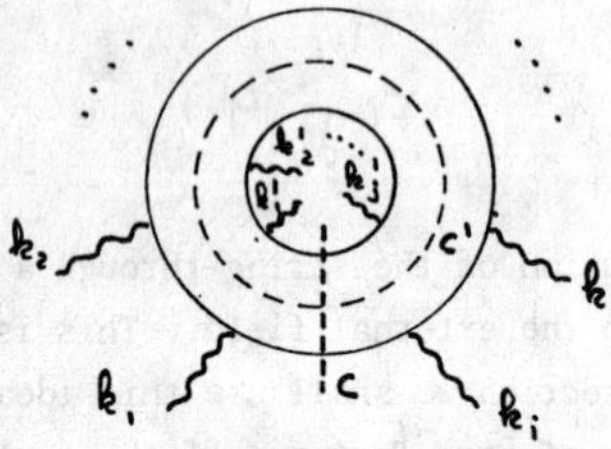

Fig 5)

The corresponding amplitude is the 1-loop amplitude of the dual resonance theory. The loop amplitude can be obtained as a trace

$$A = \mathrm{Tr}\,(VDVD \ldots \; D\tau) \tag{4.24}$$

however a projection operator over physical states must be inserted at some point in the loop. Otherwise a factorization of the surface, as the one represented by the line c in Fig. 5, might exhibit the presence of spurious states. This is a feature common to gauge theories: spurious states do not appear in tree diagrams, when the external particles are physical, but can circulate in loops, and must be explicitly projected out or subtracted. Notice that we might think of factorizing the loop amplitude by cutting the surface of evolution along line c′. This factorization would exhibit then a closed string propagating from the outer to the inner boundary (or vice-versa). A theory of the motion of a closed string can be given along exactly the same lines followed in Section 2 and 3. Only, a double set of creation and annihilation operators is required to span the Hilbert space of excitations, because the expansion of the string position and momentum is not limited any more to cosines. The classical spectrum is transverse, and transversality at the quantum level demands the existence of a massless state of spin 2 (together with d = 26)*). The loop amplitude can in fact be rewritten so as to exhibit the factorization along c′ [26]. However, it is crucial for this that only transverse states circulate in the loop[27]. The projection operator in Eq. (4.24) must be the projection operator over transverse states,

*) The spectrum of states of the closed string coincides with the spectrum of resonances appearing in the dual model proposed by Shapiro and Virasoro[10].

and, of course, we must work in 26-dimensional space-time. Then, the loop amplitude exhibits poles in the square of the sum of the outer (or inner) momenta $p^2 = (k_1 + \dots, k_i)^2 = (k_1' + \dots, k_j')^2$ corresponding to on mass shell closed strings propagating across c'. Notice that it is through the requirement that the singularities in p^2 were simple poles (poles in the Pomeron sector) and not cuts (in conflict with unitarity) that the magic number 26 made its first appearance in the theory of dual resonance models [Lovelace[28]].

OUTLOOK: THE MODELS OF NEVEU AND SCHWARTZ AND RAMOND, TOWARDS NEW FIELD THEORIES?

We wish to add to what we said about the conventional dual resonance model a brief presentation of the models of Neveu and Schwartz[9] and Ramond[11]. Remember that we can express all the dynamical variables of the conventional string [$\mathcal{P}_\mu(\sigma)$ and $x'_\mu(\sigma)$] by introducing the single field [Eq. (2.62) with $\alpha' = \frac{1}{2}$]

$$P_\mu(\sigma) = \pi \mathcal{P}_\mu(\sigma) + x'_\mu(\sigma)$$

defined over the extended range $-\pi \leq \sigma \leq \pi$. The commutation relations of P_μ with itself are

$$[P_\mu(\sigma), P_\nu(\sigma')] = 2\pi g_{\mu\nu} \delta'(\sigma - \sigma')$$

The models of Neveu and Schwartz and Ramond are obtained by supplementing the variable $P_\mu(\sigma)$ with a new variable $\Gamma_\mu(\sigma)$ (corresponding somehow to a continuous spin-degree of freedom).

For $\Gamma_\mu(\sigma)$, defined over the range $-\pi \leq \sigma \leq \pi$, one assumes anticommutation relations

$$\{\Gamma_\mu(\sigma), \Gamma_\nu(\sigma')\} = -4\pi g_{\mu\nu} \delta(\sigma - \sigma') \tag{5.1}$$

The field $\Gamma_\mu(\sigma)$ represents a sort of continuum of γ matrices, defined over the string.

The two models of Neveu-Schwartz and Ramond are differentiated by the expansions of $\Gamma_\mu(\sigma)$ into normal modes. One expands

$$\Gamma_\mu(\sigma) = i\sqrt{2} \sum_r b_{r\mu} e^{ir\sigma} \qquad r = \pm\tfrac{1}{2}, \pm\tfrac{3}{2} \dots \tag{5.2}$$

for the model of Neveu and Schwartz,

$$\Gamma_\mu(\sigma) = \sum_m \Gamma_{m\mu} e^{im\sigma} \qquad m = 0, \pm 1, \pm 2 \dots \tag{5.3}$$

for the model of Ramond.

The ground state vectors are assumed to be annihilated by all the $b_{r\mu}$ or $\Gamma_{m\mu}$ with positive index. So, we shall characterize the ground state of the model of Neveu and Schwartz by a single vector $|\phi\rangle$, with $b_r|\phi\rangle = 0$ $(r > 0)$.

In the case of the model of Ramond, however, the ground state cannot be represented by a single vector, because we must allow for a representation of the $\Gamma_{0\mu}$ matrices. We represent the ground state of the Ramond model with vectors $|u\rangle$, where u is an ordinary Dirac spinor, and define

$$\Gamma_{0\mu}|u\rangle = |\gamma_\mu u\rangle . \tag{5.4}$$

The Fock spaces of excitations H_{NS} and H_R are obtained by acting with a_m^+, b_r^+ and a_m^+, Γ_m^+ on $|\phi\rangle$ and $|u\rangle$ respectively:

$$H_{NS}: \quad \{ b^+_{\mu_1 r_1} b^+_{\mu_2 r_2} \dots a^+_{\nu_1 n_1} \dots |\phi\rangle \}, \tag{5.5}$$

$$H_R: \quad \{ \Gamma^+_{\mu_1 m_1} \Gamma^+_{\mu_2 m_2} \dots a^+_{\nu_1 n_1} \dots |u\rangle \} . \tag{5.6}$$

They clearly describe spectra consisting of an infinite number of bosons (H_{NS}) or fermions (H_R). One introduces the Virasoro gauges

$$L_n = \frac{1}{2\pi i} \oint_0 \frac{dz}{z} z^n \left\{ :P^2(z): + \frac{1}{2} : \frac{\Gamma(z)}{\sqrt{z}} \frac{d}{dz} \frac{\Gamma(z)}{\sqrt{z}} : \right\} \tag{5.7}$$

$(z = e^{in\sigma})$, and also new additional gauges

$$F_n = \frac{1}{2\pi\sqrt{2}} \oint_0 \frac{dz}{z} z^n \, \Gamma(z) P(z) \qquad (n \text{ integer}) \tag{5.8}$$

$$G_r = \frac{1}{2\pi\sqrt{2}} \oint_0 \frac{dz}{z} z^r \, \Gamma(z) P(z) \qquad (r \text{ half-integer}) \tag{5.9}$$

The new gauges are the weighted averages of a generalized Dirac operator

$$\Gamma_\mu(z) P^\mu(z) .$$

The algebra of these gauges is

$$[L_m, L_n] = (m-n) L_{m+n} + \frac{d}{8}(m^3 - m)\delta_{m,-n} , \tag{5.10}$$

$$[L_m, F_n] = \left(\frac{m}{2} - n\right) F_{m+n} \quad , \tag{5.11a}$$

$$[L_m, G_r] = \left(\frac{m}{2} - r\right) G_{m+r} \quad , \tag{5.11b}$$

$$\{F_m, F_n\} = 2L_{m+n} + \frac{d}{2}\left(m^2 - \frac{1}{4}\right)\delta_{m,-n} \quad , \tag{5.12a}$$

$$\{G_r, G_s\} = 2L_{r+s} + \frac{d}{2}\left(r^2 - \frac{1}{4}\right)\delta_{r,-s} \; .- \tag{5.12b}$$

Note the different value (d/8; d/12 in the conventional model) of the coefficient of the c-number term in Eq. (5.10). This results from an additional contribution to the c-number from the spin degrees of freedom.

One imposes on all the physical vectors the conditions that they should be annihilated by the gauges with positive index:

$$L_n |\varphi\rangle_{NS} = G_r |\varphi\rangle_{NS} = 0 \, , \tag{5.13a}$$

$$L_n |\varphi\rangle_R = F_m |\varphi\rangle_R = 0 \quad , \quad n, r, m > 0 \; .- \tag{5.13b}$$

The spectral equations are

$$L_0 |\varphi\rangle_{NS} = \frac{1}{2} |\varphi\rangle_{NS} \quad , \tag{5.14}$$

$$F_0 |\varphi\rangle_R = \left(\frac{i p \Gamma_0}{\sqrt{2}} + \sum_{m \neq 0} \frac{i \alpha_m \Gamma_{-m}}{\sqrt{2}}\right) |\varphi\rangle_R = \frac{i m}{\sqrt{2}} |\varphi\rangle_R \; .- \tag{5.15}$$

Equation (5.15) is a generalized Dirac equation for the states in the Ramond model. As in the case of the conventional model, the constraint Eqs. (5.13) to (5.15) can be interpreted as primary constraints in a suitable Lagrangian formulation of the models[29]. It would take us too long to proceed in this direction, as well to explain how the interaction is introduced in these models (see Refs. 9 and 13, Review article by J. Schwartz, and Refs. 30 to 33).

Let us only comment here on the fact that a transverse gauge can be defined in these models also, by requiring

$${}_{tr}\langle\varphi| P_+(\sigma) |\varphi\rangle_{tr} = {}_{tr}\langle\varphi| p_+ |\varphi\rangle_{tr}$$

and

$$ {}_{t_2}\langle \varphi | \Gamma_+(\sigma) | \varphi \rangle_{t_2} = 0 \ . - $$

Consistency requires then that a representation of the Lorentz group can be found within the space of transverse excitations. Considering the little group of a particle of zero transverse momentum, the non trivial generators should be given by[21])

$$ E_{Li} = \sum_n \frac{\alpha_n^i L_{-n}^{t_2}}{n} + \sum_r b_r^i G_{-r}^{t_2} $$

for the model of Neveu and Schwartz;

$$ E_{Li} = \sum_n \frac{\alpha_n^i L_{-n}^{t_2}}{n} + \sum_n \frac{\Gamma_n^i F_{-n}^{t_2}}{i\sqrt{2}} $$

for the model of Ramond.

One would expect

$$ [E_{Li}, E_{Lj}] = -2(\alpha') M^2 E_{ij} \ , \tag{5.16} $$

whereas one gets, in both cases

$$ [E_{Li}, E_{Lj}] = -2\left(L_0^{t_2} - \frac{d^{t_2}}{16}\right)E_{ij} + \left(\frac{d^{t_2}}{8} - 1\right)O_{ij} \ . - \tag{5.17} $$

We see from this equation that for both models the critical number d of space-time dimensions (the value of d for which the transverse states form a complete set of physical states) is 10 ($d^{tr} = 8$). For the masses of the ground states we obtain [comparing Eqs. (5.16) and (5.17)]:

$$ \alpha' M^2_{g.s.} = {}_{g.s.}\langle \, | L_0^{t_2} - \frac{d^{t_2}}{16} | \, \rangle_{g.s.} \ . - $$

Notice that

$$ {}_{N.S.}\langle \phi | L_0^{t_2} | \phi \rangle_{N.S.} = 0 \ ; $$

but, from Eq. (5.12a) we deduce a ground state expectation value of L_0 equal to d/16 in the model of Ramond. It follows that the values of the critical masses of the ground state bosons and fermions are $M^2_{NS} = -1/2\alpha'$ and $M^2_R = 0$ respectively.

To conclude, let me mention another possible development of the theory of dual resonance models, that has been object of recent investigations[34]. The dynamics underlying the dual models is, as we have seen, the quantum mechanics of objects with a one dimensional continuum of degrees of freedom.

For these objects a generalized wave function can be introduced, a wave functional in fact $\Psi(x_\mu(\sigma))$, which represents the amplitude for finding the string with coordinates $x_\mu(\sigma)$. In a second quantized version of the theory, the wave functional would be replaced by a generalized operator field, an operator field depending not on a point, but on a path $x_\mu(\sigma)$ (an operator valued functional), $\Phi(x_\mu(\sigma))$.

The equations leading to the spectrum of the conventional dual model can be reformulated in terms of $\Phi(x_\mu(\sigma))$ [34]. They become equivalent to the condition that $\Phi(x_\mu(\sigma))$ is independent from the choice of parametrization:

$$\int d\sigma \, h(\sigma)\, x'(\sigma) \frac{\delta}{\delta x(\sigma)} \, \Phi(x_\mu(\sigma)) = 0 ,$$

for any $h(\sigma)$ with $h(0) = h(\pi) = 0$; and to an equation for the evolution of Φ:

$$\int d\sigma \, h(\sigma) : \frac{\delta^2}{\delta x(\sigma)^2} - x'^2(\sigma) + \text{const} : \Phi(x_\mu(\sigma)) = 0 .-$$

The idea of describing the string with a quantized field $\Phi(x_\mu(\sigma))$ would seem particularly appropriate to describe the interaction among strings (three Reggeon vertex) where the number of strings is not conserved.

As we have pointed out, there exists up to now no satisfactory interpretation of the dual three Reggeon vertex as a coupling among strings. However, it is interesting to notice that the dynamics of dual models naturally leads to the idea of operators depending on a path in space-time. Path dependent operators arise also as a generalization of the bylocal operators[35]. Then, we can ask ourselves whether a connection between the two ideas exists, whether the dual nature of hadrons that manifests itself as the quantum mechanics of one dimensional extended objects and the ideas underlying current algebra and its generalizations can be related to each other. It is not clear yet if such a program can be carried out, but, if it succeeded, an interesting bridge would be laid between two theories, current algebra and duality, that have been so relevant for our understanding of strong interactions.

REFERENCES

1) G. Veneziano, Nuovo Cimento 57 A, 190 (1968).

2) K. Bardakçi and H. Ruegg, Phys. Rev. 181, 485 (1969);
H.M. Chang, Phys. Letters 28 B, 425 (1969);
H.M. Chang and Tsou S. Tsun, Phys. Letters 28 B, 485 (1969);
C. Goebel and B. Sakita, Phys. Rev. Letters 22, 257 (1969);
Z. Koba and H.B. Nielsen, Nuclear Phys. B10, 633 (1969).

3) In the following references (Refs. 4-12), we select some of the most relevant contributions to the theory of dual resonance models. For lectures and reviews on the subject, see Ref. 13.

4) S. Fubini and G. Veneziano, Nuovo Cimento 64 A, 811 (1969);
K. Bardakçi and S. Mandelstam, Phys. Rev. 184, 1640 (1969).

5) S. Fubini, D. Gordon and G. Veneziano, Phys. Letters 29 B, 679 (1969).

6) M.A. Virasoro, Phys. Rev. D1, 2933 (1970).

7) P. Di Vecchia, E. Del Giudice and S. Fubini, Ann. Phys. 70, 378 (1972).

8) R. Brower, Phys. Rev. D6, 1655 (1972);
P. Goddard and C. Thorn, Phys. Letters 40 B, 235 (1972);
J.H. Schwartz, Nuclear Phys. B46, 61 (1972).

9) A. Neveu and J.H. Schwartz, Phys. Rev. D4, 1109 (1971).

10) M.A. Virasoro, Phys. Rev. 177, 2309 (1969);
A. Shapiro, Phys. Letters 33 B, 361 (1970).

11) P. Ramond, Phys. Rev. D3, 2415 (1971).

12) Y. Nambu, Proc. Int. Conf. on Symmetries and Quark Models, Detroit, 1969 (Gordon and Breach, New York, 1970), p. 269;
H. Nielsen, Paper submitted to the XVth Int. Conf. on High Energy Physics, Kiev, 1970;
L. Susskind, Nuovo Cimento 69 A, 457 (1970).

13) G. Veneziano, Lectures given at Int. School of Subnuclear Physics "Ettore Majorana", Erice, 1970 (Academic Press, NY, 1971), p. 94;
S. Fubini, Lectures given at Ecole d'été de Physique théorique, Les Houches, 1971, to be published;
V. Alessandrini, D. Amati, M. Le Bellac and D. Olive, Phys. Rev. 1 C, No. 6 (1971);
J. Schwartz, Caltech preprint CALT-68-384 (1973), to be published in Phys. Reports;
C. Rebbi, Lectures given at the ICTP, Trieste, Nov. 1972, to be published.

14) Y. Nambu, Lectures for the Copenhagen Summer Symposium (1970), unpublished.

15) L.N. Chang and J. Mansouri, Phys. Rev. 5 D, 2535 (1972);
O. Hara, Progr. Theor. Phys. 46, 1549 (1971);
T. Goto, Progr. Theor. Phys. 46, 1560 (1971);
J. Mansouri and Y. Nambu, Phys. Letters 39 B, 375 (1972).

16) P. Goddard, G. Goldstone, C. Rebbi and C. Thorn, to be published in Nuclear Phys.

17) P.A.M. Dirac, Can. J. Math. 2, 129 (1950);
P.A.M. Dirac, Proc. Roy. Soc. A246, 326 (1958).

18) J. Douglas, Ann. Math. 40, 205 (1939).

19) S. Fubini and G. Veneziano, Nuovo Cimento 67 A, 29 (1970); and Ann. Phys. 63, 12 (1971).

20) J.L. Gervais, Nuclear Phys. B21, 192 (1970).

21) P. Goddard, C. Rebbi and C.B. Thorn, Nuovo Cimento 12 A, 426 (1972).

22) S. Fubini and P. Di Vecchia, Lectures given at the Int. School of Subnuclear Physics "Ettore Majorana", Erice, 1972, to be published, and CERN preprint TH-1542 (1972).

23) L. Brink and D. Olive, CERN preprint TH-1619 (1973).

24) M. Ademollo, A. D'Adda, R. D'Auria, P. Di Vecchia, F. Gliozzi, R. Musto, E. Napolitano, F. Nicodemi and S. Sciuto, Theory of an interacting string and dual resonance model, CERN TH preprint (1973), to be published.

25) J.L. Gervais and B. Sakita, New York City College preprint (1972).

26) E. Cremmer and J. Scherk, Nuclear Phys. B50, 222 (1972).

27) L. Brink and D. Olive, CERN preprint TH-1620 (1973);
D. Olive and J. Scherk, CERN preprint TH-1635 (1973);
L. Brink, D. Olive and J. Scherk, CERN preprint TH-1675 (1973).

28) C. Lovelace, Phys. Letters 34 B, 500 (1971).

29) Y. Iwasaki and K. Kikkawa, New York City College preprint (1973).

30) A. Neveu, J.H. Schwartz and C.B. Thorn, Phys. Letters 35 B, 529 (1971).

31) A. Neveu and J.H. Schwartz, Phys. Rev. D4, 1109 (1971).

32) C.B. Thorn, Phys. Rev. D4, 1112 (1971).

33) E. Corrigan and D. Olive, Nuovo Cimento 11 A, 749 (1972).

34) H. Fritzsch, M. Gell-Mann and C. Rebbi, unpublished.

35) H. Fritzsch and M. Gell-Mann, Proc. XVI Int. Conf. on High Energy Physics, Chicago-Batavia, 1972 (NAL, Batavia, 1973), Vol. 2, p. 135.

36) S. Mandelstam, Interacting-string picture of dual-resonance model, Univ. of California (Berkeley) preprint (1973).

D I S C U S S I O N S

CHAIRMAN: Prof. C. REBBI

Scientific Secretaries: E. Hendrick, M. Kupczynski

DISCUSSION No. 1

- *RABINOVICI:*

In what sense should the string model be regarded as a mathematical analogy and in what sense does it actually describe the space-time behaviour of a hadron?

- *REBBI:*

The model provides us with the same particle spectrum as we obtain by demanding the factorization of the intermediate states in dual resonance models. In this sense the string models serves as a mathematical analogy. We are in too early a stage to try to attack the problem of a clear space-time meaning for the string. I would not like anyone to get the impression that a hadron is a string.

- *NAHM:*

For a string with two ends you get the Veneziano model; for a closed string the Virasoro one. What do you get for different topologies, for example, a star where the strings are connected at the ends?

- *REBBI:*

Singular points create severe difficulties. It could be that such string models could be useful in the future, but for the moment there is no indication of this.

- *NAHM:*

You need additional degrees of freedom if you want to describe fermions by a string model. How can this be done?

- *REBBI:*

It is possible to add spin degrees of freedom to the string: one then obtains the models of Neveu and Schwartz and Ramond (for more details on these models, see Section 5 of the Lecture Notes). SU(3) quantum numbers can be added multiplicatively. A non-multiplicative introduction of SU(3) quantum numbers has been studied by many people, but the problem becomes more difficult, and no completely satisfactory solution has been found.

- *GENSINI:*

Can the current Veneziano model explain the power behaviour of the tail of the differential cross-section at very large values of $p_\perp$?

- *REBBI:*

Whenever the behaviour of inclusive spectra for large $p_\perp$ has been studied within the theory of dual models, one has always found an exponential cut-off in $p_\perp^2$. It must be said that the analysis of the spectra becomes harder when we go to loop diagrams. It might happen that unitarity corrections associated with dual loops generate a less damped behaviour in $p_\perp^2$. However, it may well be that the large $p_\perp$ behaviour of inclusive spectra cannot be reproduced by dual models in their present form.

- *WILCZEK:*

Is there a physical picture (as in the parton model) from which you can tell in which kinematic regions the string picture is good and in which regions it is bad?

- *REBBI:*

The string model is an interpretation of the dual model, so the question may be rephrased: where do you expect the dual models to give a good approximation to the data? Of course, if we had the right model I would be tempted to answer, "everywhere". With the present models, I would say we expect good results where the detailed structure of the lower part of the spectrum does not matter too much. We may also have to exclude regions like the large $p_\perp$ region in inclusive distributions, where other dynamical mechanisms might operate.

- *WILCZEK:*

Do you know of any way to make a connection between usual field theory and the string dual resonance model? Does the fact that the Yang-Mills theory is the zero slope limit of the dual resonance model have any relevance to this question?

- *REBBI:*

A field theoretical interpretation of the dual resonance models may be possible, but it will not be straightforward. We must remember that the counting and topology of dual diagrams and Feynman diagrams (to which we would be led by any field theory) are different. The zero slope limit is quite relevant and instructive for an analysis of the problem. Indeed, we see that in the zero slope limit, single dual diagrams give origin to many different Feynman diagrams.

- *LOSECCO:*

What can the string model tell us about unitarity in dual amplitudes?

- *REBBI:*

The string picture leads rather naturally to the dual loops. From this point of view it would suggest, as conjectured by many people, that the correct way to unitarize the model is through the dual loop expansion.

- *CHOUDHURY:*

Could you make a comment about bilocal operators and dual amplitudes? Can you impose duality conditions on the bilocal operators of the Fritzsch-Gell-Mann algebra?

- *REBBI:*

As I mentioned very briefly in my lectures, the string approach to dual models leads quite naturally to operators depending on a path in space-time. Path-dependent operators arise also in a generalization of the bilocal operators of the Fritsh and Gell-Mann. The two concepts might be related by formulating an algebra of path-dependent operators in a null plane that would reduce to the algebra of bilocals when the paths become light-like, and could embody some of the features of dual models insofar as the dependence on the path is concerned. The possibility of constructing such an algebra has been investigated by Fritsh, Gell-Mann, and myself, but at present we do not have definite results on the subject.

- *KUPCZYNSKI:*

This is a comment rather than a question. You want to derive in a rigorous way dual amplitudes, for example the Veneziano formula, which has had some success so far. However, to make possible comparison with experimental data, one has to keep only a few pole terms in the expansion of the amplitude and add the phenomenological width of the resonances, which breaks the mathematical rigor of the theory. Very often other model-dependent assumptions are needed. To obtain the hadron spectrum and the dual amplitudes theoretically requires the acceptance of additional assumptions which are far from obvious. Even in that case many models obtain tachyons. In my opinion all these difficulties make the physical interpretation quite unclear.

- *REBBI:*

People looking at the dual models often ask such questions: what about a realistic hadron spectrum, realistic amplitudes, and so on? My feeling is that it is not yet the time to ask such questions. We have already accomplished a huge amount of things. We have now some exact solutions of the superconvergence relations which seemed in 1966-67 very difficult. They are not yet adequate to explain the realistic world, but they provide a good theoretical laboratory. A lot of investigation is still needed. We should remember that we want to solve the very difficult problem of strong interactions, so we should not be too impatient.

DISCUSSION No. 2

- *WILCZEK:*

How does the dimension (d = 26) depend on the internal symmetries?

- *REBBI:*

The introduction of spin in the Neveu-Schwartz model reduces the critical number of transverse dimensions to eight. Other methods used so far to introduce symmetries, such as SU(3) or SU(2), consist of putting in the symmetries by hand, namely by putting multiplicitive factors into the dual amplitudes. Since these are non-intrinsic symmetries their introduction does not move you from d = 26.

- *WILCZEK:*

Could you elaborate on the problem of unphysical intercepts and possibilities of avoiding the problem.

- *REBBI:*

It seems necessary to have a mass zero spin-1 particle in theories of this type, because one has only transverse creation operators acting on the ground state of spin-0. For examples, in four dimensions you get only two helicities, ±1, which are characteristic of a massless spin-1 particle.

- *COLEMAN:*

What is the ground state? Is it the original zero spin particle of the dual model?

- *REBBI:*

Yes, that is correct.

- *COLEMAN:*

I am told there is a model without tachyons or ghosts. Is that correct?

- *REBBI:*

Yes, that is the model of Neveu and Schwartz. One of the gauge conditions removes the tachyon from the leading trajectory.

In the original formulation of this model one would still find a tachyon or a non-leading tracjectory of intercept ½. However, the intercept of this trajectory can be lowered and eventually it becomes negative. This possibility occurs because there is a conserved quantum number in this model. The quantum number is equivalent to G-parity, and the non-leading trajectory I was mentioning, couples only to channels of odd G-parity.

So the model of Neveu and Schwartz has a vector particle of zero mass, but no other tachyons or massless particles.

- *FRAMPTON:*

Can I take this opportunity to emphasize that the unit intercept is a general difficulty of the rubber string approach and of the operator formalism approach. Starting from a transversely vibrating string, the first excited state has spin-1 and d-2 components, where d is the space-time dimensionality, and hence must be massless. Even in a subcritical dimension where longitudinal modes are essential for completeness, the first mode needs only transverse modes (longitudinal excitations are in the second and all higher modes), so the massless vector state remains. One does not get inspiration from the subcritical situation on how to circumvent the unit intercept. Also, in the operator formalism approach no one has succeeded in finding a ghost-free Regge-behaved model without a unit intercept trajectory. Thus, a different viewpoint seems needed, and I believe that the construction of integral representations directly, exploiting the symmetries of the integrand, may be the appropriate method to obtain amplitudes for physically acceptable intercepts and mass spectrum.

- *MENDES:*

I have a comment about dual models which seem to apply to 26- or 10-dimensional worlds but not to our 4-dimensional world. I think there may be things one can learn from these models that remain invariant for changes of dimensionality. For example, if one thinks of the multiplicities and couplings of the vector mesons, one notices that the Neveu-Schwartz model has just one ρ and one ω, as it should. At the level of the ρ', where one could in principle have 16 vector mesons, I have found that after applying the gauge conditions one ends up with only two physical states, and only one of them decouples from two pions. Perhaps what we can learn about vector particles remains invariant for changes of dimensions, since a vector particle keeps the same tensor structure as a space-time vector. So regardless of the number of modes you put in the space direction, you always use them up in the extra components you need for the vector meson. This is just one example. There may be other invariant predictions worth exploring in the present unrealistic dual models.

- *REBBI:*

Yes, the comment is appropriate. Of course, there are many things about the dual model which fit the structure of the hadrons and should not be disregarded.

- *MENDES:*

Could you comment on the general pattern of changes one should expect in the spectrum of the dual models when one changes the boundary conditions at the end points of the strings? Besides changing the multiplicity as in the Virasoro model, could one also obtain, for example, non-linear trajectories by a change in the boundary conditions?

- *REBBI:*

For the case of a closed string, for example, the expansion in proper modes requires sines and cosines (instead of just cosines). Hence, there is a doubling of the modes in the spectrum. In the Neveu-Schwartz and Ramond models one introduces a new field $\Gamma_\mu(\sigma)$ with the anticommutator:

$$\{\Gamma_\mu(\sigma), \Gamma_\nu(\sigma')\} = 2g_{\mu\nu}\ \delta(\sigma - \sigma')$$

which is like a continuum of gamma-matrices over the string. We expand now in the proper modes. According to the boundary conditions one finds either

$$\Gamma_\mu(\sigma) = \sum_r b_{r\mu}\ e^{ir\sigma} \qquad r = \pm\tfrac{1}{2}, \pm\tfrac{3}{2}, \ldots$$

or

$$\Gamma_\mu(\sigma) = \sum_m \Gamma_{m\mu}\ e^{im\sigma} \qquad m = 0, \pm 1, \pm 2, \ldots .$$

In the first case one defines a non-degenerate ground state vacuum by $b_r|0\rangle$ with $r > 0$. The physical space is obtained by operating on $|0\rangle$ with $b_{-r} = b_r^+$. For the second case $\Gamma_{m\mu}|u\rangle = 0$, and the space is constructed by operating on $|u\rangle$ with $\Gamma_{m\mu}^+$. However, we have a problem with $\Gamma_{0\mu}$. We have to define a representation of $\Gamma_{0\mu}$ in the ground state

$$\Gamma_{0\mu}|u\rangle = |\gamma_\mu u\rangle .$$

So these two models, that are otherwise very similar, display very different spectra. In fact, the first is a boson model (the Neveu-Schwartz model), the second is a fermion model (the Ramond model).

Concerning the question of obtaining non-linear trajectories by an appropriate choice of boundary conditions, the answer is that it is perhaps possible, but it is a difficult problem. The difficulty is that in putting non-linear boundary conditions at the endpoints of the string, we are led to a two-dimensional field theory with a non-linear source term at the boundary, and this is hard to solve.

- *NAHM:*

I am somewhat confused by the gauging away of the longitudinal oscillation. It seems to me you should not be able to gauge away changes in the length of the string.

- *REBBI:*

In the coordinates I have used, the length of the string is not measured in space-time extension but in energy content, which is constant. I have tried to get one extra longitudinal mode coupled to the transverse modes by changing the boundary conditions, for example, by adding masses to the ends of the string. Thus one would get rid of the massless vector meson. However, if the new conditions one obtains at the boundaries cannot be linearized in a suitable gauge, it becomes extremely difficult to solve the quantum spectrum.

- *CHU:*

I have a comment concerning the question of a critical dimension of 26 in the string model and in the conventional dual models. Perhaps it should be noted that the dual model is not necessarily bound to an unrealistic dimension of space-time. In particular, the Lovelace formula for $\pi\pi$ scattering has both no ghosts and physical masses for $d \leq 4$. Perhaps if one could make a generalization of Lovelace's model to more than four pions, one could reproduce your model.

- *REBBI:*

Yes. It is going to be very interesting to listen to the lecture of Professor Frampton because he will be talking about generalizations of this sort.

DISCUSSION No. 3

- *KUPCZYNSKI:*

Your construction of the amplitude concerned only the processes $\gamma + A \to (n\gamma) + B$ where A and B denote any resonance. How can one construct other dual amplitudes?

- *REBBI:*

This is being investigated by the authors of Ref. 24. The main difficulty is the transition from the classical to the quantum mechanical quantities, which involves serious ordering problems whenever the momentum of the emitted particle is not light-like.

- *DAMMER:*

You mentioend an approach to the string model using functional integration techniques. What kind of interactions are assumed in this approach?

- *REBBI:*

There are two approaches of this type. Both make use of the idea that if one or more strings evolve through space-time they span surfaces. The quantum mechanical transition amplitudes should be obtained by summing over all possible surfaces of evolution.

- *EYLON:*

Can you find in your model the amplitudes with the space-like external particles?

- *REBBI:*

It is hard to extrapolate the amplitudes off the mass shell. The problem is to preserve the gauge identities so that no ghost couples, and to get finite results. The approach of Cremmer and Scherk*) to this problem is to deal with the propagation of a string with a hole. Taking the limit of a very small hole, it is possible to factorize the Pomeron singularity and define currents of any spin coupled to the surface for any value of p^2.

- *BOUQUET:*

Could you come back to the notion of Pomeron scattering.

- *REBBI:*

Let us consider the amplitude when the string propagates as a loop in space-time, and is coupled to some particles on the boundaries. The surface can be visualized as a closed string propagating inwards from the outer boundary, or as a cylinder in a space-time

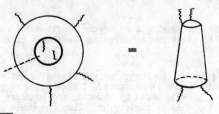

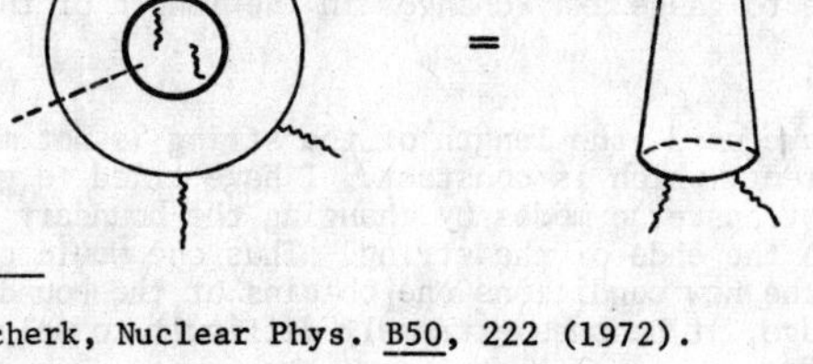

*) E. Cremmer and J. Scherk, Nuclear Phys. B50, 222 (1972).

We may cut the surface of evolution and factorize the states propagating through it. We find a spectrum of intermediate states, identical to the spectrum of quantum excitations of a closed string which defines the "Pomeron sector" of the dual model (see the Lecture Notes, at the end of Section 4).

- *WILCZEK:*

Has it been proved that the tree graphs of the dual model are the first term in an expression that in the end gives a unitary amplitude? What is the expansion parameter?

- *REBBI:*

The iteration scheme is based on perturbative unitarity. The singularity structure of the dual loop expansion has been verified to agree with the Cutkowski rules. Concerning the parameter of the expansion, one multiplies each vertex by a factor g. The width of the resonances is then proportional to g^2, so that if we had a realistic model this would fix g.

- *WILCZEK:*

So in a sense g^2 is the ratio of the width of resonance to the distance between resonances?

- *REBBI:*

Yes. A calculation based on the Veneziano amplitude (not the full D.R.M.) yields $g^2 \sim 0.1$.

- *RABINOVICI:*

Is the fact that one finds massless particles in the dual models only a mathematical technicality or can you connect it with the production of π mesons in the hadron interactions?

- *REBBI:*

I think we are forced to these particles of zero mass because of the gauge group we have in the theory. As a matter of fact, we have vector particles of zero mass and not massless π mesons. It might be that some future realistic model replaces massless vectors with massless pions.

- *NAHM:*

You told us that you tried to put masses on the end-points of the string, up to now without success. Do you think that it is just due to technical difficulties or do you think that the properties of the end-points are determined by the properties of the string?

- *REBBI:*

I do not know. From a mathematical point of view the difficulties which I met were due to the non-linearity of the boundary conditions, which made it impossible to find exact solutions or reasonable approximations for the spectrum of quantum excitations.

- *NAHM:*

You have made an analogy between the end-points of the string and the quark and anti-quark making up a meson. Then the string itself should represent the gluon field. As the gluon field should be a vector field, it is quite natural to introduce the field Γ_μ of the Neveu-Schwartz model. Having heard so much about Yang-Mills fields one may expect that Γ_μ should be given SU(3) quantum numbers, too. However, in this case there is some flow of SU(3) quantum numbers along the string. This might yield a Pomeron which has the quantum numbers of the vacuum only in the mean. Do you think such a situation is possible?

- *REBBI:*

As I said already, no entirely satisfactory dual model that embodies SU(3) quantum numbers in a non-multiplicative way exists. If a model was found that introduces a continuum distribution of SU(3) intrinsic degrees of freedom along the string, then we would have exchange of SU(3) quantum numbers in the so-called Pomeron sector of the model. Of course, it might well be that the leading trajectory in that sector still carries vacuum quantum numbers.

RECENT DEVELOPMENTS OF DUAL MODELS

P.H. Frampton

Table of Contents

RECENT DEVELOPMENTS OF DUAL MODELS

P.H. Frampton
CERN, Geneva, Switzerland

1. INTRODUCTION

1.1 Physical Born amplitudes

A principal objective in the zero-width approach to strong interactions is to obtain a realistic Born amplitude. By realistic, we imply that the Regge trajectory intercepts, and consequently the mass spectrum, should be near to the empirical values; at the same time, the amplitude should fulfil the requirements of crossing symmetry, Regge behaviour to the extent possible for a zero-width approximation, absence of imaginary-mass tachyons, and factorizability without negative-probability ghost states.

The methods of constructing such an ideal tree amplitude, that have been applied in the literature, can be listed as:

i) Operator formalism techniques;

ii) Rubber string approach;

iii) Symmetric group method.

The idea in the operator approach (i) is to realize the generalized projective algebra on a Fock space spanned by certain sets of harmonic-oscillator-like creation and annihilations operators (generally one commuting set plus additional sets which may commute or anticommute). In this way one constructs vertices such that the gauges necessary for ghost-elimination are built in from the start. This method led to the construction of the ghost-free Neveu-Schwarz model[1]. It has been further exploited in attempts to find new models by Gervais and Neveu[2,3].

The second method (ii), that of using the rubber string, makes a more physical interpretation of the internal structure of the hadrons. The string gives rise to an attractive re-interpretation of the spectrum and the absence of ghosts in the conventional model[4-6] and this has been extended to the Neveu-Schwarz model[7]. So far this method has not led to any new model, and indeed one is only beginning to obtain an understanding of the interactions[8-10] even in the known models, from this string viewpoint.

It is worth noting two general difficulties occurring in the string approach; both are associated with covariance properties. They are

a) the occurrence of massless spin-one states,

b) the special role of curious values of space-time dimensionality, $d \neq 4$.

These difficulties occur also in the operator approach. So far, the ingenious attempts[2,3] to overcome difficulty (a) in the operator formalism have not been fully successful.

We shall not discuss further the operator and string methods. Instead we shall concentrate exclusively on method (iii), the symmetric group approach.

1.2 Symmetric group approach[11,12]

This method starts from an integral representation for the four-particle tree amplitude, and then makes an extension to any number of external particles. The principle is to impose very high symmetry properties on the integrand, similar to those obtained in the earlier unphysical models.

It is to be expected that this approach will complement the other two approaches mentioned above and that the different methods may ultimately converge on an acceptable solution.

We should mention that the method is very powerful since as we shall show later we can re-derive from it essentially all the theoretically-consistent tree models that were discovered in the last several years.

We begin by making a rather detailed study of the four-meson tree amplitude.

FOUR-MESON AMPLITUDE

2.1 Veneziano model with unit intercept

The simplest proposal of a four-meson Born amplitude is to take the Euler B function of the Regge trajectory functions as suggested by Veneziano[13]. Let us therefore consider the amplitude for four scalar particles in the form

$$T = B(-\alpha_s,-\alpha_t) + B(-\alpha_t,-\alpha_u) + B(-\alpha_u,-\alpha_s) \tag{1}$$

with

$$B(-\alpha_s,-\alpha_t) = \int_0^1 dx\, x^{-\alpha_t-1}\,(1-x)^{-\alpha_s-1} \tag{2}$$

where $s = (p_1 + p_2)^2$, $t = (p_2 + p_3)^2$, $u = (p_1 + p_3)^2$, $\alpha_s = \alpha(0) + \alpha' s$

This model can be extended to a multiparticle amplitude[14]; such a generalized Veneziano amplitude is ghost-free only if $\alpha(0) = 1$, however, and this is a physically-unacceptable constraint.

The Veneziano model, which must be changed in some essential way to accommodate empirical values of the intercept, has the following properties special to the case $\alpha(0) = 1$.

i) Absence of odd daughter trajectories[13], due to a reflection property

$$R_N(\alpha_t) = (-1)^N R_N(\alpha_u) \tag{3}$$

of the residues contained in

$$B(-\alpha_s, -\alpha_t) = \sum_{N=0}^{\infty} \frac{R_N(\alpha_t)}{N-\alpha_s} \tag{4}$$

ii) Summability of three terms[15] in Eq. (1) according to

$$T = \int_{-\infty}^{\infty} dx \, |x|^{-\alpha_t - 1} \, |1-x|^{-\alpha_s - 1} \tag{5}$$

iii) The supplementary condition[13]

$$\alpha_s + \alpha_t + \alpha_u + 1 = \gamma = 0 \tag{6}$$

on the trajectory functions, when we put the external state on the internal trajectory.

iv) The phase identities[16]

$$B(-\alpha_s, -\alpha_t) - e^{\pm i\pi\alpha_t} B(-\alpha_t, -\alpha_u) - e^{\mp i\pi\alpha_s} B(-\alpha_u, -\alpha_s) = 0 \tag{7}$$

and cyclic permutations thereof. These relations are obtained by closing the contour in Eq. (5) and using Cauchy's theorem.

We now decide that the physical amplitude should maintain the first two properties but should avoid, in general, the last two. The reason for maintaining the summability condition is that at the N-particle level there is a close link between this symmetry and the ghost-eliminating gauges (see, for example, Reference 17). As we shall see, the absence of odd daughters will follow in the four-meson amplitude, from this invariance (symmetric group invariance, see later).

Concerning the phase identities, note that we can deduce from the forms

$$\begin{aligned} &B_{st} - e^{+i\pi\alpha_t} B_{tu} - e^{-i\pi\alpha_s} B_{us} \\ &+ e^{i\pi\alpha_t}\left(B_{tu} - e^{+i\pi\alpha_u} B_{us} - e^{-i\pi\alpha_t} B_{st}\right) = 0 \end{aligned} \tag{8}$$

that

$$e^{i\pi(\alpha_s + \alpha_t + \alpha_u)} = -1 \tag{9}$$

or, in other words, that there is a generalized supplementary condition

$$\alpha_s + \alpha_t + \alpha_u + 1 = \gamma = 0, \text{ modulo } 2 \tag{10}$$

In general, the physical masses will not satisfy such a condition. In a four-pion amplitude one has, for example,

$$\gamma = 3\alpha_\rho(0) - 4\alpha_\pi(0) + 1 \approx 5/2 \tag{11}$$

Therefore we must reject both properties.

2.2 No odd daughters

We now modify the integrand of the four-meson amplitude by multiplying with a general function which we choose to write as

$$A_4(-\alpha_s, -\alpha_t) = \int_0^1 dx\, x^{-\alpha_t-1} (1-x)^{-\alpha_s-1} (1-x+x^2)^{\gamma/2} \phi_4(\alpha_s, \alpha_t, \alpha_u; x) \tag{12}$$

The reason for separating out the factor[18] $(1 - x + x^2)^{\gamma/2}$ is that then ϕ_4 has much simpler analytic properties, as discussed later.

Now make the change of variables

$$x = 1 - e^{-W} \tag{13}$$

$$\nu = \frac{1}{2}(\alpha_s - \alpha_u) \tag{14}$$

to arrive at

$$A_4(-\alpha_s, -\alpha_t) = \int_0^\infty dW\, W^{-\alpha_t-1} e^{\nu W} \left[\sum_{n=0}^\infty \frac{1}{(2n+1)!}\left(\frac{W}{2}\right)^{2n}\right]^{-\alpha_t-1} \left[1 + 2\sum_{n=1}^\infty \frac{W^{2n}}{(2n)!}\right]^{\gamma/2} \phi_4(\alpha_s, \alpha_t, \alpha_u; 1-e^{-W}) \tag{15}$$

We now use, in this Laplace transform,

$$\int_0^\infty dW\, W^{-\alpha_t-1+r} e^{\nu W} = (-a\nu)^{\alpha_t - r}\, \Gamma(-\alpha_t + r) \tag{16}$$

to see that odd terms (odd r) are missing in the Regge asymptotic expansion provided that ϕ_4 satisfies

$$\phi_4(\alpha_s, \alpha_t, \alpha_u; x) = \phi_4\left(\alpha_u, \alpha_t, \alpha_s; -\frac{x}{1-x}\right) \tag{17}$$

Combining this with the requirement of crossing symmetry that

$$\phi_4(\alpha_s, \alpha_t, \alpha_u, x) = \phi_4(\alpha_t, \alpha_s, \alpha_u; 1-x) \tag{18}$$

we find that ϕ_4 satisfies invariance under an S_3 symmetric group. This is most clearly expressed by introducing a function invariant under inversion, namely

$$\beta_s(x) = x(1-x+x^2)^{-1} = \beta_s(1/x) \tag{19}$$

together with

$$\beta_t(x) = \beta_s(1-x) \tag{20}$$

$$\beta_u(x) = \beta_s\left(-\frac{x}{1-x}\right) \tag{21}$$

We can then deduce that

$$\phi_4(\alpha_s, \alpha_t, \alpha_u; x) = \Phi_4\left(\frac{\alpha_s}{\beta_s(x)}; \frac{\alpha_t}{\beta_t(x)}; \frac{\alpha_u}{\beta_u(x)}\right) \tag{22}$$

must be invariant under all 3! permutations of the argument pairs $\alpha_i, \beta_i(x)$ for $i = s,t,u$.

2.3 Analytic properties

To derive Regge behaviour for $|x| \to \infty$ at fixed t it is necessary for ϕ_4 to be analytic in Re $w > 0$ or equivalently $|1 - x| < 1$. By crossing symmetry fixed s Regge behaviour requires analyticity in $|x| < 1$. The combination gives an analyticity domain

$$\mathcal{D}(x) = (|x| < 1) \cup (|1-x| < 1) \tag{23}$$

which has much the same shape as the apparent field of view through a set of opera glasses.

This domain can be extended by consideration simultaneously of all the s,t,u Regge limits in

$$T = A_4(-\alpha_s, -\alpha_t) + A_4(-\alpha_t, -\alpha_u) + A_4(-\alpha_u, -\alpha_s) \tag{24}$$

$$= \int_{-\infty}^{\infty} dx\, |x|^{-\alpha_t - 1} |1-x|^{-\alpha_s - 1} (x - e^{i\pi/3})^{\gamma/2} (x - e^{-i\pi/3})^{\gamma/2} \phi_4(\alpha_s, \alpha_t, \alpha_u; x) \tag{25}$$

$$= \int_0^1 dx\, x^{-\alpha_t - 1} (1-x)^{-\alpha_s - 1} (x - e^{i\pi/3})^{\gamma/2} (x - e^{-i\pi/3})^{\gamma/2} \phi_4(\alpha_s, \alpha_t, \alpha_u; x)$$

$$+ \int_0^1 d\left(\frac{1}{1-x}\right) \left(\frac{1}{1-x}\right)^{-\alpha_u - 1} \left(1 - \frac{1}{1-x}\right)^{-\alpha_t - 1} \left(\frac{1}{1-x} - e^{i\pi/3}\right)^{\gamma/2} \cdot$$

$$\cdot \left(\frac{1}{1-x} - e^{-i\pi/3}\right)^{\gamma/2} \phi_4\left(\alpha_t, \alpha_u, \alpha_s, \frac{1}{1-x}\right)$$

$$+ \int_0^1 d\left(1 - \frac{1}{x}\right) \left(1 - \frac{1}{x}\right)^{-\alpha_s - 1} \left[1 - \left(1 - \frac{1}{x}\right)\right]^{-\alpha_u - 1} \left(1 - \frac{1}{x} - e^{i\pi/3}\right)^{\gamma/2} \cdot$$

$$\cdot \left(1 - \frac{1}{x} - e^{-i\pi/3}\right)^{\gamma/2} \phi_4(\alpha_u, \alpha_s, \alpha_t; 1 - 1/x) \tag{26}$$

Consideration of the Regge limits in these three terms gives the full domain of analyticity

$$\mathcal{D}(x) \cup \mathcal{D}\left(\frac{1}{1-x}\right) \cup \mathcal{D}\left(1-\frac{1}{x}\right) \tag{27}$$

which is the whole complex x-plane, including the point at infinity, except for two points, namely $x = e^{\pm 2\pi/3}$. Thus we deduce that ϕ_4 must be analytic in x except for possible poles at these points.

Since complex x-singularities have been introduced into the integrand, the unwanted phase identities and unphysical supplementary condition of the Euler B function have been successfully avoided.

2.4 Construction of ϕ_4

The construction of explicit examples of ϕ_4 is an exercise in the use of S_3, the symmetric group on three objects.

We shall make an important simplicity criterion, that the poles at $x = e^{\pm i\pi/3}$ of ϕ_4 are of low order. If we decide not to allow poles higher than second order, it follows that ϕ_4 is a linear combination

$$\phi_4 = \sum_{i=0}^{4} \lambda_i \phi_4^{(i)} \tag{28}$$

of the following five forms

$$\phi_4^{(0)} = 1 \tag{29}$$

$$\phi_4^{(1)} = \alpha_s\beta_s + \alpha_t\beta_t + \alpha_u\beta_u \tag{30}$$

$$\phi_4^{(2)} = \alpha_s\beta_t\beta_u + \alpha_t\beta_u\beta_s + \alpha_u\beta_s\beta_t \tag{31}$$

$$\phi_4^{(3)} = \alpha_s\alpha_t\beta_s\beta_t + \alpha_t\alpha_u\beta_t\beta_u + \alpha_u\alpha_s\beta_u\beta_s \tag{32}$$

$$\phi_4^{(4)} = \alpha_s^2\beta_s + \alpha_t^2\beta_t + \alpha_u^2\beta_u \tag{33}$$

We must put $\lambda_0 = 0$ to avoid a tachyon on the leading positive-intercept trajectory (we are thinking of pion-pion elastic scattering).

A more general formula for ϕ_4 is[19])

$$\begin{aligned}\phi_4 = S_1 &+ \left[\tfrac{1}{3}(2\alpha_s-\alpha_t-\alpha_u)(2\beta_s-\beta_t-\beta_u) + (\alpha_t-\alpha_u)(\beta_t-\beta_u)\right]S_2 \\ &+ \left[\tfrac{1}{3}(2\alpha_s^2-\alpha_t^2-\alpha_u^2)(2\beta_s-\beta_t-\beta_u) + (\alpha_t^2-\alpha_u^2)(\beta_t-\beta_u)\right]S_3 \\ &+ \left[\tfrac{1}{3}(2\alpha_s-\alpha_t-\alpha_u)(2\beta_s^2-\beta_t^2-\beta_u^2) + (\alpha_t-\alpha_u)(\beta_t^2-\beta_u^2)\right]S_4 \\ &+ \left[\tfrac{1}{3}(2\alpha_s^2-\alpha_t^2-\alpha_u^2)(2\beta_s^2-\beta_t^2-\beta_u^2) + (\alpha_t^2-\alpha_u^2)(\beta_t^2-\beta_u^2)\right]S_5 \\ &+ (\alpha_s-\alpha_t)(\alpha_t-\alpha_u)(\alpha_u-\alpha_s)(\beta_s-\beta_t)(\beta_t-\beta_u)(\beta_u-\beta_s)\,S_6\end{aligned} \tag{34}$$

where the six functions S_i are power series in the three quantities

$$\overline{\alpha} = \alpha_s\alpha_t + \alpha_t\alpha_u + \alpha_u\alpha_s \tag{35}$$

$$\underline{\alpha} = \alpha_s\alpha_t\alpha_u \tag{36}$$

$$Z = \beta_s\beta_t\beta_u = -x^2(1-x)^2(1-x+x^2)^{-3} \tag{37}$$

The variable Z, originally discussed by Dixon[20], is fully invariant under the S_3 group.

2.5 New four-pion amplitudes

To select from the possible candidates for ϕ_4 we consider the reality of the coupling constants. Only the amplitude corresponding to $\phi_4^{(1)}$ of Eq. (30) is consistent with this requirement. It was shown in Refs. 11 and 12 how all the levels lying lowest in mass are ghost free for the empirical intercepts.

This amplitude is explicitly, for $\alpha_\rho(0) - \alpha_\pi(0) = \frac{1}{2}$,

$$A_4 = \int_0^1 dx\; x^{-\alpha_t-1}(1-x)^{-\alpha_s-1}(1-x+x^2)^{\frac{1-\alpha_\rho(0)}{2}}\left[\alpha_s x + \alpha_t(1-x) - \alpha_u x(1-x)\right] \tag{38}$$

and it will be extended to A_N, for general N (= even integer ≥ 4) in the subsequent discussion.

Concerning the positivity requirement, we may prove[19] that A_4 satisfies the inequalities of Gribov-Pomeranchuk[21] and of Martin[22], namely

$$\frac{\partial^n}{\partial t^n}\,\mathrm{Im}\,A_4(-\alpha_s,-\alpha_t)\Big|_{t=0} \geq 0 \qquad \text{all } n = 1,2,3,\ldots\ldots \tag{39}$$

by re-writing (we take $\alpha_\rho(0) = \frac{1}{2}$ for simplicity)

$$A_4 = \int_0^1 dx\; x^{-\alpha_t-1}(1-x)^{-s}\left[\frac{1+x^3}{(1-x)(1-x^2)}\right]^{1/4}\cdot \tag{40}$$
$$\cdot\left[\alpha_s x(1-x)^{-1} + \alpha_t + (\alpha_s+\alpha_t-3/2)\,x\right]$$

In this expression, note that the factor

$$\left[\frac{1+x^3}{(1-x^2)(1-x)}\right]^{1/4} = \exp\left[\frac{1}{4}\left(-\sum_{n=1}^{\infty}(-1)^n\frac{x^{3n}}{n} + \sum_{n=1}^{\infty}\frac{x^n}{n} + \sum_{n=1}^{\infty}\frac{x^{2n}}{n}\right)\right] \tag{41}$$

has a positive-definite power expansion. The other factors in Eq. (40) are manifestly positive (for all α_s) and hence the inequalities of Eq. (39) follow. This is a non-trivial necessary condition for the ghost-free four-particle amplitude.

More important is to compare the predicted elastic $\pi\pi$ decay widths with experiment. The parent widths coincide precisely with those of the competing formula of Lovelace[23] and Shapiro[24]. Concerning the daughter resonances at low mass we note that at the ρ mass there is no ε resonance, where experimentally there is also certainly no narrow resonance[25]; at the f mass there is no ρ' p-wave resonance agreeing with its experimental absence but there is an ε' s-wave resonance at this mass with elastic width consistent with pion phase shift analysis[26].

By contrast the Lovelace-Shapiro formula has a very broad ε, a strongly-coupled ρ' and a slightly ghost-like ε'. We conclude that the present model is phenomenologically preferable at the daughter levels.

In addition to this phenomenological advantage, there is the theoretical advantage that the present model can be generalized to N-pions. We therefore now turn to the topic of multiparticle extension.

3. N-MESON AMPLITUDE

3.1 General method of multiparticle extension

We first discuss the multiparticle extension of the general four-meson amplitude

$$A_4 = \int_0^1 dx\, x^{-\alpha_t-1} (1-x)^{-\alpha_s-1} (1-x+x^2)^{\gamma/2} \phi_4(\alpha_s, \alpha_t, \alpha_u; x) \tag{42}$$

It is convenient first to change into Koba-Nielsen variables[27] z_i $(i = 1,2,3,4)$ lying in cyclic order on a circle in the complex z-plane.

We make the identification

$$x = (z_2, z_1 ; z_3, z_4) = (23)/(13) \tag{43}$$

$$(1-x) = (z_1, z_4 ; z_2, z_3) = (12)/(13) \tag{44}$$

where the notation for an anharmonic ratio is

$$(a, b; c, d) = \frac{(a-c)(b-d)}{(a-d)(b-c)} \tag{45}$$

and we have introduced the useful shorthand

$$(12) = (z_1 - z_2)(z_3 - z_4) \tag{46}$$

$$(23) = (z_2 - z_3)(z_1 - z_4) \tag{47}$$

$$(13) = (z_1 - z_3)(z_2 - z_4) \tag{48}$$

The Euler B function may be written, for $\alpha_\rho(0) \neq \alpha_\pi(0)$ in the even and odd G-parity channels

$$B_N = \int \prod_{i=1}^{N} dz_i \left[\frac{dz_a\, dz_b\, dz_c}{(z_a - z_b)(z_b - z_c)(z_c - z_a)} \right]^{-1} \prod_{i<j} (z_i - z_j)^{-2p_i \cdot p_j}$$

$$\prod_{i=1}^{N} (z_i - z_{i+1})^{\alpha_\rho(0) - 1} \left[\prod_{i<j} (z_i - z_j)^{(-1)^{i-j}} \right]^{2(\alpha_\rho(0) - \alpha_\pi(0))} \tag{49}$$

Next we should note that

$$\frac{\gamma}{2} = -\frac{1}{2}\left(\alpha_\rho(0) - 1\right) + 2\left(\alpha_\rho(0) - \alpha_\pi(0)\right) \tag{50}$$

This motivates us to derive two identities for the factor $(1 - x + x^2)$ at the $N = 4$ level.

Firstly, we compare

$$\prod_{i=1}^{4} (z_i - z_{i+1})^2 = (12)^2 (23)^2 \tag{51}$$

with the symmetrized expression*) (first introduced in Ref. 2 by Gervais and Neveu)

$$\sum_{\{q_1 q_2 q_3 q_4\}} \prod_{k=1}^{4} (z_{q_k} - z_{q_{k+1}})^{-2} = 8\left[\frac{1}{(12)^2(23)^2} + \frac{1}{(12)^2(13)^2} + \frac{1}{(23)^2(13)^2} \right] \tag{52}$$

$$= 16 \prod_{i=1}^{4} (z_i - z_{i+1})^{-2} (1 - x + x^2) \tag{53}$$

Similarly we may derive a second identity[12] by comparing

$$\prod_{i<j} (z_i - z_j)^{(-1)^{i-j}} = \frac{(13)}{(12)(23)} \tag{54}$$

with

$$\sum_{\{q_1 q_2 q_3 q_4\}} (-1)^P \prod_{k<l} (z_{q_k} - z_{q_l})^{(-1)^{k-l}} = 8\left[\frac{(13)}{(12)(23)} + \frac{(23)}{(12)(13)} + \frac{(12)}{(23)(13)} \right] \tag{55}$$

to find that there is the following re-writing

$$(1 - x + x^2) = \frac{1}{16} \prod_{i<j} (z_i - z_j)^{-(-1)^{i-j}} \sum_{\{q_1 q_2 q_3 q_4\}} (-1)^P \prod_{k<l} (z_{q_k} - z_{q_l})^{(-1)^{k-l}} \tag{56}$$

---

*) The sum in Eq. (52) and later equations is over all permutations of $\{q_1, q_2, q_3, \ldots, q_N\} = 1, 2, 3, \ldots, N$. In the present case, $N = 4$.

Now consider the factor ϕ_4. Re-written in z-variables, the S_3 invariance discussed earlier becomes the S_4 invariance of

$$\Phi_4\left(\begin{matrix}p_1\\z_1\end{matrix};\ \begin{matrix}p_2\\z_2\end{matrix};\ \begin{matrix}p_3\\z_3\end{matrix};\ \begin{matrix}p_4\\z_4\end{matrix}\right) \tag{57}$$

under all 4! permutations of the argument pairs $\{p_i,z_i\}$.

The expressions we have written are immediately generalizable to arbitrary N by extending the range of the indices. We then arrive at a multiparticle amplitude of the general form

$$A_N = \int \prod_{i=1}^{N} dz_i\,(dV_3)^{-1} \prod_{i<j}(z_i-z_j)^{-2p_i\cdot p_j}\left(S_N^{(1)}\right)^{\frac{1-\alpha_\rho(0)}{2}} \left(S_N^{(2)}\right)^{2(\alpha_\rho(0)-\alpha_\pi(0))}\Phi_N\left(\begin{matrix}p_1\\z_1\end{matrix};\ \begin{matrix}p_2\\z_2\end{matrix};\ \ldots\ldots;\ \begin{matrix}p_N\\z_N\end{matrix}\right) \tag{58}$$

in which we have defined the two symmetric-group factors

$$S_N^{(1)} = \sum_{\{q_1q_2\ldots q_N\}} \prod_{k=1}^{N}\left(z_{q_k}-z_{q_{k+1}}\right)^{-2} \tag{59}$$

$$S_N^{(2)} = \sum_{\{q_1q_2\ldots q_N\}} (-1)^P \prod_{k<\ell}\left(z_{q_k}-z_{q_\ell}\right)^{(-1)^{k-\ell}} \tag{60}$$

and ϕ_N in Eq. (58) is symmetric group invariant with respect to the N! permutations of its arguments.

3.2 Classification of the earlier (unphysical) models

The symmetric group enables us to present a unified view of four earlier different schemes. These earlier schemes arise as special cases, extrapolated to unphysical mass values, of the symmetric group approach.

i) The unit ρ - π intercept A_4 of Veneziano[13] and its A_N extension of Bardakci and others[14] corresponds to the case $\alpha_\rho(0) = \alpha_\pi(0) = 1$ and $\phi_N = 1$ in our general theory. This proposal was the first of a closed form Born amplitude, and the present models maintain the same resonance-pole producing mechanism.

ii) The $\alpha_\rho(0) = \alpha_\pi(0) = -1$ A_N model of Gervais and Neveu[2], which extends the A_4 of Mandelstam[18], is also a $\phi_N = 1$ model. This model contained the important fixed point singularities. For the negative unit intercept, however, it contained ghosts[28].

iii) The new ghost-free dual models of Gervais and Neveu[3] include one explicitly-calculated four-meson amplitude for $\alpha(0)$ arbitrary namely

$$A_4 = \int_0^1 dx\, x^{-\alpha_t-1}(1-x)^{-\alpha_s-1}\left(1-x+x^2\right)^{\gamma/2}\ {}_3F_2\left(\frac{\alpha_0-1}{6},\frac{\alpha_0+1}{6},\frac{\alpha_0+3}{6};\frac{3-\alpha_0}{2},\alpha_0-\tfrac{1}{2};y\right) \tag{61}$$

with $y = -27\beta_s(x)\beta_t(x)\beta_u(x)/4$. This amounts to a special choice of the function S_1 in Eq. (34) above. For $x = \frac{1}{2}$, we find $y = 1$ and the hypergeometric series diverges; such a singularity is inconsistent with Regge behaviour. The amplitude is important, however, since it is guaranteed by construction to sustain a no ghost theorem.

iv) Concerning the Lovelace-Shapiro formula, we find that any of the equivalent (after integration) forms

$$\phi_4 = \alpha_s\beta_s \equiv \alpha_t\beta_t \equiv \alpha_u\beta_u \equiv \alpha_s^2\beta_s + \alpha_t^2\beta_t + \alpha_u^2\beta_u \tag{62}$$

give, in A_4, the result

$$A_4 = -\frac{\Gamma(1-\alpha_s)\Gamma(1-\alpha_t)}{\Gamma(1-\alpha_s-\alpha_t)} \tag{63}$$

provided there is the unphysical supplementary condition $\gamma = 2$. More generally we may establish a linkage to the Neveu-Schwarz model by putting either of the forms

$$\prod_{i\neq j}(z_i-z_j)^{-p_i\cdot p_j}\,\phi_N = \left(S_N^{(2)}\right)^{-1}\sum_{\{q_1q_2\ldots q_N\}}(-1)^P\frac{\partial}{\partial(z_{q_1}-z_{q_2})}\frac{\partial}{\partial(z_{q_3}-z_{q_4})}\cdots\cdots\frac{\partial}{\partial(z_{q_{N-1}}-z_{q_N})}\left[\prod_{i\neq j}(z_i-z_j)^{-p_i\cdot p_j}\right] \tag{64}$$

$$\equiv \left(S_N^{(2)}\right)^{-1}\frac{1}{(z_a-z_b)}\sum_{\{q_1q_2\ldots q_{N-2}\}}(-1)^P\frac{\partial}{\partial(z_{q_1}-z_{q_2})}\frac{\partial}{\partial(z_{q_3}-z_{q_4})}\cdots\frac{\partial}{\partial(z_{q_{N-3}}-z_{q_{N-2}})}\left[\prod_{i\neq j}(z_i-z_j)^{-p_i\cdot p_j}\right] \tag{65}$$

into our general A_N and imposing $\alpha_\rho(0) = 1$, $\alpha_\pi(0) = \frac{1}{2}$. In the second form of ϕ_N, in Eq. (65), the z_a, z_b are chosen arbitrarily from the z_i and in the sum we take all permutations of $q_1, q_2, q_3, \ldots, q_{N-2} = 1,2,3,\ldots, N$ (excluding a,b).

These two forms of ϕ_N correspond to the original formulation of the Neveu-Schwarz model and to the second Fock-space reformulation respectively; the equivalence of the two forms expresses the G-gauge invariance of the theory.

We have thus classified the earlier models; we are now ready to discuss the new multiparticle amplitude for physical pions.

3.3 New N-pion amplitude

We wish now to consider the following form of the N-pion Born amplitude[29,30])

$$A_N = \int \prod_{i=1}^{N} dz_i \, (dV_3)^{-1} \sum_{\{q_1 q_2 \ldots q_N\}} (-1)^p \frac{\partial}{\partial(z_{q_1} - z_{q_2})} \frac{\partial}{\partial(z_{q_3} - z_{q_4})} \cdots \frac{\partial}{\partial(z_{q_{N-1}} - z_{q_N})} \left[\prod_{i \neq j} (z_i - z_j)^{-p_i \cdot p_j} \left(S_N^{(1)}\right)^{a/2} \right] \tag{66}$$

Here, we have treated all z-differences as independent variables.

Let us first examine the case N = 4, to see that it coincides with our earlier proposal. We may re-write

$$A_4 = \int \prod_{i=1}^{4} dz_i \, (dV_3)^{-1} \prod_{i \neq j} (z_i - z_j)^{-p_i \cdot p_j} \left(S_4^{(1)}\right)^{a/2} \sum_{\{q_1 q_2 q_3 q_4\}} (-1)^p \frac{\theta_{q_1 q_2, q_3 q_4}}{(z_{q_1} - z_{q_2})(z_{q_3} - z_{q_4})} \tag{67}$$

with

$$\begin{aligned} \theta_{q_1 q_2, q_3 q_4} = {} & (-2 p_{q_1} \cdot p_{q_2})(-2 p_{q_3} \cdot p_{q_4}) \\ & + (-2 p_{q_1} \cdot p_{q_2}) \left(S_4^{(1)}\right)^{-a/2} \frac{\partial}{\partial(z_{q_3} - z_{q_4})} \left(S_4^{(1)}\right)^{a/2} \\ & + (-2 p_{q_3} \cdot p_{q_4}) \left(S_4^{(1)}\right)^{-a/2} \frac{\partial}{\partial(z_{q_1} - z_{q_2})} \left(S_4^{(1)}\right)^{a/2} \\ & + \left(S_4^{(1)}\right)^{-a/2} \frac{\partial}{\partial(z_{q_1} - z_{q_2})} \frac{\partial}{\partial(z_{q_3} - z_{q_4})} \left(S_4^{(1)}\right)^{a/2} \end{aligned} \tag{68}$$

It is convenient to define, for general N, the modified trajectory functions

$$\hat{\alpha}_{ij} = \alpha_{ij} + \frac{a \, S_N^{(1)}\{ij\}}{S_N^{(1)}} \tag{69}$$

where $S_N^{(1)}\{ij\}$ contains only those permutations with i,j adjacent. In particular, for N = 4, this corresponds to

$$\hat{\alpha}_s = \alpha_s + a \beta_t \beta_u / 2\beta_s \tag{70}$$

$$\hat{\alpha}_t = \alpha_t + a \beta_u \beta_s / 2\beta_t \tag{71}$$

$$\hat{\alpha}_u = \alpha_u + a \beta_s \beta_t / 2\beta_u \tag{72}$$

After some algebra we then find that Eq. (67) corresponds to

$$A_4 = \int_0^1 dx\, x^{-\alpha_t - 1} (1-x)^{-\alpha_s - 1} (1-x+x^2)^{\frac{a}{2}+1} \phi_4 \tag{73}$$

with

$$\phi_4 = \hat{\alpha}_s^2 \beta_s + \hat{\alpha}_t^2 \beta_t + \hat{\alpha}_u^2 \beta_u - \frac{3a}{2} \beta_s \beta_t \beta_u \tag{74}$$

Let us agree to use the symbol ≡ (is equivalent to) with the special meaning that two expressions for ϕ_4 in Eq. (73) are equal <u>after integration</u>.

Then it is possible to prove the following equivalences

$$\phi_4 = \hat{\alpha}_s^2 \beta_s + \hat{\alpha}_t^2 \beta_t + \hat{\alpha}_u^2 \beta_u - \frac{3a}{2} \beta_s \beta_t \beta_u \equiv \hat{\alpha}_s \beta_s \equiv \hat{\alpha}_t \beta_t \equiv \hat{\alpha}_u \beta_u \tag{75}$$

$$\equiv \frac{1}{3} (\alpha_s \beta_s + \alpha_t \beta_t + \alpha_u \beta_u) \tag{76}$$

This then demonstrates that for N = 4 our general expression reduces to

$$A_4 = \int_0^1 dx\, x^{-\alpha_t - 1} (1-x)^{-\alpha_s - 1} (1-x+x^2)^{\frac{a}{2}} \cdot \frac{1}{3} (\alpha_s x + \alpha_t (1-x) - \alpha_u x (1-x)) \tag{77}$$

coinciding with our earlier Eq. (38).

Let us remark that a reformulation of our A_N may be obtained by introducing*) N complex variables ϕ_i satisfying a Grassmann algebra

$$\{\phi_i, \phi_j\}_+ = 0 \tag{78}$$

and then we may write

$$A_N = \oint \prod_{i=1}^{N} \frac{d\phi_i}{\phi_i^2} \int \prod_{i=1}^{N} dz_i \,(dV_3)^{-1} \prod_{i \neq j} (z_i - z_j - \phi_i \phi_j)^{-p_i \cdot p_j} \left[\sum_{\{q_1 q_2 \ldots q_N\}} \prod_{k=1}^{N} (z_{q_k} - z_{q_{k+1}} - \phi_{q_k} \phi_{q_{k+1}})^{-2} \right]^{a/2} \tag{79}$$

where the ϕ_i integrations are understood to be Cauchy integrals encircling the origin. The equivalence of Eq. (79) to Eq. (66) is clear once one realizes that the Grassmann algebra is simply a device for taking the first term of a Taylor series with respect to the z differences. Note that in either case, for a = 0 we arrive at the Neveu-Schwarz model for which a no-ghost theorem is well established.

---

*) Such variables have been introduced elsewhere[31,32] in discussing the Neveu-Schwarz dual-pion model.

We now make a first discussion of the properties of the new multipion amplitude, A_N.

3.4 Bootstrap consistency

We shall now check that at the internal pion pole, the six-point function A_6 factorizes correctly into two four-point fucntions.

Putting $\alpha_\rho(0) - \alpha_\pi(0) = ½$ we are led to the formula for A_6 which is

$$A_6 = \int \prod_{i=1}^{6} dz_i \, (dV_3)^{-1} \left[\prod_{i \neq j} (z_i - z_j)^{-p_i \cdot p_j} \prod_{i=1}^{6} (z_i - z_{i+1})^{-a} \prod_{i<j} (z_i - z_j)^{(-1)^{i-j}} \right]$$

$$\left[\prod_{i=1}^{6} (z_i - z_{i+1})^2 \cdot S_6^{(1)} \right]^{a/2} \tag{80}$$

$$\left[\prod_{i<j} (z_i - z_j)^{-(-1)^{i-j}} \sum_{\{q_1 q_2 \ldots q_6\}} (-1)^P \frac{\theta_{q_1 q_2, q_3 q_4, q_5 q_6}}{(z_{q_1} - z_{q_2})(z_{q_3} - z_{q_4})(z_{q_5} - z_{q_6})} \right]$$

The first square bracket has well-known factorization properties, since it is precisely the generalized Euler B function.

The second bracket also factorizes, since if we define variables $0 \leq x,w,y \leq 1$ according to the anharmonic ratios

$$x = (z_1, z_6 ; z_2, z_3) \tag{81}$$

$$w = (z_1, z_6 ; z_3, z_4) \tag{82}$$

$$y = (z_1, z_6 ; z_4, z_5) \tag{83}$$

corresponding to the (12), (123) and (56) channels respectively, then we find

$$\lim_{w \to 0} \left[\prod_{i=1}^{6} (z_i - z_{i+1})^2 \, S_6^{(1)} \right] = 96 \, (1 - x + x^2)(1 - y + y^2) \tag{84}$$

For the third square bracket, the factorization properties are more subtle. Combining terms we find

$$\lim_{\alpha_{123} \to 0} (\alpha_{123} A_6) = \int_0^1 dx \, x^{-\alpha_{12}-1} (1-x)^{-\alpha_{23}-1} (1-x+x^2)^{\frac{a}{2}+1}$$

$$\int_0^1 dy \, y^{-\alpha_{56}-1} (1-y)^{-\alpha_{45}-1} (1-y+y^2)^{\frac{a}{2}+1} \, f(x,y) \tag{85}$$

with

$$\mathcal{S}(x,y) = \Big[\beta_{12}(x)\,\beta_{56}(y)\,\overline{\Theta}_{12,34,56} + \beta_{12}(x)\,\beta_{45}(y)\,\overline{\Theta}_{12,36,45}$$
$$+ \beta_{12}(x)\,\beta_{46}(y)\,\overline{\Theta}_{12,35,46} + \beta_{23}(x)\,\beta_{56}(y)\,\overline{\Theta}_{14,23,56}$$
$$+ \beta_{23}(x)\,\beta_{45}(y)\,\overline{\Theta}_{16,23,45} + \beta_{23}(x)\,\beta_{46}(y)\,\overline{\Theta}_{15,23,46} \qquad (86)$$
$$+ \beta_{13}(x)\,\beta_{56}(y)\,\overline{\Theta}_{13,24,56} + \beta_{13}(x)\,\beta_{45}(y)\,\overline{\Theta}_{13,26,45}$$
$$+ \beta_{13}(x)\,\beta_{46}(y)\,\overline{\Theta}_{13,25,46}\Big]$$

where we defined

$$\beta_{12}(x) = (1-x)\,(1-x+x^2)^{-1} \qquad (87)$$
$$\beta_{23}(x) = \beta_{12}\,(1-x) \qquad (88)$$
$$\beta_{13}(x) = \beta_{12}\,(1-1/x) \qquad (89)$$
$$\beta_{56}(y) = (1-y)\,(1-y+y^2)^{-1} \qquad (90)$$
$$\beta_{45}(y) = \beta_{56}\,(1-y) \qquad (91)$$
$$\beta_{46}(y) = \beta_{56}\,(1-1/y) \qquad (92)$$

and

$$\overline{\Theta}_{q_1q_2,q_3q_4,q_5q_6} = \lim_{W\to 0}\left(\Theta_{q_1q_2,q_3q_4,q_5q_6}\right) \qquad (93)$$

Defining the quantities

$$f_{12} = 1 + \beta_{23}\beta_{13}/2\beta_{12} \qquad (94)$$
$$f_{23} = 1 + \beta_{12}\beta_{13}/2\beta_{23} \qquad (95)$$
$$f_{13} = 1 + \beta_{12}\beta_{23}/2\beta_{13} \qquad (96)$$
$$f_{56} = 1 + \beta_{45}\beta_{46}/2\beta_{56} \qquad (97)$$
$$f_{45} = 1 + \beta_{46}\beta_{56}/2\beta_{45} \qquad (98)$$
$$f_{46} = 1 + \beta_{45}\beta_{56}/2\beta_{46} \qquad (99)$$

and using the definition of $\hat{\alpha}_{ij}$ given earlier we find that

$$\mathcal{S}(x,y) = \hat{\alpha}_{12}\,\beta_{12}\,\hat{\alpha}_{56}\,\beta_{56}\,(1+5a/2)$$
$$+ \frac{3a}{4}\left(\hat{\alpha}_{12}\,\beta_{12}\,\hat{\phi}_4^{(2)}(456) + \hat{\phi}_4^{(2)}(123)\,\hat{\alpha}_{56}\,\beta_{56}\right)$$
$$- \frac{9a}{4}\left(\hat{\alpha}_{12}\,\beta_{12}\cdot\beta_{45}\beta_{56}\beta_{46} + \hat{\alpha}_{56}\,\beta_{56}\cdot\beta_{12}\beta_{23}\beta_{13}\right)$$
$$+ \frac{a}{8}\left(\hat{\phi}_4^{(2)}(123)\,\hat{\phi}_4^{(2)}(456)\right) + \frac{9a}{8}\left(\beta_{12}\beta_{23}\beta_{13}\cdot\beta_{45}\beta_{56}\beta_{46}\right)$$
$$- \frac{3a}{8}\left(\hat{\phi}_4^{(2)}(123)\,\beta_{45}\beta_{56}\beta_{46} + \beta_{12}\beta_{23}\beta_{13}\cdot\hat{\phi}_4^{(2)}(456)\right) \qquad (100)$$
$$\equiv \left(\hat{\alpha}_{12}^2\beta_{12} + \hat{\alpha}_{23}^2\beta_{23} + \hat{\alpha}_{13}^2\beta_{13} - \frac{3a}{2}\beta_{12}\beta_{23}\beta_{13}\right)\cdot$$
$$\left(\hat{\alpha}_{45}^2\beta_{45} + \hat{\alpha}_{56}^2\beta_{56} + \hat{\alpha}_{46}^2\beta_{46} - \frac{3a}{2}\beta_{45}\beta_{56}\beta_{46}\right) \qquad (101)$$
$$\equiv \hat{\alpha}_{12}\,\beta_{12}\cdot\hat{\alpha}_{56}\,\beta_{56} \qquad (102)$$

which confirms bootstrap consistency for any ρ intercept $\alpha_\rho(0)$ *).

3.5 Spin-lowering symmetry

It turns out that the new A_N satisfies a symmetry which is higher than the symmetric group invariance used in its construction. In particular we may re-write

$$A_N = \int \prod_{i=1}^{N} dz_i \, (dV_3)^{-1} \frac{1}{(z_a - z_b)} \sum_{\{q_1 q_2 \ldots q_{N-2}\}} (-1)^P \frac{\partial}{\partial(z_{q_1} - z_{q_2})} \frac{\partial}{\partial(z_{q_3} - z_{q_4})} \cdots \frac{\partial}{\partial(z_{q_{N-3}} - z_{q_{N-2}})} \left[\prod_{i \neq j} (z_i - z_j)^{-p_i \cdot p_j} \left(S_N^{(1)}\right)^{a/2} \right] \tag{103}$$

where z_a, z_b are chosen arbitrarily from the z_i and the sum is over all permutations of $q_1, q_2, \ldots, q_{N-2} = 1,2, \ldots, N$ (excluding a,b). We may easily express this also in terms of Grassmann variables.

At the level N = 4 we may choose inequivalently a,b = 1,2; 2,3; 1,3 to arrive at, respectively, the formulae

$$\int_0^1 dx \, x^{-\alpha_t - 1} (1-x)^{-\alpha_s - 1} (1 - x + x^2)^{\frac{a}{2}+1} \begin{cases} \hat{\alpha}_s \beta_s \\ \hat{\alpha}_t \beta_t \\ \hat{\alpha}_u \beta_u \end{cases} \tag{104}$$

which are all equal to A_4, as discussed earlier.

At the level N = 6 there are six inequivalent choices a,b = 1,4; 1,5; 1,6; 2,5; 5,6; 4,6 for considering the residue at $\alpha_{123} = 0$. In all six cases we find that ρ(x,y) defined in the previous subsection is equivalent to

$$\rho(x,y) = \hat{\alpha}_{12} \beta_{12} \cdot \hat{\alpha}_{56} \beta_{56} \tag{105}$$

as demanded by bootstrap consistency.

Note that spin-lowering symmetry is clearly associated with the tachyon-killing mechanism for the ρ trajectory since the latter is the primary function of ϕ_N which this higher symmetry describes. More remarkably, the spin-lowering symmetry in the form of the equivalences

$$\hat{\alpha}_{12} \beta_{12} \equiv \hat{\alpha}_{23} \beta_{23} \equiv \hat{\alpha}_{13} \beta_{13} \tag{106}$$

$$\hat{\alpha}_{56} \beta_{56} \equiv \hat{\alpha}_{45} \beta_{45} \equiv \hat{\alpha}_{46} \beta_{46} \tag{107}$$

---

*) A non-planar extension of A_N, by replacing in the Grassmann form $(z_i - z_j - \phi_i\phi_j)^2 \to [(z_i - z_j)(z_i^* - z_j^*) - \phi_i\phi_j\phi_i^*\phi_j^*]$ and integrating (N - 3) z-variables over the complex plane, together with 2N φ variables around the origin, is projective invariant for intercept $(1 - 2\alpha'\mu^2)$. We may anticipate by analogy with earlier models that this will become the pomeron intercept.

is responsible for the absence of an unwanted ancestor above the pion trajectory. To see this, note that spin-lowering symmetry underlies the unexpected recombination of terms in

$$\begin{aligned}
&\hat{\alpha}_{12}\beta_{12}(2p_3\cdot p_4)\hat{\alpha}_{56}\beta_{56} + \hat{\alpha}_{12}\beta_{12}(2p_3\cdot p_6)\hat{\alpha}_{45}\beta_{45} \\
&+ \hat{\alpha}_{12}\beta_{12}(2p_3\cdot p_5)\hat{\alpha}_{46}\beta_{46} + \hat{\alpha}_{23}\beta_{23}(2p_1\cdot p_4)\hat{\alpha}_{56}\beta_{56} \\
&+ \hat{\alpha}_{23}\beta_{23}(2p_1\cdot p_6)\hat{\alpha}_{45}\beta_{45} + \hat{\alpha}_{23}\beta_{23}(2p_1\cdot p_5)\hat{\alpha}_{46}\beta_{46} \\
&+ \hat{\alpha}_{13}\beta_{13}(2p_2\cdot p_4)\hat{\alpha}_{56}\beta_{56} + \hat{\alpha}_{13}\beta_{13}(2p_2\cdot p_6)\hat{\alpha}_{45}\beta_{45} + \hat{\alpha}_{13}\beta_{13}(2p_2\cdot p_5)\hat{\alpha}_{46}\beta_{46} \\
&\equiv (1-2a)\hat{\alpha}_{12}\beta_{12}\cdot\hat{\alpha}_{56}\beta_{56} \qquad (108)
\end{aligned}$$

which successfully decouples the unwanted spin-one ancestor at the pion mass.

4. CONCLUSIONS

4.1 Summary

We have indicated how the symmetric group plays a central role in the make-up of meson tree amplitudes. By exploiting this fact, we have classified earlier proposals for the Born amplitude.

We have further explicitly constructed a new four-pion and N-pion Born amplitude and have made a first discussion of its properties.

4.2 Brief experimental comment

Let us digress briefly and anticipate that we might be able to construct a realistic Born amplitude for pion-nucleon elastic scattering or for even more complex processes. It becomes a pertinent question: What type of data are expected to be useful for comparison to the theoretical prediction?

Current experiments with hadrons may be broadly classified according to their incident energy into two classes:

a) Low and intermediate energies where one studies two-body and quasi-two-body processes if possible controlling accurately the spin-states as well as all the particle momenta.

b) High and very high energy study of properties such as diffractive effects, multiparticle production, inclusive spectra and correlations.

Because the most powerful accelerating machines are the most newly acquired there is naturally an emphasis on the exciting second kind (b) of experiment at this time.

Nevertheless, the zero-width Born amplitude is likely to be the most reliable in the first kind (a) of experiment at relatively low energies; in particular, one should emphasize small momentum transfer and avoid vacuum quantum number exchange. Thus the most immediately useful data may, for example, be from pion-nucleon charge exchange in the resonance region and immediately above with small momentum transfers and with the spin maximally controlled so that details of the separate amplitudes can be fully analysed[33]. We expect that under such kinematic conditions, the unitarity corrections to the Born amplitude will be smallest and hence most reliably estimated.

4.3 Concluding remarks

It is clearly important that further consistency checks of the multipion amplitude be obtained, since at present the analysis is incomplete. One can also envisage the possible extension to include strange mesons (SU_3) and baryons. More immediately important and already sufficiently ambitious, however, is to establish the factorization properties, and if possible the absence of negative probabilities, in the multipion amplitude.

REFERENCES

1) A. Neveu and J.H. Schwarz, Nuclear Phys. B31, 86 (1971).

2) J.L. Gervais and A. Neveu, Nuclear Phys. B47, 422 (1972).

3) J.L. Gervais and A. Neveu, Orsay preprint LPTHE 73/16 (June 1973).

4) Y. Nambu, Lecture notes prepared for the Summer Institute of the Niels Bohr Institute (SINBI), 1970.

5) L.N. Chang and F. Mansouri, Phys. Rev. D5, 2535 (1972).

6) P. Goddard, J. Goldstone, C. Rebbi and C.B. Thorn, Nuclear Phys. B56, 109 (1973).

7) Y. Iwasaki and K. Kikkawa, CCNY preprints (February and March 1973).

8) J.L. Gervais and B. Sakita, Phys. Rev. Letters 30, 716 (1973).

9) S. Mandelstam, Berkeley preprint (May 1973).

10) M. Ademollo et al., CERN preprint (1973).

11) P.H. Frampton, Syracuse University preprint SU-4205-12.

12) P.H. Frampton, Phys. Rev. D, May 15 1973.

13) G. Veneziano, Nuovo Cimento 57 A, 190 (1968).

14) K. Bardakci and H. Ruegg, Phys. Rev. 181, 1884 (1969);
Chan Hong-Mo and T.S. Tsun, Phys. Letters 28 B, 485 (1969);
C.J. Goebel and B. Sakita, Phys. Rev. Letters 22, 257 (1969);
Z. Koba and H.B. Nielsen, Nuclear Phys. B10, 633 (1969);
M.A. Virasoro, Phys. Rev. Letters 22, 37 (1969).

15) D.B. Fairlie and K. Jones, Nuclear Phys. B15, 323 (1970).

16) E. Plahte, Nuovo Cimento 66 A, 713 (1970).

17) P.H. Frampton, Lecture notes on Dual Resonance Models, Part III Operator Formalism, Syracuse University preprint SU-4205-17 (1973) Section 3d. See also the references cited therein.

18) S. Mandelstam, Phys. Rev. Letters 21, 1724 (1968).

19) A.P. Balachandran and H. Rupertsberger, Syracuse University preprint SU-4205-18 (1973).

20) A.C. Dixon, Proc. Land. Math. Soc. (February 11 1904) p. 8.

21) V.N. Gribov and I. Ya Pomeranchuk, Soviet Phys. JETP 43, 208 (1962) [translation: 16, 220 (1963)].

22) A. Martin, Nuovo Cimento 42, 930 (1966).

23) C. Lovelace, Phys. Letters 28 B, 264 (1968)

24) J.A. Shaprio, Phys. Rev. 179, 1345 (1969).

25) S.D. Protopopescu et al., Berkeley LBL-787 (1972).

26) P. Estabrooks et al., CERN TH 1661 (1973).

27) Z. Koba and H.B. Nielsen, Nuclear Phys. B12, 633 (1969).

28) P.H. Frampton, CERN TH 1546 (1972).

29) P.H. Frampton, CERN TH 1698 (1973).

30) P.H. Frampton, Syracuse University preprint SU-4205-26 (1973).

31) C. Montonen, Cambridge University DAMTP 73/12 (1973).

32) D.B. Fairlie and D. Martin, Durham University preprint (1973).

33) F. Halzen and C. Michael, Phys. Letters 36 B, 367 (1971).

D I S C U S S I O N S

CHAIRMAN: Prof. P.H. Frampton

Scientific Secretaries: R. Cahalan, G. Chu

DISCUSSION No. 1

- *KUPCZYNSKI:*

What is the connection between the summability condition and ghost elimination?

- *FRAMPTON:*

In terms of Koba-Nielsen variables, the conventional model takes the form (for equal intercepts):

$$A_N = \int \prod_{i=1}^{N} dz_i \, (dV_3)^{-1} \prod_{i \neq j} (z_i - z_j)^{-p_i p_j} \prod_{i=1}^{N} (z_i - z_{i+1})^{\alpha(0)-1} .$$

For the case $\alpha(0) = 1$, we see that the integrand is symmetric group invariant under intercharge of the argument pairs $\{p_i, z_i\}$ $i = 1,2,3 \ldots, N$. This means there is a summability condition in the sense that we can sum over inequivalent permutations simply by extending the integration domain.

When we factorize the $\alpha(0) = 1$ case we write

$$A_N = \langle 0 | V_0 \, D \, V_0 \, D \, V_0 \, D \ldots D \, V_0 | 0 \rangle$$

with

$$D = (L_0 - 1)^{-1}$$

and

$$V_0 = \exp\left(\sqrt{2i} \sum_{n=1}^{\infty} \frac{a^{(n)+}}{\sqrt{n}} \cdot p\right) \cdot \exp\left(\sqrt{2i} \sum_{n=1}^{\infty} \frac{a^{(n)}}{\sqrt{n}} \cdot p\right) .$$

To factorize for $\alpha(0) \neq 1$, one method is to extend all oscillators from four to five dimensions. Putting $\lambda^2 = 1 - \alpha(0)$ we may write the $\alpha(0) \neq 1$ vertex as

$$V = V_0 V^5$$

$$V^5 = \exp\left(\lambda \sum_{n=1}^{\infty} \frac{a_5^{(n)+}}{\sqrt{n}}\right) |0\rangle_5 \; {}_5\langle 0| \exp\left(\lambda \sum_{n=1}^{\infty} \frac{a_5^{(n)}}{\sqrt{n}}\right) .$$

The crucial point is the presence of the vacuum projection on to the fifth dimension, which is essential to avoid factors other than nearest neighbours. It is easy to show that it is precisely this vacuum projection which makes the vertex transform non-covariantly under the generalized projective ghost-eliminating algebra.

Thus we are led to associate the summability condition (symmetric group invariance of the integrand of A_N for every N) with a necessary condition for the existence of the full set of ghost-eliminating gauge conditions.

- *MOEN:*

Regarding the question of factorization and absence of ghosts for realistic values of $\alpha_\rho(0)$: how does the model you propose avoid the "persistent photon" theorem?

- *FRAMPTON:*

The "persistent photon" condition has been proved only under a set of rather specialized technical assumptions and can therefore of course be avoided by violating those assumptions. Indeed there already exists in the literature a model with no ghosts and no massless vector state [I am referring to Gervais and Neveu, Orsay preprint LPTHE 73/16 (June 1973)]. It would, however, certainly be very instructive to study the mechanism by which the massive ρ meson acquires a longitudinal component in the present model.

- *KONISHI:*

Is your model dual in the sense that the resonance residues can be used to interpolate the correct Regge behaviour?

- *FRAMPTON:*

The model is semi-locally dual, to the extent that we must integrate over three resonances to obtain a curve which approximates the smooth Regge behaviour.

- *WILCZEK:*

Do you have to put in the Adler zero by hand, or is there some motivation for it in your theory?

- *FRAMPTON:*

In this model, the Adler zero is a first-order zero, arising from the constraint $\alpha_\rho(0) - \alpha_\pi(0) = \frac{1}{2}$. The daughters give a non-vanishing contribution at the Adler point. Of course, the Adler zero is for off-shell ($m_\pi^2 = 0$) scattering and does not require that the amplitude vanish, only that it be small.

I should add that the choice $\alpha_\rho(0) - \alpha_\pi(0) = \frac{1}{2}$ also simplifies the N-point function considerably.

- *VINCIARELLI:*

I would like to ask a general question. As I understand it, part of the difficulty encountered in incorporating electromagnetic and weak currents into dual models may be traced back to the harmonic nature of the potential, perhaps to the fact that this is too regular at short distances. If the good features of the theory depend only on the long-range behaviour of the potential, one may then want to try and simply modify its short-distance behaviour to improve on the off-shell properties of dual amplitudes. I would like to know whether my observation makes sense and, if it does, whether any attempts have been made on this direction?

- *FRAMPTON:*

I believe that your general observation is a sensible one. On the other hand, no attempt at including currents in a fully consistent way has been successful. The most complete work that I am aware of is by Cremmer and Scherk [Nuclear Phys. B50, 222 (1972)] who write multicurrent amplitudes to generalize single-current amplitudes proposed earlier by Rebbi and Drummond. My personal opinion is that a necessary first step is to obtain acceptable amplitudes for the strong interactions in isolation.

- *KLEINERT:*

I have a remark concerning the inclusion of currents in dual models. A few years ago I constructed a local Lagrangian with a local current minimally coupled to the electromagnetic field [Lettere Nuovo Cimento 4, 285 (1970)]. The current leads to a non-zero scaling function as long as the number of oscillators in the Veneziano model is kept finite. The form factors behave like

$$\exp\left[-q^2 \sum_{n=1}^{N} \frac{1}{n}\right]$$

and the scaling function has the form

$$F_2(x) = x_0(N)\, \delta[x - x_0(N)] .$$

In the limit $N \to \infty$, the form factors vanish faster than any exponential and at the same time

$$x_0 \to \left[\sum_{n=1}^{N} \frac{1}{n} \right]^{-1}$$

so that F_2 vanishes.

- *REBBI:*

On the question about amplitudes for currents in dual resonance models, the most complete researches up to now have been done by Cremmer and Scherk, who can write amplitudes for any number of currents with any spin. The form factors have a Gaussian behaviour which is related to the smoothness of the harmonic oscillator potential characteristic of the dual models.

Notice that early investigations of currents found a difficulty in the presence of an infinite number of modes which could generate a logarithmic divergence in the form factor exponent. In the work of Cremmer and Scherk a cut-off in the infinite sum occurs naturally. They find a scaling behaviour for deep inelastic structure functions; however, a consequence of the Gaussian behaviour of form factors is that the structure function $f(\omega)$ is identically zero for a range of ω between threshold and a definite value for ω and then becomes a non-trivial function.

- *CHU:*

You mentioned that you have checked Regge behaviour in the four-point case. Have you investigated the multi-Regge behaviour of your amplitude in the six-point case, where the analyticity properties are much more interesting? For example, one would want to verify the Steinmann relations.

- *FRAMPTON:*

The answer is no. Of course, it is very important to check the multi-Regge behaviour of the amplitude, but we have not been able to do it yet.

- *REBBI:*

Do you have an analytic proof or only numerical computations to show the absence of ghosts in your four-point function?

- *FRAMPTON:*

We have checked the absence of ghosts numerically for the first 20 levels, but no analytic proof exists. An analytic proof does exist of the positivity of Im A (elastic) in the forward direction and of its derivatives, which is a necessary (but not sufficient) condition for the absence of ghosts.

- *REBBI:*

Do you expect to find an ω in the 3π-channel of the 6π-amplitude?

- *FRAMPTON:*

We do not have complete results, but there definitely is an ω in the 3π-channel. However, it seems that the ω will appear on the π trajectory (with $m_\omega^2 - m_\rho^2 = 1/2\alpha'$) as in the model of Neveu and Schwarz.

- *MILLER:*

What does your model predict for the increase in width as a function of M^2 and can you tell us anything about the coupling of the $\rho'(1600) \to \pi\pi$ or coupling of other daughters?

- *FRAMPTON:*

I would conjecture that the behaviour of widths as a function of M^2 would be similar to that in the original Veneziano model. In answer to the second part of your question, the $\rho'(1600)$ has a $\pi\pi$ partial width of 141 MeV, while the $\varepsilon'(1250)$ has a $\pi\pi$ width of 259 MeV. The $\varepsilon(\sim 760)$ and $\rho'(1250)$ do not couple to $\pi\pi$ (all odd daughters are decoupled in my model). The widths are calculated by normalizing the ρ partial width to 112 MeV and setting $\alpha_\rho(0) = 0.48$; the widths then arise from the relative couplings of the resonances.

The present data, and $\pi\pi$ phase shifts, show that there is an s-wave under the f(1250) resonance and no p-wave. This seems to favour the new model over the Lovelace-Shapiro model.

- *NAHM:*

Is your simplicity criterion for Φ_4 just a method to make life simpler, or do you have indications that it is related to some physical feature?

- *FRAMPTON:*

The ambiguity of Φ_4 is somewhat analogous to the satellite ambiguity in the Veneziano model. It may be that one is led to unreasonable spectra if he uses a complicated Φ_4.

SECRET SYMMETRY: AN INTRODUCTION TO SPONTANEOUS SYMMETRY BREAKDOWN AND GAUGE FIELDS

S. Coleman

Table of Contents

SECRET SYMMETRY:
AN INTRODUCTION TO SPONTANEOUS SYMMETRY BREAKDOWN AND GAUGE FIELDS*

Sidney Coleman
Lyman Laboratory of Physics, Harvard University, Cambridge, Massachusetts, USA

1. INTRODUCTION

Here are some long-standing problems in particle theory:

1) How can we understand the hierarchal structure of the fundamental interactions? Are the strong, medium strong (i.e., SU(3)-breaking), electromagnetic, and weak interactions truly independent, or is there some principle that establishes connections between them?

2) How can we construct a renormalizable theory of the weak interactions, one which reproduces the low-energy successes of the Fermi theory but predicts finite higher-order corrections?

3) How can we construct a theory of electromagnetic interactions in which electromagnetic mass differences within isotopic multiplets are finite?

4) How can we reconcile Bjorken scaling in deep inelastic electroproduction with quantum field theory? The SLAC-MIT experiments seem to be telling us that the light-cone singularities in the product of two currents are canonical in structure; ordinary perturbation theory, on the other hand, tells us that the canonical structure is spoiled by logarithmic factors, which get worse and worse as we go to higher and higher orders in the perturbation expansion. Are there any theories of the strong interactions for which we can tame the logarithms, sum them up and show they are harmless?

Enormous progress has been made on all of these problems in the last few years. There now exists a large family of models of the weak and electromagnetic interactions that solve the second and third problem, and we have discovered a somewhat smaller family of models of the strong interactions that solve the fourth problem. As we shall see, the structure of these models is such that we are beginning to get ideas about the solution to the (very deep) first problem; connections are beginning to appear in unexpected places, and a optimist might say that we are on the road to the first truly unified theory of the fundamental interactions. All of these marvelous developments are based upon the ideas of spontaneous symmetry breakdown and gauge fields, the subjects of these lectures.

Honesty compels me to moderate the sales-pitch of the last paragraph

*Work supported in part by the National Science Foundation under Grant No. GP 30819X.

by pointing out that there is a fifth long-standing problem with which these theories have not yet made contact:

5) How do we explain experiments?

We can see the reason for this embarrassing lacuna if we think a little bit more closely about the second problem, constructing a renormalizable theory of the weak interactions. At the moment, there is a plethora of such theories; they all predict that higher-order weak effects are finite, and they all predict that they are small. To find which, if any, of these theories is correct requires precision measurements of higher-order weak effects (preferably purely leptonic ones, so the strong interactions don't corrupt our predictions); these are hard to come by. Phrased another way, the Fermi theory is obviously dead wrong, because it predicts infinite higher-order corrections, but it is experimentally nearly perfect, because there are few experiments for which lowest-order Fermi theory is inadequate. Likewise for electromagnetic mass differences within isotopic multiplets: to make the differences finite, we need only to tame the high-energy behaviour of self-mass integrals; to actually compute them, though, we have to know the integrals at all energies, including the low-energy region where the strong interactions are dominant (and incalculable).

These lectures are intended as an introduction to the basic ideas of spontaneous symmetry breakdown and gauge fields, not as a survey of all the work done to date, and there are some important topics that I will not discuss at all. In particular, I will not touch at all upon the important subject of model-building; indeed, in order to simplify my examples as much as possible, I will barely mention theories involving Fermions at all. This gap will be remedied by Shelly Glashow's lectures, on unified models of the weak and electromagnetic interactions. Also,although I will try and make the renormalizability of the theories we discuss plausible, I will have no time to go into the guts of the renormalization problem, and therefore will say nothing about the beautiful dimensional regularization procedure of Veltman and 't Hooft, nor about the non-Abelian generalizations of the Ward identities of quantum electrodynamics, the Slavnov identities.[1]

The organization of these lectures is as follows: Section 2 is a discussion of spontaneous symmetry breakdown, Goldstone Bosons, gauge fields, and the Higgs phenomenon in the simplest context, that of classical field theory. Section 3 shows how these ideas can be extended to quantum field theory in such a way that the classical reasoning of the previous section becomes the first term in a systematic quantum expansion. The important concept of the effective potential makes its first appearance here, and its properties are discussed at length. However, an important part of the quantization program is postponed: the quantization of gauge fields. This gaping hole in the arguments of Section 3 is filled in the next two sections. Section 4 is an introduction to functional integration as a method of quantization, and Section 5 is an application of this method to gauge fields, following the ideas of Faddeev and Popov. I have tried to make Section 4 as self-contained as possible, so it may be useful to the reader who wants to learn functional integration, even if he is uninterested in the other topics of these lectures. Section 6 takes off in a new direction and explores the asymptotic properties of gauge field theories. It includes a brief review of the renormalization group.

I have learned much from conversations with Ludwig Faddeev, Howard Georgi, Sheldon Glashow, Jeffrey Goldstone, David Gross, Benjamin Lee, David Politzer, Gerard 't Hooft, Tini Veltman, Erick Weinberg, Steven Weinberg, and Frank Wilczek. Many authors who have made major contributions to this subject (including a large subset of the above) are inadequately represented in the references at the end of these lectures, because of my eccentric choice of topics and methods of approach; to these I apologize, as I do to those whom I have omitted through ignorance.[1]

2. SECRET SYMMETRIES IN CLASSICAL FIELD THEORY

2.1 The Idea of Spontaneous Symmetry Breakdown

In general, there is no reason why an invariance of the Hamiltonian of a quantum-mechanical system should also be an invariance of the ground state of the system. Thus, for example, the nuclear forces are rotationally invariant, but this does not mean that the ground state of a nucleus is necessarily rotationally invariant (i.e., of spin zero). This is a triviality for nuclei, but it has highly non-trivial consequences if we consider systems which, unlike nuclei, are of infinite spatial extent. The standard example is the Heisenberg ferromagnet, an infinite crystalline array of spin-$\frac{1}{2}$ magnetic dipoles, with spin-spin interactions between nearest neighbors such that neighboring dipoles tend to align. Even though the Hamiltonian is rotationally invariant, the ground state is not; it is a state in which all the dipoles are aligned in some arbitrary direction, and is infinitely degenerate for an infinite ferromagnet. A little man living inside such a ferromagnet would have a hard time detecting the rotational invariance of the laws of nature; all his experiments would be corrupted by the background magnetic field. If his experimental apparatus interacted only weakly with the background field, he might detect rotational invariance as an approximate symmetry; if it interacted strongly, he might miss it altogether; in any case, he would have no reason to suspect that it was in fact an exact symmetry. Also, the little man would have no hope of detecting directly that the ground state in which he happens to find himself is in fact part of an infinitely degenerate multiplet. Since he is of finite extent (this is the technical meaning of "little"), he can only change the direction of a finite number of dipoles at a time; but to go from one ground state of the ferromagnet to another, he must change the directions of an infinite number of dipoles - an impossible task.

At least at first glance, there appears to be nothing in this picture that can not be generalized to relativistic quantum mechanics. For the Hamiltonian of a ferromagnet, we can substitute the Hamiltonian of a quantum field theory; for rotational invariance, some internal symmetry; for the ground state of the ferromagnet, the vacuum state; and for the little man, ourselves. That is to say, we conjecture that the laws of nature may possess symmetries which are not manifest to us because the vacuum state is not invariant under them.[2] This situation is usually called "spontaneous breakdown of symmetry". The terminology is slightly deceptive, because the symmetry is not really broken, merely hidden, but I will use it anyway.

We will begin by investigating spontaneous symmetry breakdown in the case of classical field theory. For simplicity, we will restrict ourselves to theories involving a set of n real scalar fields, which we will assemble into a real n-vector, $\underset{\sim}{\varphi}$, with Lagrange density[3]

$$\mathcal{L} = \tfrac{1}{2}(\partial_\mu \underset{\sim}{\varphi}) \cdot (\partial^\mu \underset{\sim}{\varphi}) - U(\underset{\sim}{\varphi}) , \tag{2.1}$$

where U is some function of the φ's, but not of their derivatives. We will treat these theories purely classically, but use quantum-mechanical language; thus, we will call the state of lowest energy "the vacuum", and refer to the quantities which characterize the spectra of small oscillations about the vacuum as "particle masses". For any of these theories, the energy density is

$$\mathcal{H} = \tfrac{1}{2}(\partial_0 \underset{\sim}{\varphi})^2 + \tfrac{1}{2}(\vec{\nabla} \underset{\sim}{\varphi})^2 + U(\underset{\sim}{\varphi}) . \tag{2.2}$$

Thus the state of lowest energy is one for which the value of $\underset{\sim}{\varphi}$ is a constant, which we denote by $\langle \underset{\sim}{\varphi} \rangle$. The value of $\langle \underset{\sim}{\varphi} \rangle$ is determined by the detailed dynamics of the particular theory under investigation, that is to say, by the location of the minimum (or minima) of the potential, U. Sticking to our policy of using quantum language, we will call $\langle \underset{\sim}{\varphi} \rangle$ "the vacuum expectation value of $\underset{\sim}{\varphi}$".

Within this class of theories, it is easy to find examples for which symmetries are either manifest or spontaneously broken. The simplest one is the theory of a single field for which the potential is

$$U = \frac{\lambda}{4!}\varphi^4 + \frac{\mu^2}{2}\varphi^2, \tag{2.3}$$

where λ is a positive number and μ^2 (despite its name) can be either positive or negative. This theory admits the symmetry

$$\varphi \to -\varphi. \tag{2.4}$$

If μ^2 is positive, the potential is as shown in Fig. 1. The vacuum is at $\langle\varphi\rangle$ equals zero, the symmetry is manifest, and μ^2 is the mass of the scalar meson. If μ^2 is negative, though, the situation is quite different; the potential is as shown in Fig. 2. In this case, it is convenient to

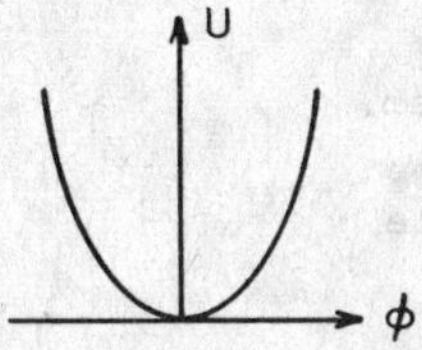

Fig. 1

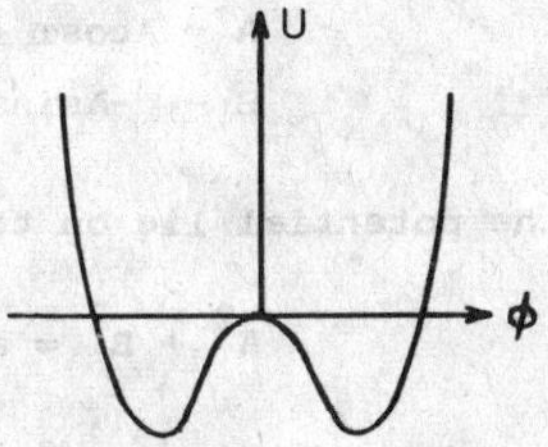

Fig. 2

introduce the quantity

$$a^2 = -6\mu^2/\lambda, \tag{2.5}$$

and to rewrite the potential as

$$U = \frac{\lambda}{4!}(\varphi^2 - a^2)^2, \tag{2.6}$$

plus an (irrelevant) constant. It is clear from this formula, and also from the figure, that the potential now has two minima, at $\varphi = \pm a$. Because of the symmetry (2.4), which one we choose as the vacuum is irrelevant to the resulting physics; however, whichever one we choose, the symmetry is spontaneously broken. Let us choose $\langle\varphi\rangle = a$. To investigate physics about the asymmetric vacuum, let us define a new field

$$\varphi' = \varphi - a. \tag{2.7}$$

In terms of the new ("shifted") field,

$$\begin{aligned} U &= \frac{\lambda}{4!}(\varphi'^2 - 2a\varphi')^2 \\ &= \frac{\lambda}{4!}\varphi'^4 - \frac{\lambda a}{6}\varphi'^3 + \frac{\lambda a^2}{6}\varphi'^2 \end{aligned} \tag{2.8}$$

We see that the true mass of the meson is $\lambda a^2/3$. Note that a cubic meson self-coupling has appeared as a result of the shift, which would make it hard to detect the hidden symmetry (2.4) directly.

2.2 Goldstone Bosons in an Abelian Model

A new phenomenon appears if we consider the spontaneous breakdown of continuous symmetries. Let us consider the theory of two scalar fields, A and B, with

$$U = \frac{\lambda}{4!}[A^2 + B^2 - a^2]^2. \qquad (2.9)$$

This theory admits a continuous group of symmetries isomorphic to the two-dimensional rotation group, SO(2):

$$\begin{aligned} A &\to A\cos\omega + B\sin\omega, \\ B &\to -A\sin\omega + B\cos\omega. \end{aligned} \qquad (2.10)$$

The minima of the potential lie on the circle

$$A^2 + B^2 = a^2. \qquad (2.11)$$

Just as before, which of these we choose as the vacuum is irrelevant, but whichever one we choose, the SO(2) internal symmetry is spontaneously broken. Let us choose

$$\langle A\rangle = a, \; \langle B\rangle = 0. \qquad (2.12)$$

As before, we shift the fields,

$$\underset{\sim}{\varphi}' = \underset{\sim}{\varphi} - \langle\underset{\sim}{\varphi}\rangle, \qquad (2.13)$$

and find

$$U = \frac{\lambda}{4!}(A'^2 + B'^2 - 2aA')^2. \qquad (2.14)$$

Expanding this, we see that the A-meson has the same mass as before, but the B-meson is massless. Such a massless spinless meson is called a Goldstone Boson;[4] for the class of theories under consideration, its appearance does not depend at all on the special form of the potential U, but is a consequence only of the spontaneous breakdown of the continuous SO(2) symmetry group (2.10).

To show this, let us introduce "angular variables",

$$\begin{aligned} A &= \rho\cos\theta, \\ B &= \rho\sin\theta. \end{aligned} \qquad (2.15)$$

In terms of these variables, (2.10) becomes

$$\rho \to \rho$$
$$\theta \to \theta + \omega, \tag{2.16}$$

and the Lagrange density becomes

$$\mathcal{L} = \tfrac{1}{2}(\partial\mu\rho)^2 + \tfrac{1}{2}\rho^2(\partial\mu\theta)^2 - U(\rho). \tag{2.17}$$

In terms of these variables, SO(2) invariance is simply the statement that U does not depend on θ. The transformation to angular variables is, of course, ill-defined at the origin, and this is reflected in the singular form of the derivative part of the Lagrange density (2.17). However, this is of no interest to us, since we wish to do perturbation expansions not about the origin, but about an assumed asymmetric vacuum. With no loss of generality, we can assume this vacuum is at $\langle\rho\rangle = a$, $\langle\theta\rangle = 0$. Introducing shifted fields as before,

$$\rho' = \rho - a,$$
$$\theta' = \theta \tag{2.18}$$

we find

$$\mathcal{L} = \tfrac{1}{2}(\partial\mu\rho')^2 + \tfrac{1}{2}(\rho' - a)^2(\partial\mu\theta')^2 - U(\rho' - a). \tag{2.19}$$

It is clear from this expression that the θ-meson is massless, just because the θ-field enters the Lagrangian only through its derivatives.

This can also be seen purely geometrically, without writing down any formulas. If the vacuum is not invariant under SO(2) rotations, then there is a curve passing through the vacuum along which the potential is constant; this is the curve of points obtained from the vacuum by SO(2) rotations - in terms of our variables, the curve of constant ρ. If we expand the potential around the vacuum, no terms can appear involving the variable that measures displacement along this curve - the θ variable. Hence we always have a massless meson.

2.3 Goldstone Bosons in the General Case

This argument can easily be generalized to the spontaneous breakdown of a general continuous internal symmetry group. I will give the generalization using somewhat more mathematical apparatus than is really necessary, in order to establish some notation that will be useful to us later on. Let us assume that we have a set of n real fields, $\underset{\sim}{\varphi}$, such that the potential is invariant under a group of transformations

$$\underset{\sim}{\varphi} \to e^{\underset{\sim}{T}_a\omega^a}\,\underset{\sim}{\varphi}, \tag{2.20}$$

where the $\underset{\sim}{T}$'s are a set of N real antisymmetric matrices, the group generators, the ω's are arbitrary real parameters, and the sum over repeated indices is implied. The associated infinitesimal transformations are

$$\delta\underset{\sim}{\varphi} = \underset{\sim}{T}_a \delta\omega^a \underset{\sim}{\varphi}. \tag{2.21}$$

Since the $\underset{\sim}{T}$'s are group generators, they obey the relations

$$[\underset{\sim}{T}_a, \underset{\sim}{T}_b] = c_{abc} \underset{\sim}{T}_c \tag{2.22}$$

where the c's are the structure constants of the group. If we choose the T's to be orthonormal (in the trace norm), then c^{abc} is completely antisymmetric. Invariance of the Lagrange density (2.1) implies that

$$U(\underset{\sim}{\varphi}) = U(e^{T_a \omega^a} \underset{\sim}{\varphi}). \tag{2.23}$$

Now let us consider the subgroup of (2.20) that leaves $\langle\underset{\sim}{\varphi}\rangle$, the minima of U, invariant. Depending on the structure of U, this may be anything from the trivial identity subgroup (all symmetries spontaneously broken) to the full group (no symmetries spontaneously broken). In any case, though, we can always choose our group generators such that this subgroup is generated by the first M generators, where N ≥ M ≥ 0. In equations,

$$\underset{\sim}{T}_a \langle\underset{\sim}{\varphi}\rangle = 0, \qquad a \leq M. \tag{2.24}$$

By definition, the remaining (N-M) generators do not leave $\langle\underset{\sim}{\varphi}\rangle$ invariant; thus we have, passing through $\langle\underset{\sim}{\varphi}\rangle$, an (N-M)-dimensional surface of constant U. Thus, by the same arguments as before, the theory must contain (N-M) massless spinless mesons, one for each spontaneously broken infinitesimal symmetry. (Note that I say "spinless", not "scalar" or "pseudoscalar". The mesons may be either scalar or pseudoscalar, depending on the parity-transformation properties of the spontaneously-broken generators; they may even have no well-defined parity at all, if parity is itself spontaneously broken, or if the original Lagrangian is not parity conserving.)

These mesons are called Goldstone Bosons, and what we have proved in the preceding paragraph is a special case of Goldstone's theorem.[4] The theorem can be proved in much greater generality: given a field theory obeying the usual axioms (Lorentz invariance, locality, Hilbert space with positive-definite inner product, etc.), if there is a local conserved current (the axiomatic version of the statement that the Lagrangian is invariant under some continuous transformation) such that the space integral of its time component does not annihilate the vacuum state, then the theory necessarily contains a massless spinless meson, with the same internal-symmetry and parity properties as the time component of the current.[5]

At first glance, Goldstone's theorem seems to be a killing blow to the idea that spontaneous breakdown (at least of continuous symmetries) is at work in the real world, for there is not a smidgen of experimental evidence for the existence of massless spinless mesons. However, there is one loophole: there do exist perfectly respectable field theories which do not obey the usual axioms. These are gauge field theories, of which quantum electrodynamics is the most familiar. There is no gauge in which quantum electrodynamics obeys all the axioms simultaneously; if we quantize in a covariant gauge, the theory contains states of negative norm, associated with the longitudinal photons; if we quantize in a gauge in which the theory has only states of positive norm, such as radiation gauge, the theory is not covariant. We will now investigate this loophole in more detail.

2.4 The Higgs Phenomenon in the Abelian Model

I will begin by reviewing the minimal-coupling prescription of ordinary quantum electrodynamics, and its connection with gauge invariance. Let $\underset{\sim}{\varphi}$ be a set of fields (not necessarily real and spinless), with dynamics determined by a Lagrange density, $\mathcal{L}(\underset{\sim}{\varphi},\partial_\mu\underset{\sim}{\varphi})$. Let $\mathcal{L}$ be invariant under a one-parameter group of transformations,

$$\underset{\sim}{\varphi} \rightarrow e^{i\underset{\sim}{Q}\omega}\underset{\sim}{\varphi}, \tag{2.25}$$

where $\underset{\sim}{Q}$ is a Hermitian matrix, called the charge matrix. (Conventionally, a set of complex basis fields of definite charge is chosen, so that Q is diagonal. However, for our purposes, it will be more convenient to choose a real set of fields, so that $i\underset{\sim}{Q}$ is a real antisymmetric matrix, like the $\underset{\sim}{T}$'s in Eq. (2.20).) The associated infinitesimal transformation is

$$\delta\underset{\sim}{\varphi} = i\underset{\sim}{Q}\underset{\sim}{\varphi}\delta\omega. \tag{2.26}$$

Now let us consider transformations of the same form as Eq. (2.26), but with $\delta\omega$ space-time dependent (gauge transformations). Our theory is not invariant under these transformations, since

$$\delta(\partial_\mu\underset{\sim}{\varphi}) = i\underset{\sim}{Q}(\partial_\mu\underset{\sim}{\varphi})\delta\omega + i\underset{\sim}{Q}\underset{\sim}{\varphi}\partial_\mu(\delta\omega) \tag{2.27}$$

and the second term spoils the invariance. We can take care of this, though, by enlarging the theory and introducing a new field, A_μ, the gauge field, that transforms according to

$$\delta A_\mu = -\frac{1}{e}\partial_\mu(\delta\omega), \tag{2.28}$$

where e is a free parameter, called the electric charge. If we now define

$$D_\mu \underset{\sim}{\varphi} = \partial_\mu \underset{\sim}{\varphi} + ie\underset{\sim}{Q}A_\mu \underset{\sim}{\varphi}, \tag{2.29}$$

then

$$\delta D_\mu \underset{\sim}{\varphi} = i\underset{\sim}{Q}\underset{\sim}{\varphi}\delta\omega, \tag{2.30}$$

and

$$\mathcal{L}(\underset{\sim}{\varphi}, D_\mu \underset{\sim}{\varphi}) \tag{2.31}$$

is gauge invariant. $D_\mu \underset{\sim}{\varphi}$ is called the gauge-covariant derivative, or sometimes just the covariant derivative. Of course, the expression (2.31) by itself can not be the total Lagrange density for a physically interesting theory; it contains no terms proportional to the derivatives of A_μ, so if we vary it with respect to A_μ we obtain, not true equations of motion, but equations of constraint. To make the gauge field a true dynamical variable, we must add a term involving derivatives; the simplest gauge-invariant choice is a term proportional to $(F_{\mu\nu})^2$, where

$$F_{\mu\nu} = \partial_\mu A_\nu - \partial_\nu A_\mu. \tag{2.32}$$

By convention, A_μ is normalized such that the final Lagrange density is

$$-\frac{1}{4}(F_{\mu\nu})^2 + \mathcal{L}(\underset{\sim}{\varphi}, D_\mu \underset{\sim}{\varphi}). \tag{2.33}$$

This is just the usual Lagrange density of minimally-coupled electrodynamics, and it has the usual physical interpretation (charged particles, massless photons, etc.), if the dynamics of the φ-fields are such that the symmetry (2.25) does not suffer spontaneous breakdown. But what happens if the symmetry is spontaneously broken, as is (2.10)?

This question is most easily answered if we use the angular variables defined by Eq. (2.15). We can avoid some tedious algebra by observing that Eq. (2.29) can be rewritten as

$$D_\mu \underset{\sim}{\varphi} = \partial_\mu \underset{\sim}{\varphi} + eA_\mu \frac{\delta \underset{\sim}{\varphi}}{\delta\omega}. \tag{2.34}$$

In this form, it can be directly applied to the angular variables. From Eq. (2.16), it follows that

$$D_\mu \rho' = \partial_\mu \rho',$$

and (2.35)

$$D_\mu \theta' = \partial_\mu \theta' + eA_\mu.$$

Applying this to Eq. (2.19) we obtain

$$\mathcal{L} = \frac{1}{4}(\partial_\mu A_\nu - \partial_\nu A_\mu)^2 + \frac{1}{2}(\partial_\mu \rho')^2 + \frac{1}{2}(\rho' - a)^2(\partial_\mu \theta + eA_\mu)^2 - U(\rho' - a). \tag{2.36}$$

It is hard to directly read off the predictions of this expression for small oscillations about the vacuum, because of the presence of quadratic cross terms, terms proportional to $A_\mu \partial^\mu \theta'$. However, these can be eliminated by introducing the new variable

$$C_\mu = A_\mu + e^{-1}\partial_\mu \theta. \tag{2.37}$$

In terms of this,

$$\mathcal{L} = -\frac{1}{4}(\partial_\mu C_\nu - \partial_\nu C_\mu)^2 + \frac{1}{2}(\partial_\mu \rho')^2 + \frac{e^2}{2}(\rho' - a)^2(C_\mu)^2 - U(\rho' - a). \tag{2.38}$$

Since the quadratic part of the Lagrangian is now in diagonal form, we can read off the eigenmodes for small vibrations about the ground state, or, in the quantum language we have been using, the particle spectrum. We see that there is a massive scalar meson associated with the ρ'-field, whose mass depends on the form of U. There is also a massive vector meson associated with the C-field, with mass given by

$$m_C^2 = e^2 a^2. \tag{2.39}$$

But the Goldstone Boson, the θ-field, has completely disappeared! This seems a little less preposterous if we count degrees of freedom. A massive vector meson has three degrees of freedom, the three spin states of a spin-one particle, while a massless vector meson has only two, the two helicity states of the photon. What has happened is that the two degrees of freedom of the massless gauge field and the one degree of freedom of the Goldstone Boson have combined together to make the three degrees of freedom of the C-field. The vector meson has eaten the Goldstone Boson and grown heavy.

This magic trick was discovered by Peter Higgs, and is called the Higgs phenomenon. (Actually, the terminology is unfair, since the phenomenon was discovered independently by several other investigators, but we will use it anyway, since it's awkward to talk of the Brout-Englert-Guralnik-Hagen-Higgs-Kibble phenomenon.[6]) We can gain further insight into the Higgs phenomenon if we remember the motivation for the minimal-coupling prescription - gauge invariance.

Gauge invariance tells us that our theory is invariant under transformations of the form

$$\theta \to \theta + \omega, \tag{2.40}$$

with ω an arbitrary function of space and time. In particular, this means we can choose ω to be minus θ, that is to say, pick our gauge in such a way that the θ-field is identically zero. The reason the Goldstone Boson disappears in the gauge-invariant theory is that it was never there in the first place; the degree of freedom that would be associated with the Goldstone Boson is a mere gauge phantom, an object that can be gauged away, like a longitudinal photon.

It is now clear how to extend the Higgs phenomenon to a general internal symmetry group, like (2.20). We merely have to add extra degrees of freedom (gauge fields) to promote the whole internal symmetry group to a gauge group. If we can do this, then we can always gauge away the degrees of freedom that would correspond to Goldstone Bosons, and kill the Goldstone Bosons before they are born. To carry out this scheme, though, we need first to develop the theory of gauge fields for general internal symmetry groups.

2.5 Yang-Mills Fields and the Higgs Phenomenon in the General Case

How do we make a general internal symmetry group a gauge group? We will follow closely our discussion of electromagnetism. We begin with a theory that is invariant under transformations of the form (2.21),

$$\delta\underset{\sim}{\varphi} = \underset{\sim}{T}_a \delta\omega^a \underset{\sim}{\varphi}. \tag{2.21}$$

Now let us consider transformations of the same form, but with $\delta\omega^a$ space-time dependent. Our theory is not invariant under these transformations, since

$$\delta(\partial_\mu \underset{\sim}{\varphi}) = \underset{\sim}{T}_a \delta\omega^a \partial_\mu \underset{\sim}{\varphi} + \underset{\sim}{T}_a (\partial_\mu \delta\omega^a) \underset{\sim}{\varphi}, \tag{2.41}$$

and the second term spoils the invariance. We will try to take care of this by introducing a set of N gauge fields, A^a_μ, one for each group generator, and defining the covariant derivatives

$$D_\mu \underset{\sim}{\varphi} = \partial_\mu \underset{\sim}{\varphi} + g\underset{\sim}{T}_a A^a_\mu \underset{\sim}{\varphi}. \tag{2.42}$$

where g, like e, is a free parameter. (For the moment, we will postpone the question of whether we can choose different g's for different gauge fields.) We wish to define the transformation properties of the gauge fields such that

$$\delta(D_\mu \underset{\sim}{\varphi}) = \underset{\sim}{T}_a \delta\omega^a D_\mu \underset{\sim}{\varphi}. \tag{2.43}$$

It is easy to see that this implies that

$$\delta A^a_\mu = c^{abc} \delta\omega^b A^c_\mu - \frac{1}{g} \partial_\mu \delta\omega^a, \tag{2.44}$$

where the c's are the structure constants of the group, defined in Eq. (2.22). (Both terms in this expression are easy to understand. The second term is a trivial generalization of the electromagnetic gauge transformation, Eq. (2.28). The first term is necessary to insure the invariance of the gauge-field couplings under space-time independent transformations; it states that, under such transformations, the gauge fields transform like the group generators. (E.g., if the gauge group is isospin, the gauge fields must form an isovector.)) It follows from Eq. (2.43) that

$$\mathcal{L}(\underset{\sim}{\varphi}, D_\mu \underset{\sim}{\varphi}) \tag{2.45}$$

is gauge invariant.

It is a bit harder to see what is the generalization of the free electromagnetic Lagrange density, $(F_{\mu\nu})^2$. The trick is to observe that, for electromagnetism

$$(D_\mu D_\nu - D_\nu D_\mu)\underset{\sim}{\varphi} = i\underset{\sim}{Q} F_{\mu\nu} \underset{\sim}{\varphi} \tag{2.46}$$

From this equation, the gauge invariance of $F_{\mu\nu}$ follows directly. In our case,

$$(D_\mu D_\nu - D_\nu D_\mu)\underset{\sim}{\varphi} = \underset{\sim}{T}_a F^a_{\mu\nu} \underset{\sim}{\varphi}, \tag{2.47}$$

where

$$F^a_{\mu\nu} = \partial_\mu A^a_\nu - \partial_\nu A^a_\mu + g c^{abc} A^b_\mu A^c_\nu. \tag{2.48}$$

From Eq. (2.46), it follows directly that $F^a_{\mu\nu}$ is, not gauge-invariant, but gauge-covariant,

$$\delta F^a_{\mu\nu} = c^{abc} \delta\omega^b F^c_{\mu\nu}. \tag{2.49}$$

However, the quadratic form $(F^a_{\mu\nu})^2$ is gauge invariant, and therefore the generalization of the electromagnetic Lagrange density (2.33) is

$$-\frac{1}{4}(F^a_{\mu\nu})^2 + \mathcal{L}(\underset{\sim}{\varphi}, D_\mu \underset{\sim}{\varphi}). \tag{2.50}$$

The first Lagrange density of this type (for the special case of the isospin group) was constructed by Yang and Mills; for this reason non-Abelian gauge fields are frequently called Yang-Mills fields.[7]

Note that for non-Abelian gauge fields, in contrast to electromagnetism, there is a non-trivial interaction even in the absence of the φ-fields, because of the non-linear form of $F^a_{\mu\nu}$. There is a good physical reason for this, which is most easily seen by going to a particular example. Let us imagine that the gauge group is isospin. Just as the photon couples to every field that carries non-zero charge, so the I_z gauge meson, for example, must couple to every field that carries non-zero I_z. But among these fields are the other two members of the isotriplet of gauge fields. (It is for precisely the same reason that gravitation is inherently non-linear; the gravitational field couples to everything that carries energy density, including the gravitational field itself.)

Now let us return to the postponed question of whether we can have different coupling constants for different gauge fields. If the gauge group is simple (like SU(2) or SU(3)), the generators of the group, and therefore the gauge fields, transform irreducibly under the action of the group; therefore they must all have the same coupling constant. However, if the gauge group is a product of simple factors (like SU(2) $\otimes$ SU(2)), then the generators of different factors never mix with each other under the action of the group, and the associated gauge fields can have different coupling constants. Thus there are as many independent coupling constants as there are simple factors in the gauge group, and Eq. (2.44), for example, should properly be written as

$$\delta A^a_\mu = c^{abc}\delta\omega^b A^c_\mu - \frac{1}{g_a}\partial_\mu \delta\omega^a . \qquad (2.44')$$

(no sum on a), where g_a can take on different values for gauge fields associated with different factor groups.

Now that we have developed the classical theory of non-Abelian gauge fields, let us apply it to spontaneous symmetry breakdown. Since the entire internal symmetry group has been promoted to a gauge group, we can always choose our gauge such that the degrees of freedom that would become Goldstone Bosons disappear. From our experience with the Abelian model, we would expect the gauge fields associated with the spontaneously broken symmetries to acquire masses. It is easy to see that the only relevant part of the Lagrange density (2.1) is the derivative term

$$\mathcal{L} = \tfrac{1}{2}(\partial_\mu \underset{\sim}{\varphi})\cdot(\partial^\mu \underset{\sim}{\varphi}) + \cdots \quad . \qquad (2.51)$$

In the presence of the gauge fields this becomes

$$\mathcal{L} = \tfrac{1}{2}(\partial_\mu \underset{\sim}{\varphi} + g_a A^a_\mu \underset{\sim}{T_a} \underset{\sim}{\varphi}) \cdot (\partial^\mu \underset{\sim}{\varphi} + g_b A^b_\mu \underset{\sim}{T_b} \underset{\sim}{\varphi}) + \cdots . \qquad (2.52)$$

When we shift the fields, this generates a mass term

$$\mathcal{L} = (g_a A^a_\mu \underset{\sim}{T_a} \langle \underset{\sim}{\varphi} \rangle) \cdot (g_b A^b_\mu \underset{\sim}{T_b} \langle \underset{\sim}{\varphi} \rangle) + \cdots . \qquad (2.53)$$

Note that gauge fields associated with symmetries that are not spontaneously broken, that is to say, those for which

$$T_a \langle \underset{\sim}{\varphi} \rangle = 0, \qquad (2.54)$$

remain massless. Thus, if we wish to have a theory of this type with a realistic particle spectrum, the entire gauge group must be spontaneously broken, except for a one-parameter subgroup. We identify this subgroup with electric charge, and the corresponding gauge field with the only observed massless vector meson, the photon.

2.6 Summary and Remarks

(1) We have discovered a large family of field theories that display spontaneous breakdown of internal symmetries. If the spontaneously broken symmetry is discrete, this causes no problems; however, if the symmetry is continuous, symmetry breakdown is associated with the appearance of Goldstone Bosons. This can be cured by coupling gauge fields to the system and promoting the internal symmetry group to a gauge group; the Goldstone Bosons then disappear and the gauge mesons acquire masses. It is pleasant to remember that, at the times of their inventions, both the theory of non-Abelian gauge fields and the theory of spontaneous symmetry breakdown were thought to be theoretically amusing but physically untenable, because both predicted unobserved massless particles, the gauge mesons and the Goldstone Bosons. It was only later that it was discovered that each of these diseases was the other's cure.

(2) Everything we have done so far has been for classical field theory. One of the main tasks before us is to see to what extent the apparatus of this section can be extended into the quantum domain. We shall see that, at least for weak couplings, it survives substantially unchanged; in particular, all of the equations we have derived can be reinterpreted as the first terms in a systematic quantum expansion.

(3) We have not touched at all on theories with Fermions. It is trivial that if we couple Fermions to the scalar-meson systems we have discussed, either directly (through Yukawa couplings) or indirectly (through gauge field couplings), then the shift in the scalar fields will induce an apparent symmetry-violating term in the Fermion part of the Lagrangian. A more

interesting question is whether spontaneous symmetry breakdown can occur in a theory without fundamental scalar fields. For example, perhaps bilinear forms in Fermi fields can develop symmetry-breaking vacuum expectation values all by themselves. I will have nothing to say about this possibility here, not because it is not important, but because so little is known about it.[8] (There is one exactly soluble model without fundamental scalars that displays the full Goldstone-Higgs phenomenon. This is the Schwinger model, quantum electrodynamics of massless Fermions in two-dimensional space-time.[9])

(4) It is important to realize that we can make the effects of spontaneous symmetry breakdown as large or as small as we want, by appropriately fudging the parameters in our models. Thus, in the real world, some of the spontaneously broken symmetries of nature may be observed as approximate symmetries in the usual sense, and others may be totally inaccessible to direct observation. Also, of course, there is no objection to exact or approximate symmetries of the usual kind coexisting with spontaneously broken symmetries. Presumably symmetries such as nucleon number conservation, neither broken nor coupled to a massless gauge meson, are of this sort.

(5) All of this is very pretty, but what does it buy us? What is the practical use of the idea of spontaneous symmetry breakdown, even by the generous standards of practicality current among high-energy theoreticians? The answer to this question will be given in the next section, when we leave classical physics and turn to quantum field theory.

3. SECRET RENORMALIZABILITY

3.1 The Order of the Arguments

We are going to plunge immediately into the study of spontaneous symmetry breakdown in quantum field theory, despite the fact that we know nothing of the properties of quantum non-Abelian gauge fields, even in the absence of spontaneous symmetry breakdown. Logically, this is not a good order in which to do things, but I would like to get to the heart of the matter as soon as possible. Thus, if you have a critical disposition, you should assume in this section that I am talking about symmetry breakdown in the presence of at most some Abelian gauge fields, and you should ignore my occasional remarks about the non-Abelian case. In any case, we will quantize non-Abelian gauge fields later on.

In this section, we will first review the elements of renormalization theory, without worrying about spontaneous symmetry breakdown. Then we will develop a formalism for handling symmetry breakdown, without worrying about renormalization. Finally, we will bring the two strands of argument together.

3.2 Renormalization Reviewed[10]

In any non-trivial quantum field theory, divergent integrals appear in the perturbation expansion for the Green's functions. Renormalization is a procedure for removing these divergences, order by order in perturbation theory, by adding extra terms, called counterterms, to the Lagrangian that defines the theory. For example, let us consider the expansion of the proper four-point-function (i.e., the off-mass-shell scattering amplitude) in the theory defined by

$$\mathcal{L} = \tfrac{1}{2}(\partial_\mu \varphi)^2 - \tfrac{1}{2}\mu^2 \varphi^2 - \frac{\lambda}{4!}\varphi^4 . \tag{3.1}$$

The first few terms in this expansion are shown in Fig. 3. All the graphs except the first correspond to divergent Feynman integrals, If we cut off

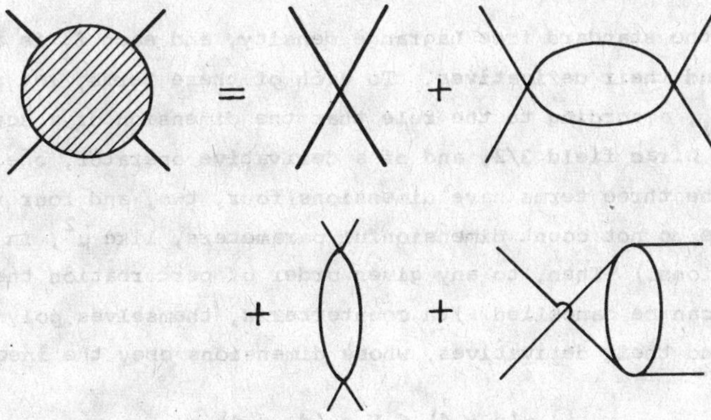

Fig. 3

the integrations at some large momentum, Λ, we obtain

$$\Gamma^{(4)} = -\lambda + a\lambda^2 \ln\Lambda + \lambda^2 f , \tag{3.2}$$

where a is a finite (i.e., cutoff-independent in the limit of large cutoff) constant and f is a finite function of the external momenta. We now change the theory, by adding an extra term (the counterterm) to $\mathcal{L}$:

$$\mathcal{L} \to \mathcal{L} - \frac{a\lambda^2}{4!}\ln\Lambda\varphi^4 . \tag{3.3}$$

The divergent term in Eq. (3.2) is now cancelled, and the Green's function is rendered finite. Of course, the extra term in the Lagrangian must be taken into account as an internal vertex when we compute to yet higher orders, but to this order at least, everything is OK.

It turns out that the obvious generalization of this idiotically simple manipulation gets rid of all the infinities for any field theory with polynomial interactions, to any order in perturbation theory. (I ask you to take this statement, and the ones that will follow it, on trust; they are true, but very difficult to prove.) Furthermore, it is possible to give a general rule for the counterterms that occur in each order of perturbation theory. For simplicity, I will begin by giving this rule and explaining its consequences for theories involving scalar (or pseudoscalar - parity conservation will not be assumed) and Dirac bispinor fields only. Let us write the Lagrange density of our theory in the form

$$\mathcal{L} = \mathcal{L}_0 + \Sigma\mathcal{L}_i, \tag{3.4}$$

where $\mathcal{L}_0$ is the standard free Lagrange density, and each $\mathcal{L}_i$ is a monomial in the fields and their derivatives. To each of these terms, let us assign a dimension, d_i, according to the rule that the dimension of a scalar field is one, of a Dirac field 3/2, and of a derivative operator, one. Thus, in Eq. (3.1), the three terms have dimensions four, two, and four respectively. (Note that we do not count dimensionful parameters, like μ^2, in computing these dimensions.) Then, to any given order of perturbation theory, all divergences can be cancelled with counterterms, themselves polynomials in the fields and their derivatives, whose dimensions obey the inequality

$$(d - 4) \leq \Sigma\, n_i(d_i - 4), \tag{3.5}$$

where n_i is the number of times $\mathcal{L}_i$ occurs in the given order.

Of course, not all counterterms allowed by the inequality (3.5) are necessary. For example, the Lagrange density (3.1) is Lorentz invariant, parity invariant, and invariant under the internal symmetry $\varphi \to -\varphi$. Thus, unless we are so foolish as to use a cutoff procedure which breaks these symmetries, we need never worry about counterterms which are not invariant under them.

Let us check (3.5) against our sample computation. For the Lagrange density (3.1), there is only one interaction, and its dimension is four. Thus, to $O(\lambda^2)$, the order to which we worked, the right-hand-side of the inequality is zero, and there are only three counterterms of appropriate dimensions and symmetry properties:

$$\mathcal{L} \to \mathcal{L} + \tfrac{1}{2}A(\partial_\mu\varphi)^2 - \tfrac{1}{2}B\varphi^2 - \frac{1}{4!}C\varphi^4, \tag{3.6}$$

where A, B, and C are cutoff-dependent. We only saw the last of these in our sample computation, but the other two are also needed in this order, to cancel the infinities in the second order self-energy (Fig. 4).

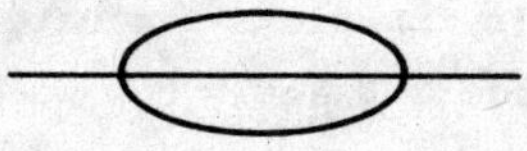

Fig. 4

But these are not only the only counterterms to second order, they are the only ones <u>to general order</u>, because no matter how many interactions we sum up, the right-hand side of the inequality is still zero. (The new interactions induced by the counterterms themselves do not affect this argument; their dimensions are also less than or equal to four.) But these three counterterms are of the same form as the three terms in the original Lagrangian; thus they can be thought of as simply readjustments of the parameters in the original theory. (More precisely, the A term can be absorbed in a rescaling of φ; the B and C terms are then corrections to the mass and coupling constant.)

A theory which has this property, for which all the counterterms induced by renormalization are of the same form as terms in the original Lagrangian, is said to be renormalizable. Phrased another way, a renormalizable theory is one for which all cutoff-dependence can be removed from the Green's functions by rescaling the fields and choosing the parameters of the theory in appropriate cutoff-dependent ways. Renormalizable theories are a very small subset of the set of all quantum field theories one can write down. (Although they may exhaust the set of theories that make sense.) For example, it is clear from our inequality (or from direct computation) that any theory involving an interaction of dimension greater than four is nonrenormalizable. However, not all theories with only interactions of dimension four or less are renormalizable. For example, the theory of mesons and nucleons interacting only through a Yukawa coupling, $\bar{\psi}\gamma_5\psi\varphi$, is not renormalizable, for this interaction induces a φ^4 counterterm, not present in the original theory. On the other hand, the same theory with both Yukawa <u>and</u> φ^4 interactions <u>is</u> renormalizable. (This is a somewhat stricter definition of renormalizability than the one in common use. Most people define renormalizable to mean that there are only a finite number of counterterms induced, whether or not they were all present in the original Lagrangian.)

I have said only that the counterterms are to be chosen to cancel the infinities. This obviously leaves them undetermined, in each order, up to finite additions. For renormalizable theories, these ambiguities are usually resolved by a set of equations, called renormalization conditions, which define the scales of fields and values of renormalized masses and coupling constants in terms of Green's functions evaluated at some conventionally chosen point in momentum space. Exactly how we choose these conventions will not be relevant to our immediate purposes. For nonrenormalizable theories, in the common sense (i.e., those with an infinite number of counterterms), there are an infinite number of free parameters, which is why these theories are commonly (and properly, I think) considered disgusting.

Until now, I have said nothing about vector fields. The rules I gave for assigning dimensions to fields were in fact derived from the high energy behaviour of free propagators, as one might expect, since these are obviously the properties that control the divergences of Feynman integrals. Thus, although the dimension of a massive vector field is one, in the normal sense of dimensional analysis, its propagator is

$$-i\frac{g_{\mu\nu} - k_\mu k_\nu/\mu^2}{k^2 - \mu^2}. \tag{3.7}$$

Because of the second term, this grows at high momentum like the propagator for the gradient of a scalar field, an object of dimension two, and our dimension-counting formula, (3.5), breaks down. In fact, most interactions of a massive vector field are nonrenormalizable. However, if the massive vector field is coupled to a conserved current, as if it were an Abelian gauge field, then we can shuffle variables to rewrite the theory in such a way that the propagator is

$$-i\frac{g_{\mu\nu} - k_\mu k_\nu/k^2}{k^2 - \mu^2}. \tag{3.8}$$

This grows just like a scalar propagator, so the dimension-counting procedure is good again. For a true (massless) Abelian gauge field, the theory may also be quantized in such a way that the propagator is of the form (3.8), (with μ^2 zero, of course). Thus, here also the dimension-counting procedure is good, as I trust you know from your experience with quantum electrodynamics. We shall see in Section 5 that this can also be done (with some complications) for non-Abelian gauge fields (but here only for the massless case).

However, even for quantum electrodynamics, dimension-counting is not sufficient to establish renormalizability. For example, Eq. (3.5) will certainly allow an $(A_\mu)^4$ counterterm (dimension four), but if we really had to introduce such a term into the Lagrangian, it would be a disaster - it would destroy gauge invariance. In QED, one shows such a term can not occur by a complicated sequence of arguments: (1) The theory is cut off in a cunning way that does not destroy gauge invariance. (2) Gauge invariance is used to establish relations between Green's functions, Ward identities. (3) The Ward identities are used to show that the possible gauge-non-invariant counterterms are not necessary. The same sequence of steps can be carried through for non-Abelian gauge theories, but the arguments are much more complicated; I will not have time to cover them in these lectures, and must refer you to the literature.[11]

3.3 Functional Methods and the Effective Potential[12]

I would now like to put aside renormalization for the moment, and begin a new line of development, one that will lead (after an orgy of formalism) to a method for treating spontaneous symmetry breakdown in quantum field theory. For simplicity, in explaining the formalism, I will restrict myself to the theory of a single scalar field, φ, whose dynamics are described by a Lagrange density, $\mathcal{L}(\varphi, \partial_\mu\varphi)$. The generalization to more complicated cases is trivial. Let us consider the effect of adding to the Lagrange density a linear coupling of φ to an external source, $J(x)$, a c-number function of space and time:

$$\mathcal{L}(\varphi,\partial_\mu\varphi) \to \mathcal{L} + J(x)\varphi(x) . \qquad (3.9)$$

The connected generating functional, $W(J)$, is defined in terms of the transition amplitude from the vacuum state in the far past to the vacuum state in the far future, in the presence of the source $J(x)$,

$$e^{iW(J)} = \langle 0^+|0^-\rangle_J . \qquad (3.10)$$

We can expand W in a functional Taylor series

$$W = \sum_n \frac{1}{n!} \int d^4x_1 \cdots d^4x_n G^{(n)}(x_1 \cdots x_n) J(x_1) \cdots J(x_n) . \qquad (3.11)$$

It is well known that the successive coefficients in this series are the connected Green's functions; $G^{(n)}$ is the sum of all connected Feynman diagrams with n external lines.

The classical field, φ_c, is defined by

$$\varphi_c(x) = \frac{\delta W}{\delta J(x)}$$

$$= \left[\frac{\langle 0^+|\varphi(x)|0^-\rangle}{\langle 0^+|0^-\rangle}\right]_J . \qquad (3.12)$$

The effective action $\Gamma(\varphi_c)$, is defined by a functional Legendre transformation

$$\Gamma(\varphi_c) = W(J) - \int d^4x J(x)\varphi_c(x). \tag{3.13}$$

From this definition, it follows directly that

$$\frac{\delta\Gamma}{\delta\varphi_c(x)} = -J(x). \tag{3.14}$$

This equation will shortly turn out to be critical in the study of spontaneous breakdown of symmetry. The effective action may be expanded in a manner similar to that of (3.11):

$$\Gamma = \sum_n \frac{1}{n!} \int d^4x_1 \cdots d^4x_n \Gamma^{(n)}(x_1 \cdots x_n)\varphi_c(x_1)\cdots\varphi_c(x_n) \tag{3.15}$$

It is possible to show that the successive coefficients in this series are the IPI Green's functions[13] (sometimes called proper vertices); $\Gamma^{(n)}$ is the sum of all IPI Feynman diagrams with n external lines. [A IPI (one-particle-irreducible) Feynman diagram is a connected diagram that cannot be disconnected by cutting a single internal line. By convention, IPI diagrams are evaluated with no propagators on the external lines.] There is an alternative way to expand the effective action: Instead of expanding in powers of φ_c, we can expand in powers of momentum (about the point where all external momenta vanish). In position space, such an expansion looks like

$$\Gamma = \int d^4x[-V(\varphi_c) + \tfrac{1}{2}(\partial_\mu\varphi_c)^2 Z(\varphi_c) + \cdots]. \tag{3.16}$$

$V(\varphi_c)$ - an ordinary function, not a functional - is called the effective potential. By comparing the expansions (3.15) and (3.16), it is easy to see that the nth derivative of V is the sum of all IPI graphs with n vanishing external momenta. In tree approximation (that is to say, neglecting all diagrams with closed loops), V is just the ordinary potential, the object we called U in Section 2.

The usual renormalization conditions of perturbation theory can be expressed in terms of the functions that occur in (3.15). For example, if we define the squared mass of the meson as the value of the inverse propagator at zero momentum, then

$$\mu^2 = \frac{d^2V}{d\varphi_c^2}\bigg|_0 . \tag{3.17a}$$

Likewise, if we define the four-point function at zero external momenta to be the coupling constant, λ, then

$$\lambda = \frac{d^2V}{d\varphi_c{}^4}\Big|_0 \,. \tag{3.17b}$$

Similarly, the standard condition for the normalization of the field becomes

$$Z(0) = 1. \tag{3.17}$$

We are now ready to apply this apparatus to the study of spontaneous symmetry breaking. Let us suppose our Lagrange density possesses an internal symmetry, like the classical field theories of Section 2. Then, spontaneous symmetry breaking occurs if the quantum field φ develops a nonzero vacuum expectation value, even when the source $J(x)$ vanishes. From Eqs. (3.12) and (3.14) this occurs if

$$\frac{\delta\Gamma}{\delta\varphi_c} = 0, \tag{3.18}$$

for some nonzero value of φ_c. Further, since we are typically only interested in cases where the vacuum expectation value is translationally invariant (that is to say, we are not interested in the spontaneous breakdown of momentum conservation), we can simplify this to

$$\frac{dV}{d\varphi_c} = 0, \tag{3.19}$$

for some nonzero value of φ_c. The value of φ_c for which the minimum occurs, which we denote by $\langle\varphi\rangle$, is the expectation value of φ in the new (asymmetric) vacuum.

To explore the properties of the spontaneously broken theory, we define a new quantum field with vanishing vacuum expectation value,

$$\varphi' = \varphi - \langle\varphi\rangle. \tag{3.20}$$

This generates a corresponding redefinition of the classical field,

$$\varphi_c' = \varphi_c - \langle\varphi\rangle, \tag{3.21}$$

from which it immediately follows that the actual mass, coupling constant, etc., are computable from equations exactly like the Eqs. (3.17), except that the derivatives are evaluated at $\langle\varphi\rangle$, rather than at zero. Thus, we have recreated the entire structure of our study of spontaneous symmetry breakdown in classical field theory. The only difference is that, instead of working with the classical potential U, we work with the effective potential V.

3.4 The Loop Expansion

Unfortunately, except for trivial models, we do not know the effective potential; to calculate it requires an infinite summation of Feynman diagrams, a task beyond our computational abilities. Thus, it is important to know a sensible approximation method for V. I shall now attempt to show that one such sensible method is the loop expansion: first summing all diagrams with no closed loops (tree graphs), then those with one closed loop, etc. Of course, each stage in this expansion also involves an infinite summation, but, as we shall see, this summation is trivial.

Let us introduce a parameter a into our Lagrange density, by defining

$$\mathcal{L}(\varphi,\partial_\mu\varphi,a) \equiv a^{-1}\mathcal{L}(\varphi,\partial_\mu\varphi). \tag{3.22}$$

We shall now show that the loop expansion is equivalent to a power-series expansion in a. Let P be the power of a associated with any graph. Then it is easy to see that

$$P = I - V, \tag{3.23}$$

where I is the number of internal lines in the graph and V is the number of vertices. This is because the propagator, being the inverse of the differential operator occuring in the quadratic terms in $\mathcal{L}$, carries a factor of a, while every vertex carries a factor of a^{-1}. (Note that it is important that we are dealing with IPI graphs, for which there are no propagators attached to external lines.) On the other hand, the number of loops, L, is given by

$$L = I - V + 1. \tag{3.24}$$

This is because the number of loops in a diagram is equal to the number of independent integration momenta; every internal line contributes one integration momentum, but every vertex contributes a δ function that reduces the number of independent momenta by one, except for one δ function that is left over for over-all energy-momentum conservation. Combining Eqs. (3.23) and (3.24), we find that

$$P = L - 1, \tag{3.25}$$

the desired result.

The point of this analysis is not that the loop expansion is a good approximation scheme because a is a small parameter; indeed, a is equal to one. (However, it is certainly no worse than ordinary perturbation theory for small coupling constants, since the set of graphs with n loops or less certainly includes, as a subset, all graphs of nth order or less in the coupling constants.) The point is, rather, since the loop expansion corresponds to expansion in a parameter that multiplies the total Lagrange density, it is unaffected by shifts of fields, and by the redefinition of the division of the Lagrangian into free and interacting parts associated with such shifts.[14]

Thus we have a systematic expansion procedure, in any order of which we can apply the methods of Section 2. Further, the first term in the expansion of V is the classical potential, U, the negative sum of all non-derivative terms in the Lagrange density. Thus, we have not only justified in the quantum world many of the classical methods of Section 2, we have justified many of the actual computations of Section 2. They should be reliable in the corresponding quantum field theories for the usual conditions under which we expect diagrams with closed loops to be negligible, that is to say, for small coupling constants.

3.5 A Sample Computation

To put some flesh on this dry formalism, let us compute the effective potential for the theory of a single scalar field with Lagrange density

$$\mathcal{L} = \tfrac{1}{2}(\partial_\mu \varphi)^2 - U(\varphi), \tag{3.26}$$

where U is a polynomial, not necessarily of renormalizable type. As stated, in the zero loop approximation,

$$V = U(\varphi_c). \tag{3.27}$$

Now let us turn to the one-loop approximation. Since the one-loop approximation does not depend on how we break the Lagrangian into free and interacting parts, let us take only the first term in (3.26) as the free Lagrange density, and all of U (including possible mass terms) as the interaction. All the one loop graphs are then shown in Fig. 5. The black dot stands for

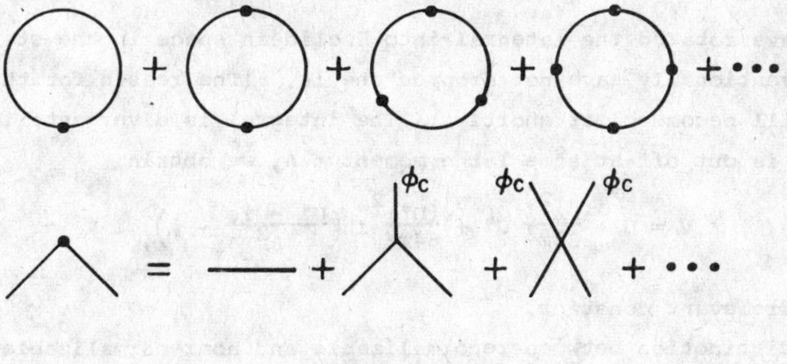

Fig. 5

a sum of terms with zero, one, two, etc. external lines, arising from terms in U of second, third, fourth, etc. order in φ. (Terms linear in φ do not contribute to IPI one-loop diagrams.) Each of these external lines carries zero external momentum and a factor of φ_c. Thus, the value of the vertex in Fig. 5 is

$$i\frac{d^2U}{d\varphi^2}\bigg|_{\varphi=\varphi_c} = iU''(\varphi_c). \tag{3.28}$$

(The *i* is just the usual *i* from Dyson's formula.) For example, if we take the U of our old Abelian model, Eq. (2.6), then

$$U''(\varphi_c) = \frac{\lambda}{6}(3\varphi_c^2 - a^2). \tag{3.29}$$

Every line carries the usual massless propagator,

$$\frac{i}{k^2 + i\epsilon}, \tag{3.30}$$

where k is the momentum going around the loop. Thus, the sum of all the graphs in Fig. 5 corrects Eq. (3.27) in the following way:

$$V = U + i\int\frac{d^4k}{(2\pi)^4}\sum_{n=1}^{\infty}\frac{1}{2n}\left(\frac{U''(\varphi_c)}{k^2 + i\epsilon}\right)^n. \tag{3.31}$$

Two factors in this expression require further explanation: (1) The *i* in front is just a reflection of the *i* in the definition of W, Eq. (3.10). (2) The 1/2n is a combinatoric factor; rotating or reflecting the n-dot graph does not lead to a new contraction in the Wick expansion, and therefore the 1/n! in Dyson's formula is incompletely cancelled.

It is easy to sum this infinite series. Aside from an irrelevant (divergent) constant, the answer is

$$V = U + \frac{1}{2}\int\frac{d^4k}{(2\pi)^4}\ln(k^2 + U''(\varphi_c) - i\epsilon), \tag{3.32}$$

where I have rotated the integral into Euclidean space in the standard way, but unconventionally have not dropped the $i\epsilon$. (The reason for this eccentricity will become clear shortly.) The integral is divergent; if the integration is cut off at some large momentum Λ, we obtain

$$V = U + \frac{\Lambda^2}{32\pi^2}U'' + \frac{(U'')^2}{64\pi^2}\ln\left(\frac{U'' - i\epsilon}{\Lambda^2} - \frac{1}{2}\right), \tag{3.33}$$

plus an irrelevant constant.

The distinction between renormalizable and nonrenormalizable interactions emerges very clearly in this computation. If U is a quartic polynomial (the renormalizable case), then we can remove all the cutoff-dependence from Eq. (3.33) by adding counterterms to the Lagrangian which are themselves at most quartic polynomials, and which can therefore be interpreted as corrections to the parameters in the original Lagrangian. For example, for our old Abelian model, we obtain in this way

$$V = \frac{\lambda}{4!}(\varphi_c^2 - a^2) + \frac{\lambda^2}{2304\pi^2}(3\varphi_c^2 - a^2)^2 \ln(3\varphi_c^2 - a^2 - i\epsilon) + b\varphi_c^4 + c\varphi_c^2, \tag{3.34}$$

where b and c are finite constants, undetermined until we state our renormalization conventions, the conditions that define the renormalized parameters of the theory, and fix the finite parts of the counterterms.[15] (Note that it is a good thing that we retained the $i\epsilon$, for the argument of the logarithm can become negative, and the $i\epsilon$ is needed to tell us the sign of the imaginary part of V. We will return to this point later.)

On the other hand, if U is of quintic order or higher (the nonrenormalizable case), the counterterms we must add are of yet higher order, and we are launched on the unending escalation of ambiguities that characterizes non-renormalizable theories. (A technical point: As in all renormalization schemes, the counterterms added in first order are to be considered as quantities of first order in the relevant expansion parameter. In our case, this is the (suppressed) loop-counting parameter, a, of Eq. (3.22). Thus, if we go to higher loops, the counterterms introduced at this stage are to be counted as one-loop internal parts, despite the fact that they are represented graphically by simple point vertices.)

3.6 The Most Important Part of This Lecture

The significant feature of the computation we have just done is that we needed to invoke no more counterterms than would have been required if there had not been spontaneous symmetry breakdown; the ultraviolet divergences of the theory respect the symmetry of the Lagrangian, even if the vacuum state does not. That this occurred in our specific computation should be no surprise; our entire formalism has been constructed so this is what happens in any computation. For $\mathcal{L}$, the Lagrange density in Eq. (3.9), is the total Lagrange density for the theory. In particular, this means that it contains all the counterterms needed to eliminate all ultraviolet divergences. None of the subsequent manipulations in Section 3.3 involve any integrations over internal momenta, and therefore none of them can introduce new ultraviolet divergences.

This point is important enough to be worth stating again in a slightly different way. We have developed the theory of spontaneous symmetry breakdown in quantum field theory in such a way that we remove all the ultraviolet divergences from the theory before we shift the fields. Before we shift the fields, everything is manifestly symmetric under the full internal symmetry group of the theory; therefore there is no way in which asymmetric counterterms can arise.

Once more, with feeling: The divergence structure of a renormalizable field theory is not affected by the occurrence of spontaneous symmetry breakdown. This simple observation is the most important part of this lecture. It is the secret of the construction of renormalizable theories of the weak interactions. These theories are apparently non-renormalizable, for they involve massive vector mesons (the W-bosons) coupled to non-conserved currents. However, this is only an appearance; in actuality, the Lagrangians of these theories involve only massless gauge fields coupled minimally to conserved currents, and are perfectly renormalizable. The mass of the vector mesons and the non-conservation of the currents are a result of spontaneous symmetry breakdown.

Likewise, we see how to construct theories in which mass differences within an isotopic multiplet are finite. We begin with a theory in which the photon is part of a set of gauge mesons that couple in an isospin-symmetric way. In such a theory, one needs only isosinglet mass counterterms. Spontaneous symmetry breakdown now occurs; the friends of the photon become massive; but there is still no need for an isospin-breaking mass counterterm.

At the end of Section 2, I asked, "What does it buy us?" We now have the wonderful answer: secret symmetry buys us secret renormalizability.

Properly, I should now go on to give detailed examples of this principle at work in specific models, but this task will be left to Glashow's lectures.

3.7 The Physical Meaning of the Effective Potential

In classical field theory, the ordinary potential, $U(\varphi)$, is an energy density; it is the energy per unit volume for that state in which the field assumes the value φ. I will now show that, in quantum field theory, the effective potential, $V(\varphi_c)$, is also an energy density; it is the expectation value of the energy per unit volume in a certain state for which the expectation value of the field is φ_c.[16] An immediate consequence of this is that, if V has several local minima, it is only the absolute minimum that corresponds to the true ground state of the theory, the state of lowest energy. As a byproduct, we will obtain the essential clue to the meaning of the mysterious imaginary part of V which appeared in our sample computation.

We begin the proof by expanding W(J), defined in Eq. (3.10), in the same way we expanded Γ in Eq. (3.16):

$$W = \int d^4x \left\{ -\mathcal{E}(J) + \tfrac{1}{2}(\partial_\mu J)^2 X(J) + \cdots \right\} \tag{3.35}$$

Now let us consider a J(x) which has a constant value, which we denote by J, throughout a box of side L, during a time T, and which goes to zero smoothly outside this space-time region. Under these conditions, for very large L and T, the first term in Eq. (3.35) is the dominant one, and

$$e^{iW} = \langle 0^+ | 0^- \rangle \approx e^{-iL^3 T \mathcal{E}(J)} . \tag{3.36}$$

What has happened physically is that, throughout the box, we have smoothly changed the Hamiltonian density of the theory:

$$\mathcal{H} \to \mathcal{H} - J\varphi \tag{3.37}$$

Thus we would expect the ground state of the theory, within the box, to go adiabatically into the ground state of the theory with the additional term in its Hamiltonian density. This state would evolve in time according to the Schroedinger equation; since it is a ground state, this means that it simply develops a phase. When we turn off the perturbation, the state goes adiabatically back to the ground state of the unperturbed theory, but the phase remains. Thus, $\mathcal{E}(J)$ is the energy per unit volume of the ground state of the perturbed Hamiltonian. (Of course, level crossing might take place. To be precise, we should say not "the ground state" but "that stationary state of the perturbed theory that is obtained from the ground state of the unperturbed theory by adiabatically turning on the perturbation".)

I will now begin an independent line of argument, which, when combined with the above observation, will yield the desired result. For notational simplicity, I will construct this argument for ordinary quantum mechanics, not for field theory, so we will speak of energies, rather than of energy densities; the proper generalization will be obvious. Let us remember the ancient Rayleigh-Ritz variational problem: To construct a state $|a\rangle$ that is a stationary state of the quadratic form

$$\langle a|H|a\rangle, \tag{3.38}$$

under the constraint that the norm of the state be one,

$$\langle a|a\rangle = 1. \tag{3.39}$$

This problem is traditionally solved by Lagrange multipliers; one introduces a Lagrange multiplier, called E, and varies without constraint the form

$$\langle a|(H - E)|a\rangle. \tag{3.40}$$

In this way one obtains

$$(H - E)|a\rangle = 0 \tag{3.41}$$

Hence, $|a\rangle$ is an eigenstate of H with energy E.

Now let us consider a slight variation of this problem. We add an extra equation of constraint

$$\langle a|A|a\rangle = A_c \tag{3.42}$$

where A is some Hermitian operator and A_c some number. We must now introduce two Lagrange multipliers, which I will call E and J, and vary without constraint

$$\langle a|(H - E - JA)|a\rangle. \tag{3.43}$$

We thus obtain

$$(H - E - JA)|a\rangle = 0 \tag{3.44}$$

Hence $|a\rangle$ is an eigenstate of the perturbed Hamiltonian, H - JA, and E is its energy. Of course, this gives us E as a function of J, and we are really interested in how things depend, not on J, but on A_c. The connection between these two quantities is easily obtained by a standard formula of first-order perturbation theory,

$$A_c = \langle a|A|a\rangle = -\frac{dE}{dJ}. \tag{3.45}$$

Hence the quantity we originally set out to make stationary is given by

$$\langle a|H|a\rangle = E + JA_c = E - J\frac{dE}{dJ}. \tag{3.46}$$

It can hardly have escaped you that (with the obvious substitution of energy densities for energies and φ for A) this is precisely the chain of manipulations that led to the definition of the effective potential. Thus we have found the physical meaning of the effective potential;

$$V(\varphi_c) = \langle a|\mathcal{H}|a\rangle, \tag{3.47}$$

for a state $|a\rangle$ such that

$$\delta\langle a|\mathcal{H}|a\rangle = 0, \tag{3.48}$$

under the constraints

$$\langle a|a\rangle = 1, \tag{3.49a}$$

and

$$\langle a|\varphi|a\rangle = \varphi_c. \tag{3.49b}$$

We can check this interpretation in another way, by reducing the four dimensions of space-time to one. The Lagrange density then becomes the Lagrangian for a particle of unit mass, φ becomes x, the position of the particle, and $U(\varphi)$ becomes $U(x)$, the potential in which the particle moves. Eq. (3.32) becomes

$$\begin{aligned} V &= U + \tfrac{1}{2}\int\frac{d\omega}{2\pi}\ln(\omega^2 + U'' - i\epsilon) \\ &= U + \tfrac{1}{2}(U'' - i\epsilon)^{\frac{1}{2}}. \end{aligned} \tag{3.50}$$

This has a direct physical interpretation: Classically, the particle sits in a minimum of the potential, and its energy is the value of the potential at the minimum. To get the first quantum correction to this picture, we approximate the potential near the minimum by a harmonic oscillator potential, and add the zero-point energy of the oscillator; this is the second term in Eq. (3.50).[17]

Once we know V is an energy density, we can understand the meaning of its imaginary part. When we follow an energy level as we change the parameters of a theory, it may often happen that, at a certain point, the energy level becomes unstable; at this moment the energy acquires a negative imaginary part, equal in magnitude to half the probability of decay per unit time. This can also be seen from our earlier discussion of $\mathcal{E}(J)$ in terms of the adiabatic turning-on of a perturbation. If the ground state of the unperturbed system adiabatically moves into an unstable state of the perturbed system, it will decay, and

$$\langle 0^+|0^-\rangle = \exp[-iL^3T\mathcal{E}(J)], \tag{3.51}$$

will be a number with modulus less than one. Of course, for a system of infinite spatial extent, one should not speak of decay probability per unit time, any more than one speaks of energy; one speaks of decay probability per unit time per unit volume, just as one speaks of energy density. Thus the imaginary part of the effective potential is to be interpreted as half a decay probability per unit time per unit volume. (Note that the $i\epsilon$ in Eq. (3.32) insures that the imaginary part is negative, as it must be if this interpretation is to be consistent.)[18]

3.8 Accidental Symmetry and Related Phenomena

Like all perturbative expansions, the loop expansion is trustworthy only for small dimensionless coupling constants. For small coupling constants, one usually expects higher terms in a perturbation expansion to be small compared to lower terms. This is indeed the case in our sample computation; for the Abelian model, for example, the zero-loop effective potential is of order λ, and the one-loop correction is of order λ^2. Nevertheless, there are important cases in which the one-loop corrections are more important than the tree graphs, and play the dominant role in determining the structure of spontaneous symmetry breaking.

This is because our theory may contain interactions that do not appear at all in the zero-loop approximation to the effective potential, such as Yukawa couplings or gauge-field couplings. We have not yet explicitly computed any graphs involving closed loops of virtual Fermions or gauge particles, but it is obvious that their magnitude depends only on the magnitude of the Yukawa or gauge coupling constants. Since these are independent parameters of the theory, it is always possible to choose them so the one-loop graphs are more important than the zero-loop graphs, even if all coupling constants are small. Thus, for example, in the Abelian gauge model of Section 2.4, closed loops of virtual photons turn out to make a contribution to V of order e^4. (See the Appendix for the computation.) This is more important than the zero-loop effective potential if e^4 is much greater than λ, which can happen even if e and λ are both much less than one.

There are even cases in which the one-loop effective potential is important whatever the relative magnitude of the dimensionless coupling constants. This is most easily explained with a specific example. Consider an SO(3) quintuplet of scalar mesons, which we denote by φ^a, where a runs from 1 to 5. The transformation properties of these fields can be most simply expressed if we assemble them into a real traceless symmetric 3 x 3 matrix, which we denote by $\underset{\sim}{\varphi}$. Under an SO(3) transformation, characterized by a rotation matrix $\underset{\sim}{R}$,

$$\underset{\sim}{\varphi} \rightarrow \underset{\sim}{R}\underset{\sim}{\varphi}\underset{\sim}{R}^T . \tag{3.52}$$

In addition, we will assume invariance under the discrete symmetry

$$\underset{\sim}{\varphi} \rightarrow -\underset{\sim}{\varphi} . \tag{3.53}$$

Thus we can only have quadratic and quartic self-couplings. The only invariant quadratic form is

$$\mathrm{Tr}\underset{\sim}{\varphi}^2 = \Sigma(\varphi^a)^2 . \tag{3.54}$$

There are apparently two possible quartic couplings, Tr φ^4 and $(\mathrm{Tr}\,\varphi^2)^2$; however, these are related by the tracelessness of φ:

$$\mathrm{Tr}\,\underset{\sim}{\varphi}^4 = \tfrac{1}{2}(\mathrm{Tr}\,\underset{\sim}{\varphi}^2)^2 = \tfrac{1}{2}[\Sigma(\varphi^a)^2]^2. \tag{3.55}$$

As is clear from the right-hand-sides of these equations, both of these terms are invariant under a larger symmetry group than SO(3), to wit, SO(5). Thus, the constraints of renormalizability (no higher than fourth-order interactions) have forced the scalar meson self-interaction, and therefore the zero-loop effective potential, to be invariant under a larger symmetry group than we started out with. This phenomenon has been dubbed accidental symmetry by Weinberg.[19] However, if the scalar mesons are coupled to a triplet of gauge fields, the gauge interaction is not forced to be (indeed, can not be) SO(5)-invariant; however, it also does not appear in the zero-loop approximation for V.

Thus, if we attempted to analyze this model in the zero-loop approximation, we would be in the soup for two reasons: (1) We would have too rich a set of vacuua - an SO(5) family instead of just an SO(3) one. (2) Even if we miraculously picked the right vacuum from this over-rich set, we would find some massless scalars that were only SO(5) Goldstone Bosons, and not SO(3) ones. (Weinberg calls these pseudo-Goldstone Bosons.) To find the right vacuum, and to give a mass to the pseudo-Goldstone Bosons, it is necessary to compute the effects of gauge-field loops.

3.9 An Alternative Method of Computation

In these lectures I have stressed a method of computation in which we first compute higher-order corrections, and then shift the fields. To be honest, I must tell you that most workers in this field prefer to do things in the other order. They rewrite the Lagrangian of the theory in terms of shifted fields

$$\underset{\sim}{\varphi} = \underset{\sim}{\varphi}' + \langle\underset{\sim}{\varphi}\rangle. \tag{3.56}$$

This gives them a Lagrangian with an extra free parameter for each spinless field (the value of the shift). These are fixed at some stage in the computation by demanding that the vacuum expectation values of the shifted fields vanish,

$$\langle 0|\varphi'|0\rangle = 0. \tag{3.57}$$

In other words, all IPI graphs with only one external line (tadpole graphs) should sum to zero.

This is just as good a way of doing things as the way I have explained; it is equivalent to computing directly the derivative of V and demanding that it vanish, without bothering to compute V first. The only reason I have developed the theory in the way I have is a pedagogical one; in the alternative method of development, it is not so easy to see that spontaneous symmetry breakdown does not lead to asymmetric counterterms. (The only case I can think of in which our method would be clearly superior would be for a theory in which V had two local minima; in this case, we would need to know the value of V in order to determine which of them was the absolute minimum, the true vacuum.)

4. FUNCTIONAL INTEGRATION (VULGARIZED)

4.1 Integration Over Infinite-Dimensional Spaces

Functional integration is a method for defining and manipulating integrals over function spaces, that is to say, over infinite-dimensional spaces, in the same way the ordinary integral calculus enables us to define and manipulate integrals over finite-dimensional spaces. It is useful in theoretical physics because it is possible to represent the generating functional of a quantum field theory as a functional integral. Such a representation has many virtues; from our point of view, the chief of these is that this makes it especially easy to see how the theory changes if we make non-linear transformations on its fundamental dynamical variables. The larger the set of physically interesting non-linear transformations, the more useful is the functional-integral representation; thus it is most useful in studying non-Abelian gauge theories.

This lecture will be devoted to explaining functional integration and its connection with field theory. Our approach will be, from a mathematical viewpoint, despicable. Nothing will be proved; everything will de done by analogy, formal manipulation of ill-defined (and sometimes divergent) quantities, and handwaving. I hope that this will at least give you an idea of what is going on and teach you to manipulate functional integrals; if you want a deeper understanding, you must go elsewhere.[20]

We begin with a very simple one-dimensional integral, the Gaussian integral,

$$\int dx\, e^{-\frac{1}{2}ax^2} = \sqrt{2\pi/a}, \tag{4.1}$$

where a is a positive real number. By analytic continuation, the formula is also true for complex a whenever the integral is defined, that is to say, whenever a has a positive real part. Eq. (4.1) can readily be generalized to n-dimensional space. In order to keep our notation from getting too awkward, we will drop the convention we have used until now of denoting vectors and matrices by wiggly underlining, and simply call a vector in such a space, x. We will denote the usual inner product of two such vectors, x and y, by (x,y). Then, if A is a real symmetric positive-definite matrix,

$$\int d^n x e^{-\frac{1}{2}(x,Ax)} = (2\pi)^{n/2}(\det A)^{-\frac{1}{2}}, \tag{4.2}$$

as can easily be seen by diagonalizing A. As before, this formula is also true if A is a complex symmetric matrix with positive-definite real part, by analytic continuation.

To keep from continually writing π's and n's, we define

$$(dx) = d^n x(2\pi)^{-n/2}. \tag{4.3}$$

Thus, Eq. (4.2) becomes

$$\int (dx) e^{-\frac{1}{2}(x,Ax)} = (\det A)^{-\frac{1}{2}}. \tag{4.4}$$

If we can integrate Gaussians, we can integrate exponentials of general quadratic forms. Let

$$Q(x) = \tfrac{1}{2}(x,Ax) + (b,x) + c, \tag{4.5}$$

where b is some vector and c is a number. Let $\bar{x}$ be the minimum of Q,

$$\bar{x} = -A^{-1}b. \tag{4.6}$$

Then

$$Q(x) = Q(\bar{x}) + \tfrac{1}{2}(x - \bar{x}, A[x - \bar{x}]), \tag{4.7}$$

and

$$Q(\bar{x}) = \tfrac{1}{2}(b,A^{-1}b) + c. \tag{4.8}$$

Whence,

$$\int (dx) e^{-Q(x)} = e^{-Q(\bar{x})}(\det A)^{-\frac{1}{2}}. \tag{4.9}$$

Once we have Eq. (4.9) , we can do the integral of any polynomial times the exponential of a quadratic form, just by differentiating with respect to b,

$$\int (dx) P(x) e^{-Q(x)} = P(-\tfrac{\partial}{\partial b})\int (dx) e^{-Q(x)}. \tag{4.10}$$

It will be convenient later to also have formulas for integrating over an n-dimensional complex vector space, not in any contour-integral sense, but merely in the sense of integrating separately over imaginary and real parts. We will denote the usual Hermitian inner product in such a space by (z^*,w), and the 2n-dimensional real integration (with appropriate factors of π inserted) by $(dz^*)(dz)$. Then if A is a positive-definite Hermitian matrix,

$$\int (dz^*)(dz)e^{-(z^*,Az)} = (\det A)^{-1}, \tag{4.11}$$

as can easily be seen by diagonalizing A. Note the change in the power of the determinant. This is because each eigenvalue of A contributes twice to the integral, once from the integration over the real part of z, and once from the integration over the imaginary part. The missing $\frac{1}{2}$ in the exponential is just a matter of convention; its effects are absorbed in the definition of $(dz^*)(dz)$. From this formula equations analogous to those we derived before follow directly; I will not bother to write them out explicitly.

Now comes the great leap of faith: There is nothing in our integration formulae that refers explicitly to the dimension of the vector space; therefore we boldly extend them to infinite-dimensional vector spaces. Let me be a bit more precise about how this is done, using Eq. (4.4) as an example. Given a quadratic form, (x,Ax), defined by a linear operator, A, on an infinite-dimensional real Hilbert space, we first restrict the form to some finite-dimensional subspace. On this finite-dimensional subspace, both sides of Eq. (4.4), the integral and the determinant, are well-defined. We then let the finite-dimensional subspace grow, until, in the limit, it becomes the whole space. More precisely, we consider an increasing sequence of finite-dimensional subspaces such that their union contains a dense set of vectors. This limit defines both the infinite-dimensional integral and the infinite determinant. It is a deep problem to determine for what operators A the limits exist and are independent of the sequence of subspaces, but it is not one I will worry about here. We will assume in our manipulations that expressions like (4.4) are well-defined whenever we need them.

The infinite-dimensional spaces we will be most concerned with will be spaces of functions, for example, the space of functions of a single real variable. This special case has an unnecessary, but traditional, special notation associated with it. The vectors in the space are traditionally denoted not by x, as we have been doing, but by some symbol that makes their nature as functions manifest, e.g., by $q(t)$, where t is the real variable. The inner product is written as

$$(q,q) = \int dt[q(t)]^2. \tag{4.12}$$

Also, a function from the vector space to the real or complex numbers is called a functional, and derivatives, like those appearing in Eq. (4.10), are called variational derivatives, and denoted by expressions like $\delta/\delta q(t)$, rather than $\partial/\partial x$. In field-theoretical applications, we will consider spaces of functions of four-dimensional space-time, usually denoted by expressions like $\varphi(x)$, where x is now not an element of the vector space but just an ordinary space-time point. In this case,

$$(\varphi,\varphi) = \int d^4x[\varphi(x)]^2, \tag{4.13}$$

and Eq. (4.4) would be written as

$$\int(d\varphi)\, e^{-\frac{1}{2}(\varphi,A\varphi)} = (\det A)^{-\frac{1}{2}}. \tag{4.14}$$

In the cases that will most concern us, A will be an integral or differential operator.

4.2 Functional Integrals and Generating Functionals

There are a large number of cases in which the generating functionals of quantum theories can be written as functional integrals. I will begin with an especially simple case, that of a single scalar field with non-derivative self-interaction. As in Section 3.3, let us write the Lagrange density for such a theory in the presence of an external c-number source, J(x),

$$\mathcal{L} = \tfrac{1}{2}(\partial_\mu\varphi)^2 - \tfrac{1}{2}\mu^2\varphi^2 + \mathcal{L}'(\varphi) + J(x)\varphi. \tag{4.15}$$

Here $\mathcal{L}'$ is the interaction, some polynomial function of φ. Let us consider this as a classical Lagrange density for a c-number field, and let us construct the classical action integral

$$S(\varphi,J) = \int d^4x\mathcal{L}. \tag{4.16}$$

S is a functional of the two c-number fields, φ and J. In Section 3.3, we also introduced the generating functional for the quantum theory, exp $[iW(J)]$, defined as the sum of all vacuum-to-vacuum graphs in the presence of the source J. I will now demonstrate the following remarkable connection between the quantum generating functional and the classical action integral:

$$e^{iW(J)} = N\int(d\varphi)\, e^{iS(\varphi,J)}, \tag{4.17}$$

where N is a normalization factor, chosen such that W vanishes when J vanishes. Eq. (4.17) is a version of Feynman's sum over histories; a quantum transition amplitude is obtained by summing over all possible classical histories of the system. As it stands, Eq. (4.17) is ill-defined, even by our sloppy standards; the integrand is an awful oscillating object, nothing like the nicely damped Gaussians of Section 4.1. This problem is remedied by stating that the generating functional on the left-hand-side of Eq.(4.17) is that of <u>Euclidean</u> Green's functions, and the functional integral is to be evaluated for fields in <u>Euclidean</u> space, that vanish at infinity.

This prescription requires some explanation: Feynman amplitudes are defined, to begin with, for real external three-momenta and real external energies. However, we can analytically continue them to imaginary energies, by simultaneously rotating all energies (internal as well as external) by $\pi/2$ in the complex energy plane. It is trivial to verify that no singularities of the Feynman integral are encountered in the course of this rotation. (The analytic continuation can also be proved without recourse to perturbation theory, but this is the easiest way to see that it is possible.) Thus we arrive at Euclidean momentum space - real three-momenta and imaginary energies. For any Euclidean momentum k_μ, we define the real variable k_4 by

$$k_0 = ik_4 \tag{4.18}$$

Thus,

$$k^2 = -k_E^2, \tag{4.19}$$

where

$$k_E^2 = \underset{\sim}{k}^2 + k_4^2, \tag{4.20}$$

the standard Euclidean square of a vector. Also,

$$d^4k = id^4k_E \tag{4.21}$$

Euclidean position-space Green's functions are defined by analytically continuing the Fourier transforms of momentum-space Green's functions. So that the Fourier exponential factor, $\exp(ik\cdot x)$, will not blow up and spoil the continuation, we must rotate x_0 through minus $\pi/2$ at the same time we rotate k_0 through plus $\pi/2$. Thus we obtain

$$x_0 = -ix_4, \tag{4.22}$$

$$d^4x = -id^4x_E, \tag{4.23}$$

etc. Thus, for example, the Feynman propagator for a free scalar field of mass μ,

$$\Delta_F(x) = \int \frac{d^4k}{(2\pi)^4} e^{-ik\cdot x} \frac{i}{k^2 - \mu^2 + i\epsilon}, \tag{4.24}$$

becomes, in Euclidean space,

$$\Delta_E(x) = -\int \frac{d^4k_E}{(2\pi)^4} e^{-ik\cdot x} \frac{1}{k_E^2 + \mu^2} . \tag{4.25}$$

Note that there is no need to retain the $i\epsilon$ in Euclidean space. It will be important to us shortly that this function obeys

$$(\Box_E^2 - \mu^2)\Delta_E(x) = \delta^{(4)}(x), \tag{4.26}$$

where

$$\Box_E^2 = \nabla^2 + \partial_4^2 . \tag{4.27}$$

Since the integrand in Eq. (4.25) has no pole, Δ_E is the unique solution to Eq. (4.26); this is in contrast to the situation in Minkowski space, where the corresponding equation has many solutions, and the $i\epsilon$ is needed to resolve the ambiguity.

Let us now turn to the verification of the functional-integral formula, Eq. (4.17). I will begin with the case of a free field, $\mathcal{L}' = 0$. The Minkowski-space generating functional is

$$e^{iW} = \exp[-\tfrac{1}{2}\int d^4x d^4y J(x)\Delta_F(x - y)J(y)]. \tag{4.28}$$

Hence, the Euclidean generating functional is

$$\exp[\tfrac{1}{2}\int d^4x_E d^4y_E J(x)\Delta_E(x - y)J(y)]. \tag{4.29}$$

This takes care of the left-hand side of Eq. (4.17). As for the right-hand side,

$$\begin{aligned} iS &= i\int d^4x(\tfrac{1}{2}[(\partial_0\varphi)^2 - (\nabla\varphi)^2 - \mu^2\varphi^2] + J\varphi) \\ &= -\int d^4x_E(\tfrac{1}{2}[(\partial_4\varphi)^2 + (\nabla\varphi)^2 + \mu^2\varphi^2] - J\varphi) . \end{aligned} \tag{4.30}$$

(I emphasize that this is not an analytic continuation, just a formal substitution. We are not proving that one well-defined object is an analytic continuation of another; we are <u>defining</u> the functional integrand.) Thus the functional integral is of the form (4.9), with

$$A = -\Box_E^2 + \mu^2, \; b = -J, \; c = 0. \tag{4.31}$$

Hence,

$$N\int(d\varphi)e^{iS} = N(\det A)^{-\frac{1}{2}}\, e^{-\frac{1}{2}(J,A^{-1}J)}\,. \tag{4.32}$$

We can now determine the normalization factor, N,

$$N = (\det A)^{\frac{1}{2}}. \tag{4.33}$$

This saves us the trouble of computing the determinant. (This is a good thing, because, in cold fact, the determinant is divergent.) Thus we obtain our final answer for the integral,

$$\exp[-\tfrac{1}{2}(J,A^{-1}J)] = \exp[\tfrac{1}{2}\int d^4x_E d^4y_E J(x)\Delta_E(x-y)J(y)]. \tag{4.34}$$

This is in agreement with Eq. (4.29); in this case, at least, the functional integral has reproduced the generating functional, as promised.

If we had attempted to evaluate the integral directly in Minkowski space, using (erroneously) the integral formulae of Section 4.1, we would have arrived at a similar result, except that A would have been the Klein-Gordon operator. We would then have been stymied, for we would not have known what Green's function to use for A^{-1}. The Euclidean calculation contains no such ambiguity; the right answer (Feynman's $i\epsilon$ rule) comes about automatically as a consequence of our prescription for continuing back into Minkowski space, after we have done the integration. Thus, Euclidean integration is not just a mathematical nicety, but is essential if we are to obtain an unambiguous answer. From now on, I will not explicitly do the continuations into Euclidean space and out again, but simply write my integrals as if they were to be done in Minkowski space, as in Eq. (4.17). You should always remember, though, that this is just a notational convention; really we are always integrating over Euclidean fields.

Now for the interacting case. Still being slapdash, I will ignore all questions of divergences, cutoffs, and renormalizations, and simply write down Dyson's formula for the generating functional:

$$e^{iW} = N'T\langle 0|\exp[i\int(\mathcal{L}'(\varphi_I) + J\varphi_I)d^4x]|0\rangle, \tag{4.35}$$

where $|0\rangle$ is the bare vacuum, φ_I is the interaction-picture field, T is the time-ordering symbol, and N' is a normalization factor, chosen as before. This can be written as

$$\begin{aligned} &N'\exp[i\int d^4y\mathcal{L}'(-i\frac{\delta}{\delta J(y)})]\langle 0|\exp[i\int J\varphi_I d^4x]|0\rangle \\ &= N'\exp[i\int d^4y\mathcal{L}'(-i\frac{\delta}{\delta J(y)})]\exp[iW_0(J)], \end{aligned} \tag{4.36}$$

where W_0 is the generating functional for the free field. Now for the functional integral. We split the action into two parts,

$$S = \int d^4x\mathcal{L}'(\varphi) + S_0(\varphi,J), \tag{4.37}$$

where S_o is the action for the free field (including the source term), the quantity denoted by S in Eq. (4.27). In the spirit of Eq. (4.11),

$$\begin{aligned} N\int(d\varphi)e^{iS} &= N\exp[i\int d^4y\mathcal{L}'(-i\frac{\delta}{\delta J(y)})]\int(d\varphi)e^{iS_0} \\ &= N\exp[i\int d^4y\mathcal{L}'(-i\frac{\delta}{\delta J(y)})]\ e^{iW_0}, \end{aligned} \tag{4.38}$$

by our preceding evaluation. Things equal to the same thing are equal to each other. Q. E. D.

This result generalizes immediately to a theory involving several scalar fields

$$\mathcal{L} = \tfrac{1}{2}(\partial_\mu\varphi^a)(\partial^\mu\varphi^a) - U + J_a(x)\varphi^a, \tag{4.39}$$

where U depends on the fields but not their derivatives, and the sum on repeated indices is implied. In this case, Eq. (4.17) becomes

$$e^{iW} = N\int\prod_a(d\varphi^a)e^{iS}. \tag{4.40}$$

Likewise, descending from four dimensions to one, we see that for the parallel system in particle mechanics,

$$L = \tfrac{1}{2}(\dot{q}^a)^2 - U(q^a) + J_a(t)q^a, \tag{4.41}$$

a similar formula applies,

$$e^{iW} = N\int\prod_a(dq^a)e^{iS}. \tag{4.42}$$

Of course, in this case, the action is just a single integral, not a quadruple one,

$$S = \int dtL. \tag{4.43}$$

We can also go backwards, from "particles" to fields, by letting the index a run over an infinite range, and identifying the q's with the Fourier components of the fields at fixed time. Thus, Eq. (4.42) is in fact more general than Eq. (4.40); it involves no conditions on the Lorentz transformation properties of the dynamical variables, merely a condition on the way in which their time derivatives enter the Lagrangian. For this reason, I will in the future use "particle" language when the discussion is general, and return to field language only for special cases.

4.3 Feynman Rules

Let us return for a moment to the case of a single scalar field with non-derivative interactions. Eq. (4.38) gives a formal expression for the functional integral in this case, but, if the interaction is non-trivial, it is impossible to turn this into an explicit closed form. However, it is perfectly feasible to evaluate it perturbatively, by expanding in powers of the interaction. Such an expansion gives the ordinary Feynman rules.

This can be seen most easily with the aid of a functional identity. I will first state and prove this identity for finite-dimensional real vector spaces, and then, as usual, extend it to function spaces. Let F(x) and G(x) be any two numerical-valued functions on a vector space; then

$$F(-i\frac{\partial}{\partial x})G(x) = G(-i\frac{\partial}{\partial y})F(y)e^{i(x,y)}\Big|_{y=0}. \tag{4.44}$$

The identity is most easily proved by Fourier analysis, that is to say, by taking F and G to be plane waves,

$$F = e^{i(a,x)}, \quad G = e^{i(b,x)}, \tag{4.45}$$

with a and b fixed vectors. Then

$$e^{(a,\partial/\partial x)}e^{i(b,x)} = e^{i(b,x+a)}. \tag{4.46}$$

and

$$e^{(b,\partial/\partial y)}e^{i(x+a,y)} = e^{i(x+a,y+b)}. \tag{4.47}$$

This proves Eq. (4.44).

Extending this to a function space, and applying it to Eq. (4.38), we obtain

$$\begin{aligned} e^{iW} &= N\exp[\tfrac{1}{2}\int d^4y\,d^4x\,\Delta_F(x-y)\,\frac{\delta}{\delta\varphi(x)}\,\frac{\delta}{\delta\varphi(y)}] \\ &\quad\times\ \exp[i\int d^4x(\mathcal{L}'(\varphi) + J\varphi)]\Big|_{\varphi=0}. \end{aligned} \tag{4.48}$$

Here I have made the obvious substitutions of J for x, φ for y, and variational derivative for ordinary derivative. Eq. (4.48) is manifestly the Feynman rules for the vacuum-to-vacuum matrix element. (If it is not manifest to you, I suggest that you compute the first few terms in the expansion for a φ^4 interaction.) Note that diagrams occur in the expansion in which two fields from the same interaction vertex are linked by a propagator; the functional integral does not normal-order the interaction for us. If we wish to treat normal-ordered interactions, we must do the normal-ordering by hand, by inserting explicit counterterms into the interaction.

This argument can immediately be extended to the general case, to give "Feynman rules" for perturbatively evaluating a functional integral of the form

$$\int \prod_a (dq^a) e^{iS}, \tag{4.49}$$

where

$$S = S_0 + S', \tag{4.50}$$

and

$$S_0 = -\tfrac{1}{2}(q^a, A_{ab} q^b). \tag{4.51}$$

Here A is a linear operator (independent of the q's) with positive-definite real part (after the rotation to imaginary time has been performed), and S' is an arbitrary polynomial functional of the q's, possibly involving source terms. Then, just as above, we can develop a diagrammatic expansion for the integral in powers of S', exactly like Feynman rules. Every power of S' is represented by a vertex, and the propagator, $D_F^{ab}(t,t')$, is the solution of

$$A_{ab} D_F^{bc}(t,t') = -i\delta_a^c \delta(t - t'). \tag{4.52}$$

Any ambiguity in solving this equation is to be resolved by rotating to imaginary time.

Note that if S' contains derivatives of the q's, these will just become derivatives of propagators in the expansion. The familiar problem of pushing time derivatives of quantum fields through a time-ordering operator, the problem that makes perturbation theory for derivative interactions such a combinatoric nightmare, has no counterpart here, for we have no time-ordering operator and no quantum fields, just an integral over c-number fields.

Thus, for any theory, if we can write the generating functional in the form (4.49), we can just read off the Feynman rules from S' in the most naive way, replacing every derivative of a field with a momentum factor, etc., without making any mistakes. Unfortunately, at the moment, the only theories for which we can write the generating functional in the form (4.49) are those without any derivatives in the interaction, so this observation is without immediate use. However, it will become very useful shortly.

4.4 Derivative Interactions

There is a large class of theories with derivative interactions for which it is possible to write a functional-integral representation of the generating functional. These are theories where the Lagrangian is no more than quadratic in time derivatives,

$$L = \tfrac{1}{2}\dot{q}^a K_{ab}\dot{q}^b + L_a \dot{q}^a - U, \tag{4.53}$$

where K, L, and U are functions of the q's. The only restriction I will place on these functions is that K be invertible, so that the equation for the canonical momenta,

$$p_a = K_{ab}\dot{q}^b + L_a, \tag{4.54}$$

can be solved for the $\dot{q}$'s and the Hamiltonian constructed,

$$H = \tfrac{1}{2}p_a (K^{-1})^{ab} p_b + \cdots \tag{4.55}$$

where the triple dots indicate terms of first and zeroth order in the p's.

For these theories, the appropriate generalization of our earlier result, Eq. (4.42), turns out to be

$$e^{iW} = N\int \prod_a (dq^a)[\det K]^{\frac{1}{2}} e^{iS}. \tag{4.56}$$

In this equation, K is to be interpreted as a linear operator on the function space, and the integral is to be interpreted in the same way our earlier (Gaussian) integrals were interpreted. Everything is to be restricted to a finite-dimensional subspace, the integral is to be done over that subspace, and the limit is to be taken. I do not know of any short argument for this formula, and have to refer you to the literature for a proof.[21] However, I can try and make it plausible to you by showing that it obeys some simple consistency checks: (1) If K is independent of the q's, and L vanishes, this reduces to the previous case. The determinant can then be pulled out of the integral and absorbed by the normalization factor, reproducing Eq. (4.42). (2) If K is independent of the q's, but L does not vanish, then, by our earlier remarks, the Feynman rules are the naive ones, with the derivative in the interaction becoming a factor of momentum at the vertex. This may be a familiar result to you if you have ever gone through the derivation of the Feynman rules for ps-pv meson-nucleon theory, or the electrodynamics of charged scalar bosons. (3) If K does depend on the q's, things are not so simple. This may be familiar to you if you followed the discussion in the literature a few years ago about the Feynman rules for chiral Lagrangians.

(4) Finally, a Lagrangian of the form (4.48) becomes one of the same form if we change coordinates. To be more precise, let us trade the q's for new variables, which we denote by $\overline{q}^a$. Then

$$L = \tfrac{1}{2}\dot{q}^a K_{ab}\dot{q}^b + \cdots = \tfrac{1}{2}\dot{\bar{q}}^a \bar{K}_{ab}\dot{\bar{q}}^b + \cdots , \qquad (4.57)$$

where

$$\bar{K}_{ab} = \frac{\partial q^d}{\partial \bar{q}^a} K_{cd} \frac{\partial q^d}{\partial \bar{q}^b} . \qquad (4.58)$$

This takes care of the transformation of the Lagrangian, but we still have to change variables in the functional integral. As always, we will figure out how to do this by going back to the finite-dimensional case. Suppose, in a finite dimensional space, we change from coordinates x to coordinates $\bar{x}$. Even though $\bar{x}$ may be a non-linear function of x, $\partial x/\partial\bar{x}$ is a linear operator (an n x n matrix, where n is the dimension of the space), and has a determinant. The change-of-variables formula is the familiar Jacobian formula,

$$(dx) = (d\bar{x})\ \det(\partial x/\partial\bar{x}) . \qquad (4.59)$$

As always, we simply extend this to the infinite-dimensional case, obtaining

$$[\det K]^{\frac{1}{2}}\prod_a(dq^a) = [\det K]^{\frac{1}{2}}\prod_a(d\bar{q}^a)\det(\partial q/\partial\bar{q}) = \prod_a(d\bar{q}^a)[\det \bar{K}]^{\frac{1}{2}}. \qquad (4.60)$$

Thus, Eq. (4.56) is independent of our choice of coordinates.

Eq. (4.56) is sometimes written in "Hamiltonian form",[22]

$$e^{iW} = N\int\prod_a(dp^a)(dq_a)e^{iS}, \qquad (4.61)$$

where S is, as usual, the integral of the Lagrangian, but the Lagrangian is written as a function of the p's and q's, considered as independent variables,

$$L = p^a\dot{q}_a - H. \qquad (4.62)$$

Formally, it is easy to see that this is equivalent to our earlier formula, by explicitly doing the integral over the p's. This is an integral of the exponential of a quadratic form, so it can be done with Eq. (4.9). We see that we get a determinant in front, just the one we need. In addition, in the exponential, the p's are replaced by their values at the point where S is stationary with respect to variations of the p's. This means that we must solve the equations

$$\dot{q}^a = \frac{\partial H}{\partial p_a} \qquad (4.63)$$

But this just reverses the standard passage from the Lagrangian to the Hamiltonian, and recreates the Lagrangian in its original form, as a function of the q's and $\dot{q}$'s.

The Hamiltonian form of the functional integral must be taken with a grain of salt. Unlike the Lagrangian form, the derivative terms do not become nicely damped exponentials when we rotate to imaginary time; they stay oscillating. Thus the Hamiltonian integral is much less well-defined than the Lagrangian one. Indeed, one can show that not even the most ingenious mathematician can make it well-defined; it is possible to find examples for which the value one assigns to (4.61) depends on whether one integrates first over the p's or first over the q's. (A simple one is $H = p^2 + q^2 + \lambda p^2 q^2$; the differences arise in perturbation theory in order λ^2.) However, there is nothing wrong with (4.61) as long as you remember that always attached to it is the rule: first integrate over the p's formally, then rotate to imaginary times.

4.5 <u>Fermi Fields</u>

Everything we have done until now has been for Bose fields. What about Fermi fields? For Bose fields, we found that the generating functional could be represented as an integral over ordinary c-number fields, the classical limits of Bose fields. By analogy, we would expect that the generating functional for a theory involving Fermi fields could be written as an integral over the classical limits of Fermi fields, <u>anticommuting</u> c-number fields. Anticommuting c-numbers are notoriously objects that make strong men quail; fortunately, we will be able to circumvent the problem of defining functional integral involving them.

Supposing we were able to define a functional integral over Fermi fields. What sort of integrals would we want to evaluate? In any theory we are interested in, the Fermi fields enter the Lagrangian at most quadratically. Thus, if we denote the Fermi field(s) by η and the conjugate field(s) by η^*, the part of the action involving Fermions is of the form

$$S_f = (\eta^*, A\eta) . \qquad (4.64)$$

where A is typically the sum of two terms: a constant term, from the free Fermion Lagrangian, and a term involving Bose fields, from the couplings to spinless mesons and/or gauge fields. For the moment, let us consider S_f to be the total action, and the Bose fields referred to above as external fields. (We can always integrate over them later; we know how to integrate over Bose fields.) If we were able to define a functional integral over Fermi fields, we would like to prove that

$$\langle 0^+ | 0^- \rangle = e^{iW} = N \int (d\eta^*)(d\eta) e^{iS_f} . \qquad (4.65)$$

Now let us consider the identical integral with η a complex Bose field. In this case, we know how to do the functional integral, by Eq. (4.11),

$$\int (d\eta^*)(d\eta)\, e^{i(\eta^*, A\eta)} = [\det(iA)]^{-1}. \tag{4.66}$$

We also know how to directly evaluate W, by perturbation theory. W is the sum of all connected Feynman graphs. For an action of the form (4.64), these are just single-closed-loop graphs, like those drawn in Fig. 5 (except that here the lines should have arrows on them, because the field is complex). What happens to the perturbation expansion if we replace Bosons by Fermions? The only difference is that there is a factor of minus one for every closed Fermi loop. Every graph that contributes to W has one and only one closed loop; therefore, W is replaced by minus W, or, equivalently, the inverse determinant in Eq. (4.66) is replaced by the determinant.

Thus, we would get the right answer if

$$\int (d\eta^*)(d\eta)\, e^{i(\eta^*, A\eta)} = \det(iA), \tag{4.67}$$

up to a constant factor, which we can always absorb in the normalization constant, N. Therefore, we <u>define</u> the left-hand side of this equation to be equal to the right-hand side. This is a poor substitute for a deep theory of integration over anticommuting c-numbers, but it does give us a compact expression (the determinant) for a sum over Fermi closed loops, and it will turn out that this is all we will need for our purposes.

4.6 <u>Ghost Fields</u>

We left the theory of derivative interactions in poor shape. It is true that we had an expression for the generating functional, Eq. (4.56), but it was not in the form of an integral of an exponential; there was a determinant sitting in front. Therefore, we could not use Eq. (4.56) to develop a diagrammatic perturbation expansion of the integral. We can now use our knowledge of Fermi fields to get the determinant up into the exponential. For, if we introduce a set of complex Fermi variables, η^a, and denote by $K^{\frac{1}{2}}$ the matrix square-root of K, then

$$[\det K]^{\frac{1}{2}} = \int (d\eta^*)(d\eta)\, e^{i(\eta^{a*}, K^{\frac{1}{2}}_{ab}\, \eta^b)}, \tag{4.68}$$

up to a multiplicative constant, which can always be absorbed in the normalization factor, N. The η's are called ghost variables (in the field-theory case, ghost fields). They are not true dynamical variables of the system, simply devices for getting a determinant up into an exponential.

Thus, the Feynman rules for the theory can be read off from an "effective Lagrangian",

$$L_{eff} = L + L_g, \tag{4.69}$$

where L_g, the ghost Lagrangian, is given by

$$L_g = \eta^{*a} K^{\frac{1}{2}}_{ab} \eta^b. \tag{4.70}$$

It is instructive to work out in detail a field-theoretic example. Let us consider the theory of a free field coupled to an external source,

$$\mathcal{L} = \tfrac{1}{2}(\partial_\mu \varphi)^2 - \tfrac{1}{2}\mu^2\varphi^2 + J\varphi. \tag{4.71}$$

Let us make a change of variables to a new field, A, defined by

$$\varphi = A + \tfrac{1}{2}gA^2. \tag{4.72}$$

where g is a constant. (This transformation is not invertible, but that shouldn't worry us; we're only going to do perturbation theory, and (4.72) is invertible near $\varphi = 0$.) In terms of A, the Lagrange density is given by

$$\mathcal{L} = \tfrac{1}{2}(\partial_\mu A)^2(1 + gA)^2 - \tfrac{1}{2}\mu^2 A^2(1 + \tfrac{1}{2}gA)^2 + JA(1 + \tfrac{1}{2}gA). \tag{4.73}$$

Thus we apparently have a very complicated interaction, with g some sort of coupling constant. Of course, this interaction is just an illusion; the vacuum-to-vacuum matrix element must be the same as in our original theory. However, this is not the answer you will get if you just read the Feynman rules naively out of (4.73). The right Feynman rules are obtained from an effective Lagrange density

$$\mathcal{L}_{eff} = \mathcal{L} + \mathcal{L}_g \tag{4.74}$$

where

$$\mathcal{L}_g = \eta^*\eta(1 + gA) \tag{4.75}$$

The unphysical nature of the ghost fields is doubly clear from this expression: (1) The ghost fields are spinless fields obeying Fermi statistics. (2) The ghost propagator has no momentum dependence; it is a constant, i.

I recommend that you compute a few things to low orders of perturbation theory, using this effective Lagrange density, to convince yourself that everything works out as it should. A good starting point is the one-point function (tadpole) to order g. This should vanish. Does it?

5. THE FEYNMAN RULES FOR GAUGE FIELD THEORIES

5.1 Troubles with Gauge Invariance

The quantization of gauge field theories is notoriously tricky. We can get an idea of the problem if we look at the simplest gauge-invariant field theory, electromagnetism.

$$\mathcal{L} = -\tfrac{1}{4}(\partial_\mu A_\nu - \partial_\nu A_\mu)^2 + \mathcal{L}' \tag{5.1}$$

Let us try and derive the Feynman propagators for A_μ by straightforwardly applying the methods of Section 4, without worrying about whether electromagnetism is in fact in the class of theories we discussed there. The computation is simplified by splitting the field into (four-dimensional) transverse and longitudinal parts

$$\begin{aligned} A_\mu &= A_\mu^T + A_\mu^L \\ &= (P_{\mu\nu}^T + P_{\mu\nu}^L)A^\nu . \end{aligned} \tag{5.2}$$

where the P's are the transverse and longitudinal projection operators; in Fourier space they are given by

$$P_{\mu\nu}^T = g_{\mu\nu} - k_\mu k_\nu/k^2 , \quad P_{\mu\nu}^L = k_\mu k_\nu/k^2 . \tag{5.3}$$

(Remember, we are secretly doing all our computations in Euclidean space, so there is no ambiguity in dividing by k^2.) Then it is easy to see that

$$S = \int d^4x[\tfrac{1}{2}(\partial_\mu A_\nu^T)^2 + \mathcal{L}'] \tag{5.4}$$

We obtain the propagators for the transverse and longitudinal parts of the field by our standard formulae; thus

$$D_{\mu\nu}^F = (g_{\mu\nu} - \frac{k_\mu k_\nu}{k^2})(-\frac{i}{k^2}) + \frac{k_\mu k_\nu}{k^2}(\frac{i}{0}) . \tag{5.5}$$

The second term is obviously unacceptable; something has gone wrong.

This debacle can be explained in two ways, either from Feynman's sum over histories or from conventional canonical quantization. (1) Sum-over-histories explanation: Feynman says that to compute a transition amplitude you must sum over all possible histories of the system. This is normally what the functional integral does for us. However, in a gauge theory, summing over all gauge fields, $A_\mu(x)$, sums over each history an infinite number of times, because fields that are connected by a gauge transformation do not represent different histories but a single history. No wonder we got divergent nonsense! (2) Canonical explanation: To canonically quantize a dynamical system, you have to find a set of initial-value variables, p's and q's, which are complete, in the sense that their values at time zero determine the values of the dynamical variables at all times. It is only in this case that the imposition of canonical commutators at time zero will determine commutators at all times and define a quantum theory. In a gauge theory, this can never be done, because you can always make a gauge transformation that vanishes at time zero but does not vanish at some other time; thus you can never find a complete set of initial-value variables. To quantize a gauge theory, you must first pick a gauge, impose some condition that eliminates the freedom to make gauge transformations. Then, if you are clever and/or lucky in your choice of gauge, you may be able to canonically quantize the theory. Of course, physical quantities are gauge-invariant, and therefore should not depend on what gauge you pick for quantization, but this always has to be proved explicitly in every particular case. We worried about none of this; no wonder we got divergent nonsense!

Both of these explanations emphasize gauge invariance as the critical feature. I personally prefer the second to the first; the injunction to sum over histories seems to me to be incomplete, for it does not tell us what measure to use when the sum is continuous, and, as we saw in our study of derivative interactions, this is not a trivial question. However, in the first part of our investigation, I will accept a quantization method invented by Faddeev and Popov, which is inspired by the first viewpoint. Later on, I will justify the Faddeev - Popov method by appealing to canonical quantization. (Please don't think I am being original in this last step; I learned the canonical justification from Faddeev.)

5.2 The Faddeev-Popov Ansatz

As usual, I will begin by discussing finite-dimensional integrals and later extend the results to function spaces. Our model of the function space of a gauge field theory will be a space of n + m real variables, which we denote collectively by z. We will also denote the first n of these variables by x, and the last m by y. The x's will be our finite-dimensional model of the gauge-independent variables (in electrodynamics, A_μ^T) and the y's of the gauge-dependent variables (A_μ^L, in electrodynamics). We will also have a model of a gauge-invariant action, a function S(z), which is independent of the y-variables,

$$\frac{\partial S}{\partial y} = 0. \tag{5.6}$$

We wish to define a (finite-dimensional model of the) generating functional that avoids the divergence problems we would encounter if we integrated over all the z's. This is easy; we just integrate over the x's only, and define

$$e^{iW} = \int (dx)\, e^{iS}. \tag{5.7}$$

(We suppress the normalization factor for the moment.) This can also be written as

$$e^{iW} = \int (dz)\, e^{iS} \delta(y). \tag{5.8}$$

Here $\delta(y)$ is an m-dimensional δ-function, normalized such that

$$\int g(y)\,(dy)\,\delta(y) = g(0), \tag{5.9}$$

for any function g. Eq. (5.8) says that we integrate along the surface y = 0. Of course, since nothing depends on the y's, we could just as well integrate along an arbitrary surface, defined by

$$y = f(x), \tag{5.10}$$

where f is an m-vector, a set of m functions of the x's. We then obtain

$$e^{iW} = \int (dz)\, e^{iS} \delta(y - f(x)). \tag{5.11}$$

We may not be given the surface in the form (5.10), but as the solution to some set of equations,

$$F(z) = 0, \tag{5.12}$$

where F is again an m-vector, a set of m functions of the z's. It is easy to rewrite the integral in a form appropriate to this description of the surface,

$$e^{iW} = \int (dz)\, e^{iS} \det(\partial F/\partial y)\, \delta(F(z)) \tag{5.13}$$

Note that, because of the presence of the δ-function, we need only evaluate the determinant on the surface. I emphasize that Eq. (5.13) defines the same expression as Eq. (5.7), and is completely independent of our choice of the functions F.

I will now state the Faddeev-Popov[23] quantization procedure. Let S be the action integral for a theory involving m fields, $\varphi^a(x)$ (not necessarily all scalar). Let S be invariant under some group of gauge transformations, parametrized by a set of n real functions, $\omega^b(x)$. For such a theory, a "gauge" is defined to be a set of n equations

$$F^b(x) = 0. \tag{5.14}$$

where the F's are functions of the φ's, possibly differential or even non-local, such that, given any $\varphi^a(x)$, there is one and only one gauge transformation that makes the Eq. (5.14) true. For electrodynamics, an example of a gauge is radiation gauge, $\nabla \cdot A = 0$. Another example is Lorentz gauge, $\partial_\mu A^\mu = 0$. (You may object that in this case the gauge transformation is not unique. This is true in Minkowski space, but remember that we are always secretly working in Euclidean space.) According to Faddeev and Popov, the theory is now quantized by declaring that

$$e^{iW} = N\int \prod_a (d\varphi^a) e^{iS} \det\left(\frac{\partial F^b}{\partial \omega^c}\right) \prod_b \delta(F^b). \tag{5.15}$$

where $F^a = 0$ is some gauge. This is the functional analogue of Eq. (5.13). The δ-function in Eq. (5.15) is a δ-function on function space, a δ-functional if you will; it obeys the equation

$$\int (d\varphi) G(\varphi) \delta(\varphi) = G(0), \tag{5.16}$$

for any functional G. We will call Eq. (5.15) the Faddeev-Popov Ansatz.

Remarks: (1) The choice of gauge in the Faddeev-Popov Ansatz is equivalent to the choice of surface in the finite-dimensional integral we discussed earlier. Thus, whether the Ansatz is true or false, it is at least self-consistent; it it independent of the choice of gauge. (2) Thus, to verify the Ansatz, it suffices to verify it for just one gauge. If it is true in one gauge, it is true in any other. (3) The gauge-independence of the Ansatz depends on the action being gauge-independent. Thus, the action can not contain source terms coupled linearly to the gauge fields. However, it can contain source terms coupled to gauge-invariant objects, like $(F^a_{\mu\nu})^2$, for example. Phrased in another way, the Ansatz only gives us gauge-invariant expressions for gauge-invariant Green's functions. Since the standard wisdom is that in a gauge theory only gauge-invariant quantities are physical observables, this is no great restriction. Also, once we have settled down in some fixed gauge, there is no objection to computing non-gauge-invariant objects, like gauge-field propagators, as a preliminary step in the computation of gauge-invariant objects. (4) I remind you that everything we are doing is on a purely formal level; we are ignoring complications that may arise as a result of ultraviolet divergences. Therefore, everything we do should be taken as merely heuristic, to be checked later by more careful analysis. The manipulation of functional integrals is more efficient than other formal methods of treating gauge theories, but it is no more rigorous.

5.3 The Application of the Ansatz

We will begin with the simplest gauge theory, electrodynamics. Since gauge transformations for this theory are parametrized by only a single function, only one equation is needed to determine a gauge. We will choose a slight generalization of the Lorentz gauge,

$$F = \partial^\mu A_\mu - f(x), \tag{5.17}$$

where f(x) is an arbitrary function. Under an infinitesimal gauge transformation, Eq. (2.28),

$$\delta F = -e^{-1}\Box^2 \delta\omega. \tag{5.18}$$

Thus,

$$\det(\delta F/\delta\omega) = \det(-e^{-1}\Box^2). \tag{5.19}$$

This is a constant and can be brought outside the integral and absorbed in the normalization. Thus we obtain

$$e^{iW} = N\int (dA)(d\psi)e^{iS}\delta(\partial^\mu A_\mu - f(x)) \tag{5.20}$$

where, to simplify notation, I have indicated by (dA) the integrals over all four components of the gauge fields, and by (dψ) the integrals over all other fields in the theory.

We still do not have the integral of an exponential, so it is hard to evaluate Eq. (5.20) perturbatively. This is easily rectified. Since the integral is independent of the function f, we can integrate it with any functional of f, G(f), without changing the integral (except perhaps for a normalization, which can always be absorbed in N). Thus,

$$\begin{aligned} e^{iW} &= N\int (dA)(d\psi)(df)e^{iS}\delta(\partial^\mu A_\mu - f)G(f) \\ &= N\int (dA)(d\psi)e^{iS}G(\partial^\mu A_\mu) \end{aligned} \tag{5.21}$$

In particular, if we choose

$$G(f) = e^{-\frac{i}{2\alpha}\int d^4x f^2}, \tag{5.22}$$

where α is some real number, we find

$$e^{iW} = N\int(dA)(d\psi)e^{i[S - \frac{1}{2\alpha}\int d^4x(\partial^\mu A_\mu)^2]}. \tag{5.23}$$

Thus the outcome of our application of the Ansatz is to replace the Lagrange density of the theory by an effective Lagrange density

$$\mathcal{L}_{eff} = \mathcal{L} - \frac{1}{2\alpha}(\partial^\mu A_\mu)^2. \tag{5.24}$$

From this it is easy to compute the electromagnetic propagator, since

$$\int d^4x(\partial^\mu A_\mu)^2 = \int d^4x(\partial^\mu A^L_\nu)^2. \tag{5.25}$$

Hence, the preposterous Eq. (5.5) is replaced by

$$D_{\mu\nu} = \frac{-i}{k^2}\left[g_{\mu\nu} - \frac{k_\mu k_\nu}{k^2} + \alpha\frac{k_\mu k_\nu}{k^2}\right] \tag{5.26}$$

Any one of these propagators should give the same results as any other in the computation of gauge-invariant objects. (I hope you know enough about electrodynamics from other sources to recognize this as a true statement.) The choice $\alpha = 1$ yields what is usually called the Feynman-gauge propagator, etc. For any choice of α, the propagator has the same high-energy behaviour as that of a scalar field, and therefore the dimension-counting formulae of Section 3.1 are applicable in computing counterterms.

Now let us turn to non-Abelian gauge fields. For rotational simplicity, we will restrict ourselves to the case where there is only one gauge-field coupling constant. We determine a gauge as in Eq. (5.17),

$$F^a = \partial^\mu A^a_\mu - f^a \tag{5.27}$$

where the f's are arbitrary functions. Under an infinitesimal gauge transformation, (2.44),

$$\delta F^a = g^{-1}[-\Box^2\delta\omega^a + gc^{abc}\partial^\mu(\delta\omega^b A^c_\mu)]. \tag{5.28}$$

In contrast to the Abelian case, here the determinant is not a constant. However, just as in Section 4.6, we can write it as an integral over a set of ghost fields, scalar fields obeying Fermi statistics,

$$\det\left(\frac{\delta F^a}{\delta\omega^b}\right) = \int(d\eta^*)(d\eta)e^{iS_g} \tag{5.29}$$

where

$$S_g = \int d^4x\mathcal{L}_g = \int d^4x(\partial_\mu\eta^{*a})(\partial^\mu\eta^a - gc^{abc}\eta^b A^c_\mu). \tag{5.30}$$

and we have chosen to absorb an overall factor of det g^{-1} into N. The ghost Lagrange density can be written in a compact way if we consider the ghosts as a set of fields that transform according to the adjoint representation of the group,

$$\mathcal{L}_g = \partial_\mu \eta^{*a} D^\mu \eta^a . \tag{5.31}$$

In this form it is clear that $\mathcal{L}_g$ is not gauge-invariant; of course, there is no reason why it should be, since it is derived from the (purposefully) non-gauge-invariant Eq. (5.27). In contrast to the example of Section 4.6, here the ghosts have a momentum-dependent propagator, that of a set of massless charged scalar fields, $i\delta^{ab}/k^2$. However, they still reveal their unphysical nature by being spinless particles obeying Fermi statistics.

The remainder of the development of the non-Abelian case is exactly the same as that of the Abelian case. Thus we arrive at the effective Lagrange density

$$\mathcal{L}_{eff} = \mathcal{L} + \mathcal{L}_g - \frac{1}{2\alpha}(\partial^\mu A^a_\mu)^2 . \tag{5.32}$$

where α is an arbitrary real number. Note that if $\mathcal{L}$ is the Lagrange density for a renormalizable field theory minimally coupled to gauge fields, every term in this expression is an interaction of renormalizable type (dimension less than or equal to four). As I explained at the end of Section 3.1, this observation is just the first step in establishing renormalizability, but it is as far as we will have time to go here.

5.4 Justification of the Ansatz

I will now justify the Faddeev-Popov Ansatz by showing that, in a particular gauge, it is equivalent to canonical quantization. The gauge is Arnowitt-Fickler[24] gauge (sometimes called axial gauge); it is defined by

$$F^a = A^a_3 = 0. \tag{5.33}$$

where the 3 indicates the third spatial component. Unlike Eq. (5.27), this is not Lorentz-covariant, so this is a terrible gauge for performing Feynman calculations; however, this is not our purpose. For simplicity, I will construct the proof for pure Yang-Mills fields, uncoupled to other fields; The generalization is straightforward.

First we must construct the Faddeev-Popov Ansatz:

$$\begin{aligned} \delta F^a &= -g^{-1}\partial_3 \delta\omega^a + c^{abc}\delta\omega^b A^c_3 \\ &= -g^{-1}\partial_3 \delta\omega^a . \end{aligned} \tag{5.34}$$

The second line follows from Eq. (5.33). Thus, the determinant is a constant, and can be absorbed in the normalization factor; in this gauge, <u>there are no ghosts</u>, even in the non-Abelian case. Thus, the Ansatz becomes

$$e^{iW} = N\int (dA)\, e^{iS} \prod_a \delta(A_3^a), \tag{5.35}$$

where (dA) indicates integration over all the A's. More explicitly,

$$(dA)\prod_a \delta(A_3^a) = \prod_a (dA_1^a)(dA_2^a)(dA_0^a). \tag{5.36}$$

I remind you that

$$S = -\tfrac{1}{4}\int d^4x (\partial_\mu A_\nu^a - \partial_\nu A_\mu^a + gc^{abc} A_\mu^b A_\nu^c)^2, \tag{5.37}$$

plus source terms, which I shall not bother to write explicitly.

(This is off the main line of the argument, but it is a point that may have been worrying you: The ghosts are fictitious particles, but they do have real poles in their propagators. Therefore, it seems that states involving ghosts might contribute to the absorptive parts of gauge-invariant Green's functions. This would be disturbing if it happened; fortunately, the existence of a ghost-free gauge shows that it does not.)

It will be convenient to rewrite the Ansatz in so-called first-order form,

$$e^{iW} = \int (dF)(dA) \prod_a \delta(A_3^a)\, e^{iS'}, \tag{5.38}$$

where

$$S' = \int d^4x [-\tfrac{1}{8}(F_{\mu\nu}^a)^2 + \tfrac{1}{4}F^{\mu\nu a}(\partial_\mu A_\nu^a - \partial_\nu A_\mu^a + gc^{abc} A_\mu^b A_\nu^c)], \tag{5.39}$$

and (dF) denotes integration over all the F's. The integral over the F's is trivial and obviously reproduces Eq. (5.35). S' is also equivalent to S in the normal sense of Lagrangian dynamics; if we vary S' with respect to the F's and A's independently, we get the same equations of motion we obtain by varying S with respect to the A's alone. That S' is a good action in both these senses is no coincidence; it is a consequence of the integration formula (4.9). If a dynamical variable appears in the action at most quadratically, and if the coefficient of the quadratic term is a constant, then integrating over the variable is the same as eliminating it from the action by using the Euler-Lagrange equations.

So much for the Ansatz; now let us turn to canonical quantization. Again, we will use S', the first-order action, and work in Arnowitt-Fickler gauge, setting A_3^a equal to zero. Let us write Eq. (5.39) in such a way that the dependence on various tensor components is explicit:

$$\begin{aligned}\mathcal{L}' = &-\tfrac{1}{2}(F^a_{\mu\nu})^2 + \tfrac{1}{2}F^{ija}(\partial_i A^a_j - \partial_j A^a_i + gc^{abc}A^b_i A^c_j)\\ &+ F^{0ia}(\partial_0 A^a_i - \partial_i A^a_0 + gc^{abc}A^b_0 A^c_i)\\ &+ F^{03a}(-\partial_3 A^a_0) + F^{i3a}(-\partial_3 A^a_i).\end{aligned} \tag{5.40}$$

where i and j run over the range 1, 2. Note the drastic simplification of the last two terms, caused by the gauge condition. We now see that canonical quantization of (5.40) is like shooting fish in a barrel: A^a_0, F^{ija}, F^{03a}, F^{13a} are constrained variables; their Euler-Lagrange equations involve no time derivatives and are, therefore, not true equations of motion but equations of constraint, fixing the constrained variables on the initial-value surface in terms of the remaining variables, A^a_i and F^{0ia}.

Let us denote the action obtained by eliminating the constrained variables by S"; it is a functional only of A^a_i and F^{0ia}. Furthermore, it is in Hamiltonian form, with the A's the canonical fields and the F's the conjugate momentum densities. Thus, we can use Eq. (4.61) to write a functional,

$$e^{iW} = N\int \pi_a (dF^{01a})(dF^{02a})(dA^a_1)(dA^a_2)e^{iS''}. \tag{5.41}$$

However, because the constrained variables enter Eq. (5.40) at most quadratically, and because the coefficients of the quadratic terms are constants, we can equally well write this as

$$\begin{aligned}e^{iW} &= N\int (dF)\pi_a (dA^a_0)(dA^a_1)(dA^a_2)e^{iS'}\\ &= N\int (dF)(dA)\pi_a \delta(A^a_3)e^{iS'}\end{aligned} \tag{5.42}$$

But this is the Faddeev-Popov Ansatz, Eq. (5.38). Q.E.D.

5.5 Concluding Remarks

(1) The chain of arguments we have just constructed shows both the power and the limitations of functional-integral methods. Functional integration is a supplement to canonical quantization, not a replacement for it. For example, when writing down the Ansatz, I could well have multiplied the integrand by some function of $(F^a_{\mu\nu})^2$. This would have been just as gauge-invariant, and just as plausible *a priori* as the original Ansatz. It would have been wrong, but there would be no way to tell this without appealing

to canonical quantization. On the other hand, once we have justified the Ansatz by canonical quantization, we can use it to pass from one gauge to another with incomparable ease. In particular, we can use it to pass from a gauge in which canonical quantization is simple to a gauge in which the Feynman rules are simple.

(2) I have said this before, but it deserves emphasis: Everything we have done in this section is purely heuristic; we have paid no attention to the problems caused by ultraviolet divergences. Properly, everything should be redone with careful attention to cutoffs, renormalizations, etc. Such careful investigations have been done[11]; the result is that the heuristic arguments have not betrayed us: These theories are renormalizable; renormalization does not spoil gauge invariance; ghost states never contribute to the absorptive parts of gauge-invariant Green's functions; etc. There is one exception: In theories in which some of the gauge transformations are chiral, the familiar Adler-Bell-Jackiw triangle anomalies can falsify our arguments. However, if the transformation properties of the Fermi fields are chosen such that there are no anomalies in the lowest-order triangle graphs with gauge currents at the vertices, then there are no anomalies anywhere, and everything is all right.[25]

(3) People are sometimes worried that the formal apparatus for treating spontaneous symmetry breakdown, explained in Section 3, is not gauge-invariant. This is true; the vacuum expectation value of a scalar field, the effective potential, indeed, even the Feynman propagators themselves, are not gauge-invariant objects. This is also irrelevant. In quantum electrodynamics, we continually do computations using non-gauge-invariant objects, like propagators, at intermediate stages. There is nothing wrong with this, as long as we are careful to express our final results in terms of gauge-invariant quantities, like masses and cross-sections. The occurrence of spontaneous symmetry breakdown does not affect this; the form of the effective potential and the location of its minimum are indeed gauge-dependent, but the values of masses and cross-sections computed with the aid of these objects are not.

(4) At the end of Section 3, I explained how many workers prefer to do computations in terms of shifted fields, defined by

$$\underset{\sim}{\varphi}' = \underset{\sim}{\varphi} - \langle \underset{\sim}{\varphi} \rangle, \tag{5.43}$$

and to determine the parameters $\langle \underset{\sim}{\varphi} \rangle$ at the end of the computation, by self-consistency. There is one awkwardness in doing things this way; the shift generates a bilinear scalar-vector coupling from the scalar Lagrange density:

$$\tfrac{1}{2}D^{\mu}\underset{\sim}{\varphi}\cdot D_{\mu}\underset{\sim}{\varphi} + \cdots = g\partial^{\mu}\varphi'\cdot A_{\mu}^{a}\underset{\sim a}{T}\langle\underset{\sim}{\varphi}\rangle + \cdots \tag{5.44}$$

This coupling causes a scalar-vector mixed propagator to appear in the Feynman rules of the theory; this is no difficulty in principle, but is an annoyance in practice. Fortunately, it is possible to cancel this term by a clever choice of gauge.[26] For our gauge condition, we choose

$$F^{a} = \partial^{\mu}A_{\mu}^{a} - f^{a}(x) - \xi\underset{\sim}{\varphi}'\cdot\underset{\sim a}{T}\langle\underset{\sim}{\varphi}\rangle. \tag{5.45}$$

where ξ is a number to be determined later. If we go through what should be by now familiar arguments, we obtain an effective Lagrange density of the form

$$\begin{aligned} \mathcal{L}_{eff} &= \mathcal{L} + \partial_{\mu}\eta^{*a}D^{\mu}\eta^{a} \\ &- g\xi\eta^{*a}\eta^{b}(\underset{\sim b}{T}\langle\underset{\sim}{\varphi}\rangle\cdot\underset{\sim a}{T}\langle\underset{\sim}{\varphi}\rangle + \underset{\sim b}{T}\varphi'\cdot\underset{\sim a}{T}\langle\varphi\rangle) \\ &- \frac{1}{2\alpha}(\partial^{\mu}A_{\mu} - \xi\underset{\sim}{\varphi}'\cdot\underset{\sim a}{T}\langle\underset{\sim}{\varphi}\rangle)^{2}. \end{aligned} \tag{5.46}$$

Hence, if we choose

$$\xi = \alpha g, \tag{5.47}$$

we can cancel the annoying cross terms. Note that the interactions in this Lagrange density are still of renormalizable type, dimension less than or equal to four.

6. ASYMPTOTIC FREEDOM

6.1 Operator Products and Deep Inelastic Electroproduction

The topic we are now going to discuss seems, at first glance, to have very little to do with the previous lectures. It is a topic in strong-interaction physics, that of reconciling the apparent scaling in the SLAC-MIT electroproduction experiments with the predictions of quantum field theory. I will begin by summarizing very briefly the standard lore on this problem.[27]

(1) The electroproduction experiments at Stanford measure the total cross sections for the process

electron + nucleon → electron + anything

which is, of course, the same thing as

virtual photon + nucleon → anything

The process is therefore described by two kinematic variables: q^2, the mass of the virtual photon, a negative number, and E, the energy of the virtual photon in the lab frame. It is convenient to trade E for the dimensionless variable

$$x = -q^2/2mE, \tag{6.1}$$

where m is the nucleon mass. Elementary kinematics restricts x to be between zero and one. The nucleons in the experiment are unpolarized, while the virtual photons can be either transverse or longitudinal; thus the cross section can be described in terms of two dimensionless invariants, $F_i(q^2,x)$, where i is 1 or 2. The F's are called structure functions; the details of their definitions will not be relevant to our immediate purposes.

As $-q^2$ increases, the F's quite rapidly lose their dependence on q^2; by $q^2 = -(2\text{ GeV})^2$, the F's appear to be functions of x alone, within experimental error. This phenomenon is called Bjorken scaling. There are two schools of thought on Bjorken scaling: (1) Bjorken scaling is a true asymptotic phenonmenon. It will persist even if the range of q^2 is increased greatly. (2) SLAC energies are too small for us to believe that we are really in the asymptotic region. Bjorken scaling is some sort of low energy epiphenomenon, and has nothing to do with true high-energy limits.

I will adopt the first position for this lecture, but you should be aware that this is just a matter of prejudice. The second position may well be correct; only future experiment can decide the question.

It will turn out to be convenient for our purposes to phrase matters in terms of the moments of the structure functions,

$$F_i^n(q^2) = \int_0^1 dx\, x^n F_i(q^2,x). \tag{6.2}$$

The problem is: Why do these moments become constants (within experimental error) as q^2 becomes large and spacelike?

(2) The operator product expansion was invented by Wilson and proved to all orders of renormalized perturbation theory by Callan and Zimmerman. It is an asymptotic expansion for the product of two local operators as the distance between them becomes small, but for our purposes it will be most convenient to express the expansion in momentum space. Let A and B be any two local operators (renormalized polynomials in canonical fields and their derivatives) and let $|a\rangle$ and $|b\rangle$ be any two states. Then,

$$\int e^{iq\cdot x} d^4x \langle a|A(x)B(-x)|b\rangle = \sum_C f_{ABC}(q)\langle a|C(0)|b\rangle. \tag{6.3}$$

as q goes to Euclidean infinity. The sum is over a complete set of local operators (all renormalized monomials in canonical fields and their derivatives). The expansion is useful because the rate of growth of the coefficient functions, the f's, is that given by naive dimensional analysis, modulo polynomials in log q^2. (This is true to any finite order in perturbation theory; we will later investigate whether these polynomials can pile up and change the asymptotic behaviour if we sum the perturbation series.) Thus, for any given A and B in any given field theory, only a finite set of operators contributes to the leading asymptotic behaviour; higher monomials give lower powers of q^2.

I emphasize that the f's are independent of the states $|a\rangle$ and $|b\rangle$. In particular, this means the operator product expansion is unaffected by the occurrence of spontaneous symmetry breakdown. This will be important to us later.

If A, B, and C are other than Lorentz scalars, f_{ABC} will have a nontrivial tensor structure. Since f is only a function of a single four-vector, q, it is a known tensor function of q times an unknown scalar function of q^2. It will be convenient to multiply this scalar function by a power of q^2 so that it becomes dimensionless. We will call the resulting dimensionless scalar function $\bar{f}_{ABC}$.

(3) By choosing $|a\rangle$ and $|b\rangle$ to be one-nucleon states, and A and B to be electromagnetic currents, we can use the operator product expansion to get an expression for the moments of the structure functions. The calculation is straightforward, and I do not want to do it in detail here; the result is of the form

$$F_i^n(q^2) = \sum_C d_{iC}^n \bar{f}_{ABC}(q^2)\langle a|C(0)|b\rangle_R. \tag{6.4}$$

Here the d's are constant coefficients, terms that are less important by powers of q^2 than the terms retained have been dropped, the subscript R indicates a reduced (scalar) matrix element, and, in any given theory, for fixed i and n, the sum runs over only a finite set of C's.

Thus the problem becomes: Why do the $\bar{f}$'s become constants (within experimental error) as q^2 becomes large and negative?

Because of the logarithmic polynomials mentioned before, this constant behaviour is not an obvious prediction of field theory. To be specific, let us consider a theory in which there is only one coupling constant, like the standard quark-vector-gluon model. In this case, a perturbative expansion of one of the $\bar{f}$'s typically yields an asymptotic expression like

$$\bar{f} = a_0 + a_{11} g^2 \log q^2 + a_{10} g^2 + a_{22} g^4 (\log q^2)^2 + \cdots \qquad (6.5)$$

where the a's are constant coefficients. Since we are interested in both large g (strong interactions) and large q^2 (asymptotic behaviour), this is worse than useless. Even for the (unrealistic) case of small g, Eq. (6.5) tells us nothing about asymptotic behaviour, for the largeness of the logarithm eventually overcomes the smallness of g. The only case in which we can predict asymptotic behaviour is free field theory ($g = 0$); in this case, the $\bar{f}$'s are indeed constants. It is for this reason that it is sometimes said that, at high negative q^2, the effects of the interactions seem to disappear, and the theory behaves as if it were free. At the moment, this may seem to you to be an excessively dramatic way of describing Bjorken scaling; nevertheless, we shall see, for a certain class of field theories, this is exactly what happens. Before we do this though, we need to develop a systematic formalism for going beyond perturbation theory and summing up the logarithms in Eq. (6.5).

6.2 Massless Field Theories and the Renormalization Group[28]

It can be shown, in any renormalizable field theory, to all orders of perturbation theory, that the asymptotic behaviour of the coefficient functions in the operator product expansion is the same as it would be in a massless field theory. By a massless field theory I mean one that has only dimensionless coupling constants in its Lagrangian; not only are masses excluded but also interactions with dimensionful coupling constants, like cubic meson self-couplings. This is very plausible; the coefficient functions depend only on a single momentum, and this momentum is going to Euclidean infinity, getting as far as it can from the mass shell, and therefore losing all memory of the masses. (I emphasize that this does not mean that the structure functions themselves are the same as they would be in a massless theory. Eq. (6.4) contains not just $\bar{f}$, but also $\langle a|C|b\rangle$, and this stays on the mass shell.)

Thus we need only analyze the behaviour of the $\bar{f}$'s in a massless theory. A massless renormalizable field theory is parametrized by a set of renormalized dimensionless coupling constants, which I will call g^a. These may be either Yukawa coupling constants, quartic meson self-interaction constants, or gauge field coupling constants. In addition, another parameter is required to complete the description - a mass, M. This extra parameter is needed to define the others (and the scale of the renormalized fields).

Let me explain why this is so, using the simplest renormalizable field theory, $\lambda\varphi^4$ theory, as an example. In the massive version of this theory, the renormalized coupling constant, λ, is usually defined as the value of $\Gamma^{(4)}$ on the mass shell, at the symmetry point, $s = t = u$. (Sometimes it is defined as the value of $\Gamma^{(4)}$ when all external momenta vanish, as in Eq. (3.17b), but this becomes the same definition when the mass vanishes.) Likewise, the renormalized field is defined as the field scaled in such a way that the derivative of $\Gamma^{(2)}$ is one on the mass shell (or, sometimes, at zero momentum). However, for a massless theory, these definitions are unworkable; all of the normal thresholds collapse on the renormalization point, and it is obviously bad policy to define λ as the value of a Green's function at the locus of an infinite number of singularities. The cure for this disease is simple; we define λ as the value of $\Gamma^{(4)}$ at some point in Euclidean space, where there are no singularities, even in the massless theory. For example, we could define λ as the value of $\Gamma^{(4)}$ at $s = t = u$, with all external momenta squared equal to $-M^2$. M can be anything; any M is as good as any other M, so long as it is not zero. Likewise, we could define the scale of the field by demanding that the derivative of $\Gamma^{(2)}$ be one when $p^2 = -M^2$. These definitions can be extended in an obvious way to more complicated theories with other kinds of couplings.

Thus, the parametrization of a massless field theory requires a mass, M. But M is arbitrary; in a given physical theory, if you change the value of M, this can always be compensated for by an appropriate change in the g's and in the scales of the renormalized fields, because the only function of M is to define these quantities. Phrased in equations, if we make a small change in M,

$$M \to M(1 + \epsilon), \qquad (6.6a)$$

where ϵ is infinitesimal, this can always be compensated for by an appropriate small change in the coupling constants

$$g^a \to g^a + \beta^a \epsilon \qquad (6.6b)$$

and a corresponding small change in the scale of renormalized operators, e.g.,

$$A(x) \to (1 + \gamma_A \epsilon) A(x). \qquad (6.6c)$$

By dimensional analysis, the β's and γ's can depend only on the g's,

$$\beta^a = \beta^a(g), \; \gamma_A = \gamma_A(g), \tag{6.7}$$

where, to simplify notation, a single g in the argument of a function stands for all the g's. If A is one of a set of operators that can mix with one another as a result of renormalization (as is the case, for example, with φ^4 and $\partial_\mu\varphi\partial^\mu\varphi$), Eq. (6.6c) should be replaced by a matrix equation. However, for simplicity, we will ignore this possible complication here.

The infinitesimal transformations (6.6) define a one-parameter group, called the renormalization group. All physical quantities must be invariant under this group. In particular, the $\bar{f}$'s must be invariant; thus

$$[M \frac{d}{dM} + \beta^a(g)\frac{\partial}{\partial g^a} + \gamma_{ABC}(g)]\bar{f}_{ABC} = 0. \tag{6.8}$$

where

$$\gamma_{ABC} = \gamma_A + \gamma_B - \gamma_C. \tag{6.9}$$

Of course, similar equations can be derived for any other object in the theory, in particular, for Green's functions. Only the γ-terms depend on the object under consideration.

Since these renormalization-group equations are exactly valid, they must be valid order-by-order in renormalized perturbation theory. Thus, from perturbation expansions of Green's functions, it is possible to deduce perturbation expansions for the β's and γ's. If this is done for the quark-vector-gluon model, for example, one finds that the power series for β begins with terms of order g^3, while those for either the quark or gluon γ begin with terms of order g^2. This is reasonable, because β reflects the effects of coupling-constant renormalization, which begin at order g^3, while γ reflects those of wave-function renormalization, which begin at order g^2.

6.3 Exact and Approximate Solutions of the Renormalization Group Equations

The differential equations of the renormalization group are a mathematical expression of a physical triviality, that the only function of the mass M is to define the renormalized coupling constants and the scale of the renormalized fields. Nevertheless, they can, in favorable circumstances, be used to obtain highly non-trivial information about the asymptotic behaviour of the theory. The basic reason for this is simple dimensional analysis; since $\bar{f}$ is dimensionless,

$$\bar{f}_{ABC} = \bar{f}_{ABC}(Q/M, g), \tag{6.10}$$

where $Q = \sqrt{-q^2}$. Thus, knowledge of the (trivial) dependence on M is equivalent to knowledge of the (non-trivial) dependence on Q.

To work this out in detail, let me assume that we know the β's and γ's exactly. Then there is a standard method[29] for solving the linear partial differential equation

$$[M\frac{\partial}{\partial M} + \beta^a(g)\frac{\partial}{\partial g^a} + \gamma(g)]\overline{f}(Q/M,g) = 0 \qquad (6.11)$$

where I have suppressed the ABC subscript for notational simplicity. The standard method goes in two steps: First, one constructs $g'^a(g,t)$, a set of functions of the g's and a single extra variable, t, defined as the solution to the ordinary differential equations

$$dg'^a/dt = \beta^a(g'), \qquad (6.12a)$$

with the boundary condition

$$g'^a(g,0) = g^a. \qquad (6.12b)$$

Then, the general solution to Eq. (6.11) is

$$\overline{f} = F(g'(g,\ln[Q/M]))\times \exp\int_0^{\ln Q/M} \gamma(g'(g,t))\,dt \qquad (6.13)$$

where F is an arbitrary function. Thus we see the power of the renormalization group; if we know everything for all g's at Q = M, then we know everything for all g's at all Q's.

Unfortunately, we do not know everything for all g's. Typically, we only know the first few terms in a power series in g. Even in this case, though, it is possible to use the renormalization group to squeeze out extra information. To show how this is done, let me return to the quark-vector-gluon model. Here I have argued that

$$\beta(g) = bg^3 + O(g^5). \qquad (6.14)$$

where b is a numerical coefficient. The only thing I will ask you to take on trust is that, if you actually do the relevant Feynman calculations, you will find that b is positive. Eq. (6.12) then becomes

$$dg'/dt = bg'^3 + O(g'^5). \qquad (6.15)$$

Now let us attempt to construct an approximate solution of this equation, for small g, by ignoring the terms of order g^5. The solution is trivially obtained by quadratures,

$$\frac{1}{g'^2} = \frac{1}{g^2} - 2bt, \tag{6.16}$$

or

$$g'^2 = \frac{g^2}{g^2 - 2bt} . \tag{6.17}$$

When can we trust this approximate solution? When t gets large and positive, the approximate g' becomes large, and the terms we have neglected become comparable to the terms we have retained. For this range of t, the approximation is garbage. On the other hand, as t becomes large and negative, the approximate g' becomes smaller and smaller, and the terms we have neglected therefore become smaller and smaller than the terms we have retained. For this range of t, the approximation is wonderful.

Now, when we plug g' into Eq. (6.13), t becomes ln (Q/M). Thus our approximation gets better and better the smaller Q is. Furthermore, we can improve on it as much as we want, simply by doing more perturbation calculations to get the higher terms in the expansions of β, γ, and F.

To phrase the whole thing more generally, an ordinary perturbation expansion, like (6.5), has two conditions for its reliability, $|g| \ll 1$ and $|\ln(Q/M)| \ll 1$. The approximation scheme I have described replaces these with a single condition, $|g'| \ll 1$. This single condition may hold in regions where the logarithm is large; in the case at hand, this includes the region of arbitrarily large negative ln(Q/M).

This is marvelous stuff; the renormalization group has tamed the logarithms in Eq. (6.5). Unfortunately, this is of no physical interest, for two reasons: (1) To start the approximation, g must be small. We are interested in strong interactions. (2) We can tame the logarithms in the region of small Q, the infrared region. We are interested in large Q, the ultraviolet region. Indeed, our whole method of approach is nonsense in the region of small Q, because, when Q is small, it is no longer sensible to neglect particle masses.

Now let us do another example, pure Yang-Mills theory for some simple Lie group. Here again there is only one coupling constant, and coupling-constant renormalization begins in order g^3, so this is hardly a new example. Everything will be exactly the same as for the quark-vector-gluon model; the only possible difference can be in the value of b, the constant in Eq. (6.15). I now announce the great discovery of the last year: b is <u>negative</u>.[30] (This is true whatever the simple Lie group.)

Thus, our previous analysis is turned on its head: Large negative t is replaced by large positive t, infrared by ultraviolet. There exists a family of renormalizable field theories for which the logarithms can be tamed in the ultraviolet region! In this region, we obtain, from lowest-order perturbation theory and the renormalization group, an approximation that gets better and better as Q gets larger and larger. Furthermore, we can improve on the approximation as much as we want, simply by doing more perturbation calculations to get the higher terms in the expansions of β, γ, and F.

6.4 Asymptotic Freedom

What we have discovered for pure Yang-Mills theory is a special case of a phenomenon called asymptotic freedom. A general renormalizable field theory is said to be asymptotically free if, for small g^a,

$$\lim_{t\to\infty} g^{a\prime}(g,t) = 0. \qquad (6.18)$$

All my remarks for pure Yang-Mills theory carry over without alteration to a general asymptotically free theory; in particular, asymptotic behaviour for large Q is exactly computable from simple perturbation theory and the renormalization group. In principle, it is simple to test whether any given field theory is asymptotically free; all one needs to do is compute the β-functions to lowest non-vanishing order, and then solve the differential equations (6.12). In practice, the test is difficult to carry out; the computation of the β-functions is straightforward, but, in the typical case, the differential equations can not be solved analytically, and one has to resort to tedious case-by-case numerical integration. Thus, although many asymptotically free theories have been discovered, and a few general theorems have been proved, we have nothing like a complete classification of asymptotically free theories. I will tell you what is known about the classification problem shortly; first, though, I would like to convince you that asymptotic freedom offers a possible explanation of Bjorken scaling.

At first thought, this is a preposterous suggestion. Asymptotic freedom is a property of field theories for small coupling constants, and Bjorken scaling is a strong-interaction effect. Nevertheless, it is possible, with a little hand-waving, to establish a connection. For simplicity, let us consider an asymptotically free theory with only one coupling constant, g. By assumption, β is negative for small positive g. Let us denote the first positive zero of β by g_1. We certainly can not compute g_1 perturbatively; if we were asked to guess, we would probably guess that g_1 was something like 1 or π or maybe even infinity (if β has no zeroes). In any case, it certainly can not be a small number; for small coupling constants, we trust perturbation theory, and perturbation theory tells us β is negative. Whatever the value of g_1, for any g less than g_1, β is negative. Therefore, if we start from such a g, and integrate

$$dg'/dt = \beta(g'), \tag{6.19}$$

g' will decrease. As we continue to integrate the equation, it will continue to decrease, until we finally reach the region of small g', where formulas like Eq. (6.17) will be valid. Thus, the asymptotic expressions derived from renormalization-group-improved perturbation theory are valid for theories defined by large coupling constants as well as small. If we are very lucky, and β has no positive zeroes, they will be valid for all values of g.

The decrease from large to small g' can be quite rapid. As an example, let us take the result of the lowest-order perturbation theory, Eq. (6.16), and imagine that it is valid for large coupling constants as well as small. (I emphasize that this is undoubtedly false; I am just using it as a simple model of rapid decrease.) For a pure Yang-Mills theory with gauge group SU(3),

$$b = -11/16\pi^2. \tag{6.20}$$

Thus, Eq. (6.16) becomes

$$\left(\frac{g'^2}{4\pi}\right)^{-1} = \left(\frac{g^2}{4\pi}\right)^{-1} + \frac{11t}{2\pi}. \tag{6.21}$$

Now let us imagine that we start out with some very large value of $g^2/4\pi$, say 10^3, at $t = 0$. Then by going to $t = 1$ (that is to say, by increasing Q by a factor of e) we arrive at $g'^2/4\pi = 2\pi/11$. From this point on, the variation is quite slow; multiplying Q by e again merely halves g', and multiplication by e^2 is required to halve it again. Thus we are led to conjecture a qualitative picture in which a very large value of g' at low momentum zooms down with lightning rapidity to a small value, and then inches its way to zero.

What sort of asymptotic behaviour do we predict, once we are in the region of small g'? To evaluate Eq. (6.13), we need to know not only g', but also γ. For small g',

$$\gamma(g') = cg'^2 + O(g'^4), \tag{6.22}$$

where c is a numerical coefficient. From Eq. (6.17), for large t,

$$g'^2 \approx -1/2bt, \tag{6.23}$$

whence,

$$\gamma \approx -c/2bt. \tag{6.24}$$

Thus, the significant variation in Eq. (6.13) comes from the upper limit in the integral; for large t,

$$\overline{f} \approx K[\ln Q/M]^{-c/2b} \tag{6.25}$$

where K is a constant. This is <u>not</u> Bjorken scaling; the moments of the structure functions are not constants, but powers of logarithms. Nevertheless, a power of a logarithm is a very slowly varying function. I have not studied the SLAC-MIT data myself, but I am told by those who have looked at them (with optimistic eyes) that they can be fit as well with powers of logarithms as with constants.

Note that, for any given model, these powers can be computed by lowest-order perturbative calculations.[31] For example, the popular colored-quark model, with a color octet of vector gluons, is an asymptotically free theory with only one coupling constant. In this model, the moments of the isospin-odd (proton minus neutron) transverse structure function have the asymptotic form

$$F^{(n)} \propto (\ln Q)^{[-.296\ln(n+2)+.051]} . \tag{6.26}$$

These are rather small powers for small moments (-0.2 for n = 0), and grow slowly with n, reaching -1 only for n = 27. Of course, since we do not know the constant coefficients of the moments, we can not reconstruct the structure functions from formulae like Eq. (6.26). However, it is easy to construct functions whose moments obey Eq. (6.26) and which display quite small deviations from scaling except for x very near to 1. (The very high moments are obviously sensitive only to the behaviour of $F(q^2,x)$ in this neighborhood.)

If we accept asymptotic freedom as the explanation of Bjorken scaling, then, whatever the field theory of the strong interactions, it must be asymptotically free. (Bjorken scaling places no restrictions on the weak and electromagnetic interactions; these are negligible in the relevant energy region.) Thus, it is important to know what field theories are asymptotically free. Here is what we know now:

(1) All pure Yang-Mills theories based on groups without Abelian factors are asymptotically free.[30]

(2) Theories of non-Abelian gauge fields and Fermi multiplets are sometimes asymptotically free and sometimes not. The Fermions make a

positive contribution to the β-function; if the theory has too many Fermions, the sign of β is reversed and asymptotic freedom is lost. "Too many" is typically a large number. For example, if the gauge group is SU(3), sixteen triplets of Fermions are not too many.[30]

(3) Much less is known about theories of non-Abelian gauge fields and scalar multiplets; typically, these theories involve a large number of quartic meson coupling constants, and this makes the investigation of the differential equations difficult. There are some theories involving scalar fields which are known to be asymptotically free.[32] At the moment, there are no known asymptotically free theories for which all the gauge mesons may be given a mass by scalar vacuum expectation values. I do not view this as a serious difficulty, for two reasons: (1) The investigation is still in its early stages; such a theory may be found next week. (2) Even if no such theory is found, we are talking about models of the strong interactions; although the couplings may become weak at large momentum, they are certainly strong at small momentum, and this is where spontaneous symmetry breakdown occurs. Therefore, symmetry breakdown might well occur non-perturbatively, as discussed in Section 2.6.

(4) Any renormalizable field theory that does not involve non-Abelian gauge fields is not asymptotically free.[33]

This last result has far reaching consequences: If we accept asymptotic freedom as the explanation for Bjorken scaling, then the field theory of the strong interactions must be asymptotically free. If it is to be asymptotically free, then it must involve non-Abelian gauge fields. Since no one has ever seen a massless hadron, these gauge fields must acquire masses. The only known mechanism by which gauge fields can acquire masses is through spontaneous symmetry breakdown. Thus, <u>the field theory of the strong interactions must be a spontaneously broken gauge field theory</u>.

This is a striking conclusion, suggestive of deep connections between the strong and weak interactions. It implies a complete reversal of the conventional wisdom of only a few years ago. We used to believe that at high (Euclidean) energies the weak interactions became strong; now we believe that the strong interactions become weak.

6.5 <u>No Conclusions</u>

I know of no way to put a proper conclusion to these lectures, because I know of no way to judge the validity of the ideas we have discussed. They are certainly ideas of great beauty, and they certainly resolve many long-standing theoretical problems, but they equally certainly have not yet quantitatively confronted experiment. Spontaneously broken gauge field theories are in the uncomfortable position of SU(3) without the Gell-Mann-

Okubo formula, or current algebra without the Adler-Weisberger relation. There are good reasons for this, which I explained in the Introduction, but still one can't help feeling nervous. It is very possible that this whole beautiful and complex structure will be swept into the dustbin of history by a thunderbolt from Batavia. All we can do is wait and see.

APPENDIX: ONE-LOOP EFFECTIVE POTENTIAL IN THE GENERAL CASE

This appendix is a computation of the one loop effective potential, V, for a general renormalizable field theory. Such a theory contains three types of interactions: spinless-meson self-interactions, Yukawa couplings of mesons and Fermions, and gauge-field interactions. We shall see that in an appropriate gauge (Landau gauge), these three types of interactions contribute additively to the effective potential in one-loop approximation; thus,

$$V = U + V_m + V_f + V_g + V_{ct}, \tag{A1}$$

where the first term is the zero loop effective potential, the next three terms are the contributions from the three types of interactions, and the last term is a quartic polynomial in φ_c, the finite residue of the renormalization counterterms, determined once we state our renormalization conditions. The method of computation will be a direct generalization of the diagrammatic summation of Section 3.[34]

(1) Spinless-meson contribution. Here the analysis is almost identical to that of Section 3; the only difference is that there may be many meson fields. Thus each internal line in the graphs of Fig. 5 carries an index a, b, etc., labeling the meson field, and the black dots represent matrices for the transition from a meson of type a to one of type b:

$$i[\underset{\sim}{U}''(\underset{\sim}{\varphi}_c)]_{ab} = i\partial^2 U/\partial\varphi^a\partial\varphi^b|_{\varphi=\varphi_c} \tag{A2}$$

In computing the graphs, we must not only integrate over the internal momentum, but also sum over the internal indices. This is equivalent to multiplying the matrices around the loop and then taking the trace. Thus, from Eq. (3.33), we obtain

$$V_m = \frac{1}{64\pi^2} \operatorname{Tr}([\underset{\sim}{U}''(\underset{\sim}{\varphi}_c)]^2 \ln \underset{\sim}{U}''(\underset{\sim}{\varphi}_c)). \tag{A3}$$

(2) Fermion contribution. Here again the graphs are almost the same as in Fig. 5; the only difference is that the internal lines are Fermion lines. (Thus, you should imagine them as carrying arrows.) The relevant term in the Lagrangian is

$$\mathcal{L} = i\bar{\psi}^a \not\partial \psi^a + \bar{\psi}^a\, m_{ab}(\underset{\sim}{\varphi})\, \psi^b + \cdots \tag{A4}$$

Here m is the sum of two terms: a constant term (the Fermion masses) and a term linear in φ (the Yukawa couplings). It can also be broken into two parts in a different way:

$$\underset{\sim}{m} = \underset{\sim}{A} + i\underset{\sim}{B}\gamma_5 . \tag{A5}$$

(I use a Hermitian γ_5.) The reality of the Lagrangian implies that A and B are Hermitian matrices. I have chosen the name $\underset{\sim}{m}$ for this matrix because $\underset{\sim}{m}(\langle \underset{\sim}{\varphi} \rangle)$ is the Fermion mass matrix, in zero-loop approximation.

The computation can be made to look like the preceding one by grouping the terms in pairs:

$$\cdots m \frac{1}{\not p} m \frac{1}{\not p} \cdots = \cdots mm^{+} \frac{1}{p^2} \cdots \tag{A6}$$

Now the only differences between the Fermion computation and the Boson one are: (1) The combinatoric factor of ½ is missing because the lines have arrows on them, and thus the graphs are not invariant under reflections. (2) This is compensated for by the fact that the odd terms in the infinite series vanish when we take the trace on Dirac indices. (3) There is an overall Fermi minus sign. Thus we obtain

$$V_f = -\frac{1}{64\pi^2} \mathrm{Tr}([\underset{\sim}{mm}^{+}(\underset{\sim}{\varphi}_c)]^2 \ln \underset{\sim}{mm}^{+}(\underset{\sim}{\varphi}_c)) . \tag{A7}$$

Note that here the trace is on Dirac indices as well as internal indices.

(3) <u>Gauge-field contribution</u>. If we work in a general gauge, the trilinear coupling between gauge fields and spinless mesons can lead to troublesome graphs of the form shown in Fig. 6. (Here the straight line is a

Fig. 6

spinless meson, and the wiggly line a gauge field.) However, if we work in Landau gauge,

$$D_{\mu\nu} = i\frac{g_{\mu\nu} - k_\mu k_\nu / k^2}{k^2} \tag{A8}$$

these graphs vanish. This is because the external meson carries zero momentum; the sum of the meson momenta is the same as the gauge-field momentum, and gives zero when we contract it with the propagator (A8). Hence we need only worry about the quadrilinear coupling

$$\mathcal{L} = \cdots + \tfrac{1}{2} A_{\mu a} A^{\mu}_{b} M^2_{ab}(\underset{\sim}{\varphi}) + \cdots, \tag{A9}$$

where

$$M^2_{ab} = g_a g_b (\underset{\sim}{T}_a \underset{\sim}{\varphi}) \cdot (\underset{\sim}{T}_b \underset{\sim}{\varphi}), \tag{A10}$$

and g_a is the coupling constant of the a-th gauge field. $\underset{\sim}{M}^2(\langle\underset{\sim}{\varphi}\rangle)$ is the gauge-meson mass-squared matrix in zero-loop approximation, whence its name.

The computation is now identical with the preceding case, except that it is now gauge fields that run around the loop. Thus,

$$V_g = \frac{3}{64\pi^2} \mathrm{Tr}([\underset{\sim}{M}^2(\underset{\sim}{\varphi}_c)]^2 \ln \underset{\sim}{M}^2(\underset{\sim}{\varphi}_c)). \tag{A11}$$

The factor of three comes from the trace of the propagator (A8).

REFERENCES AND NOTES

1. There are many excellent reviews that can be used to rectify these lapses: A. De Rujula, in Proceedings of the First International Meeting on Fundamental Physics; E. S. Abers and B. W. Lee (to appear in Physics Reports); J. R. Primack and H. R. Quinn (to appear in the Proceedings of the Santa Cruz Summer School on Particle Physics); C. H. Llewellyn Smith (CERN preprint TH-1710); M. Veltman (to appear in the Proceedings of the Bonn Conference). This last contains a meticulous history.
2. This idea goes back to the classic work of Goldstone, Nambu, and Jona-Lasinio. J. Goldstone, Nuovo Cimento 19, 15 (1961); Y. Nambu and G. Jona-Lasinio, Phys. Rev. 122, 345 (1961); 124, 246 (1961).
3. Notation: The signature of the metric tensor is (+---); $\partial_\mu = \partial/\partial x^\mu$; summation over repeated indices is always implied.
4. J. Goldstone, A. Salam, and S. Weinberg, Phys. Rev. 127, 965 (1962).
5. D. Kastler, D. Robinson, and J. Swieca, Comm. Math. Phys. 3, 151 (1966).
6. F. Englert and R. Brout, Phys. Rev. Letters 13, 321 (1964); P. Higgs, Phys. Letters 12, 132 (1964); G. Guralnik, C. Hagen and T. Kibble, Phys. Rev. Letters 13, 585 (1964); P. Higgs, Phys. Rev. 145, 1156 (1966); T. Kibble, Phys. Rev. 155, 1554 (1967).
7. C. N. Yang and R. Mills, Phys. Rev. 96, 191 (1954); R. Utiyama, Phys. Rev. 101, 1597 (1956); S. Glashow and M. Gell-Mann, Ann. Phys. (N.Y.) 15, 437 (1961).
8. For some tantalizing recent explorations, see R. Jackiw and K. Johnson (MIT preprint), and J. Cornwall and R. Norton (UCLA preprint). See also R. Jackiw, this book.
9. J. Schwinger, Phys. Rev. 128, 2425 (1962); W. Thirring and J. Wess, Ann. Phys. (N.Y.) 27, 331 (1964); J. Lowenstein and J. Swieca, Ann. Phys. (N.Y.) 68, 172 (1971). This model has another amusing feature; it possesses a continuous (non-gauge) symmetry (chirality) that is not associated with a local conserved current, because of an anomaly. This symmetry is spontaneously broken but there is no Goldstone Boson, because there is no conserved local current.
10. For more details (and references) see my lectures, "Renormalization and Symmetry" in the Proceedings of the 1971 International Summer School of Physics "Ettore Majorana" (to be published).
11. Ward identities: A. Slavnov, Theo. and Math. Phys. 10, 99 (1972). Gauge-invariant cutoff: G. 't Hooft and M. Veltman, Nucl. Phys. 44B, 189 (1973). Renormalization: G. 't Hooft, Nucl. Phys. B33, 173 (1971); B35, 167 (1971); B. Lee and J. Zinn-Justin, Phys. Rev. D5, 3121, 3137,

3155 (1972). A detailed review is Abers and Lee (Ref. 1).

12. This, and much of what follows in this section, is plagiarized from S. Coleman and E. Weinberg, Phys. Rev. D7, 1888 (1973), which contains more details, references, and applications. The effective potential was introduced by Goldstone, Salam, and Weinberg (Ref. 4), and by G. Jona-Lasinio, Nuovo Cimento 34, 1790 (1964).

13. This can most easily be proved with the aid of functional integration; see the discussion sections.

14. There is a parameter, which we have set equal to one by our choice of units, that enters the theory in the same way as a; this is h. Thus, it is sometimes said that the loop expansion is an expansion in powers of h. Y. Nambu, Phys. Letters 26B, 626 (1966).

15. The computation of the one-loop effective potential can be done along the same lines for a general renormalizable field theory, involving arbitrary numbers of spinless mesons, fermions, and gauge fields, with hardly more labor. For the interested reader, the computation is done in the Appendix. The computation may also be done with functional integrals; this method avoids the infinite summation of diagrams. See Lee and Zinn-Justin (Ref. 11) and R. Jackiw (MIT preprint).

16. This is a result of K. Symanzik, Comm. Math. Phys. 16, 48 (1970).

17. This fits in with the loop expansion being an expansion in powers of h. (See note 14) You should not allow this discussion to obscure one big difference between particle mechanics and field theory: in particle mechanics, there is no spontaneous symmetry breakdown, even for a double-welled potential of the sort shown in Fig. 2. The difference can easily be seen in the simple quadratic approximation discussed above. In particle mechanics, if we approximate the two supposed degenerate ground states by harmonic-oscillator wave functions, they have a non-zero inner product. This induces mixing which breaks the degeneracy; only one linear combination of the two states is a true ground state. It is easy to make the corresponding approximation in field theory, if we put the theory in a box of volume V. A simple computation then shows that the inner product goes to zero exponentially as the volume goes to infinity, and (at least in this approximation) the degeneracy remains.

18. This phenomenon is well-known; it occurs in the classic calculation of the effective action for electrons coupled to a constant external electromagnetic field. [W. Heisenberg and H. Euler, Z. Physik 98, 714 (1936); J. Schwinger, Phys. Rev. 82, 664 (1961).] This is real for a magnetic field, but imaginary for an electric field; the reason is that,

in the presence of a constant electric field, the vacuum is unstable and decays into electron-positron pairs.

19. S. Weinberg, Phys. Rev. Letters 29, 1698 (1972).

20. The application of functional integrals to quantum mechanics is due to Feynman, and the standard physics text is R. Feynman and A. Hibbs, Quantum Mechanics and Path Integrals (McGraw-Hill, 1965). This follows Feynman's original definition of the functional integral. This is apparently different from the definition I will give, but the two can be shown to be equivalent. A good mathematical reference is I. Gelfand and N. Vilenkin, Generalized Functions, Vol. 4 (Academic Press, 1964).

21. The clearest derivation I know of is that of K. S. Cheng, J. Math. Phys. 13, 1723 (1972). Cheng evaluates the functional integral a la Feynman (Ref. 20), and shows that it defines the same dynamics as the Schroedinger equation with an appropriate ordering of the p's and q's.

22. The Hamiltonian form first appears in Appendix B of R. P. Feynman, Phys. Rev. 84, 108 (1951). It was subsequently rediscovered many times by many authors, but I can not find any reference where its dangerous ambiguities are discussed. Indeed, one commonly finds in the literature the false statement that the Hamiltonian form is a method of quantization that is invariant under general classical canonical transformations. (A possible exception is the work of B. S. DeWitt, Rev. Mod. Phys. 29, 377 (1957). Those portions of this paper that I can understand I believe to be correct.) The remarks in the text are the product of conversations with D. Gross, C. Callan, and S. Treiman.

23. L. Faddeev and V. Popov, Phys. Letters 25B, 29 (1969) and "Perturbation Theory for Gauge Invariant Fields", Kiev lecture available in English as NAL-Thy-57.

24. R. Arnowitt and S. Fickler, Phys. Rev. 127, 1821 (1962).

25. A good review of the anomalies is R. Jackiw, in Lectures on Current Algebra and its Applications (Princeton U. Press, 1970). Implications for gauge theories are discussed in D. Gross and R. Jackiw, Phys. Rev. D6, 477 (1972) and W. A. Bardeen, in Proceedings of the XVI International Conference on High Energy Physics, Vol. 2 (NAL, 1972).

26. K. Fujikawa, B. Lee, and A. Sanda, Phys. Rev. D6, 2923 (1972).

27. So many people have made important contributions to this standard lore that a fair set of references would be longer than Sec. 6.1. A good brief review is C. G. Callan in Proceedings of the International School of Physics "E. Fermi" Course LIV (Academic Press, 1972).

28. More details, and references, can be found in my lectures, "Dilatations", in the Proceedings of the 1971 International Summer School of Physics

"Ettore Majorana" (to be published).

29. For a detailed derivation of the solution, see Ref. 28.

30. D. Gross and F. Wilczek, Phys. Rev. Letters <u>30</u>, 1343 (1973); H. D. Politzer, <u>ibid</u>., 1346. For pure Yang-Mills theory, the result was known to G. 't Hooft in the summer of 1972, but not published by him.

31. For detailed computations see D. Gross and F. Wilczek, "Asymptotically Free Gauge Theories" and H. Georgi and H. D. Politzer, "Electroproduction Scaling in an Asymptotically Free Theory of Strong Interactions" (all to appear in Phys. Rev.).

32. Gross and Wilczek, Refs. 30 and 31.

33. S. Coleman and D. Gross, Phys. Rev. Letters <u>31</u>, 851 (1973).

34. The computation can also be done by functional integration; see Ref. 15.

DISCUSSIONS

CHAIRMAN: Prof. S. Coleman

Scientific Secretaries: G. Chu, B. Calahan, E. Tomaselli, F. Wilczek

DISCUSSION No. 1

- *THIRRING:*

I have two related questions. Firstly, in the Goldstone-type theories, the non-zero vacuum expectation value must be a constant even on a cosmic scale since the motion of the earth presumably does not affect particle masses, for example. On the other hand in ferromagnets the expectation value of the magnetization exhibits domain structure (i.e. is space-dependent). Why is there no domain structure in the Goldstone-type theories?

Secondly, if one takes account of the 3° background radiation in the universe, a Bose particle of mass zero has an associated wave-length which is short by macroscopic standards. How does this affect the vacuum structure?

- *COLEMAN:*

With regard to the first question, I am told that domain structure in a ferromagnet arises from the competition between the short-range Heisenberg exchange forces which tend to align the dipoles, and the usual long-range dipole forces which tend to anti-align them. There is no analogue to these long-range anti-aligning forces in the theories I have discussed. Any lack of alignment would contribute to the positive definite kinetic energy term in the energy density $[\frac{1}{2}(\partial_0\Phi)^2 + \frac{1}{2}(\vec{\nabla}\Phi)^2]$, moving it away from its minimum. It would be amusing to construct a field theory with domain structure in the vacuum, but I do not know offhand how to do it.

It is true that the 3° background temperature gives rise to localized fluctuations of relatively small wave-lengths. However, as long as one is restricted to a finite region of space and time, which is known as "the human condition", then one cannot get from one vacuum state to another by a localized fluctuation. In the ferromagnetic analogy we can change only a finite number of dipoles. Therefore we are still left with a unique physical Hilbert space with a unique vacuum state.

- *MENDEZ:*

A general comment should be made concerning the fact that you need at least two real scalar fields in order to obtain a Goldstone boson.

- *COLEMAN:*

That is right. Goldstone's theorem applies only to continuous symmetries. The breakdown of a discrete symmetry, like a reflection in one dimension, is not associated with the appearance of a Goldstone boson.

- *WAMBACH:*

You started from a Lagrangian with a given symmetry which was hidden. By observing the real world, how can we see the hidden symmetry?

- *COLEMAN:*

This is very difficult to do operationally. Of course if we have a theory that agrees with experiment then we know what is going on. But, of course, we have not reached that stage yet.

So, how can we find the hidden symmetry experimentally? We can take our mass in the magnet, where the background field is constant in space and time. We could perform high-frequency experiments to wash out the effect of the field.

A similar thing happens in relativistic field theory. The leading asymptotic behaviour of Green's functions is insensitive to the fact that the symmetry is broken. Unfortunately, this is asymptotic behaviour in an extreme unphysical region, far off mass shell.

So for practical purposes, the answer is no; we can not do experiments that directly see the hidden symmetry.

- *DRECHSLER:*

Do you really need fundamental scalar fields in order to get spontaneous symmetry breakdown?

- *COLEMAN:*

In nature we do know of spontaneously broken gauge theories without scalar fields, although they are not relativistic field theories. Superconductors are a good example. In theories of the type we are discussing, it is a bit harder. We can try to make a theory in which the Goldstone bosons arise as bound states of, for example, a fermion and anti-fermion. This is the kind of idea used by Nambu and Jona-Lasinio. Unfortunately, their theory is not renormalizable. There are more recent attempts by Johnson and Jackiw and by Cornwall and Norton, in which they get Goldstone bosons as bound states in a theory of massless fermions exchanging vector and axial vector abelian gauge mesons, but I do not understand this work well enough to be able to judge whether the approximations they make are sensible.

- *NAHM:*

I have two questions. Firstly, for an infinite ferromagnet, there is a Hilbert space in which the ground state has a plane boundary with spin-up to the left and spin-down to the right. Is there any analogue in the theories you are discussing?

Secondly, one may create a field-free region in an infinite ferromagnet and thereby detect the rotational symmetry. Is this conceivable in the Goldstone-type field theories?

- *COLEMAN:*

Such states also exist in classical continuum field theories; the only difference is that the boundary is not sharp but continuous. I suspect these states are unstable to particle production, like those discussed in the previous question, but at the moment I have no good argument for this; it is just a suspicion.

In the second situation a region $\langle\phi\rangle = 0$ would decay not only from the boundary but also from the interior region. This is analogous to the situation in which a constant electric field in a sufficiently large region produces pairs from the vacuum. I will demonstrate this in the next lecture.

- *GROMES:*

Could you explain what happens if the potential term of the Higgs fields has more than one minimum, say two. Do you expect two different types of solutions corresponding to the two minima, or only one type?

- *COLEMAN:*

Suppose that one of the minima (at $\phi = a$) is lower than the other (at $\phi = b$). The state of lowest energy then corresponds to the constant solution at $\phi = a$. There is another constant solution at $\phi = b$. At first glance, one might think that one could build a second theory using $\phi = b$ for the ground state, and considering only states obtained from it by local operations. However, this is false: starting from $\phi = b$ we can, by local operations, change the value of ϕ to $\underline{a}$ inside some large but finite volume. We gain energy in the interior of this volume since, by assumption, $V(b) > V(a)$. However, we lose energy on the boundary, because of the $(\vec{\Delta}\phi)^2$ terms. But by making the volume sufficiently large, we can always make the interior terms more important than the boundary terms. Thus $\phi = b$ is a "false vacuum"; it is not lower in energy than any state obtained from it by local operations.

- *KLEINERT:*

Where does the extra energy go when the false vacuum collapses?

- *COLEMAN:*

In the case of the atom, a transition from the 2s "false ground state" to the 1s "true ground state" is accompanied by the production of photons. In our case, the transition is accompanied by particle production.

DISCUSSION No. 2

- *NAHM:*

In your lecture you described how to construct $\Gamma(\phi_C)$ from a current interaction $J(x)$ and $W(J)$, the sum of all connected vacuum to vacuum amplitudes. Would the situation be changed with an interaction which allows tadpoles?

- *COLEMAN:*

No. This is not a problem. The formalism has been set up to be invariant under a shift of fields, so that if there are tadpoles, we can always shift the fields to where there are no tadpoles. Since there are obviously no problems with such a special shift, there are never any problems.

- *HENDRICK:*

In your definition of renormalizability you say a theory is renormalizable if the required counter-terms correspond to terms of the original Lagrangian, so that adding the counter-terms merely results in a renormalization of the parameters of the theory. Aren't there also constraints on the renormalized parameters?

- *COLEMAN:*

For theories with gauge invariances, there are usually relations between the coefficients of the different monomials in the Lagrangian. To show the theory is renormalizable, one must show that the counter-terms do not spoil these relations. This requires more than power counting; it is done with Ward identities. I will say more (but not much) about this in a later lecture, after we have the Feynman rules for Yang-Mills fields.

DISCUSSION No. 3

- *WILCZEK:*

Recall the measure in Euclidean form for the path integrals is, for ϕ^4 theory

$$\exp - \int \{(D\phi)^2 + \mu^2\phi^2 + \lambda\phi^2\} d^4x .$$

It would seem that by adding a ϕ^6 interaction the functional integral would converge even better, but we know that ϕ^6 has many troubles compared with ϕ^4.

- *COLEMAN:*

Yes, it looks like the integral is well defined, but we know this is a lie even for $\lambda\phi^4$ because λ must be made cut-off-dependent. The answer is the integral is indeed well defined -- it is zero. You can phrase a divergence problem in two ways; either $Z_3^{-1} = \infty$, or $Z_3 = 0$. The integral is so strongly damped that it is zero, and it gives you no information. You can see this by expanding ϕ in a Fourier series and truncating at some large finite number N of modes, then letting $N \to \infty$. If you do the integrals you find an infinite product of numbers less than 1, and the product is zero.

There is nothing in what I have done to cure the divergences of quantum field theory. The path integrals are more transparent and easy to manipulate than the corresponding statements from canonical quantization, but they are no more true.

- *WILCZEK:*

The integral looks positive-definite.

- *COLEMAN:*

The integral of a positive function can be zero if the set where it does not vanish is of measure zero. Function space is so enormous that sets that look enormous by ordinary standards are in fact of measure zero.

An example: let us take an infinite-dimensional space, as the limit of a finite-dimensional space, with the measure

$$\int_{-\infty}^{\infty} \cdots \int_{-\infty}^{\infty} \prod_{1}^{n} \frac{dx_i}{\sqrt{2\pi}} e^{-\frac{1}{2}(x,x)} .$$

Then if you look at an enormous box of side N, its volume is

$$\left[\int_{-N}^{N} \frac{dx}{\sqrt{2\pi}} e^{-\frac{1}{2}(x,x)} \right]^n .$$

No matter how large N is, this approaches zero as $n \to \infty$! The measure of the whole space, of course, is 1.

- *WILCZEK:*

How do you get a non-commutativity result like the canonical commutation relations from the path integrals?

- *COLEMAN:*

You do not. There are ways of twisting and perverting the formalism so you get results that look like non-commutativity, but really the path integrals are an independent method of generating Green's functions. It is as if someone gave you an integral representation for the Legendre polynomials. It would be hard to see the connection with the algebra of angular momentum.

A pseudo-profound reason for the lack of non-commutativity in the path integral formulation is that we carry out the integrals after continuation into Euclidean space, where everything commutes because all separations are space-like.

- *RABINOVICI:*

Are the masses of Goldstone bosons and "Higgsed" vectors affected by renormalization?

- *COLEMAN:*

There is no theorem which tells you the mass of the vector particle exactly. It certainly does not vanish in perturbation theory. There will, of course, be corrections of higher order to the bare mass.

In the case of spontaneous symmetry breaking without the vector gauge particles, the Goldstone boson mass is zero. This smells like a sacred number, and in fact one can show it is exact. One can show this from the axiomatic form of field theory; this is a beautiful argument but it would take too long to reproduce it here. One can also show it from our effective potential formulation. No matter how many orders of perturbation theory one computes the potential remains symmetric, so at an asymmetric vacuum there are always directions in which the potential is flat. Excitations in this direction, of course, correspond to a zero-mass particle.

- *RABINOVICI:*

Is it possible that the "Higgsed" vector particle could become massless?

- *COLEMAN:*

All I can do is perturbative analysis. It certainly does not look like it is going to be zero. I would be very surprised if it turned out to be zero unless you could give me a good reason for it.

- *WILCZEK:*

These would have a curious effect on the former Goldstone boson.

- *COLEMAN:*

Presumably the former longitudinal part of the vector boson would resume its independent existence, if the vector meson became massless.

- *RABINOVOCI:*

What about the genuine Higgs particles?

- *COLEMAN:*

These do not have any special properties in this connection as far as I know. They have a certain mass in zeroth order, which gets renormalized in the usual way.

- NAHM:

Could you show how one gets the generating function for OPI (one-particle irreducible) Green's functions by a Legendre transformation from the generating function of the connected Green's functions?

- *COLEMAN:*

I will show you how it works. You get the connected Green's functions by putting together the OPI Green's functions in tree graphs. Let us call the generating function for OPI Green's functions $\Gamma(\phi_c)$ and consider it for a moment as the action integral for some theory. Recall that we get the tree graphs from the zeroth order of the loop expansion, so that

$$\int (D\phi_c) \exp i[\Gamma(\phi_c) + J\phi_c]a^{-1} = e^{iW[J]}$$

to lowest order in a. We can calculate the left-hand side by using the method of steepest descent, since a^{-1} is getting large. The point of stationary phase is at

$$\frac{\delta\Gamma}{\delta\phi(x)} = -J(x)$$

and hence the exponential is

$$\Gamma(\phi_c) - \phi_c \frac{\delta\Gamma}{\delta\phi_c} ,$$

which is just the inverse of a Legendre transformation. That proves it!

DISCUSSION No. 4

- *WILCZEK:*

You mentioned that trust in the Fadeev-Popov technique for gauge theories is generated by the possibility of deriving the rules alternatively from canonical quantization. Does this possibility exist also for gravitation?

- *COLEMAN:*

The Feynman rules derived by Fadeev-Popov for Yang-Mills theory can be shown to be equivalent to the rules obtained by canonical quantization by tedious but straightforward calculations, using the gauge $A_3 = 0$.

In their paper Fadeev and Popov also derived rules for quantum gravitation. Their derivation involved the same ansatz procedure I described in the lecture. To my knowledge, there is no demonstration in the literature of the equivalence of their rules to those obtained from canonical quantization. However, they are probably right because many people, particular t'Hooft and Veltman, also got them from completely different starting points and with many consistency checks. At the one-loop level the Fadeev-Popov results also agree with what Feynman and DeWitt did a decade ago.

Of course gravity, unlike Yang-Mills theory, is not renormalizable simply because it contains a dimensional coupling constant.

- *WILCZEK:*

Maybe.

- *COLEMAN:*

One challenge, for people who like interesting and difficult problems with no physical applications, is to renormalize gravity. This investigation, now that we know the Feynman rules, is easier than it would have been in the past, but still not easy.

- *WILCZEK:*

I would like to return to the question of the connection between renormalizability and path integrals, especially the so-called "non-renormalizable" ones. The "non-renormalizable" theories are those which involve an infinite number of counter-terms. The procedure of adding counter-terms looks completely nonsensical in the path integrals. What do they mean?

- *COLEMAN:*

In general, the divergence structure of the theory is completely obscured when the theory is written in path-integral form. You have summed up too much. It is impossible (at least to me) to see essentially perturbative results, such as the fact that one needs new counter-terms in each order of perturbation theory. There may be a deeper meaning, but as far as I am concerned, the path integrals are simply a convenient device for representing Feynman amplitudes obtained from canonical quantization formally, in a way that is easy to manipulate.

- *WILCZEK:*

Does this suggest to you that renormalizability is essentially an artifact of perturbation theory, or do you think it has an intrinsic meaning?

- *COLEMAN:*

I think renormalizability has a profound meaning outside of perturbation theory. My guess is there is no way to make sense out of a non-renormalizable theory. All the constructive work that has been done on theories with trivial divergence structure, super-renormalizable theories in two-dimensional models, shows that the divergence structure is given <u>exactly</u> by the divergence structure of perturbation theory. <u>Exactly</u>! I am willing to stick my neck out and say that this result will also hold in the more difficult case of merely renormalizable theories, not yet investigated. Therefore, I think that non-renormalizable theories are absolute nonsense, not merely perturbative nonsense. This statement of course is sheer chutzpah.

- *WILCZEK:*

I would like to hear your remarks about the possibility of spontaneous symmetry breaking without fundamental scalars.

- *COLEMAN:*

This would be an extraordinarly interesting thing to do. Our problem is in a sense the opposite of the problem of dual resonance modelers. We have too many theories, they have too few.

Remember the problems we started out with: the problem of non-renormalizability of the weak interactions, the problem of infinite mass differences within isotopic multiplets. These are purely theoreticians' problems. If you do not bother to calculate the corrections to the Fermi theory of weak interactions, the theory works perfectly for every problem where it is expected to apply. Theoreticians tell you the next term is infinite, but you say it is $G^2 \cdot \infty$ where G is the Fermi constant, so it is very small. The electromagnetic mass differences diverge on the high-energy end. But once you have some kind of convergence factor the major contribution is on the low-energy end. Glashow and I, nine years ago, showed how you take care of that. We made a phenomenological tadpole model which predicted electromagnetic mass differences quite well, with one free parameter which essentially sops up the divergence. So once you construct any theory which solves these theoretical problems as far as its experimental consequences go it should be the same as the old theory, but free of the theoretical difficulties. You can throw in dozens and dozens of free parameters, throw in hordes of scalar and vector mesons at will, and get many many theories all of which meet the same criteria and all of which therefore agree with experiment. I exaggerate, but there is a very large family.

On the other hand, if a theory with only Yang-Mills fields and fermions displayed spontaneous symmetry breakdown, that is if Goldstone bosons emerged as bound states, then you would have theories which were much more constrained and therefore have much more predictive power. Such a theory might even be confronted with low-energy data.

The difference is that in the theories with fundamental scalars one has many more parameters: self-couplings, masses, Yukawa couplings with the fermions. In the theories with just fermions and Yang-Mills fields one has only one coupling constant for each simple factor of the gauge group and a number of discrete choices as to what representations the fermions transform under.

What we imagine is that the Yang-Mills fields create a force which binds two fermions into a composite Goldstone boson, which the Yang-Mills then eat, like ancient gods devouring their children. If you could construct a reliable computational scheme for studying such theories, you would have a class of theories that were much more constrained. Then you might be able to calculate some low-energy phenomenon and arrive at a number. The number might be wrong, but at least we would have some way of looking on directly to experiment.

The only recent work on this problem has been done by Jackiw and Johnson, and also almost identical work by Cornwall and Norton. The model is purely abelian, so it is totally unrealistic -- just a way of getting an idea of what is going on. It consists of a massless fermion (actually the fermion is doubled to avoid anomalies -- let us forget about that) coupled to two currents, a vector and an axial vector current. Why do you need two? This is simple. The two fermions must be bound in an s-wave to become the Goldstone boson. Such a state is necessarily odd in parity. Therefore, the only kind of vector meson that can eat it is an axial vector meson. Unfortunately, the force caused by the exchange of an axial vector meson between fermion and antifermion in an s-wave state is repulsive. So you need a vector meson to glue them together.

In a totally incomprehensible approximation (at least to me) JJCN using all sorts of fancy techniques, attempt to demonstrate that this phenomenon occurs. Whatever you think of their approximation, the result they get is interesting. What happens is that the axial vector mesons and the fermion acquire a mass. There is one relation between the four constants g_A, g_V, m_A, m_F. We started with g_A, g_V and wind up with three parameters: one free parameter has seemingly appeared from out of nowhere. I understand where this free parameter has popped out of. The approximations used by JJCN are absolutely, naively scale invariant, unlike real perturbation theory. When you think of the solution as being picked by minimization of an effective potential, there is a direction in which the potential is flat -- you change the mass scale, and nothing happens. You can put the minimum anywhere along this line. The free parameter is where you put the minimum.

Other than that, there has been no progress in the grand scheme of getting everything from fermions and Yang-Mills mesons.

Based on your computation, I have a hand-waving argument that there is a symmetry breakdown even in the pure non-abelian Yang-Mills theory without fermions.

I think it is very sure that spontaneous symmetry breakdown occurs in these theories. However, what the structure of the breakdown is, or how to compute it, is still a mystery.

- *KUPCZYNSKI:*

Does not your transformations of fields give the experimentalists different things to look for? Different external particles?

- *COLEMAN:*

No. Suppose I make the change $\phi = \phi' + g^2\phi'^2$. If in one case I calculate my Green's functions for ϕ, in the other case for ϕ', I get a different result. On the other hand we know from the general analysis of the axiomatists, and more simply from just looking at graphs, that what is important is not what choice of fundamental fields you take, but the total algebra of all local observables, the fundamental fields and all functions of them. The formula which gives you your S-matrix elements (the reduction formula of Lehmann-Symanzik-Zimmerman) is independent of your choice of fundamental fields, as long as the field in question has a non-zero amplitude for creating the particle in question from the vacuum.

A good exercise is to work this out for tree graphs in free-field theory, transformed as above, so it looks interacting.

- *KUPCZYNSKI:*

How complete is the proof that the quantization and renormalization of Yang-Mills theory can be carried through consistently, preserving gauge invariance?

- *COLEMAN:*

This proof seems in good shape according to the experts. As more careful-minded people become interested in the theory, no doubt some lapses from rigour will be found.

On a formal level the Slavnov identities are sufficient to show what you want to show, that the renormalization counter-terms do not spoil the algebraic relations between the coupling constants forced on you by gauge invariance. The theory also has a regularization procedure (the dimensional regularization of t'Hooft and Veltman) which at least in theories without axially coupled fermions does not spoil the gauge invariance. This kills the possibility of anomalies in that case. For fermions it is known that there can be the Bell-Jackiw-Adler anomalies which spoil the gauge invariance of the theory. There is a paper by Bardeen which claims to show that if you arrange the theory so that these anomalies cancel on the triangle level, then they cancel from the theory entirely and you get a renormalized, gauge-invariant theory.

Finally, a large number of explicit calculations have been performed, especially on the one-loop level, and no sign of new anomalies has shown up.

- *BYERS:*

What are good references for the Fadeev-Popov technique?

- *COLEMAN:*

A lengthy and lucid discussion is given in the NAL preprint of: V.N. Popov and L.D. Fadeev, IIP 67-63. Translated and edited by B.W. Lee and D. Gordon.

DYNAMICAL SYMMETRY BREAKING

R. Jackiw

Table of Contents

DYNAMICAL SYMMETRY BREAKING

R. Jackiw

Laboratory for Nuclear Science and Department of Physics,
Massachusetts Institute of Technology, Cambridge, Massachusetts 02139.

I. Introduction

An attractive idea for field theoretic models involving massive vector mesons is that this mass is not a parameter in the Lagrangian, but arises spontaneously from a breakdown of gauge symmetry. The reason why this circumstance is desirable lies in the fact that the free meson propagator vanishes at high energy in the latter case, but not in the former. The free propagator for a massive vector meson is O(1) at large energies.[1]

$$D_o^{\mu\nu}(k) = -i(g^{\mu\nu} - \frac{k^\mu k^\nu}{\mu^2}) \frac{1}{k^2-\mu^2} \underset{k\to\infty}{\to} \frac{i}{\mu^2}\frac{k^\mu k^\nu}{k^2}$$

If there is no mass term in the Lagrangian, the free propagator is $O(k^{-2})$ in the same limit.

$$D_o^{\mu\nu}(k) = -i(g^{\mu\nu} - \frac{k^\mu k^\nu}{k^2}) \frac{1}{k^2}$$

The decrease of the free propagator at high energies in the massless case allows for a consistent perturbative expansion of the theory. This technique is not in general available, if the propagator remains constant in the high energy region, because of well-known non-renormalizable divergences. (At present it is still an open question whether non-renormalizable vector-meson theories with mass parameters in the Lagrangian are intrinsically inconsistent, or whether only perturbation theory is inapplicable.) Currently, the great activity in the theory of weak interactions is centered about just such a possibility of spontaneous mass generation for the various vector mesons that are thought to carry weak forces.[2]

The fundamental reason why an apparently massless vector meson acquires a mass has been given a decade ago by Schwinger.[3] The reason is that the vacuum polarization tensor, $\Pi^{\mu\nu}$ — the proper two point correlation function of the conserved current J^μ to which the meson couples with strength g — acquires a pole at zero momentum transfer. To see this explicitly, consider the complete vector meson propagator, $D^{\mu\nu}(q)$.

$$D^{\mu\nu}(q) = -i(g^{\mu\nu}-q^\mu q^\nu/q^2)\, D(q^2)$$
$$= -i(g^{\mu\nu}-q^\mu q^\nu/q^2) \frac{1}{q^2-q^2\Pi(q^2)}$$

$$\Pi^{\mu\nu}(q) = (g^{\mu\nu}q^2 - q^\mu q^\nu)\ i\Pi(q^2)$$

(We insist that $\Pi^{\mu\nu}(q)$ be transverse, since the current is conserved; this entails a judicious choice of seagulls.) Clearly if $\Pi(q^2)$ has a pole at $q^2=0$, the vector meson is massive, even though it is massless in the absence of interactions ($g=0$, $\Pi=0$). We shall call this the Schwinger mechanism.

In these lectures I shall discuss three different examples of the Schwinger mechanism; that is I shall present three different circumstances which lead to a pole in $\Pi(q^2)$. Two are well known, and I shall be brief. They are Schwinger's original example of 2-dimensional, massless spinor electrodynamics[3] and the currently popular "Higgs mechanism".[4] The third example, which will be treated at length, is given in a recent paper by K. Johnson and me.[5] It is based on a bound state mechanism for symmetry breaking, similar to that discussed by Johnson, Baker and Willey in their approach to massless electrodynamics,[6] and is an extension of the work of Nambu and Jona-Lasinio[7] who gave a bound state realization of the Goldstone phenomenon.

<u>Massless Spinor Electrodynamics in Two Dimensions</u>

As a first example, we consider, following Schwinger,[3] a theory in 2-dimensional space-time described by the Lagrangian

$$L = i\bar{\psi}\partial\!\!\!/\psi - \tfrac{1}{4}F^{\mu\nu}F_{\mu\nu} + eJ^\mu A_\mu$$

$$F^{\mu\nu} = \partial^\mu A^\nu - \partial^\nu A^\mu$$

$$J^\mu = \bar{\psi}\gamma^\mu\psi \tag{2.1}$$

The complete vacuum polarization tensor in this theory is given by the lowest order term.

$$\Pi^{\mu\nu} = -e^2$$  (2.2)

All other, higher order graphs vanish. The integral represented by (2.2) is evaluated unambiguously, provided transversality on $\Pi^{\mu\nu}$ is imposed, as it must be since J^μ is conserved. (The graph (2.2) is ambiguous up to constant terms, proportional to $g^{\mu\nu}$; this ambiguity is removed by the transversality condition.)

$$\Pi^{\mu\nu}(q) = \frac{ie^2}{\pi q^2}(g^{\mu\nu}q^2 - q^\mu q^\nu)$$

$$\Pi(q^2) = \frac{e^2}{\pi q^2} \tag{2.3}$$

Hence the vector meson acquires a mass μ_A.

$$\mu_A^2 = \frac{e^2}{\pi} \tag{2.4}$$

The conservation of J^μ, an essential ingredient for the unambiguous evaluation of $\Pi^{\mu\nu}$, is a consequence of the fact that we are dealing with a gauge theory, where one equation of motion is

$$\partial_\nu F^{\mu\nu} = eJ^\mu \tag{2.5}$$

Since $F^{\mu\nu}$ is antisymmetric, consistency of the theory requires that

$$\partial_\mu J^\mu = \frac{1}{e}\,\partial_\mu \partial_\nu F^{\mu\nu} = 0 \tag{2.6}$$

There exists also a symmetry principle which by Noether's theorem implies the conservation of J^μ : the transformation law $\delta\psi = i\psi$ leaves the Lagrangian invariant and leads to $\partial_\mu(\bar{\psi}\gamma^\mu\psi) = 0$. However we prefer not to invoke this property of the theory, since it requires taking quite literally the canoncial formalism, the form of the Lagrangian, and the formula for the current $J^\mu = \bar{\psi}\gamma^\mu\psi$. All these may be beset by anomalies, and may not survive in the solutions to the theory. Of course in the case of the vector current, as long as the theory exists, i.e. provided equation (2.5) is true, we know that (2.6) must hold; there are no anomalies. Thus we view current conservation not as a consequence of a symmetry principle, but as a consequence of gauge properties.

Indeed L is apparently left invariant by chiral transformations, $\delta\psi = \gamma^5\psi$ ($\gamma^5 = i\gamma^0\gamma^1$); this leads to the conservation of the axial vector current: $\partial_\mu(i\bar{\psi}\gamma^\mu\gamma^5\psi) = 0$. However, the conservation law is not maintained in the solutions to the theory;[8] there is an anomaly of the axial vector current, quite analogous to the well known triangle anomaly of 4-dimensional physics.[9] This anomaly is also responsible for the spontaneous generation of the mass. We now show this.

Consider the complete vacuum polarization tensor of the vector currents.

$$\begin{aligned} T^{\mu\nu}(q) &= \int d^2x\, e^{iqx} \langle 0|T^* J^\mu(x)\, J^\nu(0)\,|0\rangle \\ &= iT(q^2)\,[g^{\mu\nu}q^2 - q^\mu q^\nu] \end{aligned} \tag{2.7}$$

In 2 dimensions, the axial vector current, $J_5^\mu = i\bar{\psi}\gamma^\mu\gamma^5\psi$, is related to the vector current by

$$J_5^\mu = \varepsilon^{\mu\nu} J_\nu \tag{2.8}$$

since $i\gamma^\mu\gamma^5 = \varepsilon^{\mu\nu}\gamma_\nu$. Hence one can compute matrix elements of J_5^μ from those of J^μ. Specifically

$$\begin{aligned} T_5^{\mu\nu}(q) &= \int d^2x\, e^{iqx} \langle 0|\, T^* J_5^\mu(x) J^\nu(0)\,|0\rangle \\ &= \varepsilon^{\mu\mu'}\, T_{\mu'}{}^{\nu}(q) \end{aligned} \tag{2.9}$$

Observe now that

$$q_\mu T_5^{\mu\nu}(q) = iq_\mu \varepsilon^{\mu\nu} q^2 T(q^2) \neq 0 \tag{2.10}$$

Thus, in spite of the formal symmetry of the Lagrangian, the axial vector current is not conserved, and the non-conservation can be summarized by[10]

$$\partial_\mu J_5^\mu = \frac{e}{2\pi} \varepsilon_{\mu\nu} F^{\mu\nu} \tag{2.11}$$

It is easy to see that (2.11) implies the existence of a massive particle in the theory. From (2.5) and (2.8) it follows that

$$\partial^\nu \varepsilon^{\alpha\mu} F_{\mu\nu} = eJ_5^\alpha \tag{2.12a}$$

while (2.11) further yields

$$\partial_\alpha \partial^\nu \varepsilon^{\alpha\mu} F_{\mu\nu} = \frac{e^2}{2\pi} \varepsilon_{\mu\nu} F^{\mu\nu} \tag{2.12b}$$

In 2 dimensions, $F^{\mu\nu} = \varepsilon^{\mu\nu} F$, hence (2.12b) is equivalent to

$$(\Box + \frac{e^2}{\pi})F = 0 \tag{2.12c}$$

which shows that $F^{\mu\nu}$ is a free field with mass $e/\sqrt{\pi}$.[11]

There are two lessons to be learned from this simple exercise. Firstly, Noether symmetry currents, as opposed to gauge currents, need not be conserved. Secondly, the spontaneous generation of a vector meson mass is associated with the breaking of an apparent invariance of the theory; in the present case it is chiral symmetry that is broken by the anomaly. Finally note that the complete meson propagator is transverse and well behaved at high energy, in spite of the fact that it describes massive vector particles.

$$D^{\mu\nu}(q) = -i(g^{\mu\nu} - \frac{q^\mu q^\nu}{q^2}) \frac{1}{q^2 - \frac{e^2}{\pi}} \tag{2.13}$$

III. The Higgs Mechanism

The second example is the currently popular Higgs mechanism.[4] Since this subject is so much discussed these days, I shall not repeat familiar material. Nevertheless I want to show how a pole in the vacuum polarization tensor arises in this context.

A Lagrangian which exhibits the Higgs mechanism involves scalar mesons with an imaginary mass parameter, e.g.

$$L = (\partial^\mu - ieA^\mu)\phi^* (\partial_\mu + ieA_\mu)\phi - \tfrac{1}{4}F^{\mu\nu}F_{\mu\nu} + \mu^2\phi^*\phi - g(\phi^*\phi)^2 \tag{3.1}$$

Since μ^2 occurs with the wrong sign, the only way to avoid a nonsensical theory is to assume that the vaccum expectation value of ϕ is non vanishing.

$$<0|\ \phi(0)\ |0> = <0|\phi^*(0)\ |0> = \lambda \neq 0 \tag{3.2}$$

Consider now the complete vector meson propagator, to lowest order perturbation theory.

$$\begin{aligned} D^{\mu\nu}(q) &= \int d^4x\, e^{iqx} <0|\ T\, A^\mu(x)A^\nu(0)\ |0> \\ &+ ie^2\int d^4x\, d^4z\, e^{iqx} <0|\ T\, A^\mu(x)A^\nu(0)\ \phi^*(z)\phi(z)A^\alpha(z)A_\alpha(z)\ |0> \\ &= D_o^{\mu\nu}(q) + 2ie^2\, D_o^{\mu\alpha}(q)\, D_o^{\beta\nu}(q) g_{\alpha\beta} <0|\phi^*(0)\phi(0)|0> \end{aligned} \tag{3.3}$$

(In Eq. (3.3) all fields are free.) The contribution of the vacuum expectation value of ϕ to $<0|\phi^*(0)\phi(0)|0>$ is λ^2. Since the free propagator can be chosen to be transverse, the contribution of λ to (3.3) is

$$2ie^2\ \lambda^2 D_o^{\mu\alpha}(q)\ \frac{(g_{\alpha\beta}q^2-q_\alpha q_\beta)}{q^2}\ D_o^{\alpha\nu}(q)$$

Thus the vacuum polarization tensor is seen to be, to this order

$$\begin{aligned} \Pi^{\mu\nu}(q) &= i\,\frac{2e^2\lambda^2}{q^2}\,(g^{\mu\nu}q^2-q^\mu q^\nu) \\ \Pi(q^2) &= \frac{2e^2\lambda^2}{q^2} \end{aligned} \tag{3.4}$$

A pole is indeed generated, which then leads to a vector meson mass μ_A.

$$\mu_A{}^2 = 2\lambda^2 e^2 \tag{3.5}$$

Of course the same result (3.5) is obtained by the conventional method of shifting ϕ by λ. The usual method also exposes massless excitations, which however decouple.

The characteristic feature of the Higgs phenomenon is the presence of scalar fields in the Lagrangian, which acquire a symmetry breaking vacuum expectation value. For two reasons these scalar fields are an unattractive aspect of the model. Firstly, there do not exist any physical particles which would correspond to the scalar fields. Secondly, it has been observed that renormalizable non-Abelian vector meson theories can become free at high energies, when there are no scalar fields with quartic interactions.[12] When scalar fields are present, it appears difficult to implement this very desirable feature. Also it should be remarked that the Ansatz $\lambda \neq 0$ is an assumption whose validity can only be checked in low orders of perturbation theory. It is by no means obvious that the complete theory possesses solutions with this property.

It is therefore useful to inquire whether a vector meson mass can be generated in a 4-dimensional model without invoking canonical scalar fields. The exploration of such a possibility will concern us for the remainder of these lectures.

The Bound-State Model

Imagine a theory with a massless Fermion field ψ and a neutral vector meson field A^μ, interacting through a conserved current J^μ. Masslessness of the Fermion insures formal chiral symmetry in this theory. However, we assume that the Schwinger-Dyson equation for the Fermion mass operator $\Sigma(p)$ has a symmetry breaking solution, $\{\gamma^5,\Sigma(p)\}\neq 0$. It is well known that this can happen if there is a massless, bound excitation in the Fermion anti-Fermion channel. Since a Fermion mass is being generated, this zero-mass excitation couples to $\bar{\psi}\psi$ or $\bar{\psi}\gamma^5\psi$.

Ultimately we shall want the zero-mass state to combine with the massless vector meson field and generate a massive meson. This can only happen if a transition between J^μ and $\bar{\psi}\psi$ or $\bar{\psi}\gamma^5\psi$ is allowed. Therefore, as long as charge conjugation invariance is not spontaneously broken, J^μ must contain the axial vector current.

We are thus led to consider a theory described by the Lagrangian

$$L = i\bar{\psi}\not{\partial}\psi - \tfrac{1}{4}F^{\mu\nu}F_{\mu\nu} + gJ^\mu_5 A_\mu$$

$$J^\mu = i\bar{\psi}\gamma^\mu\gamma^5\psi$$

$$F^{\mu\nu} = \partial^\mu A^\nu - \partial^\nu A^\mu \tag{4.1}$$

The axial-vector anomaly occurs in the above theory by virtue of the axial vector coupling[9]. For the time being I shall ignore this; the anomaly can be eliminated by introducing additional Fermions.[13] Indeed the anomaly must be removed; otherwise the equation of motion

$$\partial_\nu F^{\mu\nu} = gJ^\mu_5 \tag{4.2}$$

is not consistent with the antisymmetry of $F^{\mu\nu}$. (It is interesting that the axial-vector current, together with the anomaly -- which to be sure must be removed -- is central to the present discussion, just as it is central in Schwinger's 2-dimensional model.) I shall return later to questions associated with the anomaly.

The theory (4.1) is chirally invariant and renormalizable, in the sense that off-mass-shell Green's functions can be computed in perturbation theory. The normalization point will not be taken at $k^2=0$ because of possible infrared divergences; rather an arbitrary value $k^2=k_o^2$ will be chosen.

The proper vertex function $\Gamma^\mu_5(p,p')$ associated with J^μ_5 satisfies a Ward-Takahashi identity

$$q_\mu\Gamma^\mu_5(p,p+q) = \gamma^5 G^{-1}(p+q) + G^{-1}(p)\gamma^5 \tag{4.3}$$

There must not be any anomalous exceptions to this equation, since the current, a source of the gauge field, must be always conserved;see (4.2). In (4.3), G(p) is the complete Fermion Green's function, which is given by the following Schwinger-Dyson equation.

$$G^{-1}(p) = -i\left(\not p - \Sigma(p)\right)$$

$$\Sigma(p) = -ig^2$$ 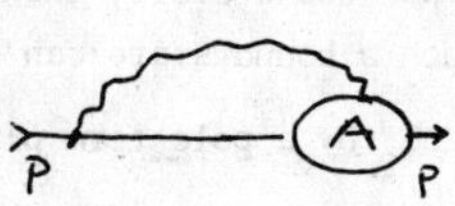

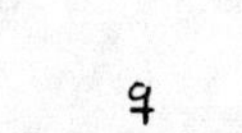 $= D^{\mu\nu}(q)$

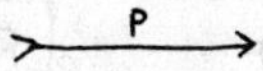 $= G(p)$

 $= i\gamma^\mu\gamma^5$

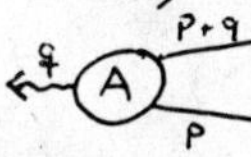 $= \Gamma_5^\mu(p, p+q)$ (4.4)

We assume a chiral symmetry breaking solution for Eq.(4.4) exists such that $\{\gamma^5;\Sigma(p)\}$ is not zero. Then (4.3) implies that $\Gamma_5^\mu(p,p+q)$ has a pole at q=0 with residue $i\Gamma_5(p)$, where

$$\Gamma_5(p) = \{\gamma^5,\Sigma(p)\} \neq 0 \tag{4.5}$$

How it can happen that $\Gamma_5^\mu(p,p+q)$ has a pole at q=0, which as I have shown guarantees a non-vanishing mass for the Fermion? To see this observe that the proper vertex function is also related to a portion of the Fermion-Fermion scattering amplitude

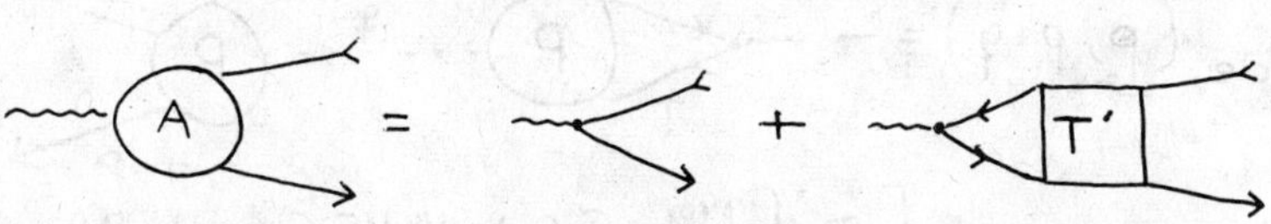

(4.6)

T' is the "one vector meson irreducible" Fermion-Fermion scattering amplitude, i.e. the full scattering amplitude T is given by

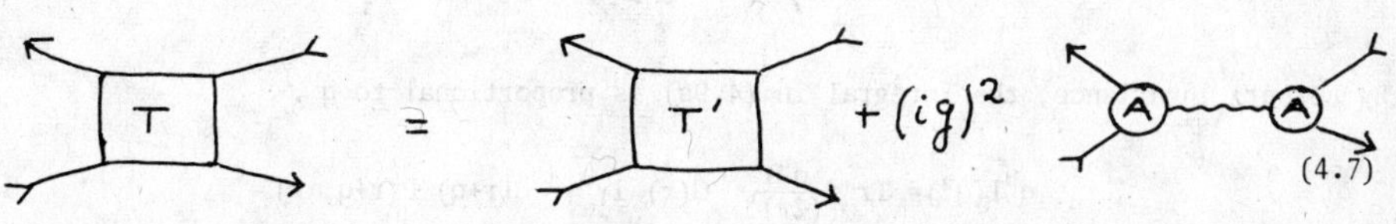

(4.7)

The reason why T' rather than T appears in (4.6) is that Γ_5^μ is the <u>proper</u> vertex function and does not contain graphs like

Our principal assumption is that the pole in $\Gamma_5^\mu(p,p+q)$ is to be attributed to a pole in T'. That is we assume that there exists a zero mass bound state in the Fermion anti-Fermion channel. In a sense this is the analog in the present theory of the assumption, made in the conventional approach to spontaneous symmetry breaking, that the vacuum expectation value of the scalar field is non-zero. First I shall work out the consequences of this zero mass bound state; then I shall turn to the much more difficult question whether or not such a bound state can exist in the theory.

Represent T' by a pole term plus a regular term R.

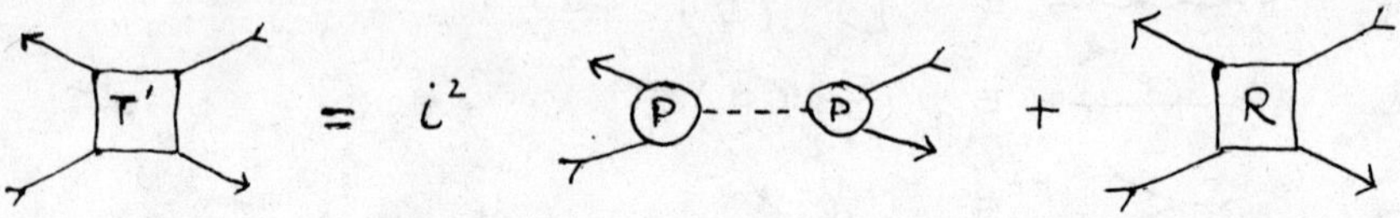

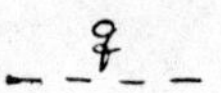

$$= \frac{i}{q^2} \tag{4.8}$$

The proper vertex which describes the coupling of the bound state to $\psi\bar{\psi}$ is represented by 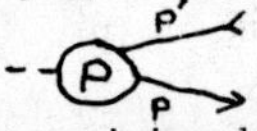 and is given by P(p,p'). (This set of circumstances is different from that envisioned for massless spinor electrodynamics. In the latter theory the electron acquires a mass spontaneously, but the Goldstone theorem is evaded due to anomalous non-conservation of the axial-vector current. In the present theory, as we have repeatedly stated, the axial vector current must be conserved, and the Goldstone theorem holds.)

Upon inserting (4.8) into (4.6), we determine the pole term in Γ_5^μ

$$\Gamma^\mu_{5\,\text{Pole}}(p,p+q) = -$$

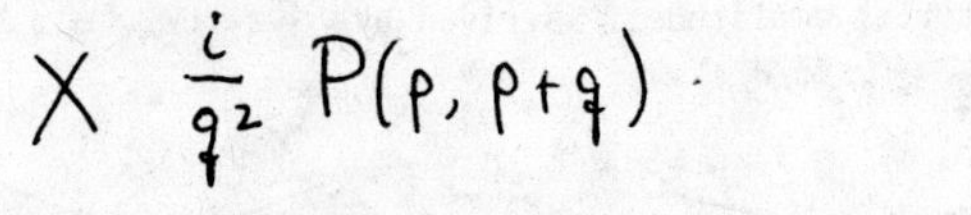

$$= -\left[-\mathrm{Tr}\int \frac{d^4r}{(2\pi)^4}\, G(r)\, i\gamma^\mu\gamma^5\, G(r+q)\, P(r+q,r)\right]$$

$$\times\ \frac{i}{q^2}\, P(p,p+q)\,. \tag{4.9a}$$

By Lorentz invariance, the integral in (4.9a) is proportional to q^μ.

$$q^\mu I(q^2) = \mathrm{Tr}\int \frac{d^4r}{(2\pi)^4}\; G(r)\; i\gamma^\mu\gamma^5\; G(r+q)\; P(r+q,\, r)$$

$$I(0) = \lambda \tag{4.9b}$$

Thus

$$\Gamma^\mu_{5\ \text{pole}}(p,p+q) = \frac{iq^\mu}{q^2} \lambda\, P(p,p+q) \qquad (4.9c)$$

and comparison with (4.3) yields

$$\Gamma_5(p) = \lambda\, P(p,p) \qquad (4.10)$$

Note that equation (4.9c) establishes the result that the singularity in Γ^μ_5 is a pole in q^2, as well as a pole in q. As a consequence $P(p,p+q)$ is ambiguous up to terms of $O(q^2)$; however, terms of $O(q)$ are well defined.

It is clear that $P(p,p+q)$ must be an odd parity vertex, and that $P(p,p)$ must be non-vanishing. Therefore the massless excitation must be a pseudoscalar, with some non-derivative coupling to the Fermion anti-Fermion channel. Also the integral (4.9b) defining λ must be non-vanishing. (It is here that charge conjugation invariance can prohibit spontaneous symmetry breaking: if $i\gamma^\mu\gamma^5$ transforms oppositely than $P(r+q,r)$ under C, then $q^\mu I(q^2) = 0$.)

Let us now examine the vacuum polarization tensor, which is given by the following Feynman-Dyson equation, apart from seagulls.

$$\Pi^{\mu\nu} = -g^2 \quad \text{[diagram with vertex A]} \qquad (4.11)$$

Since the vertex function in (4.11) has a pole, $\Pi^{\mu\nu}$ also develops this singularity, as is seen by inserting (4.9a) in (4.11)

$$\Pi^{\mu\nu}_{\text{Pole}} = g^2 \quad \text{[diagram with P P vertices]} \qquad (4.12a)$$

The integrals occurring in (4.12a) are already introduced in (4.9b). We find

$$\Pi^{\mu\nu}_{\text{pole}}(q) = g^2[q^\mu I(q^2)]\frac{i}{q^2}[-q^\nu I(q^2)]$$
$$\to -ig^2\lambda^2\, q^\mu q^\nu/q^2 \qquad (4.12b)$$

(The minus sign in the last factor in the first equation above arises as follows. The second pseudoscalar vertex in (4.12a) occurs with arguments $P(r,r+q)$, rather than $P(r+q,r)$ as in the definition $q^\mu I(q^2)$ in (4.9b). To make contact with (4.9b) the integration variable r is shifted to $r-q$, and q is replaced by $-q$.) Thus $\Pi(q^2)$, defined in (1.1) has the pole $\Pi_{\text{Pole}}(q^2) = \frac{g^2\lambda^2}{q^2}$, which indicates that the vector meson acquires a mass μ. In the approximation where only the pole term is kept, $\mu = g\lambda$. An exact theorem is $D^{-1}(0) = -g^2\lambda^2$.

The final task is to show that the zero mass pole decouples from the theory. The amplitude T' contains a massless pole; nevertheless, it is true that the full on-mass-shell scattering amplitude T does not possess this pole. To establish this important result, we combine (4.7) and (4.8)

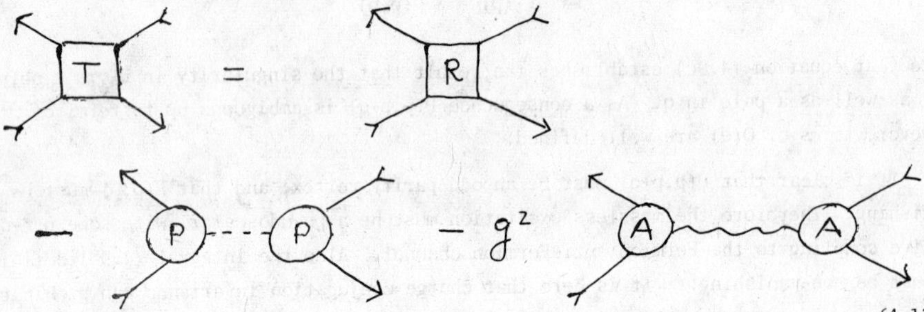

(4.13)

Only the last two terms in (4.13) contain poles at $q^2=0$. The external Fermions are on-mass-shell; thus the $q^\mu q^\nu$ term of the vector meson propagator in the last term in (4.13) gives zero contribution since the on-mass-shell vertex is transverse. Consequently $D^{\mu\nu}(q)$ may be set equal to $-i\, g^{\mu\nu} D(q^2)$. Furthermore if the left-most vertex function in the last term in (4.13) is decomposed into a regular piece and the pole term (4.9c), one sees that the pole term does not contribute: the pole is proportional to q^μ , which contracts with the right-most vertex function and annihilates it. Thus the terms in (4.13), which potentially contain a zero-mass pole, are represented by

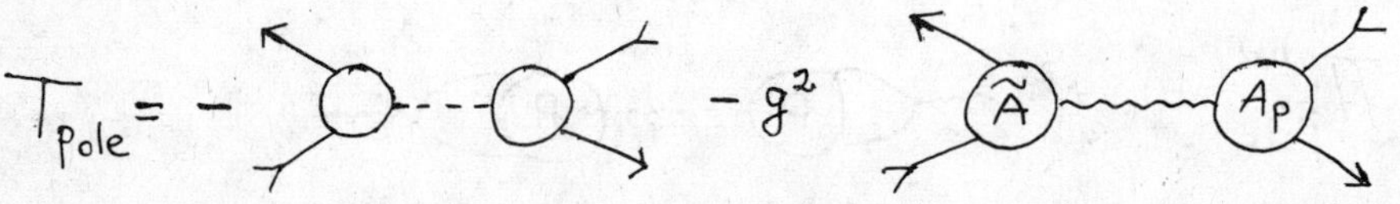

(4.14)

Here  $=\tilde{\Gamma}_5^\mu$ is the regular part of the axial vector current, and  $=\Gamma^\mu_{5\,pole}$ is the pole part given by (4.9).

We concentrate on the second term in (4.14), and set $\tilde{\Gamma}_5^\mu = \Gamma_5^\mu - \Gamma^\mu_{5\,pole}$. Since the remaining part of the graph is proportional to q^μ, Γ_5^μ does not contribute and there remains

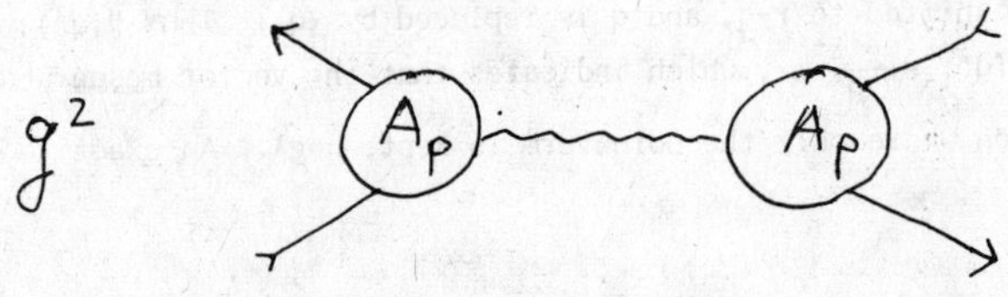

(4.15a)

When (4.9a) is inserted in (4.15a) for Γ_5^μ, the following sequence of equations now shows that the pole term of the above cancels against the first term in (4.14).

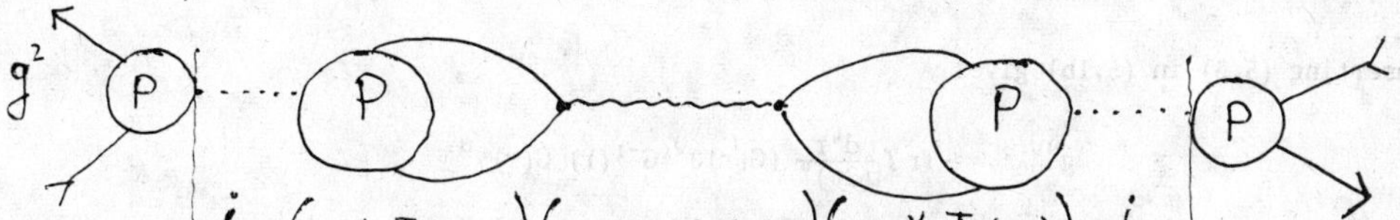

$$\frac{i}{q^2}\left(q^\mu I(q^2)\right)\left(-ig_{\mu\nu} D(q^2)\right)\left(-q^\nu I(q^2)\right)\frac{i}{q^2}$$

$$= -i\,\frac{I^2(q^2)}{q^2}\,D(q^2) \xrightarrow[q^2\to 0]{} -i\,\frac{\lambda^2 D(0)}{q^2} = \frac{i}{g^2 q^2}$$

$$=$$

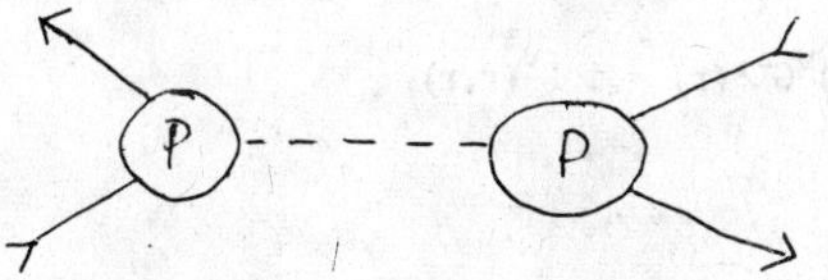

(4.15b)

The above is equal to the negative of the first term in (4.14), and T on mass shell is free from poles at $q^2=0$. This completes the discussion of our model.

A Mass Sum Rule

An interesting sum rule for the vector meson mass may be derived in our model. Let us begin with (4.9b)

$$\begin{aligned} q^\mu I(q^2) &= \mathrm{Tr}\int\frac{d^4r}{(2\pi)^4}\{G(r)i\gamma^\mu\gamma^5 G(r+q)P(r+q,r)\} \\ &= -\mathrm{Tr}\int\frac{d^4r}{(2\pi)^4}\{G(r+q)i\gamma^\mu\gamma^5 G(r)P(r,r+q)\} \end{aligned} \tag{5.1a}$$

$$\begin{aligned} g^{\mu\nu}\lambda = -\mathrm{Tr}\int\frac{d^4r}{(2\pi)^4}\{&\partial^\nu G(r)i\gamma^\mu\gamma^5 G(r)P(r,r) \\ &+ G(r)i\gamma^\mu\gamma^5 G(r)\frac{\partial}{\partial q_\nu}P(r,r+q)\big|_{q=0}\} \end{aligned} \tag{5.1b}$$

We also know from (4.3) and (4.9c) that

$$q_\mu\tilde{\Gamma}_5^\mu(p,p+q) = \gamma^5 G^{-1}(p+q) + G^{-1}(p)\gamma^5 - i\,P(p,p+q)\lambda \tag{5.2}$$

where $\tilde{\Gamma}_5^\mu(p,p+q)$ is the vertex function, <u>without</u> its pole at $q^2=0$. Hence it is true that

$$i\lambda P(p,p) = \{\gamma^5, G^{-1}(p)\}$$

$$i\lambda \frac{\partial}{\partial q_\nu} P(p,p+q)\Big|_{q=0} = \gamma^5 \partial^\nu G^{-1}(p) - \tilde{\Gamma}_5^\nu (p,p) \tag{5.3}$$

Inserting (5.3) in (5.1b) gives

$$\begin{aligned} g^{\mu\nu}\lambda^2 = -\mathrm{Tr} \int \frac{d^4r}{(2\pi)^4} \{ & G(r)\partial^\nu G^{-1}(r)\, G(r)\gamma^\mu \\ & - G(r)\partial^\nu G^{-1}(r)\, G(r)\gamma^\mu\gamma^5 G(r)\gamma^5 G^{-1}(r) \\ & + G(r)\gamma^\mu\gamma^5\, G(r)\gamma^5\partial^\nu G^{-1}(r) \\ & - G(r)\gamma^\mu\gamma^5\, G(r)\tilde{\Gamma}_5^\nu (r,r)\} \end{aligned} \tag{5.4a}$$

The middle two terms cancel against each other, while $\partial^\nu G^{-1}(r)$ may be related to the vertex function Γ^ν of the vector current.

$$\partial^\nu G^{-1}(r) = i\,\Gamma^\nu(r,r)$$

Hence

$$\begin{aligned} g^{\mu\nu}\lambda^2 = i\,\mathrm{Tr} \int \frac{d^4r}{(2\pi)^4} \{ & G(r)\,\Gamma^\nu(r,r)\, G(r)\gamma^\mu \\ & - G(r)\,\tilde{\Gamma}_5^\nu(r,r)\, G(r)\, i\gamma^\mu\gamma^5 \} \end{aligned} \tag{5.4b}$$

Recalling the representation of the vacuum polarization tensor by the Schwinger-Dyson equation (4.11), we recognize that (5.4b) is equivalent to

$$g^{\mu\nu}g^2\lambda^2 = -i[\tilde{\Pi}_A^{\mu\nu}(0) - \Pi_V^{\mu\nu}(0)] \tag{5.5}$$

$\tilde{\Pi}_A^{\mu\nu}$ is the vacuum polarization tensor associated with the non-singular, non-conserved axial vector current, while $\Pi_V^{\mu\nu}$ is the vacuum polarization of the conserved vector current, apart from seagulls.

Formally, if one ignores seagulls, $\Pi_V^{\mu\nu}(0)=0$, since the vector current is conserved. This is also seen from (5.4a) where the integrand in the first term on the right-hand side is a total divergence: $-\partial^\nu G(r)\gamma^\mu$. However, if $G(r)\to\frac{1}{r}$ for large r, $\Pi_V^{\mu\nu}(0)$ is in fact a quadratically divergent constant which cancels a similar quadratic divergence in $\tilde{\Pi}_A^{\mu\nu}(0)$. In other words, $\Pi_V^{\mu\nu}(0)$ is the seagull, which is necessary to convert the formal Feynman-Dyson equation for $\tilde{\Pi}_A^{\mu\nu}$ into a correct expression. Thus we may replace (5.5) by

$$g^{\mu\nu}g^2\lambda^2 = -i\,\tilde{\Pi}_A^{\mu\nu}(0) \tag{5.6}$$

where $\tilde{\Pi}_A^{\mu\nu}$ is now understood to include the requisite seagull.

The formula (5.6) may also be understood from (1.1) and (4.12b). We have

$$\Pi^{\mu\nu}(q) = (g^{\mu\nu}q^2 - q^\mu q^\nu)\ [i\ \tilde{\Pi}(q^2) + i\ \frac{g^2\lambda^2}{q^2}] \tag{5.7}$$

where $\tilde{\Pi}(q^2)$ is by definition regular. It is natural to identify $\tilde{\Pi}_A^{\mu\nu}$ with the regular part of (5.7). Thus the following, non-transverse expression is found

$$\tilde{\Pi}_A^{\mu\nu}(q) = i\ g^2\lambda^2 g^{\mu\nu} + (g^{\mu\nu}q^2 - q^\mu q^\nu) i\ \tilde{\Pi}(q^2) \tag{5.8}$$

and $-i\tilde{\Pi}_A^{\mu\nu}(0)=g^{\mu\nu}g^2\lambda^2$, which agrees with (5.6). Note also that $\tilde{\Pi}_A^{\mu\nu}(q)$, though non-conserved, has a conserved absorptive part.

Eq.(5.5) is also related to Weinberg's first sum rule.[14] If we identify $i\Pi_V^{\mu\nu}(0)$ with $g^2 g^{\mu\nu}$ times the Schwinger term S_V of the vector current commutator, and similarly for the Schwinger term S_A of the regular part of the axial vector current, then (5.5) reads $\lambda^2+S_A=S_V$. This is recognized as Weinberg's first sum rule, when it is recalled that the Schwinger term of the total axial vector current differs from that of the regular part by the current-pseudoscalar coupling, which in our case is λ.

VI. Problems with the Model

The specific model which I have discussed presents some problems which should be exposed. Some of these are specific to the example that I have chosen, and are trivially remedied. Others are more general.

The Lagrangian (4.1) is formally scale invariant. Consequently the emergence of mass terms appears to violate that symmetry and to lead to further massless excitations. It is clear that these potential Goldstone Bosons are spin-zero, positive parity objects. They could couple to the energy-momentum tensor, but not to a vector or axial vector meson.

However the above circumstances very likely do not occur, since scale invariance does not appear to be realized in the solutions of a field theory, in spite of the formal symmetry of the Lagrangian. The reason is not spontaneous violation, but rather the presence of anomalies, analogous to those of the axial vector current.[15] Thus we shall ignore any considerations of scale invariance.

Indeed it would appear that scalar massless mesons coupled to the energy-momentum tensor must be avoided. Such mesons would lead to a pole in the vacuum polarization tensor of two energy momentum tensors. If one then imagines coupling the energy-momentum tensor to a gravitational gauge field, the pole would produce a mass for the graviton, in a fashion entirely analogous to our previous discussions. Since massive gravity apparently is not realized in Nature, this state of affairs must be avoided.

Our model possesses two other currents which are formally conserved. These are $\bar{\psi}\gamma^\mu\gamma^5\psi^c$ and its Hermitian conjugate. Here ψ^c is the charge conjugate field $\psi_i{}^c=C_{ij}\bar{\psi}_j$; C is the charge conjugation matrix $i\gamma^2\gamma^0$ which satisfies $C=-\tilde{C}=-C^{-1}=-C^\dagger$, and which transposes the gamma matrices: $C^{-1}\gamma^\mu C=-\tilde{\gamma}^\mu, C^{-1}\gamma^5 C=\tilde{\gamma}^5$.

If these two currents are conserved in the solutions of the theory, one obtains Ward identities which at zero momentum transfer require that $G(p)\gamma^5 C+\gamma^5 C\tilde{G}(-p)=0$. By charge conjugation invariance of the theory, it is also true that $C\tilde{G}(-p)=G(p)C$. Hence the conservation of these odd currents requires massless Fermions: $\{\gamma^5,\Sigma(p)\}=0$.

Clearly this result is unacceptable, nor is its evasion with Goldstone particles satisfactory. We do not wish to deal with a theory which possesses massless, doubly charged scalars! The additional, unwanted symmetries may be disposed of by one of two devices.

Firstly, one may argue that in the solutions of the theory the currents are not conserved, again because of anomalies. The envisioned situation is analogous to the discussion of massless electrodynamics, where the electron acquires a mass spontaneously. The axial-vector current is formally conserved in that model, yet one can show that it is consistent that the equations determining the matrix elements of the axial vector current and of its divergence, admit a non-conserved solution. Unlike the anomalies of the triangle graph and of scale invariance which occur in a low order of perturbation theory, examples of this mechanism require infinite summation of graphs. I shall later show explicitly how the same mechanism which leads to a massless bound state can also generate this type of anomalous symmetry breaking.[16]

The second way of destroying the additional symmetry is by introducing vector couplings into the Lagrangian (4.1). This vector coupling can be a gauge coupling to the vector field A^μ: the interaction term in (4.1) is replaced by $\bar{\psi}\gamma^\mu(g'+ig\gamma^5)\psi A_\mu$. This leads to a parity violating theory. If one wishes to maintain parity, one can introduce an additional, massive vector field B_μ coupled to $\bar{\psi}\gamma^\mu\psi$. In the presence of a vector interaction, $\bar{\psi}\gamma^\mu\gamma^5\psi^C$ is no longer conserved. This modification has the additional attraction that scale invariance becomes explicitly broken, and one need no invoke anomalies to evade scalar Goldstone particles. Moreover, it will also be demonstrated that a vector interaction is required to produce a zero mass bound state in the weak coupling limit.

The last problem has already been mentioned: the theory is inconsistent because of the triangle anomaly.[9] In order to remove this contradiction, the number of Fermion fields must be increased.[13,17] We introduce a multiplet of n fields $\Psi = \begin{pmatrix}\psi_1\\ \vdots \\ \psi_n\end{pmatrix}$, coupled to the axial vector field by the interaction $i\bar{\Psi}\gamma^\mu\gamma^5 g\Psi A_\mu$. Provided the n x n Hermitian matrix g satisfies Tr $g^3=0$, the anomaly is absent. We shall limit the discussion to the simplest, two-Fermion case:

$$\Psi = \begin{pmatrix}\psi_1\\ \psi_2\end{pmatrix} \qquad g = \begin{pmatrix}1 & 0\\ 0 & -1\end{pmatrix} \tag{6.1}$$

The theory now possesses many conserved currents. Two, given by $\bar{\Psi}\gamma^\mu g\Psi$, and $\bar{\Psi}\gamma^\mu\Psi$, or $\bar{\psi}_i\gamma^\mu\psi_i$, i=1,2, insure the conservation of the individual species of Fermions. Fortunately this continuous symmetry is not broken even if we allow each of the Fermions to acquire a <u>different</u> mass, and a <u>different</u> coupling to the massless bound state. In addition to the Fermion number currents, the model also possesses doubly charged conserved currents as discussed in the previous subsection. These must be disposed of by one of the two methods mentioned above. Finally the axial vector current $\bar{\Psi}\gamma^{\mu}\gamma^{5}\Psi$, though beset by the triangle anomaly, can also lead to Goldstone Bosons. The reason is that the anomaly does not contribute to the Ward identity at zero momentum; while it is only the zero momentum Ward identity that is required to establish the existence of massless excitations. Thus conservation of this current must be broken by the first method discussed above, just as in massless quantum electrodynamics.

Existence of the Bound State

Now I must examine the most important question about our theory. Can the zero mass bound state, which is responsible both for the Fermion mass and for the vector meson mass, actually occur? To see what must transpire so that this bound state is present, let me return to the equation satisfied by the proper vertex function (4.6), and re-express it in Bethe-Salpeter form.

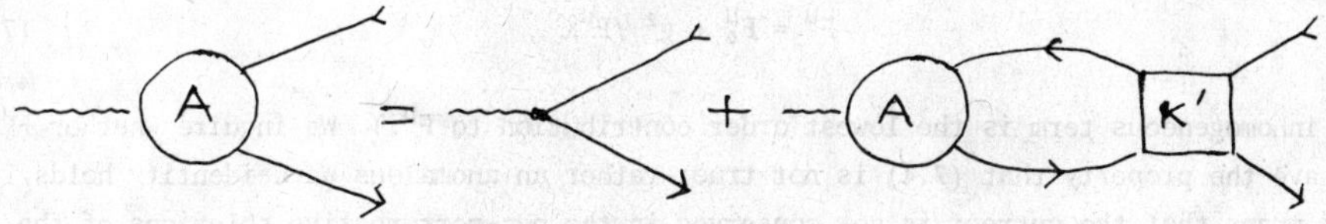

(7.1)

Here K' is the Bethe-Salpeter kernel, with one vector meson lines deleted (compare (4.7)). Decomposing Γ_5^{μ} into its regular and pole parts, we have, according to (4.9c),

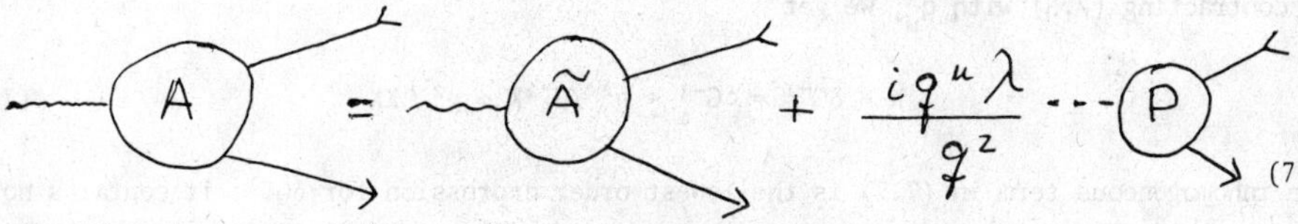

(7.2)

If (7.2) is substituted into (7.1), we find, by equating pole terms.

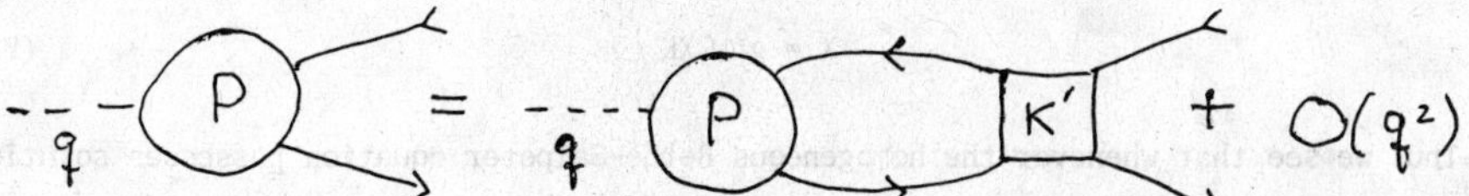

(7.3)

As is to be expected, the existence of the bound state at $q^2=0$, requires the solution of a homogeneous Bethe-Salpeter equation. The important feature of this equation is that the kernel is not Fredholm, since the theory is renormalizable and not super-renormalizable.[18] This has far reaching consequences. In particular, we shall argue that (7.3) is not an eigenvalue equation; it can be solved for arbitrary coupling constant. Consequently the solutions of (7.1) are not unique; one may always add a solution of the homogeneous equation. In particular (7.1) and (7.3) do not determine the normalization of P;

thus the Fermion mass is an arbitrary parameter in the theory.

Properties of the homogeneous Bethe-Salpeter equation may also be used to show the possibility of anomalous symmetry breaking, not by individual graphs as with the triangle anomaly and scale invariance, but rather by infinite sums of graphs summarized by the equation. (This is the symmetry breaking envisioned in the Johnson, Baker, Willey theory of massless electrodynamics.[6] It is also the mechanism which I mentioned in connection with the problems of additional, unwanted symmetries in our theory.)

Consider for example the proper vertex function F^μ associated with some current, which is formally conserved in the theory, as well as in any finite order of perturbation theory. The naive Ward identity is schematically given by

$$q_\mu F^\mu = \delta G^{-1} \tag{7.4}$$

and F^μ satisfies a Bethe-Salpeter equation of the form

$$F^\mu = F_0^\mu + g^2 \int F^\mu K \tag{7.5}$$

(The inhomogeneous term is the lowest order contribution to F^μ.) We inquire whether F^μ may have the property that (7.4) is not true; rather an anomalous Ward identity holds, which means that the current is not conserved in the non-perturbative solutions of the theory.

$$q_\mu F^\mu = X + \delta G^{-1} \tag{7.6}$$

By contracting (7.5) with q_μ, we get

$$X + \delta G^{-1} = \delta G_0^{-1} + g^2 \int G^{-1} K + g^2 \int XK \tag{7.7}$$

The inhomogeneous term in (7.7) is the lowest order expression for δG^{-1}; it contains no anomalies since the current is formally conserved. δG^{-1} satisfies its own Bethe-Salpeter equation, hence (7.7) implies that the anomaly X satisfies a homogeneous equation

$$X = g^2 \int XK \tag{7.8}$$

Thus we see that whenever the homogeneous Bethe-Salpeter equation possesses solutions; and furthermore when these solutions are not of the eigenvalue type but are always possible, which can happen when the kernel is not Fredholm, a formal symmetry can be broken. Since the normalization of X is not determined, these symmetry breaking solutions introduce new arbitrary parameters.

Let us now return to the problem of determining P, Eq.(7.3). We show that to lowest order in the coupling, solutions of (7.3) exist, for (almost) arbitrary coupling. For reasons that will emerge presently, we allow for massive vector mesons in our model, with vector coupling. (Recall that we found previously that vector couplings are needed to eliminate explicitly undesirable additional symmetries.) Thus,

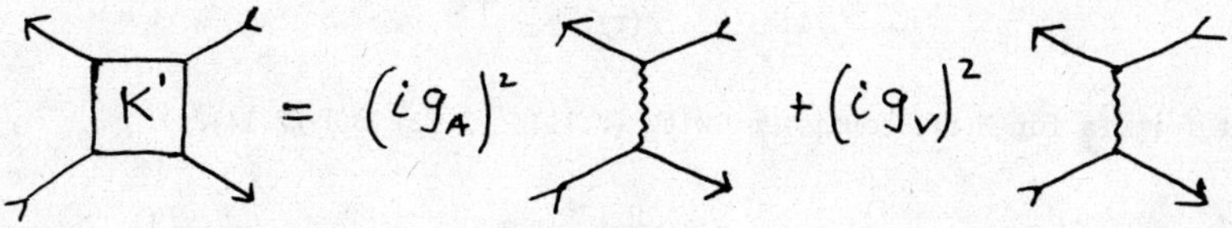

(7.9)

(The axial vector coupling constant has been called g_A; the vector, g_V.) For the propagators we take lowest order expressions, in the Landau gauge. However, it will be seen that the spontaneously induced masses are independent of coupling. Hence all mass terms are kept; they will not influence the result. We set $P(p,p)=\gamma^5 P(p^2)$, and $P(p^2)$ is found to satisfy the equation

$$P(p^2) = -3i \int \frac{d^4r}{(2\pi)^4} \frac{P(r^2)}{r^2-m^2}$$

$$\times\left[\frac{g_V^2}{(p-r)^2 - \mu_V^2} - \frac{g_A^2}{(p-r)^2 - \mu_A^2} \right] \tag{7.10}$$

The solution of this equation are known for large p^2 where μ_V and μ_A may be ignored.[19]

$$P(p^2) \underset{p\to\infty}{\to} (-p^2)^{-\varepsilon} \tag{7.11}$$

$$\varepsilon = \frac{3}{16\pi^2} [g_V^2 - g_A^2] \tag{7.12}$$

If $\varepsilon<0$, then the integral (7.10) would diverge and no solutions are possible. Clearly the only condition is that binding occurs ($g_V^2>g_A^2$); once this is achieved no further eigenvalue conditions are necessary.

The induced mass for the axial vector meson may also be computed. From the mass sum rule (5.4b) it follows that

$$g^{\mu\nu} \mu_A^2 = ig_A^2 \int \frac{d^4r}{(2\pi)^4}$$

$$\mathrm{Tr}\, \{G(r)\Gamma^\nu(r,r)\, G(r)\gamma^\mu - G(r)\tilde{\Gamma}_5^\nu(r,r)\, G(r)\, i\gamma^\mu\gamma^5 \} \tag{7.13}$$

Consistent with our lowest order approximation, we set

$$\Gamma^\nu(r,r) = \gamma^\nu \quad , \quad \tilde{\Gamma}_5^\nu(r,r) = i\, \gamma^\nu\gamma^5 \tag{7.14}$$

For G(r) we must leave a mass term, otherwise the integral in (7.13) vanishes identically

$$G(r) = \frac{i}{\not r - \Sigma(r)} \tag{7.15}$$

However it would be wrong to set $\Sigma(r)=m$, because the bare mass of the Fermion is zero,

$$\lim_{r\to\infty} \{\gamma^5, \Sigma(r)\} = 0 \tag{7.16}$$

The current formula for $\Sigma(r)$, consistent with (7.11), (7.12) and (7.14), is

$$\Sigma(p) = m\left(\frac{-p^2}{m^2}\right)^{-\varepsilon} \tag{7.17}$$

The integral (7.13) may now be evaluated. We find, for small g_V^2 and g_A^2,

$$\mu_A^2 = \frac{4}{3} m^2 \frac{1}{g_V^2/g_A^2 - 1} \tag{7.18}$$

Thus to lowest order in the coupling, our theory successfully generates masses for the Fermion and the axial vector meson. The former is an arbitrary parameter, which also determines the latter. Note μ_A^2 can become arbitrarily large as g_V^2 approaches g_A^2.

Beyond lowest order, it is of course very difficult to discuss solutions to (7.3). However a general analysis due to Johnson can be given,[20] which sheds some light on the possibility of solving that equation. Let us represent (7.3) schematically by

$$P(p^2) = g^2 \int d^4r\, K(p,r)\, P(r^2) \tag{7.19}$$

The reason that a Fredholm kernel requires an eigenvalue condition, can be understood heuristically: the validity of (7.19) requires that in some sense $g^2\int K=1$. For arbitrary g^2 this will not happen; a quantization condition on g^2 is necessary. However if $g^2\int K$ is divergent, but $g^2\int KP$ is convergent then factor of g^2 can be cancelled and plays no role in the problem (e.g. $g^2\int_1^\infty \frac{dx}{x} = g^2 \ln\infty$, $g^2\int_1^\infty \frac{dx}{x} x^{-g^2} = 1$).

Let us assume that the dominant contributions in (7.19) come from large r, because the kernel remains large in that region, and let us see whether an asymptotic solution to (7.19) for large p can be found. Upon continuing (7.19) to Euclidean space and averaging over angles, (7.19) becomes

$$\begin{aligned} P(-p^2) &= g^2\int_0^\infty dr\, r^3\, \bar{K}(p,r)\, P(-r^2) \\ &= g^2\int_0^\infty dx\, x^3\, p^4\bar{K}(p,xp)\, P(-x^2p^2) \end{aligned} \tag{7.20}$$

where the second equation follows from the first by a change of variable $r\to xp$. In the asymptotic domain of large p, $P(p^2)$ is replaced by $P_\infty(p^2)$. Also $P(-x^2p^2)$ may be set equal to $P_\infty(-x^2p^2)$, provided we assume that the region of small $x \sim 1/p$ is unimportant. Finally we assume that the kernel has the property that

$$\lim_{p\to\infty} p^4\, \bar{K}(p, xp) = f(x) \tag{7.21}$$

Then (7.20) becomes, in the asymptotic region

$$P_\infty(-p^2) = g^2\int_0^\infty dx\, x^3\, f(x)\, P_\infty(-x^2p^2) \qquad (7.22)$$

A solution of this equation can always be found.

$$P_\infty(-p^2) = (-p^2)^{-\varepsilon} \qquad (7.23)$$

provided that

$$1 = g^2\int_0^\infty dx\, f(x)\, X^{3-2\varepsilon} \qquad (7.24)$$

This is not an eigenvalue condition; rather it is an evaluation of ε in terms of g and the parameters present in f(x). The only requirement is that the answer for ε be such that the integral (7.24) converges.

The weak coupling approximation exhibits all the features which we have listed above: the kernel is non-Fredholm; (7.21) is satisfied; the infra-red region is unimportant; the significant contribution to the integral comes from the ultraviolet. Moreover, an analysis has been given of the complete Bethe-Salpeter kernel for vector meson theories without vacuum polarization.[21] All our hypotheses are valid. Thus there is some hope the requisite bound state survives beyond the weak coupling limit.

Phenomenological Lagrangians

The mechanism which we have discussed is attractive in that it dispenses with fundamental scalar fields as the agents responsible for a pole in the vacuum polarization tensor. Furthermore since the vector meson propagator has the form $-i(g^{\mu\nu} - \frac{q^\mu q^\nu}{q^2})[q^2 - q^2\Pi(q^2)]^{-1}$, its high energy behavior is no more divergent that that of the free propagator, and the theory should be renormalizable.

An obvious shortcoming of the theory, as developed so far, is that no effective method of computation has been found. Ordinary perturbation theory will never expose a bound state: to any finite order in g we shall always have massless Fermions and vector mesons.

One possible, though incomplete, approach is to describe the physical system by an effective Lagrangian. The effective Lagrangian, in tree approximation, should reproduce some of the dynamics of the complete theory. The description,necessarily limited to a low energy domain,should take into account the following features of the complete theory: (1) the excitation spectrum, (2) couplings of the states to each other, (3) the symmetries of the problem.

In the model considered by us, the excitation spectrum consists of a massive Fermion, massive axial vector meson and massless pseudoscalar meson, which, however, decouples from the theory. There may be other, massive bound states in the theory. Presumably these are important only at higher energies, and for a low energy phenomenology may be ignored.

The interactions of the theory involve Fermion-vector meson couplings, Fermion-pseudoscalar couplings, and vector meson-pseudoscalar couplings. The latter two, however, are removable by an appropriate choice of fields.

Finally, the symmetries which are to be maintained are vector and axial vector current conservation, and parity conservation. The axial current should be related to a gauge symmetry.

The above requirements lead to the following Lagrangian.

$$L = i\bar{\psi}\,\not{\partial}\psi + \tfrac{1}{2}\,\partial_\mu\phi\partial^\mu\phi - \tfrac{1}{4}F^{\mu\nu}F_{\mu\nu} + \tfrac{1}{2}\mu^2A^2$$
$$+ ig\,\bar{\psi}\gamma^\mu\gamma^5\psi\,A_\mu - m\bar{\psi}(\exp\tfrac{2g}{\mu}\gamma^5\phi)\psi - \mu A^\mu\partial_\mu\phi \tag{8.1}$$

It possesses a gauge symmetry

$$\delta\psi = \gamma^5\psi\theta$$
$$\delta\bar{\psi} = \bar{\psi}\gamma^5\theta$$
$$\delta\phi = -\frac{\mu}{g}\theta$$
$$\delta A^\mu = -\frac{1}{g}\,\partial^\mu\theta \tag{8.2}$$

It is not difficult to see that the massless field ϕ, present in (8.1), in fact decouples.[22] Change variables in (8.1)

$$\psi \to e^{-\gamma^5\frac{g}{\mu}\phi}\,\psi$$
$$\bar{\psi} \to \bar{\psi}\,e^{-\gamma^5\frac{g}{\mu}\phi}$$
$$A^\mu \to A^\mu + \frac{1}{\mu}\,\partial^\mu\phi \tag{8.3}$$

Then

$$L = i\bar{\psi}\not{\partial}\psi - m\bar{\psi}\psi - \tfrac{1}{4}F^{\mu\nu}F_{\mu\nu} + \tfrac{1}{2}\mu^2A^2 + ig\bar{\psi}\,\gamma^\mu\gamma^5A_\mu \tag{8.4}$$

This Lagrangian exhibits the physical spectrum of excitations: a massive Fermion and vector meson interacting through the axial vector current.

The phenomenological description of our theory does not lead to a renormalizable Lagrangian. This is seen either from (8.1), where the non-renormalizability resides in the non-polynomial Fermion-pseudoscalar interaction, or from (8.4) where it is the non-conservation of $i\bar{\psi}\gamma^\mu\gamma^5\psi$ that is responsible for the divergent, presumably unphysical, high-energy behavior of the theory. We do not consider the rapid growth at high energy as a defect of the fundamental theory. The phenomenological description is not meant to extend to high energies; in particular one does not expect to use the Lagrangians (8.1) or (8.4) for higher-order calculations.

IX. Conclusion

The interesting aspect of the present investigation is the demonstration, that vector particles can acquire a mass from a bound state mechanism, rather than from a vacuum expectation value of a canonical scalar field.

The present work should be extended to include internal symmetry. More importantly, an effective computational method should be developed, which bypasses ordinary perturbation theory, yet maintains renormalizability.

Especially provocative is the view, to which we have been led, that the theory is not uniquely determined by the Lagrangian. Rather, solutions of homogeneous integral equations with arbitrary constants, provide various inequivalent physical alternatives: generation of Fermion mass, anomalous non-perturbative violation of symmetry. It would obviously be most important to study this possibility further.

The physical relevance of our example is not apparent at the present time. Clearly a non-Abelian version may be used for weak interaction model building. More interesting is the possibility that the Abelian model may be relevant to pure strong interactions. A world with massless quarks and gluons which acquire their mass spontaneously without Goldstone Bosons is attractive for its economy. Moreover, we see in this picture the possibility of avoiding the problem of too much symmetry in the quark model. If chiral U(3) is spontaneously broken in the fashion described here, then the troublesome ninth Goldstone Boson disappears from the theory.

References

1) The notation is $g^{oo} = -g^{ii} = 1$, $\gamma^5 = \gamma^0\gamma^1\gamma^2\gamma^3$.

2) For a summary of the pioneering work of Salam, Schwinger, t'Hooft and Weinberg, see the review talk by B.W. Lee, Proceedings of the XVI International Conference on High Energy Physics, Batavia (1972).

3) J. Schwinger, Phys.Rev. 125, 397 (1962) and 128, 2425 (1962).

4) P.W. Higgs, Phys.Letters 12, 132 (1964), Phys.Rev.Letters 13, 508 (1964) and Phys.Rev. 145, 1156 (1966); F. Englert and R. Brout, Phys.Rev.Letters 13, 321 (1964); G.S. Guralnik, C.R. Hagen and T.W.B. Kibble, Phys.Rev.Letters 13, 585 (1964).

5) R. Jackiw and K. Johnson, Phys.Rev. in press. Some aspects of this model were previously discussed by F. Englert and R. Brout, Phys.Rev.Lett. 13, 321 (1964); F. Englert, R. Brout and M.F. Thiry, Nuovo Cimento 43, 244 (1966). Also J.M. Cornwall and R.E. Norton, U.C.L.A. preprint, have developed an example very similar to ours.

6) K. Johnson, in 9th Latin American School of Physics, Santiago, Chile (1967), I. Saavedra editor, Benjamin, New York (1968). M. Baker and K. Johnson, Phys. Rev. D3, 2516 (1971); H. Pagels, Phys.Rev.Letters 28, 1482 (1972) and Phys. Rev. in press.

7) Y. Nambu and G. Jona-Lasinio, Phys.Rev. 122, 345 (1961); see also Y. Freundlich and D. Lurie, Nucl.Phys. B19, 557 (1970).

8) K. Johnson, Phys.Letters 5, 253 (1963).

9) S. Adler in Lectures on Elementary Particles and Quantum Field Theory, S. Deser, M. Grisaru and H. Pendleton, editors, The MIT Press, Cambridge, Mass.(1970); and R. Jackiw, in Lectures on Current Algebra and its Applications, by S. Treiman, R. Jackiw, and D.J. Gross, Princeton University Press, Princeton, N.J. (1972).

10) For a recent treatment, see S.S. Shei, Phys.Rev. D6, 3469 (1972).

11) This argument was developed in collaboration with D.J. Gross.

12) G. t'Hooft, Marseilles Conference on Gauge Theories (1972); D.J. Gross and F. Wilczek, Phys.Rev.Letters 26, 1343 (1973); H.D. Politzer, Phys.Rev.Letters, 26, 1346 (1973).

13) D.J. Gross and R. Jackiw, Phys.Rev. D6, 477 (1972); C. Bouchiat, J. Iliopoulos and Ph. Meyer, Phys.Letters 38B, 519 (1972).

14) S. Weinberg, Phys.Rev. Letters 18, 507 (1967). Weinberg has emphasized a connection between the Higgs phenomenon and spectral function sum rules.

15) K. Wilson, Phys.Rev. D2, 1473, 1478 (1970); S. Coleman and R. Jackiw, Annals of Physics (N.Y.) 67, 552 (1971).

16) Such anomalous symmetry breaking was discussed by A. Maris, V. Herscovitz and G. Jacob, Phys.Rev.Letters 12, 313 (1964); as well as by the authors of ref.6.

17) One can construct two anomaly-free theories without increasing the number of Fermions, by coupling the axial vector to the real or imaginary part of $\bar{\psi}\gamma^{\mu}\gamma^{5}\psi^{c}$. These theories however do not admit a conserved vector current.

18) M. Baker, K. Johnson and B.W. Lee, Phys.Rev. 133, B209 (1964).

19) R. Willey, Phys.Rev. 153, 1364 (1967).

20) K. Johnson, unpublished.

21) K. Johnson, R. Willey and M. Baker, Phys.Rev. 163, 1699 (1967).

22) The Higgs mechanism in non-polynomial Lagrangians has been observed by S. Weinberg, Phys.Rev. 166, 1568 (1968) and L. Faddeev, unpublished.

DISCUSSIONS

CHAIRMAN: Prof. R. Jackiw

Scientific Secretaries: S. Jackson, E. Hendrick, S. Elitzur, B. Cahalan

DISCUSSION No. 1

- *GROSSER:*

In the first example of mass generation which you gave this morning -- quantum electrodynamics in two dimensions -- one has γ_5 invariance and one can therefore formally derive a conserved axial vector current. Then it turned out that the current was not conserved. Consequently, something must be wrong with this derivation. This is not surprising since nobody trusts the Noether theorem in quantum field theory. Nevertheless I would like to ask you, can one see in this simple model at what point this formal derivation is wrong?

- *JACKIW:*

I refer you to the paper of Johnson and myself. The reason is that one has to be very careful when defining the axial vector current. You have to define it by

$$J_5^\mu = i\bar{\psi}(x + \varepsilon)\gamma^\mu\gamma^5\psi(x - \varepsilon) \exp ie \int_{x-\varepsilon}^{x+\varepsilon} A^\mu \, dZ_\mu \, .$$

Then you obtain:

$$\partial\mu \, J_5^\mu = \frac{e}{2\pi} \varepsilon^{\mu\nu} F\mu\nu \, .$$

- *EYLON:*

Can we be sure that any time we find a pole in $\Pi(q^2)$ at $q^2 = 0$, its residue will be positive so that we get a positive mass?

- *JACKIW:*

A pole of $\Pi(q^2)$ means that the absorptive part of $\Pi^{\mu\nu}$ is $C\,\delta(q^2)$, and by positivity we get a positive residue C.

- *EYLON:*

If we introduce a mass for the electron in two-dimensional QED, what will happen to the photon mass?

- *JACKIW:*

Nothing will happen. The photon will remain massless, since the axial vector current is not expected to be conserved.

- *EYLON:*

So, if we look at the mass of the photon as a function of m_e will it be singular at $m_e = 0$?

- *JACKIW:*

Yes.

- *EYLON:*

What is the reason for this singularity?

- *JACKIW:*

Computing the diagram  with mass m_e for the electron propagator, we get no pole at $q^2 = 0$.

- *KUPCZYNSKI:*

Is it true that to carry out in detail your discussion about the bound states appearing in your theory, you have to renormalize your theory?

- *JACKIW:*

Yes, but to extract the information of the existence of the pole in $\Gamma_5^\mu(p, p + q)$ for $q = 0$, I need only the information that the self-energy of the Fermion $\Sigma(p)$ has a constant piece to make $\{\gamma^5, \Sigma(p)\} \neq 0$.

- *KUPCZYNSKI:*

But $\Sigma(p)$ could be infinite.

- *JACKIW:*

I shall show in the next lecture that this is not so. In this theory the Fermions are massless. With no point masses, we should have no infinities.

- *CHOUDHURY:*

If I understand correctly, it seems to me that there will be clear experimental tests of your approach in $e^+e^- \to$ hadrons. If the vacuum polarization tensor $\Pi_{\mu\nu}$ has a pole at $q^2 = 0$, then $\sigma(e^+e^- \to \text{hadrons})$ will fall off as $1/s^2$ rather than $1/s$; the latter being the expected behaviour of the usual light-cone physics?

- *JACKIW:*

In the present model the meson must always be coupled to an axial vector current, which is not relevant to the process you mentioned.

- *RABINOVICI:*

In four-dimensional QED with a massless electron, what is the mass of the photon?

- *JACKIW:*

The mass of the photon remains zero.

- *RABINOVICI:*

Is this because of gauge invariance?

- *JACKIW:*

No, this is because the photon couples to a vector current, and is a vector meson with $C = -1$. Only an axial vector meson coupled to an axial current can acquire a mass.

- *GROMES:*

For the example of massless two-dimensional QED you showed that the propagator acquires a pole for positive q^2. But the propagator also still has a pole at $q^2 = 0$. What is the meaning of this pole? Does it correspond to a particle?

- *JACKIW:*

The term $q_\mu q_\nu/q^2$ is just a gauge effect without direct physical meaning. The vector current is conserved so the $q_\mu q_\nu$ does not contribute.

- *MARCULESCU:*

You proved that for massless QED in two dimensions the appearance of a mass through the vacuum polarization tensor could be related to the axial anomaly. Do you think that a similar relationship may occur for the triangle anomaly in four dimensions?

- *JACKIW:*

It would be nice to show that the triangle anomaly generates a mass. This has never been demonstrated.

- *VILELA MENDES:*

Anomalous divergences and anomalous Ward identities are two ways of looking at the same problem. Suppose that in your example of two-dimensional QED, one insists that the divergence is not anomalous but the Ward identy for $T_5^{\mu\nu}$ is an anomalous Ward identity. Could you show how one should proceed to display spontaneous mass generation from this point of view?

- *JACKIW:*

If I have

$$\langle 0|T\, J_5^\mu\, J^\nu|0\rangle = T_5^{\mu\nu}$$

and I set

$$q^\mu\, T_5^{\mu\nu} \neq 0 ,$$

I must interpret this as the fact that the divergence of $J_5^\mu \neq 0$. It cannot be interpreted in any other way. One might think an anomalous commutator could give you similar results. However, you can exhibit $T_5^{\mu\nu}$ completely and explicitly because the theory is solvable. Then one can study all the commutators using the BJL limit. You find that there are no unexpected anomalous commutators, only an anomalous divergence.

- *VILELA MENDES:*

You must also believe the BJL limit gives the right commutators.

- *JACKIW:*

The BJL limit is the definition of the commutator.

- *VILELA MENDES:*

Is your zero-mass bound state observable?

- *JACKIW:*

No it decouples from the S-matrix.

- *TZE:*

This question is raised by Patricio Vinciarelli and myself. Would you care to compose and contrast your model with that of Z. Nambu and Jona-Lasinio?

- *JACKIW:*

I have not completed a full discussion of my model. The model of Nambu and Jona-Lasinio is based on a four-fermion interaction. Since it is non-renormalizable, they must introduce a cut-off. In our theory the interaction is $eJ_5^\mu A_\mu$, which is renormalizable. The kernels which occur in the integral equations in the two cases are quite different in terms of their high-energy behaviour. Their kernel is Fredholm, ours is non-Fredholm. In our case the Fredholm alternative ceases to hold. The bound state problem ceases to be an eigenvalue problem. In the Nambu-Jones-Lasinio case, the bound state was an eigenvalue problem and one had to fulfil very delicate conditions. We are free of constraints; the choice is made by the theoriticians. The solutions are not unique. Not everything is determined.

- *WILCZEK:*

There are theories very similar to the one you presented, but involving non-abelian groups in which the asymptotic behaviour in the deep Euclidean region need not be assumed but are calculable. It gives logs instead of powers. What is the result you expect in this case?

- *JACKIW:*

I do not know. We have not looked at non-abelian fields. However, in the abelian case, the non-perturbative behaviour has also been investigated by us. The general point is the following: I am trying to solve an equation, roughly speaking, of the form

$$P(p) = g^2 \int K(p,r)\, P(r)\, d^4r \,.$$

One can attempt to understand intuitively why it is that very frequently this is an eigenvalue equation. This is because in some sense the quantity $g^2 \int K(p,r)\, d^4r$ has to be ~ 1, since P(p) and P(r) on the two sides have the same "strength" and the only chance to balance the equation is that the quantity be 1. Now, it is clear for arbitrary g^2 that this is not going to be 1. However, if this integral of the kernel is infinite -- which is the case in the non-Fredholm-type examples -- then there is a possibility of balancing for arbitrarily small g^2. This can happen if the high-energy behaviour of the kernel is logarithmically divergent, so after putting in the P(r) the integral becomes of order $1/g^2$. This is what happens in the ladder approximation. One can even see this quite generally. I can average over angles and look at the radial equation

$$P(p^2) = g^2 \int_0^\infty r^3\, \bar{K}(p,r)\, dr\, P(r^2) \; .$$

What cancels an infinitesimal g^2 in front of the integral is a large contribution from the integrand near the upper limit. Therefore, the critical point is the behaviour of P at large p^2. To study this question I re-scale $r = px$ and get

$$P(p^2) = g^2 \int_0^\infty x^3\, dx\, P(p^2x^2)\, P^4K(p,px) \; .$$

Now let p^2 become very large and write

$$P_\infty(p^2) = g^2 \int_0^\infty x^3\, dx\, P_\infty(p^2x^2) \lim_{p\to\infty} P^4K(p,px) \; .$$

The crucial assumption is that the limit exists and is a function of x, f(x). This is indeed the case in the Bethe-Salpeter ladder approximation in which you have two Fermion propagators giving each 1/p, and a boson propagator with $1/(r - p)^2$, so you have $1/p^4$ which gives the desired limit function of x.

Therefore, the asymptotic form of the equation is

$$P_\infty(p^2) = g^2 \int_0^\infty x^3\, dx\, P_\infty(p^2x^2)\, f(x) \; .$$

The solution of this equation has the form

$$P_\infty(p^2) = c\, (p^2)^{-\varepsilon}$$

Substituting into the equation, we get

$$1 = g^2 \int_0^\infty x^3\, dx\, f(x)\, x^{-2\varepsilon} .$$

This is not an eigenvalue equation. It is a determination of ε in terms of g^2 and the various parameters in f(x). The only condition is the convergence of the integral. So all that is needed is that the Bethe-Salpeter kernel behave as it does in a scale-invariant theory, and I suspect that even in the asymptotic-free theories this kernel behaves in the same fashion. Certainly, it behaves in that way in abelian-gauge theories. In such theories it behaves, to all orders, as $1/k^4$, provided vacuum polarization is ignored.

- *NAHM:*

How much freedom do you have in choosing the non-trivial solutions of your integral equation for $P(p^2)$? Is there only one free parameter which fixes the mass scale?

- *JACKIW:*

My impression is that the only freedom one has is an over-all normalization. This includes also the trivial solution corresponding to zero normalization.

- *CHU:*

You proved that there is no pole at $q^2 = 0$ in your model for the Fermion-anti-Fermion scattering amplitude. Can you prove that the pole really decouples from the complete S-matrix?

- *JACKIW:*

I have shown the cancellation for the four-point function. I expect that in this method of using Ward identities over and over again, you can continue the cancellation to the complete S-matrix. There is an alternative way to show the decoupling of the massless excitation by writing an effective Lagrangian. In this Lagrangian there are scalar bosons corresponding to the massless excitations on which the symmetry acts non-linearly. But these scalars can be gauged away in the same way as is done in the Higgs mechanism. In this way one can see the decoupling of the massless excitations from the theory. This Lagrangian, however, is non-renormalizable and can be used only to summarize the symmetry properties of the theory, rather than for higher-order loop calculations.

DISCUSSION No. 2

- *CHU:*

Your model requires additional Fermions to cancel the anomalies of the axial vector current. So you seem to have traded in Higgs scalars for extra Fermions. Does the existence of such Fermions require that they be experimentally observed?

- *JACKIW:*

First of all, we have not traded in Higgs scalars for extra Fermions, and in fact both the present theory and the Higgs theories require the cancellation of anomalies in order to maintain the Ward identities needed for renormalizability. I do not know whether such Fermions must be observable. In a recent preprint, Susskind et al., display a two-dimensional model in which, through the Schwinger mechanism, the Fermions apparently do not appear in the asymptotic states. One may also imagine that in a strong coupling theory, it may be difficult to isolate the Fermions from their associated cloud of gauge bosons.

- *DRECHSLER:*

How many parameters does your theory end up with?

- *JACKIW:*

The original Lagrangian has two coupling constants, and the mass of the Fermion then appears as a new parameter. This appears to violate Coleman's idea of conservation of parameters in going from the bare Lagrangian to the renormalized theory. However, one may think of the new parameter as analogous to an externally imposed boundary condition as in the solution of classical problems. In addition, the choice of renormalization point also defines a new parameter.

- *WILCZEK:*

But the renormalization group equations tell you that the theory is independent of the renormalization point. Even the Yang-Mills theories, in which one is constrained to renormalize off-mass-shell, is only a one-parameter family of theories.

- *JACKIW:*

The renormalization group equations do not say the theory is independent of the subtraction points, but only how a theory with one set of renormalization conditions is related to a theory with another set.

- *LOSECCO:*

Could you compare your non-Fredholm equation with the situation in the model of Nambu and Jona-Lasinio?

- *JACKIW:*

Their model has an intrinsic cut-off, say Λ^2, so that one would have an equation of the form

$$P(p^2) = g^2 \int_0^{\Lambda^2} K(p,r)\, P(r)\, r^3 dr \ .$$

Since the r integral is now finite, the integral cannot compensate as $g^2 \to 0$ to give a result of order 1. Their equation is therefore of Fredholm type.

HADRONLIKE BEHAVIOR OF THE PHOTON

J.J. Sakurai

Table of Contents

HADRONLIKE BEHAVIOR OF THE PHOTON*)

J. J. Sakurai
Department of Physics, University of California, Los Angeles, California. USA

1. INTRODUCTION

The theme of the Summer School this year is "Laws of Hadronic Matter." It may therefore be appropriate to ask: "How many of the theoretical ideas discussed at the International School of Subnuclear Physics since its founding on this beautiful site ten years ago have stood the test of time and achieved the status of being regarded as "laws"? Apart from hadron spectroscopy where major progress has been made through SU(3) and the quark-model systematics, what have we really learned in strong interaction physics?

In two-body reactions at high energies, regardless of whether you believe in the details of Regge pole models, there is no doubt that the properties of objects that can be exchanged in the t channel have something to do with the s dependence at small values of -t. In addition, concepts such as duality and exchange degeneracy seem to be here to stay. In multibody reactions, longitudinal momentum scaling à la Feynman and Yang, proposed prior to the confirming data, and transverse momentum limitation appear to provide a good starting point, and we have developed some intuition on what happens in the target and beam fragmentation regions. In addition we are beginning to understand various aspects of multiplicity distributions in terms of short range correlation, rapidity plateaus, etc.

Perhaps you cannot yet call these systematic features "laws." Cynics may say that they are more like a set of phenomenological rules that happen to work. In any case most people now agree that there are certain features of strong interaction phenomena shared by all hadron-induced reactions.

The main purpose of this part of my lecture course is to examine to what extent these features are also shared by photoproduction and electroproduction. A parenthetical historical note is in order here. I started my 1960 Annals of Physics[1)] paper by asserting that there are some hadrons (known only as strongly interacting particles in those days) which are like the photon. In the meantime, however, the principle of "nuclear democracy" has gained wide acceptance, which states that all hadrons are created equal before the eyes of G. F. Chew and that no hadron shall enjoy a special "aristocratic" status.[2)] So nowadays it appears more appropriate to ask whether the photon behaves like hadrons than to ask whether some hadrons behave like the photon.

Hadronlike behavior of the photon was first formulated quantitatively with reference to vector-meson dominance models. As you may remember, in the sixties there was a minor industry based on the vector-meson-photon analogy equation

$$M(\gamma + A \rightarrow B) = \sum_{\rho,\omega,\phi} (e/f_V)\, M(V^{(\lambda=\pm1)} + A \rightarrow B) \qquad (1.1)$$

where A and B are hadronic states and my definition of the γ-V constant is such that we

---

*) Work supported in part by the U. S. National Science Foundation.

insert em_V^2/f_V at each γ-V junction. For obvious reasons, however, it turned out that a direct comparison of the right and left sides of (1.1) was possible only in a limited set of reactions. Meanwhile, as mentioned earlier, we have succeeded in developing some intuition on what is reasonable or unreasonable in strong interaction physics. As a result, we can examine a much wider range of photoproduction phenomena to test whether or not photons behave like hadrons.

In my Erice lectures two years ago I mentioned that evidence in favor of hadronlike behavior of the photon comes from the following five general areas.

(i) The energy dependence of the total photoproduction cross section is very similar to that of the total pion-nucleon cross section.

(ii) The photoproduction of ρ mesons exhibits features characteristic of typical elastic meson-nucleon scattering.

(iii) Two-body "inelastic" photoproduction reactions are similar to two-body meson-induced reactions.

(iv) Photon-induced inclusive reactions are similar to meson-induced inclusive reactions.

(v) The A dependence of photon-nucleus cross sections exhibits shadowing (the "Bell-Stodolsky effect").

It is probably best to start discussing what is new in each of the five areas. We then turn our attention to spacelike photons and discuss the controversial question of whether the photon still behaves like hadrons even when q^2 is spacelike.

Let me emphasize that this is an experimental talk. At the same time I attempt no comprehensive review of the existing experimental data. The data I mention are illustrative and far from exhaustive.

2. TOTAL PHOTOPRODUCTION CROSS SECTION AND VECTOR MESON PHOTOPRODUCTION

It was remarked in my Erice lectures two years ago that the total photoproduction cross section, when appropriately scaled, is very similar to the pion-nucleon cross section both in the resonance region and in the Regge asymptotic region. There is no qualitative change since that time. In quantitative details, there is now impressively precise data from Daresbury.[3] The magic formula

$$\sigma_{tot}(\gamma p) = \frac{1}{220}\,\sigma_{tot}(\pi^o p)$$
$$= \frac{1}{220}\left\{\frac{1}{2}\left[\sigma_{tot}(\pi^+ p) + \sigma_{tot}(\pi^- p)\right]\right\} \qquad (2.1)$$

still works well except in the 1520 and 1680 regions as shown in Fig. 1 where the proportionality factor is $4.6 \times 10^{-3} = 1/217$.

The total photoproduction cross section is of interest in connection with the sum rule of Stodolsky and others[4] which relates it to the vector-meson photoproduction cross sections

$$\sigma_{tot}(\gamma p) = \sum_V \left[16\pi(e/f_V)^2 \left(\frac{d\sigma'}{dt}\right)^{\gamma p\to Vp}_{t=0}\right]^{1/2} \qquad (2.2)$$

where $d\sigma'/dt$ stands for the imaginary part contribution to forward vector-meson photoproduction. If we try to saturate this sum rule using the ρ, ω, and φ contributions alone, the resulting photoproduction cross section appears to be smaller by (20-25)%. I come back to this question in the second part of my lecture course.

The reaction

$$\gamma + p \to \rho^o + p \tag{2.3}$$

is the most important exclusive channel in high-energy γp collisions. This process, whose cross section comprises a fixed percentage -- (12-15)% -- of the total photoproduction cross section, is known to exhibit all the features we associate with diffractive processes. Two years ago I already emphasized similarity of this reaction to elastic pion-nucleon scattering[5]:

(i) The differential cross section $d\sigma/dt$ is roughly independent of s.

(ii) It exhibits a forward peak with a slope parameter of (6-8) GeV^{-2}.

(iii) The phase of the forward amplitude, as measured through interference between the Bethe-Heitler pair production process and (2.3) followed by $\rho \to e^+e^-$, is predominantly imaginary.

(iv) Helicity conservation holds in the s channel.

Experimentally there is not much new for this process since my last lectures two years ago. I would like to take this opportunity to remind you of the relevance of this reaction to hadronlike behavior of the photon using an argument originally due to Knies.[6] The only "theory" used in the argument is the Schwarz inequality and unitarity, neither of which is believed to be controversial. Let us first write down

$$|i(\Phi,(T-T^+)\Psi)| = |(T\Phi,\ T\Psi)| \le ||T\Psi||\ ||T\Phi|| \tag{2.4}$$

which is a straightforward consequence of the Schwarz inequality and unitarity. Let Ψ and Φ be a $\gamma_{isov}p$ and a $\rho^o p$ ($\lambda_\rho = \pm 1$) state of the same c.m. energy moving in the same direction. We can then rewrite (2.4) as

$$16\pi(p_{\gamma p}/p_{\rho p})_{c.m.} \frac{d\sigma'}{dt}(\gamma_{isov}p \to \rho p)\Big|_{t=t_{min}} \le \sigma_{tot}(\gamma_{isov}p)\,\sigma_{tot}(\rho p) \tag{2.5}$$

where it is assumed throughout that the ρ is transverse, i.e. $\lambda_\rho = \pm 1$. It is seen from (2.4) and (2.5) that forward ρ photoproduction measures "overlap" between the final states produced in γp collisions and the final states produced in ρp collisions. In this way we can test whether photon-induced final states are indeed similar to ρ induced final states. I may add that in deriving (2.5) vector-meson dominance has not been explicitly invoked. Pure ρ dominance says that the inequality relation (2.5) is replaced by equality.

Let us put the numbers in.[7] At 9 GeV/c the total γp cross section is 120 μb, and about 85% of it is believed to be due to isovector photons. As for the hadronic ρp cross section with $\lambda_\rho = \pm 1$, we may take it to be 25 mb as determined from the A dependence of ρ photoproduction off nuclei[8] and also from ρ photoproduction off deuteron.[9] For the imaginary part contribution to forward ρ photoproduction $(d\sigma'/dt)_{t=t_{min}}$ we use 90 $\mu b/GeV^2$ even though this quantity is a little sensitive to the manner in which we extract the ρ meson cross section. Defining an "angle" Δ between the state vectors $T\Psi$ and $T\Phi$, we have

$$\frac{\text{LHS of (2.5)}}{\text{RHS of (2.5)}} = \cos^2\Delta \simeq 0.7 \quad . \tag{2.6}$$

So we see that the overlap is large. To a first approximation the photon does produce hadronic states similar to the kind produced by the ρ meson. However, the overlap is not maximal even with generous allowance for the errors. This is not surprising; some room must be left for higher-mass isovector vector mesons.

3. SINGLE PION PHOTOPRODUCTION

We now turn to single charged-pion photoproduction

$$\gamma + p(n) \to \pi^+(\pi^-) + n(p) \qquad (3.1)$$

In the early days a great deal of activities centered around a comparison between $d\sigma/dt$ of this photoproduction reaction with that of the hadronic reaction

$$\pi^- + p \to \rho^o + n \qquad (3.2)$$

after selecting the right polarization states. A more general question we should be concerned with is: Are the dynamical mechanisms responsible for (3.1) and (3.2) similar? From this point of view it is just as important to compare the effective trajectories defined by

$$\frac{d\sigma}{dt} = F(t)\ s^{2\alpha_{eff}(t)-2} \qquad (3.3)$$

as $d\sigma/dt$ themsleves. So we examine whether the effective trajectories extracted from (3.1) and (3.2) are similar.

This subject has an interesting history. Around 1967, when the π^+ photoproduction reaction was first studied at SLAC up to 18 GeV, the first obvious trend in the data (apart from the extensively discussed forward spike for very small values of $-t$) was that the product $s^2\ d\sigma/dt$ [or $(s-m_p{}^2)^2\ d\sigma/dt$] is roughly independent of s for a wide range of t. In other words, the data suggested a flat effective trajectory $\alpha_{eff} \simeq 0$ independent of t. Some theorists hastily took this as evidence for a $J = 0$ fixed pole in photoproduction.[10] That this conjecture was premature can be best seen by looking at the effective trajectory deduced from $d\sigma/dt$ of np charge exchange

$$n + p \to p + n \quad . \qquad (3.4)$$

See Fig. 2 taken from Davies et al.[11] Admire how flat α_{eff} is! In as much as most people are unwilling to tolerate a fixed pole in the purely hadronic reaction (3.4), a flat effective trajectory can hardly be taken as evidence for a fixed pole.

Very recently there has been an interesting new development in this six-year old subject. Quite generally, if we study the photoproduction reaction (3.1) with polarized photons, it is possible to separate the natural parity exchange contribution and the unnatural parity exchange contribution to (3.1) using a well-known theorem due to Stichel.[12] Even though single-pion photoproduction has been extensively studied with polarized photons for many years at low energies, it was not until last year that measurements with polarized photons were carried out by a SLAC-Tufts collaboration[13] up to 16 GeV. The effective trajectory extracted for unnatural parity exchange is shown in Fig. 3. It is noteworthy that α_{eff} in this case is no longer flat but is just what we expected from a Reggeized pion with the standard slope.

We may naturally ask: Are similar trajectories seen in the hadronic analog (3.2)?

Here $\rho_{oo}\, d\sigma/dt$, $(\rho_{11} + \rho_{1-1})\, d\sigma/dt$, and $(\rho_{11} - \rho_{1-1})\, d\sigma/dt$ receive contributions from unnatural parity, natural parity, and unnatural parity exchange, respectively. Even though the hadronic data shown in Fig. 4 taken from Estabrooks and Martin[14] are not so conclusive, there is some suggestion that unnatural and natural parity exchange follow the kind of energy dependences expected from Reggeized π exchange and Reggeized A_2 exchange, respectively.

Another aspect of two-body photoproduction which has been clarified in the past two years concerns behavior at large angles. Apart from the sharp forward spike for $-t \lesssim 0.02$ GeV^2, the π^+ photoproduction cross section roughly follows

$$\frac{d\sigma}{dt} \sim s^{-2}\, e^{3.3t} \tag{3.5}$$

up to $-t \simeq 3\ GeV^2$ and

$$\frac{d\sigma}{du} \sim s^{-3}\, e^{u} \tag{3.6}$$

in the backward direction up to $-u \simeq 1.5\ GeV^2$. What happens in between has only recently been investigated.

For some time it was known that in elastic and charge-exchange pion-nucleon scattering there is a kind of central basin in the sense that $d\sigma/dt$ tends to be relatively flat at large angles ($\theta_{cm} \simeq 90°$) with an s dependence at fixed t/s going roughly like s^{-8}. To see whether a similar behavior is exhibited in photoproduction a spectrometer experiment was carried out by the Ritson group[15] at SLAC. Their result is shown in Fig. 5 where the photoproduction data at 5 GeV are displayed together with pion-nucleon data at similar energies. The first thing to be noted is that large-angle behavior in photoproduction is remarkably similar to that in meson-nucleon scattering. Even the absolute height of the central basin is essentially the same when scaled by 220, the ratio of the pion-nucleon to the photon-proton cross section. As for the s dependence of $d\sigma/dt$ at $\theta_{cm} = 90°$ the photoproduction data suggest $s^{-7.3\pm0.4}$ to be compared with $\sim s^{-8}$ seen in typical pion-nucleon experiments. So we may infer that photons behave like hadrons even in large-angle processes.

The empirical fact that the photoproduction results closely parallel those obtained in meson-nucleon scattering naturally suggests that a common dynamical mechanism is at work in both cases. It is therefore disconcerting that the only quantitative model for this class of processes -- the parton-interchange model of Gunion, Brodsky, and Blankenbecler[16] -- predicts different s behaviors for analogous hadronic and photoproduction reactions -- $s^{-6.5}$ for the photoproduction reaction (3.1) and s^{-8} (or going down even faster) for the hadronic reaction (3.2).

4. INCLUSIVE PHOTOPRODUCTION

It was known already two years ago that inclusive reactions induced by real photons are very similar to typical hadron inclusive reactions. The principle of limited transverse momenta, which is so characteristic of all hadron-hadron collisions, is known to apply here also.

Let us concentrate on the longitudinal momentum distribution. It was conjectured by

Benecke et al.[17] that at sufficient high energies the momentum distribution of slow hadrons as measured in the rest frame of the target exhibits a limiting behavior independent of incident energies. This statement, which in their formulation is shown to follow from the principle of "limiting fragmentation," can be proved to be kinematically equivalent to Feynman scaling with respect to $x \equiv p_{||}^{(cm)}/p_{||max}^{(cm)}$ in the target fragmentation region. For hadron-hadron collisions it is also established experimentally that target fragmentation is projectile independent in the sense that, when scaled by the total cross section, the longitudinal momentum distribution becomes not only independent of the incident beam energy but also independent of the nature of the incident beam.

It is of great interest to test whether these features are shared also by photon-induced reactions. Let us look at Fig. 6 where the data points for

$$\gamma + p \rightarrow \pi^- + \text{any} \tag{4.1}$$

come from work of the DESY Streamer Chamber Group.[18] We see that for slow π^- there is indeed a general agreement between the normalized photon data points and the curves for the π^- distributions in various hadron-proton collisions.

If we are willing to rely a little more on theoretical models, even a more striking comparison between photon-induced and meson-induced reactions can be made. Those familiar with "Muellerism" may recall that factorization of various Regge pole contributions demands definite relations among invariant cross sections in the target fragmentation region. An example of this, first considered by Chan, Miettinen, and Lam,[19] is

$$f(p \overset{\gamma}{\rightarrow} \pi^-) = (\gamma_{\gamma P}/\gamma_{KP})\ f(p \overset{K^+}{\rightarrow} \pi^-)$$

$$+ [(\gamma_{\gamma f} + \gamma_{\gamma A})/4\gamma_{KM}]\ [f(p \overset{K^-}{\rightarrow} \pi^-) - f(p \overset{K^+}{\rightarrow} \pi^-)] \tag{4.2}$$

where γ_{ab} stands for the coupling constant of particle a to Regge pole b at t = 0. From the total cross section data the relevant γ's are all known (γ_{KM} may stand for $\gamma_{K\rho}$, γ_{Kf}, $\gamma_{K\omega}$, and γ_{KA_2} all assumed to be equal by exchange degeneracy). So it is possible to make a zero adjustable parameter prediction for the photon-induced reactions (4.1) in terms of the hadron-induced reactions

$$K^{\pm} + p \rightarrow \pi^- + \text{any} \tag{4.3}$$

Figure 7 shows the photoproduction data of the SLAC-Berkeley-Tufts collaboration[20] together with the predictions outlined above. We find that the theoretical predictions from the K meson data obtained by an Athens-Democritos-Liverpool-Vienna collaboration[21] agree rather well with the photoproduction data points in both magnitude and shape.

So far we have looked at pions emitted in the backward hemisphere. In the forward direction the longitudinal momentum distribution for (4.1) is known to exhibit what is known as the "leading particle effect". This, of course, fits nicely with the picture that the photon beam is to be regarded partly as a ρ^o meson beam. See Fig. 8 again taken from the data of the SLAC-Berkeley-Tufts collaboration.[20]

It may be emphasized that the highest photon energy at which inclusive photoproduction has ever been studied is 18 GeV. We are therefore still far from being able to answer the very interesting question of how a central plateau develops at high energies. When (or if?) photon beams become available at NAL, it will be interesting to compare photon-induced inclusive reactions with meson-induced inclusive reactions especially in the central region.

The last item on the 1971 list is shadowing in photon-nucleus collisions. I would like to discuss this topic of shadowing in photoproduction together with shadowing in inelastic electron scattering.

5. EXCLUSIVE ELECTROPRODUCTION

We now turn to the spacelike region and ask whether the photon continues to exhibit hadronlike behavior even when q^2 is finite. Since most people now agree that real photons behave like hadrons, it has become customary to compare electroproduction ($q^2 > 0$) with photoproduction ($q^2 = 0$).

When we compare photoproduction with electroproduction, it is important to keep in mind the role of longitudinal photons ($\lambda_\gamma = 0$) dormant at $q^2 = 0$. Obviously electroproduction of a specific final state can be very different from its photoproduction analog if there are important longitudinal contributions. Even though the longitudinal contribution to total electroproduction is only $\sim$ 20%, for certain specific channels large longitudinal-to-transverse ratios have been observed. A classical example of this is single-charged-pion electroproduction

$$e^- + p \to e^- + \pi^+ + n \tag{5.1}$$

where large longitudinal contributions are not only expected theoretically but also observed experimentally via longitudinal-transverse interference. As discussed two years ago, both the q^2 and the t dependence of this reaction can be semiquantitatively accounted for within the context of the simplest form of a ρ dominance model.[22] One may say, however, that the success of ρ dominance in this reaction for low values of t is expected on the basis of electric Born type calculations (or calculations based on OPE with a "poor man's absorption" or other more sophisticated dispersion-theoretic calculations) as long as the pion form factor is dominated by ρ.

Both in photoproduction and in electroproduction the most important exclusive channel is $\rho^o p$. As mentioned earlier, at $q^2 = 0$ the ρp channel comprises about (12-15)% of total photoproduction. How this fraction changes with q^2 has been the subject of intensive investigations by various experimental groups in the past two years or so. The main conclusion agreed by all the groups is that the fraction of ρ relative to the total definitely decreases from its photoproduction value ($q^2 = 0$) as q^2 becomes finite. An example of this is shown in Fig. 9(a) where I present the data of the SLAC Hybrid Bubble Chamber Group which studied the "muonproduction" of ρ

$$\mu^- + p \to \mu^- + \rho^o + p \tag{5.2}$$

using a 16 GeV/c muon beam at SLAC.[23] Notice that around $q^2 \simeq 1$ GeV^2 the fraction of ρ has decreased to about 1/2 of the $q^2 = 0$ value. An important question that has not yet been settled is: Does this ratio decreases monotonically to zero, or is there a tendency for this ratio to flatten out at high q^2?

What are our theoretical expectations on this vital point? If there are just transverse ρ mesons, the simplest form of ρ dominance predicts a cross section going like

$$\sigma^{(\rho)} = \sigma^{(\rho)}_{q^2=0} \; e^{bt_{min}} m_\rho^{\,4}/(q^2 + m_\rho^{\,2})^2 \tag{5.3}$$

As will be discussed in the second part of my lecture course, the total virtual-photon-proton cross section is known to go down as a single power $\sim 1/(q^2+M^2)$ with $M^2 \simeq 0.4$ GeV^2. So with just transverse ρ's, the fraction of ρ relative to the total is predicted to decrease monotonically with increasing q^2.

The actual situation is probably more involved. This is because the observed decay angular distribution of the ρ meson definitely indicates that the longitudinal contribution is important in the q^2 range investigated so far. Now the simplest ρ dominance prediction[24)] for the longitudinal-to-transverse ratio for forward ρ electroproduction is

$$R^{(\rho)} = \xi^2 q^2/m_\rho^{\,2} \tag{5.4}$$

where ξ is the ratio of the λ = 0 to the λ = ±1 total ρp cross section. Up to q^2 of about 1 GeV^2, the data[23,25)] are not inconsistent with (5.4) where the fitted value of ξ^2 is typically ∿ 0.4. If we take this form seriously up to higher values of q^2, the produced ρ's are predicted to be predominantly longitudinal, and because of the extra q^2 factor in (5.4), the fraction of ρ relative to the total should flatten out. We need, of course, much higher statistics data to study high q^2 regions.

Another aspect of ρ electroproduction which received much publicity is the q^2 dependence of the slope parameter b. In Fig. 9(b) I again present the data of the SLAC Hybrid Bubble Chamber Group.[23)] It appears from this Figure that the slope parameter (denoted by A in the Figure) does decrease with increasing q^2 for fixed W (the invariant hadronic mass of the final system). This tendency has been observed to varying degrees by other groups also. In the simplest form of diffraction picture, the slope is proportional to the square of the interaction radius. So one may say that the γp interaction radius "shrinks" as the photon becomes more virtual.

I should add that this shrinking photon effect was suggested earlier by some QED calculations of Cheng and Wu.[26)] They noted that if we study the Delbrück scattering amplitude with the external photons replaced by spacelike virtual photons, the effective transverse momentum increases as q^2 of the photon increases. They then conjectured that this tendency is general and applicable to processes like ρ electroproduction.

Even though this shrinking photon effect is still controversial, let us accept, for the sake of subsequent discussion, that the interaction radius indeed becomes smaller as the photon becomes more virtual. This has an interesting implication on π^o electroproduction, as argued by Harari.[27)] It has been known for some time that dσ/dt for π^o photoproduction has a dip at $-t \simeq 0.5$ GeV^2. Currently there are two rival explanations for this dip phenomenon. First, there is the "Nonsense School" which regards this dip as being caused by the nonsense-wrong-signature zero of the ω trajectory which goes through α = 0 at $-t \simeq 0.5$ GeV^2. Second, there is the "Radius School" which advocates that this zero is caused by a zero in the Δn = 1 helicity changing amplitude which goes like the Bessel function $J_1(R\sqrt{-t})$ with R ≃ 1 F. Harari's observation is that the two rival schools make complete different predictions for the electroproduction of π^o. According to the Nonsense School the dip is supposed to be associated with the ω trajectory; so the dip position should be exactly the same as in photoproduction even if the photon becomes virtual. On the other hand, the Radius School says that if the photon radius indeed shrinks with increasing q^2, the dip should move to a larger value of $-t$; its precise position is determined by equating $R\sqrt{-t}$ to the first zero of J_1. Typical estimates show that as q^2 is increased to 1 GeV^2, the dip position may move from $-t \simeq 0.5\ GeV^2$ to $\sim 1\ GeV^2$. I understand that an experimental study of π^o electroproduction is currently in progress at DESY to settle this interesting point.

6. INCLUSIVE ELECTROPRODUCTION

Turning now to inclusive electroproduction, the first reaction we consider is the proton distribution

$$e^- + p \rightarrow e^- + p + \text{any} \tag{6.1}$$

or equivalently

$$\gamma_{virt} + p \rightarrow p + \text{any} \tag{6.2}$$

Before the experimentalists started studying this inclusive reaction, some theorists[28] speculated, on the basis of models based on spin 1/2 constituents, that the proton may be emitted preferentially in the direction of the virtual photon. Such a behavior would be in striking contrast with the photoproduction case where the proton is known to be emitted preferentially in the backward hemisphere.

In Fig. 10 we present the invariant cross section for (6.2) at $q^2 = 1\ GeV^2$ and its photoproduction analog as a function of x with $p_\perp^2 \lesssim 0.02\ GeV^2$.[29] The first thing to be noted is that, when scaled by the total cross section, the proton distributions in the forward hemisphere are very similar. Even when q^2 is spacelike, the probability for proton emission drops rather sharply for $x \gtrsim 0.2$ just as in the photoproduction case. So the conjecture that there may be a lot of forward protons is definitely false, at least up to $q^2 \simeq 1\ GeV^2$.

If you look at Fig. 10 a little more closely, however, there is a spectacular difference between photoproduction and electroproduction just in the range $-1 < x \lesssim -0.8$. With $p_\perp^2$ fixed to be small, x is linearly related to the squared missing mass of the mesonic system emitted in the forward direction. The depletion (relative to photoproduction) of events with $-1 < x \lesssim -0.8$ means that it becomes harder to produce peripherally low-mass mesonic systems as q^2 becomes spacelike. Part of this decrease is accounted for by the disappearance of ρ^o, but more quantitative estimates show the necessity of suppressing production of highly peripheral mesonic systems <u>in general</u>, not just ρ production, to explain the observed behavior. Some people conjecture that this effect is largely due to t_{min} which depends strongly on q^2, but I have not seen detailed quantitative calculations along this line.

Let us now look at inclusive $\pi^\pm$ production

$$\gamma_{virt} + p \rightarrow \pi^\pm + \text{any} \tag{6.3}$$

Earlier discussion on this reaction was confused because some theorists were working with structure functions $\nu^3\mathcal{W}_2$ etc.,[28,30] which are rather inconvenient for the purpose of comparing electroproduction with photoproduction. The best way is, of course, to compare $(1/\sigma)E\ d\sigma/d^3p$ just as we did in the proton case. Such a comparison has been made by the

SLAC Hybrid Bubble Chamber group[23] in Fig. 11 where the dashed curves represent the photoproduction data.

The first thing to be noted in Fig. 11 is that there is no gross qualitative difference between electroproduction (actually muoproduction here) and photoproduction. The quantity $(1/\sigma)E\, d\sigma/d^3p$ depends neither on q^2 nor on W very strongly. Agreement is particularly good in the target fragmentation region ($x < 0$). If we look at details, however, there are two important differences. First, for the π^+ distribution the peak near $x = 1$ due to single π^+ production is much more pronounced for electroproduction, especially for the lower W data. This is easily explained by the growing longitudinal contribution, which, as remarked earlier, is well established to be important in this reaction. Second, for $x > 0$ the π^- distribution for electroproduction is systematically lower than the corresponding distribution for photoproduction. This tendency is more conspicuously displayed when we take the π^+/π^- ratio, as done in Fig. 12, which also shows the SLAC optical chamber data of Dakin et al.[31] who first publicized this spectacular difference. (Actually what is plotted here is the positive-to-negative hadron ratio but it is known from the other data that protons and K mesons do not make appreciable contributions.) It appears that for $0.3 \leq x < 1$ the π^+/π^- ratio grows with q^2 from ~ 1.2 at $q^2 = 0$ to something like ~ 2.2 around $q^2 = 2\ \text{GeV}^2$.

Parton enthusiasts were delighted to see this behavior for the π^+/π^- ratio because, as will be treated in the discussion session, the quark-parton model predicted this kind of tendency prior to the measurements.[32] There may, of course, be alternative explanations. The first thing we can think of is that this increase in the π^+/π^- ratio is related to the disappearance of ρ's, which produce equal numbers of positive and negative pions. However, it turns out that this alone does not explain the observed trend. Another point of interest is that the invariant mass of the recoiling baryonic system in (6.3) is typically in the range 1.5 - 2.0 GeV; in this mass range there are more N*'s (which are produced only with π^+) than Δ*'s (which can be produced with π^+ as well as with π^-). Clearly the longitudinal contribution to processes like

$$\gamma_{virt} + p \rightarrow N^{*o} + \pi^+ \tag{6.4}$$

must be studied in detail before a definite conclusion can be drawn. Quite apart from the question of the π^+/π^- ratio, it is of intrinsic interest to study the longitudinal-to-transverse ratio as a function of x in the inclusive reaction (6.3).

7. IS ω MORE RELEVANT THAN q^2?

At this point it is worth asking which is the more relevant variable for hadronlike behavior of the photon to manifest itself -- q^2 or the scaling variable $\omega \equiv 2m_p\nu/q^2$. Is the photon more likely to behave like hadrons when q^2 is low? or when ω is high? To answer this question we resort to a simple argument based on the uncertainty principle, as discussed by several authors including Gottfried and Nieh.[33]

Let us take a very naive vector-dominance picture in which the photon first converts itself into a vector state of mass m_V before it interacts with the proton target. Within the framework of the old fashioned perturbation theory in which energy conservation is violated but three-momentum is conserved, we can easily calculate in the proton rest frame the amount of energy violation involved as the photon of energy ν changes into the vector state. For $q^2 << \nu^2$ we have

$$E_\gamma - E_V \simeq (q^2 + m_V^2)/2\nu \qquad (7.1)$$

Associated with this energy difference is the time interval $\Delta\tau$ during which the hadronic vacuum polarization lasts. The longer $\Delta\tau$, the larger the "longitudinal distance" over which the photon travels in its hadronic guise. Crudely then, the longitudinal distance ℓ is given by

$$\ell = \Delta\tau \simeq 2\nu/(q^2 + m_V^2) \qquad (7.2)$$

Under certain kinematical conditions this quantity can easily become much larger than the "size" of the proton.

Observe now that this expression is similar to ω ($\equiv 2m_p\nu/q^2$). So we expect that high ω virtual photons are more likely to exhibit hadronlike behavior.

In the energy range explored at SLAC, DESY, and Cornell, high ω regions almost inevitably imply low q^2 regions. It is therefore difficult to test in a critical way whether ω is the more relevant variable than q^2. If we go to NAL or CERN II energies, this need not be the case. With 200 GeV/c muons we can comfortably explore kinematical regions with q^2 typically 5 GeV^2 and, at the same time, ω as large as 20. It would be interesting to see whether quantities like the π^+/π^- ratio in the photon fragmentation region depend more on ω than on q^2.

SHADOWING

Just to show I'm not prejudiced in favor of hadronlike behavior of the photon, I would like to conclude this part of my lecture course by mentioning phenomena which are embarrassing from my point of view.

It is well known that the A dependence of the total hadronic cross section in photon-nucleus collisions is slower than $\sim$ A. This shadowing ("Bell-Stodolsky") effect[34) was first established four years by a U. C. Santa Barbara group[35) and has subsequently been confirmed by many other groups. The physical origin of this effect was already explained in my Erice lectures two years ago. Since that time there have been two important experimental developments. First, the transition region between high-energy regions where shadowing has been observed and low-energy regions where shadowing is not expected theoretically was studied for the first time at Daresbury[36) (NINA). The amount of shadowing in the 2-3 GeV range is measured to be smaller than at higher energies just as expected. This is shown in Fig. 13 where $\sigma_{\gamma A}/A\sigma_{\gamma N}$ is plotted. Notice that no shadowing would imply $\sigma_{\gamma A}/A\sigma_{\gamma N}$ identically equal to unity.

The second important development concerns shadowing -- or rather lack of shadowing -- in electroproduction. This was studied by the SLAC-MIT collaboration (MIT for large angle data, SLAC Group A for small angle data). The striking experimental result is that for spacelike photons shadowing has not been observed in any kinematical domain -- not even at $q^2 = 0.4$ GeV^2, $\omega \simeq 30$. See Fig. 13 where the electroproduction points are taken from Kendall's Cornell talk.[37)

What about the status of our theoretical understanding of shadowing? In the late sixties there were many calculations on this based on pure ρ, ω, and φ dominance. Typically such calculations predicted too much shadowing in photon-nucleus collisions. We now know this is not surprising because pure ρ, ω, and φ dominance is deficient not only for inelastic electron-proton scattering but also for total photoproduction. Recently Schildknecht[38)] performed shadowing calculations based on generalized vector dominance which take into account higher-mass vector state contributions needed to reproduce the observed behavior of inelastic electron-proton scattering. His results are shown by the curves denoted by GVD in Fig. 13. As you can see, our understanding of shadowing is now quite satisfactory for ($q^2 = 0$) photoproduction off nuclei. However, the model fails for explaining the observed absence of shadowing in electroproduction.

I believe that this total lack of shadowing in inelastic electron scattering is difficult to understand on the basis of <u>any</u> model. As long as we can diffractively produce ρ in electroproduction, which we now know is possible, some shadowing effect is expected. Perhaps there are some subtle effects which must be taken into account when $Z\alpha$ is no longer a small quantity.

9. CONCLUSIONS

To summarize, how much progress have we made?

As far as processes involving real photons are concerned, the conjecture that photons behave like hadrons is now firmly established. In particular, since my Erice lectures two years ago, photon-induced reactions have been shown to be similar to hadron-induced reactions not only in peripheral two-body processes but also in large-angle processes and inclusive reactions. We can even be bold enough to assert that hadronlike behavior of the photon is to be regarded as one of the "laws of hadronic matter."

If we now turn to the spacelike region, we first note that some of the spectacular predictions which would demand "unhadronlike behavior" of spacelike photons -- abundance of forward protons, the average multiplicity depending only on the scaling variable ω, etc. -- were not fulfilled, at least in the q^2 range investigated so far. At the same time it has become clear that there are some important differences between photoproduction and electroproduction -- the smaller slope parameter in ρ electroproduction, the increasing π^+/π^- ratio in the photon fragmentation region, the absence of very low energy protons in the target rest frame, and the lack of shadowing.

At the present moment two views are possible.

(i) To understand the differences listed above we must invoke some "radical" ideas -- partons, pulverization, etc.

(ii) The differences can be explained in a relatively simple manner without significantly altering the fundamental notion of hadronlike behavior of the photon.

The question of which of these possibilities is realized in nature will be intensively discussed in the coming years.

ADDENDUM: MULTIPLICITY IN ELECTROPRODUCTION

After the Erice lectures were delivered, there has been an important development on the charged-hadron multiplicity in electroproduction.

At the time of the summer school the main information on the multiplicity came from the experiment of the SLAC Hybrid Bubble Chamber Group.[23] In that experiment there is a slight but statistically significant tendency for the average charged hadron multiplicity to decrease, perhaps by 10 to 15% as q^2 is raised, but this can be semiquantitatively accounted for by the known exclusive channels, the $\pi^+ n$, $\pi^o p$ (leading to single prongs) becoming more important relative to the total and the $\rho^o p$ (leading to three prongs) becoming less important. In any case it was generally conceded that the data did not cover a sufficiently wide q^2 range (the maximum value of q^2 is ~ 2 GeV^2) to examine in a critical way some of the spectacular theoretical predictions -- e.g. (a) the average pion multiplicity is a function of the scaling variable ω or ω' only[39] (Is the multiplicity at W = 9 GeV, $q^2 = 9$ GeV^2 really the same as at W = 3 GeV, $q^2 = 1$ GeV^2?) and (b) at fixed W the multiplicity increases with q^2 (Chou-Yang "pulverization,"[40] whatever that means).

Recently Berkelman and co-workers[41] at Cornell performed a counter experiment to measure the charged hadron multiplicities in inelastic electron scattering with q^2 up to ~ 10 GeV^2 and W^2 up to ~ 20 GeV^2. The important result of this experiment is that the multiplicity grows logarithmically with W^2, just as in photoproduction, but depends only weakly (if at all) on q^2. The expression

$$\bar{n} = (-0.1 \pm 0.1) + (1.3 + 0.1) \ln W^2 \qquad \text{(A.1)}$$

is claimed to give an adequate fit over the entire range. Models in which the multiplicity scales with ω or ω' are completely eliminated. A similar conclusion was also obtained by the DESY Streamer Chamber Group[42] whose data, however, go only up to $q^2 \simeq 1.5$ GeV^2.

* * *

REFERENCES

1) J. J. Sakurai, Ann. Phys. (N.Y.) 11, 1 (1960).

2) G. F. Chew, any paper.

3) T. A. Armstrong et al., Phys. Rev. D5, 1640 (1972).

4) L. Stodolsky, Phys. Rev. Letters 18, 135 (1967); P. G. O. Freund, Nuovo Cimento 44A, 411 (1966).

5) This subject was reviewed also by the 1972 Erice Lectures of E. Lohrmann.

6) G. Knies, Phys. Letters 27B, 288 (1968).

7) For a review of relevant data see, e.g., G. Wolf, Proc. 5th International Symposium on Electron and Photon Interactions at High Energies, Cornell University, Sept. 1971.

8) K. Gottfried, Proc. 5th International Symposium on Electron and Photon Interactions at High Energies, Cornell University, September 1971.

9) R. Anderson et al., Phys. Rev. D4, 3245 (1971).

10) G. C. Fox, Comments on Nuclear and Particle Phys. 3, 190 (1969).

11) M. Davies et al., Phys. Rev. Letters 29, 139 (1972).

12) P. Stichel, Z. Phys. 180, 170 (1964).

13) D. J. Sherdan et al., Phys. Rev. Letters 30, 1230 (1973).

14) R. Estabrooks and A. D. Martin, Phys. Letters 42B, 229 (1972).

15) R. L. Anderson et al., Phys. Rev. Letters 30, 627 (1973).

16) J. F. Gunion, S. J. Brodsky, and R. Blankenbecler, Phys. Letters 39B, 649 (1972).

17) J. Benecke et al., Phys. Rev. 188, 2159 (1969).

18) W. Struczinski et al., as reported by B. Wiik, Proc. 5th International Symposium on Electron and Photon Interactions at High Energies, Cornell University, Sept. 1971.

19) H.-M. Chan, H. I. Miettinen, and W. S. Lam, Phys. Letters 40B, 112 (1972). See also H. Satz and D. Schildknecht, Phys. Letters 36B, 79 (1971).

20) K. C. Moffeit et al., Phys. Rev. D5, 1603 (1972).

21) J. R. Fry et al., Nucl. Phys. B58, 408 (1973).

22) C. Iso and H. Yoshii, Ann. Phys. (N.Y.) 51, 490 (1969); J. D. Sullivan, Phys. Letters 33B, 179 (1970); H. Fraas and D. Schildknecht, Phys. Letters 35B, 72 (1971).

23) J. Ballam et al., SLAC-PUB-1163; M. Della Negra, SLAC-PUB-1242. For data of other groups see, e.g., L. Ahrens et al., Phys. Rev. Letters 31, 131 (1973).

24) J. J. Sakurai and D. Schildknecht, Phys. Letters 40B, 121 (1972). See also C. F. Cho and G. J. Gounaris, Phys. Rev. 186, 1619 (1969); H. Fraas and D. Schildknecht, Nucl. Phys. B14, 543 (1969).

25) J. T. Dakin et al., Phys. Rev. Letters 30, 142 (1973).

26) H. Cheng and T. T. Wu, Phys. Rev. 183, 1324 (1969). See also J. D. Bjorken, J. B. Kogut, and D. E. Soper, Phys. Rev. D3, 1382 (1971).

27) H. Harari, Phys. Rev. Letters 27, 1028 (1971).

28) S. D. Drell and T.-M. Yan, Phys. Rev. Letters 24, 855 (1970).

29) The photoproduction points in Fig. 10 are due to H. Burnfeindt et al., Phys. Letters 43B, 345 (1973) and W. Struczinski et al. (Reference 18) both from DESY. The electroproduction points are taken from J. C. Alder et al., Nucl. Phys. B46, 415 (1972) for $x \gtrsim 0$ (DESY) and E. Lazarus et al. (Cornell) for $x < 0$.

30) E. W. Colglazier and F. Ravndal, Phys. Rev. D7, 1537 (1973).

31) J. T. Dakin et al., Phys. Rev. Letters 29, 746 (1972).

32) J. D. Bjorken, SLAC-PUB-905, Invited talk at Tel Aviv Conference on Duality and Symmetry, April 1971.

33) K. Gottfried, Cornell University report 1969; T. H. Nieh, Phys. Rev. D1, 3161 (1970). See also B. L. Ioffe, Phys. Letters 30B, 123 (1969); V. N. Gribov, Soviet Phys.-JETP 30, 709 (1970).

34) J. S. Bell, Phys. Rev. Letters 13, 57 (1964); L. Stodolsky, Phys. Rev. Letters 18, 135 (1967).

35) D. O. Caldwell et al., Phys. Rev. Letters 23, 1256 (1969); 25, 796 (1970) (E).

36) G. R. Brooke et al., DNPL/P136 (1972).

37) H. W. Kendall, Proc. 5th International Symposium on Electron and Photon Interactions at High Energies, Cornell University, September 1971.

38) D. Schildknecht, SLAC-PUB-1230. His calculations are based on a formalism developed by S. Brodsky and J. Pumplin, Phys. Rev. 182, 1794 (1969), while the parameters used are taken from the generalized vector dominance model of J. J. Sakurai and D. Schildknecht (Reference 24).

39) S. D. Drell, D. J. Levy, and T.-M. Yan, Phys. Rev. Letters 22, 744 (1969); S. S. Shei and D. M. Tow, Phys. Rev. Letters 26, 470 (1971).

40) T. T. Chou and C. N. Yang, Phys. Rev. D4, 2005 (1971).

41) K. Berkelman et al., contributed paper to the 6th International Symposium on Electron and Photon Interactions at High Energies, Bonn, August 1973.

42) G. Wolf, private communication.

FIGURE CAPTIONS

Fig. 1. Comparison of the total γp cross section and the $\pi^o p$ cross section.

Fig. 2. Effective trajectory deduced from $n + p \to p + n$.

Fig. 3. Effective trajectory responsible for the unnatural parity exchange part of $\gamma + p \to \pi^+ + n$.

Fig. 4. Effective trajectories deduced from $\pi^- + p \to \rho^o + n$.

Fig. 5. Large angle behavior of single-pion photoproduction and meson-nucleon scattering.

Fig. 6. Inclusive π^- distributions of photon and hadron induced reactions in the target rest frame.

Fig. 7. Predictions on $\gamma + p \to \pi^- +$ any from $K^\pm + p \to \pi^- +$ any in the target fragmentation region.

Fig. 8. Inclusive π^- distribution in photoproduction compared with those of $\pi^\pm$ induced reactions.

Fig. 9. (a) The ratio of ρ electroproduction to total electroproduction.
(b) Slope parameter in ρ electroproduction.

Fig. 10. Inclusive proton distribution in photoproduction and electroproduction.

Fig. 11. Inclusive $\pi^\pm$ distributions in electroproduction. Photoproduction data are also shown (the dashed curves).

Fig. 12. The ratio of positive to negative hadrons in electroproduction.

Fig. 13. Shadowing in photoproduction and electroproduction.

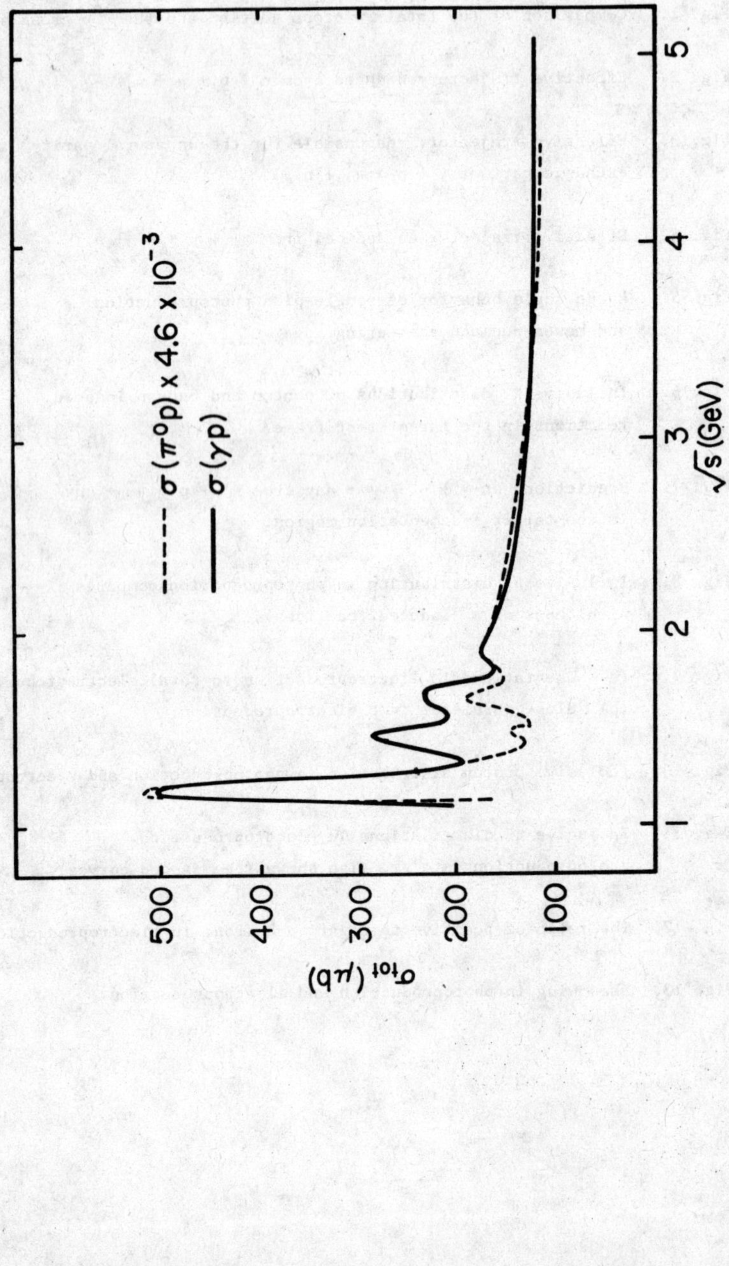

Fig.1

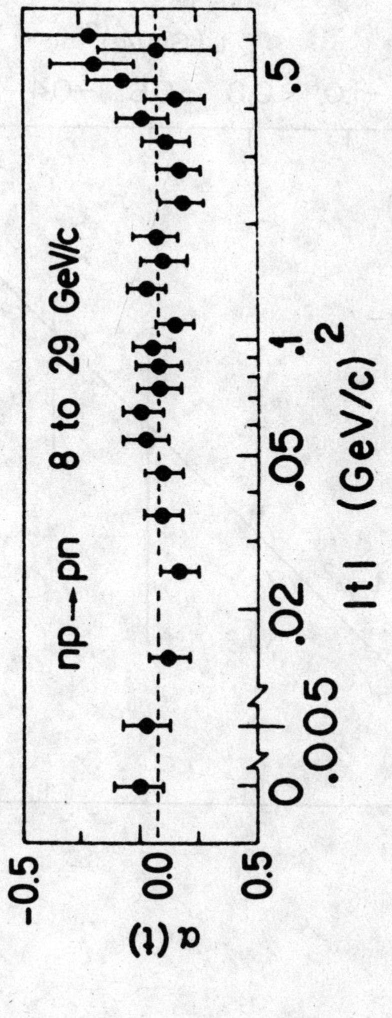

Fig. 2

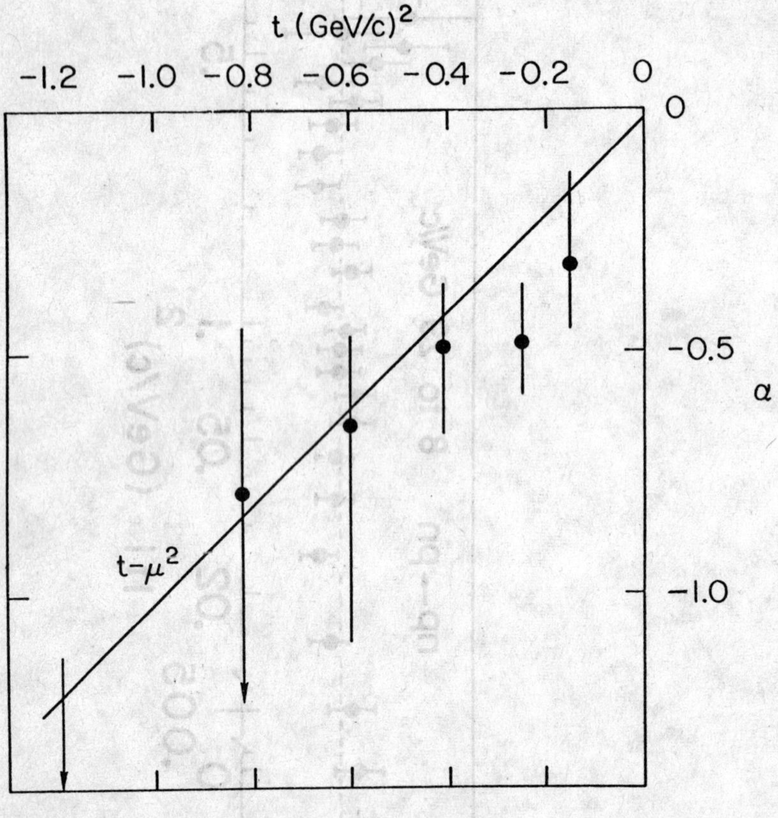

Fig. 3

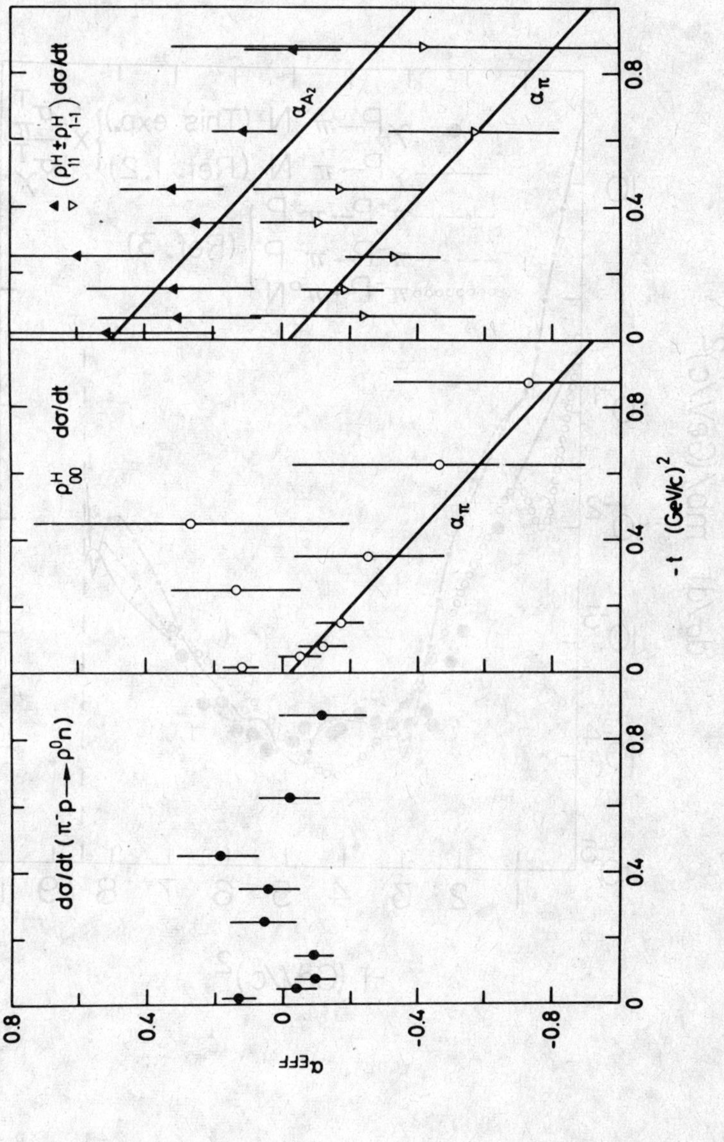
dσ/dt (π⁻p → ρ⁰n)
ρ^H_00 dσ/dt
(ρ^H_11 ± ρ^H_1-1) dσ/dt
α_π
α_A2
α_EFF
-t (GeV/c)²
0.8
0.4
0
-0.4
-0.8

Fig. 4

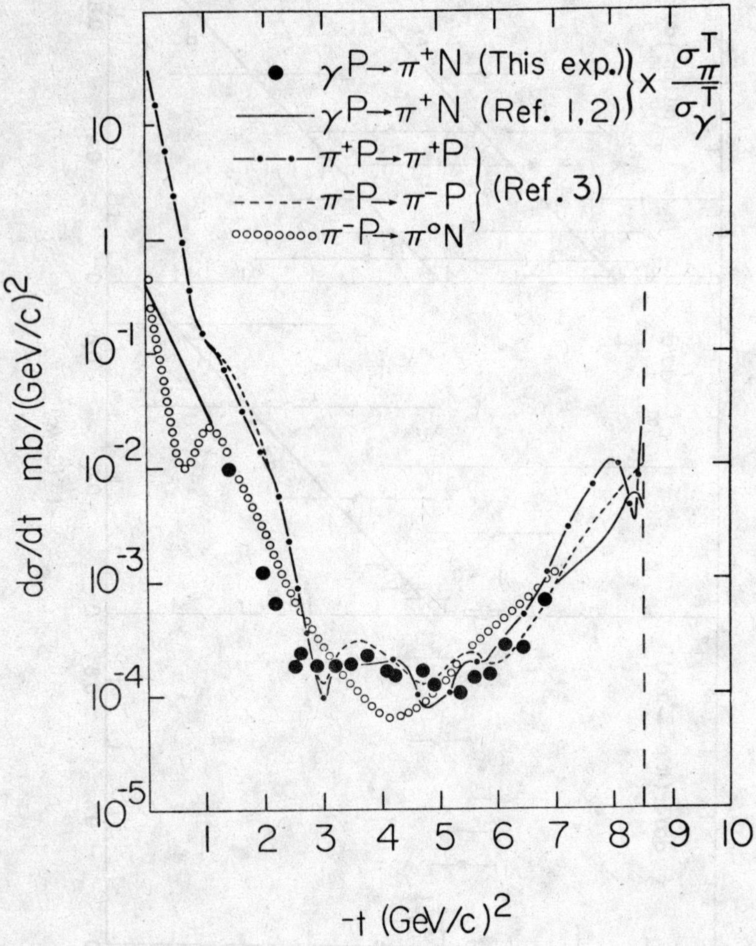
γP→π⁺N (This exp.)
γP→π⁺N (Ref. 1,2)
× σ_π^T / σ_γ^T
π⁺P→π⁺P
π⁻P→π⁻P
π⁻P→π°N
(Ref. 3)
dσ/dt mb/(GeV/c)²
10
1
10⁻¹
10⁻²
10⁻³
10⁻⁴
10⁻⁵
1 2 3 4 5 6 7 8 9 10
-t (GeV/c)²

Fig.5

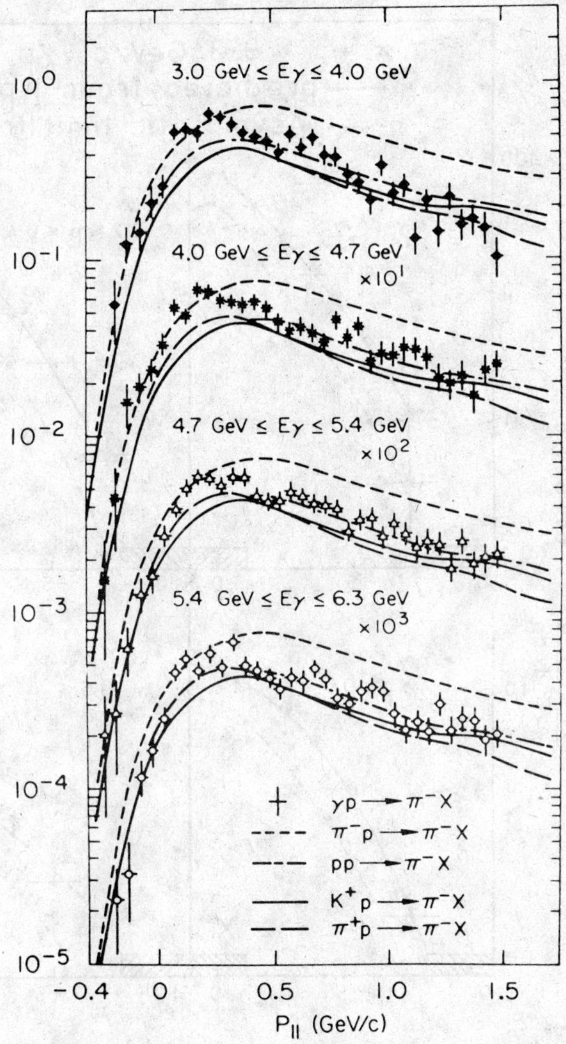
3.0 GeV ≤ Eγ ≤ 4.0 GeV
4.0 GeV ≤ Eγ ≤ 4.7 GeV
×10¹
4.7 GeV ≤ Eγ ≤ 5.4 GeV
×10²
5.4 GeV ≤ Eγ ≤ 6.3 GeV
×10³
γp → π⁻X
π⁻p → π⁻X
pp → π⁻X
K⁺p → π⁻X
π⁺p → π⁻X
1/σ_T dσ/dp_∥ (GeV/c)⁻¹
P_∥ (GeV/c)
10⁰
10⁻¹
10⁻²
10⁻³
10⁻⁴
10⁻⁵
-0.4
0
0.5
1.0
1.5

Fig.6

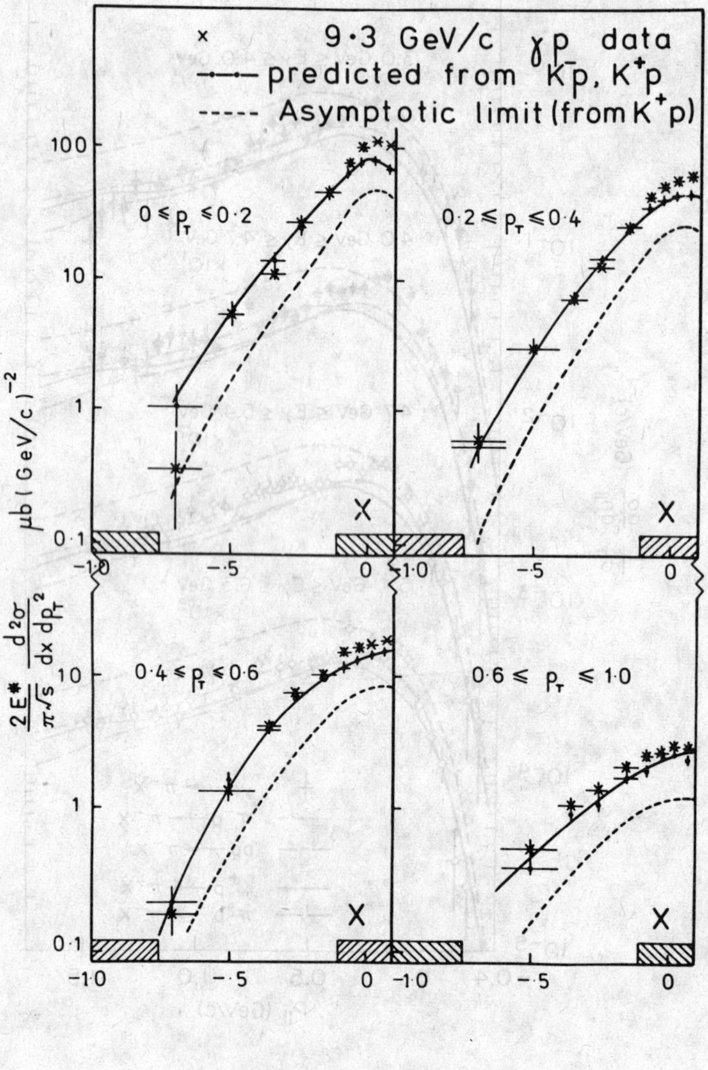
× 9·3 GeV/c γp data
predicted from K⁻p, K⁺p
---- Asymptotic limit (from K⁺p)
0 ≤ pT ≤ 0·2
0·2 ≤ pT ≤ 0·4
0·4 ≤ pT ≤ 0·6
0·6 ≤ pT ≤ 1·0
100
10
1
0·1
-1·0
-·5
0
X
μb (GeV/c)^-2
2E*/(π√s) d2σ/(dx dpT²)

Fig. 7

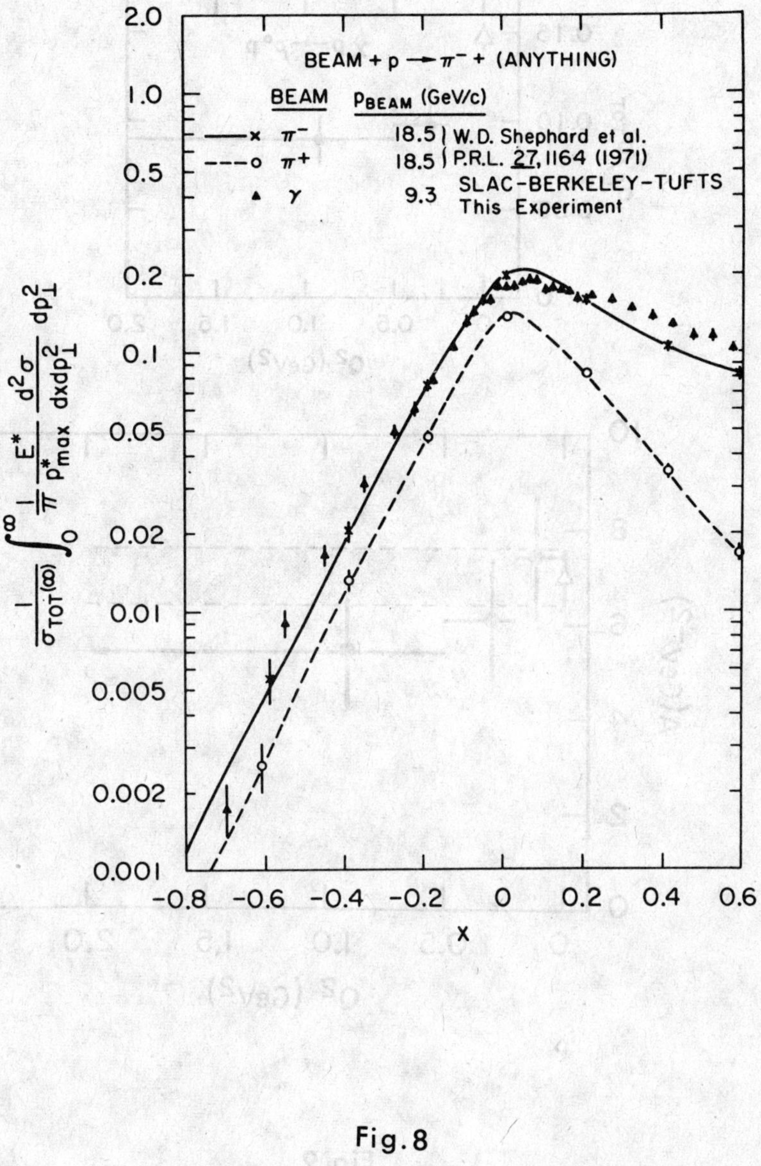
BEAM + p → π⁻ + (ANYTHING)
BEAM
p_BEAM (GeV/c)
π⁻ 18.5
π⁺ 18.5
W. D. Shephard et al.
P.R.L. 27, 1164 (1971)
γ 9.3
SLAC-BERKELEY-TUFTS
This Experiment
2.0
1.0
0.5
0.2
0.1
0.05
0.02
0.01
0.005
0.002
0.001
-0.8
-0.6
-0.4
-0.2
0
0.2
0.4
0.6
X

Fig. 8

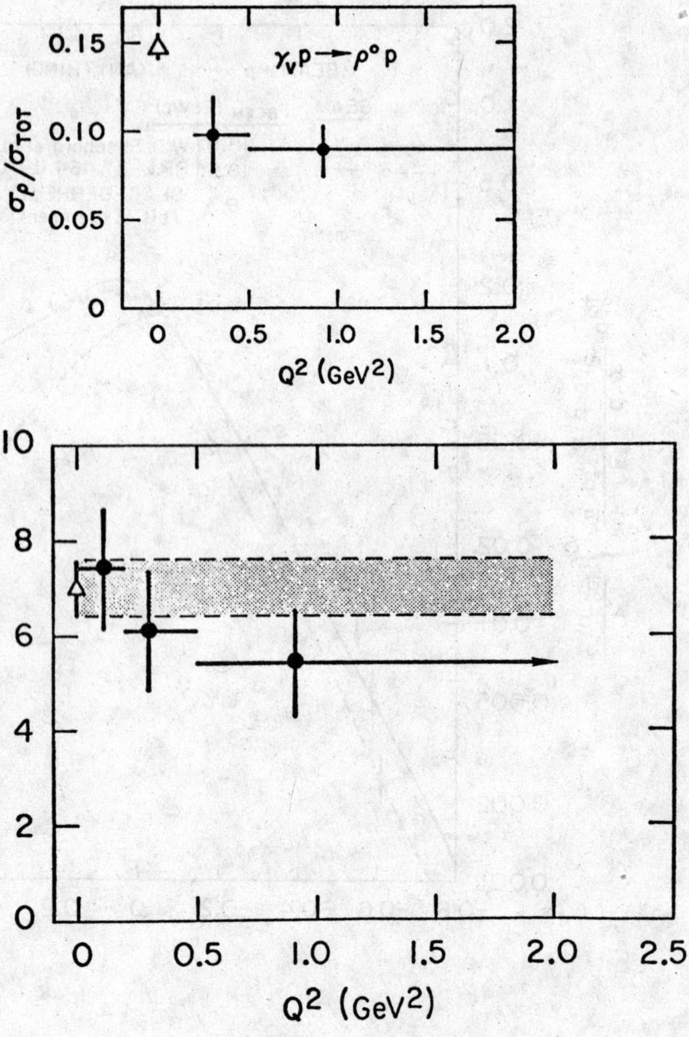
0.15
0.10
0.05
0
σ_ρ/σ_{TOT}
$\gamma_v p \to \rho^\circ p$
0
0.5
1.0
1.5
2.0
Q^2 (GeV^2)
10
8
6
4
2
0
A(GeV^{-2})
0
0.5
1.0
1.5
2.0
2.5
Q^2 (GeV^2)

Fig.9

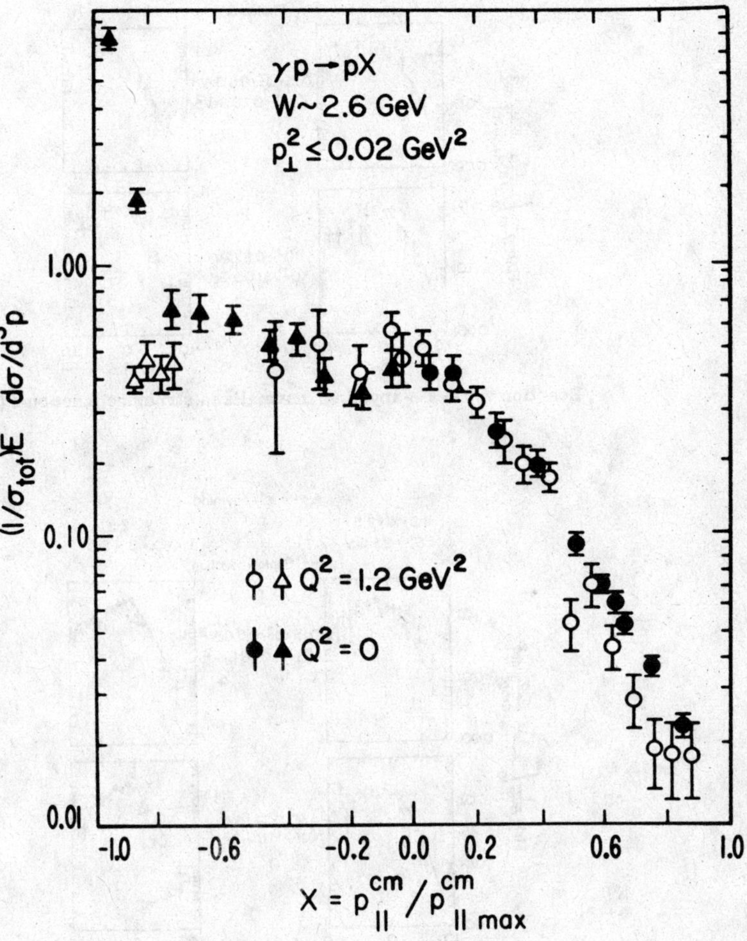

γp → pX
W ~ 2.6 GeV
$p_\perp^2 \leq 0.02$ GeV^2
$Q^2 = 1.2$ GeV^2
$Q^2 = 0$
1.00
0.10
0.01
$(1/\sigma_{tot})E\ d\sigma/d^3p$
-1.0
-0.6
-0.2
0.0
0.2
0.6
1.0
$X = p_{\parallel}^{cm} / p_{\parallel\,max}^{cm}$

Fig. 10

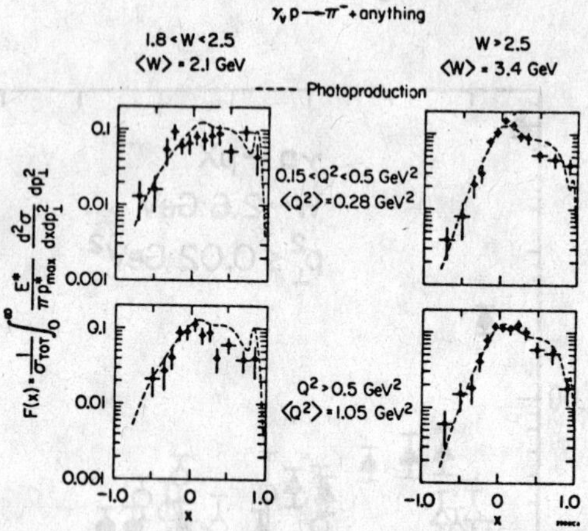

Reaction $\gamma_V p \to \pi^- +$ anything: normalized structure function F(x) vs x.

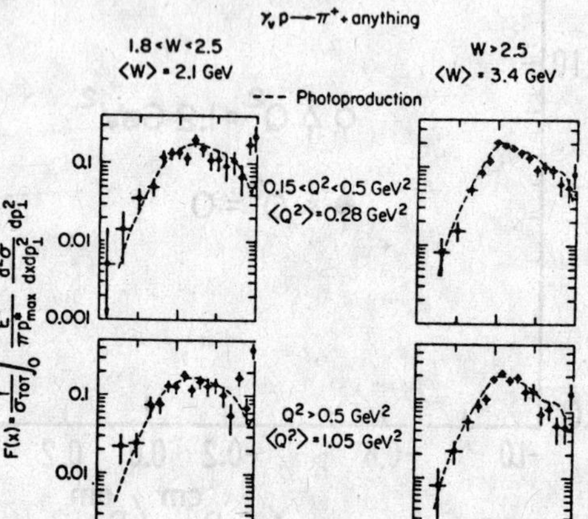

Reaction $\gamma_V p \to \pi^+ +$ anything: normalized structure function F(x) vs x.

Fig. 11

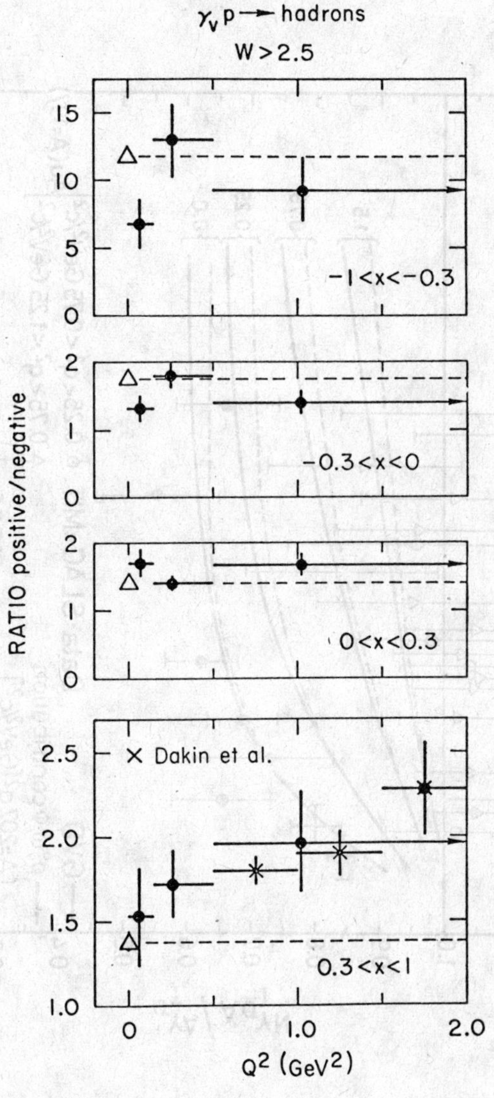
γv p → hadrons
W > 2.5
RATIO positive/negative
−1<x<−0.3
−0.3<x<0
0<x<0.3
0.3<x<1
× Dakin et al.
Q² (GeV²)
0
1.0
2.0

Fig. 12

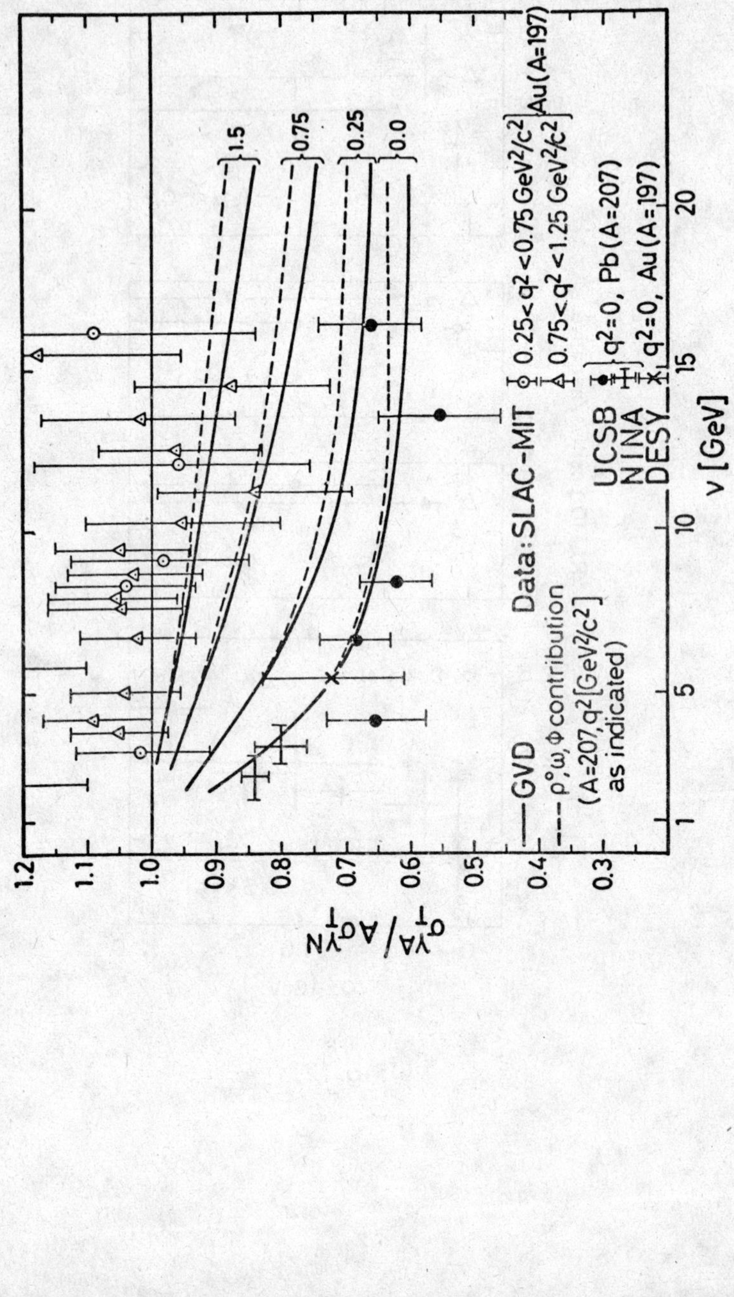

$\sigma_T^{\gamma A}/A\sigma_T^{\gamma N}$
1.2
1.1
1.0
0.9
0.8
0.7
0.6
0.5
0.4
0.3
1
5
10
15
20
ν [GeV]
1.5
0.75
0.25
0.0
— GVD
--- ρ°,ω,φ contribution
(A=207, q²[GeV²/c²] as indicated)
Data: SLAC-MIT
0.25 < q² < 0.75 GeV²/c²
0.75 < q² < 1.25 GeV²/c²
Au (A=197)
UCSB
NINA
DESY
q²=0, Pb (A=207)
q²=0, Au (A=197)

Fig. 13

D I S C U S S I O N S

CHAIRMAN: Prof. J.J. Sakurai

Scientific Secretaries: G. Chew, Y. Eylon, E. Rabinovici, D. Wright

DISCUSSION No. 1

- *EYLON:*

Using a geometrical picture of scattering, the relevant quantity governing the angular distribution is $k_f r\theta$, where k_f is the three momentum of the final particle, and r the radius of the obstacle. In usual $2 \to 2$ hadronic reactions, k_i, the three momentum of the incoming particle is about the same as k_f. Therefore an angular distribution of $e^{-(k_f r\theta)^2}$ can be expressed as $e^{-r^2(k_i k_f \theta^2)} \approx e^{-r^2 t'}$ $(t' = t - t_{min})$. In the case of "$\gamma(q^2)$"$P \to \rho P$, we get the peculiar situation where k_i is greater than the photon energy, since it is space-like. Instead of $k_i \sim k_f$, we get $k_i > k_f$, so now

$$e^{-r^2(k_f\theta)^2} \times e^{-r^2(k_f/k_i)k_i k_f \theta^2} \approx e^{-r^2(k_f/k_i)t'} .$$

Therefore in measuring the slope A of the t-distribution one finds not r^2, but $r^2 \cdot k_f/k_i$, so that the true radius is given by $r = \sqrt{k_i/k_f}\sqrt{A}$, instead of $r = \sqrt{A}$ which usually holds for real photons.

This correction factor is important in the kinematical region in which s is not much larger than q^2. It is an increasing function of q^2, and therefore although the experimentally measured A appears to shrink, the true radius r shrinks less.

- *KUPCZYNSKI:*

You have shown a formula which seems to be wrong at first sight:

$$16\pi\left(P_{\gamma p}/P_{\rho p}\right)_{cm}\left[\frac{d\sigma'}{dt}(\gamma p \to \rho p)\right]_{t=0} \le \sigma_{tot}(\gamma p)\,\sigma_{tot}(\rho p)$$

because for the $\gamma p \to \gamma p$ reaction one obtains for an arbitrary polarization η just the opposite inequality:

$$\left(\frac{d\sigma}{dt}\right)^A_{t=0}(\gamma p \to \gamma p) \ge \frac{\sigma^{A^2}_{tot}(\gamma p)}{16\pi} .$$

- *SAKURAI:*

Yes, that is correct. But $(d\sigma'/dt)$ denotes the differential cross-section divided by the factor $(1 + \eta^2)$ taking into account the possible real part of the scattering amplitude. Moreover, $(d\sigma'/dt)$, $\sigma_{tot}(\gamma p)$, and $\sigma_{tot}(\rho p)$ denote the cross-sections for the same fixed initial helicity states of the initial particles. In any case there is no contradiction. The basic physics is different. In my inequality I am saying that the overlap between the hadronic states produced in γp collisions and those produced in ρp collisions cannot be more than 100%. In your elastic scattering inequality you are saying that the forward differential cross-section may contain a non-negligible real part.

- *KUPCZYNSKI:*

To what extent is the total ρN cross-section measurable? Is it not strongly model-dependent?

- *SAKURAI:*

Crudely speaking the ρN cross-section has been obtained in two ways. First you look at the production off deuterium $\gamma d \to \rho d$. In that reaction there is a coherent very sharp peak and a part that can be regarded as a kind of double scattering contribution, and the ratio is sensitive to $\sigma_{\rho N}$. This kind of thing has been tested for πd scattering and they got the correct πN cross-section. So there is some faith in this method.

Another method is to use complex nuclei and study the A-dependence in $\gamma A \rightarrow \rho A$. You can intuitively imagine that if $\sigma_{\rho N}$ were large then the shadowing effect would be very important. I also used to think that all this was model-dependent. But the fact that the same numbers are obtained by the two entirely different methods gives impressive support to their reliability. For further reading, I suggest Gottfried's rapporteur talk at the Cornell Conference two years ago.

- *RABINOVICI:*

Could you explain how an apparently constant trajectory could be faked by the exchange of two increasing trajectories?

- *SAKURAI:*

The contribution to the cross-section from two exchanges is:

$$\beta_U(t)\; s^{2\alpha_\pi - 2} + \beta_N(t)\; s^{2\alpha_{A_2} - 2} \equiv \bar{\beta}(t)\; s^{2\alpha_{eff} - 2} \;,$$

where the right-hand side is the effective Regge behaviour. A complicated t-dependence of the β's can fake something that looks like a flat trajectory for small values of $-t$ up to $-t = 0.7$.

- *NANOPOULOS:*

Why is there a discrepancy between Blankenbeckler-Brodsky-Gunion model and the photo-production data?

- *SAKURAI:*

In BBG, $d\sigma/dt$ for π^+ photoproduction was predicted to go like $s^{-6.5}$ at $\theta_{cm} = \pi/2$. The data suggest something like $s^{-7.4}$. But this difference is not terribly significant. More relevant is the fact that BBG predict for the analogous hadronic reaction ($\pi^- + p \rightarrow \rho^0 + n$) as s-dependence that goes like s^{-8} (or even steeper). This difference is a consequence of the fact that the photon interacts, in their model, directly with a structureless parton.

- *TAYLOR:*

This question is about the interpretation of the change of slope with $|q^2$ in ρ^0-meson electroproduction being evidence for a "shrinking" photon. It is known that in ρ^0-meson photoproduction there is a significant interference between resonant and non-resonant 2π photoproduction complicating what is meant by the slope of $d\sigma/dt$ ($\gamma + P \rightarrow \rho^0 + p$). Can such an interference effect be excluded in this discussion of the $|q^2$ dependence of the slope in ρ electroproduction, and in particular, could it be that this observed change in slope with q^2 is just an interference effect?

- *SAKURAI:*

In photoproduction the ρ^0-mass distribution is asymmetric, and this may be interpreted as an interference with an s-wave $\pi\pi$ amplitude. Furthermore, the slope in $d\sigma/dt$ ($\gamma p \rightarrow \rho^0 p$) is quite mass dependent (mass = $m_{\pi^+\pi^-}$). For electroproduction, the ρ-mass distribution is more symmetric showing a more pure p-wave Breit-Wigner shape, but no data yet exists on the mass dependence of the slope. Much more data is needed, and at least now such an interference effect cannot be experimentally excluded.

- *GENSINI:*

I have two short comments and then a question to ask. The first comment is that you can obtain very simply the ρ total cross-section, if you believe in an optical picture of high-energy scattering by means of the Babinet's principle of geometrical optics. This works well for nuclei down to small A and there are no reasons why this should not work for nucleons. The second comment is that in any quark model you employ as input the asymptotic behaviour of nucleon e.m. form factors, and in this connection the dipole formula has nothing of magic in it; actually there are experimental indications that the decrease at high $-q^2$ is even steeper than that, and eminent theoreticians think it likely that it could decrease as fast as exp $[-\sqrt{-q^2}]$, saturating unitarity lower bounds! Then an argument based almost entirely on this ground has high probability of turning out wrong.

The question is: strong cuts have seldom been invoked to give a flat $\alpha_{eff}(t)$ for photoproduction and np charge exchange. Now you quoted determinations of $\alpha_{eff}(t)$ separately for natural and unnatural parity exchanges which you said were close to linear $\alpha_\rho(t)$ and $\alpha_\pi(t)$, respectively, at small values of $-t$ as expected. How large are curvatures of such effective trajectories at high values of $-t$?

- *SAKURAI:*

There is actually an indication of a small curvature at higher $-t$, similar to what is also found in πN charge exchange.

- *CHU:*

Why is there the same leading particle effect in inclusive photoproduction as in hadron-induced inclusive cross-sections?

- *SAKURAI:*

We can get a leading particle effect by considering, for example,

$$\gamma p \to \rho^0 + \text{anything}$$
$$\hookrightarrow \pi^+\pi^-$$

so there should be a leading particle effect for both π^+ and π^- inclusive distributions. But, since both are produced via ρ^0 decay, the effect should be somewhat washed out, as is borne out by the data.

- *CHU:*

In exclusive electroproduction you showed data which indicated that the fractional contribution of the $\rho^0 p$ channel decreases with $|q^2$. You explained this by saying that the $\rho^0 p$ contribution should have a factor of $1/(q^2 + m_\rho^2)^2$, and that the total cross-section should have a factor of $1/(q^2 + M_{eff}^2)$. Can you explain this?

- *SAKURAI:*

In the ρ-meson case, the $(q^2 + m_\rho^2)^2$ factor arises from naive vector dominance, for example:

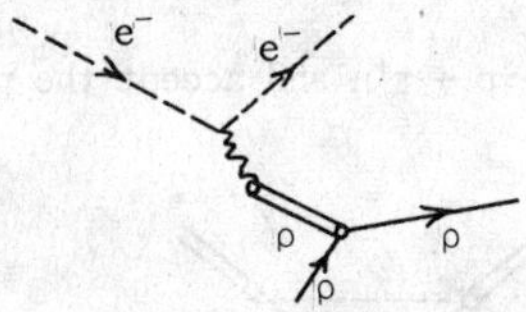

the photon couples to the hadrons via an extra ρ propagator. Apart from the well-known t_{min} effect this is adequate for transverse ρ production. However, when you talk about total electroproduction, states other than ρ may become important at higher $|q^2$. It is the combined effect of the ρ term and higher vector meson terms which produces a <u>simple</u> pole behaviour $1/(q^2 + M_{eff}^2)$.

- *WAMBACH:*

You mentioned $p_\perp^2$ distributions similar to hadronic distributions. Does an exp $(-bp_\perp^2)$ dependence which flattens out for increasing $p_\perp^2$, also show up in inclusive photo- and electroproduction?

- *SAKURAI:*

There are some measurements at SLAC of the reaction $\gamma p \to \pi^\pm + X$. They have measured distributions up to $p_\perp^2 = 4(\text{GeV}/c)^2$ and for $p_\perp^2$ up to 1 $(\text{GeV}/c)^2$ they find an exp $(-8p_\perp^2)$ dependence which flattens out to exp $(-2p^2)$. The results are very similar for π^+ and π^-. Compared to hadronic production, one observes a steeper increase for low $p_\perp^2$. These data have been taken for $x = 0.1$ to 0.4.

- *MENDES:*

If the ratio σ_ρ/σ_{tot} in electroproduction turns out to be flat in q^2 what will the theoretical implications of this fact be?

- *SAKURAI:*

In the framework of the VMD model that would mean that the longitudinal contribution is important for large q^2. There is some evidence that the longitudinal-to-transverse ratio grows (up to around 1 GeV^2) as

$$0.4\ q^2/m_\rho^2 \ .$$

Then the transverse contribution dies away as $1/(q^2)^2$, but the longitudinal part multiplied by this factor give a dependence similar to the total electroproduction cross-section. As a result you may have a flattening-out.

- *CHOUDHURY:*

If indeed the s-dependence of $d\sigma/dt$ for $\gamma p \to \pi^+ n$ and $\pi^- p \to \rho^0 n$ are different, as predicted by the parton model of Brodsky, Blackenbeckler and Gunion, will it be possible to accommodate this in some modified vector dominance model?

- *SAKURAI:*

I think it is very unlikely within the framework of the model.

- *PATKOS:*

Is it true that if you apply vector meson dominance to the incoming γ in photo-induced reactions, then you can draw the same figure as you drew for hadron-induced reactions, and so reconcile similar fixed angle asymptotic s-dependence for both of them?

- *SAKURAI:*

The main contribution to the pure hadronic reaction in the BBG model comes from the diagram

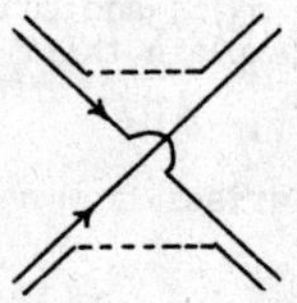

If we apply vector meson dominance for $\gamma p \to \pi^+ n$ and accept the parton picture, we have the diagram

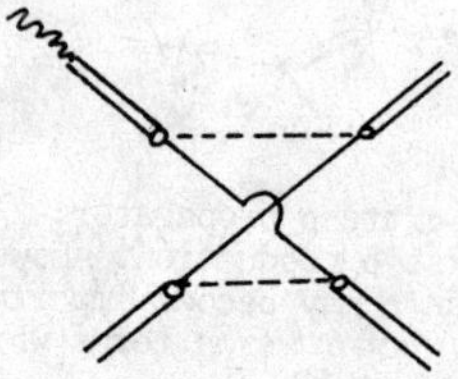

This leads to similar s-dependence of $d\sigma/dt$ for $\gamma p \to \pi^+ n$ and $\pi^- p \to \pi^0 n$, in contradiction to the prediction made by BBG.

- *VINCIARELLI:*

The presence of fixed Regge poles in photoproduction amplitudes is consistent with general principles and is perhaps suggested in some cases (such as ρ photoproduction) by experiments. Assuming that such fixed poles are really there, what do supporters of vector-meson-dominance-type models plan to do?

- *SAKURAI:*

I will think about it when there is some convincing evidence in favour of fixed poles!

DISCUSSION No. 2

- *DERMAN:*

My question is about your proof that large ω, rather than small q^2, is the relevant variable for hadron-like behaviour of the photon. Your proof seemed to be based on old-fashioned perturbation theory, i.e. finding the energy denominator for the γ - ρ^0 transition, and requiring that it be small. However, the Feynman diagram that you displayed in principle contains two time-ordered diagrams, and you only considered one of them in finding the energy denominator. Is there a good reason for ignoring the other z-type diagram, or is there some more rigorous covariant argument leading to the same result?

- *SAKURAI:*

I believe a proper treatment using covariant methods will give the same result. Did you show that it is different?

- *DERMAN:*

I have not tried yet, but the propagator just gives $1/(q^2 - m_\rho^2)$.

- *SAKURAI:*

So that makes it look like the distance from the ρ-meson pole is the relevant quantity which gives q^2 rather than ω. Actually I think the old-fashioned way is a little more relevant because it is more closely related to the concept of longitudinal distance.

- *DERMAN:*

In the expression you wrote in Greco's model for $\sigma(e^+e^- \to \text{hadrons})$, you ignored interference terms between vector meson resonances contributing to the same final state. Is this because it is the only tractable approximation or is there a better justification?

- *SAKURAI:*

If you look at the various resonances, we do not know too many. However, it appears that each seems to have its own favourite decay mode, for example, $\rho'(1600) \to \rho\varepsilon(\rho \to \pi\pi)$. In this case, ignoring interference terms may be justified, because of different final states. However, the fact that different vector mesons have different decay modes is not explained by the model.

- *BOUQUET:*

What is the physical basis for comparing the hadron production amplitude and the muon pair or quark pair production amplitude?

- *SAKURAI:*

The basis for comparison of $\sigma(e^+e^- \to \text{hadrons})$ and $\sigma(e^+e^- \to \mu^+\mu^-)$? That is a very deep question. Prior to the Frascati experiment, some people (including myself) though it would go down faster than s^{-1}. Now if you believe in scaling, naive dimensional considerations give you a $1/\text{Mass}^2$, and if you have no mass the only way to achieve this is to have a $1/s$. In quark-type models with currents constructed out of point-like spin-$\frac{1}{2}$ fields, the degree of divergence of the Schwinger term in fact implies that it goes as s^{-1}. Now what I want is to unify this point-like approach with the description based on an infinite sequence of vector mesons, and that is why I want to use $\sigma_{\mu\text{pair}}$ as a reference cross-section.

- *WILCZEK:*

What is your alternative to parton model ideas in attempting to explain the predominance of π^+ over π^- in the photon fragmentation region for semi-inclusive electroproduction?

- *SAKURAI:*

There is a large difference between photoproduction and electroproduction whenever the amplitude for longitudinal photons is large. This occurs for instance in $\gamma_{\text{virtual}} + p \to$ $\to \pi^+ + n$; in this case you can understand it because the relevant diagram is known:

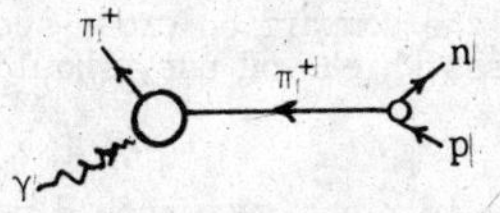

This diagram, together with its gauge-invariant partners, gives a good description of the electroproduction as well as photoproduction data.

The total electroproduction cross-section, we know from SLAC, is mainly transverse, but in certain channels the longitudinal contribution can be large.

- *WILCZEK:*

Do you have a definite mechanism in mind, or is this a pure guess? The quark-parton ideas give a definite prediction.

- *SAKURAI:*

It is a pure guess. The parton model is not a complete calculational scheme either. Some version of the parton model gives for this π^+/π^- ratio a number as high as 8.

- *NAHM:*

What about new duality for the weak current?

- *SAKURAI:*

Because of CVC there is no problem for the vector part. It is difficult to test the new duality for the axial vector part, because one does not know much about the behaviour of the A_1.

- *WAMBACH:*

In e^+e^- collisions, an experiment which will be done in the near future is the measurement of inclusive distributions. My question is, in what variables should scaling be expressed for single particle distributions since there we do not have $p_\parallel$ and $p_\perp$.

- *SAKURAI:*

In $e^+e^- \to \pi^\pm + x$, the angular distribution has to be investigated. Then one has the decomposition $A(p)(1 + \cos^2\theta) + B(p)\sin^2\theta$ where the first one is dominant for spin-$\frac{1}{2}$ constituents and the second one for spin-0 constituents. Now the question is whether $A(p)$ and $B(p)$ are analagous to longitudinal or transverse momentum distributions in hadron-hadron collisions. This question was first raised by Bjorken and Brodsky. If we assume that these are like transverse momentum distributions, their p is essentially limited, which implies a strongly increasing multiplicity ($\langle n \rangle \sim \sqrt{s}$). This does not appear to be happening. Then if $A(p)$, $B(p)$ are like longitudinal momentum distributions, one should use $x \equiv p/p_{max}$ and consider $f = d\sigma/dx$ [or $x\,(d\sigma/dx)$] distributions. Then at low energies one should have a bumpy distribution

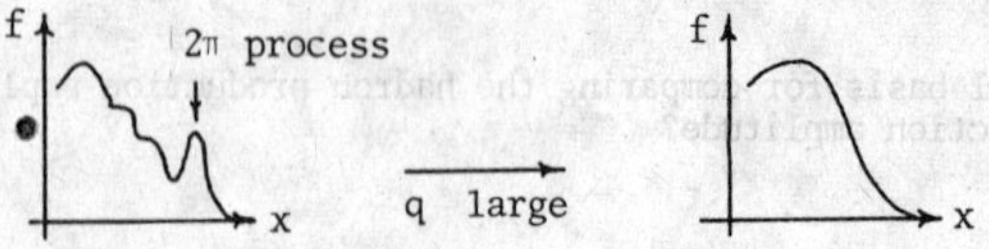

which averages out like in the duality arguments of Bloom and Gilman for high q^2.

- *KUPCZYNSKI:*

You have pointed out that the inclusive cross-sections for the reactions $\gamma p \to \pi^\pm X$ differ, namely $\sigma_{\pi^+} > \sigma_{\pi^-}$. Cannot it be simply explained by the fact that for the production of π^+, an additional channel is open, namely $\gamma p \to \pi^+ n$; whereas the π^- can only be produced in a pair?

- *SAKURAI:*

It is not as simple as that since $\gamma p \to \pi^+ n$ affects just the $x = 1$ point. However, it is important to keep in mind the constraint imposed by charge conservation: $n_+ - n_- = 1$. Now it appears that in electroproduction the average charged particle multiplicity decreases with increasing q^2. This necessarily implies via the charge conservation constraint that the positive-to-negative ratio increases.

- *KUPCZYNSKI:*

Could you explain the choice of the comparison cross-section, in particular, could you comment about the quark with low masses which you use, should they be found in experiment?

- *SAKURAI:*

If the idea of semilocal duality is right, you should use for your comparison function a threshold mass below the ρ mass, say 600 MeV, and this is precisely the mass of a quark pair in the naive non-relativistic quark model.

- *ZICHICHI:*

Kupczynski, are you satisfied with the quark being so light or not?

- *KUPCZYNSKI:*

I would be very satisfied if it would be found.

- *ZICHICHI:*

We are looking for the quarks. You know, these objects could very well exist. We have an experiment at the ISR having this in mind. The quarks could be very light, but to produce them you would have to materialize the hadronic cloud of a nucleon, which could be made of very many hadrons, say 40 pions.

People have been looking for quarks in low multiplicity events; we want to look for quarks in very high multiplicity events. It could be that the quark has very light mass, but in order to produce them you may need more than 20 GeV.

- *CHOUDHURY:*

I would like to make a comment about the asymptotic ratio

$$R = \lim_{s\to\infty} \frac{\sigma_{hadrons}}{\sigma_{\mu\mu}} \approx 2.5$$

obtained from infinite sequences of poles. This number 2.5 or 5/2 is also obtained from the parton model or the light cone current algebra. For, if one uses Han Nambu type q currents and uses the algebra of exact $SU(2)' \times SU(2)''$ rather than $SU(3)' \times SU(3)''$, then one gets $r = 5/2$ instead of $R = 2/3$. This means that in high-energy e.m. processes, the so-called quantum number "charm" is partially conserved as we know that exact charm conservation gives $R = 2/3$ and its maximal violation gives $R = 4$. So the believers in parton models or light-cone algebra will take PCCC (partially conserved charmed currents) as a physical law if $R = 5/2$, which is also obtained from Veneziano-like model as you mentioned.

- *WRIGHT:*

In the previous discussion session you stated that the contribution of the fixed pole at $J = 0$ in Compton scattering is real. Could you give a proof of that statement?

- *SAKURAI:*

There are two approaches: one by Cornwall, Corrigan and Norton based on the DGS representation, and one based on the quark-parton model. Let me comment just on the second approach. Suppose we consider Compton scattering off a structureless, charged spin-zero boson. There we have a seagull term which is purely real. For spin-$\frac{1}{2}$ quarks there is no seagull term, but if you add the usual two terms in Compton scattering, you get a term which simulates the seagull term and gives a real contribution with a q^2 independent residue. Furthermore, its s- and t-dependence are just what you expect from a $J = 0$ fixed pole.

- *RABINOVICI:*

Could you compare the average multiplicity $\langle n \rangle$ in electro- and photoproduction with $\langle n \rangle$ in hadronic reactions?

- *SAKURAI:*

I do not have the data here now. In photoproduction the usual formula $\langle n \rangle = a + b \ln s$ seems to work quite well. However, the highest data energy is only 9.3 GeV, so it is not a very stringent test of the formula. It is, however, similar to what is observed in πN scattering after you take into account the difference in initial charge.

- *RABINOVICI:*

You showed that ω is the valid parameter to describe the hadronic properties of the photon. Doing this you assumed there exists a characteristic q^2 independent mass m_V^2 representing the vector meson. Is this consistent with the infinite number of vector mesons that you need to achieve scaling?

- *SAKURAI:*

The value obtained was $\Delta E = (q^2 + m_V^2)/2\nu$. In the kind of "generalized" vector dominance models I talk about in the main lecture one can show that in deep inelastic electroproduction the important contribution is obtained for an average vector meson mass $m_V^2 \approx q^2$. For instance when $q^2 \approx m_\rho^2$, the ρ gives an important contribution. Thus $\Delta E \approx 2m_p/\omega$ and one has $\Delta \ell \sim \Delta\tau \sim 1/\Delta E \sim \omega/2m_p$, so that ω is the relevant variable.

- *RABINOVICI:*

The Greco model and your model predicted $\sigma_{hadrons}/\sigma_{\mu^+\mu^-} \xrightarrow{s\to\infty}$ const. Can you predict if this limit is approached from above or below?

- *SAKURAI:*

I cannot say anything about it because my comparison cross-section is a very naive one keeping only the leading term, namely the quark-antiquark cross-sections. If the data were infinitely better, one could try all kinds of things with the analogue of secondary trajectories, for instance.

- *EYLON:*

Do you expect q^2 duality to hold only for $e^+e^- \to$ anything, or also for inclusive annihilation $e^+e^- \to h +$ anything and even for exclusive annihilation?

- *SAKURAI:*

For inclusive, maybe. For exclusive, definitely not. At most we can have regularities in the coupling of the vector mesons to $\pi\pi$, for example, as indicated by the Veneziano model result for the $\gamma\pi\pi$ form factor. However, the actual q^2 dependence is different for inclusive and exclusive.

- *EYLON:*

If q^2 duality does hold for inclusive annihilation, and if we believe Feynman scaling in x to hold there, then the measurement of the x distribution for one energy gives us information about the x distribution for any energy; and then using q^2 duality we get information about the rates of the decays $V \to h +$ anything for all of the vector mesons.

- *SAKURAI:*

Yes, this is an interesting point.

- *EYLON:*

You argued that $\sigma(e^+e^- \to \mu^+\mu^-) \sim 1/s$ since we have no other relevant quantity with the proper dimension. Is it the range of the strong interactions which breaks this argument for hadronic cross-sections, or is there some other argument which makes total hadronic cross-sections constant?

- *SAKURAI:*

These naive dimensional arguments do not hold in the case of strong interactions. By the same argument one would say $d\sigma/dt$ for any two-body process would be $1/s^2\ f(t/s)$. This has not been observed; if anything it is like s^{-8} for πN scattering and s^{-10} for pp scattering, so for hadronic physics this is not going on.

Maybe I should make one more remark: perhaps this naive scaling is violated only for two-body processes. People have argued that it should hold for inclusive processes so that $d\sigma/dp_\perp^2 \sim (1/s^2)\ f\ (p_\perp/\sqrt{s})$, but recent data at ISR again show that the s-dependence is faster. If it fails there, you probably would not expect it to hold in exclusive processes.

- *VINCIARELLI:*

Yesterday, you said (or implied) that you might expect "generalized vector dominance" to be an exact approach to one-photon processes, while admitting that for two-photon processes it is less likely that generalized vector dominance could give the whole story. More specifically, you would allow for additional contributions, such as the ones possibly coming from fixed Regge poles, to be there in the latter case. My question is: why this difference (of principle) between one-photon and two-photon processes?

- *SAKURAI:*

Well, the fixed pole that appears because of the divergence condition shows up in two-current processes. All I am saying is that so far for one-photon processes there is no need for fixed poles.

- *CHU:*

You showed an experimental curve in which $\sigma_{e^-e^+\to hadrons}/\sigma_{e^+e^-\to\mu^+\mu^-}$ was rising as a function of s and, stated that if this effect continues to hold for higher energies, it would be a blow to the parton model people. What I do not understand is that if one is naive and notices that as s gets large, more and more hadron channels open up, it is not so surprising that the ratio should rise.

- *SAKURAI:*

What is really being measured is the vector meson density of states times the γ-V coupling constant:

$$\rho(m^2)\,\frac{1}{f_V(m^2)}\,m^2$$

Thus even with a $\rho(m^2)$ that increases with m^2, the cross-section for $e^+e^- \to$ hadrons could very well go down with s if the couplings get small fast enough.

NEW DUALITY IN ELECTROMAGNETIC INTERACTIONS

J.J. Sakurai

Table of Contents

NEW DUALITY IN ELECTROMAGNETIC INTERACTIONS*)

J. J. Sakurai
Department of Physics, University of California, Los Angeles, California, USA.

1. INTRODUCTION

In this part of my lecture course I would like to present a new approach to the electromagnetic interactions of hadrons based on "duality" with respect to the q^2 variable. Simply put, it is an attempt to unify the description based on scaling (or other asymptotic) behavior and the description based on a generalized form of vector-meson dominance. But before I start presenting the new model, let me briefly review the usual, more conventional approach to electromagnetic interactions.

Our current theoretical ideas on the electromagnetic interactions of hadrons have developed along two different lines. When q^2 is low -- spacelike or timelike -- models based on vector-meson dominance are known to be successful; on the other hand, when q^2 becomes large, it has become customary to invoke entirely different dynamical mechanisms to explain scaling or other asymptotic phenomena. As a concrete example, let us take electron-positron annihilation into hadrons. By this time everybody should be familiar with the empirical fact that below $\sqrt{s}$ = 1.1 GeV, the cross section for $e^+ + e^- \rightarrow$ hadrons is completely dominated by the ρ, ω, and ϕ peaks without any background. If we go to higher energies, however, we now know, mainly through the investigations of various Italian groups working at Frascati-Adone, that there is a large cross section for multihadron production, and most theorists nowadays tend to see in this "Frascati continuum" fundamentally different mechanisms at work -- colored quarks, light-cone dominance, etc. A similar view is also prevalent in discussing photoproduction and deep-inelastic electroproduction. For photoproduction the observed hadronlike behavior of the photon, discussed in the first part of my lecture course, can be best understood within the framework of ρ, ω, and ϕ dominance; however, when we discuss electroproduction in the scaling region ($q^2 \gtrsim 1$ GeV^2), many theorists maintain the view that we enter a fundamentally new regime in which partons -- not vector mesons -- play a dominant role.

There are two features somewhat disturbing about the usual approach outlined above. First, from a practical point of view the borderline between the scaling (or asymptotic) region and the vector-meson dominance region is not so clear-cut. As you know, there is now good evidence for a new I = 1 vector-meson state at $\sim$ 1.6 GeV,[1)] and it is plausible that there are many more vector-meson states buried in the Frascati continuum. We may legitimately ask whether at $-q^2 = (1.6)^2$ GeV$^2 \simeq 2.6$ GeV^2, we are still in the resonance region or already in the asymptotic region. Second, from a more philosophical point of view it may be a little more desirable to have continuity in the dynamics as we go from the high q^2 scaling region to the low q^2 vector-meson dominance region.

The approach I am advocating in this lecture attempts to unify the two descriptions for high-energy electromagnetic processes, which up to recent months were considered by most people to be unrelated. I would like to propose that there is some kind of "duality"

---

*) Supported in part by the U. S. National Science Foundation.

between the scaling behavior description and the vector-meson dominance description in much the same way as in pion-nucleon scattering where the well-known duality hypothesis of Dolen, Horn, and Schmidt[2)] establishes an intimate connection between asymptotic Regge behavior and low-energy s channel resonances.

Why do I prefer to call this new duality? By "new" I mean duality with respect to the variable q^2 in contrast to the old duality hypothesis where the basic variable was ν. Those familiar with old duality can understand much of what I am going to say by just replacing ν by q^2. The main results of old duality can be summarized in the following two equivalent ways:

> The Regge pole description originally designed for the high ν region works, on the average, even in the low ν region. Low energy parameters set the scale of high energy processes.

We can paraphrase the above statements, mutatis mutandis, for our new duality:

> Scaling behavior originally designed for the high q^2 region works, on the average, even in the low q^2 region. Low q^2 parameters set the scale of high q^2 processes.

In this lecture I examine to what extent the current experimental data support the above assertions in the following three inclusive processes:

$$e^+ + e^- \to \text{hadrons}, \tag{1.1a}$$

$$e^- + p \to e^- + \text{hadrons}, \tag{1.1b}$$

$$p + p \to e^+ + e^- + \text{hadrons}. \tag{1.1c}$$

I also would like to show how the ρ, ω, and ϕ mesons play an important role in building up scaling phenomena and explain, on the basis of our duality, why scaling is precocious.

To give credit to those who have indulged in this kind of speculation, let me first mention that the basic idea can be traced back to papers by Greco and collaborators,[3)] well summarized in a recent CERN preprint by Greco.[4)] While I was working on this subject a SLAC preprint by Bjorken and Kogut[5)] on "correspondence arguments" came to my attention. Their basic philosophy is also very much in line with the ideas I'm going to present; they too advocate that new kinematical regimes with high q^2 are closely linked with the old (more familiar) regimes with low q^2.

2. ELECTRON-POSITRON ANNIHILATION INTO HADRONS

For pedagogical purposes it is easiest to begin by outlining the work of Greco and collaborators[3,4)] on electron-positron annihilation into hadrons. Suppose the cross section for (1.1a) at squared photon mass $-q^2 = s$ is completely dominated by a sequence of discrete vector-meson states. Ignoring interferences among the various vector meson contributions, we can write for the total cross section of electron-positron annihilation into hadrons

$$\sigma_{had}(s) = \frac{12\pi}{s} \sum_V \frac{m_V^{\,2}\, \Gamma_V\, \Gamma(V \to e^+e^-)}{(s-m_V^{\,2})^2 + m_V^{\,2}\, \Gamma_V^{\,2}}$$

$$= \sigma_{\mu pair}(s) \sum_V \left\{ [3/(f_V^{\,2}/4\pi)] \frac{m_V^{\,2}\, \Gamma_V}{(s-m_V^{\,2})^2 + m_V^{\,2}\, \Gamma_V^{\,2}} \right\} \tag{2.1}$$

where $\sigma_{\mu pair}$ stands for the asymptotic $\mu^+\mu^-$ cross section in electron-positron annihilation:

$$\sigma_{\mu pair}(s) = 4\pi\alpha^2/3s$$
$$= [86.86/s]\ \text{nb} \quad , \quad (s \text{ in GeV}^2) \tag{2.2}$$

and my definition of the γ-V coupling constant is such that we insert em_V^2/f_V at a γ-V junction.

Now let us suppose that the above description based on a sequence of vector mesons is compatible, in a certain averaged sense, with the usual scaling behavior

$$\sigma_{had}(s) \to \text{const}/s \quad , \tag{2.3}$$

which can be derived from quark-model considerations[6] or just from consideration based on "naive dimensions." In other words, we require that the series of vector meson peaks appearing in (2.1), when suitably averaged, build up the conjectured 1/s behavior. It then follows that there must be an infinite sequence of vector mesons extending all the way to $s = \infty$. (I'm, of course, assuming here that Γ_V is not strongly s dependent; with a sufficiently complicated ρ tail anything is possible.) Furthermore there must be a particular kind of regularity in the density of the vector-meson states and the vector-meson coupling constants, viz.,[7]

$$P_V(m^2)m_V^2/f_V^2 = \text{const} \quad , \tag{2.4}$$

where $P_V(m^2)$ is the number of vector-meson states per unit squared mass interval -- for example, $1/2m_\rho^2$ in the Veneziano model to be discussed below.

To make further progress we must specify how (2.4) is satisfied. For definiteness let us assume: (i) the density function $P_V(m^2)$ for the isovector part is given by the Veneziano spectrum

$$m_n^2 = m_\rho^2(1 + 2n) \quad , \quad n = 0, 1, 2... \tag{2.5}$$

(ii) the regularity implied by (2.4) extends all the way down to the lowest-lying ρ meson region, and (iii) the isoscalar contribution is 1/3 of the isovector contribution. We can then express the asymptotic ratio R defined by

$$R \equiv \lim_{s\to\infty} \sigma_{had}(s)/\sigma_{\mu pair}(s) \tag{2.6}$$

in terms of the ρ meson coupling constant $f_\rho^2/4\pi$ as follows:

$$R = 2\pi/(f_\rho^2/4\pi) \simeq 2.5 \tag{2.7}$$

This equation is truly remarkable; it relates the asymptotic parameter R to the low-energy resonance parameter $f_\rho^2/4\pi$. Indeed, it is the first example of our new duality which asserts that low q^2 dynamics is intimately interlocked with high q^2 dynamics.

In the case of pion-nucleon scattering the connection between the Regge pole parameters and the low-lying resonance parameters can be quantitatively established by writing down a finite-energy sum rule. When I saw Greco's result (2.7) I asked myself whether a similar analyticity approach is possible in our case.[8]

From a dispersion-theorists's point of view, what the experimentalists measure in electron-positron annihilation into hadrons is the hadronic contribution to the imaginary part of the vacuum polarization amplitude[9] $\Pi_{had}(s)$:

$$\sigma_{had}(s) = 4\pi\alpha \, \text{Im} \, \Pi_{had}(s)/s \tag{2.8}$$

The vacuum polarization amplitude Π_{had} is expected to satisfy the usual cut-plane analyticity property with a cut starting at $s = 4m_\pi^2$. We may therefore write down

$$\text{Re} \, \Pi_{had}(s) = \frac{1}{\pi} \, \text{Pr.} \int \frac{\text{Im} \, \Pi_{had}(s')}{s' - s} ds' \tag{2.9}$$

However, if $\sigma_{had}(s)$ indeed goes like $1/s$, then the integral (2.9) diverges logarithmically. This, of course, reflects the well-known fact that the electric charge in pure "pointlike" electrodynamics is logarithmically divergent. Instead of considering (2.9), I propose to look at the difference between $\Pi_{had}(s)$ and some comparison amplitude $\Pi_{comp}(s)$ such that at sufficiently high energies

$$\Pi_{had}(s) \to \Pi_{comp}(s) \qquad (s \to \infty) \tag{2.10}$$

The precise form of Π_{comp} is somewhat arbitrary but if $\sigma_{had}(s)$ indeed goes like $1/s$ at asymptotic energies with no "secondary terms", it may, for instance, be chosen to be the vacuum polarization amplitude due to a pair of hypothetical "muons" of charge $\sqrt{R}$. When we look at the difference $\Pi_{had}-\Pi_{comp}$, it is possible to write a superconvergent relation as follows:

$$\int \text{Im} \, (\Pi_{had} - \Pi_{comp}) \, ds = 0 \tag{2.11}$$

Suppose $\text{Im} \, \Pi_{comp}$ does not differ appreciably from $\text{Im} \, \Pi_{had}$ for $s > s_{max}$. Equation (2.11) then leads to a "finite-energy sum rule" for the hadronic cross section

$$\int_{4m_\pi^2}^{s_{max}} s \, \sigma_{had}(s) ds \simeq \int_{s_o}^{s_{max}} s \, \sigma_{comp}(s) ds \tag{2.12}$$

where σ_{comp} is the cross section corresponding to Π_{comp} whose cut starts at s_o.

It may be mentioned that (2.12) is always valid if s_{max} is chosen to be sufficiently large. A finite-energy sum rule of this type is physically interesting only when it can be satisfied with a relatively small value of s_{max}. Empirically, below $s = 1.2 \text{ GeV}^2$ the cross section is dominated by the ρ, ω, and ϕ peaks while above 1.2 GeV^2 the multihadron contribution becomes appreciable. This motivates us to set s_{max} to be around 1.2 GeV^2. The left-hand side of (2.12) can be evaluated in terms of the parameters of the ρ, ω, and ϕ meson peaks, which are now well known from the Orsay and Novosibirsk data.[10] The integrand on the right-hand side is expressible in terms of the asymptotic form of the cross section. So with s_{max} chosen to be as low as 1.2 GeV^2, the finite-energy sum rule (2.12) embodies the view that the low-lying resonance peaks, when suitably averaged, agree with the extrapolation of the asymptotic cross section down to low energies. We may even be tempted to entertain the possibility that "asymptopia" may start already with the ρ, ω, and ϕ regions.

To predict the actual numerical value of the asymptotic ratio R, it is necessary to say something about the detailed form of σ_{comp}. The "standard" comparison cross section is R times the cross section for electron-positron annihilation into a muon pair:

$$\sigma_{comp} = R(4\pi\alpha^2/3s)(1 + 2m_\mu^2/s)(1 - 4m_\mu^2/s)^{1/2} \tag{2.13a}$$

However, in hadron production there is no reason to demand that the muon mass set the scale of the threshold. We may instead try the quark pair cross section

$$\sigma_{comp} = R(4\pi\alpha^2/3s)(1 + 2m_q^2/s)(1 - 4m_q^2/s)^{1/2} \quad (2.13b)$$

where m_q is the "quark mass." For definiteness we take the quark mass m_q to be

$$m_q = m_N/3 = 313 \text{ MeV} \quad (2.14)$$

for which there is some "empirical" evidence from hadron spectroscopy. I feel that a value for the quark-antiquark threshold much higher than $2m_N/3 = 626$ MeV would be contrary to the spirit of semilocal duality; clearly the ρ peak must be above the $q\text{-}\bar{q}$ threshold if quark pairs are to simulate the effect of the ρ meson peak. To study the sensitivity to the threshold behavior we also consider

$$\sigma_{comp} = R(4\pi\alpha^2/3s)(1 - 4m_\pi^2/s)^{3/2} \quad (2.13c)$$

whose energy dependence is appropriate for a pointlike pion pair.

In Fig. 1 we plot the integrands of (2.12), divided by $4\pi\alpha^2/3$ to conform to current usage. The solid curve is obtained from the experimental data. As for the absolute normalization of the curves corresponding to the comparison cross sections, the area under the curve in each of the three cases is adjusted to be equal to the area under the experimental curve. Our results with $s_{max} = 1.2$ GeV^2 are

$$R = 2.9\,, \quad (2.15a)$$

$$R = 5.0\,, \quad (2.15b)$$

$$R = 3.9\,, \quad (2.15c)$$

which correspond respectively to (2.13a), (2.13b), and (2.13c).

According to our speculation the comparison cross sections with R given by (2.15) should provide an adequate description of the data above $s > 1.2$ GeV^2. So, if σ_{had} indeed goes like 1/s, we predict a hadronic cross section three to five times the muon pair cross section. This is tested in Fig. 2 where the data points are taken from Silvestrini's rapporteur talk.[11)] The points above $s \simeq 5$ GeV^2 appear to be in satisfactory agreement with our expectation. (I temporarily ignore rumored data from CEA taken at $s = 25$ GeV^2.) However, the points between 1.2 GeV^2 and 4.5 GeV^2 are substantially low. Most theorists -- e.g. colored quark enthusiasts -- would like to see the data points above 5 GeV^2 lowered while I want the points between 1.2 and 4.5 GeV^2 to be raised!

I feel that the apparent disagreement seen in the 1.2 - 4.5 GeV^2 range is not completely discouraging. First of all, when the detectors cover only a limited portion of 4π, as in typical Frascati experiments, the total hadronic cross section quoted is very sensitive to the nature of final hadronic states assumed. The cross section could be higher than is indicated in Fig. 2 if low detection-efficiency channels like $\pi^o\omega$ played major roles. Second, as will be discussed later, there is reason to believe that there are still many more resonant peaks buried in the Frascati continuum. The data points are taken with typical intervals of 50 to 150 MeV; so narrow vector meson peaks may have escaped detection. (The Orsay and Novosibirsk groups would have had a hard time uncovering the ϕ meson peak if they had not known about the existence of ϕ from hadronic or photoproduction experiments.) After all, all I care about is the average cross section.

It may, of course, turn out that the rising trend suggested in Fig. 2 persists with better data and that R, in fact, increases with increasing s. Even in such a case our duality idea may still be of some value if a suitable comparison cross section can be invented.

It is hoped that experiments currently in progrss at SLAC-SPEAR will tell us a great deal about this very interesting and important question of duality in electron-positron collisions.

INELASTIC ELECTRON-PROTON SCATTERING

Let us now go to the spacelike region and consider inelastic electron-proton scattering (1.1b). I trust most people in the audience are familiar with the basic kinematics of a single-arm experiment in which only the scattered electron is detected. The double differential cross section $d^2\sigma/d\Omega dE'$ in the rest frame of the target proton can be written in two equivalent forms. The first form

$$\frac{d^2\sigma}{d\Omega dE'} = \left(\frac{d\sigma}{d\Omega}\right)_{Mott} [2W_1(q^2,\nu)\tan^2\frac{\theta}{2} + W_2(q^2,\nu)] \qquad (3.1)$$

$$(\nu \equiv E-E')$$

emphasizes similarity with the Rosenbluth formula for elastic electron-proton scattering; the second form,

$$\frac{d^2\sigma}{d\Omega dE'} = \Gamma_T[\sigma_T(q^2,\nu) + \epsilon\sigma_S(q^2,\nu)] \quad , \qquad (3.2)$$

where Γ_T, ϵ, σ_T, and σ_S respectively stand for the virtual photon flux, the polarization parameter, the transverse ($\lambda = \pm 1$) cross section and the scalar or longitudinal ($\lambda = 0$) cross section, reminds us that what is being measured is just the total cross section of virtual, spacelike photons on protons. Note the $q^2 \to 0$ limits

$$\lim_{q^2\to 0} \sigma_T(q^2,\nu) \to \sigma_{\gamma p}(\nu) \quad ,$$

$$\lim_{q^2\to 0} \sigma_S(q^2,\nu) \to \mathcal{O}(q^2) \quad , \qquad (3.3)$$

where $\sigma_{\gamma p}$ is the total photoproduction cross section.

The famous scaling hypothesis of Bjorken,[12] proposed prior to the confirming experiments,[13] states that the quantities $m_p W_1$ and νW_2, which are <u>a priori</u> functions of both q^2 and ν become functions of a single variable ω

$$\omega \equiv 2m_p\nu/q^2 \qquad (3.4)$$

as $q^2 \to \infty$ and $\nu \to \infty$ with the ratio fixed. In place of (3.4), other scaling variables have also been used, e.g.

$$\omega' = (2m_p\nu + m_p^2)/q^2 \qquad (3.5)$$

Now W_1 can be written as

$$W_1(q^2,\nu) = \frac{1}{4\pi^2\alpha}\left(1 - \frac{1}{\omega}\right)\nu\sigma_T \quad , \qquad (3.6)$$

and for $\nu^2 >> q^2$ (a good approximation for much of the SLAC-MIT data above the prominent s channel resonances) we have

$$\nu W_2 \simeq \frac{1}{4\pi^2\alpha}\left(1 - \frac{1}{\omega}\right) q^2(\sigma_T + \sigma_S) \qquad (3.7)$$

So Bjorken's scaling hypothesis demands that σ_T go like $1/q^2$ for fixed ω. When ω is high, σ_T at fixed q^2 as a function of ν, or of the hadronic invariant mass W, is known to be essentially flat, as expected from Pomeron dominance. We can therefore say that Bjorken scaling requires σ_T to go down like $1/q^2$ also for fixed W as long as the high ω condition $W^2 >> q^2$ is satisfied.

By this time it should be written in every text book that the pure ρ dominance formula for σ_T

$$\sigma_T(q^2,W^2) = [m_\rho^4/(q^2 + m_\rho^2)^2]\, \sigma_{\gamma p}(W^2) \qquad (3.8)$$

fails miserably. Clearly (3.8) is incompatible with Bjorken scaling which requires σ_T to go down like $1/q^2$. We now know that the failure of the pure ρ dominance formula is not surprising. After all, in electron-positron collisions there is copious production of higher-mass vector states (not just the tails of ρ, ω, and ϕ) up to the highest energy investigated so far. To take care of the effect due to these higher mass states we may write a spectral representation

$$\sigma_T(q^2,W^2) = \int \frac{m^4 \rho_T(m^2,W^2)}{(q^2+m^2)^2}\, dm^2 \qquad (3.9)$$

But with a sufficiently complicated spectral function we can fit anything just as with a sufficiently complicated combination of Regge poles and cuts, some phenomenologists working in strong interaction physics have succeeded in reproducing two-body data of almost any kind.

I would like to point out that an extremely simple modification of the formula (3.8) provides a semiquantitative description of the data in the high ω (diffraction) region.[14] Let us replace the m_ρ^2 in the denominator of (3.8) by m_n^2 that appears in the Veneziano spectrum (2.5) and sum over n. We have

$$\sigma_T(q^2,W^2) = \frac{8}{\pi^2}\left\{\sum_{n=0}^{\infty} \frac{m_\rho^4}{[q^2 + m_\rho^2(1+2n)]^2}\right\} \sigma_{\gamma p}(W^2) \qquad (3.10)$$

where the factor $8/\pi^2$ just normalizes σ_T at $q^2 = 0$:

$$\frac{\pi^2}{8} = 1 + \frac{1}{3^2} + \frac{1}{5^2} + \frac{1}{7^2} + \ldots . \qquad (3.11)$$

In Fig. 3 this simple, parameterless formula is compared to the SLAC-MIT transverse cross section at W = 3 GeV. Even though we don't have perfect agreement, it is much better than the old pure ρ dominance formula also shown in the same figure.

It is worth discussing a number of features of this extremely simple model. First of all, notice that the residue of the double pole at $-q^2 = m_\rho^2(1+2n)$ is independent of n. It was known for some time -- through work of Fujikawa and others[15] -- that if we are going to satisfy scaling together with Pomeron dominance in a model of this kind, the residue of the double pole must become independent of n for high-mass vector mesons.

The new point I would like to stress here is that this regularity is postulated to hold all the way down to the lowest-lying vector meson, the ρ meson itself ($n = 0$). This can actually be regarded as another example of our new duality. More of all this later.

Second, for moderate values of q^2 but with $\omega >> 1$, we can write

$$\nu W_2 \xrightarrow{\text{high } \omega} [(1+R)/\pi^4\alpha]\, m_\rho^{\;2}\, \sigma_{\gamma p} \tag{3.12}$$

where R now stands for the longitudinal-to-transverse ratio σ_S/σ_T. With R and $\sigma_{\gamma p}$ taken to be 0.2 and (100 - 130) μb, respectively, we predict $\nu W_2 \simeq 0.24 - 0.32$, in the high ω portion of the νW_2 curve, in rough agreement with the observed data.

Third, we can study how well this simple model satisfies Bjorken scaling even at relatively low values of q^2. With the usual Reggeish parametrization for $\sigma_{\gamma p}$ I have checked numerically that $\nu W_2/(1+R)$ at $\omega' = 4$ is flat within 16% between $q^2 = 0.8\ \text{GeV}^2$ and ∞. I'll come back to this subject of "precocious scaling" later on.

Fourth, suppose we try to saturate the isovector part of the photoproduction sum rule mentioned in the first part of my lecture course by the ρ meson contribution alone. This model tells us that such an attempt underestimates the total isovector photoproduction cross section by a factor of $\pi^2/8 \simeq 1.23$. Presumably a similar kind of discrepancy is expected for the isoscalar-photon channel. Experimentally it is known that the photoproduction sum rule fails by about 20-25% when we attempt to saturate it by ρ, ω, and ϕ alone,[16] and this has been considered by some rapporteurs in international conferences as an outstanding failure of vector dominance. We now see that, on the contrary, a discrepancy of about 20% is theoretically expected, hence the observed 20-25% discrepancy must be regarded as an extraordinary triumph of that version of generalized vector dominance which incorporates the duality requirement.

Fifth, this simple model relies on the vector meson spectrum (2.5) which does not skip "odd daughters". If the spacing between adjacent vector mesons is taken to be $4m_\rho^{\;2}$ rather than $2m_\rho^{\;2}$, the resulting cross section does not fit the data as well, a point noted earlier by Greco.[4] It is worth mentioning in this connection that with the spacing taken to be $4m_\rho^{\;2}$ rather than $2m_\rho^{\;2}$, Greco's model for the colliding beam cross section discussed earlier is also unsatisfactory because it leads to a value of R smaller by a factor of two compared to (2.7). Furthermore, when we attempt to construct Veneziano type models for the elastic form factor of the proton, a model which does not skip odd daughters is again favored.[17] It goes without saying that if the spacing of the isovector vector mesons is $2m_\rho^{\;2}$ rather than $4m_\rho^{\;2}$, the first ρ' is predicted to be at $\sim$ 1.25 GeV, not at 1.6 GeV. Fortunately there is some experimental evidence for diffractive photoproduction of a "B (1250)" meson with an energy-independent cross section of $\sim$ 1 μb both in a recoil-proton-detected counter experiment[18] and in laser-beam bubble chamber experiments[19]; even though its spin-parity is unknown, we are, of course, prejudiced against the diffractive photoproduction of a 1^+ object because of the Gribov-Morrison rule. Furthermore, some experts in hadron spectroscopy claim that the B mesons produced in pion-proton collisions may well have a 1^- component as well as the usual well-established 1^+ component. We should also note that SU(3) or quark-model considerations require the ω and ϕ analogs of $\rho'(1250)$ and $\rho''(1600)$.

I have discussed the simple model based on (3.10) in some detail not because I believe in its quantitative details but because it illustrates in a very simple manner the possibility of establishing a close connection between high q^2 parameters like the asymptotic value of νW_2 and low q^2 parameters like m_ρ and $\sigma_{\gamma p}$. Let us now focus our attention on the more general spectral representation (3.9). We first note that $m^4\rho_T(m^2,W^2)$ is a dimensionless quantity. Suppose we require "scale invariance" of $m^4\rho_T$ in the sense that $m^4\rho_T$ depends only on the ratio m^2/W^2 for high m^2 and W^2. At the same time, Pomeron dominance at fixed q^2 implies that $\rho_T(m^2,W^2)$ is independent of W^2. So the only possibility compatible with both Pomeron dominance and scale invariance of $m^4\rho_T$ is seen to be

$$m^4\rho_T(m^2,W^2) = \text{const.} \qquad (\text{independent of } m^2 \text{ and } W^2) \tag{3.13}$$

for high m^2. Clearly this scale invariance relation for the spectral function, when inserted in (3.9), leads to Bjorken scaling for σ_T.

A little more than a year ago, Schildknecht and I[20] constructed a generalized vector dominance model for inelastic electron-proton scattering where the weight function ρ_T is made up of the following two components: (i) δ function-like contributions at the ρ, ω, and ϕ masses whose magnitudes are determined from the vector-meson photoproduction data, and (ii) a continuum contribution satisfying (3.13) starting at $m = m_o = 1.4$ GeV. A representative result of the model is given by curve a of Fig. 4 where νW_{2T} stands for the transverse contribution to νW_2 defined by

$$\nu W_{2T} \equiv \nu W_2/(1+R) \tag{3.14}$$

The new duality hypothesis I now invoke implies that, when a suitable averaging is performed, there is no essential distinction between the ρ, ω, and ϕ regions and the higher mass region and that (3.13) must hold in some averaged sense even down to the ρ, ω, and ϕ regions. This prompts us to consider a step function approximation for $m^4\rho_T$ starting at some effective threshold m_{th}^2, which should be lower than m_ρ^2. We then obtain a <u>simple-pole</u> formula

$$\sigma_T(q^2,W^2) = [m_{th}^2/(q^2 + m_{th}^2)]\,\sigma_{\gamma p}(W^2) \tag{3.15}$$

Curve d in Fig. 4 is based on (3.15) with the effective threshold mass m_{th} fitted by Gorczyca and Schildknecht[21] to be 611 MeV.[22] Note that this value for m_{th} is below the ρ mass, just as expected. It is amusing that the shape of $m^4\rho_T$ that leads to (3.15) with $m_{th} = 611$ MeV bears resemblance to the colliding-beam comparison cross section form b multiplied by s, whose threshold starts at $\sqrt{s} = 2m_N/3$; look at Fig. 1 again. So there is some consistency between the spectral function determined from inelastic electron scattering and the colliding-beam cross section in electron-positron annihilation.

This is a good place to make comments on the "precocity" of Bjorken scaling. As is clear from (3.7), the structure function νW_2 has a kinematical zero at $q^2 = 0$. When we say that Bjorken scaling is precocious, it simply means that νW_2 at fixed ω (or ω') as a function of q^2 climbs from zero very rapidly to its asymptotic value. This is seen clearly in the data points of Fig. 4. Note also that both curve a and curve d exhibit the desired precocity property.

Very recently Gorczyca and Schildknecht[21] have attempted a quantitative estimate on the role played by the low-lying vector mesons in building up scaling phenomena. What they did was just to subtract from curve a (which fits the data) the contributions due to the ρ, ω, and ϕ, which can be computed in a noncontroversial manner using the usual double propagator rule with absolute magnitudes determined from the photoproduction ($q^2 = 0$) data. The result is shown as curve b. Note that the approach to scaling is now much slower. Thus we see that the ρ, ω, and ϕ contributions form an integral part in building up the observed scaling phenomena, a point previously emphasized by Bjorken.[23] Without them Bjorken scaling would no longer be precocious.

The question of precocity of scaling is closely related to the utility of yet another scaling variable ω_W

$$\omega_W \equiv (2m_p\nu + b^2)/(q^2 + a^2) \tag{3.16}$$

introduced by Rittenberg and Rubinstein.[24] They have observed that with[25]

$$a^2 = (0.37 - 0.42)\ \text{GeV}^2 ,$$
$$b^2 = (1.4 - 1.9)\ \text{GeV}^2 , \tag{3.17}$$

$(\omega/\omega_W)\ \nu W_2$ and W_1 (or W_1/ω_W) become independent of q^2 for fixed ω_W, not only for $q^2 \gtrsim 1$ GeV^2 but also at smaller values of q^2, even down to the photoproduction ($q^2 = 0$) point. This led them to remark: "We believe that the concept of a scaling region is a red herring since scaling, when properly defined, occurs even for physical photons."

Our new duality hypothesis, which has led to (3.15), provides a simple explanation as to why ω_W works, at least in the Pomeron dominated region. To see this let us note

$$\frac{4\pi^2\alpha}{\omega_W} W_\perp = \frac{q^2 + a^2}{2m_p\nu + b^2}\left(\nu - \frac{q^2}{2m_p}\right)\sigma_T$$

$$\xrightarrow{\text{high } \omega,\ \text{high } \nu} \frac{(q^2 + a^2)}{2m_p}\sigma_T \tag{3.18}$$

Meanwhile σ_T is well approximated by the simple pole formula (3.15). So W_1/ω_W is expected to be independent of both q^2 and ν if $a^2 = m_{th}^2$ and $\sigma_{\gamma p}$ = const. Numerically we have seen that the simple pole fit requires the effective threshold mass m_{th} to be 611 MeV; the preferred value for a according to (3.17) is

$$a = (608 - 648)\ \text{MeV} \tag{3.19}$$

The fact that the fitted value of a^2 in the approach of Rittenberg and Rubinstein is so low -- even lower than m_ρ^2 -- follows very naturally from our duality hypothesis. The success of ω_W is now understood. If you still insist on interpreting the observed precocious scaling within the framework of kindergarten quark-parton models, the only additional remark I can make is that the quark mass is as low as $m_N/3 \simeq 310$ MeV.

There are still many features of inelastic ep (en) scattering I could discuss in the context of generalized vector dominance -- the en-to-ep ratio, the reduction of the cross section in the small ω ("threshold") region, models for the longitudinal-to-transverse ratio R, etc. But in this lecture I have deliberately concentrated on those aspects which are directly relevant to our new duality hypothesis and restricted my considerations to σ_T in the diffractive region where generalized vector dominance is most likely to be successful. I would like to refer to the papers of Schildknecht and myself[20] and Schildknecht's Moriond talk[26] for discussions of the various topics not covered here.

4. LEPTON-PAIR PRODUCTION

The last process I discuss to illustrate our duality hypothesis is massive lepton-pair production in proton-proton collisions (1.1c) where we concentrate throughout on the invariant mass distribution of lepton pairs $d\sigma/dm^2$ after integration over the longitudinal and transverse momenta. This part of my talk is a progress report on work still being carried out in collaboration with a student of mine, Hank Thacker, the recipient of the Best Student Prize in the Erice School last year.

Suppose we look at the lepton-pair invariant mass distribution for (1.1c) near $m = m_{\rho,\omega}$. Then we are merely studying the inclusive hadronic reaction

$$p + p \to \rho(\omega) + \text{any} \tag{4.1}$$

where the $\rho(\omega)$ happens to decay via the leptonic mode with a branching ratio of order 10^{-5}. Everybody agrees that near the ρ mass, ρ dominance works in a trivial manner.

When the invariant mass is high, it is reasonable to expect that $d\sigma/dm^2$ exhibits a relatively simple and smooth behavior in both m^2 and s (the c.m. energy squared of the initial proton-proton system). In particular, as $m^2 \to \infty$, $s \to \infty$ with the ratio m^2/s fixed, $d\sigma/dm^2$ may satisfy a scaling law

$$\frac{d\sigma}{dm^2} = (m_o^{\alpha-4}/m^\alpha)\ F(\tau) \qquad (\tau \equiv m^2/s) \tag{4.2}$$

where a reference mass m_o has been inserted just to makϵ the scaling function $F(\tau)$ dimensionless. A special case of (4.2) with $\alpha = 4$ has received a great deal of attention because this value of α follows from considerations based on naive dimensions; historically the scaling law with $\alpha = 4$ was first publicized within the context of a parton-antiparton annihilation model by Drell and Yan.[27] However, from a more general point of view it is preferable to leave α as a parameter. For example, in a model in which the mass distribution for <u>hadronic</u> vector states scales with naive dimensions, one can show that α is, in general, not equal to 4.

Our new duality, as applied to this process, states that the scaling expression (4.2) provides an adequate "on-the-average" description of lepon-pair production even around $m^2 \simeq m^2_{\rho,\omega}$. We can then extrapolate the asymptotic formula (4.2) to the $\rho(\omega)$ meson region and express the scaling function $F(\tau)$ in terms of the integrated inclusive cross section for production of $\rho(\omega)$ mesons (4.1). This means that we can calculate $d\sigma/dm^2$ for all values of m^2 and s once we know the magnitude and the s dependence of the inclusive cross section for the hadronic reaction (4.1).

To be quantitative, the assumption that the scaling function $F(\tau)$ provides a good average description everywhere leads to

$$\overline{d\sigma/dm^2} = (m_o^{\alpha-4}/m^\alpha)\ F(\tau) \tag{4.3}$$

for all kinematically allowed values of m^2. On the other hand, near the ρ meson peak it is safe to assume

$$\left.\frac{d\sigma^{(isov)}}{dm^2}\right|_{m^2\simeq m_\rho^2} = (\Gamma_{\rho\to e^+e^-}/\Gamma_\rho)\left(\frac{d\sigma}{dm^2}\right)^{(isov)}_{had} \tag{4.4}$$

where $(d\sigma/dm^2)^{(isov)}_{had}$ is the mass distribution function for production of an I = 1 hadronic vector state (essentially all $\pi^+\pi^-$) of squared mass m^2. In the ρ meson region the average value of the lepton-pair production cross section via the isovector-photon channel is just the integral of (4.4) over the ρ meson peak divided by the squared mass interval Δm^2 over which we average $d\sigma/dm^2$:

$$\overline{\left(\frac{d\sigma^{(isov)}}{dm^2}\right)}_{m^2\simeq m_\rho^2} = (\Gamma_{\rho\to e^+e^-}/\Gamma_\rho)(1/\Delta m^2)\int_{\rho\ peak} dm^2\left(\frac{d\sigma}{dm^2}\right)^{(isov)}_{had}$$

$$= (\Gamma_{\rho\to e^+e^-}/\Gamma_\rho)(1/\Delta m^2)\, n_\rho(s)\,\sigma_{tot}(s) \tag{4.5}$$

Here the product $n_\rho(s)\,\sigma_{tot}(s)$ is the integrated inclusive cross section for ρ meson production, $n_\rho(s)$ being the ρ meson multiplicity and $\sigma_{tot}(s)$, the total proton-proton cross section. As for Δm^2, we may, for example, take it to be $2m_\rho^2$, the spacing between adjacent vector mesons in the Veneziano spectrum (2.5). Combining (4.3) and (4.5), we finally obtain

$$\left.F(\tau)\right|_{\tau=m_\rho^2/s} = (m_\rho^{\alpha}/m_o^{\alpha-4})(1/\Delta m^2)(\Gamma_{\rho\to e^+e^-}/\Gamma_\rho)\, n_\rho(s)\,\sigma_{tot}(s)$$

$$+ \text{isoscalar contribution.} \tag{4.6}$$

Thus, if we know the inclusive ρ meson cross section $n_\rho(s)\,\sigma_{tot}(s)$ for all values of s, then the isovector part of the scaling function could be determined over the entire range provided, of course, α is given.

Even without detailed data on inclusive ρ production, we can make some confident guesses about the general behavior of $n_\rho(s)$ and the corresponding function $F(\tau)$. For example, it would be quite surprising if $n_\rho(s)$ grew faster than $\ln s$ as $s\to\infty$. This means that the $\tau\to 0$ growth of the scaling function is expected to be no faster than

$$F(\tau) \sim \ln(1/\tau) \qquad (\tau\to 0) \tag{4.7}$$

In other words, $d\sigma/dm^2$ for $s\to\infty$ with m^2 fixed should grow no faster than $\ln s$. This is in contrast to the heuristic light-cone approach of Brandt and Preparata[28] and the current-algebra considerations of Sanda and Suzuki,[29] both of which predict a linear growth in s for $s\to\infty$, m^2 fixed. On the other hand, the parton-antiparton annihilation model of Drell and Yan[27] can be shown to be consistent with (4.7) for small values of τ.

The scaling formula (4.2) or (4.3) tells us that we should compare data at $m = m_\rho$, $s = s_1$ with data at $m^2 = (s_2/s_1)m_\rho^2$, $s = s_2$. Let s_1 and s_2 correspond to typical CERN PS and CERN ISR energies, respectively. Then the production cross section at ISR energies ($s \simeq 2700$ GeV^2) for lepton pairs of mass $\sim$ 6 GeV is predicted from inclusive ρ meson data taken at CERN PS energies ($s \simeq 45$ GeV^2). There have been some attempts to determine inclusive ρ meson cross section in proton-proton collisions at CERN PS energies. Preliminary bubble-chamber data of a Bonn-Hamburg-Munich collaboration[30)] appear to indicate a cross section for this process as large as a few mb. When we put the numbers in with $n_\rho(s)\,\sigma_{tot}(s)$ taken to be 4 mb at $s = 45$ GeV^2 and α set equal to 4, we predict

$$\left.\frac{d\sigma}{dm}\right|_{m=6\ \text{GeV}} \simeq 4 \times 10^{-34}\ \text{cm}^2/\text{GeV} \tag{4.8}$$

at ISR energies ($s \simeq 2700$ GeV^2) just for the isovector contribution.

Recently a CERN-Columbia-Rochester collaboration has attempted to study electron-pair production in proton-proton collisions at CERN ISR. Unfortunately they have failed to obtain a conclusively positive signature for high-mass electron pairs, and a typical upper limit currently quoted is[31)]

$$\int_{m=5\ \text{GeV}}^{\infty} \left(\frac{d\sigma}{dm}\right) dm \lesssim 10^{-34}\ \text{cm}^2 \tag{4.9}$$

provided the lepton pair angular distribution is not highly peaked in the beam direction. So Drell-Yan scaling ($\alpha = 4$) does not look too good when combined with our duality idea. It is, however, possible that the scaling law can still be made consistent with our new duality approach provided α is bigger than 4, say $\alpha = 6$. Indeed, Thacker has informed me that with $\alpha = 6$ he has succeeded in obtaining a reasonable fit to both the shape and magnitude of the mass distribution in the earlier BNL-Columbia data of Christenson et al.[32)] on muon pair production in proton-uranium-nucleus collisions; this actually is a non-trivial achievement because the only input in Thacker's calculation is a model for inclusive ρ cross section which incorporates general features of multiparticle hadronic reactions.

In any case, it is highly desirable to determine α (or to test Drell-Yan scaling in particular) in a model-independent manner before we start even talking about combining scaling with duality.

5. CONCLUSION

We have seen how our new duality provides a unified description of high q^2 phenomena and low q^2 phenomena. Throughout my lecture the basic hypothesis has been that the various regularities -- scaling, etc. -- seen in the high q^2 domain extend, in some averaged sense, all the way to the low q^2 domain. Our main results can be written as

$$\text{observables in the scaling region} = \text{functions of low } q^2 \text{ parameters} \quad . \tag{5.1}$$

Specifically, the asymptotic ratio $\sigma(e^+e^- \to \text{hadrons})/\sigma(e^+e^- \to \mu^+\mu^-)$ can be expressed in terms of the parameters of ρ, ω, and ϕ; νW_2 in the high ω region can be related to m_ρ and $\sigma_{\gamma p}$; the mass distribution of highly massive lepton pairs produced in proton-proton collisions may be computed once the s dependence of inclusive $\rho(\omega)$ cross section is known. Furthermore, our duality idea is seen to throw new light on the question of precocious scaling.

Even though there are some qualitatively encouraging signs, for more stringent quantitative tests we need better data -- better data on electron-positron annihilation into hadrons in both the Frascati-Novosibirsk region and the SLAC(SPEAR)-DESY(DORIS) region, better data on diffractive photo- and electroproduction of higher mass vector mesons, better data on massive lepton pair production especially at NAL energies, etc.

It is hoped that when I come back to Erice a few years from now, I'll be able to report on further progress along these lines.

* * *

REFERENCES

1) G. Barbarino et al., Nuovo Cimento Letters 3, 689 (1972).

2) R. Dolen, D. Horn, and C. Schmid, Phys. Rev. Letters 19, 402 (1967).

3) A. Bramón, E. Etim, and J. Greco, Phys. Letters 41B, 609 (1972).

4) M. Greco, CERN TH 1617.

5) J. D. Bjorken and J. Kogut, SLAC PUB-1213, Phys. Rev. (to be published).

6) J. D. Bjorken, Phys. Rev. 148, 1467 (1967); V. N. Gribov, B. L. Ioffe, and I. Ya. Pomeranchuk, Phys. Letters 24B, 554 (1967).

7) H. Joos and collaborators claim that an equation similar to (2.4) can be derived in their version of relativitic quark models. M. Böhm, H. Joos, and M. Kramer, DESY 72/62.

8) J. J. Sakurai, UCLA/73/TEP/76, Phys. Letters (to be published).

9) N. Cabibbo and R. Gatto, Phys. Rev. 124, 1577 (1961).

10) D. Benaksas et al., Phys. Letters 39B, 289 (1972); D. Benaksas et al., Phys. Letters 42B, 507, 511 (1972); L. M. Kurdadze et al., Phys. Letters 42B, 515 (1972).

11) V. Silvestrini, Proc. 16th International Conference on High-Energy Physics, Chicago-Batavia (1972).

12) J. D. Bjorken, Phys. Rev. 179, 1547 (1969).

13) For a summary of the famous SLAC-MIT data see, e.g., G. Miller et al., Phys. Rev. D5, 528 (1972).

14) J. J. Sakurai, UCLA/73/TEP/80.

15) K. Fujikawa, Phys. Rev. D4, 2794 (1971).

16) See, e.g., D. O. Caldwell et al., Phys. Rev. D7, 1362 (1973).

17) P. H. Frampton, Phys. Rev. 186, 1419 (1969). For comparison with recent data see P. N. Kirk et al., Phys. Rev. D8, 62 (1973).

18) R. Anderson et al., Phys. Rev. D1, 27 (1970).

19) SLAC-Berkeley-Tufts Collaboration, private communication.

20) J. J. Sakurai and D. Schildknecht, Phys. Letters 40B, 121 (1972); 41B, 489 (1972); 42B, 216 (1972).

21) B. Gorczyca and D. Schildknecht, DESY 73/28.

22) I understand that a similar fit has been considered earlier by C. Jordan (unpublished).

23) J. D. Bjorken, AIP Conference Proceedings, No. 6, Subseries No. 2, Particle Physics (Irvine Conference - 1971).

24) V. Rittenberg and H. R. Rubinstein, Phys. Letters 35B, 50 (1971).

25) F. W. Brasse et al., Nucl. Phys. B39, 521 (1972).

26) D. Schildknecht, DESY 73/21.

27) S. D. Drell and T.-M. Yan, Phys. Rev. Letters 25, 316 (1970).

28) R. A. Brandt and G. Preparata, Phys. Rev. D6, 619 (1972).

29) A. I. Sanda and M. Suzuki, Phys. Rev. D4, 141 (1971).

30) P. Söding (private communication).

31) L. di Lella and B. J. Blumenfeld (private communication).

32) J. H. Christenson et al., Phys. Rev. Letters 25, 1523 (1970).

FIGURES

Fig. 1. The ratio of the experimental colliding beam cross section to $4\pi\alpha^2/3s$ at low energies. Also shown are the corresponding ratios for the various comparison cross sections discussed in the text. The areas under the four curves are all equal when integrated up to $s = 1.2\ GeV^2$.

Fig. 2. The high-energy colliding beam cross section divided by $4\pi\alpha^2/3s$. Also shown are the corresponding ratios for the comparison cross sections whose absolute magnitudes are determined from the lower energy data of Fig. 1.

Fig. 3. Simple, parameterless formula for σ_T in inelastic electron-proton scattering. See Eq. (3.10).

Fig. 4. The transverse contribution to νW_2. Curve (a) is the prediction of Sakurai and Schildknecht.[20] Curve (d) is the simple-pole formula (3.15) fitted by Gorczyca and Schildknecht.[21] Curve (b) shows that scaling is no longer precocious when we substract from curve (a) the ρ, ω, and ϕ contributions.

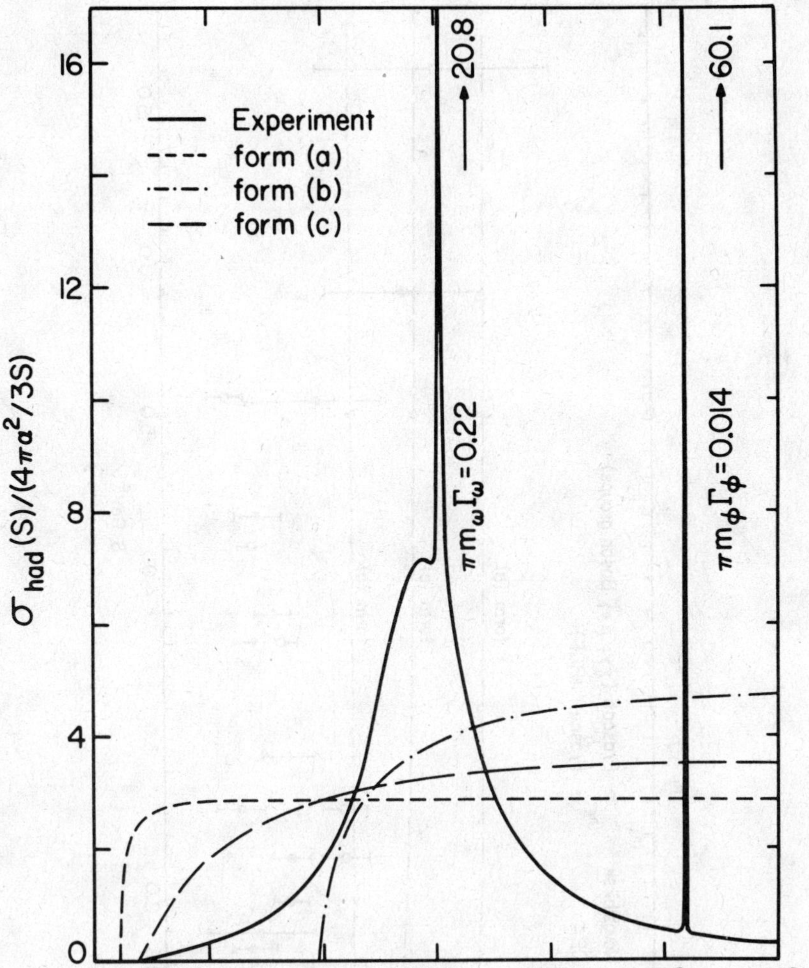
Experiment
form (a)
form (b)
form (c)
20.8
60.1
$\pi m_\omega \Gamma_\omega = 0.22$
$\pi m_\phi \Gamma_\phi = 0.014$
$\sigma_{had}(S)/(4\pi\alpha^2/3S)$
16
12
8
4
0
0.0
0.2
0.4
0.6
0.8
1.0
1.2
S GeV^2

Fig.1

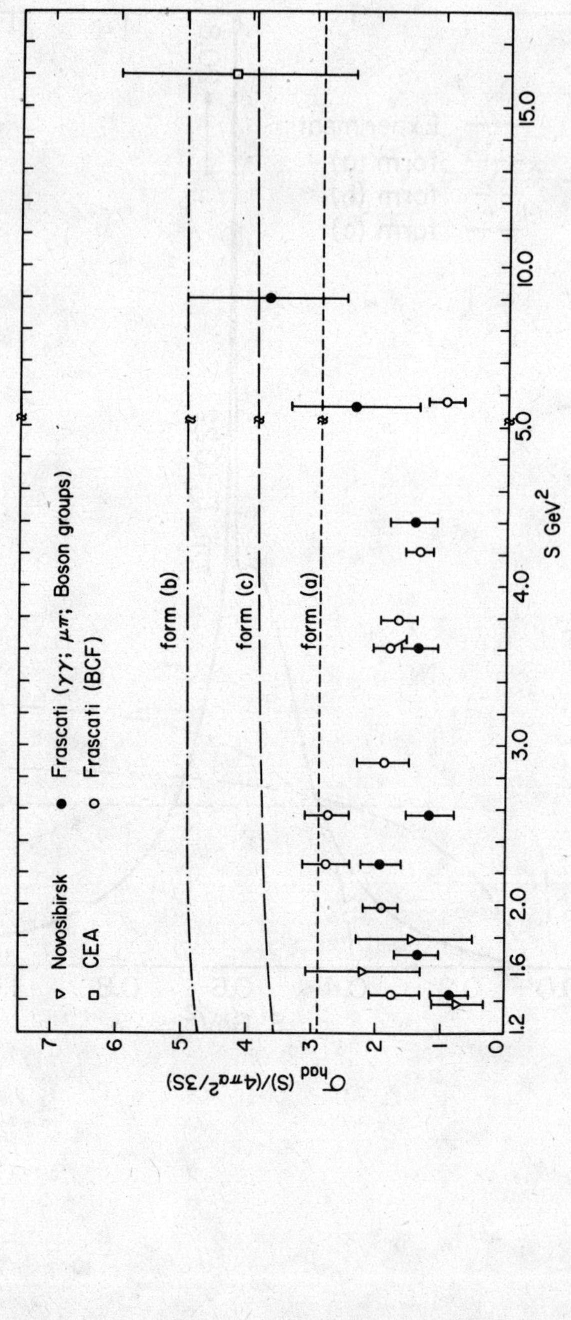

Novosibirsk
CEA
Frascati ($\gamma\gamma$; $\mu\pi$; Boson groups)
Frascati (BCF)
form (b)
form (c)
form (a)
σ_{had} (S)/($4\pi\alpha^2$/3S)
S GeV^2

Fig. 2

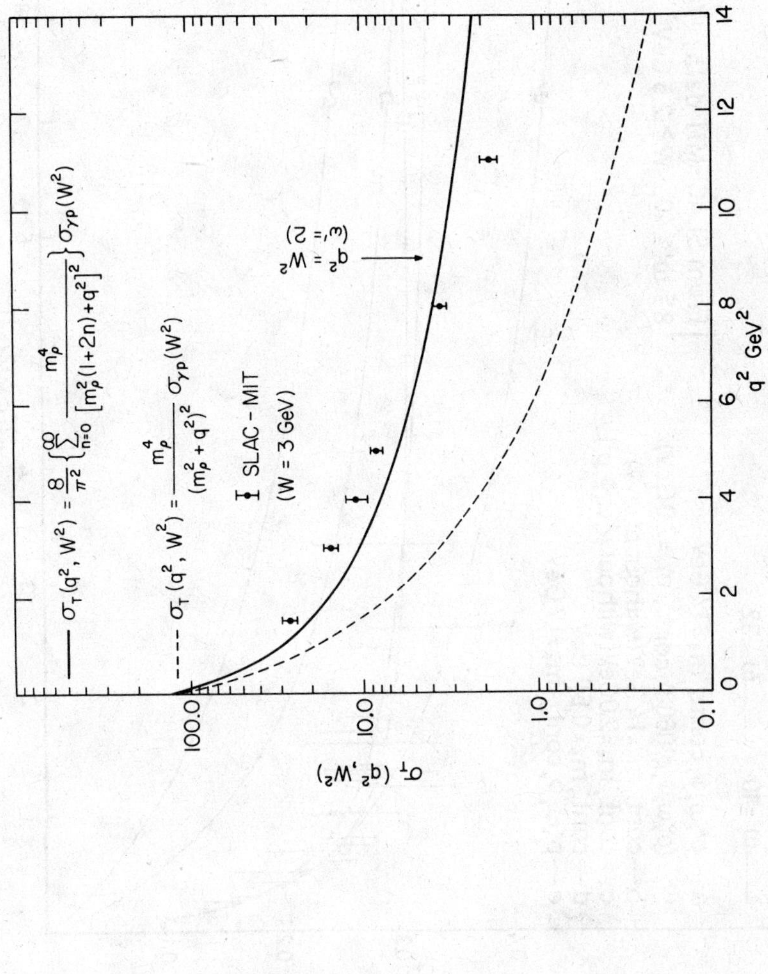

σT (q², W²) = 8/π² {Σ n=0 to ∞ m_ρ⁴ / [m_ρ²(1+2n)+q²]²} σγp(W²)
σT (q², W²) = m_ρ⁴ / (m_ρ² + q²)² σγp(W²)
SLAC - MIT
(W = 3 GeV)
q² = W²
(ω' = 2)
100.0
10.0
1.0
0.1
0
2
4
6
8
10
12
14
q² GeV²
σT (q²,W²)

Fig. 3

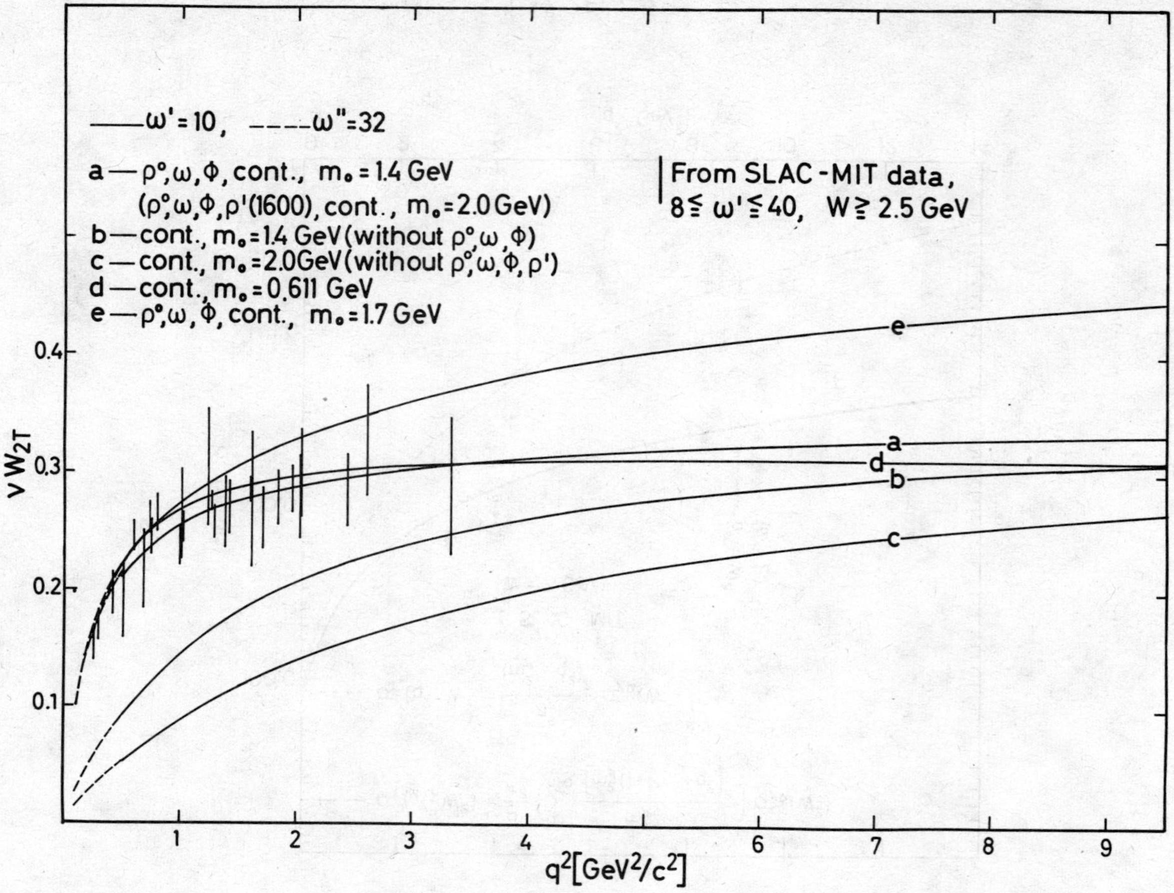
——ω' = 10, ----ω'' = 32
a — ρ°, ω, φ, cont., m₀ = 1.4 GeV
(ρ°, ω, φ, ρ'(1600), cont., m₀ = 2.0 GeV)
b — cont., m₀ = 1.4 GeV (without ρ°, ω, φ)
c — cont., m₀ = 2.0 GeV (without ρ°, ω, φ, ρ')
d — cont., m₀ = 0.611 GeV
e — ρ°, ω, φ, cont., m₀ = 1.7 GeV
From SLAC-MIT data,
8 ≦ ω' ≦ 40, W ≧ 2.5 GeV
ν W2T
0.4
0.3
0.2
0.1
1
2
3
4
6
7
8
9
q²[GeV²/c²]
e
a
d
b
c

Fig. 4

D I S C U S S I O N S

CHAIRMAN: Prof. J.J. Sakurai

Scientific Secretaries: G. Chew, Y. Eylon, E. Rabinovici, D. Wright

DISCUSSION No. 1

- *DERMAN:*

How exactly do you subtract the ρ,ω,ϕ contributions to νW_2 in electroproduction in order to demonstrate their contribution to precocious scaling, and how model-independent is the subtraction?

- *SAKURAI:*

We have done what everybody believes. It is true that for the longitudinal contribution there is some disagreement; for the transverse contribution everybody believes that the ρ,ω,ϕ contribution is given by the double propagator, and as for the magnitude we know what happens at $q^2 = 0$ from photoproduction.

- *DERMAN:*

The reason I asked was that there was once a phenomenological Lagrangian due to T.P. Lee, which produced a good fit to the longitudinal part of νW_2 for ep scattering. It was based on a vector meson dominance Lagrangian, and predicted a scaling contribution from final exclusive $n\rho^+$ channels alone. Your model gives trivially scaling contributions due to vector meson exchange, so how does that reconcile?

- *SAKURAI:*

T.D. Lee made use of <u>elementary</u> ρ^+ exchange with this naive model.

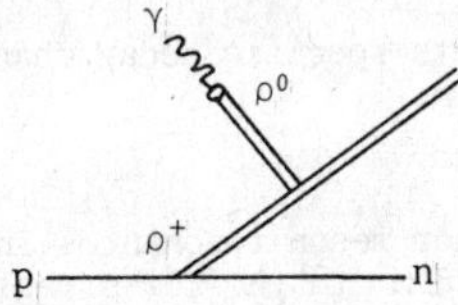

This diagram gives a scaling longitudinal contribution which reproduces the scaling curve. He then multiplied by 5 because R = 0.2 and this fits very well.

I do not think this is realistic because in realistic models ρ^+ is Reggeized etc. I tried to estimate this contribution and it seems it is much smaller than what T.D. Lee calculated.

- *DERMAN:*

Is there any evidence for the $n\rho^+$ channel scaling?

- *SAKURAI:*

I do not know.

- *WAMBACH:*

Does the rising total cross-section at the ISR also imply a rising $\sigma_{\gamma p}$ cross-section which in your model would lead to a breakdown of scaling in ep collisions?

- *SAKURAI:*

This consequence can be drawn, and if one finds in meson-baryon scattering rising total cross-section (which are supposed to be done for π and K at NAL very soon), then one should find also a rising contribution in νW_2. One could restore scaling by changing the spectral weight function a little bit, but this is ugly.

- *WAMBACH:*

Is there an idea how scaling in ep can be continued to scaling in pp collisions in the τ variable?

- *SAKURAI:*

Any connection which exists today depends on models, and there is even the possibility in certain models that one has scaling in ep and not in $pp \to e^+ + e^-X$ collisions. I think scaling in $pp \to e^+e^-X$ is a more crucial test of the quark parton model than scaling in ep collisions.

- *WAMBACH:*

For the comparison with physical data, how many vector mesons are needed in order to reproduce the observed behaviour?

- *SAKURAI:*

It depends on q^2 how many vector mesons have to be included; as a general rule one should say it is sufficient to sum up to masses of the order $|(2\text{-}3) \cdot q^2$.

- *NAHM:*

You just said that for any mass level of the Veneziano model only one combination of the mesons should couple to the photon. But this is not enough to avoid interference of heavy vector mesons, as the levels become denser and denser whereas the widths are not becoming smaller.

Now, even if for large ω you do not need interference terms, they may be necessary if one wants to formulate the new duality for small ω. Do you think that is possible?

- *SAKURAI:*

This problem also exists in hadron physics and there it has not been solved either.

- *COLEMAN:*

In a typical multichannel problem, even if resonances overlap kinematically, there is generally very little interference between them because they go into different final states.

- *SAKURAI:*

Each vector meson seems to have its specific decay channels, for some mysterious reasons, so the interference may be small.

- *EYLON:*

In $ab \to \overbrace{\ell^+\ell^-}^{M^2} + X$, you expect vector meson resonances in M^2 to average a scaling function of the variable M^2/s. Going with fixed M^2 to the low s region, do you expect the resonances in s of the ab system to average the same scaling function?

- *SAKURAI:*

This is another kind of duality, about which I do not have much to say. I did not worry about this problem, because in the pp case there is no s-channel resonance.

- *TAYLOR:*

What does your theory say about the longitudinal part of the inelastic electron-proton scattering cross-section? Does the smallness of σ_L come from some implicit transverse polarization of the vector mesons coupled to the photon?

- *SAKURAI:*

The transverse cross-section in ep inelastic scattering is given by:

$$\sigma_T = \int_{4m_\pi^2}^{\infty} \frac{m^4 \, \rho_T(m^2) \, dm^2}{(q^2 + m^2)^2} ,$$

where $m^4 \rho_T(m^2) \cong$ constant as required by scaling. The longitudinal cross-section is given by:

$$\sigma_L = \int_{4m_\pi^2}^{\infty} \frac{q^2}{m^2} \frac{m^4 \rho_L(m^2)}{(q^2+m^2)^2} dm^2$$

Taking $m^4 \rho_L(m^2) \cong$ constant, we find $R = \sigma_L/\sigma_T \sim \ln q^2$, i.e. diverges like $\ln q^2$. Experimentally $R \sim$ constant. To obtain better agreement with experiment, one can take $m^4 \rho_L(m^2)$ to be a function that decreases with m^2. Schildknecht and I have done some model calculations along this line. For details you may consult Phys. Letters 40 B, 121 (1972), and Schildknecht's Moriond Lectures.

- *GENSINI:*

1) In the Bramon-Etim-Greco model for deep inelastic annihilation, you do not have any problem with interfering states since you sum just one on the infinite sequence of vector mesons. When you turn to deep inelastic electroproduction, you actually have to sum twice before taking the imaginary part of a virtual Compton amplitude, and you can get interference between non-diagonal terms in this way. Can you please state what kind of changes do you expect to creep in when you try to account also for this aspect of your model?

2) Can you please mention some experimentally observable effects typical of this model of extended VDM, which could clearly distinguish it from any other general description of hadron-state dominance of the hadronic e.m. current?

- *SAKURAI:*

1) I use what I call the diagonal approximation. We are talking about total photoproduction, but by the optical theorem this is related to the imaginary part of the γp forward amplitude

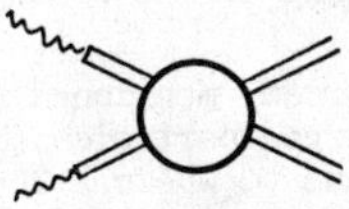

We have put the same vector state V. What happens if the V states are different? Intuition from hadron physics tells us that maybe the off-diagonal terms are not too important. If you look at pp going to $N^*(1420)$ plus proton, that is way down from pp elastic. There is an extra interesting point that somebody here might know something about. What happens to the dip mechanism when the mass changes in diffraction dissociation? If there is a dip then the diagonal approximation is quite justified. If there is no dip, maybe you should try summing over nearby states.

2) Yes, it is very important to see diffractive electroproduction of higher vector states and to study their q^2 dependence. The q^2 dependence should be slower than, say, that observed in ρ electroproduction.

- *CHOUDHURY:*

Will you make a comment on the high-energy behaviour of the central region in the deep inelastic semi-inclusive processes $ep \to ehX$ (h = hadron) from your approach? From the investigations based on the geometrical picture of Chou and Yang, the field theoretical parton model, and recently also from Regge-Mueller analysis as well as bilocal current algebra, a new behaviour is predicted in this region at very high energy. Using the language of rapidity, all these approaches predict either a "hill" or a "valley" but not a plateau. Is it possible to conjecture such a spectacular behaviour from your approach?

- *SAKURAI:*

I have not really worked in that problem. It appears that the existence of a central region is connected to the Pomeron. If as you go to the small ω region the Pomeron starts to play a less important role, then something interesting is bound to happen in the central region.

- *WILCZEK:*

We have learned to expect scaling in neutrino reactions also. Do you have any reason to expect this if your model for electroproduction scaling is true? Do you propose generalized strange vector and axial vector dominance?

- *SAKURAI:*

Yes. I expect the same qualitative behaviour, for example, hadron-like behaviour for high ω, and decreasing importance of the Pomeron contribution for low ω. But the quantitative comparison of duality results is of course more remote.

We do not know anything about A_1, A_1', A_1'', etc. What we need is an $e^+\nu$ colliding beam facility!

- *PATKOS:*

The parton model or light cone algebra predicts a non-vanishing scaling function for νR, where $R = \sigma_L/\sigma_T$. This fact very strictly constrains the possible increase of longitudinal effects in the single arm experiment. Would you compare this statement with the R's used in your models?

- *SAKURAI:*

The model I proposed is valid only for large ω. The light cone predictions were checked only in the low ω region. Therefore, it is quite possible to have deviations from the scaling law in the region of large ω. Experimentally for the high ω region there is some indication that R = constant may fit better than νR = constant (fixed ω).

- *RABINOVICI:*

Could you explain how the quark content of the proton explains the q^2 dependence of the π^+/π^- ratio in the photon fragmentation region?

- *PATKOS:*

In this comment I wish to illustrate by means of a simple example the predictive power of the parton picture and the relationship to the assumption of hadron-like behaviour by the photon.

Dakin et al., in the experiment already mentioned by Prof. Sakurai, have measured the number of positively and negatively charged particles (N^+ and N^-) produced in deep inelastic electroproduction as a function of q^2 and ν, where $X \equiv -q^2/2M\nu$.

According to the parton ideas, in the photon fragmentation region we will find the fragments of the parton that has been ejected from the proton by virtual photon. Using the factorization of this region from the proton fragmentation region as proposed by Feynman, we can write down a sum rule expressing the conservation of electric charge:

$$\left\langle \sum_i Q_i N_i \right\rangle_{\text{events}} = \left\langle Q_{\text{quark}} \right\rangle_{\text{events}} .$$

Using the normed probability functions for finding a parton of a definite type having a fraction X of the momentum of the proton $[u,\bar{u},d,\bar{d},s,\bar{s}]$, the right-hand side can be written as

$$\frac{1}{f_{ep}(x)}\left\{\frac{2}{3}\left[\frac{4}{9}\left(u(x) - \bar{u}(x)\right)\right] - \frac{1}{3}\left[\frac{1}{9}\left(d(x) - \bar{d}(x)\right)\right] - \frac{1}{3}\left[\frac{1}{9}\left(s(x) - \bar{s}(x)\right)\right]\right\} .$$

P. Hasenfratz noticed that the left-hand side is just the measured $A = N^+(q^2,\nu) - N^-(q^2,\nu)$ asymmetry, if one neglects the production of doubly charged particles. Using the explicit probability functions put forward by Weisskopf and Kuti in their model, he obtained a surprisingly good agreement with the data.

Generally the properties of A in the parton model are the following:

a) $A = A(x)$ is a scaling function;

b) at $X \simeq 0$, $A = 0$;

c) at $X \simeq 1$, $A = 2/3 = Q_u$, which is an upper limit for A.

It seems that this measurement can supply us with a very decisive test of the quark parton model.

There is an apparent contradiction if we accept the hadron-like behaviour of the photon. Then the factorization of the Regge residues enables us to write down the conservation of charge for the projectile fragmentation region alone. The sum of the charge of the projectile's fragements would give back the charge of the projectile, actually zero.

However, this latter property is valid only for large $\omega = 1/x$, around $x \simeq 0$, in the VDM of Professor Sakurai. Therefore in this respect there is no contradiction between this and the parton model. The refined VDM will work in a restricted region of x.

ABSORBED MULTIPERIPHERALISM AND RISING CROSS-SECTIONS

L. Caneschi

Table of Contents

ABSORBED MULTIPERIPHERALISM AND RISING CROSS-SECTIONS

L. Caneschi
CERN - Geneva

1. INTRODUCTION

One of the central ambitions of the theory of high energy hadronic interactions is to use the optical theorem

$$\frac{\mathrm{Im}\ A_{el}(s,0)}{\lambda^{1/2}(s,m^2,m^2)} = \sigma_{Tot}(s) = \sum_n \int d\phi_n \, |A_{2n}|^2 \qquad (1)$$

and its non-forward generalization

$$\frac{\mathrm{Im}\ A_{el}(s,t)}{\lambda^{1/2}(s,m^2,m^2)} = \sum_n \int d\phi_n \; A_{2n} \; A^*_{2'n} \qquad (2)$$

to relate the basic properties of multiparticle production processes and of two-body scattering.

The outstanding feature of multiparticle production processes is the small value of $< p_\perp^2 >$, which turns out to be about 0.1 GeV^2 (for pions) independently of s and of the multiplicity.

This feature can be interpreted in a geometrical s channel picture as a consequence of the finite size of the hadrons. The alternative (but not contradictory) way is to assume that it is a manifestation of t channel dynamics (exchanges). The very important difference between these points of view is that in the latter the observed limitation of $< p_\perp^2 >$ is only a projection in the transverse plane of the limitation of <u>invariant momentum transfers</u> t_i. Hence the longitudinal and time components of the momentum transfers are correspondingly limited. This, as we will see and as is better discussed in Refs. 1) and 2), leads to a number of qualitative and quantitative predictions, none of which is disproved by experience so far.

In this lecture we will assume therefore that

$$|A_{2n}|^2 < \prod_i f(t_i) \qquad (3)$$

where $f(t)$ is a suitably decreasing function [t^{-p} is sufficient, with p depending on the details of the model [1]]. Let us remark that (3) requires to be meaningful a definition of ordering of the final particle, since t_i is defined as

$$t_i = \left(P_a - \sum_{j=1}^{i} p_j\right)^2 = \left(P_b - \sum_{j=i+1}^{n+1} p_j\right)^2$$

In turn, (3) defines an ordering, since the preferred configuration will be the one that minimizes the momentum transfers, i.e., in the case of equal masses, the permutation of the final particles in which they are ordered following their position in rapidity.

2. LADDERS

The simplest model which obviously realizes (3) is the ladder model [3]. Let us therefore briefly summarize its features and its shortcomings. In the following, I will vastly oversimplify the mathematics in order to give a simple qualitative understanding. In the ladder model, the A_{2n} satisfy a recursion relation that can be turned into an integral equation for $\mathrm{Im}\,A_{el}$ (Fig. 1) by summation. Since the model viewed in the t channel is a sequence of two-body scattering, the t channel angular momentum β provides a simple diagonalization of the integral equation, of the form (at $t=0$):

$$A(\beta) = g\,u(\beta) + g\,u(\beta)\,A(\beta) \tag{4}$$

which, in this oversimplified treatment, can be solved algebraically to give

$$A(\beta) = g\,u(\beta)\left[1 - g\,u(\beta)\right]^{-1} \tag{5}$$

Here, $u(\beta)$ is the t channel partial wave projection of the kernel, analytic in β at the right of $2\alpha-1$, where α is the spin of the exchanged elementary particle or the intercept of the exchanged Regge trajectory. In the former case, the rightmost singularity of the kernel is an isolated pole, in the latter a logarithmic branch cut of the form

$$u(\beta) = \frac{c}{\alpha'} \int_{-\infty}^{0} \frac{dt\; f^2(t)}{\beta - \beta_c - 2\alpha' t} \tag{6}$$

In (6), $f(t)f(t')$ is a parametrization of the vertex function of Fig. 1 assumed factorizable for simplicity [4]. Assuming for the moment that $f(0) \neq 0$, it is obvious that $\mathrm{Re}\, u(\beta)$ ranges from 0 to $+\infty$ in the interval $+\infty > \beta > \beta_c$. Hence independently of the choice of g (finite and positive) in (5), $\mathrm{Im}\, A_{el}(s,0)$ is always dominated by an isolated Regge pole leading by an amount $\Delta(g) = \alpha(g) - \beta_c$. The function $\alpha(g)$ is implicitly defined by $u(\alpha) = g^{-1}$ (see Fig. 2). Hence

$$\frac{\mathrm{Im}\, A_{el}(s,0)}{\lambda^{1/2}(s,m^2,m^2)} = \sigma_{Tot}(s) = s^{\alpha(g)-1}\left(1 + O\left(s^{-\Delta}\right)\right) \tag{7}$$

Remembering now that in the ladder approximation there is a one-to-one correspondence between the g^n term in a power series expansion of σ_T and σ_n, one easily obtains

$$M_p(s) = \langle n(n-1)\ldots(n-p)\rangle = \left(\frac{d\alpha}{dg}\right)^p \ln^p s \left(1 + \frac{d^2\alpha}{dg^2}\left(\frac{d\alpha}{dg}\right)^{-2} \ln^{-1} s + \ldots\right) \tag{8}$$

and

$$\Delta_p(s) = \frac{d^p\alpha}{dg^p} \ln s + c_p \tag{9}$$

where Δ_p is the integral of the pth inclusive correlation function, and all the derivatives with respect to g are taken at the value $g = g_{physical} \neq 0$, where $\alpha(g)$ is by definition analytic. [The behaviour of the exclusive cross-sections, in principle given by

$$\left.\frac{d^n\alpha}{dg^n}\right|_{g=0} \ln^n s \quad s^{-\Delta}$$

is more delicate, since $\alpha(g)$ is analytic around $g = 0$ only in non-Reggeized models like the $\lambda\varphi^3$.]

In conclusion, we see that the ladder approximation leads very straightforwardly to the existence of an isolated Regge pole with the consequent bounded scaling of the n particle inclusive distributions and short-range correlations [these two properties are reflected in an integrated way in (8) and (9), but are trivial to derive directly from (7) and the factorization properties of the n particle inclusive distribution in the model].

3. TROUBLES

The most obvious shortcoming of the ladder approximation is the lack of s channel unitarity. As a consequence, the intercept of the leading Regge pole in the overlap function grows for large values of the coupling constant proportionally to g, thus violating the Froissart bound for g larger than some $\bar{g}$.

Furthermore, the whole scheme is hard to reconcile with the existence of diffraction scattering, i.e., with $\sigma_T(s) \underset{s\to\infty}{\to}$ const. One might say that choosing $g=\bar{g}$, it is possible to obtain $\alpha(\bar{g})=1$. Apart from the fact that it is not easy to accept that a simple phenomenon-like constant cross-sections happens because the coupling constant has a very specific value, this solution is also inconsistent. In fact if, say π-π cross-sections are asymptotically constant, i.e., correspond to $\alpha(0)=1$, their multiperipheral iteration leads to an output Regge pole with $\alpha(g) > 2\alpha(0)-1 = 1$.

This observation is the root of the several diseases associated with inelastic Pomeron exchange in the framework of a model in which an integral equation can be derived, like, at the exclusive level, the Finkelstein-Kajantie problem and, at the inclusive level, the triple Pomeron contribution. In the former kind of trouble, one observes that multiple Pomeron exchange leads to exclusive cross-sections that vanish asymptotically [like $(\ln \ln s)^{\eta}/\ln s$ thanks to the softness of the logarithmic cut] but necessarily sum to a power of s. In the latter, the contribution of the triple Pomeron region to the inclusive spectrum, of the form

$$\frac{d\sigma}{dx\,dt} \simeq (1-x)^{-2\alpha(t)+1} \qquad f(t)$$

inserted in the energy conservation sum rule

$$\sigma_T(s) = \int E \frac{d\sigma}{d^3p} d^3p$$

yields a leading contribution

$$\frac{f(0)}{\alpha'} \quad \ln\left[\frac{\alpha' \ln s}{\left[\frac{df}{dt}\right]_{t=0}} + 1\right] \tag{10}$$

that forces σ_T to diverge asymptotically, contrary to the assumption that the leading singularity is a pole at $J=1$. The source of the trouble is obviously in the fact that the iteration of a pole at $J=1$ with a cut also at $J=1$ leads to a slightly more singular behaviour at each step.

To finish the list of the too strong properties implied by the ladder approximation in its simple form discussed so far, let us remind that it leads to short-range correlations that, as is well known, force all exclusive cross-sections to decrease faster than any power of $\bar{n}$ with respect to σ_T, and this is obviously incompatible with diffraction (if $\sigma_T \to \text{const}$, $\sigma_{el} \to \text{const}$, up to a finite power of $\ln s$).

4. ABSORBED MULTIPERIPHERALISM

The difficulties presented in the previous section can somehow be circumvented without giving up a ladder structure, and therefore an integral equation simply diagonalized in t channel angular momentum by :

i) assuming that the leading trajectory intercept is $1-\varepsilon$, with the size of ε obviously related to the inelastic Pomeron coupling, or

ii) decoupling the Pomeron from inelastic channels at $t=0$. In this case the discontinuity of $u(\beta)$ at $\beta=\beta_c$ vanishes, and it is no longer necessary to have $\alpha_{out} > 2\alpha_{in}-1$ [5),6)].

Both possibilities have little phenomenological appeal, the former leading to cross-sections that decrease to zero, the second to a dip at $t=0$ of the large x inclusive p distribution in pp collisions.

An alternative point of view is to realize that the problems of the ladder approximation stem from the lack of s channel unitarity, and to try to enforce some. A traditional way to do so is to consider the ladder as an inelastic potential, rather than amplitude, and to take the matrix elements of this potential between eigenstates of elastic scattering, i.e., between distorted plane waves rather than plane waves. Thus one follows closely the absorption approach, for instance as developed by Gottfried and Jackson [7)], in the inelastic case, and assumes (Fig. 3)

$$A_{2n}(s,b) = M_{2n}(s,b)\; S_{22}^{1/2}(s,b) \tag{11}$$

Here M_{2n} is the projection of the ladder amplitude in the s channel impact parameter b, S_{22} the elastic S matrix, and the variables defining the final states are understood. We will refer to this model as the absorbed MPM (AMPM). This recipe to construct A_{2n} is only a first step towards the enforcement of s channel unitarity, but presents very attractive features like :

A) if the leading singularity is a Pomeron pole, the absorptive correction to multiparticle production generates a negative cut in the overlap function, twice as big as the Amati-Fubini-Stanghellini cut and the net result is therefore a very welcome change of sign of the AFS cut [8];

B) the cut-off properties in the momentum transfers t_i are the same for M_{2n} and A_{2n} (assuming that also elastic scattering is highly peaked in t, as it is). Hence, the properties of the multiperipheral model, like the bound $< n^p > < c_p(\ln s)^p$ proved in Ref. 1) on the basis of the inequality (3), hold also here [9].

What is spoilt is the simple structure in the t channel J plane, hence the leading Regge pole and the too strong short-range correlation properties of the ladder.

5. ASYMPTOTIC PROPERTIES

In this section we will consider the extreme asymptotic properties of the AMPM in order to check that the diseases connected with diffraction scattering have been actually disposed of. To do so, let us concentrate on the dangerous situation of multi-Pomeron exchange along the ladder [10].

We can now ask the self-consistency requirement that the three Pomeron singularities (at $J=1$, but not necessarily poles) used in the $M_{2n}(P)$, in the absorption (P') and the one obtained from the optical theorem (P") are the same. The solution is simple to find : the repeated exchange of a singularity at $J=1$ in M_{2n} generates a pole at $J=1+\Delta$ (the precise nature of the cut at $J=1$ in the kernel being irrelevant), the s channel impact parameter representation of which is

$$M(b,s) = e^{\Delta \ln s} \qquad e^{-b^2/\alpha' \ln s}$$

Thus the partial waves obtained from the shadow of the ladder saturate the unitarity bound in partial wave (for a completely absorbing disk) at a value of $b < \alpha' \Delta \ln s$. Correspondingly the factor $S^{\frac{1}{2}}(b,s)$ develops a zero that prevents A_{2n} from exceeding $\frac{1}{2}$. Hence, the only possible self-consistent solution consists of an asymptotically black disk of radius proportional to $\alpha' \Delta \ln s$ (plus possibly a grey fringe), and $\sigma_T \to \alpha'^2 \Delta^2 \ln^2 s = \frac{1}{2}\sigma_{el}$ for $s \to \infty$. The t channel J plane structure is more complicated now : at $t=0$ the leading singularity must be a triple pole (to account for the $\ln^2 s$ behaviour of σ_T). Furthermore, the diffraction peak must depend on $t \ln^2 s$, to keep $\sigma_{el} < \sigma_T$. Hence one easily guesses the resulting $((J-1)-t)^{-\frac{3}{2}}$ form [11].

This model can be studied in detail [12] to show that the exclusive sections σ_N approach a constant asymptotically, and that the inclusive distributions scale as expected, but long-range correlations are present, due to the non-factorizable nature of the leading singularity, that in particular destroys the usefulness of the inclusive Mueller approach.

Also in the $x \simeq 1$ region (that now only improperly can be referred to as the triple Regge region), the inclusive distribution scales and diverges for $x \to 1$, but absorptive effects reduce a little the type of singularity, which turns out to be

$$\frac{1}{\sigma_T(s)} \frac{d\sigma}{dx} \simeq (1-x)^{-1} \ln^{-3}(1-x)$$

and therefore is integrable [13]. Hence, as expected, the insertion of s channel unitarity corrections takes care asymptotically of the mismatch found in the energy conservation sum rule when the leading singularity is assumed to be a pole at $J=1$.

In conclusion, this asymptotic scheme shows the possibility of building a model that

i) is multiperipheral [in the sense of Ref. 1)] but has the Froissart bound built in;

ii) can accommodate in a self-consistent way diffractive phenomena, still keeping all scaling functions bounded;

iii) has scaling inclusive distributions, without factorization and with long-range correlations;

iv) has an acceptable "triple Regge" limit, without vanishing factors and with a spike in the $x \to 1$ limit.

6. APPROACH TO ASYMPTOPIA

The asymptotic treatment presented in the previous section has extreme features that certainly do not show up in present energy phenomenology. Let us therefore look for parameters still small at present energy and such that we can envisage an ISR phenomenology as a perturbative expansion of the asymptotic solution of Section 5. Two parameters immediately suggest themselves as suitable :

1) the ratio $\sigma_{el}/\sigma_{inel}$ controls the importance of absorption, i.e., of s channel Pomeron iterations. At present energy, this ratio is about 1/5, hence it is reasonable to consider absorptive effects and s channel unitarity as a perturbation;

2) the average subenergy of a π–π pair in the final state is $< s_i > \simeq$ $\simeq 600\ \text{MeV}^2$. This small value suggests that also the t channel Pomeron exchange can be treated perturbatively.

In this respect, two possible definitions of the perturbation expansion can be proposed. Consider, for instance, the π–π elastic scattering amplitude used as a kernel of a multiperipheral integral equation of the AFS type (Fig. 4). The usual multiperipheral approach would be to define Pomeron exchange as a smooth contribution that extrapolates the high energy behaviour all the way down to threshold with non-vanishing value (P in Fig. 4a), and thus separates a low energy kernel (M in Fig. 4a). Then one can solve the integral equation with M exchange and add perturbative terms in P^n. This approach turns out not to be very attractive because :

a) it lacks a phenomenological definition in terms of a classification of final states : whereas final states that present one (or more) large energy gap belong to the first (or higher) perturbative term, final states that present no large rapidity gap can contain an arbitrary number of P exchanges;

b) the sum of the perturbative series in the t channel multiple P exchange leads to a pole at $1+g$. Hence, if the zeroth order term (i.e., the solution of the iteration of the M kernel) is a pole around $J = 1$ (in order to explain in terms of multiple M exchange the existence of diffraction), the perturbative series is a series in $g \ln s/\bar{s}$. From the size of $\Delta\sigma_T$ [see Ref. 6)] $g \simeq 0.1$, hence in a perturbative approach, as described here, in which the scale s can only be the usual $s_o \simeq 1$ GeV, the perturbation parameter at ISR energy is not small enough, i.e., with this definition of P exchange at ISR there are important double P exchange contributions, even if not immediately identifiable as such from the presence of two large rapidity gaps.

An alternative way of splitting the kernel in low and high energy parts is the straightforward one of introducing a cut-off $\bar{s}$ (Fig. 4b). As a function of $\bar{s}$ the solution of the integral equation with kernel K_o will show a leading pole of intercept depending on $\bar{s}$, and if one chooses $\bar{s}$ such that the intercept is one, one can give a first order description of the large subenergy correction ΔK in terms of the solution of the iteration of K_o.

This approach is nice because the final states contributing to the zeroth, first, etc., perturbative term are easily classified, in that they show no, or one or more, rapidity gaps larger than $\ln \bar{s}/m^2$.

The events corresponding to the first order are the ones which show one large rapidity gap and, in particular, the single diffractive events. This approach leads to first order to a (less arbitrary) two-component model. Furthermore, since the zeroth order term is a factorizable pole obtained by the multiperipheral iteration of a low energy kernel, one understands why properties like factorization, approximate short-range correlations and Poisson-like distributions, hold for energies up to NAL, i.e., as long as the iterative effects connected with the rise of σ_T are not sizeable.

The predictive power of this scheme is that it connects the appearance and the abundance of large gap events with the rise of σ_T. From the triple Pomeron contribution alone, one estimates $\Delta\sigma_{inel} \simeq 2.2$ mb in the ISR range, to be compared with the experimental 3 mb. For details on this estimate, we refer the reader to Ref. 6). Other interesting subjects are the interpretation of the elastic $d\sigma/dt$ in the present framework and the comparison of our approach to alternative schemes like the one advocated by Gribov, Abarbanel and Chew, and the eikonal approach. Unfortunately, I do not have time to go into these matters here, and again the reader is referred to Ref. 6), to which the present talk can be considered as an introduction.

REFERENCES

1) A. Bassetto, L. Sertorio and M. Toller, Nuovo Cimento 11A, 447 (1972).

2) L. Caneschi, CERN preprint TH.1704, submitted to Nuclear Phys. (1973).

3) D. Amati, S. Fubini and A. Stanghellini, Nuovo Cimento 26, 896 (1962), hereafter referred to as AFS.

4) A detailed derivation of the multi-Regge integral equation which uses the notations used here can be found in L. Caneschi and A. Pignotti, Phys.Rev. 180, 1525 (1969).

5) I.J. Muzinich, F.E. Paige, T.L. Trueman and L.L. Wang, Phys.Rev. 4D, 1048 (1972), as well as in the Appendix of Ref. 6).

6) D. Amati, L. Caneschi and M. Ciafaloni, CERN preprint TH.1676, to be published in Nuclear Phys. (1973).

7) K. Gottfried and J.D. Jackson, Nuovo Cimento 34, 735 (1964).

8) L. Caneschi, Phys.Rev.Letters 23, 257 (1969).

9) This property does not imply mathematically the bounded scaling of n particle inclusive distributions, but physically this is the most reasonable way to obtain it.

10) If, as it will turn out in the following, the P singularity is not factorizable, there are technical problems to write a recursion relation also for $M_{2n} \to M_{2,n+1}$. The simplest way out is to use the non-factorizable P singularity as an asymptotic behaviour of, say, the π-π scattering kernel of the AFS original equation, thus using the factorization properties of the π to define the recursion relation.

11) J. Finkelstein and F. Zachariasen, Phys.Letters 34B, 631 (1971).

12) L. Caneschi and A. Schwimmer, Nuclear Phys. B44, 31 (1972).

13) L. Caneschi and A. Schwimmer, Nuclear Phys. B48, 519 (1972).

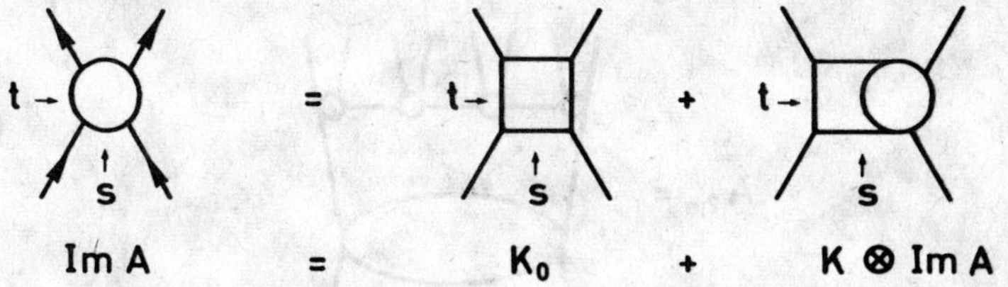

FIG. 1

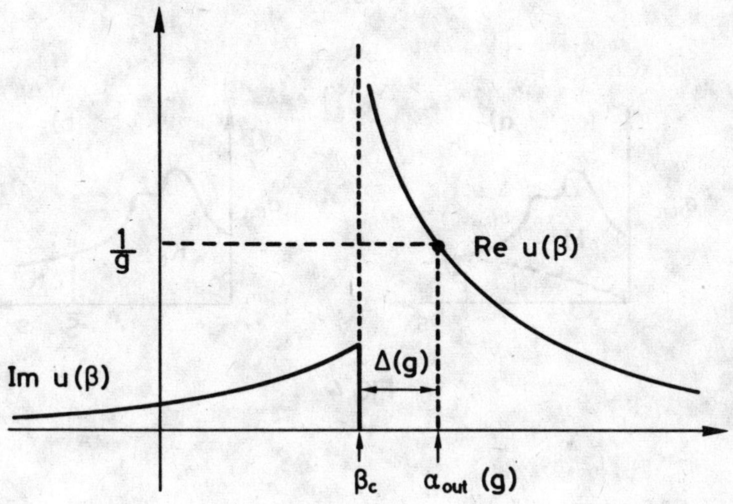

FIG. 2

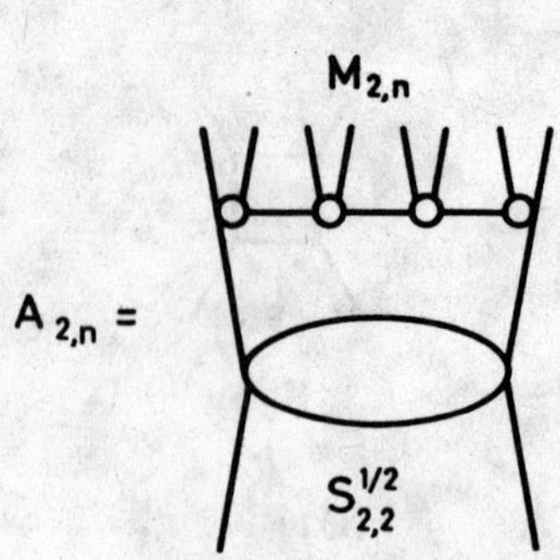

FIG. 3

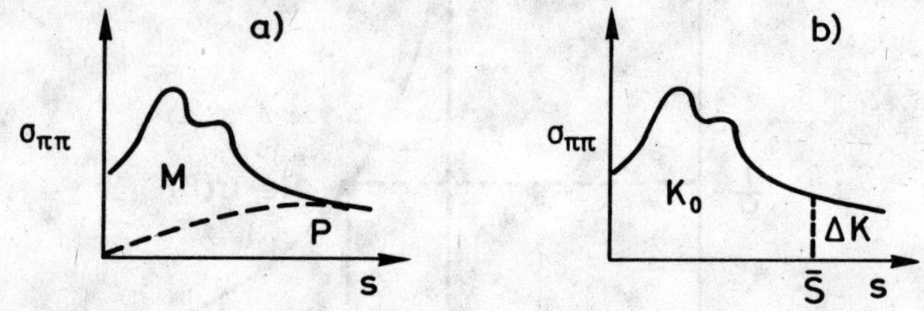

FIG. 4

D I S C U S S I O N S

CHAIRMAN: Prof. L. Caneschi

Scientific Secretaries: P. Gensini, D. Nanopoulos

DISCUSSION No. 1

- *RABINOVICI:*

On which parameters does the average subenergy in the Caneschi-Schwimmer model depend? Can any of them be determined by present experiments?

- *CANESCHI:*

The parameter which determines the average subenergy is the Pomeron inelastic coupling, that in turn determines the coefficient of the asymptotic logarithmic growth of the radius with energy.

- *RABINOVICI:*

Do you remain with low subenergies even asymptotically?

- *CANESCHI:*

Yes, you do since the average multiplicity remains logarithmic. The calculation of the average value of the subenergy is not particularly simple, because if you want a realistic model, you should make a coupled channel model in which you have this strange Pomeron and the meson trajectories both exchanged. Now the average of the subenergy is determined by how much of the leading singularity is made by the meson exchange and how much by the Pomeron exchange. In this sense, you can write a coupled channel problem in which the position of the outcome intercept is determined by an equation where a determinant found by the couplings has to be set equal to zero, and the relative weight of the couplings together with the intercepts of the input trajectories determine the average subenergy, which is asymptotically a constant.

- *RABINOVICI:*

Is there any result about inclusive cross-sections that can be computed in the multiperipheral model but not in the Regge-Muller analysis?

- *CANESCHI:*

The Muller ansatz is completely empty as long as you relate an inclusive distribution to a forward three-body amplitude nobody has seen. In order to make a phenomenological weapon out of the Muller ansatz, you have to assume that the three-to-three body amplitude has certain properties, like dominance of Regge trajectories and, in particular, if you want to make this ansatz actually tight and powerful, you have to assume that the Regge trajectories which appear in the double or single O(3,1) expansion of the Muller amplitude are the same as in two-body processes. In a very recent paper I have shown, following a method of Bassetto and Toller, that this is not the case in a multiperipheral model; in this model, in addition to the two-body J-plane singularities necessarily present, a family of "daughters" and fixed poles appears; these extra singularities are required to translate in a Muller language the multiperipheral condition of limited momentum transfers, as well as to enforce tranverse momentum conservation. These extra singularities, proportional to $\langle p_\perp^2\rangle^{-1}$, turn out to be quantitatively very important and in my opinion they prevent the possibility of constructing a sensible phenomenology starting from the Muller ansatz and expose the irrelevance of the "Ferbel plots".

- *RABINOVICI:*

How is the self-consistency achieved in the perturbative approach and how do the various quantities like $\langle n\rangle$ and σ_n change when one adds more terms in the perturbative series?

- *CANESCHI:*

The fact is that self-consistency is not achieved, and the energy-conservation sum rule is exactly the tool which brings you from the n-th to the (n+1)th order in the perturbation approach. You start from P_0 (zeroth-order Pomeron), and therefore a cross-section which is constant; then you have an inclusive distribution which can be described in this approximation by a triple P_0 coupling, and integrating it (i.e. considering the contribution of the leading particle to the energy-conservation sum rule) you obtain an overlap function which is more singular than the simple pole. Then the sum rule leads you from a P_0 to something behaving like (log s), which can call P_1, and enter it again into the scheme; you can then continue on this line and easily see that this scheme is not self-consistent. The self-consistency is achieved only when you sum over all the perturbative approach, including also absorption.

- *MOEN:*

How do you avoid a zero triple-Pomeron coupling in your scheme? To be more specific, what is the form of the triple-Pomeron contribution in your approach, and what form does the trajectory take near t = 0? Does your absorbed Pomeron factorize at t = 0?

- *CANESCHI:*

The form in the asymptotic limit is simple when you integrate over $p_\perp$, and takes the form

$$\frac{1}{\sigma}\int \frac{d^2\sigma}{dydp_\perp^2} \sim (1-x)^{-1}\,[\log(1-x)]^{-3} .$$

The strong shrinkage is due to a relation between t and $(J-1)^2$, since the J-plane singularities are given by a term

$$\left[(J-1)^2 - t\right]^{-3/2} .$$

This means that there is a relation between $\log^2$ s and t. You can see this in a simple way: since you are saturating the Froissart bound, σ_{tot} goes like $\log^2$ s and its square like $\log^4$ s, but σ_{el} can not become larger than σ_{tot} and therefore the shrinkage has to go like $\log^2$ s. At t = 0 you have a triple pole at J = 1, then as t becomes different from zero you have two complex conjugate cuts which remain at Re J = 1. There is a strange connection between this feature and the 0(2,1) invariance of the whole model: the conjugate variables to p and to J - 1 are b and the rapidity y and all the model depends only on $y^2 - b^2$.

Therefore you can consider these variables as the components of a Lorentz-metric vector and show that the whole theory is invariant and therefore only the combination I wrote can appear. From dimensional reason you can derive also the $-3/2$ exponent.

This is not a factorizable singularity: if you had a factorizable singularity of this kind you would get a mismatch in the energy-conservation sum rule of a factor $\log^2$ s instead of consistency.

- *MOEN:*

You mentioned something about the sign of the Pomeron-Pomeron cut, namely that the AFS cut-sign becomes negative when absorption is included. Can you expand on this?

You also use the finite rapidity-gap scheme of Abarbanel, which clearly leads to a positive Pomeron-Pomeron cut. How do you avoid this?

- *CANESCHI:*

As you perhaps recall in the old multiperipheral model one obtains a Pomeron pole from the sum of iterated pion exchanges. But these states do not exhaust all intermediate states, because one has forgotten at least the elastic one which produces the Amati-Fubini-Stanghellini positive cut which is known on many grounds to be wrong; on the other hand one has a phenomenologically right prescription which gives the negative sign for the cut, namely the absorptive model. As long as you stay in the t-channel the sign of the cut is positive, but in the s-channel you have to apply absorptive corrections which lead you to a negative cut. Our calculation is identical to the Abarbanel one as long as we compute in the t-channel; but we interprete this result as a "potential" rather than the final amplitude as he does.

LARGE MOMENTUM TRANSFERS AND COMPOSITENESS

L. Caneschi

Table of Contents

LARGE MOMENTUM TRANSFERS AND COMPOSITENESS

L. Caneschi - CERN - Geneva

1. INTRODUCTION

It is generally agreed that hadrons should be regarded as composite objects, the evidence in this sense being provided for instance by their electromagnetic form factors and, within the realm of purely strong interactions, by the hypothesis that particles lie on Regge trajectories.

The precise nature of this compositeness plays little rôle in determining the qualitative features of the bulk of strong interactions, the only property that really matters being the softness of the hadrons, i.e., their reluctance to absorb large momentum transfers. Therefore much attention has been recently devoted to phenomena that, on the contrary, are characterized by the presence of large momentum transfers; even if comparatively rare, they are in fact of extreme interest because they offer the possibility of investigating the details of the bound state structure.

A celebrated example of this type of reactions is provided by the SLAC deep inelastic experiments, which are readily explained if one accepts the idea of elementary constituents (partons or quarks). The small value of σ_L/σ_T seems also to suggest a spin $\frac{1}{2}$ structure for the charged components, represented by canonical fields with point-like interactions with the electromagnetic current.

Depending on whether currents are present, and whether the final state is specified (exclusive) or unspecified (inclusive), large momentum transfers can occur in four classes of reactions involving hadrons.

1) Inclusive current experiments, like the already mentioned deep inelastic scattering, or in general virtual photon reactions ($e^+e^- \rightarrow$ hadrons, μ pair production, etc.); they seem to exhibit canonical scaling.
2) Exclusive current experiments, like electromagnetic and weak form factors. These experiments show a power-like decrease in q^2.
3) Exclusive hadron experiments, like, e.g., large t two-body scattering. The exponential peak at small t seems to shallow smoothly into a power-like behaviour for these reactions and, correspondingly, the s dependence at fixed angle also suggests an inverse power.

4) Inclusive production at large transverse momentum in hadron-hadron collisions; also here an exponential p_T dependence at small p_T goes smoothly into a large p_T behaviour that can be well described by a power.

Stimulated by the SLAC results, elementary component models of the hadrons have been extensively considered recently [1)-6)]. In particular, it has been realized that the existence of a hard point-like component first detected by the interaction with currents will play a dominant rôle also in the large momentum transfer hadronic phenomena 3) and 4) [3),4)].

Actually, it might be argued that in purely hadronic events, there is always the possibility to attribute the flattening of momentum transfer distribution to a rescattering. In other words, a large momentum transfer to a particle can be given in many small bits and, of course, one cannot exclude that it is their cumulative effect that simulates a power. However, we will see that a reasonable unified picture of small and large momentum transfer events can be obtained in a framework in which elementary power-like behaviour does appear also for processes 3) and 4). Rescattering and other iterative processes are allowed and desired, but they are not expected to change the basic patterns and are not involved therefore for the qualitative aspects of large momentum transfer phenomena.

In this talk, we will exhibit the connections that are expected among the afore-mentioned four classes of phenomena in an elementary component approach, by explicitly developing a very simple picture of the hadrons as relativistic bound states of two (or more) "quarks" [7)].

In particular, we would like to stress two points that emerge from this simple model, but that we believe to be of more general validity:

A) against the widespread idea that the hard behaviour, masked by the soft structure at small momentum transfers, will eventually emerge from it, but that the two manifestations are somehow unrelated, e.g., they can be summed, we suggest that the soft and hard structures are different manifestations of the same bound state nature of the hadrons, that turn smoothly into each other;

B) the four classes of phenomena are differently sensitive to the <u>nature</u> of the components and to the <u>degree</u> of compositeness. We will see that the inclusive events 1) and 4) test especially the existence of point-like constituents, whereas the exclusive 2) and 3) depend on how much composite the hadrons are.

We cannot escape the problem, common to all quark-parton approaches, of the actual production of quarks. This problem is far from being understood and we will devote a final paragraph to some recent speculations on the matter. For the moment we will consistently assume that everything happens as if the quarks could be produced and decay without interacting.

2. FORM FACTORS AND TWO-BODY SCATTERING

We will consistently consider the hadrons as bound states of elementary constituents. The simplest bound state model that is known to reggeize is the ladder $\lambda\varphi^3$. In the same model, the electromagnetic form factor $F(q^2)$ behaves, for large q^2, like [8] $(\psi^J(q^2))/q^2$, where $\psi^J(q^2)$ is the asymptotic behaviour of the spin J bound state wave function where one of the components has large space-like mass q^2, the other being finite (Fig. 1).

In the usual ladder $\psi^J(q^2) \approx (q^2)^{-1-J}$. This result is sensitive to the degree of compositeness. For a scalar bound state of N components we expect a number of extra powers in q^2 linear in N; for instance, in $\lambda\varphi^3$ we have $\psi^0_N(q^2) \approx (q^2)^{-2N+3}$. This dependence on N has a simple physical meaning: to reconstitute the bound state when one component has received a substantial kick, some information must be transferred to the other $N-1$ components, and in the region in which q^2 is much larger than all the masses involved, each information "costs" inverse powers of q^2.

The q^2 power dependence is also sensitive to the spins of the exchanged particles and of the constituents. In particular, for a spin $\frac{1}{2}$ bound state of spin 0 and spin $\frac{1}{2}$ with the exchange of spin 0, the magnetic form factor $G_M(q^2)$ still behaves like $(q^2)^{-2}$. Even if for higher spin components we do not have in general models that explicitly reggeize, we believe this relation between the degree of compositeness and decrease of the bound state wave function to be of a rather general nature.

In this kind of model, the Regge trajectory decreases to a finite limit when $t \to -\infty$; this limit is related to the degree of compositeness in the same way as the form factor, e.g., in $\lambda\varphi^3$ $\alpha(t) \to -2N+3$. This can be understood observing that in the $t \to -\infty$ limit the lowest order term (the box diagram in the usual ladder) dominates. Therefore, in a composite model of this type, a Regge behaviour $s^{\alpha(t)}$ smoothly goes into a power behaviour at large t.

In the limit $N \to \infty$ the characteristic features of the dual model are recovered. Remark that in this limit the properties of scaling and reggeization can be accommodated, but at the price of accepting exponentially decreasing form factors (and exponential decrease with s in two-body reactions at fixed Θ). In both cases the phenomenological situation seems to favour a power behaviour that we interpret as the manifestation of finite compositeness.

In hadronic reactions that do not involve large momentum transfers, both in two-body (Fig. 2) and many-body (Fig. 3), only the soft composite nature of the hadrons can be observed, and this model is indistinguishable from any other multiperipheral (actually multi-Regge) one.

What happens when we start increasing the value of the momentum transfer is pictorially represented in Fig. 2a; the ladder shrinks into the lowest order term [i.e., as we said $\alpha(t) \to -2N+3$] and the degree of compositeness of the scattered particles also plays a rôle through the appearance of a $[\psi(t)]^2$ factor. The appearance of this factor has a very obvious physical meaning and represents the difficulty of having a composite system absorb a large momentum transfer without breaking up. In a slightly different picture (which, with our symbols, would be represented by Fig. 2b), Blankenbecler, Brodsky and Gunion [4] expect a cubic dependence on the asymptotic bound state wave function.

3. VIRTUAL γ INCLUSIVE

We have already stressed that in the composite model that we are considering, the hadrons try to minimize the number of large momentum transfer interactions. Hence for deep inelastic scattering the structure of Fig. 4, strongly reminiscent of the usual parton or multiperipheral model, will be the dominant one. From the requirement that t_1 is finite (in order not to loose a factor related to the compositeness of the first bound state vertex), we obtain as usual that $(p_1+p_2)^2 \sim -q^2$. Therefore the regular hadron picture will fill up a length of rapidity of order $\log \omega$ which, for large ω, will include a proton fragmentation region (indistinguishable in this limit from the usual hadronic one) and eventually a plateau. Hence we recover the usual results of Regge expansion in ω for large ω of W_1 and νW_2 and a contribution to the final state multiplicity of the form $g \log \omega + c$ (where g is the same coefficient that determines the $\log s$ multiplicity in hadron-hadron interactions). To this we must add the contribution coming from the two quarks 1 and 2 (whether they recombine or dress) which will, in general, be a $\tilde{n}(q^2)$.

As for the p_T distribution of the final state, let us remark that in the small ω region we expect a $\langle p_T^2 \rangle$ larger than in the purely hadronic case. In fact, when $\omega \to 1$ the minimum value of t_1 behaves like $(\omega - 1)^{-1}$ and a correspondingly large value of p_T^2 can be absorbed by the bound state wave function without sensible loss. This dependence on ω of the minimum value of t_1 will, of course, also affect the hadron spectrum in the proton fragmentation region, through the appearance of corrections of the form ω^{-1}. The necessity of these terms can be inferred from energy momentum conservation alone. In a Mueller language, they appear as daughters of the usual Pomeron (and also non-leading) trajectories, in complete analogy with the situation arising in one-particle inclusive spectra in hadron-hadron collisions.

INCLUSIVE PRODUCTION OF HADRONS OF LARGE TRANSVERSE MOMENTUM

Some recent ISR experiments [9)] have investigated the large p_T inclusive production near $\theta^{cm} \sim 90^o$. The main features of the data are:

1) the sharp exponential drop of the p_T distribution for $p_T \lesssim 1.5$ GeV turns smoothly into a much shallower behaviour that suggests a power decrease;

2) even in the ISR range, some energy dependence is observed in the data;

3) the particle composition changes substantially from the small p_T region to the large p_T one; in the latter pions do not dominate.

We have recently calculated [10)] the large p_T spectrum in the multiperipheral model with elementary particle exchange and production (Fig. 5), obtaining a satisfactory explanation of the features 1)-3). For $s \to \infty$ large p_T power-like behaviour is obviously expected and turns out to be $(p_T^2)^{-4}$ for π exchange. This power dependence in p_T is modulated, in the pionization region by a function $F(\tilde{\omega})$ of the dimensionless variable $\tilde{\omega} = \sqrt{s}/p_T$, that is related to the deep inelastic scaling functions of the model by

$$F(\tilde{\omega}) = \tilde{\omega}^{-\alpha-1} \int_1^{\tilde{\omega}} d\omega \, \omega^{\alpha} \left(1 - \frac{\omega}{\tilde{\omega}}\right)^{2\lambda+\alpha} \chi(\omega) \, \chi(\tilde{\omega} - \omega) \qquad (1)$$

Here α is the intercept of the Regge trajectory generated by the ladder, and λ is related to the spin of the exchanged particle ($\lambda = 1$ for π). The function χ is the scaling function (the analogue of νW_2 in the scalar case) of the model. This scaling law is common to parton calculations [3)] and is in reasonable agreement with the data. In particular, it offers a suitable explanation of the energy dependence: in fact, even if s is very large, the relevant dimensionless variable of the problem turns out to be (not unexpectedly) $\sqrt{s}/p_T$ at $x=0$, that has values of the order of 10 in the range considered here. Furthermore, the structure of F in terms of χ clearly shows a late asymptotism of the former with respect to the latter. Even if the function χ does not necessarily have to be identical to νW_2 for ep scattering, it is amusing to evaluate (1) with this particular choice. The resulting $p_\perp$ distribution at $\sqrt{s}=53$ GeV is shown in Fig. 6, together with the data from the CERN-Columbia-Rockefeller collaboration. In Fig. 7 we show the assumed parametrization of νW_2 (normalized to 1 at $\omega \to \infty$) and the resulting $F(\tilde{\omega})$. In Fig. 8 the predicted energy dependence of large $p_\perp$ production (solid line) is compared with the Saclay-Strasbourg data. It is seen that the choice $\chi(\omega) = \nu W_2$ seems to give slightly too early an approach to energy independence. The third feature is also very natural in a multiperipheral framework in which equally coupled particles of different masses are produced according to a universal function of $m_T^2 = \mu^2 + p_T^2$. For small p_T this favours enormously the lightest particle, the π, but for p_T^2 much larger than the hadron masses the π is just one of many possible produced particles. This feature is not shared by rescattering mechanisms.

The elementary particle exchange multiperipheral model considered so far, is, however, untenable. In fact, we always insist on the bound state nature of the hadrons and on the corresponding, e.g., multi-Regge behaviour in the appropriate (small t_i, large s_i) regions of phase space, and the elementary particle model does not fulfil them. Furthermore, in this framework the elementary π would contribute also to two-body scattering. What we want to show now is that the replacement of the elementary pion by the composite hadron discussed before does not change the results. Especially it does not change the connection between large p_T and SLAC-like scaling functions, irrespectively of the degree of compositeness, as long as we allow the parton to come out exactly as was done for the deep inelastic photon process. This connection is, of course, known since a long time. In the Regge ladder language, the argument is also simple. For small p_T Regge exchange gives results similar to elementary particle exchange and the hadron (bound state) masses fix the scale of p_T. When p_T becomes large, the response of the ladder is analogous to that discussed in the fixed angle case, i.e., the surviving diagrams are those in which the propagators containing large momentum transfers appear to the lowest order. In this case, if the particles of large p_T are partons, the dominant diagrams are of the form of Fig. 3a. We immediately recognize the parton-like calculation. The distribution is similar to the one of Ref. 10), because the relevant integration region is that in which momentum transfers, except of course the one associated to the propagator which absorbs the large p_T, remain limited. The main difference is given by the necessity to re-interpret the observed value of $p_\perp$. In fact the calculation of Ref. 10) applies in this case to the distribution of $p_\perp$ of the produced jet, that is certainly larger than the $p_\perp$ of the leading particle of the jet. If we assume, in analogy with hadron physics, that the momentum of the leading particle of the jet is a finite fraction, say one half, of the jet momentum, the following changes occur:

1) The scaling function of Fig. 7 remains unchanged, provided the variable $(\tilde{\omega}-1)$ is changed into $(\tilde{\omega}-1)/2$. Hence we are much further from asymptotism, as it is obvious.
2) The $p_\perp$ distribution at fixed s is therefore more severely cut-off by the factor $F(\tilde{\omega})$. Hence to obtain a $p_\perp$ distribution that reproduces the data of Fig. 6 we have to choose the parameter λ of Ref. 10) equals to $\frac{1}{2}$, corresponding to an asymptotic $(p_\perp^2)^{-3}$ dependence.
3) The energy variation is now much more important, as shown in Fig. 8 (dashed line).

Let us remark that the fact that the emitted large p_T particle is elementary plays a fundamental rôle (as in the deep inelastic processes). Should we emit a bound state, we would immediately loose a power of p_T^2 due to the appearance of an extra wave function $\psi(p_T^2)$ (Fig. 3b).

This shows even better the relation between deep inelastic and large p_T in this bound state framework. They are related and nearly independent of the composition of hadrons, as long as the elementary particle can be emitted. If a given mass bound state is emitted, we loose a power of the relevant variable (q^2 for deep inelastic, p_T^2 for large transverse momenta), the power depending on the compositeness character.

5. QUARK PRODUCTION PROBLEM

We have seen that the point-like constituent picture gives a reasonable comprehensive view of both soft components (Regge-like behaviour, strong scaling, etc.) and of the hard effects (the four listed before). The form factors and fixed angle scattering indicate a limited compositeness. This simple view, however, requires the elementary constituent (or its decay products) to be emitted in the inclusive hard experiments. Present indications on final state constitution seem to be inconsistent with the fragmentation of the γ into spinor states. Can we find a mechanism to avoid the point-like constituent to come out without spoiling scaling? This is a long-standing problem, on which some propositions have been made, already at the stage of deep inelastic scattering. What we want to stress is that every mechanism allowing the quarks to recombine without spoiling scaling, will work equally well for the large p_T events, i.e., will keep the relation between the two processes including the similarity of secondaries in the forward SLAC experiment with those of the large p_T region at the ISR.

The simplest reinteraction mechanisms do not succeed in avoiding the quark states to come out. In general, either they spoil scaling or they fail in filling the rapidity gap between the quark-like state and the rest ofthe secondary hadrons [11]. In order to overcome this problem it is necessary to assume that besides the short-range forces (gluons, etc.) there is a long-range one. In order to construct a formal framework to this idea, Johnson [12] introduces a singular interaction acting only at zero momentum transfer (suggested by a harmonic oscillator potential). This interaction eliminates the quark poles (outcoming quarks). The condition that the $q^2 = 0$ singularity does not generate wave functions exploding at $q^2 \to \infty$, implies a quantization of the spectrum. But the corresponding eigenfunctions for arbitrary q^2 (and in particular for large q^2) are determined fundamentally by the short-range interaction. We expect therefore the long-range interaction to play no rôle in the asymptotic property of the wave function relevant for the form factors and fixed angle scattering. Let us remark that the presence of these two forces (short-range and long-range) decouples the infinitely rising character of Regge trajectories $\alpha(t)$ for $t \to \infty$ (determined by the long-range) from that at $t \to -\infty$ (determined from the short-range) where trajectories are expected to flatten as discussed in Section 2.

The lack of dynamically acceptable formalisms which would allow to make specific predictions on final states makes it impossible for the time being to prove or disprove such a mechanism.

In the absence of a dynamical understanding of the quark recombination problem, as well as of experimental information on the photon fragmentation in deep inelastic (or e^+e^- annihilation) and on the detailed structure of the large p_T events, only wild speculations can be put forward on the matter. For instance, one could find tempting to associate the assumed linearly rising spectrum of $q\bar{q}$ bound states in this harmonic oscillator ansatz, with the spectrum of resonances of the dual model, and therefore to expect an analogy between their decay products. The final states of the γ fragmentation region could be similar to the states that saturate the unitarity relation for the discontinuity of a quantum number (non-Pomeron) exchange reaction: in this case the contribution of the photon fragmentation to the multiplicity in deep inelastic (and the whole multiplicity in e^+e^- annihilation) could be given by, e.g.,

$$\tilde{n}(Q^2) = \frac{\bar{n}(K^-p)\,\sigma(K^-p) - \bar{n}(K^+p)\,\sigma(K^+p)}{\sigma(K^-p) - \sigma(K^+p)}$$

In the realm of these bold speculations, one might even remark that after all the multiperipheral model gives a satisfactory generation mechanism of non-Pomeron trajectories, and argue that this multiplicity should be logarithmic, but with a smaller coefficient (related to the crossing matrix for the projection of a ladder on t channel quantum numbers $\neq 0$; typically $\frac{1}{2}$ for isospin 1 exchanges) and conclude that the complete multiplicity distribution in deep inelastic scattering could be $\bar{n}(q^2,\omega) \approx g(\log\omega + \frac{1}{2}\log(-q^2))$, corresponding to an inclusive distribution showing in the $\omega \to \infty$, $q^2 \to -\infty$ limit two plateaus, the photon fragmentation one [of length $\log(-q^2)$] being lower than the usual $\log\omega$ hadronic one.

In the framework of this trend of speculations, one could argue also that $\langle p_T \rangle$ in the photon fragmentation could be comparable to the hadronic one for large ω, and that the e^+e^- annihilation final states could exhibit the same p_T limitation, after an event-by-event rotation to account for the finite (one) total angular momentum. Remark, however, that in the dual model the energy independence of $\langle p_T \rangle$ stems from a very subtle coherence of the s channel resonances, that could easily be spoiled by summing them with different weights or phases. Hence, we feel that these last speculations on $\langle p_T \rangle$ are on an even weaker basis than the previous ones on the average multiplicity.

A more elegant possibility to obtain the parton recombination, also exploiting a singular interaction at $t = 0$, has been proposed by Casher, Kogut and Susskind [13], in the framework of an infrared divergent field theoretical model.

The root of the problem is to let the "forbidden" quantum number, (charm or triality) flow from one quark to another distant in rapidity space, and the usual field theoretical attempts would solve it by having this quantum number propagated all the way through quark lines, i.e., by a short range interaction. But the time necessary to initiate, i.e., a multiperipheral chain of this sort is proportional to Q^2 by time dilatation, and if q is sufficiently large the two quarks are so far away from each **other** in configuration space that their chains cannot possibly overlap, hence a violation of scaling if we insist that they do so. If one could have a long range interaction, like vector exchange, to effectively carry the forbidden quantum number, the problem could be solved. This is exactly what happens in the case of a polarization current, in which two opposite charges $+$ and $-$ are separated in space. If the vacuum is sufficiently polarizable it starts producing $+-$ pairs and in the final situation everything happens as if the charge $+$ had flown back to join the charge $-$, even if the actual interaction happens through a neutral photon. A graphical representation of the phenomenon is given by Figs. 10 and 11. The net result is hopefully a uniform rapidity distribution of "dipoles", in our case zero triality states, hence a logarithmic behaviour of $\tilde{n}(q^2)$. The coefficient g_p of $\ln q^2$ does not necessarily have to be the same as the coefficient g_s of $\ln s$ in the multiperipheral multiplicity. Hence the complete multiplicity, e.g., in events with large $p_\perp$ or in electroproduction, is expected to be

$$\tilde{n}(s, Q^2) = g_s \ln s + (g_p - g_s) \ln Q^2$$

Intuitively g_p is expected to be large, and if $(g_p - g_s) > 0$ one can explain the observed increase of the multiplicity with $p_\perp$ in the ISR data of the Pisa-Stony Brook collaboration [9].

In order to make sure that no parton escapes, the process of vacuum polarization has to occur with unit probability, hence the need of an (infrared) divergent field theoretical model. An explicitly solvable example of this kind is given by a two-dimensional quantum field theory of spinors interacting through a massless gauge vector field [14], and in Ref. 13) it is shown that in such a model the light-cone properties are the same as in a full spinor theory, even though the asymptotic states are saturated by bosons only. It is hoped that the divergence that ensures this mechanism is present also in four dimensions if the interaction associated with the coupling of the forbidden quantum number with the hypothetical vector field is sufficiently strong [14]. The limitation of the transverse momentum of the jet products, that obviously cannot be investigated in the two-dimensional model, is also shown to be plausible in Ref. 13), but the arguments adduced are not conclusive.

6. CONCLUSIONS

The points that we wanted to stress are the smoothness of the transition between the soft and the hard regime and the close connection that **exists** among the exclusive large momentum transfer experiments, all sensitive to the bound state wave function, and on the other hand, the inclusive ones that only depend on the possibility of the point-like constituents to be emitted, or, alternatively, to recombine along a common scheme such as the one discussed in the last paragraph.

It is also clear in this framework why the inclusive large p_T production is easier than a corresponding exclusive one (e.g., elastic scattering at $t \simeq -p_T^2$) by several orders of magnitude. For the production of large p_T hadrons we can actually envisage a hierarchy, depending on whether one or both "jets" of large p_T particles are restricted to have finite mass (e.g., to consist of only one particle) or not. In the former case, the bound state wave function of the produced particle appears, in the latter it does not. Also in the former case, the multiplicity should show the $\log(s/p_T^2)$ behaviour expected in elementary particle multiperipheralism; in the latter, an additional $\tilde{n}(p_T^2)$ identical to the one defined for the photon fragmentation in deep inelastic scattering is expected to contribute.

REFERENCES

1) R.P. Feynman, Phys.Rev.Letters 23, 1415 (1969).

2) J.D. Bjorken and E. Paschos, Phys.Rev. 185, 1975 (1969).

3) S.M. Berman, J.D. Bjorken and J.B. Kogut, Phys.Rev. D4, 3388 (1971).

4) J.F. Gunion, S.J. Brodsky and R. Blankenbecler, Phys.Rev. D6, 2652 (1972).

5) S.D. Drell and T.D. Lee, Phys.Rev. D5, 1738 (1972).

6) P.V. Landshoff and J.C. Polkinghorne, Cambridge preprint DAMTP 72/43 (1972).

7) In this sense, our language is particularly close to the one of Ref. 5).

8) D. Amati, R. Iengo, H.R. Rubinstein, G. Veneziano and M. Virasoro, Phys.Letters 27B, 38 (1968);
D. Amati, L. Caneschi and R. Iengo, Nuovo Cimento 58A, 783 (1968);
M. Ciafaloni and P. Menotti, Phys.Rev. 173, 1575 (1968);
J.S. Bell and F. Zachariasen, Phys.Rev. 170, 1541 (1968).

9) CERN-Columbia-Rockefeller collaboration, Saclay-Strasbourg collaboration as reported by M. Jacob, and Pisa-Stony Brook collaboration; CERN preprint TH.1683 (1973).

10) D. Amati, L. Caneschi and M. Testa, Phys.Letters 43B, 186 (1973).

11) J. Kogut, D.K. Sinclair and L. Susskind, Princeton preprint COO 2220-6 (1973).

12) K. Johnson, Phys.Rev. D6, 1101 (1972).

13) A. Casher, J. Kogut and L. Susskind, Tel-Aviv University preprint, 373-73, June 1973.

14) J. Schwinger, Phys.Rev. 128, 2425 (1962).

FIGURE CAPTIONS

Figure 1 : Elastic form factor.

Figure 2 : Two-body scattering at small t.

a) Two-body scattering at large t.

b) Two-body scattering at large t in the exchange model of Ref. 4).

Figure 3 : Production processes at small p_T.

a) Production process at large p_T in the inclusive situation, i.e., with a jet of mass proportional to p_T^2.

b) Production process at large p_T in the exclusive situation, i.e., one particle or a jet of finite mass has large p_T.

Figure 4 : Deep inelastic electron scattering.

Figure 5 : Large p_T production in the elementary exchange multi-peripheral model.

Figure 6 : Large $p_\perp$ π^0 production. Data from the CERN-Columbia-Rockfeller collaboration.

Figure 7 : The functions $\chi(\omega)$ and $F(\tilde{\omega})$.

Figure 8 : Energy dependence of the large $p_\perp$ $\pi^\pm$ yield. Data from the Saclay-Strasbourg collaboration.

Figure 9 : Time evolution of the large mass parton anti-parton system in the vacuum polarization model of Ref. 13).

In all figures, the following symbols are used:

quark (or quark system) with small q^2 gluon

gluon

large q^2 quark

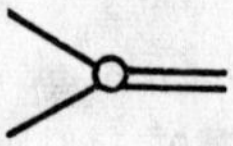

bound state wave function for small values of the constituent masses

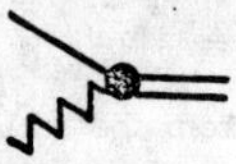

asymptotic bound state wave function for one of the components much off-shell.

current

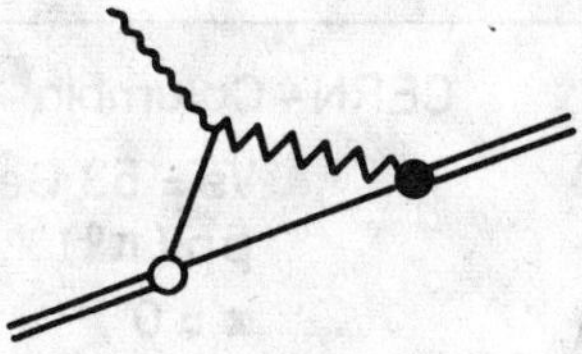

FIG.1

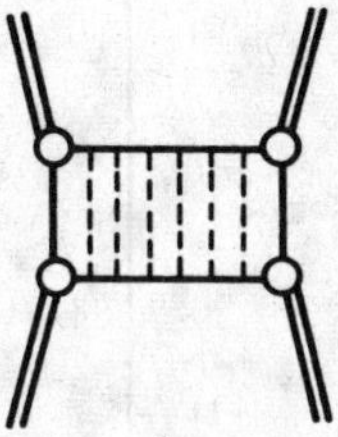

FIG.2

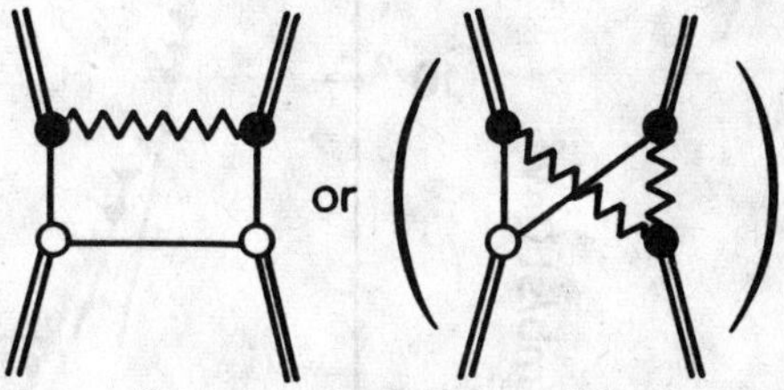

2a

2b

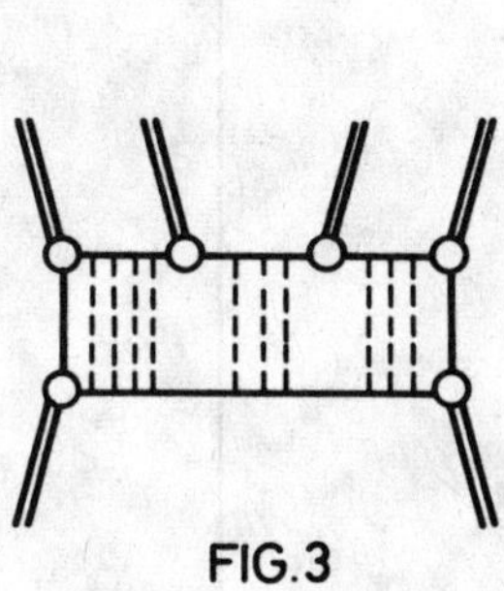

FIG.3

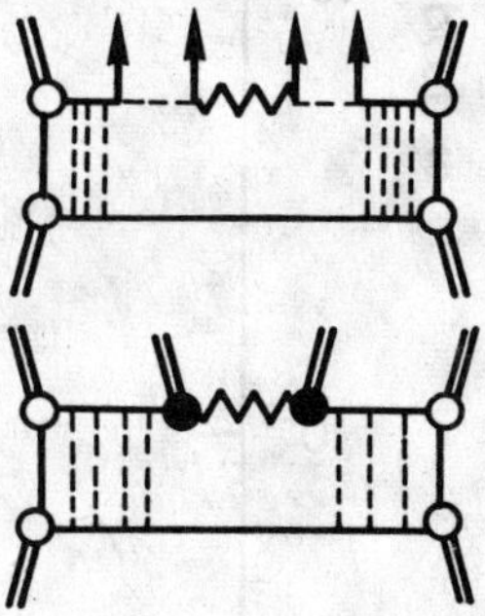

3a

3b

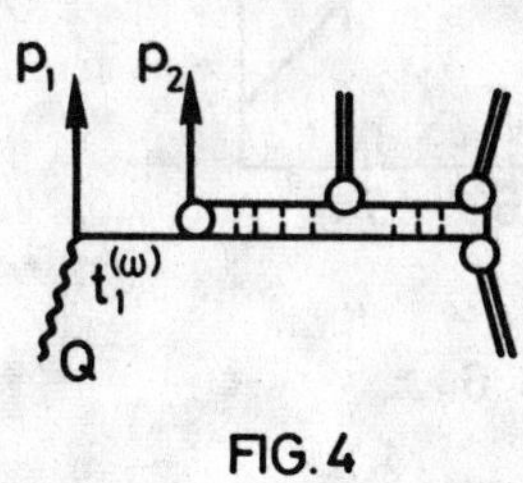

FIG.4

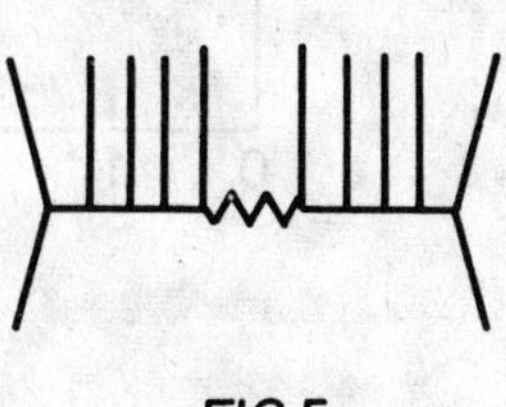

FIG.5

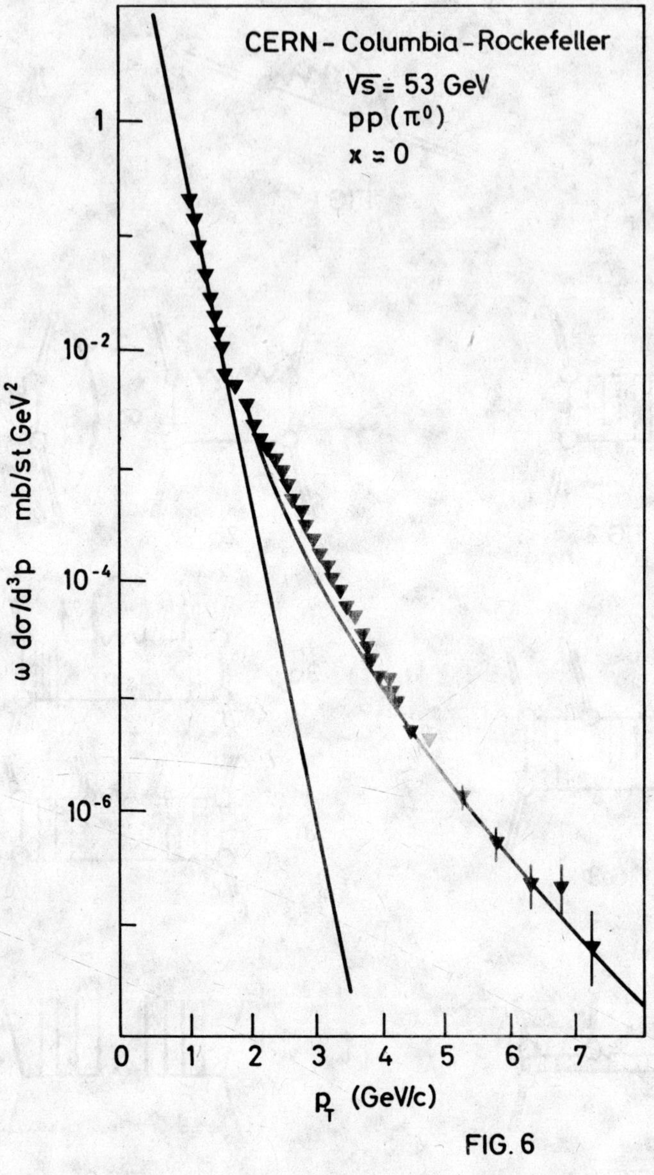
CERN - Columbia - Rockefeller
√s = 53 GeV
pp (π⁰)
x = 0
1
10⁻²
10⁻⁴
10⁻⁶
ω dσ/d³p mb/st GeV²
0
1
2
3
4
5
6
7
p_T (GeV/c)

FIG. 6

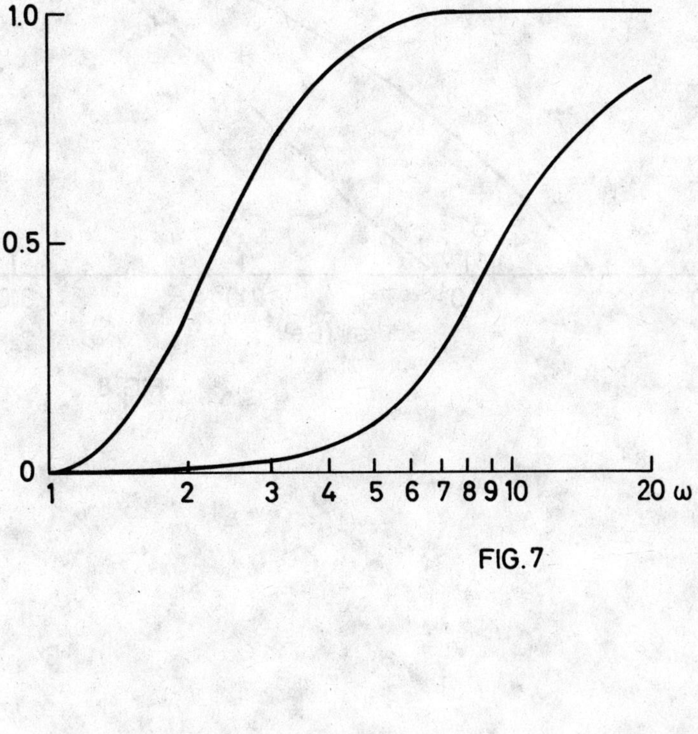

FIG. 7

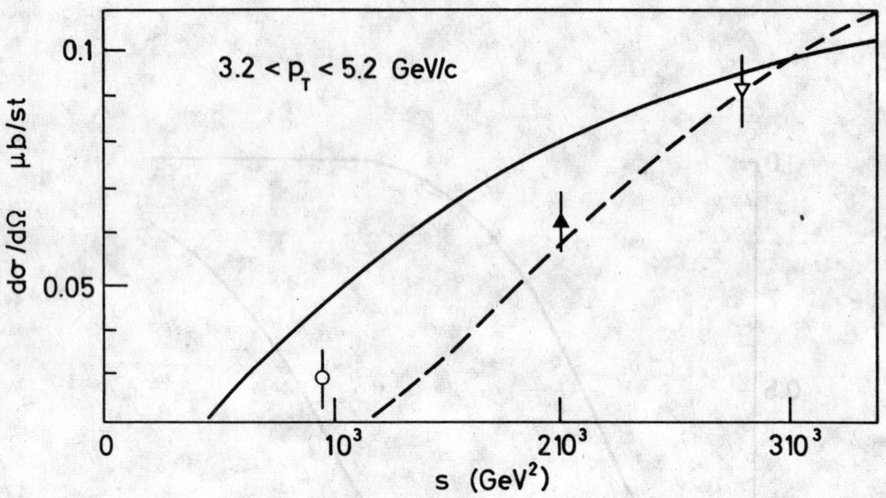

FIG. 8

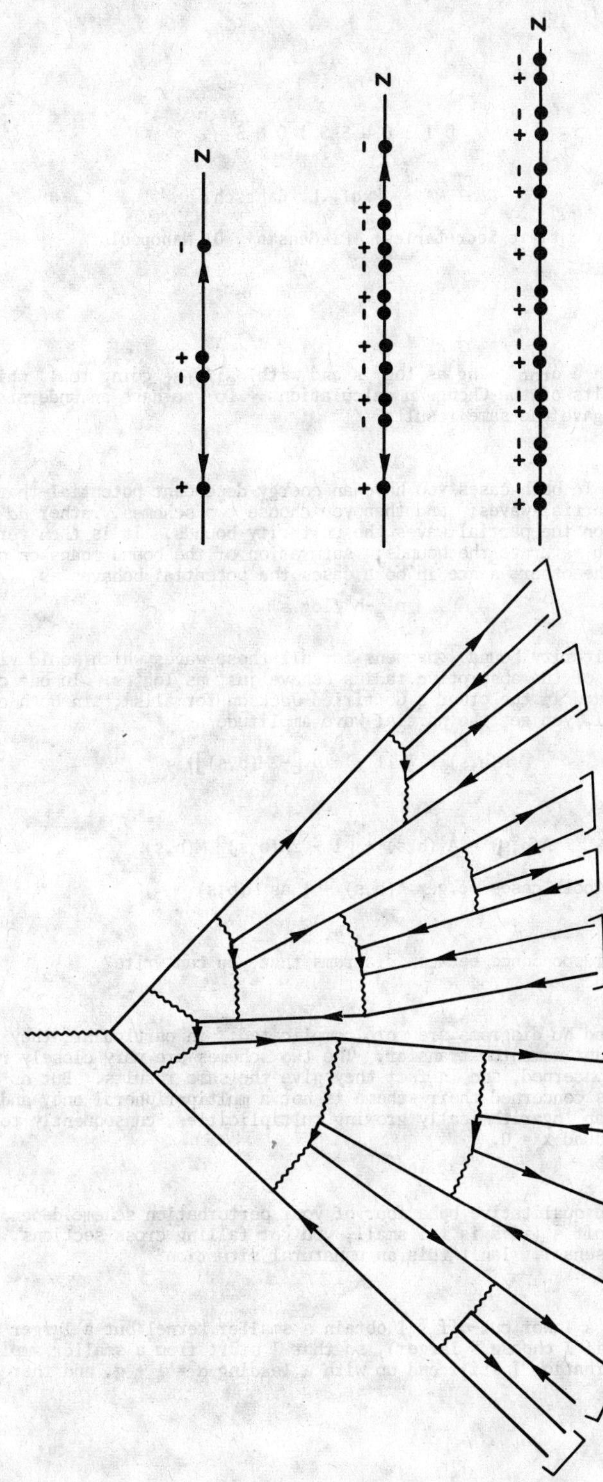

FIG.9

DISCUSSIONS

CHAIRMAN: Prof. L. Caneschi

Scientific Secretaries: P. Gensini, D. Nanopoulos

DISCUSSION No. 1

- *HENDRICK:*

You ended up with a σ_{tot} going as $\log^2 s$ and with σ_{el}/σ_{tot} going to $\frac{1}{2}$, which are identical to the results of the Cheng-Wu calculations. Do you have an understanding of why these two approaches gave the same results?

- *CANESCHI:*

Of course I do. In both cases you have an energy-dependent potential that violates unitarity bounds in partial waves; and then you choose two schemes, rather *ad hoc*, which are known to enforce on the partial waves the unitarity bounds. It is then very natural that in both cases you saturate the bounds; saturation of the bound means on one side $\sigma_{el} = \sigma_{inel}$, and on the other, since in both cases the potential behaves as

$$s^p \, e^{-b^2/\log s} ,$$

the cut-off at the unitarity bound $\frac{1}{2}$ happens for all those waves which would violate it and this makes the square of the absorptive radius behave just as $\log^2 s$. In one case we have an eikonal formalism and in the other a Gottfried-Jackson formalism; in both cases, once you know the potential, you get the partial wave amplitude as

$$A(b,s) = \tfrac{1}{2} \left\{1 - \exp\left[-2M(b,s)\right]\right\}$$

in the first case, and

$$A(b,s) = A^2(b,s) + \left[1 - 2A(b,s)\right] M(b,s)$$

in the second, and in both cases we get $A(b,s) \to \frac{1}{2}$ as $M(b,s) \to \infty$.

- *HENDRICK:*

Is there any correspondence between diagrams that you can write?

- *CANESCHI:*

No. The Cheng and Wu diagrams are more complicated: in particular, they do violate limitations on invariant momentum transfer. The two schemes are very closely related as far as unitarity is concerned, and in fact they give the same results. But as far as particle production is concerned their scheme is not a multiperipheral one, and in particular, leads to more than logarithmically growing multiplicities, consequently to a violation of bounded scaling around $x = 0$.

- *WILCZEK:*

It seems that the qualitative behaviour of your perturbation scheme depends sensitively on the choice of cut-off $\bar{s}$ (if $\bar{s}$ is too small, you get falling cross-sections, and if $\bar{s}$ is too large you get nonsense). Isn't this an unnatural situation?

- *CANESCHI:*

No. If I choose a lower cut-off $\bar{s}$ I obtain a smaller kernel but a larger perturbative term (and conversely if I choose $\bar{s}$ larger), so that I start from a smaller amplitude; but applying larger perturbation I still end up with a leading $\alpha = 1 + g$, and there is no internal inconsistency.

- *WILCZEK:*

What is (roughly) the empirical value of $\bar{s}$?

- *CANESCHI:*

It is hard to say. If you want to make a connection between the raising of the cross-section and the appearance of the peak at x = 1 in the proton spectrum, then you should choose an $\bar{s}$ corresponding to a gap in rapidity of the order of 2 or 3 units.

- *NAHM:*

Amplitudes which give rise to cross-sections growing like $\log^2 s$ have to satisfy very restrictive conditions, for example, infinitely many zeros near t = 0 and scaling as an entire function of $t \log^2 s$. Are these conditions satisfied by your model?

- *CANESCHI:*

Certainly. These restrictions are just consequences of having a completely absorptive disk, the radius of which increases logarithmically. Then the Bessel transform of $\theta(c \cdot \log s - b)$ is just $J_0(\sqrt{t} \log s)$, which gives you both infinite series of zeros and the required connection between the t- and s-dependence.

- *WAMBACH:*

You mentioned that a cut-off in t implies also a cut-off in $p_\perp$. However, looking in phase-space the two prescriptions populate different kinematical regions. Experimentally a cut-off in $p_\perp$ seems to be favoured. Could you comment on this?

- *CANESCHI:*

That is correct, and that has upset me very much. A decisive test would be to check if transverse momentum is conserved locally, which is a typical feature of the multiperipheral model.

- *WAMBACH:*

You mentioned the behaviour of the σ_n as a function of s to go to a constant limit. Is this limit reached from above or below?

- *CANESCHI:*

Only in the Pomeron exchange model considered with Schwimmer is it $d\sigma_n(s)/ds > 0$ always, but the introduction of a dominant low-energy kernel leads to a decrease at finite s.

- *GENSINI:*

Is there any other internally self-consistent model which predicts rising cross-sections?

- *CANESCHI:*

There is another model on the market which is very fashionable and has been compared with the ISR data, namely the Gribov Reggeon calculus. There is an essential difference between the two mechanisms: one produces a saturation of the Froissart bound with in-definitely increasing total cross-section, whereas in the Gribov scheme the leading singularity is a pole, accompanied by a cut with negative residue.

What distinguishes the two schemes is the behaviour of the elastic cross-section. In the Gribov scheme the total cross-section rises as $\sigma_{tot} = a - b/\log s$, while the elastic cross-section goes down as $b/\log s$, being given as a Pomeron-Pomeron cut.

You must also note that in the Gribov approach the rise in σ_{tot} is numerically related to the decrease in σ_{el}, so that if both quantities increase this scheme is ruled out. Actually you can have a temporary situation where both cross-sections grow, but a condition for this to happen is that the impact parameter representation has a minimum at b = 0. Hence in this scheme one will get into trouble fitting both an increase of 4 mb in σ_{tot} and an increase of 1 mb in σ_{el} over the whole ISR energy range.

- *WAMBACH:*

Can you explain the charge excess at large $p_\perp$ in your model?

- *CANESCHI:*

In the elementary particle exchange model this result is very natural, since the charge excess also is a function of $s/p_\perp^2$ only. In the bound state approach when two constituents are produced at large $p_\perp$ they tend to radiate filling up all the rapidity gap. However, the charged particles will be mostly distributed at large $p_\perp$, since the charge flows along the lines of the outgoing charged constituents, while particles radiated in the central region of p will tend mostly to be neutral.

- *PATKOS:*

Treating the hadron-hadron elastic scattering in the composite model in ladder approximation at fixed angle and high energy, you arrive at a box diagram:

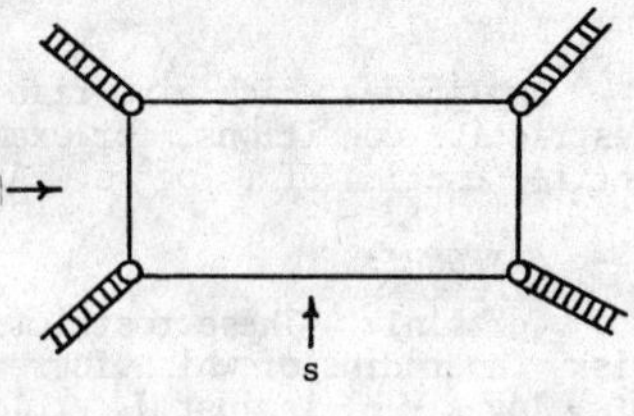

This diagram, if the constituents are quarks in the $\pi^+\pi^+$ channel contradicts the Rosner-Harari-type duality diagrams. Second, in the model of Blankenbecler, Brodski and Gunion, the dominant diagram is the quark exchange diagram:

Can you explain the reason for this difference?

- *CANESCHI:*

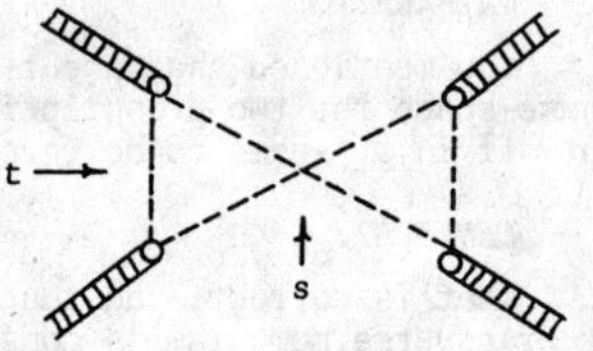

I guess that nobody can give a reason why the Rosner-Harari duality had to be valid at large angles. In the Blankenbecler-Brodski-Gunion model three hadronic wave functions are in the asymptotic region $q^2 \to \infty$, and therefore in our model the contribution coming from their diagram is negligible compared to that of the box diagram.

- *PATKOS:*

What will be your scaling variable at $x \neq 0$?

- *CANESCHI:*

The one-particle distribution function at $x \neq 0$ will have the form:

$$G(p_c \cdot p_a/p_\perp^2,\ p_c \cdot p_b/p_\perp^2) \ .$$

For fixed x, G depends only on the ratio $s/p_\perp^2$.

Therefore, the distribution for $x \neq 0$ approaches faster the limiting function than for $x = 0$. This can give interesting experimental implications. We could expect a local transient minimum in the large $p_\perp$ distribution around $x = 0$.

The simple physical explanation for the change of the scaling variable is the following: at $x = 0$, in the c.m. frame we have a symmetric situation so that fragments of both a and b will have a mass of order $\sqrt{s}$. So the only dimensionless quantity will be $p_\perp/\sqrt{s}$.

If $x \neq 0$ then the dimensionless quantities which can be formed from the energies of the final state bodies will depend essentially only on $p_\perp^2/xs$.

ALGEBRA OF CURRENTS AND REGGE COUPLINGS

H. Kleinert

Table of Contents

ALGEBRA OF CURRENTS AND REGGE COUPLINGS

Hagen Kleinert,
Freie Universität Berlin, Berlin, W. Germany.

Introduction

In particle physics there exist two basic sets of observable quantities: local currents of any spin J and Regge couplings. The first set is measured in electromagnetic and weak interactions, either via the exchange of single photons or W mesons or via deep-inelastic electron and neutrino scattering. The second set is observed rather directly in purely hadronic interactions at high energies, in particular in those reactions in which diffraction effects are absent (like charge exchange scattering etc.). In these lectures I shall report on an attempt [1] at obtaining a joint algebraic understanding of all these observables together. My presentation is divided into two parts. I shall show that

I. Algebraic saturation schemes of chiral SU(2) X SU(2) can naturally be extended by Regge couplings to form a larger algebra. In particular, isospin T_a , axial charge X_a, and the residues of ρ_a , A_{1a} , π_a , and f trajectories form the "superalgebra" SU(2) X SU(2) X O(5).

II. The "superalgebra" is closely related to the algebra of bilocal form factors of Fritzsch and Gell-Mann (ABFF) [2]

$$[F_a(\underset{\sim}{k}, z), F_b(\underset{\sim}{k}', z')] = i\varepsilon_{abc} F_c(\underset{\sim}{k}+\underset{\sim}{k}', z+z') \qquad (1)$$

Symbolically one may say :

$$\frac{\text{SUPERALGEBRA}}{\text{ABFF}} = \frac{\text{SUPERCONVERGENCE}}{\text{CURRENT ALGEBRA}} \qquad (2)$$

I. From the Chiral to the "Superalgebra"

Given any chiral saturation scheme on a set of resonances [3,4,5] $\alpha, \beta, \gamma, \ldots$ whose mass^2 values cover some intervall (0,N) [+]. Let $[T_a]_{\beta\alpha}$, $[X_a]_{\beta\alpha}$ denote the matrices of isospin and axial charge between those resonances in the infinite momentum frame, respectively. Then T_a, X_a satisfy the well-known commutation rules of SU(2) X SU(2). The saturation scheme can be tested by assuming PCAC that relates $[X_a]_{\beta\alpha}$ to the pionic decay widths $\Gamma(\beta \to \alpha + \pi_a)$. In fact, the relevant coupling is

$$\langle \beta p' | j_a^\pi | \alpha p \rangle \Big|_{q^2 = m_\pi^2} \overset{PCAC}{\approx} \frac{1}{f_\pi} \langle \beta p' | \partial A_a | \alpha p \rangle \Big|_{q^2=0} = \frac{i}{f_\pi} (m_\beta^2 - m_\alpha^2) \left[X_a \right]_{\beta\alpha} \tag{3}$$

i.e., it is directly given by the commutator

$$[m_a^2]_{\beta\alpha} \equiv i\, [m^2, X_a]_{\beta\alpha} \tag{4}$$

of the mass^2 matrix $[m^2]_{\beta\alpha} \equiv m_\alpha^2 \delta_{\beta\alpha}$ with $[X_a]_{\beta\alpha}$. This matrix will be the crucial quantity for the extension of the charge algebra. The basic technical tool consists in the saturation of appropriate finite-energy sum rules (FESR) for the scattering amplitudes

$$\pi_a \alpha \longrightarrow \pi_b \beta \tag{5}$$

$$\pi_a \pi_{a'} \alpha \rightarrow \pi_b \beta \tag{6}$$

$$A\, A'\, \alpha \rightarrow B\, B' \beta \tag{7}$$

in terms of the resonances $\alpha, \beta, \gamma, \ldots$ of the saturation scheme.

1. Applying FESR to (5) and using the pionic couplings m_a^2 as an input for the resonance contributions we find from the isospin even amplitude that the commutator $[X_a, m_b^2]$ has no isospin I_t=2 part [++] and that the I_t=0 part

$$[m_4^2]_{\beta\alpha}\, \delta_{ab} \equiv i\, [X_a, m_b^2] \tag{8}$$

is, up to a factor, equal to the residue R^f of the f trajectory [+++] [6,7]

+) For simplicity, let us count masses in units of m_o = 1 GeV.

++) This paricular statement is well-known [3].

+++) We neglect contributions of Regge daughters. Also we assume resonances to be dual to the f trajectory undisturbed by diffractive effects.

$$[m_4^2]_{\beta\alpha} \approx c^f N^{\alpha_f(0)} [R^f]_{\beta\alpha} \tag{9}$$

As saturation schemes increase, $[m_4^2]_{\beta\alpha}$ will diverge like $N^{\alpha_f(0)} \sim N^{\frac{1}{2}}$. We recall that the information of $[X_a, m_b^2]$ having no $I_t = 2$ part amounts to the statement that m^2 can be written as

$$[m^2]_{\beta\alpha} = [m_o^2]_{\beta\alpha} + [m_4^2]_{\beta\alpha} \tag{10}$$

with m_o^2 being an SU(2) X SU(2) invariant and m_4^2 forming the isosinglet part of a $(\frac{1}{2},\frac{1}{2})$ representation [3]. Thus the chiral mass splitting matrix is proportional to the couplings of the f trajectory.

2. From the isospin odd amplitude (5) we find, by the same token, that the commutator between m_a^2 and m_b^2 is proportional to the residue R^ρ of the ρ trajectory [6,7]

$$\frac{1}{2i} \varepsilon_{abc} [m_a^2, m_b^2]_{\beta\alpha} \approx c^\rho N^{\alpha_\rho(0)+1} [R_c^\rho]_{\beta\alpha} \tag{11}$$

Notice that R^ρ as a commutator of two $(\frac{1}{2},\frac{1}{2})$ representations transforms like $(0,1) \pm (1,0)$. Its chiral partner +) is the commutator $-i[m_a^2, m_4^2]$. By considering an appropriate double FESR of the process (6) this can be shown to be proportional to the coupling R^{A_1} of the A_1 trajectory.

$$-i [m_a^2, m_4^2]_{\beta\alpha} \approx c^{A_1} N^{\alpha_{A_1}(0)+1} [R_a^{A_1}]_{\beta\alpha} \tag{12}$$

3. As a final step for closing the algebra consider now the process (7) at large subenergies $(p_A + p_\alpha)^2, (p_{B'} + p_\beta)^2$ and at zero momentum transfers $(p_A - p_B)^2 = (p_{A'} - p_{B'})^2 = (p_\alpha - p_\beta)^2 = 0$.
Then the scattering amplitude is presumably dominated by Regge particle scattering

$$R^i \alpha \rightarrow R^j \beta \tag{13}$$

where R^i, R^j are the leading trajectories.

+) Since

$$[X_a, \frac{1}{2i} \varepsilon_{bef} [m_e^2, m_f^2]] = i\varepsilon_{abc} (-i[m_c^2, m_4^2])$$

Also for this amplitude we may write FESR and saturate it with the resonances $\alpha, \beta, \gamma, \ldots$ of the saturation scheme. We make the standard assumption that at high CM energies M the amplitude (13) is again dominated by the exchange of some trajectory R^k. It will then behave like

$$\approx g^{ji}_{k} (M^2)^{\alpha_k(0)-\alpha_j(0)-\alpha_i(0)} \tag{14}$$

where g^{ji}_{k} is proportional to the triple- Regge coupling between R^j, R^i, and R^k. Performing the FESR integral we find [8]

$$[R^j, R^i] \approx g^{ji}_{k} N^{\alpha_k(0)-\alpha_j(0)-\alpha_i(0)+1} R^k \tag{15}$$

In this way, R^{ρ}_{a}, $R^{A_1}_{a}$ and R^{ρ}_{a}, R^{π}_{a} can be shown[+)] to commute like SU(2) X SU(2):

$$[R^{\rho}_{a}, R^{\rho}_{b}] = g^{\rho\rho}_{\rho} N^{\frac{1}{2}} i\varepsilon_{abc} R^{\rho}_{c}$$

$$[R^{\rho}_{a}, R^{A_1}_{b}] = g^{\rho A_1}_{A_1} N^{\frac{1}{2}} i\varepsilon_{abc} R^{A_1}_{c}$$

$$[R^{A_1}_{a}, R^{A_1}_{b}] = g^{A_1 A_1}_{\rho} N^{\frac{3}{2}} i\varepsilon_{abc} R^{\rho}_{c} \tag{16}$$

with similar commutators for R^{ρ}_{a}, R^{π}_{a}. In addition one finds:

$$[R^{\rho}_{a}, R^{f}] = 0$$

$$[R^{A_1}_{a}, R^{f}] = g^{A_1 f}_{\pi} N^{\frac{1}{2}} i R^{\pi}_{a}$$

$$[R^{A_1}_{a}, R^{\pi}_{b}] = -g^{A_1 \pi}_{f} N^{\frac{3}{2}} i\delta_{ab} R^{f}$$

$$[R^{\pi}_{a}, R^{f}] = -g^{\pi f}_{A_1} N^{\frac{1}{2}} i R^{A_1}_{a} \tag{17}$$

These are indeed the commutators of O(5).

+) Here we have used $\alpha_{\rho}(0) \sim \alpha_{f}(0) \sim \frac{1}{2}$; $\alpha_{\pi}(0) \sim \alpha_{A_1}(0) \sim 0$.

Let us now make the assumption that the pion does not reggeize much at t=0. Then we can identify

$$R^{\pi}_{a} \approx \frac{1}{f_\pi} m_a^2 \tag{18}$$

This closes the algebra.

Note that the results (9) ,(11),(12),(18) make many of the commutators (16) and (17) be automatically fulfilled. It can easily be shown that the algebra closes if one only enforces:

$$\left[[m_a^2, m_b^2], m_c^2\right] \propto N^2(\delta_{ac} m_b^2 - \delta_{bc} m_a^2) \tag{19}$$

In order to see that we are dealing in fact with the algebra of SU(2) X SU(2) X O(5) we introduce renormalized Regge couplings ρ_a, A_{1a}, π_a, f which satisfy simple commutation rules

$$[\rho_a, \rho_b] = i\varepsilon_{abc}\, \rho_c \tag{20}$$

Then $T_a = T_a - \rho_a$, $X_a = X_a - A_{1a}$ form an SU(2) X SU(2) that completely decouples from the Regge O(5).

For the construction of physical saturation schemes see again the literature [1]. Let us just mention that the old chiral scheme [3] containing the mesons π, A_1, ρ, σ in representations $(\frac{1}{2},\frac{1}{2})$ and (1,0) ± (0,1) allows for a natural extension to the whole "supergroup" without adding more states. It can be viewed as the representation (0,0) X <u>10</u> of SU(2) X SU(2) X O(5).

Among many results of this group theoretic approach let us only mention a few:

1. The algebra enforces certain consistency relations among the triple Regge couplings via the Jacobi identity:

$$g^{\rho\rho}{}_{\rho} = g^{\rho\pi}{}_{\pi} = g^{\rho A_1}{}_{A_1}$$

$$g^{A_1 f}{}_{\pi}\, g^{A_1\pi}{}_{f} = g^{\rho\rho}{}_{\rho}\, g^{A_1 A_1}{}_{\rho} \tag{21}$$

$$g^{\pi f}{}_{A_1}\, g^{\pi A_1}{}_{f} = g^{\rho\rho}{}_{\rho}\, g^{\pi\pi}{}_{\rho}$$

2. With c^{g}, c^{f}, c^{π} being known, triple-Regge couplings can be predicted from saturation schemes.
3. The mass spectrum is highly restricted since m_4^2 is not only a chiral $(\frac{1}{2},\frac{1}{2})$ representation but a generator of the larger group O(5).

II. Connection with the Algebra of Bilocal Form Factors

If we are willing to accept the light cone philosophy, structure functions $F(\underset{\sim}{k},\xi)$ of deeply inelastic Compton amplitudes +) (at $t=-\underset{\sim}{k}^2$, $q^2 \to q^2 \to \infty$, $\xi = -\frac{q^2}{2\nu}$ fixed) are related to the bilocal form factors $F(\underset{\sim}{k},z)$ via

$$F^{S}_{A}(\underset{\sim}{k},z) = \int_{-1}^{1} e^{i\xi z} F^{S}_{A}(\underset{\sim}{k},\xi)\,d\xi \tag{22}$$

Here S,A denote symmetric and antisymmetric structure functions (in ξ). The important observation is now the following:

One can introduce signatured bilocal form factors

$$F^{\pm}(\underset{\sim}{k},z) = 2\int_{0}^{1} e^{i\xi z} F^{A}_{S}(\underset{\sim}{k},\xi)\,d\xi \tag{23}$$

Then the coefficients $F^{J\pm}(\underset{\sim}{k})$ of a power series expansion

$$F^{\pm}(\underset{\sim}{k},z) = \sum_{J=1}^{\infty} \frac{(iz)^{J-1}}{(J-1)!} F^{J\pm}(\underset{\sim}{k}) \tag{24}$$

are equal to the Khuri type of partial wave amplitudes of the Compton amplitude $T^{\pm}(k,\xi)$ whose absorptive part is $F^{\pm}(k,\xi)$. The proof follows directly ++) by expanding the dispersion relation

$$T^{\pm}(\underset{\sim}{k},\xi) = \xi^{-n+1}\int_{0}^{1} \xi'^{\,n-1} \frac{F^{\pm}(\underset{\sim}{k},\xi')}{\xi'-\xi+i\varepsilon}\,d\xi' + P_{n-1}(\underset{\sim}{k},\xi^{-1}) \tag{25}$$

in powers of ξ^{-1} outside the circle $|\xi|=1$. Thus $F^{J\pm}(\underset{\sim}{k})$ can be continued analytically in J and is expected to possess Regge poles

$$F^{J\pm}(\underset{\sim}{k}) = \frac{R^{i}(\underset{\sim}{k})}{J-\alpha_i(\underset{\sim}{k})} \tag{26}$$

+) $\underset{\sim}{k}$ is the transverse momentum in the infinite momentum frame.

++) The short proof presented here demonstrates this only for $J \geq n$. It can, however, easily be extended to all J [1].

Moreover, as a consequence of (1), $F^{J\pm}(\underset{\sim}{k})$ satisfy the "algebra of signatured form factors" [1])

$$\left[F_a^{J,\eta}(\underset{\sim}{k}),\, F_b^{J',\eta'}(\underset{\sim}{k}')\right] = i\varepsilon_{abc}\, F_c^{J+J'-1,-\eta\eta'}(\underset{\sim}{k}+\underset{\sim}{k}') \tag{27}$$

Now our relation (2) can easily be proved. First we observe that we can use the method of de Alfaro et al.[+)] and continue (27) in $\underset{\sim}{k},\underset{\sim}{k}'$ until $-\underset{\sim}{k}^2$, $-\underset{\sim}{k}'^2$ hit meson poles at m^2, m'^2 of spin J, J'. What follows are standard superconvergence relations[+)] for these meson couplings.
The new feature of our algebra (27) is that we can also do the converse. Keep k, k' fixed and continue J,J' to Regge poles! This leads to

$$\left[F_a^{J,(-)^J}(0),\, R_b^{i}(\underset{\sim}{k})\right] = i\varepsilon_{abc}\, R_c^{i,J-1}(\underset{\sim}{k}) \tag{28}$$

$$\left[R_a^{i}(\underset{\sim}{k}),\, R_b^{j}(\underset{\sim}{k}')\right] = 0 \tag{29}$$

Here $R^{i,J-1}(\underset{\sim}{k})$ denotes the $(J-1)^{st}$ daughter of $R^i(\underset{\sim}{k})$. Thus charges of normal parity are "daughter lowering operators" by J-1 units while Regge couplings are superconvergent.
The second result appears, at first sight, to contradict our algebra (15) found previously. However, we know that sum rules in the infinite momentum frame are only valid if the so called class II graphs[+)] can be neglected. This is, in fact, ensured if the corresponding scattering amplitude is asymptotically superconvergent. Thus the commutator (27) will hold if the leading Regge trajectory has a value[++)]

$$\alpha_k(\underset{\sim}{k}+\underset{\sim}{k}') < 1 \tag{30}$$

This imposes no problem for integer J, J' : For any $\underset{\sim}{k},\underset{\sim}{k}'$, $\alpha_k(k+k') < 1$.[+++)] If one, however, continues analytically in J,J' to $\alpha_i(k)$, $\alpha_j(k')$, this inequality may be violated easily at $k=k'=0$. Then FESR have to be used and the commutator (15) results. For general momentum transfers it has to

---

+) See any of the text books on current algebra by B.Renner or S.Adler and R.Dashen.

++) Note that for brevity we write $\alpha(\underset{\sim}{k})$ rather than the conventional form $\alpha(-\underset{\sim}{k}^2) = \alpha_0 + \alpha'(-\underset{\sim}{k}^2)$.

+++) In the infinite momentum frame all momentum transfers are spacelike.

be replaced by [1]

$$[R^i(\underline{k}), R^j(\underline{k}')] = i g^{ij}{}_k(\underline{k},\underline{k}') N^{\alpha_k(\underline{k}+\underline{k}')-\alpha_i(\underline{k})-\alpha_j(\underline{k}')+1} R^k(\underline{k}+\underline{k}') \tag{31}$$

This completes the proof of our symbolic relationship (2) and shows, in addition, how the Regge algebra (15) is extended to arbitrary momentum transfers $\underline{k},\underline{k}'$.

Finally let us remark that on the basis of (26) signatured form factors possess Regge behaviour for large z, but only if $|z| \to \infty$ in the upper half plane [1]. The original bilocal form factors $F(\underline{k},z)$, on the other hand, don't, contrary to what can be found in the literature.

Outlook

The algebraic approach to physical observables has yielded some interesting new results. Currents and Regge couplings appear on the same footing. We hope that physically meaningful solutions of the combined "superalgebra" may contribute to a deeper understanding of the structure of particle interactions.

References

1) H.Kleinert, Bilocal Form Factors and Regge Couplings, Berlin Preprint, March 1973, Nuclear Physics B (in press)

2) H.Fritzsch and M.Gell-Mann, Caltech Preprints 1972

3) S. Weinberg, Phys.Rev. 177, 2604 (1969)

4) C.Boldrighini, F.Buccella, E.Celeghini, E.Sorace and L.Triolo, Nucl.Phys. B22, 651 (1970)

5) F.Buccella, H.Kleinert, C.A.Savoy et al., Nuovo Cimento 69A, 133 (1970)

6) H.Kleinert, Phys.Letters B39, 511 (1972) and Fortschr. Phys. 21, 377(1973)

7) H.Kleinert and L.R.Ram Mohan, Nucl.Phys. B52, 253 (1973)

8) H.Kleinert, Lettere Nuovo Cimento 6, 583 (1973)

D I S C U S S I O N S

CHAIRMAN: Prof. H. Kleinert

Scientific Secretary: S. Kitakado

DISCUSSION No. 1

- *BUCCELLA:*

As you have a higher algebra than SU(2) × SU(2), I want to know if this gives some information on the mixing operator that describes the transformation from constituent to current quarks.

If you start from the fact that there exists SU(6), does this algebraic condition give further restrictions?

- *KLEINERT:*

Yes. In the old language, SU(2) × SU(2) scheme, m_4^2 had only the property of forming a $(\frac{1}{2},\frac{1}{2})$ representation. Now it is a generator of a bigger group 0(5). This restricts the matrix element considerably. You get many more selection rules, more than just for a $(\frac{1}{2},\frac{1}{2})$ operator. Certainly, this puts constraints on any mixing operator. In which form precisely, I have not yet investigated.

If you want to start with SU(6) you also have to extend my algebra by SU(3). This is to be studied in the future.

- *GENSINI:*

Do you have ρ-A_2, π-B, or f-ω exchange degeneracy?

- *KLEINERT:*

Exchange degeneracy has been shown to be related via local duality to the absence of both exotic Regge exchanges in the t-channel and exotic resonances in the s-channel. Since my algebra has these properties, I certainly predict exchange degeneracy; for example, of ρ and f. The specific trajectories you are asking for are not contained in my algebra. For this it would have to be extended. Then ρ,ω; A_1,B; π,η; A_2,f trajectories form the group SU(4) and you obtain the correct degeneracies among these trajectories.

- *GENSINI:*

The second question is: when you look at the expansion of the bilocal currents you get a set of local tensors of higher and higher rank, each of which can be thought as decomposable in a "tower" of Poincaré group representations. All these towers can be thought of as forming a whole Toller pole family, as in the Cabibbo and Testa approach; you can then try to arrive at a Regge-coupling algebra from this side instead of yours. Can you please comment on the parallel, or on the differences between your approach and theirs?

- *KLEINERT:*

My algebra of signatured form factors consists of the matrix elements of these currents in the infinite momentum frame with an additional signature treatment I explained. These are the only commutators I am willing to believe in the presence of interactions among quarks. In this frame, however, only the leading trajectory of your Toller families survive and $F^{J,\pm}(\varepsilon)$ describes only exchange of spin J. This is the first difference of my approach to that of Cabibbo and Testa. In addition, they are making some vector meson dominance assumptions. This causes them to miss out on the divergence properties of the commutators. They do not notice that triple Regge behaviour forces the commutator of two ρ trajectories to blow up like $N^{\frac{1}{2}}$ as the saturation scheme goes to infinity.

- *DRECHSLER:*

How can your algebra be derived for non-vanishing values of the masses $-k'^2$, $-k^2$, $-(k + k')^2$ of the Reggeons and the momentum transfer, respectively?

- *KLEINERT:*

For this argument I have to refer you to the original paper. Basically one starts with an amplitude

$$T^{J',J}(p',q',p,q) = -i \int e^{iqx} \langle p' | \overbrace{j^{++\cdots+}}^{J'}(x), \overbrace{j(0)^{++\cdots+}}^{J} | p \rangle$$

and derives finite energy sum rules at $q'^{+} = q^{+} = 0$ in the variable q^{-}, and Im $T^{J',J}$ is analytic in q^{-} and Im $T^{J',J}$ behaves asymptotically like $(q^{-})^{\alpha_k(k+k')-J'-J}$. If signature is introduced properly in $T^{J',J}$, this gives rise to the right-hand side

$$g_k^{ji}(k,k')\, N^{\alpha_k(k+k')-\alpha_j(k')-\alpha_i(k)}$$

I referred to in the lecture.

- *SALVINI:*

What is the experimental test you wish for your representation? What should experimentalists do?

- *KLEINERT:*

Experimentalists themselves cannot give me an answer. There have to be phenomenological people fitting the data. What I predict is the coupling of the ρ (or A_1, π, f) trajectory to different resonances.

- *SALVINI:*

Can you suggest one or more experiments? Is there any new experiment you strongly wish for this verification?

- *KLEINERT:*

When we plot the cross-section of resonance production at fixed q^2 as a function of the missing mass squared, the areas under the different resonance bumps will be determined by the matrix elements of the Regge residues ρNR_i, A_1NR_i, πNR_i, etc. where N is the target nucleon and R are the different resonances. My superalgebra predicts these matrix elements. Thus all I need are analyses of production experiments in the resonance region. Charge-exchange reactions are preferable in order to get rid of diffractive effects.

FROM CONSTITUENT TO CURRENT QUARKS BREAKING SU(6)W

F. Buccella

Table of Contents

FROM CONSTITUENT TO CURRENT QUARKS BREAKING $SU(6)_W$

F. Buccella
Istituto di Fisica, Università di Roma, Italy.

The work I am going to describe has been done in collaboration with F. Nicoló, A. Pugliese and C.A. Savoy[1,2].

As you can find in the Erice Proceedings of 1970 [3], the problem of the saturation of chiral $SU(3) \otimes SU(3)$ seems to have a rather elegant solution; in the $P_z = \infty$ frame the axial charges Q_5^α are obtained with a unitary transformation on the corresponding $SU(6)_W$ generators $A(\sigma_z\lambda^\alpha/2)$:

$$Q_5^\alpha = UA\left(\sigma_z \frac{\lambda^\alpha}{2}\right) U^\dagger \ , \tag{1}$$

with

$$U = 1 + iZ - \frac{Z^2}{2} + \ldots \ , \tag{2}$$

and

$$Z = (\vec{W} \times \vec{M})_z \tag{3}$$

has simple transformation properties under the classification group $SU(6) \otimes O(3)$: $\vec{W}$ is a member of the <u>35</u> and $\vec{M}$ is a vector under $O(3)$.

Melosh[4] has given a rather suggestive interpretation of the operator U: in the free quark model it transforms the charges obtained from the densities $\bar{\psi}\gamma_z\gamma_5\lambda^i\psi$ into symmetries of the Dirac Hamiltonian. Still more interesting, the transformation properties under $SU(6) \otimes L_z$ of the generator of U are the same as Z *).

This result makes still more appealing the form of the axial charges described by Eqs. (1)-(3); however, since it has been derived in a situation where $SU(6)_W$ is a symmetry, it is worth establishing if the unitary operator of Ref. 3 takes into account correctly $SU(6)_W$ breaking.

This problem may be studied in connection with the Weinberg equations[5]:

$$\left[Q_5^{1+i2}, \left[Q_5^{1+i2}, m^2\right]\right] = 0$$
$$\left[Q_5^{1+i2}, \left[Q_5^{1+i2}, 2m(J_x + iJ_y)\right]\right] = 0 \tag{4}$$

*) In Ref. 4 $SU(6)_W$ is considered rather than $SU(6)_S$: for baryons the two SU(6) coincide; for mesons, $\vec{W}$ defined in Eq. (3), is a <u>35</u> under both SU(6) [more exactly there it is just a generator of $SU(6)_W$].

From Eq. (4) one derives:

$$\begin{aligned}
&\left\langle p,\ h = \tfrac{1}{2}\ |U^{\dagger}m^2U|\ p,\ h = \tfrac{1}{2}\right\rangle &&= \left\langle \Delta,\ h = \tfrac{1}{2}\ |U^{\dagger}m^2U|\ \Delta,\ h = \tfrac{1}{2}\right\rangle \\
&\left\langle \pi,\ h = 0\ |U^{\dagger}m^2U|\ \pi,\ h = 0\right\rangle &&= \left\langle \rho,\ h = 0\ |U^{\dagger}m^2U|\ \rho,\ h = 0\right\rangle \\
&\left\langle \rho,\ h = 1\ |U^{\dagger}m^2U|\ \rho,\ h = 1\right\rangle &&= \left\langle \omega,\ h = 1\ |U^{\dagger}m^2U|\ \omega,\ h = 1\right\rangle \\
&\left\langle p,\ h = \tfrac{1}{2}\ |U^{\dagger}m(J_x + iJ_y)U|\ p,\ h = -\tfrac{1}{2}\right\rangle &&= \tfrac{1}{2}\left\langle \Delta,\ h = \tfrac{1}{2}\ |U^{\dagger}m(J_x + iJ_y)U|\ \Delta,\ h = -\tfrac{1}{2}\right\rangle .
\end{aligned} \tag{5}$$

If there is no unitary transformation (U = 1) then Eq. (5) should imply:

$$\begin{aligned} m_{\Delta} &= m_N \\ m_{\rho} &= m_{\omega} = m_{\pi}\ , \end{aligned} \tag{6}$$

which are the predictions of $SU(6)_W$ symmetry. More generally Eq. (5) give rise to sum rules such as:

$$\sum_a |\langle p,\ h = \tfrac{1}{2}\ |U^{\dagger}|\ p_a,\ h = \tfrac{1}{2}\rangle|^2\ m^2_{p_a} = \sum_b |\langle \Delta,\ h = \tfrac{1}{2}\ |U^{\dagger}|\ \Delta_b,\ h = \tfrac{1}{2}\rangle|^2\ m^2_{\Delta_b}\ . \tag{7}$$

From Eq. (7) and the analogous ones that can easily be derived from Eqs. (5), one gets, up to second order in the expansion in Z,

$$\begin{aligned}
m^2_{\Delta} - m^2_N &= \frac{3}{2}\left(1 - \frac{3}{5}\left|\frac{G_A}{G_V}\right|\right)\left(m^2_{70,L=1} - m^2_{56,L=0}\right) \\
m^2_{\rho} - m^2_{\pi} &= 2\left(1 - G_{\rho\pi}\right)\left(m^2_{35,L=1} - m^2_{35,L=0}\right) \\
m^2_{\rho} &= m^2_{\omega} \\
m_{\Delta} - m_N &= \frac{3}{2}\left(1 - \frac{3}{5}\left|\frac{G_A}{G_V}\right|\right)\left(m_{70,L=1} - m_{56,L=0}\right) .
\end{aligned} \tag{8}$$

These results have the right sign and order of magnitude.

It is worth stressing that it is the choice of 70,L = 1, rather than 56,L = 1, as the representation connected by Z to the states of the 56,L = 0 which gives rise to the correct sign of the Δ-N mass difference. A similar role is played for mesons by the fact that no L = 1 state has the right quantum numbers to be mixed with the ρ with zero helicity.

So it is reasonable to think that the transformation of Ref. 3 not only has the group properties suggested by the interpretation of Ref. 4, but also is needed to take into account the breaking of $SU(6)_W$. The specific form of the unitary transformation in the free-quark model has inspired[6] the making of more precise assumptions for the U operator defined in Eqs. (2) and (3). In fact it is rather appealing to assume.

$$U = \Pi_j U_j\ , \tag{9}$$

where

$$U_j = \exp\left[i\ \frac{\vec{\sigma}_j \times \vec{k}_j}{2k_{j\perp}}\ \theta\left(\frac{k_{j\perp}}{m_j}\right)\right] \tag{10}$$

and j stands for each one of the quarks in the quark-antiquark pair for mesons and each one of the three quarks for baryons: in both cases, since one is at $p_z = \infty$, $p_x = p_y = 0$, the transverse parts of the $\vec{k}_j$, which are the only parts appearing in Eq. (10), obey the constraints:

$$\sum_j k_j^x = 0$$
$$\sum_j k_j^y = 0 .$$

From Eqs. (1), (9) and (10) one obtains easily:

$$Q_5^\alpha = \sum_j \frac{\lambda_j^\alpha}{2} \left\{ \sigma_j^z \cos\theta \left(\frac{k_{j\perp}}{m} \right) + \frac{\sigma_j^x k_j^x + \sigma_j^y k_j^y}{k_{j\perp}} \sin\theta \left(\frac{k_{j\perp}}{m} \right) \right\} . \qquad (11)$$

It is easy to see that the axial charges defined in Eq. (11) transform as a 35 under the classification group*); moreover they obey the selection rule $\Delta L_z = 0$ for $\Delta\vec{L}$ = even and $\Delta L_z = \pm 1$ for $\Delta\vec{L}$ = odd: these properties allow us to recover, at all orders in the mixing, the results previously found[3] (it is worth recalling among these the transverse decay of the B meson and the ratio D/F = $\frac{3}{2}$ for the baryon octet $\frac{1}{2}^+$). They are also in agreement with the actual knowledge about the experimental phases of the resonant amplitudes for $\pi N \to \pi\Delta$ [8]**) and finally imply the validity of the Johnson-Treiman relations as soon as one is allowed to consider altogether the contributions of states belonging to an SU(6)⊗O(3) supermultiplet.

An attractive feature of Eq. (11) is that about the same value is required for the diagonal matrix element of $\cos\theta\ (k_\perp/m)$ in the ground state ($\sim 1/\sqrt{2}$) to fit the corresponding diagonal matrix elements of the axial charges, both for baryons and for mesons.

There is finally a very interesting feature, which shows up for baryons in the symmetric harmonic quark model: exactly for asymptotic values of the radial quantum number n, but practically already at rather low values, the contributions of all the states with given n sum up to give the Cabibbo, Horwitz and Ne'eman pattern for the factorized couplings of the t-channel exchanges.

In our work discussed here last year[9], where a set of axial charges for baryons obeying Eq. (4) was proposed, the property just described was shown to apply to all the supermultiplets except the ground state: the price one has to pay there is to have a spectrum which is not realized in Nature (56 ⊕ 70 for any L). Here, instead, one arrives at the situation with 56 and 70 coupled to the ground state in such a way as to have CHN in the crossed channel only asymptotically (indeed for n = 0 we have only 56; for n = 1 only 70; for n = 2 more 56 than 70; for n = 3 more 70 than 56 and so on).

It is reasonable to think that this property is not a consequence of the specific use of the harmonic oscillator quark model (which, however, is rather appealing considering the actual experimental situation for baryonic states), but of the fact that the interference between the three terms in Eq. (11) becomes less important for the transitions to higher states so that one recovers the results of Ref. 9, where, in fact, only one quark is responsible for these transitions.

---

*) This property has been abstracted from the free quark model, in order to make predictions for the matrix elements of the axial charges[7].

**) It is a pleasure to thank D. Faiman for a very interesting discussion on the subject.

The absence of t-channel exotic exchanges at high energy, which follows also from Eq. (11), is what is needed to satisfy Eq. (4) and ensures the convergence of the sum rules (5).

* * *

REFERENCES

1) F. Buccella, F. Nicoló and A. Pugliese, Nuovo Cimento Letters 8, 244 (1973).

2) F. Buccella and C.A. Savoy, to be published in Nuovo Cimento Letters.

3) F. Buccella, *in* Elementary Processes at High Energy, Proc. Erice School 1970 (Ed. A. Zichichi) (Academic Press, N.Y., 1971), p. 510.
F. Buccella, E. Celeghini, H. Kleinert, C.A. Savoy and E. Sorace, Nuovo Cimento 69 A, 133 (1970).

4) H.J. Melosh, California Institute of Technology, Thesis (1973), unpublished.

5) S. Weinberg, Phys. Rev. 177, 2604 (1969), and Phys. Rev. Letters 22, 1023 (1969).

6) C.A. Savoy, Nuovo Cimento Letters 7, 841 (1973).

7) F.J. Gilman and M. Kluger, Phys. Rev. Letters 30, 518 (1973).
A.J.G. Hey and J. Wevers, Phys. Letters 44 B, 263 (1973).
A.J.G. Hey, J.L. Rosner and J. Weyers, Current quarks, constituent quarks, and symmetries of resonance decays, CERN preprint TH.1659 (1973), to be published in Nuclear Phys. B.
F.J. Gilman, M. Kugler and S. Meshkov, Phys. Letters 45 B , 481 (1973).

8) A.H. Rosenfeld, see these Proceedings.

9) F. Buccella, to be published in Proc. Erice School, 1972 (Ed. A. Zichichi).
F. Buccella, F. Nicoló and A. Pugliese, Nuovo Cimento 12 A, 640 (1972).

D I S C U S S I O N S

CHAIRMAN: Dr. F. Buccella

Scientific Secretaries: M. Kupczynski, D.K. Choudhury

DISCUSSION No. 1

- *WILCZEK:*

The Melosh transformation translates the information on the unitary transformation U into physical predictions. What are the assumptions people have made on U other than the obvious ones of P- and C-evenness, SU(3) invariance, etc.?

- *BUCCELLA:*

The assumption made for the unitary transformation is that it can be developed as a power series

$$U = 1 + i\,Z - \frac{Z^2}{2} + \ldots$$

with

$$Z = (\vec{W} \times \vec{M})_z$$

where $\vec{W}$ transforms according to a $\underline{35}$ irreducible representation of SU(6), and $\vec{M}$ is a vector under 0(3). Thus one can derive the measurable predictions for the transitions $\underline{70}\ \ell = 1 \to \underline{56}\ \ell = 0$ and $\underline{56}\ \ell = 2 \to \underline{56}\ \ell = 0$, $\underline{35}\ \ell = 1 \to \underline{35}\ \ell = 0$ in reasonable agreement with experiment. Good predictions come out also for the renormalization of the charges in the $\underline{56}\ \ell = 0$ multiplet. Further restrictions on U come from the Weinberg equation

$$[Q_5^+, [Q_5^+, m^2]] = 0 .$$

One finds that the $\vec{M}_i$ commute and are just the coordinates of the harmonic oscillator. Finally, if one makes the assumption that the unitary operator is $U = U_1 U_2$ where

$$U_i = \exp\left[i\,\frac{\vec{\sigma}_i \times \vec{K}_i}{m}\, f\left(\frac{K_i}{m}\right)\right]$$

for mesons, and $U = U_1 U_2 U_3$ where

$$U_i = \exp\left[i\,\frac{\vec{\sigma}_i \times \vec{K}_i}{m}\, g\left(\frac{K_i}{m}\right)\right]$$

for baryons and

$$\sum_i K_{i_\perp} = 0$$

there are a lot of predictions which can be derived for arbitrary functions f and g, like the transverse decay of the B meson, the Johnson-Treiman relations for the total meson-nucleon cross-sections, Regge couplings identical to those of the Cabibbo, Horowitz and Ne'eman model, anti-W behaviour for the Δn = odd transitions and the W for the Δn = even ones. All the axial couplings are expressed in terms of one parameter. The assumption f = g = 1 is in reasonable agreement with the experiment.

- *WILCZEK:*

We have become accustomed to think about momentum distributions of quarks from electroproduction experiments. Does Melosh have anything to say about that?

- *PASCHOS (Comment):*

In deep inelastic scattering there is also an analogous problem of current and constituent quarks. On the one hand, the parton model visualizes the proton at infinite momentum built up of quarks (finite or infinite in number). On the other hand, the light-cone algebra considers the currents to be of the form $\bar{q}\,\gamma_\mu\,\lambda_j\,q$. Consequently, one desires to have a transformation which takes you from one picture to the other. It is my understanding that such a transformation is not presently available.

- *ZICHICHI:*

The trouble which I have with many models presented at this School, is that they look like a trick to me. An example is the story about constituent and current quarks. The current quarks are those objects which build the currents, and which for magic reasons behave in such a way as to produce simple objects, the currents. The constituent quarks are not simple objects, but when the physical states in those theories are needed, they are built in such a way as to fit the experiment. This I cannot digest. Think of the analogy between current electrons which build up $\bar{\psi}\gamma_\mu\psi$ and constituent electrons.

- *BUCCELLA:*

Some years ago one could think that there was no reason for the transformation between current and constituent quarks. However, we have good classification groups SU(6) × O(3) which should be an approximate symmetry of the theory. One cannot have this symmetry with a local current as shown by Coleman. If one wants to connect the hadron classification with an algebra of charges, one is led to this unitary transformation. Since the unitary transformation is not local, Coleman "no-go" theorem does not hold.

- *WILCZEK:*

Is the parameter m fixed in Melosh's theory either rigorously or intuitively?

- *BUCCELLA:*

In Melosh transformation this mass is put in by hand and is 1/3 of the proton mass.

- *KUPCZYNSKI:*

You said that to understand the SU(6) × O(3) calssification of hadrons, one has to use Melosh transformation and constituent quarks. I would like to mention the following. The SU(6) × O(3) symmetry is quite well understood for slowly moving hadrons when one can consider the non-relativistic bound state wave functions of hadrons. In this non-relativistic theory, the constituent quarks appear which are believed to have large masses, and it is for these quarks that experimentalists hunt for many years. In the current algebra approach the hadronic states are considered as appropriate superpositions of the free $q\bar{q}$- or qqq-states (with masses being, for example, 1/3 of the proton mass). By applying the Melosh transformation, the interaction between quarks cannot be introduced, so the "constituent quarks" in this approach are still the free quarks. Such a formal approach is quite justified by its successes in comparison with the experiments, but the name "constituent quarks" is quite misleading. To have real constituent quarks one should consider the relativistic bound state equations. However, until the discovery of real quarks, such investigations are perhaps too early.

- *BUCCELLA:*

The non-relativistic understanding of the SU(6) × O(3) symmetry is not satisfactory, since we want to have Poincaré invariance in the theory.

- *KUPCZYNSKI:*

But in your model you defined your hadron states only in the infinite momentum frame. How do your hadron states look like in the finite momentum frame?

- *BUCCELLA:*

The Melosh transformation can also be used in the finite momentum frame.

- *NAHM:*

Can you explain why you have $SU(6)_W$ for some multiplets, and anti-$SU(6)_W$ for others?

- *BUCCELLA:*

Our unitary transformation consists of repeated application

$$(\vec{\sigma} \times \vec{K})_z = i(\sigma^{1+i2}\ K^{1-i2} - \sigma^{1-i2}\ K^{1+i2})\ .$$

Thus, if you flip one spin you get $\Delta L_z = \pm 1$, for example, for 70 $\ell = 1 \rightarrow 56$ $\ell = 0$ you have anti-$SU(6)_W$, whereas for 56 $\ell = 2 \rightarrow 56$ $\ell = 0$ you have $SU(6)_W$.

- *NAHM:*

Melosh's transformation contains an arc tan in the exponent, yours does not. Why?

- *BUCCELLA:*

The exponent of arc tan (K) gives one term proportional to cos arc tan (K), which goes very smoothly from 1 to 0 for $K \rightarrow \infty$, plus one term proportional to sin arc tan (K), which gives transitions. Experiments require a rapidly varying function f(K). Thus one should not have a function arc tan x.

- *WAMBACH:*

As far as I understand, the U-transformation connects the proton with all $\ell = 0, 1, 2, \ldots$ representations and higher multiplets. Therefore, my question is how well does the saturation scheme for the $m_p - m_\Delta$ mass difference converge if at all?

- *BUCCELLA:*

Higher terms in the saturation scheme which correspond to higher multiplets ($\ell > 0$) have not been investigated. Therefore, we do not know how well the mass differences converge to finite values.

- *WAMBACH:*

In your scheme, you start from an exact SU(3) × SU(3), then induce $SU(6)_W$ breaking through the U-transformation. Why do you not take into account the rather large breaking of SU(3) × SU(3)?

- *BUCCELLA:*

I am not considering spontaneous broken SU(3) × SU(3), but rather the collinear subgroup of SU(3) × SU(3) (generator is σ_z, not γ_5). Hence, the question on the connection between the chiral SU(3) × SU(3) and the $SU(6)_W \times O(3)$ symmetry of hadrons cannot be answered in the framework of the U-transformation. To my knowledge, nobody has tried up to now to interpret $SU(6)_W$ as a chiral group.

- *KIM:*

When you apply the unitary operator of Melosh on a three-quarks state, say proton, one gets a three-quark state and many particle states. Is it reasonable to interpret the three-quark state as valence quark and the others as core-quarks in parton language?

- *BUCCELLA:*

Yes. However, in the present approach one can connect, in general, any arbitrary configuration of constituent quarks to current quarks. For example, one can consider the constituent quark system to consist of three quarks with lots of gluons not taking any role in weak and electromagnetic interactions. Success of the scheme, however, depends to a large extent on spin-$\frac{1}{2}$ nature of the current and constituent quarks. As you know, such models predict a large transverse cross-section compared with the longitudinal cross-seciton in electroproduction processes in good agreement with experiment.

- *CHOUDHURY:*

As far as I understand, the scheme of Melosh eventually introduces an angle α such that SU(6) result $g_A/g_V = 5/3$ is modified to $g_A/g_V = -5/3 \cos\alpha$. Instead of fitting α from some decay width, could you imagine any internal symmetry scheme to determine α? It seems to me quite important, since as far as decay widths are concerned, one can get reasonably good results by introducing the so-called "recoil effect" or through some phenomenological form factors [and thereby modifying the usual SU(6) result].

- *BUCCELLA:*

In the work of Melosh

$$\cos \alpha = \frac{m}{\sqrt{m^2 + k_{\perp 2}}} = \frac{1}{\sqrt{2}} .$$

It would be very interesting to succeed in understanding the reason for this value of cos α. It is important to stress, however, that the renormalization of g_A/g_V is connected to the leakage to the other states.

- *PATKOS:*

When you use the Melosh transformation for baryons, the symmetrized form of transformation comes as a result of your theory, or is it put as an input?

- *BUCCELLA:*

If we use the symmetric representation $(56,0^+)$ for the classification of baryons in the constituent quark picture, then only the symmetric part of the most general transformation appears. This is consistent with usual classification schemes.

- *PATKOS:*

If you want to use the Melosh transformation for actual calculations, you need the value of the parameters m (the mass of the quark) and $k_\perp$ (the eigenvalue of the operator $k_\perp$). You said previously that one takes for $k_\perp$ the $\langle k_\perp \rangle$ of hadrons measured in inelastic reactions. What is the value of m?

- *BUCCELLA:*

Actually in all calculations, $k_\perp$ and m appear in ratio $k_\perp/m$. So m remains as a parameter.

PART B

REVIEW LECTURES AND SEMINARS ON EXPERIMENTAL TOPICS

THE CURRENT STATUS OF MESON SPECTROSCOPY

David H. Miller

Table of Contents

THE CURRENT STATUS OF MESON SPECTROSCOPY

David H. Miller*)
Purdue University and CERN**)

1. INTRODUCTION

This report is a review of the current status of Meson Spectroscopy and is basically composed of results published between January 1972 and June 1973. Some historical perspective is included, however, together with indications of directions this subject will take in the future.

The number of invited talks and topical conferences on Meson Spectroscopy have been numerous over the past year. There has been a conference on $\bar{p}p$ reactions[1], the Philadelphi Meson Conference[2], the Batavia Conference and a recent $\pi\pi$ conference held in Tallahassee, Florida. All except the latter have their proceedings in print and widely available. Of necessity some of the material will be the same as covered in these conferences and reviewed by Diebold[3] and Leith[4]. There are some new results since Batavia and much of the data submitted to that conference has now been published.

The latest April 1973 Particle Data Group tables[5] have also been utilised and should be consulted for a more extensive bibliography on the subject covered. Since the heaviest confirmed mesons have masses $\sim$ 1.7 GeV, the term "high mass meson" will be used to cover the mass region > 1.6 GeV.

The author is very grateful to those people who allowed him to include pre-publication results in this report and who took the time and trouble to send or discuss their results with him.

2. REVIEW OF THE CURRENT SITUATION

2.1 Models

Experimental results are the key to how nature behaves, but one would be foolish to ignore the predictions of theoretical models which have either had some success or seem logical. The simplest model to use which gives an idea of the complexity we might expect is the quark model in which one uses a $q\bar{q}$ pair to form a meson. Various calculations[6] have been done using a simple non-relativistic calculation and including spin orbit splitting to predict which meson states one should expect. Fig. 1 shows the results of such a calculation and as can be seen many states could exist at masses < 2 GeV. Two specific predictions are that no exotic states should exist, that is states that cannot be made of a $q\bar{q}$ pair. This seems to be true experimentally in that for example no doubly charged mesons have been seen or mesons with abnormal C.

---

*) John Simon Guggenheim Fellow.

**) This report was prepared whilst the author was a visiting Scientist at CERN on leave of absence from Purdue University W. Lafayette, Indiana.

Regge theory tells us to expect resonances to lie on trajectories which could be linear on an M^2 versus J plot (fig. 2). Unfortunately, M^2 and J are not known for all resonances and the g and $K^*(1760)$ are the highest masses and spins known, so although most mesons can be accomodated on reasonable trajectories one would like more data. One prediction is that if trajectories are linear, many high mass mesons should exist getting closer and closer together in mass. For example at masses $\sim$ 3.5 GeV the separation becomes less than one pion mass.

The SU_3 symmetry predicts that mesons will occur in octets and singlets with mixing between the two singlets. The well established nonets are shown in fig. 3 and there are many mesons left over to place in other nonets. One can then check the mass splittings, mixing angles and two-body decay modes. This has been done by several authors[7], who find good agreement with predictions although in many cases the data is poor.

One can also look at other types of resonances, e.g. the baryon resonances and assume that the mesons will have similar complexity. An examination of the baryon table in the PDG tables reveals that unlike the Michelin guide we actually have four star resonances with spins to 7/2 and three star resonances with spins to 11/2 and masses to 3 GeV. Unfortunately, most of this data comes from phase shift analyses which are a much more difficult proposition using mesons. One reason that so much interest was generated by the original CERN missing mass experiment was that not only did the states observed fall on linear trajectories but the widths were narrow so that one did not need a phase shift analysis to resolve them. Since that experiment[8] is in disrepute we are left with the possibility of large widths and many overlapping states.

2.2 Experimental problems

The main problem is always lack of good data. Sometimes, however, as in the case of the A_1, A_3 where the data is good we still do not completely understand the situation. Listed below are a few problems some of which will be discussed in this report in detail.

(i) The relation of diffractive phenomena to meson states.

(ii) The existence of narrow states with particular reference to the 1 GeV mass region.

(iii) The observation of high mass mesons.

(iv) The determination of J^{PC} and branching ratios, to fill the SU_3 nonets.

These are fairly general categories but in fact each meson tabulated in the PDG tables has its own problems. In fig. 4 is a list of these mesons with some comments on their status and an indication of their typical production cross sections in the decay mode normally studied and correcting for unobserved events where possible.

Most experiments from which this data comes have between 5-30 events/μb and do not analyse all possible channels. The total cross sections of all channels analysed in which resonances occur is usually 10-15 mb. Since there are several confirmed resonances with production cross sections of a few microbarns it is clear that many resonances could exist that have not been observed yet. Since most resonances are first observed as mass bumps, the difficulty is obvious, if we wish to observe a narrow 1μb effect at 5 standard deviations with a 30 event/μb experiment the background has to be ∿ 36 events. This means that one needs a channel whose cross section is ∿ 100μb for an experiment with a 30 MeV FWHM resolution. Indeed we do see such mesons as the f' produced in $K^-p \to \Lambda f'$ with $f' \to K\bar{K}$ where the background is very low, if a similar meson decayed to 4π for example, we would not as yet have seen it. Similarly, the effective events/μb can be drastically reduced in certain channels. As an illustration a recent experiment studied $\pi^+d \to p_s p\, K^o_s K^o_s \pi^o$ to establish the charge conjugation of the D and E mesons[9]. To get a unique fit to such a final state, we would like to see the spectator proton and both K decays. The probability for this for uncorrelated K decays is (1/9 * .4). The spectrum obtained is shown in fig. 5 with a total of 40 unique events from an 18 event/μb experiment and the result is inconclusive. Another example is the $A_2^\pm$ with a claim for an ωππ decay mode of ∿ 10%. The $A_2^\pm$ have been studied extensively but the decay $A_2^\pm \to \omega\pi^\pm\pi^o$ occurs with two pi zero's and cannot be selected uniquely. So experimentally we have looked for only particular resonances at cross sections of a few micro barns and there are severe problems in looking for mesons with properties which depress the signal to noise, such as:

(i) Broad states $\Gamma \to 150$ MeV
(ii) Many decay modes
(iii) Decay modes involving multineutrals or γ rays
(iv) Small production cross section
(v) Production in high multiplicity processes

There are also problems in finding multiple states in the same mass region because of the difficulty of using phase shift analysis. All the above is true even if the meson states decay freely, however, interference effects can also occur and cause problems.

2.2.1 Importance of diffractive N^* production

One example of such a process with a large cross section which occurs in the same final state as meson production is the diffraction dissociation of the nucleon into Nπ and Nππ. Particularly in the case of the reaction KN → KπN and πN → ππN this background complicates the analysis of the ππ, Kπ system. The experimental situation is that the events in the diffractive N^* fall in the very forward region of the cos $\theta_{\pi\pi}$ distribution and in ππ mass regions which move to increasingly high mass as the beam momentum increases. Since the spin of observed Kπ and ππ resonances increase with mass they also populate predominantly the same region of cos $\theta_{\pi\pi}$ and this causes problems. For example, in the K^*(1760) decay into Kπ the published moments of the ππ system for two very similar sets of data were quite different because in one an attempt was made to reduce the diffractive background (see sect. 4.3.2).

Interference effects have been directly observed by W. Michael[10] in the reaction $\pi^+p \to \pi^+p\pi^o$ at 2.67 GeV. In this experiment the width and mass of the ρ are found to vary as a function of position on the Dalitz plot. The major variation occurs when the ρ overlaps with the $p\pi^o$ diffractively produced N^*. This is shown in fig. 6. At higher incident momenta this overlap occurs at and above 1.6 GeV and if similar effects occur could explain some of the problems with the $g \to \pi\pi$ and $K^*(1760) \to K\pi$ data. The diffractive N^* production has a cross section of a few hundred microbarns and the $g \to \pi\pi$ a few tens of microbarns. It is possible that the two amplitudes given by the diagrams

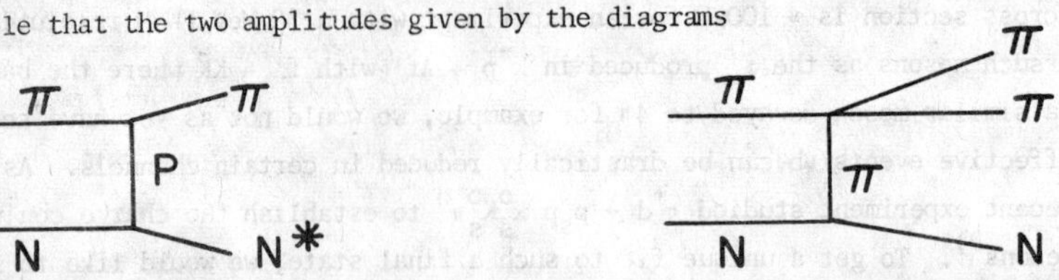

interfere with differing phase depending on the incident beam momentum and distort the properties of the $\pi\pi$ resonance. It is possible if high mass mesons exist that their production amplitude could be enhanced by such a mechanism.

2.3 Speculation

One suspects that the models are correct and that as yet we have not isolated all the states that exist. Nearly all our information on new mesons has come from bubble chambers with typical experiments in the 5 → 30 event/μb range with mass resolutions of 30 MeV FWHM. An important reason why more states have not appeared could be because of the restrictive situation in which we look for them. That is in general we look using π, K beams at reactions dominated by the exchange of a single trajectory. For example:

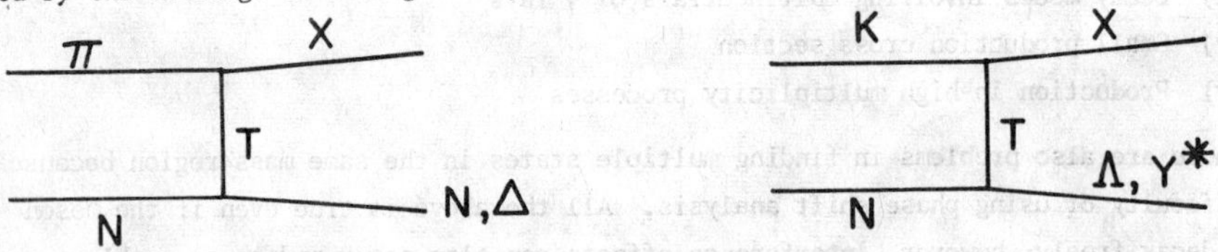

and similar ones for mesons with strangeness. The cross section for such processes will depend on the $B\bar{B}T$ coupling at the lower vertex and the $XT\pi(K)$ coupling at the upper vertex.

It is not unreasonable to put in the couplings at each vertex to just the lowest members of the trajectory in order to estimate production cross sections. In fact, one might put in the suitable coupling corresponding to the on mass shell decay X → π, K (+ lowest member of a trajectory) for example ππ, πK, πω etc. This would mean that any meson not having a sizeable branching ratio into (π + lowest member of a trajectory) or (K + lowest member of a trajectory) has a small production cross section. One might expect that high mass mesons that decay preferentially by single pion emission or states that have small partial widths for two-body decays involving a π or a K would be difficult to produce in the processes normally studied. Selection of events corresponding to diagrams of the type

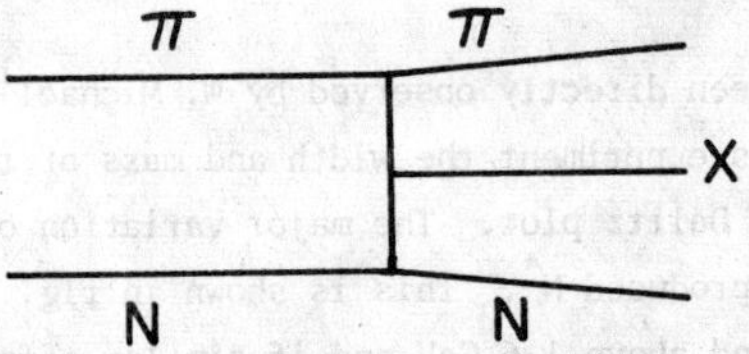

could yield new states of this type. In fact, selection of different production mechanisms may be the only way states with similar masses and widths can be separated.

There are some observations which support such a theory, for example:

(i) The M meson in $K^-p \to K^-pM$.

(ii) A four pi enhancement seen in $K^+p \to K^op4\pi$.

(iii) The ρ' produced in $\gamma p \to 4\pi p$ maybe has a dominant $\rho\varepsilon$ decay and is not produced strongly in $\pi N \to \rho' N$.

(iv) The many effects claimed in $p\bar{p}$ annihilations which are not produced in the single exchange region.

(v) Mesons like the f' which are produced in a small number of reactions.

(vi) Mesons seen mainly as decay products of a higher mass meson e.g. δ, B1. This is presumably because the BBT coupling is weak for their direct production.

2.4 Production cross sections

The above is also borne out by an examination of what we know about existing states.

Fig. 4 is an attempt to collect together the production characteristics of mesons. The results are taken from bubble chamber experiments since this is where most mesons have first been seen. A typical cross section is given and the incident energy. The main use for this table is to indicate what cross sections we would expect for states not yet observed. The table does not in general indicate the current state of our knowledge about particular states. Those mesons which are narrow on small backgrounds, the vector states and mesons like the ρ, A_2 have all been studied much more extensively using various techniques. Although at M > 1.3 only specific decay modes have been studied for the $A_2^{\pm}$ and the g^o.

The following conclusions can be drawn about the cross sections:

(i) Hypercharge exchange cross sections are about a factor of 10 lower than charge exchange reactions e.g. $K^-p \to \Lambda\rho$, f:$\pi^-p \to n\rho$, f.

(ii) The cross section for a Regge Recurrence is about a factor of 10 lower than for the preceding members e.g. ρ, g: ω, $\omega(1675)$.

An extrapolation of these cross sections above M = 1.8 GeV predicts production cross sections of a few microbarns for production by non strangeness changing interactions and < 1 μb for strangeness changing reactions.

Some mesons like the Ø and f' because of their couplings are only seen in the hypercharge exchange reactions. Others like the δ, D, E, B, are not produced in peripheral processes.

One would predict that experiments in the 500 event/μb range should be able to observe some states above the g, $K^*(1760)$. As an example, the missing singlet for the 3^- nonet the Ø(1820) should be seen in $K^-p \to Y$Ø with a cross section of 1 μb.

2.5 General Conclusions

Existing experiments have explored mass regions up to 1.8 GeV in particular ways. Above 1.8 GeV the experiments as yet do not have sufficient sensitivity. Below 1.8 GeV many state could exist that are not produced in the reactions we study or are hidden under the ones observed.

3. EXPERIMENTAL RESULTS IN THE REGION M < 1.7 GeV

3.1 The 1 GeV mass region

Each mass region has its own personality and in this case there is a penchant for the observation of narrow states, particularly in the neutral final state. Lamb and co-workers[11) have published evidence for possible states at 940, 950, 1033, 1150 MeV in the missing mass spectrum recoiling against a neutron in the reaction $\pi^- p \to n + MM$ making a crude selection on the final charged multiplicity. Confirming evidence is needed before classifying these states particularly since Binnie and co-workers[12) with a similar experiment only see the η'. Holloway et al.,[13) also claim an effect but overall the situation is still confused.

Two other observations are more amenable to further study since they occur in well defined final states. The first is the result that in the reaction $K^-p \to K^-p\pi^+\pi^-\gamma$ at 3.9 and 4.6 GeV[14) an enhancement is seen which is compatible with the η' in all respects except that no $\rho\gamma$ is observed in the $\pi^+\pi^-\gamma$ whereas the η' they observe in $K^-p \to \Lambda\eta'$ has a $\rho\gamma$ decay consistent with previous observations. The evidence for the M is shown in fig. 7 and the final numbers show a three standard deviation difference between the $\pi^+\pi^-\gamma$ decay mode of the M and η'. One point is that the reaction in which the M is observed is much less peripheral a cut of cos $\theta^* > 0.0$ between initial proton and final state K^-p is used whereas a cut of cos $\theta^* > .8$ is used for the $K^-p \to \Lambda\eta'$. It is possible final state interactions are a problem even though the η' has a small natural width or it could be that in quite a different production mechanism a new state is excited.

The δ is another intriguing state which has been observed off and on. In cases where a partial decay mode is inferred, $\pi\eta$ is always favoured. One way it is observed is in the decays of the D and the E in $\bar{p}p$ annihilations which are covered later in this report. It is sufficient to say here that the observation and a J^P determination are plagued by high backgrounds and reflections of the ω. It was also observed clearly several years ago in

$K^-p \to \Lambda\pi\delta$[15)

$K^-n \to \Lambda\delta$[16)

where the decay $\pi\eta$ ($\eta \to$ all neutrals) was observed. What is still lacking is on observation in such reactions of the decay

$$\delta^\pm \to \pi^\pm\eta \to \pi^\pm\pi^+\pi^-\pi^o$$

and then sufficient data to determine J^P. As can be seen in fig. 8, the evidence for its existence is quite good and its production cross section $\sim$ 3μb. This is one case where a lot of observations can be tied together by a fairly easy experiment. As was noted by Astier[17), the $K\bar{K}$ threshold enhancement could be related to a virtual $\pi\eta$ state below $K\bar{K}$ threshold for which the δ would be the ideal candidate. One would then expect to observe it in the above production mechanisms produced by K exchange and only weakly in πN collisions involving η exchange. The D for example, which decays via $K\bar{K}$ with the $K\bar{K}$ in the I = 1 threshold effect also decays via $\pi\delta$ in its $\pi\pi\eta$ decay mode. So everything holds together very nicely except one would still like to see a clean decay of the $\delta^\pm \to \pi^\pm(\pi^+\pi^-\pi^o)$.

3.2 The A_2 meson

Now that the dust has finally settled the consensus is that the A_2 is not split and behaves as a simple Breit Wigner resonance. This result has come from a series of high resolution experiments covering a wide energy range. The mass spectra from these experiments is shown in fig. 9. These experiments are

- 7 GeV π^+p data of Berkeley[18) — fig. 9a
- Northeastern Stonybrook results[19) — fig. 9b
- The $K\bar{K}$ mass spectrum of CERN-Munich[20) — fig. 9c
- The $K\bar{K}$ mass spectrum of Foley et al.[21) — fig. 9d
- The mass spectrum of Ankenbrandt et al.[22) — fig. 9e
- The $\pi\eta$ mass spectrum of Key et al.[23) — fig. 9f
- The missing mass spectrum of Conforto et al.[24) — fig. 9g
- The A_2^- mass spectrum of Binnie et al.[25) — fig. 9h
- The A_2^o mass spectrum of Holloway et al.[26) — fig. 9i

All of these experiments find no evidence for a split and prefer a single Breit Wigner over a dipole. This is an impressive series of experiments which have all refuted a previous observation. Unfortunately, we have not learnt much about the A_2 from all the high statistics and effort, since a recent report has found a new two-body decay with a magnitude of nearly 10%. This is a result of Defoix et al.[27) from the reaction $\bar{p}p \to 3\pi^+3\pi^-\pi^o$ where in the $\omega 2\pi^+2\pi^-$ final state a $\omega\pi^+\pi^-$ enhancement is seen at 1.315 GeV which decays via an intermediate state at 1.04 GeV which they call B_1. The relation to the A_2 is inferred from the mass and width, and the decay $A_2 \to B_1\pi \to \omega\pi\pi$ is claimed, the evidence is shown in fig. 10a). The complication in such a high multiplicity final state is that the combinatorial background is very high. This means that resonant signals sit on very high backgrounds and that spurious correlations could be generated by the selection techniques and the restricted phase space.

There are nine $\pi^+\pi^-\pi^o$ combinations which cause problems in selecting the true events. The same authors in the same channel with the same events also find both D and $E \to \pi\delta(\to\pi\eta)$ where the δ is an enhancement at 972 MeV. The D, E and A_2^o lie within 130 MeV of each other and the δ and B_1 within 70 MeV so the analysis depends crucially on selecting the ω and η correctly. It would have made a more consistent picture if the $\omega\pi\pi$ decay of the A_2 had proceeded through $B\pi$, where the B is the normal one observed. Because of energy conservation this would have to be through the Breit Wigner tail of the B. A recent paper by Cohen et al.[28) studying B production in $\pi^-n \to p\pi^-B^-$ find their data implies a large π A_2 coupling to the B. Since the A_2 is produced easily in reactions which do not suffer the same analysis problems confirmation of the effect should not be hard. This illustrates the remark made before that since in the counter experiments a single decay mode was looked for no information is available on other modes. Some preliminary confirmation, however, is coming from Bubble chamber experiments of the $\omega\pi\pi$ decay. Flatté[29) has shown a possible confirmation from the $\Delta^{++} A_2^o$ final state but the significance is low (see fig. 10b). Better evidence comes from a 3.9 GeV π^-p experiment of Chaloupka et al.[32) who see a bump (fig. 10c) in $\pi^+\pi^-\pi^+$ MM which would include the $\omega\pi^-\pi^o$ final state. They obtain a branching ratio of 10 ± 5% in agreement with the other experiments. More data is obviously needed to confirm this new decay mode.

An observation of A_2 production at 25 and 40 GeV/c will be included with a discussion of A_1 and A_3.

3.3 D and E

New results have been published on these mesons from $\bar{p}p$ annihilations on both the $K\bar{K}\pi$ decay modes and the $\pi\pi\eta$ decay modes. Defoix et al.[31] analyse the same events the $A_2^o \rightarrow \omega\pi\pi$ decay was observed in, that is $\bar{p}p \rightarrow 3\pi^+3\pi^-\pi^o$ at .72 GeV. The $\pi^+\pi^-\pi^o$ mass spectrum with 9 combinations/event all of which fall at masses less than 1 GeV is shown in fig. 11a. η production is only a few percent of the channel which is dominated by ω production. Selection of the η by a mass cut alone is not sufficient since the chance of a real ω event giving a mass combination in the η region is very high. The $(5\pi)^o$ mass spectrum is shown in fig. 11b for events with at least one 3π combination in the η region and the $\eta\pi^\pm$ mass spectrum is also shown. The D and E are contained in the former and the δ in the latter since the final conclusions of the paper are to see the decays

$$D, E \rightarrow \pi^\pm\delta^\mp \rightarrow \eta\pi^+\pi^-$$

In order to improve the η selection, the authors try several techniques one of which is to show that the D and E mass regions have correlated η and δ signals whereas adjacent regions do not. Perhaps more striking visually is an attempt to select η by using the 3π Dalitz plot variables to enhance the η signal and reduce the ω background. The result of this is shown in fig. 11c which also reveals that the final state involving the D is

$$\bar{p}p \rightarrow D\rho^o$$

which agrees with the analysis of the $D \rightarrow K\bar{K}\pi$ in the same data[32]. The final numbers from this analysis are shown in fig. 11d. The analysis is obviously very tricky but the properties of the D seem to be stable in all experiments[33] which is encouraging. Duboc et al.[34] in $\bar{p}p$ annihilations analyse 3, 4, 5 and 6-body final states containing a $K\bar{K}$ pair. The D is observed in nearly all channels in $(K\bar{K}\pi)^o$ (fig. 12) and always decays with the $K\bar{K}$ pair in the I = 1 $K\bar{K}$ threshold enhancement. With some assumptions the D is found to have $J^{PC} = 1^{++}$. The evidence for the E is somewhat weaker (fig. 12), but once again when it is observed its properties appear to be consistent. Definitive measurements of the D and E spin parity would be very nice to have, in experiments with smaller background and well defined production processes.

3.4 Diffractive production of mesons

This is one area of meson resonances which does not suffer from lack of events, however, as always there are still problems. An overall review of diffraction dissociation has recently been given by D.W. G.S. Leith[4]. This section will deal mainly with the $\pi \rightarrow 3\pi$ diffraction process. The production of the 3π system has been studied extensively over a wide range of incident momentum and partial wave decompositions of the 3π system have been made using techniques developed by G. Ascoli and co-workers[35]. This analysis has been carried out up to 3π masses of $\sim$ 1.9 GeV using data from a compilation[36] of bubble chamber data of the reaction

$$\pi^- p \rightarrow \pi^- \pi^+ \pi^- p$$

in the momentum region 5 $\rightarrow$ 25 GeV/c and also on the same reaction at 25 and 40 GeV/c using data from the CERN-IHEP boson spectrometer[37]. The mass spectra for these two sets of data is shown in fig. 13 and reveals the broad A_1 with some A_2 on the high side and the A_3 at a mass $\sim$ 1.6 GeV. The results of the partial wave analysis for the CERN-IHEP data is shown in fig. 14 and they agree with the bubble chamber data extremely well. The main results are:

1) The A_1 enhancement is a 1^+ s-wave $\rho\pi$ system
2) The A_2 enhancement is a 2^+ d-wave $\rho\pi$ system
3) The A_3 enhancement is a 2^- s-wave $f\pi$ system
4) Interference between partial waves is observed and relative phases are determined
5) Only the A_2 relative to the 1^+p and 1^+s shows the expected behaviour for a resonance
6) The A_3 is well fitted by a Breit-Wigner shape with M = 1.66 GeV and Γ = .30 GeV
7) The $\pi\eta$ branching ratio of the A_2 at 40 GeV/c is the same as at lower energies
8) The energy dependence of the A_1, A_2, A_3 production cross section is very similar
9) All observed waves at the high momenta are produced by natural parity exchange $J^P = (0^+, 1^-, 2^+, ---)$.

The results shown in fig. 14 from the CERN-IHEP experiment were obtained with an apparatus which detects the recoil proton and has a large solid angle for secondaries in the forward arm. This gave a high acceptance for the reaction $\pi^- p \rightarrow p\pi^-\pi^-\pi^+$ to masses above the A_3 with the requirement that two of the pions be momentum analysed in addition to the recoil proton triggers. In the cases where three pions were momentum analysed a 1C fit to a missing π^o was possible.

Fig. 14a shows the intensities for each of the partial waves used in the fit. Other fits were done with up to 15 partial waves but all but the ones shown were consistent with zero intensity and finally not used in the fit.

Fig. 14b shows the results for the A_1 and the very flat relative phases with respect to the other waves with slowly varying intensities. The overall shape of the A_1 cannot be fitted with a simple Breit-Wigner.

The results for the A_3 shown in fig. 14c are similar except a Breit-Wigner fit to the intensity is possible.

The A_2 does exhibit the expected resonant behaviour as shown in fig. 14d which indicates that the method works correctly.

Another interesting result is that the A_1, A_2, A_3 cross sections have a similar behaviour as a function of P_{LAB}. This was unexpected since the A_2 is not in the diffractive series. In addition to the $\pi\rho$ the $\pi\eta$ decay was also observed as shown in fig. 15. A clean η signal is seen which is correlated mainly with A_2 production. The number of events 52 is small and the correction for the acceptance high (~ 2) but the branching ratio of $\eta\pi^-/\rho\pi$ $= 0.22 \pm .03$ is in good agreement with the accepted value. Further confirmation of the $A_2 \to \eta\pi$ comes from the decay which yields a 2^+ wave in a pure $|M| = 1$ state produced by natural parity exchange ($\rho_{11} + \rho_{1-1} = .97 \pm .06$), for the $\pi\rho$ decay $\rho_{11} + \rho_{1-1} = .93 \pm .04$. The fall off for A_1, A_2, A_3 is close to $p^{-0.5}$ at these high momenta. The A_1, A_3 are produced in $|M| = 0$ states with ρ_{00} being close to 1.

Another very interesting result which is a little more tentative, is that in the analysis of the CERN-IHEP data the 2^+p $f\pi$ system behaves in a resonant manner. The actual amplitude has a resonant shape (fig. 16a) with a mass ~ 1.75 GeV but more importantly the relative phases move through 90° (fig. 16b). This result has been checked in detail and the solution remains stable. The magnitude of the effect is about 10% of the total 3π amplitude in that region.

So what conclusions can be drawn. The A_1, A_3 unlike the A_2 do not seem to behave as resonances so the question remains as to how they should be assigned. Whether as just kinematic effects of a Deck mechanism or connected in some way with SU3 multiplets. The other surprise is that the A_2 cross section falls at the same rate as the A_1 and A_3. The results from this experiment are very interesting although there are still problems.

A new result from G. Thompson[38)] using the same analysis program obtained from Illinois, but on the reaction

$$\pi^+ p \to \pi^+ \pi^+ \pi^- p$$

also finds an interesting effect. The mass spectrum of the 3π system after removing the Δ^{++} is shown in fig. 17a and the number of events is more than the compilation of ref. 36 with the advantage of being at a single momentum 13 GeV/c. The analysis yields similar results for the intensities of the partial waves which are shown in fig. 17c. One difference is that in the A_3 region there appears to be relative phase changes of 90°. The most significant one is the 2^- relative to the 1^+p which is shown in fig. 17d. This result conflicts with the π^- data and the reasons as yet are not clear. The behaviour of the A_1 and A_2 shown in fig. 17e are in good agreement with the CERN-IHEP results.

Apart from this result, the phase shift analysis is giving consistent results and is clearly an important way to analyse the data. It may be the only way in which resonances can be disentangled if their main decay is to 3π. The interpretation of the A_1, A_3, however, is still not resolved.

A recent paper by Cohen et al.[39)] compares $\pi^+ p \to \pi^+ \rho p$ with $\pi^- n \to \pi^- \rho^- p$ and find similar mass spectra in the A_1 region (see fig. 18). Since the $\pi^- \rho^-$ is exotic and the data can be fitted with a Reggeised Deck calculation the authors are in favour of a kinematic interpretation of the A_1.

The situation with the $K\pi\pi$ system is very similar except it lacks the detailed partial wave analysis carried out on the 3π system. There have been various reviews of the existing data[40)] and the situation has not changed on the non-charge exchange data. Werner et al.[41)] have looked for the Q produced by charge exchange with a compilation of 100 events/μb.

No clear Q is observed (see fig. 19) for the I = 1/2 or I = 3/2 $K\pi\pi$ systems. This data also does not agree with a Reggeised Deck calculation. Q production in the u channel is observed by Firestone[42] with similar characteristics to the normal diffractive production (see fig. 20). These observations have not clarified the overall situation which is still ambiguous.

3.5 The ρ' meson

This meson was first observed about two years ago and since that time has been studied in e^+e^- colliding beams at Adone and in photoproduction mainly at SLAC. The only certain decay mode is $\rho' \to 4\pi$ and all experiments agree that the quantum numbers are the same as the ρ that is $J^{PC} = 1^{--}$.

The initial data from Adone shown in fig. 21 come from experiments studying hadron production in e^+e^- colliding beams at total centre of mass energies up to 2.4 GeV. The first published result[43] on $e^+e^- \to 4\pi$ consisted of 29 events in the energy range 1.2 → 2.4. Although statistically weak the deviation from predictions was interpreted[44] as indicating the existence of a ρ' with mass ∿ 1.6 GeV. A new analysis[45] has since been published of 23 events in the energy range 1.5 → 1.7 and this analysis finds the data is compatible with a 100% intermediate $\rho\varepsilon$ final state, i.e. $e^+e^- \to \rho^o\varepsilon^o\ (\pi^+\pi^-)(\pi^+\pi^-)$ where M_ε = .8 GeV and Γ_ε = .3 GeV. This data is shown in fig. 22. One important point stressed in the paper and by Silvestrini at Batavia is that the one intermediate photon hypothesis for e^+e^- interactions immediately gives $J^P = 1^-$ and the decay into 4π gives C = -1. A description of the whole experiment has just been published[46] and fig. 23 shows the energy dependence for the $e^+e^- \to \pi^+\pi^-\pi^+\pi^-$ and for events including $e^+e^- \to \pi^+\pi^-\pi^o\pi^o$ which if the $\rho\varepsilon^o$ decay is 100% should exhibit the same variation with a factor of 2 depression in cross section. There are experimental biases in the data and it is statistically weak but the main conclusions are convincing.

The other reaction in which the $\rho' \to 4\pi$ is clearly seen is $\gamma p \to 4\pi p$ and recent data has come from a bremsstrahlung beam and a polarized photon beam. The data from the 9.3 GeV polarized photon experiment comes from a bubble chamber experiment[47] at SLAC and is shown in fig. 24a.

In the final state $\gamma p \to \pi^+\pi^-\pi^+\pi^- p$ there is a broad enhancement in the 4π mass distribution at ∿ 1.5 GeV which is most clearly seen when the strong Δ^{++} production is removed. There is substantial ρ production and the dominant decay of the enhancement is $\rho\pi^+\pi^-$ and no $\rho^o\rho^o$ is observed. The decay of the ρ', is analysed in the helicity frame by looking at the vector $Q = P^+_{\pi 1} + P^+_{\pi 2}$. Strong $\sin^2\theta$ and $\cos^2\psi$ components are found as shown in fig. 24b. The amount of the $\sin^2\theta\ \cos^2\psi$ component is shown in fig. 24b as a function of 4π mass. These distributions are characteristic of the decay of an s-channel helicity conserving vector meson produced in a natural parity exchange[48] and the data therefore are consistent with $J^P = 1^-$. From the observed ratio of $\pi^+\pi^-\pi^+\pi^-$: $\pi^+\pi^-$MM isospin 2 for the $\pi^+\pi^-$ pair not in the ρ is ruled out and from the lack of a $\rho\rho$ decay I = 1 for the odd $\pi^+\pi^-$ pair is ruled out and, therefore, I = 0 is preferred. This gives an overall isospin for the ρ' of 1. A J^P analysis was carried out assuming s-channel helicity conservation and that the ρ' decayed to $\rho + \varepsilon$ with the lowest allowed angular momentum between the ρ and ε for a given J^P. With these assumptions the data rules out all but $J^P = 1^-$, the fit to 1^-+ phase space being shown in fig. 24. The mass is found to be 1.43 ± .05 GeV and Γ = .65 ± .1 GeV and

$$\rho^- \rightarrow \frac{\pi^+\pi^-}{\rho^0\pi^+\pi^-} < .2, \quad \frac{K^+K^-}{\rho^0\pi^+\pi^-} < 0.04$$

The data from the bremsstrahlung beam[49] covering the energy range 4.5 → 18 GeV and taken in a streamer chamber yields similar results. The 4π mass spectrum is shown in fig. 25a and the $\rho\pi^+\pi^-$ together with the (non ρ) $\pi^+\pi^-$ spectra in fig. 25b. These clearly show a preferred ρππ decay. Once again the data are consistent with an s-channel helicity conserving meson with $J^P = 1^-$ decaying to ρ + ε. A mass of 1.62 ± .03 and Γ = .31 ± .07 is found. These values depend critically on the parametrization used and acceptable fits with M as low as 1.2 GeV can be obtained. Comparison with other data yields

$$\rho' \rightarrow \frac{\pi^+\pi^-}{4\pi^\pm} < .14$$

The other interesting fact is that the production cross section is energy independent as shown in fig. 26.

There have been some other indirect analysis of the ρ'. D. Mortara[50] has put forward the proposal that the ρ' is really the decay of the ρ → 4π far out on the Breit-Wigner tail of the ρ. His calculation of this effect fitted to the SLAC data is shown in fig. 27. An argument against this model is that put forward in a paper by Y. Eisenberg et al.[51] who say that in such a case apart from phase space effects, the ratio of ρ/ρ' in every experiment should be the same. Their data on $\pi^+p \rightarrow \Delta^{++}\rho(\rho')$ sets limits on ρ' production which are in disagreement with the ρ/ρ' ratio of 10:1 found in γp experiments. This paper also sets a limit on the ρ' → 2π mode which appears to be incorrect. This is important since the PDG tables contain this result in the meson table. Apart from arguments against the technique used there is a conflicting experimental result from the phase shift analysis of the CERN-Munich Group[52] who use a one-pion exchange model in $\pi^-p \rightarrow \pi^-\pi^+n$ to find the ππ phase shifts. They find clear evidence within the framework of their assumptions for a resonance in the p wave with a mass of 1590 ± 20 and Γ = 180 ± 50 and an elasticity of 25% ± 5%. The result of this analysis is shown in fig. 28. The problem with the analysis of Eisenberg is probably that at such a low momentum and high t one-pion exchange is no larger valid, the minimum t at the ρ' mass is .3 at 5 GeV/c for $\pi^+p \rightarrow \Delta^{++}\ \rho'$. No g is observed in the data either which also indicates pion exchange is not good. The ρ' has some properties very reminiscent of the A_1, A_3. It is produced in a diffractive way with constant cross section and is a threshold effect in the ρε system. It would be very nice to observe the 2π decay directly and ρ' production in other processes and determine its parameters. Although convincing the current data is still very meagre.

. EXPERIMENTAL RESULTS ON HIGH MASS AND HIGH SPIN MESONS

4.1 Missing Mass Experiments

This is a separate topic since such experiments have been very important although the sum total of what we have learnt about meson resonances from them is small. The first most important experiment using this technique was the original CERN experiment[53] claiming to see the R.S.T.U. This was followed by a new experiment at CERN which saw further peaks at masses between 2.6 and 3.6 GeV[54]. These results are now in doubt following the Northeastern Stonybrook[55,56,57,58] experiment which has failed to observe narrow structures. A somewhat different experiment because of the high energies has been carried out at Serpukhov and has produced results on the A_1, A_2, A_3[59,60]. All the experiments use the reaction $\pi^- p \to pX^-$ and detect the final state proton and look at the X^- mass spectrum. In addition, most of the experiments have used a forward spectrometer arm to detect charged particles from the X^- decay. This has meant that data has been obtained on $\pi^- p \to p\pi^- X^o$ where the detected π^- is relatively fast forward. With the CERN-IHEP experiment it was possible to select events of the type $\pi^- p \to p\pi^- \pi^+ \pi^-$. In all cases a small fraction of the beam were K^- and the corresponding K^- induced reactions were studied. The various mass spectra obtained are shown in figs 29 → 32 and in addition, the parameters obtained by the various authors for resonance production. The conclusions that can be drawn are:

1) The R,S,T,U, as claimed by the CERN experiment do not appear to exist[55,56] (figs 29 and 30).

2) Only the R is observed in the X^- spectrum from Northeastern Stonybrook[55] and is consistent with a mixture of known effects (g, A_3) (fig. 29).

3) The Q, $K^*(1420)$ and L are observed in the $K^- p$ reaction[58] (fig. 29).

4) An enhancement whose mass and width are consistent with the D is observed in the neutral spectrum[57] (fig. 29).

5) The CERN-IHEP experiment does not observe any high mass state with $\frac{\Delta\sigma}{\Delta t} > 3\mu b/(GeV/c)^2$ (fig. 31).

6) The Q and L are observed[60] by CERN-IHEP (fig. 32).

7) The K^*890 and K^*1420 are observed and analysed at 25 and 40 GeV/c[60] (fig. 32).

What have we learnt overall about mesons that is useful? From the pure missing mass experiments we have learnt little except that important claims should be verified.

The observation of the D (which is also seen by CERN-IHEP) is interesting but it is possible that the f and A_2^o are somehow producing this bump.

Interesting results have come from the exclusive channels in the CERN-IHEP experiments.

Bubble chamber experiments are approaching the numbers of events in the missing mass experiments although not within the same t-region. A recent result by G. Thompson et al.[61] from $\pi^+ p \to pX^+$ at 13 GeV obtained by compiling all the topologies measured yields the spectrum shown in fig. 33 and there is a striking similarity to the Northeastern Stonybrook spectrum as of course there should be. Only the A_1 region and the R region show deviations from a smooth curve. The difference is that in the individual channels there is copious resonance production which is not observed and cannot be fully analysed in the missing mass experiments.

4.2 Formation Experiments

These experiments study $\bar{p}p$ interactions as a function of centre of mass energy looking for direct resonance formation.

This process would then give a peak in one or more partial cross sections as the centre of mass energy passed through the X mass. The relation to incident momentum and centre of mass energy is shown in fig. 34. The specific final states $\pi^+\pi^-$ and K^+K^- can also be analysed in terms of an angular decomposition to look for effects in specific partial waves. The latest large scale review of all these processes is contained in the proceeding of the Chexbres meeting[1).

The current experimental situation is that no resonance effects have been unambiguously observed using this technique. Many claims to see resonant behaviour have either been withdrawn or not substantiated in similar experiments and no effect has been confirmed separately in independent experiments except the broad bumps seen in the total cross sections of $\bar{p}p$ and $\bar{p}n$ by Abrams[62)]. Recent papers have come to the following conclusions:

1) Donald et al.[63)] withdraw their claim for a $K_1^o K_1^o \omega$ effect in the T-region.

2) Chapman et al.[64)] find no evidence for direct channel resonance production in the mass range 2.294 → 2.5 GeV in states with at least one visible K_1^o.

3) Oh et al.[65)] no longer see strong evidence for a narrow effect at a momentum of 1.8 GeV. They also see no evidence for a claim by Chapman et al.[66)] to see an effect at 1.76 GeV.

4) Bacon et al.[67)] find little evidence for a $\rho\rho\pi$ effect previously seen by Kalbfleisch et al.[68)] or for resonant behaviour in the $\bar{p}p \to \pi^+\pi^-$ and K^+K^- reactions. Although the angular distribution, of the $\pi^+\pi^-$ system contain evidence for high spin components it is not needed to involve narrow resonances to fit these distributions although it is possible[69)]. Confirming evidence for this comes from the CERN-Munich group[70)] who look at the reaction in the one-pion exchange region and thus study the inverse reaction $\pi\pi \to p\bar{p}$.

Their mass spectrum is quite smooth (see fig. 35) with no hint of any resonant structure. This absence of narrow resonant effects is also born out by a very high statistics counter experiment done with the Rutgers Annihilation Spectrometer[71)]. They have used an apparatus with an incident $\bar{p}$ beam and 32 detectors to measure multiplicities. They confirm the broad bumps seen in ref. 62 at 2193 and 2339 MeV in the total cross section and find that the positions and widths of these structures are invariant to changes in the multiplicity and degree of non-peripherality. Their data is shown in fig. 36. In summary, there now seems to be no firm evidence for high mass bosons produced in $\bar{p}p$ formation and one awaits new experiments. One such experiment at CERN[72)] is collecting data on $\bar{p}p \to \pi^+\pi^-$, K^+K^-. This experiment already has ∿ 40,000 events and is now accumulating a similar number with a polarised target.

4.3 The 3^- nonet

The highest mass and highest spin mesons which are known to exist are members of the 3^- nonet, these are the g and the $K^*(1760)$. There are, however, unresolved difficulties with both mesons and in addition the other members of the nonet need to be established.

A problem with the study of the higher mass and spin mesons is that they occur in mass regions where several states could exist and each state may have several decay modes. To be sure that one has isolated a single state one really requires consistency of M, Γ and J^P in each mode and perhaps the stability of the branching ratios in different experiments.

4.3.1 The g-meson

The non-strange member of the octet, the g-meson has been known for a long time and in both $\pi^+\pi^-$ and $\pi^\pm\pi^o$ yields $J^P = 3^-$. The other enhancements observed in the region occurs mainly in 4π and it has not been clear if the g and the 4π enhancement correspond to a single resonance and even assuming that they are the branching ratios have only been poorly determined.

One crucial question is whether the differing experimental results published reveal the existence of more than one structure. A secondary problem is then to sort out the various branching ratios (e.g. $\rho\rho$, πA_2, $\pi\omega$) that contribute to the 4π final state. An earlier review was published by J. Bartsch et al.[73], with the majority of data coming from experiments with incident momenta < 10 GeV/c. The results quoted in this paper are shown in figs 37 and 38. Since that publication there have been many other results published[74-85] and there are some other high statistics data which is unpublished[86]. If we take the positive results first, they can be summarized as follows:

1) $g \to 2\pi$ is well established both in $\pi^\mp\pi^o$ and $\pi^-\pi^+$.
2) There is no evidence for an I = 0 object going to $\pi^+\pi^-(\pi^o\pi^o)$.
3) The 4π enhancement is well established.
4) The $\pi\omega$ enhancement is well established.
5) The $\pi\pi$ decay has $J^P = 3^-$.
6) The $\pi\omega$ decay is consistent with $J^P = 3^-$.

These points are clearly illustrated in figs 37-41. Fig. 37 shows the compiled results of Bartsch et al.[73] for 2π and 4π and the varying parameters found. Their 8 GeV π^+p results are shown in fig. 38 and are consistent with a single state. Results published since then are shown in fig. 39 and consist mainly of bubble chamber data except for fig. 40a from the CERN-Munich group which in a high statistics study of $\pi N \to \pi\pi N$ confirmed $J^P = 3^-$ for the $g \to \pi\pi$. The other results show 2π and 4π enhancements of varying size and parameters.

One quite different result reported initially by Barnham et al., and confirmed by Holmes[84] is shown in fig. 40 and is a 4π enhancement with no corresponding 2π enhancement. The evidence is not overwhelming and the data could accomodate perhaps a 10 → 20% $\pi\pi$ decay.

Fig. 41 shows data from the Purdue Group for $\pi^+p \to pg^+$ and $\pi^+p \to \Delta^{++}g^o$ at 13 GeV/c[86]. The spectra in figs 41a and 41b show the combined data for channels in which the g decays. Both the g^+ and g^o have impressive signals and similar masses and widths (there is a small contamination from the $A_3 \to \pi^+\pi^o\pi^o$ in the g^+ spectrum which does not affect the statements made). A more detailed look reveals a problem, that is, only a narrow $\pi^-\pi^o$ effect is seen whilst a broad $\pi^+\pi^-$ effect is seen (figs 41c and 41d). On a more positive note a clear $g^+ \to \pi\omega$ is observed (fig. 41e) and no I = 0 state is observed. This latter statement comes from comparing $\Delta^{++}\pi^+\pi^-$ to Δ^{++} (all neutrals) and multiplying the former by a Clebsch-Gordon coefficient of ½. A normalization to the f^o makes a nice cross check of any possible biases. As can be seen in figs 41d and 41f, no $\pi^o\pi^o$ (or $4\pi^o$) state is observed beyond the f. So what are the conclusions. The easiest thing would be to say several states exist. As evidence one could give that some experiments see no $\pi\omega$[73,83], some see little or no $\pi\pi$[80,86], masses and widths tend to vary. Perhaps this is too easy, if one throws in fluctuations, centrifugal barrier effects, possible ρ' production etc., perhaps one can force consistency. Other observations which have to be fitted into a overall picture are possible $\pi^+\pi^+\pi^-\pi^-$ states[87]. Once again we will have to wait and see.

4.3.2 The $K^*(1760)$

The strange partner of the g has been seen in several experiments recently and has similar properties to the g already discovered except that not so much data exists. There are two effects observed both in Kπ and Kππ which because of the nearness in mass and width and because the reactions involve charge exchange (therefore excluding the L) could be the same object. This meson was first reported by Carmony et al.[88], who see a clear Kππ signal and a clear but difficult to interpret Kπ signal. The Kπ mass spectrum fig. 42 from their experiment $K^+n \to K^+\pi^-p$ shows a clear dip after the $K^*(1420)$ followed by a sharp rise. The structure which follows appears to be more complicated than just a single resonance. The Kππ spectrum now with more data than published originally gives a very clear peak. The Kπ spectrum published by Firestone et al.[89], at the same time is very similar with perhaps less structure after the sharp rise. They, however, do not see a corresponding Kππ enhancement. A recent paper by Aguilar Benitez et al.[90] in lower energy K^-p data does see a well separated single peak in Kπ but once again little Kππ has been observed. An analysis of the moments in all these experiments indicates that the effect has spin 3 and therefore could be the strange partner to the g. Some new data has also been reported recently[91] on Kπ and Kππ systems produced in $K^+p \to \Delta^{++}K^*$ at high energies. This data shown in fig. 43 appears to be very similar to the original observation with a clear Kππ peak and a more complicated Kπ structure. There are clearly then, discrepancies between experiments which have to be resolved. Some of these problems may be due to strong N^* diffractive background in some of the channels studied and others may be purely statistical fluctuations, but clearly resonant structures exist.

An analysis of the 3^- nonet has been carried out recently and agreement is found using SU3 with the current experimental quantities. Observation of the singlet member Ø (1820) and better branching ratios for the $K^*(1760)$ and g are clearly desirable, however, before definitive comparison to SU3 predictions can be made.

4.4 Very high mass mesons

There have been observations of peaks in mass spectra in the mass range 2.5 → 3.6 GeV, however, no single effect has been confirmed in independent experiments. One source of observations has been from the CERN Boson Spectrometer group who see enhancements in the X^- system in $\pi^- p \to p X^-$ [54]. The other main claims are from $\bar{p}p$ annihilations into multi-pion final states where for example Alexander et al.[92] found enhancements in the $(3\pi^+ 3\pi^-)$ and $(4\pi^+ 4\pi^-)$ systems produced in

$$\bar{p}p \to X^o \pi^o \quad \text{at 6.9 GeV/c.}$$

These effects may be resonances but until clear consistent results are obtained in several experiments the situation remains in doubt.

5. SOME NEW RESULTS ON VARIOUS TOPICS

5.1 ρ-ω interference

A very pretty high statistics result on ρ-ω interference has come from Diebold and co-workers[93]. The data consists of more than 500 000 events of the type $\pi N \to \pi\pi N$ obtained with the Argonne effective mass spectrometer. This is one effect which although observed in bubble chambers needs the high statistics and good resolution which the spectrometer technique provides. The reactions studied were

$$\pi^- p \to \pi^- \pi^+ n$$

$$\pi^+ n \to \pi^+ \pi^- p$$

at 3, 4 and 6 GeV/c. If one uses a coherent sum of ρ and ω amplitudes to describe the mass spectrum then for the π^- reaction one gets $|A\rho + A\omega|^2$ whereas for the π^+ reaction one gets $|A\rho - A\omega|^2$. The mass spectra obtained are shown in fig. 44a and show clearly the constructive interference in one case and the destructive interference in the other. In fig. 44b is shown the difference between the mass spectra (= 4 Re $A\rho$ $A\omega$) and the normalized difference for the different incident momenta. A fit using the 4 GeV/c data and assuming $\emptyset(\pi^+) = \emptyset(\pi^-) + 180^o$ and complete coherence yields an overall phase angle of $\emptyset = 26 \pm 12^o$.

5.2 The ω width

There are two new results on the ω width. The first from Brown et al.[95], using a recoil neutron spectrometer find a natural width of 7.7 ± 1.0 MeV. This width is somewhat less than observed by other experiments but it does depend on an accurate knowledge of the resolution since the observed peak has a FWHM of 14.5 MeV.

The other result is from e^+e^- colliding beams[96] and yields a width of 9.1 ± .8 which is not inconsistent with the above result or with the PDG average of 9.8 ± .5.

5.3 Coherent production of mesons

There are some new results[97] presented at the Washington APS meeting by the University of Seattle group on $\pi^- d$ interactions at 15 GeV/c. Their data from ∿ 1/3 of their experiment show clear production of the A_1, A_3 in the reaction $\pi^- d \rightarrow 3\pi d$ and in addition they have some evidence for a πg threshold effect. This is shown in fig. 45. The evidence is still rather weak and if it exists one would have expected that it would have shown up in the much higher statistics experiments using $\pi p \rightarrow 3\pi p$. It will be interesting to see if higher mass threshold enhancements are confirmed.

5.4 The δ revisited

In the same experiment at Argonne that studied the $A_2 \rightarrow \eta\pi$ a result was obtained on δ production at 4.5 GeV/c with the $\delta \rightarrow \eta\pi$[98]. The evidence for this is shown in fig. 46 where a fit gives a 3.7 σ effect with M = 980 ± 11 MeV, $\Gamma = 60^{+50}_{-30}$ MeV and a σ_{TOT} of 1.7 ± .8μb based on a $\frac{d\sigma}{dt} \sim e^{12t}$ (a 10% change occurs for e^{7t}). This compares with the original CERN experiment that found $\sigma_{TOT} \sim$ 5μb with 4 GeV/c $\pi^- p$[99].

5.5 The f'

The f' continues to show up in expected places although the number of events is never very large. This is one meson which is observed clearly because it is produced in final states where the background is very low. Fig. 47 shows the mass spectra from the only papers published concerning the f' since 1969. These are $K^- p \rightarrow \Lambda f'$ at 3.9 and 4.6 GeV[14] and $K^+ p \rightarrow K^+ p f'$ at 10 GeV/c[100] and $K^- p \rightarrow \Lambda f'$ at 3.95 GeV/c[101]. In all cases spin $\gtrsim$ 2 is favoured but more data for a J^P analysis is clearly needed.

5.6 Search for exotics

No doubly charged mesons have been observed or exotics of the second kind with the wrong quantum numbers for a $\bar{q}q$ system. Not many published results exist on specific searches for such objects although most people examine their data for indications of the existence of exotics. Implicitly, therefore, production cross sections must be less than a few microbarns. One recent paper by Cohen et al.[102] gives results for the reaction $\pi^- d \rightarrow pp X^{--}$ in a 7 event/μb exposure and set an upper limit of 9 μb on exotic production.

Theoretical predictions in general, favour the production of such objects through a baryon - anti-baryon production coupling. A recent paper by Lipkin[103] suggests looking for double charged mesons in Δ exchange reactions decaying to 3 or 4 pions. Faiman et al.[104] suggest similar reactions but predict baryon anti-baryon decays such as $p\bar{\Sigma}^+$, $p\bar{\Xi}^+$. Experimental data on such processes is either non existent or meagre and it is obviously interesting to explore such reactions.

6. FUTURE DIRECTIONS

It is clear that to solve some of the problems in meson spectroscopy one would like to have more data. Up until now except for a few isolated instances 5 → 30 events/μb has been a typical experiment. One would like to improve this to 300 → 1000 events/μb. There are three new spectrometer systems under construction which should be able to produce data at this level.

6.1 Spectrometers

6.1.1 Omega

This spectrometer is being built at CERN by a large team of people under the direction of A. Michelini[105]. A layout is shown in fig. 48 and the parameters of the apparatus in fig. 49. This apparatus is working well and the first experiments started taking real data in May. Previously test data has been taken when Ω was only partly complete with one coil and analysis of that data revealed that the mass resolutions etc. agree with the design predictions. One unique feature of this apparatus is that the spark chambers are scanned by plumbicons a system which works exceedingly well. The approved experiments for this apparatus are shown in fig. 50 and represent a large fraction (if not all) of the Ω running at PS energies before the West Area at CERN is shut-down to prepare for the SPS. It appears, therefore, that Ω will not answer many of the current questions in Meson Spectroscopy but may unearth a new set at higher energies.

6.1.2 MPS

This spectrometer under construction at B.N.L.[106] is very similar to the Ω system in philosophy. That is a large magnet filled with detectors to give a large solid angle efficiency for secondary particles. In detail they differ considerably as can be seen from figs 51 and 52. The other main difference is that the apparatus will have available a Medium Energy Separated Beam and eventually a high energy beam. These is also no enforced shut-down as in the case of Ω and in principle the apparatus will be able to produce Physics in the energy range < 25 GeV as long as the interest justifies it. The approved program is shown in fig. 53 and has some overlap with Ω.

6.1.3 LASS

This spectrometer is being built by David Leith and co-workers at SLAC[107] and is quite a different design to achieve the same ends of large solid angle acceptance for secondaries. The apparatus has the unique feature of a large solenoïd magnet to obtain the large solid angle for secondaries. This apparatus is further from completion than the previous two and the experimental program is not so fixed. Because it will be at SLAC, however, it will have the advantage of being able to do unique photon, muon and electron physics, both with and without a polarised target.

6.1.4 Other Spectrometers

I have included the above three because in principle they will be very successful large general purpose facilities. Many other spectrometers exist for example that of Diebold at Argonne[108] and CERN-Munich at CERN[109]. Both of these are working devices and have shown what a powerful technique can do with specific problems.

6.1.5 Summary

If the spectrometers are successful the next few years could be very exciting as data on known states is increased by factors of 10-100 and the search for new effects with cross sections below 1 μb is pursued. In addition, all these spectrometers can use polarised targets opening up new areas of amplitude analysis in meson production. One thing which they will not do is to look with mass resolutions $\sim$ 1 MeV. It is possible that very fine structure exists which has not been observed and that could lead to a deeper understanding of the fundamental interactions.

6.2 Bubble Chambers

World-wide there is still an enormous amount of data being produced from Bubble Chambers. It still has the advantages of 4π detection efficiency, vertex detection and ability to detect decays. It is still very hard, however, to gain a factor of 10 and go over the 100 events/μb. There are, however, a few very large scale experiments under way, particularly in K^- interactions below 10 GeV.

One development which should improve this is that of rapid cycling techniques. The 15" chamber at SLAC has operated very well in the 30 cycles/sec. range and the 40" up to 10 c.p.s. These is also an approved program at the Rutherford Laboratory to build a rapid cycling chamber. If these chambers can be triggered efficiently 500 events/μb does seem reasonably easy to achieve.

One should not close without mourning the imminent demise of the 82" chamber at SLAC which is planned to be mothballed at the end of the year. First as the 72" and now the 82" this chamber has produced data which has contributed more to our knowledge or resonant states than any other device. At the time it was built it was a far sighted and great technical achievement and in over a decade of continuous use has proved itself beyond measure.

7. SUMMARY AND CONCLUSIONS

An asterisk in column 4 of fig. 4 indicates those mesons where sufficient data has accumulated to study production mechanisms, density matrix for their decay etc. There are 18 such states, of these four may not be resonances, but we can add perhaps five that are resonances with not enough data for such studies. Unfortunately, in this case knowledge does not accumulate from a lot of small experiments only confusion results. So we have to wait for new definite results, which could take another decade. There are all the missing states to find, branching ratios to determine (in many cases now only known to 10 or 20%) and electromagnetic decays. Exotics are interesting since if they do not exist it is a very strong constraint on the way nature behaves. One would really like to probe a level of a 100 less in cross section with a 100 times better resolution.

Some experiments are clear since we know of mesons which exist and all that is required is more events. In the high mass region one needs at least 500 events/μb and even then if widths are increasing as well as the number of states there will be problems. Better partial wave analysis is needed of systems such as 3π, $\pi\omega$ to see if many partial waves resonate. One would also like to look in many different reactions at differing energies for new phenomena. Polarised targets with a complete amplitude analysis will also be very valuable.

Many of the problems are clear so that hopefully the next five years will bring a significant advance in our knowledge of mesons.

REFERENCES

1) CERN 72-10 Proceedings of the Symposium on Nucleon-Antinucleon Annihilations held in Chexbres, Switzerland, 27-29 March 1972.

2) A.I.P. Conference Proceeding Number 8. Experimental Meson Spectroscopy edited by A.Rosenfeld and K.-W. Lai, 1972.

3) R. Diebold Review of Meson Resonances, Proceedings of the XVI Rochester Conference, Batavia, 1972.

4) D.W.G.S. Leith Review of Diffraction Dissociation, Proceedings of the XVI Rochester Conference, Batavia 1972.

5) Review of Modern Physics 45 (1973) 51.

6) An early paper is by R. Dalitz in Meson Spectroscopy. Edited by C. Baltay and A.H. Rosenfeld, published by Benjamin Inc. 1968.

7) E. Flaminio et al., B.N.L. 14572.
Recent papers on the 3^- nonet are: R.H. Graham and T.S. Yoon, Phys. Rev. D6 (1972) 336.
A. Bramon and M. Greco, Nuovo Cimento 10A (1972) 521.

8) A review of the early results is M.N. Foccaci, Phys. Rev. Letters 17 (1966) 890.

9) P.L. Hoch Thesis LBL 1051.

10) W. Michael Phys. Rev. D7 (1973) 1985.

11) D.L. Cheshire et al., Phys. Rev. Letters 28 (1972) 520.
R.W. Jacobel et al., Phys. Rev. Letters 29 (1972) 671.
A.F. Garfinkel et al., Phys. Rev. Letters 29 (1972) 1477.

12) D.M. Binnie et al., Phys. Letters 39B (1972) 275.

13) L. Holloway et al., University of Illinois preprint COO 1195 241.

14) M. Aguilar-Benitez et al., Phys. Rev. D6 (1972) 29.

15) R. Ammar et al., Phys. Rev. Letters 21 (1968) 1832.
V.E. Barnes et al., Phys. Rev. Letters 23 (1969) 610.

16) D.H. Miller et al., Phys. Letters 29B (1969) 255.

17) A. Astier et al., Phys. Letters 25B (1967) 294.

18) M. Alston-Garnyost et al., Phys. Letters 33B (1970) 607.

19) D. Bowen et al., Phys. Rev. Letters 26 (1971) 1663.

20) G. Grayer et al., Phys. Letters 34B (1971) 333.

21) K.J. Foley et al., Phys. Rev. D6 (1972) 747.

22) C.M. Ankenbrandt et al., Phys. Rev. Letters 29 (1972) 1688.

23) A.W. Key et al., Phys. Rev. Letters 30 (1973) 503.

24) G. Conforto et al., University of Wisconsin preprint UW COO 354.

25) D.M. Binnie et al., Phys. Letters 36B (1971) 257.

26) L.E. Holloway et al., University of Illinois preprint 1195, December 1972, 250.

27) C. Defoix et al., Phys. Letters 43B (1973) 141.

28) D. Cohen et al., University of Rochester preprint COO 3065, February 1973, 39.

29) Proceedings of XVI Rochester Conference, Batavia, V1 P144.

30) V. Chaloupka et al., Phys. Letters 44B (1973) 211.

31) C. Defoix et al., Nuclear Phys. B44 (1972) 125.

32) B. Lorstad et al., Nuclear Phys. B14 (1969) 63.

33) The Analysis from the $\bar{p}p$ data agree with the few observations of the D in πN interactions e.g.
O. Dahl et al., Phys. Rev. 163 (1967) 1377.
J. Campbell et al., Phys. Rev. Letters 22 (1969) 1204.

34) J. Duboc et al., Nuclear Physics B46 (1972) 429.

35) These techniques are described in ref. 36 and 37. The assumptions are that the 3π system decays independently of the nucleon in the final state and that the decay is through $\pi\rho$, πf and $\pi\varepsilon$ final states with $J \leq 4$. Δ production is removed and the effect of this on the 3π system taken into account.

36) G. Ascoli et al., Phys. Rev. D7 (1973) 669.
G. Ascoli et al., Phys. Rev. Letters 26 (1971) 929.

37) These results are taken from a series of papers either submitted or to be submitted to Nuclear Physics B by the CERN-IHEP Boson Spectrometer Group:

CERN	: G. Damgard, W. Kienzle, C. Lechanoine
IHEP	: Y. Antipov, L. Landsberg, A. Lebedev, F. Yotch
University of Geneva	: R. Baud, R. Busnello, M. Kienzle-Focacci, P. Lecomte, M. Martin
University of Illinois	: G. Ascoli, R. Sard
University of Munich	: R. Klanner, A. Weitsch
Academy of Science Tbilisi	: V. Roinishvoli

38) This is a result from a first J^P analysis of a 30 event/μb π^+p experiment.
G. Thompson private communication.

39) D. Cohen et al., Phys. Rev. Letters 28 (1972) 1601.

40) A recent review is by M. Bowler in the Proceedings of European Physical Society meeting on Mesons Resonances and related Electromagnetic Phenomena, edited by R.H. Dalitz and A. Zichichi. A compilation of events for $K^+p \to Q^+p$ is contained in H.H. Bingham et al., Nucl. Phys. B48 (1972) 589.

41) B. Werner et al., Phys. Rev. D7 (1973) 1275.

42) A. Firestone, Nucl. Phys. B47 (1972) 348.

43) G. Barbarino et al., Lett. al Nuovo Cimento, 3 (1972) 689.

44) A. Bramon and M. Greco, Lett. al Nuovo Cimento, 3 (1972) 693.

45) F. Ceradini et al., Phys. Letters 43B (1973) 341.

46) M. Grilli et al., Nuovo Cimento 13A (1973) 593.

47) H.H. Bingham et al., Phys. Lett. 41B (1972) 635.

48) J. Ballam et al., Phys. Rev. D5 (1972) 545.

49) M. Davier et al., SLAC publ. 1205 submitted to Nucl. Phys. B

50) D. Mortara, University of Illinois preprint COO, October 1972, 195-249.

51) Y. Eisenberg et al., Phys. Let. 43B (1973) 149.

52) B. Hyams et al., Max Plank Institute preprint MPI-PAE/Exp. EI.28 (sub. to Nucl. Phys. B).

53) An early review of all results is M.N. Foccacci et al., Phys. Rev. Letters 17 (1966) 890. A recent re-analysis has revealed that the quoted cross sections for the S.T.U. are too large by a factor of 4.
W. Kienzle to be published in Phys. Rev. D.

54) R. Baud et al., Phys. Letters 30B (1969) 129.
R. Baud et al., Phys. Letters 31B (1970) 549.

55) D. Bowen et al., Phys. Rev. Letters 29 (1972) 890.

56) D. Bowen et al., Phys. Rev. Letters 30 (1973) 332.

57) R. Thun et al., Phys. Rev. Letters 28 (1972) 1733.

58) H.R. Bleiden et al., Phys. Letters 39B (1972) 668.

59) Y.M. Antipov et al., Phys. Letters 40B (1972) 147.

60) See ref. 37.

61) G. Thompson et al., Purdue University Preprint.

62) R.J. Abrams et al., Phys. Rev. D1 (1970) 1917.

63) R.A. Donald et al., Phys. Letters 40B (1972) 586.

64) J.W. Chapman et al., Nucl. Phys. B42 (1972) 1.

65) B.Y. Oh et al., Nucl. Phys. B51 (1973) 57. This paper together with
P.S. Eastman et al., Nucl. Phys. B51 (1973) 29, and
Z. Ming Ma et al., Nucl. Phys. B51 (1973) 77, give extensive results on the momentum region 1.51 → 2.9 GeV.

66) J.W. Chapman et al., Phys. Rev. D4 (1971) 1275.

67) T.C. Bacon et al., Phys. Rev. D7 (1973) 577.

68) G. Kalbfleisch et al., Phys. Letters 29B (1969) 259.

69) T. Fields et al., Phys. Letters 40B (1972) 503.

70) G. Grayer et al., Phys. Letters 39B (1972) 563.

71) J. Alspector et al., Phys. Rev. Letters 30 (1973) 511.

72) This is the experiment of Q.M.C., RHEL, DNPL, Liverpool with approximately 2000 events at each of 20 momenta with a hydrogen target. The current run with a polarised target is expected to yield ∿ 50 000 events. The preliminary analysis indicates that at 2 GeV J = 5 is needed to fit the angular distributions although it is not clear yet if resonances are necessary (A. Astbury private communication).

73) J. Bartsch et al., Nucl. Phys. B22 (1970) 109.

74) Results are included which are listed in the PDG tables for the years 1970, 71, 72 and published results in 1973.

75) N. Armenise et al., Letter al Nuovo Cimento IV (1970) 199.

76) K.W.J. Barnham et al., Phys. Rev. Letters 24 (1970) 1083.

77) C. Caso et al., Letter al Nuovo Cimento III (1970) 707.

78) S. Kramer et al., Phys. Rev. Letters 25 (1970) 396.

79) P.H. Stuntebeck et al., Phys. Letters 32B (1970) 391.

80) J. Ballam et al., Phys. Rev. D3 (1971) 2606.

81) G. Grayer et al., Phys. Letters 35B (1971) 610.

82) J.A.J. Matthews et al., Nucl. Phys. B33 (1971) 1.

83) N. Armenise et al., Letter al Nuovo Cimento 4 (1972) 205.

84) R. Holmes et al., Phys. Rev. D6 (1972) 3336.

85) N.M. Cason et al., Phys. Rev. D7 (1973) 1971.

86) This data is from a 30 event/μb $\pi^+ p$ experiment by the Purdue Group of G. Thompson, T. Mulera, J. Gaidos, D.H. Miller, R.B. Willmann.

87) Recent papers claiming a ρρ effect at M - 1.7 GeV are J. Clayton et al., Nucl. Phys. B47 (1972) 81. H. Braun et al., Nucl. Phys. B30 (1972) 213.

88) D.D. Carmony et al., Phys. Rev. Letters 27 (1971) 1160.

89) A. Firestone et al., Phys. Letters 36B (1971) 513.

90) M. Aguilar-Benitez et al., Phys. Rev. Letters 30 (1973) 672.

91) These are preliminary results from the Birmingham-Bruxelles-CERN-LPNHE(Paris)-Mons-Saclay Collaboration. This data was presented at an open TCC meeting at CERN and private communication.

92) G. Alexander et al., Nucl. Phys. B45 (1972) 29.

93) D.S. Ayres et al., ANL/HEP 7320 and ANL/HEP 7318.

94) B.N. Ratcliff et al., Phys. Letters 38B (1972) 345.

95) R.M. Brown et al., Phys. Letters 42B (1972) 117.

96) D. Benaksas et al., Phys. Letters 42B (1972) 507.

97) K. Morirasu et al., University of Washington Visual technique Laboratory HEP 19.

98) This data was kindly supplied to me by R. Prepost. The other participants and the experimental apparatus is the same as for ref. 23, 24.

99) See reference 14.

100) D.C. Colley et al., Nucl. Phys. B50 (1972) 1.

101) I. Videau et al., Phys. Letters 41B (1972) 213.

102) D. Cohen et al., Nucl. Phys. B53 (1973) 1.

103) H.J. Lipkin, Phys. Rev. D7 (1973) 2262.

104) D. Faiman et al., Phys. Letters 43B (1973) 307.

105) The whole project is described in CERN NP 68-1 and many other internal reports. A recent review of spectrometers was given by A. Michelini at the Frascati Instrumentation Conference 1973 (NP OM 197).

106) The details included in this report were taken from E.D. Platner et al., a paper submitted to the Frascati Instrumentation Conference 1973.

107) This apparatus is discussed in several SLAC Internal reports.

108) This apparatus has already shown what good physics it can produce, (Ref. 94). The technical aspects have been discussed most recently in ANL/HEP 7314.

109) This spectrometer has concentrated mainly on $\pi p \to \pi\pi N$, $K\bar{K}N$, $p\bar{p}N$ for example as in ref. 20, 70, 81, 52 and in detailed $\pi\pi$ phase shift analysis.

Table I. I^G (J^P) of mesons from $\bar{q}q$ model. For the distinction between abnormal J^P and abnormal C, see text below Eq. (5). K mesons share the same values of J^P as the I = 0 and 1 states shown, but are not eigenstates of G. The middle column, which gathers together (J^P) [N or A] CP, is a redundant intermediate step intended to make the table easier to read.

Parity	$\bar{q}q$ State CP −	$\bar{q}q$ State CP +	(J^P) [ ] CP Normal or abnormal	$I^G(J^P)C_n$	Examples and comments
Parity −	$^1S_0$		$(0^-)_A-$	$0^+(0^-)+$	η, η'
				$1^-(0^-)+$	π
		$^3S_1$	$(1^-)_N+$	$0^-(1^-)-$	ω, ϕ
				$1^+(1^-)-$	ρ
Parity +	$^1P_1$		$(1^+)_A-$	$0^-(1^+)-$	
				$1^+(1^+)-$	B
		$^3P_0$	$(0^+)_N+$	$0^+(0^+)+$	ϵ, S*
				$1^-(0^+)+$	$\pi_N(1016)$
		$^3P_1$	$(1^+)_A+$	$0^+(1^+)+$	
				$1^-(1^+)+$	A1
		$^3P_2$	$(2^+)_N+$	$0^+(2^+)+$	f, f'
				$1^-(2^+)+$	A2
Parity −	$^1D_2$		$(2^-)_A-$	$0^+(2^-)+$	Regge recurrence
				$1^-(2^-)+$	of $^1S_0$, 0^-
		$^3D_1$	$(1^-)_N+$	same as $^3S_1$	
		$^3D_2$	$(2^-)_A+$	$0^-(2^-)-$	Regge recurrence of top abnormal-C state
				$1^+(2^-)-$	below: $(J^P)C_n = (0^-)-$
		$^3D_3$	$(3^-)_N+$	$J > 2$	
Parity +	$^1F_3$		$(3^+)_A-$	$J > 2$	
		$^3F_2$	$(2^+)_N+$	same as $^3P_2$	
		$^3F_3$	$(3^+)_A+$	$J > 2$	
		$^3F_4$	$(4^+)_N+$	etc.	

ABNORMAL C STATES THAT CANNOT COME FROM $\bar{q}q$ MODEL

Abnormal C states Have no $\bar{q}q$ model	$(0^-)_A+$	$0^-(0^-)-$	All except $J^P = 0^-$ are J^P = normal, CP = −1
		$1^+(0^-)-$	
	$(1^-)_N-$	$0^+(1^-)+$	
		$1^-(1^-)+$	
	$(0^+)_N-$	$0^-(0^+)-$	
		$1^+(0^+)-$	
	$(2^+)_N-$	$0^-(2^+)-$	
		$1^+(2^+)-$	
	$(3^-)_N-$	$0^+(3^-)+$	
		$1^-(3^-)+$	

Figure 1. A list of possible $q\bar{q}$ states (Ref. 5).

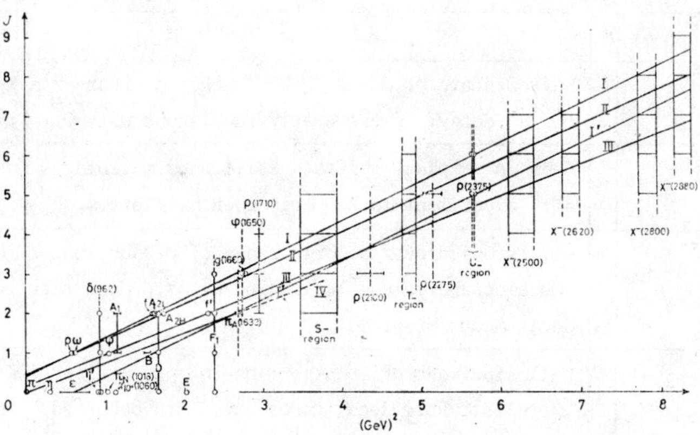

Fig. 1.

Figure 2. Possible Regge trajectories for known states taken from M. Roos Lett. al Nuovo Cim. 3 (1970) 257.

Established Nonets, and octet-singlet mixing angles from Appendix IIB, Eq. (2'). Of the two isosinglets, the "mainly octet" one is written first, followed by a semi-colon.

$(J^P)C_n$	Nonet members	$\theta_{lin.}$	$\theta_{quadr.}$
$(0^-)+$	π, K, η; η'	24 ± 1°	10 ± 1°
$(1^-)-$	ρ, K^*, ϕ; ω	36 ± 1°	39 ± 1°
$(2^+)+$	A_2, $K_N(1420)$, f'; f	29 ± 2°	31 ± 2°

Figure 3. Established nonets (Ref. 5).

STATUS	NAME	$I^G J^P C$	DYNAMICS	COMMENTS	MAIN REACTION IN WHICH MESON IS OBSERVED
	π(140)	$1^- 0^-$ +	*		
	η(549)	$0^+ 0^-$ +	*	Narrow state produced in low background for $\pi^-\pi^+\pi^o$ decay. η assymmetry consistent with 0.	$K^- p \to \Lambda\eta$ $\pi p \to \Delta\eta$
	ε(600)	$0^+ 0^+$ +		Seen in ππ phase shifts. Still need matching $\pi^o\pi^o$ experiment to help pin down parameters.	$\pi^- p \to \pi^-\pi^+ n$
	ρ(770)	$1^+ 1^-$ -	*	Well studied over a large range of production energies and in e^+e^- colliding beams. Some rare decay modes observed and limits set.	$\pi N \to \pi\pi N$
	ω(784)	$0^- 1^-$ -	*	Well studied over a large range of production energies. Rare decay modes have been observed. Width is fairly well measured.	$K^- p \to \Lambda\omega$ $\pi N \to \omega N$ $\pi N \to \omega\Delta$
→	M(940)			Only seen as a bump in a neutron missing mass experiment not confirmed in a similar experiment	$\pi^- p \to n(MM)$
→	M(953)			Mass and width consistent with η' but $\pi^+\pi^-\gamma$ decay has no ργ contribution. Needs confirmation	$K^- p \to K^- p\pi\pi\gamma$
	η'(958)	$0^+ 0^-$ +	*	Good data in a few experiments with detailed Dalitz plot fit. Spin 2 is still allowed but highly unlikely. Some rare decay modes observed	$K^- p \to \Lambda\eta'$ $\pi p \to \Delta\eta'$
	δ(970)	$1^- 0^+$ +		Situation still uncertain. Seen in D decay and weakly in KN interactions. Not clear if all observations correspond to same state.	$D \to \pi\delta$ $\delta \to \pi\eta$ $K^- p \to \Lambda\pi\delta$
→	H(990)	0^-A -		If something is confirmed with this mass it will called the H.	$\pi N \to HN$
	S^*(1000)	$0^+ 0^+$ +		Seen as strong $K\bar{K}$ threshold effect and in ππ phase shifts.	$\pi N \to K\bar{K}N$
	Ø)1019)	$0^- 1^-$ -	*	Well studied in KN interactions and e^+e^- colliding beams. Some rare decay modes known.	$K^- p \to \Lambda Ø$
→	M(1033)			Only seen in a neutron missing mass experiment	$\pi^- p \to X^- n$
→	B_1(1040)			Seen in $A_2^o \to \pi B_1 \to \pi\pi\omega$. Needs confirmation.	$\bar{p}p \to$

Figure 4a A list of all non-strange mesons confirmed or claimed

STATUS	NAME	$I^G J^P C$	DYNAMICS	COMMENTS	MAIN REACTION IN WHICH MESON IS OBSERVED
→	$\eta_N(1080)$	$0^+(N)$ +		Seen in $\pi^- p \to \pi^+\pi^- n$ predominantly at backward decay angles. Not clear if a resonance or interference between non-resonant waves.	$\pi^- p \to \pi^- \pi^+ n$
	$A_1(1100)$	$1^- 1^+$ +	*	Broad $\pi\rho$ enhancement in a single J^P 1^+. Interpretation uncertain. Lots of data.	$\pi^\pm N \to (3\pi)^\pm N$
→	M(1150)			Seen in one neutron missing mass experiment.	$\pi^- p \to nX^-$
→	$A_{1.5}(1170)$			There have been many claims for a 3π bump between A_1 and A_2. Not confirmed.	$\pi N \to 3\pi N$
	B(1235)	$1^+ 1^+$ -	*	Well established in many experiments. $\pi\omega$ only decay seen. Need partial wave analysis to see if other waves resonate.	$\pi^\pm N \to N(\pi\omega)^\pm$
	F(1270)	$0^+ 2^+$ +	*	Studied over a wide range of incident energy 4π, $K\bar{K}$ decays not well studied.	$\pi p \to \pi\pi n$
	D(1285)	$0^+ A$ +		Mainly seen in $\bar{p}p$ annihilations and a few πN experiments. $\pi\delta$, 4π decays still need verification.	$\bar{p}p$
	$A_2(1310)$	$1^- 2^+$ +	*	$A_2^\pm$ well studied to see if split. $\rho\pi$, $\eta\pi$, $K\bar{K}$ modes well studied. A_2^o not so well studied. Possible new $\omega\pi\pi$ decay mode observed.	$\pi N \to A_2 N$
	E(1420)	$0^+ A$ +		Seen mainly in $\bar{p}p$ annihilations. Evidence is still weak and much data is needed to determine branching ratios J^P.	$\bar{p}p \to$
→	X(1430)			No satisfactory observation usually 3 standard deviation effects.	
→	X(1440)			Not clearly seen. One expt. has cross section .1 μb ∿ 10 events.	
	f'(1514)	$0^+(2^+)$ +	*	Only seen in $K\bar{K}$. Very poor limits on other modes. Total world data < 100 events.	$K^- p \to \Lambda f'$
	$F_1(1540)$			Only seen in $K^* K$ and in $\bar{p}p$ annihilations.	$\bar{p}p \to$
	ρ'(1600)	$1^+ 1^-$ -	*	Directly seen in e^+e^- colliding beams and $\gamma p \to 4\pi p$. Seen in some $\pi\pi$ phase shift analysis. Resonance interpretation needs to be confirmed and $\pi\pi$ decay.	$\gamma p \to 4\pi p$ $e^+e^- \to 4\pi$

Figure 4a Continued

STATUS	NAME	$I^G J^P C$	DYNAMICS	COMMENTS	MAIN REACTION IN WHICH MESON IS OBSERVED
	$A_3(1640)$	$1^- 2^-$ +	*	Broad πf in 2^-. Interpretation uncertain. Lots of data.	$\pi^\pm N \to (3\pi)^\pm N$
	$\omega(1675)$	0^-N -		Only clear decay is $\pi\rho$. Not too much data exists although it is seen in many experiments.	$\pi n \to p\pi^o\rho^o$ $\pi p \to \Delta^{++}\pi^o\rho^o$
	g(1680)	$1^+ 3^-$ -	*	2π decay well established to be 3^-. 4π enhancement is well established. Not clear if several states exist and how 4π is made up of $\rho\rho$, πA_2 etc.	$\pi N \to gN$
→	X(1690)			Not seen since 1969, needs further confirmation	
→	X(1795)			Seen as a $\bar{p}n$ bound state decaying to four or more pions. Needs confirmation.	$\bar{p}n \to$
→	$\eta/\rho(1830)$	G = +		Seen in $\bar{p}p$ annihilations in 4π decay on a large background.	$\bar{p}p \to$
→	$\omega/\pi(1830)$	G = -		Seen in $\bar{p}p$ annihilations.	$\bar{p}p \to$
→	S(1930)			Many claims no consistency.	
→	$\rho(2100)$			Few claims no compelling evidence.	
→	T(2200)			Many claims from $\rho\rho\pi$ to $A_2\omega$. There is a broad effect in σ_{TOTAL} $\bar{p}p$.	
→	$\rho(2275)$			No evidence it exists.	
→	U(2360)			Broad bumps in $\bar{p}p$ σ_{TOTAL} but no consistent results for a single meson.	
→	$N\bar{N}(2375$			Seen in $\bar{N}N$ total cross section not confirmed as a resonant effect.	
→	X_{3600}^{2500} (2500 ↓ 3600)			Many claims in missing mass experiments and from $\bar{p}p$ annihilations but no one experiment has enough data and no observation has been duplicated.	$\bar{p}p \to$ $\pi^- p \to X^- p$

Figure 4a Continued

STATUS	NAME	$I^G J^P C$	DYNAMICS	COMMENTS	MAIN REACTION IN WHICH MESON IS OBSERVED
	K(494)	$\frac{1}{2}0^-$	*		
	$K^*(892)$	$\frac{1}{2}1^-$	*	Very well studied over wide range of energies. Mass difference between 0 and ± known.	$K^-n \to K^*N$
	κ	$\frac{1}{2}0^+$		Possibly δ_0^1 is resonant somewhere in which case the resonance will be called κ.	$KN \to K\pi N$
→	$K_A(1175)$	3/2		No evidence it exists.	
→	$K_A(1265)$	3/2		No evidence it exists.	
	Q	$\frac{1}{2}1^+$	*	Broad bump in $K\pi\pi$ with $J^P = 1^+$ mainly $K^*\pi$ but some $K\rho$ and $K\pi\pi$. Interpretation still uncertain but can be fitted with two resonances.	$K^\pm N \to (K\pi\pi)^\pm N$
	$K_N(1420)$	$\frac{1}{2}2^+$	*	Studied in many experiments. Most strong decays known.	$KN \to K^+N$
→	$K_N(1660)$	$\frac{1}{2}$		Observations not definite or consistent.	
→	$K_N(1760)$	$\frac{1}{2}$		Clear evidence for its existence but some discrepancies over $K\pi/K\pi\pi$ ratio.	$K^\pm N \to K^{*o}N$
	L(1770)	$\frac{1}{2}$ A		Strange analog of the A_3 which is mainly $K^*(1420)\pi$. Other decay modes uncertain.	$K^\pm N \to K^{*\pm}N$
→	$K_N(1850)$			Should be classified with $K^*(1760)$ experimental data the same interpretation similar.	$K^+n \to K^{*o}p$
→	$K^*(2200)$			Seen in Antihyperon-nucleon system, needs further data.	
→	$K^*(2800)$			No clear evidence for its existence.	

Figure 4b A list of all mesons with strangeness confirmed or claimed

MESON	REACTION	CROSS SECTION MICROBARNS	SIGNAL / BACKGROUND	ENERGY
η	$K^-p \to \Lambda\eta$	25	26/0	3.9, 4.6
	$\pi p \to \Delta\eta$	76	30/0	5
ε	$\pi N \to \pi\pi N$	not studied directly $\sim$ 10% of the ρ		
ρ	$\pi N \to \pi\pi N$	500	5000/500	3 → 13
	$K^-p \to \Lambda\rho$	70	260/26	3.9, 4.6
ω	$K^-p \to \Lambda\omega$	70	260/10	3.9, 4.6
	$\pi N \to \Delta\omega$	200	2000/100	4
M(940)	$\pi^-p \to nMM$		55/100	2.4
M(953)	$K^-p \to K^-p\pi\pi\gamma$		50/40	3.9, 4.6
η'	$K^-p \to \Lambda\eta'$	15	80/8	3.9, 4.6
	$\pi p \to \Delta\eta'$	17	25/3	5
δ	$\bar{p}p \to D(\text{decay})$	250	130/1200	.7
	$K^-p \to \Lambda\pi\delta$	5	20/4	5.5
	$\pi n \to D(\text{decay})$	44	40/100	2.7
H		Too uncertain for comment		
S^*	$\pi N \to K\bar{K}N$	Threshold effect in $K\bar{K}$ system and a pole in $\pi\pi$ phase shift analysis.		
Ø	$K^-p \to \Lambda Ø$	25	300/6	3.9, 4.6
M(1033)	$\pi^-p \to nMM$		60/80	2.4
M(1033)	$\pi^-p \to nMM$		60/80	2.4
B_1	$p\bar{p} \to A_2(\text{decay})$	300	1000/5000	.7
$\eta_N(1080)$	$\pi p \to \pi\pi n$	20	100/100	4
A_1	$\pi p \to 3\pi p$	150	5000/500	4 → 40

Figure 4c Typical cross sections and signal to background ratios for non-strange meson production.

MESON	REACTION	CROSS SECTION MICROBARNS	SIGNAL / BACKGROUND	ENERGY
M(1150)	$\pi^- p \rightarrow nMM$		65/150	2.4
$A_{1.5}$	$\pi p \rightarrow 3\pi p$	80	60/300	7
B	$\pi p \rightarrow pB$	40	1200/700	7
F	$\pi p \rightarrow \pi\pi N$	200	2000/700	5 → 15
	$K^- p \rightarrow \Lambda f$	50	300/100	3.9, 4.6
D(δπ)	$\bar{p}p \rightarrow D{+}n\pi$	250	130/1200	.7
	$\pi n \rightarrow p\,D^o$	44	30/60	2.7
D (KKπ)	$\bar{p}p \rightarrow D +$	40	30/50	.7
	$\pi p \rightarrow Dn$	10	15/30	2.4 → 4
A_2	$\pi p \rightarrow A_2\ P$	50	$K\bar{K}.\eta\pi$ $\frac{300}{50}$	4 → 40
			$\pi\rho$ $\frac{1000}{1500}$	
E	$\bar{p}p \rightarrow E +$	60	100/200	.7
	$\pi p \rightarrow nE$	44	30/60	2.4 → 4
X (1430)				
X (1440)				
f'	$K^- p \rightarrow \Lambda f'$	10	40/70	3.9, 4.6
F_1	$\bar{p}p \rightarrow F\pi$	50	108/180	.7
ρ'	$\gamma p \rightarrow 4\pi p$	1	100/40	4.5 → 18
A_3	$\pi p \rightarrow 3\pi p$	40	1000/1000	4 → 40
ω(1675)	$\pi n \rightarrow p\omega$	19	100/300	8
g	$\pi p \rightarrow gN$	40	800/1600	13

Figure 4c Continued

MESON	REACTION	CROSS SECTION MICROBARNS	$\frac{\text{SIGNAL}}{\text{BACKGROUND}}$	ENERGY
$K^*(892)$	$\pi p \to K^*\Lambda$ $Kp \to NK^*$ $Kp \to K^*\Delta$	50 600 700	800/80 3000/300 4000/400	6 3.9, 4.6 7.3
κ	$Kp \to K\pi(\Delta,N)$	Possible resonance in $\delta_o^1\Gamma$ and m not certain		
$K_A(1175)$	Exotic Existence very doubtful.			
$K_A(1265)$	$\bar{p}p \to K_A K\pi\pi$		15/28	3.6
Q	$KN \to QN$ $\pi p \to \Lambda Q$	200 5	3000/700 25/70	4 → 14 6
$K_N(1420)$	$\pi p \to \Lambda K^*$ $Kp \to K^*\Delta$ $Kp \to K^*N$	15 70 200	250/80 400/400 1400/2500	6 7.3 3.9, 4.6
$K_N(1660)$	$KN \to K^*N$	20	25/30	5
$K_N(1760)$	$Kn \to K^*p$	20	160/160	9
L(1770)	$Kp \to K^*p$	15	300/1400	12
$K_N(1850)$	$Kn \to K^*p$	15	200/500	12
$K^*(2200)$	$K^+p \to (\Lambda p)$ + $K^+n \to K\pi p$	4 4	20/30 25/25	9 9
$K^*(2800)$	One observation of a bump at the top end of phase space.			

Figure 4d Typical cross sections and signal to background ratios for the production of mesons with strangeness

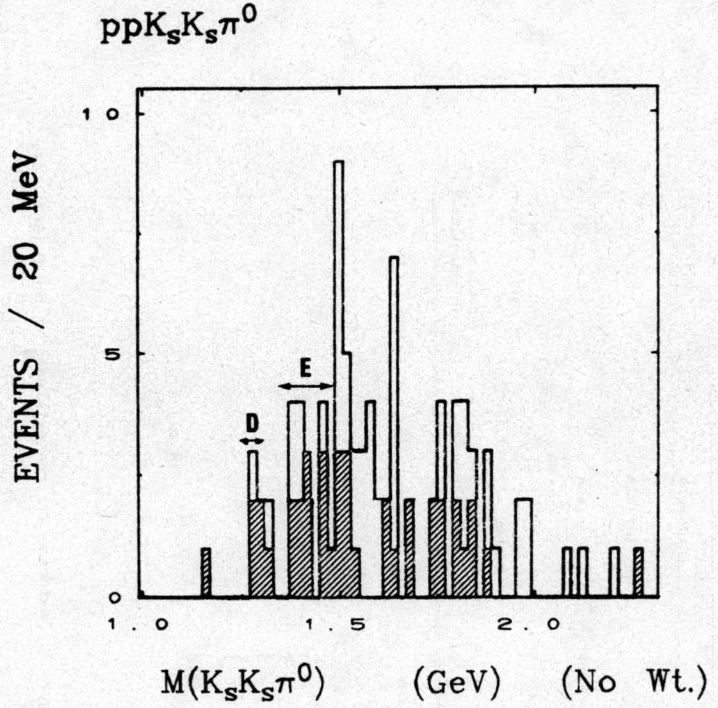

Figure 5. $K_SK_S\pi^0$ spectrum from π^+d interactions at 2.7 → 4.2 GeV (ref. 9).

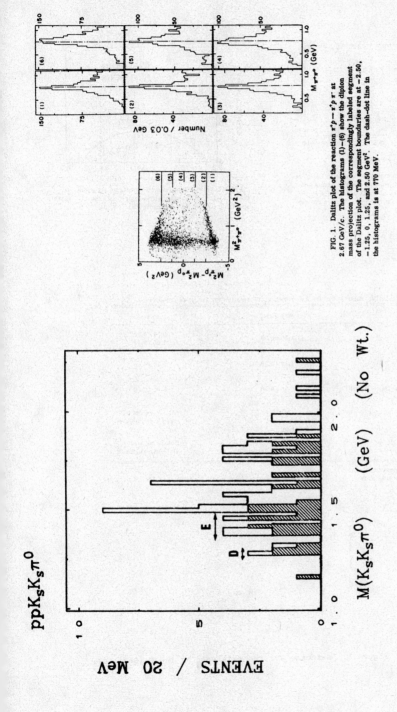

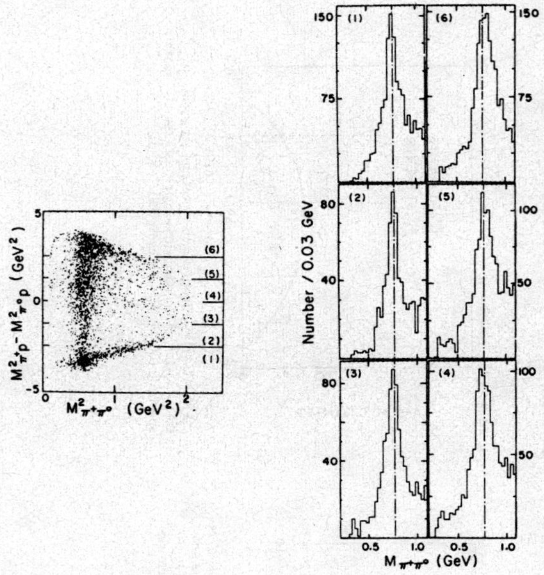

FIG. 1. Dalitz plot of the reaction $\pi^+p \to \pi^+p\pi^-$ at 2.67 GeV/c. The histograms (1)–(6) show the dipion mass projection of the correspondingly labeled segment of the Dalitz plot. The segment boundaries are at −2.50, −1.25, 0, 1.25, and 2.50 GeV^2. The dash-dot line in the histograms is at 770 MeV.

Figure 6. The effect of diffractive N^* production on the ρ mass and width (ref. 10).

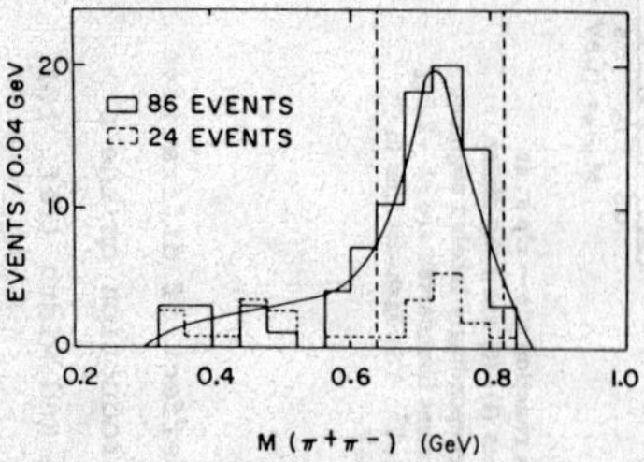

FIG. 34. $\pi^+\pi^-$ mass projection for events in the η' region. The dotted curve represents the $\pi^+\pi^-$ mass projection of the adjacent control bands normalized to the otal number of events in the background under the resonance.

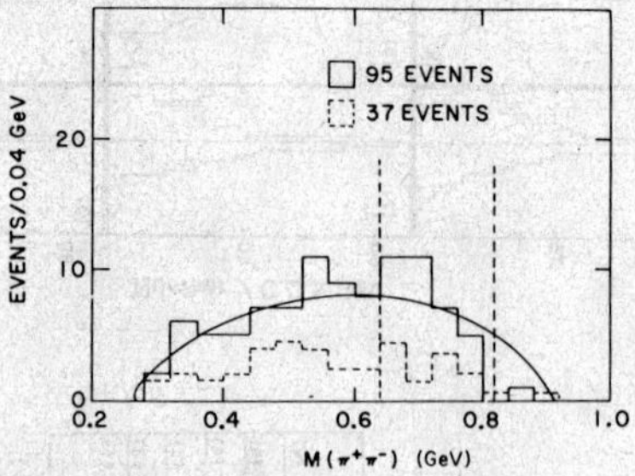

FIG. 47. $\pi^+\pi^-$ effective-mass spectrum for events in the M region. The dotted curve represents the $\pi^+\pi^-$ mass spectrum for the adjacent background region.

TABLE XIV. Summary of the experimental information for the η'-M comparison.

	K^-p sample	Λ sample
$\frac{\pi^+\pi^-\gamma}{\eta_N\pi^+\pi^-}$	1.20 ± 0.30	0.54 ± 0.10^a 1.05 ± 0.16^b
$\frac{\rho_{out}\gamma}{\rho_{in}\gamma}$	1.12 ± 0.38	0.10 ± 0.08
$\frac{\rho^0\gamma}{\pi^+\pi^-\gamma}$ from Dalitz plot fit	0.05 ± 0.10	0.94 ± 0.20
$\eta_N\pi^+\pi^-$ Dalitz plot	0^{-+} and 2^{-+} (mix) favored but 1^{+-} not ruled out	All J^{PC} except 0^{-+} and 2^{-+} (mix) ruled out
$\pi\pi$ helicity decay angular distribution	ρ_{out} is flat. ρ_{in} has $\sim 3\sigma$ for $\sin^2\theta$	Consistent with $\sin^2\theta$ in ρ region

[a]This experiment.
[b]Reference 8, p. 95.

Figure 7. Evidence for the existence of the M (Ref. 14).

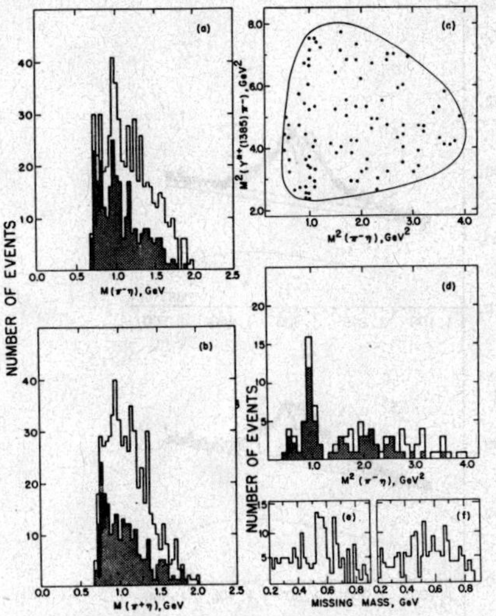

FIG. 1. (a) The $\pi^-\eta$ invariant-mass distribution for 571 events fitting Reaction (1), in which the η is a missing neutral with a missing mass within ±50 MeV of the η mass. There were 274 events which had the square of the four-momentum transfer to the $\pi^-\eta$ system $\Delta^2(\pi^-\eta) \leq 1.5$ (GeV/c)^2, and they are shown cross-hatched. Of these latter events, 31 represent the decay $X^0 \rightarrow \pi^+ + \pi^- + \eta$ and are shown completely blackened. (b) The $\pi^+\eta$ invariant-mass distribution for the same events as in (a) and with similar cuts. There were 231 events with $\Delta^2(\pi^+\eta) \leq 1.5$ (GeV/c)^2. (c) Dalitz plot of the three-body final state for 96 events representing Reaction (2). The $Y_1{}^*$ was selected to have a $\Lambda\pi^+$ invariant mass between 1335 and 1435 MeV. (d) Projection of the Dalitz plot of (c) onto the axis representing the square of the $\pi^-\eta$ invariant mass. The cross hatched and blackened areas have the same significance as in (a) and (b). (e) Mass of the missing neutral, M^0, for a subset of the 8500 two-prong + V events interpreted as Reaction (3), and having a squared four-momentum transfer to the π^-M^0 system $\Delta^2(\pi^-M^0) \leq 1.5$ (GeV/c)^2. The π^-M^0 invariant mass was selected to lie between 940 and 1040 MeV. (f) Similar to (e) except that the π^-M^0 invariant mass was selected to lie between either 890 and 940 MeV or between 1040 and 1090 MeV. All plots in this figure, except (e) and (f), are made using quantities kinematically fitted to Reaction (1).

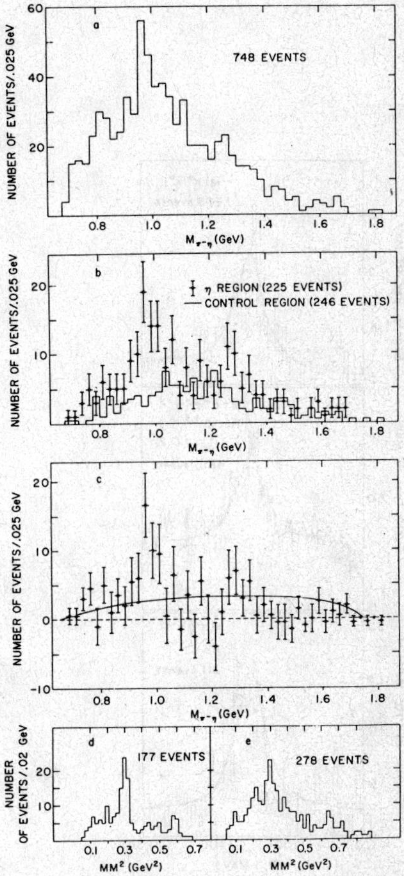

FIG. 2. (a) π^-MM spectrum from Reaction (1) with MM in the η region and the $\eta^*(960)$ events removed. (b) Same as (a) but with $M_{\Lambda\pi^+}$ in the $\Sigma^+(1385)$ region. The histogram with (without) the error bars corresponds to those events with MM in the η region (control region). Events in the control region are normalized to the number of background events in the η region. (c) $\pi^-\eta$ spectrum after the histogram without the error bars is subtracted from the one with error bars in (b). The solid curve corresponds to the phase space for the final state $\Sigma^+(1385)\pi^-\eta$ and is normalized to the total number of events. (d), (e) MM^2 spectrum from Reaction (1) for $\Sigma^+(1385)$ events with π^-MM in the δ region (control region).

Figure 8. Production of the δ in the reaction $K^-p \rightarrow \Lambda\pi\delta$ (Ref. 15).

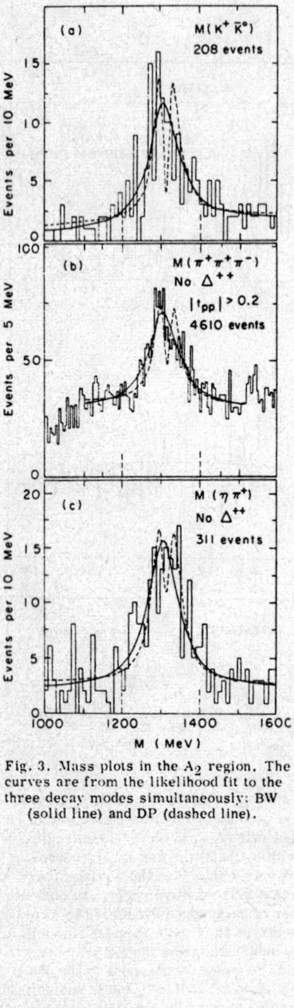

Fig. 3. Mass plots in the A_2 region. The curves are from the likelihood fit to the three decay modes simultaneously: BW (solid line) and DP (dashed line).

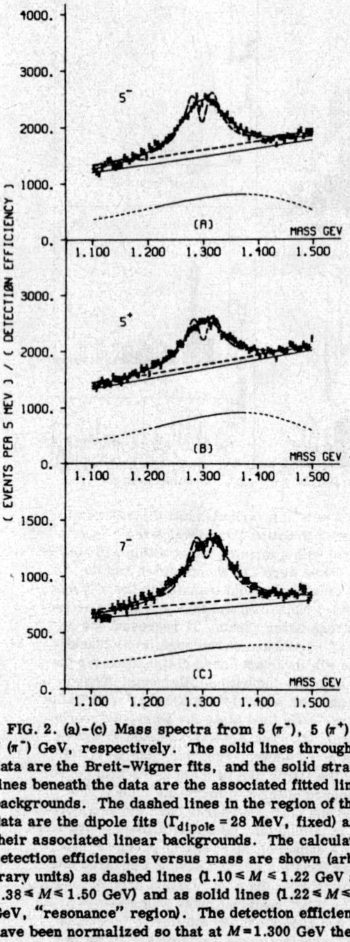

FIG. 2. (a)-(c) Mass spectra from 5 (π^-), 5 (π^+), and 7 (π^-) GeV, respectively. The solid lines through the data are the Breit-Wigner fits, and the solid straight lines beneath the data are the associated fitted linear backgrounds. The dashed lines in the region of the data are the dipole fits ($\Gamma_{dipole} = 28$ MeV, fixed) and their associated linear backgrounds. The calculated detection efficiencies versus mass are shown (arbitrary units) as dashed lines ($1.10 \leq M \leq 1.22$ GeV and $1.38 \leq M \leq 1.50$ GeV) and as solid lines ($1.22 \leq M \leq 1.38$ GeV, "resonance" region). The detection efficiencies have been normalized so that at $M = 1.300$ GeV the ordinates on the graphs indicate the actual number of events detected in the experiment per 5-MeV bin.

Figure 9. a) (on the left) A_2^+ mass spectra from $\pi^+p \rightarrow p\, A_2^+$ at 7 GeV/c (Ref. 18).

b) $A_2^\pm$ mass spectra from $\pi^\pm p \rightarrow A_2^\pm p$ at 5 and 7 GeV (Ref. 19).

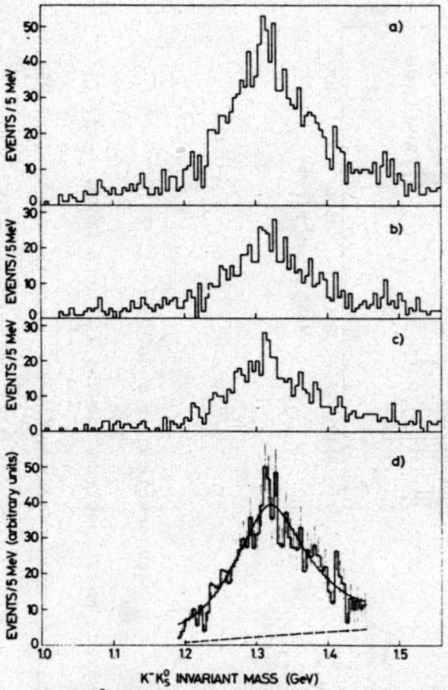

Fig. 5. $K^-K^0_S$ invariant mass: a) all $|t|$. unweighted: b) $|t| < 0.2$. unweighted: c) $0.2 < |t| < 0.7$ $(GeV/c)^2$. unweighted: d) weighted $0.0 < |t| < 0.7$ $(GeV/c)^2$. Full line: Breit-Wigner type fit of the form

$$F(m) = \text{linear background} + \frac{2}{\pi}\,\frac{m - m_0 - \Gamma(m)}{(m^2 - m_0^2)^2 + m^2\Gamma^2_{(m)}}$$

$$\Gamma(m) = \Gamma_0\left(\frac{q}{q_0}\right)^5 \frac{9 + 3R^2q_0^2 + R^4q_0^4}{9 + 3R^2q^2 + R^4q^4}$$

[6] giving $m_0 = 1321 \pm 3$ MeV and $\Gamma_0 = 123 \pm 7$ MeV with $P_{\chi^2} = 32\%$. Dashed line: fitted background.

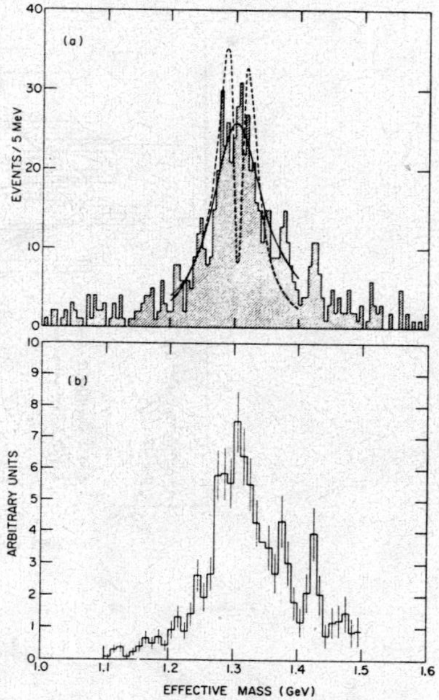

FIG. 8. The K^0K^- effective-mass spectrum for those events having a proton recoil; (a) uncorrected for acceptance, and (b) corrected for the acceptance of our apparatus. The A_2 signal at 1.3 GeV is produced with very little background. The curves are explained in the text. The mass scale is uncertain by ≈1 MeV.

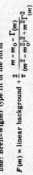

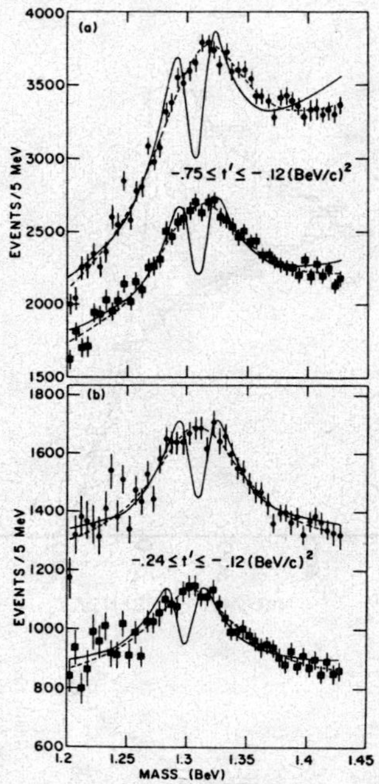

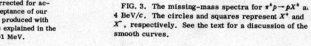

FIG. 3. The missing-mass spectra for $\pi^{\pm}p \to pX^{\pm}$ at 4 BeV/c. The circles and squares represent X^+ and X^-, respectively. See the text for a discussion of the smooth curves.

Figure 9 c) (on the left) A_2^- mass spectrum in the K^-K^0 mode from π^-p at 17.2 GeV (Ref. 20).

d) K^0K^- mass spectrum at 20.3 GeV (Ref. 21).

e) Spectrum from Ref. 22 at 4 GeV/c.

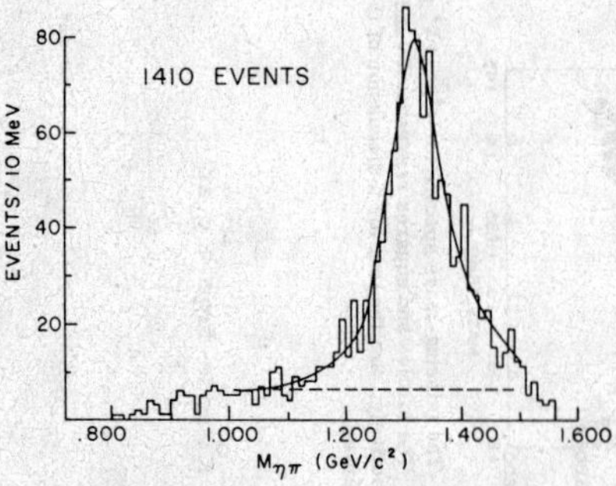

FIG. 2. Invariant mass of $\eta\pi^-$ from events fitting the reaction of Eq. (3). The solid curve shows the result of the single *D*-wave Breit-Wigner fit with parameters described in the text. The dashed curve indicates the fitted background.

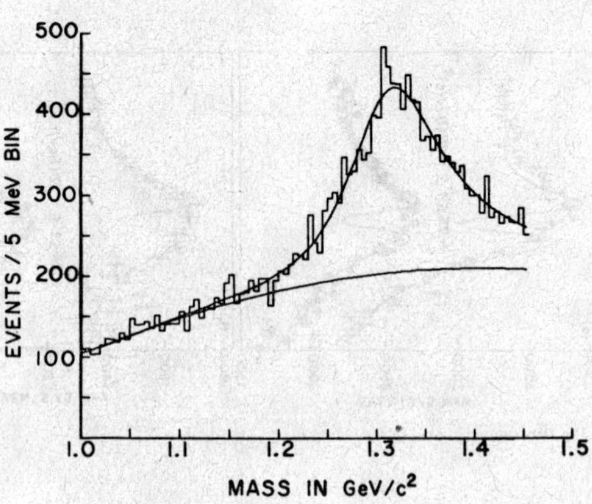

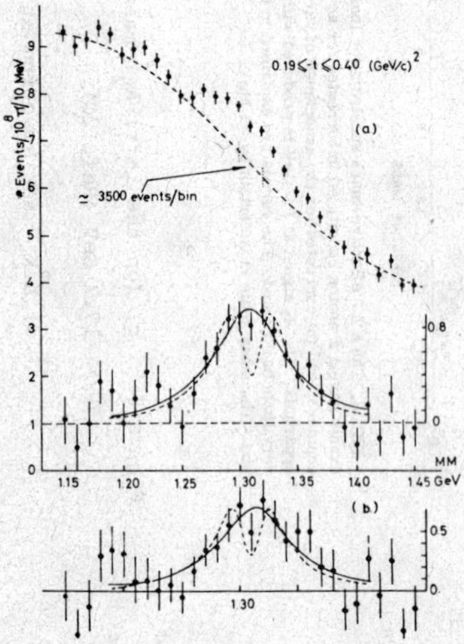

Fig. 3(a). Missing mass spectrum in the low *t* region and the predicted background. The lower figure shows the spectrum after background subtraction. Best fits to a dipole with $\Gamma = 28$ and a Breit-Wigner are also shown. In these fits the normalization of the background was allowed to vary.
Fig. 3(b). Missing mass spectrum in the *t* region examined in the CERN boson spectrometer experiment at 2.6 GeV/*c* after background subtraction.

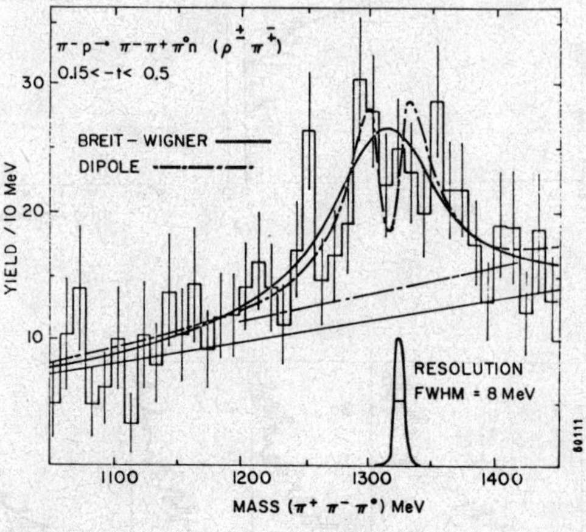

Figure 9 f) (top left) The $A_2^- \to \pi^-\eta$ mass spectrum at 6 GeV/c (Ref. 23).

g) (top right) The missing mass spectrum for the A_2^- at 6 GeV/c (Ref. 24).

h) (lower left) The missing mass spectrum for A_2^- near threshold (Ref.25).

i) (lower right) The A_2^0 mass spectrum at 4.5 GeV/c (Ref. 26).

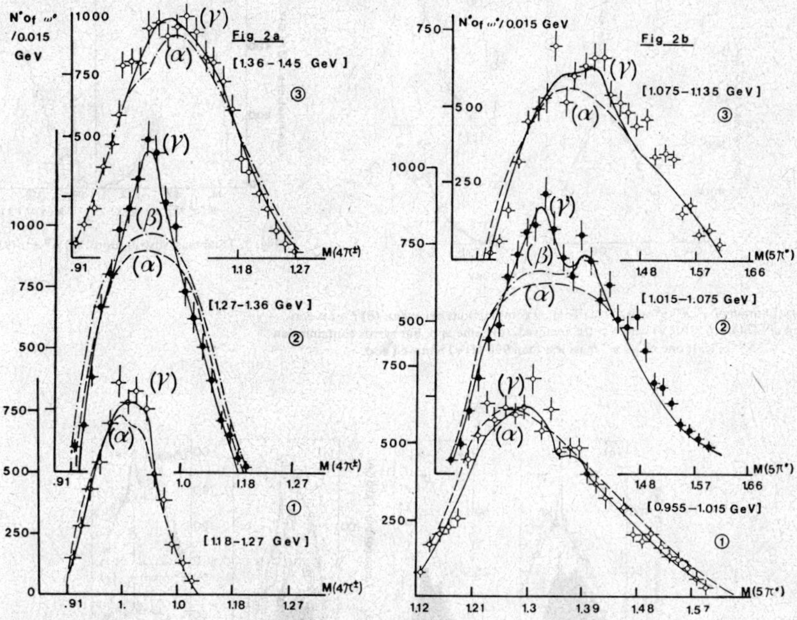

Fig. 2. Curves and normalization are described in fig. 1, except for β which takes into account two simple $(\omega^\circ\pi\pi)$ effects respectively at 1315 and 1405 MeV (without sequential decay).
a distributions of $\omega^\circ\pi^\pm$ masses contained in $\omega^\circ\pi^+\pi^-$ combinations in the three successive mass intervals indicated in figure.
b distributions of $\omega^\circ\pi^+\pi^-$ masses containing at least one $\omega^\circ\pi^\pm$ combination in three successive mass intervals.

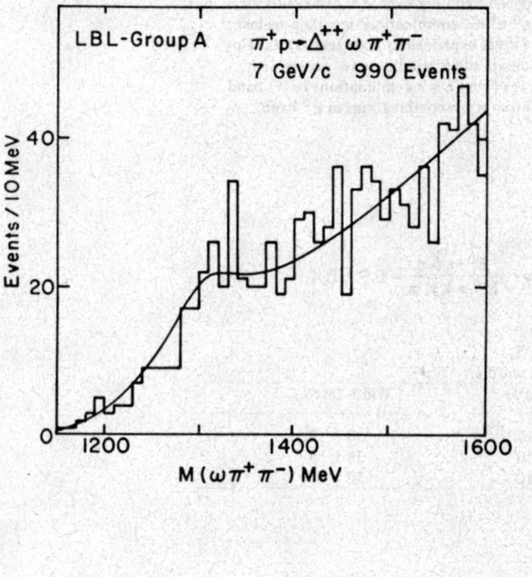

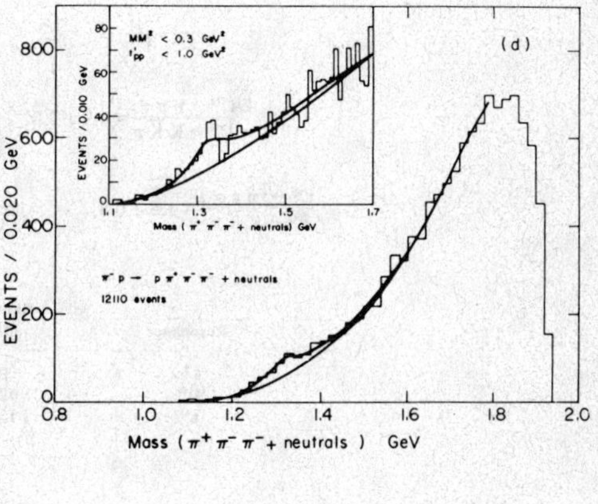

Figure 10 a) (top) Evidence for the $A_2^0 \to \omega\pi\pi$ (ref. 27);

b) (lower left) Supporting evidence for $A_2^0 \to \omega\pi\pi$ (ref. 29);

c) (lower right) Evidence for $A_2^- \to \omega\pi\pi$ (ref. 30).

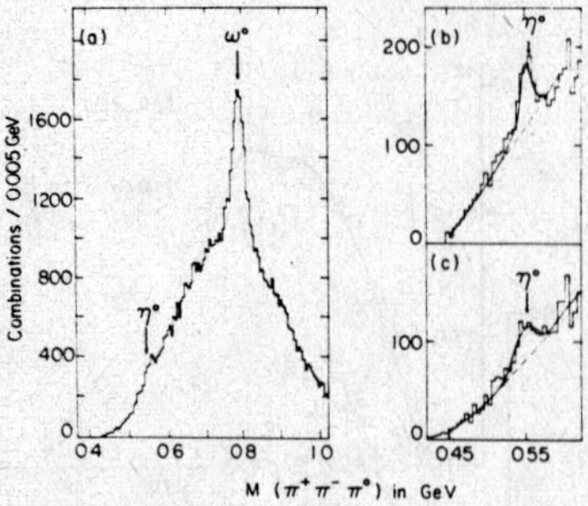

Fig. 2. Distribution of $\pi^+ \pi^- \pi^0$ masses. (a) Total, 9 combinations per event. (b) Events containing an ω^0 (744 – 820 MeV) with $\lambda > 0.5$ removed. (c) Same as b, but events containing an X^0 (at least one $\eta^0 \pi^+ \pi^-$ mass less than 990 MeV) removed also.

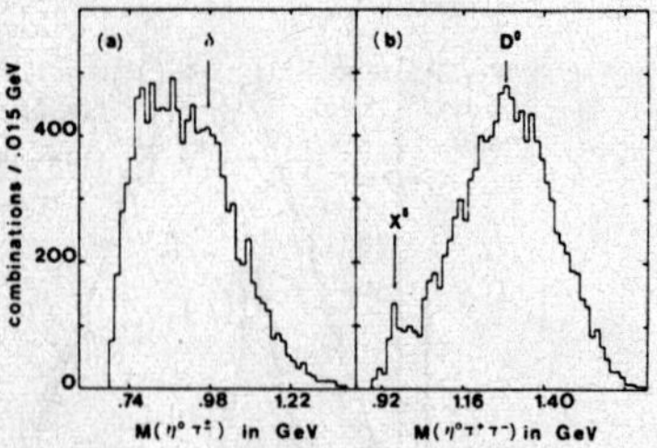

Fig. 3. Total mass distributions. (a) $\eta^0 \pi^\pm$. (b) $\eta^0 \pi^+ \pi^-$.

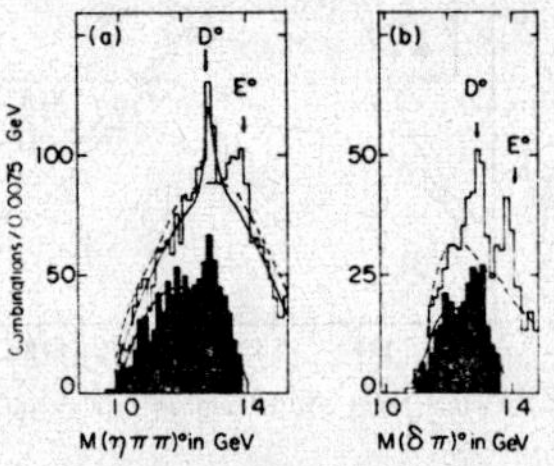

Fig. 12. Distribution of $\eta^0 \pi^+ \pi^-$ masses after estimator selection, each $(5\pi)^0$ combination being counted only once. Shaded events are for $\eta^0 \pi^+ \pi^-$ combinations recoiling against ρ^0. Curves explained in text (sect.4). (a) Total $\eta^0 \pi^+ \pi^-$. (b) $\delta^\pm \pi^+$.

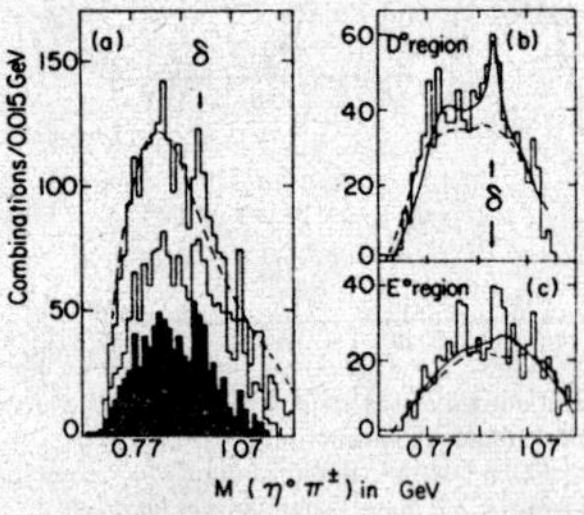

Fig. 13. Distribution of $\eta^0 \pi^\pm$ masses after estimator selection, each $(4\pi)^\pm$ combination being counted only once. Shaded events are for $\eta^0 \pi^+ \pi^-$ combinations recoiling against ρ^0. Curves explained in text (sect. 3). (a) Upper curve: total. Middle curve: events with D^0. (b) From $\eta \pi \pi$ combinations in D^0 band (c) From $\eta \pi \pi$ combinations in E^0 band.

$$\frac{D^0 \to \eta \pi \pi}{D^0 \to K \overline{K} \pi} = 4.9 \pm 2.0,$$

$$\frac{E^0 \to (\eta \pi \pi)}{E^0 \to K \overline{K} \pi} = 1.5 \pm 0.8, \qquad \text{including} \quad \frac{E^0 \to \delta \pi}{E^0 \to K\overline{K} \pi} = 0.9 \pm 0.4.$$

Table 2
Masses and widths.

Resonance	Mass (MeV)	Width (MeV)
$\delta^\pm$	972 ± 10	30 ± 5
D^0	1292 ± 10	28 ± 5
E^0	1398 ± 10	50 ± 10

Figure 11. Evidence for D and E production in $\bar{p}p \to 3\pi^+ 3\pi^- \pi^0$ annihilations and final results from the fit. The D is consistent with a 100% $\pi\delta$ decay (Ref. 31). (11 a is top left. 11 b top right. 11 c is the middle.)

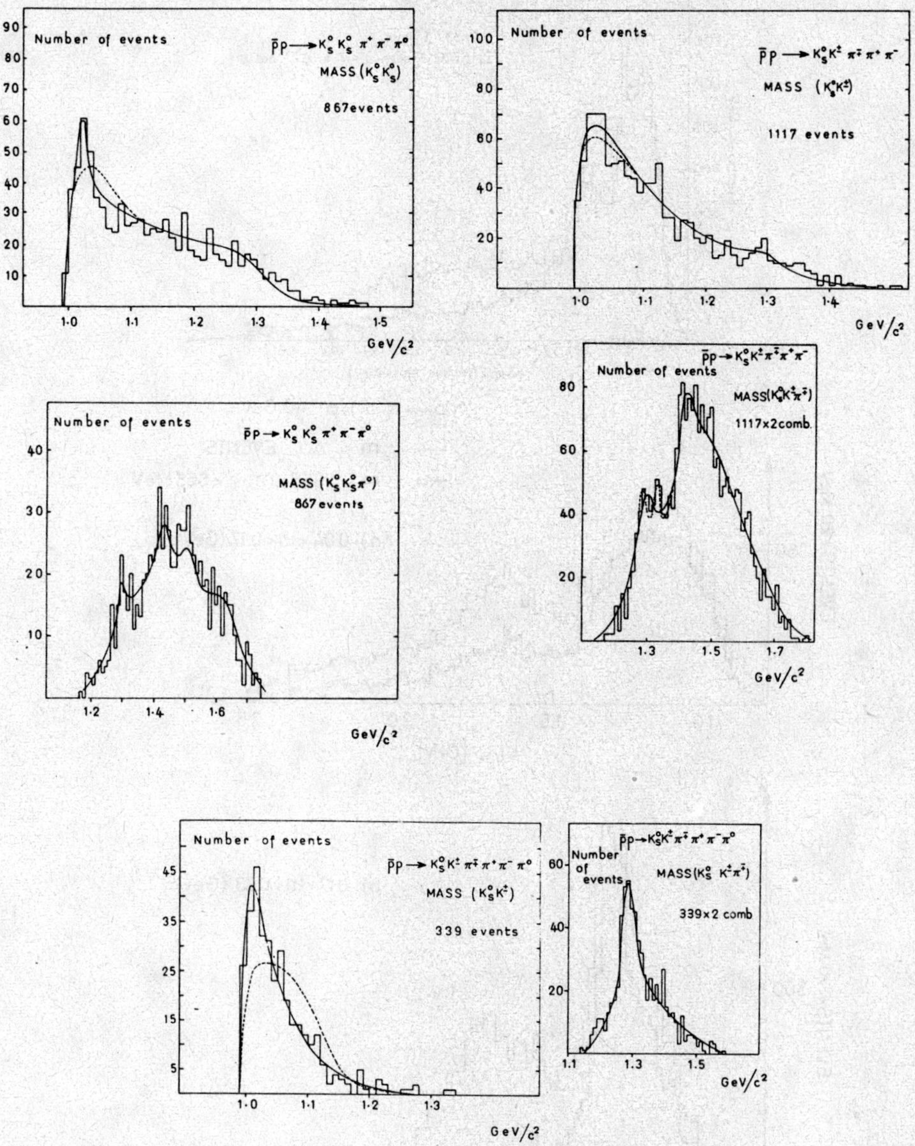

Figure 12. Evidence for the $K\bar{K}\pi$ decays of the D and E in $\bar{p}p \rightarrow K\bar{K}\pi$ (+ $n\pi$) at 1.2 GeV (Ref. 34).

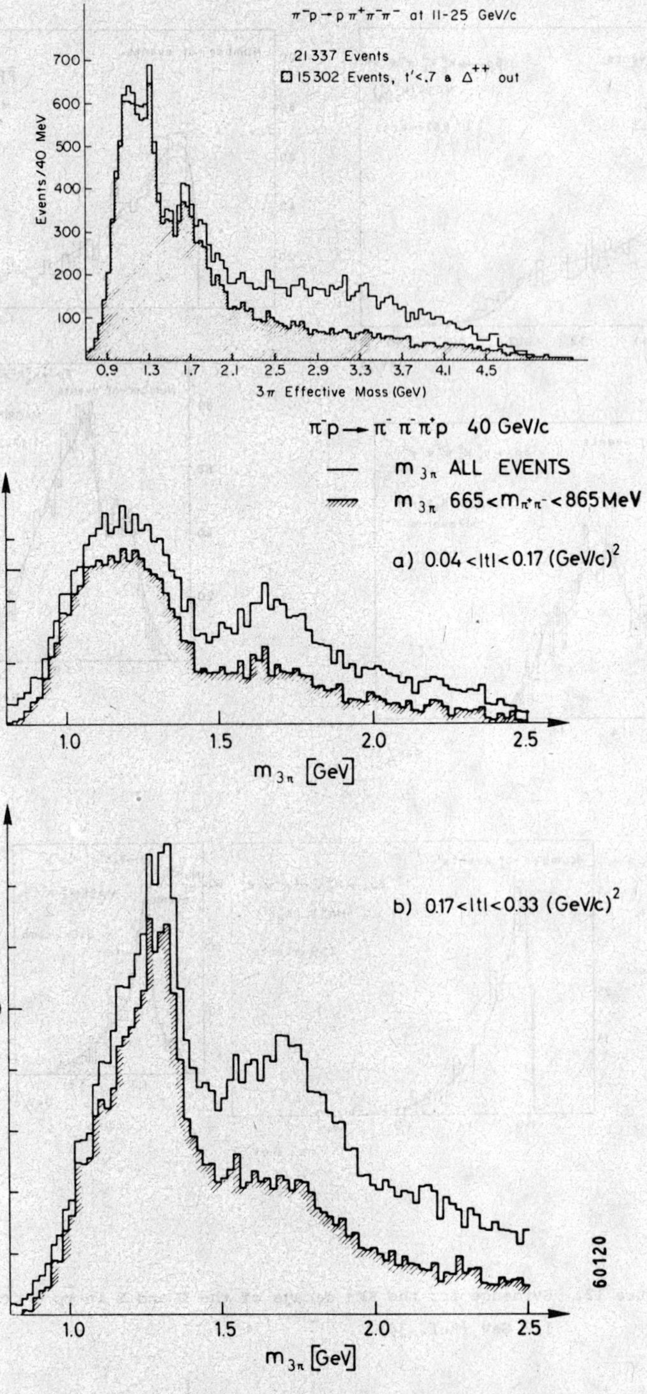

Figure 13 The 3π mass spectra from Ref. 36 (top) and Ref. 37 (bottom)

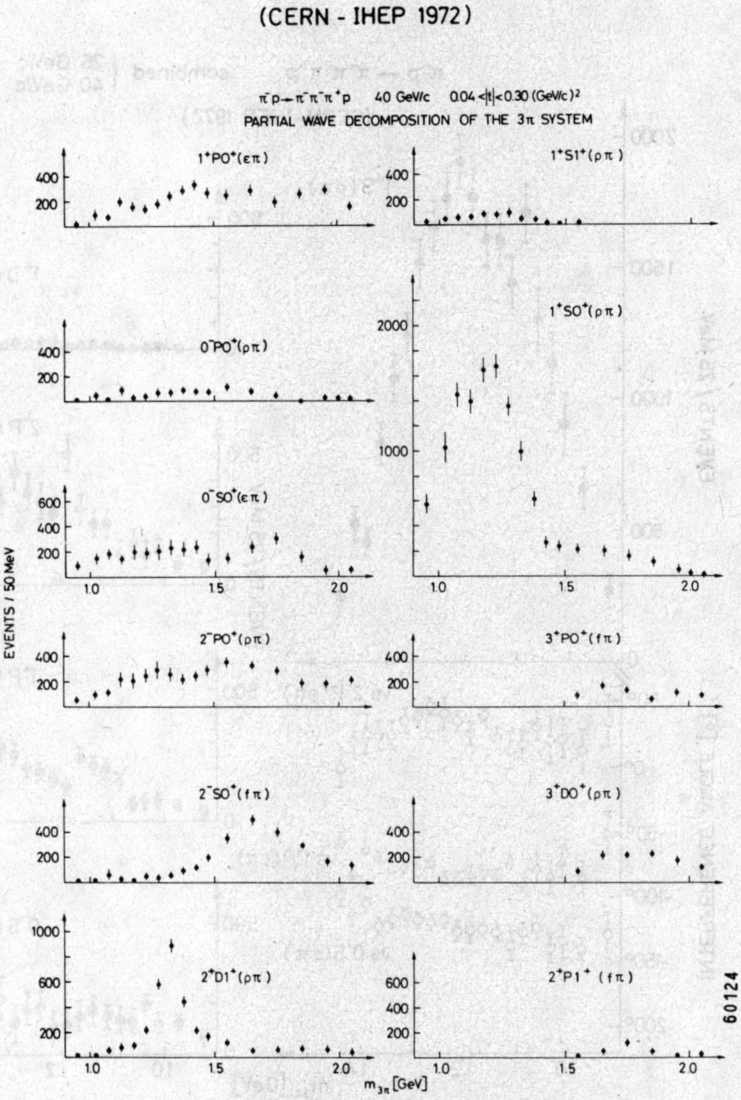

Fig. 5

Figure 14a) Results for the partial wave decomposition of the 3π system (Ref. 37)

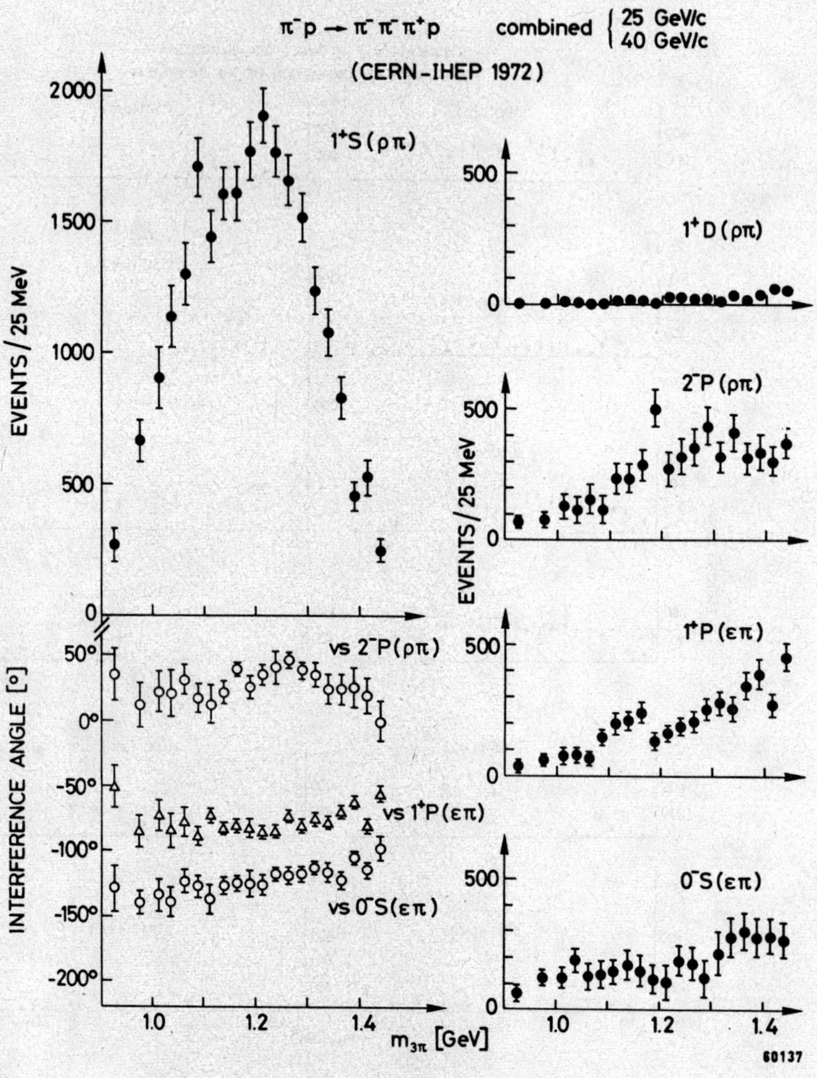

Figure 14b) Results for the A_1 intensity and relative phases (Ref. 37)

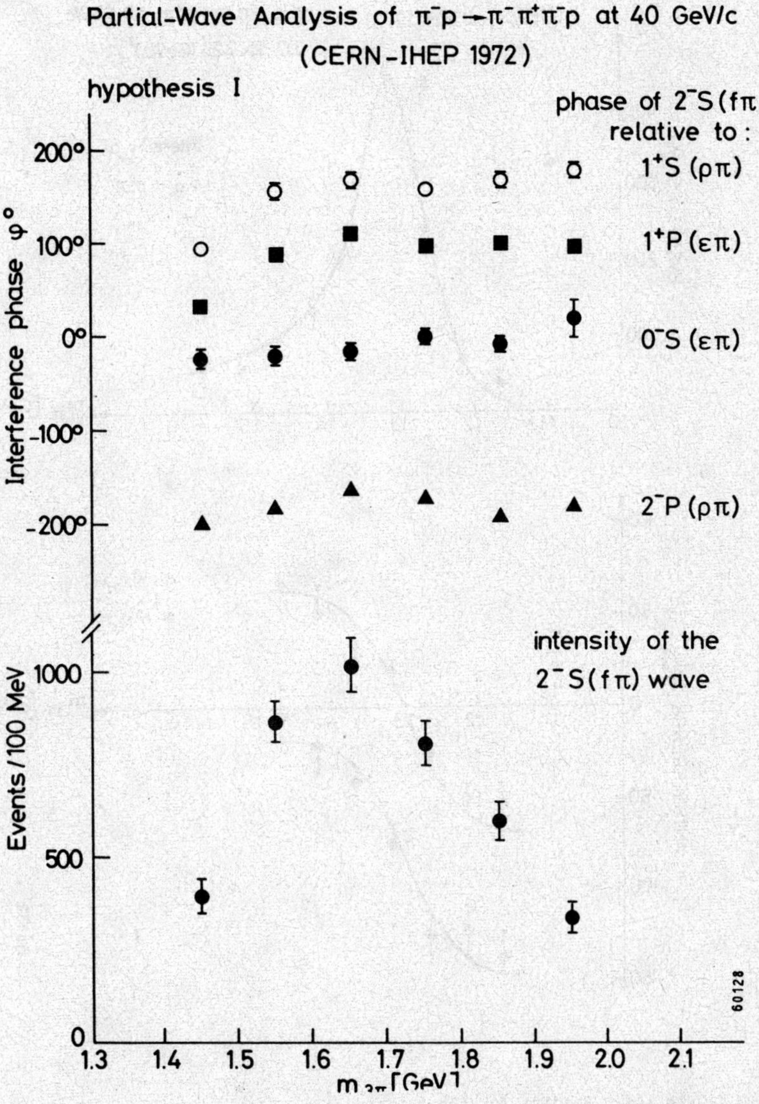

Figure 14c Results for the A_3 intensity and relative phases (Ref. 37).

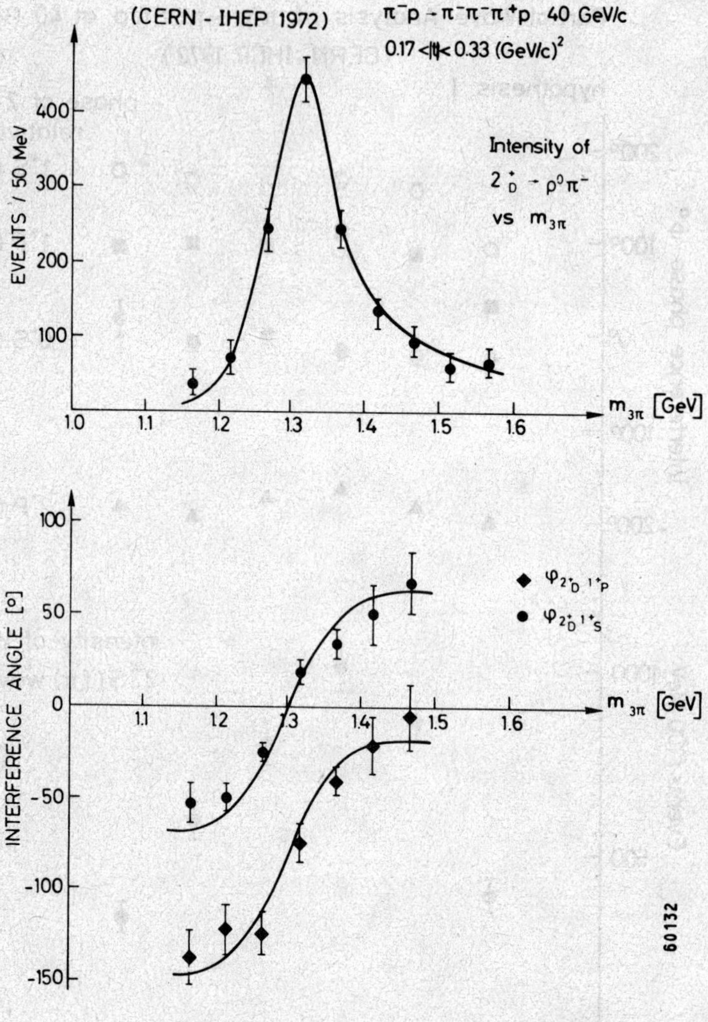

Figure 14d. Results for the A_2 intensity and relative phases (Ref. 37).

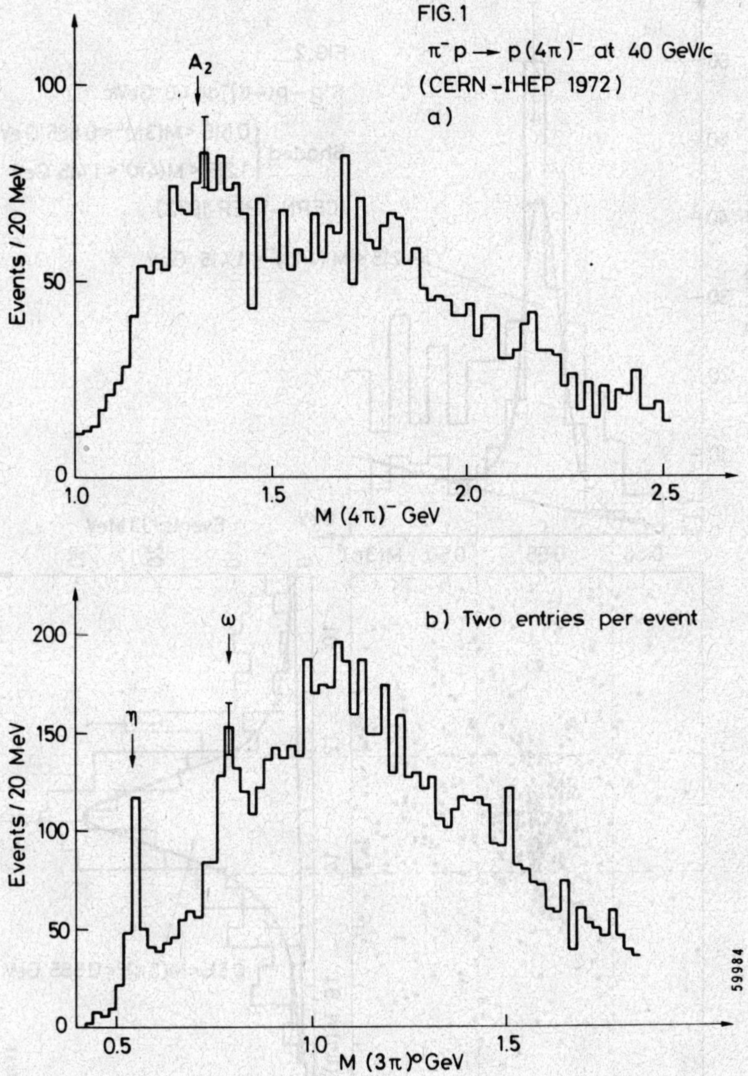

Figure 15 a) Data showing the $\pi^-\eta$ decay of the A_2^- at 40 GeV/c (Ref. 37).

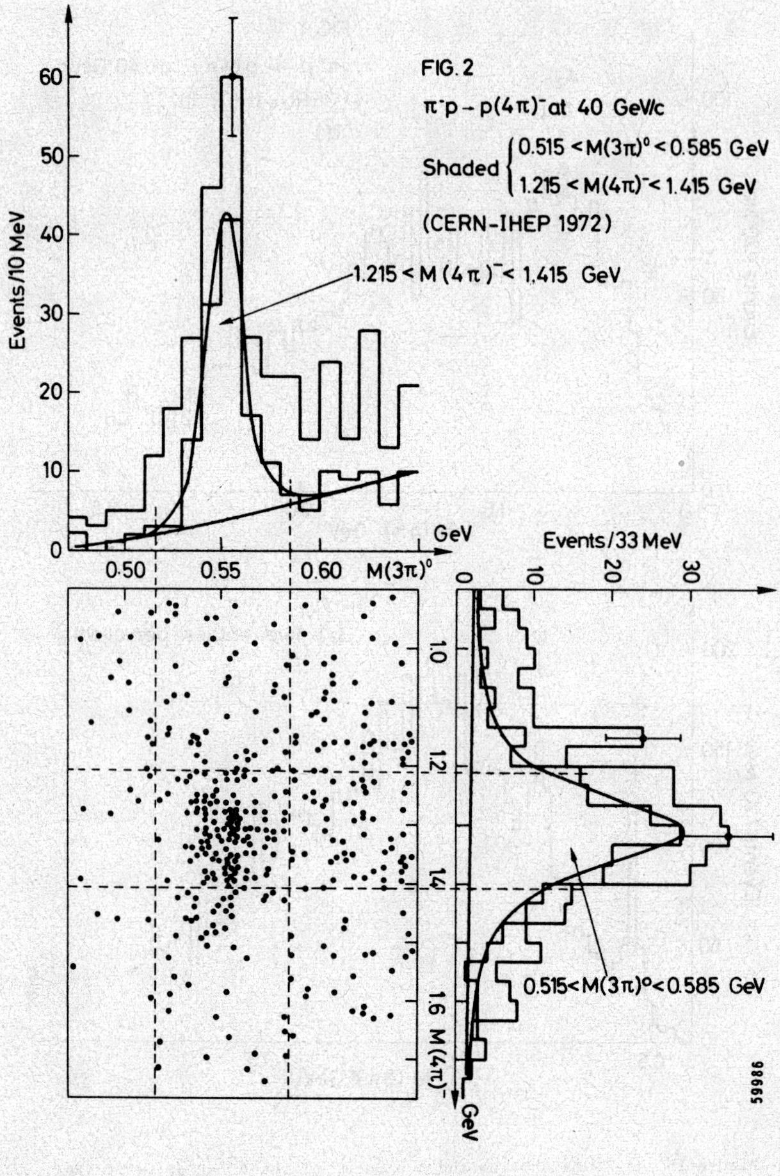

Figure 15b) The correlation of the η with the A_2^-.

FIG. 3

$\pi^- p \rightarrow p(4\pi)^-$ at 40 GeV/c

A_2 Decay Angles ($\cos\vartheta, \varphi$)

(CERN-IHEP 1972)

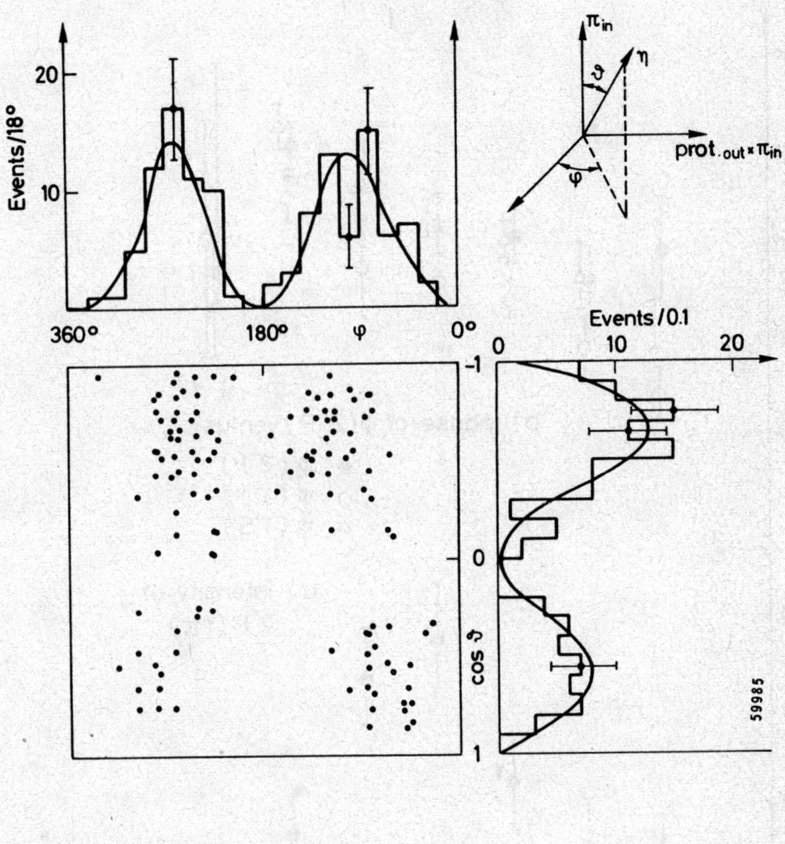

Figure 15c) Angular distributions for the $\pi^-\eta$ decay of the A_2^- (ref. 37).

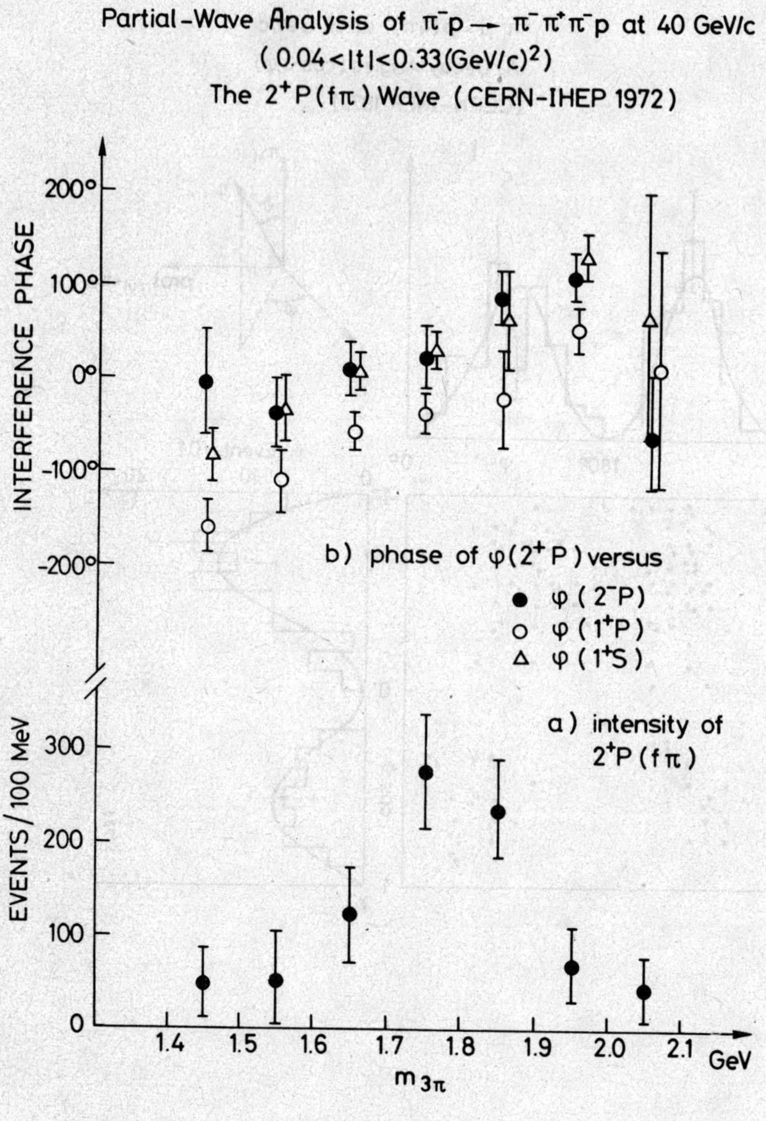

Figure 16 Results for the $2^+p(f\pi)$ wave showing possible resonant behaviour (Ref. 37)

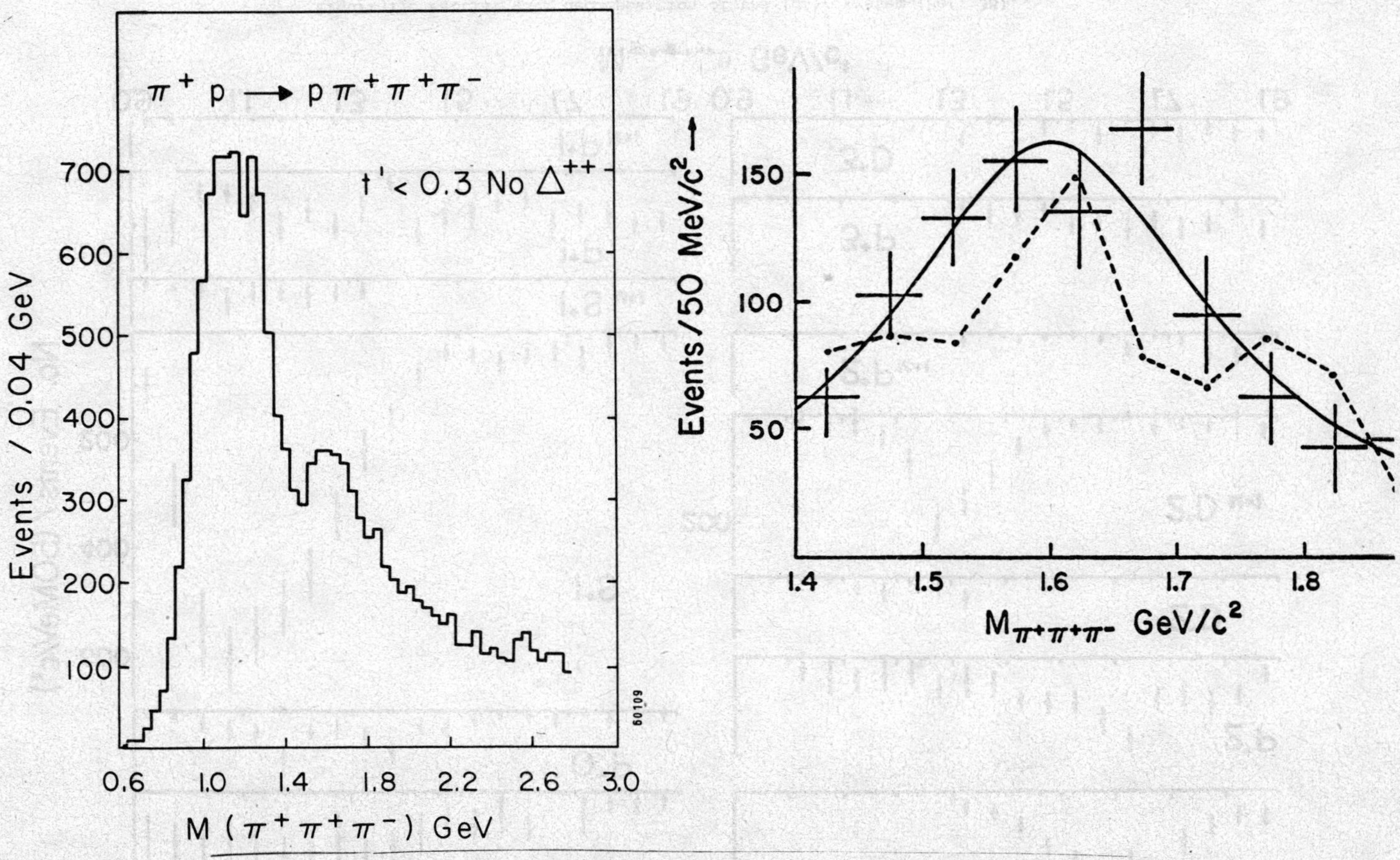

Figure 17a (left) $(3\pi)^+$ mass spectrum at 13 GeV/c (Ref. 38)

b (right) The A_3 intensity (Ref. 38).

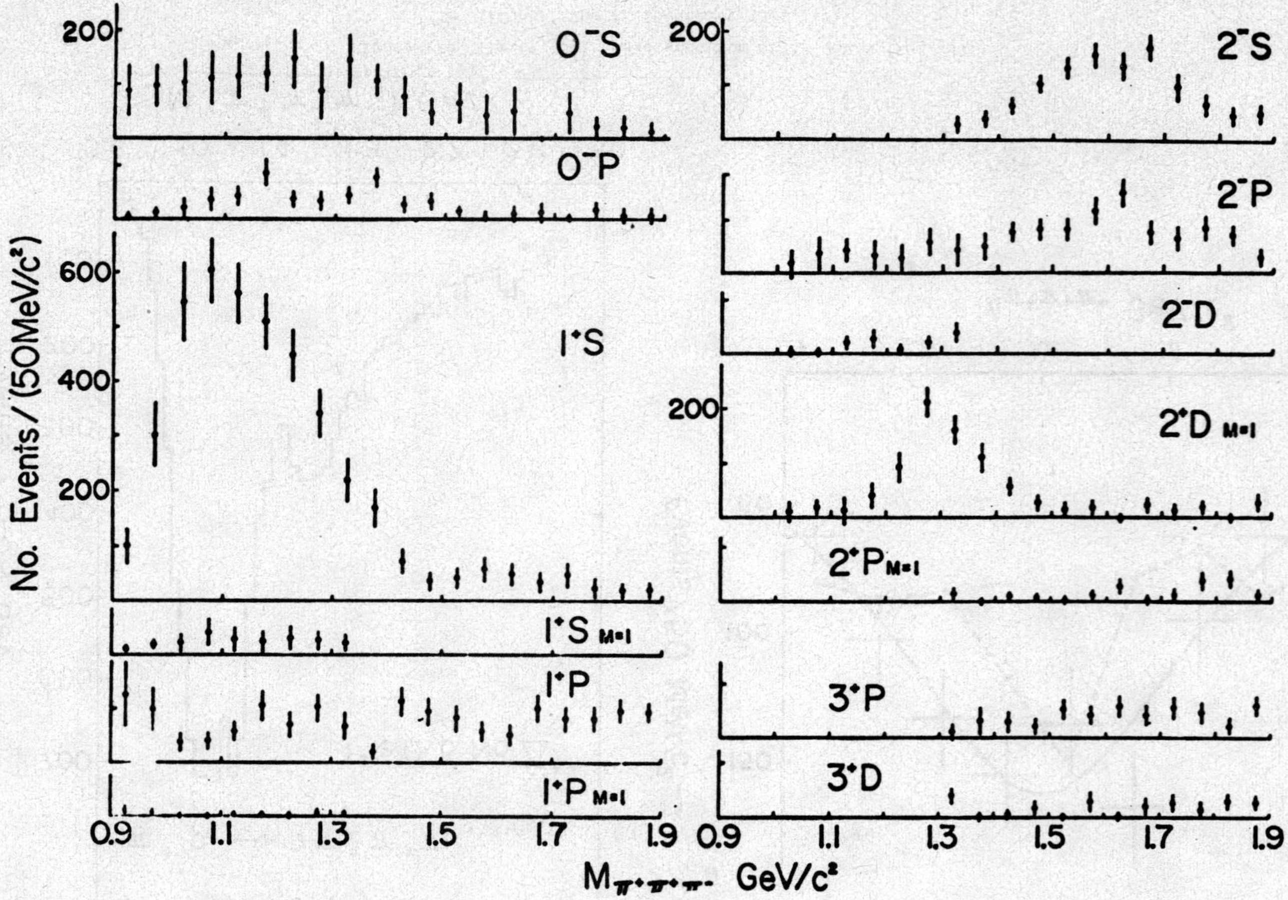

Figure 17c Partial wave decomposition of the $(3\pi)^+$ system (Ref. 38).

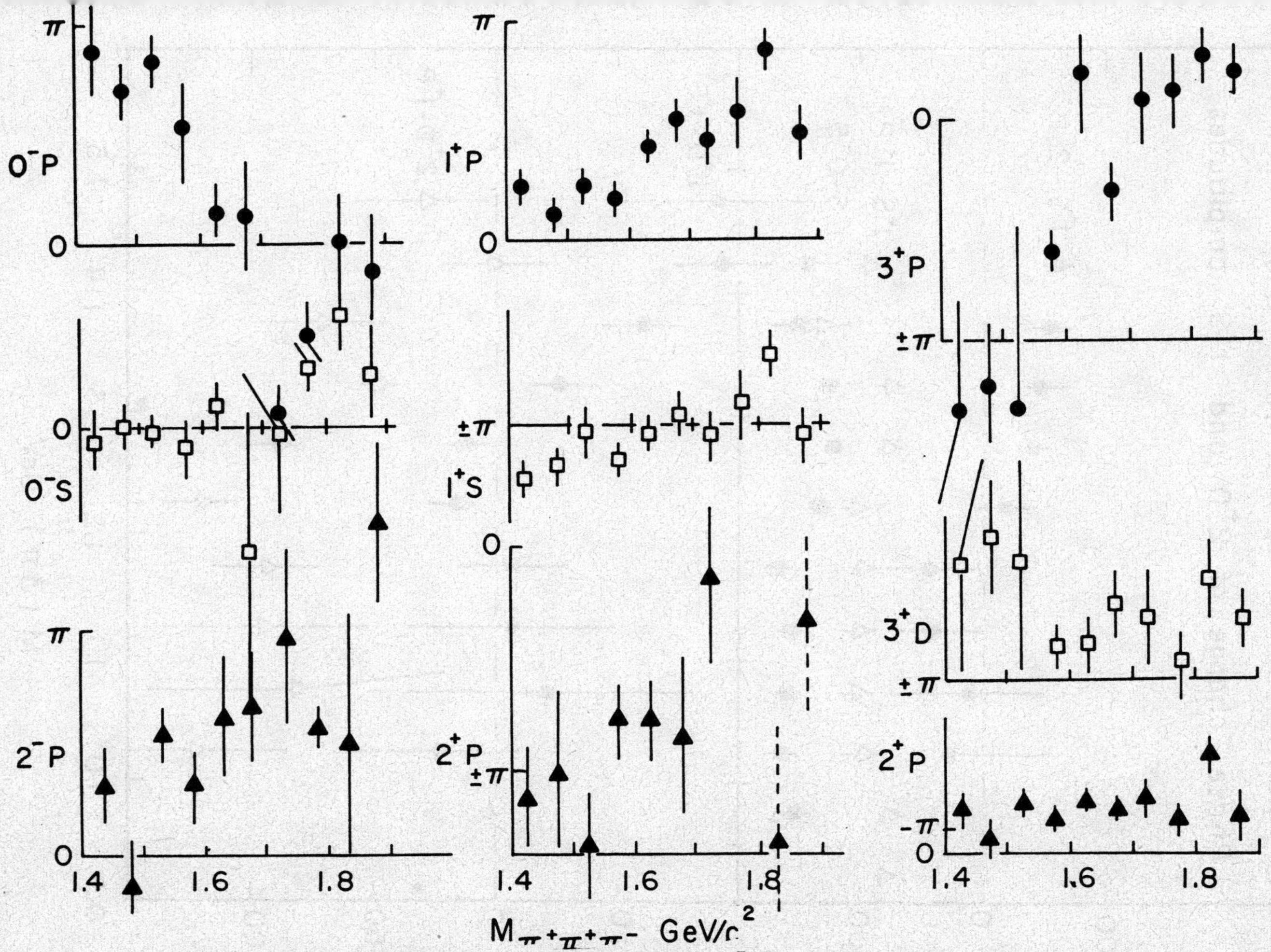

Figure 17d Relative phase variation between the $2^{-}S$ A_3 and other waves (Ref. 38).

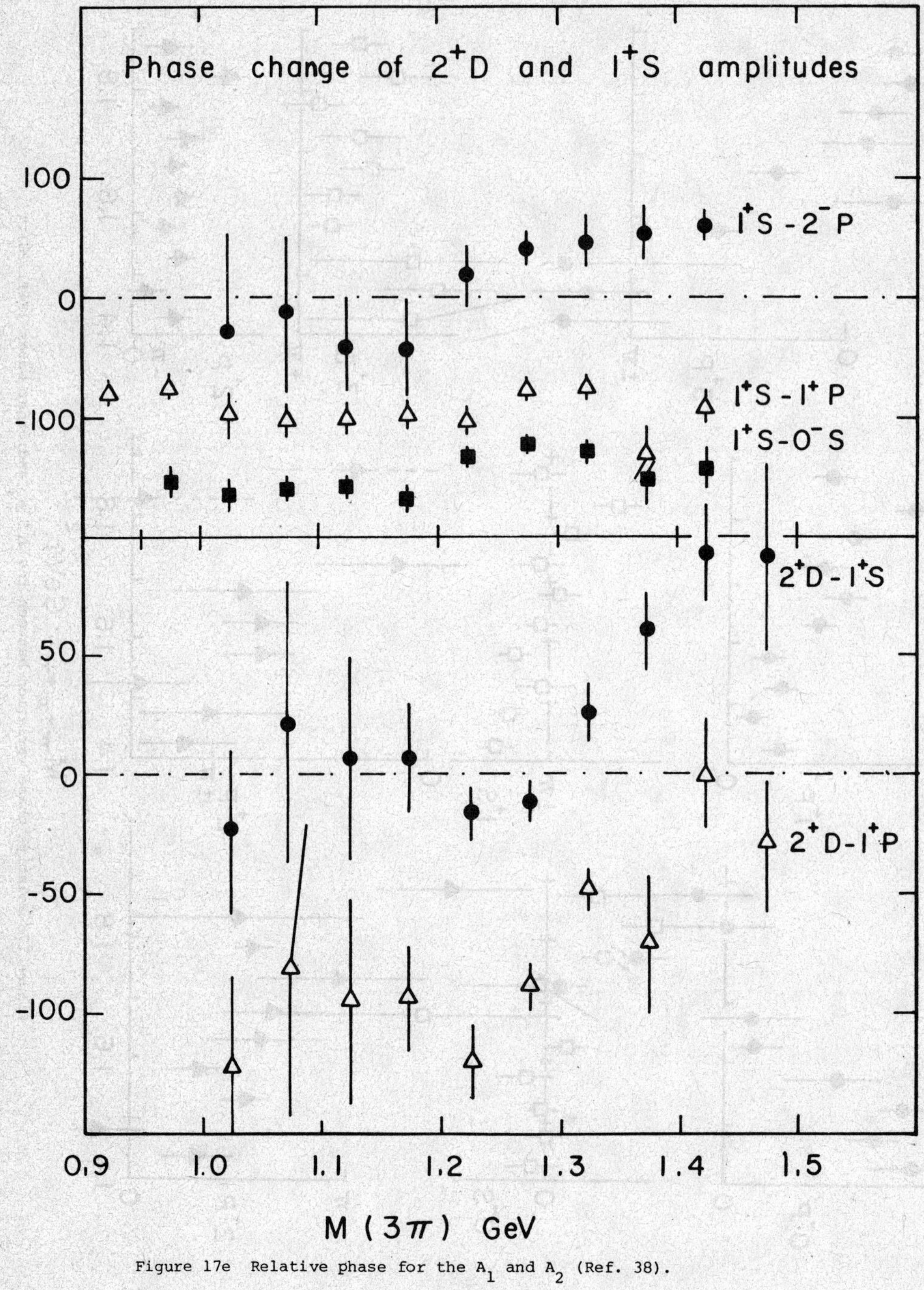

Figure 17e Relative phase for the A_1 and A_2 (Ref. 38).

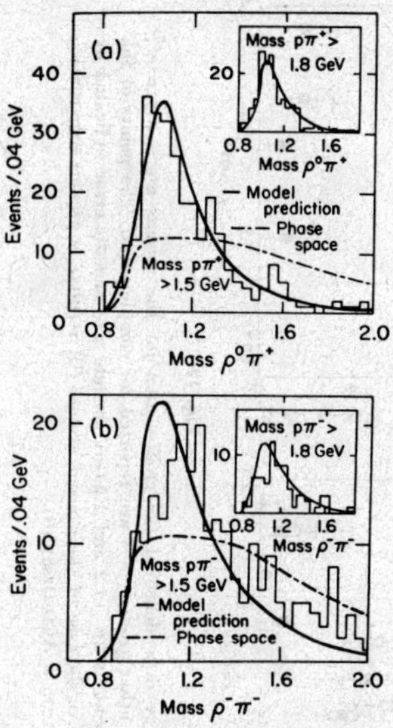

FIG. 3. Invariant mass distributions for those $\rho\pi$ events fitted by the model described in the text. See text for complete details concerning all selections made. (a) Reaction (1); (b) Reaction (2); the insets differ from the main figures only in the $p\pi$ mass cutoffs employed. There are 10.7 events/μb in (b).

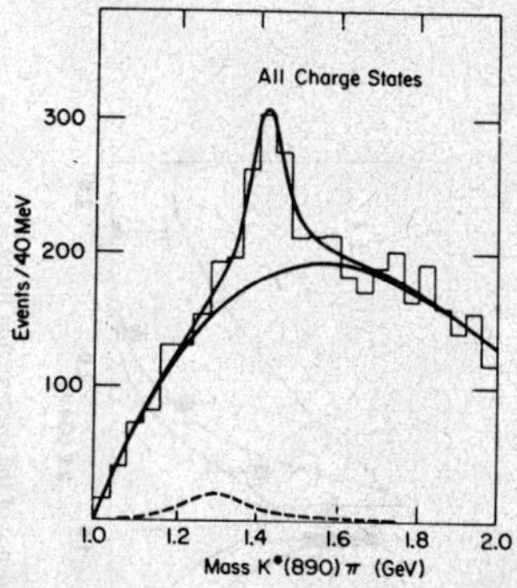

FIG. 1. The $[K^*(890)-\pi]$ mass spectrum for all the data from the sources listed in Table I. The solid curves represent the results of a fit to a third-order background polynomial plus $K^*(1420)$.

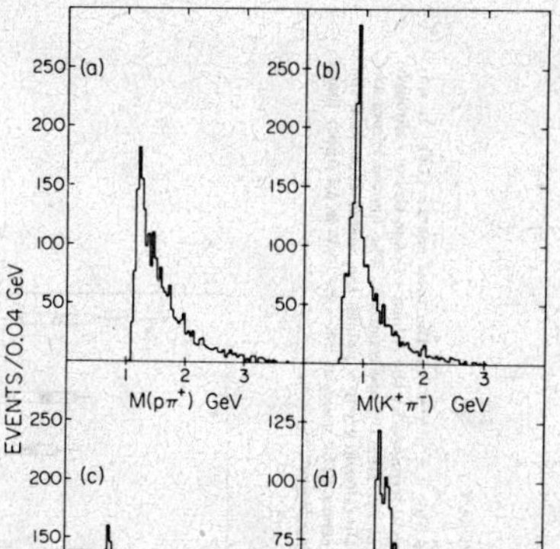

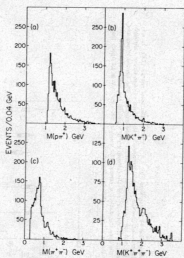

Fig. 3. Mass distributions for the events in the backward peak of reaction (3): (a) $M(p\pi^+)$, (b) $M(K^+\pi^-)$, (c) $M(\pi^+\pi^-)$ and (d) $M(K^+\pi^+\pi^-)$.

Figure 18. (top left) Evidence for "A_1" production in π^-, (Ref. 39).

Figure 19. (top right) Null evidence for Q production in charge exchange reactions (Ref. 41).

Figure 20. (bottom) Evidence for u channel Q production (Ref. 42).

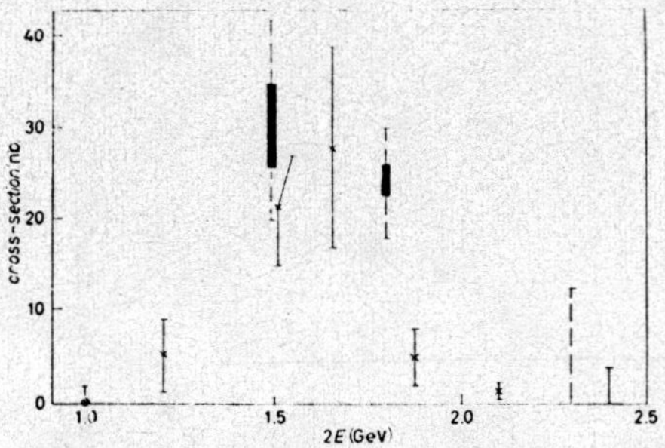

Fig. 1. – Experimental cross-section of the process $e^+e^- \to \pi^+\pi^-\pi^+\pi^-$ *vs.* the total energy ($2E$). In addition to the results of this experiment, also the values determined by the Adone « Boson Group » through an overall fit of their data ([11]) and the ACO result at 990 MeV ([10]) are reported. The quoted errors are only statistical for the present experiment. For the Boson group's points the boxes represent the systematic uncertainty and the bars the statistical errors. The point at $2E = 2.4$ GeV is an upper limit (see Table I): ○ ACO, ■ Boson group, × this experiment.

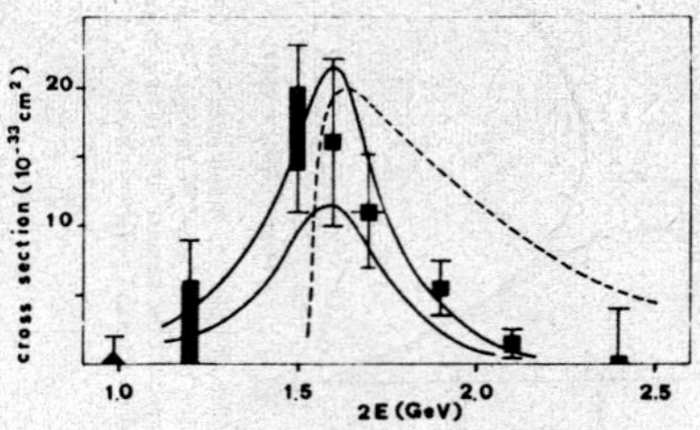

Fig. 4. Energy dependence of the cross section for process $e^+e^- \to \rho^\circ\epsilon^\circ \to \pi^+\pi^-\pi^+\pi^-$. The curves are taken from ref. [2] (═) and ref. [8] (----). The dotted curve taken from ref. [8] corresponds to that with the steepest energy dependence for $2E > 1.6$ GeV.

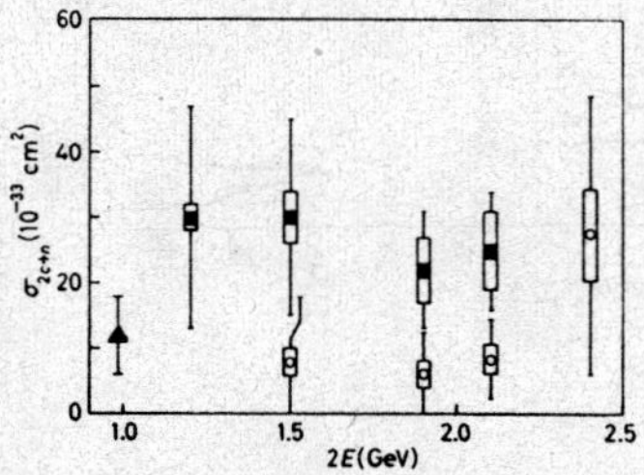

Fig. 17. – Energy dependence of the cross-section $\sigma_{2c+n} = \sigma(\pi^+\pi^-\pi^0) + \sigma(\pi^+\pi^-\pi^0\pi^0)$. The open boxes indicate systematic errors. In our case these errors correspond to assuming only one of the two quoted reactions present at a time. The lower limit corresponds to considering only the process $e^+e^- \to \pi^+\pi^-\pi^0\pi^0$: ■ present experiment, ○ γγ group ([12]), ▲ Orsay ([3b]) (only $\pi^+\pi^-\pi^0\pi^0$).

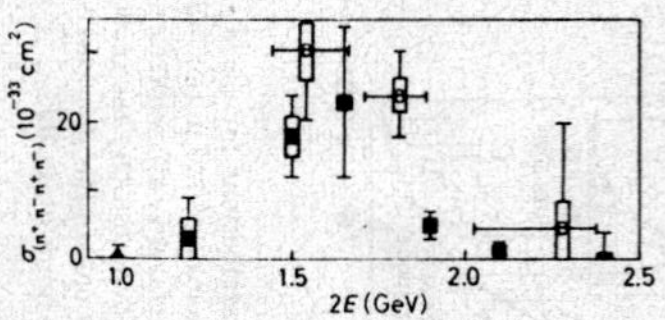

Fig. 18.

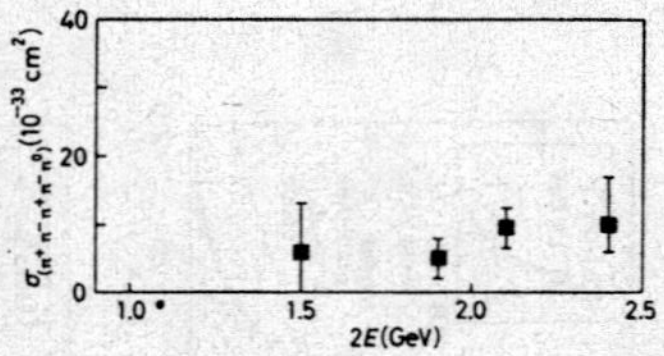

Fig. 19.

Fig. 18. – Energy dependence of the cross-section for the process $e^+e^- \to \pi^+\pi^-\pi^+\pi^-$. Also data from other groups ([3b,11]) are reported for completeness. The points of the present experiment at $2E = 1.2$ and 1.5 GeV include a systematic error indicated by an open box, as in the case of the « boson group » ([11]) results: ■ present experiment, B boson group ([11]), ▲ ACO-Orsay ([3b]).

Figure 21. (top left) ρ' production (Ref. 43)

Figure 22. (top right) ρ' production (Ref. 45).

Figure 23. (bottom) Summary of ρ' production (Ref. 46)

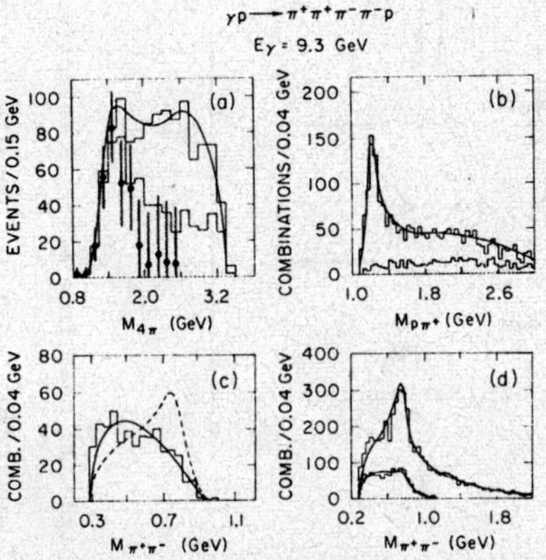

Fig. 1. Reaction $\gamma p \to 2\pi^+2\pi^- p$. (a) Four pion mass spectrum. The shaded histogram has events with Δ^{++} removed ($M_{p\pi^+} > 1.32$ GeV). The points are Π (see eq. (2)) corrected for efficiency. The curve is from a maximum-likelihood fit to the channel. (b) π^+p mass distribution. The shaded events are for $M_{4\pi} < 1.7$ GeV. The curve is from the maximum-likelihood fit. (c) $\pi^+\pi^-$ mass distribution ($M_{4\pi} < 1.7$ GeV) for $\pi^+\pi^-$ pairs opposite a ρ^0. The dotted (solid) curve shows the distribution expected for $\rho^0\rho^0(\rho^0\pi^+\pi^-$ + phase space). (d) $\pi^+\pi^-$ mass distribution. The shaded events are for $M_{4\pi} < 1.7$ GeV. The curve is from the maximum-likelihood fit.

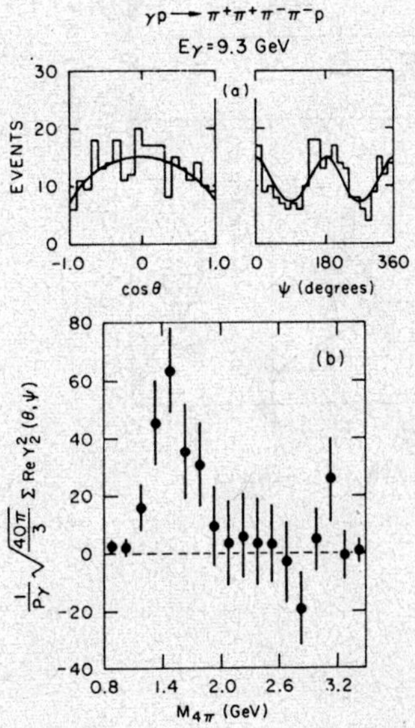

Fig. 2. (a) Distribution of the angles θ and ψ for $M_{4\pi} < 1.7$ GeV. The curve is from the maximum-likelihood fit. (b) Π uncorrected for analyzer efficiency.

Figure 24. ρ' production using a 9.3 GeV polarized photon beam (Ref. 47).

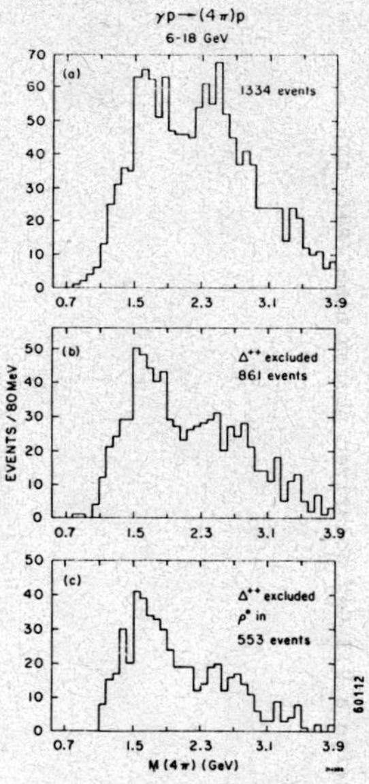

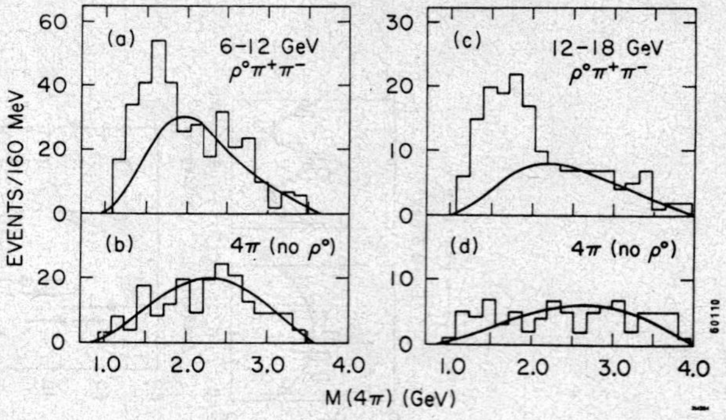

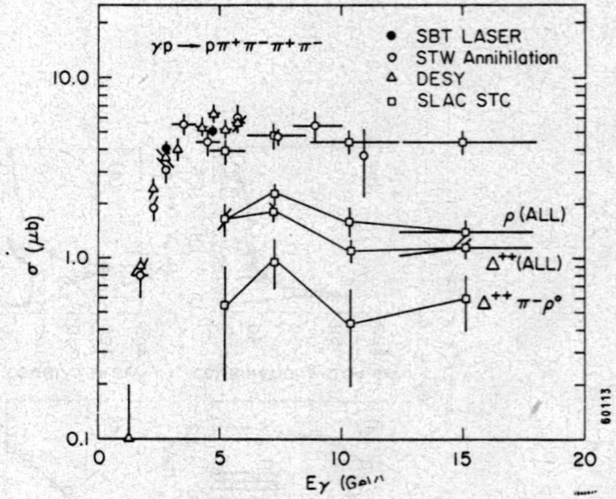

Figure 25. ρ' production using a bremsstrahlung beam from 4.5 → 18 GeV/c (Ref. 49).

Figure 26. (bottom) Variation of the ρ' production cross section with incident energy (Ref. 49).

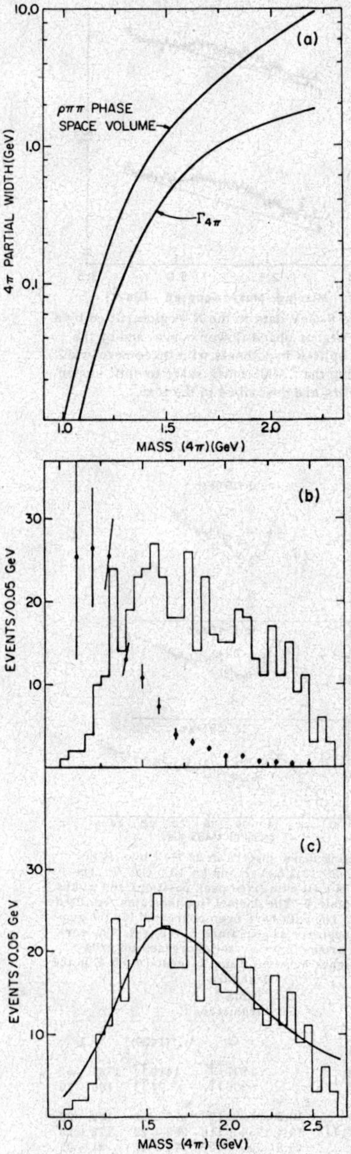

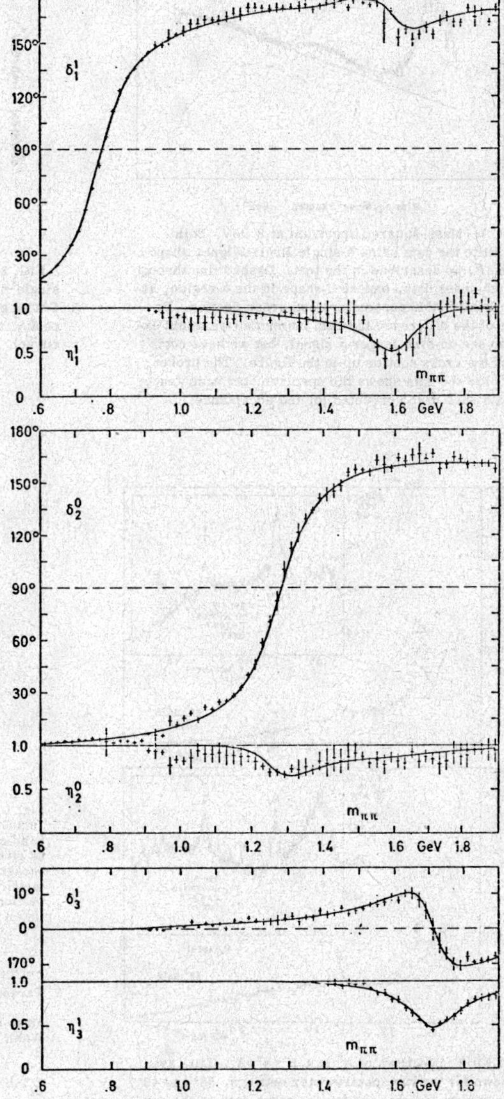

Figure 27. (left) The photoproduction data fitted to a model assuming the 4π decay is the tail of the ρ (Ref. 50).

Figure 28. (right) Evidence for the existence of the $\rho' \to \pi^+\pi^-$ from a phase shift analysis (Ref. 52).

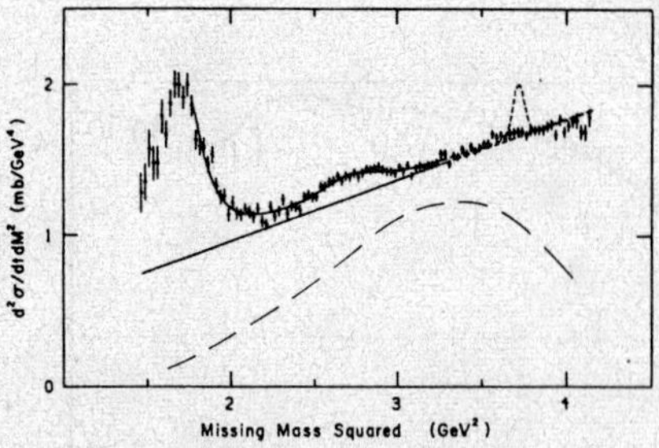

FIG. 1. Mass-squared spectrum at 8 GeV. Solid line, fit to the data using a single Breit-Wigner shape for the R, as described in the text. Dashed line through the high-mass data, expected shape in the S region, assuming the CMMS parameters and cross section. Because of the difference in beam momenta, we would expect to see an even larger S signal, but we have not scaled the cross section up in the figure. The broken line below the data shows the spectrometer acceptance; the data have been corrected for the acceptance.

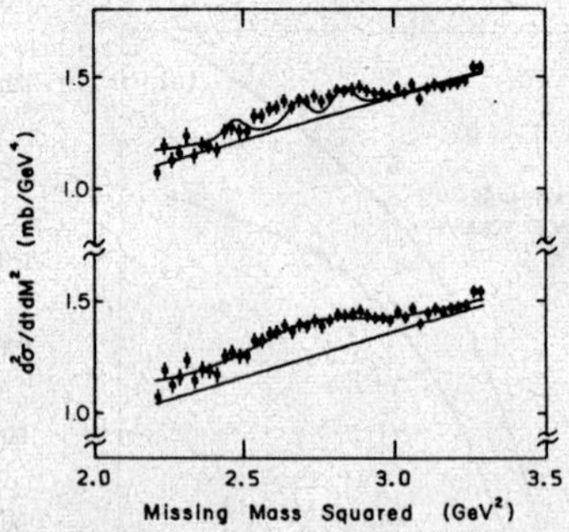

FIG. 2. The 8-GeV data in the R region, fitted by a single Breit-Wigner shape (lower curve) and by the CMMS group split-R hypothesis with three zero-width peaks, allowing the CMMS mass scale to shift (upper curve). The fits are described in the text.

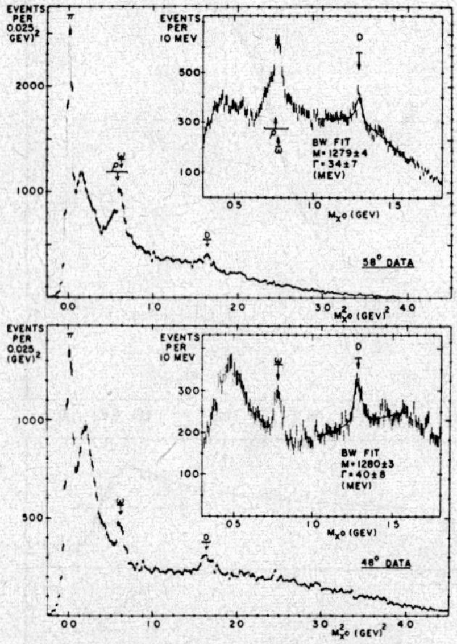

FIG. 2. Spectrum of X^0 in $\pi^- p \to \pi^- p X^0$. Data are shown for the two spectrometer settings, 58° and 48°. Mass plots (insets) are for the same data as the corresponding plots of X^0 mass squared. The fits of the peak at 1280 MeV are described in the text. Both fits have χ^2 probabilities larger than 40%. Bars above the arrows pointing to peaks in the spectra represent intrinsic widths folded with experimental resolution.

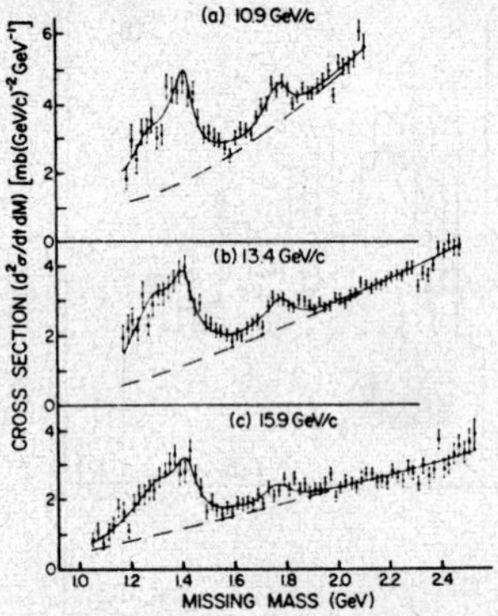

Fig. 1. Missing mass spectra in 20 MeV bins at (a) 10.9 GeV/c, (b) 13.4 GeV/c and (c) 15.9 GeV/c. The solid curve is a fit with fixed peak positions and widths as given in table 2. The dashed line indicates the fitted background. The data have been corrected for the geometrical acceptance as explained in the text. The correction is 1 around 2.0 GeV and increases towards lower and higher masses, being largest (about 4) in the Q-region

Table 2
Fit parameters †

Parameter		Q	"K*(1420)"	L
Mass (MeV)		1297††	1410††	1767 ± 6
Width (MeV)		226††	77††	100 ± 26
$\frac{d}{dt}$ (μb/GeV2)†††	10.9	615 ± 128	272 ± 58	162 ± 29
	13.4	754 ± 110	203 ± 29	122 ± 23
	15.9	505 ± 122	174 ± 41	81 ± 23
Slope B (GeV^{-2})		11.5 ± 1.9	6.4 ± 1.9	9.3 ± 2.1

† All errors are statistical only.
†† These parameters were held fixed in the fits.
††† Average differential cross section in the interval $0.12 < |t| < 0.40$ GeV^2.

Figure 29 a) (upper left) non observation of the S (Ref. 55)

b) (upper right) Evidence against the split R (Ref. 55)

c) (left) Evidence for the D (Ref. 57)

d) (right) $K^- p \to pX^-$ showing Q, K^*(1420), L production (Ref. 58).

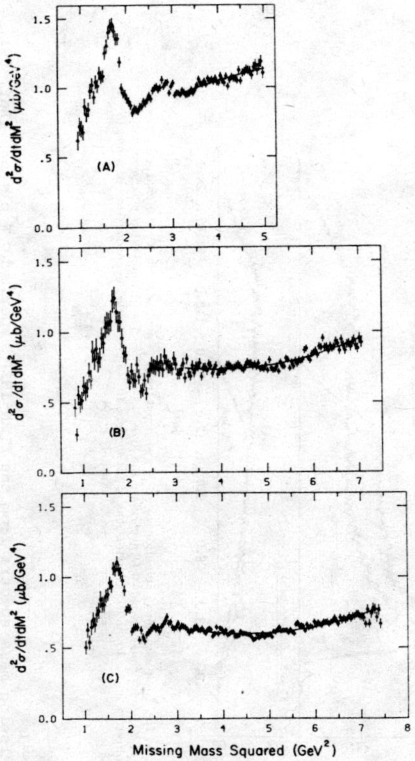

FIG. 1. (a)–(c) Missing-mass spectrum for $\pi^- + p \rightarrow (MM)^- + p$ at 11, 13.4, and 16 GeV, respectively, for $0.2 \leq |t| \leq 0.3$ GeV^2. The quadratic background fits are shown. For details, see text.

TABLE II. S, T, and U cross sections.

Beam momentum (GeV)	S	$\langle d\sigma/dt \rangle^a$ (μb/GeV^2) T	U	Assumed width	Mass interval (GeV^2)	χ^2	Degrees of freedom
11	3.2 ± 2.1	4.3 ± 3.8		0	3.2–5.0	43	40
13.4	2.2 ± 2.7	1.2 ± 2.1	−2.6 ± 2.0	0	3.2–6.1	86	66
16	1.1 ± 2.2	−4.2 ± 1.8	4.9 ± 2.1	0	3.2–6.1	68	66
11	11.6 ± 4.5	5.2 ± 5.9		b	3.2–5.0	38	40
13.4	3.6 ± 5.4	1.5 ± 3.0	−6.4 ± 4.5	b	3.2–6.1	86	66
16	2.9 ± 3.8	−6.3 ± 2.4	7.0 ± 3.9	b	3.2–6.1	68	66
12	8.8 ± 2.5[c]	7.3 ± 2.5[c]	10.5 ± 2.5[c]				

[a] For t interval $0.2 \leq |t| \leq 0.3$ GeV^2. Errors shown are statistical only. For a discussion of possible systematic errors, see footnote 10.

[b] $\Gamma_S = 35$, $\Gamma_T = 13$, $\Gamma_U = 30$ MeV.

[c] CMMS results. For the CMMS experiment, for the S and T, $0.22 \leq |t| \leq 0.36$ GeV^2; and for the U, $0.28 \leq |t| \leq 0.36$ GeV^2. These are the renormalized cross sections. See Ref. 3. The normalization factor of ~4 is only given approximately in Ref. 3, and no errors are attached. We have used a factor of 4.0 here.

TABLE III. Peak signal-to-background ratios.

	Beam momentum (GeV)	S	T	U
NU/SUNY[a]	11	0.057 ± 0.022	0.036 ± 0.040	...
	13.4	0.026 ± 0.039	0.015 ± 0.030	−0.046 ± 0.032
	16	0.020 ± 0.026	−0.050 ± 0.020	0.048 ± 0.029
	Average	0.039 ± 0.015	−0.020 ± 0.015	0.010 ± 0.021
CMMS[b]	12	0.08 ± 0.02	0.08 ± 0.02	0.16 ± 0.03

[a] Ratios are for maximum-width cross-section estimates ($\Gamma_S = 35$, $\Gamma_T = 13$, $\Gamma_U = 30$ MeV).

[b] We have assigned an error based on the cross section accuracy. The errors in Ref. 1 may include normalization errors as well as statistical accuracy, and thus overestimate the errors in the ratios. These ratios are adjusted to an all-charged background. See text.

Figure 30. Missing mass spectra showing the non observation of any states above M = 1.8 GeV. In particular the S, T and U are not observed (Ref. 56)

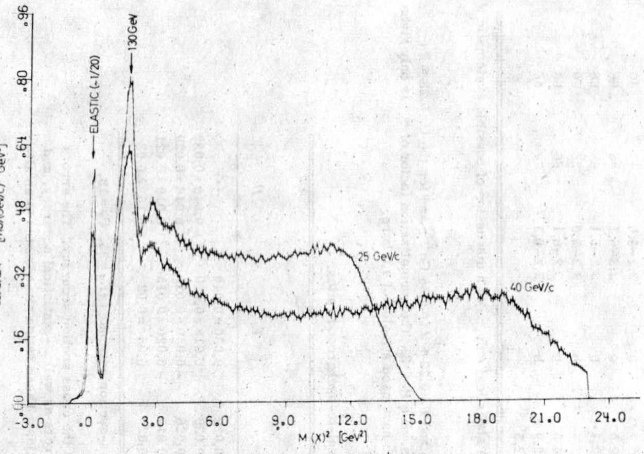

Fig. 2. X^- mass spectrum at 25 GeV/c and 40 GeV/c ($0.17 < |t| < 0.35$ GeV/c^2), all decay modes. The elastic peak has been scaled down by an arbitrary factor. Bin size: 0.10 GeV^2. In all figures "dt" means the Δt interval indicated.

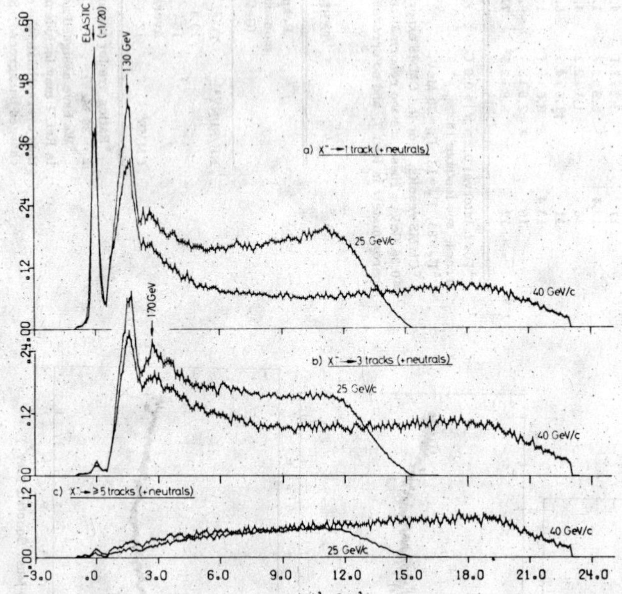

Fig. 3. Same as fig. 2 but with decay selection: a) $X^- \to$ 1 charged track (+ neutrals); b) $X^- \to$ 3 charged tracks (+ neutrals); c) $X^- \to \geq 5$ charged tracks (+ neutrals). Bin size: 0.10 GeV^2. The little elastic peak seen in the higher multiplicities is due to small imperfections of the system as, for example, secondary interactions in the target, etc.

Figure 31. Missing mass spectra from the CERN-IHEP experiment $\pi^- p \to pX^-$ (Ref. 59).

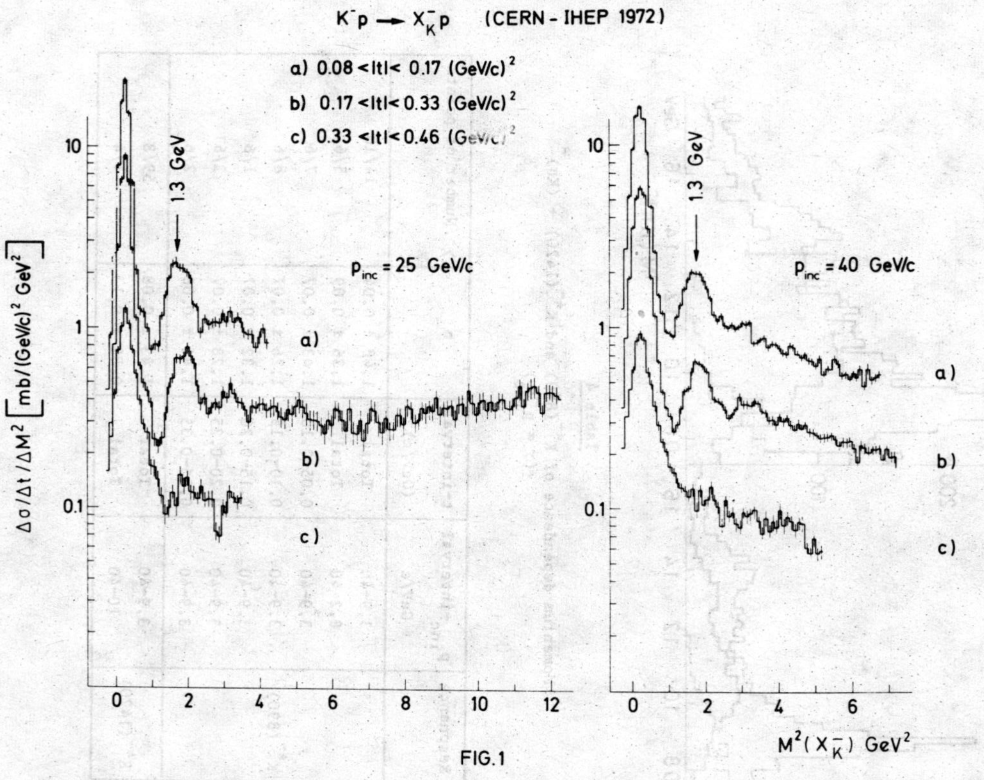

Figure 32a Missing mass spectra for the reaction $K^-p \to pX^-$ (Ref. 37).

FIG. 2 $K^-p \to (K\pi)^-p$ at 25 and 40 GeV/c (CERN-IHEP 1972)

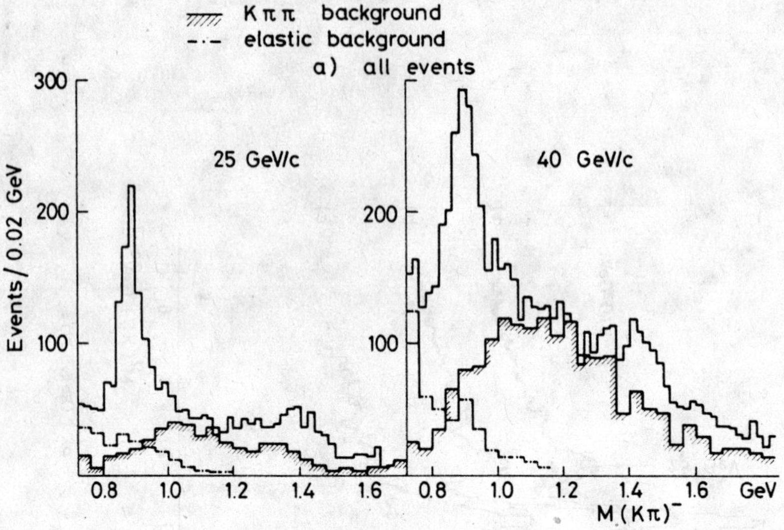

Table 4

Momentum dependence of $K^{*-}(890)$ and $K^{*-}(1420) \to (K\pi)^-$
($\sigma \propto p_{inc}^{-n}$)

Resonance	p_{inc}-interval GeV/c	t-interval $(GeV/c)^2$	n	χ^2/Number of points
$K^{*-}(890)$	3.9-40	Total	1.48 ± 0.04	14/10
	8.2-40	Total	1.36 ± 0.09	5/6
	3.9-40	0.05-0.10	1.43 ± 0.07	7/6
	3.9-40	0.10-0.15	1.26 ± 0.07	8/6
	3.9-40	0.15-0.20	1.32 ± 0.07	1/6
	3.9-40	0.20-0.25	1.28 ± 0.07	3/6
	3.9-40	0.25-0.35	1.46 ± 0.06	2/6
$K^{*-}(1420)$	3.9-40	Total	1.43 ± 0.08	59/8
	10-40	Total	0.82 ± 0.17	1/4

Figure 32b Results on the reaction $K^-p \to p\,(K\pi)^-$ (Ref. 37).

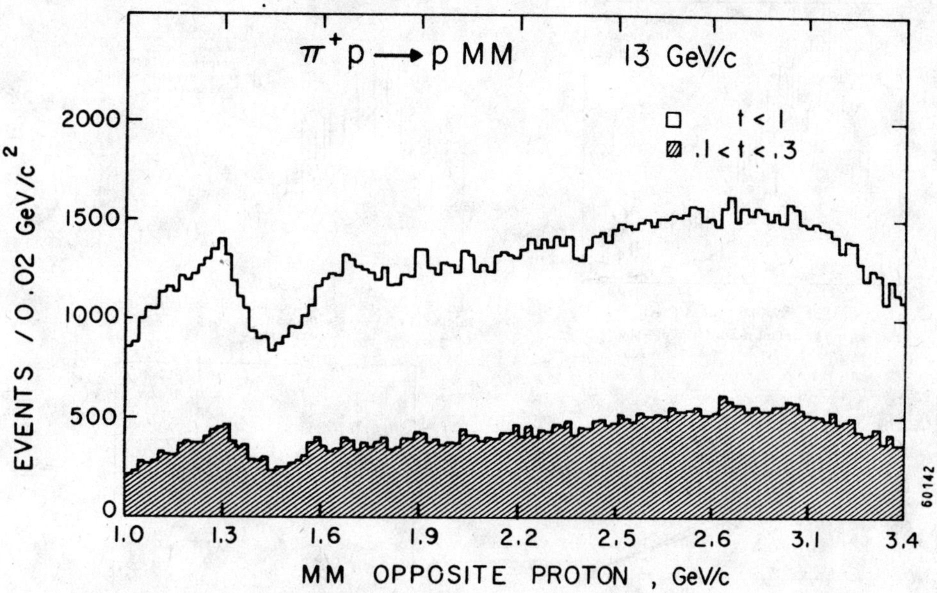

E^*	P	E^*	P
1.8791	.1000	2.2893	1.6000
1.8869	.2000	2.3246	1.7000
1.8997	.3000	2.3599	1.8000
1.9168	.4000	2.3951	1.9000
1.9379	.5000	2.4301	2.0000
1.9622	.6000	2.4650	2.1000
1.9892	.7000	2.4996	2.2000
2.0184	.8000	2.5340	2.3000
2.0494	.9000	2.5682	2.4000
2.0817	1.0000	2.6021	2.5000
2.1150	1.1000	2.6357	2.6000
2.1491	1.2000	2.6691	2.7000
2.1837	1.3000	2.7021	2.8000
2.2187	1.4000	2.7349	2.9000
2.2539	1.5000	2.7674	3.0000

Figure 33. Missing mass spectrum compiled from all the final states with a proton measured in a π^+p experiment at 13 GeV/c (Ref. 61).

Figure 34. Variation of centre of mass energy with momentum for $\bar{p}p$ interactions

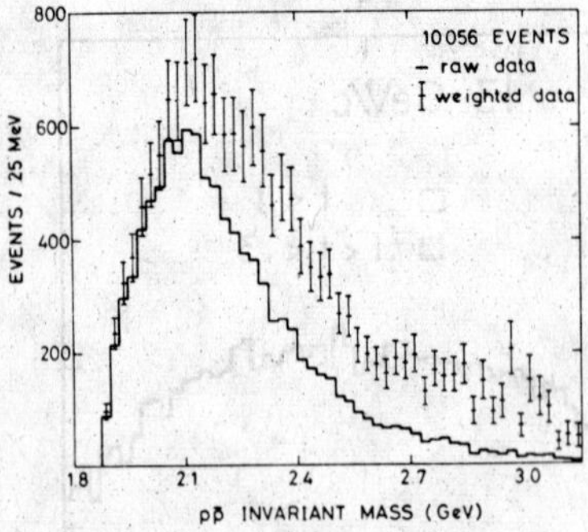

Fig. 2b. pp̄ invariant mass for $\pi^- p \rightarrow p\bar{p}n$.

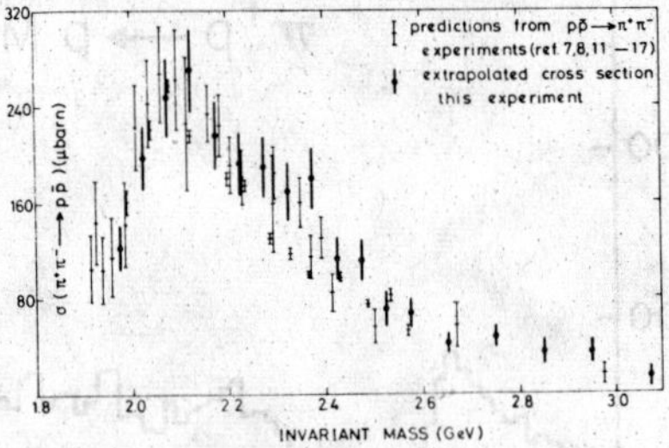

Fig. 4. Comparison of extrapolated on-shell cross-section of $\pi^+\pi^- \rightarrow p\bar{p}$ with prediction of $p\bar{p} \rightarrow \pi^+\pi^-$ experiments.

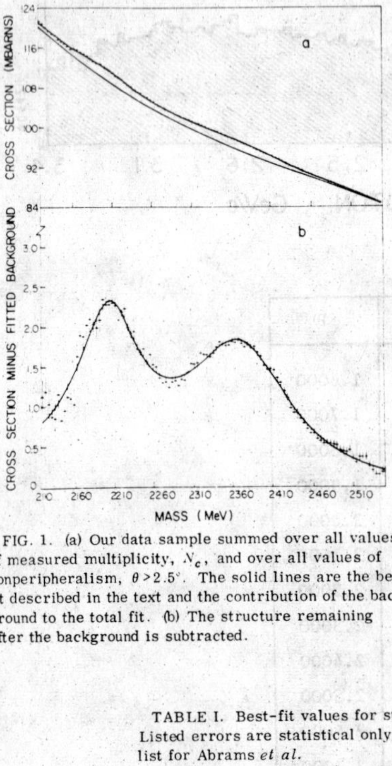

FIG. 1. (a) Our data sample summed over all values of measured multiplicity, N_c, and over all values of nonperipheralism, $\theta > 2.5°$. The solid lines are the best fit described in the text and the contribution of the background to the total fit. (b) The structure remaining after the background is subtracted.

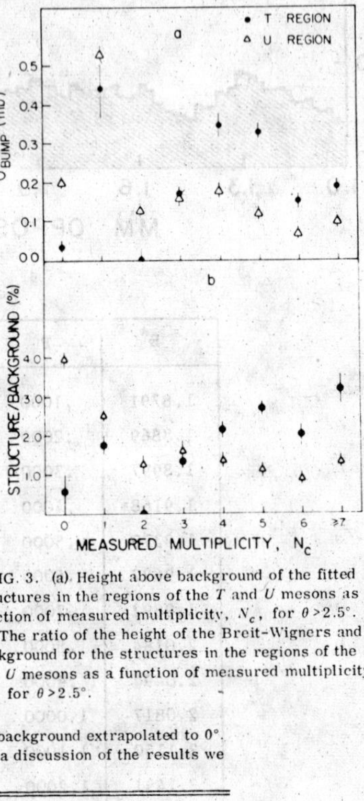

FIG. 3. (a) Height above background of the fitted structures in the regions of the T and U mesons as a function of measured multiplicity, N_c, for $\theta > 2.5°$. (b) The ratio of the height of the Breit-Wigners and the background for the structures in the regions of the T and U mesons as a function of measured multiplicity, N_c, for $\theta > 2.5°$.

TABLE I. Best-fit values for structures above background extrapolated to 0°. Listed errors are statistical only. See Ref. 9 for a discussion of the results we list for Abrams *et al.*

	T region			*U* region		
	Mass (MeV)	Width (MeV)	Height (mb)	Mass (MeV)	Width (MeV)	Height (mb)
Rutgers	$2192.7^{+1.2}_{-1.5}$	$97.6^{+8.1}_{-5.9}$	$2.32^{+0.13}_{-0.08}$	$2359.4^{+1.4}_{-1.2}$	$164.9^{+17.9}_{-7.6}$	$2.06^{+0.20}_{-0.12}$
Abrams *et al.* (our fit)	2187 ± 3	56 ± 8	1.85 ± 0.25	2363 ± 2	171 ± 10	2.52 ± 0.28

Figure 35. (top) The mass spectrum of the $p\bar{p}$ system from the reaction $\pi^- p \rightarrow p\bar{p}n$ and a comparison to the reaction $p\bar{p} \rightarrow \pi^+\pi^-$ (Ref. 70).

Figure 36. Data from a high statistics counter experiment confirming broad structures in the $p\bar{p}$ total cross section (Ref. 71).

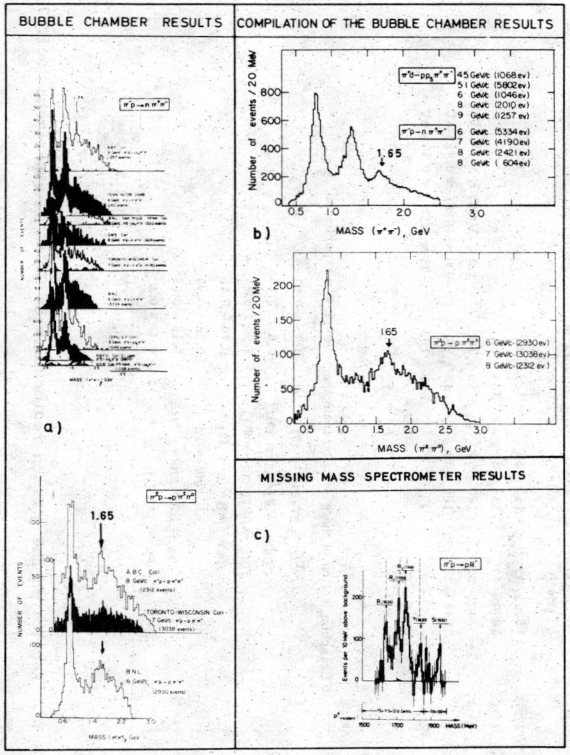

Fig. 2. (a) Effective-mass distributions of the $(2\pi)^o$ and $(2\pi)^\pm$ systems at various energies, as compiled by French [14]. (b) Compilations of the $(2\pi)^o$ and of the $(2\pi)^\pm$ mass distributions of fig. 2a. (c) Multi-pion mass distribution obtained with the missing-mass spectrometer at CERN [12].

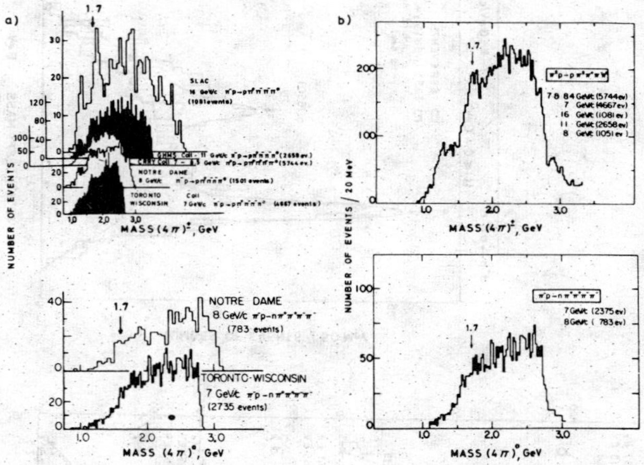

Fig. 3. (a) Effective mass distributions of the $(4\pi)^\pm$ and $(4\pi)^o$ systems at various energies as compiled by French [14]. (b) Compilations of the $(4\pi)^\pm$ and $(4\pi)^o$ mass distributions of fig. 3a.

Table 1
Mass and width of the mesons in the R-region as deduced from bubble chamber experiments.

	M_o(MeV)	Γ(MeV)	I^G	J^P
$g^\pm \to \pi^o\pi^\pm$	1640 ± 20	120 ± 30	1^+	3^- ?
$g^o \to \pi^+\pi^-$	1680 ± 15	200 ± 50	$?^+$	-
$A_3^\pm \to (3\pi)^\pm$	1648 ± 10	100 ± 20	1^-	
$A_3^o \to (3\pi)^o$	1651 ± 10	130 ± 20	0^- ?	
$\rho^\pm \to (4\pi)^\pm$	1700 ± 15	120 ± 30	1^+	
$R_\omega \to \omega\pi^+\pi^-$	1670 ± 15	50 ± 15	$?^-$	-
$\pi(1840)$	≈ 1840	~	$?^-$	-
$\rho(1840)$	≈ 1840	~	$?^+$	-
$\rho(1630)^- \to \omega\pi^-$	≈ 1630	~ 60	$?^-$	-

Figure 37. Results from a compilation of results on the g meson region (Ref. 73).

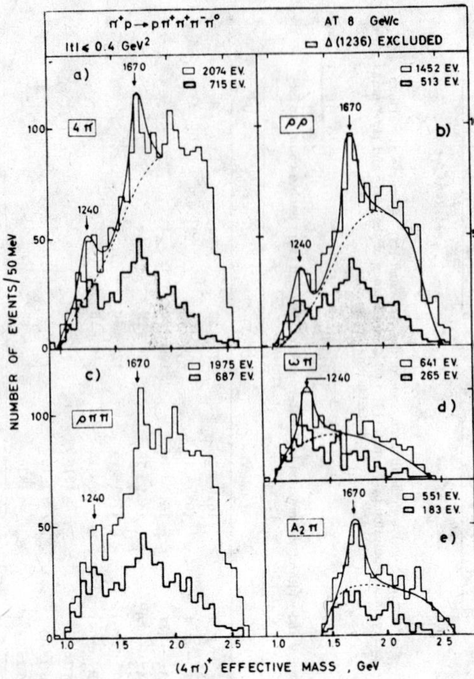

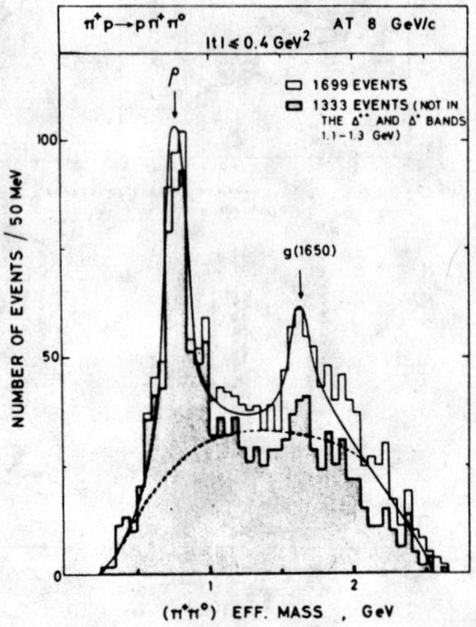

Table 3
2π and 4π enhancements in the g-meson mass region from the 8 GeV/c π^+p experiment.

	$M_o \pm \Delta M_o$ (MeV)	$\Gamma \pm \Delta\Gamma$ (MeV)	Number of events	Conditions
$(2\pi)^+$	1650 ± 35	180 ± 30	122 ± 20	
$(4\pi)^+$	1680 ± 40	135 ± 30	144 ± 20	
$(A_2\pi)^+$	1640 ± 20	180 ± 30	90 ± 15	$1.2 \leq M(3\pi) \leq 1.4$ GeV
$(\rho\rho)^+$	1689 ± 20	160 ± 30	102 ± 20	$0.6 \leq M(2\pi) \leq 0.82$ GeV
$\omega\pi^+$	-	-	≤ 5 ± 10	$0.65 \leq M(3\pi) \leq 0.85$ GeV

Table 4
Branching ratios of the g^+-meson from the 8 GeV/c π^+p experiment.

$\sigma(g \to 2\pi)/\sigma(g \to 4\pi)$	0.8 ± 0.15
$\sigma(g \to A_2\pi)/\sigma(g \to 4\pi)$	0.6 ± 0.15
$\sigma(g \to \rho 2\pi)/\sigma(g \to 4\pi)$	1.0 ± 0.15
$\sigma(g \to \rho\rho)/\sigma(g \to 4\pi)$	0.70 ± 0.15
$\sigma(g \to K\bar{K})/\sigma(g \to 2\pi)$	0.08 ± 0.03
$\sigma(g \to K\bar{K}\pi)/\sigma(g \to 2\pi)$	0.10 ± 0.03
$\sigma(g \to 2\pi)/\sigma_{all}$	0.4 ± 0.1

Figure 38. Results on the g meson region from an 8 GeV π^+ experiment (Ref. 73).

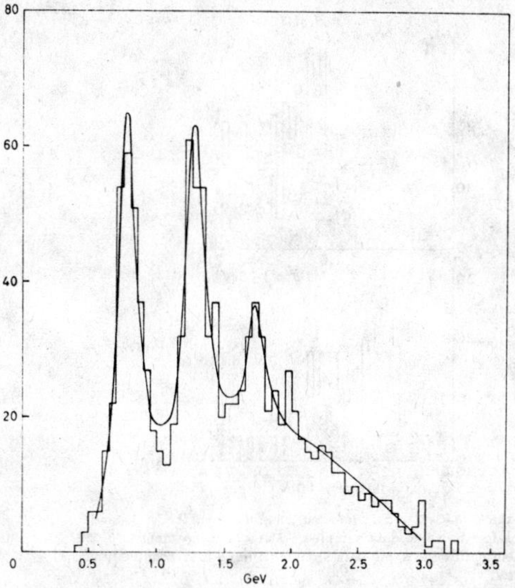

Fig. 2. Dipion mass spectrum with $t' < 0.04$ (GeV)², $\pi^+ n \rightarrow p\pi^+\pi^-$ reaction, 1109 events.

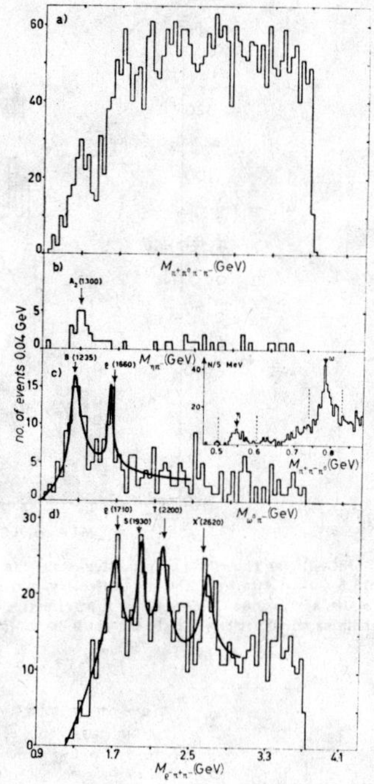

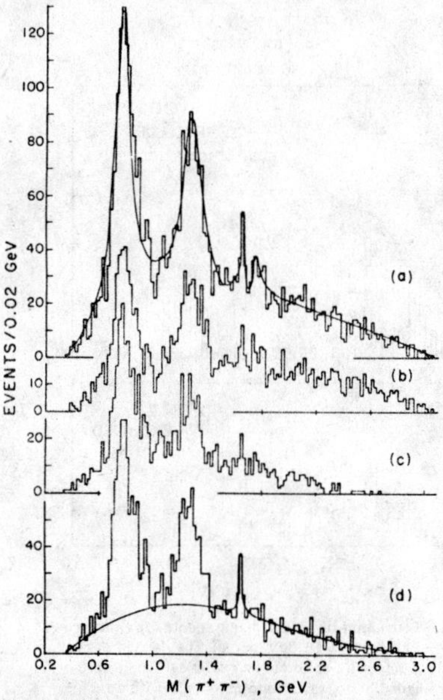

Fig. 1. Effective mass distributions for a) events of reaction (1) and reaction (2) combined, b) events of reaction (1) only, c) events of reaction (2) only, and d) events of the combined sample for which $|t - t_{min}| \leq 0.15 (GeV/c)^2$. In a) the fit in the g° region is shown for two coherent Breit-Wigner resonances, with the background described in table 1. In d), the fit shown is for a single Breit-Wigner resonance plus incoherent background.

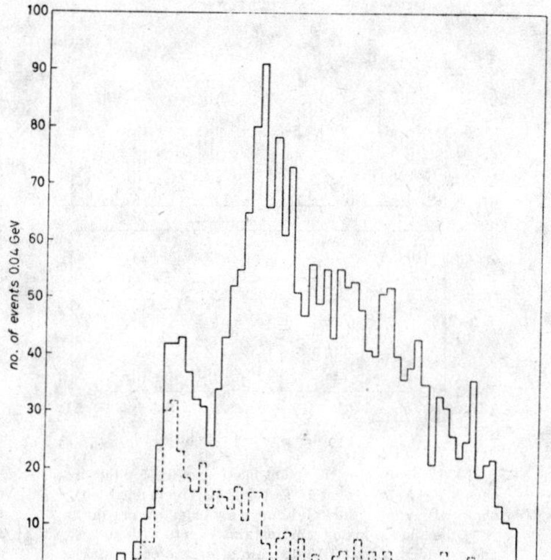

$M(\pi^+\pi^-\pi^-\pi^0)$ from reaction (1). Dashed line: $M(\omega\pi)$ (433 eV).

Figure 39. Recent results on the g meson region.

(a) (top left) $\pi^+ n \rightarrow p\pi^+\pi^-$ at 9 GeV/c (Ref. 75)

(b) (top right) $\pi^- p \rightarrow p4\pi$ at 11.2 GeV/c (Ref. 77).

(c) (bottom left) $\pi N \rightarrow N\pi\pi$ at 5.4 and 8 GeV/c (Ref. 79)

(d) (bottom right) $\pi^- p \rightarrow p4\pi$ at 9 GeV/c (Ref. 83).

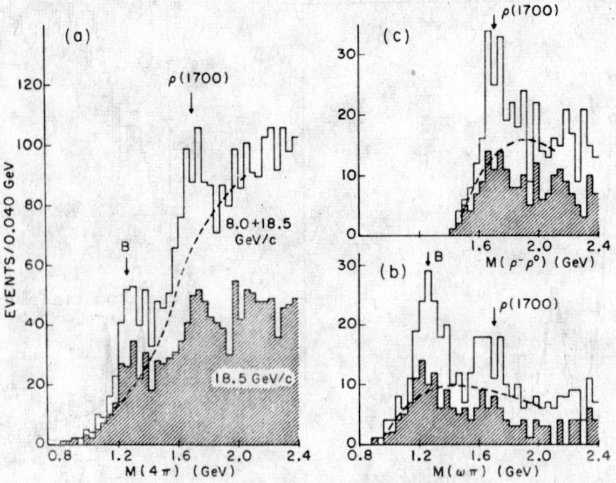

FIG. 1. (a) The $\pi^+\pi^-\pi^-\pi^0$ effective-mass distribution for the 18.5-GeV/c data (shaded), and the combined 8.0- and 18.5-GeV/c data. (b) The $\omega^0\pi^-$ effective-mass distributions for the same data samples. The ω^0 requirement is $0.720 \leq M(\pi^+\pi^-\pi^0) \leq 0.84$ GeV. (c) The $\rho^-\rho^0$ effective-mass distribution for the same data samples. Here it is required that both $\pi\pi$ combinations satisfy the cut $0.660 \leq M(\pi\pi) \leq 0.860$ GeV.

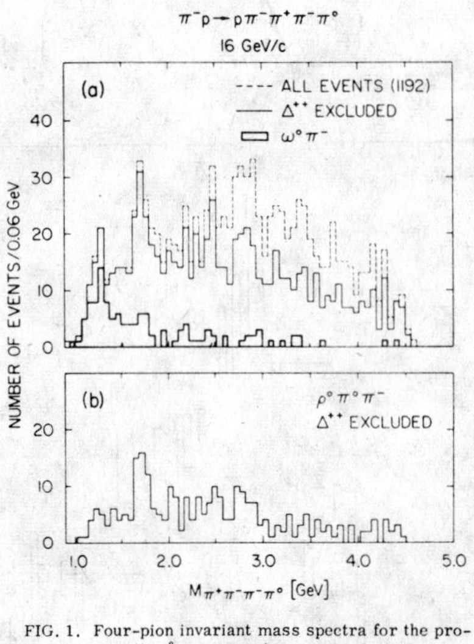

FIG. 1. Four-pion invariant mass spectra for the process $\pi^-p \to p\pi^-\pi^+\pi^-\pi^0$ at 16 GeV/c. (a) The dashed curve shows all events; the solid curve excludes events in the Δ^{++}; the shaded histogram contains events with an ω^0, and excludes Δ^{++}. (b) Histogram of events containing an associated ρ^0.

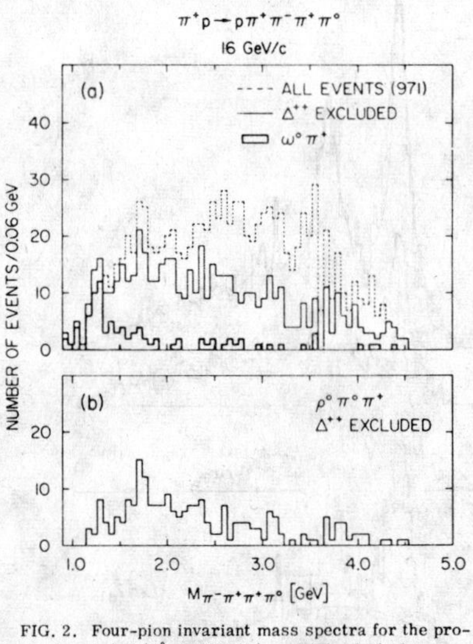

FIG. 2. Four-pion invariant mass spectra for the process $\pi^+p \to p\pi^+\pi^-\pi^+\pi^0$ at 16 GeV/c. (a) The dashed curve shows all events; the solid curve excludes events in the Δ^{++}; the shaded histogram contains events with an ω^0, and excludes Δ^{++}. (b) Histogram of events containing an associated ρ^0.

Figure 39 (e) (top) $\pi^- p \to p4\pi$ at 8 and 18.5 GeV (Ref. 85)

(f) (bottom) $\pi^{\pm} p \to p4\pi$ at 16 GeV/c (Ref. 80).

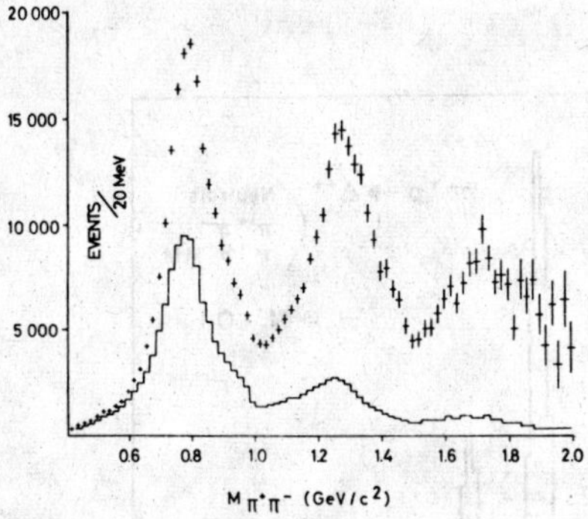

Fig. 1. Dipion mass distribution of observed events with $|t| < 0.2$ (GeV/c)^2 (lower curve), and corrected for acceptance losses (upper curve) using a fit with $l_{max} = 6$, $m_{max} = 1$.

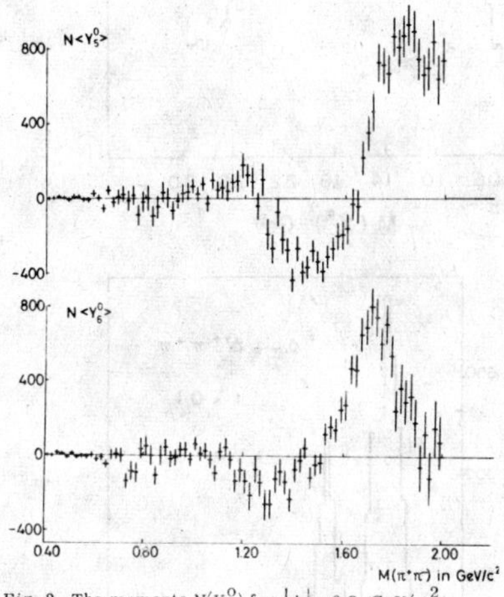

Fig. 2. The moments $N\langle Y_l^0\rangle$ for $|t| < 0.2$ (GeV/c)^2 ($l_{max} = 6$, $m_{max} = 1$).

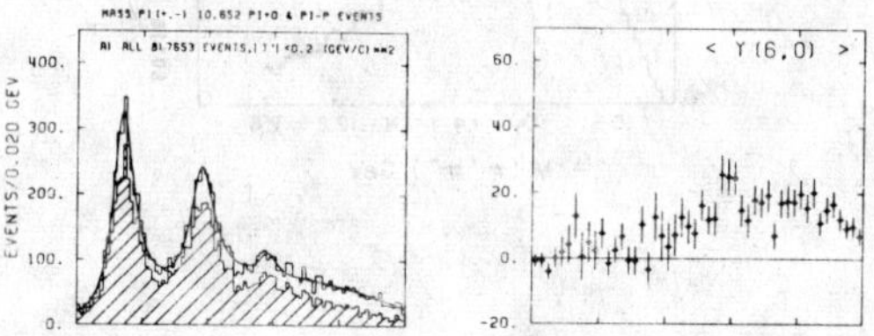

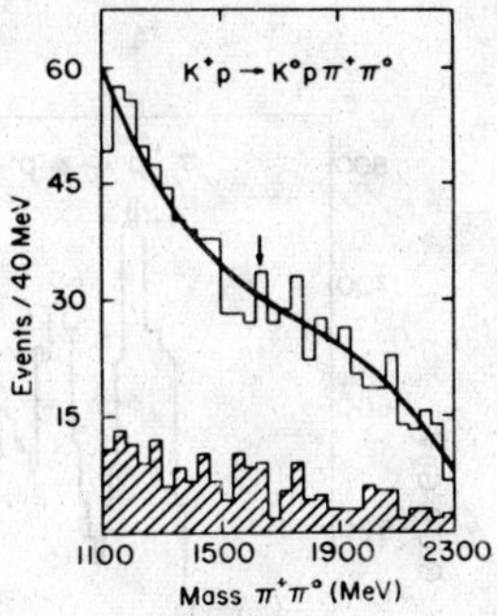

FIG. 2. Combined $\pi^+\pi^0$ mass spectrum; cross-hatched events are from the 12.7-GeV/c experiment alone. The curve represents a polynomial fit to the combined data.

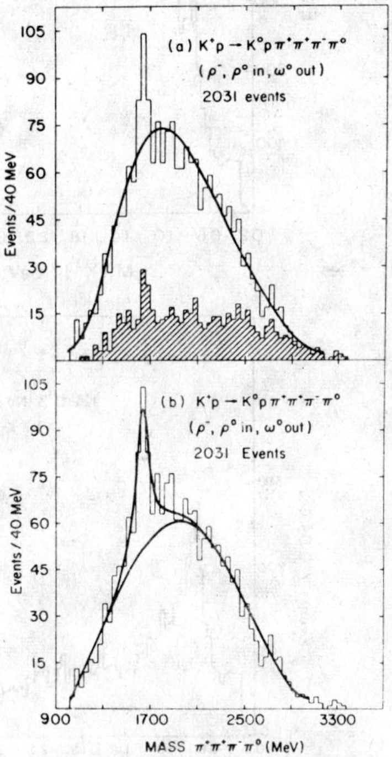

FIG. 1. (a) Combined $\pi^+\pi^+\pi^-\pi^0$ mass spectrum; cross-hatched events are from the 12.7-GeV/c experiment alone. Equivalent $\rho^{-,0}$ selections and ω^0 antiselection have been applied to both data samples (see footnote 8). The curve represents a polynomial fit to the combined data. (b) Combined $\pi^+\pi^+\pi^-\pi^0$ mass spectrum. The curves represent a fit of the data to a Breit-Wigner resonance shape and a polynomial representation of the background.

Figure 40 (a) (left) Results showing the 3^- behaviour of the g (Ref. 81, 82).

(b) (right) Results on the g from K^+p interactions showing a small branching ratio for the 2π mode (Ref. 84).

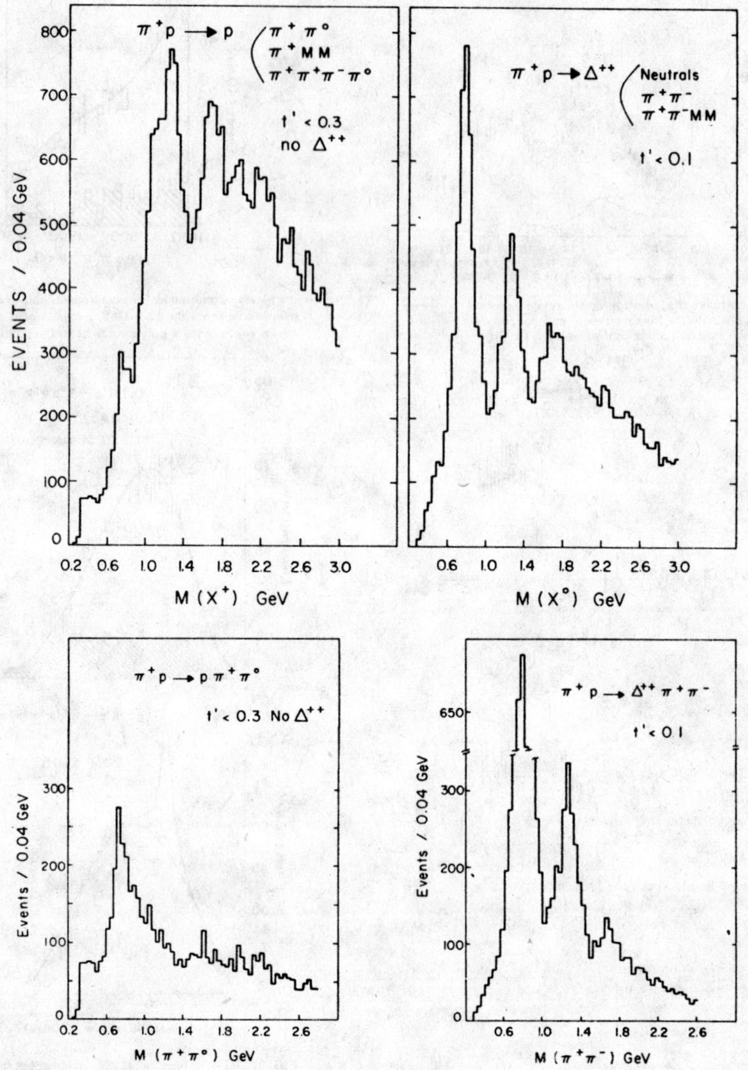

Figure 41a) Preliminary data on the g-meson from a 13 GeV/c π^+p experiment (ref. 86).

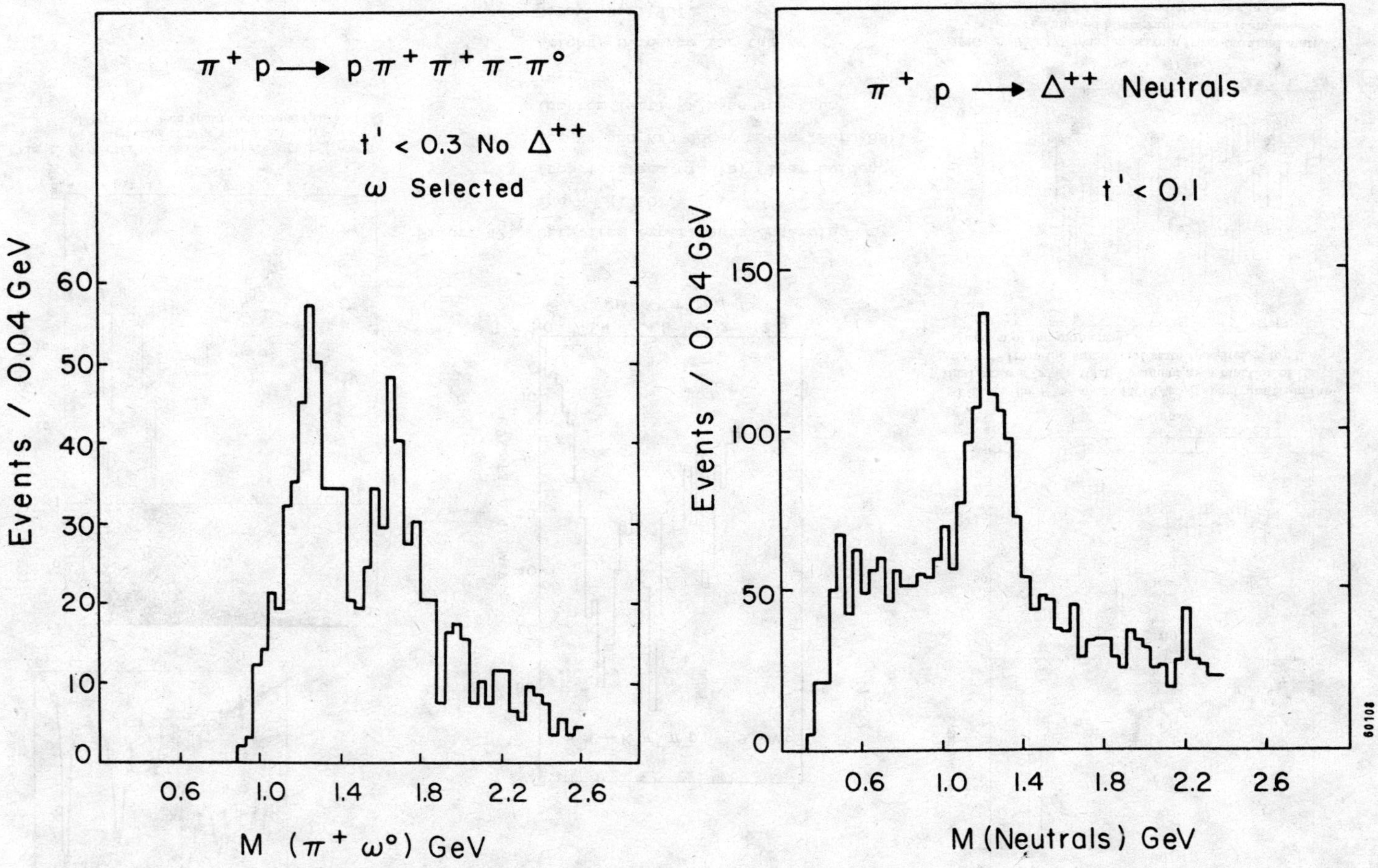

Figure 41b) Data showing a clear $\omega\pi$ and no $\pi^o\pi^o$ decay of the g (ref. 86).

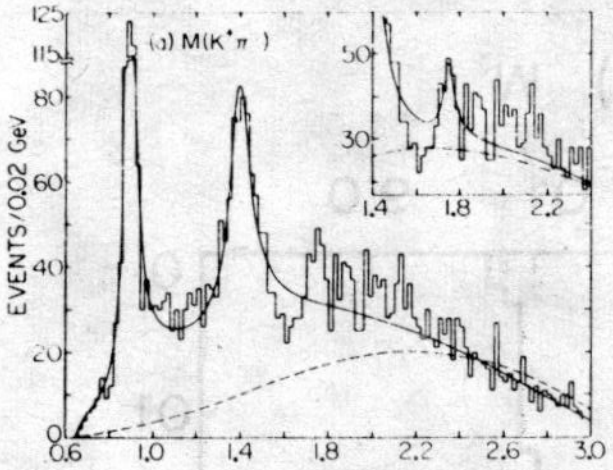

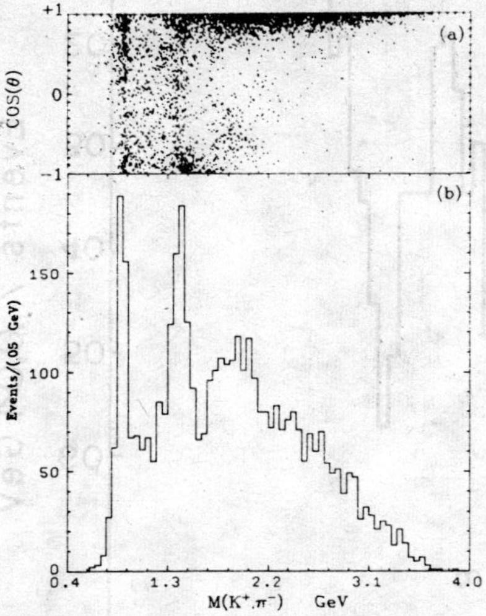

Fig. 1. (a) Scatter plot of cos θ versus $M(K^+\pi^-)$; (b) distribution of $M(K^+\pi^-)$. A cut at $|t'| < 0.2$ $(GeV/c)^2$ has been applied to the data shown here.

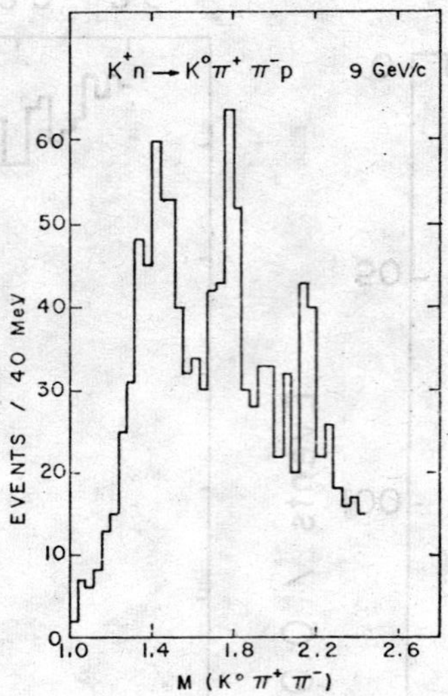

Figure 42. Data from experiments observing the K^* (1760).

(top left and middle) Clear evidence for Kπ and Kππ decay modes (Ref. 88).

(bottom left) Kπ mass spectrum (Ref. 89).

(right) Evidence for the Kπ decay (Ref. 90).

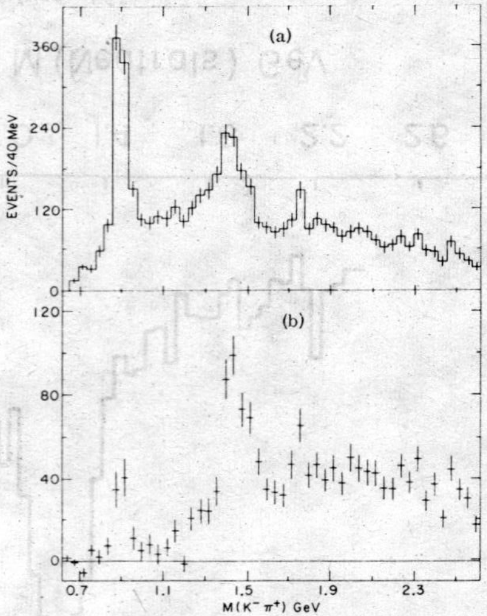

FIG. 1. (a) $K^-\pi^+$ mass $[M(K^-\pi^+)]$ spectrum from the final state $K^-\pi^+n$. (b) Y_2^0 moment as a function of $M(K^-\pi^+)$ from the same final state (see Ref. 8 for the definition of the moments).

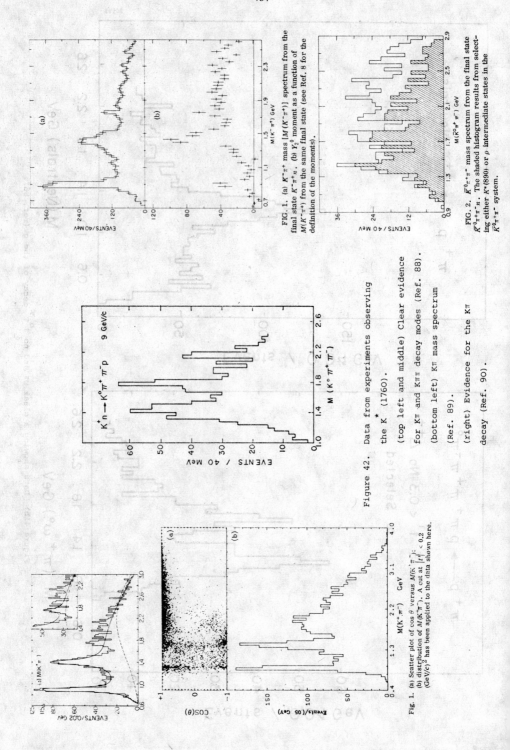

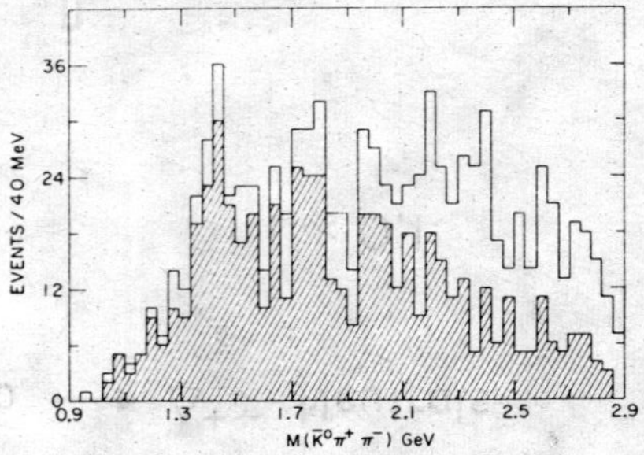

FIG. 2. $\bar{K}^0\pi^+\pi^-$ mass spectrum from the final state $\bar{K}^0\pi^+\pi^-n$. The shaded histogram results from selecting either $K^*(890)$ or ρ intermediate states in the $\bar{K}^0\pi^+\pi^-$ system.

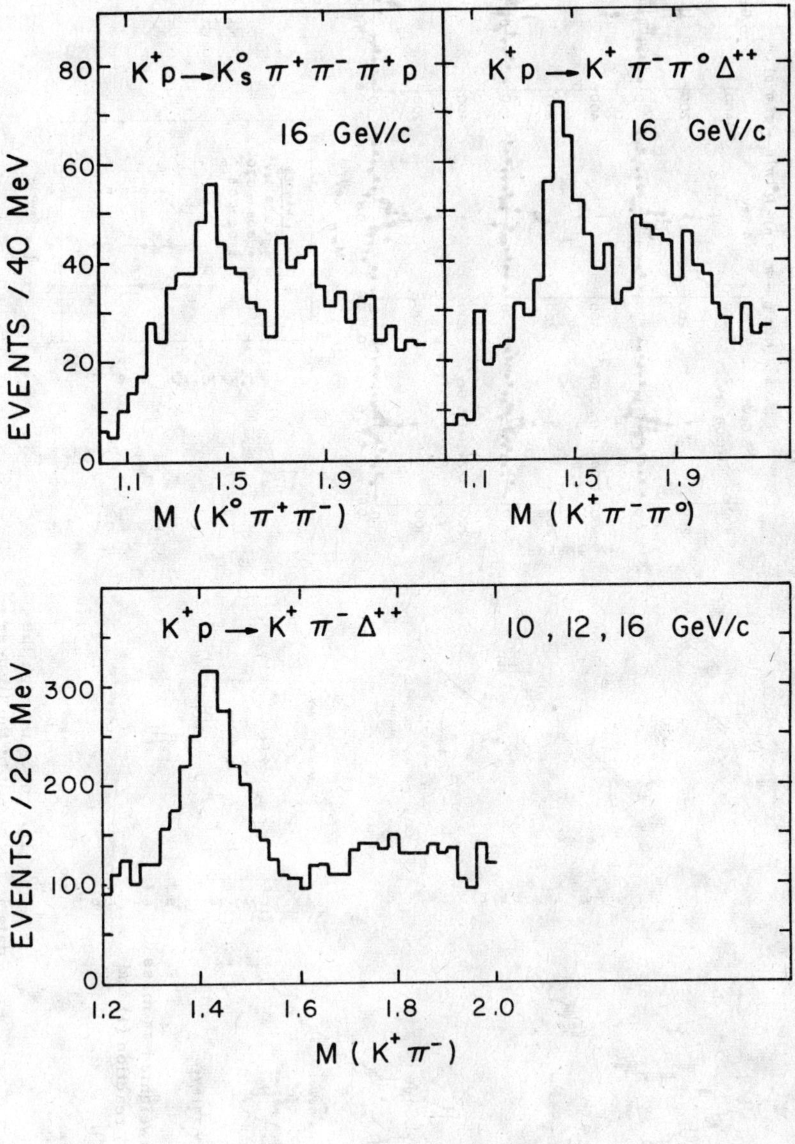

Figure 43. Preliminary evidence for K^*(1760) production in the reaction $K^+p \to \Delta^{++}K^*$(1760) (Ref. 91).

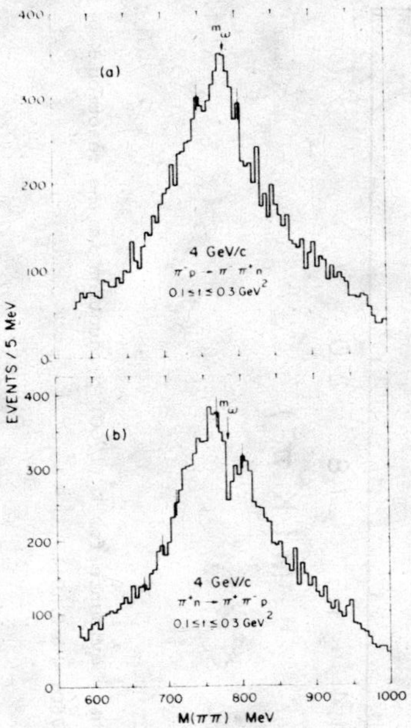

Fig. 1 The unweighted ππ mass spectra for (a) reaction (1) and (b) reaction (2).

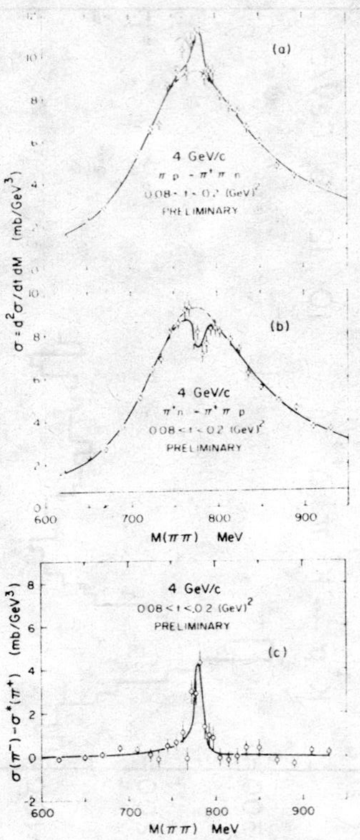

The curve through the data is the result of the fit described in the text, the dashed curve is the fitted ρ shape and the thin line the fitted background.

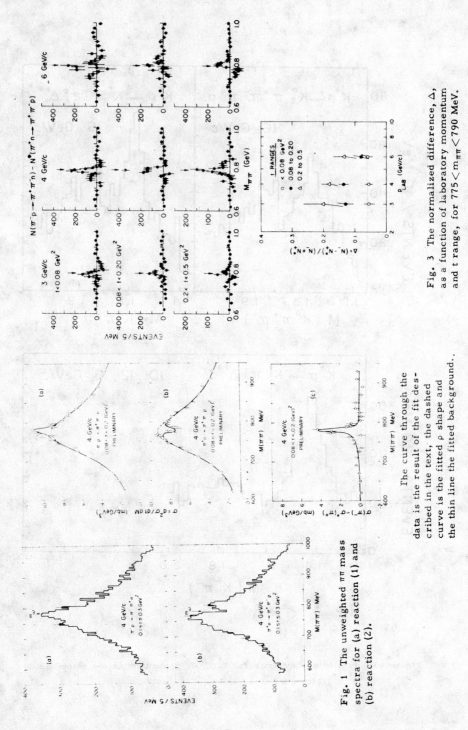

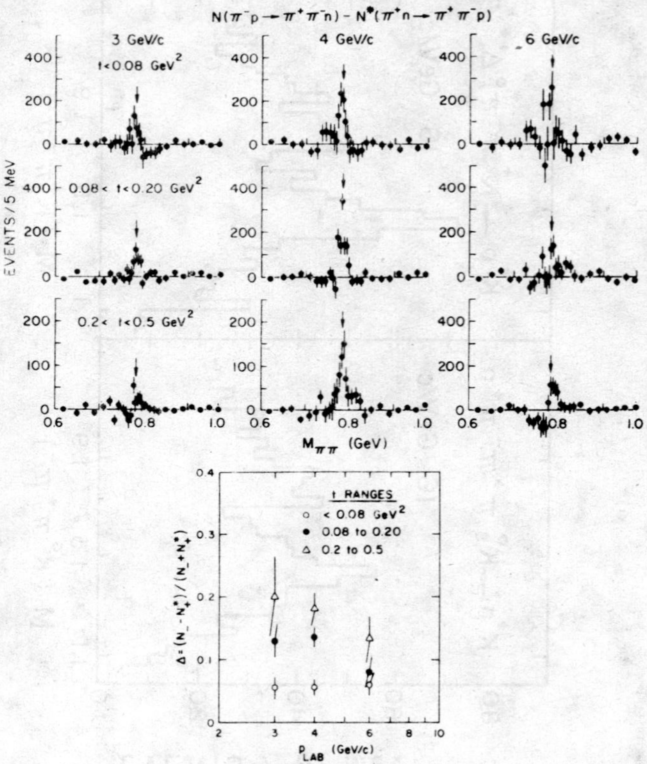

Fig. 3 The normalized difference, Δ, as a function of laboratory momentum and t range, for $775 < m_{\pi\pi} < 790$ MeV.

Figure 44. Observation of ρ-ω interference in a high statistics spectrometer experiment (Ref. 93).

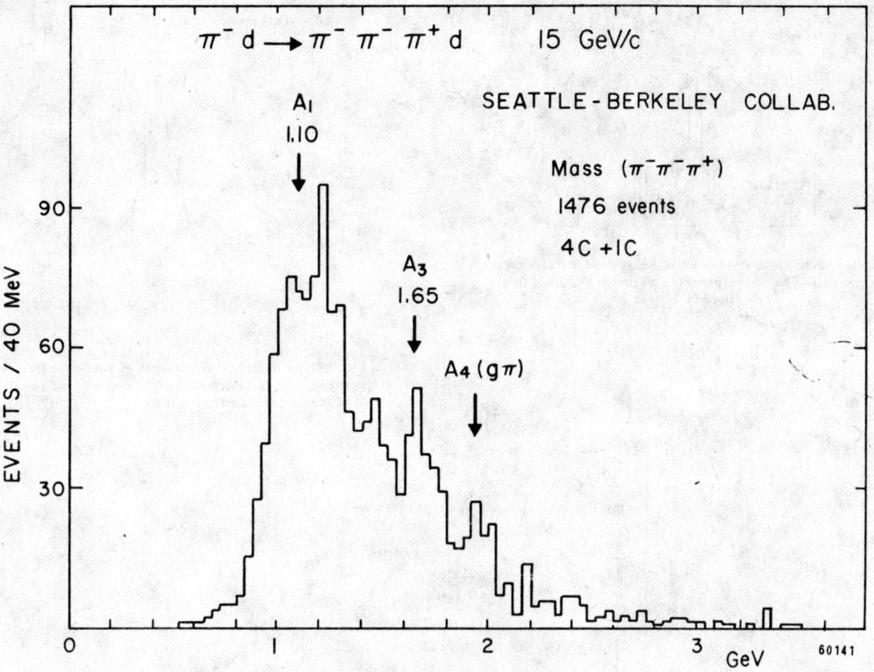

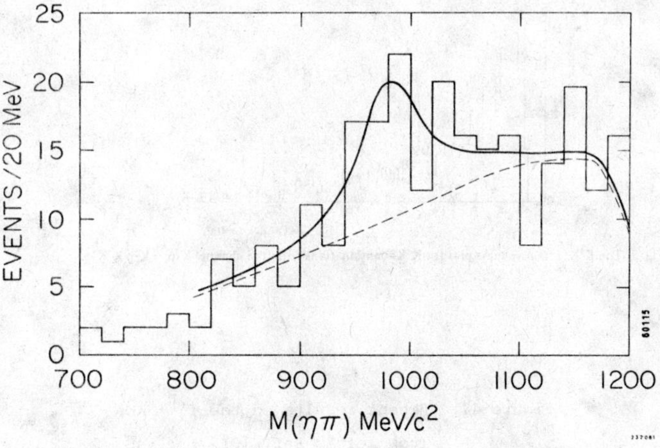

Figure 45 (top) The 3π mass spectrum from $\pi^- d \rightarrow 3\pi d$ at 15 GeV/c

Figure 46 (bottom) The $\pi^- \eta$ mass spectrum in the δ region (Ref.98) from the reaction $\pi^- p \rightarrow p\pi^- \eta$ at 4.5 GeV/c.

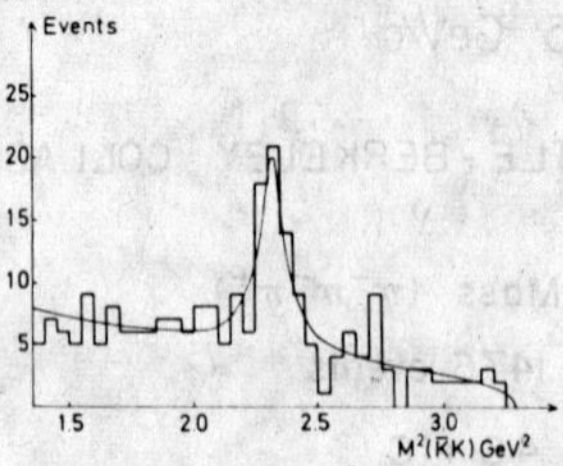

Fig. 1. Effective mass distribution of $\bar{K}K$ systems in reactions $K^- p \to \Lambda \bar{K}K$. The curve represents a maximum likelihood fit described in the text.

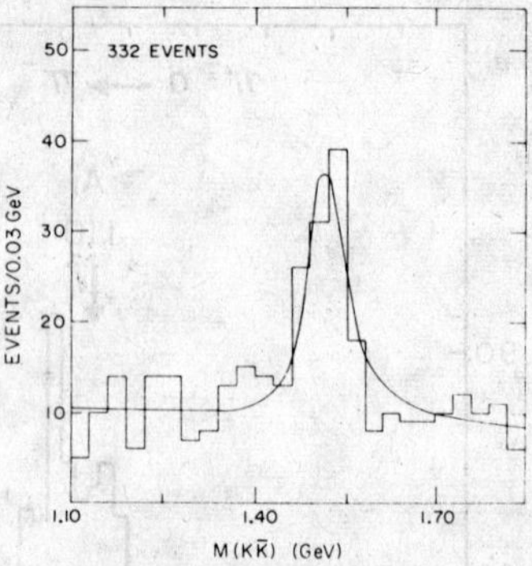

FIG. 68. $K\bar{K}$ effective distribution at 4.6 GeV/c from reactions 5, 8, 11, 12, and 14. The curve is the result of the fit described in the text.

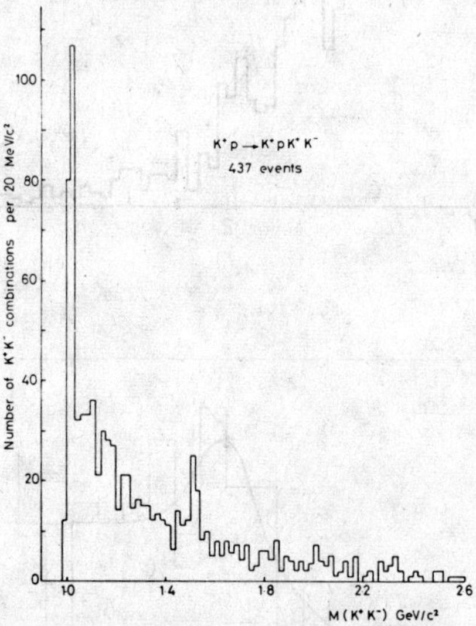

Fig. 1. The K^+K^- effective mass plot for K^+K^- combinations from the channel $K^+p \to K^+p\,K^+K^-$

Figure 47. Recent results on the $f' \to K\bar{K}$.

(top left) Ref. 101

(top right) Ref. 14

(bottom) Ref. 100

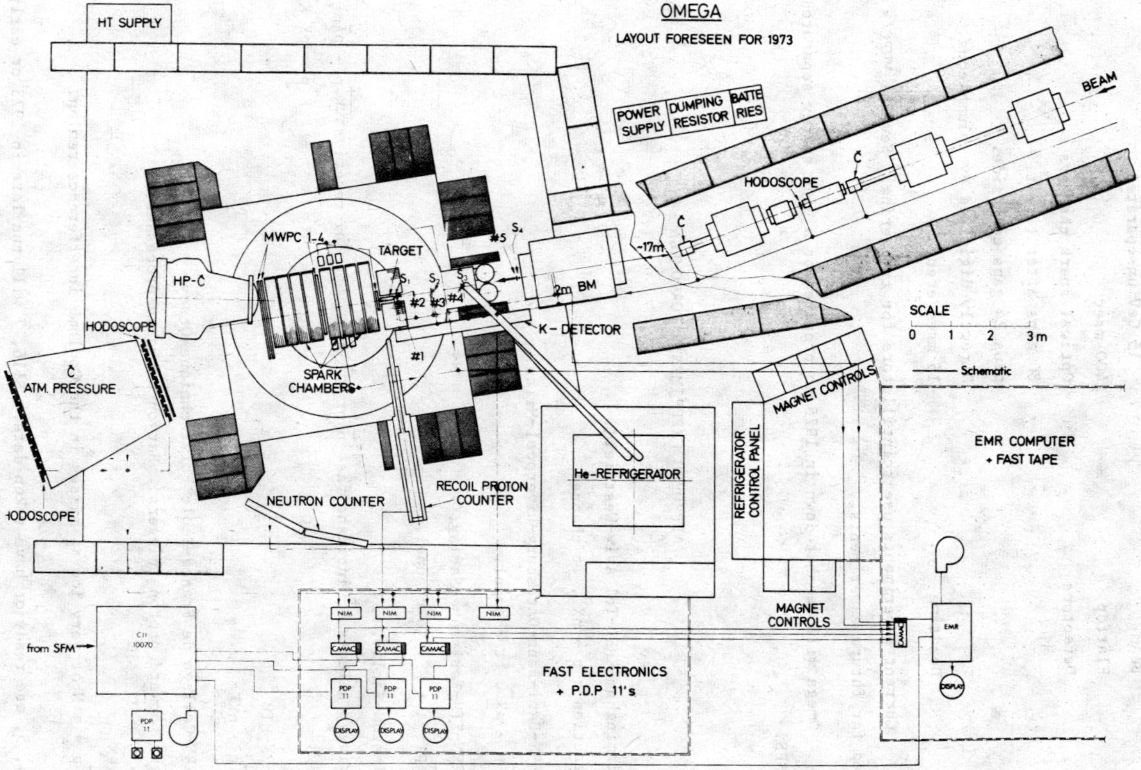

Figure 48 A plan view of the Omega Spectrometer

TECHNICAL DATA FOR THE OMEGA SPECTROMETER

Magnet	Superconducting 6 m x 6 m x 1.5 m
Field	19 k gauss
Beam	15 GeV unseparated
Flattop	400 msec
Detectors	Optical spark chambers 80 gaps after target plus 24 gaps each side. Directly digitized by plumbicons.
Deadtime	15 msec/event

Comments Current experiments use scintillators for triggering. Several MWPC's are being added for future experiments.

Omega will be shut down in 1975 to prepare for a high energy separated beam from the SPS.

APPROVED EXPERIMENTS FOR OMEGA

S112 Birmingham-RHEL-Tel Aviv-Westfield.
Reaction $\pi^- p \rightarrow nX^o$. t channel.

S113 CERN-Bari-Bonn-Daresbury-Liverpool-Milan.
$\pi^- p \rightarrow pX^-$. t channel.

S114 CERN-ETH-Karlsruhe-Saclay.
$\pi^- p \rightarrow \Lambda + K^*$. u channel.

S115 Glasgow-Saclay.
Trigger selects K^+ incident fast $\bar{p}$ or K^- out. Special reactions $K^+p \rightarrow p\bar{\Lambda}p$
$K^+p \rightarrow pK^+K^-K^+$.

S116 CERN-ETH.
$K^-p \rightarrow nKK\pi$

S117 CERN-College de France-Ecole Polytechnique-Orsay.
Fast forward proton triggers. Baryon exchange processes.

S133 CERN
$\pi^- p \rightarrow \pi\pi N$ at very low $\pi\pi$ masses in the Coulomb interference region.

S112, 3, 4, 5, are ready or have taken data. S116, 7 will run late in 1973 or early 1974.

Figure 49. Some technical details of Omega.

Figure 50. List of approved experiments for Omega.

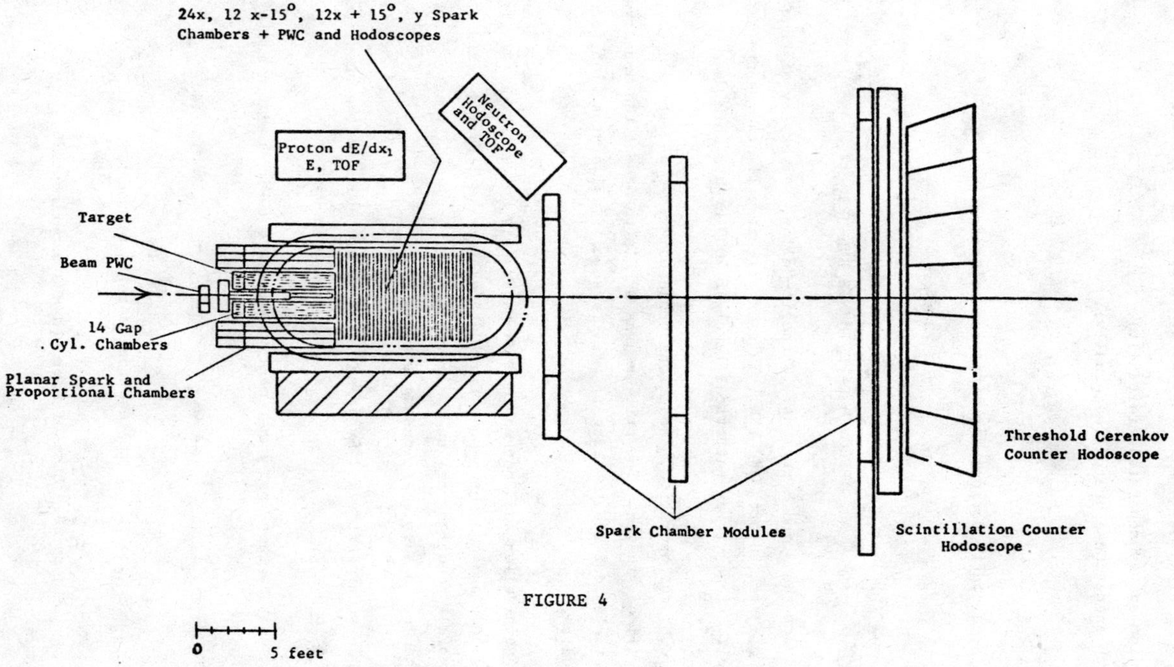

FIGURE 4

Figure 51. A typical configuration of the BNL-MPS spectrometer.

TECHNICAL DATA FOR THE BNL-MPS

Magnet	C type 6' x 15' x 4' Conventional
Field	10 kgauss
Detectors	Wire spark chambers with digitized read out. Plus several MWPC's giving close to 4π detection efficiency.
Beam	Medium energy separated (initially). High energy unseparated (eventually).
Deadtime	10 - 15 msec.
Flattop	1.4 sec.

Flexible target area configurations for different triggers.

APPROVED EXPERIMENTS BNL-MPS

AGS # 557. Study of the δ and R regions. Carnegie-Mellon, University of Pennsylvania, University of Massachusetts and BNL. This experiment will study $\pi^- p \to X^- p$ with a proton recoil trigger to a statistical level of 4000 events/μb.

AGS # 594. BNL-CCNY collaboration. Reactions produced by $\pi^\pm K^\pm p^\pm$ with a detectable neutral vee decay will be studied. The trigger uses MWPC's to detect the appearance of two charged particles outside the target. All types of final states with forward K^0 or Λ will be studied.

AGS # 596. Carnegie-Mellon and Southeastern Massachusetts University. This is a study of several two-body and quasi-two-body final states to make detailed tests of exchange models. Typical reactions are $\bar{p}p \to \pi^\pm \pi^\pm$, $\pi^- p \to p\pi^-$ produced by baryon exchange and exotic exchanges in such reactions as $K^- p \to \pi^+ \Sigma^-$.

AGS # 601. Brandeis University and Syracuse University. Study of the reaction $\bar{p}p \to \bar{V}^0 V^0$ + neutrals at 6 GeV/c.

It is expected that the first experiments will start taking data in early 1974.

Figure 52. Some technical details of the MPS.

Figure 53. List of approved MPS experiments.

D I S C U S S I O N S

CHAIRMAN: Prof. D.H. Miller

Scientific Secretary: L. Svensson

DISCUSSION No. 1

- *NAHM:*

You told us that one should look at unconventional places for the production of new mesons. What about large $p_\perp$? Of course, the crosss-section is very small, but if one gets as many δ's as π's it might be sensible to look there.

- *MILLER:*

Indeed, if one explores a process with small t-values, one apparently sees only mesons coupling to π,K plus the lowest member of a trajectory, and where the coupling of the lower vertex is strong. One condition for alternative experiments, because of the low cross-section, is that one should be able to trigger effectively. This has been possible in the u-channel; for example, the old BNL-Collins group experiment which saw meson production, but only in the missing mass. Another important process for 1^- states is e^+e^- annihilation, but it has its problems too. For example, one has not observed the $\rho'(1250)$ which should lie on a daughter trajectory below A_2. This could be because the ρ' decays into $\pi^0\omega$, and in this case one would have two neutrals in the final states, and together with the expected angular distribution for the decay makes the current experiments insensitive to its detection.

In summary, my hope is that different processes will excite different states, but that the data outside of the small t-region in π,K reactions are very meagre.

- *LOSECCO:*

What is the experimental situation with respect to C-violation in η decay?

- *MILLER:*

There is a recent result of Wonyong Lee which is very close to zero (perhaps $\frac{1}{2} \pm \frac{1}{2}\%$). There is also unpublished data from the Lipman group at the Rutherford Laboratory which also gives a number close to zero (0.3 ± 0.3%, I think). This latter number is a private communication which should not be quoted. In this case numbers from individual experiments are probably better than the PDG world average.

- *KUPCZYNSKI:*

Can the results on the D, E, A_2 from $\bar{p}p$ annihilations depend on the particular model used to overcome the combinatorial background?

- *MILLER:*

The technique is certainly difficult since in addition to the 9 $\pi^+\pi^-\pi^0$ combinations, there are also 15 $(\pi\pi)^{+-0}$ combinations. In addition, at higher energies where the phase space is less restrictive, the cross-sections for the observed states decreases very fast. My own feeling is that the D, E (possibly the A_2) have these decays through intermediate states near 1 GeV, but to determine their properties we need cleaner experiments.

- *EYLON:*

Can we expect to find those high mass resonances which are below the 5 μb limit, by performing the experiments at higher energies where the fact that the resonances are heavy has no kinematical importance?

- *MILLER:*

In all cases we know production cross-sections fall off as p_{lab} increases once one is some way above threshold, and it would be very unrealistic to expect cross-sections for higher mass mesons to increase at higher energies. Kinematically the situation could be clearer; for example, no final state interactions, but the cross-section will have decreased.

- *WEBER:*

Dual theoreticians and Regge phenomenologists tend to draw their Regge trajectories with a ruler. In view of what you have said about the uncertainty of high mass mesons, is there any evidence at all from the meson spectrum for linear meson trajectories?

- *MILLER:*

You can certainly draw a straight line through the ρ and the g. Exchange degeneracy holds quite well and the f and A_2 also lie on this line. The spin parity for lower-lying mesons (D, E, et.) have yet to be established, so little can be said about them. I would like to see some high mass resonances established before we can say whether the trajectories are straight for any considerable mass interval.

DISCUSSION No. 2

- *BUCCELLA:*

In the Ascoli analysis one does not consider the fact that the S-matrix is not diagonal in the orbital angular momentum; in fact it is possible for $\rho\pi$ in the A_1 region elements, connecting the 1^+s to the 1^+d-wave. This consideration applies also to $f\pi$ in the A_3 region. In particular, pure s-wave is in contradiction with the angular distribution $(1 + \cos^2 \theta)$ observed for the direction of the two pions forming a ρ with respect to the other pions. In fact in a CERN experiment, they claim to have pure 1^+d-wave. I am particularly interested in this, since in a theory based on the transformation from current to constituent quarks the $A_1 \to \rho\pi$ comes out with the angular distribution $(1 + 3 \cos^2 \theta)$, but the phase between the transverse and longitudinal modes is the opposite to the one predicted by pure d-wave.

- *MILLER:*

I do not really know what to say. This is a statement that may be perfectly true. I am not familiar with the analysis and I have not seen detailed angular distributions and Dalitz plots of the analysis. In the analysis I was talking about this morning, the simplest set of 10 waves are used which seem to fit the data, and certainly in their analysis the 1^+s is the only thing contributing to the A_1. One thing is that the A_1 sits right at the beginning of phase space and whether other waves are important enough to be really sensitive to interference effects could be a problem. If I remember correctly, if one allows a d-wave decay for the A_1, the fit always gives an answer consistent with zero.

- *GENSINI:*

The $2^{-}S0^{+}$ wave showed for the CERN-IHEP results seems to have a FWHM of about 400 MeV. Now, with such a width you cannot expect any longer a "normal" B-W behaviour of the phase and in general you expect the phase over the resonance to be much less than π. Some of the phase differences you mentioned (I recall at least one) showed some change of order $\pi/2$. Therefore I think the A_3 and A_4 should be on the same footing.

- *MILLER:*

I agree one needs to be careful with wide states, but in the A_3 region the relative phases only change by at most 20° which is very small.

- *NAHM:*

I am totally confused about the A_1. Professor Sakurai just used it to explain scaling for the weak current. Professor Kleinert included it into his $O(5) \times SU(2) \times SU(2)$ algebra. Whereas, the experimentalist say there is no significant evidence either for an A_1 resonance or for an A_1 Regge exchange. Could you classify this situation?

- *MILLER:*

I would be famous if I could. Actually one would like to have a 1^+ nonet, and certainly the B exists and seems to have stable properties. However, nobody has extracted the other members in an unambiguous way, and perhaps there is something in the large $\pi\rho$ enhancement which is resonant.

- *ROSENFELD:*

There is a paper of Wright which explains, by a suitably situated pole, the bump in the cross-section of the A_1 region and the smooth phase.

- *MILLER:*

Does that calculation conflict with the Deck diagrams which seem capable of reproducing the data?

- *ROSENFELD:*

No, it does not conflict.

- *WILCZEK:*

Is there any outstanding clear experimental discrepancies between quark model predictions and meson spectroscopy?

- *MILLER:*

No. There are missing states, but they have not been explored with sufficient sensitivity yet. In addition, if you take the SU(3) predictions, the current state of the branching ratios are in good agreement, but on the other hand, I do not think they really have been tested very well since the experimental errors are large.

- *RABINOVICI:*

What is the experimental status of the existence of the σ and ε mesons which appears in multiparticle production models due to the closeness of its mass to the average subenergies between neighbouring produced particles?

- *MILLER:*

This is something I am not going to talk about because I am really not familiar with the current state of the $\pi\pi$ and $K\pi$ phase shift for these scalar mesons. There was a Conference in Tallahassee which I did not attend this year, and so I am a little out of touch.

- *ZICHICHI:*

This topic was reviewed at this School last year. There are not very much new data on the subject since then.

- *ROSENFELD:*

After the recent Tallahassee Conference the situation is still as in PDG table. Two sorts of solutions fit the $\pi^+\pi^-$ scattering data; both have a narrow S^* at 997 MeV, but one set has a pole close enough to the real axis to be called an ε, the other set has no pole. Thus the pole for Protopescu's "ε" solution is of (660 ± 100) - i(320 ± 20) MeV, where 320 MeV corresponds (vaguely) for $\Gamma/2$, not Γ.

As for the $K\pi$ s-wave phase shift, there seems to be no solution which has a 90° phase shift below 1300 MeV, so your κ must be massive or far from the real axis.

DISCUSSION No. 3

- *KUPCZYNSKI:*

In the phase shift analysis of the CERN-Munich data, the fitted contribution of the ρ' in the total cross-section is large. How good is the refit of the data neglecting the ρ' from the beginning?

- *MILLER:*

If you do a complete refit without including the ρ', you get something which fits the data but the χ^2 is very bad. CERN-Munich do find that they need a pole in the p-wave and it comes out with the right mass for the ρ'. Their statement is that although many phase shift solutions could exist, it is very unlikely that a p-wave resonance will not be needed.

- *NAHM:*

You have shown us one solution for the p-wave, and there have been different ones. What are the values for the ρ' mass and elasticity in the other solutions?

- *MILLER:*

This is the only fit that has been done in detail, but there are ambiguities in this region. Their feeling is that it does not matter which solution you choose, you will always get this pole in the p-wave. I do not know the values for the elasticities, but I would think that the mass has to be approximately the same for different solutions as it has to fit the behaviour of the moments over the relevant mass region.

- *YU:*

In the pion-nucleon scattering how does the production of pions that results from the decay of meson resonances compare to that coming from baryon resonances?

- *MILLER:*

You mean in quasi-two-body processes, not in formation; you do not mean just πN going into a N^*? That is a very general question, but let me give you one answer. If you look at $\pi^+p \to p\pi^+\pi^+\pi^-$ then the Δ^{++} is roughly 30% of the total, and in the 3π J^P analysis you have to make a cut and take that out.

When I talked about $\pi^+p \to p\pi^+\pi^0$ the diffractive N^*-type bump is $\sim$ 300 μb (typical cross-section in 10-13 GeV for g signal is $\sim$ 10 μb), and reflecting this into the meson system gives roughly equal contributions in the g region because the spread of the N^* is over several hundred MeV. The question is, however, too general to answer unless you have a special point.

- *TAYLOR:*

The question really is a philosophical one. What do you consider is the minimal of information necessary to establish the existence of a meson resonance?

- *MILLER:*

I think I did that by the way I classified the meson resonances. Certainly it is necessary to have corroborative experimental evidence of good statistics. Having a phase shift go through 90° and/or a complete spin-parity analysis is also convincing. It must be pointed out that simple mass enhancements can be deceptive as in the case of A_1. These are perfectly rigorous statements one can make "theoretically", but they have not been too useful on a practical basis. Most well-studied states were first seen as mass enhancements and could be associated with a single J^P because they were produced in a single-exchange process.

- *GENSINI:*

Concerning again the ρ' seen in 4π photoproduction, I would like to ask if any signal is seen connected to the g(1700) state, which according to the Gribov-Morrison sum rule ought also to be produced peripherally, and how the ρ'(1600) is differentiated from such a background?

- *MILLER:*

Probably A. Rosenfeld will add a comment since he has done some of these experiments and has been closer to them than I have. As far as I am concerned, you do not see a 3-signal, and from the angular correlations of the break-up in the decay you can prove from the polarized photon data and from the bremsstrahlung data that the ρ' is consistent with a 1^- object.

- *ROSENFELD:*

I cannot give any number. What we did was that we took advantage of the fact that the beam was polarized and wrote down the simple matrix element of Zemack for production of mesons with spin-1 and 3 decaying into 4π, and found that we could completely explain the spectrum with 1^- decaying into ρε. For 3^- we could not see anything at all, but to turn that into an answer, not more than 10% is a little difficult without a piece of paper.

- *ROSENFELD (Comment):*

Let me now try to explain John Wright's idea in terms of the facts that Miller has presented. He told us that the A_1(1050) meson decayed into $\rho\pi$(s-wave), and that the magnitude of $\rho\pi$ 1^+ signal did not look like a B-W resonance, and its phase was almost constant. Wright suggested that the A_1 really decays through $\varepsilon\pi$ and $S^*\pi$ in a p-wave, and that the apparent $\rho\pi$ channel just results from a final state interaction

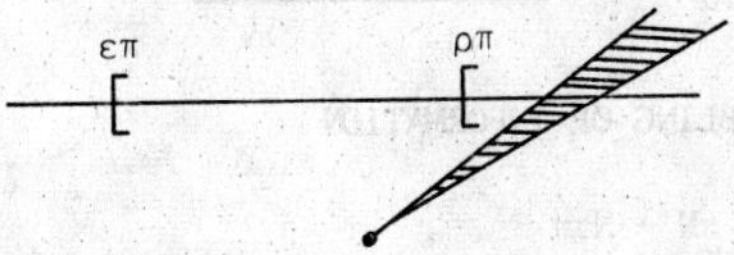

Then consider an A_1 pole in the $1^+(\varepsilon\pi + S^*\pi)$ W-plane. If the pole is located, as sketched, on a sheet (III or IV) which communicates with the real world via the $S^*\pi$ branch point, then it will not produce its full effect until $S^*\pi$ threshold (~ 1120 MeV) and the phase (which is exactly the shadowed triangle in the sketch) will vary little with energy. I should warn you that when I last heard this idea discussed, it was still just a conjecture. I do not know if Wright has yet actually done a proper K-matrix fit of "$\rho\pi$" data and really explained the amplitude and phase.

ALMOST EVERYTHING ABOUT BARYON RESONANCES

A.H. Rosenfeld

Table of Contents

ALMOST EVERYTHING ABOUT BARYON RESONANCES

Arthur H. Rosenfeld

Lawrence Berkeley Laboratory, University of California 94720, USA

LECTURE I.

1. INTRODUCTION. RECENT DOUBLING OF INFORMATION

The title for these talks is both too ostentatious and too short, but in one line I could not add that I shall confine myself to Non-Strange Baryon Resonances which I'll call N*s. I'll show that the interesting information on most of these states has roughly doubled in the last year or so, permitting some interesting tests of SU(6) and the Quark Model.

Until roughly last year, almost all quantitative information on N* resonances came from Elastic Partial - Wave Analyses ($\pi N \to \pi N$), which are still called "Elastic Phase Shift Analyses" or "EPSA" for short. Two excellent recent examples are the work of Almehed and Lovelace (CERN-72)[1] and of Ayed, Bareyre and Lemoigne (Saclay 72)[2]. Some of the photon couplings had been measured in photo-production experiments ($\gamma N \to \pi N$)[3], and one could not find any Argand Diagrams for $N^* \to \Delta\pi$, $N\rho$, etc.

But suddenly there are several computer programs capable of doing a partial-wave analysis of

$$\pi N \to I(J^P) \to N\pi\pi \tag{1}$$

in terms of the "Isobar Model", i.e.

$$N\pi\pi = \Delta\pi + N\rho + N\varepsilon + \ldots \tag{2}$$

At first it seems scandalous that such information comes so late, but remember the situation in "EPSA". Resonances were not disentangled until good polarization data became available about 10 years ago. The equivalent to polarization in $N\pi\pi$ analyses is the interference between the various resonance bands on an $N\pi\pi$ Dalitz plot. So one has to analyse the whole Dalitz plot, using about 10,000 events at a single energy. That takes big programs, both experimental and computational.

To convince you that the available information has really doubled recently, I present Table 1. Actually, more interesting than the new partial widths for $\Delta\pi$, $N\rho$, ..., is the fact that the sign of the amplitude is now known for each channel. And we'll be able to compare the imaginary part of the newly-found pole position with the new partial widths.

So when I say that we now know "almost everything" about N^*s I mean that the only obvious reaction that has not been measured and analysed is $\pi N \to N\pi\pi$ with a polarized target. That would be the most efficient way to find the couplings, but in fact, if the forth-coming unpolarized analyses are really unique, the problem may be solved before the polarization experiments are done.

PARTIAL WAVE ANALYSIS OF $\pi N \rightarrow N\pi\pi$

A total of five charge channels can be fitted :-

$$\pi^- p \rightarrow n\pi^-\pi^+, \quad p\pi^-\pi^o, \quad n\pi^o\pi^o, \tag{3}$$

$$\pi^+ p \rightarrow p\pi^+\pi^o, \quad n\pi^+\pi^+. \tag{4}$$

At $E_{c.m.} = 1520$ MeV ~ 13 mb of the total inelastic $\pi^- p$ cross section of 15 mb is accounted for by reactions (3) while at $E_{c.m.} = 1700$ MeV the numbers are respectively 21-22 mb out of ~ 25 mb. Reaction (4) is also large.

Thus to restate more quantitatively my introductory comment, we conclude that if the $N\pi\pi$ final states can be understood we will have an essentially complete description of πN scattering at these energies. We will then be in a position to attempt a multichannel analysis of the πN reactions with the added knowledge that no further new experimental information will become available (although, of course, the present inelastic partial wave amplitudes may be somewhat modified in light of new results; e.g., polarization measurements in the inelastic reactions).

In general two methods have been followed : isobar model analyses of the whole final state[5,6], and straight, quasi-two-body partial-wave analysis of specific reactions, e.g., $\pi N \rightarrow \Delta\pi$ (Ref. 7) which have been isolated by applying judicious cuts to the data to select this final state.

2.1. Isobar model

Groups at Oxford[8,9], Saclay[10] and LBL/SLAC[11] have used this technique, differing mainly in their methods of fitting the data. The method itself consists of writing the transition amplitude for reaching a given $N\pi\pi$ final state as a coherent sum of quasi-two-body processes as indicated in Fig. 1. The transition matrix is then written in an LS representation as[5,6]

$$T(W,w_1,w_2,\theta,\phi) = \sum_{IJLL'S\ell} A^{IJLL'S\ell} \times C^{I}_{X}{}^{JLL'S\ell}(w_1, w_2, \theta, \phi) F^{\ell}(w_1,w_2), \tag{5}$$

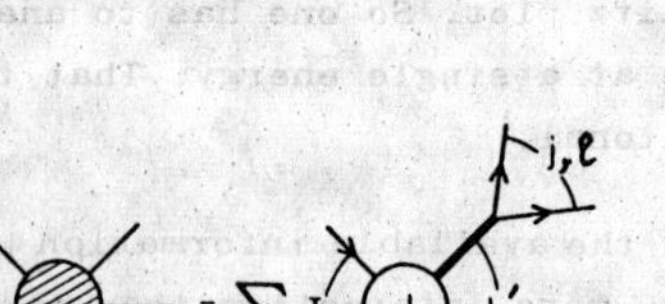

FIG. 1--The isobar model.

Table 1. Recent doubling of information available on a typical resonance: F_{35}

a. Entry in 1972 Particle Data Booklet.
b. Extra information in 1973 edition, or to appear in 1974

State	$I(J^P)$	Mass (MeV)	Width (MeV)	Partial Widths Mode	Partial Widths (MeV)
a. 1972 Entries					
$\Delta(1890)$	$3/2(5/2^+)$ F_{35}	1840 to 1920	135 to 350	$N\pi$	50
				$N\pi\pi$	large
b. 1973 or 1974 Additions					
Pole at $1824 - i\frac{282}{2}$				$N\pi\pi$ { [$\Delta\pi$	55]
				[$N\rho$	219]
				$N\gamma$	.03
Breit-Wigner "Refit" :-		1907	324	Sum	324

where : $W, w_1 w_2, \theta, \phi$ are the kinematic variables required to completely specify the reaction; C^I is the product of isospin Clebsch-Gordon coefficients to reach different charge final states; $X^{JLL'S\ell}$ contains all factors related to the angular momentum decompositions, including barrier factors; $F^\ell(w_1, w_2)$ is the Final State enhancement factor e.g., a Breit-Wigner or Watson final state interaction factor,[12], where ℓ is the orbital angular momentum in the decay of the isobar. The variable parameters, the <u>partial wave amplitudes</u>, $A^{IJLL'S\ell}$ are assumed to be dependent only on the total c.m. energy W. The differential cross section is

$$d^4\sigma(W, w_1, w_2, \theta, \phi) \propto |T(W, w_1, w_2, \theta, \phi)|^2 \tag{6}$$

The data are fitted in a variety of manners by the different groups, always treating <u>each c.m. energy independently</u> - Oxford[8,9] fit invariant mass and angular projections of the data in $\pi^{\pm}p$ collisions for $1300 < E < 1500$; Saclay[10] fit moments of the angular distribution for several zones on the Dalitz plot. The old analysis covered π^-p and π^+p separately: π^-p ($1390 < E < 1580$) and π^+p ($1650 < E < 1970$); the new analysis uses π^+p <u>and</u> π^-p, starting at 1390 MeV ; LBL/SLAC[11] make maximum likelihood fits to $\pi^{\pm}p$ reactions for $1300 < E < 1970$ (ie., to all the kinematic variables).

This isobar-model approach is optimistic in that one hopes to fit the whole reaction, making maximum use of all interference effects associated with the overlap of the various resonance bands. However, as we shall see this has proved to be possible (for at least 10,000 events at each energy) and provides us with an immense amount of information.

2.2. Single-Channel Partial-Wave analyses (e.g. $\pi N \to \Delta\pi$).

The LBL/SLAC[13] collaboration, and an LBL/UC Riverside[14] group, have used this technique to analyse specifically

$$\pi N \to \pi \Delta \tag{7}$$

After applying cuts, one assumes that one has a pure sample of reactions (7) and then performs fits to the production angular distribution of the Δ and sometimes also its density matrix elements, in terms of the partial wave amplitudes, usually with an energy dependent formalism. [These single-channel analyses throw away so much interference information that it is no longer possible to get unique fits at a single energy.] The major advantage of the single-channel analysis is that the formalism is easier to handle, whereas the great dangers lie in the assumption of a pure sample and in the energy-dependent parametrizations one uses. Furthermore, it is impossible to relate in phase, reactions such as

$$\pi N \to N\rho \tag{8}$$

to reaction (7) because the regions of interference which would define the phase are specifically removed from consideration. I feel that this method provides useful information but only on the large unambiguous partial waves present, and one should be much more skeptical of small effects[13].

2.3. Justification of the Isobar Model.

We think that the Isobar Model is a theoretically adequate way to analyse the data, and Smadja[15] estimates that the approximations involved affect our amplitudes by < 5%, i.e. $\delta T < .025$. This is tiny compared to our stated accuracy of $\delta T \approx 0.1$ (see sect. 5.2), or even our statistical error $\delta T(\text{stat.}) \approx 0.03$ at a single energy. Nevertheless the theoretical approximations are interesting, so I'll outline them.

a) Even if only one final-state resonance were involved, we don't know precisely how to write the final state enhancement factor. We all use Watson's[12] $e^{i\delta} \sin\delta$ because it's simple and consistent with observation, but it is not unique[16] and it could be more complicated.

b) Consider our case, when resonances overlap. For example, consider $\pi^- p \to S_{11} \to \Delta^-\pi^+(L'=0) + n\rho^\circ(L'=0)$. For the first term by definition we include the factor $e^{i\delta} \sin\delta$ for the N and π composing the Δ. For the second ($N\rho$) term we do not provide for any Δ. But of course there is some probability that the n and the π^- from the ρ will be in an $I = \frac{3}{2}$ p-wave. This is what can introduce the 5% error in our results. The error is proportional to the overlap $\langle \Delta\pi | N\rho \rangle$ between $\Delta\pi$ and $N\rho$ wave functions which in any case we have to calculate to compute the cross section, Eqs. (5) and (6).

2.4. Isobar Model Analyses of 3π and K2π Production.

The Illinois groups of Ascoli and Kruse have pioneered the isobar analysis of the 3π subsystem produced in the reaction πp → p3π, according to the model

$$3\pi \rightarrow I(J^P) \rightarrow \rho\,\pi + \varepsilon\,\pi. \tag{9}$$

Here $I(J^P) = 1\,(1^+)$ corresponds to the partial wave associated with the A1 meson, $1\,(2^+)$ corresponds to the A2, etc. I need not discuss the interesting results; you can read David Miller's contribution to these same proceedings. But I can mention that Ed Ronat at LBL is fitting 7 GeV $\pi^+ p \rightarrow p(3\pi)^+$ events with the LBL/SLAC[11] programs and seems qualitatively to be confirming the results obtained with the Ascoli program, which uses a rather different parametrization. It is hard to write and debug these complicated programs (see Sect. 5), so this confirmation should be a relief to all concerned. The Nππ fits of Eq. (2) require up to 60 isobar-model waves, and so need ≳ 10,000 events at each energy ; the 3π fits of Eq. (9) need only a dozen waves, and Ascoli and Kruse achieved their first fits with only ~ 15,000 events spread over half a dozen energies.

In addition to 3π systems, isobar-model programs are now being used on Kππ (the problem of the Q- "meson") and on NKπ (the question of a Z_1^* "resonance").[17].

3. CONNECTION a) BETWEEN ENERGIES AND b) TO ELASTIC PARTIAL WAVES

Before we proceed, we should define some notation. Because there are pion beams, and no ρ or Δ beams, we call the πN channel "number 1", and the reaction πN → πN we call "elastic". For each incoming partial wave, $I(J^P)$, we define the T-matrix by

$$\left.\begin{array}{l}\sigma(\pi N \rightarrow N\pi)\\ \sigma(\pi N \rightarrow N\pi\pi)\\ \sigma(\pi N \rightarrow N\gamma)\end{array}\right\} = 4\pi\lambda\!\!\!^{-2}(J+\tfrac{1}{2})\left\{\begin{array}{l}|T_{11}|^2\\ |T_{1\Delta}+T_{1\rho}+T_{1\varepsilon}+\dots|^2\\ |T_{1\gamma}|^2\end{array}\right\} \tag{10}$$

The values of the magnitude and phase of $T_{11}\,(I,\,J^P,\,E_i)$ are already known from "EPSA", and we want to take advantage of this valuable and expensive information.

After a single-energy Nππ fit we know the magnitude and relative phase of the inelastic terms $T_{1\Delta}$, $T_{1\rho}$, ... but one crucial overall phase is still free, and must be tied to that of T_{11}.

In addition, before we present an Argand Diagram, we want to impose two constraints :

1. Continuity in Energy.
2. Unitarity simultaneously on all the elements of the T-matrix. Specifi-

cally the S-matrix, $S = 1 + 2iT$, must be unitary and symmetric.

We can tie the overall inelastic phase to the EPSA phase, and impose these two constraints by means of an energy-dependent, multi-channel K-matrix fit simultaneously to the T_{11} amplitudes from EPSA and the off-diagonal T_{1j} from our own fits. More details are given in Sect. 4 of Lecture II - here I want merely to outline the battle plan. For the moment let me just say that if we write the matrix equation for each J^P

$$S = \frac{1+iK}{1-iK} = 1 + 2iT, \tag{11}$$

then if K is real and symmetric, S will be unitary and symmetric (i.e. will satisfy unitarity, and time-reversal invariance). Solving (11) for the T-matrix for each J^P we have

$$T = \frac{K}{1-iK}, \tag{12}$$

so we can parametrize T in terms of a real, symmetric matrix. Moreover we shall write K as a sum of factorizable poles (corresponding to a sum of resonances in T) plus a non-factorizable background (linear in c.m. energy E) :

$$K_{ij} = \sum_{R=1}^{3} \frac{\gamma_i \gamma_j}{E_R - E} + B_{ij} + C_{ij}\, E. \tag{13}$$

For each J^P we then get K-matrix parameters from our fit to all available amplitudes (typically three, but for D_{13} we need five) at 20 different energies.

From the K-matrix parameters we can extract smooth Argand diagrams. This procedure is summarized in Fig. 2. [As you can see, we go on even futher, but that is reserved for Lecture II, Sect. 6]. You should now have enough of an outline to understand the Argand plots at the end of this section.

The final reaction, $\pi N \rightarrow N\gamma$ (inverse photoproduction) has such a small cross section ($\alpha \propto e^2$) that unitarity is no help, and we do not include it in the K-matrix fit. Instead the K-matrix parameters from the hadronic reactions are used as starting values for a final energy-dependent K-matrix fit to photoproduction. This is discussed in Lecture II, Sect. 1.

4. COUNTING AND NAMING THE WAVES

How many isobar model waves can be fed by a single incoming πN partial wave, e.g. D_{13}? If you peek ahead at Fig. 9, you will see that the answer is at least 5.

a. D_{13} can feed two $\Delta\pi$ waves, i.e. the Δ can be produced in a D-wave ($L = L' = 2$ in the notation of Fig. 1), or even more likely in an S-wave ($L' = 0$, $j_\Delta = \frac{3}{2}$, $\underline{L}' + \underline{j}_\Delta = \underline{J} = \frac{3}{2}$). We call these waves ΔDD_{13} and ΔDS_{13}

Input :- 10,000 $N\pi\pi$ events at one energy E_i.

↓

TRIANGLE - RUMBLE

↓

Output :- { About 24 partial wave amplitudes $T_\alpha(E_i)$, i.e one energy point ready for 24 different Argand plots. }

Sorting :- From now on each incoming partial wave is treated separately.

Input : { Amplitudes $T_{1\alpha}$ for 20 different energies, several channels, e.g.

Channel 1 :- $\pi N \rightarrow F_{15} \rightarrow N\pi$ from CERN or Saclay.

2 :- $\rightarrow \Delta\pi$

3 :- $\rightarrow N\rho$ } from Triangle - Rumble

4 :- etc., see Figs 7 - 13. }

↓

K-MATRIX (J^P) FIT

↓

Output :→ / Input :← { K-matrix parameters and smooth Argand Plots $T^{J^P}_{\alpha\beta}(E)$ }

↓ (left branch)

BREIT - WIGNER REFIT

↓

Output: B.W. Parameters near E_R for

$$T_{\alpha\beta} = \frac{\frac{1}{2}\sqrt{\Gamma_\alpha \Gamma_\beta}}{E_R - E - i\Gamma/2} + B_{\alpha\beta} + C_{\alpha\beta}\, E.$$

↓

POLE - HUNT →

↓ (right branch)

POLE - HUNT

↓

{ Output : Pole Position

$m_{Real} - i\,\frac{\Gamma}{2}$

Compare direct values with those obtained indirectly from BW Refit. }

Fig. 2. Sequence of extracting Argand Plots and parameters for Resonances.

and in fact find some of both amplitudes.

b. D_{13} can in principle feed three ρN waves. The spin of the ρ ($j = 1$ in fig. 1) can couple with the spin of the nucleon to form $S = \frac{1}{2}$ or $\frac{3}{2}$. If $S = \frac{1}{2}$, L' can take only one value, and we call the wave $\rho_1\ DD_{13}$. If $S = \frac{3}{2}$, L' can be a D- or an S-wave, and we write $\rho_3\ DD_{13}$ and $\rho_3\ DS_{13}$. In Fig. 9 we report evidence for only the last of these three waves.

c. We give the name "ε" to an s-wave dipion $[I(J^P) = 0(0^+)]$. Then an incoming $\pi N\ D_{13}$ wave can feed only $\varepsilon\ DP_{13}$.

So, if we include the πN channel, D_{13} could be coupled to 7 decay channels, and we find we need 5 of them (but 2,3 is more typical).

If we confine our analysis to F waves or less ($L \leq 3$, $L' \leq 3$) we find that 14 incoming waves (7 with $I = \frac{1}{2}$ — S_{11} through F_{17} — plus 7 with $I = \frac{3}{2}$) can in principle feed 60 inelastic waves. Our program searches for all 60 complex amplitudes (119 real numbers) but we find a need for only about half of them, and in fact in the region at or below 1520 MeV, only for a quarter of them.

5. CHECKS OF THE LBL/SLAC PROGRAMS

5.1 The LBL/SLAC Analysis

For the rest of this Lecture I shall concentrate on the LBL/SLAC $N\pi\pi$ analysis, which is the only one which has presented Argand plots of all channels at 18 energies from 1300 to 2000 MeV. It is well documented. The most recent publication is by Cashmore[18], in the Proceedings of the 1973 Purdue Conference. See also Refs. 6 and 11.

This analysis has the following advantages :

i) It spans the c.m. energy range $1300 < E < 2000$ except for a 100 MeV gap $1540 < E < 1650$, where the data are still being analysed by Saclay.

ii) It utilizes the data in the most efficient manner, making a simultaneous max. likelihood fit[11,19] to the three major channels at each energy

$$\begin{aligned} \pi^- p &\rightarrow n\pi^-\pi^+ \\ \pi^- p &\rightarrow p\pi^-\pi^0 \\ \pi^+ p &\rightarrow p\pi^+\pi^0 \end{aligned} \qquad (14)$$

iii) We obtain excellent agreement with the inelastic reaction cross sections predicted by elastic phase shift analyses (EPSA). (We used 1970 solutions, by now, unfortunately, obsolete).

iv) From the single-energy fits at each energy we have been able to establish two continuous solutions over the full energy range, thus producing

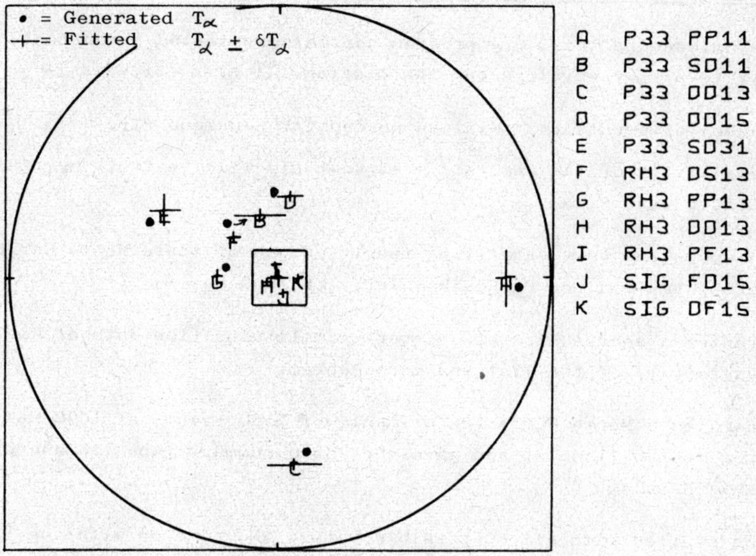

XBL 726-934

Fig. 3 - Results of a fit to 7500 Monte Carlo events generated at 1650 MeV to test Triangle/Rumble. This is Figure 8 of Herndon's thesis[20].

Argand diagrams for the first time. We hope to resolve these solutions as soon as the amplitudes in the middle of the Saclay gap become available.

5.2 Monte Carlo tests and sensitivity limits

Before unleashing this big program on data, we tested it on Monte Carlo events. This is a fine way to debug the program. It also forcasts :-

a. The number of events that will be needed for a unique fit.

b. The sensitivity of the analysis - at what $|T|$ will we fail to find a Monte Carlo wave ?

Fig. 3 shows the very satisfying result of one of these Monte Carlo tests. Larry Miller played God (or Prince Rainier). As such he :-

a) Invented a secret list of eleven amplitudes (the dots of Fig. 3). Even the length (11) of the list was kept secret.

b) Generated a Monte Carlo "experiment" of 7500 events at 1690 MeV, corresponding to the 11 dots, and gave the "data summary tape" to another student, David Herndon.

Herndon started with all 60 possible waves, and came up with the 11 crosses on Fig. 3, plus 13 more "noise" waves, as big as the four smallest secret waves, but all inside the box shown on the figure, whose half-side is $\pm.05$. We conclude that our signal : noise is better than 1:1 only for $|T|>.05$.

Given the extra uncertainties and systematic errors of real data (where our model can also not be perfect), we prefer to quote a "sensitivity" of $|T| \sim 0.1$.

The same experiment fails when tried with only 2500 instead of 7500 events (we find many solutions); and works poorly with 5000 events (several solutions). Hence our slogan that we need ∿ 10,000 events at each energy in the region of 1690 where we have to consider 60 waves.

5.3 Sign checks

Our programs can be internally consistent, pass the Monte Carlo tests of Section 5.2, and still have a wrong sign or sign convention, e.g. for a Clebsch-Gordan coefficient or a D-function. So we decided to put the same data through all the programs available, from Oxford[8,9], Saclay[10] and LBL/SLAC. None of the results agreed in all waves! We found a bug or a misunderstanding in both our own program and our version of Saclay's old program; Oxford won. Now that independent programs agree, we tend to believe them.

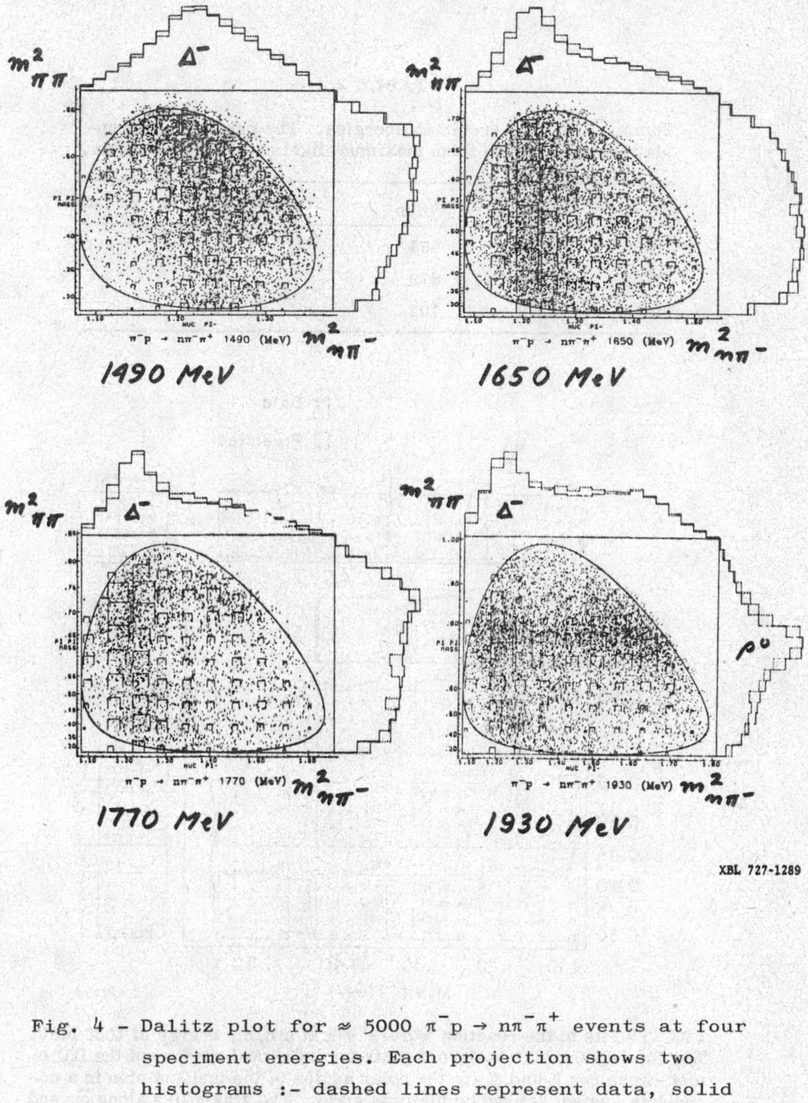

Fig. 4 - Dalitz plot for ≈ 5000 $\pi^- p \to n\pi^-\pi^+$ events at four speciment energies. Each projection shows two histograms :- dashed lines represent data, solid lines are predictions from the fit. This is Fig. 1.1 of Ref. 11.

TABLE 2

Summary of χ^2 at specimen energies. The predicted bin populations are derived from maximum likelihood fits to the data.

$E_{c.m.}$	χ^2	N_{bins}	Number of Partial Waves
1530	790	681	15
1690	1086	679	20
1970	2372	702	24

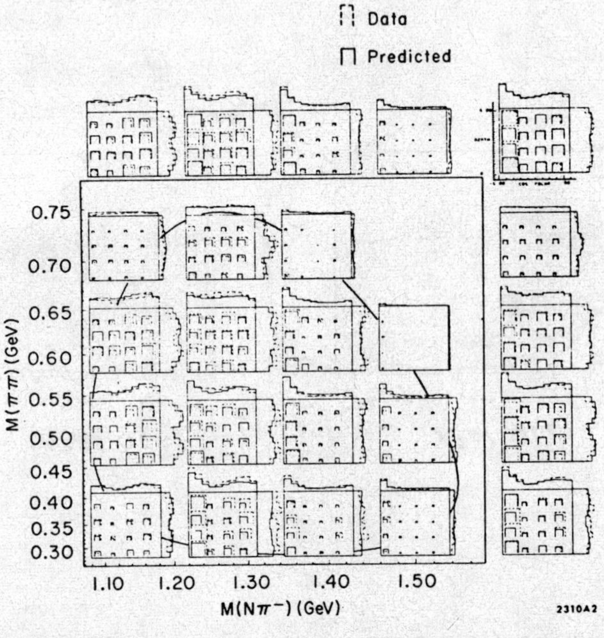

FIG. 5 -Fits to the reaction $\pi^-p \to \pi^+\pi^-n$ at a c.m. energy of 1690 MeV. The figure contains cos θ vs ϕ plots for individual regions of the Dalitz plot where cos θ and ϕ are the polar angles of the incident pion in a co-ordinate system defined by the final state. The z axis lies along $\vec{p}_N$ and the y axis lies along $\vec{p}_{\pi^-} \times \vec{p}_{\pi^+}$. The plots outside the Dalitz plot are the sums of the corresponding plots within the boundary.

5.4 Adequacy of fits

The satisfactory quality of our fits is illustrated by Table 2 and by Figs. 4, 5, and 6. Table 2 represents the χ^2 at 3 energies in our analysis, the ratio χ^2/N being excellent at lower energies but deteriorating as the energy increases. There are enormous variations of structure within the data at a given energy and in general the model reproduces them well, as can be seen in Fig. 4 (a standard Dalitz plot) and in Fig. 5, which shows our 4-D representation of the fit to $n\pi^+\pi^-$ at 1690 MeV.

Figure 5 consists of 2-D plots of the angular variables, cos θ and ϕ for individual regions of the Dalitz plot. We can also use our partial wave amplitudes to predict the cross section for

$$\pi^- p \rightarrow n\pi^0\pi^0 \tag{15}$$

and the good agreement with the experimental results is demonstrated in Fig. 6a. Finally, and this will be continually apparent throughout this talk, we have excellent agreement with the 1970 EPSA predictions.

The remaining point that must be addressed is the question of uniqueness of the solutions. For energies below 1540 MeV we are fairly certain of a unique solution because many random starting values always lead to one final solution. For energies greater than 1650 MeV we cannot be certain. We obtain several solutions at each energy from which we have identified the present solutions by requiring reasonable agreement with EPSA predictions and continuity of the solution at the adjacent energy point. This continuity, in modulus and phase, is vitally important because it allows us to show Argand diagrams.

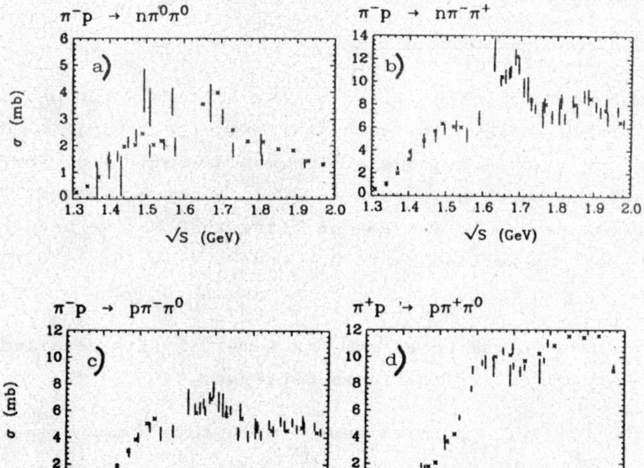

FIG. 6 -Single pion production cross sections. Data points are indicated by | and the predictions from our partial wave amplitudes by x.

RESULTS: ARGAND DIAGRAMS FOR 1972 SOLUTION ONLY

6.1 1972 vs. 1973 Solutions

Finally I can present some Argand Diagrams, but first a warning. As mentioned in Sect 5.1, paragraphs i) and iv), we are hampered by a gap in the available data between 1540 and 1650 MeV. Note on the figures following (7 through 12) that energy points M,N,O appear only on the elastic Argand plots, and note the energy gap in the inelastic T and $|T|^2$ plots.

We have found two solutions, which bridge this gap in different ways. One solution is old; we presented it at the 1972 Batavia Conference, and it is the only one for which I have Argand plots.

The other "1973" solution is still being explored. We found it only after considerable prodding and help by distraught theorists, mainly Gilman and Faiman. It has three more waves (ρ_1 PP_{11} - which Saclay decided independently should be included - ΔSP_{11}, and ΔFF_{15}), and a higher likelihood. All the big amplitudes are similar in both solutions, except for ΔPP_{11}, which is crucial for bridging the 1540 - 1650 MeV energy gap. The 1972 ΔPP_{11} amplitude moves fast at 1650 MeV (see Fig. 7), and (by continuity) also in the gap. So point R at 1690 MeV is nearly 180° out of phase with point K at 1520 MeV. The 1973 amplitude is motionless at 1650 (!), so continuity keeps the phase for the 1688 MeV region the same as at 1520. But PP_{11} is such a large wave, on both sides of the gap, that it influences all others. So the 180° difference between the 1972 and 1973 PP_{11} solutions produces a similar change in all other waves. I don't think we'll clear up this ambiguity until Saclay reports amplitudes (or events) in the gap.

6.2 Comments on selected Argand plots

On the Argand plots of Figs. 7,...11, the letters A through Z are the results of each single-energy fit, with statistical errors $\delta T \approx .03$ (twice the size of a letter). The magnitude of T comes directly from the fit. The phase has been calculated as discussed in Sect. 3, by a K-matrix fit to one or two large partial waves. In the region below 1540 MeV we used P_{11}; above the 1540 to 1650 MeV gap we tied on to D_{15} and F_{15}; in the 1920 region we rely on F_{35}.

The smooth Argand curves come from the K-matrix fits detailed in Lecture II. More details are given in the figure captions.

I include the following partial waves, with some comments on each:

- Fig. 7 (P_{11}). I have already mentioned the overwhelming importance of P_{11} in bridging the phase across the gap. Before this analysis EPSA told us that for the 1470 resonance $x_{el} = \Gamma(\text{elastic})/\Gamma(\text{total})$ was about 50 %; for 1780, $x_{el} \approx 15$ %. Now we see that the inelasticity at 1470 is due both to Δ PP_{11} ($x_\Delta \approx 30$ %) and εPS_{11} ($x_\varepsilon \approx 25$ %). These estimates (and the signs) for many

Fig.7

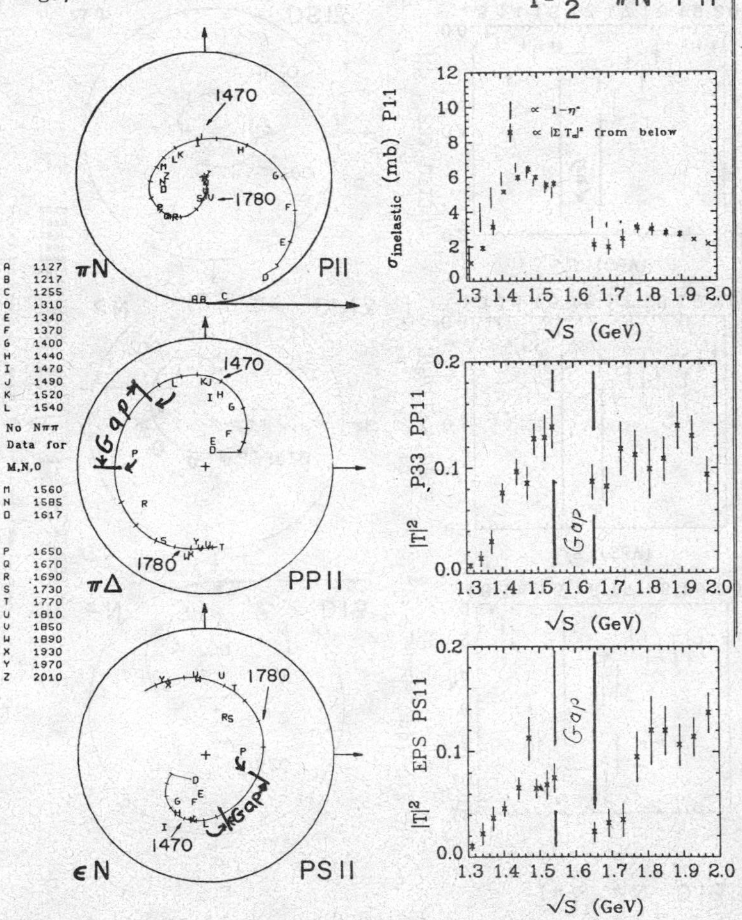

Fig.8

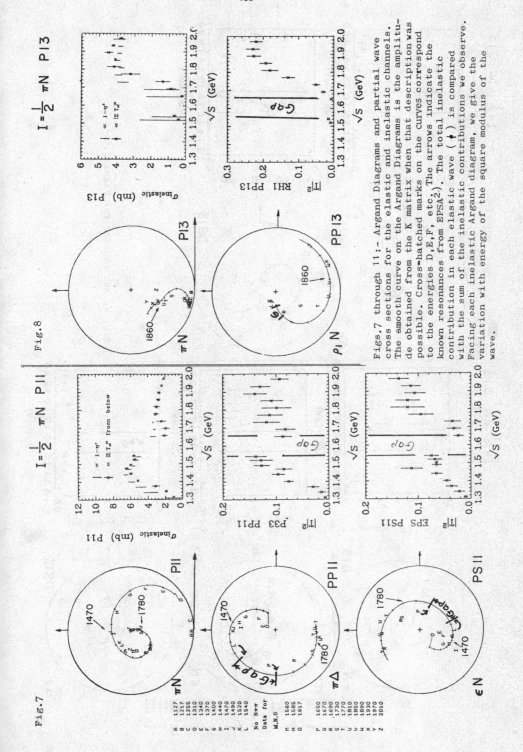

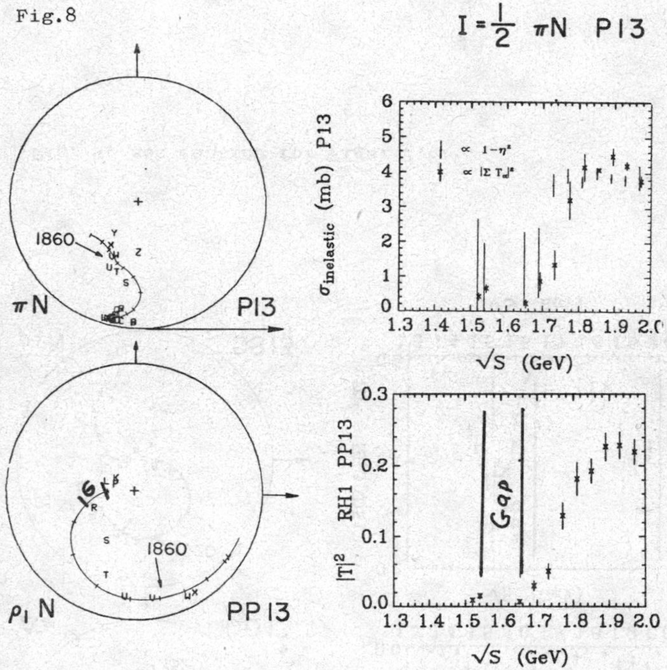

Figs.7 through 11:- Argand Diagrams and partial wave cross sections for the elastic and inelastic channels. The smooth curve on the Argand Diagrams is the amplitude obtained from the K matrix when that description was possible. Cross-hatched marks on the curves correspond to the energies D,E,F, etc. The arrows indicate the known resonances from EPSA[2)]. The total inelastic contribution in each elastic wave (ɸ) is compared with the sum of the inelastic contributions we observe. Facing each inelastic Argand diagram, we give the variation with energy of the square modulus of the wave.

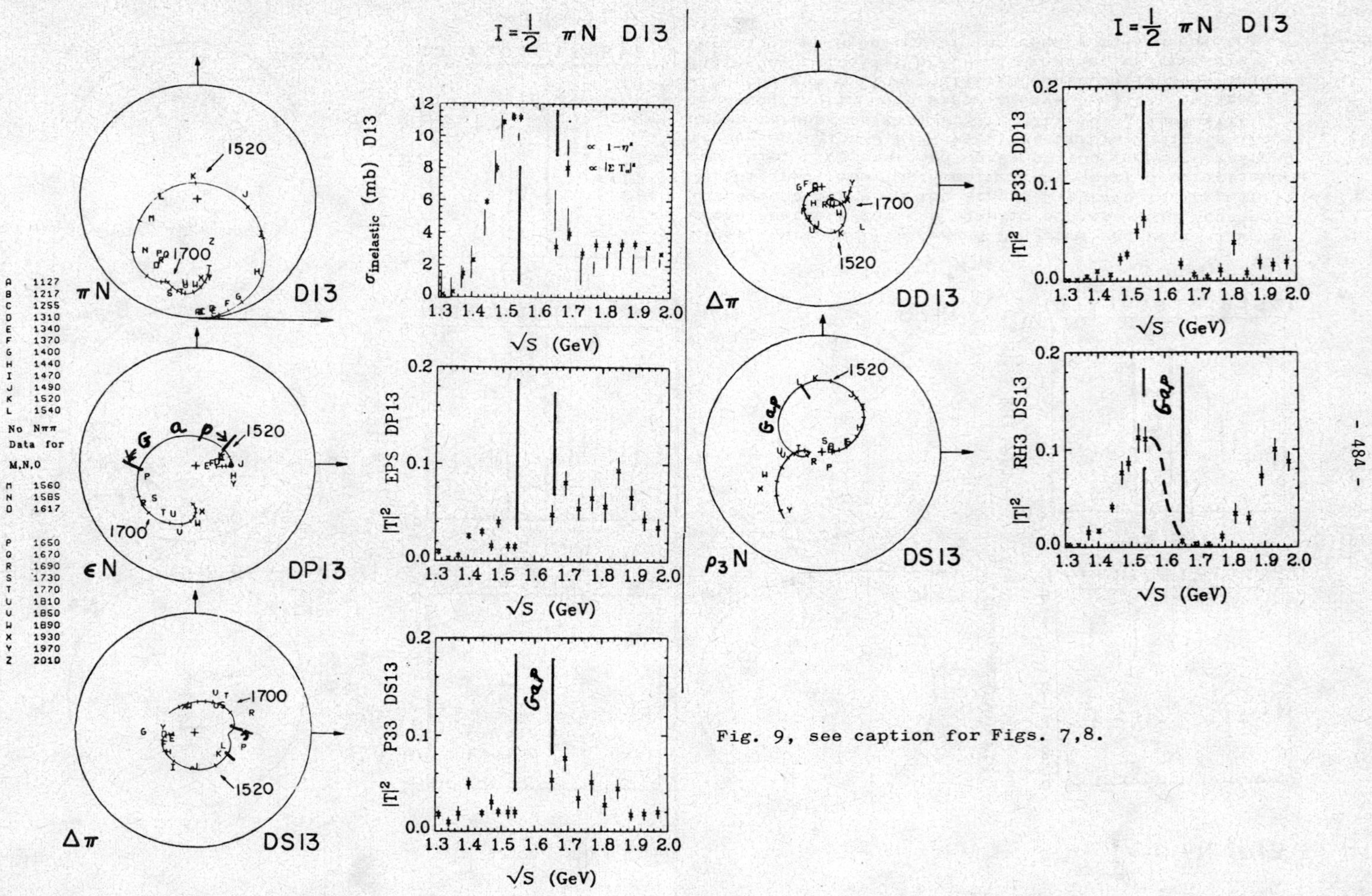

Fig. 9, see caption for Figs. 7,8.

Fig. 10

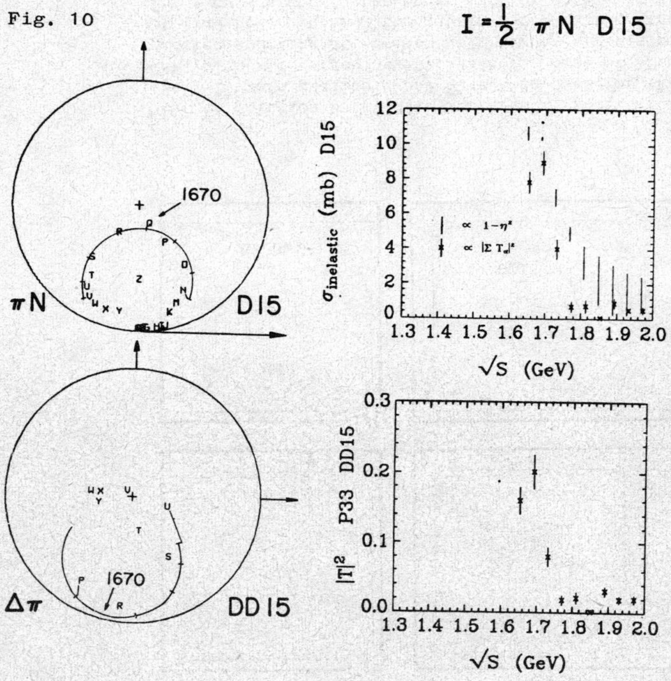

Fig. 11

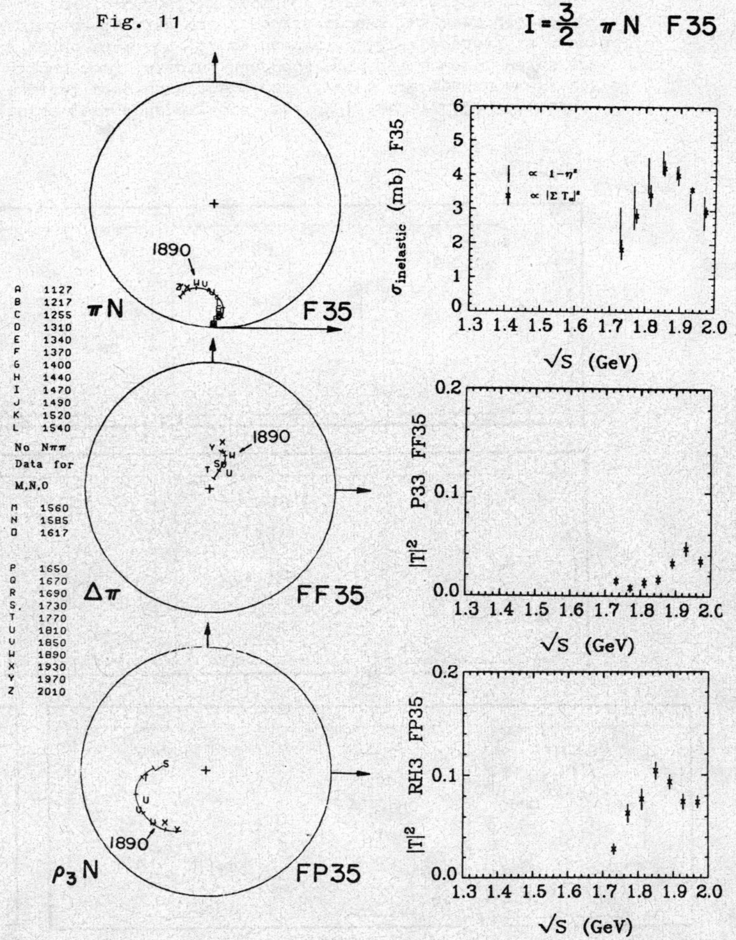

Figs. 10,11, see caption for Figs. 7,8.

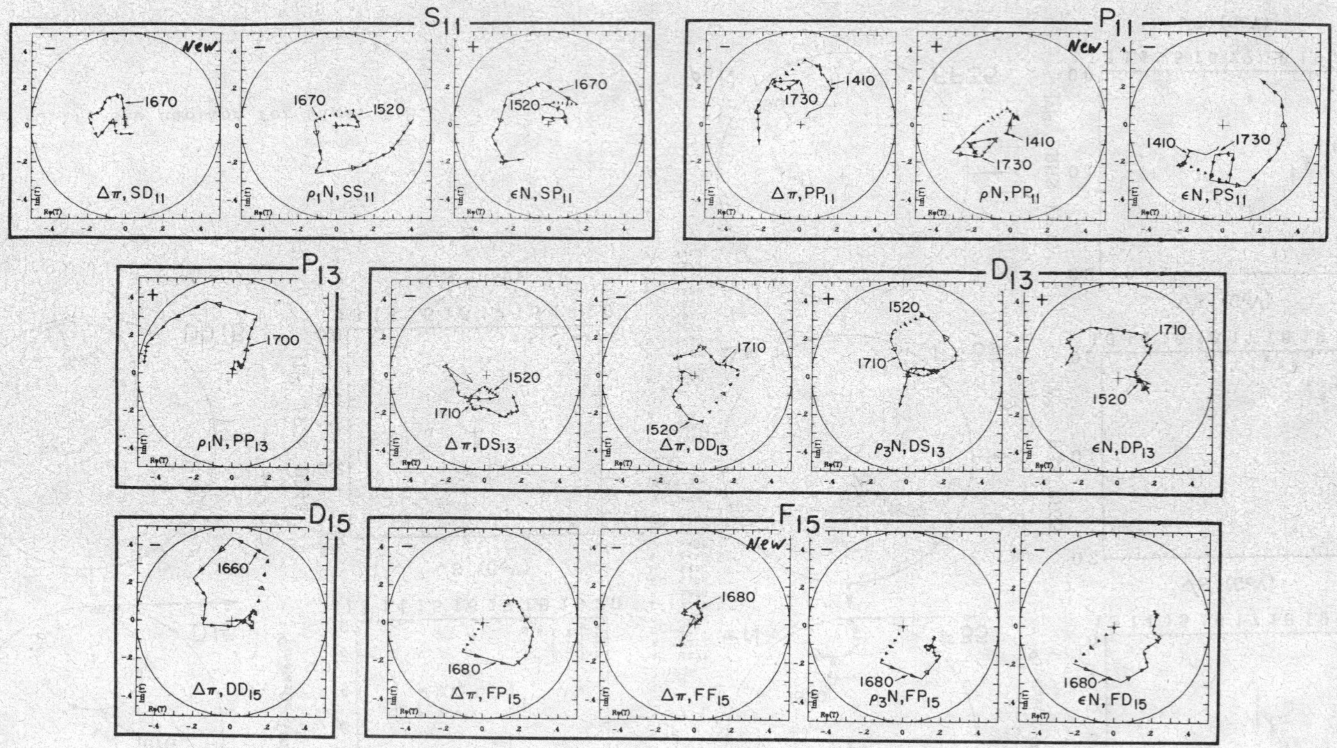

XBL741-2048

Fig. 12 Solution "B" (1973), $N\pi\pi$ Argand Plots. For these more modern plots, the 1973 Saclay EPSA solutions (Ref. 2) were available and have been used for the nominal resonance energies. Arrows are spaced every 20 MeV, with wide arrows every 100 MeV: base of wide arrows mark integral hundreds of MeV. Lower-ℓ waves are plotted starting at $\sqrt{s}$=1400 MeV; higher-ℓ waves only where they were first needed. Last arrowhead is always at 1940 MeV. To show the gap in our data the straight line joining the five gap arrows has been deleted. The + or - signs to the upper left of each circle show how to transform from our sign conventions to the "Baryon-first" convention.

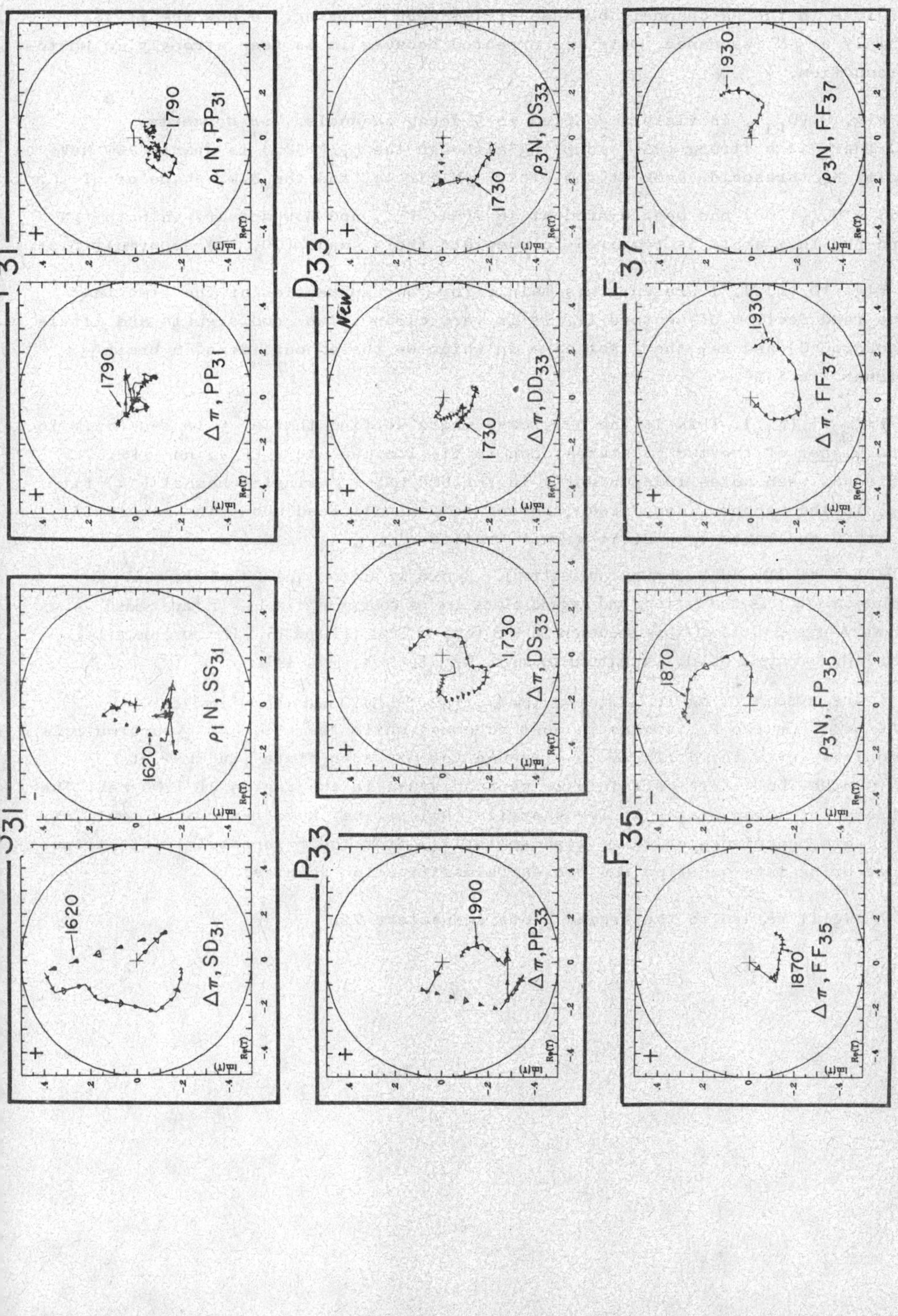
P31
Δπ, PP31
1790
ρ1 N, PP31
1790
S31
Δπ, SD31
1620
ρ1 N, SS31
1620
D33
New
Δπ, DS33
1730
Δπ, DD33
1730
ρ3N, DS33
1730
P33
Δπ, PP33
1900
F37
Δπ, FF37
1930
ρ3N, FF37
1930
F35
Δπ, FF35
1870
ρ3N, FP35
1870

reactions are in Table 2 of Lecture II.

- Fig. 8 (P_{13}). Here is an example of an inelastic resonance which is barely visible in the πN channel, because of its weak coupling. We now see it is mainly a ρN resonance; this was expected because it is seen strongly in photoproduction.

- Fig. 9 (D_{13}), is visibly coupled to 5 decay channels. Two comments:-
i) There is a strong ρDS_{13} coupling although the $D_{13}(1520)$ is nearly 200 MeV below $N\rho$ threshold. Look at the clear ρN circle, and the B.W. shape of $|T_{\rho N}|^2$.
ii) A $D_{13}(1700)$ has been hinted at in EPSA [1,2], and now appears in both εN and $\Delta\pi$. This state is required to complete the N^* and $\Delta(70, 1^-)$ supermultiplet.

- Fig. 10 (D_{15}). I present this mainly for your inspection at the time that you read Sect. 6 of Lecture II. It is very clean (two good signals and little background) and was the first case in which we tested our ideas on Breit - Wigner "refits".

- Fig. 11 (F_{35}). This is the only case where we find that a Δ is seen only in the higher of the two L' states open to it, i.e. we see ΔFF_{35}, no ΔFP_{35}. This has been noted independently in the UCR/LBL [14] single-channel ($\Delta\pi$) fit. F_{35} is the second, less "clean", resonance you will read about in Lecture II, Sect. 6 and Table 4, on Breit-Wigner "refits".

- NOTE ADDED JAN. 1974 (during publication). By now we are convinced that the 1973 Solution ("B") is the better, and Argand plots are of course available. I have added summary Argand plots of this solution in the form of Figs. 12 and 13. For more details see A.H. Rosenfeld et al., submitted to Phys. Rev. Letters, Feb. 1974.

The presence of two P_{11} states at low energies ($\sim$ 1470 and $\sim$1750) implies the need for two P_{33} states in most schemes, while the $[56, L=2^+]$ supermultiplet requires yet a third. There is evidence for one such state in EPSA at $\sim$1900 MeV but there certainly is no such state in the region of 1700 MeV. The absence of these states at low energies, unless they have remarkably small πN, $\pi\Delta$, etc, couplings, or large mass splittings from their supermultiplet partners, must bring into question the present classification schemes.

We'll return to the Argand Plots in Lecture II.

REFERENCES FOR LECTURE I.

1) S. ALMEHED and C. LOVELACE Nucl. Phys. B40, 157 (1972).

2) R. AYED, P. BAREYRE, Y. LEMOIGNE, contributed paper to the XVI International Conference on High Energy Physics, Batavia (1972).

3) For complete photoproduction references see the Particle Data Group's latest annual Review of Particle Properties e.g. reference 4.

4) Particle Data Group, Rev. Mod. Phys. 45, N°2 Part II, (1973).

5) B. DELER and G. VALLADAS, Nuovo Cimento 45A, 559 (1966).

6) R.J. CASHMORE, D.J. HERNDON, and P. SÖDING, LBL-543 (unpubl. 1973).

7) A.D. BRODY and A. KERNAN, Phys. Rev. 182, 1785 (1969).

8) M.G. BOWLER and R.J. CASHMORE, Nucl. Phys. B17, 331 (1970).

9) W. CHINOWSKY, J.H. MULVEY, and D.H. SAXON, Phys. Rev. D2, 1790 (1970).

10) For the latest in a long series of Saclay papers, see J. DOLBEAU and F.TRIANTIS, Paper Submitted to the 2nd Aix-en-Provence Conference on Elementary Particles, 1973. This is based on an entirely new computer program which yields both the phase and magnitude of the amplitudes. For older analyses of magnitude only, see several theses:-
NGUYEN THUC DIEM, Thesis, note CEA N-1602, Paris (1972);
P. CHAVANON, Collège de France, Thesis (1971).

11) D.J. HERNDON, R. LONGACRE, L.R. MILLER, A.H. ROSENFELD, G. SMADJA, P. SÖDING, R.J. CASHMORE, D.W.G.S. LEITH, LBL-1065/SLAC-PUB-1108 (1972), submitted to Phys. Rev. as LBL 1065 (Rev.) Sept. 1973.

12) K.M. WATSON, Phys. Rev. 88, 1163 (1952).

13) A.D.BRODY, R.J.CASHMORE, A.KERNAN, D.W.G.S.LEITH, B.G.LEVI, B.C.SHEN, D.J. HERNDON, L.R. PRICE, A.H. ROSENFELD, P. SÖDING, Phys. Letters 34B, 665 (1971).

14) V. MEHTANI, S.Y. FUNG, A. KERNAN, T.L.SCHALK, Y. WILLIAMSON, R.W.BIRGE, G.E. KALMUS and W. MICHAEL, Phys. Rev. Letters 29, 1634 (1972);

15) G. SMADJA, "Resonances that Overlap" LBL-382 (unpublished 1971).

16) R. OMNES, Nuovo Cimento 8, 316 (1958).

17) A. BERTHON, L. MONTANET, E. PAUL, P. SAETRE, A. YAMAGUCHI (CERN) and G. BURGUN, E. LESQUOY, A. MULLER, E. PAULI, F.A. TRIANTIS, S. ZYLBERAJCH (CEN, Saclay); CERN Preprint D.Ph.II/PHYS 73-31, submitted to the IInd Aix-en-Provence Conference on Elementary particles, 1973.

18) R.J. CASHMORE, SLAC-PUB - 1257 and Proc. Purdue Conference on Baryon Resonances (1973).

19) L.R. MILLER, Ph.D. Thesis, LBL-38 (unpublished).

20) D.J. HERNDON, Thesis, LBL-544 (unpublished, 1972).

D I S C U S S I O N S

CHAIRMAN: Prof. A.H. Rosenfeld

Scientific Secretaries: R.T. Ross, M. Leneke

DISCUSSION No. 1

- *EYLON:*

Could data on high spin N^* states be obtained more easily from deep inelastic electroproduction, than from πp formation experiments?

- *ROSENFELD:*

Unless they decayed via a chain of high-spin resonances, such states could be seen easily in πp formation experiments, since the production cross-section is given by:

$$\sigma = 4\pi\lambda\!\!\!^{-2}(J + \tfrac{1}{2})\ \Gamma_{\pi N}/\Gamma\ .$$

$\Gamma_{\pi N}$ would be small due to the high J, but also Γ would be small since πN will be the dominant channel. Hence this state will show up as a clear resonance spike in the elastic cross-section. Further, there are major experimental difficulties involved in obtaining N^* production information from electroproduction, and also the cross-section for electroproduction goes down as something like $\sigma \sim 1/q^{12}$ (similar to that of the elastic cross-section), so these N^*'s cannot yet be studied by high q^2 electroproduction.

- *EYLON:*

Is it possible to measure the decays of nucleon resonances into nucleon + electron pair?

- *ROSENFELD:*

The branching ratios for normal radiative decays of the known N^*'s are very small ($\sim 10^{-3}$), and decays into Ne^+e^-, that is, involving "heavy γ's" will be even much smaller than the $N\gamma$ decay mode which is must barely measurable now.

- *ZICHICHI:*

If high mass N^*'s with narrow widths exist (as suggested by the Veneziano model), the electromagnetic decays of these states may become competitive, and hence decays into Ne^+e^- may have an observable branching ratio.

- *NAHM:*

In your isobar model you use only the Δ and no higher N^*'s. Did you exclude them *a priori*, or do you have some bounds on their contribution?

- *ROSENFELD:*

One gets a good fit over some 700 bins (with $\chi^2 \sim 1.3$ per bin), and this indicates that the model is good. However, we have left out some final states where our model proved to be bad. For example, the reaction $\pi^+p \to p\pi^+\pi^0$ obtained a good fit, but $\pi^+p \to n\pi^+\pi^+$ could not be fitted with the model.

In the $n\pi^+\pi^+$ reaction there are no ρ resonances in the I = 2 combination ($\pi^+\pi^+$). Also the Δ is only singly charged and has no dominant Clebsch-Gordan coefficient. There is also an indication that the $N^*(1520)$ contributes to this final state.

On the whole, however, Nature seems to be reasonable, and a simple model is sufficient.

- *PARSONS:*

Can you comment further on why we never find ρ_1 and ρ_3 contributions together in the same resonance decay, and why one finds waves with higher angular momentum dominating where centrifugal barriers would lead one to expect the waves with the lower orbital angular momentum to dominate?

- *ROSENFELD:*

The point raised in the first part of this question is an embarrassment to the analysis. For instance, in the process $\pi N \to \Delta\pi + \rho_1 N + \rho_3 N$ where the $\rho_3 N$ contribution was negligible, there was an attempt to put the values for $\Delta\pi$ and $\rho_1 N$ back into the analysis and to try and generate a $\rho_3 N$ contribution, but this has not been successful.

The answer to the second point is that this occurs only rarely (1 in every 15 cases) and in this case we just have to suppose that some process favours the higher orbital angular momentum.

- *PARSONS:*

Are the ρ_1 and ρ_3 simply related to the ρtransverse and ρlongitudinal, if so what is the relationship?

- *ROSENFELD:*

Yes. The two representations are connected by a 3 × 3 matrix of the form:

$$\begin{pmatrix} \gamma_{\lambda=1/2} \\ \gamma_{\lambda=3/2} \\ \gamma_{\lambda=\text{longitudinal}} \end{pmatrix} = \begin{pmatrix} 3 \times 3 \\ \text{matrix} \end{pmatrix} \begin{pmatrix} \rho_1 \ L \\ \rho_3 \ L \\ \rho_3 \ L+2 \end{pmatrix}$$

where the representation on the left is in terms of the total γ helicity (as used to describe the γN state in photoproduction) and that on the right is the form used in our analysis; ρ_1 L+2 being the higher orbital angular momentum state of the ρ_3 contribution.

- *PARSONS:*

Is there any difficulty in the isobar model in the formation of $\pi N \to \pi\Delta \to \pi\pi N$, with the ρ interference of the two π's in the final state?

- *ROSENFELD:*

I can only repeat that in the $E^* = 1520$ region our χ^2 are $\sim$ 800/200 degrees of freedom, and then get worse (1100 at 1200 MeV, 2400 at 2000 MeV) apparently because of OPE and higher mass isobars, and so the difficulty you mention must not be our most serious one.

- *MONTANET:*

Could you elaborate a little more on the basic assumption you made in your analysis, namely the dominance of final state interactions and its validity.

- *ROSENFELD:*

The model we used assumes that we can add the different amplitude ($\Delta\pi$, ρN ...) with a relative phase and modulus which varies only slowly, apart from Watson's factors corresponding to the two-body final states. This is the simplest parametrization that we can use and we believe that the quality of our results are *a posteriori* justification of the method.

Moreover, I agree that with the new generation of experiments which bring us detailed information on the three-body final states, theoreticians should pay more attention to the problems involved in this method. We certainly intend to compare results with the predictions of OPE, duality, and Veneziano amplitudes, for instance.

- *MILLER:*

You said that you use the N(1520) to give a reference phase. Do you put in a circle on the Argand diagram for this state? Secondly, when you are certain that a state exists, have you tried to re-fit, smoothing out the behaviour on the Argand diagrams? The experimental points you showed had quite a ragged behaviour on the Argand diagram.

- *ROSENFELD:*

We simply use the Saclay phase shifts for the P_{11} reference phase. In reply to your second point, yes we have done this and in tomorrow's lecture I will talk about K-matrix fits and energy-dependent fits.

LECTURE II: FROM ARGAND DIAGRAMS TO PHYSICS

A.H. Rosenfeld

Table of Contents

LECTURE II: FROM ARGAND DIAGRAMS TO PHYSICS

In this lecture I shall discuss some physics which we can learn from the Argand plots explained and presented in Lecture I.

This lecture has separately numbered figures, tables, and references, because I want to use part of the lecture as a contribution to the forthcoming (Sept. 73) conference at Aix-en-Provence.

1. COMPARISON OF $\gamma P \to \pi N$ AND $\pi N \to \rho N$ WITH THE QUARK MODEL

In sections 2 and 3 of this lecture I shall take up various higher-symmetry tests of $\Delta\pi$ amplitudes; first I want to dispose of γN and ρN.

SU(3) tests of (Vector Meson) × (Nucleon) would need analyses such as $K^- p \to K^*N$ or ρY, and I am not aware of any relevant results.

γN and ρN can be compared either:-
a) directly, using the notions of vector dominance, or
b) separately, each with the quark model or SU(6).

For a) we must transform our partial-wave amplitudes $T(J^P, L, L')$ with a 3×3 matrix into helicity amplitudes which are conventional for photon reactions: call them $T(\lambda = \frac{1}{2})$, $T(\lambda = \frac{3}{2})$, and the unobservable T for longitudinal photons. In our analysis we have yet to propagate all the errors through this transformation, so I shall say no more at present.

I take up next the direct comparisons, b).

1.1 γN vs. the Quark Model

At LBL, Moorhouse and Oberlack[1)] have recently done a partial-wave analysis of photoproduction, and have found really encouraging agreement with the quark model. Fig. II-1 is just a photograph of their summary table.

To make a stringent comparison with the quark model we may take only the larger couplings of the prominent resonances, and only those where quark model predictions are "starred" in Fig. 1. We find seven such cases, underlined in Fig. 1, where the experimental sign is sure and which theoretically depend on, and only on, the Clebsch-Gordan coefficients of the quark model (the same in either the "relativistic" or "nonrelativistic" models. These 7 signs all agree! About 7 more signs which are less certain also agree [but not those for $P_{11}(1470)$] and in general all 33 magnitudes agree within a factor of 3. I remind you that the chance of random agreement of say 10 signs is $2^{-10} \approx 10^{-3}$, so I consider this to be an impressive systematic test of the quark model.

Fig. II-1. Comparison of pion photoproduction amplitudes with Quark Model. This a reproduction of Table 1 of Moorhouse and Oberlack (Ref. 1).

Average resonance couplings from seven fits to the data compared with quark-model predictions. The result from the partial wave analysis is an average over seven fits and the error is the spread over the seven fits; directly underneath the partial wave analysis result we give the quark-model result for the usual assignment of the resonance to an $\{SU6\}L$, [SU3, $2S+1$] multiplet. An asterisk denotes that the quark-model result does not involve a difference of two terms. Table 1a comprises resonances assigned to the $\{56\}L = 0^+$ and $\{70\}L = 1^-$ multiplets and table 1b the $\{56\}L = 2^+$, $\{56\}_2 L = 0^+$ and $\{70\}_2 L = 0^+$ multiplets where the suffix denotes radial excitation. In table 1b we also give quark-model results for some resonances for which we do not have partial wave results since they are outside our data range. $A_{1/2}$ and $A_{3/2}$ denote decays through helicity-1/2 and helicity-3/2 states, respectively, and superscripts + and 0 denote decays of charge +1 and charge 0 particles respectively. Units are $GeV^{-1/2} \times 10^{-3}$.

Table 1a

	N*(mass) [SU3, $2S_{quark}$ +1] J^P	$A^+_{1/2}$	$A^+_{3/2}$	$A^0_{1/2}$	$A^0_{3/2}$
$\{56\}L = 0^+$	$p_{33}(1230)$	−142± 6	−259±16		
	[10, 4] $3/2^+$	-108^*	-187^*	etc	
$\{70\}L = 1^-$	$s_{11}(1545)$	53±20		−48±21	
	[8, 2] $1/2^-$	156		−108	
	$d_{13}(1512)$	−26±15	194±31	−85±14	−124±13
	[8, 2] $3/2^-$	−34	109^*	−31	-109^*
	$s_{31}(1620)$	90±76			
	[10, 2] $1/2^-$	47			
	$d_{33}(1635)$	68±42	22±52		
	[10, 2] $3/2^-$	88	84^*		
	$s_{11}(1690)$	66±42		−72±66	
	[8, 4] $1/2^-$	0		−30	
	$d_{13}(1700)$	3±?	20±?	−28±?	27±?
	[8, 4] $3/2^-$	0^*	0^*	10^*	-40^*
	$d_{15}(1670)$	11±12	21±20	10±40	−35±14
	[8, 4] $5/2^-$	0^*	0^*	38^*	-53^*

Table 1b

	N*(mass) [SU3, $2S_{quark}$ +1] J^P	$A^+_{1/2}$	$A^+_{3/2}$	$A^0_{1/2}$	$A^0_{3/2}$
$\{56\}L = 2^+$	[8, 2] $3/2^+$	−11	30	30	0^*
	$f_{15}(1690)$	−8±4	100±12	17±14	−5±18
	[8, 2] $5/2^+$	−10	60^*	30^*	0^*
	[10, 4] $1/2^+$	−30			
	[10, 4] $3/2^+$	−30	50		
	$f_{35}(1870)$	−60±?	−100±?		
	[10, 4] $5/2^-$	−20	−90		
	$f_{37}(1950)$	−133±46	−100±41		
	[10, 4] $7/2^+$	-50^*	-70^*		
$\{56\}_2 L = 0^+$	$p_{11}(1470)$	−55±28		2±25	
	[8, 2] $1/2^+$	27		−18	
$\{70\}_2 L = 0^+$	$p_{11}(1750)$	26±28		27±22	
	[8, 2] $1/2^+$	−40		10	

1.2 ρN vs. the Quark Model

Very recently, Moorhouse and Parsons[2] have made the same quark model comparison for the photon's heavy relative, the rho meson, using our amplitudes for $N \to I(J^P) \to \rho N$. Alas, neither the length of the table nor its contents are quite so impressive, so I shall not reproduce it. There are only 3 "starred" predictions, and they are indeed satisfied by our amplitudes. In addition there is an unstarred prediction for $\rho_3 FP_{35}$ which does not agree with our solution. Another starred prediction awaits the bridging of the Saclay gap. Tune in later for more details.

2. SU(3) TESTS: $\pi N \to \Delta\pi$ vs $\bar{K}N \to \Sigma(1385)\pi$.

It is well known that one of the major triumphs of SU(3) has been the agreement between "isoscalar coefficients" c_i and experimental signs for amplitudes $T_{1\alpha}$ for reactions like

$$K^-p \to (\tfrac{3}{2}^- \text{ nonet}) \to \bar{K}N; \qquad T_{11} \propto g_1^2\, c_1 c_1$$
$$\to \Sigma\pi; \qquad T_{12} \propto g_1^2\, c_1 c_2 \qquad (1)$$
$$\to \Lambda\pi; \qquad T_{13} \propto g_1^2\, c_1 c_3$$

According to SU(3) these are all examples of elastic scattering with a coupling constant g(actually g_D and g_F but let's ignore that annoyance), so the coefficients c_i are just generalizations of Clebsch-Gordan coefficients. So far, about six SU(3) multiplets are established (2 nonets, 2 octets, 2 decuplets) wich satisfy about 20 sign checks. I repeat that this cannot just be good luck because $2^{-20} \approx 10^{-6}$.

Now we can compare our amplitudes for $\pi^-p \to (\frac{3}{2}^-, \frac{5}{2}^-$, and $\frac{5}{2}^+) \to \Delta\pi$ with a CHS analysis[3] of $\bar{K}^-p \to$(same)$\to \Sigma(1385)\pi$. The relative signs of the two $J=\frac{5}{2}$ waves agrees with SU(3); the sign for DD_{13} if we chose our 1973 solution.

3. COMPARISON OF $\Delta\pi$ AMPLITUDES WITH SU(6) AND QUARK MODEL

In SU(6) the nucleon and Δ belong to the same $\underline{56}$ supermultiplet, so elastic scattering generalizes to reactions like

$$\pi N \to (\underline{70},\ L^P = 1^-) \to N\pi; \qquad T_{11} \propto g_1^2\, c_1^2$$
$$\text{and} \qquad \to \Delta\pi; \qquad T_{12} \propto g_1^2 c_1 c_2. \qquad (2)$$

We have to "waste" one reaction to define the overall phase (and thus the sign of c_2/c_1), but then the other reactions via the same $\underline{70} \to \underline{56} \times \underline{\pi}$ serve as sign checks. Unfortunately, just as in SU(3) there were really two cou - plings ($g = g_F + g_D$), so in SU(6) there are again two for the $\Delta\pi$ case , this time because of the fact that a resonance can decay into $\Delta\pi$ via two different values of L' (e.g. $D_{13} \to \Delta DD_{13}$ and ΔDS_{13}, see Sect. 4a. of Lecture I).

It is the open choice of the relative sign of the two L' couplings which leads to the two alternative columns of Table 1, labelled either:

Faiman-Rosner	→ "$SU(6)_W$"	"Anti-$SU(6)_W$"
Gilman-Kugler-Meshkov	→"$(8,1)_0 - (1,8)_0$"	"$(3,\bar{3})_1 - (\bar{3},3)_{-1}$"

The Faiman-Rosner names are old, based on the argument about the two different values of L'. The group-theoretical names come from the transformations studied by Melosh[4] who showed that indeed there are two couplings, with those two transformation properties. Note on Table 1, however, that the Quark Model does not put up with this ambivalence -- it predicts a unique column, which corresponds to the Anti-SU(6) choice.

Table II-1

Signs of the amplitudes for $\pi N \rightarrow N^* \rightarrow \pi\Delta$ for N^*'s in the 70 L = 1 and 56 L = 2. Products of the theoretical and experimental signs for decays through the $(8,1)_0 - (1,8)_0$ and $(3,\overline{3})_1 - (\overline{3},3)_{-1}$ terms are presented, with the overall phase chosen so that DD13(1520) is positive. Signs which are independent of which term dominates are denoted by a "*". Experiment and theory agree if within the 70 L = 1 or 56 L = 2 decays all the signs in a column agree.[a)]

		Faiman-Rosner[5)] → SU(6)$_w$ Gilman-Kugler-Meshkov[6)] → $(8,1)_0 - (1,8)_0$	Anti-SU(6)$_w$ $(3,\overline{3})_1 - (\overline{3},3)_{-1}$	Moorhouse[2)] and Parsons quark model	
70 L=1 → 56 L=0	DD13(1520)	+*	+*	+*	
	DS13(1520)	-	+	+	
					Energy gap[a)]
	SD31(1640)	+	-	?[b)]	
	DS33(1690)	+	-	?[b)]	
	DS13(1700)	+	- ?[c)]	?[b)]	
	DD15(1670)	-*	-*	-*	
56 L=2 → 56 L=0	FP15(1688)	-	+	+	Only disagreement
	FF35(1880)	-*	-*	-*	
	FF37(1950)	-*	-*	-*	

a) Because of experimental inability so far to bridge 100 MeV gap between 1520 region and 1688 region, signs so far need not check across this gap.
b) Moorhouse's " ?" means he feels the experiment is uncertain.
c) ? in Anti-SU(6)$_w$ column means we feel experiment is uncertain.

Table II - 1 is taken from Gilman et al.[6)] (but Faiman and Rosner give the same prediction); it gives the product of theoretical signs with our experimental signs for our 1972 solution. (For our 1973 solution change the two signs below the energy gap with respect to the 7 signs above). We see that the theorists badly need our 1973 solution (in fact they helped us find it). Only for that solution can one find a complete column of minus signs, by choosing Anti-SU(6) for 70 decays and Straight-SU(6) for 56 decays. As usual, tune in again after Saclay helps us bridge the gap.

4. ESTIMATING RESONANCE PARAMETERS "BY EYE", WITHOUT A K-MATRIX.

We are through with the glamorous problem of signs and higher symmetries. The rest of this lecture deals with a more pedestrian question: "Is there a reliable way to parametrize a resonance?" We shall exploit our multichannel amplitudes to find two consistent descriptions, which are compared in Table 4. But be careful: some partial widths will differ by a factor two, depending on which description you chose.

Before we go on to Fancy Method I (K-matrix fits and T-matrix poles), I present in Table 2 the most conventional method of all --- "eye-ball" fits to the Argand plots of Lecture 1 according to the following recipe:-

1) We look at all Argand plots coupled to a given incoming partial wave (e.g. all three P_{11} channels of Fig. 7 of Lecture I), giving the most weight to the ones which look most resonant, and pick an energy where they all simultaneously seem to have the greatest speed. This is called the resonance energy. We then draw semi-circles through the points near the greatest speed, and estimate the radius r of each circle. Then $r = \sqrt{X_{el} X_{\alpha}}$ (see below).

2) We get help from Elastic Phase Shift Analyses (EPSA, Refs. 7 and 8) in two ways:

a) The Argand plots have already had their phases set to agree with some resonance seen strongly in EPSA -- e.g. P_{11} near 1520 MeV (see Lect.I, Sect. 6.2).

b) We use the EPSA values of Γ_{tot} and X_e in order to calculate X_{inel}. The numbers appearing in Table 2 in each inelastic channel are $\sqrt{X_{el} X_{inel}}$ and (below that) Γ_{inel}; the final column corresponds to the sum of the branching fractions for the given resonance. It should be noted that X_{inel} are very sensitive to variation in the X_{el}, the elastic branching fraction.

Finally, one might note that in many cases all decay modes of the resonance are essentially accounted for ($\sum X_i \approx 1$).

Table II-2

Resonance couplings estimated by eye for $N\pi\pi$ channels, with help from EPSA[7,8]. Each entry contains the partial wave considered, the amplitude at resonance and the partial width in MeV.

Resonance	E (MeV)	Γ_{tot} (MeV)	x_{el}	$\pi\Delta$	$\pi\Delta$	$N\rho_3$	$N\rho_1$	$N\epsilon$	Σx_i
P11	1440	236	.52 124	PP11 +.29 36				PS11 -.25 28	.80
D13	1520	119	.57 68	DS13 -.27 15	DD13 -.21 9.0	DS13 +.31 20.0			.94
S31	1630	160	.32 51	SD31 -.325 52			SS31 +.307 47		.95
D15	1670	141	.40 56	DD15 -.46 75					.93
F15	1690	133	0.6 80	FP15 +.31 21		FP15 +.27 16		FD15 +.24 13	.97
D33	1670	207	0.16	DS33 +.37 172					.99
S11	1700	148	0.50 74				SS11 +.19 11	SP11 -.35 35	.81
D13	1730	130	0.10 13	DS13 +.11 17				DP13 -.29 109	1.07
P11	1750	183	.15 28	PP11 -.345 140				PS11 +.21 52	1.21
P13	1850	250	.25 63				PP13 -.44 195		1.03
F35	1890	260	.15 40		FF35 +.10 16	FP35 -.29 140			.75
F37	1930	230	.40 92	FF37 +.25 36		FF37 -.25 36			.71

MULTICHANNEL K-MATRIX AND T-MATRIX FITS

Our ability to account for all of the πN inelasticity in many partial waves indicates hat we are now in the position to perform multichannel fits, exploiting the constraints of unitarity to their fullest possible extent, in attempting to understand the πN interaction.

For this purpose we used a K matrix to parametrize our T-matrix elements obtained from our isobar model fitting program.

It is well known that for a partial wave which is coupled to several particle states, a real K matrix can be related to the Argand amplitudes by[10)]

$$T_{ij} - K_{ij} = i \sum_{l} T_{il} Q_l K_{lj}, \tag{3}$$

where the Argand amplitude is related to T by

$$A_{ij} = Q_i^{1/2} T_{ij} Q_j^{1/2} \tag{4}$$

and Q is a diagonal matrix corresponding to the c. m. momentum of the particles in each channel.

We now can make a reduced K-matrix equation by putting in the barrier penetration factors. We let

$$\begin{aligned} K_{ij} &= B_i^{1/2} k_{ij} B_j^{1/2}, \\ T_{ij} &= B_i^{1/2} \tau_{ij} B_j^{1/2}, \end{aligned} \tag{5}$$

where B is the Blatt-Weisskopf[11)] barrier factor. Thus Eq. (3) becomes

$$\tau_{ij} - k_{ij} = i \sum_{l} \tau_{il} Q_l B_l k_{lj} \tag{6}$$

and Eq. (4) becomes

$$A_{ij} = Q_i^{1/2} B_i^{1/2} \tau_{ij} Q_j^{1/2} B_j^{1/2}. \tag{7}$$

In order to extend this prescription to isobars which do not have a fixed mass, we replace $Q_l B_l$ by their weighted average value $\overline{Q_l B_l}$, where $\overline{Q_l B_l}$ is defined by the integration of $Q_l B_l$ over a normalized Dalitz plot projection of this isobar's diparticle mass[12)].

Because the isobars are not an othogonal set, they have an overlap with respect to one another. So off-diagonal terms will enter into the momentum matrix. Thus Eq. (6) becomes

$$\tau_{ij} - k_{ij} = i \sum_{lm} \tau_{il} \Delta_{lm} k_{mj}, \tag{8}$$

where Δ_{lm} for the diagonal terms are

$$\Delta_{11} = \overline{Q_1 B_1} \qquad (9)$$

and the off-diagonal terms are related to the overlaps between the isobar states.

The reduced K matrix is then parametrized by simple factorizable poles and linear background terms which are not factorizable. These parameters are then adjusted to fit the Argand amplitudes. The resulting K-matrix parameters yielded "ridiculous" values for the masses and partial widths of the resonances (i.e., if interpreted literally they correspond to resonances which are shifted by ~ 100 MeV from their nominal value and have much greater widths than expected from inspection of the Argand diagrams). This is not surprising, the K-matrix is merely a good way to parametrize the T-matrix in terms of real numbers, and K-matrix pole positions and residues even change along with the number of channels considered.

5.1 Poles of the T-matrix

If we have a good representation of the Argand diagrams, this implies that we have a comparatively good description of the T matrix as a function of energy. In order to identify resonances and their properties we now search the T matrix for poles in the complex energy plane and determine the residues at these poles. The motivations for this procedure are:

i) we expect the pole positions and residues in the T matrix to be independent of our parametrization of the T matrix, providing, of course, that it is good. This expectation stems from the work on the P33(1236) resonance.[13,14] and investigation of our own[12];

ii) we expect the pole position and residue to be closely related to the Breit-Wigner parameters but the pole position does not equal M_0, $1/2\ \Gamma_0$, the conventional Breit-Wigner parameters, and the residues are not necessarily equivalent to the widths. We expect these equalities to become very poor when we either have large backgrounds or wide resonances.

The results of these investigations are contained in Table 3, where we give the real and imaginary parts of the pole position together with the residues of the $\tau_{\alpha\alpha}$ matrix scaled by $2 \times \overline{Q_\alpha B_\alpha}$ calculated at an energy $E = \text{Real}(E_{pole})$. These will correspond to the partial widths, and the residues of the τ matrix correspond to the couplings. Several comments about these results are in order:

i) often the pole positions are a long way from the position one might expect, e.g., F35, F37, or P13;

ii) $1/2\ \Sigma|\Gamma_i| \neq -\text{Im}(E_{pole})$ in many cases (where Γ_i is given by $\Gamma_i = 2\times\overline{Q_i B_i}\ (\text{res}_i)^2$). However, it should be noted that even a pure Breit-Wigner will not have this property. The way we have defined Γ_i gives the closest agreement with the equality for a pure Breit-Wigner[12].

If the background becomes large, the disagreement becomes worse.

iii) The last point is further emphasized by the fact that the residues have large phases even after taking into account the phases associated with the kinematical factors.

Table II-3

T-matrix poles and residues. Partial widths Γ_i are calculated by $\Gamma_i = 2 \times \overline{Q_i B_i}$ (evaluated at E = Re E(pole)) $(\text{residue}_i)^2$. Entries for Γ_i are $\Gamma_{real} / \Gamma_{imag.} / |\Gamma|$ in MeV.

Wave	Pole	$\Gamma_{\pi N}$	$\Gamma_{\pi \Delta_L}$	$\Gamma_{\pi \Delta_{L'}}$	$\Gamma_{N\rho_3}$	$\Gamma_{N\rho_1}$	$\Gamma_{N\epsilon}$	Other channel		$\Gamma_{tot} = \Sigma \lvert\Gamma_i\rvert$
S11	$1503-i\frac{65}{2}$	7 -6 9				6 0 6	23 35 42	2 10 10	(ηN)	67
	$1652-i\frac{100}{2}$	26 -37 45				-3 -9 9	-2 -4 5	5 -32 32	(ηN)	91
P11	$1385-i\frac{235}{2}$	36 -109 115	21 -25 33				5 -5 7			155
	$1724-i\frac{283}{2}$	-39 -115 122	47 5 47				31 -56 64			233
P13	$1728-i\frac{159}{2}$	1 -25 25				42 -73 84				109
D13	$1514-i\frac{142}{2}$	88 13 89	5 36 36	3 14 14	34 6 34		-3 0 3			176
	$1647-i\frac{117}{2}$	5 -15 16	-8 -22 24	0 -2 2	-1 4 4		-57 -32 65			111
D15	$1666-i\frac{159}{2}$	68 -14 69	91 -10 92							161
F15	$1672-i\frac{155}{2}$	99 -17 101	5 11 12		33 -27 42		15 -16 21			182
S31	$1600-i\frac{79}{2}$	-3 -20 20	22 16 27			-5 102 102				149
D33	$1657-i\frac{109}{2}$	7 -4 8	-9 -49 50		36 9 37					95
F35	$1824-i\frac{282}{2}$	36 -26 44	19 -18 26		-20 -105 107					177
F37	$1866-i\frac{255}{2}$	33 26 42	-5 29 29		-21 4 22			41 -31 51	(Junk)	144

The implication of these statements is that it is not easy (and sometimes impossible) to relate pole parameters to the parameters of the Breit-Wigner amplitude which we normally discuss. This point will be demonstrated more in the following sections.

It does appear, however, that these pole parameters are unique (if calculated in the same manner with equally good fit to data), and thus it will be necessary for any future theories to present the results on resonances in terms of the properties of the corresponding second sheet poles (or whichever sheet is appropriate in the specific multichannel problem).

6. UNITARY (BREIT-WIGNER AND BACKGROUND) FITS TO THE T-MATRIX

In order to better estimate the conventional Breit-Wigner parameters, we assume that in the region of a pole our T-matrix amplitude can be described as a Breit-Wigner plus unitary background:

$$T(\text{refit}) = T^{BW} + T^{Bkgd}, \tag{10}$$

where

$$T^{BW}_{ij} = \frac{\frac{1}{2}\,\Gamma_i\Gamma_j}{E_R - E - \frac{i}{2}\sum_k \gamma_k^2 Q_k}, \tag{11}$$

$$\Gamma_j = Q_j^{\frac{1}{2}}\,\gamma_j\,e^{i\theta j}, \tag{12}$$

and the background S-matrix ($S = 1 + 2iT$) is separately unitary,

$$S^{Bkgd}_{ij} = \delta_{ij} + 2i\,Q_i^{\frac{1}{2}}\,T^{Bkgd}_{ij}\,Q_j\ . \tag{13}$$

As in Eqs. (3) and (4), T^{Bkgd} is parameterized as a K-matrix (this time a linear function of E) and the BW phase θ_j is adjusted by a matrix unitarity constraint [15]

$$S^{Bkgd}\,\Gamma^* = \Gamma. \tag{14}$$

Unfortunately a general multichannel solution of Eq. (14) is not possible, so we added Eq. (14) as an additional chi-square term and fitted θ_j as a polynomial in E [12].

One can see that as the number of channels increases the number of parameters rises sharply. Because of this limitation, these fits are time-consuming.

Once the parameters of the B.W. refit are found, we can recalculate T(refit) via Eq.(10) and again hunt for its poles. We find that this (indirect) pole is close to that of the original amplitude, if (and only if) we have imposed the unitarity constraint (14). This agreement must mean that, near a resonance, Eq. (10) is a good approximation. Thus we have found a self-consistent way to parameterize a resonance, but we repeat our earlier warnings about the differences between pole parameters and BW parameters, both of which are plotted in Table 4:

1) The 2 sets of parameters do not (and cannot) always agree.

Table II-4.

Comparison of resonance parameters from (a) coupling estimate and elastic phase shift analysis; (b) poles of the T-matrix; and (c) unitary (Breit-Wigner + background) fit.

	M	Γ_{tot}	πN	$\pi\Delta$	$\pi\Delta$	$N\rho$	$N\epsilon$	
D15	1670	141	56	75				Elastic/coupling estimate [a]
	1666	159	69	92				T-matrix pole [b]
	1692	176	71	105				Unitary (BW + background) [c]
F15	1690	133	80	21		16	13	Elastic/coupling estimate [a]
	1672	155	101	12		42	21	T-matrix pole [b]
	1682	153	88	15		33	17	Unitary (BW + background) [c]
F35	1890	260	40	16		140		Elastic/coupling estimate [a]
	1824	282	44	26		107		T-matrix pole [b]
	1907	324	51	55		219		Unitary (BW + background) [c]
D13	1520	119	68	15	9	20		Elastic/coupling estimate [a]
	1514	142	89	36	14	34		T-matrix pole [b]
P13	1850	250	63			195		Elastic/coupling estimate [a]
	1728	159	25			84		T-matrix pole [b]

a) using results from elastic analyses[4)] (Breit-Wigner and background fit to elastic Argand diagram) together with "eye-ball" estimates of coupling from Argand diagram;

b) T-matrix pole quantities from the K-matrix parametrization;

c) unitary (Breit-Wigner plus background) refit to smooth Argand diagrams from K-matrix parameters.

2) The BW parameters depend on the form chosen for the background, while of course the parameters of the pole itself should be stable against changes in the form of the background.

7. RESULTS FROM THE DIFFERENT RESONANCE PARAMETERISATIONS

In Table 4 we have compared the various parameters obtained for the resonances by the several methods discussed above:

We think the lessons of this table are clear:

i) for clear narrow resonances, e.g., D13, F15, D15, one obtains reasonable qualitative agreement although quantitatively there are factors of 2(or more) disagreement in partial widths;

ii) for wide resonances, e.g., P13, F35, the displacement of M (conventional) and Real (E_{pole}) can be on the order of 100 MeV.

The above observations mean that one should be wary of using quoted resonance parameters without checking their origin and, further, the partial widths are only reliable to factors of ~ 2.

8. CONCLUSIONS

1) We have measured 23 couplings in sign and magnitude and this will be an important testing ground for any new theories.

2) It is possible to obtain good representation of the Argand diagrams in all channels and then extract the pole structure of the T matrix. This has been done for all the resonances we observe with $E < 2000$ MeV.

3) It is not possible in general to relate the pole parameters unambiguously to the parameters of Breit-Wigner. In order to obtain such quantities it is necessary to make a fit to the data with a unitary model resonance plus background. We have obtained such for three pronounced resonances.

4) The uniqueness of the pole parameters indicated in analyses of elastic P33 amplitude seems to be present in the inelastic waves we have considered.

5) The various theoretical calculations are consistent with our results only for our 1973 continuation across the energy gap. It is clearly essential to obtain partial wave amplitudes in this region as soon as possible.

9. ACKNOWLEDGEMENTS.

I want to thank Frixos Triantis for carefully reading these lectures and suggesting improvements. I am grateful to DPhPE at Saclay, and particularly to Mme. M. Thellier, for typing this manuscript.

REFERENCES FOR LECTURE II.

1) R.G. Moorhouse and H. Oberlack, Phys. Letters 43B, 44 (1973); R.G. Moorhouse, H. Oberlack and A.H. Rosenfeld, LBL-1590 (submitted to Phys. Rev. (July 1973)).

2) R.G. Moorhouse and N. Parsons, Comparison of the Quark Model with SU(3) Inelastic Amplitudes University of Glasgow, G128QQ. (submitted to Nucl. Phys. B. (July 1973)).

3) CERN-Heidelberg-Saclay Collaboration, J. Prevost et al. DPhPE 73-07, June 1973.

4) H. Melosh IV, California Institute of Technology preprint, June 1972 (unpublished).

5) D. Faiman and J. Rosner, Ref. TH-1636 CERN (1973).

6) F. Gilman, M. Kugler, and S. Meshkov, SLAC Report SLAC-PUB-1235 (1973).

7) R. Ayed, P. Bareyre, Y. Lemoigne, contributed paper to the XVI International Conference on High Energy Physics, Batavia, Illinois (1972).

8) S. Almehed and C. Lovelace, Nucl. Phys. B40, 157 (1972).

9) Particle Data Group, Phys. Letters 39B, 1 (1972).

10) H. Pilkuhn, The Interactions of Hadrons (John Wiley, New York 1967), Ch. 8.

11) J. M. Blatt and V. F. Weisskopf, Theoretical Nuclear Physics (John Wiley, New York, 1952).

12) R. S. Longacre (Ph. D. thesis), Lawrence Laboratory Report LBL-948 (1973), unpublished.

13) T. A. Lasinski and A. Barbaro-Galtieri, Phys. Letters 39B, 1 (1972).

14) J. S. Ball, R. R. Campbell, P. S. Lee, and J. L. Shaw, Phys. Rev. Letters 28, 1143 (1972).

15) C. J. Goebel and K. W McVoy, Phys. Rev. 164, 1932 (1967).

DISCUSSIONS

CHAIRMAN: Prof. A.H. Rosenfeld

Scientific Secretaries: R.T. Ross, M. Leneke

DISCUSSION No. 1

- *ROSS:*

Is the variation of results found for resonance position and widths [for example, in FF35(1890)] due just to the different definitions in the various methods. Or, for example, if you had much better statistics would the different methods give closer results? Also are the values you obtain influenced by the way you parametrize the background terms?

- *ROSENFELD:*

I am sure that if the data were better the signal would show up better above background, so things would improve slightly. But I think there is just an intrinsic disagreement between the position of pole and something that is not going to be a perfect circle on the Argand plot, but that we approximate by a circle.

As far as changing the background parametrization is concerned, I do not know. One experimental problem is that of the large amount of computer time required, so we do not try all possibilities. However, putting a quadratic term in the background or parametrizing in terms of a different variable could perhaps change things a bit.

- *RABINOVICI:*

What is known experimentally about the density of resonances per unit mass as a function of the mass?

- *CHU:*

Hagedorn has published plots in support of his thermodynamic model, which show that the density of resonances per unit mass rises approximately exponentially in mass, up to the highest mass region studied where of course the density falls due to lack of experimental evidence.

- *RABINOVICI:*

Is there experimental evidence for another Roper resonance and if so, is the slope of $d\sigma/dt$ in its diffractive production known? Is there an explanation of the fact that the diffractive slope of the Roper resonance is different from the other N^* production slopes?

- *ROSENFELD:*

From the most recent Saclay solutions (Ayed et al.) there is some indication of the existence of a second Roper resonance, but this state is not required by the 1972 CERN solution.

As to the slope of $d\sigma/dt$: since the quantum numbers of the P_{11} Roper resonance are $J^P = \frac{1}{2}^+$ (identical with the proton), I guess that they should be peripherally produced very strongly, hence the very large slope. However, the data are not yet good enough to distinguish slopes for the rumoured two different Roper twins.

- *GENSINI:*

You mentioned in your lectures results on only P-, D-, and F-waves. Did your analysis see any new features in S_{11}- and S_{31}-waves?

- *ROSENFELD:*

No, but this is not the best technique for looking for new resonant states or new features of partial waves. Such discoveries are best made in very high statistics πN elastic scattering experiments.

- *NAHM:*

SU(6) predictions and experiment disagree on one sign which has no "star". How reliable is that sign?

- *ROSENFELD:*

The stars have been placed by theoreticians; they mean that there are no particular difficulties with configuration mixing or similar problems.

- *PARSONS:*

The sign for this "bad" resonance depends on the relative size of a quark recoil term and a normal recoil term. This size depends on the wave function. We used four-dimensional harmonic oscillator wave functions, but for a different interaction one might get a different result.

- *ROSENFELD:*

Let me make one remark. For all of the resonances which have stars, the three groups of theoreticians (Moorehouse and Parsons; Gilman et al; Faiman and Rosner) get the same predictions from different approaches. Therefore one may have confidence in the "starred" predictions.

- *EYLON:*

I did not understand how one can determine experimentally the relative phase of resonant amplitudes.

- *ROSENFELD:*

We look for the interference between two adjacent resonances by studying the behaviour of the amplitudes in the region in which the two resonances overlap. In some cases it is convenient to use the phase of some intermediate resonance as a reference.

- *WILCZEK:*

Suppose you have charmed baryons that decay weakly (with narrow widths corresponding to lifetime $\sim 10^{-13}$ sec) but with large phase space (suggesting decays predominantly into multiparticle final states). How heavy would these particles have to be in order to have missed detection?

- *ROSENFELD:*

Of course charmed states could not be seen in formation, since such a state must be produced in association with another charmed state. However, the detection of similar (uncharmed) states still presents difficulties such as:

1) The limit of detection depends on the product of production cross-section x width. These limits are tabulated in many papers.

2) One usually deals with two-body final states. Therefore partial widths are important.

3) In our bubble chamber experiment, the energy bins were of width 40 MeV, so a very narrow resonance may be diluted by binning.

As far as charmed states are concerned it is worth while remembering a comment made by Prof. Miller earlier in this School: at momenta of a few 100 GeV/c, particles with these lifetimes would leave measurable tracks in, for example, a bubble chamber and so these particles would be easily identified.

- *HENDRICK:*

You mentioned that the sign agreement between experiment and the SU(6) predictions seem to be quite good for γN resonances and not good for ρN resonances. Assuming that this trend continues as experimental data comes in, you might consider this to be a failure of the vector dominance model (VDM). It seems, however, that such evidence would not merely rule out VDM but also raise the question of why SU(6) works so well in predicting the γN signs, but makes poor predictions for ρN signs.

- *ROSENFELD:*

Certainly, what you say would be a reasonable implication. However, we feel that the techniques of relating γN and ρN coupling constants via VDM are not totally reliable since they are complicated by different kinematics. The success of SU(6) predictions, say for the $\Delta\pi$ decays certainly means we cannot say SU(6) is good only for electromagnetic decays!

- *YELLIN:*

Does your maximum-likelihood program come out with statistical errors on the parameters, and could these errors account for the difference between the two methods of getting resonance parameters for FF35?

- *ROSENFELD:*

The statistical errors are calculated in the usual way. But then these errors are multiplied by a factor of 4 to account for model differences. This is the recipe used in other πN phase shift analyses. The statistical errors do not account for the variation of results, although they may hide some of the divergencies at the present level of statistics. However, in future experiments I believe the differences will still remain and are just associated with fundamental differences in the definition of these quantities. On the Argand plot the circles were not too bad, but we have to decide on a suitable set of parameters to give to phenomenologists.

- *ZICHICHI:*

I have fundamental doubts on all these matters when I see a circle that is small compared with the background. Given a set of data produced by Monte Carlo, what is the probability that you produce resonances with these shapes on an Argand diagram?

- *ROSENFELD:*

This is not a case of looking for small signals buried in a sea of background. They are pretty good circles.

- *ZICHICHI:*

I do not agree. I still believe this could be caused by statistical fluctuations. Has anyone tried to analyse Monte-Carlo-produced events in a pole-hunting search?

- *ROSENFELD:*

No, this has never been done.

- *TING:*

I too am worried. Consider a very high mass state, say πp at 100 GeV with a width of 5 GeV. Your method is clearly useless in this case. All the resonances you consider, tie in with a specific phenomenological model. Is there any clean way to sort out this problem?

- *ROSENFELD:*

No. But I am convinced that a lot of these states are well-defined resonances; we just have to decide how to parametrize them properly.

- *MONTANET:*

I should point out that we did an analysis similar to your πN, on $K^+p \to K^0 p\pi^+$ using some 100,000 events at low energy. We have the same problem with final states like $\Delta^{++}K^0$ and K^*p, but of course we do not expect poles in K^+p. And we do succeed in fitting all the characteristics of this reaction with only a background type of term and no poles.

- *ROSENFELD:*

This is a very good example. There has been strong psychological pressure to find a K^+p pole for the last four years, but no has been able to find any. This then gives us confidence in the poles found in our πN analysis.

- *YELLIN:*

What do you mean when you say you fit a Breit-Wigner "with unitarity"?

- *ROSENFELD:*

The K-matrix solution is certainly unitary. Instead of fitting to only the channels that have been experimentally observed, we fit to all the Argand diagrams that are given by the K-matrix solution, and if the fit is good the new parameterization for the amplitude will also have to give a unitary S-matrix.

- *YELLIN:*

The T-matrix resonances have poles at complex masses with complex residues. The quark model gives particles at real mass with real coupling constants. But in the K-matrix, poles are at real energy with real residues. This property makes it tempting to consider the quark model as applying to the K-matrix rather than to the T-matrix. Has any theoretician considered this?

- *ROSENFELD:*

I do not see why Nature should recognize the K-matrix, since the K-matrix is just a mathematical trick to parametrize a complex unitary S-matrix in terms of real quantities.

- *WAMBACH:*

You introduced your program which fits a lot of resonances simultaneouly. What are the future prospects following this way? Do the available or potential computer facilities allow one to extend such a program to higher energies, for example, up to 3 GeV?

- *ROSENFELD:*

Concerning computer costs, we have been spending 5000 dollars per month for several years, and I am reluctant to keep this up until we see how much interest this work attracts. Of course, higher quality new data [e.g. $\pi N \rightarrow \pi N$ with "A" and "R" measurements, or πN(polarized) $\rightarrow N\pi\pi$] can be analysed with less computer cost.

- *WAMBACH:*

Can you comment on the possibility of constructing a phase shift analysis program that can use the parameters of a Breit-Wigner fit in an iterative way for say the K-matrix fit, and hence obtain some unique solution as the iteration converges.

- *ROSENFELD:*

One could certainly construct a big program to handle all events at all energies simultaneously, and then one could do energy-dependent fits with various parametrizations, but this was not quite what you had in mind.

- *WAMBACH:*

No, the idea is to build an iterative scheme. We know there should be resonances and that these should be fitted with Breit-Wigners.

- *ROSENFELD:*

I do not agree! Chew would say we know there are poles in the S-matrix in the complex plane, but they probably will <u>not</u> look much like Breit-Wigners. Only a one-channel resonance should be well described with a Breit-Wigner (like the P33 for example). As far as your iterative scheme is concerned, I do not think we should spend lots of time and money on such a scheme. It would be better if some theoreticians got together and sorted out what the parametrization of resonances should be.

- *YU:*

Would you comment on the impact on phase-shift analysis of the method of optimized polynomial expansion of Ciulli and Cutosky and Deo. What have we learned about baryon resonances using this method?

- *ROSENFELD:*

This is an interesting point. Cutosky et al., developed the ACE method perhaps a bit late in the game. I believe it is the most powerful method available, but so far it has only confirmed results of previous analysis; no striking new discoveries have been made by this method.

MESON DAUGHTERS - REALITY OR FICTION?

L. Montanet

Table of Contents

MESON DAUGHTERS - REALITY OR FICTION?

L. Montanet, CERN, Geneva

1. REGGE TRAJECTORIES FOR MESONS

It is well known that a Regge trajectory corresponds to a physical particle when it crosses alternate integral values of spin J (the energy squared S being positive). Thus, Regge trajectories connect particles which differ in spin by two units, all other quantum numbers (baryon number B, hypercharge Y, Isospin I, parity P) being constant.

The mass separation of two particles on the same trajectory depends on the "slope" of this trajectory, for which we have no a priori information if we except the linear extrapolations which can be attempted from the variations of some differential cross sections with energy - (for instance, the slope of the ρ trajectory can be deduced from a fit to the charge exchange differential cross section: $\pi^- p \to \pi^o n$ measured between 4 and 20 GeV/c).

In general, these slopes are found to be close to 1 GeV^{-2}, but the extrapolation method is limited to the very few cases where a single trajectory dominates the reaction. Moreover, we do not know to what extent the linear extrapolation is valid, if at all.

A direct determination of the Regge trajectories in the region where S is positive requires the knowledge of at least two particles with identical quantum members but for their spins, which must satisfy $\Delta J = 2$. The only well established example for mesons is the ρ - g doublet[1]. There are however several other "candidates", and before discussing the plausibility of the "daughters", let us review briefly the status of the "parents".

Figs 1, 2 and 3 are the Chew-Frautschi[2] plots (Real part of the spin α against $S = m^2$) for isovector, isodoublet and isoscalar mesons, respectively.

In fig. 1, we see that the ρ - g trajectory has an intercept at S = 0 approximately equal to the value 0.5, deduced, as already mentioned, from the analysis of πN charge exchange scattering experiments. Another remarkable feature of this trajectory is that the A_2 particle lies very close to it. If we assume that the "A_2 trajectory" has the same slope than the ρ - g one (no reliable 4^+ particle has yet been observed and the 0^+ member of the A_2 trajectory may very well have a negative value for S, therefore not corresponding to a physical particle), we infer that these two trajectories of opposite signature coïncide almost exactly. In terms of Regge pole theory, this means that there is no significant "exchange force" since the forces generating even signature trajectories depend on the sum of the t and u channel singularities, whereas for negative signature, it is the difference which matters[3]. The absence of u-channel exchange for mesons can be explained by the absence of exotic state in the quark model.

In addition to the ρ, A_2, g mesons, only two other isovector mesons are well known: the π and the B-(respectively $J^P = 0^-$ and 1^+). This is obviously not enough to build up with confidence another trajectory. In any case, we know that the mass of the π mesons (and of the pseudoscalar nonet in general) may be misleading for the classification under consideration.

It may be however interesting to note that two "candidates" for isovector mesons, the A_1, with $J^P = 1^+$, and the F_1 with $J^P = 2^-$ [4], fall on a "π trajectory" nearly parallel to the ρ - g trajectory.

But, after the analysis of the 3π system by Ascoli et al.[5], it is not clear that the A_1 is a $J^P = 1^+$ $\rho\pi$ resonant system. As concerns the F_1, the experimental evidence for its existence and its quantum numbers is still too limited to draw definite conclusions. The F_1 is essentially observed as a $K^* K$ enhancement in four-body annihilations

$$\bar{p}p \to K\bar{K}\pi\pi,$$

after the abundant double K^* production has been removed.

On the other hand, the B ($J^P = 1^+$) and A_3 ($J^P = 2^-$) could be associated in a unique degenerated trajectory, but the comments we have made for the A_1 are also valid for the $J^P = 2^-$ $f\pi$ enhancement asociated with the A_3 [5].

The $\delta(970)$ is in such a controversial experimental situation[1] that it seems premature to comment on its trajectory.

The situation of the isodoublets, the K^*, (fig. 2) presents several similarities with the isovectors. If we assume that the $K^*(1760)$ observed by Carmony et al.[6] decaying in particular into $K\pi$, is a $J^P = 3^-$ particle, we can build a 1^-, 3^- trajectory which has, as the ρ - g one, an intercept at S = 0 close to 0.5 and which embodies the $J^P = 2^+$ $K^*(1400)$.

The only other $I = \frac{1}{2}$ meson which is well known is the $J^P = 0^-$ K(494) particle and from there, one can draw trajectories which go through one of the two $J^P = 1^+$ candidates and through the $J^P = 2^-$ K(1770), the L meson, but these particles have to be understood better before drawing firm conclusions.

The Chew-Frautschi plot of the isoscalar particles (fig. 3) is more populated, if not better ordered. The $\omega(1675)$, first observed by Armenise et al.[7], is a I = 0 ω-object which seems to be natural spin-parity assignment.

Although its spin is not yet known, it seems natural to interpret this resonance as a recurrence of $\omega(784)$[8]. We can then draw an ω-trajectory which embodies the $J^P = 2^+$ f^o particle.

Similarly for the ω, but even less convincing, is the evidence for a recurrence at 1830 MeV[9], which suggests a trajectory close to the $J^P = 2^+$ f'.

The analogy between the ω-f on one hand, and the ϕ-f' on the other hand, can be taken as a result of the closeness of the mixing angles of the 1^- and 2^- nonets.

If we ignore the pseudo-scalar resonances, the η and η', for which the mixing is abnormal, we are left with only one isoscalar resonance of spin different from zero, the D-mesons: as long as we do not know the other isoscalar partner of the $J^P = 1^+$ nonet, it is difficult to incorporate the D in our present analysis.

The remaining isoscalars, in particular the $J^P = 0^-$ and 0^+ candidates, will be discussed later.

2. REGGE DAUGHTER TRAJECTORIES

When spin and reaction of unequal mass particles are introduced in the simple Regge pole theory, the resulting scattering amplitudes exhibit singularities at t = 0. P.D. Collins[3] shows that the helicity amplitude can be written:

$$A_{H_t}(s,t) = -\beta_H(t)\ \frac{e^{-i\pi\alpha} + \sigma}{2\sin\pi\alpha}\ f_H(\alpha)\ \xi_{\lambda\lambda'}(z_t)\ \times$$

$$\left\{ \left(\frac{S}{4S_o}\right)^{\alpha-M} + \Delta(\alpha-M)\, 4S_o \left(\frac{S}{4S_o}\right)^{\alpha-M-1} + \left[\frac{(\alpha-M)(\alpha-M-1)}{2}\,(4S_o\Delta)^2 + a_1(\alpha)\left(\frac{q_{t_{13}}\, q_{t_{24}}}{S_o}\right)^2\right]\left(\frac{S}{4S_o}\right)^{\alpha-M-2} + \cdots \right\}$$

where:

$$\beta_H(t) = 16\pi\,(2\alpha+1)\, t^{\delta}\, K_{\lambda\lambda'}(t)\ \gamma_H(t)\ (-1)^{(\lambda-\lambda'-|\lambda-\lambda'|)/2}$$

$$\lambda = \lambda_1-\lambda_3,\quad \lambda' = \lambda_2-\lambda_4,\quad M = \max\{|\lambda|,|\lambda'|\}$$

$$N = \min\{|\lambda|,|\lambda'|\}$$

$$\delta = -\tfrac{1}{2}(M-N)$$

$K_{\lambda\lambda'}(t)$ is a threshold kinematical factor defined in ref. 3, $\gamma_H(t)$ a reduced residue free of kinematic singularities,

$$f_H(\alpha) = \frac{(2\alpha)!}{[(\alpha+M)!\,(\alpha-M)!\,(\alpha+N)!\,(\alpha-N)!]^{\frac{1}{2}}},$$

$$\xi_{\lambda\lambda'}(z) = \left(\frac{1-z}{2}\right)^{|\lambda-\lambda'|/2}\left(\frac{1+z}{2}\right)^{|\lambda+\lambda'|/2}$$

$$\Delta = \frac{1}{2t}\left[t^2 - t\,(m_1^2+m_2^2+m_3^2+m_4^2) + (m_1^2-m_3^2)(m_2^2-m_4^2)\right]$$

It is this last term which introduces a singularity at t = 0 for unequal mass particle reactions. More precisely, the term of order $\left(\frac{S}{4S_o}\right)^{\alpha-M-n}$ has a singularity t^{-n}.

To cancel this unwanted singularity, several suggestions have been made[3,10]. One of them is to introduce additive Regge poles, the "daughters", which have residues which cancel the singularities produced by the "parent" trajectory. From the expansion given above for the amplitude, it is clear that one needs a series of daughters with

$$\alpha_n(o) = \alpha(o)-n,\quad n = 1,2,3$$

and with residues which are related one to each other. For instance, for n = 1,

$$\beta_1(t) \underset{t\to o}{\to} -\beta(o)\,(\alpha_o-M)\,4S_o\,\frac{(m_1^2-m_3^2)(m_2^2-m_4^2)}{2t}$$

One should however remark that these constraints on the daughters are only valid at t = 0: everywhere else, the daughter trajectories may have a behaviour independent of their parents.

Until now, the arguments invoked for the introduction of daughters were of kinematical nature. The presence of daughters has also been suggested for dynamical reasons. The need for a ρ' conspiring with the ρ in the πN finite energy sum rule[11], the "cross over" (change of sign of the difference $\frac{d\sigma}{dt}(\bar{A}B) - \frac{d\sigma}{dt}(AB)$) for πN, KN processes at $t \sim -0.2\ \mathrm{GeV}^2$ [12], the non-vanishing neutron polarisation in the $\pi^- p \to \pi^o n$ charge exchange cross section (since, in this case, we have a priori only ρ exchange, and that a single Regge pole gives the same phase to all amplitudes, this single pole fit must be wrong), are several aspects of this new role of the daughters. However, all these arguments are only suggestive, and many other ways have been proposed to cure the weaknesses of the Regge pole theory.

3. VENEZIANO'S DAUGHTERS

The idea that there is an equivalence between the crossed channel Regge poles and resonances, the duality, has produced an impressive set of predictions. If we subtract the Pomeranchon from the amplitudes to which it can contribute, the duality tells us that we can fit the remaining amplitudes with Regge pole trajectories which are dual to non-exotic resonances. As we have already noticed, this leads to quadruple degenerescence of Regge trajectories: consider for instance the $\pi^+\pi^-$ scattering; apart from the Pomeranchon, only ρ and f can be exchanged. Since there is no I = 2 exotic resonance, the ρ and f exchanges must be exchange degenerate and must cancel each other. Introducing now the $K\bar{K}$ scattering, for which we have in addition to the ρ and f, the ω and A_2 exchanges, and for which again there is no exotic resonance, we have a quadruple degenerescence between ρ, f, ω and A_2 exchanges.

The simplest analytic function which satisfies most of the requirements of this pole duality has been proposed by Veneziano[13]. Introducing the Euler's Gamma function $\Gamma(1-\alpha(s))$ which has the desired property of having poles when $\alpha(s)$ is a positive integer, we can tentatively write the amplitude for $\pi\pi$ scattering, i.e.:

$$A(s,t) = \Gamma(1-\alpha(s))\,\Gamma(1-\alpha(t))$$

since we require an identical behaviour for s and t-channels (and no pole in u-channel). This amplitude would have however double poles each time that both (s) and (t) are integer. We remove these double poles by introducing a denominator with $\Gamma(1-\alpha(s) - \Gamma(t))$ and we get the Veneziano amplitude:

$$V(s,t) = \frac{\Gamma(1-\alpha(s))\,\Gamma(1-\alpha(t))}{\Gamma(1-\alpha(s) - \alpha(t))}$$

This function reduces to the expected Regge behaviour for large s values provided $\alpha(s)$ is a linear function of s: $\alpha(s) = \alpha(o) + \alpha's$, the slope α' being equivalent to the normalisation $\frac{1}{S_o}$ which appears in the Regge pole amplitude (hence the value of $\sim 1\ \mathrm{GeV}^2$ usually taken for S_o).

At a pole $\alpha(s_r) = J$, the Gamma function has a residue:

$$\frac{(-1)^J}{(J-1)!\ \alpha!}$$

and the Gamma functions which depend on (t): $\frac{\Gamma(1-\alpha(t))}{\Gamma(1-J-\alpha(t))}$ can be written as a sum of Legendre polynomials in z, the highest term being $P_J(z)$. The pole at $s = s_r$ corresponds therefore to a degenerate sequence of resonances of spin J, J-1, ---0. We find therefore that the Veneziano amplitude leads automatically to the introduction of daughters. In figs 4 to 7, we show how the Chew-Frautschi plot is populated by Veneziano daughters generated by some of the known parents.

However, it is known that Veneziano amplitudes have only zero width resonances. A direct comparison with experiment is therefore not possible and one should not expect that all the predictions provided by this simple model are verified: if the daughters exist, they may be displaced in mass, there widths and branching ratios cannot be predicted in a reliable way.

In conclusion, if several theoretical arguments can be put forward to suggest the existence of daughter trajectories, one should not take these predictions too literally when analysing experimental data, keeping simply in mind that "below" classical resonances like the ρ, f, A_2 ... several resonances of low spin may be present, at least in some of the formation or production processes examined. It seems to be the case for the baryons observed in formation experiments, although their ordering into "daughters" is still a completely open problem. For the mesons, the experimental situation is even more confused. This is partly due to the fact that we cannot perform "formation" experiments (except for $M > 2m_p$, with proton-antiproton annihilations). Still, several "candidates" have been suggested recently as daughters of the ρ, the g, the A_2 .... We shall now briefly review this situation.

THE G = +1, NATURAL PARITY SYSTEM

This is the meson system which has been most extensively studied, in particular recently with the introduction of detailed $\pi\pi$ phase-shift analysis[14], using the one-pion exchange approximation or assuming the dominance of final state interactions in $\bar{p}p$ annihilations[15]. The four pion system has been studied in photo-production experiments[16], in e^+e^- collisions[17] and also in $\bar{p}p$ annihilations[18].

For a general review of the $\pi\pi$ scattering, see ref. 19.

It is generally agreed that the I = 0 $\pi\pi$ S-wave phase-shift δ_0^0 slowly increases up to $\sim 90^0$ at m $\sim$ 850 MeV, then quickly reaches 180^0 at m $\sim$ 990 MeV and then 270^0 in the 1100 MeV mass region. Fig. 8 reproduces the results of $\pi\pi$ phase shift analysis based on OPE model, by Protopopescu and Hyams, and those of Frenkiel based on the analysis of the $\pi^+\pi^-$ system produced in $\bar{p}p$ annihilations at rest into $\omega^0\pi^+\pi^-$.

Fig. 9 shows the Argand plot for the $\pi\pi$ S-wave[22] the S-wave moves along the unitary circle in an anticlockwise direction up to the $K\bar{K}$ threshold, then becomes inelastic up to $\sim$ 1100 MeV where it moves again along the unitary circle in the f^0 mass region.

The anal; ic structure of this S-wave amplitude (Protopopescu and Hyams) contains various poles in the complex energy plane. One pole seems to be present in all the fits obtained with a reasonable χ^2: the S^*, with $M - i\Gamma/2 = (1000 - 20i)$ MeV. The recent results of Hyams et al., which are deduced from the analysis of the $\pi^+\pi^-$ system alone M = 1007 ± 20 MeV, Γ=30 ± 10 MeV are in fair agreement with the old results of Aguilar et al.,[20], who observed the $K_1^o K_1^o$ spectrum produced in $\bar{p}p$ annihilations:

$$\bar{p}p \to K_1^o K_1^o \pi^+\pi^-$$

and found for the S^* : M = 1048 ± 10 Γ = 35 ± 10 MeV.

It is therefore very likely that a relatively narrow $1^G J^P = 0^+0^+$ resonance, strongly coupled to the $K_1^o K_1^o$ system, exists.

The other poles observed by Hyams and Protopopescu are far away from the real axis, and therefore their interpretation in terms of resonance is doubtful: Protopopescu finds a second pole in the second Riemann sheet at $M-i\Gamma/2 = (660 \pm 100) - i(320 \pm 70)$ MeV. The poles of T_o^o nearest to the real axis of the complex energy phase are, in the analysis of Hyams et al., at $M-i\Gamma/2 = 1049 - 250i$ and $1537 - 233i$ MeV.

In the high mass region (above m = 1 GeV), the phase-shift analysis becomes more and more complex since several partial waves with unknown inelasticities must be included. Moreover, more and more channels are opened and the K-matrix formalism is consequently more and more involved. The results of Hyams clearly indicate that the $\pi\pi$ S-wave is important in the f^o mass region, being essentially elastic: This may explain the old known difficulties met with the angular decay distribution of the $f^o \to \pi^+\pi^-$. However, much more detailed analysis, based on more complete data (including all opened channels) are still necessary before we can conclude in the presence (or absence) of an S-wave resonance in the 1200 MeV mass region.

We see that, for the time being, the only well established scalar resonance is the S^* strongly coupled to the $K\bar{K}$ system, and being therefore a plausible daughter of the ϕ meson. In terms of the naive $\bar{q}q$ model for mesons, the S^* would be, like the ϕ, a $\lambda\bar{\lambda}$ object. The $\varepsilon(\sim 700)$ and $\varepsilon'(\sim 1250)$ are too far from the real axis to be considered, yet, as well established resonances.

Let us now turn to the ρ', ρ''. Fig. 4 shows that Veneziano's approach would predict the existence of a ρ' and of a ρ'' with mass in the neighbourhood of the f^o and g meson, respectively. Three classes of reactions have been used for searches on these vector mesons:

- e^+e^- collisions and photoproduction
- $\pi^\pm p$ interactions
- $\bar{p}p$ annihilations.

The e^+e^- collisions offer one of the most attractive way of studying the production of vector mesons. It is well known that around $S = m_\rho^2$, m_ω^2 and m_ϕ^2, the e^+e^- annihilations into $\pi^+\pi^-$, $\pi^+\pi^-\pi^o$ and $K\bar{K}$, respectively, can be very well represented by simple Breit-Wigner form factors or by more refined formulas like the Gounaris-Sakurai one[23]. Unfortunately, the statistics are still rather poor to allow definite conclusions. The results obtained on the dipion production[24], the pion form factor, show that the cross-section is somewhat larger than expected from the extrapolation of the ρ, ω, ϕ above 1 GeV. But the most striking results come from the multiparticle production, the annihilations

$$e^+e^- \to \pi^+\pi^-\pi^o\pi^o$$

exhibiting an enhancement (fig. 10) around 1300 MeV (only 2 σ.d.) and the annihilations

$$e^+e^- \to \pi^+\pi^+\pi^-\pi^-$$

exhibiting another one (fig. 11) around 1600 MeV[25] - (M $\sim$ 1650 MeV, $\Gamma \sim$ 400 MeV).

Photoproduction of $\pi^+\pi^-$ off carbon at 6 GeV/c (fig. 12)[26] adds evidence for an excess of events (excess with respect to the tail of the ρ) in the 1200-1800 MeV mass-region.

Four-pion photoproduction has been observed in two experiments, one using a streamer chamber[27], the other one the SLAC 82 inc. bubble chamber[28]. This five-body final state is extremely difficult to analyse, even at the relatively high energy used for these experiments (6 to 18 GeV). Baryon resonances production is important and may affect the (4π) system. Once the Δ^{++} region has been excluded, the (4π) spectrum shows an excess of $\rho^o\pi^+\pi^-$ events in the 1600 MeV mass region which does not seem to be simply due to peripheral phase space mechanism (fig. 13). Its internal structure indicate the presence of ρ^o, but not of $\rho^o\rho^o$ in the 4π final state. It is more difficult to exclude the $A1\pi$ final state. In the bubble chamber experiment, a polarized photon beam was used and a spin-parity analysis can be performed: $J^P = 1^-$ is strongly favoured over other spin-parity assignments. The isospin is indicated by the fact that the $\pi\pi$ system outside the ρ^o seems to be I = 0 (R = $\pi^o\pi^o/\pi^+\pi^-$ compatible with 1/2). A maximum likelihood fit gives for the mass and width of this ρ'': M = 1450 $\pm$ 50 MeV, Γ = 650 $\pm$ 100 MeV and the ratio $\gamma p \to p\rho''$ to $\gamma p \to p\rho$ at 9 GeV/c is found to be 0.10.

The production of vector mesons in $\pi^\pm N$ interactions has been analysed several times in great details. It is well known that in these interactions, the ρ, the f and the g mesons are observed in the dipion system. The 4π system is not so well known: there are some indications that the g meson ($J^P = 3^-$) has an important 4π decay mode[1]; the $\omega^o\pi$ system is under analysis in two new high statistics recent experiments[29]: in addition to a $J^P = 1^+$ B meson resonance at $\sim$ 1230 MeV, a $J^P = 3^-$ contribution probably due to the g meson is observed around 1600 MeV. $J^P = 1^-$ may not be excluded in the 1250 MeV mass region.

Unless detailed spin parity analysis are performed, the presence of a $\rho' \to \pi^+\pi^-$ is difficult to establish in the 1200 MeV mass region, the $J^P = 2^+$ f^o being probably largely dominant. The same remark is valid for the g-meson region ($\sim$ 1600 MeV) and even extends to charged dipion states in this last case. It may explain why attempts to put forward $\rho' \to \pi\pi$ production have usually been unsuccessful[30]. However, in their detailed phase-shift analysis of the dipion system, Hyams et al.,[14] found that, in addition to the $\rho(778)$, they had to introduce a second pole to interpret the I = 1, J = 1 wave: the inelasticity η falls to 0.5 in the 1600 MeV region, the phase δ_1^1 showing a drop in the same energy region. These facts are best explained in terms of a ρ'' resonance with the following parameters:

M = 1590 $\pm$ 20 MeV, Γ = 180 $\pm$ 50 MeV,

and an elasticity $\chi = 0.25 \pm 0.05$ (fig. 14). This elasticity is compatible with the upper limit of 20% given by Bingham et al.[28], although the masses, and even more the widths quoted for the ρ'' by these two groups show some contradiction.

Four reactions bring interesting information on the production of vector meson in proton-antiproton annihilations at rest: two of them correspond to the "formation" of ρ'' at a fixed total energy: $s = (2Mp)^2$:

$$\bar{p}p \to \rho'' \to \pi^+\pi^-$$

$$\bar{p}p \to \rho'' \to \pi^+\pi^-\pi^+\pi^-$$

This assumption is not irrealistic if we remember that $\bar{p}p$ annihilations at rest occur essentially in S-states[31]: ${}^1S_o$ ($J^{PC} = 0^{-+}$) and ${}^3S_1$ ($J^{PC} = 1^{--}$) and if we remark that the mass and width of the ρ'' as measured in e^+e^- collisions or in photoproduction is compatible, with the total energy available in $\bar{p}p$ annihilations.

The other two reactions correspond to the "production" of ρ'' in association with a recoiling pion:

$$\bar{p}p \to \rho''\pi \to 3\pi$$

$$\bar{p}p \to \rho''\pi \to (\omega^o\pi)\pi$$

In principle, we could also add the (4π) decay:

$$\bar{p}p \to \rho''\pi \to (4\pi)\pi$$

but in practice this 5π final state reaction is more difficult to analyse.

$\bar{p}p \to \pi^+\pi^-$: the rate for this two-body annihilation channel is known to be small: $R = 0.37\%$. One should notice that the only possible initial state is ${}^3S_1$, which has the quantum numbers of the ρ.

$\bar{p}p \to 2\pi^+2\pi^-$ [32]. Diaz et al., have shown that this final state is dominated by the process,

$$({}^3S_1)\ \bar{p}p \to \rho^o\pi^+\pi^-,$$

most of this final state corresponding to a dipion (outside the ρ^o) in an S-wave (for which the authors use a parameterisation identical to the one already discussed[15], see fig. 8). The rate for this $\rho^o\ \delta^o_o$ final state is found to be:

$$R = (0.98 \pm 0.20)\%$$

(see figs 15, 16, 17 and table 1).

From these observations, one can conclude that the tail of the ρ'' ($m = 1.88$ GeV) has the following properties:

1) $\pi^+\pi^-/(4\pi+2\pi) = 0.20$. This is in fair agreement with the results of Hyams[14] and Bingham[28].

2) the 4π decay is dominated by the $\rho^o\pi^+\pi^-$ final state where the dipion is predominantly S-wave.

In particular, $A_1\pi$ and $A_2\pi$ (which are allowed decays for $J^{PC} = 1^{--}$) and $\rho^o\rho^o$ (which uld be forbidden) are small or completely negligible.

The production processes offer the possibility of measuring, in principle, the mass of e ρ''. It has been known for a long time[33] that $\bar{p}p$ annihilations at rest into $\pi^+\pi^-\pi^o$ are ninated by P-wave dipion final states but that even sophisticated ρ form factors do not count for the details of the Dalitz-plot distributions. In a more recent analysis, Besliu al.[34] have shown that the introduction of a ρ'' with M $\sim$ 1400 and $\Gamma \sim$ 400 MeV would greatly lp in the understanding of the results. These results, however, cannot be compared to the nous Lovelace analysis of $\bar{p}n \to \pi^+\pi^-\pi^-$ [35] where the initial state is ^1S_o. The results of sliu on $\bar{p}p$ annihilation at rest refer to $^3S_1$ initial state and suggest that $\rho\pi$ and $\rho''\pi$ from is initial state are in the ratio 1/8. If this ρ'' observed in the annihilation $\bar{p}p \to \rho''\pi$ the same object than the one observed in formation reactions with the decay ratio $2\pi/4\pi = 1/4$, should observe the process: $\bar{p}p \to \rho''\pi \to (4\pi)\pi$ with the rate R = 2%. The five-pion annihila-n is abundant and the rate expected for the above process is not in contradiction with the ta. However, a detailed analysis of this channel is made difficult by the important back-ound due to combinational effects.

Annihilations at rest into $\omega^o\pi^+\pi^-$ seem to bring a completely different type of formation[15]. As for the five pion annihilations mentioned above, the background behind e ω^o signal is large and makes difficult a detailed analysis (fig. 18). Moreover, the $\pi^+\pi^-$ final state is found to be rather complex, the ^1S_o initial state contributing to 30% ch the final state $\omega^o\rho^o$, and the $^3S_1$ initial state containing an important production with e dipion in an S-wave (32%), $\omega^o f^o$ (23%) and $B\pi$ (41%). But the interpretation of the Dalitz-ot and of the polarization of the ω^o needs the introduction of a ρ' (13%) with the mass and dth: M = 1256 ± 10 MeV, Γ = 129 ± 20 MeV (see table 1 and fig. 19).

Again this object cannot be compared with the ρ' introduced by Lovelace[35] since it is entially coupled to $^3S_1$ initial state whereas Lovelace's ρ' is associated with ^1S_o initial te. Moreover, its mass and width are not compatible with those of the ρ'' observed in $\bar{p}p$ rmation" experiments, nor with that of the ρ'' observed in photoproduction and in e^+e^- lisions. Finally, this $\rho' \to \omega^o\pi$ does not seem to have a $\pi\pi$ decay mode: this mode has been rched for in $\bar{p}p \to 3\pi$ annihilations, and we also have the negative results of the $\pi^+\pi^-$ se-shift of Hyams[14].

This new ρ' needs obviously more experimental support before introducing it with confi-ce in a theory. This may not be easy if it is not coupled to the $\pi\pi$ system. But it is arkable that the (weak) evidence for a $\rho' \to \pi^+\pi^-\pi^o\pi^o$ from e^+e^- collisions (fig. 10) peaks 1250 MeV with a width of $\sim$ 200 MeV, in agreement with the $\rho' \to \omega^o\pi^\pm$ of Frenkiel, the ay mode observed in e^+e^- collisions being compatible with an $\omega\pi$ decay.

An attractive explanation of these $\rho(770)$, $\rho'(1260)$, $\rho''(1600)$ vector recurrence has been posed by Renard[36] who suggests that they represent the inelastic effects of ρ propagator ch incorporates in a simple way the effects of an energy dependent width and of a strong lasticity:

$$\Delta\rho\ (s) = (m^2\rho - s)\ [1 - (m^2\rho - s)\ G\ (s)] - im\ \Gamma(s)$$

with, for $s > m^2$, $G(s) = \Sigma\, G_i(s)$

$$G_i(s) = \frac{A_i}{m_\rho^2\,(1+\delta)} \; \frac{\gamma_i\,(s_i - s)}{(s_i - s)^2 + \gamma_i^2} ,$$

and for $s < m_\rho^2$ $\quad G(s) = -\frac{\delta}{m_\rho^2}$

$i = 1,2 \ldots n$ denotes the hadronic decay channels.

δ is the finite width correction factor determined experimentally: $\delta = 0.13 \pm 0.08$.

S_i, A_i, γ_i are the threshold, the coupling and cut-off parameters corresponding to channel i.

However, it has been shown by Eisenberg[30] that this "explanation" does not seem to be valid since the production ratio ρ/ρ' seems to depend on the production mechanism.

In terms of the "naive" $q\bar{q}$ model for mesons, two vector mesons are expected to occur in the 1-2 GeV mass region, one corresponding to the orbital excitation L = 2, the isovector member of the $^3D_1$ nonet, and the radialy excited state of the ρ (n = 2 of $^3S_1$). One may speculate that the $^3D_1$ is more likely the $\rho'(1260)$, and the radialy excited one the $\rho''(1600)$, which would decay into $\rho\varepsilon$, the ε being one of the scalar objects discussed before.

We shall encounter another example of this de-excitation by emission of a scalar meson when we discuss the pseudo-scalar particles.

Coming back to fig. 4, we see that, although the experimental information is still incomplete, it is not unrealistic to imagine that the observations discussed above are related with the ρ' and ρ'' predicted by the Veneziano model. Fig. 20 summarizes the experimental situation for these vector resonances.

THE G = -1 NATURAL PARITY SYSTEM

As discussed by Miller[1], the isovector $J^P = 0^+$ candidate, the $\delta(970)$, is still in a poor experimental situation. If its mass is below the $K\bar{K}$ threshold, it may produce in the I = 1 $K\bar{K}$ system a threshold enhancement similar to the effect noticed in the $K_1^o K_1^o$ system, which is generally attributed to the I = 0 channel (the S^*), but could be due also to I = 1. This is indeed supported by the observation of such a threshold enhancement in the $K^oK^\pm$ system by Astier[37], in $\bar{p}p \to K^oK^\pm\pi^\mp$, but the analysis of the three-body annihilations: $\bar{p}p \to \eta^o\pi^+\pi^-$ does not support the existence of a resonance in the $\eta\pi$ system (this negative result can of course be interpreted in terms of a higher limit for the $\eta\pi$ decay mode of the δ).

Experiments with high statistics on the $\eta\pi$ and $K\bar{K}$ systems allowing a detailed analysis of the coupled channels are still needed to clarify the situation of the δ.

Referring to fig. 5, we see that another $J^P = 0^+$, I = 1 resonance can be predicted at the mass of the A_2 meson. The best spin-parity analysis of the A_2 mass region has been done by Antipov, Ascoli[38] but this analysis is not relevant for our purpose since a 3π system cannot have spin-parity 0^+. We are therefore left with the information collected on the $\eta\pi$

and $K\bar{K}$ system, still rather limited as concerns the spin-parity analysis. It is generally assumed that the all enhancement observed in the $K\bar{K}$ and $\eta\pi$ systems is due to the 2^+ state.

Taking advantage of the fact that $\bar{p}p$ annihilations at rest occur in well defined quantum number states, Espigat[39)] and Chung[40)] have analysed the $\eta\pi$ system as observed in the reaction $\bar{p}p \to \eta^0\pi^+\pi^-$, the first authors using the final state interaction model, the second ones a Veneziano amplitude. Of the 3 sets of quantum numbers tried for the $\eta\pi$ system in the A_2 mass region, $J^P = 0^+$, 1^- and 2^+, Espigat can reject 1^- but cannot distinguish between 0^+ and 2^+, both states being possibly present. For the singlet $\bar{p}p$ contribution, which is the amplitude of interest here, Chung et al., have used the expression given by Baacke et al.[41)] for the similar process:

$$\eta \to \eta\pi\pi$$

$$A = \beta_o F_o + \beta_1 F_1 + \beta_2 F_2$$

with
$$F_o = \frac{\Gamma(1-\alpha_s)\Gamma(1-\alpha_t)}{\Gamma(1-\alpha_s-\alpha_t)} + \frac{\Gamma(1-\alpha_u)\Gamma(1-\alpha_t)}{\Gamma(1-\alpha_u-\alpha_t)} + \frac{\Gamma(1-\alpha_s)\Gamma(1-\alpha_u)}{\Gamma(1-\alpha_s-\alpha_u)}$$

$$F_1 = \frac{\Gamma(1-\alpha_s)\Gamma(1-\alpha_t)}{\Gamma(2-\alpha_s-\alpha_t)} + \frac{\Gamma(1-\alpha_u)\Gamma(1-\alpha_t)}{\Gamma(2-\alpha_u-\alpha_t)}$$

$$F_2 = \frac{\Gamma(1-\alpha_s)\Gamma(1-\alpha_u)}{\Gamma(2-\alpha_s-\alpha_u)}$$

α_s and α_u correspond to the A_2 trajectory in the $\eta\pi$ channel and α_t to the f (assumed degenerate with the ρ) trajectory in the $\pi\pi$ or $\eta\eta$ channel. The trajectory functions have been fixed at

$$\alpha_x = 0.20 + 1.06\ x + i\ 0.18\ (x - x_o)^{\frac{1}{2}} \quad \text{for } A_2$$

and
$$\alpha_x = 0.39 + 1.06\ x + i\ 0.13\ (x - x_o)^{\frac{1}{2}} \quad \text{for } f^o,\ x_o \text{ being the x-channel threshold.}$$

The fit is very good and gives for the singlet contribution $\beta_o = 1$ (fixed), $\beta_1 = 1.9 \pm 0.2$ and $\beta_2 = 1.9 \pm 0.3$.

It is interesting to study the pole content of this solution for the $\eta\pi$ system.

Using the property of the Gamma functions:

$$\Gamma(x) = \frac{1}{x}\ \Gamma(1 + x) \quad ,$$

we get, at the first pole ($\alpha_s = 1$):

$$F_o = \frac{-\alpha_t}{1-\alpha_s}\ \Gamma(2-\alpha_s) + \frac{\Gamma(1-\alpha_u)\Gamma(1-\alpha_t)}{\Gamma(1-\alpha_u-\alpha_t)} - \frac{\alpha_u\ \Gamma(2-\alpha_s)}{1-\alpha_s}$$

$$F_1 = \frac{\Gamma(2-\alpha_s)}{1-\alpha_s} + \frac{\Gamma(1-\alpha_u)\ (1-\alpha_t)}{\Gamma(2-\alpha_u-\alpha_t)}$$

$$F_2 = \frac{\Gamma(2-\alpha_s)}{1-\alpha_s}$$

At the pole $\alpha_s = 1$, the residue can therefore be written:

$$R_1 = [\beta_1 + \beta_2 - \beta_o (\alpha_t + \alpha_u)] \quad \Gamma(2-\alpha_s)$$

$\Gamma(2-\alpha_s)$ depends on s and therefore affects the shape of the resonance but not the angular distributions

$$\alpha_t + \alpha_u = 0.59 + i\,0.18\,(u - u_o)^{\frac{1}{2}} + i\,0.13\,(t-t_o)^{\frac{1}{2}}$$

at $s = m_\rho^2$, R_1 is of the order:

$$R_1 \sim (\beta_1 + \beta_2 - 3.7\,\beta_o) \quad \Gamma(2-\alpha_s)$$

With the values found for β_1 and β_2, we see that

$$R_1 \sim 0$$

In other words, the data does not require an $\eta\pi$ $J^P = 0^+$ resonance at the mass on the ρ.

At the second pole, $\alpha_s = 2$:

$$F_o = \frac{-\alpha_t(1+\alpha_t)\Gamma(3-\alpha_s)}{2-\alpha_s} - \frac{\alpha_u(1+\alpha_u)\Gamma(3-\alpha_s)}{2-\alpha_s} + \frac{\Gamma(1-\alpha_u)\Gamma(1-\alpha_t)}{(1-\alpha_u-\alpha_t)}$$

$$F_1 = \frac{\alpha_t\,\Gamma(3-\alpha_s)}{2-\alpha_s} + \frac{\Gamma(1-\alpha_u)\Gamma(1-\alpha_t)}{\Gamma(2-\alpha_u-\alpha_t)}$$

$$F_2 = \frac{\alpha_u\,\Gamma(3-\alpha_s)}{2-\alpha_s}$$

the residue can be written:

$$R_2 = \{\beta_o\,[-\alpha_t\,(1+\alpha_t) - \alpha_u\,(1+\alpha_u)] + \beta_1\alpha_t + \beta_2\alpha_u\}\;\Gamma(3-\alpha_s)$$

The best fit gives $\beta_1 = \beta_2 = \beta$

$$R_2 = \{(\alpha_t + \alpha_u)(\beta - \beta_o) - \beta_o(\alpha_t^2 + \alpha_u^2)\} \quad \Gamma(3-\alpha_s)$$

Since $t \sim \frac{1}{2}\,(M^2-s)(1 + \cos\theta)$

$$u \sim \tfrac{1}{2}\,(M^2-s)(1 - \cos\theta)$$

(relations valid in the limit: $M^2 >> M_\eta^2 + 2M_\pi^2$) the degenerescence of the trajectories and the experimental result $\beta_1 = \beta_2$ imply that this pole does not contain P-waves, as expected from the quark model. On the other hand, with the values found for β_1 and β_2, both D and S waves are present and an explicit calculation shows that S-waves are indeed important ($\sim$ 30%).

This example shows that a detailed analysis of the $\eta\pi$ (or $K\bar{K}$) system in the A_2 mass region may lead to some interesting information on a possible δ' daughter of the A_2. The results obtained above are interesting, but give only very limited indications on the $\eta\pi$ system. They are based on too limited experimental data (the "η" has a large background: 30%, the statistics are marginal, the $\eta\pi$ spectrum is cut-off by phase-space, the presence of $^3S_1$ initial state increases the difficulty of the analysis ...). Further experiments are badly needed to clarify the situation.

. THE UNNATURAL PARITY SYSTEM

There is still much too limited information on the axial vector and 2^- systems to allow any discussion here.

The pseudo-scalar case offers a more interesting situation, although the large violation of the "ideal mixing" of the basic nonet tends to obscure the picture.

Our experimental information has not progressed in the last five years on possible high mass $I^GJ^P = 0^+0^-$ objects. The E-meson is still a reasonable candidate, but $J^P = 1^+$ is not excluded and all its decay modes (K^*K, $\delta\pi$, $\eta\pi\pi$) are not yet well established. Moreover, it has essentially been observed, for the time being, in a single reaction:

$$\bar{p}p \rightarrow E^o\pi^+\pi^-.$$

It may be interesting to remark that this reaction is a good example of a possible de-excitation mechanism by emission of a scalar, as already mentioned when discussing the ρ'':

$$0^- \rightarrow 0^- 0^+.$$

Indeed, it is found[41] that the $\bar{p}p$ initial state contributing to the production of E is $'S_o$ ($J^{PC} = 0^{-+}$) and not $^3S_1$ (1^{--}), and that the dipion emitted together with the E has the properties of a scalar object.

Moreover, if the $\delta\pi$ decay of the E is confirmed to be one of the main decay modes, this decay is another example of the same process:

$$0^- \rightarrow 0^- 0^+.$$

Even if we consider the $'S_o$ $\bar{p}p$ and the E-meson as possible pseudo-scalar candidates, we do not know how to associate them with the states predicted on fig. 6.

THE I = ½ MESONS

The phase-shift analysis performed on the $K\pi$ system have shown striking similarities with the $\pi\pi$ system. Before the analysis of Protopopescu and of the CERN-Munich group, one possible solution for the $\pi\pi$ S-wave phase shift δ_o^o was a rapid increase in the vicinity of the ρ-mass, suggesting the presence of a "narrow" E. This solution is now rejected in favour of a smooth slowly increasing δ_o^o in the ρ mass, up to $\sim$ 1 GeV.

The same situation seemed to happen to the S-wave $K\pi$ phase-shift: two solutions were in presence in the $K^*(892)$ mass region, one of them suggesting the presence of a "narrow" κ.

A recent analysis of A. Barbaro-Galtieri et al.[42] shows that the smooth slowly increasing solution is preferred in the K^* mass region. This analysis is based on the study of the reaction

$$K^+p \rightarrow \Delta^{++} K^+\pi^- \quad \text{at 12 GeV/c}$$

and is made for the $K\pi$ mass region:

$$0.79 < M(K\pi) < 1.0 \text{ GeV}$$

Both energy independent and energy dependent analysis reject the "narrow" 0^+ resonance hypothesis, unless it is much narrower than the experimental resolution ($\Gamma < 7$ MeV). It seems therefore that any $J^P = 0^+$ $S = 1$ resonance must be looked for above a mass of 1GeV.

8. CONCLUSIONS

Although there are many indications which speak in favour of the presence of meson resonances of low spin in the 1-2 GeV mass region, we are still far from being able to introduce enough correlation between these indications to draw a clear picture. We are far from being able to prove or disprove the existence of meson daughter trajectories.

Apart from the $\pi\pi$, and, to a certain extent, the 3π systems, the meson states are still badly known. The data are statistically too limited, the methods of analysis too primitive. More coupled channels analysis and complete spin-parity analysis are necessary.

Still, contrarily to current opinion, interesting observations have been made on meson spectroscopy in the last few years. For the time being, these indications suggest the existence of a daughter to the ϕ: the S^*, a daughter to the g, the ρ'', may be a daughter to the A_2, the ρ', weakly coupled to $\pi\pi$.

The A_2 may have another daughter with $I^G = 1^-$ $J^P = 0^+$ (a δ') but no such $I^G = 1^-$ daughter is observed with $J^P = 1^-$, as expected from the quark model. The poles detected for the I^G $I^P = 0^+$ 0^+ $\pi\pi$ amplitude are too far from the real axis of the complex energy plane to be interpreted in terms of resonance, the ε' ($\sim$ 1250 MeV) being however in a slightly better position than the ε($\sim$ 700 MeV). It remains to transform these indications into well established facts. Then we shall be in a good position to understand, hopefully, their relations with Regge and Veneziano theories and, ultimately, with a satisfactory theory of strong interactions.

REFERENCES

1) D.H. Miller, "Meson spectroscopy". Erice International School of Subnuclear Physics, 1973.

2) G. Chew, S. Frautschi, Phys. Rev. Lett. 8 (1962) 41.

3) P. Collins, Phys. Reports IC (1971) 103.

4) M. Aguilar et al., Phys. Lett. 29B (1968) 379, Nucl. Phys. B14 (1969) 195.
J. Duboc et al., Phys. Lett. 34B (1971) 343.

5) G. Ascoli et al., Philadelphia Conference Proceedings, 1972.

6) Carmony et al., Phys. Rev. Lett. 27 (1971) 1160.

7) N. Armenise et al., Phys. Lett. 26B (1968) 336.

8) J.A. Matthews et al., Nuovo Cimento Lett. 1 (1971) 361.

9) J.A. Danysz et al., Nuovo Cimento 51A (1967) 801.

10) P.D. Collins and E.J. Squires - "Regge Poles in Particle Physics", Spinger Tracts in Modern Physics, Vol. 45 (1968) 1.

11) R. Dolen, D. Horn, C. Schmid, Phys. Rev. 166 (1968) 1768.

12) Shu Yvan Chu, B. Desai, D. Roy, Phys. Rev. 187 (1969) 1896.

13) G. Veneziano, Nuovo Cimento 57A (1968) 190.

14) Baton et al., Phys. Lett. 33B (1970) 528.
P. Baillon et al., Phys. Lett. 38B (1972) 585.
Carroll et al., Phys. Rev. Lett. 28 (1972) 318.
B. Hyams et al., MPI-PAE/Exp. EL28 (1973) (to be published in Nucl. Phys. B).
S. Protopopescu et al., Phys. Rev. D7 (1973) 1279, 1425.

15) P. Frenkiel et al., Nucl. Phys. 47B (1972) 61.

16) Davier et al., Paper presented at the 1972 Batavia Conference.
H. Bingham et al., Phys. Lett. 41B (1972) 635.

17) Silvestrini, rapporteur's talk at the 1972 Batavia Conference.

18) J. Diaz et al., Nucl. Phys. B16 (1970) 239.

19) J. Peterson Phys. Reports 2C (1971) 157.

20) M. Aguilar et al., Phys. Lett. 29B (1969) 241.

21) C. Lovelace, CERN preprint TH 1041 (1969).

22) R. Diebold, ANL/HEP 7254 (1972), rapporteur's talk at the 1972 Batavia Conference.

23) G. Gounaris, J. Sakurai, Phys. Rev. Lett. 21 (1968) 244,
F. Renard, Nucl. Phys. 15B (1970) 118.

24) V. Allès-Borelli et al., Phys. Lett. 40B (1972) 433,
V. Balakin et al., Phys. Lett. 41B (1972) 205.

25) B. Bartoli et al., Phys. Rev. D6 (1972) 2374.
G. Bacci et al., Phys. Lett. 38B (1972) 551.
G. Barbarino et al., Let. Nuovo Cimento 3 (1972) 689.
G. Cosme et al., Phys. Lett. 40B (1972) 685.
L. Kurdadze et al., Phys. Lett. 42B (1972) 515.

26) Alvensleben et al., Phys. Rev. Lett. 26 (1971) 273.

27) M. Davier et al., SLAC preprint (1972)

28) H. Bingham et al., Phys. Lett. 41B (1972) 635.

29) The 7 GeV $\pi^+ p \to p\pi^+\omega^o$ studied at LRL by Ott (paper 798, Batavia Conference, 1972) and the 3.9 GeV $\pi^- p \to p\pi^-\omega^o$ studied at CERN by V. Chaloupka (private communication).

30) Y. Eisenberg et al., Phys. Lett. 43B (1973) 149.

31) For a discussion of this point, see R. Bizzarri in the Chexbres Symposium on $\bar{p}p$ annihilations. CERN yellow report 72-10.

32) J. Diaz et al., Nucl. Phys. B16 (1970) 239.

33) M. Foster et al., Nucl. Phys. B6 (1967) 107.

34) C. Besliu et al., An. de Fisica 67 (1971) 383.

35) C. Lovelace, Phys. Lett. 28B (1968) 264.

36) F. Renard, talk given at the VIII Rencontre de Moriond, 1973.

37) A. Astier et al., Phys. Lett. 25B (1967) 294.

38) Antipov, Ascoli et al., 1972, Philadelphia Conference on Meson Spectroscopy, AIP, New York.

39) P. Espigat et al., Nucl. Phys. B36 (1972) 93.

40) S.U. Chung et al., Nucl. Phys. B31 (1971) 261.

41) P. Baillon et al., Nuovo Cimenta 50A (1967) 393.

42) A. Barbaro-Galtieri et al., LBL 1972.

TABLE 1

$\bar{p}p$ annihilation rates for some final states

Final state	Initial State J^{PC}	R:(%)
$\pi^+\pi^-$	1^{--}	0.38
$\rho^o\rho^o \rightarrow 2\pi^+2\pi^-$	0^{-+}	0.12
$A_2\pi \rightarrow 2\pi^+2\pi^-$	0^{-+}	1.4
$A_2\pi \rightarrow 2\pi^+2\pi^-$	1^{--}	0.56
$\rho^of^o \rightarrow 2\pi^+2\pi^-$	1^{--}	0.90
$\rho^o\delta^o_o \rightarrow 2\pi^+2\pi^-$	1^{--}	0.90
$\omega^o\rho^o \rightarrow 2\pi^+2\pi^-\pi^o$	0^{-+}	4.3
$\omega^o\delta^o_o \rightarrow 2\pi^+2\pi^-\pi^o$	1^{--}	4.6
$\omega^of^o \rightarrow 2\pi^+2\pi^-\pi^o$	1^{--}	3.3
$B\pi \rightarrow 2\pi^+2\pi^-\pi^o$	1^{--}	6.0
$\rho'\pi \rightarrow 2\pi^+2\pi^-\pi^o$	1^{--}	1.9

FIGURE CAPTIONS

Fig. 1 Chew-Low plot for isovector mesons

Fig. 2 Chew-Low plot for isodoublet mesons

Fig. 3 Chew-Low plot for isoscalar mesons

Fig. 4 Veneziano's daughters for G = +1 natural parity mesons

Fig. 5 Veneziano's daughters for G = -1 natural parity mesons

Fig. 6 Veneziano's daughters for G = +1 unnatural parity mesons

Fig. 7 Veneziano's daughters for G = -1 unnatural parity mesons

Fig. 8 δ_o^o $\pi\pi$ phase shift analysis of Hyams, Protopopescu (ref. 14) and Frenkiel (ref. 15)

Fig. 9 Argand plot for the $\pi\pi$ S-wave (ref. 22)

Fig. 10 Yield of e^+e^- annihilations into $\pi^+\pi^-\pi^o\pi^o$ (results of Cosme, Bacci and Kurdadze (ref. 25))

Fig. 11 Yield of e^+e^- annihilations into $\pi^+\pi^-\pi^+\pi^-$ (ref. 22)

Fig. 12 Mass spectrum for $\pi^+\pi^-$ photoproduction off carbon at 6 GeV/c (ref. 26)

Fig. 13 Photoproduction of $\pi^+\pi^-\pi^+\pi^-$ (ref. 16)

Fig. 14 $\pi\pi$ phase-shift analysis of Hyams et al., (ref. 14) for I = 1 P and F waves, and for I = 0 D-wave

Fig. 15 $\bar{p}p \rightarrow 2\pi^+2\pi^-$. $(\pi^+\pi^-)$ and $(\pi^\pm\pi^\pm)$ spectra with the contributions of $'S_o\ A_2\pi$ and $^3S_1\ \rho\pi\pi$ final states

Fig. 16a) Distributions of the angular part A_1^2 and A_7^2 of the amplitude $(^1S_o\ \rho^o\rho^o)$ and $(^3S_1\ \rho^o\delta_o^2)$

b) Cosine of the angle between unlike-charged pions and like-charged pions

c),d),e) Angular distributions showing the relative importance of $(^1S_o\ A_2\pi)$ and $(^3S_1\ \rho\pi\pi)$ amplitudes (see ref. 32).

Fig. 17 $\pi^+\pi^-$ mass distribution produced in association with a ρ^o in $\bar{p}p \rightarrow \rho^o\pi^+\pi^-$. Fit B assumes phase-space for the S-wave dipion. Fit A assumes a phase-shift δ_o^o as described in fig. 8

Fig. 18 $\bar{p}p \rightarrow 2\pi^+2\pi^-\pi^o$. $(3\pi)^o$ spectrum

Fig. 19 $\bar{p}p \rightarrow \omega^o\pi^+\pi^-$. $(\omega\pi^\pm)$ spectrum with a selection favouring $J^P = 1^-$ $(\omega\pi)$ final state

Fig. 20 Summary of the experimental situation of the ρ' and ρ''.

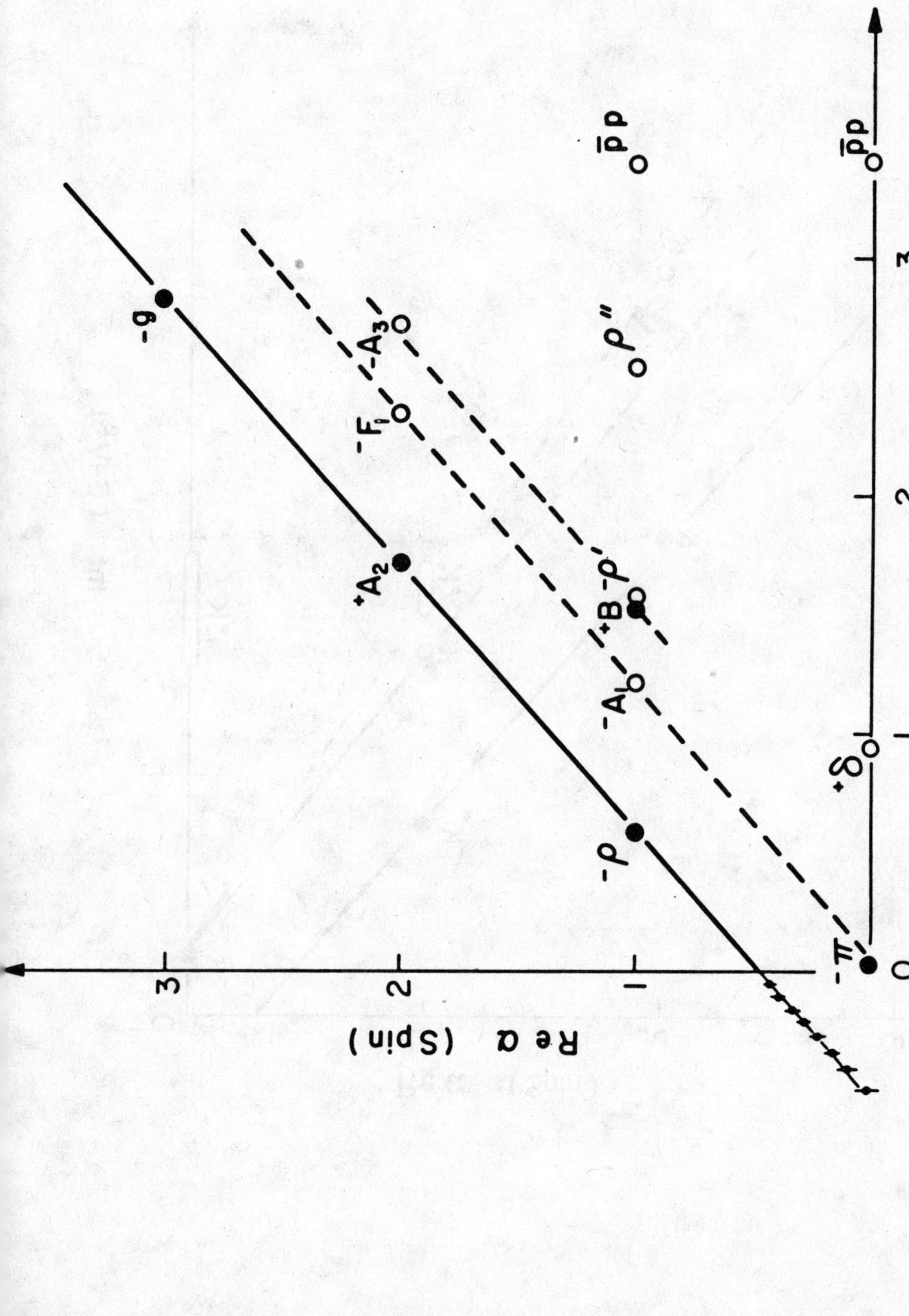
Re α (Spin)
S = m² (GeV²)
0
1
2
3
π
ρ
δ
A1
B
ρ'
A2
F1
A3
g
ρ''
p̄p

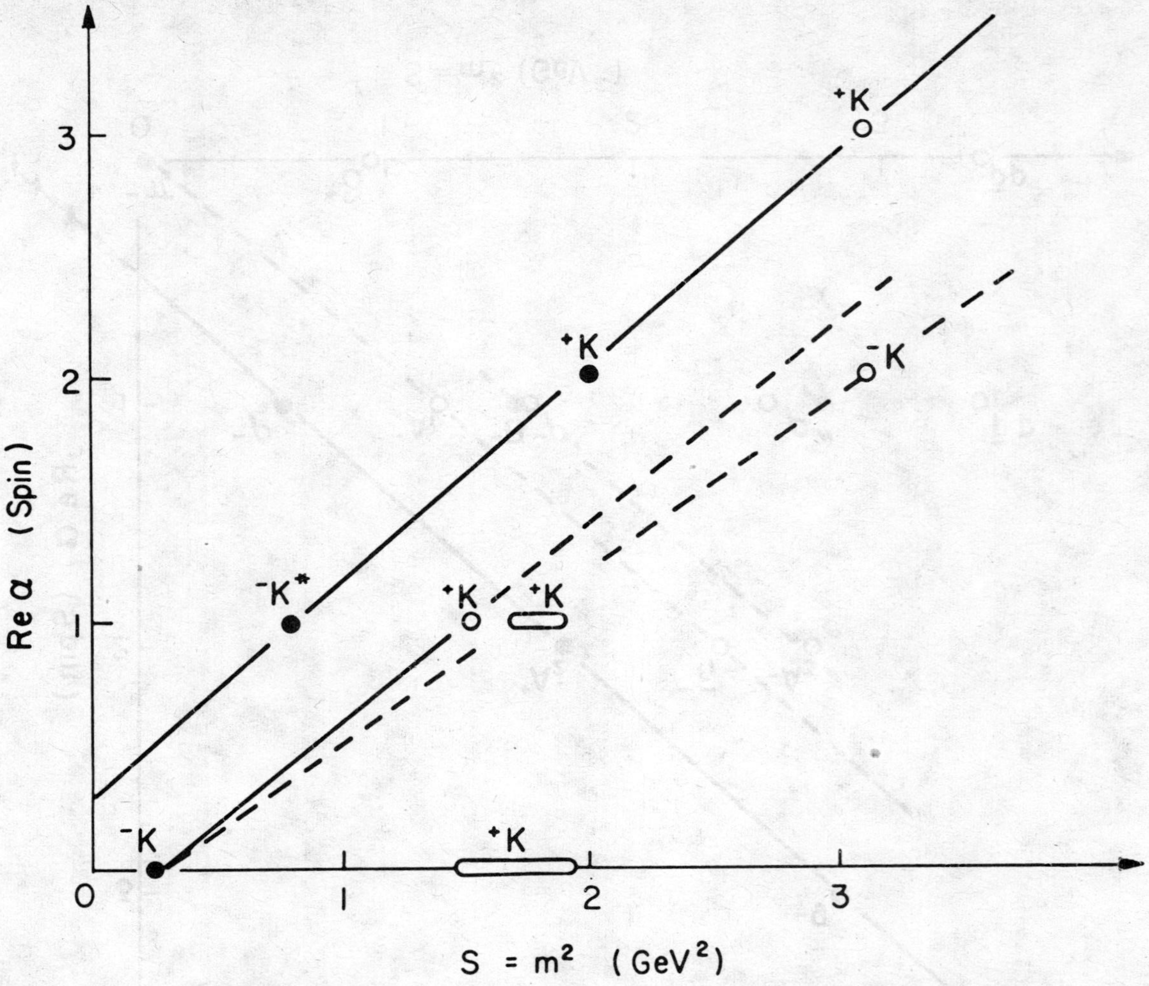

Fig. 2

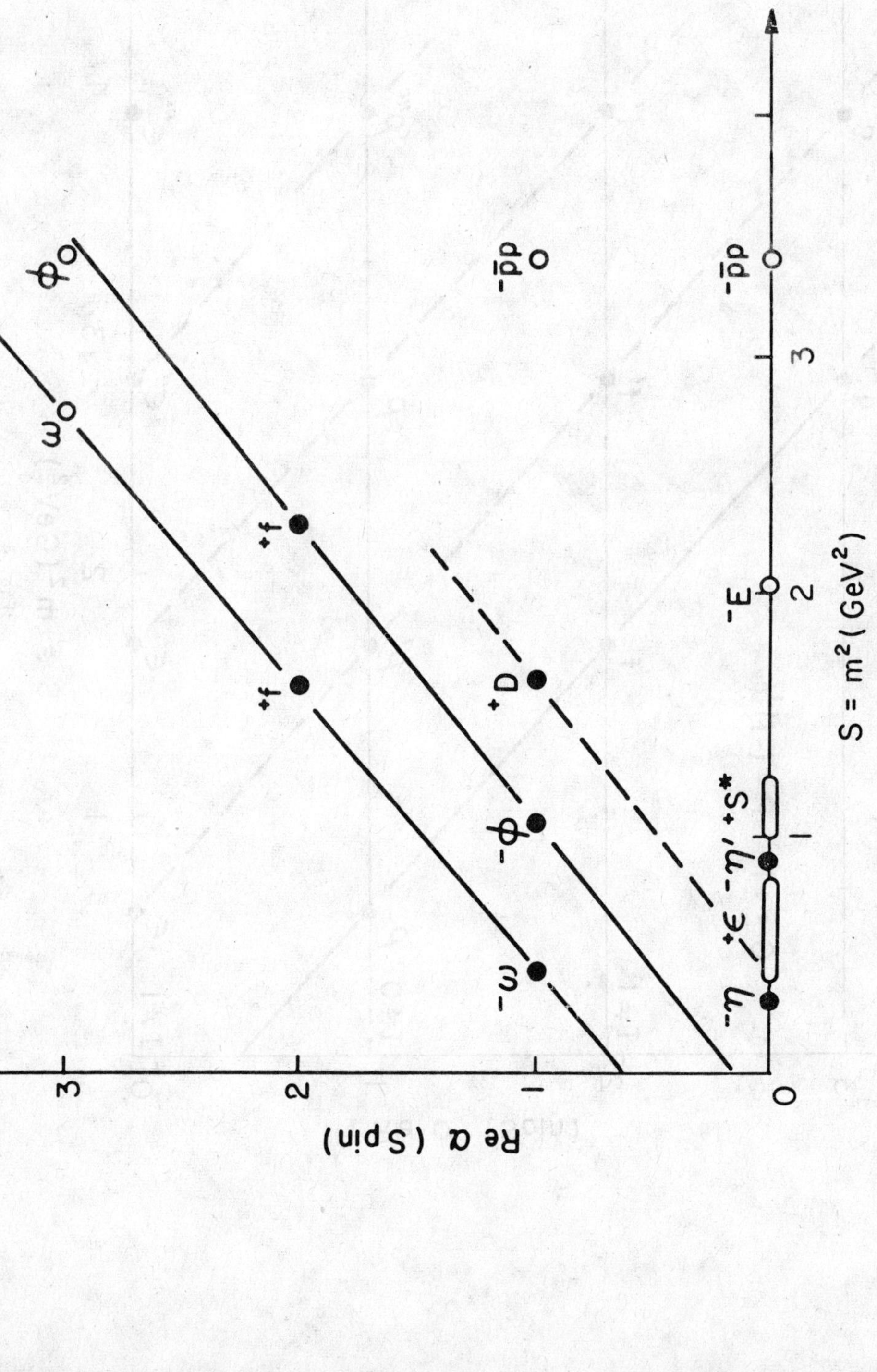

Fig. 3

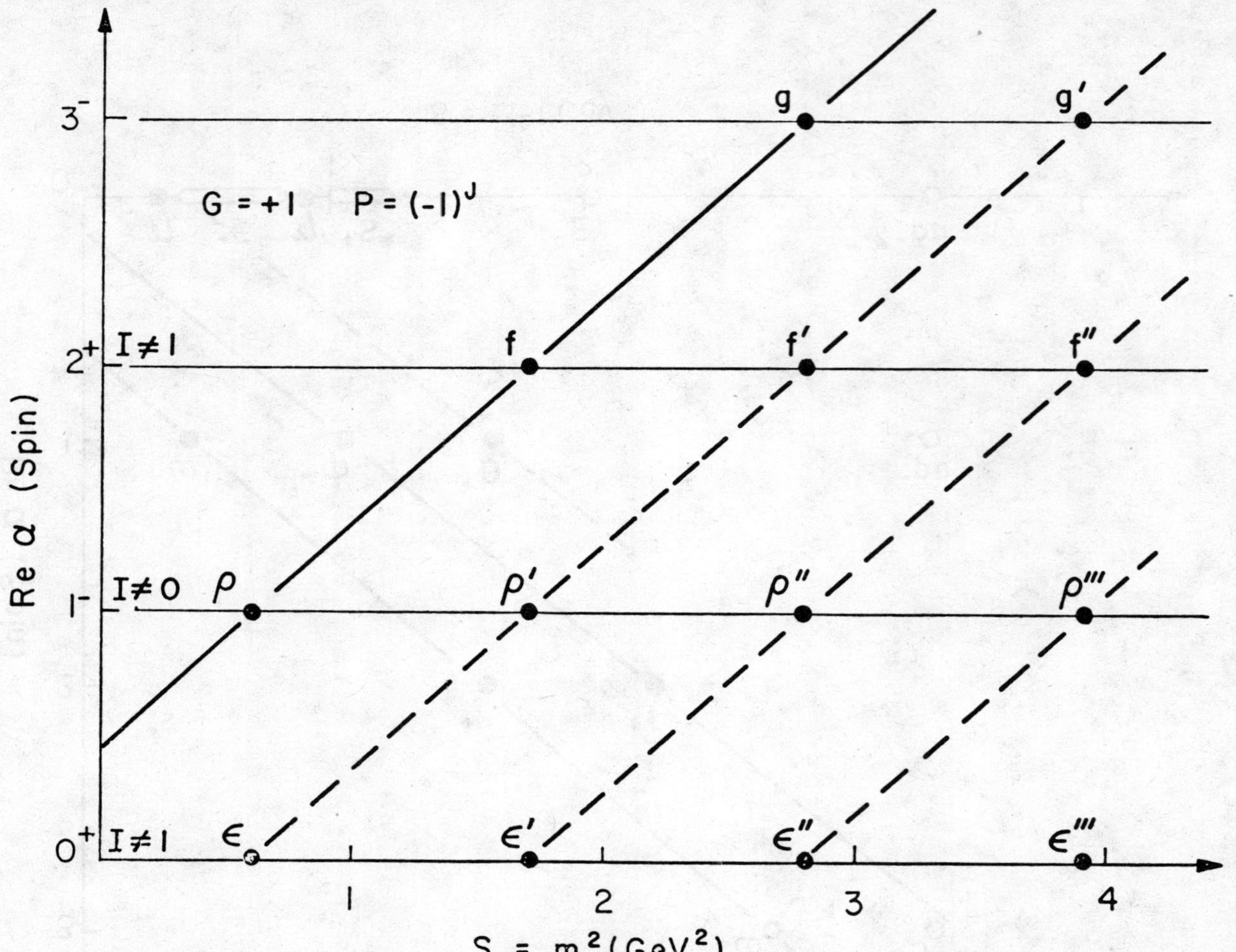

Fig. 4

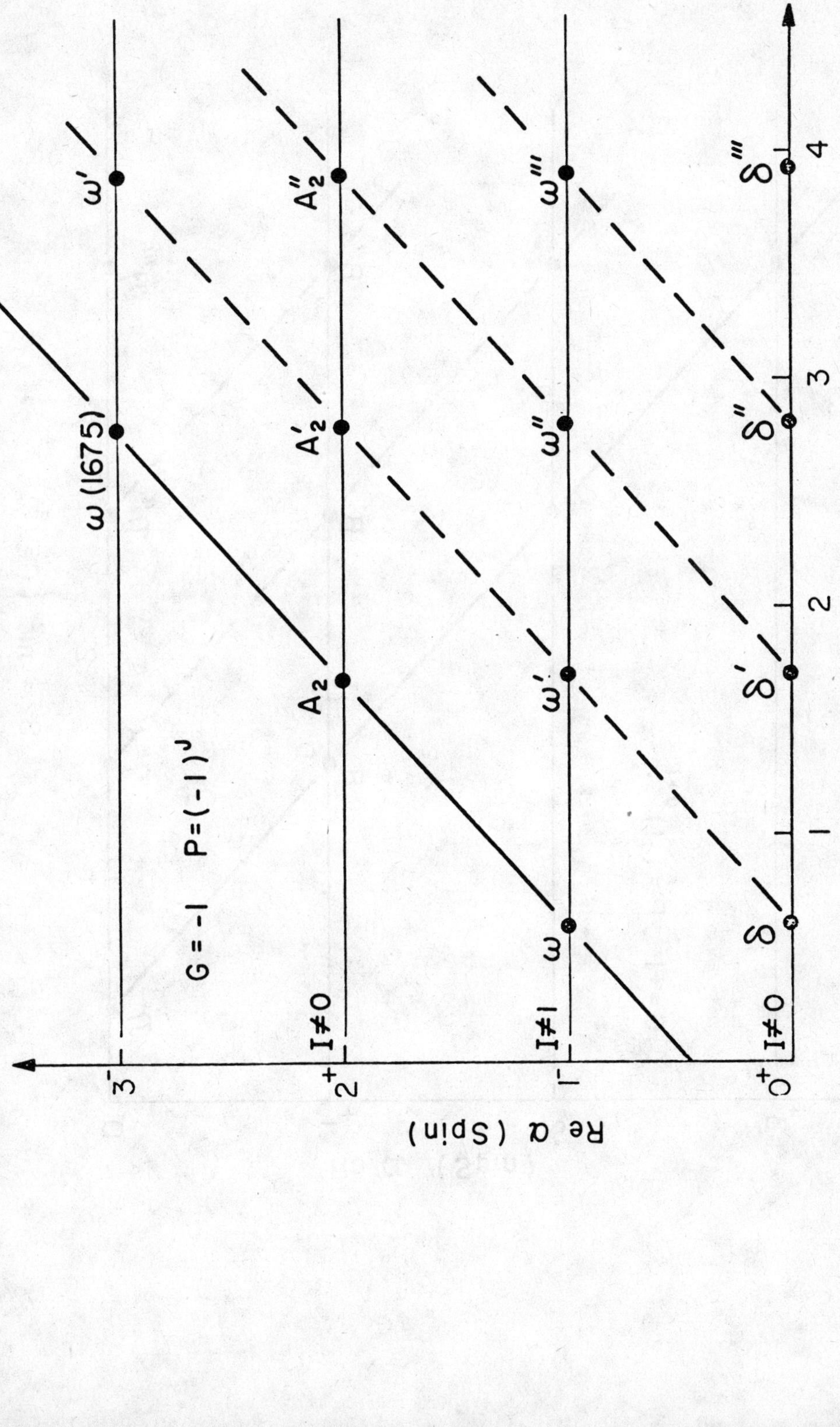

Re α (Spin)
S = m² (GeV²)
G = -1   P=(-1)^J
0⁺
1⁻
2⁺
3⁻
I≠0
I≠1
I≠0
ω
ω'
ω''
ω'''
A₂
A₂'
A₂''
ω (1675)
ω'
δ
δ'
δ''
δ'''
1
2
3
4

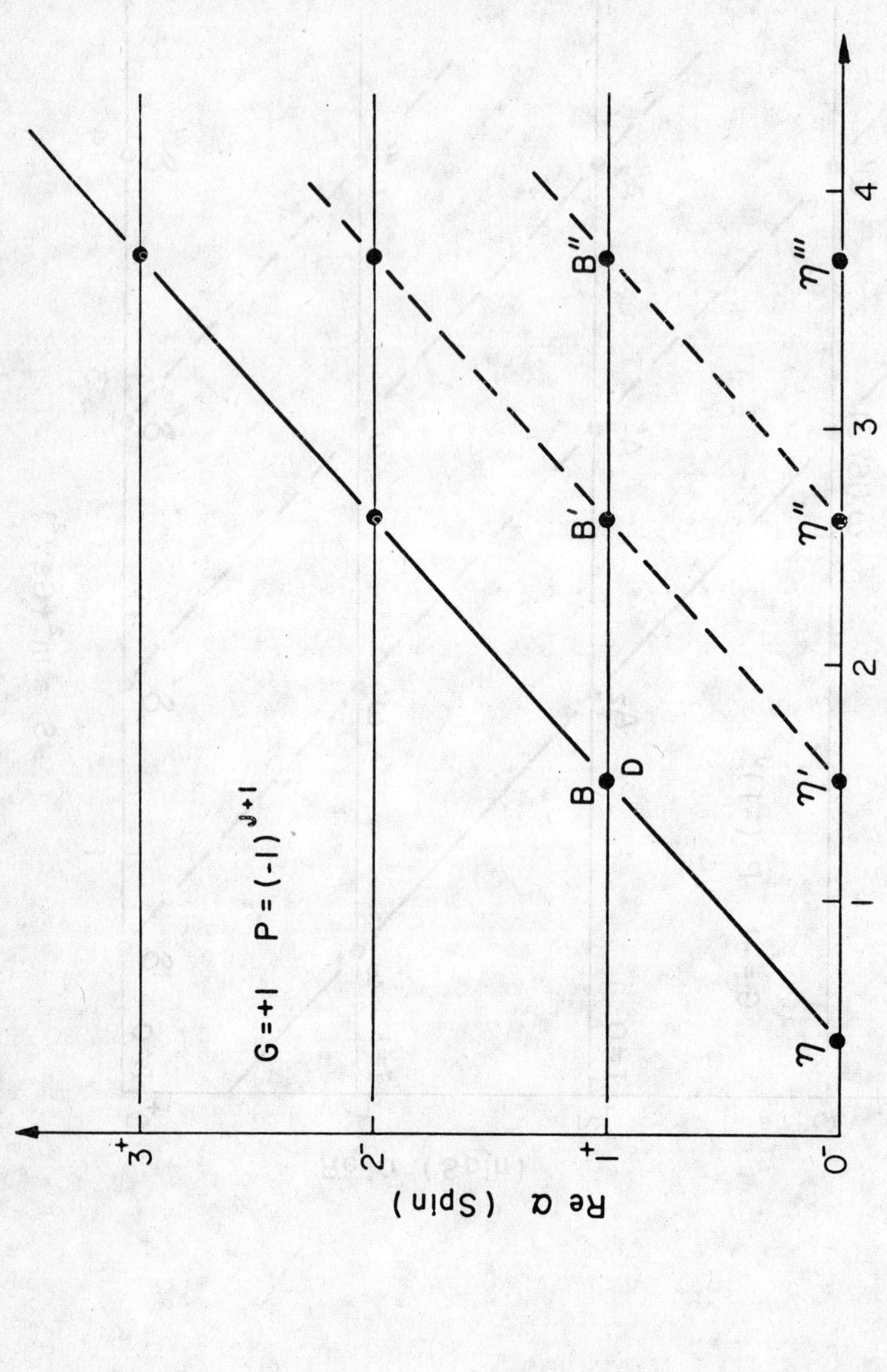
G=+1 P=(-1)^{J+1}
Re α (Spin)
0⁻
1⁺
2⁻
3⁺
S = m² (GeV²)
1
2
3
4
η
η'
η''
η'''
B
D
B'
B''

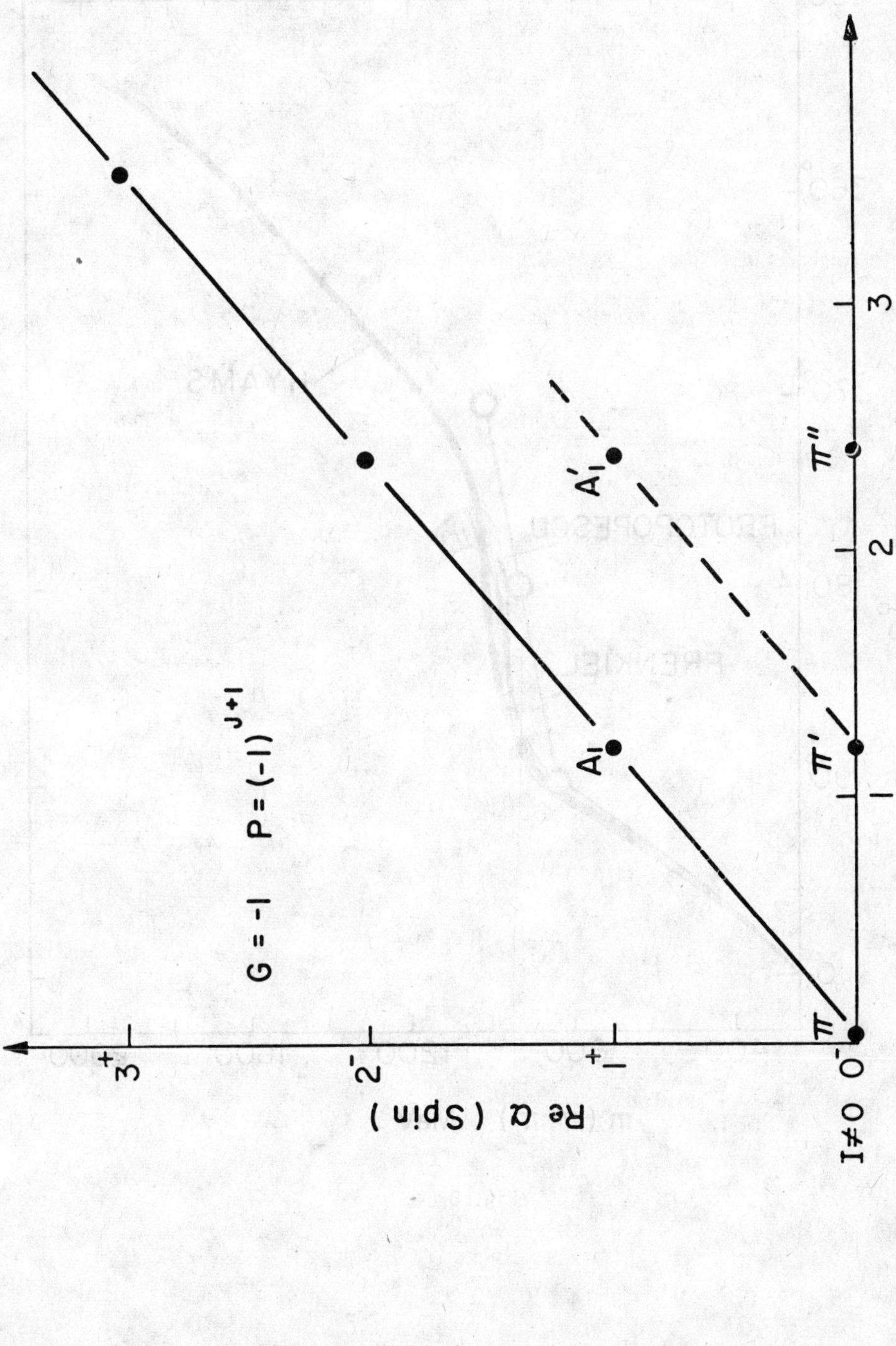
G = -1   P = (-1)^{J+1}
Re α (Spin)
I≠0
0⁻
1⁺
2⁻
3⁺
1
2
3
S = m² (GeV²)
π
π'
π''
A₁
A₁'

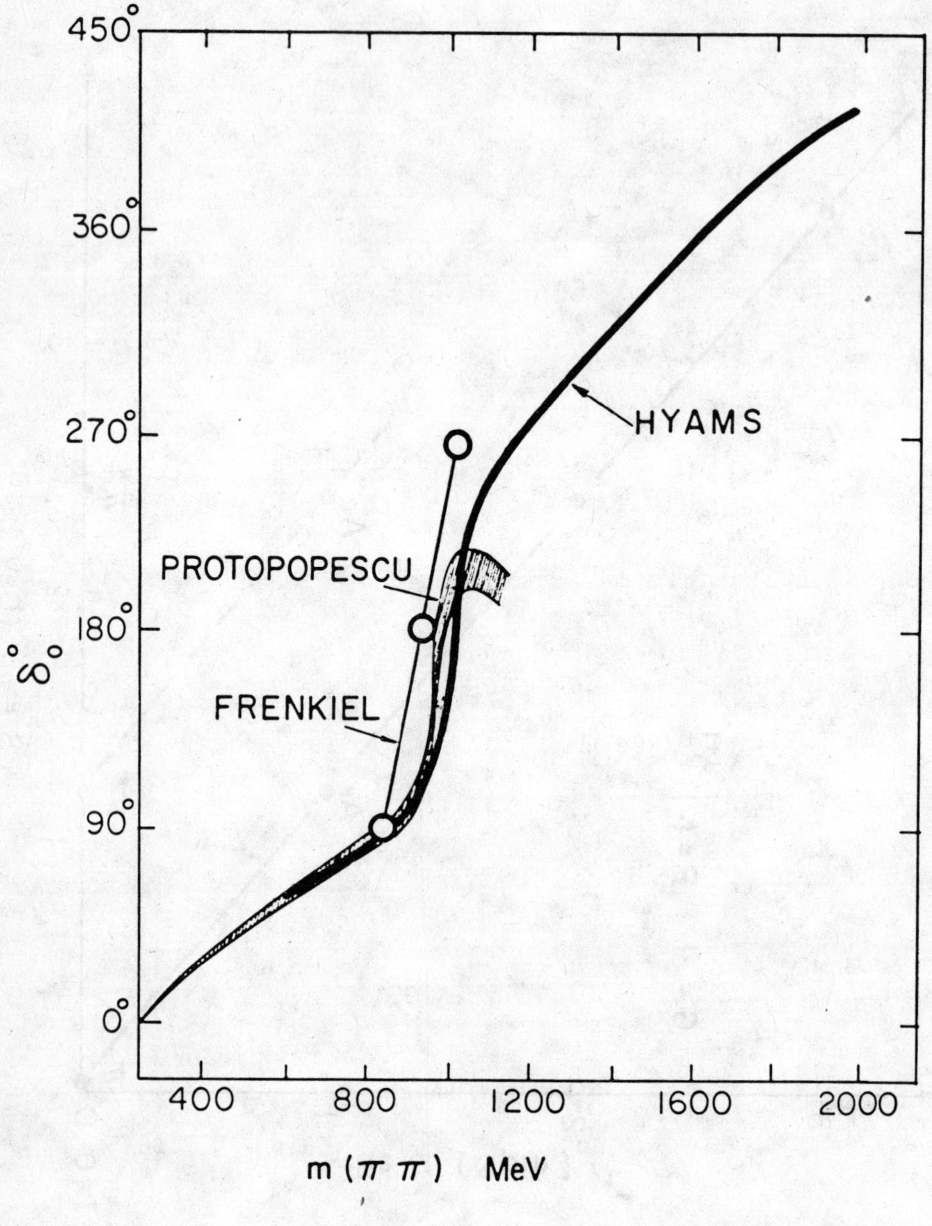
450°
360°
270°
180°
90°
0°
δ°
HYAMS
PROTOPOPESCU
FRENKIEL
400
800
1200
1600
2000
m (π π) MeV

Fig. 8

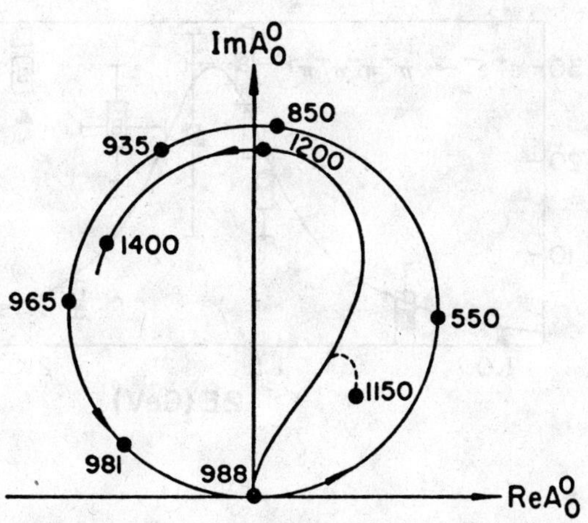
ImA_0^0
850
935
1200
1400
965
550
1150
981
988
ReA_0^0

Fig. 9

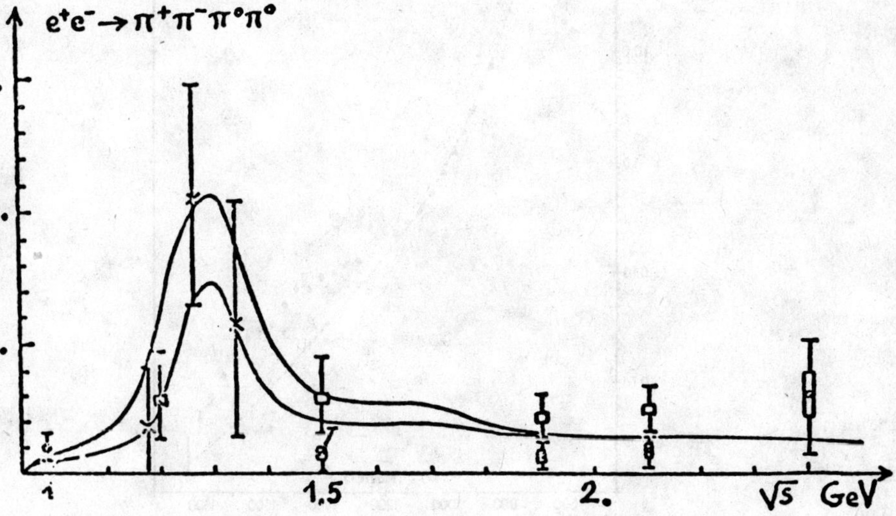
$e^+e^- \to \pi^+\pi^-\pi^0\pi^0$
15.
10.
5.
1
1,5
2.
√s GeV

Fig. 10

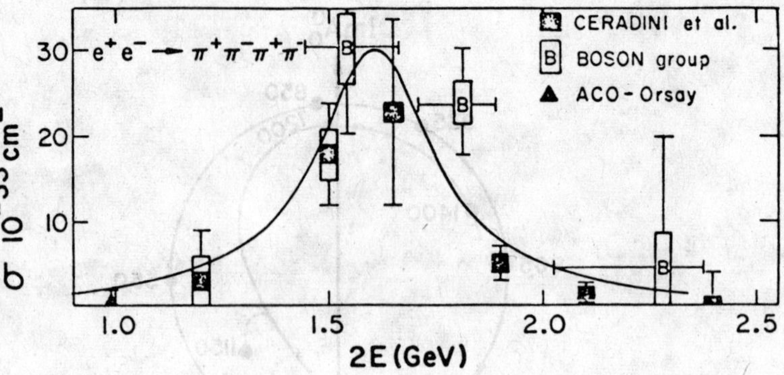

CERADINI et al.
BOSON group
ACO-Orsay
$e^+e^- \rightarrow \pi^+\pi^-\pi^+\pi^-$
σ $10^{-33}cm^2$
2E (GeV)

Fig. 11

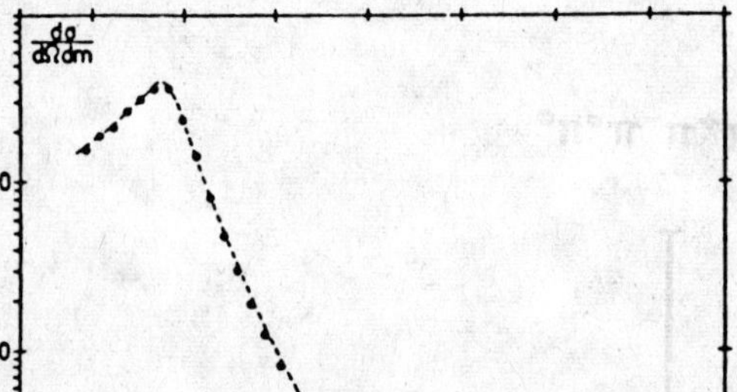

$\frac{d\sigma}{d\Omega dm}$
$M_{\pi\pi}$ (MeV/c²)

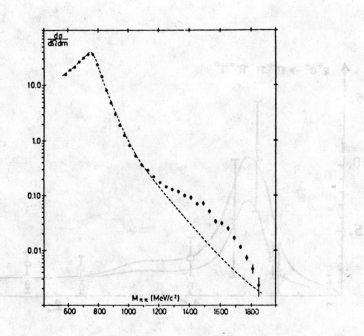

Fig. 12

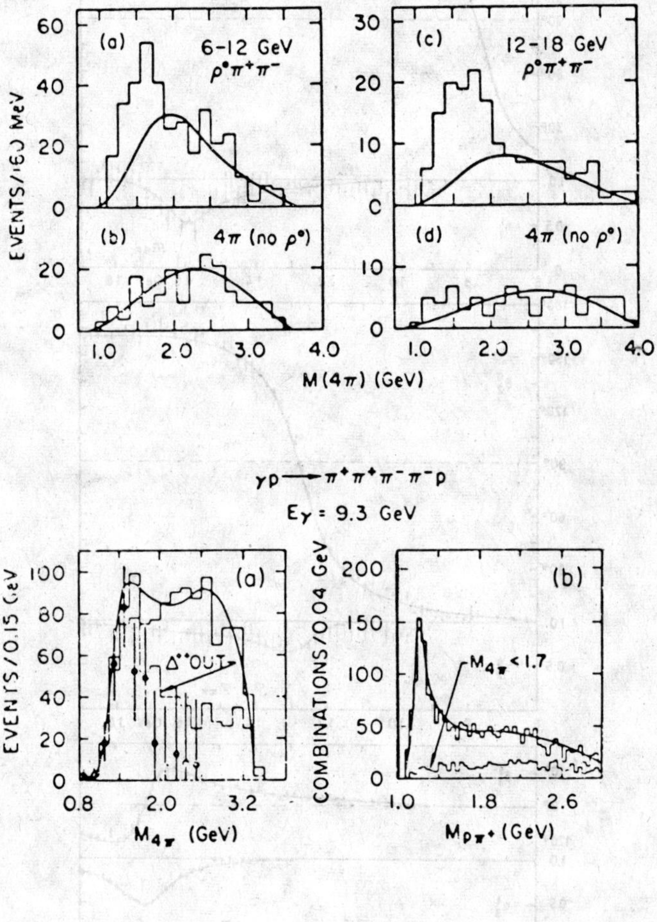
(a)
6-12 GeV
ρ°π⁺π⁻
(b)
4π (no ρ°)
(c)
12-18 GeV
ρ°π⁺π⁻
(d)
4π (no ρ°)
EVENTS/160 MeV
M(4π) (GeV)
γp → π⁺π⁺π⁻π⁻p
Eγ = 9.3 GeV
(a)
Δ⁺⁺OUT
EVENTS/0.15 GeV
M4π (GeV)
(b)
M4π < 1.7
COMBINATIONS/0.04 GeV
Mpπ⁺ (GeV)

Fig. 13

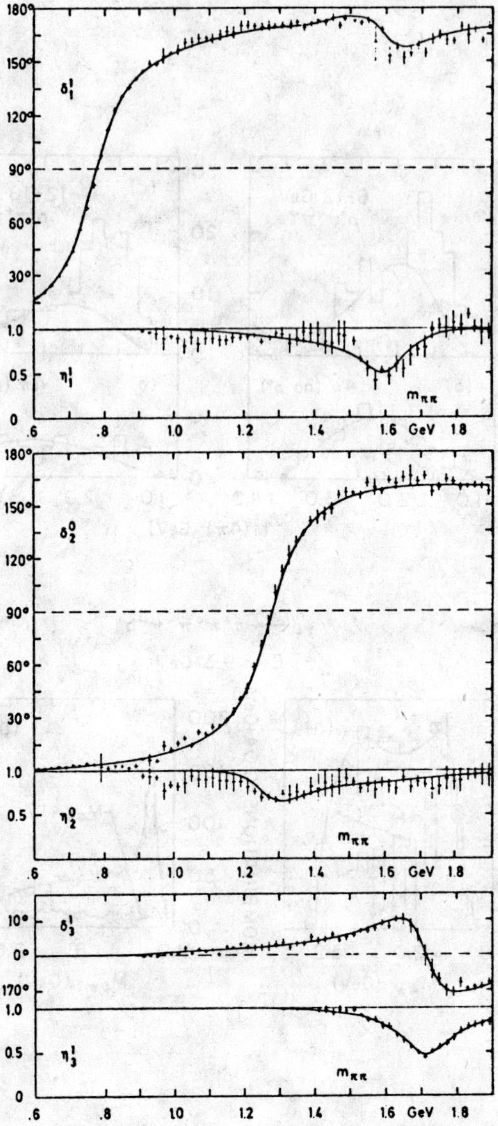
180°
150°
120°
90°
60°
30°
δ_1^1
1.0
0.5
0
η_1^1
$m_{\pi\pi}$
.6
.8
1.0
1.2
1.4
1.6
GeV
1.8
180°
150°
120°
90°
60°
30°
δ_2^0
1.0
0.5
η_2^0
$m_{\pi\pi}$
10°
0°
170°
δ_3^1
1.0
0.5
0
η_3^1
$m_{\pi\pi}$

Fig. 14

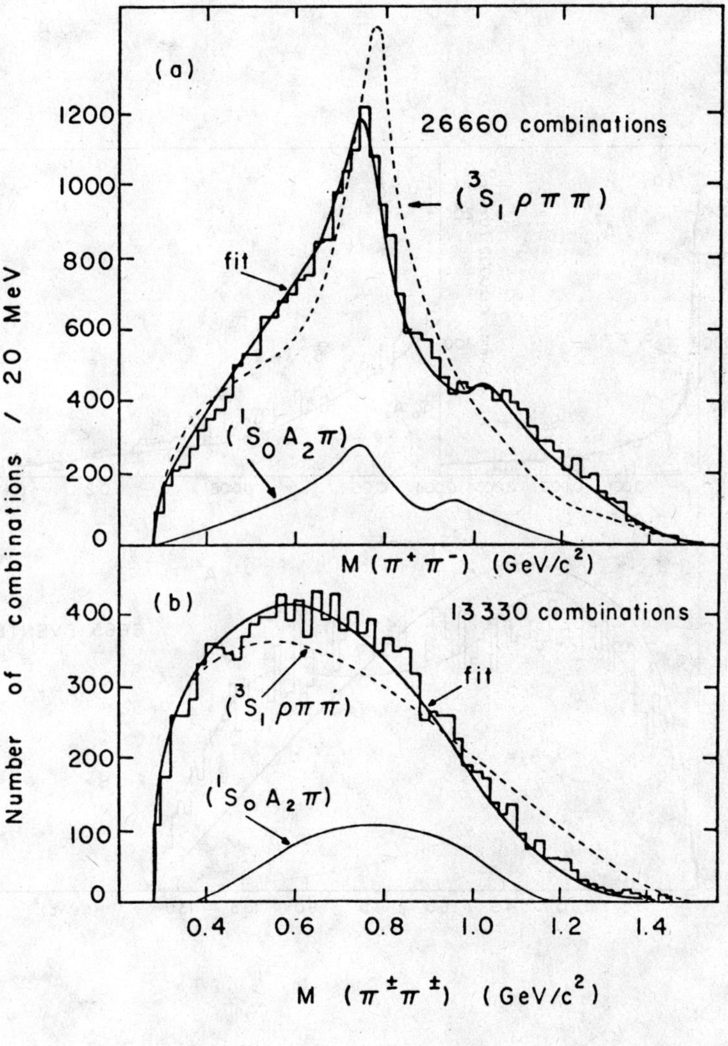
(a)
26660 combinations
($^3S_1\,\rho\pi\pi$)
fit
($^1S_0\,A_2\pi$)
1200
1000
800
600
400
200
0
M ($\pi^+\pi^-$) (GeV/c^2)
(b)
13330 combinations
($^3S_1\,\rho\pi\pi$)
fit
($^1S_0\,A_2\pi$)
400
300
200
100
0
0.4
0.6
0.8
1.0
1.2
1.4
Number of combinations / 20 MeV
M ($\pi^\pm\pi^\pm$) (GeV/c^2)

Fig. 15

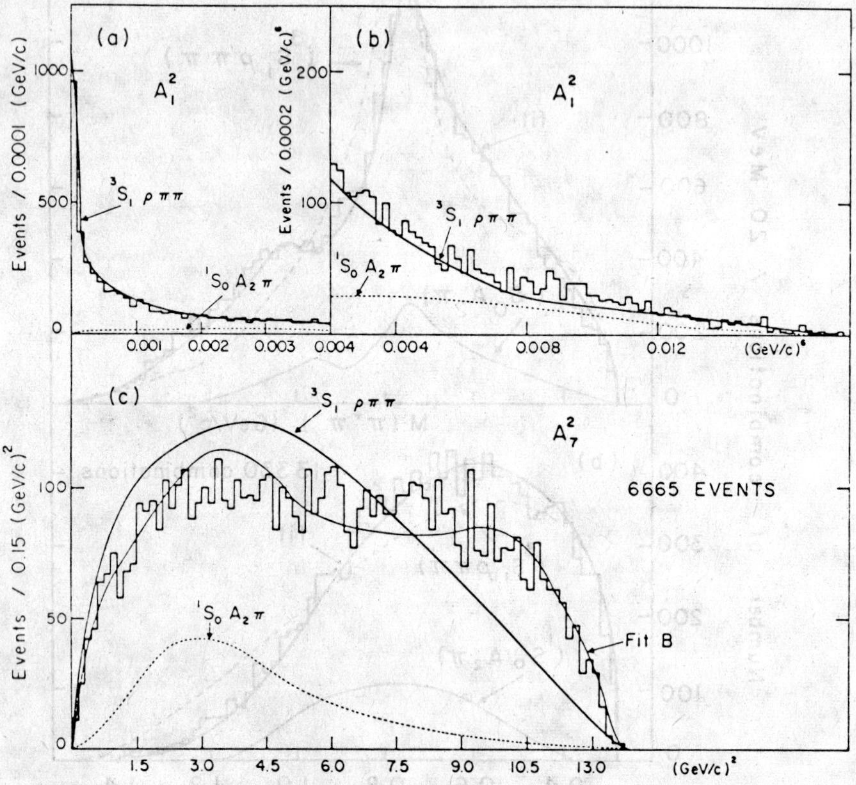

(a)
A_1^2
Events / 0.0001 (GeV/c)
$^3S_1\ \rho\pi\pi$
$^1S_0\ A_2\pi$
1000
500
0
0.001
0.002
0.003
0.004
(b)
A_1^2
Events / 0.0002 (GeV/c)^6
200
100
$^3S_1\ \rho\pi\pi$
$^1S_0\ A_2\pi$
0.004
0.008
0.012
(GeV/c)^6
(c)
$^3S_1\ \rho\pi\pi$
A_7^2
6665 EVENTS
Events / 0.15 (GeV/c)^2
100
50
0
$^1S_0\ A_2\pi$
Fit B
1.5
3.0
4.5
6.0
7.5
9.0
10.5
13.0
(GeV/c)^2

Fig. 16a

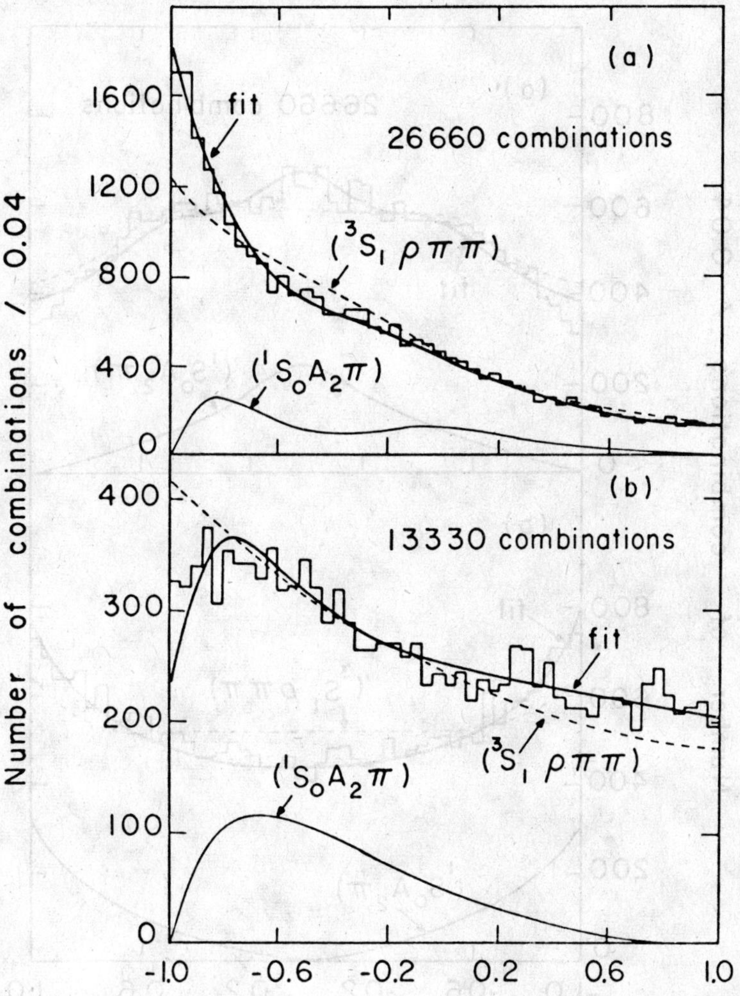
Number of combinations / 0.04
(a)
26660 combinations
fit
$(^3S_1\ \rho\pi\pi)$
$(^1S_0 A_2\pi)$
1600
1200
800
400
0
(b)
13330 combinations
fit
$(^3S_1\ \rho\pi\pi)$
$(^1S_0 A_2\pi)$
400
300
200
100
0
-1.0
-0.6
-0.2
0.2
0.6
1.0

Fig. 16b

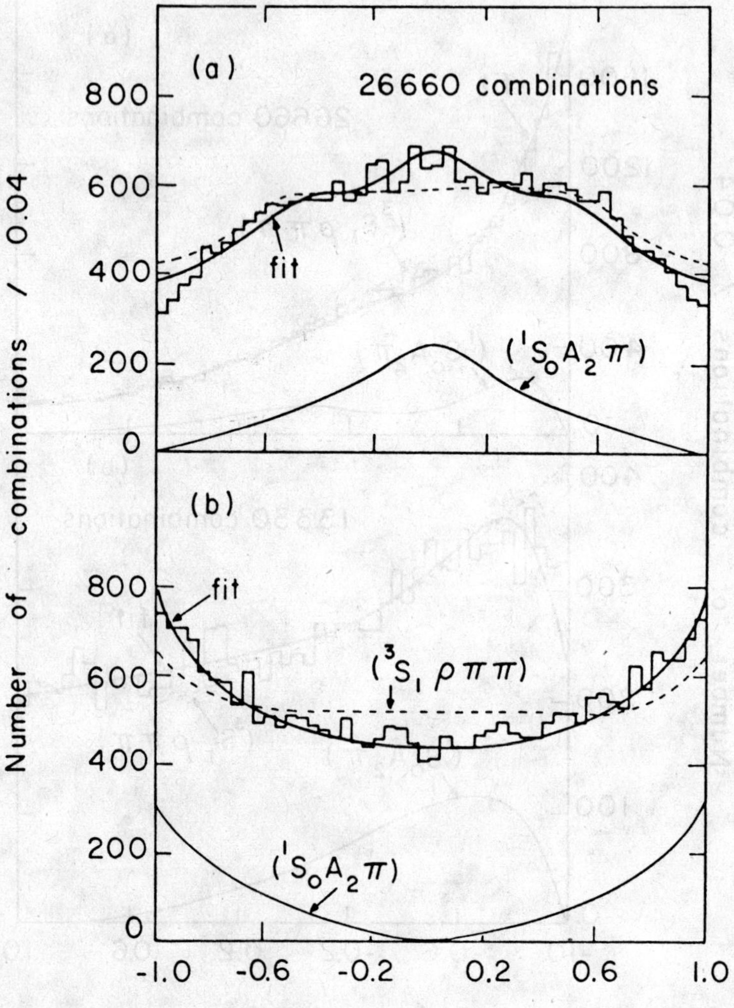
(a)
26660 combinations
800
600
400
200
0
fit
($^1S_0 A_2 \pi$)
(b)
800
600
400
200
0
fit
($^3S_1 \rho\pi\pi$)
($^1S_0 A_2 \pi$)
Number of combinations / 0.04
-1.0
-0.6
-0.2
0.2
0.6
1.0

Fig. 16c

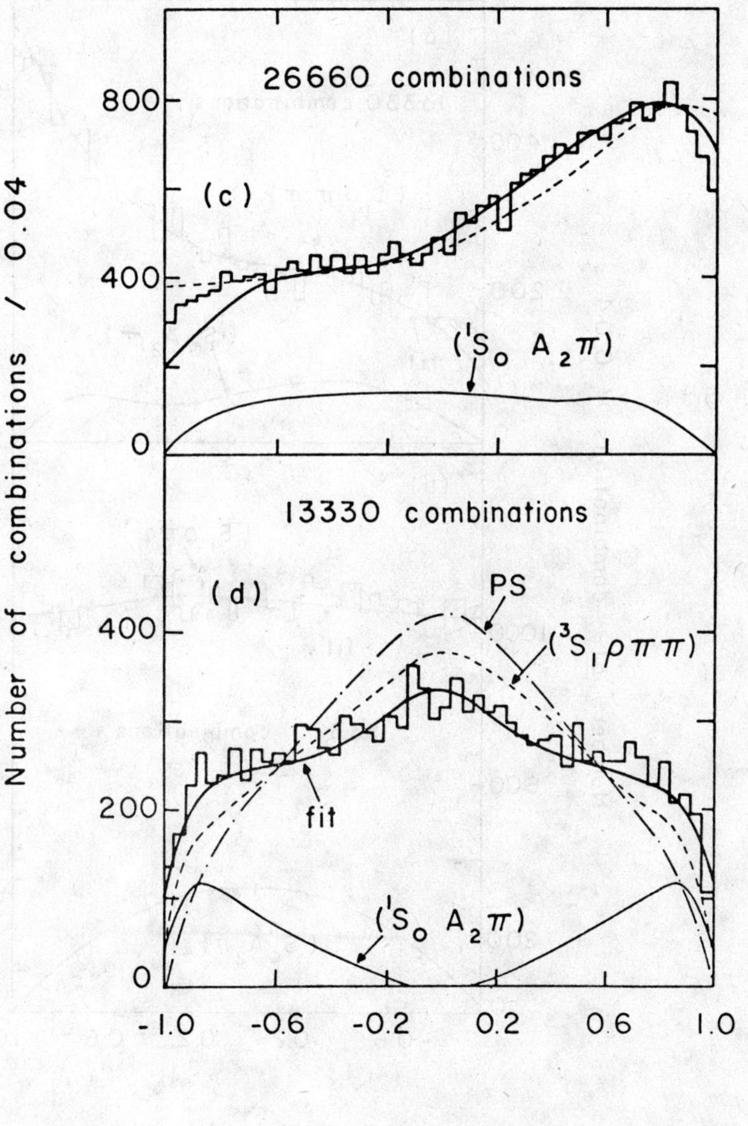
26660 combinations
(c)
800
400
0
($^1S_0$ $A_2\pi$)
13330 combinations
(d)
PS
($^3S_1$ $\rho\pi\pi$)
fit
($^1S_0$ $A_2\pi$)
400
200
0
-1.0
-0.6
-0.2
0.2
0.6
1.0
Number of combinations / 0.04

Fig. 16d

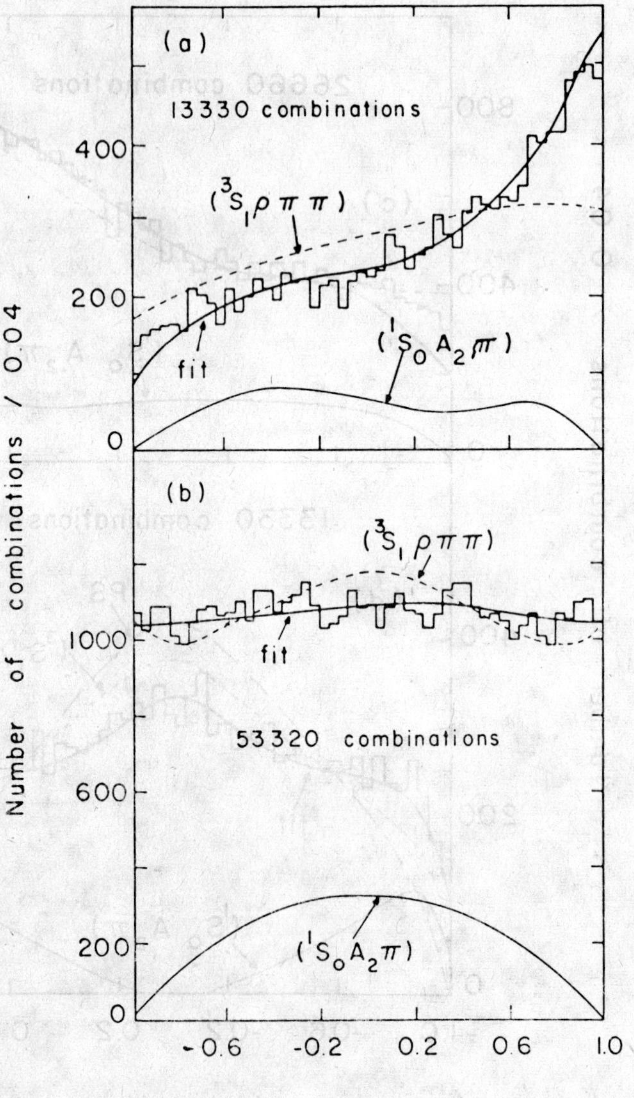
(a)
13330 combinations
($^3S_1\rho\pi\pi$)
fit
($^1S_0A_2\pi$)
(b)
($^3S_1\rho\pi\pi$)
fit
53320 combinations
($^1S_0A_2\pi$)
Number of combinations / 0.04
400
200
0
1000
600
200
0
-0.6
-0.2
0.2
0.6
1.0

Fig. 16e

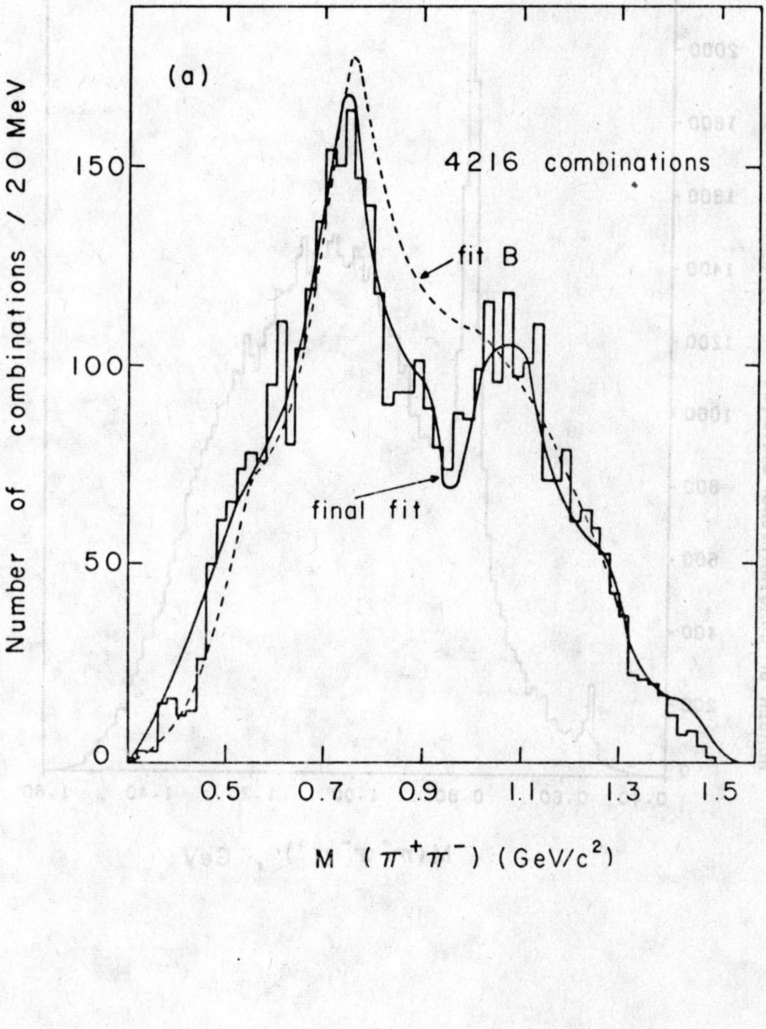
(a)
4216 combinations
fit B
final fit
Number of combinations / 20 MeV
150
100
50
0
0.5
0.7
0.9
1.1
1.3
1.5
M $(\pi^+\pi^-)$ (GeV/c^2)

Fig. 17

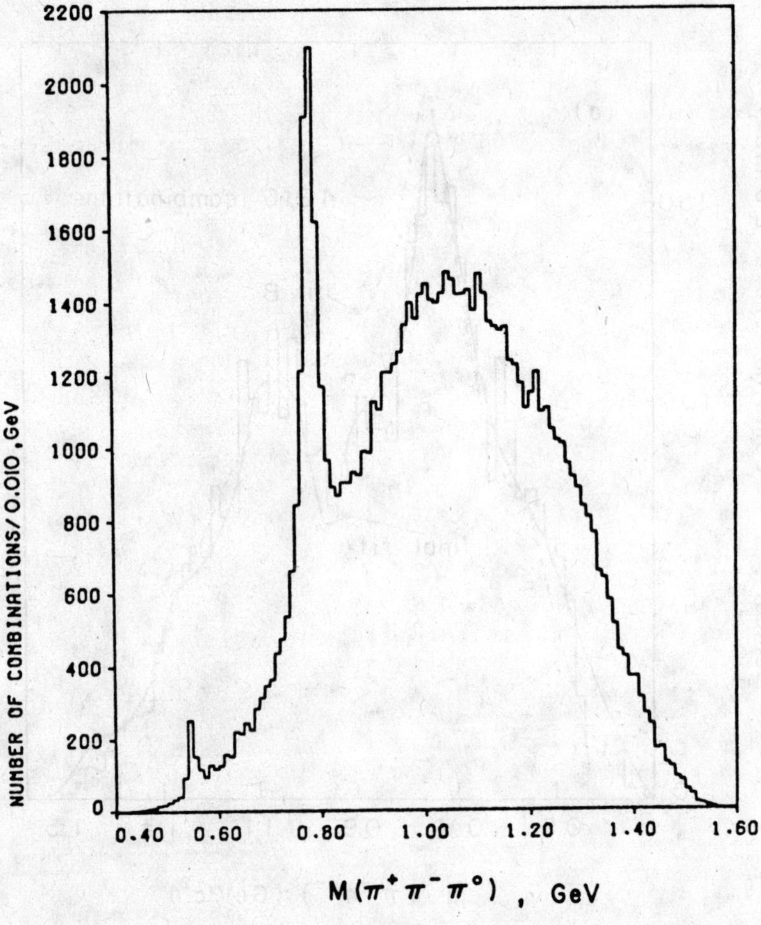
2200
2000
1800
1600
1400
1200
1000
800
600
400
200
0
NUMBER OF COMBINATIONS/0.010,GeV
0.40
0.60
0.80
1.00
1.20
1.40
1.60
M(π⁺π⁻π°) , GeV

Fig. 18

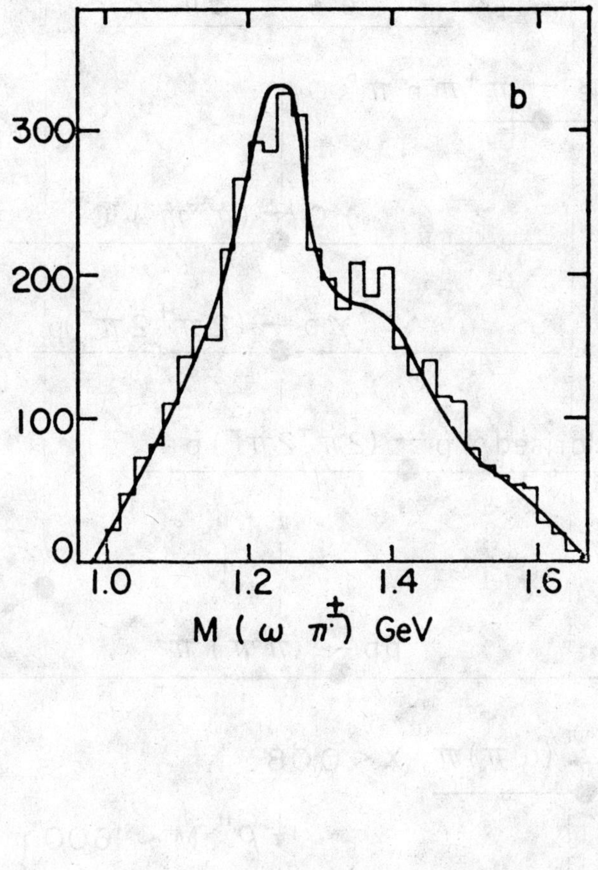

Fig. 19

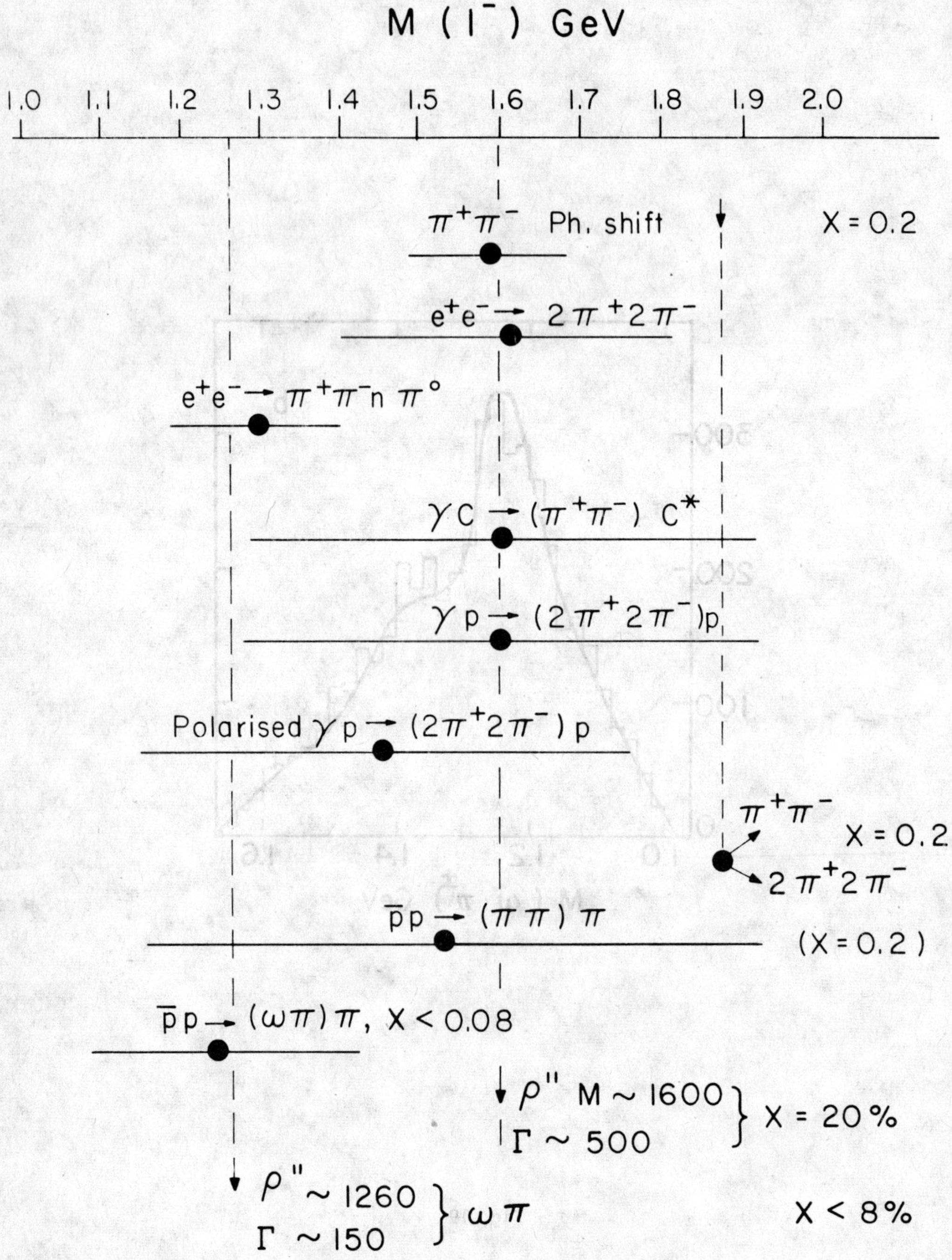
M (1⁻) GeV
1.0 1.1 1.2 1.3 1.4 1.5 1.6 1.7 1.8 1.9 2.0
π⁺π⁻ Ph. shift
X = 0.2
e⁺e⁻ → 2π⁺2π⁻
e⁺e⁻ → π⁺π⁻n π°
γC → (π⁺π⁻) C*
γp → (2π⁺2π⁻)p
Polarised γp → (2π⁺2π⁻) p
π⁺π⁻
X = 0.2
2π⁺2π⁻
p̄p → (ππ) π
(X = 0.2)
p̄p → (ωπ) π, X < 0.08
ρ'' M ~ 1600
Γ ~ 500
X = 20%
ρ'' ~ 1260
Γ ~ 150
ωπ
X < 8%

Fig. 20

D I S C U S S I O N S

CHAIRMAN: Prof. L. Montanet

Scientific Secretary: M.N. Tugulea

DISCUSSION No. 1

- *WAMBACH:*

Concerning the Münich analysis of $\pi\pi$ scattering I have the following questions:

1) Can one use the isobar model discussed by Rosenfeld here?

2) Why do they use the K-matrix parametrization?

- *MONTANET:*

1) In principle, the Chew-Low extrapolation is a good method to study $\pi\pi$ scattering but, of course, there are kinematical singularities in the region $0 < t < \mu^2$. The isobar model implies other assumptions and in the case of the CERN-Munich experiment the OPE model with absorption is reasonable.

2) The authors prefer K-matrix parametrization since it is simpler to look for several resonances in the same partial wave.

- *CORDS:*

You said that for the fit of the $\pi\pi$ partial waves by the CERN-Münich group, 36 parameters have been used and only 13 spherical harmonic moments could be measured per mass bin. How could they avoid running into ambiguities with so many parameters?

- *MONTANET:*

The 36 parameters refer to the energy-dependent analysis which refers to 705 data points. The ambiguities arise when you perform an energy-independent analysis (11 parameters for 13 data points at each energy). To avoid these ambiguities the authors used a method proven to be useful in the CERN I πN phase-shift analysis, i.e. they combine the energy-independent χ^2 fit with a χ^2 coming from the comparison of the results with those of the energy-dependent analysis.

- *MENDES:*

In Miller's lectures there was definite evidence that A_1 and A_3 did not show resonant behaviour, however they were included in your "parents" table (although with question marks). Should I consider these effects as possible resonance candidates, or should I forget about them altogether?

- *MONTANET:*

Ascoli has shown that the so-called A_1 and A_3 resonances do not have the expected behaviour for a reasonable Breit-Wigner resonance when he performed his analysis. If we want to know more about the A_1 (and the A_3) we have to look for other production reactions. Another example is the "Q bump", which may be better understood if observed in hypercharge exchange reactions ($\pi^- p \to \Lambda K^0 \pi^+ \pi^-$).

- *KUPCZYNSKI:*

What further assumptions, besides the OPE model, were used to describe the $\pi^- p$ interactions?

- *MONTANET:*

Since we have not performed, for the time being, an experiment with a polarized target, we have to introduce amplitudes to take into account the presence of moments which are forbidden in the OPE model.

The CERN-Münich group introduced a model, similar to Williams' one, but in a less restrictive way and their results give a parametrization still compatible with the prescription of Williams.

- *CHU:*

Concerning A_1, the region around 1000 MeV must necessarily contain a contribution from the Deck effect since it is near the $\pi\rho$ threshold. Therefore, if there is an A_1 with a mass of around 1000 MeV, the data will be a superposition of the A_1 and the Deck effect. One would certainly not expect the $\pi\rho$ enhancement to exhibit pure resonance structure. The question is: can the Ascoli data be explained by a combination of Deck mechanism and A_1 resonance production?

- *MONTANET:*

No, I do not believe that the Ascoli data allows for the presence of a normal Breit-Wigner resonance behaviour. It would be nice to be able to observe the three-pion system in a reaction where there can be no Deck effect, e.g. in $\bar{p}p \to \pi^+\pi^-\pi^+\pi^-$, or in backward scattering for instance. However, it is very difficult to observe a $\pi\rho$ resonance since both the ρ and presumed A_1 are fairly broad.

- *BYERS:*

I would like to ask you about the Q. This is an interesting question since by SU_3 it is related to the A_1.

- *MONTANET:*

The situation is similar to that of A_1. Unfortunately, there are not enough statistics to give us a definitive spin-parity analysis.

In general, there are two difficulties when doing such experiments: one, which is purely experimental, is to insure that your data are clean, unbiased, and statistically significant; the second one is to appreciate the reliability of the "model" usually necessary to interpret the data. If there was not this last difficulty, I think that the experiment which observes the K^*, $J^P = 1^+$, at M = 1270 MeV, with the two coherent decays, namely:

$$\bar{p}p \to K^0K^*(1270) \ , \quad K^*(1270) \to K^*\pi K\rho$$

would effectively solve the problem. Unfortunately, this depends once more on the validity of the final-state interaction model, and that is why I believe these results have to be confirmed by a completely different experiment, for instance, by a statistically significant result in hypercharge exchange reactions such as $\pi^-p \to \Lambda K^*(1300)$.

- *ZICHICHI:*

If we could have only one experiment, but a good one from the experimental point of view, then we could solve your second problem too and it would be enough to understand the 1^+K^* problem.

DISCUSSION No. 2

- *NAHM:*

I want to ask you about the 12 poles found in a fit to the $\pi\pi$ scattering: are these poles randomly scattered, or does one have a clear cut distinction between one class of three reasonable poles and nine poles which are nonsense?

- *MONTANET:*

There is no problem with the S^*, but there is no clear separation between the ε', the ε'' pole and other poles found in the partial wave amplitude. They are just further away from the real axis.

- *ROSENFELD:*

There may be no conflict at all between the ρ'' parameters of CERN-Münich from $\rho''(1590 \pm 20 \to 2\pi)$, where they report $\Gamma = 180 \pm 20$ MeV and the wider $\rho''(1600 \to 4\pi)$ with $\Gamma \sim 500$ MeV. CERN-Münich see

$$d\sigma \propto \left| \frac{1}{\varepsilon_{\rho''} - \Gamma} \right|^2 d(\text{2-body})$$

We see

$$d\sigma \propto \left| \frac{1}{\varepsilon_{\rho''} - \Gamma} \right|^2 d(\text{4-body}) \; .$$

d(2-body) is easy: it is $q(\text{decay})/\sqrt{s} \to 1$. d(4-body) is a diverging puzzle. Even if the ρ(750) in the final state had zero width, d(4-body) would still increase with mass as fast as d(3-body), which is the area of a Dalitz plot $\propto s^2 = m^4(\rho'')$ (a very high power).

So the $\rho'' \to 4\pi$ spectrum is skewed to high mass, and in fact the 4π mode probably robs the 2π mode at 1600-1800 MeV, and so skews the 2π mode to low mass. Thus the spectra and width for 2π and 4π will be different.

- *GENSINI:*

I would like to make two comments on the topics touched by Prof. Montanet in his lecture. The first one concerns extrapolation of partial wave data onto unphysical sheets: first of all no analysis of this kind has ever tried to establish the domain in which, given a set of data with errors, a meaningful search for singularities can be accomplished. There are mathematical theorems and known methods for this, but they seem to be little known by phenomenologists. I would also like to remind you that the unphysical sheets have a much richer structure than is usually assumed in these analyses; that is to say, there are also cuts around, and many of the poles seen in the s_0-wave would fake these cuts. For instance, in the s_0-wave we expect a ρρ cut with threshold energy (1.55 - i0.15) GeV, very close to the found ε″.

The second comment concerns the first and strongest theoretical rationale for an s-wave low-energy resonance in the ππ system, namely the 50% discrepancy between the prediction of Adler-Weisberger relation for ππ scattering and the saturation of its right-hand side with known states over, say, the mass of the ρ. A recent, still unpublished, computation by Kellett, Paver, Verseguassi and myself, employing the more powerful form of dispersion relations developed by Sorin Ciulli and his collaborators, shows that the recent data by Protopopescu et al., together with information on the very low-energy region coming from recent experiments on $K_{\ell 4}$ lead to a very good saturation indeed of Adler-Weisberger relation even if no clear ε signal is displayed by the data (the latter are well represented even by a single S^* pole formula).

- *KLEINERT:*

Let me remark that the use of backward dispersion relation allows us to differentiate between even and odd partial wave resonances in ππ scattering. Thus, since $P_\ell(\cos\theta = 1) = (-1)^\ell$ one can show that both have to contribute with equal strength to the Adler-Weisberger sum rule.

Thus knowing that the $\ell = 1$ meson fills up half of the sum, we conclude that the remaining 50% are due to event partial waves. Since f contributes only 7%, the bulk will be due to the phase shift in the ε region.

- *CHU:*

You showed an experiment in which the f′(1500) was not observed as a 2π resonance. The Rosenfeld tables indicate that the f′(1500) has only been observed in $K\bar{K}$. Is there any other evidence for the f′(1500) in ππ or πω channels?

- *MONTANET:*

No, there is not. Of course, the width as presently quoted in the Particle Data Tables is only 40 MeV, and with the mass resolution of present experiments which search for f′(1500) in 2π channels, it is possible that we could have missed it if it has a small branching ratio into 2π.

- *FRAMPTON:*

On the question of theoretical guidance on interpretation of daughter states in ππ scattering, the most detailed theoretical predictions on masses and widths are probably coming from dual models. An important general prediction of dual models, which is consistent with the experimental data presented here, is the complete absence of odd daughters strongly coupled to ππ; this implies absence of ε(760), ρ′(1250), ε″(1600), f′(1600) in the two-pion channel. (Note, however, that daughter states with these masses and quantum numbers may couple in four-pion channels.)

For the even daughters ε(1250) and ρ′(1600) the elastic partial widths are predicted, by the simplest dual model, as 260 MeV and 140 MeV, respectively, in the two-pion channel.

- *KLEINERT:*

If you have no ε and only a 260 MeV wide ε', you must have a tremendous f width in order to balance even and odd partial waves as I just explained.

- *FRAMPTON:*

The f width is consistent with experiment in the model.

MULTIPARTICLE PRODUCTION PROCESSES AT HIGH ENERGY

Edmond L. Berger

Table of Contents

MULTIPARTICLE PRODUCTION PROCESSES AT HIGH ENERGY

Edmond L. Berger
CERN - Geneva

1. INTRODUCTION

1.1. Qualitative summary

Operations of the NAL accelerator and of the CERN Intersecting Storage Rings have provided new data which extend our knowledge of many facets of hadronic production from the centre-of-mass energy of $\sqrt{s} = 8$ GeV, available previously, to $\sqrt{s} = 62$ GeV. Events of the past year have changed our outlook on high multiplicity hadronic reactions considerably. New directions have been established in theoretical and experimental research by the observation of rising proton-proton total cross-sections, of inelastic diffraction scattering of fast protons, and of reasonably strong yet short-range inclusive correlations among copious secondaries produced with relatively small momentum in the over-all collision centre-of-mass system (the "central region"). In these lectures, I attempt to examine critically these new data and corresponding theoretical developments. The focus here is restricted to recent progress. This is neither the place, nor does it seems to be the right moment, for yet another comprehensive review. For subjects not treated here, and for other points of view, the articles listed under reference 1) may be consulted profitably.

I concentrate almost entirely on the inclusive approach to multiparticle physics, because as yet too little high energy information of a more differential character is available. The desirability of greater insight into the topological structure of individual events is very apparent, however.

After a very brief survey of kinematics, notation, and sum rules, I catalogue some empirical regularities of multihadron production whose validity is confirmed and extended by measurements at NAL and the ISR. Next, a list is given of the past year's new observations which must be incorporated into an over-all phenomenological scheme. After this introduction, in Section 2, I discuss the concept of dominant short-range order as an asymptotic unifying theme. The implications of this hypothesis for single particle inclusive spectra are compared with available data, particularly with regard to energy independence of a central plateau in rapidity space.

---

*) On leave from Argonne National Laboratory.

Two more explicit phenomenological frameworks are then described in Section 3 as a basis for making further guesses. One of these is the Amati-Bertocchi-Fubini-Stanghellini-Tonin model [2] which is, in fact, the prototype of models from which the hypothesis of short-range-order is abstracted. The second is the fragmentation [3] concept, in which both short and long-range structure is found. These view-points suggest radically different topological structure of individual multiparticle events, but may be made to agree in their "predictions" of an asymptotic energy independent central plateau in the single particle inclusive rapidity distribution. From the inclusive approach, two-particle spectra can discriminate between these models, as discussed in Section 4, where data are also examined in detail. The dominant short-range order alternative is favoured by data in the central region, but more precise experimental investigations of the energy dependence of two-particle spectra is clearly called for. In contrast to one year ago, when the experimental picture was consistent with either fragmentation or dominant short-range order models, correlation data appear to favour the latter alternative. Several features of data suggest, however, that there is significant cluster formation, an aspect of hadronic phenomena emphasized in the fragmentation approach. One way to imitate data is thus to imagine a multiperipheral-like emission [4] of clusters of pions. A mean of four pions per cluster [4] is suggested by the correlation data. Cluster formation and its parametrization are described in detail in Sections 3.2 and 3.3.

Evidence for long-range correlations is present also in ISR and NAL data. A striking example is the inelastic diffractive peak observed in the inclusive proton longitudinal momentum spectrum [5]. Further experimental study is also necessary here, but present indications support a peak which is energy independent at fixed $x = 2p_L/\sqrt{s}$. Here p_L is the final proton's longitudinal momentum in the centre-of-mass frame. Data and theoretical understanding are described in Section 5. As energy increases, states of ever-increasing mass are excited diffractively. Multiplicity rises accordingly. The cross-section in this inelastic diffractive peak may rise logarithmically, and thus be associated with the observed growth with s of the total cross-section. This association is discussed quantitatively in terms of triple-Regge models. On the basis of this approach, predictions are made for the single and two-particle inclusive spectra in rapidity space expected for particles recoiling from the inelastic fast proton.

1.2. <u>Notation, variables and sum rules</u>

Although necessary, this section can be kept short because several treatments are published [1),6)] on the choice variables and no great advances in the subject are evident. I specialize to proton-proton collisions, since much of the NAL and all the ISR data are of this type.

In an inclusive single particle process,

$$pp \rightarrow hX$$

one is interested in the differential cross-section for production of particle h, with four-momentum $(E,\vec{p})$, regardless of whatever else ("X") accompanies h. An implicit or, even, explicit sum is made over states with various quantum numbers which comprise X. The Lorentz-invariant inclusive cross-section, denoted

$$f(s,\vec{p}) = E\frac{d\sigma}{d^3p} \qquad (1.1)$$

is a function of the total energy $s = (p_a+p_b)^2$ and of the momentum $\vec{p}$ of produced hadron h. Here p_a and p_b are four-momenta of the incident protons in the collision. The factor E appears in (1.1) in order to (partially) remove inessential kinematical effects.

In a two-particle inclusive process, $pp \rightarrow h_1h_2X$, one studies

$$\frac{E_1E_2\, d^2\sigma}{d^3p_1\, d^3p_2} = G(s,\vec{p}_1,\vec{p}_2)\,. \qquad (1.2)$$

According to the definition of $f(s,\vec{p})$. one count is registered in a counter subtending element d^3p/E of phase space for each time a hadron of type h falls into the counter. It is then clear that the integral of f over all phase space gives the <u>mean</u> multiplicity of hadrons of type h, times the cross-section.

$$\int f(s,\vec{p})\,\frac{d^3p}{E} = \int \frac{d\sigma}{d^3p}\,d^3p = \langle n_h\rangle\,\sigma \qquad (1.3)$$

In (1.3), termed the "multiplicity sum rule", we may employ either the total, the inelastic, or some other cross-section σ ; the meanings of $< n_h >$ change accordingly. Obviously, an alternative expression for $< n_h >$ is

$$\sigma \langle n_h \rangle = \sum n \sigma_n \tag{1.4}$$

where σ_n denotes the partial cross-section for production of n and only n hadrons of type h.

Another useful sum rule is that expressing energy conservation. Since $\sigma^{-1} f(s,\vec{p})$ can be thought of as a number density, the product $\sigma^{-1} E_h f_h(s,\vec{p})$ (d^3p/E) is the fraction of energy deposited in phase-space element (d^3p/E) by hadrons of type h. Upon summing over hadrons of all types and integrating over all phase space, we obtain

$$\sum_h \int E_h f_h(s,\vec{p}) \left(d^3p/E \right) = \sqrt{s}\, \sigma \tag{1.5}$$

Expression (1.5) is especially relevant in attempts to assess the implications of a rising cross-section for the s dependence of inclusive cross-sections. This is treated in Sections 2.4 and 5.

The reader can easily supply other sum rules expressing conservation of charge, baryon number, and random other quantities.

For a single particle process, with spin ignored, there are three independent kinematic variables necessary for a specification of $pp \to hX$. Three popular choices, dictated by differing physics considerations, are :

1) s , $|\vec{p}_T|$, p_L

2) s , $|\vec{p}_T|$, y

3) s , t , M_X^2

In all three sets, $s = (p_a+p_b)^2$. In sets 1 and 2, an attempt is made to exploit the observed fact that the distribution in transverse *) momentum $\vec{p}_T$ is a rapidly decreasing and roughly energy-independent function. Thus, in some sense, all the physics goes on in the single longitudinal dimension. In set 1, p_L is the longitudinal momentum in some chosen rest-frame (e.g., centre-of-mass, stationary target frame, beam frame, etc.).

The Feynman scaling variable is defined as

$$x = 2p_L/\sqrt{s} \quad , \tag{1.6}$$

where here p_L is the centre-of-mass value.

In set 2, rapidity **)

$$y = \frac{1}{2} \log \left[\frac{E+p_L}{E-p_L} \right] = \log \left[\frac{E+p_L}{m_T} \right] \tag{1.7}$$

with

$$m_T^2 = p_T^2 + m_h^2 \tag{1.8}$$

One advantage of rapidity as the choice of longitudinal variable is well known. Under a longitudinal Lorentz transformation from one rest frame to another,

$$y \longrightarrow y' + \text{constant}$$

Thus, distributions in y are easily shifted from one frame to another.

Inverting (1.7), we write

$$E = m_T \cosh y \; ; \tag{1.9}$$

---

*) Transverse is defined with respect to the longitudinal axis given by the incident particle direction in the over-all centre-of-mass frame.

**) Throughout this paper, log means natural logarithm, i.e., to the base e.

$$p_L = m_T \sinh y . \qquad (1.10)$$

The full kinematic range of x is obviously

$$-1 < x < 1$$

independent of s, whereas

$$-\tfrac{1}{2} Y < y < \tfrac{1}{2} Y \; ; \; Y = \log\left(\frac{s}{m_p^2}\right) \qquad (1.11)$$

For future reference, a table is given here of various momenta, energies, and the corresponding full interval Y in rapidity.

Momentum on a stationary target in GeV/c	Momenta of ISR colliding beams in GeV/c	s $(GeV)^2$	$\sqrt{s}$ GeV	Y/2
30		58.2		2.1
70		133.4		2.5
100		189.8		2.7
200		377.8		3.1
300		565.8		3.3
	11		23.2	3.2
	15.4		30.4	3.5
	22		44.4	3.9
	26.7		52.7	4.0

The third set (s,t,M_x^2) will be used primarily in Section 5.

Expressed in terms of the three variable sets, the inclusive cross-section is

$$E\frac{d\sigma}{d^3p} = \frac{\bar{x}\, d\sigma}{\pi\, dx\, dp_T^2} = \frac{d\sigma}{\pi\, dy\, dp_T^2} = \frac{d\sigma}{\pi\, dt\, d(M^2/s)} \qquad (1.12)$$

$$(\bar{x} = 2E/\sqrt{s})$$

Because it expands the region of phase space in which $p_L^{(cm)} \approx 0$, y is a good variable for displaying pion inclusive spectra, whose cross-section is largest at small $y^{(cm)}$ (or small x). By contrast, the physics of proton inclusive spectra is such that $< x_p > \approx 0.5$; and $E d\sigma/d^3p$ peaks near $x_p \simeq 1$, as will be discussed in Section 5. Thus plots in terms of x seem to be better when final proton spectra are displayed.

1.3. Ancient history

Having disposed of kinematics, we turn to revered old facts whose range of validity is extended by observations at NAL and ISR.

1.3.1. As discussed eloquently elsewhere [7], measurements of the total [8] and elastic [9] pp cross-sections at ISR have produced their surprises. However, it remains true that inelastic production accounts for the bulk of the total cross-section, with a roughly energy independent fraction of 80% :

$$\sigma_{inel} \simeq 0.8\, \sigma_{tot}$$

1.3.2. Although larger at large p_T than perhaps expected, the distribution in transverse momentum $d\sigma/dp_T^2$ is largest at small p_T, and falls precipitously as p_T increases. A very crude parametrization [10] over the range $0.1 < p_T < 1\ (\mathrm{GeV}/c)^2$ is

$$E\, d\sigma/d^3p \sim \exp(-6\, p_T) \tag{1.13}$$

Surely this form is not appropriate all the way to $p_T = 0$ [28]. Moreover, the parameter 6 is subject to some variation with x, and with the type of hadron observed.

For $p_T > 2$, the scaling form

$$E\, d\sigma/d^3p \sim p_T^{-8}\, g(p_T/\sqrt{s}) \tag{1.14}$$

is suggested by some theories [11] and appears to fit data [12], with

$$g_{\pi^0}(p_T/\sqrt{s}) \sim \exp\left[26\, p_T/\sqrt{s}\right] \tag{1.15}$$

Nevertheless, the integrated cross-section for $p_T > 1$ (GeV/c) is on the order of tenths-of-millibarns. Thus, to a good approximation throughout the ISR range, the mean $< p_T >$ has the "small" value ≈ 350 MeV and is independent of energy. Transverse momenta are "limited" for the great bulk of σ_{inel}.

Some recent data are shown in Fig. 1.

1.3.3. An interesting compilation of mean multiplicities from pp collisions has been published [13] by the CERN-Bologna group ; it is reproduced here as Fig. 2. Representative fits made to the mean number of charged hadrons per collision give [13),14)]

$$\langle n_{ch} \rangle \approx 1.4 \log s - 1. \tag{1.16}$$

$$\approx 1.5\, s^{0.3} \tag{1.17}$$

$$\approx 1.9 \log s - 3.8 + \frac{6.4}{s^{1/2}}. \tag{1.18}$$

Although these functional forms differ, they show that $< n_{ch} >$ grows much less rapidly than the maximum possible rate of growth $(< n_{ch} >_{max} \simeq \frac{2}{3}\sqrt{s})$. With "minor corrections" for differing leading particle energy fractions, multiplicity data from πp, Kp, and $\bar{p}p$ reactions fall on roughly the same curve as the pp data [15].

1.3.4. Inasmuch as the energy available does not go into transverse motion (point 1.3.2.) nor fully into particle production (point 1.3.3.), its influence is most obvious in altering the longitudinal momentum dimension. To first approximation, the mean centre-of-mass longitudinal momentum grows in proportion to $\sqrt{s}$:

$$\langle p_L \rangle = k_h \sqrt{s}\,. \tag{1.19}$$

The constant of proportionality k_n differs for different hadrons h. Statement (1.19) is related, of course, to the hypothesis of Feynman scaling. If

$$f(x, p_T^2, \sqrt{s}) \rightarrow f(x, p_T^2) \tag{1.20}$$

then, obviously,

$$\langle x_h \rangle = k_h \tag{1.21}$$

For final protons, $k \simeq \frac{1}{2}$. This relatively large value of k_p emphasizes a result known for years from cosmic ray data that the final baryons carry away on the average a substantial fraction of the collision energy.

1.3.5. Scaling in the fragmentation regions. For the sake of definiteness, I limit the fragmentation region to the kinematic region $|x| > 0.1$. Because

$$x \simeq \exp\left[-(y_{end} - y)\right] , \tag{1.22}$$

fragmentation covers roughly two units of rapidity from the kinematic edges of the y plot.

In this fragmentation region, as discussed at length in earlier reviews [1], inclusive distributions of $\pi^\pm$ in pp collisions show a remarkable independence of s from $p_{lab} = 10$ to 30 GeV/c, when plotted versus the scaling variable x. Thus, for $pp \rightarrow \pi^\pm X$, whereas $Ed\sigma/d^3p = f(s,x,p_T^2)$ is in general a function of three independent variables, the dependence on s seems to drop out. ISR data extend the region of experimental validity of this statement to $\sqrt{s} \cong 53$ GeV. I defer further discussion of new data to Section 2.2.

Not all single particle inclusive spectra scale in the 10-30 GeV/c interval, to be sure. Some (e.g., $pp \rightarrow KX$) rise and others fall with s. A systematic experimental study of the question of approach to scaling through the Serpukhov and NAL energy ranges would be very valuable.

1.4. New developments

As a result of data accummulated during the past year, new items to be included in the grand design are : (1) the rise with s of the total cross-section [8] ; (2) the behaviour of the elastic differential cross-section [9] $d\sigma/dt$ as a function of s and of momentum transfer t ; (3) scaling and non-scaling behaviour of single particle inclusive spectra $(a+b\rightarrow h+X)$; (4) the dependence of these spectra on momentum of the observed hadron, including the emergence of a plateau in rapidity [16] ; (5) correlations [17)-19)] seen between hadrons in two-particle inclusive processes $(a+b\rightarrow h_1+h_2+X)$; (6) the behaviour of the cross-section [20] σ_n for n charged particles (prongs) as a function of n and s ; (7) the remarkably large hadronic inclusive cross-section at large transverse momentum [12] p_T ; and, last but far from least important, (8) salient features gleaned from more differential, exclusive analyses of events.

This itemization is arbitrary to a certain extent. Category (7) could well be included in (4). It is separated off just to point out that there may be different physics dominating the low and high p_T parts of data. Next, the nebulous subject of "diffraction dissociation" could well be listed by itself, but I include it in (3) and (4) ; in later pages I will treat it separately. The listing is also by experimental topics, rather than according to such theoretical concepts as short-range order, long-range order, cluster formation, diffraction, factorization, particle exchange, fragmentation and so forth. However, some of these concepts emerge as simplifying themes and are emphasized throughout the text.

Elastic and total pp cross-sections are discussed in detail by others [7], so I can dispense with my categories (1) and (2), and concentrate on inelastic multibody phenomena. The growth with s of σ_{tot} will of course still affect our discussion.

Because it is still unclear how to integrate large p_T phenomena into the whole picture, and because of lack of time, I do not treat this topic here. Multiplicity distributions have generated a vast literature of their own during the past year [21]. I will avoid increasing this volume. There is little to add. It seems necessary now to develop a view-point in which an explanation of both the inclusive spectra and $\sigma_n(s)$ emerge together.

2. SHORT-RANGE ORDER AND SINGLE-PARTICLE SPECTRA

2.1. Statement of the hypothesis of short-range order

Although I risk biasing the presentation, I think it is useful to introduce the hypothesis of short-range order [22] at this point, because it serves well as a putative asymptotic unifying theme. It is a principle which can be abstracted from a broad class, but by no means from all reasonable models. Furthermore, as will be discussed, it is consistent with much experimental data from ISR, in particular data on the behaviour of pions which are relatively slow in the over-all centre-of-mass frame. On the other hand, there is at least one class of events which does not seem to satisfy the hypothesis ; this is the inelastic diffraction class which is treated in Section 5. It is important to keep in mind from the start that the approximate domain of applicability of the hypothesis must be specified carefully.

Beginning on a purely intuitive level, it is perhaps easy to swallow the casual remark that if particles are far enough apart in phase-space, then they behave in an uncorrelated manner. Such a remark is true enough of gas molecules in a room. Can it also be true of hadrons, where for them rapidity difference is the measure of separation ? It is surely not true of fairly low multiplicity events, where momentum conservation, if nothing else, provides strong correlation. Nevertheless, various models, the prototype of which is the multiperipheral ansatz [2] and, more recently, inclusive Regge (Mueller) analysis [23], suggest the validity of a short-range order (SRO) hypothesis for inclusive spectra at asymptotic energies, as defined below.

We begin with an n particle <u>inclusive</u> rapidity distribution

$$\frac{d^n \sigma_{ab}}{dy_1 \cdots dy_n} \equiv f_{ab}^{(n)}(y_1, y_2, \ldots, y_n ; y_a, y_b) \tag{2.1}$$

As usual, y_a and y_b are centre-of-mass rapidities of the incident (beam) particles : $|y_a| = |y_b| = Y/2$. By Lorentz invariance, $f^{(n)}$ is actually function of rapidity <u>differences</u> [e.g., $(y_i - y_j)$] not individual rapidity variables. The number n may be 1,2,..., etc.

Suppose we group rapidity variables into two sets

$$(y_a ; y_1, \ldots y_k)$$

and

$$(y_{k+1}, \dots y_n ; y_b) ,$$

where y_k is the largest rapidity of the first set, and y_{k+1} the smallest of the second set.

The hypothesis of SRO states that if the rapidity difference $|y_{k+1} - y_k|$ is large enough, then $f^n(y_j)$ does not depend on this difference and factorizes into a product of terms :

$$f^{(n)}(y_j) \longrightarrow g_a^{(k)}(y_a ; y_1, \dots, y_k)\, g_b^{(\ )}(y_{k+1}, \dots, y_n ; y_b) \tag{2.2}$$

The functions g_a and g_b are not specified generally, but in strong versions of SRO, each factor is itself a separately measurable inclusive cross-section.

$$\sigma_{ab} f_{ab}^{(n)} \longrightarrow f_{ab}^{(k)}(y_a ; y_1, \dots, y_k ; y_b)\, f_{ab}^{(n-k)}(y_a ; y_{k+1}, \dots, y_n ; y_b) \tag{2.3}$$

The strong version carries an additional postulate about the functional form and normalization of the factors. The strong version may not agree with data, whereas the weaker form has experimental support.

To have a measure of distance in y space, we introduce an energy independent correlation length λ. We can then rephrase the hypothesis : if $|y_k - y_{k+1}| \gg \lambda$, then (2.2) follows.

The hypothesis itself does not provide a scale of distance but models agree that $\lambda \approx 2$. This value is found from inclusive Regge analysis [23], since $\lambda \equiv (\alpha_P(0) - \alpha_R(0))^{-1} \simeq 2$. In models in which behaviour at small rapidity differences is dominated by resonance-like behaviour (fragmentation models, cluster formation models, etc.), $\lambda \simeq 2$ is dictated essentially by kinematics, and by the mean transverse momentum (or Q value) of the decay (cf. Section 3.3). Data from ISR are consistent with the value $\lambda \simeq 2$ (Section 4).

The SRO hypothesis suggests a division of the inclusive rapidity axis into three regions :

1) target fragmentation, length λ about y_b ;
2) projectile fragmentation, length λ about y_a ;
3) central region of length $Y - 2\lambda$.

Particles within a distance λ from y_a (or y_b) are possibly highly correlated among themselves. Central region particles should be uncorrelated from those near y_a and/or y_b.

Existence of a central region requires $Y = y_b - y_a > 4$, if $\lambda \simeq 2$. Thus, it is meaningful to discuss the possibility of a physically distinct central region only if $p_{lab} > 30$ GeV/c (cf. the Table in Section 1.2).

The hypothesis of SRO is primarily a statement about (the absence of) correlations. As such, its predictions are most useful at the level of two and more particle inclusive processes. However, the hypothesis makes testable assertions also about single-particle inclusive spectra in various regions of phase space. It is well to examine how well these agree with data before proceeding to more complex situations.

2.2. Implications of SRO for single-particle spectra in the fragmentation regions

The single-particle inclusive distribution $E d\sigma/d^3p = \pi^{-1} d^2\sigma/dy dp_T^2$ is a function of three variables y, p_T^2 and s. The dependence on s can be reexpressed as a dependence on $Y = \log(s/m_p^2) = y_b - y_a$. Thus, in terms of p_T^2 and rapidity differences, the three free variables are p_T^2, $(y - y_a)$, and $(y - y_b)$.

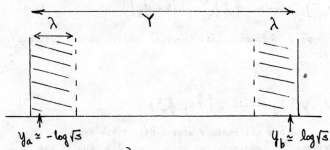

If y is in the fragmentation region of b, then $|y - y_b|$ is small and the difference $|y - y_a| \approx |y - Y|$ is large, provided that s is high enough. According to the SRO hypothesis, dependence on $|y - y_a|$ drops out, and

$$f(y, s, p_T^2) \longrightarrow f(p_T^2, |y - y_b|) \tag{2.4}$$

Because $x \simeq \exp[-(y_b - y)]$, a function of $(y - y_b)$, only, is a function of x only. Thus, SRO gives

$$f(x, s, p_T^2) \longrightarrow f(x, p_T^2) \tag{2.5}$$

Expressions (2.4) and (2.5) are alternative statements of the famous <u>scaling</u> law of ABFST [2] and Wilson [22], rediscovered by Feynman [24], Yang [3] and Mueller [23].

The hypothesis also predicts an energy independent flat central plateau in y for single-particle spectra, as discussed in the next subsection. Observation of scaling in the fragmentation regions and/or an s independent central plateau does not prove SRO, of course. There are models, without dominant SRO in which both predictions emerge naturally. The observation is a first consistency check. Note, furthermore, that no energy scale is provided. For different produced hadrons h, we are not told a priori the energy value at which non-asymptotic terms should be negligible. For the latter, we need more detailed models.

For comparison of scaling predictions with data, I will confine myself to spectra with $p_T < 1$ (GeV/c)^2. At larger p_T, an increase of f with s is observed in the ISR range.

Beginning with the fragmentation regions, and restricting attention to $0.2 < x < 0.8$ for the moment, we observe in Figs. 3.a-c that for all practical purposes $\bar{p}$, π^-, and K^- inclusive distributions are all energy independent (scale) within the ISR range [25]. Data on π^+, K^+ and p spectra show similar behaviour [26]. Energy independence for pions $(\pi^\pm)$ seems to extend down to much lower energy [27], beginning somewhere in the 10-30 GeV/c PS range. The PS/ISR comparison demonstrates that K^+ spectra rise somewhat as energy is increased, K^- and $\bar{p}$ spectra grow even more dramatically, and the p rate falls. As remarked above, these statements apply to the region $0 < p_T <$ < 1 GeV/c and $0.2 < x < 0.8$.

For $x > 0.8$, where only protons (and, perhaps neutrons) are observed, we encounter the intriguing question of whether the leading proton peak scales. ISR data [5], confined to a few p_T values, are shown in Fig. 4. This topic will be treated again in Section 5.

2.3. Single-particle spectra in the central region

When both $|y_a - y| \gg \lambda$ and $|y_b - y| \gg \lambda$, the SRO hypothesis assures us that $f(y - y_a, y - y_b, p_T^2)$ can depend only on p_T^2. Thus, an energy independent, flat rapidity distribution is predicted in the central region.

$$\frac{d\sigma}{dy\,dp_T^2} \longrightarrow f(p_T^2) \tag{2.6}$$

One immediate implication is that of an asymptotic log growth of $< n >$ inasmuch as

$$\langle n \rangle \sigma = \int \frac{d\sigma}{dy\,dp_T^2}\, dy\, dp_T^2 \longrightarrow \log s \int f(p_T^2)\, dp_T^2 + A \tag{2.7}$$

Constant A arises from the s independent contribution of the fragmentation regions to the integral.

The coefficient of log s is given by the plateau height divided by cross-section σ. This log s growth is characteristic of an effectively one-dimensional phase space, the single-dimension arising because p_T is sharply limited. An incremental increase of energy goes into expanding the longitudinal dimension in the uniform manner specified by SRO. Note that the log prediction is not necessarily expected to hold below $p_{lab} \approx 50$ GeV/c, i.e., until there is reason to have a plateau in y.

Experimental tests of (2.6) and, more generally, the question of the existence of an energy independent plateau in y, are beset with unnecessary confusion. Three types of data exist. By far the best are data on particles of well defined type (e.g., π, p, K, $\bar{p}$, etc.) whose y and p_T values are both specified. The two-dimensional distribution $d^2\sigma/dy\,dp_T^2$ is given as a function of s. Second, some data on particles of a specific type are given only in the integrated form $d\sigma/dy$ versus s. Finally, the most difficult data to interpret are those in which the distribution $d\sigma/d\eta$, for all charged particles combined, is given as a function of s. Variable $\eta = \log \tan(\theta/2)$ is an approximation to y, and involves a necessary average over p_T. Angle θ is the ISR frame or centre-of-mass frame scattering angle of the charged particle.

Bubble chamber data on $d\sigma/d\eta$ for π^- production at 200 GeV/c at NAL [28] are compared with 28.5 GeV/c data [29] in Fig. 5. An approximate plateau covering $|\eta| < 1$ is seen in the NAL data, whereas no such plateau is visible for $p_{lab} < 30$ GeV/c. At least a 30% increase of $d\sigma/d\eta$ at $\eta = 0$ is seen from 28.5 to 200 GeV/c.

A summary of some ISR data on the s and y dependences of $d^2\sigma/dp_T^2dy$ for $pp \to \pi^+X$ is given in Fig. 6. Qualitatively similar results are true for $pp \to \pi^-X$. The British-Scandinavian [30] (BS) and Saclay-Strasbourg [31] (SS) data at $y = 0$ are consistent. The BS data at $\sqrt{s} = 30.6$ and 52.8 GeV show that the invariant cross-section at fixed p_T and fixed s increases by $\approx 15\%$ as y changes from 1 to 0. Thus, even at $\sqrt{s} = 52.8$ GeV, there is not a flat plateau, but, rather, a bulge in $d\sigma/dy$ near $y = 0$. The magnitudes of the slopes in y decrease, however, from 0.18 ± 0.01 for π^- and 0.15 ± 0.01 for π^+ at $\sqrt{s} = 30.6$ GeV to 0.13 ± 0.01 and 0.12 ± 0.01 at $\sqrt{s} = 52.8$ GeV, consistent with a gradual approach to a flat plateau in the central region at infinite energies, for small p_T (< 1.0 GeV/c). This aspect of the SRO prediction is therefore supported for pions.

Turning next to s dependence, two observations may be made. For fixed p_T, it appears from Fig. 6 that data at different y and $Y = \log s$ are consistent with lying on a universal curve which is a function of the difference $y_{proj} = (y_{cm} + \frac{1}{2}Y)$. If this were true, then knowledge of the y dependence at fixed s would give us directly the s dependence at fixed y_{cm}. The 15% increase of $d\sigma/dy$ from $y = |1|$ to $y = 0$ then translates into an 8% rise of the "plateau height" $d\sigma/dy|_{y=0}$ from $\sqrt{s} = 30.6$ to $\sqrt{s} = 52.8$ GeV. However, experimental uncertainties in energy calibration of flux monitors are unfortunately sufficiently large that this rise cannot be asserted strongly. The BS group [30] cautions therefore that the π production cross-section, integrated over the measured range of p_T and over the centre-of-mass system rapidity range 0 to 0.5, shows an increase between the two energies of $8 \pm 7\%$. Saclay-Strasbourg data [31] on $\pi^\pm$ production show perhaps a 10% rise of $Ed\sigma/d^3p$ at $x = 0$ and $0.3 < p_T < 0.9$ GeV/c as $\sqrt{s}$ changes from 22 to 30 GeV. Over the range

$\sqrt{s} = 30$ to $\sqrt{s} = 53$, no further increase is observed within the 5% systematic error set by limitations on luminosity measurements.

Thus, the picture which emerges from measurements of s dependence of $d\sigma/dy$ for $\pi^{\pm}$ near $y = 0$ and $0.1 < p_T < 1$ GeV/c is one of a factor of 2 or so rise from the 20-30 GeV/c BNL-PS energy range to the ISR. Within the ISR range of $\sqrt{s} = 30$ to $\sqrt{s} = 53$ GeV, the data are consistent with an energy independent inclusive cross-section $d\sigma/dydp_T^2$ at $y = 0$ and $0.1 < p_T < 1$ GeV/c, as would be predicted by the hypothesis of short-range order.

Data on antiproton production are given in Fig. 7. The cross-section integrated over $0 < y_{cms} < 0.5$ for fixed p_T shows an average increase from $\sqrt{s} = 30.6$ to 52.8 GeV of 25 ± 7% for $\bar{p}$. At fixed s and fixed p_T, $d\sigma/dydp_T^2$ increases by 42 ± 8% as y_{cms} changes from $|1|$ to 0. Thus, neither a plateau nor s independence is apparent for $\bar{p}$ production even in the ISR energy range. Validity of an SRO hypothesis for $\bar{p}$ inclusive production would require energies perhaps an order of magnitude higher than available at ISR. A comparison of $Ed\sigma/d^3p$ for $\bar{p}$ and π^- production at $y = 0$ and $p_T = 04.$ GeV/c gives $(\bar{p}/\pi) \simeq 1/15$.

For proton production, $pp \to pX$, the BS group [30] reports an energy independent flat inclusive distribution in y from $1 > |y_{cm}| > 0$, and $0.3 < p_T < 1.0$ GeV/c.

The final figure on the plateau situation is Fig. 8. These data from the Pisa-Stony Brook ISR collaboration [32] are plots at different energies of the distribution $\sigma_{inel}^{-1} d\sigma/d\eta$ summed over all p_T and over all types of charged particles (including background). As suggested by the P-SB authors [32], the η and energy dependences shown by these data cannot be taken entirely at face value because of the background problem and neglect of certain experimental corrections. However, in Section 4, I discuss correlation data at different energies published by the same group. The s dependence of these correlation data is given experimentally in terms of the s dependence of the single particle spectra. Because s dependence of the correlation data is a much more crucial test of SRO than single particle spectra, it is important to see what s dependence is present, for whatever reason, in the P-SB single particle spectra.

Figure 8 shows good scaling of the distribution $\sigma^{-1} d\sigma/d\eta$ in the fragmentation regions. However, the very rough plateau in the central interval $|\eta| < 1$ rises with $\sqrt{s}$ by 20% or so over the ISR energy range. Combining this with the 10% rise of σ_{inel}, we see a net increase of $\approx 30\%$ for $d\sigma/d\eta|_{y\approx 0}$, which is a far greater increase than is seen by BS or SS.

The P-SB data are, of course, of a very different type from those of the BS and SS groups. The P-SB results include production at both larger p_T and at smaller p_T. Directly produced charged particles as well as converted neutrals are accepted. It is unlikely that energy dependent large p_T data, with small cross-section, can account for the discrepancy in s dependence. Curious s dependent, background-related phenomena at low p_T, where cross-sections are large, could be the problem.

To summarize, the BS and SS collaborations present what seems to be compelling evidence for the existence of a central, energy independent plateau in y, for pions, extending over at least $-1 < y < +1$ at ISR energies. The rise with s of other inclusive spectra in the central region (e.g., $pp \rightarrow \bar{p}X$) suggests that there are still important threshold effects in the ISR range for production of heavy secondaries. The strong energy dependence of $d\sigma/d\eta$ seen by the P-SB group is of concern for conclusions about inclusive correlations, discussed in Section 4.

2.4. Implications of a rising total cross-section

The total cross-section is a function of s, or, equivalently, of (y_a-y_b) : $\sigma_{tot}(y_a-y_b)$. Applying the SRO hypothesis naively to this function, we expect that

$$\sigma_{tot} \rightarrow \text{constant} \tag{2.8}$$

inasmuch as $|y_a-y_b| \gg \lambda$. Thus, the rise of σ_{tot} over the ISR range shows a violation of SRO or, at least, that there are important non-asymptotic effects in this energy region.

It may be argued that the SRO statement should really be made for the ratio

$$\sigma_{tot}^{-1} \, d^n\sigma/dy_1 \cdots dy_n \, ,$$

which is, after all, a density. Such a procedure would be suggested by some proponents of gas/liquid analogies [33]. On the other hand, it is not clear that dynamical models from which the SRO hypothesis is abstracted support the suggestion that the ratio is the appropriate quantity. For example, we may take the inclusive Regge (Mueller) approach [23], and introduce cuts or other singularities to reproduce the rise in σ_{tot}. Then such effects will presumably be present also in inclusive spectra. However, these dependences on s generally need not cancel out in the ratios $\sigma^{-1} d^{(n)}\sigma/dy_1 dy_2, \ldots, dy_n$.

I will consistently apply the SRO hypothesis to inclusive cross-sections and not to ratios. Purely empirically, the 10% rise of σ_{tot} over the ISR range can then be used to estimate the extent to which we should expect the constant plateau and the scaling predictions of SRO to be valid in this energy range.

Sum rules give us further insight, but they do not solve the problem. As we will see below, it is possible for most inclusive spectra to scale perfectly, simultaneously with a rising σ_{inel}. It is a very intriguing <u>experimental</u> question to ascertain precisely where the rise of σ_{inel} really comes from. Theoretical models suggest different possibilities.

Rewritten in terms of other variables, the energy-conservation sum rule (1.5) takes the forms

$$2\sigma_{inel} = \pi \sum_h \int dx \int dp_T^2 \, f_h(s, x, p_T^2) \tag{2.9}$$

and

$$\sqrt{s}\,\sigma_{inel} = \sum_h \int dy \int dp_T^2 \, m_T \cosh y \, \frac{d\sigma_h}{dy\,dp_T^2} \tag{2.10}$$

We note that if all hadronic spectra scale perfectly $[f_h = f_h(x, p_T^2)$ in the integrand of (2.9)], and if none has singular behaviour near the end points of integration $(|x| \to 1)$, then Eq. (2.9) requires σ_{inel} = constant. Thus, the rise of σ_{inel} over the ISR range demands that at least for some h, f_h has residual non-scaling dependence on s, and/or that some f_h behave in singular fashion as $|x| \to 1$. As an example of the latter possibility, the triple-Pomeron Regge formalism discussed in Section 5 gives

$$f_p(x) \propto (1-x)^{-1} \tag{2.11}$$

as $x \to 1$, with an upper cut-off $x_c = 1-M_c^2/s$. With this form

$$\int^{x_c} f_p(x)\, dx \propto \log s \tag{2.12}$$

which is then the basis for a theoretical suggestion [34] that the rise of σ_{inel} comes from inelastic diffraction. The experimental credibility of this idea is examined in Section 5.

The sum rule clearly allows that some spectra (e.g., π spectra in the fragmentation region) may scale perfectly. As discussed above (Section 2.2), π and K data in the region $|x| \gtrsim 0.1$ do seem to satisfy this possibility. On the other hand, the pion spectra may rise with s for $|y| < 1$. It is worth while to see how much of the rise of σ_{inel} can be associated directly with a rise of the central plateau height for pions and other hadrons.

Beginning with Eq. (2.10), let us make the simplifying assumption (criticized later) that there is a full pion plateau, for $-Y/2 < y < Y/2$. Thus,

$$\frac{d\sigma}{dy\, dp_T^2} = h(p_T^2) \tag{2.13}$$

$$\sqrt{s}\, \sigma_{inel} = \frac{\langle m_T \rangle}{m_p} \sqrt{s} \int dp_T^2\, h(p_T^2) \equiv \frac{\langle m_T \rangle}{m_p} \sqrt{s}\, H \tag{2.14}$$

and

$$\frac{\Delta \sigma_{inel}}{\sigma_{inel}} = \frac{\langle m_T \rangle}{m_p} \frac{\Delta H}{H} \left(\frac{H}{\sigma_{inel}} \right) \tag{2.15}$$

Experimentally, $H\sigma_{inel}^{-1} \simeq 0.8$ to 1.0 for π^-, and $<m_T> = <\sqrt{m^2 + p_T^2}>$ $\simeq 350$ MeV. Taking equal contributions from π^+, π^- and π^0, I multiply (2.15) by 3, and find

$$\left(\frac{\Delta\sigma_{inel}}{\sigma_{inel}}\right)_\pi \simeq \frac{\Delta H}{H} \qquad (2.16)$$

Thus, an 8% rise of the pion plateau height H implies directly an 8% rise of σ_{inel}. A more careful analysis should be done, of course, in which actual data are integrated over y, not the full plateau form I used. This will reduce the right-hand side of (2.16) by a correction factor which I estimate to be $\gtrsim 0.5$. A net rise is left for σ_{inel} due to central pion production perhaps as large as 8%, based on the British-Scandinavian numbers, of 8±7%, (and perhaps zero, also !, such are the experimental errors).

For the rise of σ_{inel} attributable to antibaryon production in the central region, a similar exercise gives

$$\left(\frac{\Delta\sigma_{inel}}{\sigma_{inel}}\right)_{\bar{B}} \gtrsim \left(\frac{\Delta H}{H}\right)_{\bar{p}}\left(\frac{H}{\sigma_{inel}}\right)_{\bar{p}} \qquad (2.17)$$

On the right-hand side I have included a factor of 2 to take into account unseen $\bar{n}$ production. This is cancelled by other numerical factors in the computation. As described above, experiment gives $(\Delta H/H)_{\bar{p}} \simeq 25\pm7\%$ from $\sqrt{s} = 30$ to 53 GeV, and $H_{\bar{p}} \simeq H_{\pi^-}/15$. Thus, from antibaryon production, I obtain

$$\left(\frac{\Delta\sigma_{inel}}{\sigma_{inel}}\right)_{\bar{B}} \approx \left(\frac{\Delta H}{H}\right)_{\bar{p}}\left(\frac{H_{\bar{p}}}{H_{\pi^-}}\right)\left(\frac{H_{\pi^-}}{\sigma_{inel}}\right) \gtrsim 2\% \qquad (2.18)$$

from $\sqrt{s} = 30$ to 53 GeV. This is a 1 mb, or so, increase - and so may account for about 1/3 of the observed rise of σ_{inel}. It would not be correct to multiply this number by 2 on the grounds that $\bar{p}$'s are produced as baryon/antibaryon pairs. Data show no increase of the proton flux at y = 0, meaning, in this context, that an increase with s of the proton's $d\sigma/dy$ due to production of p from $(\bar{p}p)$ pair formation is offset by a decrease coming from some other source.

It is not easy to sum over all produced hadrons in the central region to obtain the net increase of σ_{inel} due to central production. For example, one must decide how to treat K^o, Λ, and η production. Are they effectively included already inasmuch as we have looked at inclusive spectra of π's, p's, $\bar{p}$'s, etc., into which these other particles decay ? Such questions aside, the sum rule argument certainly supports an increase of σ_{inel} over the ISR range. All observed inclusive spectra in the central region, except p, are consistent with an increase with s over the ISR range (the p spectra are constant). In the fragmentation regions, all spectra are essentially s independent over the ISR range. The sum rule (2.9), (2.10) will then show a net increase of σ_{inel} over this same range. From central production alone, I estimate an increase $1\text{mb} < \sigma_{inel} < 3\text{mb}$.

3. TWO EXPLICIT MODELS

Crucial predictions of the SRO hypothesis appear only at the level of two-particle inclusive spectra and inclusive correlations. Indeed, the implications of SRO for single-particle spectra are not unique to models with only short-range order. Correlations are treated in Section 4, but first, to provide a more concrete idea of the short-range order concept it is useful to introduce a specific, if idealized, model embodying short-range order and, further, to contrast such a model with the fragmentation view-point [3], which has both long-range and short-range structure.

3.1. ABFST reviewed

We begin with the traditional Amati-Bertocchi-Fubini-Stanghellini-Tonin multiperipheral model [2], in which the amplitude for $pp \rightarrow (n+2)$ hadron final state is sketched below. The inclusive spectra are obtained from this model after an appropriate integration and sum is made over unobserved particles.

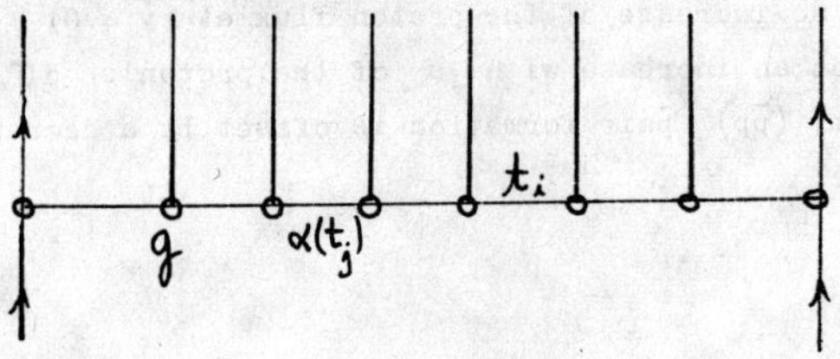

The amplitude is large only when all momentum transfers t_i along the multiperipheral chain are small. For our present purposes it does not matter very much what specific particles or Reggeons $\alpha(t_i)$ are exchanged.

Analytically, the amplitude is represented as

$$A(2 \to n+2) = g^n \prod_{i=0}^{n} s_{ij}^{\alpha(t_i)} f_i(t_i) \tag{3.1}$$

Here $f(t_i)$ falls off rapidly as $|t_i|$ grows.

Detailed analyses may be found elsewhere [35] but here I will simply catalogue some predictions of the approach.

(a) $$\langle n \rangle = g^2 \log s \tag{3.2}$$

(b) $$\sigma_n = \sigma_0 e^{-\langle n \rangle} \langle n \rangle^n / n! \quad \text{(Poisson form)} \tag{3.3}$$

(c) $\dfrac{d\sigma}{dy}$ is s independent and flat. (3.4)

(d) $\dfrac{d\sigma_n}{dy}$ The single-particle spectrum from an n particle final state is s independent and flat for each n. (3.5)

The sum $\sum \sigma_n = \sigma_{inel}$ is an energy independent constant. What emerges is a picture in which particles are expected to be produced with a roughly uniform spacing in y. Each event has the plateau structure of the inclusive sum, except, of course, for statistical fluctuations. The density of particles along the rapidity axis, $\langle n \rangle / Y$, is given by g^2.

3.2. Cluster formation

In the summary above, I imagined that individual hadrons are emitted along the multiperipheral chain. This need not be the case. Indeed, if all exchanges are pions, then obviously G parity forbids emission of single pions. That the objects emitted should be resonances or, at least, clusters of several hadrons is theoretically possible and, moreover, seems phenomenologically required.

Three facts argue against independent emission of single hadrons along the chain.

(a) Important resonances are known to exist. There is no reason to suppose that they are not formed in multiparticle events. Duality arguments, in which resonance formation is presumed to be represented by a sum of exchange terms, are not quantitatively reliable at small invariant mass. Resonance-like effects should be present explicitly in a production model.

(b) The parametrizations of $< n_{ch} >$ given in Section 1.3 suggest that for the total number of hadrons

$$\langle n \rangle \sim b \log s$$

with $b \sim 2$ to 3. Thus the mean spacing Δy of particles along the rapidity axis is only $\Delta y \sim 0.3$ to 0.5. Because

$$\langle M_{ij} \rangle \simeq \langle m_T \rangle \exp\left(\tfrac{1}{2}\Delta y\right) \tag{3.6}$$

the mean invariant mass of two produced hadrons is of order 500 MeV, well within the resonance region. Thus, strong correlations should be expected among produced hadrons, not independent emission.

(c) A direct measure of correlation is the parameter

$$f_2 = \langle n(n-1) \rangle - \langle n \rangle^2 \tag{3.7}$$

Shown in Fig. 9 is a compilation of data for negative tracks (i.e., $n = n_-$). Note that f_2 grows increasingly positive for $p_{lab} \gtrsim 50$ GeV/c. The independent emission Poisson distribution, Eq. (3.3), gives $f_2 \equiv 0$, for all s. Moreover, energy-momentum related effects in an independent emission model would depress f_2 to negative values [42]. Thus, the positive f_2 suggests clustering, as discussed more explicitly below.

We imagine, therefore, that a possible <u>imitation</u> of nature is one in which clusters of hadrons are emitted along a multiperipheral chain [4]. The mean number of clusters $< N >$ grows as

$$\langle N \rangle = g^2 \log s \tag{3.8}$$

In an (over) idealized model, the number of clusters produced may follow a Poisson law

$$\sigma_N = \sigma_0 \left[\exp(-\langle N\rangle)\right] \langle N\rangle^N / N! \tag{3.9}$$

Thus, if there is an average of $\langle K \rangle$ hadrons per cluster, the net mean number $\langle n \rangle$ of produced hadrons in the final state is

$$\langle n\rangle = \langle K\rangle\langle N\rangle = \langle K\rangle\, g^2 \log s \tag{3.10}$$

and

$$f_2 = \langle n(n-1)\rangle - \langle n\rangle^2 = \frac{\langle K(K-1)\rangle}{\langle K\rangle} \langle n\rangle \tag{3.11}$$

Parameter f_2 will be naturally positive if $\langle K(K-1) \rangle > 0$.

According to our multiperipheral ansatz, the distribution of clusters in y space $\rho(y_c) = 1/\sigma\,(d\sigma/dy_c)$ will be asymptotically uniform and s independent. To pass from this remark to one about the final distribution of hadrons requires a study of the distribution in rapidity of hadrons from cluster decay.

3.3. Cluster decay

Resonance behaviour implies many things - Breit-Wigner line shape, phases, spin-related decay characteristics and so forth. What I shall most like to emphasize, however, is the idea that decay particles from resonances occupy neighbouring regions of rapidity. Since the parent resonance invariant mass is relatively low, decay particles have small Δy [cf. Eq. (3.6)]. For my purposes, then, the idea of resonance is solely that cross-section is enhanced at small invariant mass, and therefore that small Δy is also enhanced. Clearly this effect is strongest where quantum numbers allow resonance formation.

Although some clustering very likely results physically from resonance formation, in order to stress the proximity of particles in rapidity, and to avoid other connotations of the term resonance, I use the word <u>cluster</u> to label the phenomenon of closer than average spacing of particles in rapidity space. The search of cluster formation in individual multiparticle events at NAL and ISR energies is only now beginning [36),37)]. Efforts in this direction deserve the strongest encouragement.

The parametrization of clusters and of their expected contribution to inclusive processes is straightforward. Begin by defining an average decay distribution in the cluster rest frame $D(\vec{q})$. Next, a transformation of variables is made to a set including rest-frame rapidity, namely the set (q,y,ϕ) where q is the magnitude of three momentum in the cluster rest frame. We easily obtain exactly

$$\frac{dy}{\cosh^2 y} = \frac{q\, d\cos\theta}{2\omega} \tag{3.12}$$

and thus

$$\frac{d^3q}{\omega} = \frac{dq^2\, d\phi\, dy}{\cosh^2 y} \tag{3.13}$$

with $\omega^2 = \mu^2 + q^2$; μ is the decay particle's mass. If one is interested only in the final decay rapidity distribution, he may integrate so as to derive [36)]

$$\frac{dD}{dy} = (\cosh y)^{-2} \int dq^2\, d\phi\, D(\vec{q}) \, . \tag{3.14}$$

In general, D in Eq. (3.14) may depend on $\cos\theta$ (i.e., on y). However, let us make an assumption that the cluster decay distribution is isotropic in the cluster rest frame : $D(\vec{q}) \rightarrow D(q)$. Then setting $r = q/\langle q \rangle$, we get [36)]

$$\frac{dD}{dy} = \frac{2\pi \langle q \rangle^2}{\cosh^2 y} \int_L^\infty dr\, D(r) \, , \tag{3.15}$$

with lower limit

$$L = \left(\frac{\mu \sinh y}{\langle q \rangle} \right)^2 . \tag{3.16}$$

In (3.16), $< q >$ is determined by the energy available (Q value) of the cluster decay. Empirically, the observed $< p_T >$ determines this scale, and for pions, $\mu/< q >$ is small. For small y, then $L \to 0$, and the decay spectrum

$$\frac{dD}{dy} \propto (\cosh y)^{-2} \tag{3.17}$$

The function $\cosh^{-2} y$ is in turn well approximated numerically by a Gaussian $\exp(-y^2/2\delta^2)$ whose dispersion $\delta \approx 0.9$. Corrections owing to $L \neq 0$ in general reduce δ by a small amount, but are relevant only for larger y where the Gaussian has fallen substantially from its peak. Dispersion δ is independent of both cluster mass and multiplicity.

Properly normalized to (integrate to) unity, the isotropic decay distribution in rapidity of a single cluster may then be given as

$$g(y) \simeq \frac{1}{\delta\sqrt{2\pi}} \exp(-y^2/2\delta^2), \tag{3.18}$$

or as

$$g(y) \simeq 0.5 \cosh^{-2} y \tag{3.19}$$

(Note that the two approximations differ by $\approx$ 10% at $y = 0$.)

Suppose, next, we have some production model which specifies the dynamical distribution of clusters in y space. Call this distribution $\rho(y_c) = 1/\sigma\,(d\sigma/dy_c)$. Obviously, the inclusive rapidity distribution for decay particles is then

$$\frac{1}{\sigma}\frac{d\sigma_h}{dy} = \int dy_c\, \rho(y_c)\, n_h(y_c)\, g(y-y_c) \tag{3.20}$$

Function $n(y_c)$ is just the average multiplicity of decay particles of the type h which come from a cluster located at y_c. Note that

$$\langle n_h \rangle \equiv \sigma^{-1} \int \frac{d\sigma_h}{dy}\, dy = \int dy_c\, \rho(y_c)\, n_h(y_c) \tag{3.21}$$

In very similar fashion, we write the two-particle inclusive y distribution for identical hadrons h as

$$\frac{1}{\sigma}\frac{d^2\sigma}{dy_1 dy_2} = \int dy_c\, \rho(y_c) \langle n_h(n_h-1)\rangle_{y_c}\, g(y_1-y_c)\, g(y_2-y_c)$$

$$+ \int dy_{c1}\, dy_{c2}\, \rho(y_{c1})\, \rho(y_{c2})\, g(y_1-y_{c1})\, g(y_2-y_{c2}) \times n_h(y_{c1})\, n_h(y_{c2}) \tag{3.22}$$

There are two terms, corresponding to situations in which both particles decay from the same cluster and in which they arise from different clusters. In the integrand of the first term as the right-hand side of (3.22), $\langle n_h(n_h-1)\rangle_{y_c}$ stands for twice the mean number of pairs of hadrons of type h in the decay cluster of a cluster located at y_c. In this integrand, the simplifying approximation is made that two decays from a given cluster are independent, i.e.,

$$g_2(y_1, y_2 ; y_c) = g(y_1, y_c)\, g(y_2, y_c) .$$

Surely this guess must be corrected at least for energy momentum and charge conservation effects in a more serious treatment [4]. In the two cluster term, the second integral on the right of (3.22), I have explicitly made the assumption that the distribution for two cluster formation $\rho_2(y_{c1}, y_{c2})$ factors to give

$$\rho_2 = \rho(y_{c1})\, \rho(y_{c2}) ,$$

as would be the case for independent production of clusters. Again, this guess would be modified in more careful treatment of the effects of energy momentum conservation in, say, a Monte Carlo computation [4]. The second integral therefore gives $\sigma^{-2}(d\sigma_h/dy_1)(d\sigma_h/dy_2)$.

Defining a "correlation function"

$$C(y_1, y_2) = \sigma^{-1}\frac{d^2\sigma}{dy_1 dy_2} - \sigma^{-2}\frac{d\sigma}{dy_1}\frac{d\sigma}{dy_2} , \tag{3.23}$$

we see that $C(y_1,y_2)$ is given by the first integral on the right-hand side of Eq. (3.22)

$$C(y_1,y_2) = \int dy_c\, \rho(y_c) \langle n_h(n_h-1)\rangle_{y_c}\, g(y_1-y_c)\, g(y_2-y_c) \quad (3.24)$$

The integral

$$f_2 \equiv \iint C(y_1,y_2)\, dy_1\, dy_2 = \int dy_c\, \rho(y_c) \langle n_h(n_h-1)\rangle_{y_c} \quad (3.25)$$

$$\equiv \langle n_h(n_h-1)\rangle^2 - \langle n_h\rangle^2$$

The above arguments are reasonably general, independent of the forms chosen for $\rho(y_c)$, $n_h(y_c)$ and $< n_h(n_h-1) >_{y_c}$. The only (and serious) approximations involve cavalier treatment of kinematic correlations connected with energy momentum and charge conservation, as I indicated above. I now turn to specific models for the rapidity distribution and multiplicity content of clusters, to see what these imply for $\sigma^{-1} d\sigma/dy$, $C(y_1,y_2)$, f_2 and so forth.

3.4. Multiperipheral emission of clusters *)

For exercise, we can take both $n(y_c)$ and $\rho(y_c)$ as (both s and y_c independent) constants : $< K >$ and B respectively, whereupon

$$\frac{d\sigma_h}{dy} = \sigma \langle K\rangle B \int_{-kY/2}^{kY/2} g(y-y')\, dy'$$

$$= \frac{\sigma\langle K\rangle B}{2} \left\{ \tanh\left(y+\frac{kY}{2}\right) - \tanh\left(y-\frac{kY}{2}\right)\right\} \quad (3.26)$$

---

*) Much of the work I discuss in this section and in Sections 5.6-5.7 was done in a long-range collaboration with G.C. Fox during the past year. See also Ref. 4).

As can readily be seen, when $y \ll Y$, $d\sigma_h/dy \to \sigma \langle K \rangle B$, which is just the energy independent flat distribution in rapidity, indicated by data and to which so many models aspire.

$$\langle n_h \rangle \sigma \simeq \sigma \langle K \rangle B \log s \tag{3.27}$$

With this same multiperipheral-like model, $\langle n_h(n_h-1) \rangle = \langle K(K-1) \rangle$ is independent of y_c and s. We obtain from (3.24)

$$C(y_1, y_2) = \frac{B\langle K(K-1)\rangle}{2\delta\sqrt{\pi}} \exp\left[\frac{-(y_1-y_2)^2}{4\delta^2}\right] \tag{3.28}$$

An alternative form is

$$C(y_1, y_2) = \frac{\langle K(K-1)\rangle}{2\delta\sqrt{\pi}\,\langle K\rangle\,\sigma} \left.\frac{d\sigma}{dy}\right|_{y=0} \exp\left[\frac{-(y_1-y_2)^2}{4\delta^2}\right] \tag{3.29}$$

The correlation function falls off in a Gaussian fashion as a function of $|(y_1-y_2)|$. It is independent of energy s, if the single particle plateau height is itself energy independent. The distance λ over which $C(y_1-y_2)$ falls to e^{-1} of its peak value may be defined to be a "correlation length"

$$\lambda = 2\delta \simeq 1.8 \tag{3.30}$$

This correlation length is essentially identical to that given in inclusive Regge (Mueller) analysis [23),38)]. In this latter approach, $\lambda = (\alpha_P - \alpha_R(0))^{-1} \simeq 2$ if $\alpha_R(0) \simeq \frac{1}{2}$. The two approaches may differ, however, if the quantum numbers of the two-particle system in question are exotic. In this case, it is not altogether clear whether inclusive Regge analysis is applicable ; if it is, then $\alpha_R(0) < 0$ and $\lambda < 1$. It will be noted that the Gaussian form in (3.29) is valid only for small $|y_1-y_2|$, where neglect of the pion mass implicit in (3.17) is justified. On the other hand, the exponential form for C predicted by Mueller analysis [38)] is correct only at large $|y_1-y_2|$. The two approaches are in a sense complementary. The relevant data are all at small $|\Delta y|$.

The integral of $C(y_1,y_2)$ [Eq. (3.29)] over all phase space gives

$$f_2^{(h)} = B\langle K(K-1)\rangle Y \quad \left(\simeq \frac{\langle K(K-1)\rangle}{\langle K\rangle}\langle n_h\rangle\right) \tag{3.31}$$

where $Y = \log s$. The coefficient of $\log s$ is positive as long as the number of particles per cluster $N \geq 2$.

Taking $f_2^{(-)} \simeq (0.7\pm0.1)\langle n_-\rangle$ as a crude parametrization of data [20] on negative particle production for $p_{lab} > 50$ GeV/c, I find that hadronic clusters produced in the multiperipheral chain contain roughly 1.7 negative hadrons on the average. More detailed analysis (Section 4.3) shows this is a slight overestimate.

Before leaving the multiperipheral approach and turning to an alternative explanation of the scaling plateau, I should call attention to the fact that, as I have developed it here, the multiperipheral cluster emission model cannot describe a possible diffractive component in the data. Thus, in the model all $\sigma_n(s) \to 0$ as $s \to \infty$. This is equivalent to the statement that there are no Pomeron exchanges in our multiperipheral chain. Data on the s dependence on σ_n are consistent with this possibility, but it may also turn out that $\sigma_n(s) \to C_n$, with all least some $C_n > 0$ as $s \to \infty$. A stronger indication of the presence of a diffractive component is the observation of an inelastic diffractive peak in the proton x spectrum near $x = 1$ (cf. Section 5). By introducing an additional set of graphs in which one Pomeron is present in the chain of exchanges, we can reproduce both the inelastic peak and obtain $\sigma_n(s) \to C_n > 0$. This point is developed in Section 5. For the present, we can regard the model discussed in the present subsection as a model for the non-diffractive component of multiparticle data. For discussion of more theoretical points associated with the multiperipheral cluster emission approach, the papers of Hamer and Peierls may be consulted [39].

3.5. Fragmentation models

The constant assumptions $\rho(y_c) \to B$ and $n(y_c) \to \langle K\rangle$ made in Section 3.4 are not the only way to generate a plateau in y for the single particle rapidity spectrum. As a second example, I may cite the nova fragmentation approach [3),40] in which production of a cluster of mass M is postulated to have production cross-section $d\sigma/dM \sim \sigma_0/M^2$, and decay multiplicity $n(M) \sim K_0 M$. Only one or at most two clusters are produced per event.

For simplicity I will use these formulas over the full interval $1.0 < M < \sqrt{s}$. Thus

$$\sigma_{inel} = \int (d\sigma/dM)\, dM = \sigma_0 \tag{3.32}$$

$$\langle n \rangle = \tfrac{1}{2} K_0 \log s$$

The kinematic relationship between y_c and M is $|y_c| = \log(\sqrt{s}/M)$. Therefore,

$$\rho(y_c) = \frac{1}{\sigma}\frac{d\sigma}{dM}\frac{dM}{dy_c} = \frac{1}{M} \tag{3.33}$$

and

$$n(y_c) = K_0 M = K_0 \sqrt{s}\, \exp(-|y_c|) . \tag{3.34}$$

The important result is that the product

$$n(y_c)\, \rho(y_c) = \text{Constant} = K_0 \tag{3.35}$$

This guarantees the single particle inclusive plateau, as in Eq. (3.26).

Although both models can be made to agree on the existence of an asymptotic s independent central plateau, the fragmentation and multi-peripheral cluster models are seen to differ in a spectacular way when one looks at their very different expectations for the evolution of dynamics with energy. The differences are seen most clearly upon examination of idealized individual event structure.

Below $p_{lab} = 30$ GeV/c, the two models are in essential agreement. Both expect one or two clusters to be produced per event. Indeed, low multiplicity, low mass clusters seem clearly present in data [41], but it is possible that these effects are explained as a reflection of leading particle effects. At these low energies, presence of clustering may also be inferred indirectly from successful fits with cluster-dominated models to two particle inclusive spectra [40], in contrast to the failure of independent emission models [42].

As energy grows, and phase space opens, the fragmentation approach restricts the number of clusters to 2. In the multiperipheral approach, more clusters pop out of the vacuum. Their number grows in proportion to log s, $< N > \simeq B \log s$. Based on fits [4] to data,

$$B \simeq 0.5 \text{ to } 0.75 \tag{3.36}$$

Thus, the mean distance between clusters in the multiperiphal approach is

$$|\Delta Y_c| \simeq 1.3 \text{ to } 2$$

This interval is comparable to the intrinsic typical spread of $\Delta y = 2$ for decay particles from a single cluster.

To fulfil the requirement that $< n > \sim \log s$, the typical multiplicity (mass) in a cluster of the fragmentation type must grow with s. In our multiperipheral approach, the mean number of hadrons per cluster ($\simeq 4$) is energy independent. For $p_{lab} \gtrsim 100$ GeV/c, typical events in the two models are sketched below in rapidity y space.

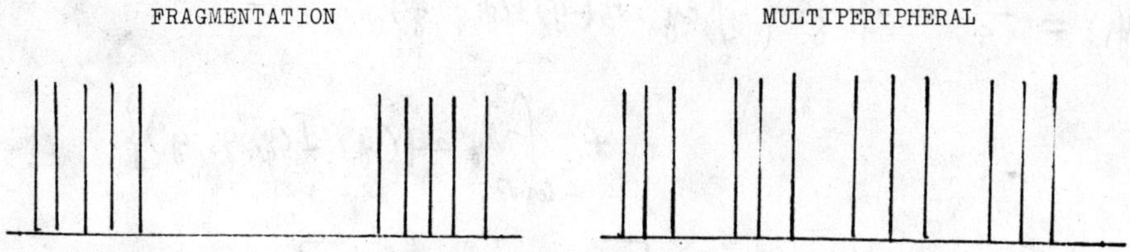

Each vertical bar marks the position in y of a charged track.

New techniques of data analysis must be developed to help in identifying patterns of these types in individual events. One possibility is the measurement of the dispersion δ^1 of tracks in y space [36]. In Monte Carlo simulations, this parameter shows good promise in distinguishing between the two broad classes discussed above [4),36),37]. Moreover, within the multiperipheral class of models, one finds

$$\langle \delta^1 \rangle = c \log s + d \tag{3.37}$$

The value c is sensitive to the number of particles per cluster [4),37)].

At the conventional inclusive level, discrimination between the (two-cluster) fragmentation and (multicluster) multiperipheral view-points is possible by studying certain features of the two-particle rapidity distribution [40),43)]. The <u>magnitude</u> of predicted cross-sections at $y_1 \approx y_2 \approx 0$ as well as <u>s dependence</u> are very different in the two cases, with data seeming to favour the first example.

In the fragmentation case, with $n(M) = K_o M$, as stated above, we have $n(M)(n(M)-1) \sim K_o^2 M^2$. Thus, the product

$$\rho(y_c) \langle n_h(n_h-1) \rangle_{y_c} \sim K_o^2 M = K_o^2 \sqrt{s}\, \exp(-y_c) \tag{3.38}$$

For the single fragmentation component, with one cluster per event, the correlation function is

$$C_1(y_1,y_2) = \frac{K_o^2}{4} + \frac{K_o^2\sqrt{s}}{4\pi\delta^2}\left\{\int_0^{\log\sqrt{s}} dy' \exp(-y')\, I(y_1,y_2,y') + \int_{-\log\sqrt{s}}^{0} dy' \exp(y')\, I(y_1,y_2,y')\right\}$$

$$I(y_1,y_2,y') = \exp\left[\frac{-(y_1-y')^2-(y_2-y')^2}{2\delta_o^2}\right]$$

After some algebraic manipulation, we can re-express this as

$$C_1(y_1,y_2) = \frac{K_o^2}{4} + \sqrt{s}\, \exp\left[\frac{-(y_1-y_2)^2}{4\delta^2}\right] F\big((y_1+y_2), \log s\big) \tag{3.39}$$

The function $F(0,\log s)$ is approximately constant. A second contribution $C_2(y_1,y_2)$ with similar structure arises from the double fragmentation component, having two clusters per event.

We observe several features of this expression.

(A) For (y_1+y_2) fixed, it has the same Gaussian fall-off in (y_1-y_2) as does the multiperipheral example (3.29).

(B) However, it displays explicit dependence on (y_1+y_2) which the multi-peripheral-like example did not. It is not a function just of the difference (y_1-y_2).

(C) It shows a growth with s proportional to $\sqrt{s}$ at $y_1 = y_2 = 0$, when $K/2$, the height of the single-particle inclusive spectrum in the model is constant. Again, this is in contrast with (3.29) which shows no s dependence.

Data [17)-19)] seem to favour (3.29), although there is some doubt (cf. Section 4). In fact, the hint from data that correlations are s independent in the central region and a function only of the difference (y_1-y_2) is confirmation of much deeper ideas than those which are apparent in either of the crude examples treated here. We return now to the short-range order hypothesis which gives these effects in a model independent manner.

4. INCLUSIVE CORRELATIONS

4.1. Apology

The measurement of correlation is after all a major part of what physics is all about. Short of an article on the subject by itself, I could not attempt a complete treatment even of all the two-body correlation measurements presently made or desirable for theoretical reasons. I will therefore only mention in passing data on neutral-charged particle correlations in multiplicity distributions [44)]. Likewise, owing to scanty data, I will not treat the intriguing question of inclusive correlations between hadrons as a function of angle ϕ between their respective transverse momentum vectors. Experimental hints [17)] that such correlations are positive and of long-range in y deserve detailed investigation. Since I have set aside any treatment of large p_T phenomena, neither will I speculate here on the character of events which give hadrons with large p_T [11)]. Collective phenomena, such as clustering, require new methods for statistical analysis of data [36),37),41)].

My focus will be restricted essentially to inclusive correlations in rapidity between two hadrons. There are important theoretical expectations and relevant data on this interesting subject [45].

4.2. Two-particle rapidity correlations

The two-particle inclusive cross-section is a priori a function of five independent variables : s, y_1, y_2, $|p_{T1}|$, $|p_{T2}|$ and ϕ, where ϕ is the angle between the two transverse vectors $\vec{p}_{T1}$ and $\vec{p}_{T2}$. I will henceforth neglect the ϕ and p_T dependences, concentrating only on y dependences. Therefore,

$$\frac{d^2\sigma}{dy_1\,dy_2} = G(y_1, y_2, s) \qquad (4.1)$$

In this section, I will discuss in detail only correlations observed between hadrons which are both in the central region. Correlations found when one or the other hadron may be outside this central region are summarized well by Sens [1]. Obvious long-range correlations present when one or both protons have $|x| > 0.8$ are treated in Section 5.

If both observed hadrons are in the central region, then $|y_i - y_a| > \lambda$ and $|y_i - y_b| > \lambda$. As before, y_a and y_b are the rapidities of the incident protons. According to the SRO hypothesis, in this central region, G cannot depend on $|y_i - y_a|$ or $|y_i - y_b|$. The only remaining rapidity variable is $|y_1 - y_2|$. The hypothesis predicts therefore that

$$\frac{d^2\sigma}{dy_1\,dy_2} \longrightarrow g(y_1 - y_2) \qquad (4.2)$$

In other words, the two-particle distribution should tend to an energy independent function of Δy. [If, further, it were kinematically possible to have $(\Delta y) > \lambda$, with both y_i still in the central region, then $g \to$ const. This latter prediction seems to have little practical significance even at ISR energies.] The hypothesis does not predict the magnitude of $g(y_1 - y_2)$ nor its functional form. For this, we need explicit SRO model, such as that discussed in Section 3.4.

An alternative to the simple prediction (4.2) is provided by fragmentation models, in which one finds both long-range and short-range structure. Explicit fragmentation predictions were developed in Section 3.5, where, we saw that a $\sqrt{s}$ growth of $G(y_1,y_2,s)$ was expected, in contrast to (4.2). Furthermore, in fragmentation models, dependence on the variables $|y_i-y_a|$ and $|y_i-y_b|$ does not drop out.

I now turn to an examination of available correlation data from NAL[19], and from the Pisa-Stony Brook[18] and CERN-Hamburg-Vienna[17] ISR collaborations. All these data are expressed in terms of

$$R(y_1,y_2) = \frac{C(y_1,y_2)}{\left(\frac{1}{\sigma_{inel}}\right)^2 \frac{d\sigma}{dy_1}\frac{d\sigma}{dy_2}} \tag{4.3}$$

This ratio is simpler to determine experimentally than $C(y_1,y_2)$. To the extent that $\sigma^{-1}d\sigma/dy$ is independent of s and y, the denominator is a constant, and conclusions about $R(s,y_1,y_2)$ bear directly on $C(s,y_1,y_2)$.

The existence of a scaling plateau is best established for pion production at fixed small p_T : $0.1 < p_T < 1.0$ GeV/c. Thus, inclusive correlations between <u>pions</u> ($\pi^+\pi^-$, $\pi^+\pi^+$, $\pi^-\pi^-$, $\pi^\pm\pi^0$) in the central region, with $0.1 < p_{Ti} < 1.0$ GeV/c for both pions would offer the greatest opportunity for a clean investigation of new phenomena. Unfortunately, all present central region correlation data are averaged over all p_T. Moreover, in the Pisa-Stony Brook experiment[18], correlations are observed between any pair of charged particles, without regard to sign ($\pm$) and particle type ($\pi, K, \bar{p}$, etc.). In the CHV study[17], correlations between a γ ray and a charged track are studied. Finally, in the NAL and ISR experiments, the approximate rapidity variable $\eta = \log(\tan\theta/2)$ rather than y is used for charged particles. For these reasons, results are not as precise as one might like.

In Fig. 10 results of the CHV collaboration[17] are shown for $\sqrt{s} = 22$, 30, 45 and 53 GeV. Presented is

$$R(y_\gamma, y_{ch}) = \frac{\sigma_{inel}\, d^2\sigma/dy_\gamma dy_{ch}}{(d\sigma/dy_\gamma)(d\sigma/dy_{ch})} - 1 \tag{4.4}$$

versus y_γ for $y_{ch} \approx 0$ (open circles) and $y_{ch} \approx -2.5$ (dark circles). The point $y_{ch} \approx -2.5$ falls outside the central region. Thus those predictions of SRO developed above do not apply to the data shown in dark circles.

Concentrating on the data for which $y_{ch} \approx 0$, we see that $R(0,0)$ is positive (≈ 0.65), R is essentially independent of s from $\sqrt{s} = 23$ to 53, for all $|y_\gamma| < 3$. The function $R(y_\gamma, 0)$ is well fitted by a Gaussian [cf. Eq. (3.28)] with correlation length $\lambda \approx 2$.

A compendium of Pisa-Stony Brook charged-charged correlation data[18] appears in Fig. 11 for $\sqrt{s} = 23$ and 63 GeV. The function $R(\eta_1, \eta_2)$ is plotted versus η_2 for several fixed values of η_1. For small $|\eta_i|$ it is clear that $R(\eta_1, \eta_2)$ is largest at $\eta_1 \approx \eta_2$, and that the maximum value $R(\eta_1 \approx \eta_2) \approx 0.65$ is s independent. An energy independent function of $(\eta_1 - \eta_2)$ is certainly in accord with dominant features of these data, but some deviations are notable even at $|\eta_i| < 2$. In particular, while the average correlation length λ is ≈ 2, the plot with $\eta_1 \approx 0$ shows that this length is effectively larger at $\sqrt{s} = 63$ than it is at $\sqrt{s} = 23$. Some dependence on $|\eta_i - \eta_{ab}|$ (i.e., on $|\eta - \frac{1}{2}\log s|$) may also be discerned in the plots.

Data on charged-charged correlations from an NAL bubble chamber experiment[19] at 200 GeV/c are portrayed in Fig. 12. These results are plotted versus $|\eta_1 - \eta_2|$ for two fixed values of η_1. The maximum occurs at $\eta_1 \cong \eta_2$, where $R = 0.6 \pm 0.1$. A correlation length of 2 is also appropriate here. Similar results for $\pi^- \pi^-$ correlations show $R(0,0) \simeq 0.3$ with large errors.

It would be heartening to be able to conclude from these data that the principle of short range order is in perfect accord with data on two-particle inclusive spectra, and that the fragmentation alternative in the central region is dead. However, the crucial issue of energy dependence is still unresolved. As will be recalled, the CHV and P-SB groups observe a healthy 20-30% rise of their $\sigma^{-1} d\sigma/dy$ over the ISR energy range. Therefore, their true correlation functions

$$C = R\left(\sigma^{-1} d\sigma/dy_1\right)\left(\sigma^{-1} d\sigma/dy_2\right)$$

rise by 50% or more. In a sense, this rise splits the difference between the zero growth and $\sqrt{s}$ growth predictions of pure SRO and pure fragmentation, respectively [Eqs. (3.29) and (3.39)].

Partisans of either view can have their way. One may speculate that whatever causes the rise of $\sigma^{-1} d\sigma/dy$ and $\sigma^{-1} d\sigma^2/dy_1 dy_2$ in the CHV and PSB experiments properly cancels out in the ratio $R(\eta_1,\eta_2)$, and therefore, that R rather than C represents unbiased reality. This will remain a <u>speculation</u> until clean data on <u>pion-pion</u> central region rapidity correlations become available, with p_T of both pions in the range $0.1 < p_T < 1.0$ GeV/c (where the pion scaling plateau is established).

A summary of rapidity correlation data in the central region follows.

(a) Strong positive correlations are observed ; $R(0,0) = 0.6$ to 0.7 for the charged-charged configuration.

(b) The correlation length λ is roughly energy independent ; $\lambda \simeq 2$.

(c) The magnitude and shape of $R(\eta_1,\eta_2)$ is roughly the same for γ charged and charged-charged correlations.

(d) For $|\eta_1-\eta_2| < 2$ and $|\eta_i| < 2$ $R(\eta_1,\eta_2)$ shows dominant dependence on the difference $|\eta_1-\eta_2|$, in accord with expectations of pure SRO.

(e) $R(0,0)$ is energy independent. With uncomfortable reservations, this suggests that $C(0,0)$ may be approaching the s independent limit predicted by pure SRO.

4.3. <u>Implications for models</u>

The issue of s dependence of $C(0,0)$ is crucial for two-fireball models [46]. The particular fragmentation example discussed in Section 3.5 provides a $\sqrt{s}$ growth of $C(0,0)$. While somewhat weaker [e.g., $(\log s)^2$ or $s^{1/4}$] behaviour might be obtained in other versions of the two fireball approach, growth with s seems inescapable. Qualitatively, the reason is simple. Since a fireball by definition occupies a finite spread in rapidity, its centre y_F must be located reasonably near $y = 0$, in order that we observe its decay particles in the region near $y = 0$. In a one or two fireball event, the kinematic relationship $y_F \simeq \log(\sqrt{s}/M_F)$ holds, where M_F is the fireball mass. Therefore, the fireball mass $M_F \propto \sqrt{s}$. Inasmuch as multiplicity grows with mass $(n \gtrsim \log M)$, the fireball providing particles near $y \simeq 0$ will have multiplicity of order $(\log s)^p$ or s^q, where powers p and q depend upon assumptions made about fireball decay. Given a hadron near $y = 0$, the chance of finding another is proportional to the multiplicity of the fireball from which the particles come. Thus

$$C(0,0) \simeq f(s) , \tag{4.5}$$

where function f(s) grows with s.

Observation of a scaling C(0,0) would seem therefore to eliminate the entire two fireball class, not just the specific realization discussed in Section 3.5.

There is, of course, other circumstantial evidence against two fireball models. For example, it has been argued that with two fireball models it is impossible to obtain a simultaneous numerical fit to both the plateau $d\sigma/dy$ for pions and the NAL prong cross-section data [47]. A successful fit to σ_n vs. n at large n forces a dip in $d\sigma/dy$ near $y = 0$. A judgement here depends, however, on the cleverness of the person fitting data. I know I can overcome this particular objection, and several others, but the resulting model is not what everyone would consider most natural [48]. In particular, the use of non-asymptotic terms and of elongated (multiperipheral-like) decay distributions in the nova or two fireball approach begins to beg the issue of what a fireball or nova is meant to be.

In the minds of many people, a multiperipheral or short-range order dominated model is emerging as the most natural explanation at hand for describing phenomena in the central region. The approach easily provides an energy independent plateau for single particle spectra, energy independent short-range correlations, and prong cross-sections σ_n which fall off as $(n!)^{-1}$ for $n > \langle n \rangle$, all in qualitative agreement with data. If the objects emitted along the chain are hadronic clusters (cf. Sections 3.2 and 3.4), then many quantitative aspects of data are also reproducible. Diffractive effects, discussed in Section 5, are easily accommodated by inclusion of graphs in which a Pomeron link is present in the multiperipheral chain. Geoffrey Fox and I have performed a specific calculation along these lines [4]. For details of the parametrization, consult Ref. 4). Spectra were evaluated with a Monte-Carlo program in which all energy-momentum and other constraints are fully respected. A sample of results is presented in Figs. 13, 14, 15 and 16.

Charged particle multiplicity data [20] from 50 to 300 GeV/c are well reproduced (Fig. 13). Rapidity distributions at several energies [49] are shown in Fig. 14. Scaling in the fragmentation regions and the eventual appearance of a plateau at ISR energies are apparent. The general shape and energy dependence (including the rise at $y = 0$ versus energy) are in accord with data [28),29),49),30),31)]. Our results for $R(y_1, y_2)$ are given in Fig. 15 (a,b). We see that R is energy independent over the ISR range for $y_1 \approx 0$. For a value of y_1 which is outside the plateau, R rises with energy to its limiting form as a function of $(y_1 - y_2)$ only. These features all agree with data discussed earlier in this section, as does the rough shape and, in particular the zero in R seen at $y_1 \approx 0, |y_2| = 3$.

Although diffractive effects can in principle and do affect correlations in the central region (cf. Section 5.7), our analytic calculations and Monte-Carlo results both indicate that essentially all correlation in the central region is of a short-range character. Thus,

$$R(0,0) \simeq 0.1 + R_{SR}(0,0) \tag{4.6}$$

Here,

$$R_{SR}(0,0) \simeq C_{SR}(0,0) \Big/ \left(\frac{1}{\sigma_{SR}} \frac{d\sigma_{SR}}{dy} \right)^2_{y=0} \tag{4.7}$$

with C_{SR} given by Eq. (3.29). The constant term (0.1) in Eq. (4.6) is the <u>net</u> long-range contribution. It is associated with diffractive effects, as discussed in Section 5.7. [Its magnitude is controlled, of course, by the net cross-section σ_D associated with single inelastic diffraction. Our results for σ_D agree well with information available from NAL and ISR on the magnitude and shape of $E d\sigma/d^3p$ for $pp \to pX$ at large x_p ; cf. Section 5.]

From Eqs. (4.6) and (3.29) we deduce that

$$\frac{\langle K(K-1) \rangle}{2\sqrt{\pi}\, \delta \langle K \rangle} \simeq 0.55 \left(\frac{1}{\sigma_{SR}} \frac{d\sigma_{SR}}{dy} \right)_{y \simeq 0} \tag{4.8}$$

An average $\langle K \rangle = 4$ pions (of all charges) per cluster decay fits both the correlation data and correlation moment f_2. Other more complicated correlation data [50] from ISR have also been examined, and so far, we find good agreement. (In our approach, $\langle K \rangle$ is s independent.)

The dispersion of tracks in rapidity provides a good model independent measure of the extent of cluster formation in individual events [36]. Its measurement would provide an independent check on the approximate validity of the idea of multiperipheral emission of clusters.

For each event, we define

$$\overline{y} = \sum_{i=1}^{\ell} y_i / \ell$$

$$\delta^1_{ch}(y) = \left\{ \frac{1}{\ell - 1} \sum_{i=1}^{\ell} (y_i - \overline{y})^2 \right\}^{1/2} \tag{4.9}$$

where the sum runs over the ℓ particles obtained by removing the leading particle (farthest away in rapidity) from an $(\ell+1)$ charged particle final state. Our model, as in all multiperipheral models, predicts a logarithmic rise in $\delta^1_{ch}(y)$. The prediction is shown for two multiplicities in Fig. 16 as a function of p_{lab}. At 300 GeV/c, we calculated δ from this cluster model, from a simple multiperipheral model with production of single π's (not in clusters), and from the nova model. Both the nova model and the cluster model agree qualitatively with the experimental dispersions at this energy [37]. [The Monte Carlo program indicates that these models would be distinguishable at 300 GeV/c if true rapidity and not the approximation $-\log(\tan\theta_{proj}/2)$ were used.] Data on δ^1 at ISR energies should be forthcoming soon from the Pisa-Stony Brook group. It will be instructive to see how the experimental results compare with predictions of Fig. 16.

Data on δ^1_n vs. n will provide an independent estimate of the mean number of particles per cluster, which can be compared with the value $\langle K \rangle = 4$ we obtained from inclusive rapidity correlation data. It is not obvious a priori that the two estimates should agree, because δ^1_n measures clustering seen in individual events, of fixed multiplicity n, whereas the magnitude and shape of $C(y_1, y_2)$ are influenced in part by the over-all multiplicity distribution σ_n vs. n. [This last point is evident from the relationship

$$f_2 = \langle n(n-1) \rangle - \langle n \rangle^2 = \iint C(y_1, y_2)\, dy_1\, dy_2 \, .]$$

Although I am admittedly too close to the effort to be entirely objective, this is one of the few analyses I know of which attempts a description of all known aspects of inclusive data. For $p_{lab} \gtrsim 50$ GeV/c, we tried to accommodate knowledge of the x, y, p_T and s dependences of single particle spectra, the prong distributions [20] $\sigma_n(s)$, the systematic change with multiplicity [28),51] n of $d\sigma_n/dy$, and the two-particle inclusive correlation data. Thus, our successful description of the inclusive correlation data was not tailored to this purpose alone.

To be sure, obvious improvements are needed on several fronts, but the reasonable agreement between our model and experiment for most important attributes of multiparticle data at small p_T provides support for the input clustering and multiperipheral-like matrix element. Having in hand a simple framework which successfully imitates the more striking features of NAL and ISR results allows us to estimate the dynamical significance of other observations, in particular those not interpretable directly by fully inclusive (Mueller) analysis [23].

At $p_{lab} \leq 30$ GeV/c, where the nova model is particularly successful [40], our model and the nova model are very similar. At these energies, we produce only one or two clusters, just as in the nova model. As s increases, the number of clusters grows in our model, rather than remaining fixed at one or two.

The evolution of dynamics with increasing energy may be portrayed as in the following crude sketch.

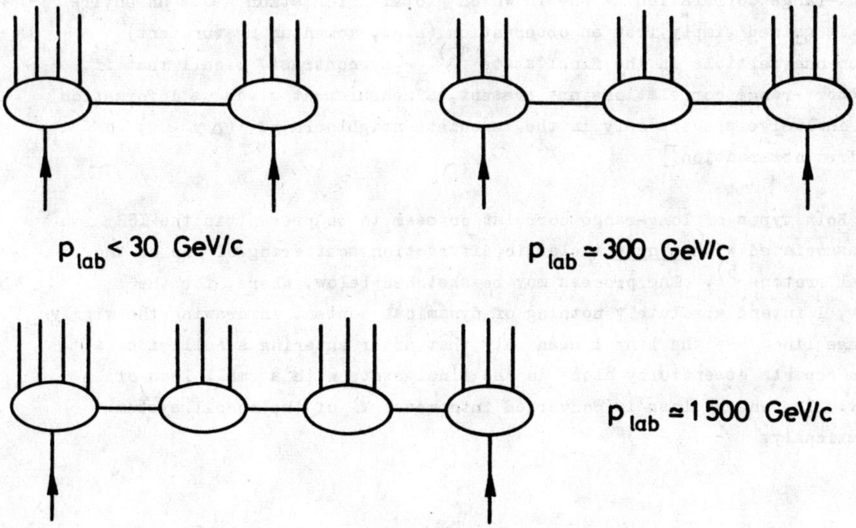

As I have drawn them, clusters always decay into 3-4 particles. However, a broader distribution about the mean of 4 is certainly to be expected. The end clusters, in fact, are sometimes to be replaced by a single diffractively scattered proton. The ratio of diffractive to non-diffractive scattering is an important parameter, which may be determined by careful fits to the inclusive proton spectrum $Ed\sigma/d^3p$ (cf. Section 5). Data in the fragmentation regions are observed to reach their asymptotic form at relatively low p_{lab}. In the model, this would mean that the clusters at the end of the chain attain their asymptotic limiting character at low s. Further increases in energy allow these two end-clusters to separate in y space. New clusters are generated in the intervening space.

. INELASTIC DIFFRACTION SCATTERING

5.1. Qualitative remarks ; long-range correlations

In the last sections, I emphasized the rôle and magnitude of short-range correlations among hadrons in the central region of rapidity. There appears to be evidence in ISR data also for important long-range effects. These can manifest themselves in several ways. Obvious long-range correlation would be present if the inclusive correlation function is enhanced when the distance Δy between two hadrons is large. This is an inclusive long-range correlation between two hadrons. A second and more general type of long-range correlation is one in which global information about an entire event is gained simply from an observation (e.g., momentum measurement) made on one particle in the final state [52]. [By contrast, recall that if only short-range correlations are present, a measurement gives us information about inclusive spectra only in the immediate neighbourhood $(\Delta y \leq 2)$ of the first observation.]

Both types of long-range correlation seem to be present in the ISR data associated with single inelastic diffraction scattering of one of the initial protons [5]. The process may be sketched below, where, for the moment, I intend absolutely nothing of dynamical content in drawing the wiggly exchange line. By the line I mean only that after entering a collision, a proton departs essentially alone in the final state with a small loss of energy. The energy loss is converted into mass M of the recoil system. Kinematically,

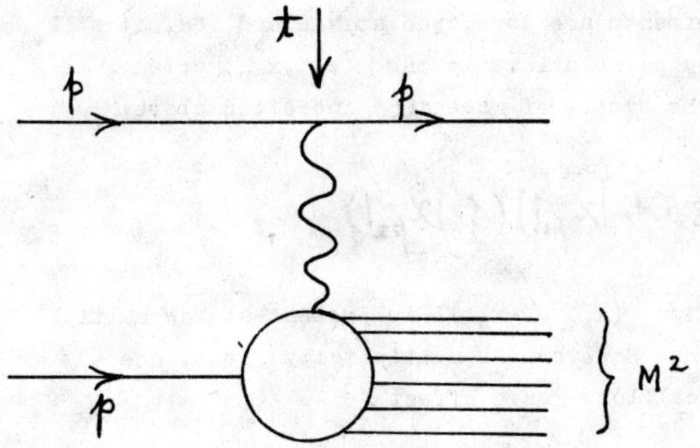

$$M^2 \simeq s\,(1-x_p) \tag{5.1}$$

The recoil system is the result of "dissociation" or fragmentation of the other proton.

Use of the word "diffraction" is in some ways unfortunate, since it has had many imprecise meanings to many people over the past 10-20 years. However, in the NAL and ISR ranges, it does seem finally possible to discus a component of inelastic data which shows an energy (in)dependence similar to that of elastic scattering data. Thus, an attempt at more precise definition can now be justified, perhaps.

Along with single inelastic diffraction, it may be kept in mind that a process such as sketched below probably also contributes at some level. As

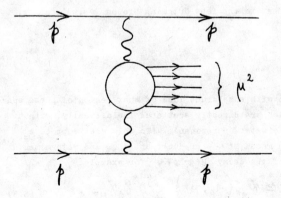

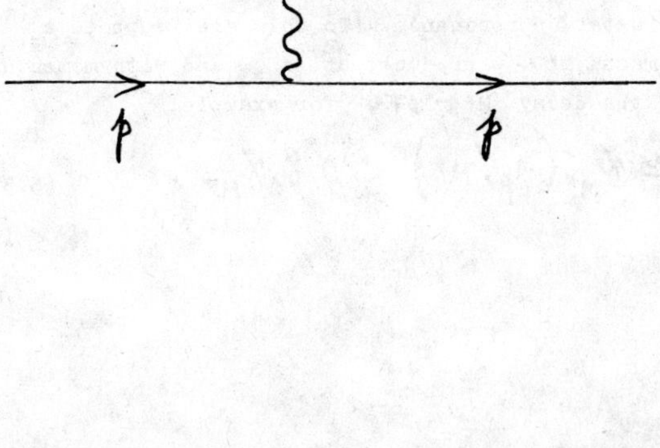

double-arm spectrometer experiments are developed at NAL and ISR, it will be extremely interesting to study correlations in the (x_{p1}, x_{p2}) plane, as a function of multiplicity of the system of mass μ indicated above.

$$\mu^2 \simeq s(1-|x_{p1}|)(1-|x_{p2}|) \tag{5.2}$$

Obviously very large correlation $(x_{p1} = -x_{p2})$ is present at low multiplicity, owing to momentum conservation. However, as multiplicity rises, one may or may not see a residual dynamical long-range effect [53].

Finally, a third process is double dissociation, in which two clusters are formed, in some appropriate diffractive sense.

Whether this process can be defined precisely and kinematically isolated is not yet clear.

I will confine my remarks here to <u>single</u> inelastic diffraction, or single "diffraction dissociation", as defined above.

In the following pages, I first summarize present data on single inelastic diffraction and then discuss theoretical interpretations. The variables I use are mass-squared M^2 of the recoil system and invariant momentum transfer t from the incident to the fast outgoing proton, as well as x, defined in Eq. (5.1).

5.2. Summary of data

It is useful to begin with some crude idea of what region of phase space is populated by protons which are directly scattered inelastically, without having "decayed" from some cluster or resonance. To this end, suppose a resonance-like object, with mass M^* is produced at x_{M^*} and with momentum transfer t. As a result of the decay $M^* \to p\pi$, for example,

$$x_p \simeq x_{M^*}(m_p/M^*) < 0.8\, x_{M^*} . \tag{5.3}$$

Thus, the chance of seeing direct scattering is enhanced if one looks at $x_p > 0.8$, and small t. The notion of "diffractive" scattering adds the requirements that the differential cross-section $d^2\sigma/dx\,dt$ should fall off reasonably sharply in t, and have a dependence on energy which is more or less as independent of s as that of $d\sigma_{el}/dt$.

Relevant data have been obtained by the Rutgers [54] and Columbia-Stony Brook [55] groups at NAL, the NAL pp bubble chamber collaborations [56],[57],[58] at 100, 200 and 300 GeV/c, and by the CERN-Holland-Lancaster-Manchester [5] and Aachen-CERN-Harvard-Genova-Torino [53] groups at the ISR. The reaction $\pi^- p \to pX$ at 205 GeV/c has also been studied by the NAL-LBL collaboration [59]. The picture is far from complete, but nevertheless very intriguing. Several points can be made.

5.2.1. x dependence

At fixed s, there is a sharp forward peak for $x > 0.9$ in the invariant $Ed\sigma/d^3p$ plotted versus x. This peak is present for a broad range of p_T values. At the larger p_T values a valley appears between the peak near $x \approx 1$ and a broad secondary maximum in the range $x < 0.8$. These points are illustrated in Fig. 17.

5.2.2. Scaling in x

At some fixed values of p_T it has been possible to obtain data at several values of $\sqrt{s}$. Because of a non-zero lower bound on the scattering angle, fixed by the apparatus, the p_T values accessible at all $\sqrt{s}$ are unfortunately rather large, where cross-section and physical interest are not at their greatest. However, present measurements at ISR are consistent with an s independent $Ed\sigma/d^3p$ at (all) fixed x and fixed p_T. These data are illustrated in Fig. 4. At lower energies [54] both scaling and non-scaling components can be identified.

The scaling properties of $Ed\sigma/d^3p$ for $x > 0.8$ are of great interest, as will be discussed below. More precise data over the widest possible s and p_T range are awaited anxiously.

5.2.3. Mass dependence

Mass dependence mixes s and x dependences. Because

$$d\sigma/dt\,d(M^2/s) = d\sigma/dt\,dx \qquad (5.4)$$

s independence of $E d\sigma/d^3p$ at fixed x, implies that

$$s\, d\sigma / dM^2$$

is s independent at fixed (M^2/s). That is, $d\sigma/dM^2 = s^{-1}g(M^2/s)$.

The traditional view of fixed mass diffraction holds that a particular mass region or resonance [e.g., N*(1688)] is produced with an energy independent cross-section

$$\frac{d\sigma}{dM^*} = \text{constant}$$

Note that perfect scaling in x all the way to the kinematic limit

$$x_{Max} = 1 - m_p^2/s$$

may be incompatible with this tradition. Experiments with excellent mass resolution are needed to examine carefully the s and M^2 dependences of cross-section in the region $M^2 \lesssim 5\ \text{GeV}^2$. Spectra from NAL and ISR are shown in Figs. 18 and 19, respectively.

5.2.4. t dependence

Data on the t dependence of inelastic diffraction are available at several energies. Bubble chamber data [57] from 200 GeV/c pp collisions are shown in Fig. 20. The slope β in a parametrization of the form

$$\frac{d^2\sigma}{dM^2 dt} \sim \exp(\beta t) \qquad (5.5)$$

varies smoothly with M^2, as shown in the Table below [57]

M^2	β
<5	9.1 ± 0.7
5 - 10	8.0 ± 1.1
10 - 25	6.1 ± 0.7
25 - 50	5.8 ± 0.7

These data show no evidence of any turn-over, dip, or vanishing of $d\sigma/dt$ near $t = 0$. Thus, these bubble chamber results, which should be reliable at small $|t|$, contradict some NAL counter data [55] at 200 GeV/c, in which a dip in $d\sigma/dt\, dM^2$ is seen at small $|t|$, for $M^2 < 20\ GeV^2$. Because the t dependence at small $|t|$ is very interesting theoretically (see below), more experimental work seems in order. Owing to the minimum scattering angle limitation mentioned above, ISR data are confined to a range of t values with $|t| \gtrsim 0.1\ (GeV/c)^2$.

For pp scattering, the slope β_{el} of $d\sigma_{el}/dt$ is about $\beta_{el} \simeq 12\ (GeV/c)^{-2}$. Combining this with $d\sigma_D/dt \simeq \exp(-6|t|)$, and using factorization, we see that there appears to be no t suppression associated with the coupling :

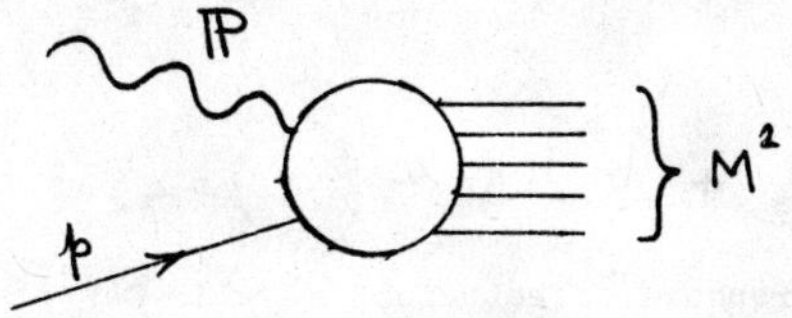

The double dissociation process would appear to suffer from no strong t dependence, therefore, and may grow very rapidly in importance as s increases [60].

5.2.5. Cross-section

The cross-section for single inelastic diffraction is easy to estimate, but it is hard to define precisely. In analyses of NAL bubble chamber results, different groups make different assumptions about the "background" to substract from the observed missing-mass distributions $d\sigma/dM^2$, in order to reduce these to proper cross-sections $d\sigma/dM^2$, differential in mass-squared of a cluster recoiling from a free proton. Next, there is the question of how high up in M^2 one should integrate. These ambiguities result in different answers [56),57),58),61)]. Nevertheless, most groups agree on a cross-section somewhere in the interval 5 to 8 mb, for energies in the NAL and ISR range. At $\sqrt{s} = 23.2$ and $\sqrt{s} = 30.4$ GeV, estimates by the CHLM collaboration [61)] give $\sigma_D = 5.4 \pm 1.0$ and 5.0 ± 1.0 mb, respectively. These numbers result from summing cross-section for $|x| > 0.94$, after an extrapolation is made in t with the help of NAL data.

5.2.6. Multiplicity

The structure of the recoil cluster and the manner in which its decay particles populate phase space are two areas well worth investigation. The mean charged multiplicity $n(M_x)$ has been obtained from NAL data on $pp \to pX$ as a function of missing mass [56),57),58)]. A compilation is given in Fig. 21. It would be valuable to investigate the t dependence also : $n(M_x, t)$.

Observe that $n(M_x^2)$ appears not to depend on energy s. It grows slowly with M_x^2. Two curves are drawn on Fig. 21 to provide a basis for comparison. Neither curve is in any sense a best fit. I took the coefficient b = 1.4 in the parametrization

$$\langle n(M^2) \rangle = a_M + b \log M^2 \qquad (5.6)$$

directly from my fit to the s dependence of over-all pp charged multiplicity [14]

$$\langle n(s) \rangle = a_s + b \log s$$

No attempt is made to reproduce threshold effects for $M^2 < 10$ GeV^2. As will be seen, the rate of growth of $< n(M^2) >$ with $\log M^2$ is consistent with being the same as the rate of growth of $< n(s) >$ versus $\log s$ [62].

The linear curve [58]

$$\langle n(M^2) \rangle = 0.6\, M + 2 \qquad (5.7)$$

is not an unreasonable parametrization at low M, but begins to fail for $M^2 > 50$ GeV^2.

Theoretical interpretation of these multiplicity data is not altogether direct. The nova approach [40] is predicated on a linear increase of multiplicity with mass of a properly defined recoil cluster. In fits to low energy data $[p_{lab} \leq 30$ (GeV/c)$]$ with this model, the form

$$\langle n_{ch}(M) \rangle \simeq 1.4\,(M - m_p) \qquad (5.8)$$

was found appropriate (for charge multiplicity) [40]. The rather small coefficient (0.6) of the linear term found in NAL data would appear to contradict nova model parametrizations, even at low M. This objection is not as strong as it might be, however, because the missing mass in NAL experiments is not identical to recoil cluster mass of the model. Likewise, direct interpretation of the logarithmic fit [Eq. (5.6)] in terms of multiperipheral-like models is ambiguous inasmuch as a unique or clean exchange situation is not obtained in the data unless appropriate selections are made on M^2 and t.

On a purely qualitative level, the fact that the same coefficient $b \simeq 1.4$ is found in fits $< n(s) > \sim b \log s$ and $< n(M^2) > \sim b \log M^2$ suggests that what is important for determining asymptotic multiplicity is the energy available Q in whatever system or subsystem one is dealing with [63]. One might hazard the guess that this remark is both model independent, and invariant to changes of parametrizations of n_{ch} vs. Q.

5.2.7. Correlations

The actual distribution in phase space of particles which decay from the single dissociation cluster is intriguing to contemplate. There are, of course, some limitations imposed purely by kinematics. If the fast proton has a given x_p, then the mass of the cluster is $M^2 = s(1-x_p)$. The position in rapidity of the centre of the cluster is

$$|y_c| = \log(\sqrt{s}/M) = \frac{1}{2}\log(1-|x_p|) . \tag{5.9}$$

Thus, if $x_p \simeq 0.95$, then the recoil cluster is centred at $y_c \simeq 3/2$.

The manner in which decay particles distribute themselves about y_c is a question of dynamics. Two (extreme) alternatives are provided by the isotropic cluster assumption, on the one hand, and an elongated bremsstrahlung picture on the other. If the decay is isotropic, then the distribution in y has dispersion of about one unit [cf. Eq. (3.18)]. A bremsstrahlung picture allows a roughly uniform spread of pions extending from $y = -\log M + y_c$ to $y = +\log M - y_c$.

Taking $s = 2000\ \text{GeV}^2$ and $x = 0.95$, again, $\log M = 2.3$. The two extreme pictures would predict very different behaviour at this value of s, for $x_p \simeq 0.95$.

With $x_p \geq 0.99$, and $s = 2000\ \text{GeV}^2$, $M^2 = 20\ \text{GeV}^2$ and $\log M \cong 1.5$. Thus, for $x_p \gtrsim 0.99$, the two view points lead to pretty much the same (largely kinematic) conclusion at ISR energies : a tight cluster moving directly opposite the fast proton.

A sample of available data is shown in Fig. 22. The CHLM group has a series of counters placed in various locations about an ISR intersection region enabling them to record a charged particle signal in coincidence with the fast proton in their small angle spectrometer [50]. The auxiliary charged particle counters record only a rough angle measurement, not momentum, charge, or mass. As shown in Fig. 22, detection of a charged particle in or near the central region (counters at angles $90\pm14^\circ$ and $117.5\pm12.5^\circ$) has a considerable effect on the shape of $f_p(x)$. The inclusive cross-section shows a dip rather than a peak as $x \to 1$.

These data therefore give direct evidence for negative long-range two-particle inclusive correlations. They suggest also that events providing a proton with large x $(|x| > 0.95)$ are events in which the remaining produced hadrons are to be found primarily in the opposite hemisphere. This result, partly of kinematic origin, is evidence for the second type of collective long-range correlation mentioned in the opening paragraphs of this section.

More correlation data obtained by the CHLM group are published [50], and other groups are also undertaking analyses of correlations associated with diffractive events [53],[64]. As data become more detailed, their full understanding will undoubtedly require development of phenomenological models realistic enough to accommodate most gross features of data. These correlation data will then probe further features of models and help in identification of new dynamical aspects heretofore buried.

5.3. Theoretical games

Perhaps the most obvious theoretical guess to make is that the system of mass M recoiling from the fast proton is produced with an s independent cross-section $d\sigma/dM^2 = f(M^2)$. Attempting a power-behaved structure for $f(M^2)$, we write

$$\frac{d\sigma}{dM^2} = (M^2)^{-\lambda} \tag{5.10}$$

The total cross-section associated with this phenomenon, which I label σ_D, is then

$$\sigma_D = \int_{M_0^2}^{M_c^2} M^{-2\lambda}\, dM^2 \tag{5.11}$$

where M_o and M_c are the lower and upper limits some dynamical model would assign for the region of validity of (5.10).

Taking $M_c = \alpha s$, we find the following choices for the s dependence of σ_D :

$$\text{constant} - s^{1-\lambda} \qquad \lambda > 1$$

$$\log (s/\text{constant}) \qquad \lambda = 1$$

$$s^{1-\lambda} \qquad \lambda < 1$$

The possibility $\lambda < 1$ is ruled out asymptotically inasmuch as it implies a violation of the Froissart bound.

Because of the kinematic relationship

$$\frac{d\sigma}{dt\, d(M^2/s)} = \frac{d\sigma}{dt\, dx}$$

(5.10) implies

$$\frac{d\sigma}{dx} \propto s\, M^{-2\lambda} = s^{1-\lambda}\, (1-x)^{-\lambda} \qquad (5.12)$$

Perfect scaling in x of the proton inclusive spectrum then requires $\lambda = 1$. The choice $\lambda = \frac{3}{2}$ made in diffractive excitation models [3], produces a non-scaling form

$$\frac{d\sigma}{dx} \propto \frac{1}{\sqrt{s}}\, \frac{1}{(1-x)^{3/2}} \qquad (5.13)$$

whose falling s dependence seems in disagreement with ISR data, unless resolution problems are playing curious tricks. (For similar reasons, any $\lambda > 1$ is inappropriate.) Thus, the ISR results are most simply interpreted as requiring $\lambda = 1$. This latter choice is, in fact, that suggested by the triple-Pomeron Regge formalism, as I shall now describe. However, it is perhaps useful to remark that the expressions $d\sigma/dM^2 \propto M^{-2}$, $d\sigma/dx \propto (1-x)^{-1}$, and $\sigma_D \propto \log s$ could well turn out to be in good agreement with data, while the simple triple-Regge formalism may have to be abandoned.

5.4. Triple-Regge limit

Specific Regge based predictions can be made for single-particle inclusive spectra near the "phase-space boundary", where $|x| \to 1$. Inasmuch as only protons have substantial cross-section in this region, I limit my remarks to them : $pp \to pX$.

Consider the diagram drawn in Fig. 23a in the limit $s \to \infty$, small fixed $t = (p_c - p_b)^2$, and, for the moment, fixed $M_x^2 = (p_a + p_b - p_c)^2$. An ordinary Regge limit is appropriate here as indicated in 23b. The cross-section is

$$\frac{d^2\sigma}{dt\, dM^2} = \sum_j \frac{1}{s^2} \beta_j(t, M^2) \left(\frac{s}{M^2}\right)^{2\alpha_j(t)}, \tag{5.14}$$

where a sum is made over the different trajectories $\alpha_j(t)$ which can be exchanged.

We now sum and integrate over the various contributions in missing mass M_x. The procedure is equivalent to standing Fig. 23b on its head, and then sewing together the unobserved missing mass channels, as illustrated in Fig. 23c. Summation in the X channel at fixed M_x^2 is accomplished in the usual fashion of unitarity sums. We obtain the Reggeon particle $(\alpha_j a)$ total cross-section

$$\beta_j(t, M^2) \sim \sigma_{aj}(M^2). \tag{5.15}$$

If M^2 is large, with (s/M^2) kept large, a Regge expansion in the X channel gives

$$\beta_j(t, M^2) \propto \gamma_P(t)\, M^{2\alpha_P(0)} + \gamma_R\, M^{2\alpha_R(0)} \tag{5.16}$$

where $\alpha_P(0)$ is the $t = 0$ intercept of the Pomeron trajectory, and $\alpha_R(0)$ the intercept of secondary trajectories (ρ, ω, A_2, f). This is illustrated in Fig. 23d. The end result is Fig. 23e. For the moment I assume that poles are the only singularities we need to consider. Intercept $\alpha_P(0)$ is ≤ 1 ; with $\alpha_P(0) \equiv 1$, we obtain constant total cross-sections for all processes. The complications and requirements of rising σ_{tot} will be addressed subsequently.

Imagine that M_x^2 and (s/M^2) are both large enough to justify our keeping only the leading Pomeron contributions. Assembling pieces, we obtain

$$\frac{d\sigma}{dt\, d(M^2/s)} = \gamma_{P}(t)\, (s/M^2)^{2\alpha_{P}(t) - \alpha_{P}(0)} \tag{5.17}$$

Here $\gamma_P(t)$ is proportional to the function which describes the coupling of three Pomerons (!). Transforming variables to the proton's scaled longitudinal momentum $x = 2p_L/\sqrt{s} \cong 1 - M^2/s$, we rewrite (5.17) as

$$E\frac{d\sigma}{d^3p} = \frac{d^2\sigma}{\pi\, dt\, d(M^2/s)} = \gamma_{P}(t)\,(1-x)^{-1-2\alpha' t} \tag{5.18}$$

Because of the requirement $s \gg M^2 \gg M_{th}^2$, this formula is meant to be valid in a small region bounded by $(1-x) \ll 1$ or $x > x_o$ and $x \ll (1-M_c^2/s)$.

Before proceeding to a comparison of (5.18) with data, let us catalogue other important triple Regge contributions ; taking $\alpha_R(0) = \frac{1}{2}$ in (5.16)

Graph f(x) at t = 0

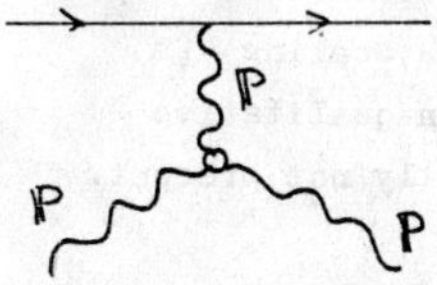

$$\frac{1}{1-x}$$

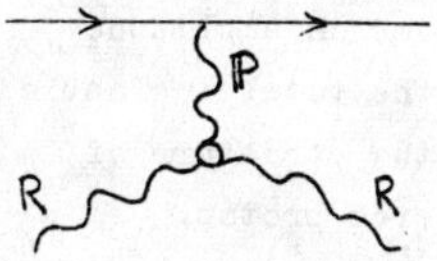

$$(1-x)^{1-2\alpha_R(t)} \longrightarrow \text{constant} \tag{5.19}$$

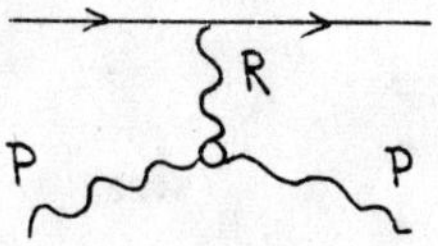

$$\frac{1}{\sqrt{s}}\,(1-x)^{-3/2} \tag{5.20}$$

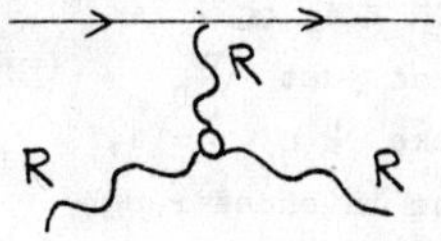

$$\frac{1}{\sqrt{s}}\,(1-x)^{-1/2} \tag{5.21}$$

Only the first two contributions survive for all x as $s \to \infty$.

It will be particularly instructive to compare expressions (5.18) with detailed data from NAL and ISR. Fits will require keeping some of the non-Pomeron contributions (5.19)-(5.21) to be sure. Expression (5.17) is particularly interesting in that it specifies both the s dependence at fixed M^2 and the M^2 dependence at fixed s, in terms of the same ratio $(s/M^2)^{2\alpha(t)-\alpha(0)}$.

At fixed ISR energy, for $x > 0.9$, analyses of experimental M^2 dependence by members of the CERN-Holland-Lancaster-Manchester group [65] give reasonable fits to (5.17) with $\alpha_P(t) \approx 1 + 0.2\,t$. To be sure, experimental errors allow some variation of intercept $\alpha_P(0)$ about 1 and of slope in the vicinity of 0.2. The value $\alpha(0) \approx 1$ for $x > 0.9$ should be contrasted with $\alpha(0) \simeq 0.5$ which is obtained for $0.3 < x < 0.8$; this latter region is evidently dominated by the RRP term.

5.5. Implications of the triple-Pomeron term

Since the formula with $\alpha_P(0) = 1$ is at least consistent with the M^2 dependence of data, let us take it seriously and contemplate some of its implications. Expression (5.18) shows directly the prediction of a scaling (s independent) peak in the proton x spectrum near $x_p = 1$, in qualitative agreement with data. The apparent infinity as $x \to 1$ is really not present. The formula ceases to be valid there.

Other consequences of a finite coupling of the triple-Pomeron amplitude are also interesting. We consider first its contribution to the total inelastic cross-section, and then proceed to a possible description of the structure of inelastic events associated with the fast diffractively-scattered proton.

The cross-section σ_D is given by

$$\sigma_D = \int \frac{d^2\sigma}{dt\,dM^2}\,dt\,dM^2 \qquad (5.22)$$

where the integral extends over the region in t and $M_o^2 < M^2 < M_c^2 \propto s$ in which the PPP is dominant. In principle, $[< n_p >]_D$ and not just σ_D should appear on the left-hand side of (5.22). However, I take $< n_p > = 1$, consistent with our interpretation of expression (5.17) in the intended range of integration. Converted to x, and with the PPP term displayed explicitly, (5.22) becomes

$$\sigma_D = \int dt \int_{x_0}^{x_c} \gamma_P(t)\,(1-x)^{-1-2\alpha' t}\,dx \qquad (5.23)$$

As before, x_0 = constant, and $x_c = 1-(M_c^2/s)$.

If $\gamma_P(0)$ is finite and $\alpha'_P = 0$, we find

$$\sigma_D \propto \log\left(\frac{1-x_0}{1-x_c}\right) = \log\left(\frac{s(1-x_0)}{M_c^2}\right). \qquad (5.24)$$

This form of the PPP coupling therefore provides a log s increase of the single inelastic diffractive cross-section σ_D.

Before proceeding further, I should like to illustrate how this log s increase comes about. In the top half of Fig. 24, I show the region of integration in x from fixed x_0 to x_1, at $s = s_1$; and from x_0 to x_2, at $s = s_2 > s_1$. Here $x_i = 1 - M_c^2/s_i$, with M_c^2 fixed to be independent of s. The increase $\Delta\sigma_D$ as s is changed from s_1 to s_2 arises because at higher s, we integrate up closer to $x = 1$ along the scaling curve $f_p(x)$. However, it is important to keep in mind that this extra cross-section really comes from high values of excitation mass M^2. As shown in the lower part of Fig. 24, the equivalent region of integration in M^2 is from the fixed lower limit M_c^2 to a variable upper limit. The correspondence between the x and M^2 plots is simple. As energy is increased, more and more of the M^2 plot is squeezed up in x above our fixed lower cut-off $x = x_0$.

Unless compensating terms are hidden, the log s increase of σ_D will result in a component of σ_{inel} which rises as log s. Such a growth is not, in itself, theoretically inadmissible [34] nor experimentally contradicted, as discussed in more detail below. However, once the PPP coupling is admitted, then graphs are also allowed in which an arbitrarily large number of Pomerons are exchanged. As shown some years ago [66] with an exchange of n Pomerons, we derive a cross-section

$$\sigma_n \sim (\log s)^{n-1}$$

For $n > 3$, an asymptotic violation of the Froissart bound results.

Because the Froissart bound is by no means saturated in the ISR range, a local growth of σ_D faster than $(\log s)^2$ is presumably acceptable. However, the problem of principle arises sooner or later. Considerable theoretical effort has gone into the general question of Pomeron couplings [67].

The choices $\alpha' \neq 0$ and/or $\gamma_P(0) = 0$ in Eq. (5.23) modify the argument. Taking $\gamma_P(t) = \exp(bt)$ and $\alpha' \neq 0$, we find

$$\sigma_D \propto \log\left[b + 2\alpha' \log(s/M_c^2)\right]. \tag{5.25}$$

This expression gives a slow $\log(\log s)$ increase asymptotically, but, to the extent that $(2\alpha'/b)\log s \ll 1$, we obtain here also a direct logarithmic increase

$$\sigma_D \propto \frac{2\alpha'}{b} \log s \tag{5.26}$$

over some range of s. With $\alpha' \simeq 0.3$ and $b \simeq 6$, the appropriate range is $\log(s/M_c^2) \ll 10$. With $M_c^2 \simeq 10\ \mathrm{GeV}^2$, $s \lesssim 100\ \mathrm{GeV}^2$ is acceptable. The theoretical problem of asymptotic violation of the Froissart bound is still present.

If $\alpha' \neq 0$ and $\gamma_P(0) = 0$, the cross-section σ_D also rises, but now to a constant asymptotic limit. Taking $\gamma_P(t) = t\exp(bt)$, we find easily that

$$\sigma_D \propto -\int_{x_0}^{x_c} \frac{d\log(1-x)}{\left[b - 2\alpha' \log(1-x)\right]^2} \tag{5.27}$$

$$= K_1 - \frac{K_2}{1 + \frac{2\alpha'}{b}\log(s/M_c^2)}$$

Vanishing of the triple-Pomeron coupling $\gamma_P(0) = 0$, thus prevents an unbounded increase of σ_D and avoids eventual violation of the Froissart bound.

Data, however, seem to show no evidence for a zero at $t = 0$, at least at 200 GeV/c (Fig. 20). It seems, therefore, that $\gamma_{PPP}(t=0) \neq 0$, and one must look to some other method to achieve a consistent theoretical picture in which the asymptotic bound $\sigma_D \lesssim (\log s)^2$ is satisfied. Much effort in this direction is expected during the coming months.

The logarithmic rise shown in (5.24) suggests that we may attempt to "blame the rise" of σ_{tot} observed in the ISR range on σ_D. As pointed out by various people [34], the numbers do in fact seem to work out, in the following sense.

A fit by Capella and collaborators [34] to CHLM ISR data at $\sqrt{s} = 23$ and 30 GeV, where the covered x and t ranges are reasonably large, gives

$$E \frac{d\sigma}{d^3p} = \frac{2 \exp(4.65\, t)}{(1-x)} \tag{5.28}$$

under the assumption that $\alpha'_P = 0$. Using the form

$$E \frac{d\sigma}{d^3p} = \frac{A_D \exp(B_D t)}{(1-x)} \tag{5.29}$$

my eyeball fit to these data (Figs. 4 and 17) gives $A_D = 1.6$ to 2.0, $B_D = 4.5$ to 6, consistent with Capella's numbers. Thus, the predicted change $\Delta\sigma_D$ in inelastic diffractive cross-section from $s = s_1$ to $s = s_2$ is easily seen to be

$$\Delta\sigma_D = 2\pi \frac{A_D}{B_D} \log\left(\frac{s_2}{s_1}\right) . \tag{5.30}$$

The factor 2 in (5.30) arises from the fact that either proton may be diffractively scattered. There are two graphs. I ignore diffractive scattering of both protons simultaneously. With $\sqrt{s_1} = 23$, $\sqrt{s_2} = 53$, I find

$$\Delta\sigma_D = (3 \pm 0.7) \text{ mb}. \tag{5.31}$$

Adding to this the experimental increase $\sigma_{el} \simeq 1$ mb, from the elastic channel, we see that diffractive inelastic events may indeed fully account for the observed experimental rise of σ_{inel}. However interesting, this conjecture is of course by no means proved. First, it may be stressed that, purely experimentally, it has not yet been possible to ascertain whether $\Delta\sigma_D > 0$. The experimental numbers [61] $\sigma_D = 5.4\pm1$ and 5.0 ± 1 mb at $\sqrt{s} = 23.2$ and 33.4 show no trend because of the large quoted errors. From $\sqrt{s} = 23.2$ GeV to $\sqrt{s} = 30.4$ GeV, the predicted rise of σ_D based on (5.30) is $\Delta\sigma_D = (1.0\pm0.3)$ mb, which is consistent with the data, in view of the large experimental error. Direct experimental observation of a rise of σ_D would be much more convincing. Second, our previous discussion (Section 2.4) showed that the rise of σ_{inel} is at least partially accounted for by the increase of inclusive spectra in the central region. In this sense, (5.31) is too large.

Obviously, data which should be forthcoming during the coming months from NAL and ISR on the t and x structure of inelastic diffraction will allow a much more careful separation and determination of the various triple-Regge terms, including PPP. How much of $\Delta\sigma_{inel}$ is due to $\Delta\sigma_D$ will then be much better known.

5.6. Multiperipheral model for diffractive events *)

The suggestion that the exchanged Pomeron couples as a simple pole, consistent with data on $E d\sigma/d^3p$, leads to the possibility of constructing a model for the full diffractive event [69]. Indeed, because of the simple exchange assumption, the amplitude $\mathbb{P}\, p \rightarrow$ anything in a multiperipheral-like approach

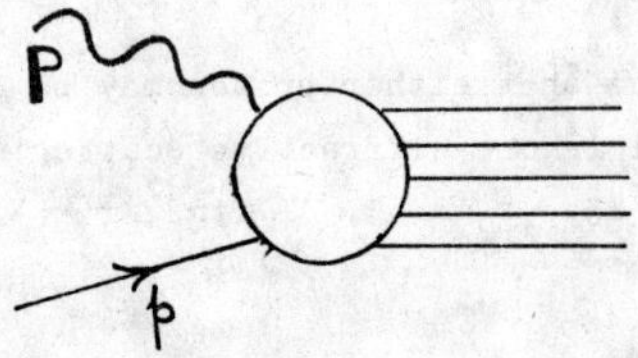

*) The work discussed in this and the next section was done in collaboration with G.C. Fox during the past year, Ref. 4). Similar work was done by E. Rabinovici (private communication from H. Harari).

should bear much similarity to that for $pp \rightarrow$ anything. For example, in $\mathbb{P} p \rightarrow M^2$, the produced hadrons should populate a rough plateau in rapidity space of width $\log M^2$ and centred at

$$|y_c| = \log(\sqrt{s}/M) .$$

Therefore, at $s = 1000\ \mathrm{GeV}^2$ an inelastic diffractive event - in which $x_p = 0.95$ (i.e., $M^2 = 50\ \mathrm{GeV}^2$) should appear in y space as shown below.

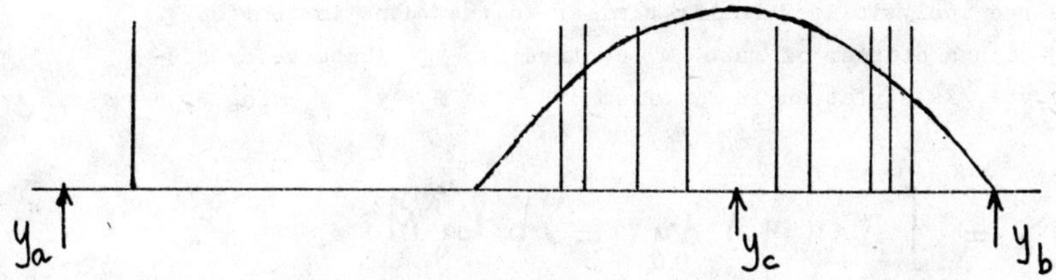

Given are both vertical bars, suggesting a single event, and a smooth curve simulating the distribution of tracks in y which might arise after a sum is taken over many events with $x_p \cong 0.95$. At energy s, it would be of great interest to measure the experimental quantity [68]

$$\left(\frac{d\sigma}{dx_p\, dt_p}\right)^{-1} \left(\frac{d^2\sigma}{dx_p\, dt_p\, dy}\right) \tag{5.32}$$

where, here, x_p and t_p are variables connected with the fast proton, and y is the rapidity of a pion from the system which has recoiled from the fast proton. In the model we are discussing, (5.32) should look similar to $d\sigma/dy$ for pions from $pp \rightarrow \pi X$, measured at the lower energy $s_L = M^2 = s(1-x_p)$.

Lacking experimental data to the contrary, we continue to speculate. The multiplicity $n(M^2)$ in the model should be given by the same functional form which fits the total multiplicity $n(s)$. In particular, if

$$n(s) = a + b \log s \tag{5.33}$$

we expect

$$n(M^2) = a' + b \log M^2 \tag{5.34}$$

Here coefficient b is the same in both (5.33) and (5.34), but the constant term may be different. Such a conjecture is at least consistent with NAL data, Fig. 21, discussed above.

It is now a simple matter to integrate over M^2 (equivalently, over x_p) to obtain the full distribution in rapidity space for pions from inelastic diffractive events. By this I mean that we no longer single out the proton, but ask rather just about the π spectrum which our Pomeron exchange model predicts.

The required analysis is formally similar to that done in Section 3. Here, we begin with a cluster of mass M centred at y_c whose decay distribution $D(y,y_c)$ is a plateau in y from $y_c - \log M < y < y_c + \log M$.

$$\langle n(M^2)\rangle = \int D(y,y_c)\,dy = b \log M^2 .$$

The kinematic relationship between M and y_c is

$$|y_c| = \log(\sqrt{s}/M) \tag{5.35}$$

or

$$M^2 = s \exp(-2|y_c|) . \tag{5.36}$$

According to our PPP formalism, the cluster is produced with a density distribution

$$\rho(y_c) = \frac{1}{\sigma_D^{(1)}}\frac{d\sigma_D^{(1)}}{dy_c} = \frac{M^2}{\sigma_D^{(1)}}\frac{d\sigma_D^{(1)}}{dM^2} = \frac{2K}{\sigma_D^{(1)}} . \tag{5.37}$$

Constant K is fixed by

$$\sigma_D^{(1)} = K\int_{M_0^2}^{M_c^2} M^{-2}\, dM^2 \tag{5.38}$$

Setting $M_c^2 = s$ and $M_0^2 = 1$ we find $\sigma_D^{(1)} = K \log s$. The inclusive y distribution for pions in conjunction with a proton which is inelastically diffractively scattered (into the hemisphere $x_p < 0$) is then

$$\frac{d\sigma^{(1)}}{dy} = \sigma_D^{(1)} \int \rho(y_c)\, dy_c\, D(y, y_c) = 2\sigma_D^{(1)} b \int I(y_c)\, dy_c \tag{5.39}$$

where $I(y_c) = 1$ if $\log\sqrt{s} > y > 2y_c - \log\sqrt{s}$ and is zero otherwise. Thus,

$$\frac{d\sigma^{(1)}}{dy} = K b \left(y + \log\sqrt{s}\right) \tag{5.40}$$

This triangle shaped distribution is shown by the shaded area in the sketch below.

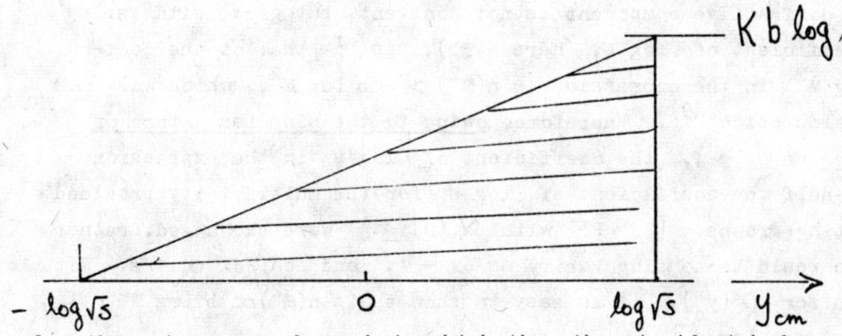

Obviously, there is a second graph in which the other incident hadron is diffractively scattered (into the forward hemisphere). Pions produced by the latter process provide a distribution

$$\frac{d\sigma^{(2)}}{dy} = K b \left(\log\sqrt{s} - y\right) \tag{5.41}$$

which peaks at $y = -\log\sqrt{s}$. The sum of both contributions is a plateau in y, extending (ideally !) from $y = -\log\sqrt{s}$ to $y = +\log\sqrt{s}$ and whose height rises in proportion to $\log s$:

$$\frac{d\sigma_D}{dy} = K b \log s = \frac{1}{2}\sigma_D b \tag{5.42}$$

The net diffractive pion multiplicity $< n_D >$ is given by

$$\langle n_D \rangle = \sigma_D^{-1} \int K b \log s \, dy = \sigma_D^{-1} K b (\log s)^2$$

where

$$\sigma_D = \sigma_D^{(1)} + \sigma_D^{(2)} = 2 \sigma_D^{(1)} .$$

Thus,

$$\langle n_D \rangle = \tfrac{1}{2} b \log s \tag{5.43}$$

Three features of this last expression are notable. First, the multiplicity of the diffractive component is not constant, but grows with s. Second, the coefficient of log s, here $(\frac{1}{2}b)$, is $\frac{1}{2}$ that of the coefficient of $\log M^2$ in the expression $< n(M^2) > = b \log M^2$, which was input to our calculation [69]. Therefore, owing to the singular nature of the PPP coupling as $x \to 1$, the coefficient of log s in the expression $< n_D >$ is one-half the coefficient of log s for the multiplicity provided by non-diffracting graphs. [If P' with $\alpha(0) \simeq \frac{1}{2}$ were exchanged, rather than P, there would be no singularity as $x \to 1$, and $d\sigma/dM^2 \propto 1/s$. With this assumption for $\rho(y_c)$, it is easy to show that $n(M^2) \propto b \log M^2$ results in $n(s) \propto b \log s$.]

Third, and last, the net multiplicity arising from both diffractive and non-diffractive graphs has the form

$$\langle n \rangle = \langle n_{ND} \rangle \frac{\sigma_{ND}}{\sigma} + \langle n_D \rangle \frac{\sigma_D}{\sigma} \tag{5.44}$$

Here $\sigma = \sigma_{ND} + \sigma_D$. If we were to take our model seriously as $s \to \infty$, we would find

$$\langle n \rangle_\infty \propto \langle n_D \rangle \sim \frac{1}{2} b \log s \tag{5.45}$$

whereas, at low energy, where $\sigma_{ND} \gg \sigma_D$

$$\langle n \rangle \sim \langle n_{ND} \rangle \sim b \log s \tag{5.46}$$

This would imply that the measured coefficient b_{exp} of $\log s$ in fits of the form

$$\langle n \rangle \sim b_{exp} \log s$$

should decrease as s increases. This is clearly in disagreement with data, which if anything show an increase of b_{exp} as s increases [13),14)]. However, it was at least implicit from the start that the model should be invoked only in a limited region [34),69)] (ISR range ?) where $\sigma_D \ll \sigma_{ND}$. At energies at which $\sigma_D \approx \sigma_{ND}$, we would have to take seriously into account the prospect of non-negligible multi-Pomeron exchange. Thus I intend all my remarks here about a simple multiperipheral model with factorized Pomeron exchange to be restricted in validity to a region of energy encompassing NAL and ISR. The really asymptotic properties of the model have not been worked out satisfactorily [*)]. (In this sense, and owing directly to the rising nature of σ_{tot} with s, the ISR is not an asymptotic machine !).

Besides s dependence, several other aspects of our treatment are obviously overly naïve. First, the pure plateau assumption for the cluster decay distribution needs to be modified with rounded-off fragmentation edges. Second, use of the PPP term for the full range $s \geq M^2 \geq 1\ \mathrm{GeV}^2$ is outrageous. Corrections to both, however, do not change the essential lessons of the discussion :

(a) that pions from the diffractively produced cluster may well reach the middle [**)] of the rapidity plot, and

(b) their multiplicity grows with s, perhaps logarithmically.

These results suggest that clean kinematic separation of a full "diffractive component" is virtually impossible. Of course, an enriched part of the component may be had by selecting events with one $|x_p| > 0.95$ (say), but this a biased sample which may not be trustworthy for reaching general conclusions.

---

*) For discussions of a multiperipheral "perturbative" approach to hadron production, see Refs. 34).

**) Note that the centre is reached whenever $(y_c - \log M) < 0$; that is, whenever $M > s^{1/4}$. At $s = 1000\ \mathrm{GeV}^2$, and $x_p = 0.9$, $M = 10$ GeV, which is greater than $s^{1/4} = 5.6$.

Our argument showed that diffractive events lead to a log s increase of the height of the central plateau for $pp \to \pi X$: $d\sigma_D/dy = \frac{1}{2}\sigma_D b$. It is worth evaluating numerically how great this increase is over the ISR range. We start from $d\sigma/dy = d\sigma_{ND}/dy + d\sigma_D/dy$. Taking differences, we find

$$\Delta\left(\frac{d\sigma}{dy}\right) = \frac{1}{2}(\Delta\sigma_D)\, b \tag{5.47}$$

because $d\sigma_{ND}/dy$ is, by hypothesis, energy independent. For pions of fixed charge (i.e.) π^-, $b \simeq 0.7$ to 1.0. Thus, using Eq. (5.47),

$$\Delta\left(\frac{d\sigma}{dy}\right)\Big|_{y\simeq 0} \simeq 1.2 \text{ mb}$$

is the expected increase from $\sqrt{s} = 22$ to $\sqrt{s} = 53$ GeV. Data from NAL at 200 GeV/c give [28]

$$\frac{d\sigma}{dy}\Big|_{y\simeq 0} \simeq 26 \text{ mb}$$

Thus, our diffractive contribution provides a rise of ≈4.5% in the plateau height (integrated over p_T^2). This number is to be compared with the increase of (8±7)% quoted by the British-Scandinavian group [16], and so may fully account perhaps for any experimental rise. Insofar as s dependence is concerned, the remaining distribution $d\sigma_{ND}/dy$ would be in "perfect" agreement with the predictions of the short-range order hypothesis.

The British-Scandinavian data [16] showed also that the y distribution was not flat at fixed $\sqrt{s} = 53$ GeV. Whether pions from the diffractive cluster could cause the observed bulge at $y_{cm} \approx 0$ is not obvious, but, in any case, requires a less approximate treatment than can be given purely analytically [4]. [It is very unlikely that the diffractive component could explain the rise of the $\bar{p}$ inclusive rate at $y = 0$.]

5.7. Correlations in the diffractive component

That pions from the diffractive component may reach $y = 0$, and the idea that long-range correlations are intimately associated with diffraction make it imperative that we examine the structure of rapidity correlations among pions produced in a diffractive exchange situation. I turn to this subject now and derive the form of $d^2\sigma_D/dy_1 dy_2$, where y_1 and y_2 denote rapidities of pions which have come from a system which recoils from a fast inelastic proton scattered diffractively.

As we discussed above, if x_p is fixed, the recoil system of mass $M = [s(1-x_p)]^{\frac{1}{2}}$ behaves in our model just as would a final state of $\sqrt{s} = M$ in an over-all pp collision. Thus, if we fix x_p, and we look only at pions from the recoil system, we expect to see only short-range rapidity correlations, whose correlation function $C_{x_p}(y_1,y_2)$ has exactly the same form as that in (3.29) [so long, of course, as both y_i are in the allowed kinematic range $y_c - \log M < y < y_c + \log M$; $y_c = \log(\sqrt{s}/M)$]. Explicitly,

$$\frac{1}{\sigma_x} \frac{d^2\sigma_x}{dy_1 dy_2} = \frac{1}{\sigma_x^2} \frac{d\sigma_x}{dy} \frac{d\sigma_x}{dy_2} + C_x(y_1,y_2), \quad (5.48)$$

where $C_x(y_1,y_2)$ is given by Eq. (3.28). Here, cross-section σ_x is just the inclusive cross-section given by the triple-Pomeron formula

$$\sigma_x = f_p(x,t)$$

$1/\sigma_x(d\sigma_x/dy_1)$ = plateau of height b extending over the interval $y_c - \log M < y_1 < y_c + \log M$. Expression (5.48) applies for fixed $(M^2(x),t)$ of the fast proton.

We ask next for the full form of correlations among pions in the diffractive component, after we integrate over x_p. As will be derived below, this function has both short-range and long-range pieces.

Formally, it is clear that

$$\frac{1}{\sigma_D^{(1)}} \frac{d^2\sigma_D^{(1)}}{dy_1 dy_2} = \int dy_c\, \rho(y_c)\, D_2(y_1,y_2;y_c) \quad (5.49)$$

where $\rho(y_c)$ is given by (5.37) and $D_2(y_1,y_2,y_c)$ is provided by (5.48). Inserting the first term of (5.48) into (5.49), we obtain the contribution

$$\frac{1}{\sigma_D^{(1)}} 2Kb^2 \int dy_c\, I_2(y_1,y_2;y_c)$$

where $I_2 = 1$ if both $2y_c < y_1 + \log\sqrt{s}$ and $2y_c < y_2 + \log\sqrt{s}$, but $I_2 = 0$ otherwise. The result is

$$\frac{Kb^2}{\sigma_D^{(1)}}\left(\log\sqrt{s} + y_L\right) \tag{5.50}$$

where y_L is the lesser of (y_1, y_2). Similarly, the integral over $C_x(y_1, y_2)$, [when (5.48) is inserted in (5.49)] gives

$$\frac{K}{\sigma_D^{(1)}}\, C(y_1, y_2)\left[\log\sqrt{s} + y_L\right] . \tag{5.51}$$

Recall that $\sigma_D^{(1)} = K \log s$ defines K. Thus,

$$\frac{1}{\sigma_D^{(1)}}\;\frac{d^2\sigma_D^{(1)}}{dy_1\,dy_2} = \frac{1}{2}\left[1 + \frac{y_L}{\log\sqrt{s}}\right]\left[b^2 + C(y_1, y_2)\right] \tag{5.52}$$

The second diffractive graph, whose fast proton heads towards $x > 0$, gives

$$\frac{1}{\sigma_D^{(2)}}\;\frac{d^2\sigma^{(2)}}{dy_1\,dy_2} = \frac{1}{2}\left[1 - \frac{y_G}{\log\sqrt{s}}\right]\left[b^2 + C(y_1, y_2)\right] \tag{5.53}$$

where y_G is the greater of (y_1, y_2).

The total two-pion inclusive spectrum from the diffractive component is therefore

$$\frac{1}{\sigma_D}\;\frac{d^2\sigma_D}{dy_1\,dy_2} = \frac{1}{2}\left[1 - \frac{|y_1 - y_2|}{\log s}\right]\left(b^2 + C(y_1, y_2)\right) \tag{5.54}$$

Setting $C = 0$, just for the moment, I sketch below $d^2\sigma_D/dy_1dy_2$ versus y_1 for $y_2 = 0$

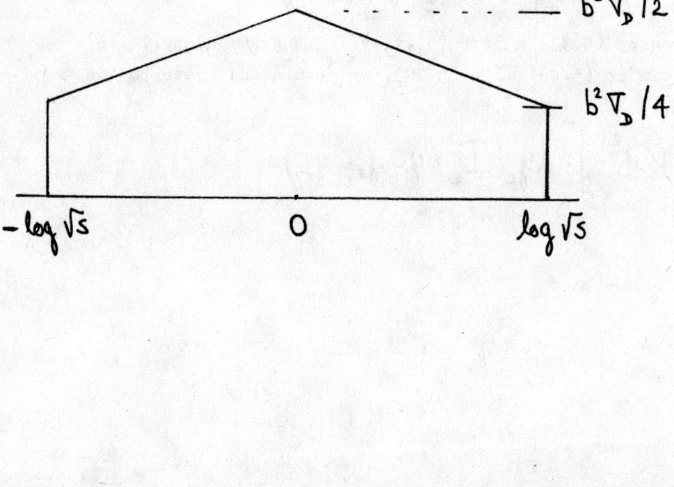

Defining a diffractive component correlation function $C_D(y_1,y_2)$ by

$$C_D(y_1,y_2) = \frac{1}{\sigma_D}\frac{d^2\sigma_D}{dy_1 dy_2} - \frac{1}{\sigma_D^2}\frac{d\sigma_D}{dy_1}\frac{d\sigma_D}{dy_2}$$

I find

$$C_D(y_1,y_2) = \frac{1}{2}\left[\frac{1}{2} - \frac{|y_1-y_2|}{\log s}\right] b^2 + \frac{1}{2}\left[1 - \frac{|y_1-y_2|}{\log s}\right] C_{ND}(y_1,y_2) \qquad (5.55)$$

The term C_{ND} appearing in (5.55) is exactly the function C of Eq. (3.29). I have now added the subscript ND (for non-diffractive) to distinguish it from C_D. The constant b appearing in (5.55) is the coefficient of $\log s$ in the expansion

$$\langle n_{ND}\rangle = a + b \log s$$

Equivalently, it is the full height of the central plateau in the single-particle inclusive spectrum

$$b = \frac{1}{\sigma_{ND}}\frac{d\sigma_{ND}}{dy}$$

It will be noted that C_D has both a long-range part, given by the first term in (5.55), and a short-range piece. At $y_1 = y_2 = 0$, the long-range part is $1/4\ b^2$ (≈ 0.5 to 1 for charged-charged correlations). The short-range piece at $y_1 = y_2 = 0$ is $0.5\ C_{ND}(0,0)$. To evaluate this latter term, we note that

$$C_{ND}(0,0) \cong b^2 R_{ND}(0,0)$$

where

$$R(y_1,y_2) = C(y_1,y_2) \Big/ \left[\frac{1}{\sigma^2}\frac{d\sigma}{dy_1}\frac{d\sigma}{dy_2}\right]$$

Experimentally, $R(0,0) \simeq 0.6$ (which I interpret as largely due to non-diffractive effects). Thus, the long-range and short-range parts of $C_D(0,0)$ are roughly equal in numerical value.

It is interesting to evaluate the integral f_2^D defined as

$$f_2^D = \iint C_D(y_1, y_2)\, dy_1\, dy_2 \tag{5.56}$$

Simple arithmetic gives

$$f_2^D = \frac{b^2}{12}(\log s)^2 = \frac{1}{3}\langle n_D\rangle^2 = \frac{1}{12}\langle n_{ND}\rangle^2 \tag{5.57}$$

Note that the correlation moment f_2^D grows as $(\log s)^2$, not as $(\log s)$. Therefore, our diffractive component shows long-range correlations both in the intuitive sense, discussed in the opening paragraphs of Section 5.1, and in the formal sense [70] as well.

It may seem somewhat surprising that we obtain long-range correlation from a pure multiperipheral model. After all the cluster recoiling from the fast proton has pure short-range correlation structure, internally, and the exchanged Pomeron is a pure factorized pole in our model. Thus, at a superficial glance, one might expect the pure short-range answer $f_2^D \sim O(\log s)$. However, on closer inspection, one sees that the $f_2^D \sim (\log s)^2$ result is related directly to the $f_p(x) \sim (1-x)^{-1}$ singularity structure of the exchange. A typical non-diffractive multiperipheral situation would arise if we were to replace the exchanged Pomeron by an f^o or ρ trajectory. In this case, the proton x distribution is non-singular as $x \to 1$, and we calculate $f_2^{ND} \sim O(\log s)$. The long-range effect does not arise simply from our having integrated over a spectrum of masses, but over a quite particular spectrum [which has behaviour $d\sigma/dM^2 \sim (M^2)^{-1}$].

It is intriguing that the multiperipheral approach can reproduce both long-range and short-range correlation effects in data. Only detailed analyses will verify whether the approach is successful quantitatively ; present steps in this direction are promising [4].

The full correlation function observed experimentally receives contributions from both diffractive and non-diffractive graphs. Ignoring interference effects between these two types of processes *), we may write formally

*) Clearly this would need justification, in a more complete treatment, inasmuch as diffractive and non-diffractive graphs populate overlapping regions of phase space.

$$C(y_1, y_2) = \alpha_D C_D + \alpha_{ND} C_{ND}$$

$$+ \alpha_D \alpha_{ND} \left[\frac{1}{\sigma_{ND}} \frac{d\sigma_{ND}}{dy_1} - \frac{1}{\sigma_D} \frac{d\sigma_D}{dy_1} \right] \times (1 \rightarrow 2) \tag{5.58}$$

Here $\alpha_D = \sigma_D / \sigma_{inel}$; $\alpha_{ND} = \sigma_{ND} / \sigma_{inel}$; $\alpha_D + \alpha_{ND} = 1$. The last term in (5.58) produces an apparent long-range correlation effect in the central region $(y_1 \simeq y_2 \simeq 0)$, as long as

$$\left[\frac{1}{\sigma_{ND}} \frac{d\sigma_{ND}}{dy_i} - \frac{1}{\sigma_D} \frac{d\sigma_D}{dy_i} \right] \neq 0$$

In a sense, this apparent long-range effect is fictitious, since it does not come directly from dynamics, but rather from the existence of two contributing different dynamical mechanisms [71]. That the presence of two mechanisms (each of which may be purely of short-range character) can result in a term in $C(y_1, y_2)$ which has apparent long-range structure, points up the danger of expressing data always in terms of $C(y_1, y_2)$ [or $R(y_1, y_2)$]. Interpretations based on $C(y_1, y_2)$ may be less reliable than those based directly on $\sigma^{-1} \, d^2\sigma / dy_1 dy_2$.

The maximum size of the "fictitious" long-range term in the central region is controlled by

$$\alpha_D \alpha_{ND} \left| \frac{1}{\sigma_{ND}} \frac{d\sigma_{ND}}{dy} - \frac{1}{\sigma_D} \frac{d\sigma_D}{dy} \right|_{y \simeq 0} \tag{5.59}$$

Models in which $d\sigma_D / dy |_{y \simeq 0} \simeq 0$ produce very large apparent long-range effects. For example, using their concept of diffraction, Pirilä and Pokorski [72] have estimated that the fictitious long-range term accounts for one-half of the experimentally observed correlation at $y_1 = y_2 = 0$. This estimate is directly tied to their assumption that their entire diffractive component of ≈ 7 mb gives no pions at $y = 0$ (i.e., $d\sigma_D / dy = 0$).

In the multiperipheral-like model I have been discussing here,

$$\frac{1}{\sigma_D}\frac{d\sigma_D}{dy} \simeq \frac{1}{2}\frac{1}{\sigma_{ND}}\frac{d\sigma_{ND}}{dy} \tag{5.60}$$

Thus, the difference in (5.59) is only 1/4 the value it has in models such as Pokorski and Pirilä's. Combining (5.56), (5.58) and (5.55) with subsequent discussion, I find that our multiperipheral model gives

$$C(0,0) = \left(1 - \frac{\alpha_D}{2}\right) C_{ND}(0,0) + \frac{\alpha_D b^2}{4}(1+\alpha_{ND}) \tag{5.61}$$

Upon forming

$$R(0,0) \simeq C(0,0)/b^2$$

I obtain

$$R(0,0) = \left(1 - \frac{\alpha_D}{2}\right) R_{ND}(0,0) + \frac{\alpha_D}{4}(1+\alpha_{ND}) \tag{5.62}$$

The diffraction associated long-range correction term $\alpha_D(1+\alpha_{ND})/4 \simeq 0.09$ thus provides a very small net long-range effect at $y_1 \simeq y_2 \simeq 0$. Most of the observed charged-charged correlation $[R(0,0) \simeq 0.6]$ is therefore interpreted as a pure short-range effect in the multiperipheral approach I have developed here [73].

The same term $0.25\alpha_D(1+\alpha_{ND})$ which appears in (5.62) will be present in all two-pion correlations, independent of charges. Thus, this term, whose magnitude (≈ 0.1) is insignificant for charged-charged correlations, is relatively more important, for example, for $(\pi^-\pi^-)$ correlations. This points up again that experimental study of $(\pi^-\pi^-)$ and $(\pi^+\pi^-)$ correlations in y, at fixed p_T, will greatly improve our present picture, which unfortunately must be based solely on data for charged-charged correlations in $\log\tan(\theta/2)$, averaged over p_T.

REFERENCES

1) G. Giacomelli - Rapporteur's talk, Chicago-NAL International Conference (1972) ;

D.R.O. Morrison - Review given at the Royal Society Discussion Meeting on Proton-Proton Scattering (March, 1973) ;

A.N. Diddens - Lectures given at IV Seminar on Theoretical Physics, GIFT, Barcelona (1973) ;

M. Jacob - Rapporteur's talk, Chicago-NAL International Conference (1972) ;

J.C. Sens - Invited Review, Conference on Recent Advances in Particle Physics, New York Academy of Sciences (1973) ;

D. Horn - Physics Reports 4C, 1 (1972) ;

W.R. Frazer et al. - Revs.Modern.Phys. 44, 284 (1972) ;

K. Gottfried - CERN Academic Training Lectures, CERN Preprint TH. 1615 (1973) ;

M. Jacob - 1973 CERN School of Physics, Ebeltoft, and 1973 Louvain Summer Institute Lectures, CERN Preprint TH. 1683 (1973) ;

H.M. Chan - Lectures given at IV Seminar on Theoretical Physics, GIFT Barcelona (1973) ;

J.D. Jackson - Summary talk, International Conference on High Energy Collisions, Vanderbilt (1973) ;

A. Mueller - Parallel Session Summary, Chicago-NAL International Conference (1972) ;

L. Foà - Invited review, Italian Physical Society Meeting, Cagliari (1972).

2) D. Amati, S. Fubini and A. Stanghellini - Nuovo Cimento 26, 896 (1962) ;

L. Bertocchi, S. Fubini and M. Tonin - Nuovo Cimento 25, 626 (1962).

3) J. Benecke, T. Chou, C.N. Yang and E. Yen - Phys.Rev. 188, 2159 (1969) ;

R. Hwa - Phys.Rev.Letters 26, 1143 (1971) ;

M. Jacob and R. Slansky - Phys.Rev. D5, 1847 (1972) ;

R. Adair - Phys.Rev. 172, 1370 (1968) ; D5, 1105 (1972).

4) E.L. Berger and G.C. Fox - CERN Preprint TH. 1700 (1973), to be published in Physics Letters.

5) CERN-Holland-Lancaster-Manchester ISR Collaboration, M.G. Albrow et al. - Nuclear Phys. B51, 388 (1972) ; Nuclear Phys. B54, 6 (1973) ; Phys. Letters 42B, 279 (1972).

6) L. Van Hove - Physics Reports 1C, 347 (1971).

7) U. Amaldi - 1973 Erice School Lectures ;

M. Jacob - CERN Preprint TH. 1683 (1973).

8) CERN-Rome ISR Collaboration, U. Amaldi et al. - Phys.Letters 44B, 112 (1973) ;

Pisa-Stony Brook ISR Collaboration, S.R. Amendolia et al. - Phys.Letters 44B, 119 (1973).

9) Aachen-CERN-Harvard-Genova-Torino ISR Collaboration, M. Holder et al. - Phys.Letters 35B, 355 (1971) ; 36B, 400 (1971) ;

G. Barbiellini et al. - Phys.Letters 39B, 663 (1972) ;

CERN-Rome ISR Collaboration, U. Amaldi et al. - Phys.Letters 36B, 504 (1971).

10) CERN-Bologna ISR Collaboration, A. Bertin et al. - Phys.Letters 42B, 493 (1972) ;

CERN-Holland-Lancaster-Manchester ISR Collaboration, M.G. Albrow et al. - Phys.Letters 42B, 279 (1972) ;

Saclay-Strasbourg ISR Collaboration - Phys.Letters 41B, 547 (1972).

11) S.M. Berman, J.D. Bjorken and J.B. Kogut - Phys.Rev. D4, 3388 (1971) ;

R. Blankenbecler, S.J. Brodsky and J.F. Gunion - Phys.Letters 42B, 461 (1972) ;

P.V. Landshoff and J.C. Polkinghorne - Phys.Letters 45B, 361 (1973) ;

D. Amati, L. Caneschi and M. Testa - Phys.Letters 43B, 186 (1973) ;

E.L. Berger and D. Branson - Phys.Letters 45B, 57 (1973).

12) CERN-Columbia-Rockefeller ISR Collaboration, F.W. Büsser et al. - CERN Report (1973), submitted to Phys.Letters.

Other large p_T ISR data :

Saclay-Strasbourg ISR Collaboration, M. Banner et al. - Phys.Letters 44B, 537 (1973) ;

British-Scandinavian ISR Collaboration, B. Alper et al. - Phys.Letters 44B, 521 (1973).

13) CERN-Bologna ISR Collaboration, M. Antinucci et al. - Nuovo Cimento Letters 6, 121 (1973).

14) E.L. Berger - Phys.Rev.Letters 29, 887 (1972).

15) France-Soviet Union and CERN-Soviet Union Serpukhov Collaborations, V.V. Ammosov et al. - Nuclear Phys. B58, 77 (1973).

16) British-Scandinavian ISR Collaboration - CERN Reports (1973), submitted to Phys.Letters.

17) CERN-Hamburg-Vienna ISR Collaboration, H. Dibon et al. - Phys.Letters 44B, 313 (1973).

18) Pisa-Stony Brook ISR Collaboration, S.R. Amendolia - CERN Report (1973), submitted to Phys.Letters.

19) Argonne-NAL-Iowa-Maryland-Michigan State 205 GeV/c Collaboration, as reported by J. Whitmore - International Conference on High Energy Particle Collisions, Vanderbilt (1973).

20) Multiplicity data in pp collisions :

13-28.5 GeV/c : E.L. Berger, B.Y. Oh, Z.-M. Ma and G.A. Smith, to be published ;

50 and 69 GeV/c : France-Soviet Union Mirabelle Collaboration - paper submitted to XVI International Conference on High Energy Physics Chicago-NAL (1972) ;

100 GeV/c : J.W. Chapman et al. - Phys.Rev.Letters 29, 1686 (1972) ;
200 GeV/c : G. Charlton et al. - Phys.Rev.Letters 29, 515 (1972) ;
300 GeV/c : F.T. Dao et al. - Phys.Rev.Letters 29, 1627 (1972).

21) Recent discussions include :
A. Wroblewski - Lectures given at XIII Cracow School, Zakopane (1973) ;

Z. Koba - 1973 CERN School of Physics, Ebeltoft ;
W.R. Frazer - Conference on Recent Advances in Particle Physics, New York Academy of Sciences (1973).

22) K. Wilson - Acta Phys.Austriaca 17, 37 (1963) ;
A. Krzywicki - Nuclear Phys. B58, 633 (1973).

23) A.H. Mueller - Phys.Rev. D2, 2963 (1972).

24) R.P. Feynman - Phys.Rev.Letters 23, 1415 (1969).

25) CERN-Holland-Lancaster-Manchester ISR Collaboration, M.G. Albrow et al. Nuclear Phys. B56, 333 (1973).

26) For additional references to the scaling behaviour of single-particle spectra, see the articles by Diddens, Foà, Giacomelli and Jacob, Ref. 1), as well as the proceedings of the IV International Conference on High Energy Collisions, Oxford (1972).

27) J.V. Allaby et al. - Contribution to IVth International Conference on High Energy Collisions, Oxford (1972).

28) Argonne-NAL 205 GeV/c Collaboration, Y. Cho et al. - NAL-Pub-73/26-EXP ; ANL/HEP 7316 (1973), submitted to Phys.Rev.Letters.

29) W.H. Sims et al. - Nuclear Phys. B41, 317 (1972).

30) British-Scandinavian ISR Collaboration - Ref. 16).

31) Saclay-Strasbourg ISR Collaboration, M. Banner et al. - Phys.Letters 41B, 547 (1972).

32) Pisa-Stony Brook ISR Collaboration, G. Bellettini et al. - Phys.Letters 45B, 69 (1973).

33) J.D. Bjorken - in "Particles and Fields" (1971), Ed. by A.C. Mellissinos and P. Slattery (AIP, New York) ;
R. Arnold - Argonne Lectures ANL/HEP 7317 (1973).

34) A. Capella and M.-S. Chen - Stanford Report SLAC-PUB-1252 ;
W.R. Frazer, D.R. Snider and C.-I. Tan - San Diego Preprint UCSD-10P10-127 (1973).
G.F. Chew - Berkeley Preprints LBL-1556 and LBL-1701 (1973).
D. Amati, L. Caneschi and M. Ciafaloni - CERN Preprint TH. 1676 (1973) ;
M. Bishari and J. Koplic - Berkeley Preprints (1973), and Phys.Letters 44B, 175 (1973).

35) C. De Tar - Phys.Rev. D3, 128 (1971) ;

S. Fubini - in "Strong Interactions and High Energy Physics", Ed. by R.G. Moorhouse (Oliver and Boyd, Edinburgh, 1964) ;

W. Frazer et al. - Ref. 1).

36) E.L. Berger, G.C. Fox and A. Krzywicki - Phys.Letters 43B, 132 (1973) ;

E.L. Berger and A. Krzywicki - Phys.Letters 36B, 380 (1971).

37) F.T. Dao et al. - Phys.Letters 45B, 73 (1973).

Data on δ should also be forthcoming soon from the Pisa-Stony Brook ISR Collaboration.

38) C. Quigg - in Proceedings of the Vanderbilt Conference on High Energy Collisions (1973).

39) C.J. Hamer and R. Peierls - Brookhaven Reports (1973).

40) E.L. Berger and M. Jacob - Phys.Rev. D6, 1930 (1972) ;

E.L. Berger, M. Jacob and R. Slansky - Phys.Rev. D6, 2580 (1972).

41) Vanderbilt-Brookhaven Collaboration, W. Burdett et al. - Nuclear Phys. B48, 13 (1972) ;

Vanderbilt-Yale Collaboration, J. Hanlon et al.- Report Yale 3075-45 (July 1973).

42) D. Sivers and G.H. Thomas - Phys.Rev. D6, 1961 (1972).

43) C. Quigg, J.-M. Wang and C.N. Yang - Phys.Rev.Letters 28, 1290 (1972).

44) For a summary of data and possible theoretical interpretations, see :

E.L. Berger, D. Horn and G.H. Thomas - Phys.Rev. D7, 1412 (1973) ;

D. Horn and A. Schwimmer, Nuclear Phys. B52, 627 (1973).

45) For other theoretical discussions of correlations covering some material which I do not treat here, consult :

M. Jacob - Ref. 1) ;

C. Quigg - Ref. 38) ; and

M. Le Bellac, R. Peccei and L. Caneschi - in "Multiparticle Phenomena and Inclusive Reactions", Seventh Rencontre de Moriond (1972), Ed. by J. Tran Thanh Van.

46) The long history of the one and two fireball approach is hard to trace completely. Some references, prior to those in Ref. 3), include :

M. Good and W.D. Walker - Phys.Rev. 120, 1857 (1960) ;

G. Cocconi - Phys.Rev. 111, 1699 (1958) ;

P. Ciok et al. - Nuovo Cimento 10, 741 (1958).

See also :

E.L. Feinberg - Physics Reports 5C, 238 (1972).

47) M. Le Bellac and J.L. Meunier - Phys.Letters 43B, 127 (1973).

48) E.L. Berger - Argonne Report ANL/HEP 7239 (1972), submitted to 1972 Chicago-NAL Conference.

49) Michigan-Rochester-NAL Collaboration data at 103 GeV/c as presented in Fig. 5 of :

J. Whitmore - International Conference on High Energy Collisions, Vanderbilt (1973).

50) CERN-Holland-Lancaster-Manchester ISR Collaboration, M. Albrow et al. - Phys.Letters 44B, 207 and 518 (1973).

51) Pisa-Stony Brook ISR Collaboration, G. Bellettini - XVI International Conference on High Energy Physics, Chicago-Batavia, Vol. 1, 279 (1972).

52) This point has also been emphasized by :

M. Jacob - Ref. 1).

53) Preliminary data on this question have been obtained by :

Aachen-CERN-Harvard-Genova-Torino ISR Collaboration, G. Goldhaber - Contribution to the Chicago-NAL International Conference (1972).

54) F. Sannes et al. - Phys.Rev.Letters 30, 766 (1973).

55) Columbia-Stony Brook Collaboration, S. Childress et al. - Contribution to Vanderbilt Conference (1973) ;

S. Childress et al. - Report submitted to Phys.Rev.Letters (1973) ; and

R. Schamberger et al.

56) University of Rochester - University of Michigan 100 GeV Bubble Chamber Collaboration, C.M. Bromberg et al. - Report Nr. UMBC 72-14 and UR 416 (January 1973).

57) Argonne-NAL 205 GeV/c Bubble Chamber Collaboration, S.J. Barish et al. - Argonne Report ANL/HEP 7338 (1973) ; M. Derrick et al. - Argonne Report ANL/HEP 7332 (1973).

58) NAL-UCLA 303 GeV/c Bubble Chamber Collaboration, F.T. Dao et al. - Phys.Letters 45B, 399 and 402 (1973).

59) NAL-LBL 205 GeV/c Bubble Chamber Collaboration, F.C. Winkelman et al. - Report Nr. LBL-2113 (1973).

60) This observation has also been made by K. Kajantie, private communication.

61) J.C. Sens - Ref. 1).

62) A similar conclusion is reached upon comparing data on $\langle n(M_x^2) \rangle$ from $\pi^- p \to pX$ at 205 GeV/c, with $\langle n(s) \rangle$ from $\pi^- p \to X$. NAL-LBL Collaboration, Ref. 59).

63) As remarked by K. Gottfried, the relatively low multiplicity observed in collisions of hadrons with nuclei is also consistent with this suggestion.

64) Pisa-Stony Brook ISR Collaboration.

65) CERN-Holland-Lancaster-Manchester ISR Collaboration, M.G. Albrow et al. - CERN Report (July 1973), submitted to Nuclear Phys.

66) J. Finkelstein and K. Kajantie - Phys.Letters 26B, 305 (1968).

67) See, for example, the papers listed in Ref. 34) and other papers cited in those articles. A recent paper is that of :

J.L. Cardy and A.R. White - CERN Preprint TH. 1726 (1973).

68) A similar request has been made by C. Quigg, private communication.

69) The factor $(\frac{1}{2})$ has also been derived by :

D. Snider and W. Frazer - Phys.Letters 45B, 136 (1973).

70) A. Mueller - Phys.Rev. D4, 150 (1971).

71) The idea of two distinct components has received much attention especially in attempts to fit multiplicity distributions. See, for example :

E.L. Berger - Ref. 14) ;

C. Quigg and J.D. Jackson - NAL Report NAL-THY-93 (1972) ;

K. Fialkowski and H. Miettinen - Phys.Letters 43B, 61 (1973) ;

W. Frazer, R. Peccei, S. Pinsky and C.-I. Tan - San Diego Preprint (1972) ;

H. Harari and E. Rabinovici - Phys.Letters 43B, 49 (1973) ;

L. Van Hove - Phys.Letters 43B, 65 (1973) ; and

J. Lach and E. Malamud - Phys.Letters 44B, 474 (1973).

72) P. Pirilä and S. Pokorski - Phys.Letters 43B, 502 (1973).

73) The use of clusters in one form or another to interpret or to imitate positive inclusive correlations has recently become very popular. Clusters are discussed phenomenologically in my Refs. 4), 36) and 39). Recent articles which came to my attention after I gave these lectures at Erice are by :

F. Hayot and A. Morel - Saclay Preprint D.Ph.T/73/58 (1973).;

W. Schmidt-Parzefall - CERN Report (1973), submitted to Phys.Letters ;

P. Pirilä and S. Pokorski - CERN Preprint TH.1686 (1973) ;

A. Białas, K. Fialkowski and K. Zalewski - Phys.Letters B45, 337 (1973) ;

J. Kripfganz, G. Ranft and J. Ranft - Nuclear Phys. B56, 205 (1973).

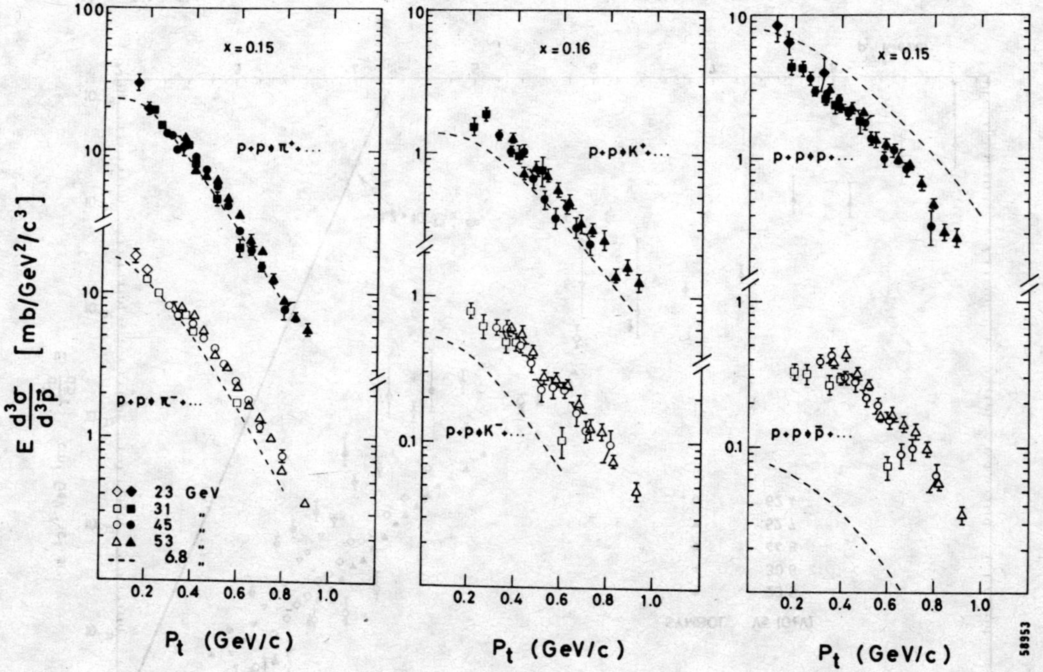

Fig. 1a The invariant inclusive cross-sections $Ed\sigma/d^3p$ are plotted versus p_T at fixed x for the production of $\pi^{\pm}$, $K^{\pm}$, p and $\bar{p}$. ISR data from the Argonne-CERN-Bologna collaboration, Ref. 10). The dashed lines represent interpolations through data at p_{lab} = 24 GeV/c and $x \approx 0.16$ [Ref. 27)].

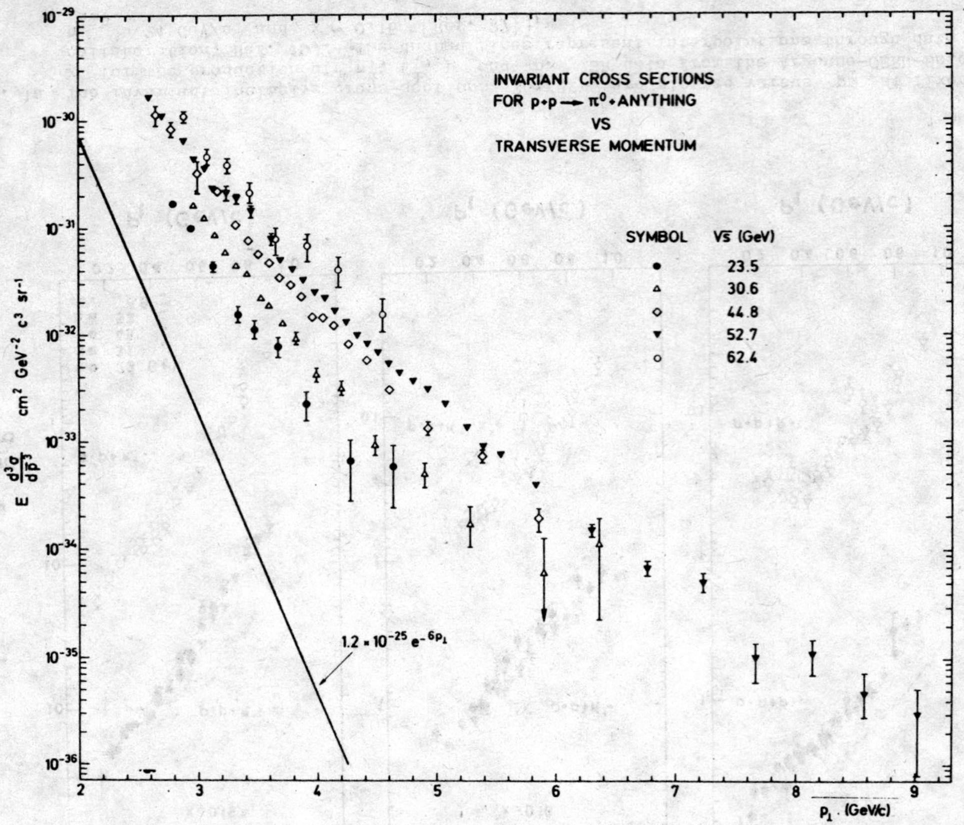

Fig. 1b The transverse momentum dependence at $x \approx 0$ of the invariant inclusive cross-section $Ed\sigma/d^3p$ at five centre-of-mass energies. Data of the CERN-Columbia-Rockefeller collaboration, Ref. 12). Shown for comparison is the extrapolation to large p_T of a fit to data in the interval $p_T < 1$ GeV/c.

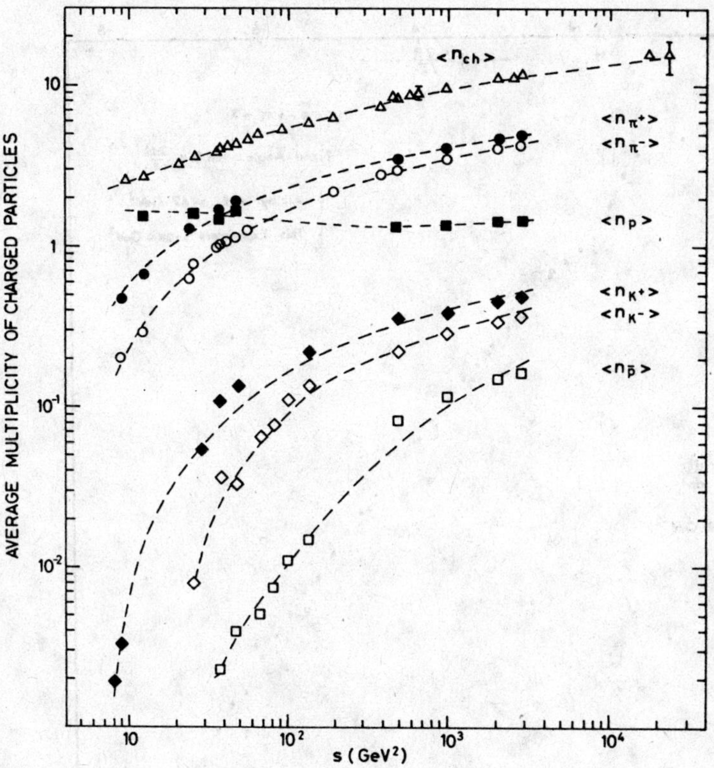

Fig. 2 The average multiplicities of π^-, π^+, K^-, K^+, $\bar{p}$ and p observed in pp interactions are plotted as a function of s. From Ref. 13).

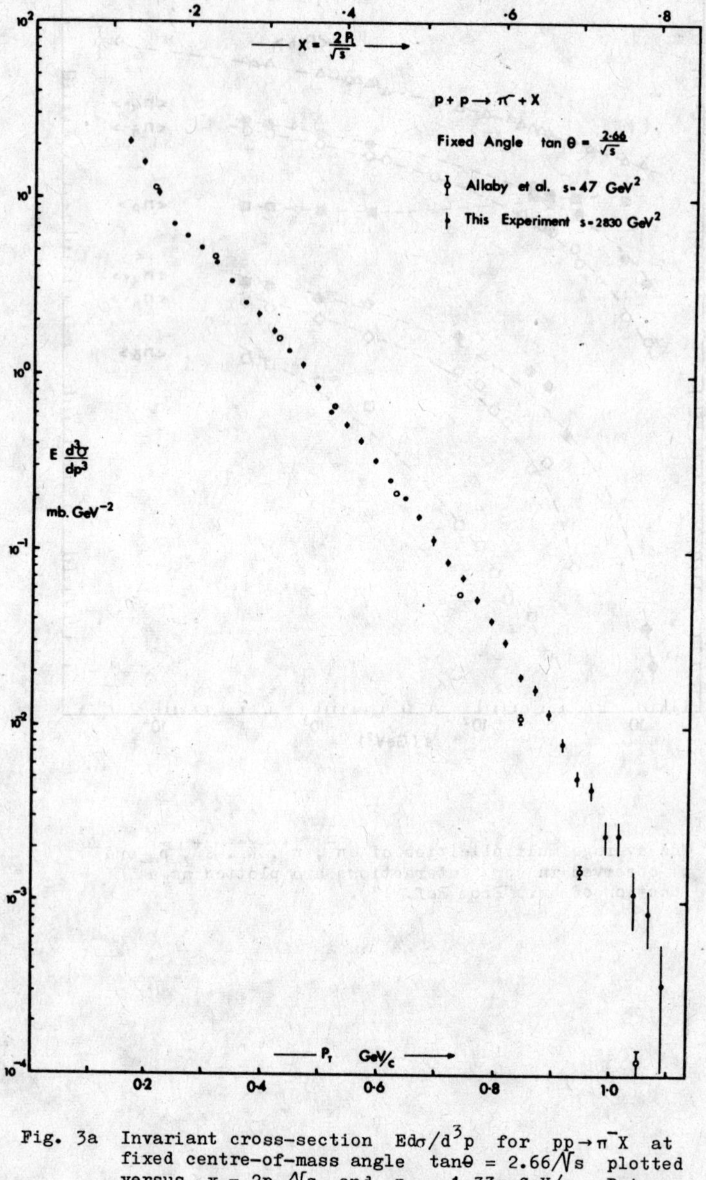

Fig. 3a Invariant cross-section $Ed\sigma/d^3p$ for $pp \rightarrow \pi^- X$ at fixed centre-of-mass angle $\tan\Theta = 2.66/\sqrt{s}$ plotted versus $x = 2p_L/\sqrt{s}$ and $p_T = 1.33x$ GeV/c. Data from the CHLM experiment at $s = 2830$ GeV^2 [solid points, Ref. 25)] are compared with data at $s = 47$ GeV^2 [open circles, Ref. 27)].

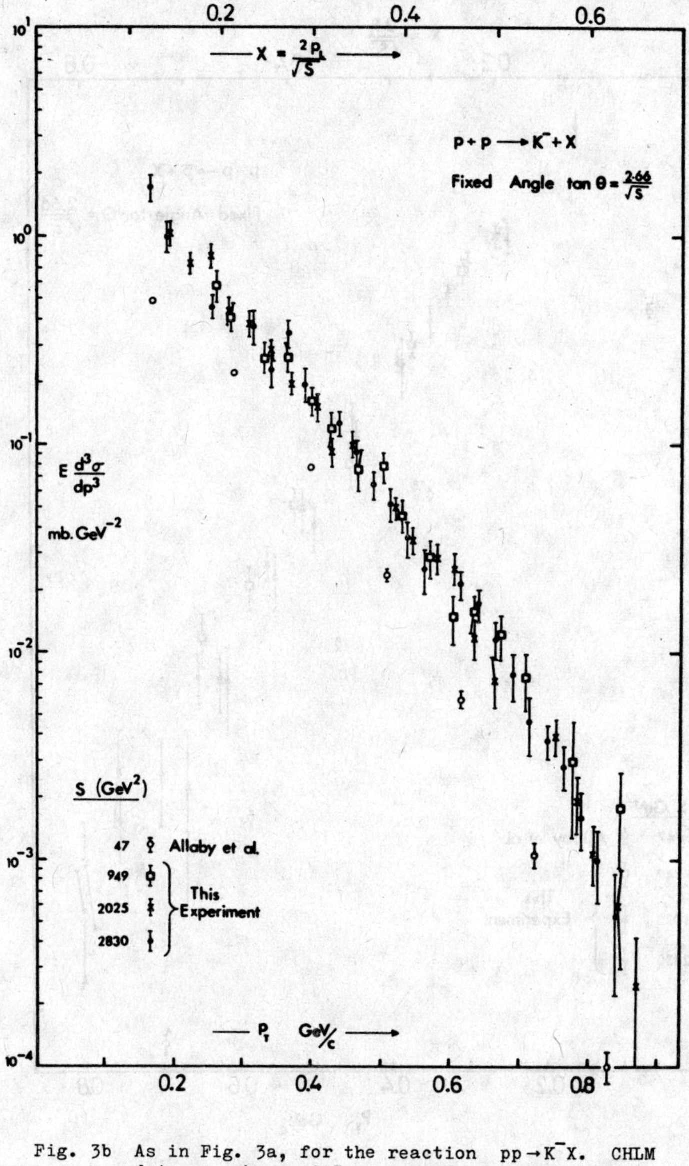

Fig. 3b As in Fig. 3a, for the reaction $pp \rightarrow K^- X$. CHLM data are shown at 3 centre-of-mass energies.

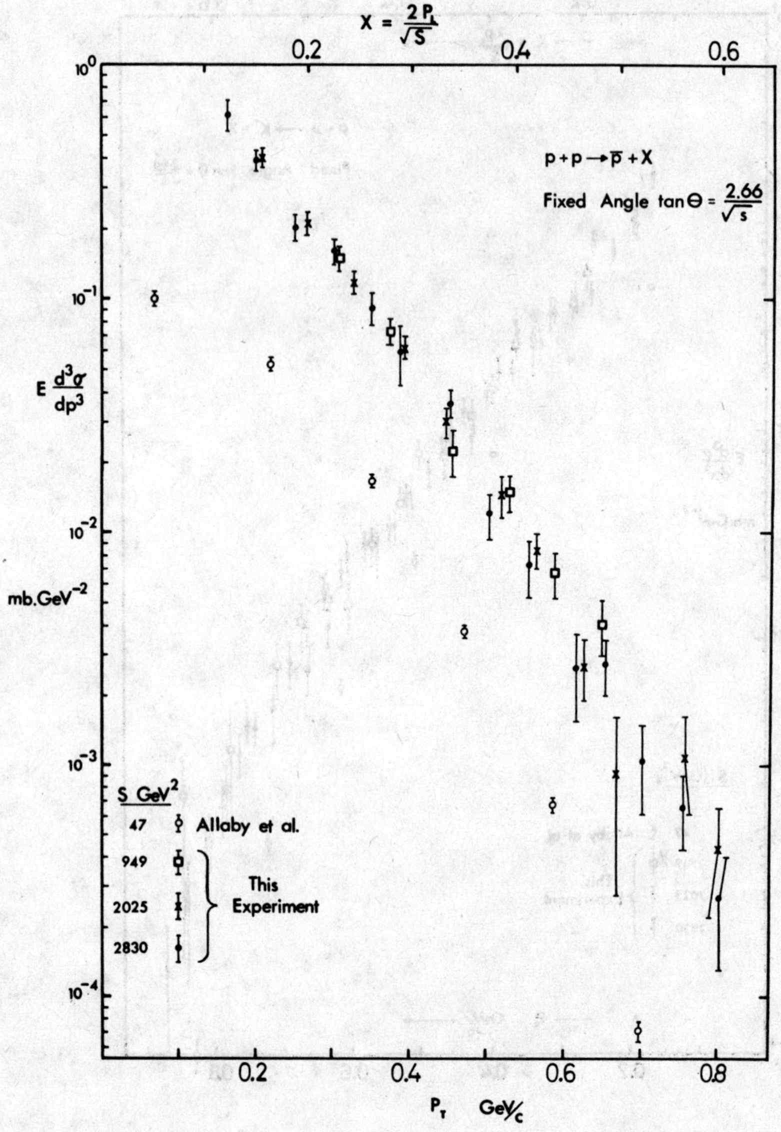

Fig. 3c As in Fig. 3b, for the reaction $pp \to \bar{p}X$.

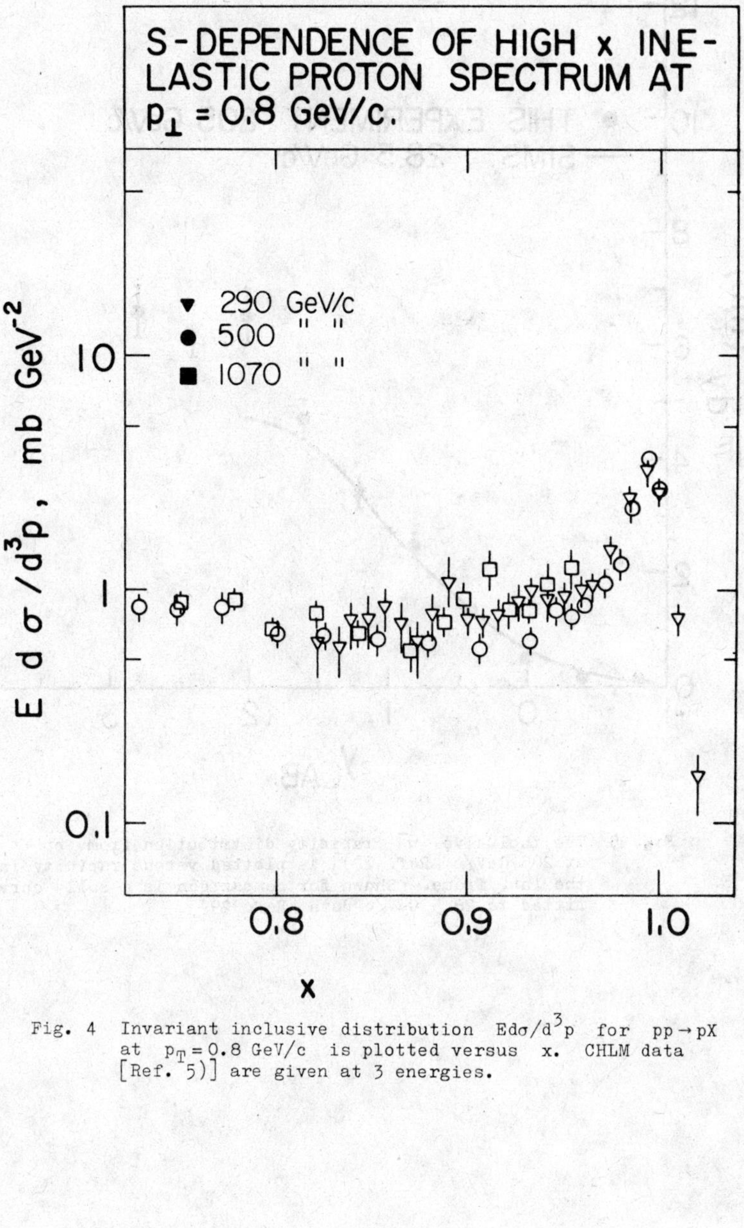

Fig. 4 Invariant inclusive distribution $Ed\sigma/d^3p$ for $pp \to pX$ at $p_T = 0.8$ GeV/c is plotted versus x. CHLM data [Ref. 5)] are given at 3 energies.

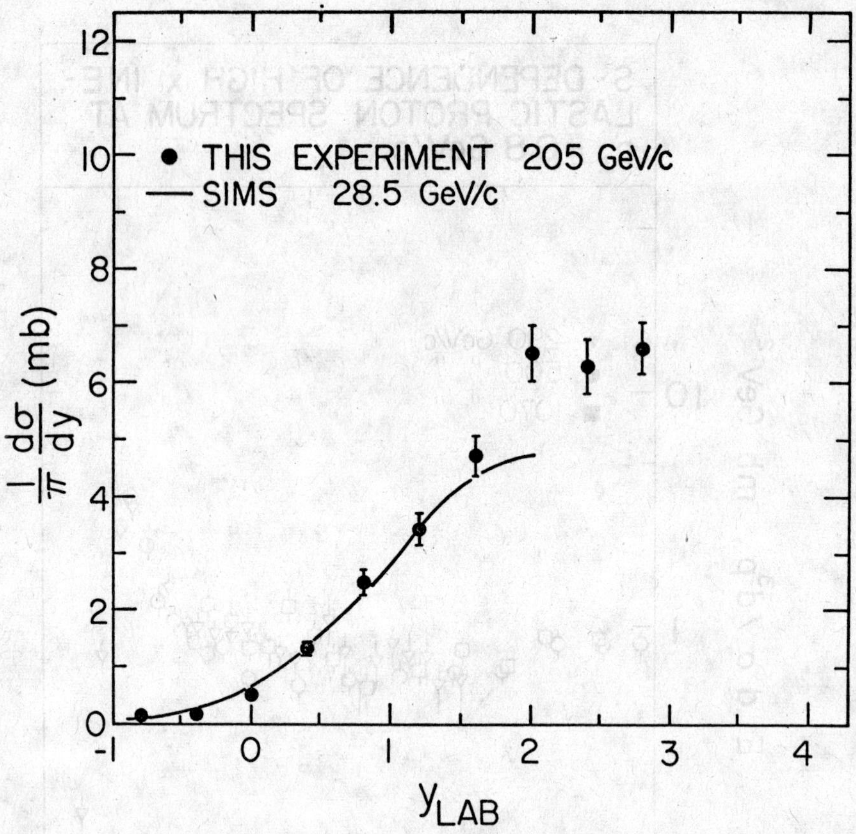

Fig. 5 The inclusive π^- rapidity distribution from $pp \to \pi^- X$ at 205 GeV/c [Ref. 28)] is plotted versus rapidity in the lab. frame. Shown for comparison is a solid curve fitted to 28.5 GeV/c data [Ref. 29)].

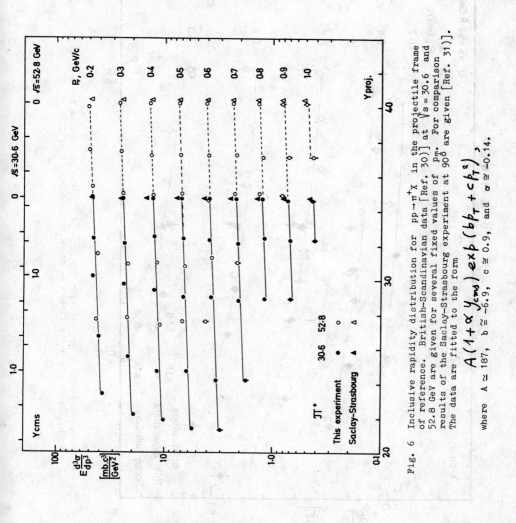

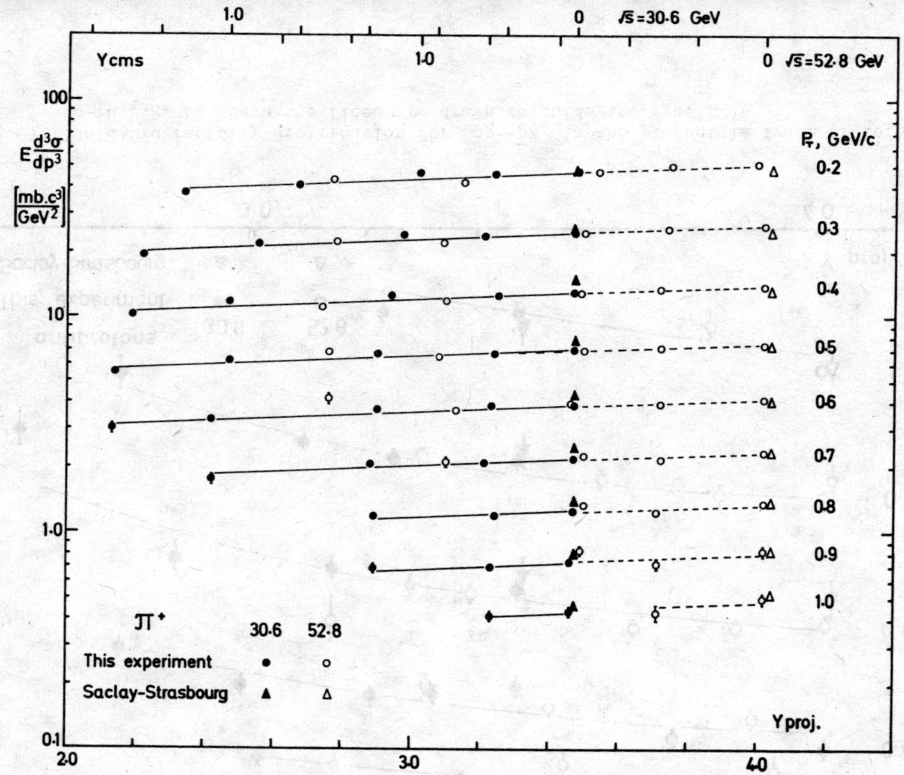

Fig. 6 Inclusive rapidity distribution for $pp \to \pi^+ X$ in the projectile frame of reference. British-Scandinavian data [Ref. 30)] at $\sqrt{s} = 30.6$ and 52.8 GeV are given for several fixed values of p_T. For comparison results of the Saclay-Strasbourg experiment at 90° are given [Ref. 31)]. The data are fitted to the form

$$A(1+\alpha\, y_{cms})\exp(b p_T + c p_T^2),$$

where $A \simeq 187$, $b \cong -6.9$, $c \cong 0.9$, and $\alpha \cong -0.14$.

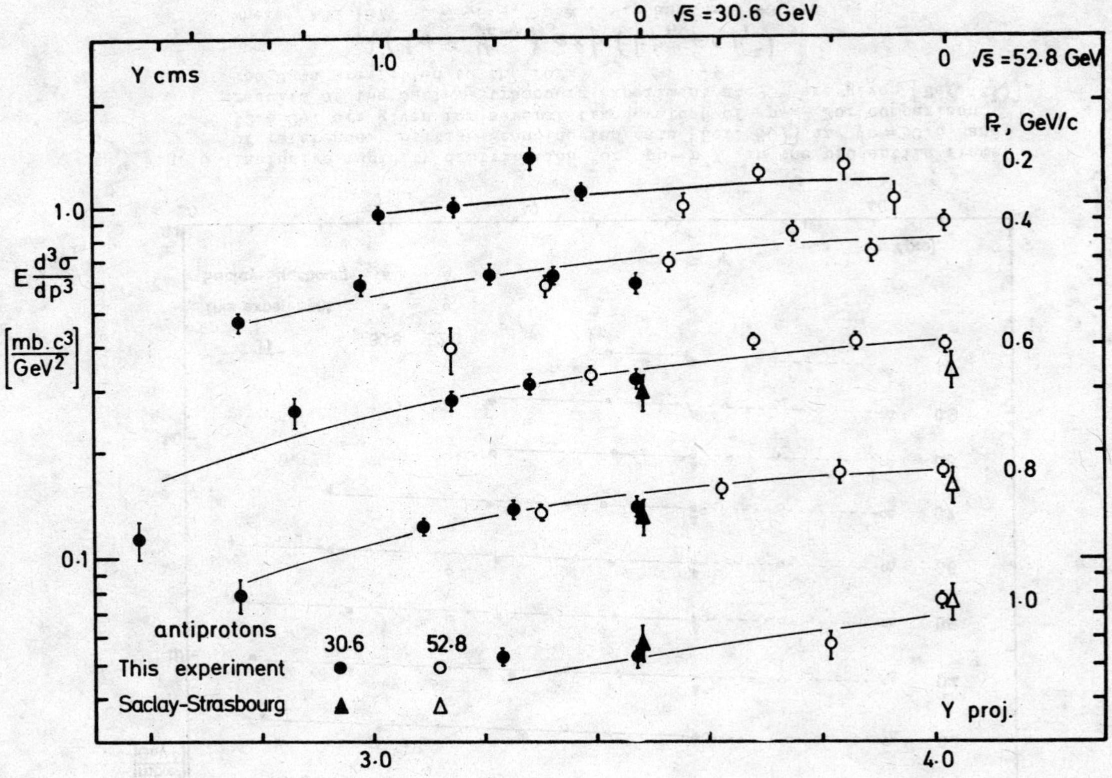

Fig. 7 Inclusive rapidity distribution for $pp \to \bar{p}X$ in the projectile frame of reference, as in Fig. 6. Here the lines are drawn to guide the eye.

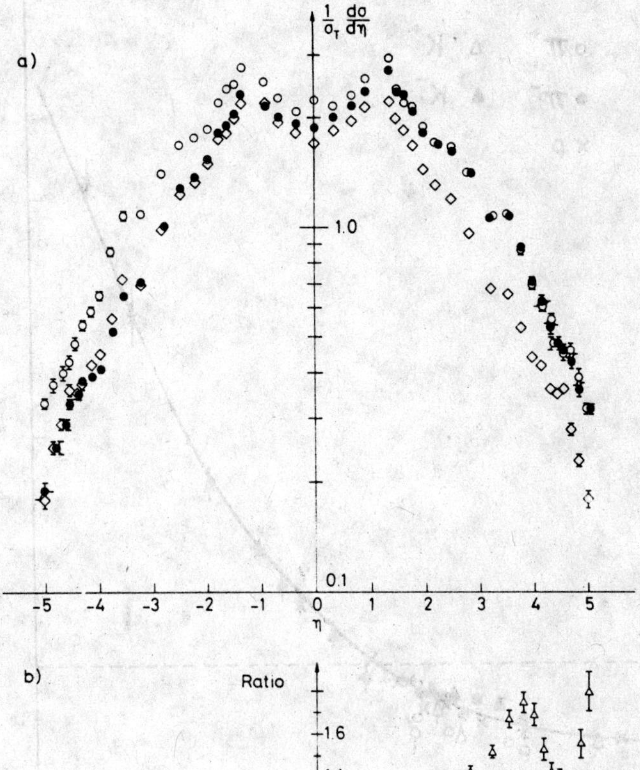

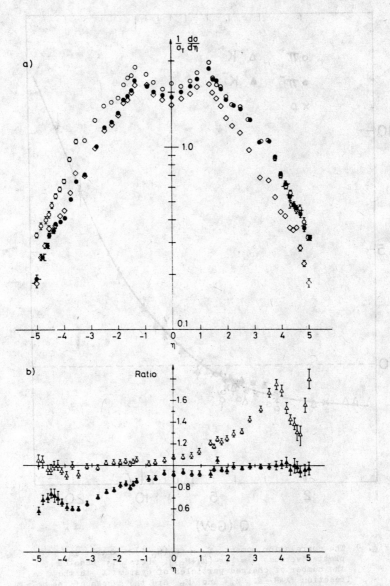

Fig. 8 Pisa-Stony Brook [Ref. 32)] single-particle inclusive density distributions $(1/\sigma_T)(d\sigma^{(1)}/d\eta)$.

a) Comparison of $(1/d\sigma_T)(d\sigma^{(1)}/d\eta)$ for three runs :
(A) $p_1=p_2=15.4$ GeV/c; ◇
(B) $p_1=p_2=26.6$ GeV/c; o
(C) $p_1=15.4$ GeV/c, $p_2=26.6$ GeV/c; ●

b) Ratios of $(1/\sigma_T)(d\sigma^{(1)}/d\eta)$ for asymmetric energy to $(1/\sigma_T)(d\sigma^{(1)}/d\eta)$ same energies are plotted versus η.
△ : run (C)/run (A)
▲ : run (C)/run (B).

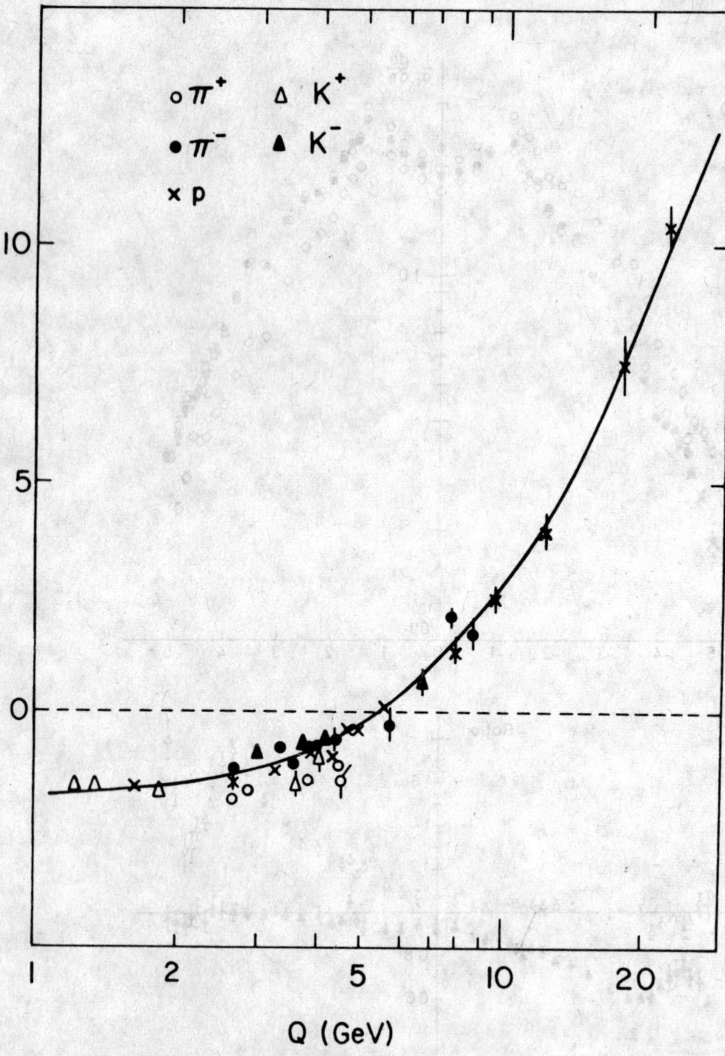

Fig. 9 The correlation moment $f_2 = \langle n_c(n_c-1)\rangle - \langle n_c\rangle^2$ is plotted versus $Q = [\sqrt{s}-(M_A+M_B)]$. Here n_c denotes the number of charged particles of system X in the reaction $A+B \rightarrow X$. M_A and M_B are the masses of the incident particles. Data are shown for the indicated choices of particle A ($\pi^{\pm}, K^{\pm}$, and p); in all cases B = proton. This figure is taken from Ref. 15).

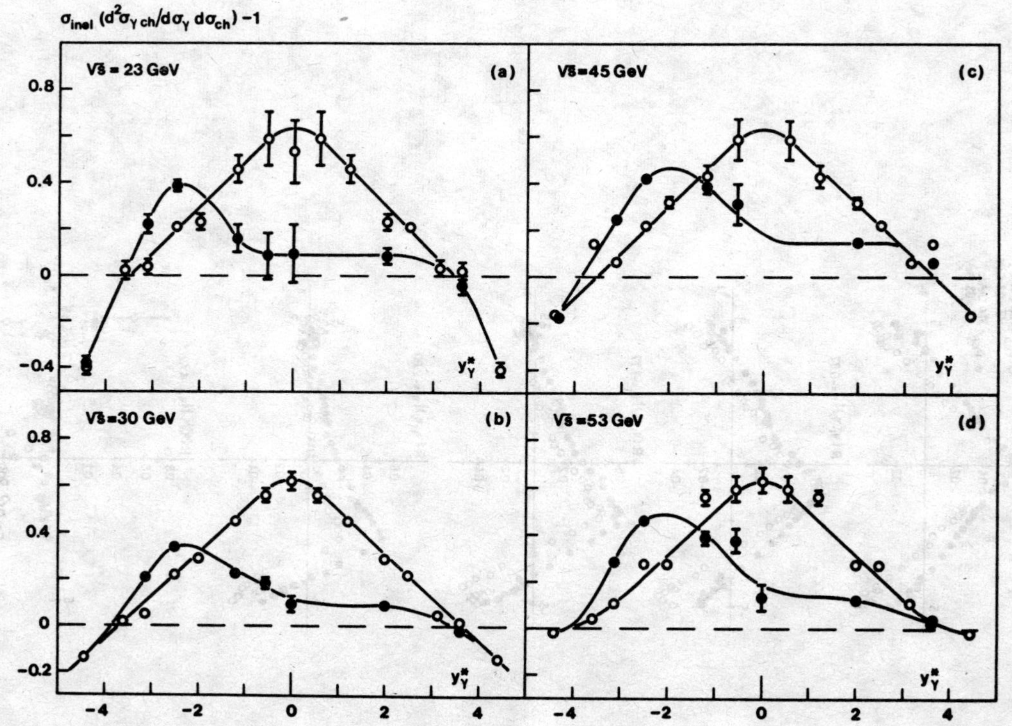

Fig. 10 For the reaction $pp \to \gamma hX$, the reduced correlation function $R(y_\gamma, y_{ch})$ is plotted versus centre-of-mass gamma rapidity y_γ for two fixed values of approximate centre-of-mass rapidity y_{ch} of charged hadron h. Open circles give CHV data [Ref. 17)] with $y_{ch} \approx 0$ and full circles for $y_{ch} \approx -2.5$. Results are presented at four energies, a) $\sqrt{s} = 23$, b) $\sqrt{s} = 30$, c) $\sqrt{s} = 45$ and d) $\sqrt{s} = 53$ GeV.

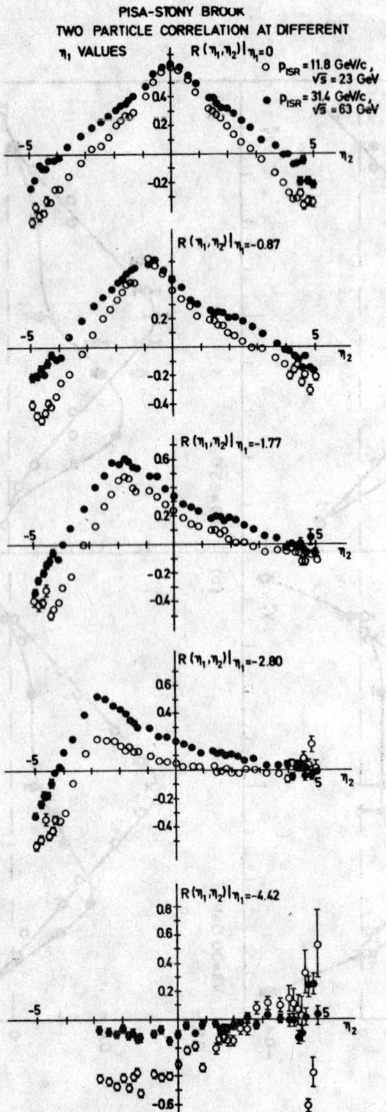

Fig. 11 Reduced charged particle correlation functions $R(\eta_1, \eta_2)$ for several fixed η_1 are plotted versus η_2. Open circles $\sqrt{s}$ = 23 GeV. Dark circles $\sqrt{s}$ = 62 GeV. Errors shown are statistical only; from Ref. 18).

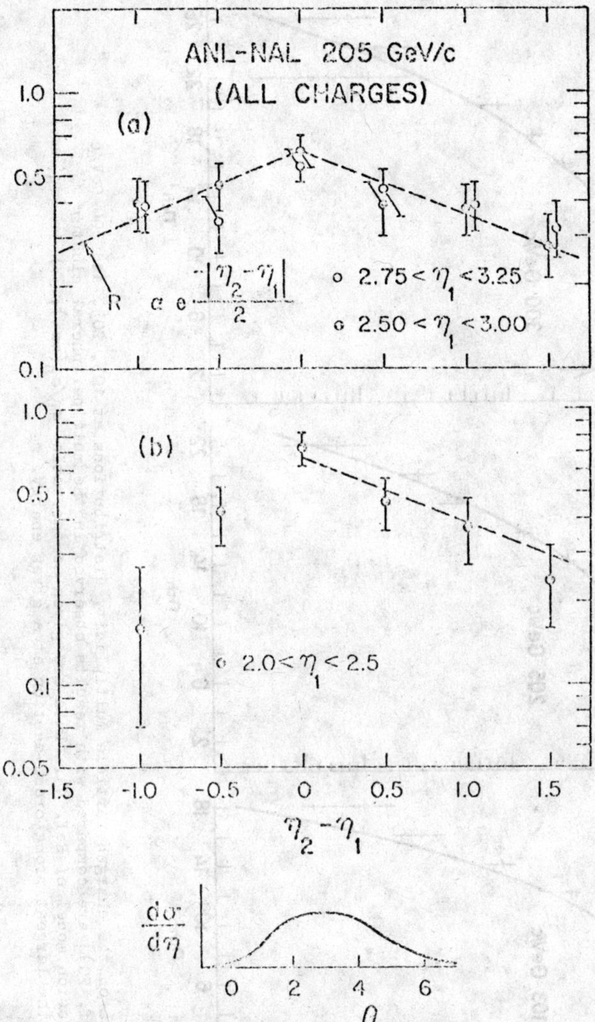

Fig. 12 The reduced charged particle correlation function $R(\eta_1, \eta_2)$ at 205 GeV/c is plotted versus $(\eta_1 - \eta_2)$ for three selections on η_1, defined in the lab. rest frame. The zero value of centre-of-mass longitudinal momentum corresponds to $\eta \approx 3$. In a), the two selections place η_1 in the central plateau interval. In b), η_1 is near the edge of the rapidity plateau. Data from Ref. 19).

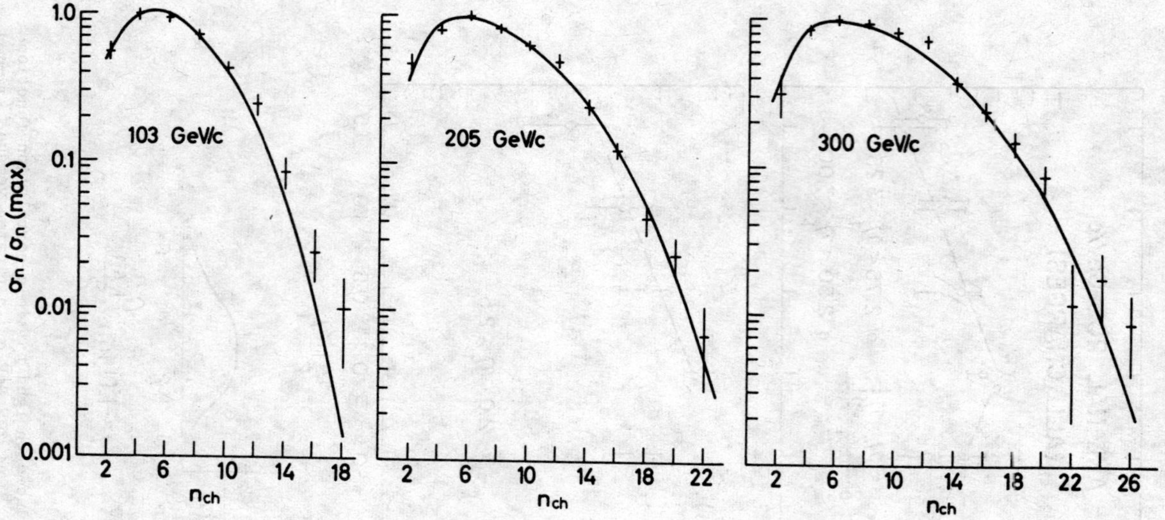

Fig. 13 Data on the charged particle multiplicity distributions at 103, 205, and 300 GeV/c [Ref. 20)] are compared with results observed in the multiperipheral cluster emission model of Ref. 4). Data are given as the ratio $\sigma_n/\sigma_n(\max)$, where $\sigma_n(\max)$ is the largest prong cross-section at a given energy.

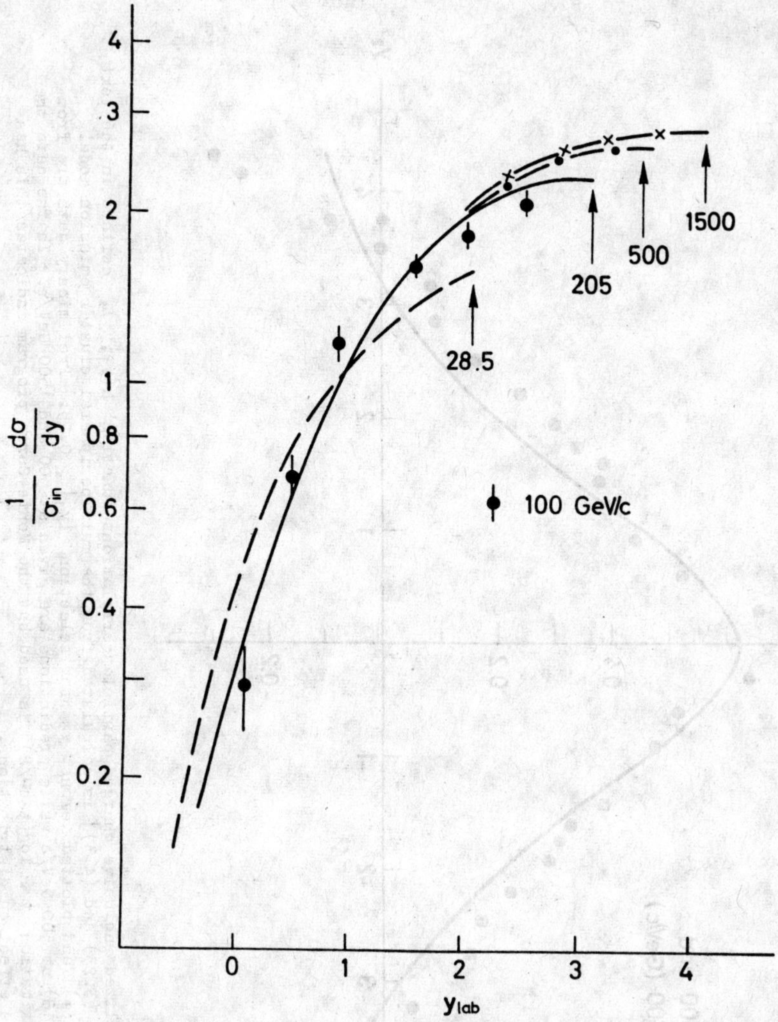

Fig. 14 Single pion inclusive rapidity distribution $\sigma_{inel}^{-1} d\sigma/dy$ versus y_{lab} for 28.5, 205, 500 and 1500 GeV/c. Theoretical curves are normalized absolutely and summed over all charges. Data at 100 GeV/c [Ref. 49)] are $1.5 \times [d\sigma/dy(\pi^+) + d\sigma/dy(\pi^-)]/\sigma$. Theoretical curves are computed from a multiperipheral cluster emission model, Ref. 4).

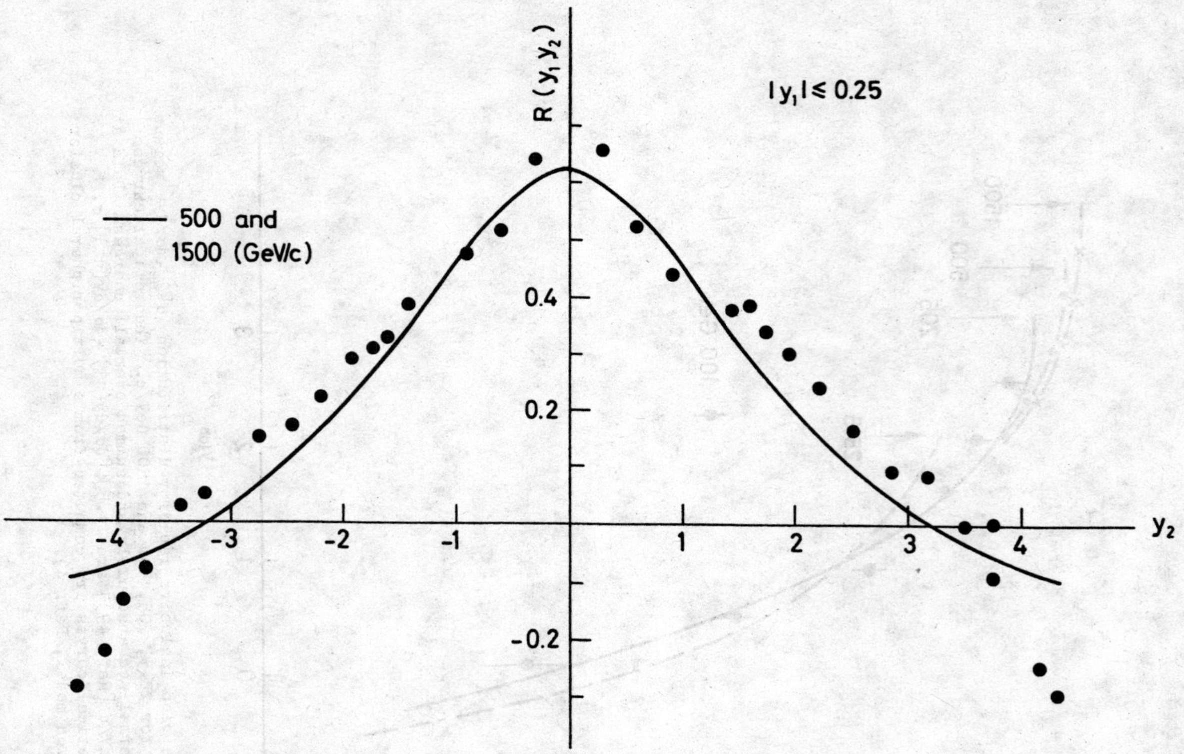

Fig. 15a Normalized inclusive charged particle correlations function $R(y_1,y_2)$, defined in the text [Eqs. (3.23) and (4.3)] is calculated from the multiperipheral cluster emission model [Ref. 4)] and plotted versus y_2 for selection $|y_1| \leq 0.25$. Preliminary data are from Ref. 18) at 500 GeV/c while predictions are given at 500 and 1500 GeV/c. In the data the approximation $y \approx \log(\tan\theta/2)$ is used but the Monte Carlo program indicates this has little effect on distributions.

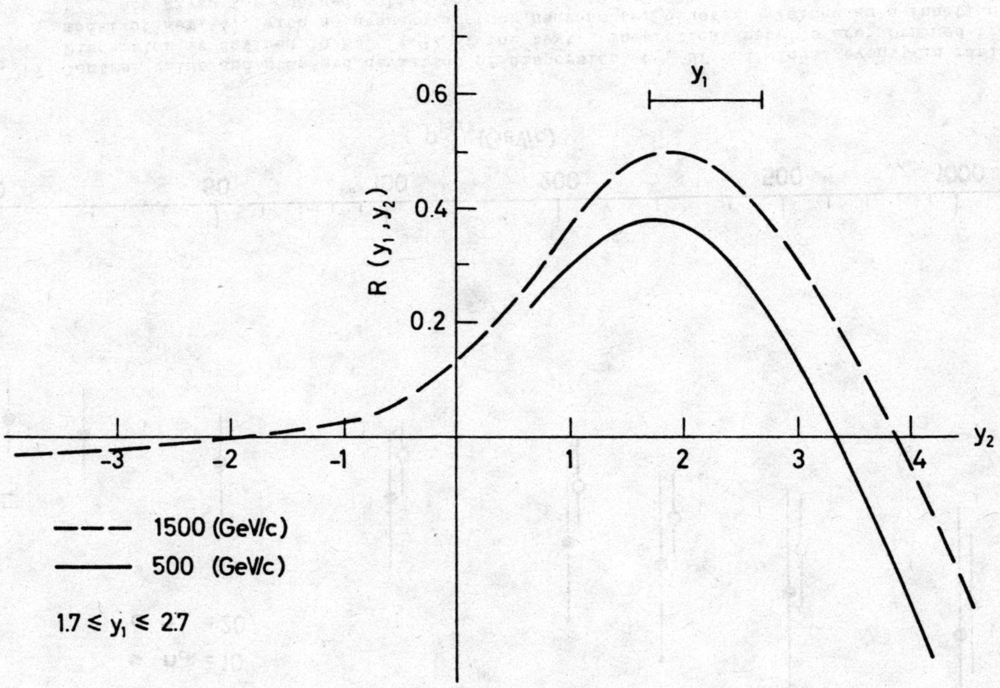

Fig. 15b As in Fig. 15a, but for the selection $1.7 \leq |y_1| \leq 2.7$.

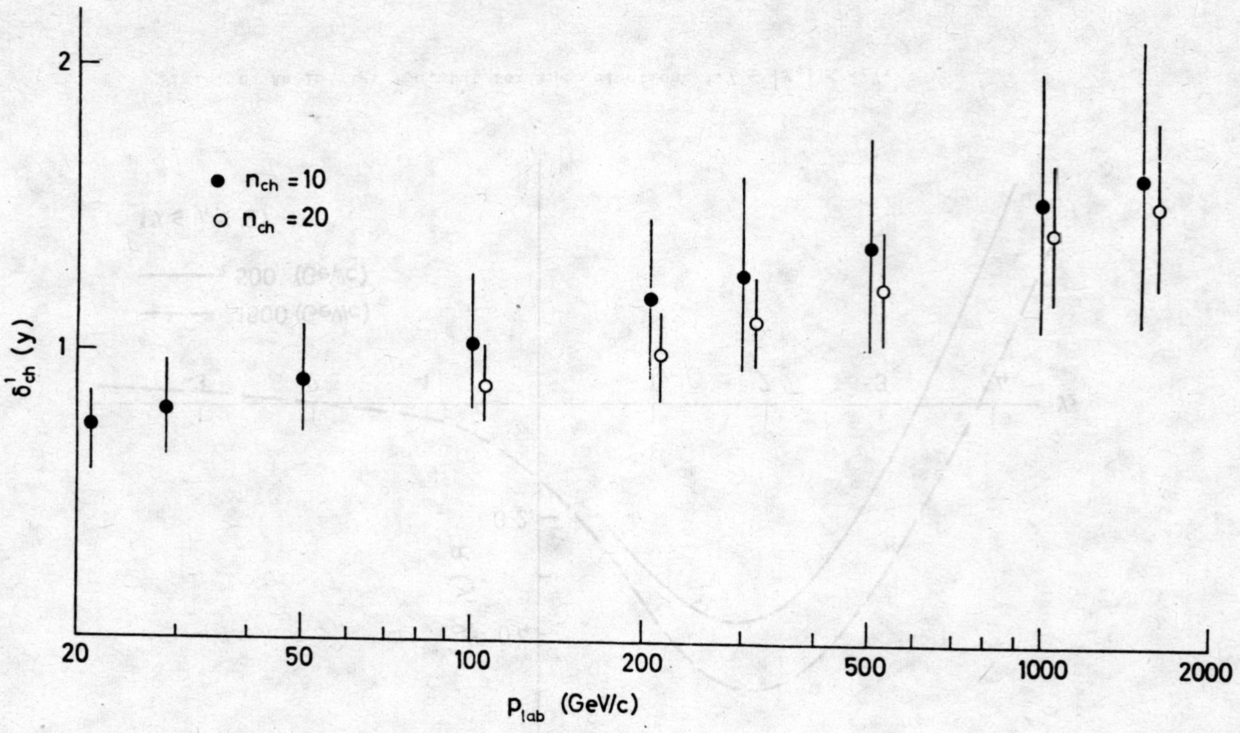

Fig. 16 Central value and standard deviation of dispersion δ^1 of individual events in rapidity. Dispersion is defined in Eq. (4.9) of the text. Theoretical results are obtained from the model of Ref. 4), with an average of four hadrons per cluster. Values as a function of p_{lab} are given for charged multiplicity n_{ch} = 10 and 20.

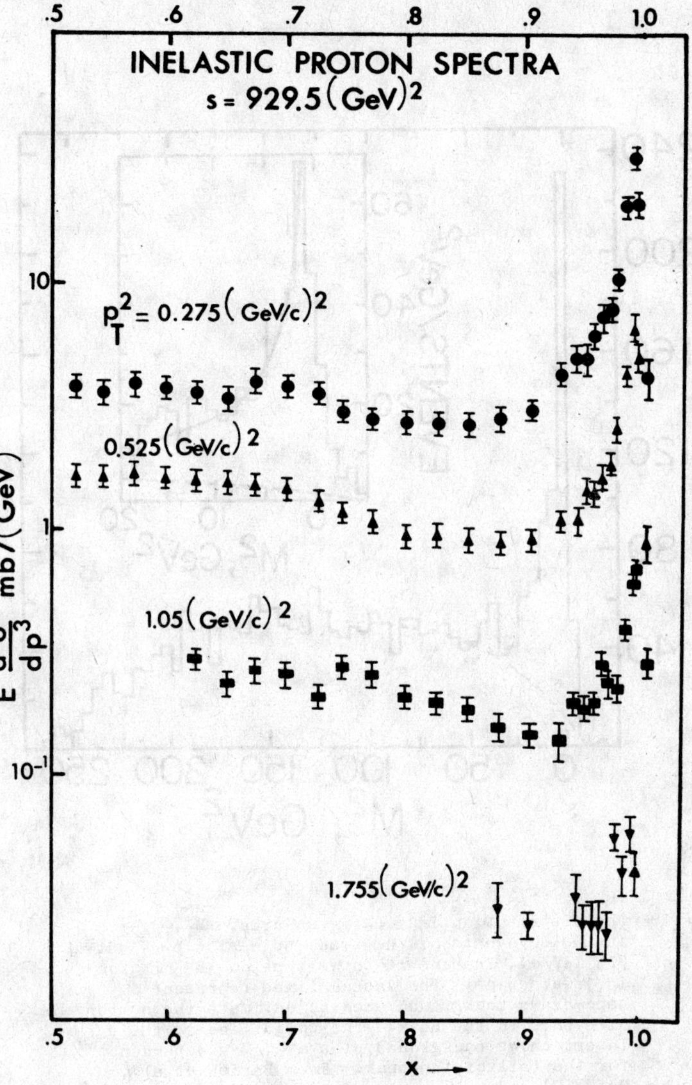

Fig. 17 The CHLM invariant inclusive spectrum $Ed\sigma/d^3p$ for $pp \to pX$ at $s = 929.5\ (GeV)^2$ is plotted as a function of x for four different fixed values of p_T. Ref. 5).

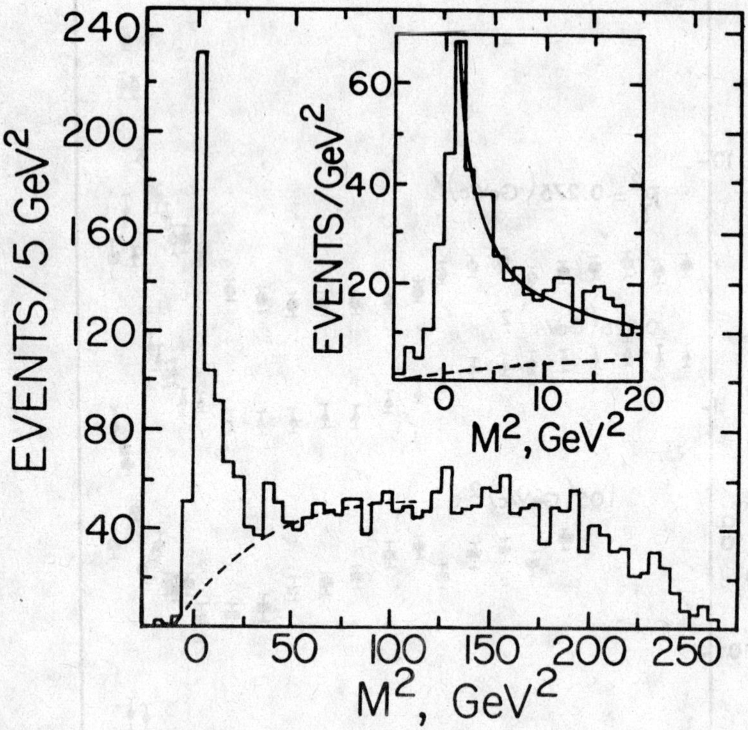

Fig. 18 The distribution in missing squared, M^2, for inelastic events of the type $pp \rightarrow$ slow $p + X$ at 205 GeV/c. The insert shows the low M^2 region in in 1 GeV^2 bins. The dashed lines represent a hand-drawn background used to estimate the number of events in the peak. The solid line on the insert shows background plus a $1/M^2$ dependence for the tail of the peak. From Barish et al., Ref. 57).

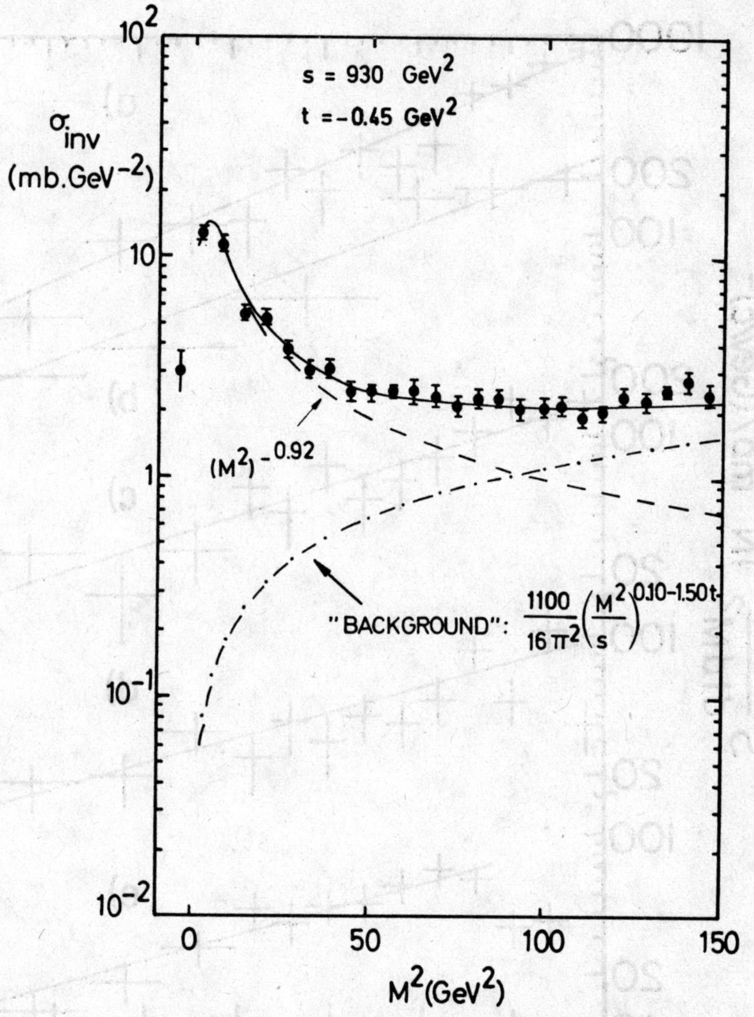

Fig. 19 The invariant inclusive cross-section $(s/\pi)d^2\sigma/dt\,dM^2$ measured by the CHLM group, Ref. 65), for the reaction $pp \rightarrow pX$ at $s = 930$ GeV^2 and $t = -0.45$ GeV^2 is plotted versus M^2, where M is the mass of system X and t is the square of the momentum transfer to the final proton. The solid curve results from a triple-Regge fit to the data, as discussed in Ref. 65).

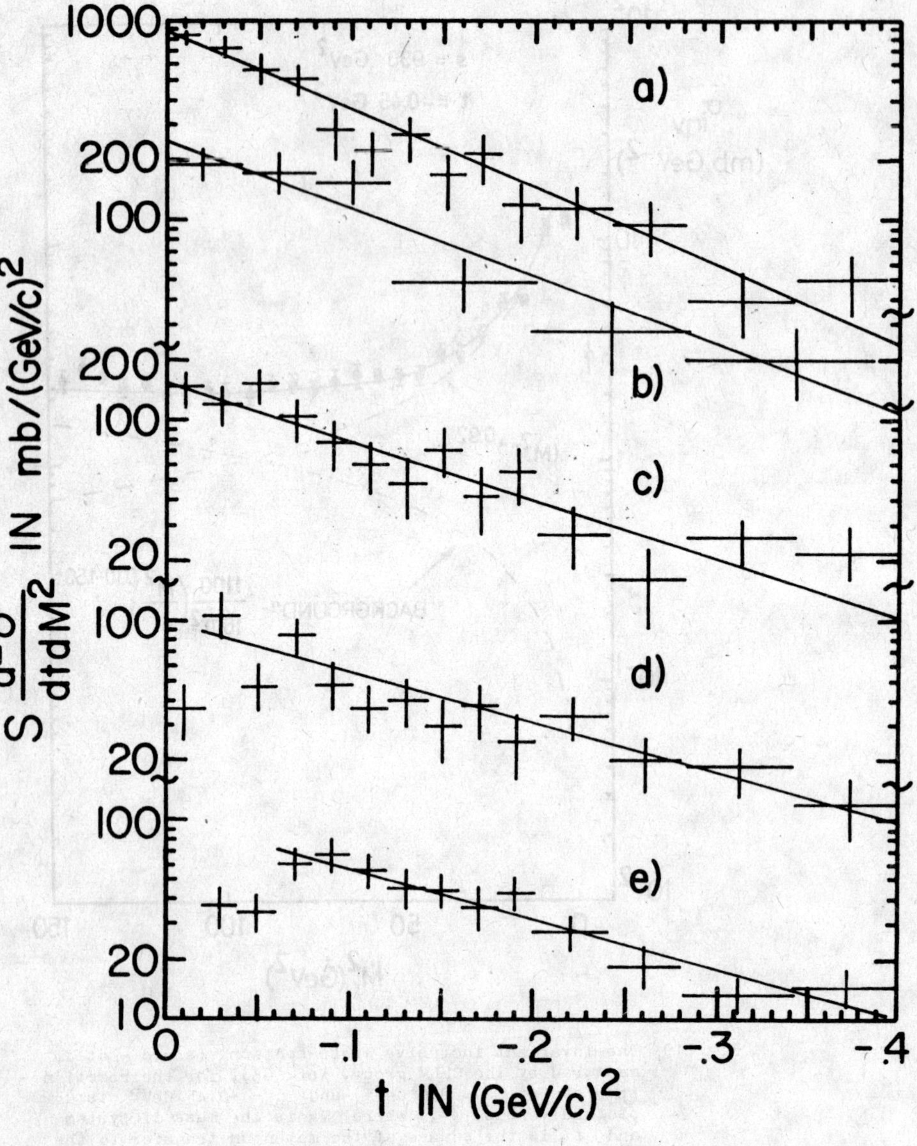

Fig. 20 Invariant cross-section $s(d^2\sigma/dt\ dM^2)$ versus four-momentum transfer squared t for all events from $pp \to pX$ at 205 GeV/c. a) for $M^2 < 5$ GeV^2, b) $5 \leq M^2 < 10$ GeV^2, c) $10 \leq M^2 < 25$ GeV^2, d) $25 \leq M^2 < 50$ GeV^2 and e) $50 \leq M^2 < 100$ GeV^2. The lines drawn show fits of the form Ae^{Bt} to the data (see Table in Section 5.2.4.) [Ref. 57].

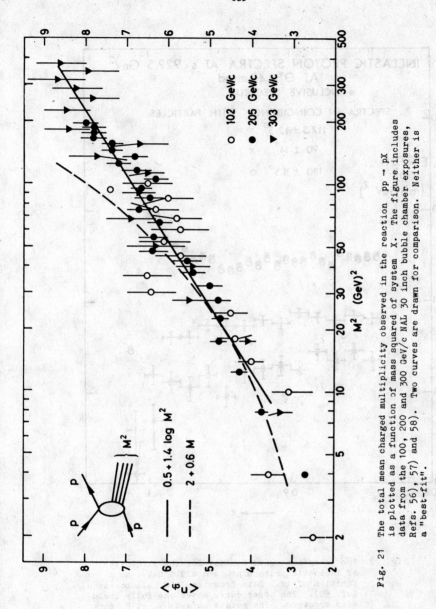

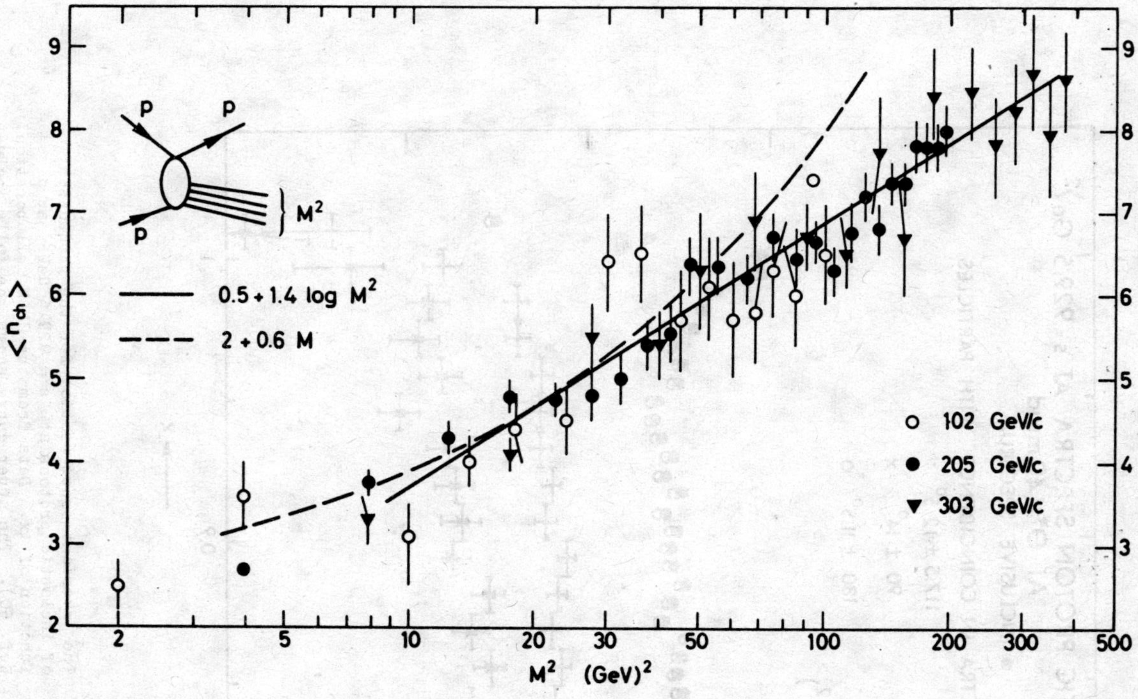

Fig. 21 The total mean charged multiplicity observed in the reaction $pp \rightarrow pX$ is plotted as a function of mass squared of system X. The figure includes data from the 100, 200 and 300 GeV/c NAL 30 inch bubble chamber exposures, Refs. 56), 57) and 58). Two curves are drawn for comparison. Neither is a "best-fit".

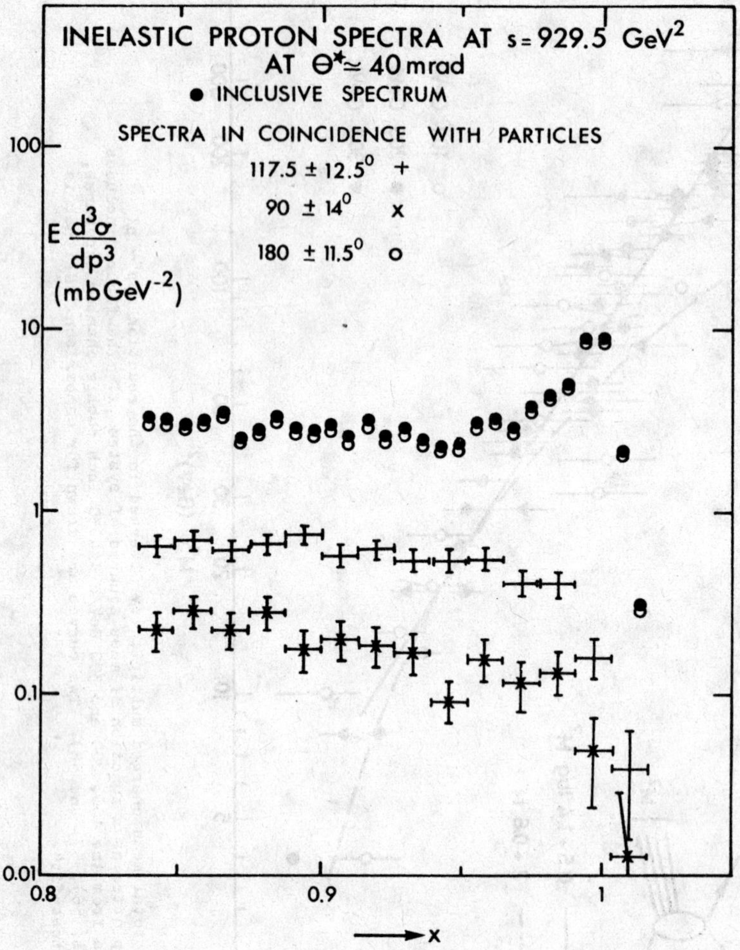

Fig. 22 The invariant cross-section for the production of inelastic protons at $s = 929.5\ GeV^2$ as a function of x. Data from the CHLM Collaboration, Ref. 50). The upper curve shows the fully inclusive spectrum. The points marked x, +, 0 show the spectrum in coincidence with one <u>or more</u> charged particles registered in telescopes placed at the indicated angular positions.

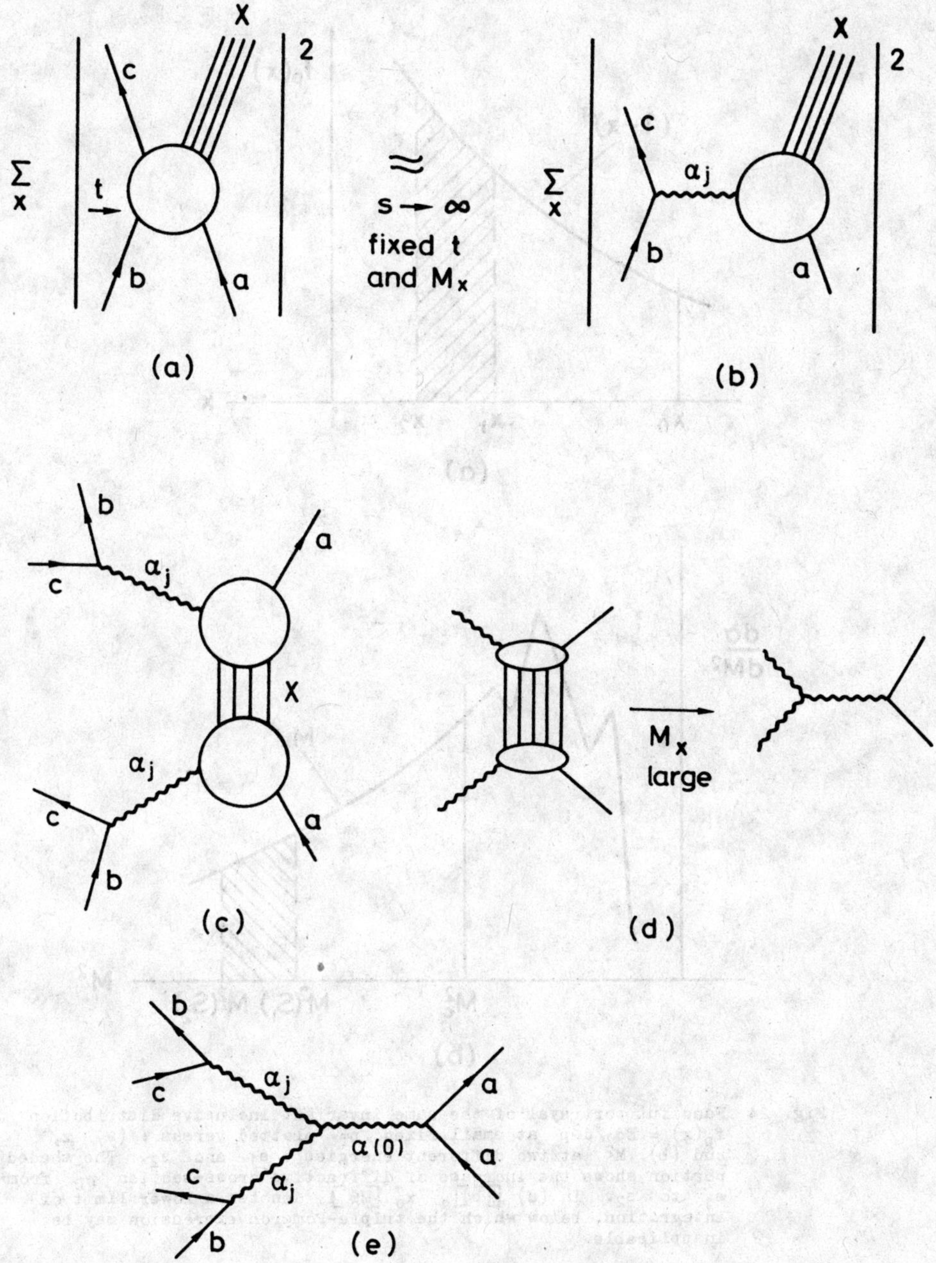

Fig. 23 Sequence of diagrams illustrating various Regge expansions made in passing to the triple-Regge graph shown in (e). The corresponding equations are found in Section 5.4.

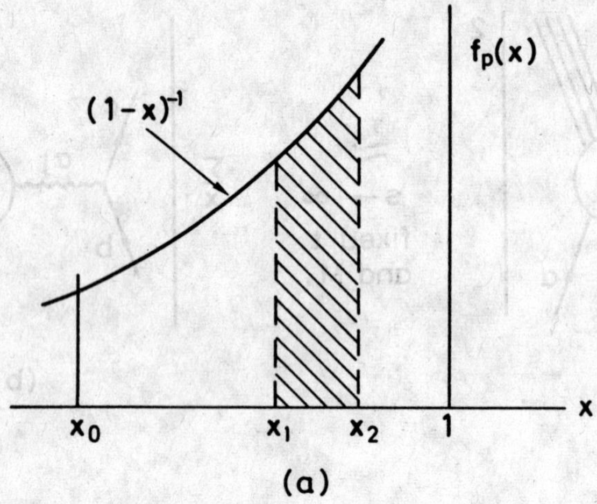

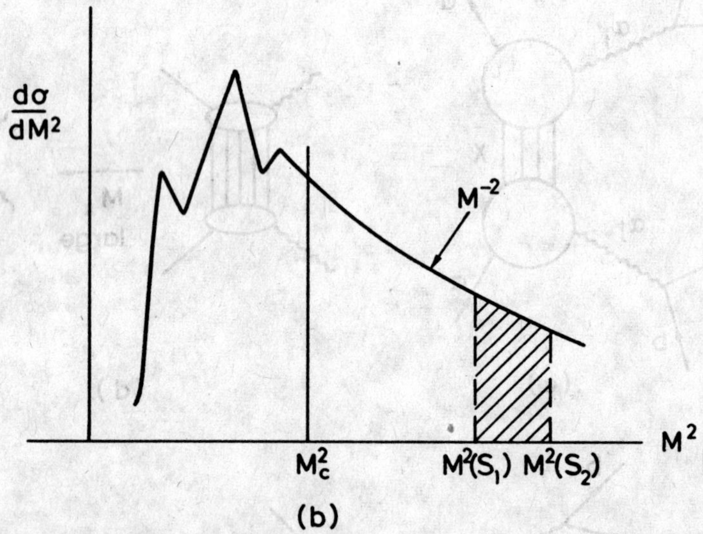

Fig. 24 Fanciful portrayal of the same invariant inclusive distribution $f_p(x) = Ed\sigma/d^3p$ at small fixed p_T plotted versus : (a) x, and (b) M^2 at two different energies : s_1 and s_2. The shaded portion shows the increase of diffractive cross-section σ_D from s_1 to s_2. In (a) [(b)], x_o [M_c^2] denotes a lower limit of integration, below which the triple-Pomeron expression may be inapplicable.

D I S C U S S I O N S

CHAIRMAN: **Prof. E.L. Berger**

Scientific Secretaries: Y. Eylon, W. Nahm, E. Rabinovici, W.Y. Yu

DISCUSSION No. 1

- *YU:*

With regard to the rise with energy in the central plateau, you have mentioned conflicting experimental results. On the one hand, the Pisa-Stony Brook experiment and the British-Scandinavian experiment showed a rise; on the other hand, the Saclay-Strasbourg experiment does not. Now if I remember correctly, the British-Scandinavian experiment has much better statistics than the Saclay-Strasbourg experiment. What would you like us to believe in?

- *BERGER:*

My impression is that within errors the British-Scandinavian and Saclay-Strasbourg experiments are consistent. If you look at the Saclay-Strasbourg data within their errors you can put a flat straight line; within their errors you can also put a line which increases by 8 ± 7% over the ISR energy range. The Pisa-Stony Brook data are not yet reliable on the question of s dependence in the central region.

- *YU:*

Concerning the transverse momentum dependence of the rise in the central plateau, I think it would be difficult to infer any conclusion on the fact of the data of the British-Scandinavian experiment you showed. I think there is a suggestion that as we go to smaller transverse momentum, the rate of the rise increases for the following reason. The rate of the rise for the Pisa-Stony Brook experiment is much larger than that of the British-Scandinavian experiment. But when a larger momentum cut-off is imposed in the Pisa-Stony Brook experiment, the increase is brought down to the level of the British-Scandinavian experiment. Maybe the cause of the larger rate of rise in the central plateau for the Pisa-Stony Brook experiment is due to small transverse momentum events.

- *BERGER:*

I would hesitate to make such a speculation. There are perhaps other difficulties with the Pisa-Stony Brook data, which are under intense study.

- *KEPHART:*

For low momentum, you have a hard time disentangling secondary particles produced, δ-rays, etc. from the primary production spectrum. Probably the 40% rise which you showed is more like a 20% rise in the actual spectrum.

- *AMALDI:*

I have been told recently by the Pisa-Stony Brook group that they have now better understood the problem of δ-rays produced. After this correction, the increase is compatible with the 20% measured by the British-Scandinavian group.

- *YU:*

Would you think it would be more attractive to attribute the rise in the total cross-section to the rise in the central plateau, than to attribute it to the proton spectrum near x = 1 in the manner of Capella and others? In other words, would it be aesthetically more elegant to allow non-scaling inclusive spectra than to have an inclusive spectrum which is singular at the kinematic boundary?

- *BERGER:*

I would prefer to side-step the question of aesthetics and stay with empirics. We can take the single particle spectrum of the proton, integrate it, and see to what extent it is compatible with an increase of the total cross-section. As discussed in Lecture 3, from the data alone, we cannot see an increase in the cross-section. If we fit the data to a specific model, then that specific model (namely the triple Regge model) will give an increase in the associated cross-section which grows like log s. On the other hand, using our energy sum rule, we see that any particle's inclusive spectrum can play a role in the increase in the total cross-section. We know the $\bar{p}$ spectrum rises in the central region, and if we ask how much rise in the total cross-section can be blamed on anti-baryons, I think the numbers do not support more than 1 mb. Thirdly, we have to worry about the problem of double counting. If we know (as taking a specific model would teach us) that the rise in the total cross-section is associated with the inelastic diffractive component, then we have to worry about the particles produced in the recoil cluster of this inelastic diffractive proton. Those particles can populate the central region. If the cross-section associated with the inelastic diffractive proton is rising, then so will the inclusive cross-section of particles produced by the recoil cluster. So we have to be very careful as to where we place the blame. I will try to work out the numbers in the third lecture.

- *RABINOVICI:*

You excluded multiperipheral single particle production models by showing that the average subenergies are small. Duality taught us that one can try to interpolate some quantities in the resonance region in terms of exchanges. Why do you exclude this possibility?

- *BERGER:*

Duality is not always quantitatively reliable. It works quantitatively for ρ exchange but fails for π and baryon exchanges. So I would prefer a specific production model in which the cross-section at low subenergies is represented by resonances or what I call clusters.

- *RABINOVICI:*

In increasing the masses of the clusters you decreased the density in rapidity and increased the average subenergies. On the other hand energy conservation decreases the rapidity at your disposal, thus decreasing the average subenergies. Could you compare the different effects?

- *BERGER:*

When numerical calculations are done, it appears that in the cluster model the average subenergies increase enough with respect to the one-particle production model. See for example, CERN preprint Ref. TH.1700 (1973) by Berger and Fox, published in Physics Letters.

- *RABINOVICI:*

Once one choses the production mechanism to be described in terms of a Poisson distribution, and one assumes σ_{tot} = const. (up to log s), the value of the Regge intercept determines the coupling constant through the Chew-Pignotti relation $2\alpha - 2 + g^2 = 0$. Thus a natural g would be 1 or 2. On the other hand you showed that the true $\langle n \rangle$ is 2 ln s, so it seems $\langle K \rangle$ can not be much higher than 2.

- *BERGER:*

For these questions, let me refer you to a recent Brookhaven report by Hamer and Peierls. I believe that K = 4 particles per cluster and a coupling $g^2 \approx 1$ are consistent.

- *NAHM:*

1) You said that the 15% increase in the inclusive cross-section over the plateau region may be a low-energy effect. Have you any experimental indication for that?

2) Which value do you get for the multiplicity K of particles coming from a cluster?

- *BERGER:*

1) In order to test this idea, one has to wait for NAL DATA. At PS energies one has no plateau.

2) 4.

- *WEBBER:*

If the rising total cross-section is connected with a $(1 - x)^{-1}$ behaviour in the single particle distribution for observed protons near x = 1 as you suggested, shouldn't a large proportion of this effect come from the elastic cross-section, in contradiction to the data shown earlier by Prof. Amaldi where the inelastic and elastic cross-sections seem to rise equally?

- *BERGER:*

No. You do not assume a behaviour like $(1 - x)^{-1}$ all the way to x = 1 but only to $x \leq 1 - M_C^2/s$ where M_C^2 is taken large enough to justify Pomeron dominance in inelastic production. This question will be discussed more fully in my third lecture. Elastic events are excluded.

DISCUSSION No. 2

- *DAY:*

I have become concerned about the variables used. The models have all been developed in terms of rapidity, while at least some of the data is in terms of $\eta = \log \tan \theta/2$. In those cases where the data are in terms of η, have the theoretical distributions been rewritten in terms of the same variable? Doesn't the change of variable change theoretical distributions?

- *BERGER:*

The data, of course, cannot be corrected to a distribution in rapidity since that would require knowledge of the momentum of the hadron which usually is not measured. However, under certain conditions the two variables are essentially the same. The theories can be rewritten in terms of η. A gaussian distribution in y would transform into gaussian distribution in η with roughly the same width. A flat rapidity distribution in the central region would develop a dip of about 20% when expressed in terms of η.

- *WAMBACH:*

Can you explain in your model what is exchanged and what is produced? What are the differences between this model and that of Ranft (1972) which also considered the multiple production of clusters in the thermodynamical model?

- *BERGER:*

It is important to specify what is exchanged when one wants to express the output trajectory in terms of the input trajectory.

Hamer and Peierls, in a multiperipheral multicluster model, use as an input $\alpha_\pi = 0$, $g^2 = 0.7$ to obtain the output singularity at $\alpha_P = 1$.

In answer to the second part of your question, I have not read the paper you mentioned.

- *WAMBACH:*

There are data from the CHLM Collaboration on the correlation between the pionization and fragmentation region in which they found that the normalization cross-section σ_0 in the distribution

$$\left[\frac{\sigma_0 (d\sigma/dy_1 dy_2)}{(d\sigma/dy_1)(d\sigma/dy_2)} - 1\right] = R_{12}(y_1, y_2)$$

should be the total cross-section and not the inelastic one. Are your statements on $C_{12}(y_1, y_2)$ affected by this result?

- *BERGER:*

It is not so important to distinguish between σ_{tot} and σ_{inel} as they both have the same energy dependence; only the numerical value of R will be changed. If one is interested in checking predictions of a specific model, one must choose the cross-section dictated by the model. For instance, in a Mueller-Regge analysis one should see if $R_{12}(y_1, y_2)$ tends to zero (for large rapidity separations) when normalized by σ_{tot}.

- *WAMBACH:*

Can you compare the energy behaviour of σ_{inel} and σ_{tot}?

- *AMALDI:*

Our results give a rise of (10 ± 2)% in both cases.

- *NAHM:*

If you have a slowly increasing $C(y_1,y_2)$, this might mean that the correct model should lie somewhere in between the multiperipheral model and the nova model.

- *BERGER:*

I do not believe that there is absolute truth in any of the models I have presented. Nature is much more imaginative than any of us here today.

- *RABINOVICI:*

You calculated $\langle K \rangle$ using the experimental data on R(0,0). The formula you used contained also $\langle K(K - 1) \rangle$. What information about this quantity had you used to obtain $\langle K \rangle$.

- *BERGER:*

One must invent a distribution in K in order to determine the ratio $\langle K(K - 1) \rangle / \langle K \rangle$. Results are not very sensitive to a choice from among reasonable distributions: uniform, Poisson, etc.

- *RABINOVICI:*

Evaluating $C(y_1,y_2)$ in the cluster model you assumed the factorization $d\sigma/dy_1 \cdot d\sigma/dy_2$, even if $\langle K \rangle$ equals 4. Is this assumption kinematically valid?

- *BERGER:*

In order to get a closed analytic form one neglects such effects as energy conservation, charge conservation, etc. In doing explicit fits to the data one must do Monte-Carlo calculations which take account of the effects mentioned.

- *RABINOVICI:*

I wonder if the fact you mentioned, that the correlation length λ does not change from 21 GeV up to ISR energies, is not due to the coincidence between the values of λ obtained by Berger and Krzywicki from kinematical arguments, and the value predicted in a Mueller-Regge analysis.

- *BERGER:*

The Mueller-Regge gives

$$C(y_1,y_2) \propto e^{-\frac{1}{2}|y_1-y_2|}$$

for large $|y_1 - y_2|$ with both y's in the plateau region. Our results this morning are restricted to small rapidity separations. The correlation length I found is indeed due to kinematical assumptions, namely, isotropic decay of the clusters.

- *RABINOVICI:*

You showed data on correlations between π^- and γ (which are actually π^-, π^0 correlations) and data on charged particle correlations. I think one can not expect them to be the same due to charge conservation correlations.

- *BERGER:*

You are correct. If we examine the 200 GeV/c bubble chamber data from NAL we find that R(0,0) for $\pi^-\pi^-$ and $\pi^+\pi^+$ is 0.3 ± 0.1 and not 0.6

- *ZICHICHI:*

Production of resonances will induce correlations. Has anybody calculated the correlations resulting from the production of known resonances? If these can account for the present data, why should one introduce new entities such as clusters?

- *BERGER:*

I think that all the correlation information or low subenergy behaviour of scattering amplitudes can be explained in terms of known resonances. If one would assume only ρ production one would obtain the right correlation length. However, R(0,0), calculated in such a model, is only half the experimental number. Production of pairs of known resonances such as ρ's would give the desired correlations. The introduction of the cluster concept allows us to perform calculations without dealing with the detailed structure and decay modes of resonances. No calculations using only resonances have been done.

- *ZICHICHI:*

This would be very interesting work for young theorists.

- *EYLON:*

You have decomposed the double inclusive distribution into two parts. In the first part the two detected particles came from the same cluster. In the second part they came from two different clusters. In the exact expression for the second part, one should include the double distribution of clusters $\rho(y_{C_1},y_{C_2})$. Assuming independent emission of clusters, you have factorized it to $\rho(y_{C_1})\cdot\rho(y_{C_2})$ and the result was that the first part alone gives the correlation function $C(y_1,y_2)$. However, in the fragmentation model, when there are only two clusters, their emission is no longer independent. Therefore, why have you identified in this model the first part with the correlation function?

- *BERGER:*

In principle, it is important to include such energy-momentum correlation in any numerical calculation. However, we have excluded them from the calculations in the lecture in order to get a simple analytic result.

- *EYLON:*

Due to the left-right symmetry in pp collision, $C(y_1,y_2)$ should be symmetric under the reflecton of y_1 and y_2 together, around the centre of the rapidity axis. In the fragmentation model we got $C(y_1,y_2) \propto \sqrt{s}\ e^{-\frac{1}{2}(y_1+y_2)}$ which does not satisfy the left-right symmetry, unless we shall restrict outselves to the c.m. frame and take $\sqrt{s}\ e^{-\frac{1}{2}|y_1+y_2|}$

- *BERGER:*

We should add the absolute value sign.

- *CHOUDHURY:*

If the single particle distribution in the central region does not scale but rather grows with s, I do not understand why this is inconsistent with short-range order. To me, it only means that Feynman's hypothesis of scaling is wrong in the central region. I think only if one finds a discontinuity at x = 0, one should believe inconsistency with short-range order.

- *BERGER:*

For pure short-range order one should have asymptotically a rapidity independent, energy independent plateau. The ≈ 8% increase at ISR energies shows either a violation of pure S.R.O., or it is a non-asymptotic effect. The data show no discontinuity at x = 0.

- *PATKOS:*

Prof. Berthelot said in his lecture that at Serpukhov energies the charged topological cross-sections follow a Poisson distribution for particle pairs. In the light of the multiperipheral model this should mean that at this energy the clusters contain in average those particles. This seems to indicate an increase of $\langle K \rangle$ from the Serpukhov energies to ISR energies.

- *BERGER:*

It is possible that $\langle K \rangle$ grows from PS energies to ISR energies, but I do not think that this is a necessary consequence of the data. In pp collisions the Poisson fit is rather good at 50 GeV and not good at 20 GeV. With four particles per cluster, you can get an excellent fit from 21 GeV on up. The point is that energy-momentum conservation and non-asymptotic effects are quite important in the over-all multiplicity distribution at these energies.

- *HENDRICK:*

You mentioned that you tried to make a two-fireball fit to the rapidity distribution and that it failed. Could you say what you tried to fit and why the fit failed?

- *BERGER:*

Let me instead refer to the work of Le Bellac and his collaborators at Nice. They tried to get a sensible fit with a Hwa-type two-fireball diffractive model. They found that when they cut off the number of particles per fireball in order to fit the NAL multiplicity data, they were forced to get a dip in the middle of the plateau. If one makes the fireballs sufficiently heavy to populate the central region, one obtains non-negligible prong cross-sections σ_n with n up to 30 prongs, which is not compatible with the NAL data. Sufficiently complicated models of this type can be invented, and will fit both σ_n and give a plateau; however, every piece of data seems to be treated in an *ad hoc* way.

- *BRANDT:*

Is there any experimental evidence for the multiple-cluster formation per event, which is introduced to reflect all the experimental data (target and projectile fragmentation and central plateau)?

If not, is there not the possibility of an anisotropic decay of target and projectile clusters with high masses which could as well lead to a plateau in the central part, perhaps with double Pomeron contribution to explain the strong correlations for the central part?

- *BERGER:*

In princple, you can try anisotropy, but then you are begging the issue what a fireball really is. Below 30 GeV a two-fireball model describes the data correctly. Now if you move up and try to fit the data at ISR energies you have "right wing" or "left wing" choices. You may keep two fireballs and give them an elongated decay distribution covering all the rapidity axis. There is interference in the central area where the decays of the two fireballs overlap. You really end up with decays which look multiperipheral. The left wing choice allows for more clusters, and this is the road I took.

- *BRANDT:*

But you do not see the clusters experimentally, or do you?

- *BERGER:*

Any inclusive distribution includes so much averaging that no detail of any model can be proved. One can, however, look at individual events and check if any clustering is seen. One possible parameter which allows us to do this is the dispersion of the rapidity distri bution of the charged tracks in a single event, after one has taken out the track which is farthest from the mean rapidity. Fox, Krzywicki and I have done some calculations with this parameter. You see the following. Events with one fireball yield a dispersion which is identically 1. For the most naive multiperipheral model without clustering you get a curve which rises logarithmically to 2.5 at the highest ISR energy $\sqrt{s} \cong 62$. The curves cross for 300 GeV/c. With clusters you get intermediate curves. Preliminary data from the Pisa-Stony brook experiment are consistent with clusters of about four particles each.

I fully agree with you that one has to look in a much more differential way at the data to test the details of models [Phys. Letters 43 B, 132 (1973)].

- *BRANDT:*

How can you get the growth of σ_{tot} by introducing the x = 1 singularity for the proton-spectrum, where the inelastic peak of diffractively scattered protons does not seem to grow at ISR energies?

- *BERGER:*

A good question, but it is treated in my third lecture.

- *MOEN:*

What reasons other than experimental are there for choosing $\lambda = 2$?

- *BERGER:*

Most models give this value for the correlation length λ. In Regge models, for example, $\lambda = (\alpha_P - \alpha_\rho)^{-1} \approx 2$. Also in a fragmentation model the typical size of a cluster in rapidity space is 2 units.

DISCUSSION No. 3

- *WILCZEK:*

As we heard from Amaldi, the behaviour of total pp cross-sections indicates increasing rather than decreasing opacity, unlike what you expect from a Regge pole. Also the simple fact of rising cross-sections prohibits a simple pole description of the Pomeron. In view of this, what is the justification for using the simple Pomeron pole idea at all, especially in attempts to explain rising cross-sections? More generally, how can you do Mueller analysis if the Pomeron singularity is complicated?

- *BERGER:*

A Pomeron pole with intercept 1 gives rise to factorization and scaling and thus "explains" some experimental data. One can only hope that cuts introduced in order to satisfy the Froissart bound can be treated as a second-order effect in so far as factorization and scaling at these low energies are concerned. That hope is a bit pious at this point, but we can expect that in the coming year it will get a more rigorous foundation.

- *RABINOVICI:*

In what sense can we speak of a collision of a Pomeron with a proton? Are we on the same ground as when discussing virtual photon-proton collisions, or is the situation here much worse? If one does have some confidence in the picture, can we learn from ep collisions something on the Pomeron-proton cross-section for large negative t?

- *BERGER:*

We are much worse off than in the virtual photon case. But if you want to imagine that the Pomeron couples like a vector meson, then I guess you can learn something from virtual photon scattering.

- *RABINOVICI:*

In addition I have a comment. The model is asymptotically unstable against shrinking of the Pomeron. An attempt to put $\alpha' = 0$ will change all the results obtained; σ_T will go like ln (ln s) and not like ln s; $\langle n \rangle$ will grow like A ln s and not like A/2 ln s; the plateau will disappear and so on. For a limited range of energy this effect can be masked by an appropriate slope of the form factor $\gamma(t)$.

- *BERGER:*

I agree. The model I discussed has the wrong asymptotic behaviour, as multi-Pomeron contributions would violate the Froissart bound. On the other hand, as you point out, over a limited range of energies an effective slope can yield all the things I have discussed.

- *AMALDI:*

Do you know which is the impact parameter distribution of the absorption due to the various successive terms due to the iteration of the Pomeron?

- *BERGER:*

This is a very interesting problem, but I do not know the answer. I refer you to Caneschi's lecture.

- *MENDES:*

Concerning the violation of the Froissart bound by the multi-Pomeron contributions, I have two questions:

1) Does the result mean that multi-Pomeron couplings are long-range effects, or does it simply mean that our way of constructing such amplitudes is incorrect?
2) Could it be that if you were able to sum all the $(\ln s)^n$ contributions you would regain an $(\ln s)^2$ behaviour?

- *BERGER:*

If you include a Pomeron exchange somewhere in the chain you get long-range correlations, e.g. an f_2 which increases faster than ln s. Concerning your second question: if one allows the Pomeron to couple like a factorized particle, one can write down amplitudes with many Pomerons exchanged. That yields an expansion of σ as a sum over arbitrarily high powers of ln s. You may treat this as a perturbation series. It may be possible to sum this and get no violation. It may also be that what we have written down as a single Pomeron term is not correct, that one should include some cut structure to describe unitarity effects. This would alter the structure of the series and might bring you back to a cross-section which grows like ln s or $(\ln s)^2$.

- *MENDES:*

What does your cluster short-range order model predict concerning large transverse momentum secondaries? Are the predictions comparable to those of the old AFS multiperipheral model?

- *BERGER:*

Not necessarily. In the usual version of the model, which Amati, Caneschi, and Testa used, one obtains only two particles with opposite large transvers momenta. The multicluster model should yield a jet of about four hadrons with large transverse momentum.

- *EYLON:*

From your picture of the particle distribution due to the diffractive component, it seems that the usual argument that the diffractive and the non-diffractive parts populate different regions of phase space is no longer true. What is, therefore, the justification for neglecting the interference between those two components?

- *BERGER:*

There is no justification. One should include interference, if possible.

- *EYLON:*

Taking $x > 0.8$, we are sure that the proton did not come from the decay of a cluster of small mass. Is this condition sufficient also to exclude events in which the proton comes from the decay of a very heavy object?

- *BERGER:*

Yes. The heavy object should have a very small rapidity, and the rapidity of a decay proton should be only one or two units away.

- *EYLON:*

Let us take an exaggerated example. Consider a pp collision which yields a cluster at rest in the c.m. frame. Now let this cluster decay to πp. Then one could have $x \approx 1$.

- *BERGER:*

This may be kinematically possible, but as you have said it is highly improbable.

- *YU:*

Concerning the rise in the total cross-section, I would like to bring up for discussion an alternative speculation that it can be attributed to the rise in the central plateau. Let us consider the multiplicity sum rule $\int (d\sigma/dy)\, dy = \langle n \rangle\, \sigma$. On the left-hand side the width of the plateau increases as log s and the height is rising 8% if we take the data of the British-Scandinavian experiment. On the right-hand side the total cross-section is rising as log s, and if we take the average multiplicity as also increasing as log s, we see that the numbers show that the rise in the central plateau is manifestly able to explain the rise in the total cross-section. The weak point of the argument is that the multiplicity can increase so fast that it kills off all the increase from the central plateau. My speculation is that it does not. I have the feeling and I believe this is shared by some people, that scaling is approached earlier in the fragmentation region than in the pionization region. We know that at accelerator energies the fragmentation contribution dominates and pionization is small, whereas at ISR energies the reverse is probably true. In a particular set-up of the Pisa-Stony Brook group, the cross-section for "single diffraction" events is small and is constant with respect to energy. This is a private communication from a member of the group. The data are unpublished.

- *BERGER:*

There are members of the Pisa-Stony Brook group here, and I guess they will tell you they do not yet know the single diffraction cross-section from their experiment.

Amaldi presented the problem very eloquently. Where do we really place the blame for the rise of the cross-section? Do we blame it on central events or on peripheral events? This multiplicity sum rule indicates that if the plateau rises with energy then $\sigma\langle n\rangle$ also rises. Unfortunately one does not know how to attribute the rise to the two terms in the product. That is why this particular sum rule is at the present time not used for quantitative calculations. If you could quantitatively separate the two terms, people would be very happy. But this has not been done. I agree that it is an intriguing speculation. What I showed this morning is that you can quantitatively associate the rise of the cross-section with the diffractive part. I did not prove that this is the only alternative. Indeed, in our earlier exercise we saw that a rise of σ_{inel} of $\approx$ 1 mb over the ISR range is due to antibaryon production in the central region.

- *FRAMPTON:*

I would like to make a distinction between two things: one theoretical and one phenomenological. On the theoretical side if we start by assuming the Pomeron is a factorizable pole of intercept 1, insertion of the triple-Pomeron expression into the energy conservation relation may lead to total cross-section increasing like log s for $s \to \infty$. We must then iterate this procedure for consistency, putting back log s into the inclusive formula. Eventually this generates by iteration arbitrarily high powers of log s violating unitarity. This is the theoretical inconsistency.

On the other hand, you choose to invent a phenomenological model with diffractive component increasing like log s as you are fully entitled to do.

I do not believe, however, that the first step of a theoretical argument, that by iteration leads to an inconsistency, can be used to motivate the phenomenological model. I believe that it is better to regard the two things as completely separate and distinct issues.

ELASTIC AND INELASTIC PROCESSES AT THE INTERSECTING STORAGE RINGS THE EXPERIMENTS AND THEIR IMPACT PARAMETER DESCRIPTION

Ugo Amaldi

Table of Contents

ELASTIC AND INELASTIC PROCESSES AT THE INTERSECTING STORAGE RINGS THE EXPERIMENTS AND THEIR IMPACT PARAMETER DESCRIPTION

Ugo Amaldi

CERN, Geneva, Switzerland and
Istituto Superiore di Sanità, Roma, Italy.

1. INTRODUCTION

The approach adopted in this series of four lectures has been chosen on purpose to be complementary to the discussion of "Multiparticle production processes at high energy" presented at this School by E.L. Berger. I shall move from the experimental information availa on proton-proton elastic scattering to its impact parameter description. This, in my opinio represents a simple and physically intuitive framework apt to discuss the implications of th known behaviour of elastic scattering on the dynamics of the inelastic processes. During th discussion of these implications, various classes of inelastic phenomena will be examined as possible dynamical sources of the increasing average opaqueness observed in the ISR energy range, which has its most spectacular manifestation in the 10% increase of the total cross-section. At present, no definite conclusion can be drawn, but it is hoped that this "s-channel" point of view will be a useful complement of the discussion given by E.L. Berger or the phenomenology of the inelastic processes, based mainly on the "t-channel" approach.

2. ISR RESULTS ON PROTON-PROTON ELASTIC SCATTERING

2.1 A general outlook

In a contribution to last year's School I discussed the "ISR results on proton-proton elastic scattering and total cross-sections" from an experimental point of view, presenting at the same time the many possible interpretations of the observed phenomena[1]. In the conclusions of this paper, which will be denoted by I in the following, the main results obtained in this field at the ISR up to February 1973 were summarized as follows:

i) The logarithmic slope of the elastic differential cross-section changes by about 2 GeV^{-2} around a momentum transfer $|t| \simeq 0.15$, so that at the highest ISR energies ($\sqrt{s}$ = 53 GeV) the slope equals (13.1 ± 0.3) GeV^{-2} for $0.01 \leq |t| \leq 0.15$ GeV^2 and (10.8 ± 0.2) GeV^{-2} for $0.17 \leq |t| \leq 0.31$ GeV^2.

ii) The forward slope b ($|t| < 0.15$ GeV^2) increases with energy. A behaviour proportional to ln s is compatible with the existing ISR data, and is confirmed by the results of the jet experiment recently performed at NAL[2]. If a ln s behaviour is assumed, b increases by (8 ± 2)% in the "standard" ISR energy range, by which I mean $23.5 \leq \sqrt{s} \leq 53$

iii) The differential elastic cross-section dσ/dt shows a clear minimum at $|t| \simeq 1.4$ GeV^2. Since the systematic uncertainties on the absolute value of the cross-section are quite large, at that time it was believed that for $|t| \gtrsim 0.2$ GeV^2 the cross-section di not vary very much in passing from $\sqrt{s}$ = 31 to 53 GeV. In the same energy interval the position of the minimum displaces by (0.08 ± 0.11) GeV^2 towards smaller momentum transf passing from (1.45 ± 0.10) GeV^2 at $\sqrt{s}$ = 31 GeV to (1.37 ± 0.03) GeV^2 at $\sqrt{s}$ = 53 GeV.

iv) The real part of the nuclear amplitude in the forward direction is very small[3] (ρ = Re/Im = 0.025 ± 0.035 for $23 \leq \sqrt{s} \leq 31$ GeV).

v) The total proton-proton cross-section σ_t increases by (10 ± 2)% in the ISR energy [3-5]

vi) The elastic cross-section σ_{el} increases by (12 ± 4)% in the ISR energy range if a 10% increase of the forwa·d slope b is assumed. Indeed, one has (apart from small corrections due to the change of slope around 0.15 GeV²)

$$\sigma_{el} \propto \frac{\sigma_t^2}{b} \tag{1}$$

so that a slower increase of b with energy implies a faster increase of the computed value of σ_{el}.

vii) The inelastic cross-section $\sigma_{in} = \sigma_t - \sigma_{el}$ increases in the ISR energy range by ∿ 10%, passing from (32.3 ± 0.4) mb to (35.6 ± 0.5) mb.

Some of the above points will now be discussed, together with the relevant new information known at the end of July 1973.

2.2 More about the forward slope and the real part

As far as the forward slope is concerned [point (ii) above], the quoted data obtained at NAL with the hydrogen-jet technique[2] point to the conclusion stated above: the forward slope increases with energy proportionally to ln s. A fit to the data in the energy range $75 \leq s \leq 2800$ GeV² gives $b = 8.32 + 2\alpha' \ln s$, with $\alpha' = 0.275$ GeV^{-2} [6]. The estimated error is $\Delta\alpha' \simeq 0.02$ GeV^{-2}.

Passing to the measurement of the real part [point (iv) above] in the experiment performed by the CERN-Roma collaboration[3], elastic scattering in the Coulomb region has been detected and two quantities have been derived at each energy: the total cross-section σ_t and the ratio ρ between the real and the imaginary parts of the nuclear amplitude in the forward direction.

Later experiments[4,5] lead to independent values of the total cross-sections, which now can be fed into the fit to the measured angular distributions, together with their errors, so as to leave only one free parameter: the value of ρ. Note that this is the usual procedure adopted in all previous experiments in which elastic data in the Coulomb regi·· ·ve been measured. The results are summarized in Table 1.

Table 1

Ratio ρ of the real to the imaginary parts of the forward nuclear amplitude obtained by the CERN-Roma collaboration

$\sqrt{s}$ (GeV)	σ_t a) (mb)	ρ b)
23	39.15 ± 0.4	0.00 ± 0.03
31	40.6 ± 0.5	+0.02 ± 0.04

a) This value is the average of the values measured in Refs. 4 and 5.

b) Systematic and statistical errors are combined together in the quoted standard deviations.

For comparison, let us recall that the values obtained by leaving both ρ and σ_t as free parameters in the fit were $\rho = +0.02 \pm 0.05$ and $\rho = +0.03 \pm 0.06$, respectively[3]. The new values are plotted in Fig. 1 together with lower energy data.

2.3 More about proton-proton total cross-sections

The experiments on total cross-sections by the CERN-Roma collaboration[4] and by the Pisa-Stony Brook collaboration[5] have been discussed in I and, more recently, by A.N. Diddens[7] and by G. Bellettini[8]. Here only the principle of the methods will be outlined and few relevant comments and new information added.

For the illustration of the Van der Meer method, used in both experiments to measure the luminosity L of the ISR, we refer to I. At present it is sufficient to recall that during the measurements of L, the two ISR beams are vertically displaced in small and precisely-known steps, and that the main source of error on the measured luminosity comes from the uncertainty in the absolute knowledge of the size of these steps. At the time of publication of the two papers[4,5], a ±2% scale error, equal at all momenta, was attributed to the step size, mainly because the ISR group had detected a disagreement of this order between the step size computed from the known properties of the magnets, which are used to perform the vertical displacements, and the step size measured by means of pick-up electrodes. Since then, the ISR group has performed very accurate measurements of the step size by means of mechanical devices ("shavers") and found agreement with the computed vertical displacements within (0.4 ± 0.5)%. The Pisa-Stony Brook collaboration has continued working on the measurement of the vertical beam positions by means of spark chambers and has found agreement with the ISR "nominal" step within (1.7 ± 0.5)%[8]. In conclusion, recent measurements have confirmed that there is no indication of an energy-dependent effect on the displacement scale and, very probably, the scale error is smaller than ±2%. Other important experimental remarks, which confirm the validity and the accuracy of the Van der Meer method, are discussed by G. Bellettini in Ref. 8.

The CERN-Roma collaboration has obtained the total proton-proton cross-section by measuring elastic scattering and applying the optical theorem. The steps of the procedure are the following:

a) Measurement of the elastic differential rate $\Delta R_{el}/\Delta\Omega$ in a known solid angle $\Delta\Omega$ and in the vertical plane around $\theta = 6$ mrad, which corresponds to a momentum transfer $t = p^2\theta^2$, where p is the ISR momentum.

b) Determination of the elastic differential cross-section by means of the luminosity L measured with the Van der Meer method:

$$\left(\frac{d\sigma}{d\Omega}\right)_t = \frac{1}{L}\,\frac{\Delta R_{el}}{\Delta\Omega}\,. \tag{2}$$

c) Extrapolation of this cross-section to zero angle assuming an exponential behaviour in t, and thus using the factor $\varepsilon = e^{+bt}$. The slope b was taken from the measurements discussed above.

d) Application of the optical theorem with $\rho = 0$

$$\sigma_t = \frac{4\pi\hbar}{p}\sqrt{\left(\frac{d\sigma}{d\Omega}\right)_{t=0}} = \frac{4\pi\hbar}{p}\,\frac{1}{\sqrt{L}}\sqrt{\frac{\Delta R_{el}}{\Delta\Omega}}\ \varepsilon. \quad (3)$$

This method has been applied at the four standard ISR energies with the results collected in Table 2, together with the values of the elastic cross-section σ_{el} and of the inelastic one $\sigma_{in} = \sigma_t - \sigma_{el}$.

Table 2

Elastic differential and total cross-sections measured by the CERN-Roma collaboration[4)]

$\sqrt{s}$ (GeV)	$\lvert t\rvert$ $(GeV^2) \times 10^{-3}$	$(d\sigma/dt)_t$ (mb/GeV^2)	b [a)] (GeV^{-2})	$\varepsilon = e^{b\lvert t\rvert}$ Extrapolation factor[b)]	$(d\sigma/dt)_{t=0}$ (mb/GeV^2)	σ_t (mb)	σ_{el} [c)] (mb)	σ_{in} (mb)
23.4	8.1	71.0 ± 1.5	11.8	1.10 ± 0.004	78.1 ± 1.7	39.1 ± 0.4	6.8 ± 0.2	32.3 ± 0.4
30.5	13.5	71.0 ± 1.6	12.3	1.18 ± 0.007	83.8 ± 1.9	40.5 ± 0.5	7.0 ± 0.2	33.5 ± 0.4
44.8	17.5	73.8 ± 1.7	12.8	1.25 ± 0.010	92.3 ± 2.2	42.5 ± 0.5	7.5 ± 0.3	35.0 ± 0.5
52.8	23.0	70.6 ± 1.8	13.1	1.35 ± 0.014	95.4 ± 2.6	43.2 ± 0.6	7.6 ± 0.3	35.6 ± 0.5

a) The values of the forward slope b are obtained by fitting the Serpukhov and the ISR results on the hypothesis that $b \propto \ln s$.

b) The extrapolation factor ε increases with energy because the elastic scattering events are measured at a fixed angle, and thus at a momentum transfer $|t|$ which increases by a factor of about three when the energy is increa The error on ε is obtained by assuming a very generous error $\Delta b = \pm 0.5\ GeV^{-2}$ on the interpolated values of b. The fitted errors are half of the error chosen.

c) The elastic cross-section σ_{el} is obtained from the quoted values of $(d\sigma/dt)_{t=0}$ and the shape of the differenti cross-section as measured by the CERN-Roma and the Aachen-CERN-Genova-Harvard-Torino collaborations.

Note that the values of $d\sigma/dt$ measured at the four energies are practically constant. This is already a qualitative indication that the forward differential cross-section increases sizeably in the ISR energy range, because the momentum transfer corresponding to a fixed scattering angle increases by a factor of three when the centre-of-mass energy passes from 23.4 GeV to 52.8 GeV. Note also that the increase in the value of σ_t, computed by means of Eq. (3), depends only upon the square root of the extrapolation factor ε, so that also if the same value of b is used at the four energies, the corresponding increase of σ_t would be reduced very little. In fact, it would be $\Delta\sigma_t = (3.8 \pm 0.7)$ mb, instead of $\Delta\sigma_t = (4.1 \pm 0.7)$ mb. A last remark concerning the choice of $\rho = 0$ in the application of the optical theorem. As discussed above at the two lower energies, the measured average value of ρ is $+0.01 \pm 0.025$ is certainly compatible with zero. Anyway, also the effect of a sizeable real part on σ_t is very small: for $-0.5 \leq \rho \leq +0.20$ one has a variation of σ_t which is $|\Delta\sigma_t| \leq 0.2$ mb, much smaller than the errors quoted in Table 2.

The Pisa-Stony Brook collaboration has measured the total interaction rate R_t in intersection region number 6 of the ISR and obtained directly the total cross-section σ_t by using the luminosity measured with the Van der Meer method:

$$\sigma_t = \frac{R_t}{L}\ . \quad (4)$$

The system of about 500 scintillation counters which detect the particles due to beam-beam interactions, cover about 90% of the full solid angle, mainly because of the unavoidable "holes" in the forward direction due to the presence of the vacuum pipes in which the stored beams circulate. For this reason the total rate R_t is obtained from the measured rate R_m by application of a correction factor η, so that

$$\sigma_t = \frac{\eta R_m}{L} . \qquad (5)$$

The correction factor is mainly due to events which have some charged tracks at small angle inside the "hole" due to the vacuum pipes. The events of this type can be elastic or quasi-elastic. The elastic differential cross-section is well known and can be integrated to obtain the rate lost in the holes. The corresponding cross-section varies from 0.5 mb to 2.0 mb in the ISR energy range. This increase in the losses is an obvious consequence of the fact that, increasing the energy, the flux in a hole of fixed angular aperture increases because the angular distribution is an approximately universal function of the momentum transfer and not of the angle.

Table 3

Total cross-sections measured by the Pisa-Stony Brook collaboration[5)]

$\sqrt{s}$ (GeV)	Detected cross-section R_m/L (mb)	Increment for elastic losses (mb)	Increment for inelastic losses (mb)	η	σ_t (mb)
23.4	38.66 ± 0.79	0.54 ± 0.10	0.10 ± 0.02	1.017 ± 0.003	39.30 ± 0.79
30.5	39.93 ± 0.81	0.75 ± 0.10	0.17 ± 0.04	1.023 ± 0.004	40.85 ± 0.82
44.8	40.69 ± 0.84	1.54 ± 0.15	0.34 ± 0.10	1.047 ± 0.006	42.57 ± 0.86
52.8	40.53 ± 0.83	1.95 ± 0.20	0.50 ± 0.12	1.060 ± 0.008	42.98 ± 0.84

The increment due to the inelastic losses is obtained by studying the quasi-elastic events as a function of the maximum angle of the tracks in one hemisphere[7)]. This is a very small correction which contributes very little to the observed increase of σ_t.

As discussed in detail by G. Bellettini[8)], the Pisa-Stony Brook collaboration has checked the luminosity measurements to a level of ±3% by means of a system of counters and spark chambers that measure the vertical profiles of the two beams in the crossing region[9)]. More recently, the group has performed other measurements adding lead converters in front of some of the hodoscopes, so as to see also converted gamma rays and increase the detection efficiency. The results obtained on σ_t agree with the published values. A further check has been obtained by measuring σ_t with two beams of unequal energies. Finally, new measurements have been performed at (31.5 - 31.5) GeV/c. The preliminary result is

$$\sqrt{s} = 63 \text{ GeV} ; \quad \sigma_t = (43.3 \pm 0.8) \text{ mb} .$$

The values obtained in the two experiments are plotted in Fig. 2 versus s, together with the two points obtained by the CERN-Roma Collaboration by detecting elastic scattering in the Coulomb region[3]. The agreement between the three sets of data is very good. In particular, the fact that the two measurements based on the Van der Meer method agree, represents a very important check of the reliability of this method, because the luminosity L enters linearly in Eq. (5), but under a square root in Eq. (3). Any energy-dependent error in the method would thus lead to different effects in the two experiments. For instance, if the whole increase in σ_t, as measured by the Pisa-Stony Brook collaboration, was due to an energy-dependent error of -10% in the luminosity measurements, the expected effect on the CERN-Roma experiment would only be a +5% increase of σ_t.

2.4 A new method for measuring cross-sections at the ISR

This observation leads to the suggestion of a third method for measuring σ_t independent of the value for the luminosity. By eliminating L between Eqs. (3) and (5) one obtains

$$\sigma_t = \left(\frac{4\pi\hbar}{p}\right)^2 \frac{\varepsilon}{\eta} \frac{(\Delta R_{el}/\Delta\Omega)}{R_m} , \qquad (6)$$

which shows that σ_t can be obtained by measuring simultaneously the elastic scattering rate at a small angle and the interaction rate R_m on a large fraction of the whole solid angle.

To study the possibility and the difficulties of the method, the CERN-Roma and the Pisa-Stony Brook collaborations have applied it to the data collected in the two different interaction regions (I6 and I8), where the experiments were mounted, during the last months of 1972. To this end the following hypotheses have been made: i) the luminosities in I6 and I8 are the same; ii) the beams are vertically centred in both intersection regions.

Hypothesis (i) can be checked by comparing the vertical effective heights of the source measured simultaneously in various intersection regions. Table 4 shows part of the data collected by K. Potter

Table 4

Comparison of h_{eff} (mm) in various intersections

Date	p (GeV/c)	Intersection region				
		I2	I4	I5	I6	I8
11/12/72	22.6	2.98	2.95	2.90	2.99	2.94
18/12/72	15.4	4.58	4.66	4.40	4.60	4.56
19/12/72	15.4	3.67	3.66	3.44	3.68	3.70

While the values measured in I5 (odd intersection) are about 4% lower, the measurements in the even intersections are consistent with a maximum spread of ±1%. The ratio between the effective heights in two intersection regions is not expected to vary with momentum, and introduces a scale error which can generously be estimated to be ±2%, i.e. ±0.8 mb on σ_t.

To satisfy hypothesis (ii), only runs in which the beams had been carefully centred both in I6 and I8 have been used. Residual centring errors are estimated to introduce ±1% errors in the measured rates R_m and ΔR_{el}.

Since the runs in I6 (CERN-Roma) and I8 (Pisa-Stony Brook) in general were not simultaneous, the total rate R_m has been corrected with a factor κ accounting for the decay of the luminosity, which was carefully followed in the Pisa-Stony Brook experiment. In the accepted runs the factor κ never differed from 1 by more than 6% and its estimated error is ±1%.

Seven runs satisfied the conditions stated above, one at (15.4 + 15.4) GeV and two for each of the other standard ISR energies. The results of the average of these runs are collected in Table 5.

Table 5

Total cross-section obtained in the CERN-Roma-Pisa-Stony Brook exercise

p GeV/c	ε a)	η b)	σ_t mb
11.8	1.10	1.017	40.5 ± 1.0
15.4	1.18	1.023	40.3 ± 1.0
22.6	1.25	1.047	42.6 ± 1.1
26.6	1.35	1.060	43.1 ± 1.2

a) From Table 2
b) From Table 3

The errors on σ_t have been obtained by quadratic combinations of the point-to-point errors appearing in Table 6.

The observed spread of the two measurements available at 11.8, 22.6, and 26.6 GeV/c are ±2.4%, ±1.25%, and ±0.5%, respectively, in good agreement with the expectation obtained from the last row of Table 5. The results of the exercise are in excellent agreement with the measurements based on the Van der Meer luminosity (Tables 2 and 3). This conclusion is not a surprise, if the very good agreement between the results of the CERN-Roma and the Pisa-Stony Brook experiments is considered a simultaneous check of the fact that (i) the luminosity measurements are correct, and (ii) the extrapolation factors ε and η are correctly estimated. On the other hand, one can imagine combinations of errors in the quantities appearing in Eqs. (3) and (5), which would lead to agreement within the errors between the results obtained by the two groups, but disagreement with total cross-sections computed by means of Eq. (6). [Note, for example, that η is in the numerator of Eq. (5) and in the denominator of Eq. (6).]

Table 6

Point-to-point errors in the CERN-Roma-Pisa-Stony Brook exercise

Source of error	p = 11.8	15.4	22.6	26.6
Subtraction of inelastic background in ΔR_{el} a)	0.5%	0.7%	1%	1.5%
Knowledge of b ($\Delta b = 0.5\ GeV^{-2}$ b))	0.4%	0.6%	0.7%	1.0%
Centring in I8	1%	1%	1%	1%
Centring in I6	1%	1%	1%	1%
Knowledge of κ	1%	1%	1%	1%
Knowledge of η	0.3%	0.4%	0.6%	0.7%
Over-all error on σ_t	±2%	±2.2%	±2.4%	±2.7%

a) This is the same error estimated for this subtraction in the CERN-Roma experiment.

b) As in Table 2.

Table 6 shows that important contributions to the relatively large errors obtained in this exercise come from the centring errors and from the knowledge of the luminosity decay factor κ. These sources of errors would disappear in an experiment in which R_m and $\Delta R_{el}/\Delta\Omega$ are measured simultaneously in the same intersection region. In this case also the percentage subtraction of the inelastic background to obtain ΔR_{el} would be greatly reduced, because the large detectors measuring R_m could be used to veto part of the inelastic rate in the telescopes measuring elastic scattering. In conclusion, one could reasonably aim with this method at errors on σ_t definitely smaller than 2% at the maximum ISR momenta. Work along these lines is in progress.

2.5 Energy dependence of the inelastic cross-section

The total, inelastic and elastic proton-proton cross-sections are plotted versus the laboratory momentum in Fig. 3. The figure shows that the behaviour of the total cross-section σ_t, which is practically flat between 30 GeV and 150 GeV and then increases, is obtained as the sum of a decreasing elastic cross-section σ_{el} and of an inelastic cross-section σ_{in} which steadily increases starting from threshold. Taking also into account the fact that at high energies the elastic cross-section is mainly the shadow of the inelastic processes, the following remark by D. Morrison is relevant[10]. "The important experimental result is the continuous increase of the inelastic cross-section with energy". At low energies the elastic cross-section decreases rapidly (also because of the potential scattering contribution which tends to disappear) and hence the total cross-section, being the sum of σ_{el} and σ_{in}, decreases also. At higher energies the elastic cross-section becomes essentially diffractive, and the continuous slow rise of σ_{in} causes a corresponding rise in σ_{el}. As a consequence, $\sigma_t = \sigma_{el} + \sigma_{in}$ also rises.

In the laboratory momentum range 6-1500 GeV/c, the continuous rise of the inelastic cross-section can be fitted with various simple expressions[10]:

$$\sigma_{in} = (26.22 \pm 0.27)\, s^{0.037\pm0.002}\ \text{mb} \lessapprox s^{0.04}\ , \tag{7}$$

$$\sigma_{in} = \left[(23.36 \pm 0.073)(\ln s)^{0.173\pm0.03} + (0.69 \pm 0.35)\right] \text{mb}\ , \tag{8}$$

where s is measured in GeV².

The above remark makes the analytical behaviour of the total cross-section less surprising and points our attention to the inelastic processes which are responsible for the continuous increase of the inelastic cross-section.

2.6 Behaviour of the differential cross-section: a summary of information

Combining all the results obtained on elastic scattering and total cross-sections at the ISR, I have drawn Fig. 4, where the behaviour of the elastic differential cross-section is plotted versus the momentum transfer at $\sqrt{s}$ = 2.35 GeV and $\sqrt{s}$ = 53 GeV. The forward differential cross-sections differ by 20% (see Table 2), because of the 10% increase of the total cross-section and because the contribution of the real part to $(d\sigma/dt)_{t=0}$ is negligible (less than a few per cent). At the two extreme ISR energies the forward slopes differ by about 10% (Table 2), so that there is a cross-over of the two differential cross-sections at $|t| \simeq 0.15$ GeV². At larger momentum transfers the data obtained by the Aachen-CERN-Genova-Harvard-Torino collaboration indicate very clearly, at least in my opinion, that there is a continuous shrinkage of the differential cross-section also for $t \gtrsim 0.15$ GeV². The data at all energies and all momentum transfers can then be made consistent only with the assumption that the position of the minimum at $\sqrt{s}$ = 53 GeV is at a smaller momentum transfer than at $\sqrt{s}$ = 23 GeV. The minimum displaces inward by (10 ± 3)%. Note that the differential cross-section at the secondary diffraction maximum for $\sqrt{s}$ = 23 GeV has not been measured and the curve drawn is obtained by extrapolating the data available at $\sqrt{s}$ = 30.5 GeV.

The curve of Fig. 4 (together with the independent information that the forward real part is compatible with zero for $23.5 \lesssim \sqrt{s} \lesssim 30.5$) pictorially summarizes all the information available both on elastic scattering and total cross-sections in the ISR energy range, and will be the starting point of the discussion of Section 4.

TOTAL CROSS-SECTIONS ABOVE THE ISR ENERGY RANGE

3.1 Cosmic-ray data

At present, only cosmic-ray data can give an indication of the behaviour of the total proton-proton cross-section above 2000 GeV of laboratory energy. Most of the measurements are now a few years old and are summarized in Fig. 5. These "direct" measurements of the inelastic cross-section have been performed in the Russian Proton Satellites and in the Echo Lake experiment[11]. The energy spanned in these experiments is essentially within the ISR energy range and, after the results discussed in the last section, the weak indication that some of the cosmic-ray experiments had a rising cross-section is by now put on a firm experimental basis. "Indirect" measurements attempt to observe unaccompanied hadrons at different depths in the atmosphere and to compare their spectrum with the primary proton spectrum. Since the experimentally-measured "unaccompanied" flux of protons has to be considered an upper limit to the flux of protons which have not interacted with air nuclei, the measured attenuation gives a lower limit for the inelastic cross-section σ_{in}(p-air). To be reliable, the method needs: i) an accurate knowledge of the primary spectrum, ii) clear experimental separation of "unaccompanied" protons from the protons which have interacted and produced a small shower, and iii) a reliable way of computing the total proton-proton (or better proton-nucleon) cross-section from the measured values of σ_{in}(p-air).

The differential primary spectrum has been accurately measured for energies $E \lesssim 2000$ GeV [12] and in the range $50 \le E \le 2000$ GeV is well represented by the power law

$$\frac{dN_p}{dE} = (8.6 \pm 0.8) \times 10^{-6}\, E^{-2.75\pm0.03}\ \frac{\text{protons}}{\text{m}^2\ \text{sr s GeV}} . \tag{9}$$

Above 1000 GeV the integral spectrum has been measured in the Russian Proton Satellites. The results indicate that the power in Eq. (9) changes from 2.7 to about 3.2 [13]; however, it is generally felt that a second measurement is needed before definitely concluding that there is a real change in the spectrum slope above about 1000 GeV.

To measure the hadron energy, total ionization calorimeters are used, together with air shower arrays to detect showers accompanying the hadrons. The definitions of "unaccompanied" hadrons are not the same in different experiments and thus non-coincident results are to be expected. However, the method aims to establish a lower limit to the cross-section σ_{in}(p-air), and the condition to obtain a limit is only that no unaccompanied proton is missed. Of course the lower limit will be closer to the true value of the cross-section if the apparatus is such that a large fraction of the hadrons interacting in the atmosphere above the apparatus are rejected.

Since about 1970 the spectra of unaccompanied hadrons measured at mountain level have been showing some unusual behaviour (see, for instance, the papers of Ref. 14). In 1972, before the ISR results, Yodh, Pal and Trefil concluded[15], on the basis of an analysis of a selected sample of the existing data, that the cross-section σ_{in}(p-air) increases with energy up to $E \simeq 20{,}000$ GeV and deduced the energy behaviour of the proton-nucleon cross-section σ_t by applying the Glauber theory to the interactions of the incoming protons with air nuclei. The result was stated in the form

$$\sigma_t \ge 38.8 + 0.4\,[\ln(s/100)]^2 \quad (s \text{ measured in GeV}^2) . \tag{10}$$

More recently Wdowczyk and Zujewska[16], using another set of data for the spectrum of the unaccompanied protons at mountain level, came to the conclusion that "there is no evidence to suggest a change in the magnitude of the inelastic proton-proton cross-section up to 50,000 GeV".

The different conclusions do not come from the choice of the primary spectrum. In fact, both analyses agree on the use, as primary proton flux, of the spectrum of Eq. (9) extrapolated with the same power law from 2,000 GeV to about 50,000 GeV. In doing so they discard the information obtained with the Russian Proton Satellites, which indicates a steeper spectrum in this energy range and would tend to give a much flatter proton-nucleon cross-section. On the other hand, different "unaccompanied" spectra are used in Refs. 15 and 16. Both spectra used in the two analyses come from measurements performed at Chacaltaya (5200 m a.s.l.), Bolivia, under 550 g/cm^2 of air. They are plotted in Fig. 6 as spectrum A (the one used by Yodh, Pal and Trefil and measured by Kaneko et al.[17]) and as spectrum B (the one used by Wdowczyk and Zujewska and measured by Murakami et al.[14,18]). The continuous line is calculated by attenuating the primary spectrum of Eq. (9) with an interaction length of protons in air (λ = 86 g/cm^2) equal to its value at lower energies. The two spectra have approximately the same slope but differ in absolute value by about a factor of four. They have been obtained with the apparatuses schematically shown in Fig. 7. In both experiments the existence of an energetic hadron is established by a burst in one, and only one, of the shielded detectors, which measures the energy of the hadron itself. Moreover, in spectrum A it is required that one, and only one, unshielded detector, directly above the shielded counter which recorded the burst, records a signal. The pulse-height of the unshielded detector must be less or equal to that due to two minimum ionizing particles. In spectrum B a three-counter telescope must record an event directly above the shielded detector which registered the burst and no pulse-height cut is made on the data.

Recently, Yodh, Pal and Trefil have discussed these triggering requirements[19] and came to the conclusion that in spectrum B accompanied events are registered which are correctly rejected in arrangement A. A typical event of this type is the one shown in the upper part of Fig. 7: two particles come together and, while one crosses the three counters of a vertical hodoscope, the second one crosses only the bottom counter of the same hodoscope. The pulse-height requirement eliminates this event in arrangement A. This is certainly a worrying effect, but to my knowledge no detailed numerical calculation has definitely confirmed that this effect explains the factor of four by which the two spectra disagree. However, in Ref. 19 it is stated that an effect of this order of magnitude has been found in attempting to trigger a cloud-chamber-calorimeter-spark-chamber array. The conclusion of Yodh, Pal and Trefil is that "spectrum A is the correct flux and the original observation that there is evidence for rising proton-proton total cross-sections is on a firm experimental basis". The lower bound derived from these cosmic-ray data[19] is compared with the ISR measurements in Fig. 8.

3.2 Generalities on rising total cross-sections

At present, cosmic-ray data do not show convincingly that the proton-proton total cross-section continues to increase up to 20,000 GeV, but indicate that this has to be considered an open possibility. It is then worth while discussing the general properties of indefinitely rising total cross-sections. This has already been done in I, where the physical meaning of the famous Froissart bound[20]

$$\sigma_t \leq \frac{\pi}{m_\pi^2} \left[\ln \left(\frac{s}{\tilde{s}} \right) \right]^2 \tag{11}$$

has been discussed. The starting point was the usual impact parameter description for the small-angle and high-energy elastic amplitude for scalar particles ($\hbar = 1$)

$$F(s, t) = \frac{1}{\pi} \int d^2a \, e^{i\vec{q}\cdot\vec{a}} f(s, a) \, . \tag{12}$$

In this equation $|\vec{q}| = \sqrt{-t}$ is the bi-momentum transfer (real in the physical region), $\vec{a}$ is the bivector "impact parameter" and f(s, a) is a "partial wave" amplitude corresponding to an orbital angular momentum $\ell \simeq ak$ perpendicular to the momenta of the colliding hadrons, where k is the wave number. The normalization of the amplitude F(s, t) is such that

$$\frac{d\sigma}{dt} = \pi |F(s, t)|^2 , \tag{13}$$

and

$$\sigma_t = 4\pi \, \text{Im} \, F(s, 0) = 8\pi \int \text{Im} \, f(s, a) \, a \, da \, . \tag{14}$$

The bound (11) is a consequence of the fact that the unitarity condition (i.e. probability conservation) and the analyticity properties of the amplitude (which follow from micro-causality) impose very strong limitations on the dependence of Im f(s, a) upon s and a. The unitarity condition is automatically satisfied if one writes

$$f(s, a) = \frac{i}{2} \left[1 - e^{2i\delta(s,a)} \right] , \tag{15}$$

where the phase (eikonal) δ(s, a) is in general a complex quantity

$$\delta(s, a) = \delta_R(s, a) + i \, \delta_I(s, a) \, . \tag{16}$$

On the hypothesis that at very high energy elastic scattering is essentially diffractive (i.e. is due to the absorption of the incoming wave caused by the many open inelastic channels), one has $\delta_R \simeq 0$ and

$$\text{Im} \, f(s, a) = \frac{1 - e^{-2\delta_I(s,a)}}{2} \tag{17}$$

For $\delta_I \to \infty$ the partial wave is completely absorbed and one has

$$\text{from unitarity:} \qquad \text{Im } f(s, a) \leq \frac{1}{2} \ . \tag{18}$$

At the same time it can be shown (see Ref. 1) that

$$\text{from analyticity:} \quad \text{Im } f(s, a) \leq e^{-a/a_0} \left(\frac{s}{\tilde{s}}\right)^{\beta} \left[\text{with } a_0 > \frac{1}{2m_\pi} \simeq 0.7\text{fm}; \quad \beta \leq 1\right] \ . \tag{19}$$

Then the maximum value of σ_t in Eq. (14) is obtained when the partial wave amplitude Im f(s, a) follows the limits (18) and (19) with the behaviour shown in the following figure:

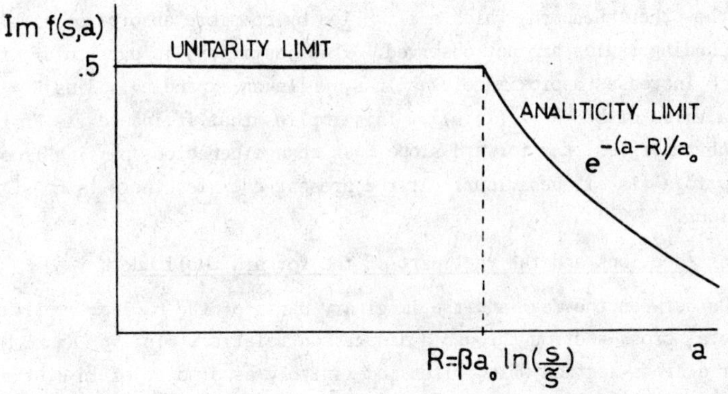

In conclusion, when the Froissart bound is diffractively saturated the partial waves up to a radius $R = \beta a_0 \ln (s/\tilde{s})$ are fully absorbed. The radius of this black disc increases at maximum as ln s, so that the total cross-section cannot increase faster than $(\ln s)^2$, as indicated in Eq. (11). At the same time the "fringe" has a constant width a_0 and its area increases as ln s.

On the hypothesis of purely diffractive scattering, the total cross-section can be written in the form

$$\sigma_t = 4\pi \int_0^{\infty} \left[1 - e^{-\Omega(a)}\right] a \, da \ , \tag{20}$$

where the "opaqueness" $\Omega(a) = 2\delta_I$ at the impact parameter a has been introduced. At the same time the unitarity condition (to be discussed in Section 4.3) implies

$$\sigma_{el} = 2\pi \int \left[1 - e^{-\Omega(a)}\right]^2 a \, da \tag{21}$$

$$\sigma_{in} = 2\pi \int \left[1 - e^{-2\Omega(a)}\right] a \, da \ . \tag{22}$$

We say that the Froissart bound is saturated when $e^{-\Omega(a)} \ll 1$ [i.e. when $\Omega(a) \gg 1$] for $a \lesssim R$. Then $\sigma_{el} \simeq \sigma_{in} \simeq \sigma_t/2 \simeq \pi R^2$. This is certainly not the case in the ISR energy range, since $\sigma_{el}/\sigma_t \simeq 0.17$ (see Table 2) so that we conclude that the rising cross-section in the ISR range has nothing to do with the saturation of the Froissart bound. This observation can be made more precise on the "diffractive" hypothesis [at sufficiently high energies the elastic amplitude F(s, t) is essentially imaginary] by inverting Eq. (12) and by computing first Im f(s, a) and then $\Omega(a)$. The results of a calculation of this type is presented in Fig. 9 [21]. The decrease of the central value of the opaqueness is uncertain owing to the errors of the input data, but the definite conclusion is reached that the opaqueness increases around a = 1 fm. The central absorption $[1 - e^{-\Omega(0)}]$ is about 75% in the ISR energy range, and $\Omega(a)$ is too small to give a flat behaviour of the partial wave amplitude Im f(s, a), as would be necessary in order to observe the Froissart régime.

Since the Froissart bound is not saturated in the ISR energy range, it comes as no surprise that the other phenomena which are implied by complete absorption within a logarithmically expanding radius are not observed. This justifies, in particular, the fact that the slope b increases approximately as ln s, while an expanding radius $R \propto \ln s$ would require a value of $b \propto R^2 \propto (\ln s)^2$. This implies that if indeed the Froissart régime sets up at higher energies, the forward slope must change behaviour passing from an approximate (ln s) law to a $(\ln s)^2$ behaviour. At the present energies there is no sign of this changing behaviour.

3.3 Rising cross-sections and the real part of the forward amplitude

The ratio ρ between the real and the imaginary parts of the nuclear amplitude is related to the total cross-section through a dispersion relation. In 1965 Khuri and Kinoshita[22] proved that for a cross-section which rises indefinitely as a power of ln s, the real part ρ approaches zero at infinite energy from above. For the scattering of scalar particles, the arguments can be easily presented by considering the relativistically normalized scattering amplitude A(E) as a function of the energy E in the laboratory. [A(E) is the amplitude for which dispersion relations are normally written and is related to the total cross-section through the relation $\sigma_t = 4\pi \,\mathrm{Im}\, A/E$.] A cross-section which increases as $(\ln s)^\nu$, i.e. as $(\ln E)^\nu$, then asymptotically implies

$$A(E) \propto i\, E\, (\ln E)^\nu \,. \tag{23}$$

This expression does not satisfy crossing symmetry, which requires on the real energy axis $A(-E) = A^*(E)$. In fact when E is changed into -E, ln E becomes $\ln|E| + i\pi$. This defect can easily be cured by writing

$$A \propto iE\left(\ln E - \frac{i\pi}{2}\right)^\nu \simeq iE(\ln E)^\nu + \frac{\pi\nu}{2}\, E(\ln E)^{\nu-1} \,, \tag{24}$$

which asymptotically implies

$$\rho = \frac{\mathrm{Re}\, A}{\mathrm{Im}\, A} = + \frac{\pi\nu}{2}\, \frac{1}{\ln E} \tag{25}$$

As anticipated, the real part goes to zero from positive values, and its value depends upon the power ν. The derivation presented above is valid for the scattering of scalar particles, but it is also applicable to the sum of the proton-proton and antiproton-proton amplitudes (which is the even-signature amplitude). If the odd-signature amplitude goes to zero rapidly enough, Eq. (25) applies separately to the proton-proton and proton-antiproton amplitudes.

It would be useful to have a more physical justification of the results represented by Eq. (25). The following one, even if not completely satisfactory, is the best I have found. When the Froissart bound is saturated, the imaginary part of each partial wave amplitude f(s, a) for any fixed impact parameter increases from zero to the maximum allowed value $\frac{1}{2}$ with the power law of Eq. (19). When the energy is large enough, according to the argument of the previous subsection, the partial wave is completely absorbed and its real part is zero. This follows from the fact that the real phase shift of a fully absorbed wave is zero because there is no outgoing wave whose phase could be compared with the phase of the incoming wave. On the other hand at low energy, when Im f(s, a) starts to be different from zero, in order not to violate the causality condition, the point representing the amplitude in an Argand diagram must move in the usual counter-clockwise direction, as shown in the following figure[23].

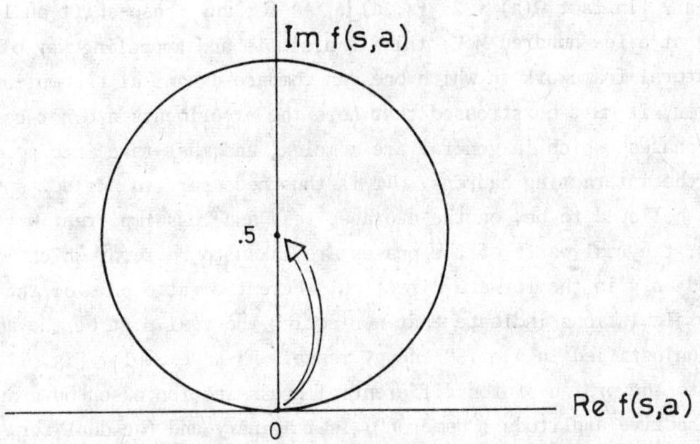

As a consequence, the partial waves which are within the completely absorbing disc of radius $R = \beta a_0 \ln (s/\tilde{s})$ sit at the centre of the Argand diagram and contribute mainly to Im F(t =0), while the partial waves which are in the fringe of radius R and constant width a_0 add essentially positive contributions to Re F(t = 0). Since the area of the disc is $\pi R^2 \propto (\ln s)^2$ and the area of the fringe is $2\pi R a_0 \propto \ln s$, the ratio of the real part (positive and coming essentially from the fringe) to the imaginary part (due mainly to the absorbing disc) is proportional to $2\pi R a_0/\pi R^2 \propto 1/\ln s \propto 1/\ln E$, as in Eq. (25).

After the ISR experiments, various dispersion relation calculations of the forward real part have been published, in which indefinitely rising cross-sections were introduced[24-26]. The continuous curve of Fig. 1 represents the results obtained by Bartel and Diddens[7,25], in the hypothesis of a cross-section rising as $(\ln s)^2$. The experimental points of the CERN-Roma collaboration agree better with these calculations, which predict that the real part crosses zero at $E \simeq 300$ GeV, than with the results of previous calculations in which it was assumed that asymptotically: $\sigma_t^{pp} = \sigma_t^{\bar{p}p} = 40$ mb (dashed line of Fig. 1). Bartel and Diddens have also computed ρ assuming that, above a certain value of the laboratory energy E, the total cross-section becomes constant. The results are presented in Fig. 10 and indicate that measurement of the real part at the maximum ISR energy is sensitive to the behaviour of the total proton-proton cross-section up to energies of the order of 10,000 GeV. However, more recently Joynson and von Schlippe[26)] have pointed out that, from a measurement of ρ alone, it is not easy to distinguish between the following two possibilities: i) cross-sections which satisfy $\sigma_t^{pp}/\sigma_t^{\bar{p}p} \to 1$ and go to a constant value at infinite energy; ii) indefinitely rising cross-sections which are compatible with $\sigma_t^{pp}/\sigma_t^{\bar{p}p} \to 1$ but for which $\sigma_t^{pp} - \sigma_t^{\bar{p}p} \to$ constant.

4. PROTON-PROTON OPAQUENESS AND INELASTIC PROCESSES

4.1 Energy dependence of the proton-proton opaqueness

On the hypothesis that the scattering amplitude is purely imaginary, the momentum-transfer behaviour of the proton-proton elastic cross-section is completely determined by the impact-parameter dependence of the opaqueness $\Omega(a)$. This procedure is nothing else than a phase-shift analysis performed at high energy using the hypothesis that the phases are purely imaginary [in fact $\Omega(a) = 2\delta_I(s, a)$]. As for any phase-shift analysis of proton-proton scattering at a few hundred MeV, this is a simple and appealing way of representing the data and a natural framework in which one can compare dynamical assumptions with experiments. However, it must be stressed that here the experiments are not complete enough to determine the phases, which in general are complex, and must take into account the spinor nature of the interacting hadrons. It is thus necessary to neglect spin-effects, which however are believed to be, on the average, less and less important with increasing energy. Moreover, the real parts of the phases are taken to be zero, which is experimentally justified only in the forward direction, where the ratio ρ is of the order of few per cent. Model calculations indicate that neglecting the real part of the amplitude is probably not too unjustified in the ISR energy range. As an example, Fig. 11 shows the results of a fit to the proton-proton differential cross-section based on a dual model, in which the diffractive amplitude ("Pomeron") is imaginary and the dual term is real[27)]. While at PS energies ($s \simeq 50$ GeV^2) the real contribution dominates above $|t| \simeq 0.8$ GeV^2, in the ISR energy range ($s \simeq 2000$ GeV^2) it simply fills the diffraction minimum.

The opaqueness profiles of Fig. 9 show that at small impact parameters the absorption either decreases or, perhaps, remains constant passing from $\sqrt{s} = 20$ GeV to 50 GeV. At large impact parameter the increase is definite and large. This contrasts with the expectation derived from the simplest Regge picture, in which elastic scattering is dominated by the exchange of a Pomeron which is a pole in the J-plane and whose trajectory has non-zero slope in t, thus explaining the energy dependence of the forward slope parameter b. According to this picture the central absorption decreases with energy, while the radius increases in such a way as to produce a constant total cross-section (Fig. 12). Of course a more complicated structure of the Pomeron in the J-plane (cuts) can fit the observed behaviour of $\Omega(a)$. Various Regge models which have been proposed in order to explain the energy behaviour of σ_t and b have already been reviewed in I, and will not be discussed here any further[28)].

Figure 9 contains other important information: the average impact parameter of the increase in the opacity in the ISR energy range is of the order of 1 fm, which is larger than the average impact parameter of the opaqueness at the minimum energy of the ISR ($\sim$ 0.5 fm). In other words, the increase in the opaqueness when $\sqrt{s}$ passes from $\sim$ 20 GeV to $\sim$ 50 GeV is more peripheral than the opaqueness itself. This is quite natural, since Fig. 4 shows that the increase in the elastic cross-section comes mainly from the small momentum transfers, which corresponds to large impact parameters. In conclusion, the dynamics of the inelastic processes must be such that i) the inelastic cross-section has the continuously increasing behaviour shown in Fig. 3; ii) in the ISR energy range the increase in the opaqueness produced by their shadows must be more peripheral, in impact parameter space, than the opaqueness itself. To discuss the implications of these two statements let us first summarize the physics of the inelastic processes, as observed at the ISR and at NAL.

4.2 Inelastic processes in the ISR energy range

Complete experimental reviews have recently been prepared by Sens[29] and Diddens[7], a very physical description has been given by Morrison[10], while the more theoretical aspects of the inelastic processes have been thoroughly discussed by Jacob[29] and, at this School, by Berger[30]. Here the presentation will be short.

Table 7 [29] summarizes the experiments performed on inelastic processes at the ISR. In Fig. 20 the various experiments are localized around the rings.

At present the main conclusions of the study of the inelastic processes at the ISR and at NAL can be summarized as follows.

i) The multiplicity of the produced particles grows slowly with s and most of the particles are pions. Energy dependences of the type ln s and $s^{\frac{1}{4}}$ fit the data qualitatively.

ii) The cross-section σ_n for producing n-particles first increases with energy and then slowly decreases. The threshold of each σ_n increases with n.

iii) The average transverse momentum of $\langle p_T \rangle$ is of the order of 0.4 GeV/c for all energies and particles, and the longitudinal momentum is roughly proportional to $s^{\frac{1}{2}}$.

iv) Typical ISR invariant cross-sections for the hadron species h,

$$f_h(s, x, p_T) = E \frac{d\sigma_h}{d^3p} \tag{26}$$

at p_T = 0.4 GeV/c are shown in Fig. 21 versus the Feynman variable x [10]. The protons show a "leading particle" effect, while the other spectra drop rapidly with x. When plotted versus the rapidity in the centre of mass y, the invariant cross-section displays a "plateau" around y = 0. A compilation of the data[31] is presented in Fig. 22 (the variable here is y_{lab}). All the particles show a "plateau", with the exception of antiprotons, and the invariant cross-section does not depend upon s, i.e. it scales, within the experimental errors (±10 - 15%).

v) Recent data of the BS Collaboration show that the "plateaus", integrated over p_T and at a given energy, are not flat in y [32] (the π^+ spectra are plotted in Fig. 23). The π-plateau is found to increase by (8 ± 7)% when the energy passes from $\sqrt{s}$ = 30.6 to 52.8 GeV. Data of the SS Collaboration show no increase within 5% errors. The antiproton "plateau" increases by (25 ± 7)% [32].

Table 7

ISR experiments on inelastic processes

Intersection Region	Group	Abbreviation	Apparatus	Experiment
I1	CERN-Columbia-Rockefeller	CCR	Fig. 13	π^0 spectra at 90° for $p_T < 9$ GeV/c. Correlations associated with large p_T.
I1	Saclay-Strasbourg	SS	Fig. 14	Charged particles and π^0 around 90°.
I2	CERN-Holland-Lancaster-Manchester	CHLM	Fig. 15	Particle spectra at small angles. Diffraction dissociation; correlations.
I2	British-Scandinavian	BS	Fig. 16	Particle spectra $30° \leq \theta \leq 90°$ for $p_T \leq 3$ GeV/c.
I2	Argonne-Bologna-CERN	ABC	Magnetic spectrometer around ~ 100 mrad	Particle spectra at medium angles.
I4	CERN-Hamburg-Orsay Vienna	CHOV	Lead-glass Čerenkov counters	π^0 spectra.
I4	CERN-Hamburg-Vienna	CHV	Čerenkov and scintillation counters	Charged particle/photon correlations.
I6	Aachen-CERN-Genova-Harvard-Torino	ACGHT	Fig. 17	Diffraction dissociation.
I8	Pisa-Stony Brook	PSB	Figs. 18, 19	Single-particle spectra without momentum analysis. Correlations and large p_T.

vi) On the average, the colliding protons keep about 40% of their energy and their spectrum is quite flat. However, a peak is observed in the invariant cross-section for $x \simeq 1$ (Figs. 21 and 24), after subtraction of the elastic event, usually attributed to the production of high masses in the "opposite" hemisphere. The cross-section for this process (5-6 mb) is of the same order of magnitude as the elastic one, and is approximately constant in the NAL and ISR energy range. (This point is discussed in detail in all review papers quoted above and in particular by J.C. Sens.) The process is normally described as the "diffraction" production of large masses, similar to the production of isolated proton isobars well known at a few GeV of laboratory energy. The relation between the missing mass M and the Feynman variable x is

$$x \simeq 1 - \frac{M^2}{s} \ . \tag{27}$$

Since diffraction production is present for $x > 0.95$ (Fig. 24), masses as large as 10 GeV should be excited at the top ISR energies. The invariant cross-section seems to scale with energy, and a recent detailed analysis of the CHLM Collaboration[33] that the s, t, and M^2 dependence of the data at $\sqrt{s} = 23.5$ and 30.5 GeV can be well represented by a formula which is suggested by the "triple Pomeron" definition of the phenomenen[30]:

$$f = E\frac{d\sigma}{d^3p} = \frac{s}{\pi}\frac{d^2\sigma}{dM^2dt} = G(t)\left(\frac{M^2}{s}\right)^{1-2\alpha(t)} \ , \tag{28}$$

where

$$\alpha(t) = 1 + 0.2\ t \ .$$

Recent data obtained at NAL by the Columbia-Stony Brook Collaboration seem to require more complicated parametrizations[34]

vii) In the CCR, SS, and BS experiments, gamma-rays and charged particles of large transverse momenta have been obsered. The invariant cross-section above $p_T \simeq 2$ GeV/c deviates from the exponential fall-off characteristic of low transverse momenta. The argument is discussed at this School by L. Caneschi.

viii) The results of the measurements of particle correlations by the PSB and CHV Collaborations are evidence for the hypothesis of dominant short-range order (SRO) in rapidity. In particular they favour a multiperipheral model in which clusters are produced that decay with a mean multiplicity in the ISR energy range of the order of four[30].

At present the above phenomena are best described in the framework of the so-called two-component models[35]. The largest component, which accounts for $\sim$ 80% of the inelastic cross-section, is characterized by SRO correlations in rapidity space. The amount of correlations indicates a certain clustering of the emitted particles. The smallest component, the diffractive one, explains $\sim$ 20% of σ_{in} and can be physically defined as the one which leads to events with large rapidity gaps. It presents long-range correlations and its general features seem at present reasonably well described, within the experimental errors, by the "triple-Pomeron" formula of Eq. (28). Moreover, there are events that are characterized by the production of particles with large transferse momenta. These events could also be explained in the framework of a conventional multiperipheral model[36], but at present the data available on correlations seem to indicate that they require at least the production of two clusters with opposite and large momenta. Needless to say, strong interactions are so complicated that these two (or three) components are certainly related one to the other. In particular, if the mechanism of diffraction production of large masses is similar to elastic scattering, from an s-channel point of view diffraction dissociation can be computed as the shadow of the non-diffractive processes[37]. The two components are also strongly related in the t-channel Mueller-Regge approach, the shadow of the SRO process being described as exchange of the Pomeron, whose triple coupling originates the scaling part of the diffractive events[30]. Based on this point of view, perturbative approaches have been developed which make the hypothesis that is zeroth order only SRO process are present, and then generate from this contribution the diffractive component[38].

The experimental data on the energy dependence of the two (or three) components are at present not accurate enough to establish if and how much each of them contributes to the measured increase of the inelastic cross-section in the ISR energy range $[\Delta\sigma_{in} = (3.3 \pm 0.7)\text{mb}]$ E. Berger gives estimates of the various contributions[30], and I shall refer to them in the following. Of course only accurate experimental data can settle the question; for this reason the last section is devoted to a short presentation of the relevant experiments under way or planned at the ISR.

In the next sections I shall summarize what has been done in connecting elastic and inelastic processes at ISR energies. As anticipated in the Introduction, at present no definite conclusion can be drawn. However, I consider the approach useful because it raises some simple and physical questions on the properties of the various classes of inelastic processes, and in particular of the two main components. The answers to these questions will undoubtly improve our understanding of high-energy hadrodynamics.

4.3 Elastic scattering as the shadow of the inelastic processes

In the large, but limited, ISR energy range the increase in the opaqueness is more peripheral than the absorption itself. The first question is, which is the most peripheral between the two inelastic components? From a geometrical point of view the diffraction component is more peripheral, because it is associated with lower average multiplicity and thus with glancing collisions. In the same picture the pionization (SRO) component, which corresponds to larger multiplicities, is more "central" because in head-on collisions more energy can be transformed into the masses of the produced particles. The opposite description holds in the multiperipheral model, where larger multiplicities correspond to longer exchanged chains, and thus to larger impact parameters[39]. The same conclusion would be reached from an optical point of view, in which diffraction production, as diffraction scattering, is the shadow of the non-diffractive processes: since diffraction phenomena are larger where the absorption is larger, diffraction production and diffraction scattering take place, on the average, at impact parameters which are smaller than the average impact parameter of the processes which cause the absorption. Some of these arguments have to be reconsidered when there is a strong reabsorption of the incoming and outgoing particles. In fact the (negative) absorption corrections are more central than the amplitude to which they apply, so that by introducing them the peripherality increases. This discussion indicates that "intuitive" arguments must be regarded with caution and calls for an at least qualitative understanding of the opaqueness profile at a fixed energy in terms of the known inelastic processes. This would already be a good starting point towards understanding the observed increasing opaqueness.

The unitarity condition in momentum space reads:

$$4\pi \text{ Im } F(t) = \sum_{\text{el.states } m} \langle f|T^{+}|m\rangle \langle m|T|i\rangle + \sum_{\text{inel.states } n} \langle f|T^{+}|n\rangle \langle n|T|i\rangle , \tag{29}$$

where the sums over intermediate elastic and inelastic channels have been separated. The tree terms are the total, elastic and inelastic overlap functions[40]:

$$h_t(t) = 4\pi \text{ Im } F(t)$$

$$h_{el}(t) = \sum_{\text{el.states}} \langle f|T^{+}|m\rangle \langle m|T|i\rangle \tag{30}$$

$$h_{in}(t) = \sum_{\text{incl.sta:es}} \langle f|T^{+}|n\rangle \langle n|T|i\rangle ,$$

and are so normalized that

$$h_t(0) = \sigma_t \; ; \quad h_{el}(0) = \sigma_{el} \; ; \quad h_{in}(0) = \sigma_{in} \; . \tag{31}$$

On the hypothesis of a purely imaginary amplitude, for each value of t the function $F(t) = \text{Im } F(t)$ can be obtained from the measured differential cross-section [see Eq. (13)] and the elastic overlap integral computed by integrating over angular variables. By subtraction one eventually obtains the inelastic overlap function. This procedure has been applied to proton-proton data by de Groot and Miettinen[41,42] and the results are shown in Fig. 25. The inelastic overlap changes sign around 0.6 GeV². This is a very general property which was first noticed by Zachariasen[43] and has a very important consequence: the phases of the production amplitudes cannot be neglected in computing $h_{in}(t)$ because the zero of $h_{in}(t)$ is sensitive not only to the absolute values but also to the phases. Thus experimental information on the absolute values of the production matrix elements are not enough to compute diffraction scattering; a model which provides the phases is also needed. The physical reason for this is that the phases are related to the position in space at which the particles are produced[41,44], and of course this position is relevant in computing the elastic shadow.

No calculation exists at present of the inelastic overlap function in the framework of a two-component model of the inelastic processes. A phenomenological approach along the lines suggested by Van Hove has been used by de Groot and Miettinen, and the results are shown in Fig. 25. The inelastic overlap function is taken as a superposition of a central part (in impact parameter space) proportional to $J_1(R\sqrt{-t})$, which they attribute to the non-diffractive production, and a more peripheral contribution proportional to $J_0(R\sqrt{-t})$, which they suggest to be due to the diffractive processes. Both terms contain a modulating function of the momentum transfer, which takes into account that the non-diffractive (diffractive) production does not happen in a black disc (in a ring) of radius R, as implied by the use of the Bessel function J_1 (J_0), but in interaction volumes which have smooth edges. The fit is very good and indicates that the elastic data at a fixed energy can be fitted using for $h_{in}(t)$ the sum of two contributions of simple analytical form. The most peripheral one (which has been chosen to contribute an inelastic cross-section of about 6 mb) is responsible for the forward peak in the differential cross-section[45] and would thus be connected very naturally with the increasing cross-section. However, at this stage both the relevance of the fit and the identification of the more peripheral contribution with diffractive production is a matter of opinion. A better understanding can come in going from momentum transfer to impact parameter space, where unitarity is diagonal.

By Fourier-transforming the amplitude F(t) [Eq. (12)] and the inelastic overlap function

$$G_{in}(a) = \frac{1}{(4\pi)^2} \int e^{-iq\,a}\, h_{in}(t = q^2)\, d^2q \ , \tag{32}$$

the unitarity condition assumes the very simple diagonal expression:

$$\mathrm{Im}\, f(a) = |f(a)|^2 + G_{in}(a) \ , \tag{33}$$

which relates the inelastic overlap function in impact parameter space to the elastic amplitude at the <u>same</u> impact parameter (this is a consequence of angular momentum conservation). On the diffractive hypothesis Eq. (33) can be solved for f(a). This can easily be done by writing

$$G_{in}(a) = \frac{\left[1 - e^{-2\Omega(a)}\right]}{4} \ . \tag{34}$$

Then the expression already introduced

$$f(a) = \frac{\left[1 - e^{-\Omega(a)}\right]}{2} \tag{35}$$

satisfies Eq. (33). Integrating Eq. (33) on the impact parameters one gets $\sigma_t = \sigma_{el} + \sigma_{in}$, where the cross-sections have the expressions anticipated in Eqs. (20), (21), and (22).

The inelastic overlap function $G_{in}(a)$ is easily computed from the values of the opaqueness $\Omega(a)$ plotted in Fig. 9. The results are represented by the continous curve of Fig. 26, together with the outputs of three other similar calculations[46-48]. Note that a straight line in Fig. 26 represent a Gaussian overlap function in impact parameter space. It is no surprise that from the data one obtains an overlap function which is not very different from a Gaussian, since the experimental ratio of the elastic to the total cross-section is ~ 0.17, and Van Hove has shown many years ago that for a Gaussian fully absorbing at the centre this ratio is 0.185. Three of the four curves plotted in Fig. 26 show a variation of slope around $a^2 = 3$ fm^2. The agreement between the three analyses has to be considered reasonable good in view of the fact that the input data are taken from experiments which have to be normalized one to the other, and that Henyey et al. use a particular prescription to obtain a real part different from zero at all momentum transfers while I and Miettinen and Pirila take the real part to be zero.

The calculation of Henzi and Valin gives an overlap function which is much more similar to a Gaussian[47]. At present I can only attribute the discrepancy, with respect to the other three calculations, to an unknown difference between the cross-section data used as input.

The lower curve of Fig. 26 represents the increment of the inelastic overlap function in the ISR energy range obtained from the values of $\Delta\Omega$ plotted in the lower part of Fig. 9. Note that the uncertainty for small values of a^2 has very little influence on the increment $\Delta\sigma_{in}$ of the inelastic cross-section, whose explicit form is [Eqs. (22) and (34)]:

$$\Delta\sigma_{in} = 8\pi \int_0^\infty \Delta G_{in}(a) a \, da = 4\pi \int_0^\infty \Delta G_{in}(a) \, da^2 \, . \qquad (36)$$

It is interesting to remark that, above $a^2 = 2.5$ fm^2, the increment ΔG_{in} is approximatively proportional to G_{in} with a proportionality factor equal to ~ 4. It is impossible to assess at present the meaning of this fact. It could be accidental, or it could indicate that the inelastic mechanism, which causes the increase of the total cross-section, has an overlap function which, being proportional to ΔG_{in} at all impact parameters, is peripheral and corresponds to an integrated cross-section about 4 times larger than σ_{in}, i.e. to $\sigma_{in} \simeq 14$ mb at $\sqrt{s} = 53$ GeV. No obvious identification of such a mechanism is possible at present[49].

Equally difficult is the problem of recognizing in $G_{in}(a)$ the sum of two main contributions, one due to SRO processes and the other to diffraction dissociation. To proceed further, it is necessary to go to specific models.

4.4 Models of the inelastic processes and of the increasing absorption

As discussed by Berger[30], various papers have appeared[38,50,51] which attribute the increasing behaviour of the inelastic cross-section to the diffraction component. The ISR data summarized in Section 4.2 indicate that for $t \simeq 0$ the invariant cross-section $f(x, t)$ behaves as $1/(1 - x)$, where x is the Feynman variable [Eqs. (27) and (28)]. To compute the inelastic diffraction cross-section, $f(x, t)$ has to be integrated up to the kinematical limit $x_{max} = 1 - M_0^2/s$ [see Eq. (27)] where M_0^2 is the minimum value of the mass that is supposed to be produced through the triple-Pomeron mechanism, which leads to Eq. (27). The integration gives a cross-section which increases as $\ln(s/M_0^2)$, so that the increment of the inelastic cross-section for $s_1 < s < s_2$ is proportional to $\ln(s_1/s_2)$ in the small shrinkage limit. The proportional factor can be obtained by fitting Eq. (28) to the measured spectra. As a result one finds that this process could contribute as much as $\Delta\sigma_{in} = (1.7 \pm 0.7)$ mb. This figure and its error are my personal best estimates, taking also into account the numbers obtained by various people who have played this type of game. The increment of the diffractive inelastic cross-section does not seem to saturate the observed increase $[\Delta\sigma_{in} = (3.3 \pm 0.7)\text{mb}]$. However, by stretching the errors, one could try to attribute a large fraction of the increasing inelastic cross-section to this mechanism. The following question arises: which is the impact parameter distribution of the inelastic overlap function due to proton diffraction dissociation? Sakai and White came to the conclusion that this overlap function is central or peripheral according to whether one assumes s-channel or t-channel helicity conservation in the diffraction process[52]. The impact parameter distributions of the various overlap functions are plotted in Fig. 27 for $\sqrt{s} = 30.5$ GeV in the case of t-channel helicity conservation. It is seen that the diffractive component has a maximum around 0.8 fm, while the non-diffraction contribution is peaked at $b = 0$. The opposite happens when s-channel helicity conservation is assumed. Since t-channel helicity conservation fits the data on proton diffraction dissociation in isolated resonances obtained at PS energies[53], it seems reasonable to assume the same to be true also for the production of the very large masses observed at the ISR. If the increment of the overlap function has the same peripheral nature as the overlap function itself, a qualitative understanding of the phenomenon can be obtained. No quantitative calculations of the increment of the overlap function exist at present, but in the framework of the Mueller-Regge approach Amati, Caneschi and Ciafaloni[51] have paid attention to the impact parameter distribution of the inelastic overlap functions. Two extreme possibilities have been considered. First, the non-diffractive (pionization) process contribute a central overlap function, while the triple-Pomeron contribution is more peripheral. This picture is considered unnatural by the authors because in a multiperipheral scheme the diffraction component, which gives rise to events with relatively small multiplicities, is more central than the pionization component. As a second possibility, they consider a central diffraction component and a more peripheral pionization. Thus the tail of G(b) (Fig. 26) is due to pionization, which is supposed to be described by the exchange of a Pomeron pole plus a cut, whose contribution disappears with energy and contributes about 0.7 mb to $\Delta\sigma_{in}$. This second model, although in agreement with multiperipheral views, is less transparent than the former one. Indeed the over-all peripheral increase of the inelastic overlap function is interpreted as due to the sum of a central increase (due to diffraction) and of a typical Regge behaviour (due to pionization). This second contribution is "typical" because it gives the well-known Regge phenomenology (Fig. 12): a decreasing central absorption together with an expanding radius. In the model the central decrease of the absorption compensate for the increase due to diffraction (which is also supposed to be central) so that the over-all increase around 1 fm is a manifestation of the expanding radius of the pionization part, which is not cancelled by the diffraction contribution.

The "triple-Pomeron" approaches, in which the increasing cross-section is related to the peak observed in the inclusive protons spectra, assume that an increase of the diffractive cross-section does not influence the other contributions, so that it leads to an equal rise of the total inelastic cross-section. Recent work by Blankenbecler has shown that there are reasons to believe that this could not be the case[54]. Under quite general circumstances the rise of a diffractive cross-section can induce, because of the consequent increased absorption implied by unitarity, a twice as large decrease of the pionization component, so that the total inelastic cross-section decreases. This should happen when the driving mechanism is diffractive, i.e. it is imaginary, but does not apply when the production process is due to the exchange of real particles.

An approach which is based on the exchange of real particles has been advocated, for completely different reasons, by Henyey, Hong Tuan and Kane[46] to explain the increasing absorption at large impact parameters. They feel that it is unreasonable to assume that the peripheral component is caused by diffraction dissociation alone, because diffraction is the shadow of other scattering so that if there is no other scattering there cannot be diffractive scattering either. The suggested mechanism consists in the virtual dissociation of one of the protons into a system of two (or more) particles (which occupies more area than the original proton) followed by interaction of the fragments with the other proton. The tail appearing in the overlap function for a $\gtrsim$ 2 fm (Fig. 26) is attributed to this dissociation phenomenon which gives essentially rise to a Deck effect. Qualitative arguments suggest that this tail could increase with energy, because the phase space enlarges and more dissociative states can be produced. The main drawback of a crude model based on these ideas and proposed in the same paper by Henyey et al.[46] is in the very small cross-section which it predicts for the true diffraction dissociation events (less than 2 mb), definitely smaller than the value measured at NAL and ISR energies.

Let us now pass to a different approach. Since the finding of the rising cross-section no explicit model of the inelastic processes has been proposed in which the increasing absorption is attributed to the pionization component. Probably this is due to the fact that from a t-channel point of view a model of this type can be stated in very simple words, which however hide many mysteries: The Pomeron intercept is larger than 1 (let us say equal to $1 + \varepsilon$). In such a model the cross-section would increase as a power of the energy ($\sigma \propto s^{\varepsilon}$) and, in order not to violate the Froissart bound, absorptive corrections have to be introduced. The absorbed multiperipheral model of Finkelstein and Zachariasen[55] saturates Froissart bound ($\sigma_t \propto \ln^2 s$) and in this regime departs from the short-range order characteristic of the non-absorbed multiperipheral model[39]. However, as already remarked, at ISR energies this regime is not reached at all and the effect of the absorptive corrections are probably small, so that short-range order can be a reasonable approximation in describing multiparticle production, in agreement with ISR data. As a first approximation one can neglect absorption altogether, and choose the parameter ε equal to 0.04, as deduced by the fit to the inelastic cross-section given in Eq. (7). Then it is easy to see that an increment of the elastic partial wave amplitude Im f(a), which has the correct impact parameter distribution, is obtained. In fact, to fit the elastic slope, a Pomeron of intercept 1 must give a decrease of Im f(a) at a = 0 equal to $\sim$ 10% together with an expansion of the impact parameter scale by ∿ 5%. For an intercept at 1.04, the factor $s^{0.04}$ in front of the amplitude compensates the decrease at a = 0, leading to an increment of the amplitude which is practically zero for a = 0, and has the correct peripheral behaviour[56]. This model would predict an increase of the plateau of the single particle rapidity distribution dσ/dy which is also of the order of 10% in the ISR energy range, not in contradiction with the ISR data presented in Section 4.2.

It seems that no difficulty arises in attributing the observed increment of the absorption in the ISR energy range to the SRO component. However, no consideration has yet been given to the basic problem: does the shadow of the SRO component reproduce the observed inelastic overlap function at a fixed (and high) energy? Short-range order is a natural feature of multiperipheralism, so that this is a reasonable framework in which to try and answer this question. It is quite well known that the standard multiperipheral model of Amati-Bertocchi-Fubini-Stanghellini and Tonin is not satisfactory from this point of view. In particular the forward logarithmic slope of the elastic cross-section is too large, i.e. the overlap function is too much spread out in impact parameter space, and the shrinkage is too fast with energy[57]. Recently Henyey has studied multiperipheral models in impact parameter space[58] coming to the conclusion that this difficulty can be overcome by introducing the hypothesis that (i) clusters of particles are produced along the multiperipheral chain and (ii) the cluster size increases with energy. This result agrees with the conclusion reach in Ref. 57 and with indications deduced from the study of correlations in the ISR energy range[30]. However more calculations are needed, with particular attention to the effect of the phases of the inelastic amplitudes.

The rise of the proton-proton total cross-section can be explained as a threshold effect. From this point of view it is necessary to justify the very high value of the observed threshold ($s \simeq 100$ GeV^2). In the diffraction model described at the beginning of this section this is obtained through the requirement $s \gg M^2$, where M is the "large" diffractively-produced mass. It has also been suggested[59] that a large threshold effect could be due to the onset of the production of baryon-antibaryon pairs, which increases markedly in the ISR energy range (Section 4.2). From a multiperipheral point of view the large threshold for this phenomenon is due to the fact that the production of a baryon-antibaryon pair along a multiperipheral chain absorbs an amount of rapidity $\Delta y \simeq 3$, which is large with respect to the total rapidity range available at $s \simeq 100$ GeV^2 ($Y \simeq 4.5$). Based on the measured inclusive distributions, various estimates of the value of $\Delta\sigma_{in}$ due to this phenomenon have been made. They range from $\Delta\sigma_{in} \simeq 1$ mb [30] to 5-6 mb [59]. The result is model dependent because the observed 25% increase of the antiproton inclusive distribution can be obtained either with a constant cross-section of the events in which at least one antiproton is produced and an increasing number of antiprotons per event, or with an increasing cross-section and an almost constant antiproton multiplicity per event.

While it seems that this phenomenon could even account for the whole observed increase of the inelastic cross-section, I feel that the impact parameter distribution of its overlap function has not the correct peripheral behaviour to explain the known experimental facts without the intervention of some other mechanism. The qualitative argument goes as follows. The production of particles along a multiperipheral chain is equivalent to the production at the vertices of a random walk in impact parameter space. The radius R of the interaction is proportional to the step size Δa and to $\sqrt{n}$, where n is the number of steps and equals the average multiplicity, if particles are produced singly along a chain. When a proton-antiproton pair (and in general a baryon-antibaryon pair) is produced, the rapidity interval available to the other particles is greatly reduced and thus also the number of steps. This implies that the radius of these events is smaller than the average radius of the SRO component, in disagreement with the property required for the mechanism which causes the increase of the total cross-section. The argument is admittedly only qualitative and unfortunately no detailed calculation exists. Moreover it is not at all conclusive because the difficulty can be overcome if a second mechanism acts at the same time (for instance, if the SRO component has an overlap function which behaves as a Regge pole[60)] or if the central part of the antibaryon overlap function is absorbed at small impact parameters).

The production of particles of large transverse momenta is another phenomenon which sets in at the ISR. A model has been proposed[61)] which considers it as a manifestation of the existence of a "hard" collision between parton pairs, which also causes the increase of the total cross-section. The increment of the inelastic cross-section is estimated to be of the right order of magnitude on the basis of the observed behaviour of the single particle spectra at large transverse momenta measured at the ISR[61)] and with cosmic rays[62)]. It is my impression that these estimates are too optimistic. Moreover no calculation exists of the impact parameters distribution of the overlap inelastic function due to such a process, but a simple-minded geometrical approach would suggest that these "hard" collisions are more central than the average "soft" inelastic events. However, as remarked in connection with the antibaryon model, the simultaneous presence of another mechanism could give to the overall increment of the inelastic overlap function the needed peripheral nature.

Eikonal models are the last to be discussed here in spite of the fact that the most well known of them [the Cheng and Wu model[63)]] was the first to consider in recent times, but prior to the publication of the ISR results, the possibility of indefinitely rising hadron-hadron cross-sections. Eikonal models have been already discussed in I and here the presentation will be short. They aim to derive the high energy properties of the eikonal, whose imaginary part is the opaqueness $\Omega(a)$. The approach is based on the perturbation expansion of some field theory and the consequent summing of the graphs which are dominant in the high-energy limit[63,64)]. The quite complete attempt by Cheng and Wu uses quantum electrodynamics with massive photons as starting field theory. They concluded that in such a theory [which satisfies unitarity and analyticity but has difficulties with energy conservation[65)]] the imaginary part of the eikonal (i.e. the opaqueness) increases as a power of s, so that the Froissart bound is saturated at infinite energy. Since this result is obtained by summing an infinite set of graphs, it is not easy to attribute the rising cross-section to a particular physical mechanism. However by considering the structure of the summed series, it is seen that in this model the imaginary "potential" which acts between the increasing particles is mainly due to the shadow of the inelastic processes in which slow particles are produced. This potential becomes larger and larger with increasing energy and causes the complete absorption of the central partial wave and thus the saturation of the Froissart bound. Since a similar effect is found in other field theories, Cheng and Wu inferred that this probably is a general feature of any local field theory and, in particular, of high-energy hadrodynamics. However it has also been shown that singular "potentials" do not necessarily lead to saturation of Froissart bound[38].

Cheng, Walker and Wu[66] proposed a model which has the asymptotic features now described and fits the existing data for $\pi^{\pm}$, $K^{\pm}$ and $p^{\pm}$ scattering on protons. In the model the opaqueness increases as a power of s and for its impact parameter dependence a simple but arbitrary form is taken, so that

$$\Omega(a) \propto s^{c}\, e^{-\lambda(a^2+a_0^2)^{1/2}} \qquad (c > 0) \tag{37}$$

where λ is the same for all channels, while the parameter a_0, has different values for $\pi^{\pm}p$ $K^{\pm}p$ and $p^{\pm}p$ scattering. The model is made crossing symmetric by introducing a suitable phase factor in the eikonal, so that the real part of the nuclear amplitude comes out automatically at all momentum transfers. Its value for $t = 0$ [24] agrees with the data available and with dispersion relation calculations, once the fourteen parameters of the model are fixed in such a way that a very good fit to the six total cross-section is obtained above $s \simeq 50$ GeV². The last figure of Ref. 66 show that with the same parameters a reasonably good fit is obtained to the elastic differential proton-proton cross-section for $30.5 \leq \sqrt{s} \leq 53$ GeV. This fact at first sight seems to indicate that Eq. (37) is a good representation of the impact parameter distribution of the opaqueness and of its increment in the ISR energy range. However this conclusion is based on an inaccurate comparison of the data with the output of the model and is not justified because we have seen that a detailed analysis of the experimental data shows that the increment $\Delta\Omega(a)$ of the opaqueness is more peripheral than the opaqueness $\Omega(a)$, while Eq. (37) implies that the increment $\Delta\Omega(a)$ has the same impact parameter distribution of $\Omega(a)$ [67]. This criticism applies only to the parametrization chosen in the fit by Cheng, Walker and Wu [in which the s- and a-dependence of $\Omega(a)$ are factorized] and , of course, has nothing to say about the validity of the Cheng and Wu model, which is an interesting possibility still open.

FUTURE ISR EXPERIMENTS

In two years of running the ISR have provided a large amount of new and unexpected information and, as usually, have suggested many more questions than answers. Clearly the point of view adopted in this discussion represents only a very particular cut through the complexity of the experimental data. However it may be useful to list briefly the experiments, either under way or planned at the ISR, which should shed some light on the problems touched upon.

At the beginning of 1974 the Aachen-CERN-Genova-Harvard-Torino collaboration will perform a measurement of proton-proton elastic scattering at small angles ($\theta \gtrsim 3$ mrad) with a new apparatus consisting of proportional chambers. By extrapolating to $\theta = 0$ and using the Van der Meer method, they will obtain a new measurement of the total cross-section. The CERN-Roma and Pisa-Stony Brook collaborations are proposing an experiment in intersection region 8, based on the new method for measuring total cross-sections described in Section 2.4. This experiment should be running in the first half of 1974. Measurements of the real part up to the maximum ISR energies are planned by the CERN-Roma collaboration to start at the beginning of 1975. Passing to inelastic processes, the Split Field Magnet is now working and will provide data on the diffraction production of isolated resonances and on two-body correlations. The CERN-Holland-Lancaster-Manchester collaboration is continuing the study of diffraction processes with improved resolution and, by having beams of different energies in the two rings, will measure differential cross-sections for $t \geq 0.13$ GeV^2 and $500 \leq s \lesssim 2{,}000$ GeV . The British-Scandinavian-MIT collaboration will accurately measure the shape of the "plateau" around $y = 0$ and aims at enough accuracy to detect a 5% rise through the ISR energy range. Correlations, with particular accent on large transverse momenta, will be provided by the streamer chamber run by the Aachen-CERN-München collaboration. Large transverse momenta are under study by the Saclay Group and the CERN-Columbia-Rockefeller collaboration, who are also proposing to construct a large solenoid which will provide a longitudinal magnetic field and the possibility of measuring the momenta of particles emmitted at 90° within a large solid angle. Correlations between gamma rays of large transverse momentum and charged particles are measured by the Pisa-Stony Brook collaboration.

I am grateful to E.D. Berger, L. Caneschi, H. Miettinen, A. Mueller and P. Pirila for enlightning discussions.

REFERENCES AND FOOTNOTES

1) U. Amaldi, "ISR results on proton-proton elastic scattering and total cross-section", *in* Highlights in particle physics (ed. A. Zichichi) (Editori Compositori, Bologna, 1973), and NP Internal Report 73-5. Referred to as I in the text.

2) V. Bartenev, A. Kunetsov, B. Morozov, V. Nikitin, Y. Pilipenko, V. Popov, L. Zolin, R.A. Carrigan, Jr., E. Malamud, R. Yamada, R.L. Cool, K. Goulianos, I. Hung Chiang, A.C. Melissinos, D. Gross and S.L. Olsen, Phys. Rev. Letters 31, 1088 (1973).

3) U. Amaldi, R. Biancastelli, C. Bosio, G. Matthiae, J.V. Allaby, W. Bartel, M.M. Block, G. Cocconi, A.N. Diddens, R.W. Dobinson, J. Litt and A.M. Wetherell, Phys. Letters B43, 231 (1973).

4) U. Amaldi, R. Biancastelli, C. Bosio, G. Matthiae, J.V. Allaby, W. Bartel, G. Cocconi, A.N. Diddens, R.W. Dobinson and A.M. Wetherell, Phys. Letters 44 B, 212 (1973).

5) S.R. Amendolia, G. Bellettini, P.L. Braccini, C. Bradaschia, R. Castaldi, V. Cavasinni, C. Cerri, T. Del Prete, L. Foà, P. Giromini, P. Laurelli, A. Menzione, L. Ristori, G. Sanguinetti, M. Valdata, G. Finocchiaro, P. Grannis, D. Green. R. Mustard and R. Thun, Phys. Letters 44 B, 213 (1973).

6) U. Amaldi, Proc. 2nd Int. Conf. on Elementary Particles, Aix-en-Provence, 1973, J. Phys. (France) Suppl. 10, C1-241 (1973).

7) A.N. Diddens, "Very high energy hadron experiments", Lectures given at the 4th Seminar on Theoretical Physics, organized by GIFT, Barcelona, April, 1973.

8) G. Bellettini, "Total proton-proton cross-section at the ISR", 8e Rencontre de Moriond, Méribel-Les-Allues, France, 1973 and Int. Conf. on High Energy Collisions, Stony Brook, 1973.

9) S.R. Amendolia, G. Bellettini, P.L. Braccini, C. Bradaschia, R. Castaldi, V. Cavasinni, C. Cerri, T. Del Prete, L. Foà, P. Giromini, P. Laurelli, A. Menzione, L. Ristori, G. Sanguinetti, M. Valdata, G. Finocchiaro, P. Grannis, D. Green, R. Mustard and R. Thun, Pisa-Stony Brook collaboration, Proc. Int. Conf. on Instrumentation for High-Energy Physics, Frascati, 1973.

10) D.R.O. Morrison, "Review of inelastic proton-proton reactions", Preprint CERN/d-Ph II/PHYS 73-11, to be published in Proc. Roy. Soc.

11) Russian Proton Satellietes: V.V. Akimov et al., Acta Phys. Hungar. 29, Suppl., 1, 517 (1970).
Echo Lake Experiment: A.E. Bussian et al., Proc. 12th Int. Conf. on Cosmic Rays, Hobart, 1971 (University of Tasmania, Hobart, 1971), Vol. 3, p. 1194.

12) M.J. Ryan, J.F. Ormes and V.K. Balasubrahmanyan, Phys. Rev. Letters 28, 985 (1972).

13) N.L. Grigorov, Yu.V. Gubin, I.D. Rapoport, I.A. Savenko, B.M. Yakovlev, V.V. Akimov and V.E. Nesterov, Proc. 12th Int. Conf. on Cosmic Rays, Hobart, 1971 (University of Tasmania, Hobart, 1971), Vol. 5, p. 1746.

14) K. Murakami, K. MacKeown, C. Yokoyama, C. Aguirre, A. Trepp, G.R. Mejia, J. Pacheco, K. Kamata, Y. Toyoda, T. Meeda, K. Suga, K. Uchino and M. La Pointe, Proc. 6th Interamerican Seminar on Cosmic Rays, La Paz, 1970 (Universidad Mayor de San andres, la Paz, 1971), Vol. 3, p. 695, and P.I. Mackeown, *ibid,* p. 684.
G.B. Yodh, J.R. Wayland and Y. Pal, Proc. 6th Interamerican Seminar on Cosmic Rays, La Paz, 1970 (Universidad Mayor de San Andres, La Paz, 1971), Vol. 3, p. 706.

15) G.B. Yodh, Y. Pal and J.S. Trefil, Phys. Rev. Letters 28, 1005 (1972).

16) J. Wdowczyk and E. Zujewska, J. Phys. A (GB), 6, L9 (1973).

17) T. Kaneko, C. Yokoyama, K. Nishi, G. Mejia, A. Trepp, C. Aquirre, L. Murakami, K. Kamata, K. Suga, Y. Toyoda, Proc. 12th Int. Conf. on Cosmic Rays, Hobart, 1971 (University of Tasmania, Hobart, 1971), Vol. 3, p. 945. No data are contained in this abstract.

18) K. Kamata et al., Suppl. Acta Phys. Hungar. 3, 49 (1970).

19) G.B. Yodh, Y. Pal and J.S. Trefil, "On the evidence for rapidly rising p-p total cross-sections from cosmic-ray data", University of Maryland, Tech. Rep. No. 73-114 1973, to be published in Phys. Rev.

20) M. Froissart, Phys. Rev. 123, 1053 (1961);
A. Martin, Nuovo Cimento 42, 930 (1966).

21) U. Amaldi, Proc. 2nd Int. Conf. on Elementary Particles, Aix-en-Provence, 1973 J. Phys. (France), Suppl. 10, C1-241 (1973).

22) N.N. Khuri and T. Kinoshita, Phys. Rev. 137 B, 720 (1965).

23) A similar observation has been made by T.T. Chou and Cheng Ning Yang, Proc. 2nd Int. Conf. on High-Energy Physics and Nuclear Structure, Rehovoth, 1967 (ed. G. Alexander) (North Holland, Amsterdam, 1967).

24) C. Bourrely and J. Fisher, Nuclear Phys. B61, 513 (1973).
P. Kroll, Nuovo Cimento Letters 7, 745 (1973).
H. Cheng, J.K. Walker and T.T. Wu, Phys. Letters 44 B, 283 (1973).

25) W. Bartel and A.N. Diddens, NP Internal Report 73-4 (1973).

26) D.W. Joynson and W. von Schlippe, "The real part of the pp amplitude and rising cross-sections", Westfield College, preprint, 1973.

27) D.D. Coon, U.P. Sukhatme and J. Tran Thanh Van, Phys. Letters 45 B, 287 (1973).

28) For a recent fit see: C. Pajares and D. Schiff, Lett. Nuovo Cimento, 8, 237 (1973).

29) J.C. Sens, "Review of recent results from the European Organization for Nuclear Research Intersecting Storage Rings", Invited paper at the Conference on Recent Advances in Particle Physics, New York, 1973.
M. Jacob, "Multi-body phenomena in strong interactions", CERN-JINR School of Physics, Ebeltoft, Denmark, 1973, Summer Institute on Particle Interaction at Very High Energies, Louvain, Belgium, 1973, CERN preprint TH.1683.

30) E.L. Berger, this School and CERN preprint TH.1737.

31) M. Antinucci, A. Bertin, G. Capiluppi, G. Giacomelli, A.M. Rossi, G. Vannini and A. Bussière, "Particle production at medium angles at the CERN ISR", presented at Int. Conf. on New Results from Experiments on High-Energy collisions, Vanderbilt University, Nashville, 1973.

32) B. Alper, H. Boggild, P. Booth, F. Bulos, L.J. Carroll, G. von Dardel, G. Damgaard, B. Duff, N.N. Jackson, G. Jarlskog, L. Jønsson, A. Klovning, L. Leistam, E. Lillethun, G. Lynch, M. Prentice, D. Quarries and J.M. Weiss, "Large angle inclusive production of charged pions at the CERN ISR with transverse momenta less than 1.0 GeV/c", "Inclusive production of protons, antiprotons and kaons and particle composition at small x at the CERN ISR", to be published.

33) M.G. Albrow, A. Bagchus, D.P. Barber, A. Bogaerts, B. Bosnjakovic, J.R. Brooks, A.B. Clegg, F.C. Erné, C.N.P. Gee, D.H. Locke, F.K. Loebinger, P.G. Murphy, A. Rudge, J.C. Sens and F. van der Veen, "Missing-mass spectra in p-p inelastic scattering at total energies of 23 and 31 GeV", to be published in Nuclear Phys.

34) This point is fully discussed by L. Foà, Proc. 2nd Aix-en-Provence Int. Conf. on Elementary Particles, 1973, J. Phys. (France), Suppl. 10, C1-317 (1973).

35) K.G. Wilson, Cornell preprint, CLNS-131 (1970);
M. Bander, Phys. Rev. D2, 164 (1972).
W.R. Frazer, R.D. Peccei, S.S. Pinsky and C.I. Tan, Phys. Rev. D7 2647 (1973).
K. Fiałkowski, Phys. Letters 41 B, 379 (1972).
H. Harari and E. Rabinovici, Phys. Letters 43 B, 49 (1973).
K. Fiałkowski and H. Miettinen, Phys. Letters 43 B, 61 (1973).
L. Van Hove, Phys. Letters 43 B, 65 (1973).
C. Quigg and T.D. Jackson, NAL-THY-93 (1972).

36) D. Amati, L. Caneschi and M. Testa, Phys. Letters 43 B, 186 (1973).

37) A very interesting attempt in this direction has been recently performed using the uncorrelated jet model as non-diffractive process: A. Białas and A. Kotanski, "Diffractive dissociations and shadow scattering", TPJU-8/73, June 1973, Submitted to Acta Phys. Polon. B.

38) A comprehensive presentation of this point of view is given by A. Mueller, Proc. 2nd Aix-en-Provence Int. Conf. on Elementary Particles, 1973, J. Phys. (France) Suppl. 10, C1-307 (1973).

39) On the contrary the multiperipheral absorbed model by Finkelstein and Zachariasen has the geometrical behaviour: L. Caneschi and A. Schwimmer, Nuclear Phys. B44, 31 (1972).

40) L. Van Hove, Nuovo Cimento 28, 798 (1963); Rev. Mod. Phys. 36, 655 (1964).

41) D.H. de Groot and H.I. Miettinen, "Shadow approach to diffraction scattering", Proc. 8e Rencontre de Moriond, Meribel-les-Allues, France 1973, and report RL-73-003.

42) H.I. Miettinen, Proc. 2nd Aix-en-Provence Int. Conf. on Elementary Particles 1973, J. Phys. (France) Suppl. 10, C1-263 (1973).

43) F. Zachariasen, Phys. Reports 2 C, 1 (1971).

44) Z. Koba and M. Namiki, Nuclear Phys. B8, 413 (1969).

45) For a similar approach in the Regge framework see: V. Barger, R.J.N. Phillips and K. Geer, Nuclear Phys. B47, 29 (1972).

46) F.S. Henyey, R. Hong Tuan and G.L. Kane, "Impact parameter study of high-energy elastic scattering", UMHE 73-18.

47) R. Henzi and P. Valin, "Overlap function analysis and pp elastic scattering data at ISR energies and at large momentum transfer", 2nd Aix-en-Provence Int. Conf. on Elementary Particles, paper 246 (1973).
R. Henzi and P. Valin, "On the energy dependence of the overlap function from a recent analysis of the pp elastic scattering data at ISR energies", 2nd Aix-en-Provence Int. Conf. on Elementary Particles, paper 450 (1973).

48) H.I. Miettinen and P. Pirila, private communication and to be published.

49) Recently the tail beyond $a^2 \simeq 7\ fm^2$ has been attributed to two-pion exchange: J.W. Alcock, N. Cottingham and C. Michael, CERN preprint TH-1765 (1973).

50) A. Capella and M. Shin Chen, Phys. Rev. D8, 2097 (1973).
M. Bishari and J. Koplik, Phys. Letters 44 B, 175 (1973).
G.F. Chew, Berkeley preprint, LBL-1556 (1973) and Phys. Letters 44 B, 169 (1973).
W.R. Frazer and D.R. Snider, Phys. Letters 45 B, 136 (1973).
T.L. Neff, Phys. Letters 45 B, 349 (1973).

51) D. Amati, L. Cansechi and M. Ciafaloni, Nuclear Phys. B62, 173 (1973).

52) N. Sakai and J.N.J. White, Nuclear Phys. B59, 511 (1973).

53) N. Sakai and J.N.J. White, "The helicity structure of diffraction dissociation and the role of absorption", preprint, June 1973, to be published.

54) R. Blankenbecler, Phys. Rev. Letters 31, 964 (1973).

55) J. Finkelstein and F. Zachariasen, Phys. Letters 34 B, 631 (1971).

56) L. Caneschi and M. Ciafaloni, Proc. 2nd Aix-en-Provence Int. Conf. on Elementary Particles, 1973, J. Phys. (France) Suppl. 10, C1-268 (1973).

57) For very complete discussions see: R.C. Hwa, Phys. Rev. D8, 1331 (1973); C.J. Hamer and R.F. Peierls, Phys. Rev. D8, 1358 (1973).

58) F.S. Henyey, Phys. Letters 45 B, 363 (1973) and 45 B, 469 (1973).

59) D. Sivers and F. von Hippel, Argonne preprint ANL/HEP 7323.
M. Suzuko, Univ. of California, Berkeley preprint (1973).
T.K. Gaisser and C.I. Tan, Brookhaven preprint BNL-18070 (1973).
J. Koplik, Berkeley APS Meeting, August 13-17, 1973, and LBL-2175.
G.F. Chew, Proc. Int. Conf. on High-Energy Collisions, Stony Brook, 23-24 August, 1973.

60) T.K. Gaisser and Chung-I Tan, BNL 18070 (1973).

61) G. Frye, J. Kogut and L. Susskind, Phys. Letters 40 B, 469 (1972).
A. Casher, S. Nussinov and L. Susskind, Phys. Letters 44 B, 511 (1973).

62) D. Cline, F. Halzen and J. Luthe, Phys. Rev. Letters 31, 491 (1972).

63) H. Cheng and T.T. Wu, Phys. Rev. Letters 24, 1456 (1970); Phys. Letters 34 B, 647 (1971); Phys. Letters 36 B, 357 (1971).

64) S.J. Chang and T.M. Yan, Phys. Rev. Letters 25, 1586 (1970); Phys. Rev. D4, 537 (1971).

65) V. Abramovskii, O. Kancheli and V. Gribov, Proc. 15th Conf. on High-Energy Physics, Batavia, 1972, Vol 1, p. 389.

66) H. Cheng, J.K. Walker and T.T. Wu, Phys. Letters 44 B, 97 (1973).

67) The presence of a real part different from zero at all momentum transfers does not change the above conclusion. The implications of the same model for production processes have been discussed in: H. Cheng and T.T. Wu, Phys. Letters 45 B, 367 (1973).

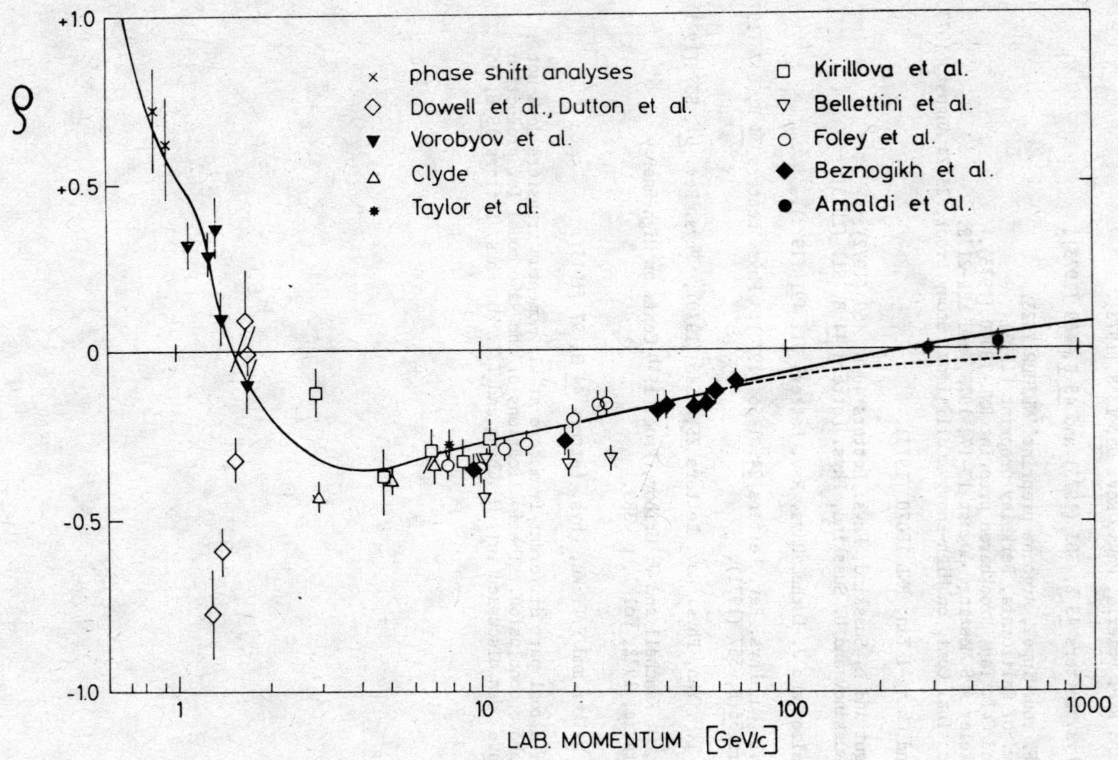

Fig. 1 The ratio ρ between the real and the imaginary parts of the forward nuclear amplitude. The CERN-Roma points (Amaldi et al.) have been obtained by using for σ_t the values measured in Ref. 4 and 5 (see Table 1). The dotted line is the result of a dispersion calculation in which it is assumed that the common value of the proton-proton and antiproton-proton total cross section is 40 mb, while in the calculation of the continuous line the asymptotic choice $\sigma_t \; \alpha(\log s)^2$ has been made (this point is discussed in Section 3.3). Preliminary results of the USSR-USA jet gas experiment, which is running at NAL, indicate indeed that the real part becomes positive for $p_L \simeq 300$ GeV/c. [Note added in proof: ρ crosses zero at $p_L = (280 \pm 60)$ GeV/c: V. Bartenev et al., Phys. Rev. Letters 31, 1367 (1973)]

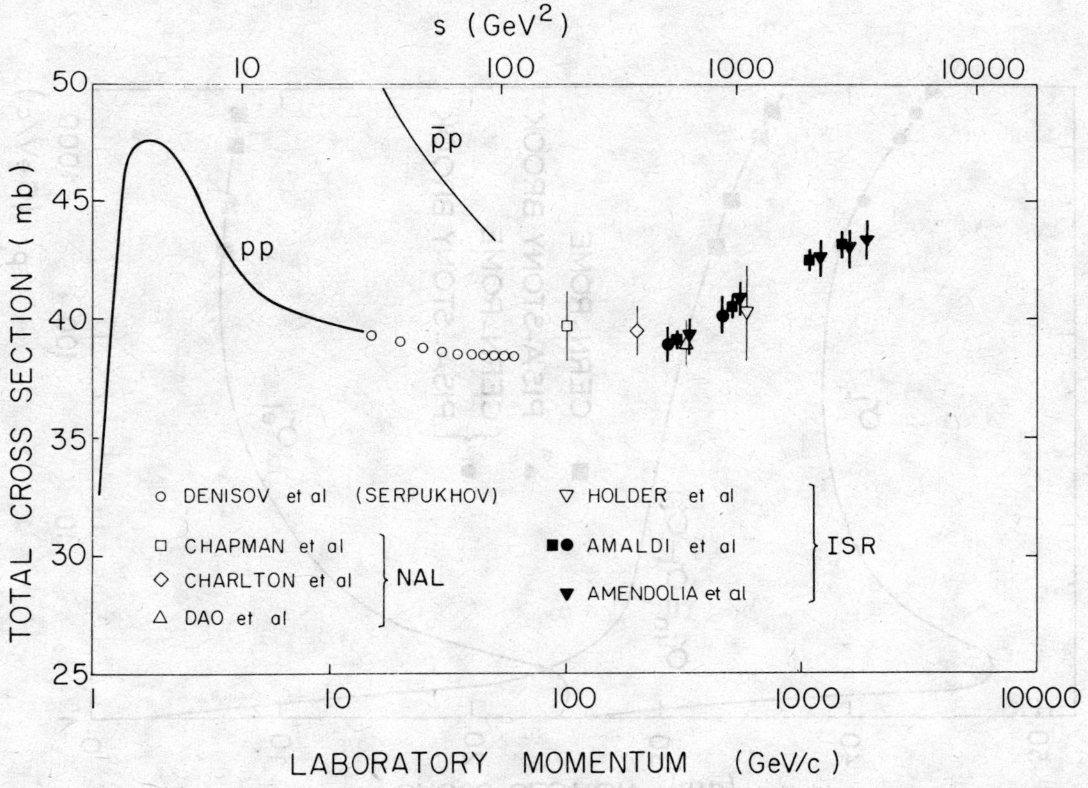

Fig. 2 Total proton-proton cross-sections measured at the ISR by the CERN-Roma collaboration (Amaldi et al. Ref. 4) and by the Pisa Stony Brook collaboration (Amendolia et al. Ref. 5). For more recent data of the Pisa-Stony Brook collaboration see Bellettini's contribution to the Stony Brook Conference (Ref. 8).

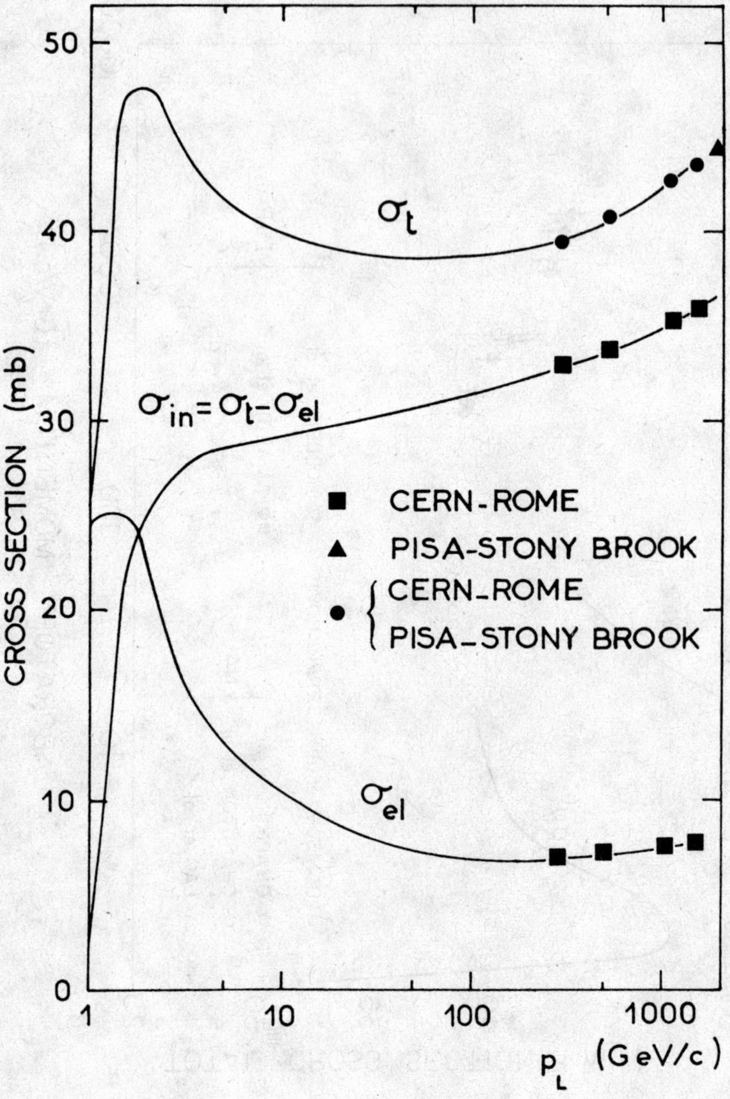

Fig. 3 Total, inelastic and elastic proton-proton cross-sections versus the laboratory momentum (Ref. 10).

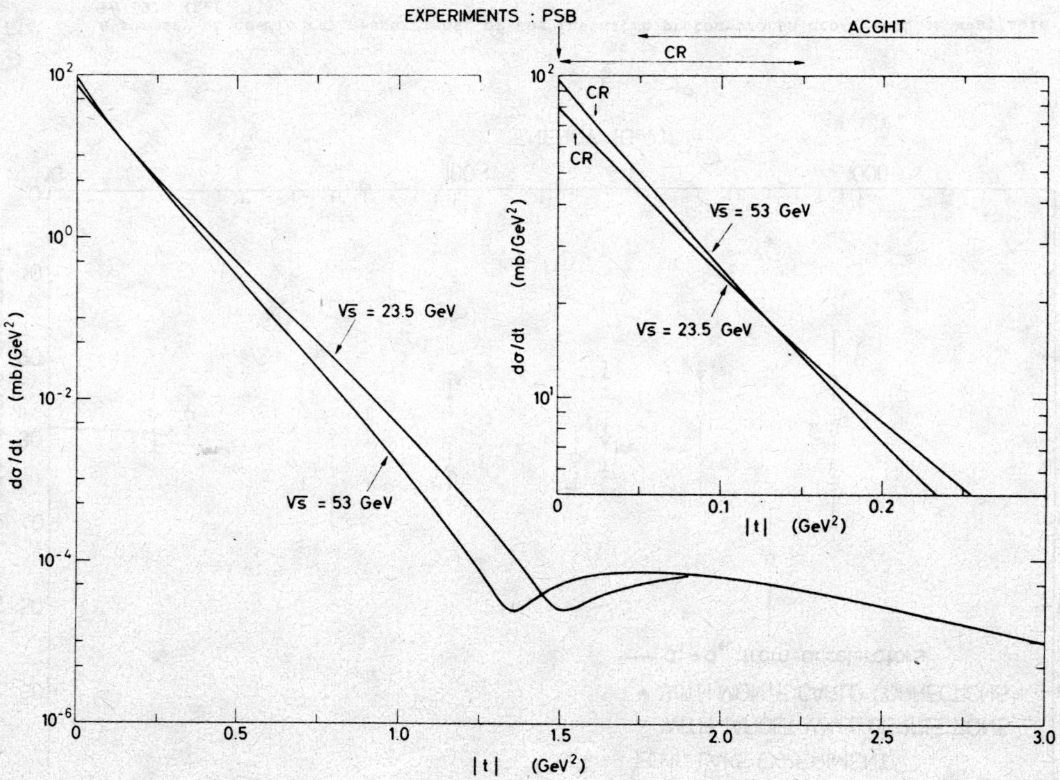

Fig. 4 A personal compilation of the information available on elastic scattering and total cross-sections at the two extreme ISR energies. While the forward behaviour is on firm basis, in drawing the behaviour at larger momentum transfers it has been assumed that the position of the minimum moves towards smaller values of $|t|$ as the energy is increased. This fact is strongly indicated by the data obtained by the Aachen-CERN-Geneva-Harvard-Torino collaboration, but on the other hand has not been definitely demonstrated.

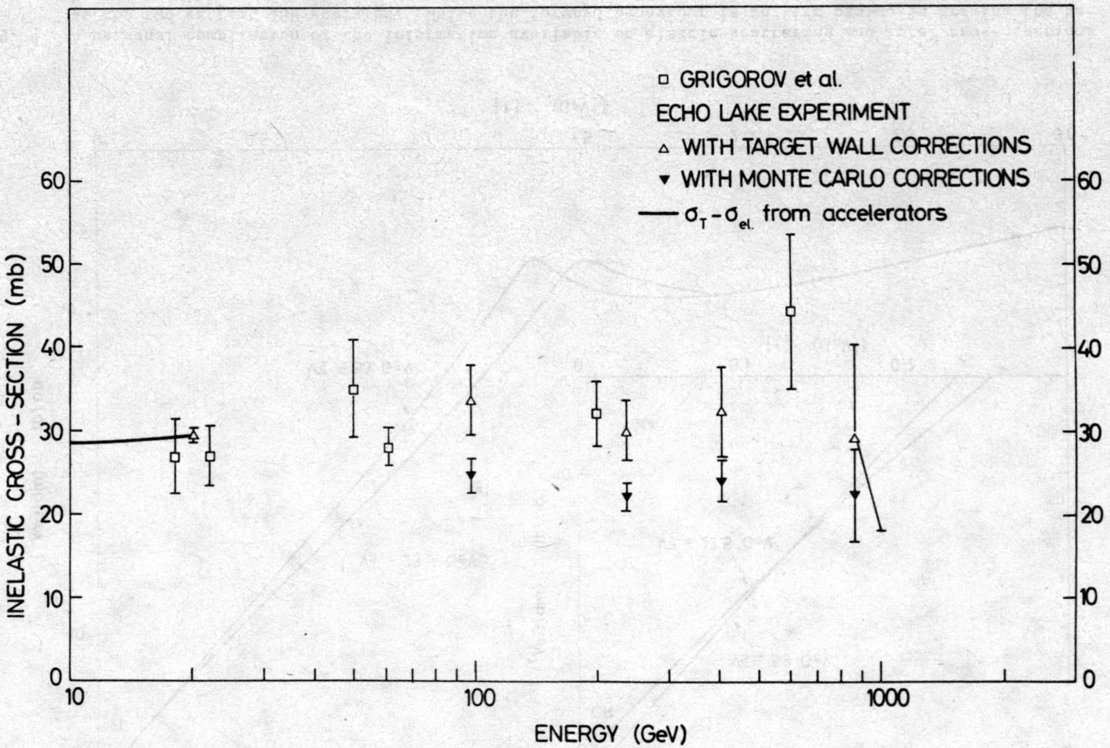

Fig. 5 A summary of cosmic ray measurements of the inelastic proton-proton cross-section available in 1971 (Ref. 11).

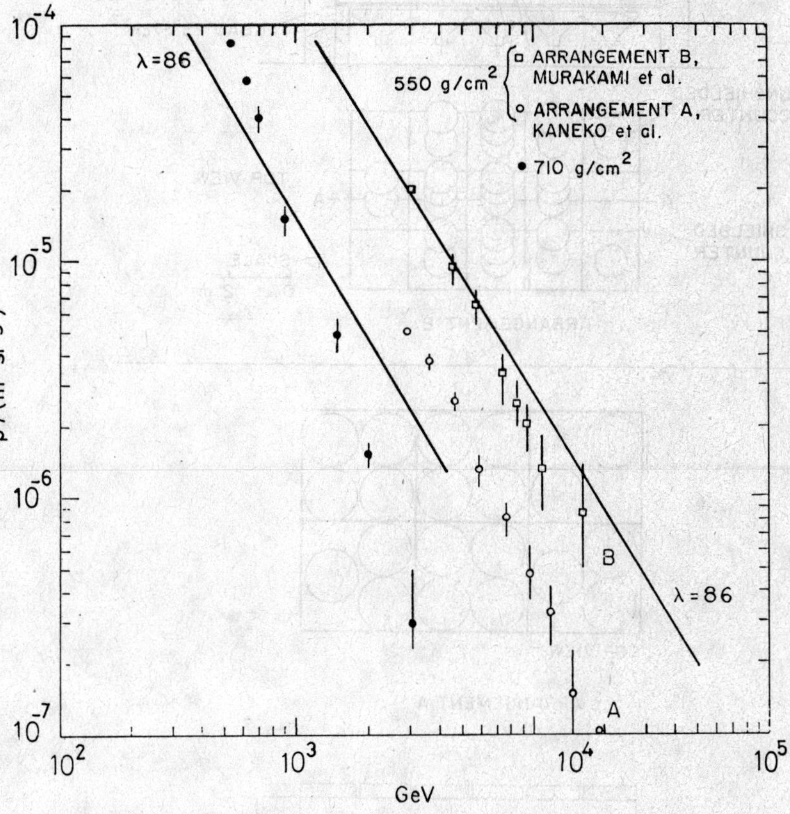

Fig. 6 The spectra of "unaccompanied" protons measured at mountain level (550 g/cm^2) and at sea level (710 g/cm^2), are compared with the expectation based on the attenuation of the primary spectrum with an interaction length of protons in air (λ = 86 g/cm^2) equal to its value at accelerator energies.

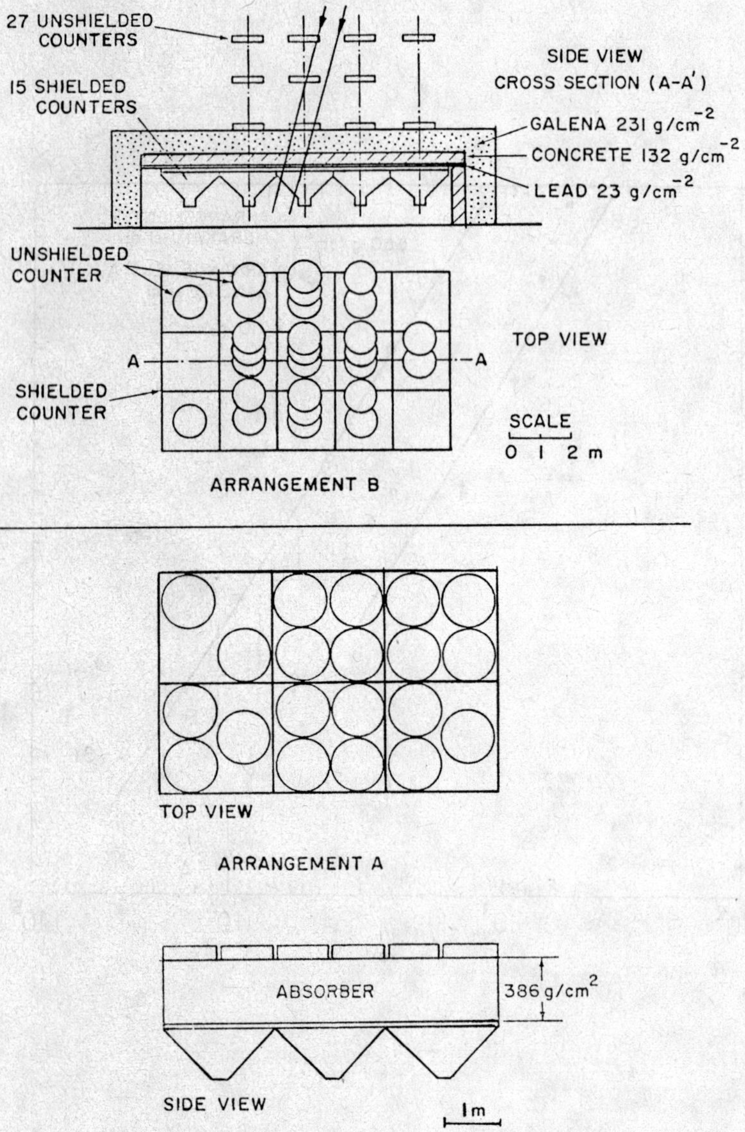

Fig. 7 The arrangements used to measure the spectra plotted in Fig. 6. The difference of scale between the two drawings should not induce in error: the shielded counters of arrangement A are nine counters among the fifteen of arrangement B. The two arrangements differ essentially for the disposition of the unshielded counters and the triggering requirements.

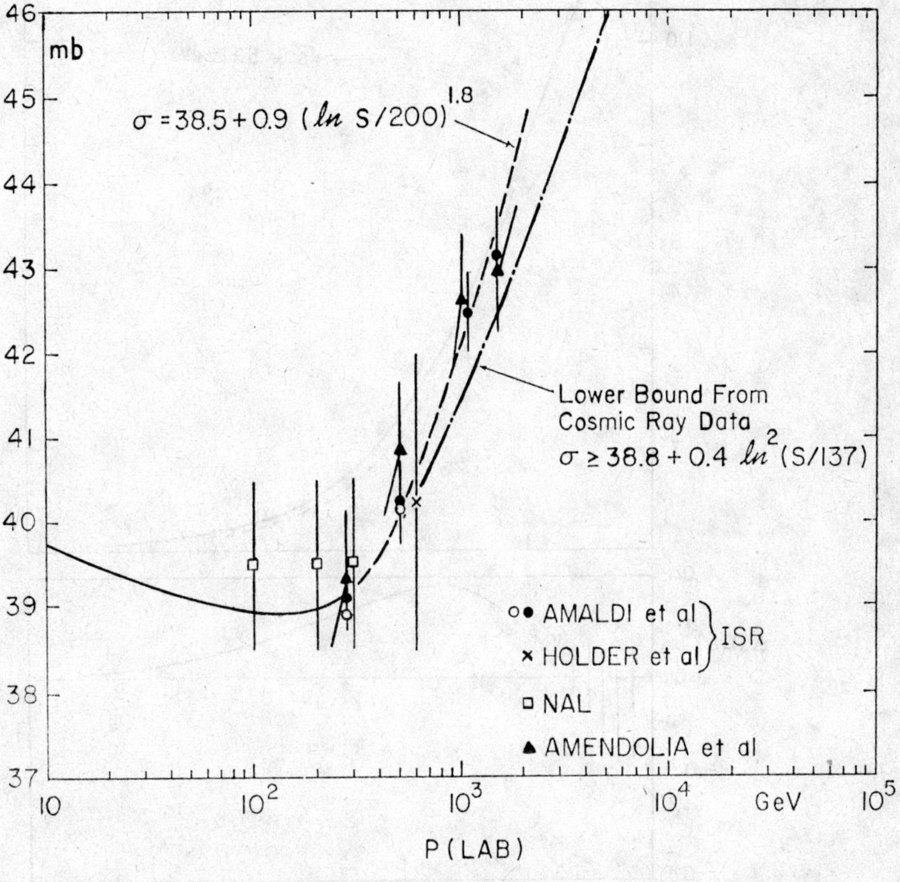

Fig. 8 The lower bound from cosmic ray data (Ref. 19) is compared with the ISR results. The energy "scale" 137 GeV^2 is not well determined by the data and is taken from the mnemonic fit to the accelerator and ISR proton-proton cross-sections given in I.

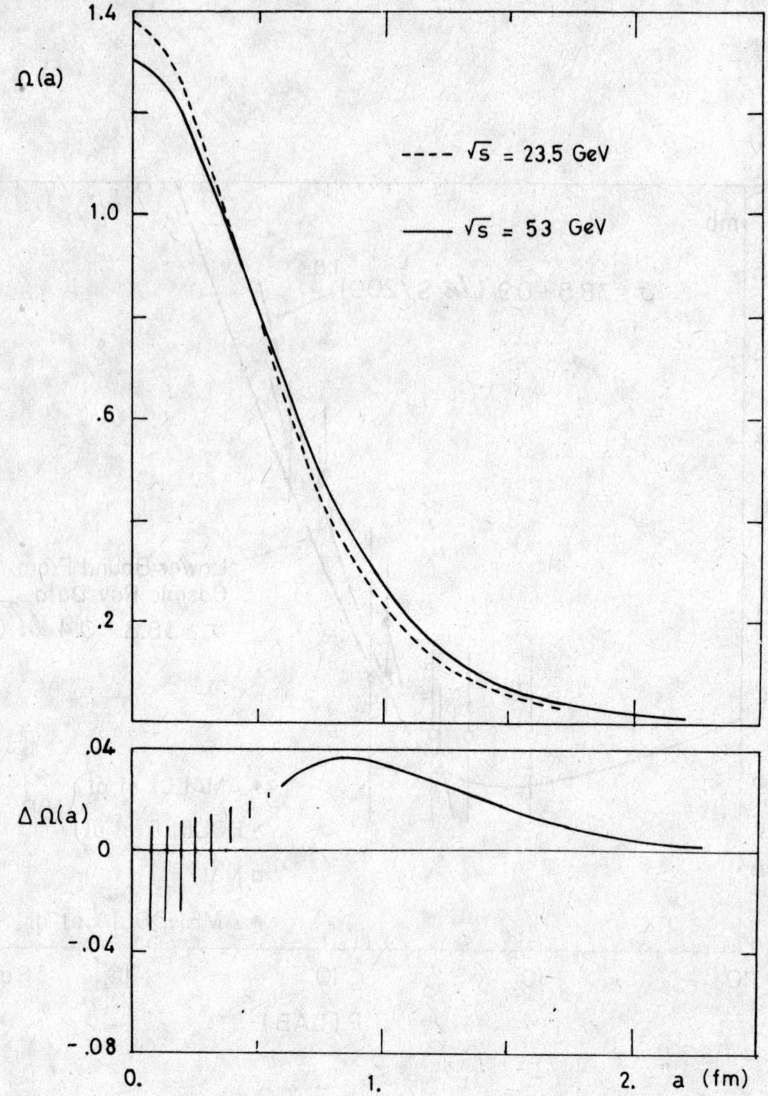

Fig. 9 Proton-proton opaqueness as a function of the impact parameter at the two extreme ISR energies. $\Omega(a)$ is computed from the differential cross-sections plotted in Fig. 4. Uncertainties in the data influence $\Omega(a)$ mainly at small impact parameters ($a \lesssim 0.5$ fm). The lower part of the figure shows the increment of the opaqueness in the ISR energy range. The quoted uncertainties in the data reflect in a range of possible values for $\Delta\Omega(a)$, for small values of a, qualitatively represented in the figure by the vertical bars

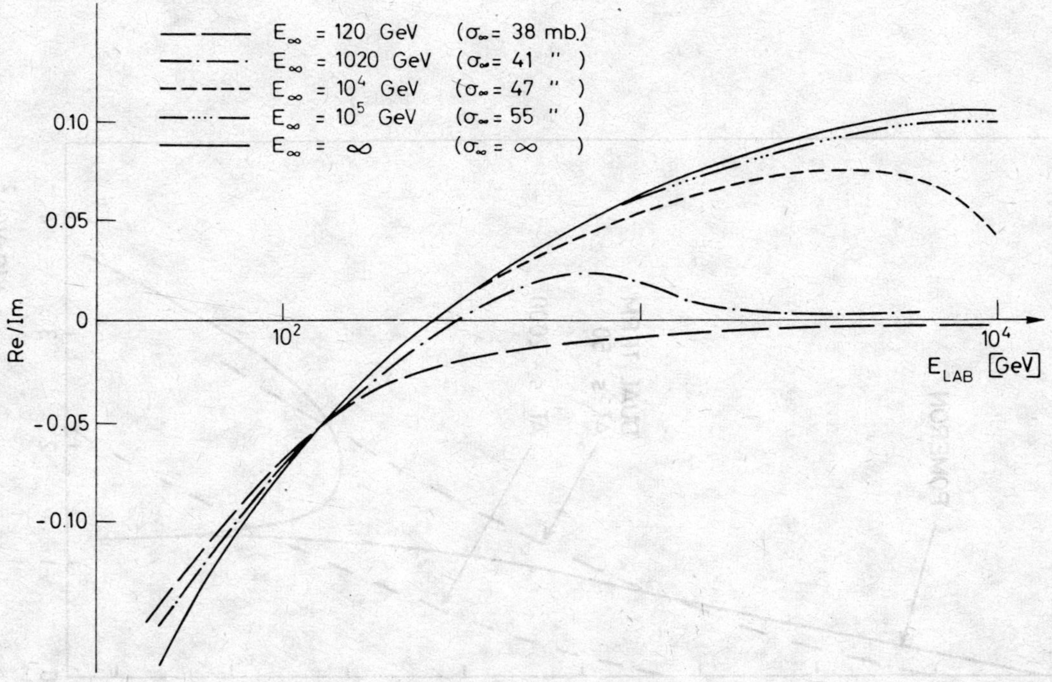

Fig. 10 Results of a dispersion relation calculation of the ratio ρ of the real to the imaginary part of the nuclear amplitude (Ref. 25). The various curves refer to various hypothesis on the high energy behaviour of σ_t^{pp}. The continuous one is computed in the asymptotic hypothesis that the difference between σ_t^{pp} and $\sigma_t^{\bar{p}p}$ decreases as a power of s and that σ_t^{pp} $\alpha(\ln s)^2$. The other curves have been obtained by assuming that, above a certain energy, the proton-proton cross-section flattens.

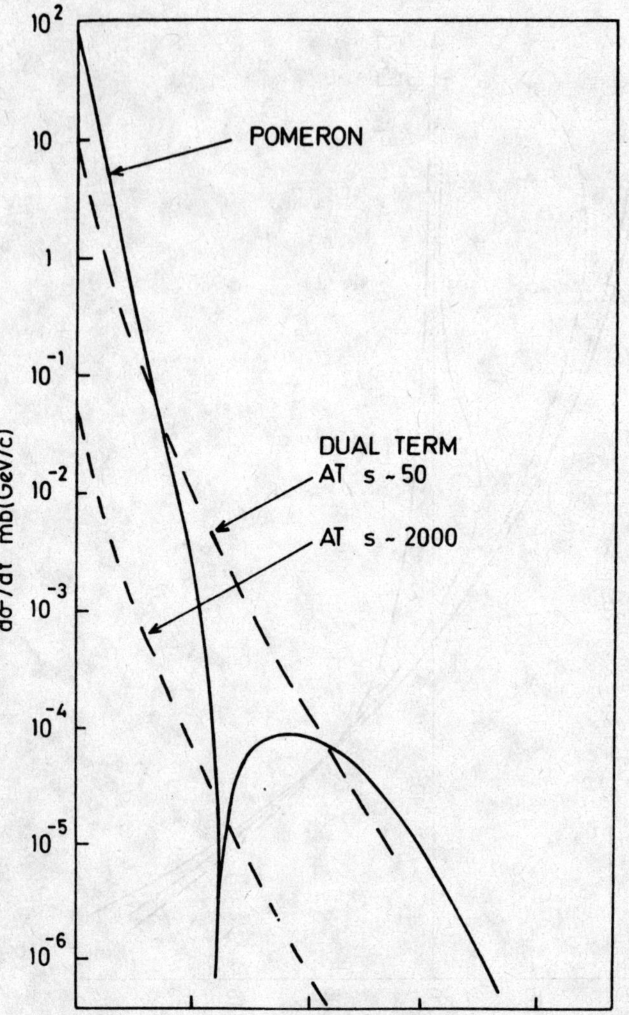

Fig. 11 Fit to the proton-proton differential cross-section in the dual model of Coon, Sukhatme and Tran Thanh Van (Ref. 27). At ISR energies the real dual contribution is negligible with respect to the diffractive pomeron exchange, but in the region of the diffraction minimum.

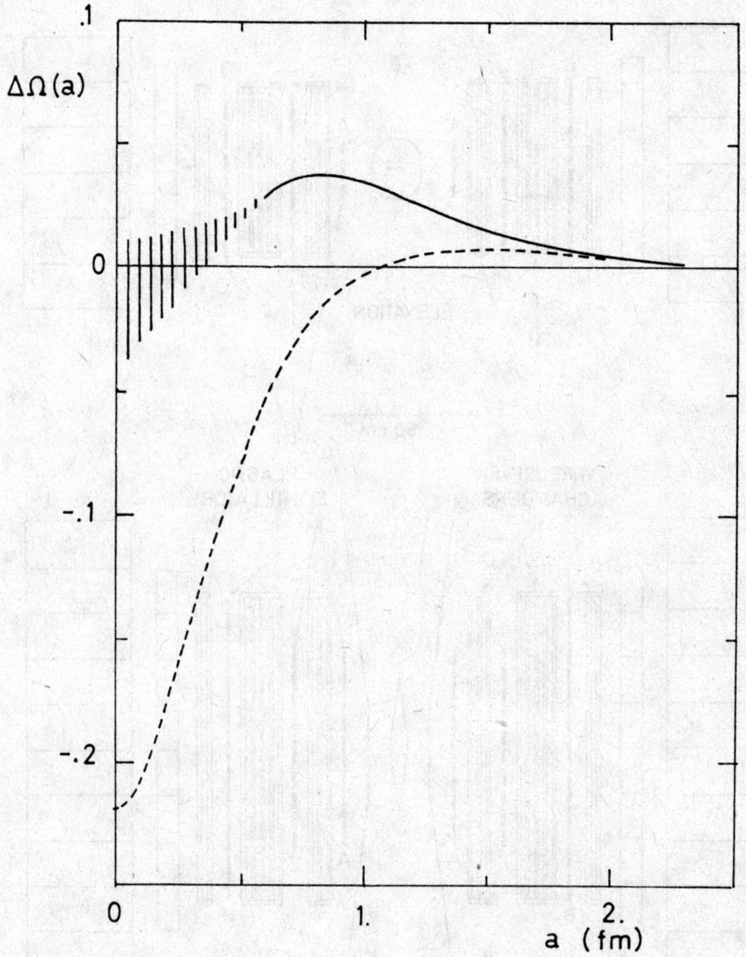

Fig. 12 The "experimental" increment of the opaqueness ΔΩ(a) (continuous curve) is compared with the expected behaviour (dashed curve) due to the exchange of a pomeron pole, whose trajectory is such to explain the 8% increase of the forward slope b, but with a constant total cross-section. In performing this comparison it is assumed that the pomeron exchange does not "eikonalizes", i.e. that the Regge formula represents, as usually, the scattering amplitude and not the opaqueness.

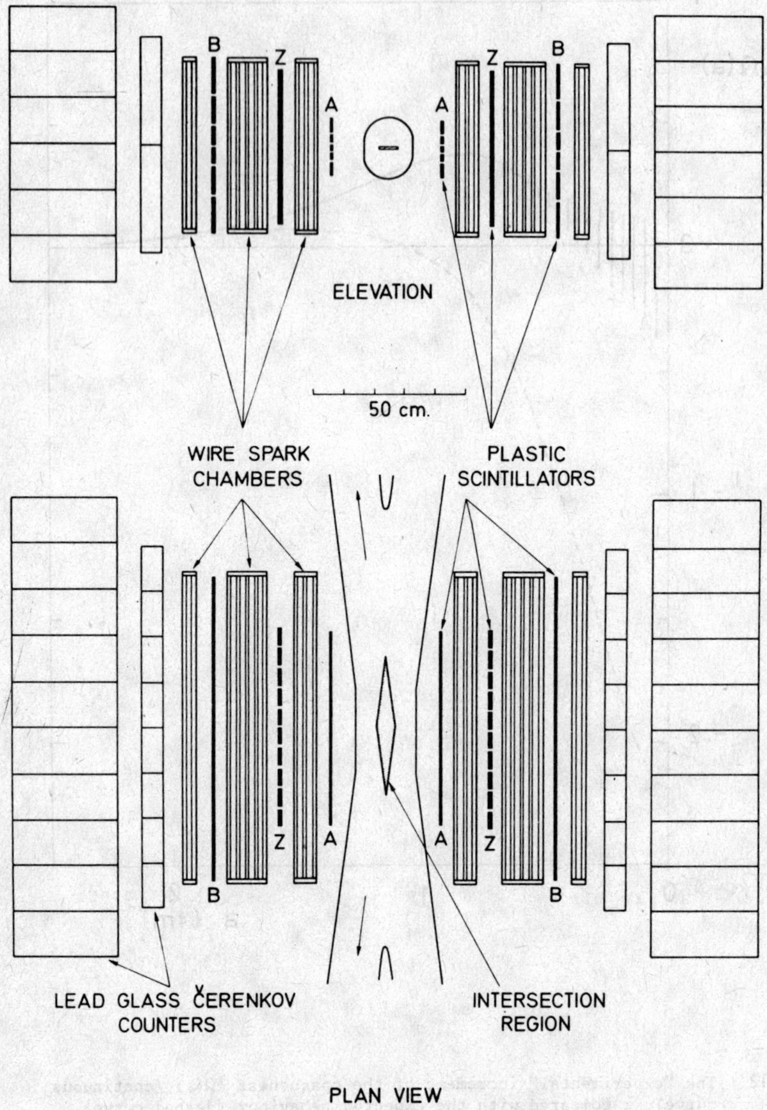

Fig. 13 The apparatus of the CERN-Colombia-Rockefeller collaboration. Gamma rays are detected by Čerenkov counters around 90° and the position of the charged particles are measured by means of wire spark chambers.

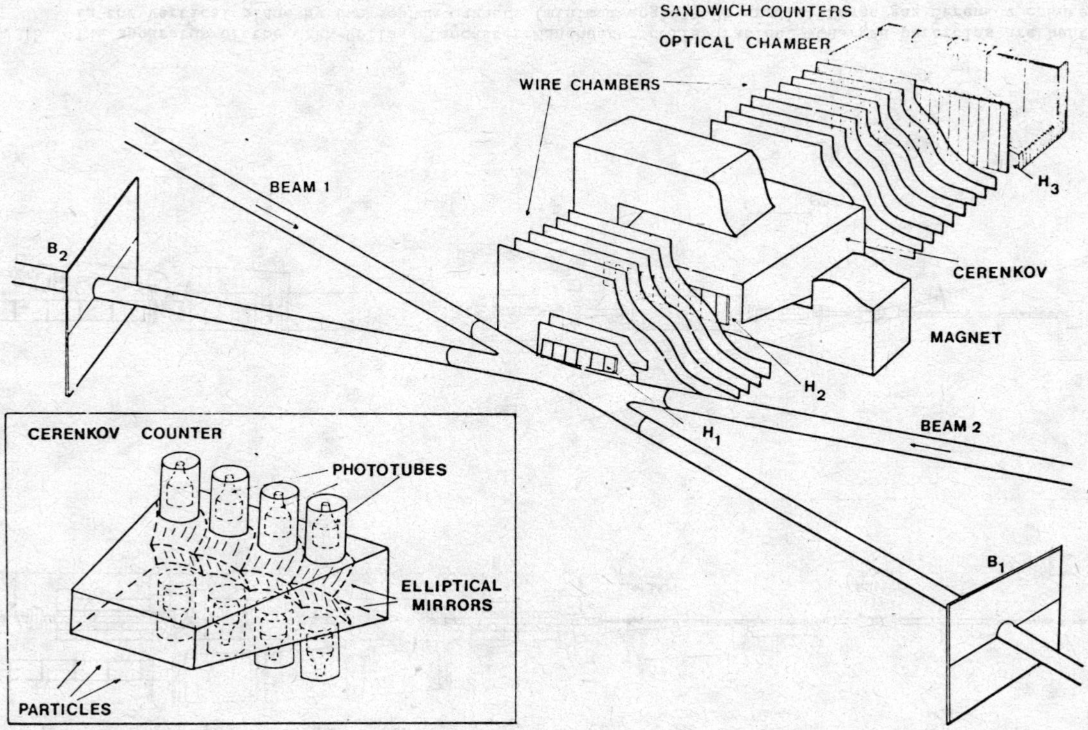

Fig. 14 Expanded view of the apparatus of the Saclay-Strasbourg collaboration. The momenta of the charged particles are measured through the deflection in the magnetic field, while the gamma rays are detected through their conversion in electron pairs in the vacuum chamber and in hodoscope H_1.

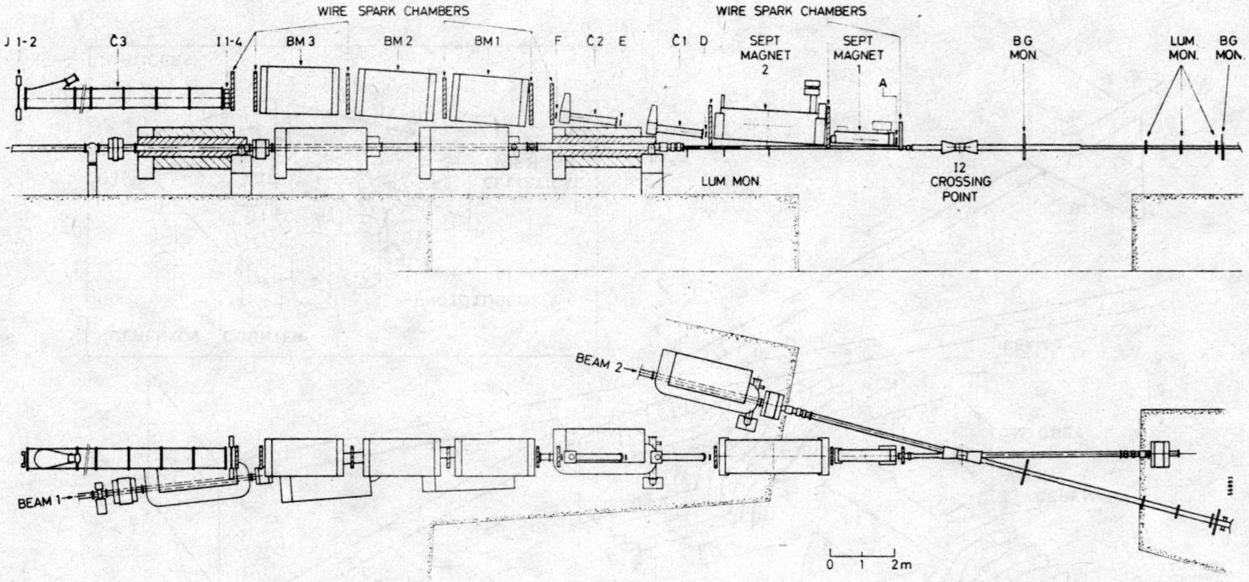

Fig. 15 The apparatus of the CERN-Holland-Lancaster-Manchester collaboration. Charged particles are bent in the vertical plane by two septum magnets (minimum angle ∿ 25 mrad). Three gas Čerenkov counters and three bending magnets determine the nature and momentum of the charged particles.

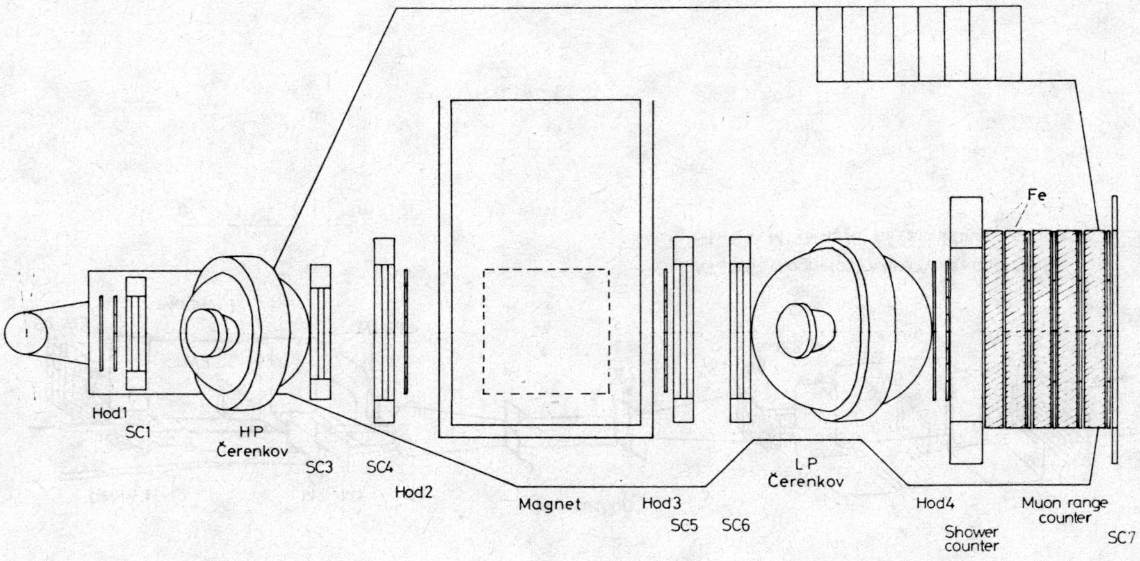

Fig. 16 Horizontal projection of the apparatus of the British-Scandinavian collaboration. It consists of a large aperture bending magnet, two Čerenkov counters, two sets of wire chambers and a muon detector. The platform can be rotated between 30° and 90°.

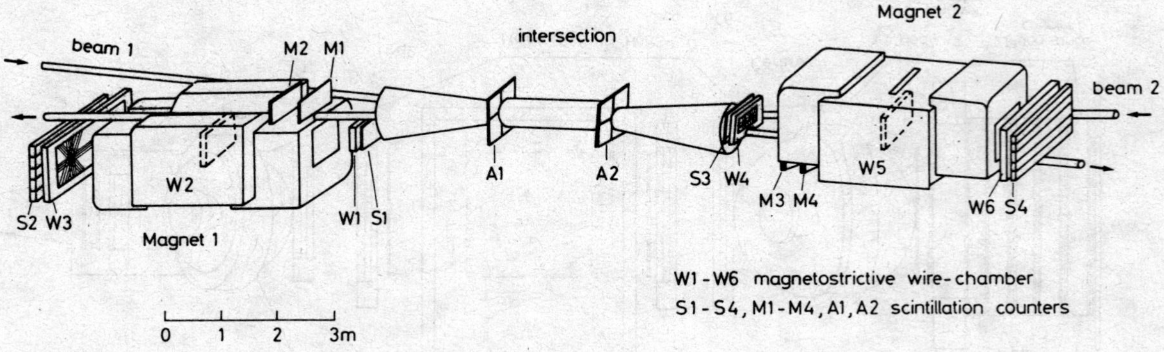

Fig. 17 The apparatus of the Aachen-CERN-Genova-Harvard-Torino collaboration, used for measuring elastic scattering at large momentum transfers and for detecting diffractive events.

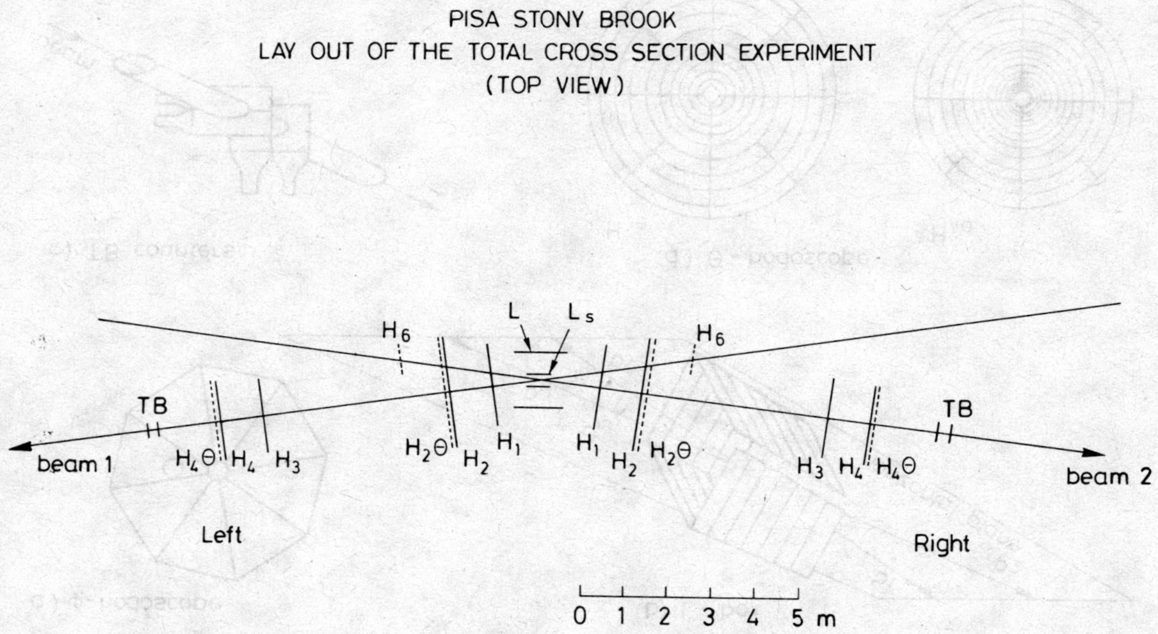

Fig. 18 Sketch of the apparatus of the Pisa Stony Brook collaboration. It consists of about 500 counters, which cover 90% of the full solid angle.

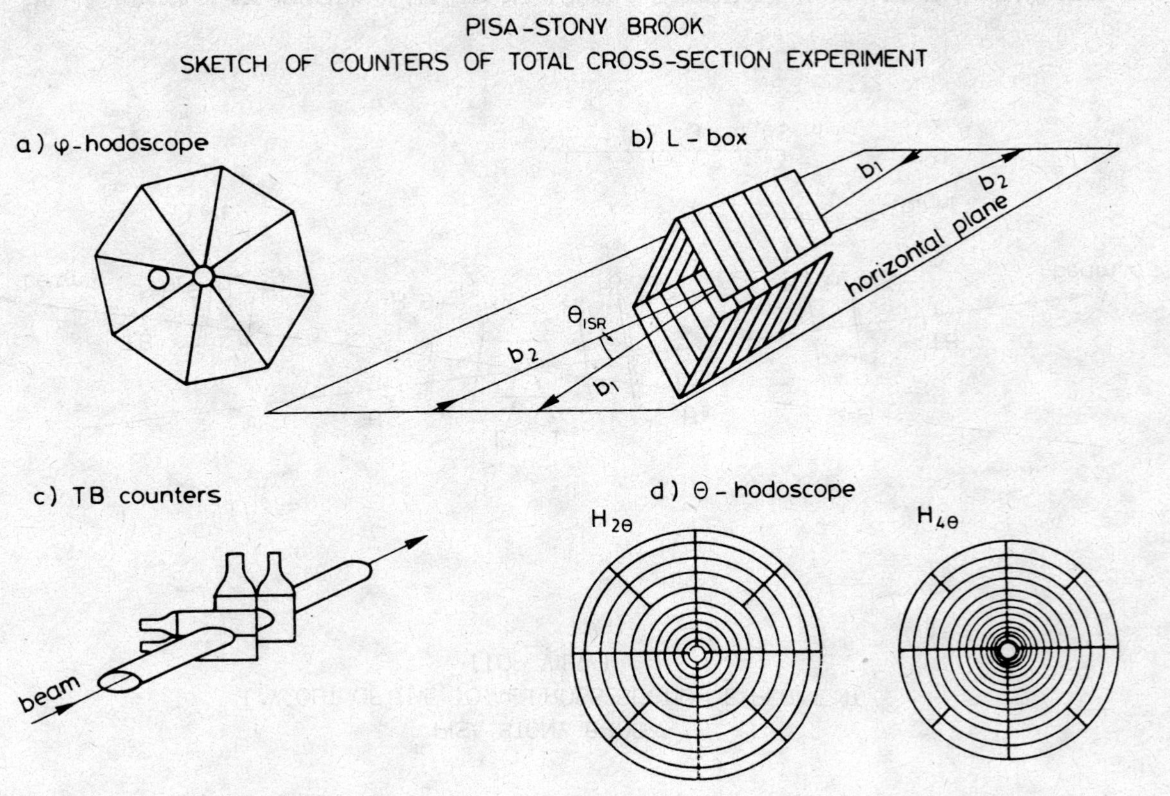

Fig. 19 Details of the counter hodoscopes of the Pisa-Stony Brook experiment.

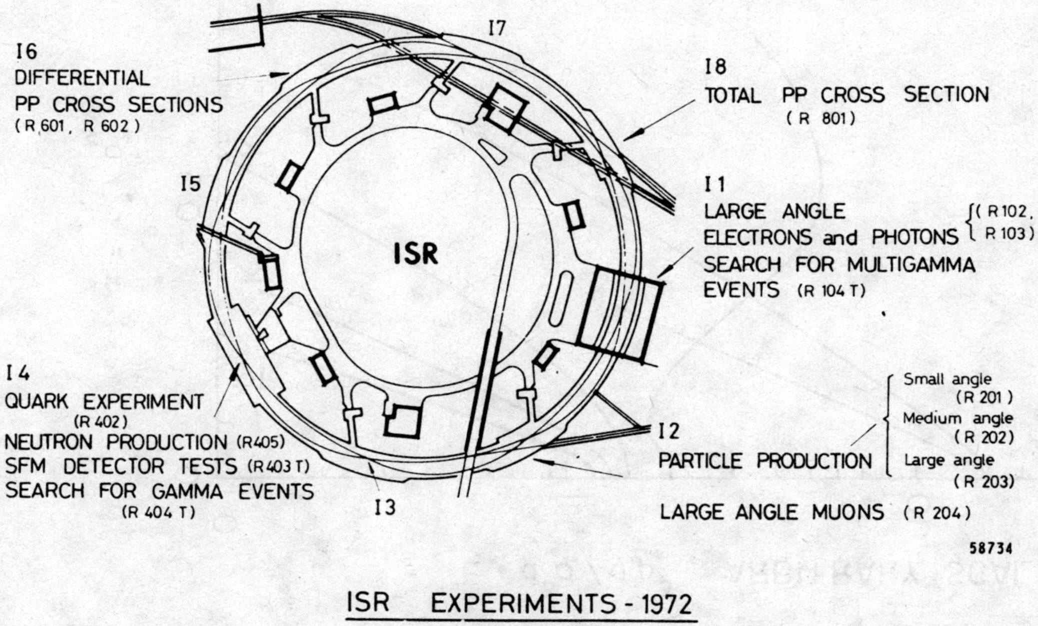

Fig. 20 The experiments performed during 1972 are localized around the rings.

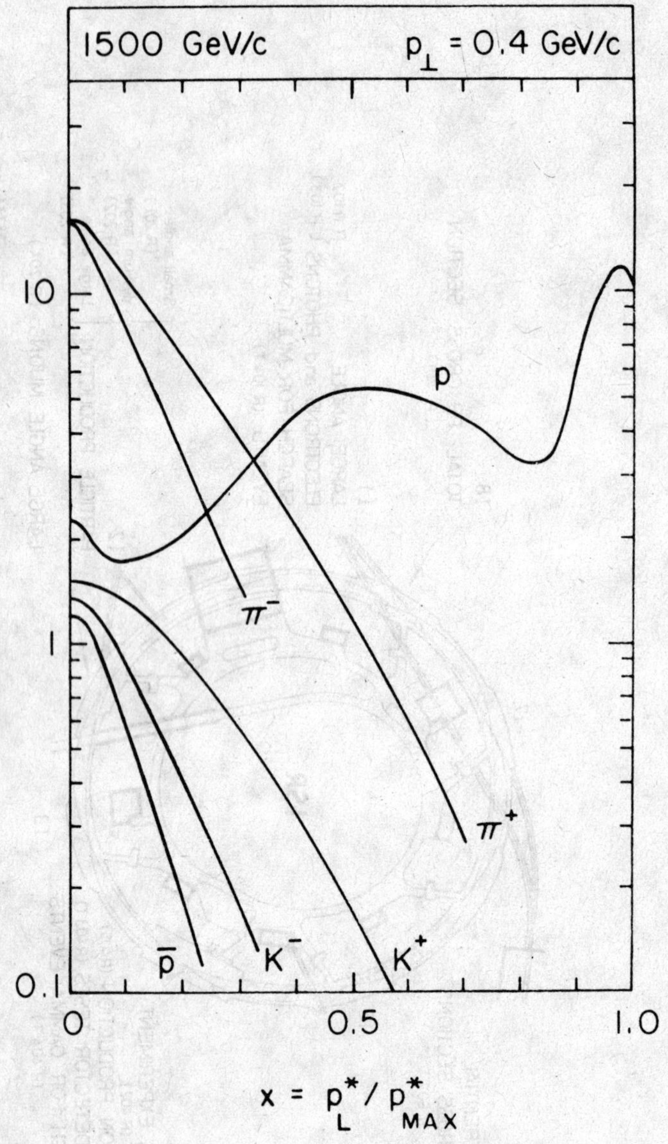

Fig. 21 The invariant cross section is plotted at $\sqrt{s} \simeq 53$ GeV for various particles versus the Feynman variable $x = p_L^*/p_{max}^*$, where p_L^* and p_{max}^* are the longitudinal momentum and the maximum momentum of each particles in the centre-of-mass system. The scale is close to mb/GeV^2. (The figure is taken from Ref. 10).

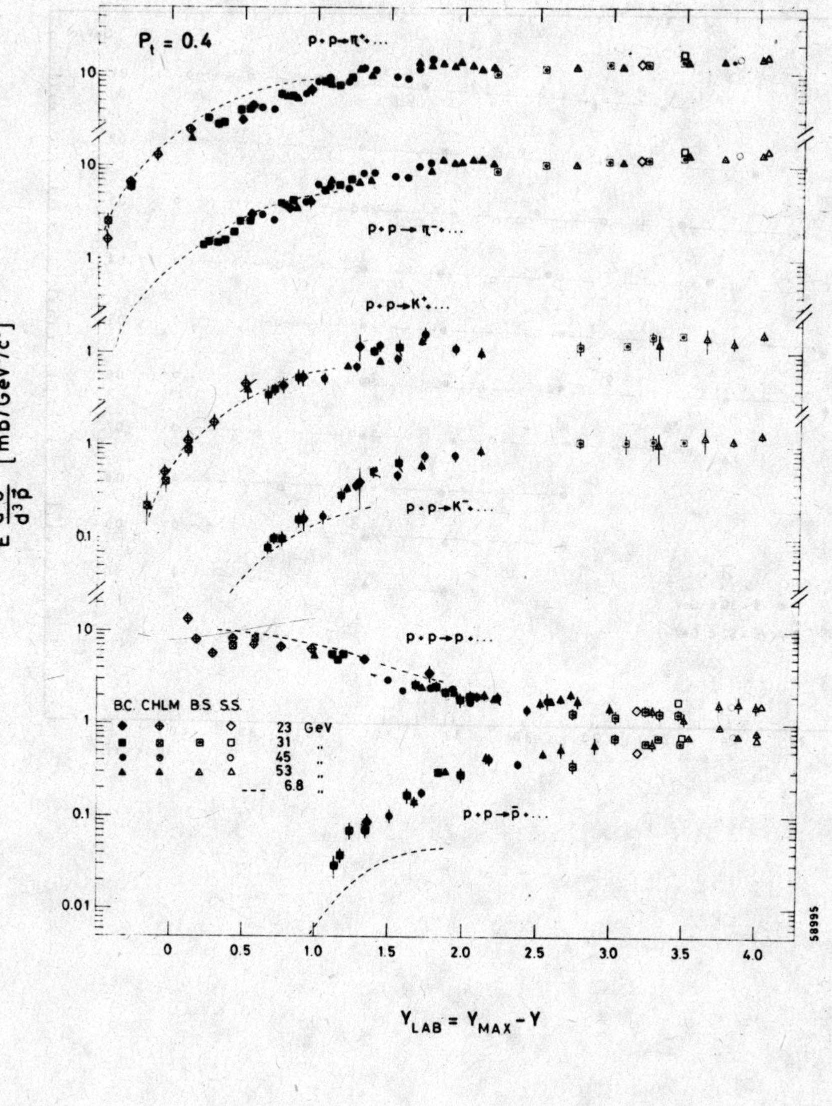

Fig. 22 Compilation of data on invariant cross-sections from the ISR (Ref. 31) at $p_t = 0.4$ GeV/c. The approximate scaling from PS to ISR energies is apparent for all particles (with the exception of the antiprotons). The errors are typically 15%.

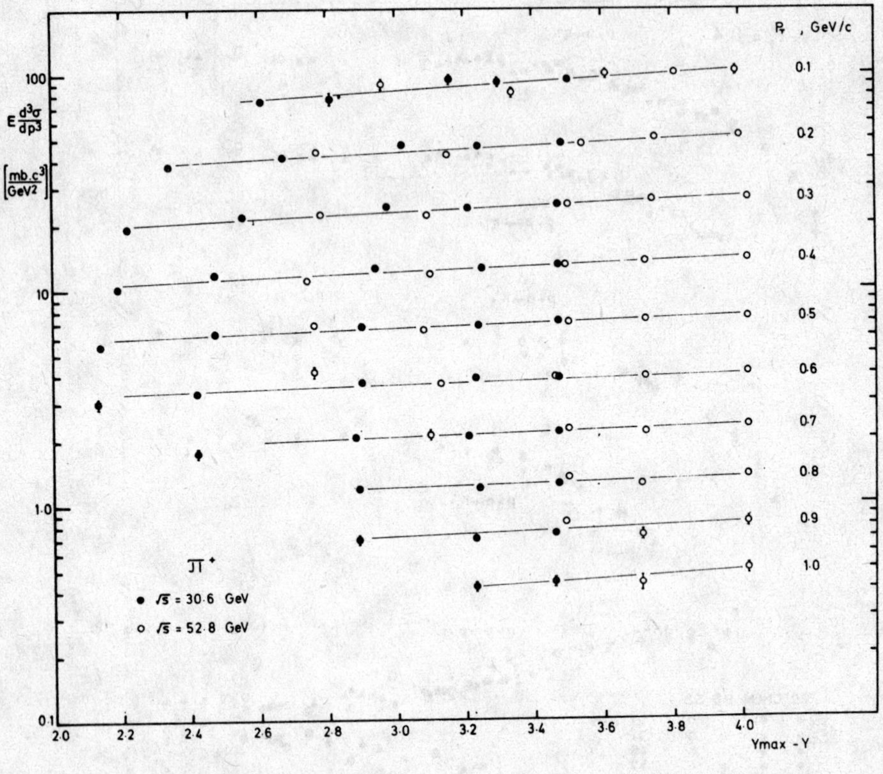

Fig. 23 Recent results from the British-Scandinavian collaboration on the behaviour of the plateau for π^+ versus $y_{LAB} = y_{max} - y$. These measurements show an increase of the plateau of (8 ± 7)% in the ISR energy range.

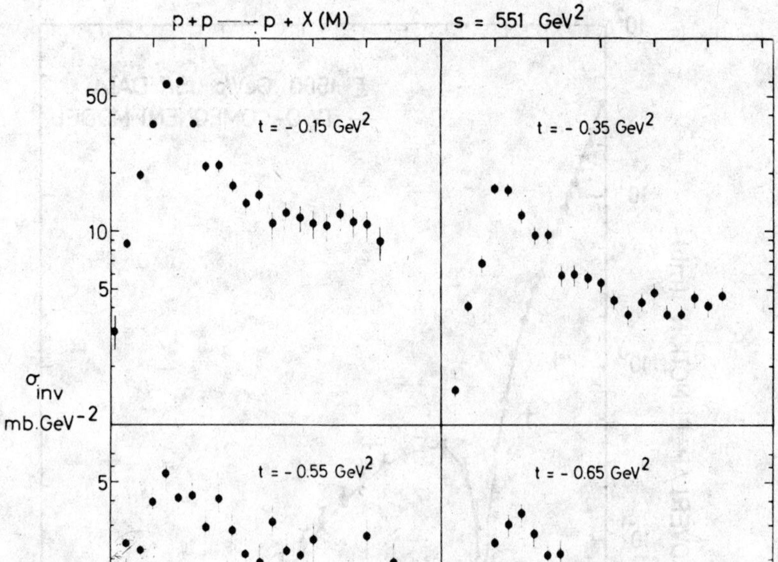

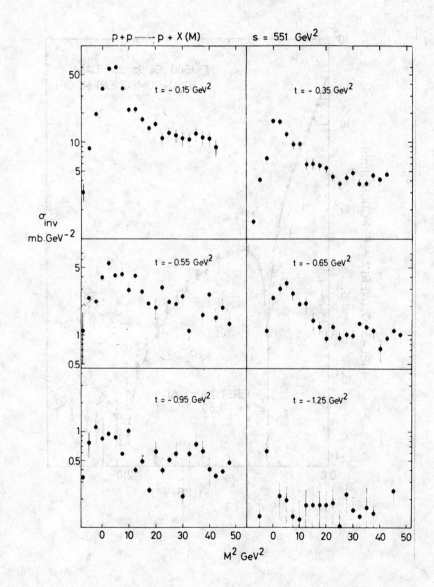

Fig. 24 Results of the CERN-Holland-Lancaster-Manchester collaboration on the diffractive production of large masses. The elastic events have been subtracted from the sample by requiring that there is no collinear charged particle in the opposite hemisphere.

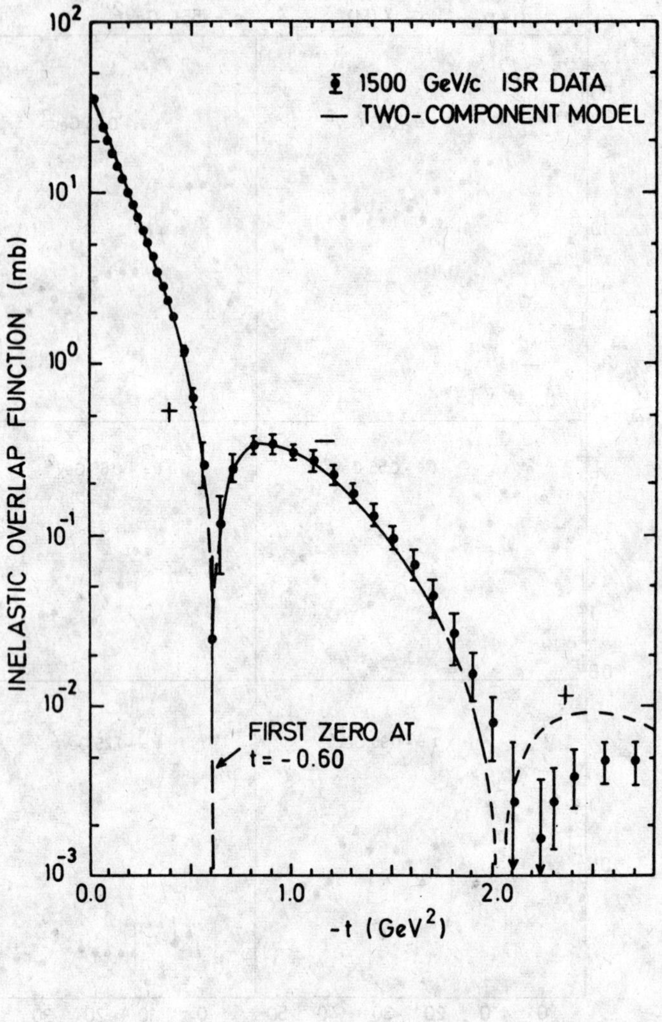

Fig. 25 The "experimental" inelastic overlap integral at $\sqrt{s} \simeq 53$ GeV in momentum transfer space is compared with the parametrization of de Groot and Miettinen (Ref. 41), which contains a central and a peripheral component.

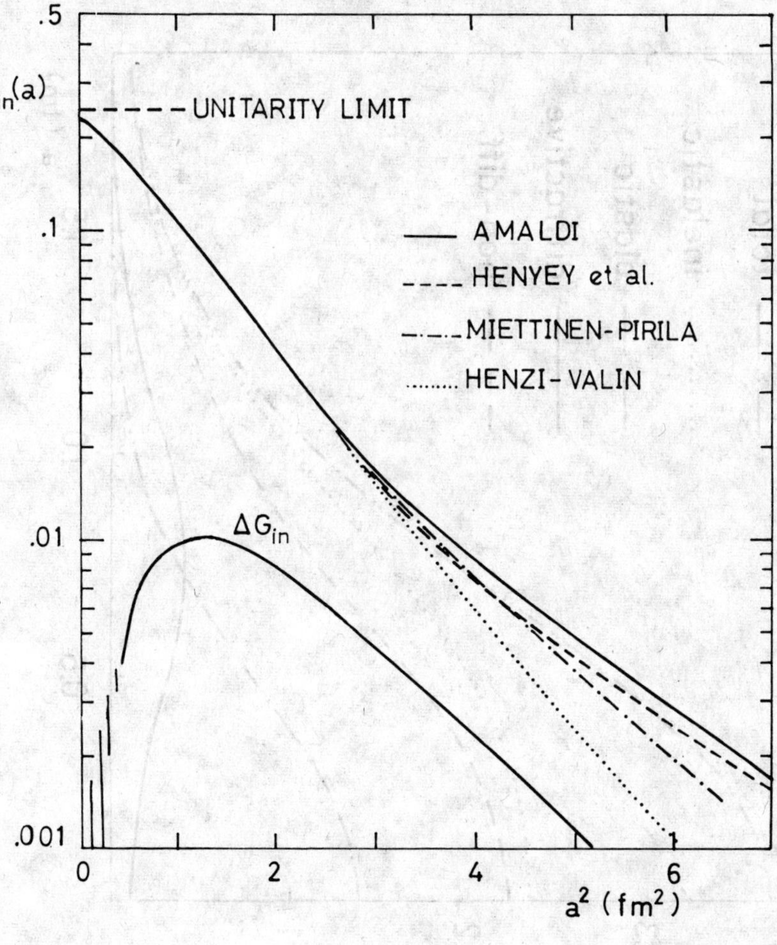

Fig. 26 The continuous curve represents the inelastic overlap function at $\sqrt{s}$ = 53 GeV computed from the values of the proton-proton opaqueness plotted in Fig. 9. The dashed, dashed-dotted and dotted lines represent the results of similar calculations reported in Refs. 46, 47 and 48. The lower continuous curve is the increment ΔG_{in} of the inelastic overlap function in the ISR energy range, taken from fig 9. For very small values of a^2 the behaviour of ΔG_{in} is affected by the experimental uncertainties, but above $a^2 \simeq 0.5$ cm^2 it is on good experimental basis.

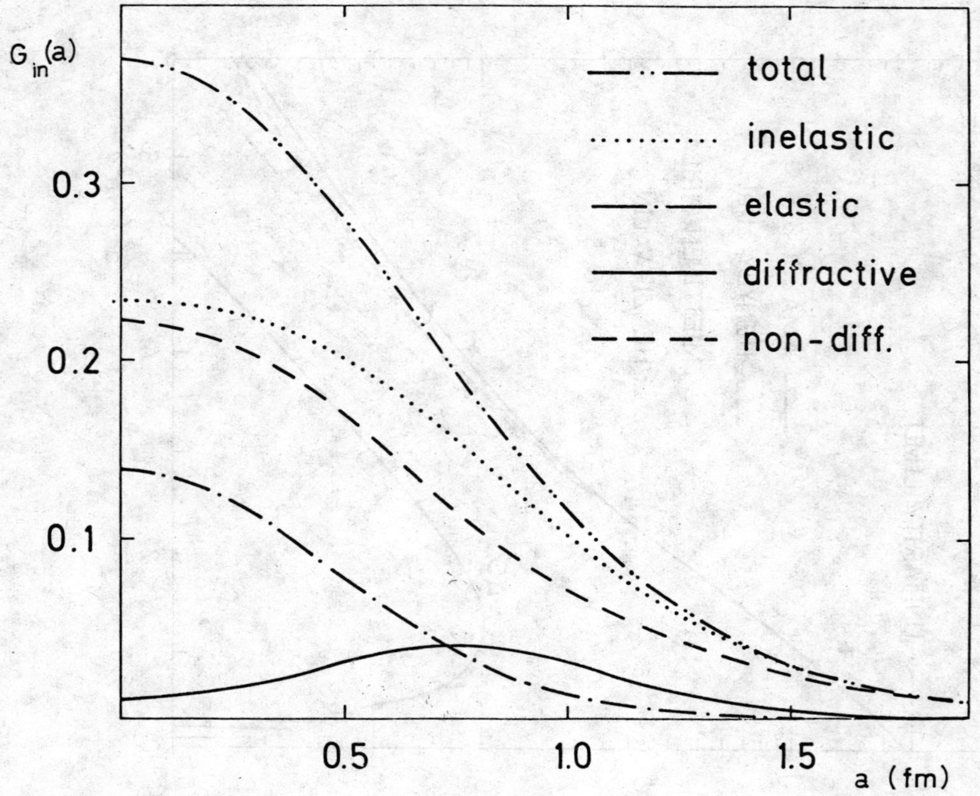

Fig. 27 Overlap functions at $\sqrt{s}$ = 30.5 GeV computed from the experimental data by Sakai and White (Ref. 52) in the hypothesis that in diffractive dissociation t-channel helicity is conserved. The continuous curve is deduced from the measured momentum dependence of the invariant cross-section for the protons of large x, as measured by the CHLM collaboration.

DISCUSSIONS

CHAIRMAN: Prof. U. Amaldi

Scientific Secretaries: A. Bouquet, P. Gensini, A. Patkos

DISCUSSION No. 1

- *YU:*

If you draw a smooth curve through the Serpukhov and ISR data for the total cross-section, you get a concave curve. Now consider only the ISR points. What kind of curve will result? This would be interesting if one wants to speculate whether the cross-section will continue to rise or it becomes finally constant.

- *AMALDI:*

In the ISR energy range the four points averaged over the two experiments are consistent with a linear dependence on log s. I shall treat some of the speculations you mentioned in one of my next lectures.

- *MOEN:*

Is it possible to parametrize the total zero-crossing you get as a - b $(\log s)^{-1}$? Does it work?

- *AMALDI:*

Yes, it can be done with a good fit. Certainly a Regge-type parametrization with a negative cut contribution, which gives a cross-section which goes to a constant value from below, is consistent with the data.

- *HENDRICK:*

Are your data consistent with a cross-over of ρ = Re f/Im f at $\sqrt{s}$ about 16 or 17 GeV?

- *AMALDI:*

Yes. In this connection I would like to mention that there have been recent computations of dispersion relations (with various hypotheses for the behaviour of the pp total cross-section at high energies) all of which say that, if σ_{tot} increases enough, the real part must cross zero where our data points are.

- *HENDRICK:*

Can you say if it continues increasing after it crosses zero?

- *AMALDI:*

No experimental data exist, but as you know on the theoretical side the Khuri-Kinoshita theorem says that if the cross-section goes up fast enough, then ρ goes to zero from above as $(\log s)^{-1}$. This point will be covered in one of my lectures.

- *RABINOVICI:*

What are the plans to measure topological cross-sections $\sigma_n(s)$ at the ISR and look for the Wilson dip?

- *AMALDI:*

There are now essentially three ways to collect information on multiparticle production at the ISR. There is the Pisa-Stony Brook experiment, which however cannot measure momentum even if it collects information on a very large solid angle. There is the split-field magnet facility on which various experiments have been approved. It can measure momenta, but covers only a smaller solid angle. Then there is an experiment which is already on the machine employing a streamer chamber. For the moment it has no magnetic field but will probably see tracks in a better way than with proportional chambers.

It will, however, take quite a long time before all these experiments will be able to give some data on topological cross-sections.

- *TING:*

I have a remark on the split-field magnet facility, since our group has an approved experiment at the SFM together with the British-Scandinavian Collaboration. One can measure two- and three-particle correlations, but the identification of the final particles can be accomplished only for momenta below 1 GeV/c. Below this momentum one can easily measure $\pi\pi$, πK, KK, and $\bar{p}p$ correlations in the central region.

- *MELLISINOS:*

The NAL small angle pp data obtained using the gas jet target by the Soviet-USA Collaboration are still preliminary. They cover the range of s from ~ 20 to ~ 800 $(GeV)^2$. At this time the data on the forward peak slope parameter b are compatible with Serpukhov and ISR data. They do not exclude an energy dependence of b linear in ln s.

Similarly, the data on the ratio ρ of the real to the imaginary part of the forward amplitude indicate that possibly ρ becomes positive within the domain of NAL energies.

- *BYERS:*

The cross-section actually measured by the Pisa-Stony Brook Collaboration seems constant and the rise comes from corrections applied to take into account events in the beam pipe region. Would you tell us about those corrections?

- *AMALDI:*

In both experiments the measured cross-section is not the total cross-section and in the Pisa-Stony Brook case the total cross-section is obtained as the sum of the directly observed one and of two increments. The elastic correction is known with high precision. By definition, elastic events have two particles and if they go into the pipes, as the counters surround the pipes, they are not detected. To compute this correction the Pisa-Stony Brook Collaboration takes the CERN-Rome values for the elastic cross-section and the angular distribution measured by Aachen et al. The correction increases with energy because the opening angle of the pipes is fixed and this corresponds to increasing the minimum measured $|t|$ as the energy increases. The increment of the inelastic losses in the ISR range is only 0.4 mb. To compute these losses they subtract the elastic events seen outside the pipes from the total number of events and plot the remaining number of events versus the angle Θ_{max} of the track which forms the largest angle with the pipe axis. An extrapolation to $\Theta_{max} = 0$ is then done to know the number of inelastic events which go inside the solid angle covered by the pipes, which their apparatus therefore does not see.

- *DRECHSLER:*

Is it fair to say that all total cross-section measurements, in particular the Pisa-Stony Brook experiment using absorption techniques, are dependent on the behaviour of the differential cross-section at small values of momentum transfer and its energy variation? The extrapolation to $t = 0$ is indeed model dependent.

- *AMALDI:*

In principle this is true, in practice this model dependence is negligible. For example, in the CERN-Rome experiment a variation of 10% (± 1.3 GeV^{-2}) of the slope parameter would give a variation of the total cross-section at (26.5 + 26.5) GeV of ± 0.6 mb, equal to the quoted error of the measurement. Remember also that the dependence on luminosity of the two determinations is different and from this point of view the agreement between the two is clearly a very remarkable one.

- *TING:*

Before results from your experiment and the Pisa-Stony Brook experiment were published, I heard that ISR gave constant cross-sections. Now nobody speaks about that any longer. What did happen?

- *AMALDI:*

The Aachen-CERN-Genova-Harvard-Torino Collaboration presented at Batavia values of the pp total cross-section at the three highest ISR energies, compatible with a constant behaviour. More precisely, the values were 38.5 ± 1.0 mb, 38.0 ± 1.0 mb, and 38.5 ± 1.0 mb. These results were obtained towards the end of 1971, when various problems connected with the application of the Van der Meer method were not yet well understood.

The CERN-Rome experiment also presented results at Batavia, which were indicating an increase of about 2 mb but with large errors (1.0 ÷ 1.5 mb at the extreme points). After Batavia a 4% error was discovered in the program which computes the Van der Meer displacement, so that all the quoted cross-sections have to be increased by 2% (+0.8 mb). Then the Pisa-Stony Brook Collaboration presented data in a seminar at CERN showing a similar increase with the same size of the errors, and giving also a very suggestive high value at (31.5 + 31.5) GeV. If one plots on the same graph the results of these three "first generation" experiments, they do not actually disagree outside the large errors, and on the whole they give an indication of a rising cross-section. The values I have presented here are instead the results of much more refined second-generation experiments, based on the experience previously gained. In particular, it must be said that the apparatus of the CERN-Rome Collaboration has been completely changed. Consequently the errors have been very much reduced, essentially because the problem of the cross-talks between the intersections has been fully appreciated and controlled. Did I answer your question?

- *TING:*

No. Do ACGHT have new data or do they stick to their old data?

- *AMALDI:*

They do not have new data, but are now in the process of debugging a new second generation experiment in interaction region 6. They will have very probably results before the end of 1973.

- *KUPCZYNSKI:*

In the Coulomb interference method you fit the elastic differential cross-section with three parameters: the total cross-section, the scale, and the ratio ρ of the real to the imaginary part of the forward elastic amplitude. Have you checked how much is changed the value for σ_{tot} if you change the value of ρ and how big is the scale error?

- *AMALDI:*

If we change ρ from -0.05 to +0.20 the total cross-section changes by less than 0.2 mb. There is no scale error in the Coulomb interference measurement; in the Van der Meer method the value of the scale error published is ±0.8 mb and was mainly due to the luminosity measurements. Further checks of the diplacement have reduced it to less than ±0.5 mb.

- *KUPCZYNSKI:*

I have also a general comment: in the different methods of the evaluation of the cross-sections in ISR the following assumptions were introduced, which could be untrue:

1) The Coulomb interference formula, which at high energies may require modifications.
2) The optical theorem, which may turn out to be violated as pointed out by Eberhard [Nuclear Phys. B48, 333 (1972)] and myself (Phys. Letters B, to be published).
3) The assumptions that $d\sigma/dt$ does not start to decrease in the unobserved forward region, a fact which could cause an over-estimation of the additive corrections to σ_{tot}.
4) The ratio ρ can change significantly in the high-energy region.

Of course the fact that using all the conventional assumptions one obtains the rising total cross-sections is very important, even if it could turn out that the usual assumptions 1 to 4 are no longer true.

DISCUSSION No. 2

- *TING:*

I would like to mention an interesting experiment we are planning at NAL to measure pp total and particle production cross-sections and multiplicities at $s \simeq 16000$ GeV^2 or $p_L \simeq 8000$ GeV/c.

We take an extracted 10 GeV proton beam from the booster (87 pulses, 2 ns long each with intensity of about 10^{10} protons/pulse). We guide the 10 GeV beam to collide head on with the extracted 400 GeV beam. The 400 GeV beam has 1113 pulses, 2 ns long each with again 10^{10} protons/pulse. We then take 87 pulses in parasitic mode: the total rate is $\sim$ 1200 events/day in an interaction region of 1 m × 1 mm × 1 mm (the beam has a cross-section of 1 mm), with 3×10^4 pulses per day and a total cross-section of 40 mb. Remember that, extrapolating the formula quoted by Dr. Amaldi as a fit to accelerator and cosmic-ray data, we can expect a cross-section of 55 mb at the c.m. energy of this experiment.

- *NAHM:*

The cosmic-ray data for σ_{pp}^{tot} have been parametrized with a term $\log^2 (s/137)$. However, an energy scale of the order of 100 GeV is probably unphysical and can only indicate that non-leading terms are important over the measured energy range. Does one know anything about the question whether σ_{tot} rises as log s or $(\log s)^2$?

- *AMALDI:*

The only thing I can say about the increase of σ_{tot} is that over the ISR range it is compatible with a log s increase. The formula I showed you is only an easy way for everybody to remember how to compute total pp cross-sections from 20 to 2000 GeV/c laboratory momentum, which are accurate up to some tenth of a millibarn, but it has no deeper meaning at all. If this increase has some physical meaning the scale has probably to be smaller than that. Then the increasing behaviour becomes dominant only at the ISR energies because below it is hidden by the decreasing probability for some other phenomenon. On the other hand we are still far from saturating the Froissart bound at the present energies, and an increase in σ_{tot} even steeper than $(\log s)^2$, like a power, will not violate it.

- *RABINOVICI:*

I would like to comment on Dr. Nahm's remark. The high s_0 number in the $(\log s/s_0)^2$ fit can have physical meaning, if for example it is the threshold for the Pomeron-proton cross-section as Prof. Berger told us in his lecture.

- *AMALDI:*

This is not my understanding of it. If we would fit with this formula only one side of the energy dependence of σ_{tot}, then you could be right. But we are using it for the whole s-region. So this explicit number is just to help our memory.

- *MILLER:*

I would like to know whether in the Pisa-Stony Brook experiment, where in principle one can look for the topology of the events, are there any evidence for the increasing contribution of multiprong events or for the converse: the decreasing cross-section of the few prong events?

- *AMALDI:*

In the last few months the Pisa-Stony Brook Collaboration has made a big step forward. They can distinguish the multiparticle and quasi-elastic events from one another. However, they have not yet numbers on this very important piece of information. It will be of course very important to know whether the cross-section is increasing because the cross-section of quasi-elastic events (diffractive type) or that of SRO-type of events increases.

- *HENDRICK:*

Is there any evidence that Pomeranchuk-type theorem for differential cross-sections is close to being satisfied?

- *AMALDI:*

In the forward direction they are satisfied within 10-20%, as at lower energies.

- *WAMBACH:*

Do you know whether one can say something on the form log s or (log s)^2 for the increase of the total cross-section by using the Martin-Yndurain bounds?

- *AMALDI:*

As far as I know, with the present data they cannot add new information.

- *ZICHICHI:*

I would like to make a remark on the cosmic-ray experiment described this morning. You said that the solution of "unaccompanied hadron" is based on a dE/dx cut at a value of ~ 2. As you probably know we have recently done an accurate measurement of dE/dx using high-energy protons (14-20 GeV). The dE/dx distribution, as observed with a 2.5 cm plastic scintillator, shows an anomalous Landau tail due to strong interactions. This tail is energy-dependent as expected from the energy dependence of the collision cross-section. Therefore a fixed dE/dx cut using a thick plastic scintillator is known to be energy-dependent. The question is: has this point been investigated by the authors of the cosmic-ray experiment or by the other specialists you have mentioned?

- *AMALDI:*

Not as far as I know

- *KUPCZYNSKI:*

My comment is connected with the questions by Drs. Nahm and Wambach. The upper bounds on the rise of the total cross-section cannot say how σ_{tot} grows to eventually saturate them. We have to remember that the rise of σ_{tot} means the rise of the range R of the strong interactions as the energy increases. Of course it cannot be excluded *a priori* that $R \to \infty$ as $s \to \infty$, but in my opinion the speculations in that direction are premature.

DISCUSSION No. 3

- *RABINOVICI:*

I wish to comment about a model by Caneschi and Schwimmer which was made about a year ago before the ISR results were known. This model, although a M.P.M., is a diffractive model as the particles exchanged along the chain are Pomerons, and a large number of events are produced at small impact parameter instead of the usual M.P.M., which gives the largest production at large impact parameter.

- *AMALDI:*

I understood from Caneschi that this happens only when there is absorption in the model, in particular when the absorption is complete in the centre. Experimentally we have 80% absorption. Therefore the model can apply at much higher energies and solve the problem, but does not apply in the present energy range.

- *RABINOVICI:*

The log^2 s behaviour of the Froissart bound follows only from the assumption of a power behaviour of the impact parameter amplitude. I would like to know what sacred principle of physics would be violated by an exponential behaviour?

- *AMALDI:*

The best answer I got was from Prof. Van Hove. His point is the following: an exponential in the complex plane is a crazy function, exponential in one direction and oscillating in another. Since most people believe in simple analyticity properties, a power behaviour is preferable as it has no essential singularity at infinite complex energy. According to Van Hove, it is not true that axiomatists have shown that exponential behaviour is forbidden by the causality principle without further assumptions. Indeed, they do assume a certain smoothness of the function. A violation of the Froissart bound should thus lead to the conclusion not that micro-causality is wrong but that the specific smoothness assumptions do not apply.

- *RABINOVICI:*

Ball and Zachariasen produced a type of model requested by Prof. Amaldi, using a model for inelastic scattering. They constructed an s-channel equation for the elastic shadow in terms of the inelastic shadow. For small t they have an explicit solution whose main features are a ln s growth for σ_{tot}, $\sigma_{el} \sim$ const. and a behaviour of the elastic amplitude like $J_1(R\sqrt{-t})/\sqrt{-t}$. This is a self-consistent model without absorption which predicts the total cross-section to rise.

- *KUPCZYNSKI:*

You have shown the fit of the upper limit of the inelastic contributions of the elastic pp scattering using the generalized optical theorem. Elastic two-particle contributions contain the overlap of the two-particle amplitudes for different t, integrated over the whole range of the internal momentum transfer. What approximations did Miettinen use to obtain his fit?

- *AMALDI:*

They calculated the overlap functions assuming the elastic amplitudes to be imaginary and obtained them from data for $d\sigma/dt_{elastic}$.

- *HENDRICK:*

You said that we can reduce by a factor 2 the Froissart bound in the diffractive scattering case. You take Im f = $\frac{1}{2}$ instead of 1. Could you explain that?

- *AMALDI:*

The Froissart bound was derived by mathematical majoration. Unitarity says that the partial wave amplitude should be inside a circle of radius $\frac{1}{2}$ (Argand plot). On the circle lie points of zero absorption. On the contrary, absorption corresponds to being inside the circle. We believe that at high energy elastic scattering is essentially diffractive so that elastic scattering is just the shadow of inelastic scattering. This corresponds to points at the centre of the unitarity circle, where Im f = $\frac{1}{2}$.

- *KUPCZYNSKI:*

When de Groot and Miettinen use the extrapolation formula, are they using the exponential formula or just a functional fit to the data?

- *AMALDI:*

They use the data.

- *WILCZEK:*

The Chou-Yang formula gives a good fit for elastic pp scattering slope at s = 2800 GeV^2, the input being the proton form factor and the only parameter being adjusted by the total cross-section. A small change in the total cross-section gives a small change of the parameter but a large change in the height of the second maximum. The data seem inconsistent with this. Do you think possible to reconcile this theory with experiment, or do you think the agreement at s = 2800 is fortuitous?

- *AMALDI:*

Chou and Yang recently turned the problem around and calculated the form factor from the proton-proton elastic cross-section. They found a remarkable agreement with the electromagnetic form factor. Their model, I think, is even too good and indicates a remarkable connection between the density of charged matter seen by the photon and the density of strong interacting matter in pp elastic scattering.

- *CHU:*

Have the experimentalists been able to measure the ratio σ_{el}/σ_{tot} in the cosmic-ray experiments?

- *AMALDI:*

No. It is very difficult to measure the σ_{el} component since for such high energies an elastically scattered proton will be nearly indistinguishable from an unscattered proton. The total cross-section is calculated by measuring σ_{inel} and then using the Glauber model to relate σ_{inel} to σ_{tot}.

DISCUSSION No. 4

- *LOSECCO:*

The Cheng, Walker and Wu fit to hadron-hadron data seems to indicate a common asymptotic behaviour for all cross-sections. The Bartel and Diddens fit you showed had a different high-energy behaviour. How do the two fits differ?

- *AMALDI:*

The Bartel and Diddens fit uses 16 parameters. The Cheng, Walker and Wu fit requires a generalized Pomeranchuk theorem in which the cross-sections must all rise to the same value and therefore it is a 14-parameters fit. The behaviour of ρ in the ISR range does not differ very much from one fit to the other: for instance for a cross-section rising as 0.5 mb $(\log s/s_0)^2$ ρ is always in the range $0.06 \div 0.09$, independent of the other parameters.

- *TING:*

Suppose that at NAL one measures the K^+p total cross-section and finds out that it is decreasing or flat, what will be the implications?

- *AMALDI:*

I do not know. To get a more complete information I would suggest, for instance, to measure the differential elastic cross-section and try to evaluate from it the inelastic contributions. It may well be that one could then find a decreasing absorption with an increasing radius. As far as pp scattering is concerned, ISR data are completely consistent with a wide range of asymptotic behaviour for $\sigma_{tot}(s)$, as predicted by different models; for instance, a pole-cut model, formulated before the ISR data were obtained and fitted to the rise in the Serpukhov data, fits quite well also the ISR data.

- *CANESCHI:*

You said you can fit your σ_{tot} data also with a Pomeron plus a negative cut. What happens to ρ in such a case?

- *AMALDI:*

I showed you the results of Bartel and Diddens. They have assumed σ_{tot} to stay constant after a set of values of s. It is seen that if the cross-section flattens more than one decade higher in laboratory momentum (for instance above 10^4 GeV/c), the effect on the real part is negligible.

- *LOSECCO:*

When does the Cheng, Walker and Wu prediction $\sigma_{el}/\sigma_{tot} = \frac{1}{2}$ set in?

- *AMALDI:*

Very far away. At our energies the ratio is only about 17%, but I can not give you any precise number.

- *TAYLOR:*

I can supply a number for this ratio of the elastic scattering cross-section to the total cross-section for proton-proton scattering as a function of energy in the Cheng, Walker and Wu impact picture. Their model predicts that at very high s, $\sigma_{el}/\sigma_{tot} \to \frac{1}{2}$. This limit is, however, very slowly approached: at ISR energies $\sigma_{el}/\sigma_{tot} \approx 0.17$ and at $s \simeq 10^6$ GeV^2, the ratio is only ≈ 0.25.

- *EYLON:*

Is there any chance to have experimental data about exclusive two-body reactions like $pp \to pN^*$?

- *AMALDI:*

There are experiments set up to measure $pp \to p\Delta$ in intersection region 6 at the ISR and $pp \to pN^*$ in the split-field magnet facility, and there are also plans to measure $pp \to \Delta\Delta$ in the same apparatus of the ACGHT Collaboration.

- *MELISSINOS:*

Such experiments have been done at NAL with the recoil technique and there are data concerning production of $N^*(1420)$, with a t slope of about 20 GeV^{-2}, and of $N^*(1680)$ with a slope similar to that of the elastic forward peak, in addition to the broad bump at higher masses shown by Berger.

- *EYLON:*

Were those experiments made at different values of the energy, such that we could study the energy dependence of the exclusive cross-sections in order to check Bjorken-Kogut's assumptions?

- *AMALDI:*

We expect to have more precise data in a short time, but with a simple-minded approach we expect no energy dependence, since those are diffractive processes to isolated isobars and this is roughly consistent with present data.

- *EYLON:*

But this might violate scaling in x, since the reaction $pp \to p + X(M^2)$ will have, according to Bjorken and Kogut, a cross-section $\sigma_X(s) \propto f_p(1 - M^2/s)$, where f_p is the proton scaling function.

- *EYLON:*

In the discussion held this morning, during the lecture one could get the wrong impression that the log s increase of the number of partons inside the proton is a new feature of pp scattering. In the simplest parton model for deep inelastic scattering one finds that the density of partons in x is given by $\nu W_2(x)/x$; since the data seem to indicate $\nu W_2 \xrightarrow[x\to 0]{}$ const., one finds a divergent distribution in x. This means that the rapidity distribution is constant, although its range increases like ln s.

- *AMALDI:*

In fact I intended to make it clear that those ideas and the figures those authors plugged in were taken from deep inelastic scattering description in the parton model.

- *KLEINERT:*

I am somewhat confused by Eylon's remark. The parton distribution you see in deep inelastic electroproduction is the longitudinal momentum distribution inside the proton, and you naively expect their average number to be constant; but what you mention when you speak about parton distribution in rapidity space in pp collisions is their distribution inside the "fireball" formed by the two colliding particles, where we can very well have an average number of elementary constituents increasing logarithmically with the c.m. energy of the pp system.

- *ZICHICHI:*

This is a physical picture I like more, but Caneschi will lecture on this subject anyhow, so let us wait for him to explain us these problems.

- *RABINOVICI:*

Some cosmic-ray data are claimed to show that the average transverse momentum of produced pions is an increasing function of energy. What do the ISR data indicate?

- *BERGER:*

In the ISR range and up to the ISR range the average transverse momentum is not substantially increasing.

- *HENDRICK:*

In these fits you described, the authors use as input an extrapolation of the total cross-section data out to infinite s, and using dispersion relations, get as output the behaviour of ρ out to infinity. Since these fits use analyticity and crossing they should reproduce the Khuri-Kinoshita behaviour of $\rho \to C/\ln s$ as $s \to \infty$. Is that correct?

- *AMALDI:*

Yes, that is correct.

- *HENDRICK:*

Then if you use as input, say, an extrapolation of the total cross-section behaviour seen at the ISR, you should have ρ increasing for a certain range of s, then turning around and going to zero. What is the s-point at which ρ is maximum in such a fit?

- *AMALDI:*

It is at $s \simeq 5.10^4$ GeV^2.

- *HENDRICK:*

What input parameters is this point sensitive to? That is, for certain inputs will the position in s of this maximum change?

- *AMALDI:*

I am not sure.

- *MELISSINOS:*

Don't push your expectations too high, because if ρ is positive, its effects are much harder to be measured than if it is negative.

HIGH ENERGY pp ELASTIC SCATTERING AND THE CHOU-YANG FORMULA

N. Byers

Table of Contents

HIGH ENERGY pp ELASTIC SCATTERING AND THE CHOU-YANG FORMULA

Nina Byers
Department of Theoretical Physics, University of Oxford,
Oxford, England

Introduction

This lecture is a report of a remarkably good fit to pp elastic scattering data at $\sqrt{s} = 53$GeV. It includes a summary of presently available information concerning the energy dependence of the pp elastic scattering data taken over the ISR range of energies $31 \le \sqrt{s} \le 53$GeV, a discussion of the Chou-Yang formula, a parton model interpretation of this formula, and some concluding remarks.

I. Summary of ISR Data

Measurements of pp elastic scattering at the ISR[1)] revealed three new features:

(i) log $(d\sigma/dt)$ is not simply a linear function of t in the small t range;

(ii) $d\sigma/dt$ has a minimum at $-t \cong 1.4(\mathrm{GeV/c})^2$;

(iii) $d\sigma/dt$ has a secondary maximum at $-t \cong 1.8(\mathrm{GeV/c})^2$.

Data on the behavior of $d\sigma/dt$ both as function of energy and t, in the small t region, are summarized in Table I where we quote published values of the slope b of $\ln(d\sigma/dt)$ (obtained by straight line fits to data at various energies in the specified t intervals).

Table I

Values of slope parameter $b = d/dt\,(\ln(d\sigma/dt))$ at three ISR energies.

$\sqrt{s}$ GeV	range of $\lvert t\rvert (\mathrm{GeV/c})^2$	$b(\mathrm{GeV/c})^{-2}$	Ref.
31	0.015 - 0.055	13.0 ± 0.7	2
	0.046 - 0.090	11.9 ± 0.3	3
	0.138 - 0.240	10.9 ± 0.2	3
45	0.01 - 0.05	12.6 ± 0.4	4
	0.046 - 0.089	12.9 ± 0.2	3
	0.136 - 0.239	10.8 ± 0.2	3
53	0.01 - 0.05	13.1 ± 0.3	4
	0.06 - 0.112	12.4 ± 0.3	3
	0.168 - 0.308	10.8 ± 0.2	3

The variation of b as function of t interval shown in Table I has been fitted by a "break in slope"; i.e., with $b \approx 13$ for $0 \le -t < 0.1(\mathrm{GeV/c})^2$ and

$b \approx 11$ for $-t \gtrsim 0.1(\text{GeV/c})^2$. The data at $\sqrt{s} = 53\text{GeV}$ are displayed in Figure 1 where one sees that these data are also compatible with a curved line. If ln $(d\sigma/dt)$ is curved in the small t region, values of b obtained from straight line fits to data will vary according as to what interval of t is used. Therefore, if one wishes to examine the variation of forward slope with energy most reliably one should, in my view, compare b values obtained from data in the same range of t. This is done in Table I and one sees that the data indicate that for $31 \leq \sqrt{s} \leq 53\text{GeV}$ the shrinkage observed[5)] in the range $6.5 \leq \sqrt{s} \leq 12\text{GeV}$ appears to have slowed down appreciably or stopped.

Data over the full range of t measured at 53GeV are displayed in Figure 2. The solid curves in Figures 1 and 2 show the Chou-Yang[6)] fit of Kac[7)] to these data. This will be discussed below. As regards the behavior with energy of the larger t data, the dip at $-t \approx 1.4(\text{GeV/c})^2$ and the second maximum at $-t \approx 1.8(\text{GeV/c})^2$ both appear in the data taken for $31 \leq \sqrt{s} \leq 53\text{GeV}$[8)]. The absolute normalization of the data in this region is not known. If the data in the larger t region are smoothly joined onto those at small t where the absolute normalization has been measured, the preliminary data of the ACGHT collaboration indicate that height of the secondary maximum is, up to about a factor 2, the same throughout the range $31 < \sqrt{s} \leq 53$. Calculations show that the height of the secondary maximum is highly model dependent and very sensitive to small changes in parameters. Models which fit the small t data yield secondary maxima whose heights can differ by orders of magnitude[7)]. The position of the dip is relatively stable against small changes because it is a diffraction zero determined mainly by the opacity and size of the absorbing material. However, the height of the secondary maximum varies enormously. For example, the Chou-Yang model which fits the 53GeV data shown in Figures 1 and 2 predicts that a 10% rise in $d\sigma/dt$ at $t = 0$, corresponding to a 5% rise in the total cross section σ_T, is accompanied by a 250% rise in $d\sigma/dt$ at $-t \approx 1.8(\text{GeV/c})^2$. It is, therefore, of great interest to learn from experiment about the energy variation of the height of the second maximum, since σ_T is reported[8)] to rise from 41 to 43mb over the range $31 \leq \sqrt{s} \leq 53\text{GeV}$.

I shall conclude this first section with some remarks concerning the Kac fit of the Chou-Yang formula to the 53GeV data[7)]. Kac had only one free parameter. It is remarkable that he obtained so good a fit to both the small t and larger t data. If this is not fortuitous, one would expect:

(i) if σ_T increases by 5%, the height of the secondary maximum should increase by a factor of about 2.5;

(ii) a second minimum should be seen.

The position of the second minimum can, at present, be predicted only up to an accuracy of about 15% and is expected to occur for $-t \approx 5(\text{GeV/c})^2$.

Unfortunately it may be extremely difficult to observe because the order of magnitude of the cross section after the minimum may not rise above 10^{-9}mb(GeV/c)$^{-2}$.

II. The Chou-Yang Formula

The Chou-Yang formula is a one parameter relation between two sets of data -- measured electromagnetic form factor and measured elastic scattering.[6] Neglecting spin, Chou and Yang conjectured that the high energy limit of pp elastic scattering would be given by

$$\frac{d\sigma}{dt} = \pi |F|^2 \qquad (1)$$

where

$$F = i\ (1 - e^{-\Delta})_{\otimes} \qquad (2)$$

with

$$\Delta = A\ G(t)^2 \qquad (3)$$

The real parameter A is to be determined by data; G(t) is the measured electromagnetic form factor normalized to G(0) = 1. The symbol $\otimes$ in (2) means that $(1 - e^{-\Delta})$ is to be expanded as a power series in Δ and all products of Δ replaced by convolutions of Δ; for example, Δ^2 replaced by $\Delta \otimes \Delta$ where

$$\Delta \otimes \Delta = \int \frac{d^2\kappa'}{2\pi}\ \Delta(-|\vec{\kappa}' - \vec{\kappa}|^2)\ \Delta(-|\vec{\kappa}'|^2) \qquad (4)$$

and $\vec{\kappa}$ is a two dimensional vector with magnitude $|\vec{\kappa}|^2 = -t$. The parameter A and therefore Δ has dimension (momentum)$^{-2}$. Consequently $\Delta \otimes \Delta$ has the same dimension as Δ. The dimensionless parameter that governs the convergence of the series is the product $A\beta$ where β is the scale of t over which $\Delta(t)$ varies appreciably. For example, if $G(t)^2 = f(t/\beta)$, the scattering amplitude F has the form

$$F = i\ A\ \mathcal{F}(A\beta;\ t/\beta) \qquad (5)$$

where $\mathcal{F}$ is a dimensionless function of the dimensionless parameter $A\beta$ and dimensionless variable t/β. Generally, since $\Delta = AG(t)^2$ is a rapidly decreasing function of -t, the larger -t the more important the higher order terms in the expansion.

For the case of interest here, Kac finds $A \simeq 12$(GeV/c)$^{-2}$ and since G(t) approximately follows the dipole form $G_D(t) = (1 - t/0.71)^{-2}$ we have $A\beta \approx 8$; however, since $\Delta \sim (1 - t/0.71)^{-4}$ the expansion parameter is more nearly $A\beta/4$ -- i.e., of order one. The contributions to F(t) from the first four terms are displayed in Table II.

Table II

Contributions to the Kac amplitude[7] from the first four terms in the expansion of F.

$-t(GeV/c)^2$	0.0	0.1	0.5	1.8
Δ	11.89	6.84	1.40	0.08
$-(2!)^{-1}\Delta\otimes\Delta$	-3.47	-2.48	-1.28	-0.19
$(3!)^{-1}\Delta\otimes\Delta\otimes\Delta$	0.99	0.88	0.54	0.14
$-(4!)^{-1}\Delta\otimes\Delta\otimes\Delta\otimes\Delta$	-0.24	-0.23	-0.16	-0.06

Note that even at $t = 0$ the leading term is not a very good approximation to F and that, owing to strong cancellations, for $-t \gtrsim 0.5$ many terms must be taken into account. Fortunately the series can be summed. Using Parseval's Theorem, we can write

$$F(t) = i \int \frac{d^2b}{2\pi} (1 - e^{-A\rho}) e^{i\vec{k}\cdot\vec{b}} \tag{6}$$

where

$$\rho(b) = \int \frac{d^2k}{2\pi} \, G(-|\vec{k}|^2)^2 e^{-i\vec{k}\cdot\vec{b}} \tag{7}$$

Thus one can calculate F by parameterizing the data for G(t) and performing the Fourier integrals. Actually they are one dimensional Hankel transforms since

$$\int_0^{2\pi} \frac{d\varphi}{2\pi} e^{ib|\vec{k}|\cos\varphi} = J_0(b\sqrt{-t}) \tag{8}$$

The form (6) for F shows that the Chou-Yang formula is an optical model which relates the scattering amplitude to a transmission factor $e^{-A\rho}$ for an absorbing disc with transverse dimension $\vec{b}$. From this viewpoint A measures the absorption cross section and ρ the density of absorber (in "gm"/cm^2).

The form (6) also may be regarded as an eikonal approximation to the partial wave expansion for F: viz.,

$$F = (i/2k^2) \sum_{\ell=0}^{\infty} (2\ell+1)(1-e^{2i\delta_\ell}) P_\ell(\cos\theta) \tag{9}$$

where for small θ and large ℓ we use the approximation

$$P_\ell(\cos\theta) \simeq J_0(b\sqrt{-t})$$

with $b = (\ell+\frac{1}{2})/k$ and $t = -4k^2\sin^2\theta/2$. If, for large CM momentum k, the phase shifts δ_ℓ become pure imaginary and the S-matrix elements $e^{2i\delta_\ell}$ smooth functions of b, the partial wave sum is approximately given by the integral (6). This approximation would apply for small angles; i.e., $t/s \ll -1$.

III. Parton Model Interpretation of Chou-Yang Formula

Glauber's nuclear multiple scattering analysis[9] shows that the nuclear optical model may be regarded as the small angle, high energy eikonal approximation to multiple nucleon-nucleon encounters. Similarly, the Chou-Yang formula may be viewed as an optical model of multiple parton-parton encounters. The Chou-Yang formula is a certain limit of a Glauber multiple scattering series.[10] This limit is one for which:

(i) the number of constituents $\sqrt{N}$ of the proton, which we here indiscriminately call partons, approaches infinite (i.e., 1/N effects may be neglected);

(ii) the parton-parton scattering is short range and absorptive so that the parton-parton partial wave scattering amplitudes $\alpha_\ell = 1-e^{2i\delta_\ell}$ may be approximated by

$$\alpha(b) = (A/N)2\pi\delta^2(\vec{b}) \tag{10}$$

with $b = (\ell+\frac{1}{2})/k$ as before;

(iii) parton correlations may be neglected so that, e.g., a double scattering contribution is given approximately by the square of that due to a single scattering[10];

(iv) the transverse distribution of partons (in the infinite momentum frame) is the same as the charge distribution.

As regards assumption (i): if N is finite, the transmission factor $S(b) = e^{-A\rho}$ in (6) has a finite number of terms rather than the infinity of terms in the exponential. However if $N \gtrsim 10$, (6) yields a good approximation for not too large values of t. As regards assumption (ii): the parton-parton scattering amplitude f(t) is related to α(b) by

$$f(t) = i\int\frac{d^2b}{2\pi}\,\alpha(b)\,e^{i\vec{K}\cdot\vec{b}} \tag{11}$$

If α(b) is given by (10), the parton-parton scattering is pure imaginary and point-like; i.e., f(t) = iA/N. If f(t), on the other hand, varies with t; i.e., if it is given by

$$f(t) = i(A/N)\,g(t) \qquad \text{with} \qquad g(0) = 1 \tag{12}$$

then, if assumptions (i), (iii), and (iv) are valid, Δ in the Chou-Yang formula would be given by

$$\Delta = A\,G(t)^2\,g(t). \tag{13}$$

Since the form (1) can be solved for Δ,[6] if Δ is compared with accurately measured $G(t)^2$ values one may find that $g(t) \neq 1$. In particular, if Δ changes sign this would, in this parton model, be unambiguous evidence for a non-pointlike parton-parton scattering amplitude.

Since f(0) = i(A/N), we can use the optical theorem and Kac value of A to obtain a measure of the parton-parton total cross section σ_T; viz., $4\pi A/N = \sigma_T$ and with $A = 12(\text{GeV}/c)^{-2}$ we get

$$N\sigma_T \text{ (parton-parton)} = 4\pi A = 59\text{mb} \tag{14}$$

A remark regarding the number N should be made here. It was introduced in Reference 10 where the Chou-Yang formula was heuristically discussed as the high energy limit of a Glauber multiple scattering series. The treatment followed a non-relativistic theory of nucleus-nucleus scattering where the number of scattering centers in the target and projectile is fixed. In our case, however, we do not expect the number of partons in a proton to be a fixed number and, presumably, N or $\sqrt{N}$ is an average number. In the parton-quark model of Feynman,[11] $\sqrt{N}$ may be estimated from integrals of scaling functions measured in deep inelastic lepton scattering.

The exponential dependence of F on the density function $\rho(b)$ in (6) arises owing to assumption (iii); $\rho(b)$ is the convolution of the probability $D_\infty(\vec{b})$ for finding a parton at transverse distance $\vec{b}$ in the frame where the proton has momentum $P\to\infty$ along z direction: viz.,

$$\rho(b) = \int \frac{d^2b'}{2\pi}\, D_\infty(\vec{b}'-\vec{b})\, D_\infty(\vec{b}'). \tag{15}$$

In Glauber's theory, $D(\vec{b})$ for a nucleus is

$$D(\vec{b}) = \frac{2\pi}{n}\sum_{i=1}^{n}\int \prod_{j=1}^{n} d^3r_j\, \delta^2(\vec{r}_i-\vec{b})\, \left|\psi(r_1, r_2, \ldots, r_n)\right|^2 \tag{16}$$

where $\vec{r}_i$ is the two-dimensional component of the three-vector $\underline{r}_i$ transverse to the collision axis and Ψ is a nuclear wave function. Therefore one does not expect to see D(b) exponentiated. This occurs in (6) as a consequence of (iii) and neglect of 1/N effects.[10]

Assumption (iv) is used to evaluated $D_\infty(b)$; viz.,

$$D_\infty(b) = \int \frac{d^2\kappa}{2\pi}\, e^{-i\vec{\kappa}\cdot\vec{b}}\, G(-|\vec{\kappa}|^2) \tag{17}$$

where G(t) is measured in electron scattering. This assumes that all the constituents of the proton that interact strongly (absorptively here) are transversely distributed as is the charge.

In this section, as in the original Chou-Yang conjecture, I have neglected the spin and isospin of the proton. These are taken into account and discussed in the next section.

IV. Is $G = G_E$ or G_M or what?

Since the proton has spin (and isospin) there are two (four) form factors. For the proton, the two form factors are G_M and G_E. Which should we use for G? Either G_E or G_M or some linear combination of these could be used. This is certainly a problem in principle. In practice, however, the problem does not need to be solved because, at the present level of accuracy, measurements[12] show that for the proton $G_E = G_M$ (G_M is normalized so that $G_M(0) = 1$) for $0 \le -t \lesssim 2(\text{GeV/c})^2$. For $-t \gtrsim 2\ (\text{GeV/c})^2$, G_E

appears to decrease more rapidly than G_M. For comparison with pp scattering data for $-t \lesssim 2(\mathrm{GeV}/c)^2$ without much ambiguity, therefore, one may take $G = G_M$. It is of interest to note here that in the same range $0 \le -t \lesssim 2(\mathrm{GeV}/c)^2$, measurement of the neutron form factor yields $G_M^{(n)} = G_M^{(p)}$ to the accuracy of the measurement.[12] Of course, $G_E^{(n)} \neq G_E^{(p)}$ since $G_E^{(n)} = 0$. To the extent we may take $G_E^{(n)} = 0$ for small t (measurements give $G_E^{(n)}(t) \approx (-t/4M^2)\mu_n G_M(t)$), we can express these equalities by setting the isoscalar and isovector form factors all equal; i.e., take

$$G_E^{(s)} = G_E^{(v)} = G_M^{(s)} = G_M^{(v)} = G \tag{18}$$

where

$$G_E^{(p)} = \tfrac{1}{2}\left(G_E^{(s)} + G_E^{(v)}\right), \quad \mu_p G_M^{(p)} = \tfrac{1}{2}\left(\mu_s G_M^{(s)} + \mu_v G_M^{(v)}\right)$$

$$G_E^{(n)} = \tfrac{1}{2}\left(G_E^{(s)} - G_E^{(v)}\right), \quad \mu_n G_M^{(n)} = \tfrac{1}{2}\left(\mu_s G_M^{(s)} - \mu_v G_M^{(v)}\right) \tag{19}$$

with $\mu_s = 0.88$ and $\mu_v = 4.70$. For $-t \gtrsim 2\,(\mathrm{GeV}/c)^2$, (18) is not a good approximation; $G_E^{(p)}$ falls faster than $G_M^{(p)}$ and $-G_E^{(n)} \approx G_M^{(n)}$. Consequently for $-t \gtrsim 2(\mathrm{GeV}/c)^2$ there will be differences in the Chou-Yang differential cross section according as to which form factor is used for G.

However, in the range $0 \le -t < 2(\mathrm{GeV}/c)^2$ we can test the Chou-Yang conjecture without much ambiguity using (18). Note that this gives precisely the same differential cross sections for pp, $\bar{p}p$, np, $\bar{n}p$, nn and $\bar{n}n$ elastic scattering.

V. Comparison With Data

Kac[7] computed the Chou-Yang formula using G. Shaw's semiphenomenlogical fit to the $G_M^p(t)$ data; viz.,

$$G(t) = [(1 - t/0.43)(1 - t/1.43)]^{-1} . \tag{20}$$

This form takes account of the well-known[13] deviations of G_M^p from the dipole form $G_D = (1 - t/0.71)^{-2}$ for $0 \le -t \lesssim 2(\mathrm{GeV}/c)^2$. Kac's fit to the p-p data is shown in Figures 1 and 2. It is a remarkably good fit which was achieved with only the one free parameter A. If instead of Shaw's form for G(t) the dipole form G_D is used, the fit is not so good. In Figure 3 is displayed the Chou-Yang formula computed with $G = G_D$ and A chosen to fit the small t data.

New data[12] on $G_M(t)$ at small t indicate that the three parameter formula

$$G = [(1 - t/0.382)(1 - t/2.1)(1-t/14.0)]^{-1} \tag{21}$$

is a better fit to the data over the range $0 \le -t \lesssim 6(\mathrm{GeV}/c)^2$; see Figure 4. Shaw's fit to earlier data is shown in Figure 5. Since, as is indicated in Figure 5, the data fall below G_D for $-t \gtrsim 5$ and Shaw's form (20) does not, Shaw's form for G(t) should not be used to calculate $d\sigma/dt$ for $-t > 2(\mathrm{GeV}/c)^2$. If one uses (21) instead, one obtains a second minimum in $d\sigma/dt$ at $-t \approx 5.8(\mathrm{GeV}/c)^2$. (Shaw's form gives a second minimum at $-t \approx 4.3$

$(\mathrm{GeV}/c)^2$.) However, in the small t region, Shaw's G(t) fits the pp data better than (21). This is probably due to the singularity structure of (20) which has poles nearer the ρ and ρ' mass squared. On the scale of a graph like those shown in Figs. 2 and 3, the fits using (20) and (21) are indistinguishable for $0 \leq -t < 2(\mathrm{GeV}/c)^2$. For $-t \gtrsim 2(\mathrm{GeV}/c)^2$ they are different and (21) seems to yield better agreement with the data.

VI. Summary and Conclusions

Presently available data in the small t region indicate that the shape of $d\sigma/dt$ does not change much over the range $31 \leq \sqrt{s} \leq 53\mathrm{GeV}$ and its magnitude rises by ~10%. If the Chou-Yang fit of Kac to the 53GeV preliminary data is not fortuitous, one would expect the height of the secondary maximum to change by about a factor of ~2.5 if $d\sigma/dt$ at $t = 0$ changes by 10%. This corresponds to about a 5% change in A. Such a change produces very small changes in the shape of $d\sigma/dt$ near $t = 0$.

If there is energy dependence, i.e., if the fitting parameter A changes with energy, one should allow A to have an imaginary as well as real part. If $\mathrm{Im}\,A/\mathrm{Re}A \approx 1\%$, the effect is merely to fill in the minimum in Kac curve (Fig. 2); if $\mathrm{Im}\,A/\mathrm{Re}A \approx 10\%$, the minimum goes over into a shoulder and the dip - secondary peak structure disappears.

The fit to the 53GeV data with A real predicts a second minimum at $-t \approx 5(\mathrm{GeV}/c)^2$. This may, however, be very difficult to observe at present.

Assuming A is real, one may invert the Chou-Yang formula and solve for Δ from the pp scattering data.[6] If the parton model interpretation is correct, the ratio $R(t) = \Delta/G^2$ gives a measure of the t dependence of parton-parton scattering amplitudes. Since we do not know precisely what to use for G, it is difficult at present to extract R(t). One may, of course, evaluate R(t) for various choices of G(t). If R(t) changes sign, one has unambiguous evidence for non-pointlike parton-parton scattering amplitudes. More generally, it is of interest to extract Δ from pp scattering data. The result one gets depends only on the assumptions of spin independence (equation 1) and that the amplitude is pure imaginary (equation 2).

The apparent success of the Chou-Yang fit to pp data lends further interest to comparison of πp, Kp, etc. elastic data with their corresponding Chou-Yang formalae.[6]

If indeed this relation between electromagnetic form factors and elastic scattering holds, one may infer similar relations between diffractive production of resonant states and electroproduction form factors.[10] To check these relations, one needs accurate electroproduction data at small t. It is likely that one cannot neglect spin in production processes. The problem of how to include spin in these considerations has not yet been solved.

References

1) G. Giacomelli report to 16th International Conference on High Energy Physics, Batavia, Illinois (1972).
2) U. Amaldi et al., Physics Lett. 36B, 504 (1971).
3) G. Barbiellini et al., Physics Lett. 39B, 663 (1972).
4) U. Amaldi et al., Physics Lett. 44B (1973).
5) C.G. Beznogikh et al., Physics Lett. 30B, 274 (1969); Kh.M.Chernev et al., Physics Lett. 36B, 266 (1971).
6) T.T. Chou and C.N. Yang, Proceedings of Conference on High Energy Physics and Nuclear Structure, ed. G. Alexander, North-Holland Publishing Co., Amsterdam (1967), pp.348-360; Phys.Rev.Lett. 20, 1213 (1968).
7) Maxwell Kac, Oxford Preprint 33/73, 1973.
8) U. Amaldi, Lectures given at XIth International School of Subnuclear Physics, Erice (1973).
9) R.J. Glauber, Proceedings of Conference on High Energy Physics and Nuclear Structure, ed. G. Alexander, North-Holland Publishing Co., Amsterdam (1967), pp.311-338.
10) N. Byers and S. Frautschi, "Quanta", ed. Y. Nambu, Univ. of Chicago Press (1969) pp.367-393.
11) R.P. Feynman, "Photon-Hadron Interactions", W.A. Benjamin, Inc. (1972); see page 155.
12) W. Bartel et al., Nuclear Physics B58, 429 (1973). [In our text here we use the normalization $G_M(0) = 1.$]
13) T. Massam and A. Zichichi, Lett.al Nuovo Cimento 1, 387 (1969). This paper contains references to earlier studies.

Figure Captions

1) The fit of Kac[7] to the 53GeV pp elastic scattering differential cross section in the small t region. The data points are from Ref. 1.
2) The fit of Kac[7] to the 53GeV pp elastic scattering differential cross section. The data are taken from the ACGHT report to the Batavia Conference[1].
3) Comparison of Kac differential cross section (B) with that calculated using for G the dipole form $G_D = (1-t/0.71)^{-2}$ normalized to fit the data near t = 0 (A). This figure is taken from Ref. 7.
4) Comparison of measured values of $G_M(t)$ with $G_D(t) = (1 - t/0.71)^{-2}$. Data are taken from Ref. 9 (DESY) and a recent preprint of P.N. Kirk et al. submitted for publication to Physical Review (SLAC). The curve is a three parameter fit to these data; see text.
5) Comparison of the G(t) used by Kac with early data. This figure also is reproduced from Ref. 7.

Fig. 1

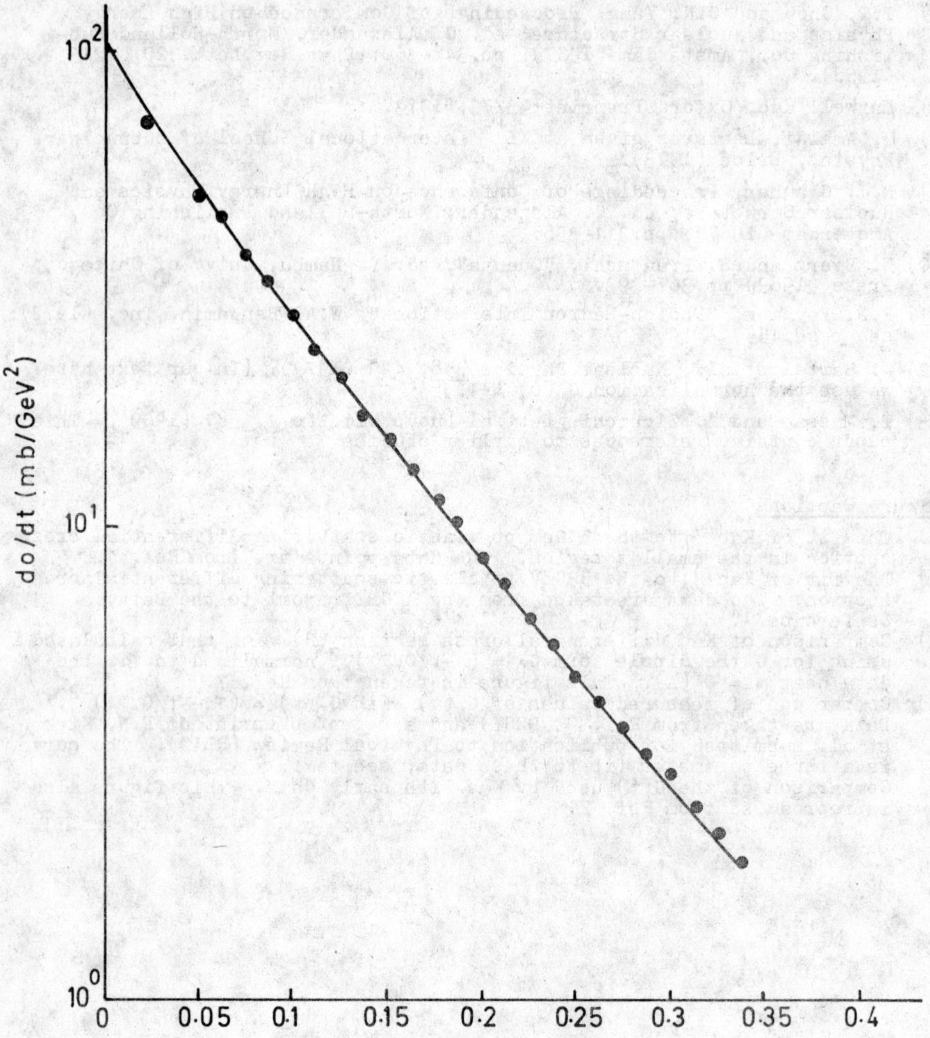

Fig. 2

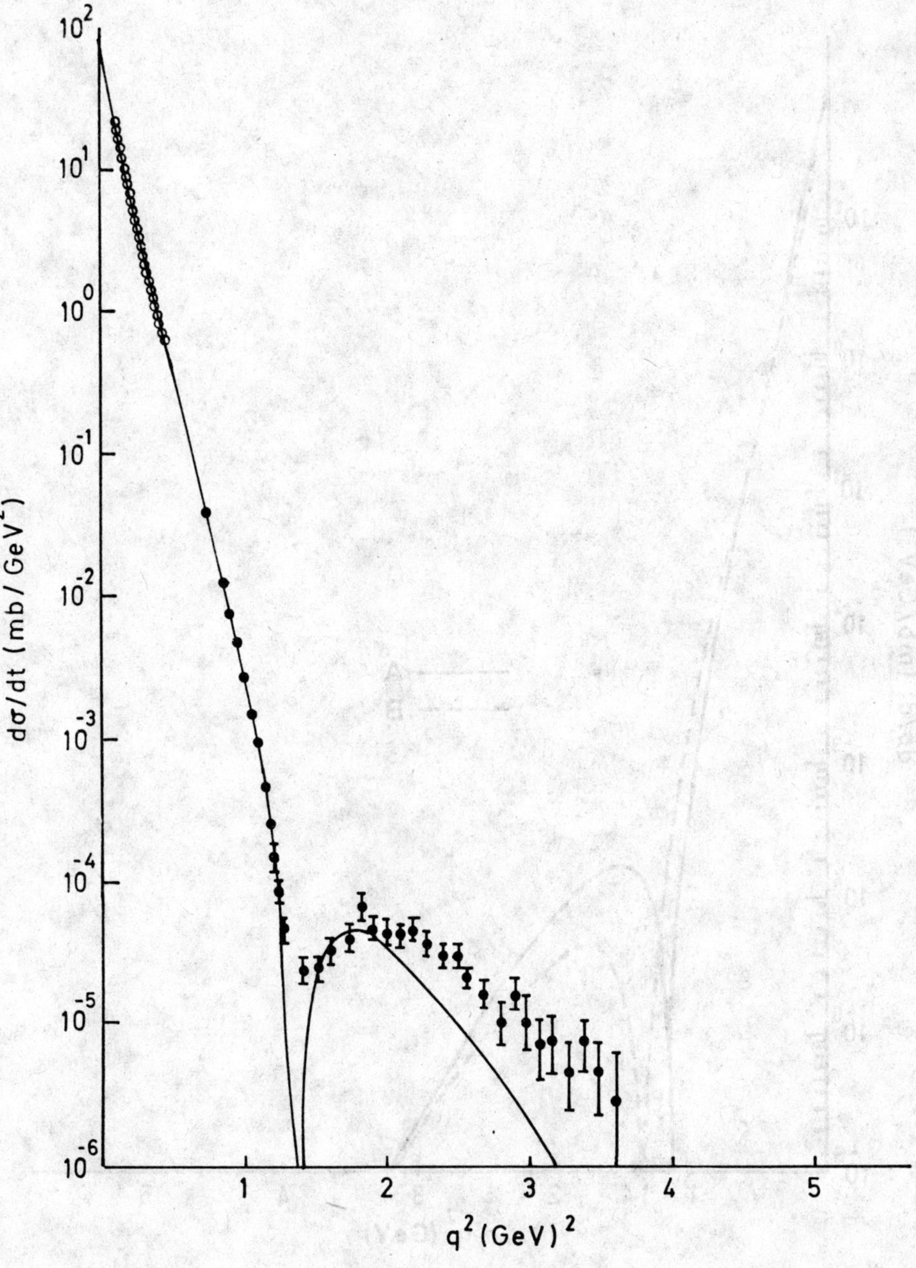
10^2
10^1
10^0
10^{-1}
10^{-2}
10^{-3}
10^{-4}
10^{-5}
10^{-6}
$d\sigma/dt\ (mb/GeV^2)$
1
2
3
4
5
$q^2\ (GeV)^2$

Fig. 3

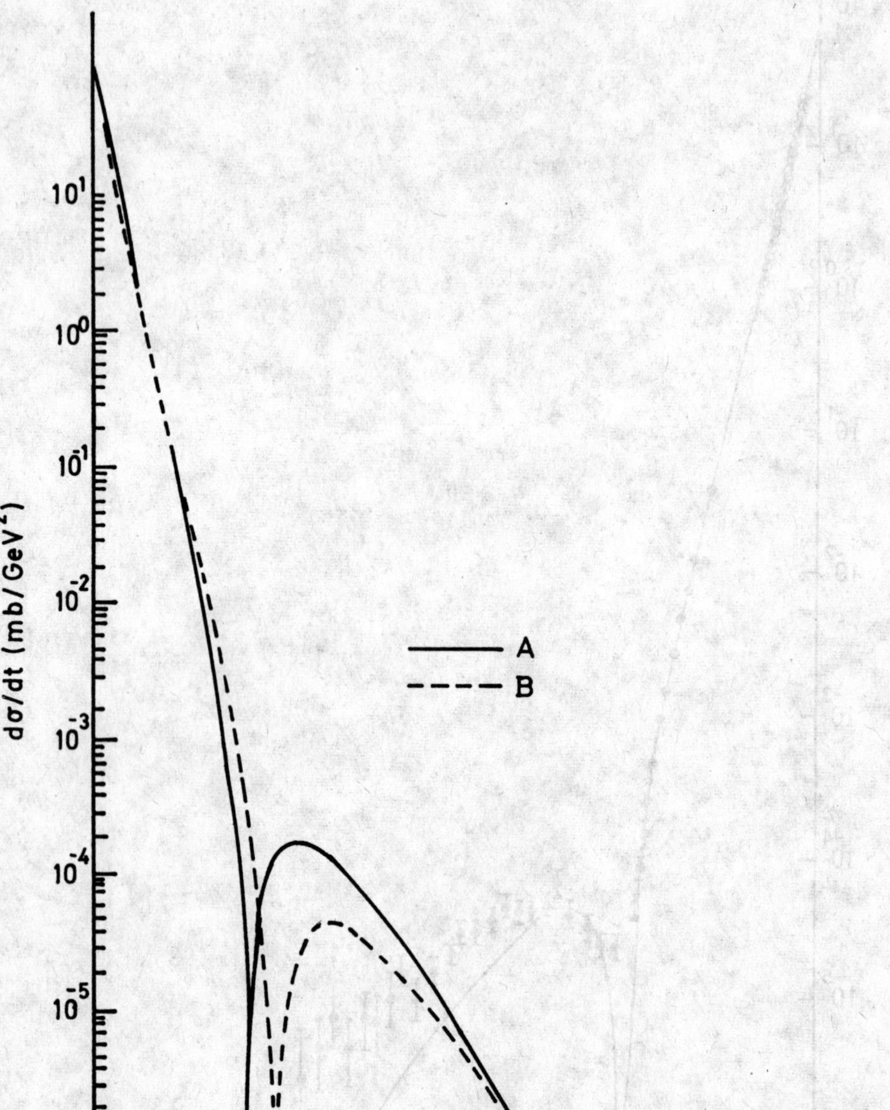

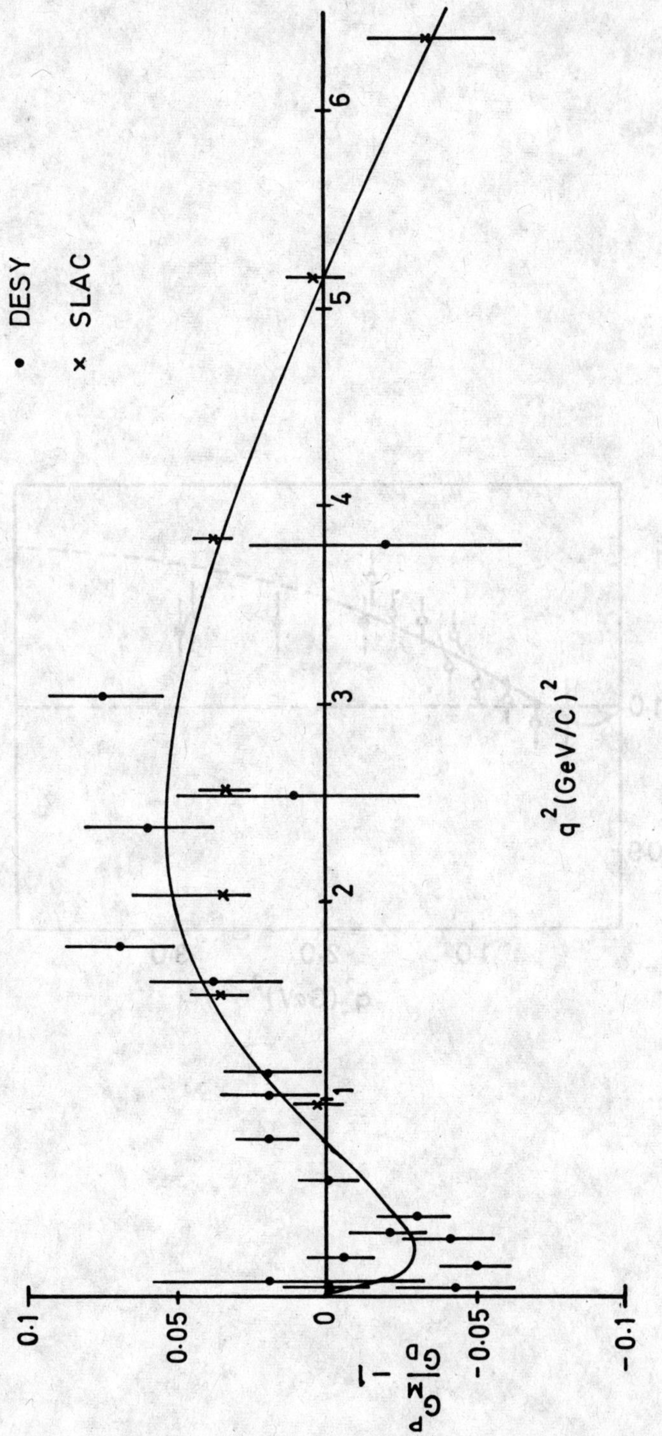

Fig. 4

Fig. 5

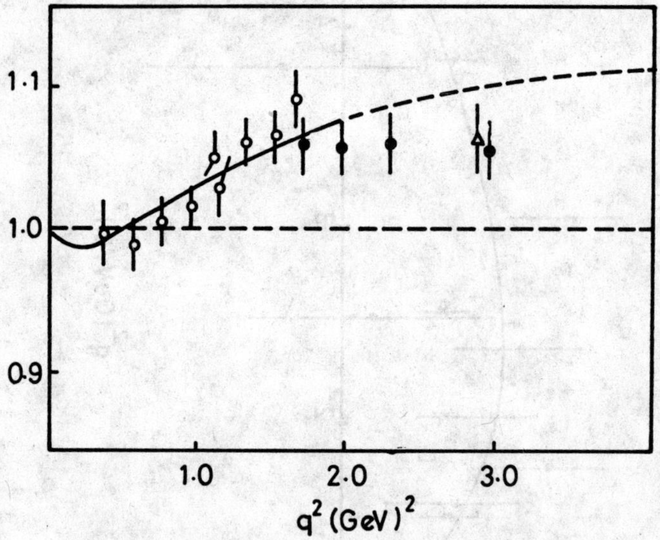

ELECTROMAGNETIC INTERACTIONS

S.C.C. Ting

Table of Contents

ELECTROMAGNETIC INTERACTIONS

Samuel C. C. Ting,
Deutsches Elektronen-Synchrotron, Hamburg, Germany, and
Laboratory for Nuclear Science, Massachusetts Institute of Technology,
Cambridge, Massachusetts, U.S.A.

1. INTRODUCTION

These lectures will consist of three parts: A. The interaction between photons and electrons or muons. - The study of quantum electrodynamics. B. The interaction between real photons and vector mesons. - The questions on vector dominance. And C. The interaction between virtual photons and hadrons. - The question on scaling and structure functions.

2. EXPERIMENTS ON QUANTUM ELECTRODYNAMICS

On the high q^2 test of quantum electrodynamics most of the experiments are done with e^+,e^- storage rings. Most of the low q^2 test of QED has been reviewed in many international conferences and will not be discussed here. The Delbrück scattering results from DESY tests QED to very high orders and shall be discussed in detail.

None of the experiments have found any deviation from the predictions of quantum electrodynamics:

2.1 Study of $e^+ + e^- \to e^+ + e^-$: to first order two diagrams dominate:

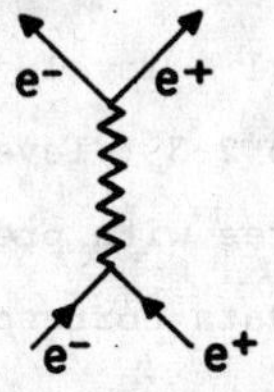

$q^2=4E^2>0$
"time like"

+

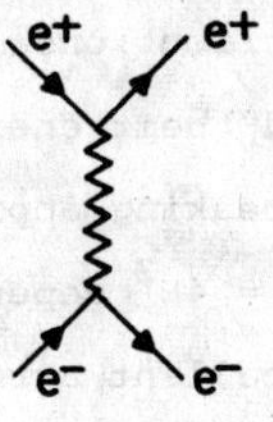

$q^2=-4E^2\sin^2\theta/2<0$
"space like"

For scattering angles $\theta < 150^o$ - the "space like" diagrams dominate and the "time like" contributions are only few per cent.

To compare with experimental results the theoretical cross sections were corrected for radiative effects according to a formula of Tavernier[1)]

This calculation uses the "peaking approximation" which predicts that the azimuthal angle between scattered particles differ from 180° by $\Delta\phi = 0$.

The experiment at the CEA[2)] uses 2 GeV electron and positron beams collided head-on in a straight section of a bypass to the synchrotron. The measurement of luminosity was based on the event rate from $e^+ + e^- \to e^+ + e^- + 2\gamma$. This process is dominated by low q^2 where the validity of QED has been verified.

Figure 1 shows the detector of this experiment. It is similar to other set ups of all the first generation non-magnetic detectors in e^+,e^- storage rings. It uses spark chambers to measure coplanarity of two body events and range energy relations to identify particles.

Figure 2 shows the result of the CEA experiment compared with the predictions of QED. They find with 230 events, a σexpt./σQED = .88±·10. One way of determining the significance of this result is to assign any possible deviation from QED to a heavy photon of mass Λ with either a positive (Λ+) or a negative (Λ-) metric. This model leads to a modification of the photon propagator by

$$F(q^2) = 1 \pm \frac{q^2}{q^2-\Lambda^2}$$

with 95 % confidence they find $\Lambda_+ > 12$ GeV, $\Lambda_- > 4.5$ GeV. If $\Lambda^2 >> 4E^2$, we find $F(q^2) = 1 \pm q^2/\Lambda\pm^2$, which is the conventional parametrization. Then the CEA results yield, with 95 % confidence, $\Lambda_+ > 12$ GeV, $\Lambda_- > 6$ GeV.

A precise and important experiment was done by the CERN-Bologna-Frascati group[3)] at ADONE, where high statistics were obtained in $e^+e^- \to e^+e^-$ in the total c.m. energy range 1.6 GeV to 2.0 GeV. This experiment compared the angular distributions, collinearity, acoplanarity, and absolute cross sections and their energy dependence with QED to ≲1 % level. This experiment shows that peaking approximation can be applied to 2-3 % level. Figure 3 shows the $S = 4E^2$ dependence of e^+e^- events compares with prediction of QED. As seen, excellent agreements were observed. The data corresponds to their results in s = 1.44 to 9.0 GeV^2.

Taking into account experimental correction factors, the data in Fig. 3 yields an energy dependence $\sigma = (1.00\pm0.02)\ S^{-(199\pm.02)}$ in good agreement with $\sigma = \frac{1}{s}$ from QED.

2.2 Study of $e^+ + e^- \to \mu^+ + \mu^-$

In this reaction only the time-like diagram contributes: comparing this reaction with $e^+ + e^- \to e^+ + e^-$ where the space-like photon dominates enables one test crossing symmetry. Or, if one assumes cross symmetry comparing $e^+ + e^- \to \mu^+ \mu^-$ with $e^+ e^- \to e^+ e^-$ checks μe universality.

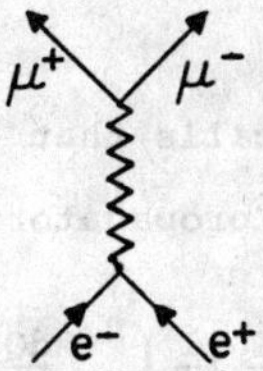

$$q^2 = 4E^2$$

Two experiments were done at ADONE. The Conversi group's[4)] result is shown in Figure 4. Where R is the ratio between corrected number of $e^+e^- \to \mu^+\mu^-$ and $e^+e^- \to e^-e^+$ events from the same apparatus, plotted as a function of 2E, after normalization to the corresponding theoretical cross-sections integrated over solid angle. The dashed line responds to estimated systematic uncertainties, ignoring radiative corrections, with

$$R = \frac{[(e^+e^- \to \mu^-\mu^+)/(e^+e^- \to e^+e^-)]\text{exp}}{[(e^+e^- \to \mu^+\mu^-)/(e^+e^- \to e^+e^-)]\text{QED}}$$ and 400 $\mu^-\mu^+$ events,

and 7192 e^-e^+ events. As seen, the results agree with predictions of QED.

Quantitatively we can compare this result with the hypothesis of negative metric[5)] "heavy photon" of mass m_Γ with

$$\frac{1}{q^2} \to \frac{1}{q^2} - \frac{1}{q^2 - m_\Gamma{}^2}$$

The best fit of Figure 4 yields $m_\Gamma > \sim 10\ m_p$.

Alternatively we can view this experiment as a direct comparison between pure time-like process at $q^2 = 2.2 - 4.4\ \text{GeV}^2$, and essentially space-like process at $-3.5 < q^2 < -0.45$.

This will provide a check of crossing symmetry in QED with the measured $R = .96 \pm 0.065$ in good agreement with $R = 1$ from crossing.

A similar experiment was done by the CERN-Bologna-Frascati group[6)] which compares the $e^+e^- \to e^+e^-$ $(-3.4 < q^2 < -.38\ \text{GeV}^2)$ with $e^+e^- \to \mu^+\mu^-$ $(2.56 < q^2 < 4.0\ \text{GeV}^2)$. The result shows $\mu\mu\gamma$ and $ee\gamma$ vertices are the same in agreement with other checks of μe universality. Figure 5 shows the result of this measurement. The flatness of this distribution shows that the muon behaves like electrons in the $2.56 < q^2 < 4.0\ \mathrm{GeV}^2$ region.

An interesting experiment on $e^+e^- \to \mu^+\mu^-$ at mass of ϕ was done at Orsay[7], where they observed a deviation from the $\frac{1}{s}$ dependence of the cross section predicted from QED. A good fit to the data require one takes into account of $\gamma \to \phi \to \gamma$ transitions. Thus this experiment provides a first evidence on vacuum polarization effect.

To see this more clearly, one recalls that the Källén-Lehmann representation of the complete photon propagator.

$$D'_{\mu\nu}(K) = \frac{\delta_{\mu\nu}}{K^2} + \left(\delta_{\mu\nu} - \frac{K_\mu K_\nu}{K^2}\right)\frac{1}{\pi}\int_0^\infty \frac{dQ}{Q}\,\frac{\mathrm{Im}\,\pi(Q)}{Q^2+K^2-i\varepsilon}$$

where $\mathrm{Im}\pi(4E^2) = \frac{E^2}{\pi\alpha}[\Sigma_f\,\sigma^f]$ with $\Sigma_f\,\sigma^f$ is the total annihilation cross section for $e^+e^- \to f$. This propagator can be tested if one chooses a pure QED process like $e^+e^- \to \phi \to \mu^+\mu^-$.

The $\sigma'(e^+e^- \to \mu^+\mu^-)$ taking into account of modification due to ϕ mesons is

$$\sigma'(e^+e^- \to \mu^+\mu^-) = \left|1 - \frac{3B}{\alpha}\,\frac{M_\phi\,\Gamma_\phi}{M_\phi^2 - 4E^2 - iM_\phi\Gamma_\phi}\right|^2 \sigma(e^+e^- \to \mu^+\mu^-)$$

with Γ_ϕ is the ϕ width, B = leptonic branching ratio $\approx 3 \times 10^{-4}$. We expect a 12 % deviation when $2E = M_\phi \pm \Gamma_\phi/2$.

The experiment was done at the Orsay storage ring with luminosity measurement by $e^+e^- \to e^+e^- + 2\gamma$ to ±5 % level and energy calibration done by detecting $e^+e^- \to \phi \to K^+K^-$. The result is shown in Figure 6. Comparing the data with σ' formula we obtain $B = (2.93 \pm 0.96) \times 10^{-4}$ in good agreement with direct result of $B = (3.01 \pm 0.12) \times 10^{-4}$. This is a direct observation of hadronic vacuum polarization.

2.3 Delbrück Scattering

A very important experiment was done recently at DESY[8] on elastic scattering of photons in the Coulomb field of nuclei via virtual electron-positron pairs. It is one of the nonlinear processes in QED which are a direct consequence of vacuum polarization. They are characterized by closed fermion loops and are forbidden via Maxwell's classical electrodynamics as a result of the linear form of its field equations and the principle of superposition The lowest order diagram is 6 order

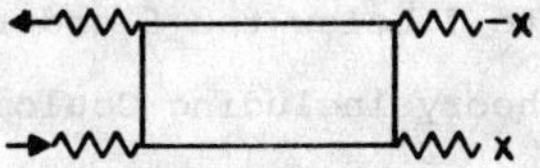

$\alpha^6 Z^4$ the next order being the 10th order

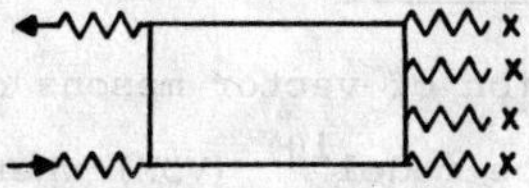

$\alpha^{10} Z^8$

It has been pointed out by Cheng and Wu[8)] that in the GeV energy region the higher (>6) order diagrams make a large contribution to the measured cross sections.

By performing this experiment one checks the validity of QED to higher orders and particularly the expansion of $Z\alpha$ when Z is large.

The experiment was done with a photon beam from 1 to 7.3 GeV. Scattered photons were detected in the angular range from 1 to 3 mrad. with Cu.Au. Au. U. targets. Figure 7 shows the experimental set up of this experiment. A well collimated bremsstrahlung beam from DESY with beam divergence of ±.15 mrad, sport size 6 x 6 mm^2 hits the scattering target. The scattered photons were converted into e^+e^- pairs and measured by a pair spectrometer. The momentum resolution is ~1 %, space resolution on the pair converter was ± 3 mm. The angular resolution was ±.25 mrad given by combined effect of reconstruction errors, beam spot size and beam divergence. Photon energies were measured from maximum E_0 down to 0.75 E_0 with 1 % resolution.

Of all the backgrounds the following 3 are most serious.

i) Compton scattering on electrons which can be calculated exactly and found to 16 % of Delbrück scattering.

ii) secondary photons from showers

iii) photon splitting

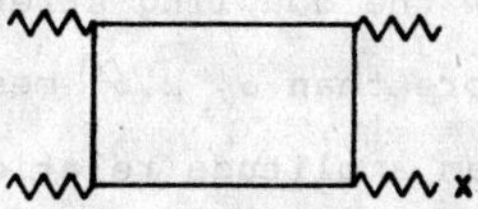

Process ii) and iii) are inelastic and do not contribute to the counting rate at the edge of the bremsstrahlung spectrum. By using 3 % wide energy band just below E_0 one rejects most of these backgrounds.

Figures 8 and 9 show the result on Delbrück cross section for Au and

U targets, compared to theoretical prediction with and without Coulomb correction. The measured values are a factor 2 to 7 below the 6 orders Born approximations and are in good agreement with theory including Coulomb correction.

3. INTERACTION OF PHOTON WITH VECTOR MESONS

The physics of photoproduction of vector mesons can be visualized through the Vector Meson Dominance model[10] (VDM) where one relates the electromagnetic current of hadrons with the fields of ρ,ω,ϕ,V etc. via the relation

$$J_\mu(X) = -\left(\frac{m_\rho^2}{2\gamma_\rho}\rho_\mu(X) + \frac{m_\omega^2}{2\gamma_\omega}\omega_\mu(X) + \frac{m_\phi^2}{2\gamma_\phi}\phi_\mu(X) + \frac{m_V^2}{2\gamma_V}V_\mu(X) + \ldots\right)$$

where m_V is the vector meson mass and γ_V is the γ - v coupling strength

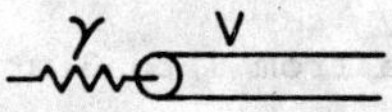

$$\frac{e\, m_V^2}{2\gamma_V} = \sqrt{\pi\alpha}\,\frac{m_V^2}{\gamma_V}$$

This model has proven to be a useful guide in understanding of many of the photon induced reactions. For example, one can relate the compton scattering amplitude to transverse vector meson photoproduction amplitude:

$$A_{\gamma p\to\gamma p} = \sum_V \frac{\sqrt{\pi\alpha}}{\gamma_V} A^{Trans}_{\gamma p\to\gamma p} \qquad (1)$$

We shall go over in detail some of the problems involved in checking the validity of (1).

We make the following observations:

i) One needs to know the coupling strength γ_V.

ii) It may require more than ρ, ω,ϕ mesons

iii) Equation (1) is an amplitude relation involving $A_{\gamma p\to Vp}$ and $A_{\gamma p\to\gamma p}$

3.1 Measurement of γ_V

The measurement of coupling constant γ_V has been carried out from DESY[11], ORSAY[12], NOVOSIBIRSK[13].

The Orsay results agree with the earlier DESY measurements and are listed below:

	ρ	ω	φ
$\frac{\gamma_V^2}{4\pi}$	0.64±0.05	4.8±0.5	2.8±0.2

There are also very precise measurements on the forbidden decays of $\phi, \omega \rightarrow 2\pi$ by the DESY-MIT group[14] (Fig. 10). ω and φ are 3π resonances with I = 0. It decays into 2π via electromagnetic transitions of the type

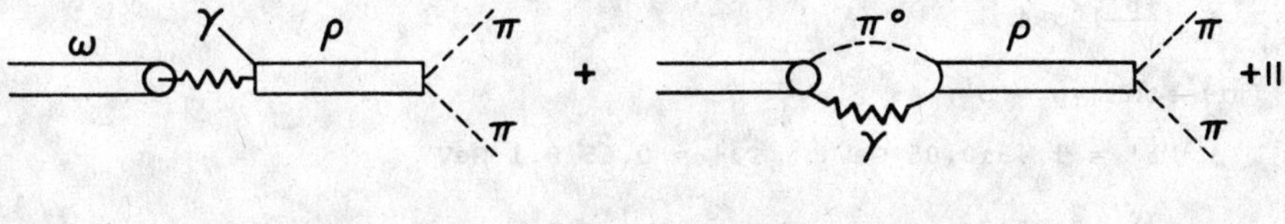

The DESY-MIT experiment measured both the production and decays of these mesons with the same apparatus and studied the non-resonant background by doing the experiment on many nuclear targets.

Their results are: $\Gamma\omega\rightarrow 2\pi/\Gamma\omega\rightarrow all$ = (1.22±0.30) %, $\Gamma\phi\rightarrow 2\pi/\Gamma\phi\rightarrow all = 2.7\times 10^{-4}$ with 95 % confidence.

3.2 <u>Search for new vector mesons</u>

The first indication on the possible existence of new vector mesons at 1.6 GeV comes from the work of H. Alvensleben[15] et al. where they studied the reaction $\gamma + C \rightarrow C + \pi^+ + \pi^-$ and found a strong enhancement at mass 1.6 GeV.Figure 11 shows the result of this measurement.

Recently at Frascati[16] a 4π enhancement were observed at the same mass region from $e^+ + e^- \rightarrow 2\pi^+ + 2\pi^-$. Figure 12 shows this result. The Frascati observation implies that the enhancement is in a 1^- state.

The 4π enhancement was also observed in the Berkeley-SLAC collaboration (Figure 13) analyzing the reaction

$$\gamma + p \rightarrow p + 2\pi^+ + 2\pi^-$$

with a 9.3 GeV, $\frac{\Delta p}{p}$ = ±3 % linearly polarized monochromatic back scattered laser beam. The experiment established that the 4π state is reached via decay into ρ^{o} + σ where σ denotes a isospin zero s-wave $\pi^{+}\pi^{-}$ system.

It should be noted that neither the $e^{+}e^{-}\to 4\pi$ nor the $\gamma + p \to p + 4\pi$ experiment can decide whether the ρ' is a resonance or a continuum. A possible way of distinction is to measure the phase of ρ' via $\gamma + p \to p\rho'$ ($\rho' \to e^{-}e^{+}$) interfering with QED pairs. Such an experiment is being performed currently at DESY.

If we interpret the ρ' being a vector meson then we have the following parameters:

i) From $e^{+} + e^{-} \to$ pions

$M\rho' = 1.6$ GeV, $\Gamma\rho' = 0.3\text{-}0.4$ GeV

$$\left(\frac{\gamma\rho'}{\gamma\rho}\right)^{2} \approx 4$$

ii) From $\gamma p \to p + 4\pi$

$M\rho' = 1.43 \pm 0.05$ GeV, $\Gamma\rho' = 0.65 \pm 0.1$ GeV

$$\left(\frac{\gamma\rho'}{\gamma\ p}\right)^{2} \approx 6 \pm 2$$

There are many other experiments looking for new vector mesons[18),19)]. So far the experiment has been not successful and are limited by statistics.

Two large scale experiments are being planned at Brookhaven ($p+p\to e^{+}e^{-}+X$) and at DESY ($\gamma+p\to e^{+}e^{-}+X$) to search for new $e^{+}e^{-}$ enhancements with sensitivity down to $\sigma B \approx 10^{-38}$ cm^{2}, and mass resolution ΔM ±5 MeV in the mass region $1.0<m<6$ GeV.

To see how such an experiment is done let us look into the BNL experiment. In this experiment one uses a slow extracted 30 GeV external proton beam with intensity of 4 x 10^{12} per pulse. Since the invariant particle production cross section $\propto e^{-6P_{\perp}*}$ and independent of $P_{\parallel}*$. The maximum yield is at $P_{\perp}* = P_{\parallel}* = 0$. I.e. at rest in the c.m. system. Since particles are produced at rest, most likely it does not carry a polarization; let us look into 90° decays.

$P_{\perp}* = \frac{m^{*}}{2}$, $P_{\parallel}* = 0$

We have 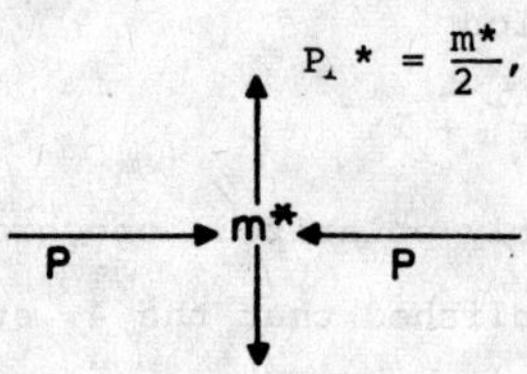

$E^{*} = 7.5$ GeV, m* produced at

rest decay into e^+e^- each with $P_\perp^* = \frac{m^*}{2} \cdot P_{//}^* = 0$.

Transferred into laboratory system we have $P_\perp = P_\perp^*$; $P_{//} = \frac{m^*}{2} \cdot \frac{30}{7.5} = 2m^*$

Thus the e^+ or e^- has a laboratory emission angle of

$$\tan\theta = \frac{\frac{m^*}{2}}{2m^*} = \frac{1}{4} \quad \underline{\text{independent of the mass } m^*}$$

A pair spectrometer with opening angle of 30° and vertical bending to decouple P and θ is the best way to search for e^+e^- enhancement: the mass search is done by magnet exciation alone, the high incident proton beam enables us to search $V^o \to e^-e^+$ down to 10^{-38} cm^2 level.

The DESY experiment is done with a 7.5 GeV photon beam, measuring $\gamma+p\to p+e^+e^-$. It has the added advantage that one can study the phase of new e^+e^- enhancement by measuring the interference with QED pairs.

3.3 Study of the amplitude of $A\gamma p\to\gamma p$ and $A\gamma p\to vp$

3.3.1 Study of $A\gamma p\to\gamma p$

Let us now return to equation (1) and look into the physics of measuring the amplitude of compton scattering $A\gamma p\to\gamma p$. Beside the VDM equation (1) the measurement of $A\gamma p\to\gamma p$ will provide us with a direct check of the Kramers-Kronig relation[20),21)]

The measurement was carried out by the DESY-MIT group last year at DESY. This being a very difficult and important experiment, we will go into some detail of the physics, the techniques and the analysis involved.

The Kramers-Kronig relation[20)] was first derived more than 40 years ago from the causality principle. Gell-Mann, Goldberger, and Thirring[21)] obtained the same result from field theoretical considerations. It relates the real part of the forward Compton scattering amplitude, Re f_1, to the total hadronic photon nucleon cross section σ_T, via the relation

$$\text{Re } f_1\ (k) = \frac{-\alpha}{M} + \frac{k^2}{2\pi^2} P \int_{k\pi}^{\infty} \frac{dk'}{k'^2-k^2}\, \sigma_T\ (k') \qquad (2)$$

where P denotes the Cauchy-principal value of the integral, k_π is the one pion threshold energy, α is the fine structure constant, and M is the mass of the proton. The explicit evaluation of equation (2) has been carried out by Gilman and Damashek[22)] using the known total photon nucleon cross sections

σ_T. The purpose of this experiment is to measure Re f_1 directly and to compare it with the prediction of equation (2), thereby to check the validity of dispersion relations for photons.

The classical way to study the phase of the amplitude of πp and pp scattering is to measure the interference between the elastic scattering amplitude and the Coulomb amplitude. For Compton scattering the scattered photon, being a neutral particle, does not interfere with the Coulomb field. To study the Compton amplitude we consider the case where the scattered photon is "almost real", i.e., we study the asymmetric pair distribution from the reaction

$$\gamma + p \to p + \gamma \text{ (virtual)} \qquad (3)$$
$$\hookrightarrow e^+e^-$$

where the invariant mass of the pair is almost zero. To first order the amplitude for the reaction $\gamma + p \to p + e^+ + e^-$ is

$$A = A_C(\gamma) + A_{BH}(2\gamma)$$

where $A_C(\gamma)$ is the Compton amplitude with the scattered γ decaying into e^+e^-. $A_{BH}(2\gamma)$ is the ordinary Bethe-Heitler amplitude which behaves like two photons under charge conjugation. It follows from charge conjugation invarianc that

$$2d\sigma_{int} \equiv |A(e^+,e^-)|^2 - A(e^-,e^+)|^2 = 4 \text{ Re } A_C(\gamma)A_{BH}(2\gamma)$$

is odd under exchange of e^+,e^-. Following Gell-Mann, Goldberger, and Thirring, we write the forward Compton amplitude

$$A_C = f_1 \vec{\varepsilon} \cdot \vec{\varepsilon}' + i f_2 \vec{\sigma}\cdot\vec{\varepsilon} \times \vec{\varepsilon}'$$

where $\vec{\varepsilon}$ and $\vec{\varepsilon}'$ are the polarization vectors of initial and final photons, respectively. If we average over nucleon spins in the amplitude, we are left with f_1. At $t = 0$, or in forward direction, the imaginary part of f_1, Im f_1, is Im f_1 (k) = $(k/4\pi)\sigma_T(k)$, and the real part of f_1 is related to the Im f_1 via equation (2).

Since the interference cross section $d\sigma_{int}$ is asymmetrical under interchange of 4-momenta of electron (P_-) and positron (P_+), it can be observed by taking the difference between two yields, N_+ and N_-, for settings of opposite polarity of an asymmetric detector. The resultant yield is then independent of the Bethe-Heitler and Compton yields. The contribution from higher order diagrams has been estimated by Brodsky[23)] and been found to be small compared to the interference term, $d\sigma_{int}$, and has been neglected.

To compare the results with the dispersion relation calculation directly, the following facts should be noted:

i) The scattered photon has to be "almost real", i.e., the electron-positron pair's invariant mass should be close to zero.

ii) The momentum transfer to the recoil proton has to be very small. $-t \leq m_{\pi}^2$

iii) The incident photon energy should be well above the nuclear resonance region.

These conditions are the main experimental difficulties. One is forced to study zero opening angle electron-pairs near the forward direction, where the single arm electron rate is greater than the pair rate by a factor $2x10^4$. The accidental coincidence rate becomes a serious problem.

Figure 14a shows the schematic of the double arm spectrometer, where MA, MD_1, and MD_2 are large aperture dipole magnets. RF, RB, R1, R2, R3 etc. are scintillation counters, RC and LC are threshold Cerenkov counters and LS and RS are lead lucite shower counters used to distinguish pions from electrons. QM is a specially built small quantameter which provides the clearance for forward e^+e^- to enter the spectrometer.

The incident beam with k_{max} = 3 GeV and an intensity of $2x10^9$ equivalent quanta per second had a sport size of 6x6 mm^2. The beam was defined by two adjustable collimators and cleaned of charged particles by bending magnets.

In order to minimize accidentals it was desirable to run the experiment at low intensities. The large acceptance of the spectrometer allowed us to do so and still obtain a sufficient event rate. We had Δk = 1 GeV, Δt = 0.025 $(GeV/c)^2$ (FWHM), $\langle k \rangle$ = 2.2 GeV and $\langle t \rangle$ = -0.027 $(GeV/c)^2$.

The acceptance of the spectrometer was defined by scintillation counters alone. To reduce dead time of the system to the percent level, the "hottest" counters LF, RF, LB, RB were split into hodoscopes. The rates in each hodoscope were about 300 kHz. Additional hodoscopes at locations R1, R2, R3 were used to determine the kinematics of each event.

To reduce the background from the reaction

$$\gamma + p \longrightarrow X + \pi^{o} \quad \pi^{o} \longrightarrow e^{+}e^{-} \tag{4}$$

which peaks strongly at zero opening angles, the mass acceptance of the spectrometer was designed to peak at 25 MeV with Δm = 20 MeV (FWHM).

To control the accidentals, the triggering time of each counter was measured separately by a time-of-flight system. Thus we could determine time distributions of accidentals in each arm separately ("single arm accidentals") as well as distributions of double arm accidentals. Because of the large acceptance of the spectrometer, neither first nor second order transport equations could be used; instead, a three dimensional magnetic field mapping of the spectrometer was made and quadratic interpolation methods were used to calculate the acceptance. The accuracy of this calculation was further confirmed by floating wire measurements.

During the experiment many checks were made to monitor the functioning of the spectrometer. We list the following examples:

i) The spectrometer polarities were reversed every 3 hours. A proton resonance meter was used to ensure that the magnetic field returned to 1 part in 10^4 of the designed value. The single arm rates were recorded continuously and found to be reproducible to 1 %.

ii) To ensure that the acceptance of the spectrometer was really defined by counters, we systematically moved the shielding away from the accepted region by 2.5 cm. To an accuracy of 1 % no detectable effects were found.

iii) Target out rates were measured and were found to be approximately 3 %.

iv) The stability of the quantameter, the hydrogen target and the incident beam direction were monitored by a pair of spectrometers looking at the H_2 target (Fig. 14a). The rate was constant to 1 % with respect to the flux measured by the quantameter.

Half the data were taken at each spectrometer polarity. These yields are denoted as N_+ and N_-. 14b shows the time-of-flight spectrum of all the detected electron pairs. As seen, most of the events are contained within

1.2 ns. The accidentals outside the peak were equal for both polarities. They do not contribute to the observed asymmetry $N_+ - N_-$. The smallness of the accidental background in Fig. 14b enables one to subtract the accidentals with confidence. The validity of the background treatment was checked by analyzing the asymmetric yield $N_+ - N_-$ without background subtraction. Our final result was not significantly changed. Fig. 14c shows the total event distribution $N_+ + N_-$ as a function of mass.

The solid line in Fig. 14c is a theoretical yield of e^+e^- pairs. For pair mass below 20 MeV, a systematic uncertainty of ±30 % comes mainly from uncertainty in the cross section for reaction (4). For invariant mass greater than 20 MeV the contribution comes almost completely from QED pairs, for which the systematic error is small. The agreement between theory and experiment indicates that all systematic uncertainties have been understood and taken into account. The spectrum shown in Fig. 15a can be compared directly with the predictions of dispersion relations. The solid line in Fig. 15a is a dispersion relation calculation following Gilman and Damashek. The data of Fig. 15a yield a value for the real part of the Compton amplitude of -12.3±2.4μb·GeV.

As seen in Fig. 15b, this value agrees well with the dispersion calculation as a function of k.

3.3.2 Study of $A\gamma p \to vp$

Let us look into the measurement of $A\gamma p \to vp$ of Equation (1). Up to now most of the measurements are of the type

$$\gamma + p \to V^o + P, \quad V^o \to \text{hadrons}$$

to obtain the cross section for photoproduction of vector mesons and checks of vector dominance are made by square both sides of equation (1). The comparison on both sides of Equation (1) are difficult to make because the cross section from $\gamma + p \to V^o$ (↳hadrons) + P are model dependent on the mass m_V, width Γ_V, the shape of the resonance R (m), and the analytical form of the nonresonant backgrounds assumed.

For $\gamma p \to \rho p$: In using the reaction $\gamma + p \to \pi^+\pi^- + p$ to obtain $\gamma + p \to \rho^o + p$ cross section one commonly assumes two kinds of nonresonant backgrounds.

a) The Soding type background with photon split into 2π and one π elastically scatters off nucleon. The 2π are in a p state and interfere with the $\rho \to 2\pi$ amplitude to produce the observed mass spectrum. The difficulty with this model being on how to avoid the double counting problem of distinguishing the 2π in the p state from $\rho \to 2\pi$ amplitude.

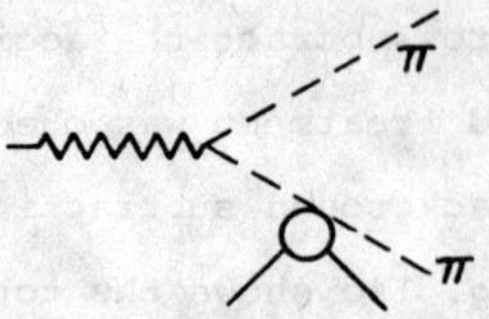

b) The other more phenomenological approach is to assume the $\gamma+p\to\pi\pi p$ spectrum are all from $\rho \to 2\pi$ and the $\rho \to 2\pi$ spectrum has a $(\frac{m_\rho}{m_{\pi\pi}})^{n(t)}$ factor in it. The two approaches can yield cross sections which differ from each other by 30 % or more[24).]

For narrow resonances like $\gamma + p \to \omega + p$ and $\gamma + p \to \phi + p$ the background problems are much less (<10 %) and a reasonable model independent analysis of the data can be made and the data yields:

For $\gamma + p \to \omega + p$: The total cross section can be separated into contributions σ^N, σ^U form natural and unnatural parity exchanges in t channel.

$$\sigma^{N,U} = \frac{1}{2}\,(1 \pm P_\sigma)\ .$$

The data[24)] (Fig. 16) shows that σ^U decreases rapidly ($E\gamma^{-2}$) with increasing energy while σ^N is approximately constant.

For $\gamma + p \to \phi + p$: This reaction is thought to proceed only by Pomeron exchange in the t-channel. Fig. 17 shows the differential cross sections[24),25)] and the observed slope is $\sim$4 GeV^{-2} smaller than the ρ, and ω production slopes. The integrated cross section is shown in Fig. 18. As seen, the production cross section may increase slightly with energy.

Finally we discuss the "best way" to study photoproduction cross section and the photoproduction phase of vector mesons:

With the reaction

$$\gamma + p \to p + V^0$$
$$\qquad\qquad\quad \hookrightarrow e^+e^-$$

measuring e^+,e^- and recoil protons will enable us to reject all the inelastic contributions of the type $\gamma + p \to \rho + N^*$. As discussed earlier, the total amplitude comes from $A = A_{BH}(2\gamma) + A_{vp}(\gamma)$ with $|A_{BH}(2\gamma)|^2 \sim \frac{1}{\theta^8}$, $|A_{vp}|^2 \sim \frac{1}{\theta^4}$

where θ is the opening angle in the laboratory with respect to incident direction. When $\theta \to$ large $|A_{BH}(2\gamma)|^2 \to$ small and can be calculated exactly, and there are no other backgrounds. The interference term $|A(e^+,e^-)|^2 - |A(e^-,e^+)|^2 = 4A_{BH}$ Re A_{vp} measures the phase of $A_{\gamma p} \to vp$. Thus for a wide enhancement like ρ' the phase measurement is the only way to distinguish it from a real resonance.

4. ELECTROPRODUCTION

The electroproduction data will be discussed in terms of the standard variables:

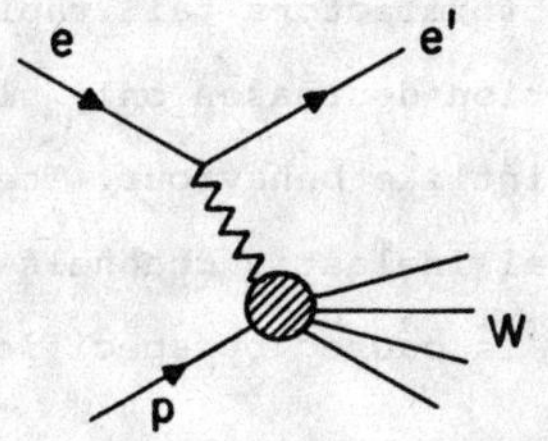

$e = (\vec{e},E)$, $e' = (\vec{e}',E')$	four momenta of the incoming and scattered electrons
$\theta =$	e' scattering angle in the Lab. system
$p = (0,m_p)$	target proton
$-Q^2 = (e-e')^2 = -4EE' \sin^2 \frac{\theta}{2}$	mass squared of virtual photon
$\nu = E - E'$	Lab. energy of virtual photon
$W^2 = (e-e'+p)^2 = 2m_p\nu + m_p^2 - Q^2$	mass squared of outgoing hadron system

For virtual photons $(-Q^2) < 0$ the photon polarization vector has a transverse as well as a longitudinal component. The differential cross section for electroproduction, $d^2\sigma/dQ^2dW$, can be expressed in terms of the cross sections σ_T and σ_L for scattering of transverse and longitudinal photons:

$$\frac{d^2\sigma}{dQ^2dW} = \frac{\pi}{EE'}\frac{W}{m_p}\Gamma_T \quad \{\sigma_T(Q^2,W) + \varepsilon\ \sigma_L(Q^2,W)\}$$

with the transverse flux, Γ_T, being defined as

$$\Gamma_T = \frac{\alpha}{E\pi^2}\frac{E'}{E}\frac{W^2 - m_p^2}{m_p\ Q^2}\frac{1}{1-\varepsilon}$$

and $\varepsilon = \left\{1 + 2(Q^2 + \nu^2)\, Q^{-2}\, tg^2 \frac{\theta}{2}\right\}^{-1}$

In terms of structure functions W_1, W_2 we have

$$\frac{d\sigma}{dWdQ^2} = \sigma_{mott} \{W_2 + 2 \tan^2 \frac{\theta}{2} W_1\} \quad \text{with } \sigma_T \approx \frac{\nu W_2}{Q_2} \underset{Q^2 \to 0}{\to} \sigma_{\gamma p}(s)$$

Much of the experimental work at electron accelerators is now concentrated on the study of inelastic electron and muon scattering in an effort to understand the cross section observed in the deep inelastic region[26),27)]. While the nucleon electromagnetic formfactors fall rapidly with Q^2 the total inelastic ep scattering cross section decreases only slowly with Q^2 (see Fig. 19) and appears to have a pointlike behaviour. There are now experimenters to measure individual inelastic channels in order to see whether the observed Q^2 behaviour is caused by specific final states.

4.1 Electroproduction of vector mesons

There are two main interests in the electroproduction of ρ^o mesons:

i) VDM relates $\gamma_V + p \to \rho + p$ with the virtual compton scattering $\gamma_V + p \to \gamma_V + p$ and through optical theorem to the total inelastic cross sections, and

ii) It has been suggested[28)] that the hadronic interaction radius of the photon shrinks as Q^2 increases.

Figure 20 shows the result for $e + p \to e + p + \pi^+ + \pi^-$ as a function of W in different Q^2 intervals[29)]. Qualitatively the same W dependence is observed in photoproduction[30)]. The Q^2 dependence of the cross section for $ep\pi^+\pi^-$ is approximately the same as that of total inelastic cross section (Fig. 21).

To study the cross section for $e + p \to e + p + \rho^o$ from $e + p \to e + p + \pi^+\pi^-$ one encounters same uncertainties as for photoproduction. And again in the absence of any reliable theory, one tries to fit the $\pi^-\pi^+$ spectrum with a relativistic Breit-Wigner with a p wave width, multiplying by a generalized mass shift factor

$$\frac{(m_\rho^2+Q^2-t)^2}{(m_{\pi\pi}^2+Q^2-t)^2}$$

No attempts to solve the problem of $\rho-\omega$ interference has yet been made, and Fig. 22 summarizes the various measurements of the diffractive slope for $\gamma_V + p \to (\rho,\omega) + p$[31] plotted against the dimensionless inverse lifetime parameter $\frac{Q^2+m_\rho^2}{2M\nu}$. The data indicate that the slope decreases to less than half of its photoproduction value.

A decrease in slope has been predicted for electroproduction as the photon makes the transition from $Q^2 = 0$ where it has an effective hadronic "size" determined by its virtual vector meson components, to large space like Q^2, where the life time $\Delta\tau = (E\rho - \nu)^{-1} = \frac{2\nu}{(Q^2+m_V^2)}$ of the vector meson state becomes short and the photon-proton interaction becomes more pointlike.

4.2 Multiplicities of Charged Hadrons

The data (Fig. 23) from 16 GeV SLAC track chambers[32] shows that the average multiplicity scales with $\omega' = 1 + s/Q^2$ for $|Q^2| > 1$ GeV.

4.3 Inclusive Particle Spectra

Several experiments have been focused on the Q^2 variation of inclusive π, k, p momentum and angular spectra. The results are usually presented in terms of the transverse momentum P_T and the Feynman variable $x = P^*_{/\!/}/P^*_{max}$ where $P^*_{/\!/}$ and P^*_{max} are the longitudinal and maximum possible momentum in the cms for the particle in question. All quantities refer to the process

$$\gamma_{virtual} + P \to \text{hadrons}$$

At high energies we expect to find three distinct x regions with qualitatively different behaviour:

$x \approx -1$ particles from target fragmentation

$x \approx +1$ particles from beam fragmentation

$x \approx 0$ central region with particles coming - in terms of the multiperipheral model - from the middle rungs.

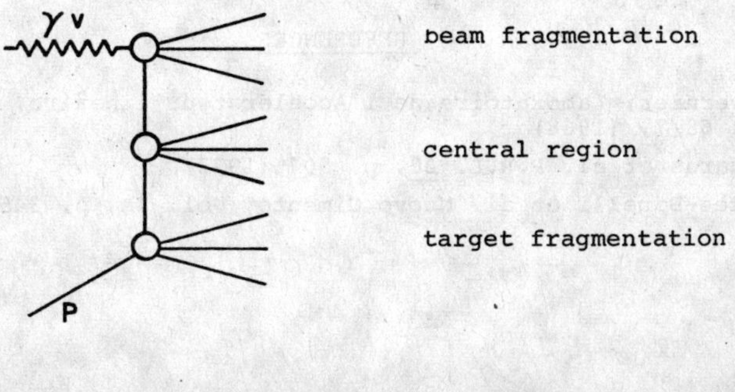

With this picture in mind we expect e.g. the pion distribution in the target region to be independent of the nature of the beam particle. Applied to electroproduction, at large energies the pion distribution for $x < 0$ should not depend on Q^2.

4.3.1 <u>π Spectra:</u> Fig. 24, 25, 26 showed the π^+, π^- distribution integrated over P_T for given Q^2 intervals for DESY and SLAC energies, and we observe that:

i) for $X < 0.3$ the data is independent of Q^2

ii) for $X > 0.3$ the data is decreasing with increasing Q^2.

The $X > 0.3$ data is namely from two body channels like ρ^o production and $\gamma_V p \to \pi^+ n$, $\pi^+ \Delta^o$, $\pi^+ N^*$ (1520)

4.3.2 <u>K Spectra:</u> The limited data on the inclusive K^+ spectra in the forward direction increases with Q^2 due to two body reactions[33] $\gamma_V p \to K^+ (\frac{\Lambda}{\Sigma})$, $K^+ Y^+$. No systematic study of $K^{\pm}$ spectra as a function of X has yet been made.

4.3.3. <u>Proton Spectra:</u> The yield at backward going protons $X < .5$ is shown in Fig. 27. The backward peak at $Q^2 = 0$ from $\gamma p \to (\rho\omega) + p$ disappears with increasing Q^2.

The model of Drell, Levy and Yan[34] assuming spin 1/2 partons inside protons thus predicts the forward proton yield should increase with Q^2 is not yet observed at DESY energies (Fig. 28).

* * *

ACKNOWLEDGEMENT

I wish to thank Dr. T. McCorriston, Dr. M. Della-Negra and Prof. A. Zichichi for many interesting conversations and for having provided me with useful data.

* * *

REFERENCES

1) S. Tavernier, Laboratoire de l'Accelerateur lineaire, Orsay, Report No. Rl 68/7, (1968)

2) R. Madaras et al. P.R.L. <u>30</u>, p. 507 (1973)

3) V. Alles-Borelli et al. Nuovo Cimento, Vol. 7A, p. 345 (1972)

4) M. Bernardini et al., Submitted to Physics Letters, 1973.
B. Borgia et al. Nuovo Cimento letters, Vol. 3, p. 115 (1972)

5) T. D. Lee and G. C. Wick, Nucl. Phys. 9B, 209 (1969)

6) V. Alles-Borelli et al., Nuovo Cimento, Vol. 7, p. 330 (1972)

7) J. E. Augustin et al., P.R.L., Vol. 30, p. 462 (1973)

8) G. Jarlskog et al., DESY 73/4 (1973)

9 H. Cheng and T. T. Wu, P.R.L. 22, 666 (1969) and DESY 71/69

10) J. J. Sakurai, Ann. Phys. 11, 1 (1960)

11) H. Alvensleben et al., P.R.L. 27, 444 (1971)

12) J. le Francois, proceedings 1971, international symposium on Electron and Photon Interactions at High Energies, Cornell University, ed, by N. Mistry, p. 52

13) V. J. Auslander et al., P.L. 25B, 433 (1967)

14) H. Alvensleben et al. P.R.L. 27, 888 (1971)
H. Alvensleben et al. P.R.L. 28, 66 (1972)

15) H. Alvensleben et al. P.R.L. 26, 273 (1971)

16) G. Bacci et al., P.L. 38B, 551 (1972), G. Barbarino et al., lettre al Nuovo Cimento 3, 689 (1972)

17) H. H. Bingham et al., SLAC-Pub.-1113 and LBL-1085 (1972)

18) M. Bernardini et al. P.L. 44B, p. 393 (1973)

19) G. Barbiellini et al., I.N.F.N. report 420 (1973)

20) R. Kronig, J. Opt. Soc. Am. 12, 547 (1926), H. A. Kramers, Atti, Congr. intern. fisici, Como 2, 545 (1927)

21) M. Gell-Mann, M. C. Goldberger, W. Thirring, P.R. 95, 1612 (1954)

22) M. Damashek, F. J. Gilman, P.R. D1, 1319 (1970)
Also: T. A. Armstrong et al. P.R. D5, 1640 (1972)

23) S. J. Brodsky, J. G. Gillespie, P. R. 173, 1011 (1968)

24) J. Ballam et al., SLAC-pub-1143 (1972) and ref. therein

25) H. Alvensleben et al., P.R.L. 28, 66 (1972)

26) G. Wolf, DESY 72/61, E. D. Bloom et al. P.R.L. 23, 930 (1969)
M. Breidenbach et al. P.R.L. 23, 935 (1969)
G. Miller et al. P.R. D5, 528 (1972)

27) W. Albrecht et al. Nucl. Phys. B27, 615 (1971)

28) H. Cheng and T. T. Wu, P.R. 183, 1324 (1969); J. D. Bjorken, J. Kogut and D. Soper, P.R. D1, 1382 (1971), H. T. Nieh, P.L. B38, 100 (1972)

29) V. Eckardt et al., Proceedings of High Energy Physics Conference, Chicago 1972

30) ABBHHM Collaboration, P.R. 175, 1669 (1968)

31) L. Ahrens et al. (LNS 227, May 1973)

32) J. Ballam et al., M. Della-Negra, private communication

33) G. Wolf, DESY 72/61

34) S. D. Drell, D. J. Levy and T. M. Yan, P.R.L. 22, 744 (1969)

* * *

FIGURE CAPTIONS

Fig. 1 Layout of the by-pass on-line detector, showing a typical event.

Fig. 2 Experimental angular distribution of Bhabha scattering events compared with the normalized prediction of QED.

Fig. 3 S-dependence of $e^+e^- \to e^-e^+$ compared with QED, data from the CERN-Bologna-Frascati group.

Fig. 4 Results from Conversi group on $e^+e^- \to \mu^+\mu^-$ vs $e^+e^- \to e^-e^+$

Fig. 5 Results from the Zichichi group $\mu^+\mu^-/e^+e^-$ ratio accepted in the apparatus at various energies.

Fig. 6a Best fit of $e^+e^- \to \phi \to \mu^-\mu^+$ with the Kallen-Lehmann representation

Fig. 6b The excitation curve for $\phi \to K^-K^+$ for calibration

Fig. 7 Experimental set up of the DESY Delbrück experiment.

Fig. 8 Measured differential cross sections for Delbruck scattering vs t for Au compared with theory.

Fig. 9 Measured differential cross section for Delbrück scattering for U at small t

Fig. 10 Cross section of $\gamma+A \to A+\pi^-\pi^+$ from DESY/MIT. The solid line is fit with $\rho\omega$ interference. The other curves are a) no ω contribution, b) and c), d) backgrounds. See (Ref. 14)

Fig. 11 Mass spectra for $\gamma+c \to c+\pi^-\pi^+$ enhancement at 1.6 GeV is clearly seen.

Fig. 12 Cross section for $e^+e^- \to 2\pi^+2\pi^-$ for Ref. (16)

Fig. 13 $\gamma+p \to p + 2\pi^+2\pi^-$ at 9.3 GeV, 4π mass spectrum. The shaded histogram has events with Δ^{++} removed

Fig. 14a Plan view of the spectrometer

Fig. 14b Difference between arrival times of e^- and e^+ for all detected pairs.

Fig. 14c Distribution of all events as a function of mass. The solid line is the theoretical prediction. The size of the vertical bars indicate the theoretical uncertainties. For m>20 MeV QED dominates and the theoretical uncertainties are negligible.

Fig. 15a Distribution of interference events as a function of mass. The solid line is the prediction of $d\sigma_{int}$

Fig. 15b Comparison of experimental results with dispersion relation prediction.

Fig. 16 Reaction $\gamma+p \to p\omega$. Total cross section as a function of the incident photon energy. The points labeled ABBHHM (P.R. <u>175</u>, 1669 (1968)) and SLAC Annihilation Beam (P.R. <u>D5</u>, 15 (1972)) are earlier work. The full and dashed curves give the contributions of a diffractive process and OPE respectively.

Fig. 17 Reaction $\gamma+p \to \phi+p$

Fig. 18 Reaction $\gamma+p \to \phi+p$. Total cross section and exponential slope A of the differential cross section as function of incident photon energy.

Fig. 19 Total electroproduction cross section $\sigma_\tau^{tot} + \varepsilon\sigma_\tau^{tot}$, as a function of Q^2 for various W. Figure taken from G. Wolf, DESY 72/61.

Fig. 20 Total cross section for the reaction $ep \to ep\pi^+\pi^-$ as a function of the total hadron mass W for different Q^2 intervals, Ref. 26, ref.29, ref. 30.

Fig. 21 Reaction $ep \to ep + \pi^+\pi^-$; Q^2 dependence of the cross section for different W intervals (ref. 29). The $Q^2 = 0$ points are from Ref. 30. Fig. taken from Ref. 29

Fig. 22 Measured slope of the t distribution for ρ^o and ω production plotted against the dimensionless inverse life time parameter $\frac{Q^2+m_p^2}{2m_\nu}$ Figure from Ref. 31.

Fig. 23 Average charged multiplicity vs ln W' from SLAC track chamber data.

Fig. 24 The normalized π^- yield for $e\ p \to e\pi^- + \ldots$ at W = 2.6 GeV and $Q^2=0$, and at $2.0<W<2.7$ GeV for different Q^2. The curve shows yield at $Q^2=0$ when $\gamma p \to p\rho^o$ are removed (Ref. 33, Fig. 28)

Fig. 25 Normalized π^- yield from 16 GeV SLAC track chamber data (Ref. 32)

Fig. 26 Normalized π^+ yield from 16 GeV SLAC track chamber data (Ref. 32)

Fig. 27 The normalized invariant cross sections for inclusive proton production for $x<-.5$ data from Ref. 32

Fig. 28 The normalized invariant cross sections for inclusive proton production at $P^2<0.02$ GeV^2, W = 2.63 GeV, $Q^2 = 0$ and $Q^2 = 1.16$ (Fig. from Ref. 33)

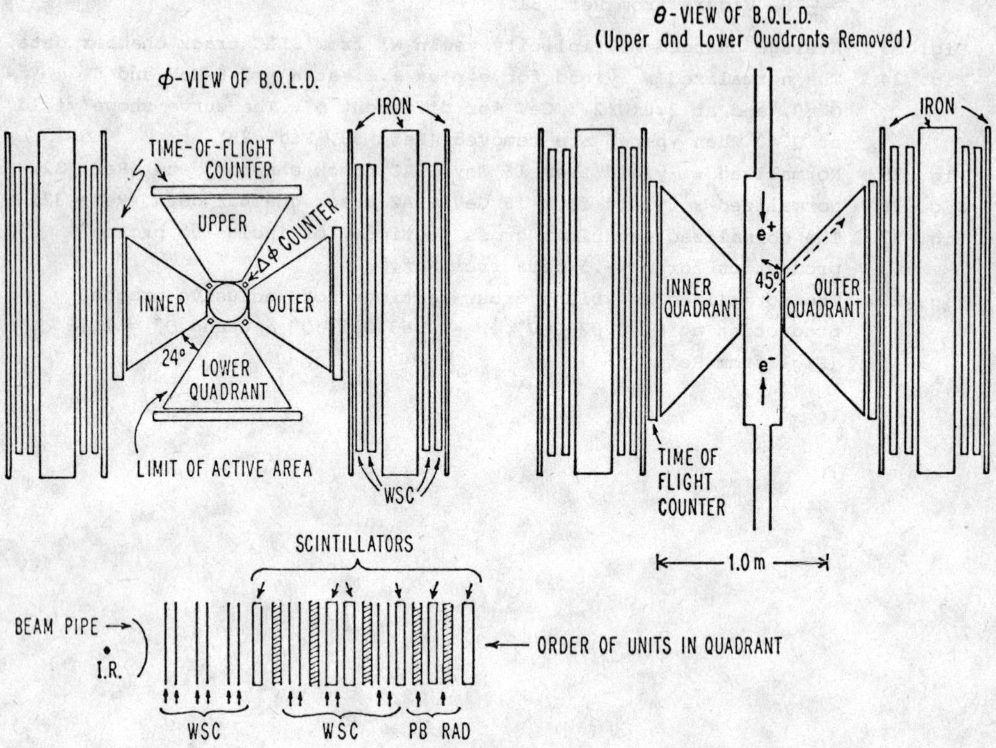
ϕ-VIEW OF B.O.L.D.
TIME-OF-FLIGHT COUNTER
UPPER
Δϕ COUNTER
INNER
OUTER
24°
LOWER QUADRANT
LIMIT OF ACTIVE AREA
IRON
WSC
θ-VIEW OF B.O.L.D.
(Upper and Lower Quadrants Removed)
IRON
e+
45°
INNER QUADRANT
OUTER QUADRANT
e-
TIME OF FLIGHT COUNTER
1.0 m
SCINTILLATORS
BEAM PIPE
I.R.
ORDER OF UNITS IN QUADRANT
WSC
WSC
PB RAD

Fig. 1

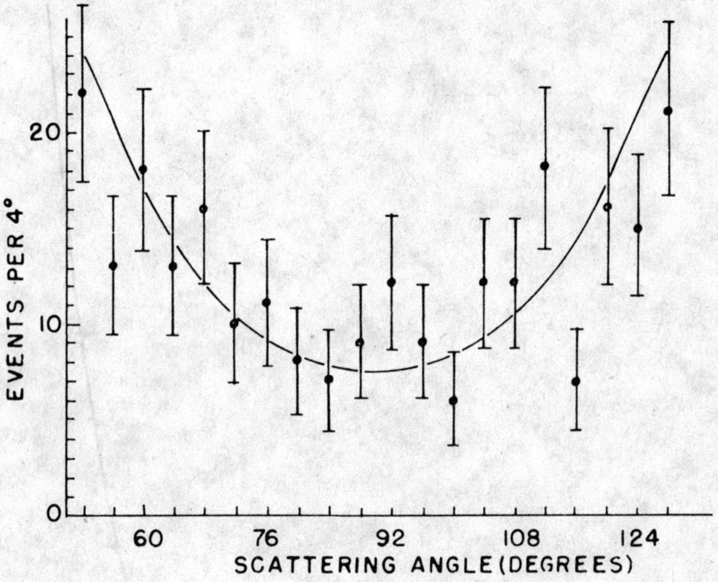

Fig. 2

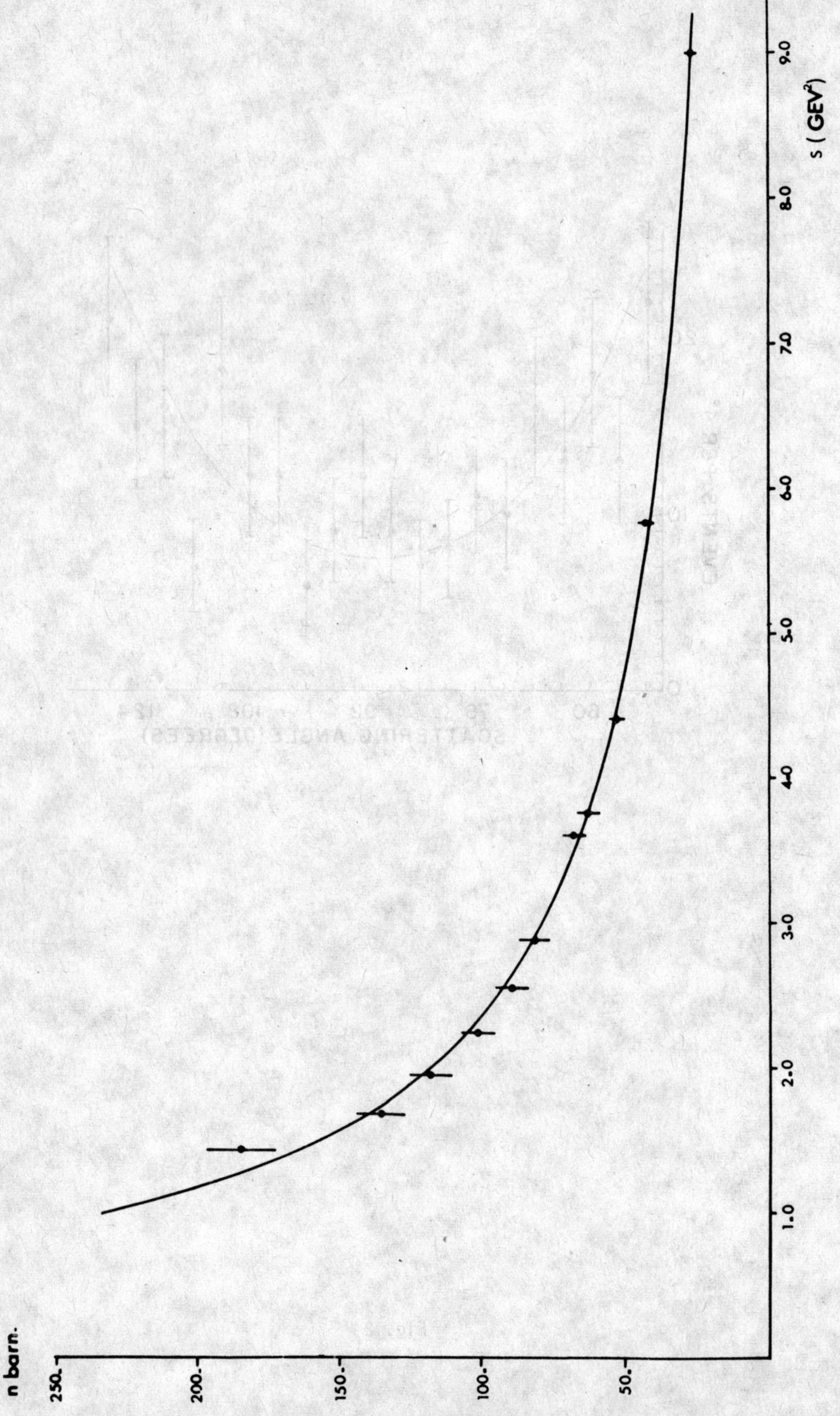

Fig. 3

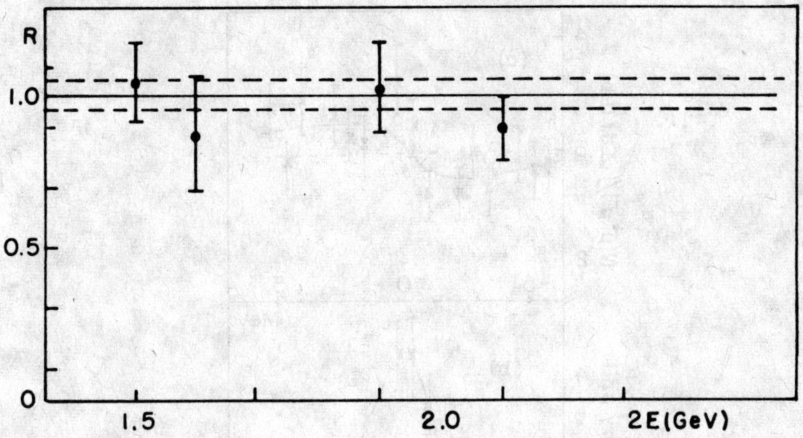
R
1.0
0.5
0
1.5
2.0
2E(GeV)

Fig. 4

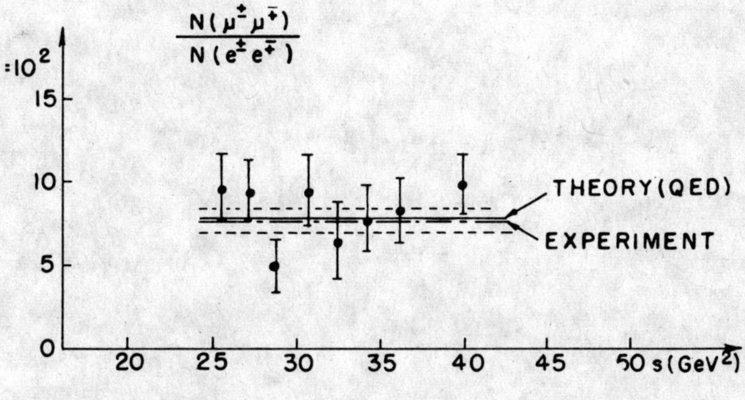
$\frac{N(\mu^{\pm}\mu^{\mp})}{N(e^{\pm}e^{\mp})}$
×10²
15
10
5
0
20
25
30
35
40
45
50 s(GeV²)
THEORY(QED)
EXPERIMENT

Fig. 5

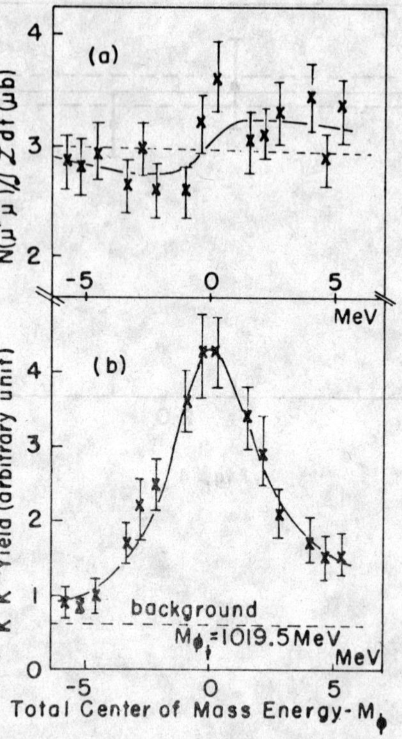

(a)
N(μ⁺μ⁻)/∫𝓛dt (μb)
4
3
2
-5
0
5
MeV
(b)
K⁺K⁻ Yield (arbitrary unit)
4
3
2
1
0
background
Mφ=1019.5 MeV
MeV
-5
0
5
Total Center of Mass Energy-Mφ

Fig. 6

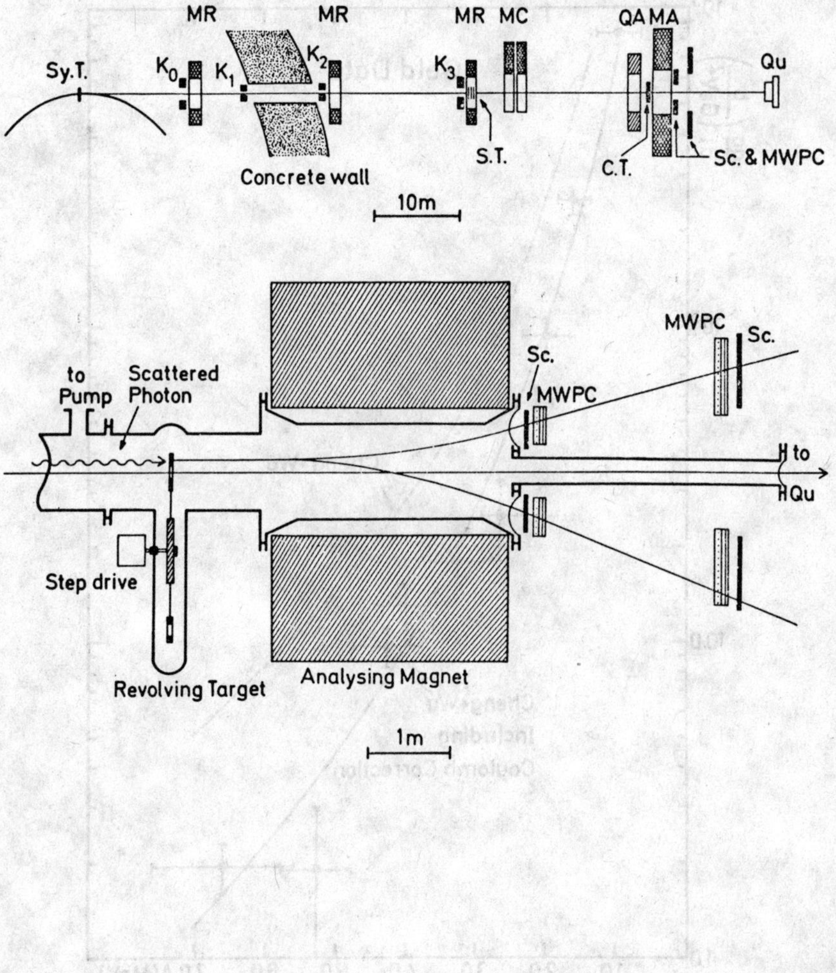
MR
MR
MR
MC
QA
MA
Sy.T.
K0
K1
K2
K3
Qu
S.T.
C.T.
Sc. & MWPC
Concrete wall
10m
MWPC
Sc.
Sc.
MWPC
to
Pump
Scattered
Photon
to
Qu
Step drive
Revolving Target
Analysing Magnet
1m

Fig. 7

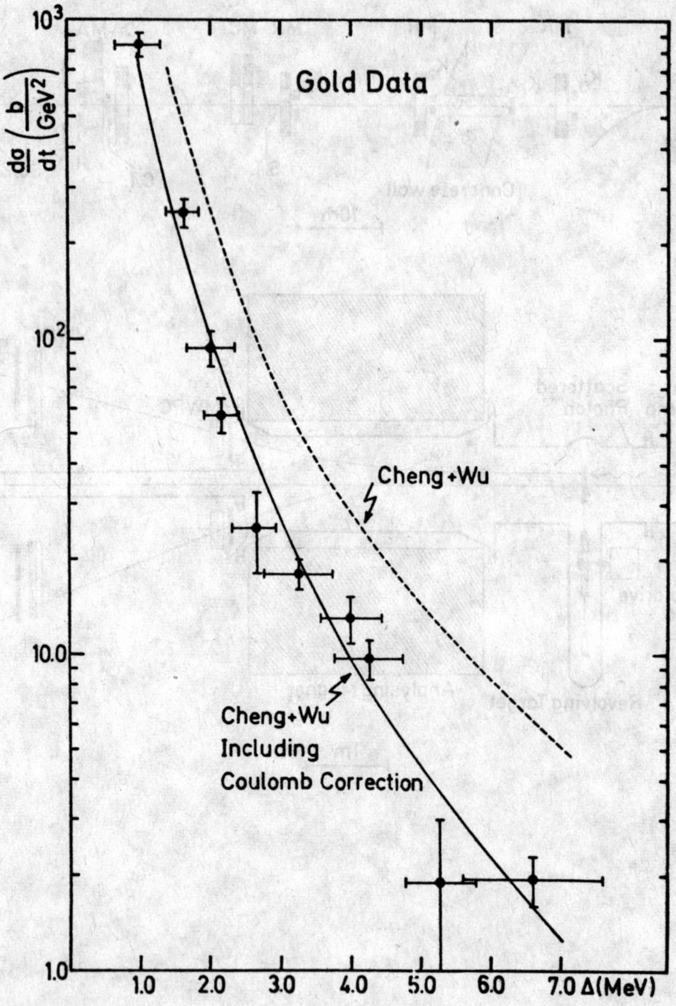
Gold Data
dσ/dt (b/GeV²)
10³
10²
10.0
1.0
1.0
2.0
3.0
4.0
5.0
6.0
7.0 Δ(MeV)
Cheng+Wu
Cheng+Wu
Including
Coulomb Correction

Fig. 8

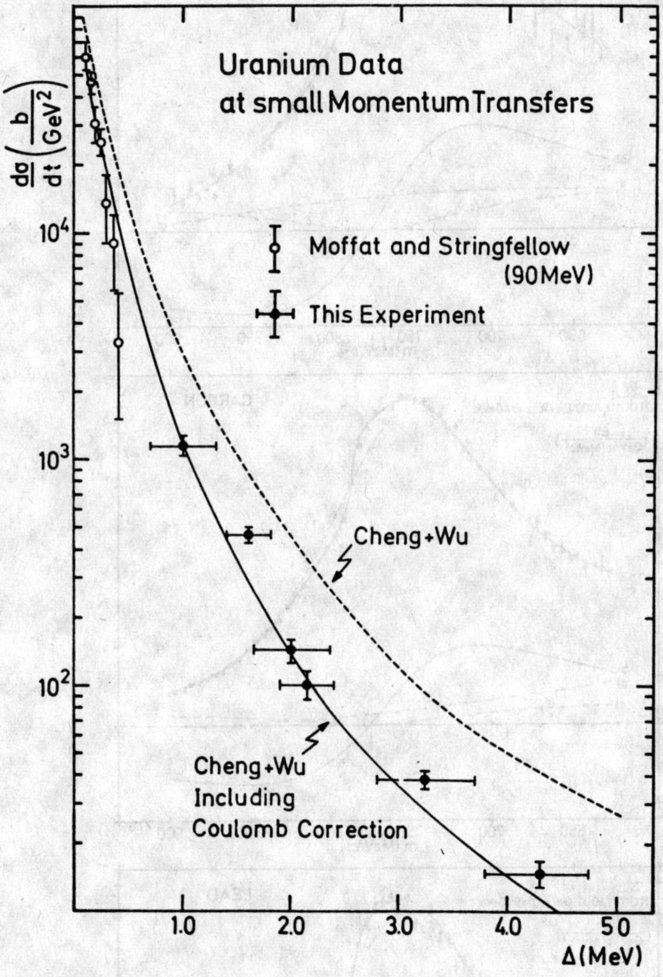
Uranium Data
at small Momentum Transfers
Moffat and Stringfellow
(90MeV)
This Experiment
Cheng+Wu
Cheng+Wu
Including
Coulomb Correction
$\frac{d\sigma}{dt}\left(\frac{b}{GeV^2}\right)$
10^4
10^3
10^2
1.0
2.0
3.0
4.0
5.0
Δ(MeV)

Fig. 9

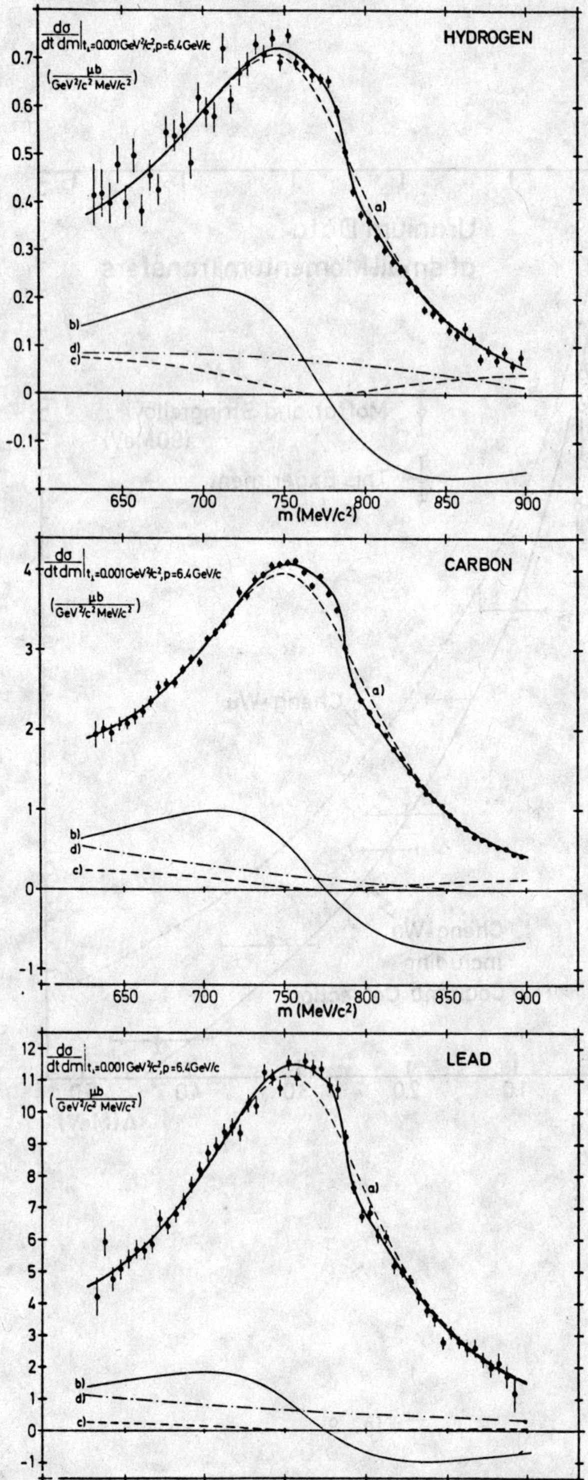
HYDROGEN
CARBON
LEAD
dσ/dt dm | t⊥=0.001GeV²/c², p=6.4GeV/c
(μb / GeV²/c² MeV/c²)
a)
b)
c)
d)
m (MeV/c²)
650
700
750
800
850
900

Fig. 10

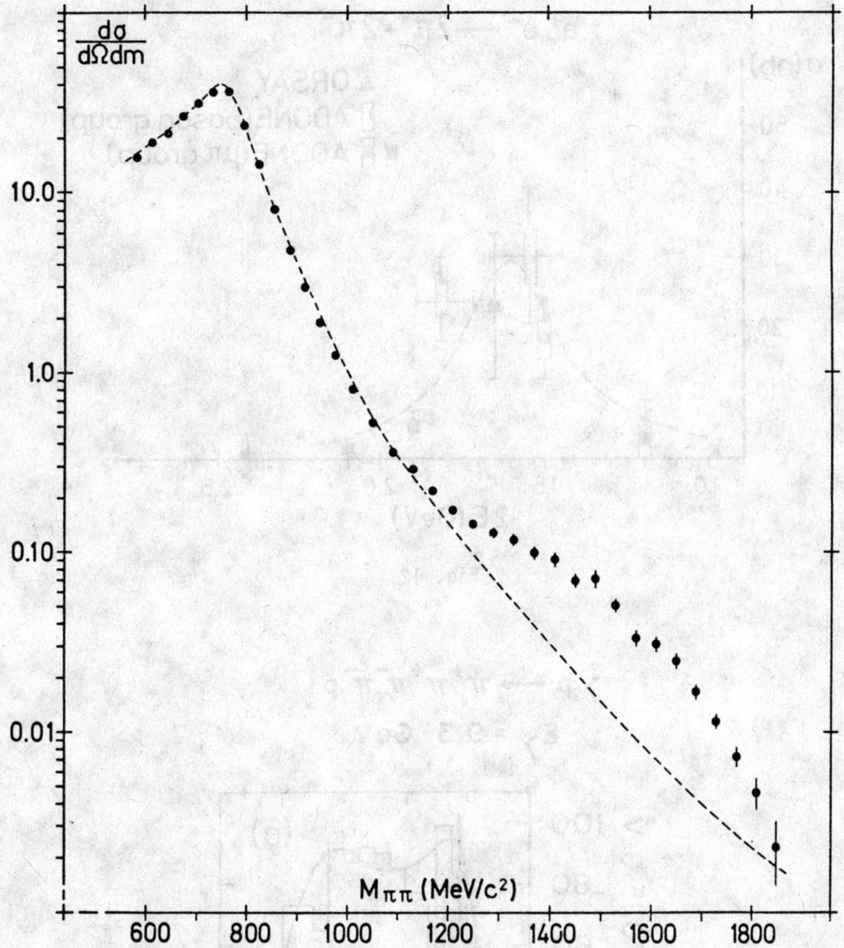
dσ
dΩdm
10.0
1.0
0.10
0.01
M ππ (MeV/c²)
600
800
1000
1200
1400
1600
1800

Fig. 11

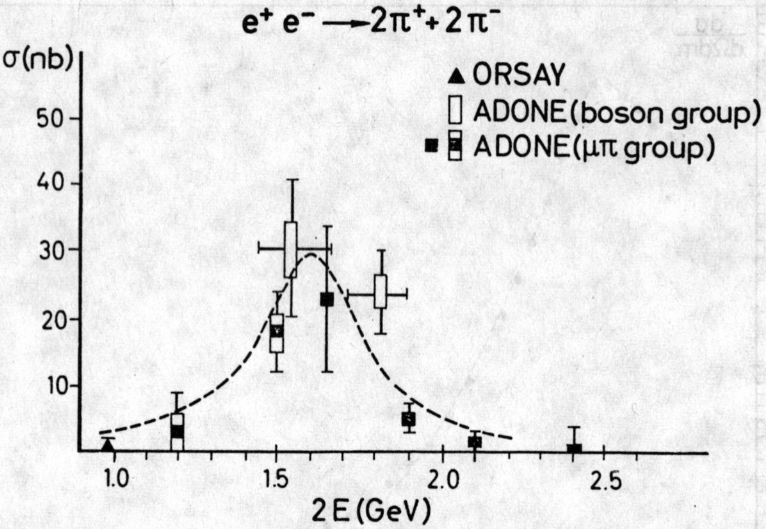
e+ e- → 2π+ + 2π-
σ(nb)
50
40
30
20
10
▲ ORSAY
ADONE (boson group)
ADONE (μπ group)
1.0
1.5
2.0
2.5
2E (GeV)

Fig. 12

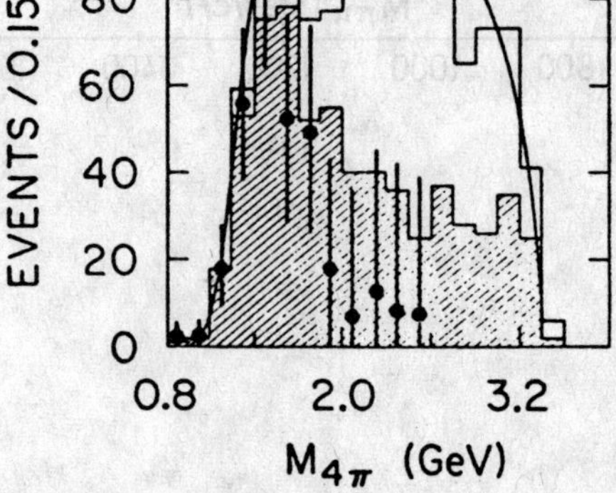
γp → π+π+π-π-p
Eγ = 9.3 GeV
EVENTS / 0.15 GeV
100
80
60
40
20
0
(a)
0.8
2.0
3.2
M4π (GeV)

Fig. 13

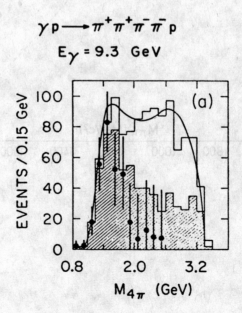

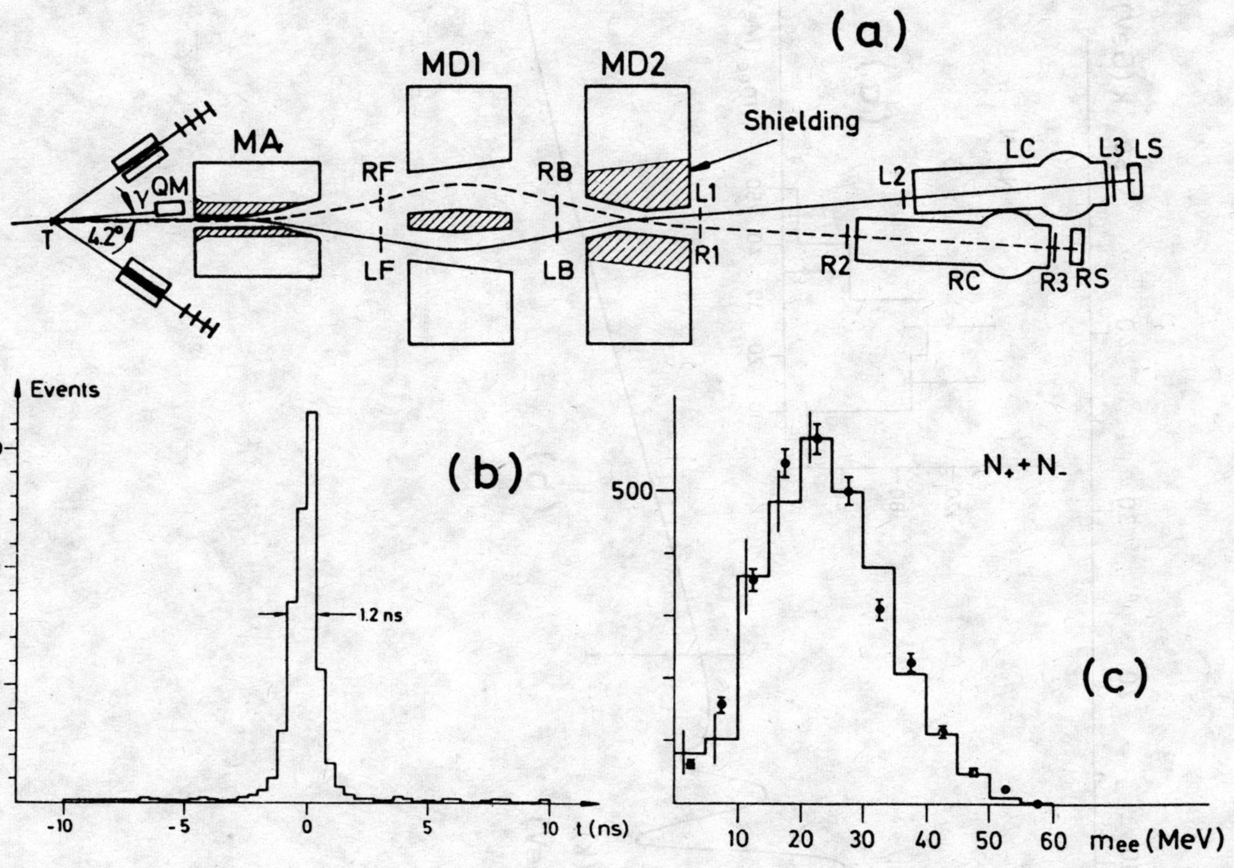

(a)
MA
MD1
MD2
Shielding
T
4.2°
γ
QM
RF
LF
RB
LB
L1
R1
L2
R2
LC
RC
L3
LS
R3
RS
Events
1500
1000
500
(b)
1.2 ns
-10
-5
0
5
10
t (ns)
500
N₊ + N₋
(c)
10
20
30
40
50
60
mee (MeV)

Fig. 14

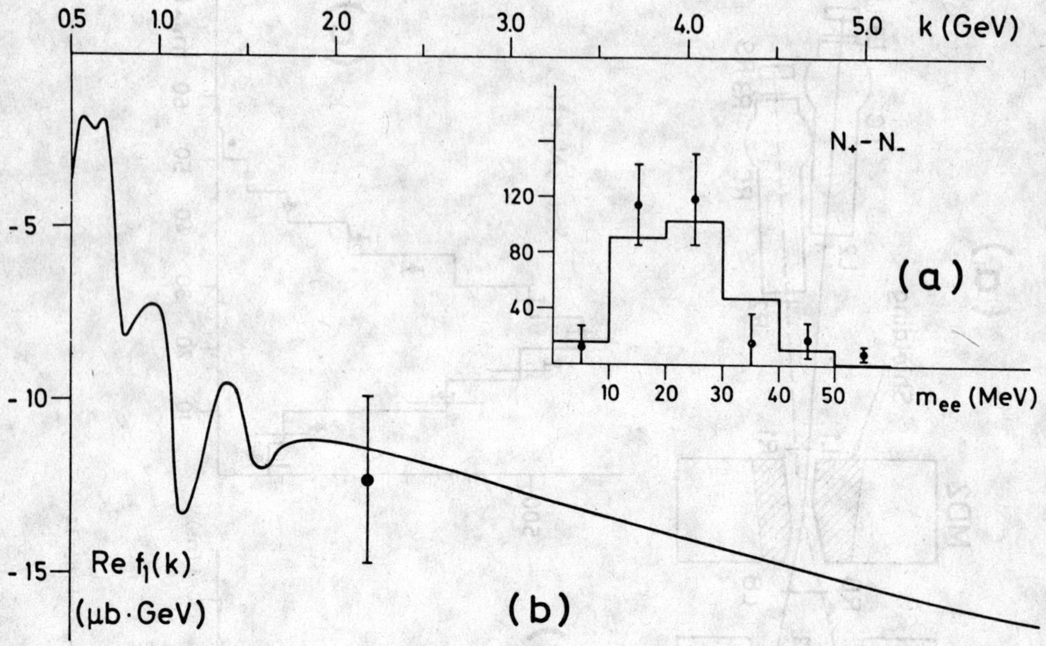
k (GeV)
0.5
1.0
2.0
3.0
4.0
5.0
Re f₁(k)
(μb · GeV)
- 5
- 10
- 15
(b)
N₊ - N₋
120
80
40
10
20
30
40
50
m_ee (MeV)
(a)

Fig. 15

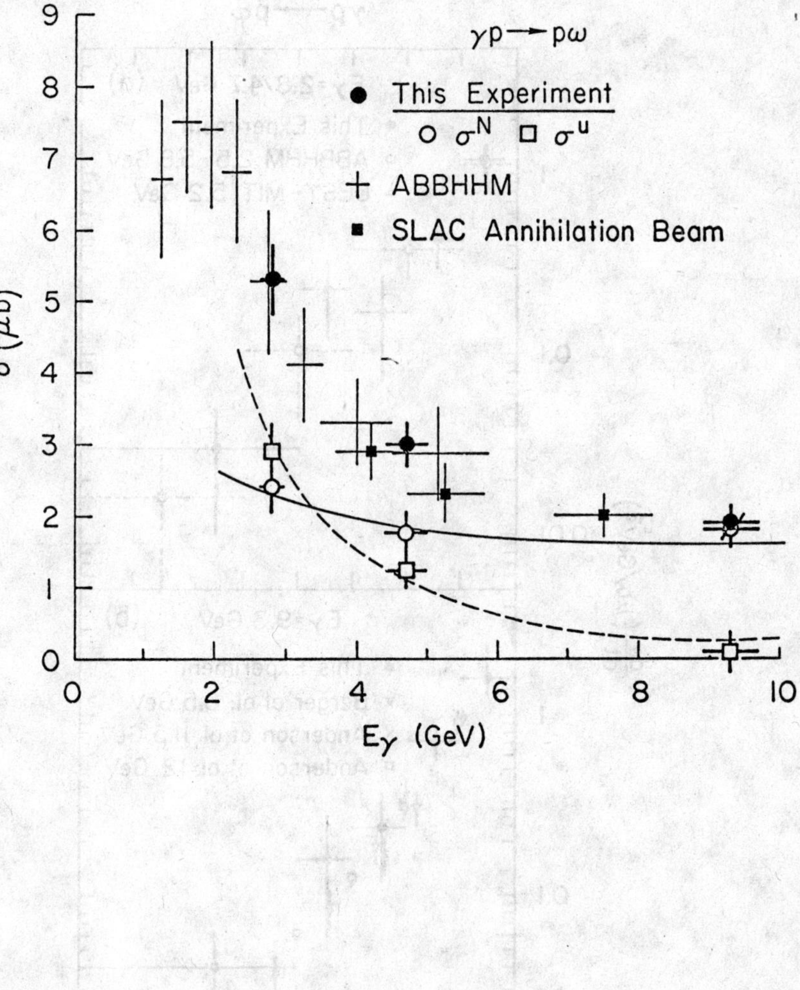
γp → pω
This Experiment
σ^N σ^u
ABBHHM
SLAC Annihilation Beam
σ (μb)
E_γ (GeV)
0
1
2
3
4
5
6
7
8
9
0
2
4
6
8
10

Fig. 16

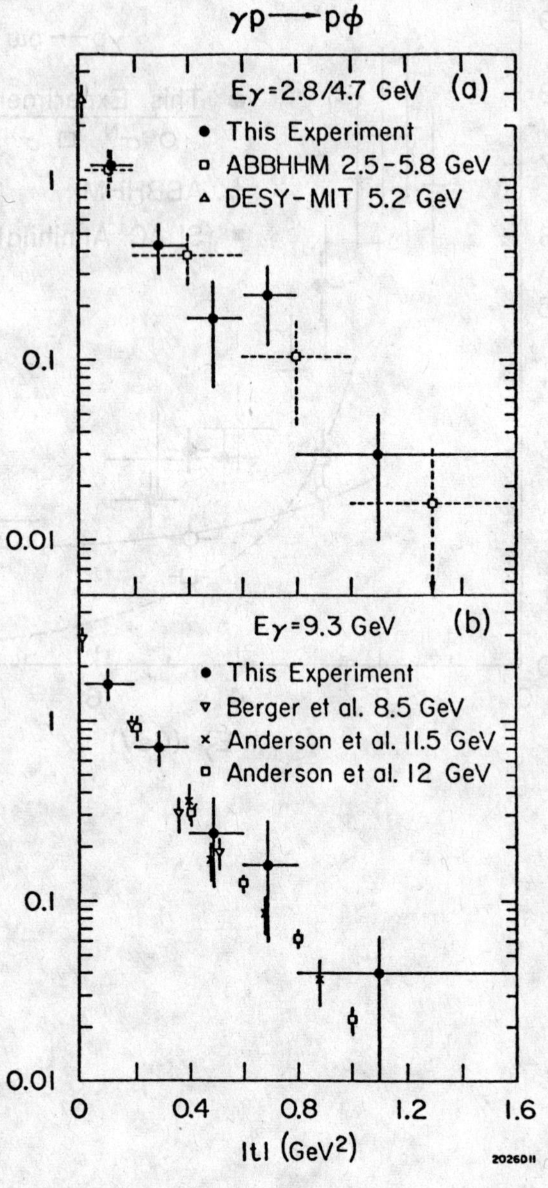
γp ⟶ pφ
Eγ=2.8/4.7 GeV
(a)
This Experiment
ABBHHM 2.5-5.8 GeV
DESY-MIT 5.2 GeV
Eγ=9.3 GeV
(b)
This Experiment
Berger et al. 8.5 GeV
Anderson et al. 11.5 GeV
Anderson et al. 12 GeV
dσ/dt (μb/GeV²)
1
0.1
0.01
0
0.4
0.8
1.2
1.6
|t| (GeV²)
2026D11

Fig. 17

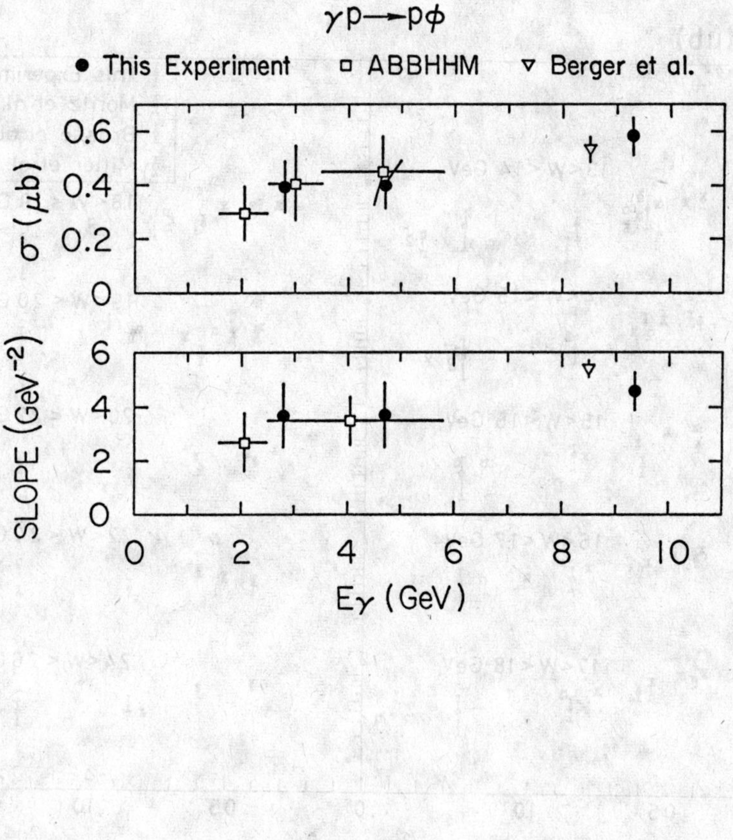

γp→pφ
• This Experiment
□ ABBHHM
▽ Berger et al.
σ (μb)
0.6
0.4
0.2
0
SLOPE (GeV^{-2})
6
4
2
0
0
2
4
6
8
10
Eγ (GeV)

Fig. 18

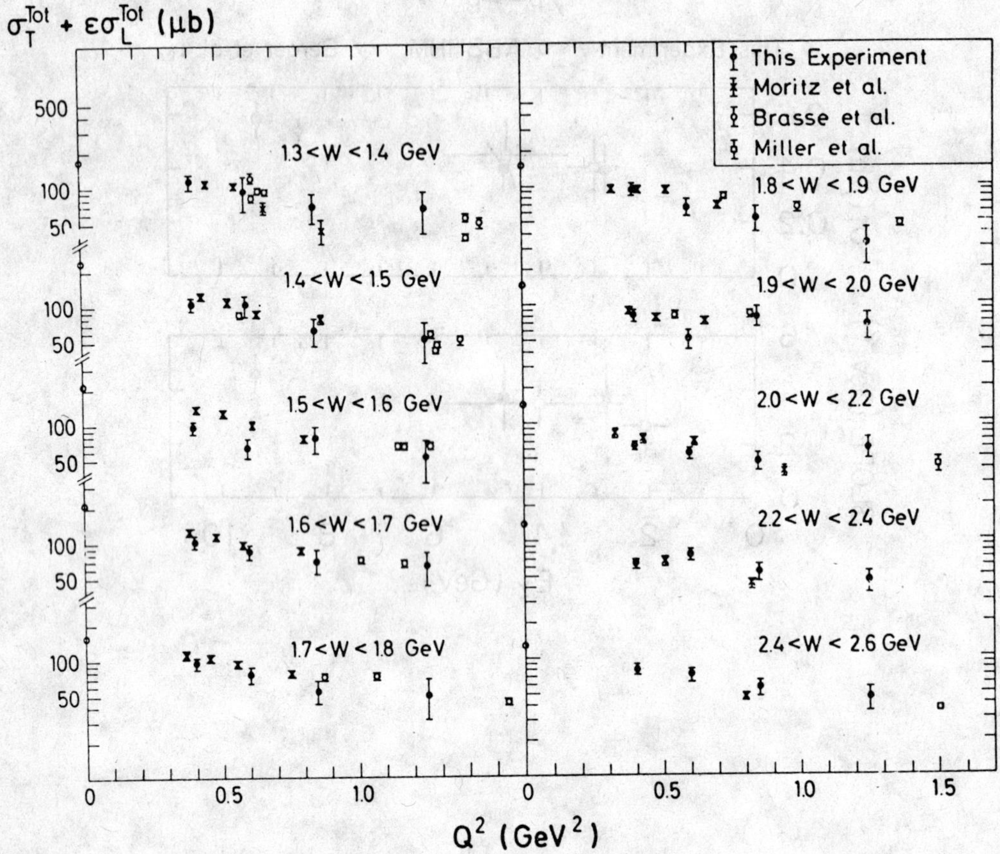

$\sigma_T^{Tot} + \varepsilon\sigma_L^{Tot}$ (μb)
This Experiment
Moritz et al.
Brasse et al.
Miller et al.
1.3 < W < 1.4 GeV
1.4 < W < 1.5 GeV
1.5 < W < 1.6 GeV
1.6 < W < 1.7 GeV
1.7 < W < 1.8 GeV
1.8 < W < 1.9 GeV
1.9 < W < 2.0 GeV
2.0 < W < 2.2 GeV
2.2 < W < 2.4 GeV
2.4 < W < 2.6 GeV
500
100
50
0
0.5
1.0
1.5
Q^2 (GeV^2)

Fig. 19

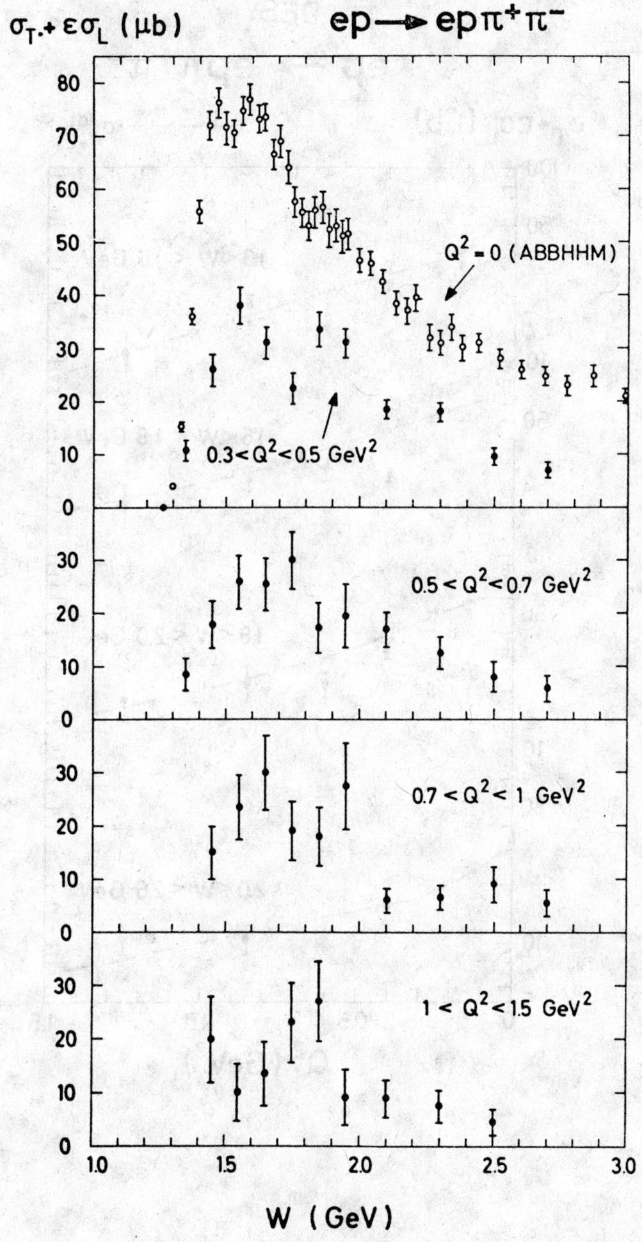
σT.+ εσL (μb)
ep⟶epπ+π−
Q2 = 0 (ABBHHM)
0.3 < Q2 < 0.5 GeV2
0.5 < Q2 < 0.7 GeV2
0.7 < Q2 < 1 GeV2
1 < Q2 < 1.5 GeV2
W (GeV)

Fig. 20

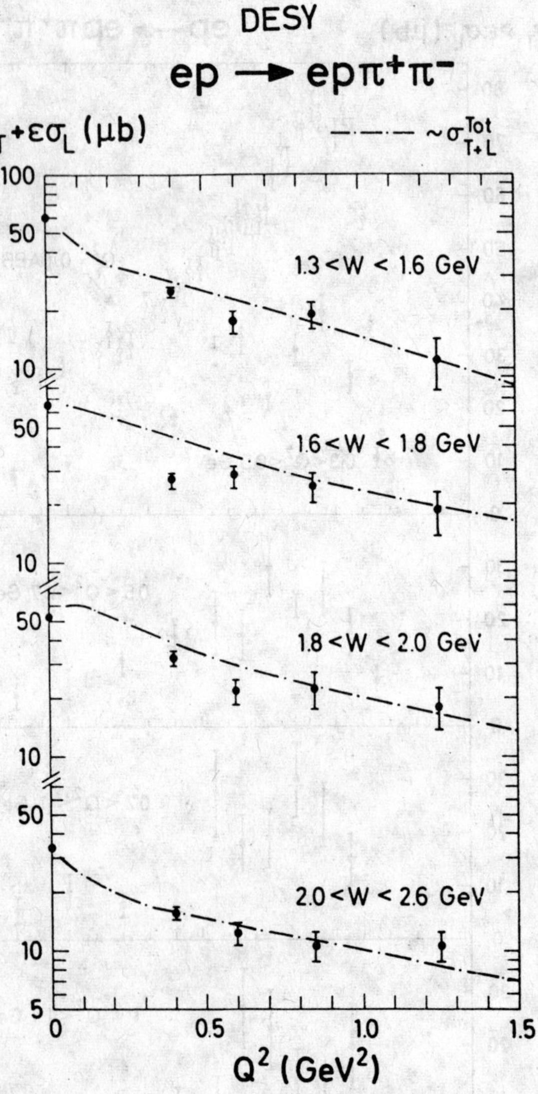
DESY
ep → epπ⁺π⁻
σT+εσL (μb)
~σ Tot T+L
100
50
10
1.3 < W < 1.6 GeV
1.6 < W < 1.8 GeV
1.8 < W < 2.0 GeV
2.0 < W < 2.6 GeV
5
0
0.5
1.0
1.5
Q² (GeV²)

Fig. 21

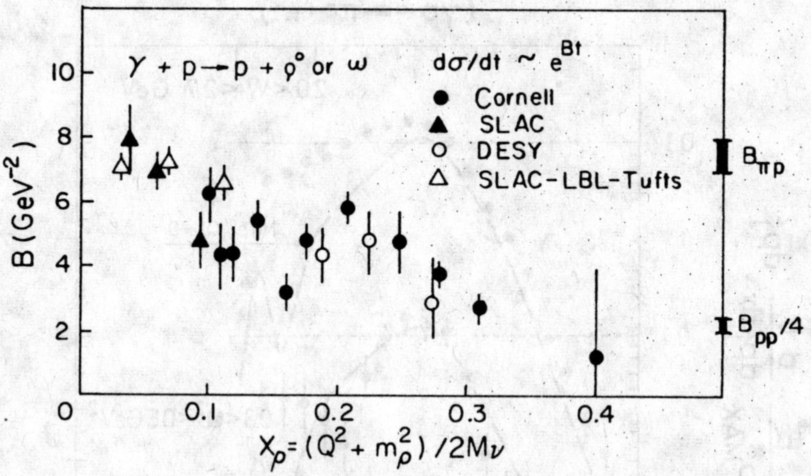
γ + p → p + ρ° or ω
dσ/dt ~ e^Bt
Cornell
SLAC
DESY
SLAC-LBL-Tufts
B (GeV^-2)
Bπp
Bpp/4
Xρ = (Q² + mρ²) /2Mν

Fig. 22

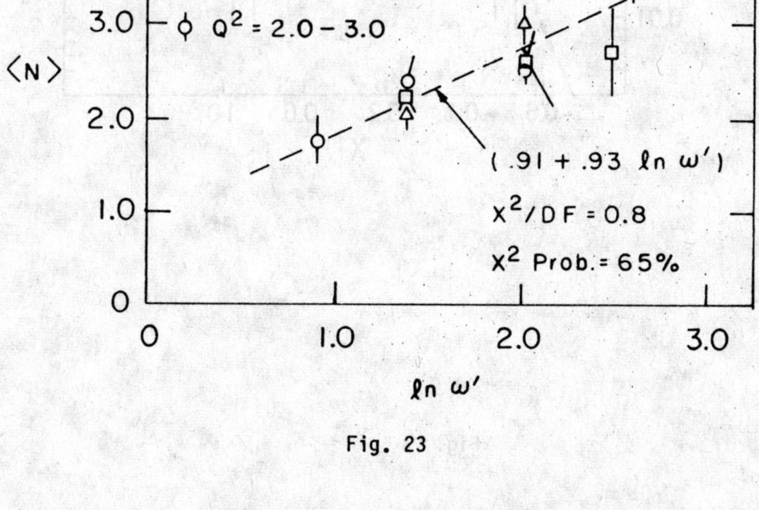
AVERAGE CHARGED MULTIPLICITY vs ln ω'
Q² = 1.0 − 1.25
Q² = 1.25 − 2.0
Q² = 2.0 − 3.0
⟨N⟩
(.91 + .93 ln ω')
X²/DF = 0.8
X² Prob. = 65%
ln ω'

Fig. 23

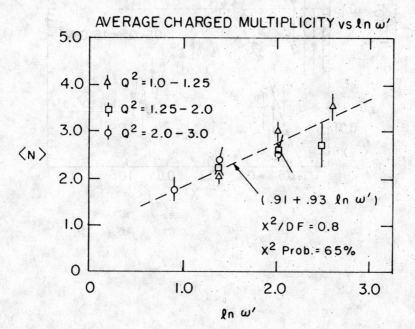

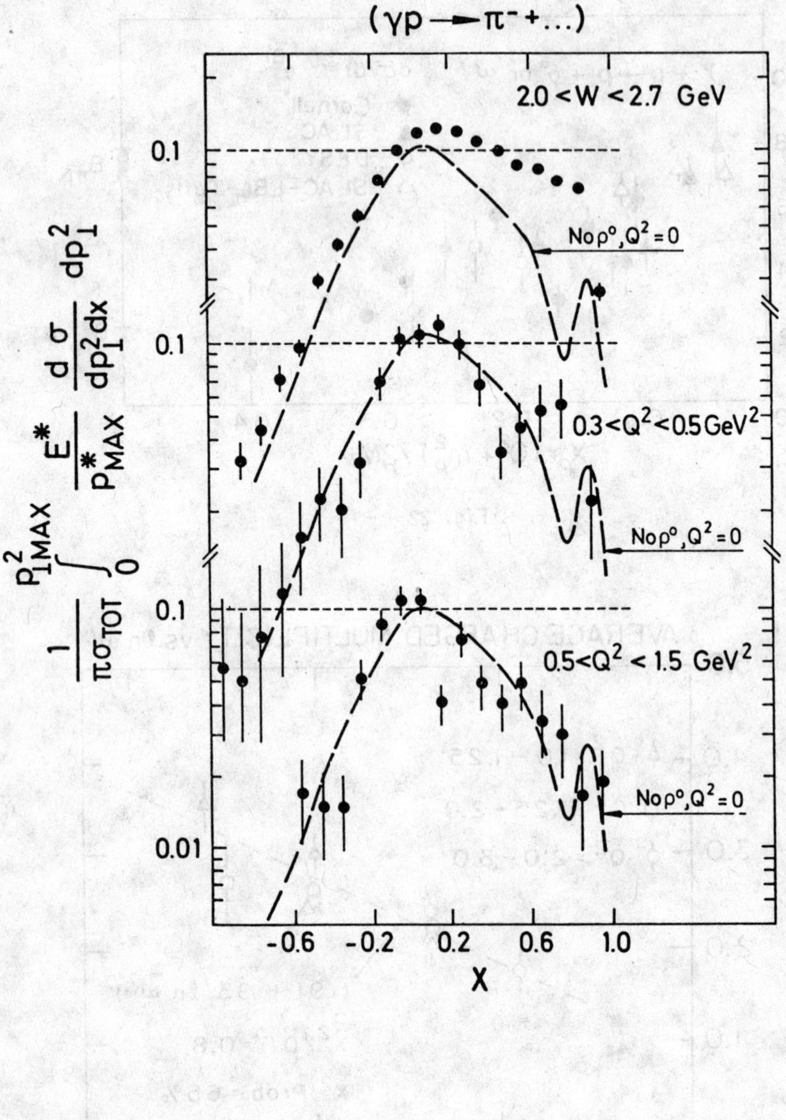

(γp → π⁻ + ...)
2.0 < W < 2.7 GeV
No ρ°, Q² = 0
0.3 < Q² < 0.5 GeV²
No ρ°, Q² = 0
0.5 < Q² < 1.5 GeV²
No ρ°, Q² = 0
0.1
0.1
0.1
0.01
-0.6
-0.2
0.2
0.6
1.0
X

Fig. 24

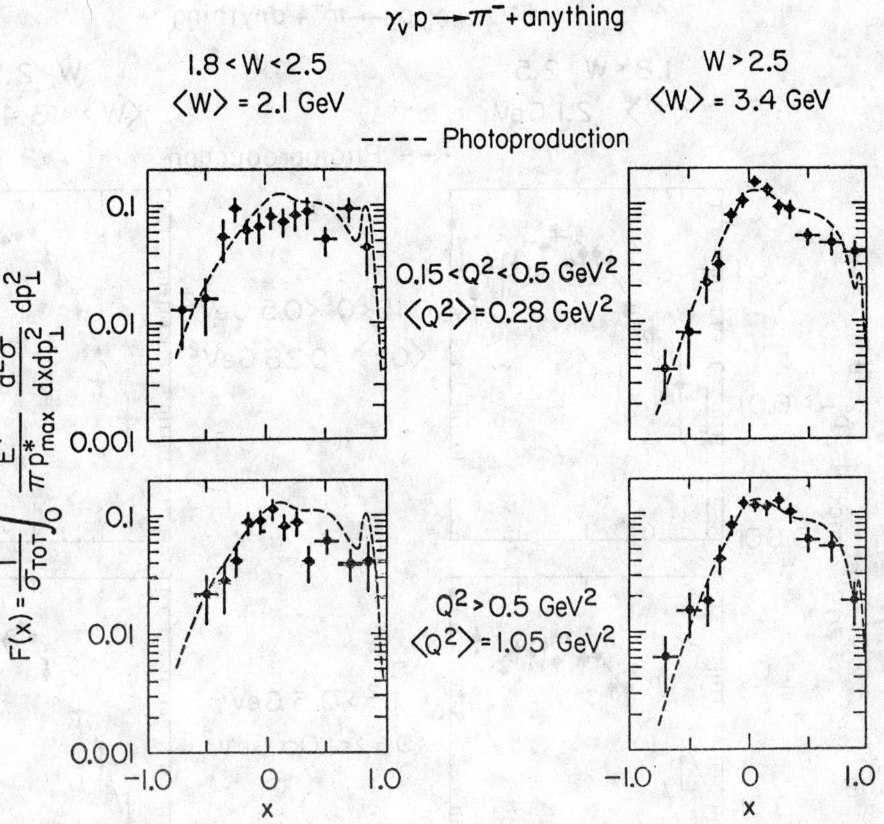

$\gamma_v p \rightarrow \pi^- +$ anything
1.8 < W < 2.5
⟨W⟩ = 2.1 GeV
W > 2.5
⟨W⟩ = 3.4 GeV
---- Photoproduction
0.15 < Q^2 < 0.5 GeV^2
⟨Q^2⟩ = 0.28 GeV^2
Q^2 > 0.5 GeV^2
⟨Q^2⟩ = 1.05 GeV^2
$F(x) = \frac{1}{\sigma_{TOT}} \int_0^\infty \frac{E^*}{\pi p^*_{max}} \frac{d^2\sigma}{dx dp_\perp^2} dp_\perp^2$
0.1
0.01
0.001
-1.0
0
1.0
x

Fig. 25

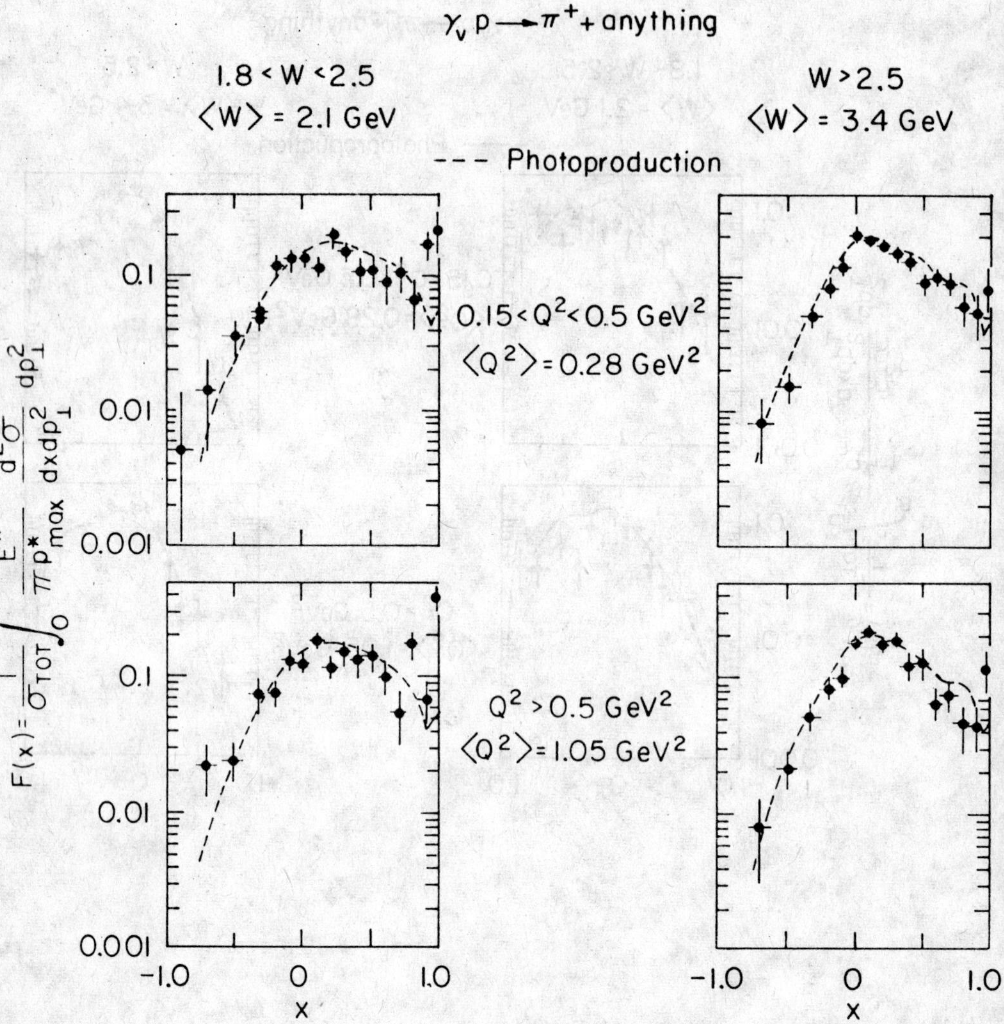

$\gamma_v p \rightarrow \pi^+ +$ anything
1.8 < W < 2.5
⟨W⟩ = 2.1 GeV
W > 2.5
⟨W⟩ = 3.4 GeV
--- Photoproduction
0.15 < Q² < 0.5 GeV²
⟨Q²⟩ = 0.28 GeV²
Q² > 0.5 GeV²
⟨Q²⟩ = 1.05 GeV²
$F(x) = \frac{1}{\sigma_{TOT}} \int_0^\infty \frac{E^*}{\pi p^*_{max}} \frac{d^2\sigma}{dx dp_\perp^2} dp_\perp^2$
0.1
0.01
0.001
-1.0
0
1.0
X

Fig. 26

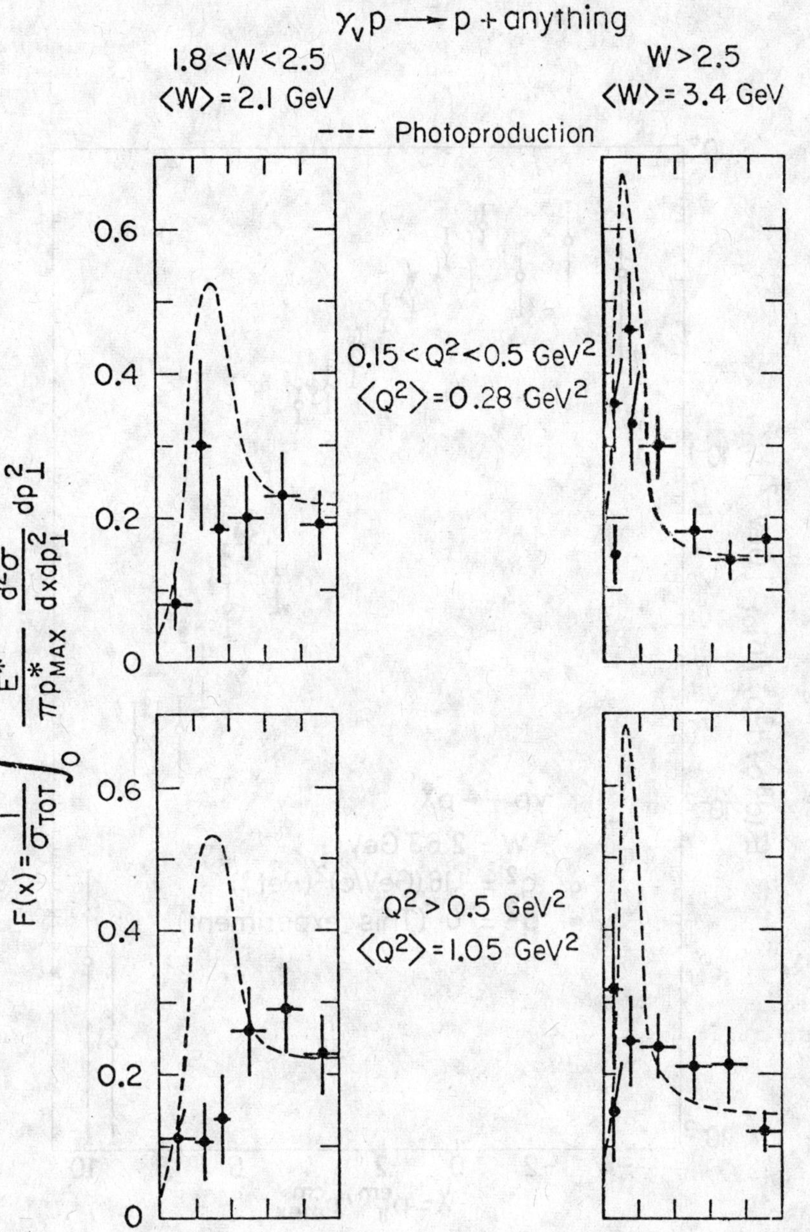

$\gamma_v p \longrightarrow p + \text{anything}$
$1.8 < W < 2.5$
$\langle W \rangle = 2.1$ GeV
$W > 2.5$
$\langle W \rangle = 3.4$ GeV
--- Photoproduction
$0.15 < Q^2 < 0.5$ GeV^2
$\langle Q^2 \rangle = 0.28$ GeV^2
$Q^2 > 0.5$ GeV^2
$\langle Q^2 \rangle = 1.05$ GeV^2
$F(x) = \frac{1}{\sigma_{TOT}} \int_0^\infty \frac{E^*}{\pi p^*_{MAX}} \frac{d^2\sigma}{dx dp_\perp^2} dp_\perp^2$
0.6
0.4
0.2
0
-1.0
-0.5
x

Fig. 27

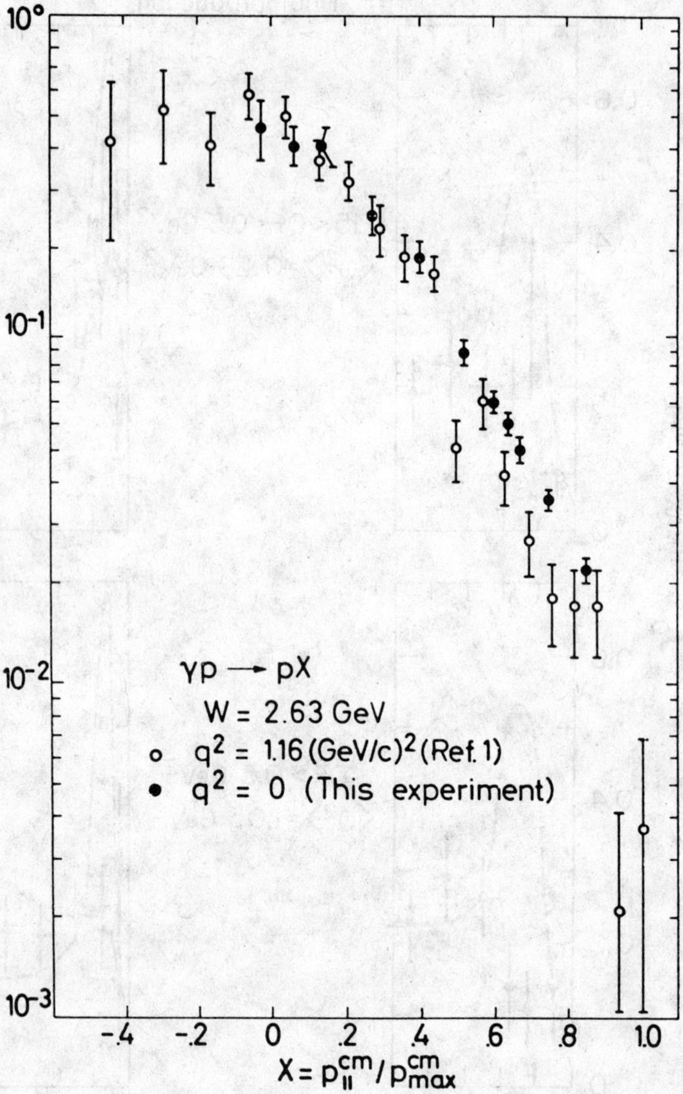
10°
10-1
10-2
10-3
E (d³σ/d³p) / σtot [GeV-2]
γp → pX
W = 2.63 GeV
q² = 1.16 (GeV/c)² (Ref. 1)
q² = 0 (This experiment)
-.4
-.2
0
.2
.4
.6
.8
1.0
X = p∥cm / pmaxcm

Fig. 28

D I S C U S S I O N S

CHAIRMAN: Prof. S.C.C. Ting

Scientific Secretaries: S.B. Berger, J.K. Taylor, G. Chu

DISCUSSION No. 1

- *YELLIN:*

In order to avoid disagreement between theory and experiment and the latest Delbrück scattering experiment, to what order in perturbation theory is one forced to go? Are terms like 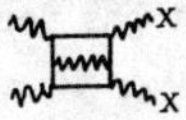 included?

- *TING:*

10th order is not enough, but it is not clear from the calculations of Cheng and Wu what order is enough because all terms are summed together. Terms like the one you mentioned are not included.

- *NAHM:*

Could you explain the Coulomb correction in the calculation of Cheng and Wu?

- *TING:*

It comprises all diagrams with more than two virtual photon lines with the Coulomb field.

- *CHU:*

Did the calculation of Cheng and Wu for Delbrück scattering take into account form factors in the nucleus?

- *TING:*

No. The reason was that they were making a comparison to the experimental data where the maximum momentum transfer q was only $\simeq$ 5 MeV ($\simeq 10^{-11}$ cm^{-1}). Thus if the form factor is parametrized as

$$F(q^2) = 1 - \frac{1}{6} \langle r \rangle^2 q^2 + O(q^4)$$

and we note that for the nuclei $\langle r \rangle \simeq 5 \times 10^{-13}$ cm, it is clear that $F(q^2) \simeq 1$ for the experiment.

- *HENDRICK:*

1) You mean they do not take into account the spatial distribution of the field of the nucleus?

2) Then how does this calculation differ from the Delbrück scattering by Schwinger around 1953?

- *TING:*

No.

- *GLASHOW:*

I think I can clarify this. Schwinger's original calculation was only to first order in α and first order in αZ, considering the field of the nucleus to be a constant field. The recent calculation of Cheng and Wu is to the first order in α and all orders of αZ, using for the field of the nucleus a Coulomb field. This is a remarkable calculation.

- *ARONSON:*

Can you comment on the result of the recent Cornell experiment on photoproduction of μ pairs? The result is supposed to be anomalously large compared to theory.

- *TING:*

This result has nothing to do with QED because the BH diagrams are very small in the kinematical region considered. The result is 5-10 times larger than a parton model calculation of Bjorken and Paschos. Bjorken told me this is an order of magnitude-type calculation for a single parton amplitude.

- *TAYLOR:*

Could you elaborate on the physics of the Källen-Lehman representation of the photon propagators, and in particular, what is the meaning of the off-diagonal term? Has this vacuum polarization effect been measured for the ρ and ω?

- *TING:*

I cannot elaborate on the Källen-Lehman representation of the photon propagator, but I can answer the second part of your question. This vacuum polarization effect has not been measured in the ρ-ω region because the effect is small.

- *AMALDI:*

The vacuum polarization for the ρ would be a 10% effect and would be difficult to measure. In the ϕ-meson mass region the polarization of the vacuum is a 12% effect and is accessible to experiment.

- *ZICHICHI:*

The experiment described by Prof. Ting does not prove that the Källen-Lehman representation is right, but it just shows that the vacuum polarization effect exists. One can produce resonances via the first part of the diagram

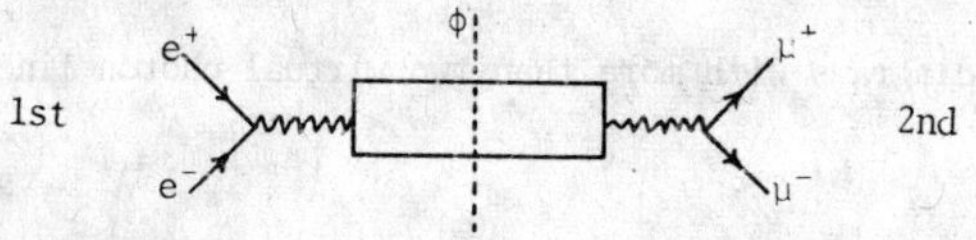

and know that the ϕ decays into two leptons via the second part of the diagram. There are very good experiments on $e^+e^- \to \phi$ and $\phi \to \mu^+\mu^-$. It would be a tremendous accident of Nature if the two parts did not join together.

- *EYLON:*

The optical theorem relates Im (γp) to the total γp cross-section and not to $\sigma(\gamma p \to \to$ hadrons only), which is the measured quantity. What is the experimental difference between these two cross-sections?

- *TING:*

You can look at this equation in two ways. First as an exact relation valid to all orders in α, or as a relation which is also valid if you take the first order in α. If you do the latter you should include only in σ the processes $\gamma + p \to p +$ hadrons and not, for instance, $\gamma + p \to p + \gamma$ or $p +$ pair, because this would be of higher order. And I guess this is the sense in which you use it.

- *EYLON:*

Do you have data about the dependence of the ρ photoproduction slope b on energy?

- *TING:*

The slope b varies from 6 $(\mathrm{GeV}/c)^2$ to 8 $(\mathrm{GeV}/c)^2$ depending on how you subtract the background. Furthermore, if you plot the slope b as a function of $M_{\pi\pi}$, the mass of the $\pi^+\pi^-$ system, there is a strong variation of b. For $M_{\pi\pi} \simeq 200\ \mathrm{MeV}/c^2$, $b \simeq 15\ (\mathrm{GeV}/c)^{-2}$, and for $M_{\pi\pi} \sim 760\ \mathrm{MeV}/c^2$, $b \simeq 6\text{-}8\ (\mathrm{GeV}/c)^{-2}$. Within these uncertainties, it is meaningless to talk about the shrinking or non-shrinking of the slope b with energy.

- *MILLER:*

In your BNL experiment you said you could reach a sensitivity of $10^{-38}\ \mathrm{cm}^2$. I wonder what is the signal to background ratio?

- *TING:*

The question is: in the experiment to measure e^+e^- pairs from the reaction:

$$p + p \rightarrow V + X, \quad V \rightarrow e^+e^-$$

what is the background? The mass of the electron-positron system, M_{ee}, we are going to observe is between $2 \leq M_{ee} \leq 6$ GeV/c^2, and the mass resolution is ±5 MeV/c^2. To estimate the background, one uses an old measurement done by a Columbia group with protons incident on uranium going to $\mu^+\mu^- + X$. In this experiment, the incident energy is the same and the mass resolution is about 2 GeV/c^2. The sensitivity of our experiment is about a factor of 1000 times more than the Columbia experiment.

- *PATKOS:*

My question concerns the possible background of the search for heavy vector mesons at the BNL experiment in pp scattering. I would like to know how the e^+e^- pairs coming from π^0 decay having large transverse momenta (whose presence was demonstrated at the ISR) can be discarded?

- *TING:*

This is a very good question. The answer is yes, we have looked into this. The results from the ISR have shown that the inclusive cross-section for producing π^0's in pp collisions has an excess at large $p_\perp$ for $p_\perp > 2$ GeV/c from a pure exponential behaviour, and therefore there are a lot of energetic π^0's. The π^0 can decay into a Dalitz pair $\pi^0 \rightarrow \gamma + e^+e^-$, with a branching ratio of 1/80. External γ-ray conversion may be eliminated by making the target window thin enough. The Dalitz decay is always there and is dealt with by placing two special magnets near the target to bend out low-energy electrons or positrons, but does not affect the high-energy electrons or positrons. By placing counters after the magnet, these events may be eliminated by a triple coincidence veto.

- *BERGER:*

Do you know of any measurement of two-photon decays of tensor mesons?

- *TING:*

I do not know of any.

- *ZICHICHI:*

It has been attempted; for instance, the $f_0 \rightarrow 2\gamma$ has been investigated and wide limits on the branching ratio have been established.

- *THIRRING:*

As an experimentalist, could you briefly tell us what is the evidence for, or against, vector dominance in the sense that all interactions of photons with hadrons go via intermediate ρ, ϕ, or ω?

- *TING:*

The question is, how well one can visualize the interaction of photons with hadronic matter as going via a diagram of the type:

For $q^2 < 0$ this is a very bad approximation. For $q^2 = 0$, a subclass of experiments seem to give roughly a 30% confirmation of the vector dominance model if you use ρ, ω, and ϕ. But the approximation is not true at all if you compare the process $\gamma + p \rightarrow \pi^+ + n$, for example. For $q^2 > 0$ there is really not very much information at this moment. For a detailed comparison you compare the amplitude: $A(\gamma p \rightarrow \gamma p)$ to $\Sigma_V(1/\gamma_V)\, A(\gamma + p \rightarrow V + p)$. To make this comparison you must first know the coupling constants γ_ρ, γ_ω, and γ_ϕ. Secondly, you must know the photoproduction amplitudes of the vector mesons. These amplitudes are not well known, and for example, for the ρ meson the amplitude is known only to 30% or 40% and is strongly dependent on the treatment of the background. In one model

$$\left.\frac{d\sigma}{dt}\right|_{t=t_{min}} (\gamma p \rightarrow \rho p) = 105\ \mu b$$

at 2.7 GeV and in another model = 148 μb for the same experiment, and it strongly depends on how you subtract background.

- *NAHM:*

If the ρ' is not a resonance but a dynamical effect, is there any reason why the SLAC linearly polarized beam measurement and the Adone storage ring measurement of the photon-ρ' coupling constant should yield the same result?

- *TING:*

The agreement is only to an accuracy of ±50% which is not so significant.

- *NAHM:*

But if there is no reason why the measured coupling constants should be the same, then there is the possibility, with a more accurate measurement of the photon-ρ' coupling constant, of distinguishing whether the ρ' is a resonance, or a dynamical effect.

- *TING:*

The way a resonance is defined is having a phase shift going through 90°. Therefore, to decide whether the ρ' is a resonance or not is to measure the interference of the leptonic decay of the $\rho' \to e^+e^-$ with the Bethe-Heitler amplitude as a function of the mass of the e^+e^- pair and see if the relative phase goes through 90°

- *WILCZEK:*

The Cheng-Wu calculation of Delbrück scattering you mentioned was to all orders in αZ, and lowest order in α. It does not involve the selective summation of diagrams. It does not therefore tell us if this summation really reproduces the high-energy behaviour of QED. Do you know of any way of testing their predictions to higher-energy QED experimentally?

- *GLASHOW:*

This is a lowest-order electromagnetic calcualtion to all orders in an external electromagnetic field, and is adequate to describe the low-energy experiment described by Prof. Ting. When one does experiments with 500 GeV photons, then one can ask more serious questions about taking into account higher-order corrections.

- *WILCZEK:*

I just wanted to ask an experimentalist if he had an idea of testing the high-energy predictions.

- *TING:*

No.

- *VILELA MENDES:*

You said that the measured branching ratio of the decay $\omega \to 2\pi$ was ten times the expected value. Could you please elaborate on this?

- *TING:*

The ω meson is 3π resonance and therefore it cannot decay to 2π directly, but it can decay to 2π via the diagram:

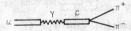

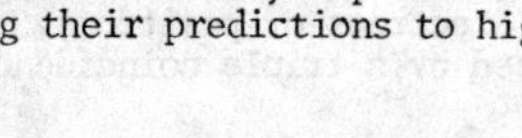

since the ω and ρ masses are very close, and the width of the ρ is of order 100 MeV; and since there is a virtual photon this is an α^2 process, and therefore the branching ratio is $\Gamma_{\omega\to 2\pi} \cong \alpha^2 \cdot 100$ MeV $\simeq 10$ keV. The measured number is a factor of 10 larger. This is, of course, not the whole story, because you can have diagrams of the type

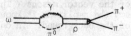

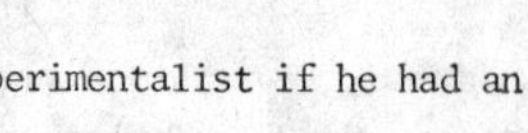

This diagram is difficult to calculate.

- *VILELA MENDES:*

Are there more elaborate models to calculate this?

- *GLASHOW:*

Yes, I have a model where the number is of the right size. One wants a mixing between ρ^0 and ω^0 which is of the typical size of electromagnetic mass differences of the order of several MeV. The mixing can be obtained by the diagram above, or ones involving intermediate vector meson states and photons.

- *NAHM:*

How can you connect the cross-section for $\gamma_V + p \to V^0 + p$ to the total cross-section for virtual photons, if vector meson dominance connects this process to one where one has a space-like incoming photon and an outgoing time-like photon?

- *TING:*

Vector meson dominance connects a process with $q^2 > 0$ to one with $q^2 = 0$. The hypothesis is that this extrapolation is smooth. To obtain the total cross-section one has to continue further to $q^2 < 0$. Of course, the results of this extrapolation are uncertain, since it is known that vector meson dominance is bad for $q^2 < 0$.

- *LOSECCO:*

How does the parton model predict more protons in electroproduction than in photoproduction in the photon fragmentation region?

- *DERMAN:*

The parton model is only valid in the region $-q^2 = Q^2 >> M^2$ and does not apply to photoproduction. Only when $Q^2 >> M^2$ do we have the picture where point-like baryon number carrying partons are kicked out of the proton.

- *WILCZEK:*

Would you comment on the predominance of π^+ over π^- in the photon fragmentation region $\gamma_V + p \to \pi^\pm + X$? Is it evidence for the parton model?

- *TING:*

No, after you subtract for the two-body decay reaction $\gamma_V + p \to \pi^\pm + \Delta^{0++}$, there is no predominance of π^+ or π^- within experimental errors.

- *EYLON:*

Are the SLAC experiments limited to $Q^2 \leq 3$ GeV^2 because of the small cross-sections for $Q^2 > 3$ GeV^2, or because these events cannot be detected for technical reasons?

- *TING:*

Because of the small cross-section.

- *EYLON (Comment):*

Those who believe in the shrinkage of the radius of the virtual photon usually predict that r will decrease as a function of q^2, and therefore the slope b will decrease as a function of q^2. You have shown the decrease of b as a function of $M/4\tau$ which is an "ω"-type variable, and thus indicated that ω is probably a more relevant variable than q^2. In Prof. Sakurai's lecture, I claimed that at least part of the decrease in b is due to a kinematical effect. The kinematical factor was a function of ω, and perhaps the new data you have presented indicate the importance of this kinematical effect.

- *KUPCZYNSKI:*

Can you tell us how one finds the total cross-section for ρp scattering?

- *TING:*

What one must do is study the process $\gamma + A \to \rho^0 + A$ (where A is an atom) for a large number (15) of different nuclei. According to vector dominance, the incident photon is converted into a ρ^0 inside the nucleus, and the ρ^0 subsequently undergoes multiple scattering with the nucleons inside the nucleus. Then, by using the Glauber model, we can compute the $\rho^0 + p$ total cross-section.

- *MELISSINOS (Comment):*

This procedure is very delicate. I am thinking of an experiment which was designed to study the A_1 + p cross-section via the conversion $\pi \rightarrow A_1$. This experiment produced apparently nonsensical results.

- *TING:*

I agree that the method is delicate. But if the experimentalist is careful, one can do it definitely. In our experiment at DESY, we were able to obtain what I think is a reliable ρp total cross-section with a 10% uncertainty.

- *THIRRING (Comment):*

I have the following comment regarding scaling of multiplicities. Whenever you have a smooth function of two variables you can, at least locally, find coordinates such that it depends only on one variable and not on the other (unless you are in a minimum or a maximum where it is flat in any direction). Therefore, I cannot admire scaling as a miracle of nature, unless you can tell me what is so particular about the variable on which she does not depend. You showed evidence for scaling in $\omega = 1 + s/Q^2$, and I wonder whether one can explain the physical significance of this variable.

- *TING:*

I do not know.

THEORETICAL INTERPRETATION OF NEUTRINO EXPERIMENTS

E.A. Paschos

Table of Contents

THEORETICAL INTERPRETATION OF NEUTRINO EXPERIMENTS

E. A. Paschos

National Accelerator Laboratory,*] Batavia, Illinois 60510

I. INTRODUCTION

In the last year new experimental results with neutrino beams have been obtained in several laboratories and their theoretical interpretation is the subject of this article. Experiments with neutrino beams are capable of investigating the structure of hadrons and the nature of weak interactions. The first part of this article deals with deep inelastic scattering. It is shown that the scaling phenomenon for the weak structure functions implies the linear rise of the total cross sections and provides bounds for the ratio $\sigma(\bar{\nu} N \to \mu^+ x)/\sigma(\nu N \to \mu^- x)$. Recent determinations of this ratio indicate that it is near the lowest bound allowed by scaling, which, in turn, has several consequences. A comparison of such detailed information with theoretical expectations reveals remarkable consistency between theory and experiment.

The second part deals with neutral currents. Experimental search for neutral currents over the last year established not only new bounds for several processes, but also credible candidates for some of the reactions. A survey of existing results is presented within the context of gauge models of leptonic and semileptonic interactions.

II. TOTAL CROSS SECTIONS

The processes that we are dealing with are shown schematically in Fig. 1. The process is described in the laboratory frame. An incident neutrino with energy E hits a nucleon at rest, leading to a final muon with energy E' and a final hadronic state with momentum P_n. When we sum over all final hadronic states, the process depends on three kinematic variables:

E : incident energy

$\nu = E - E'$: energy transfer

$q^2 = -Q^2 = -4EE'\sin^2\theta/2$: square of the momentum transfer.

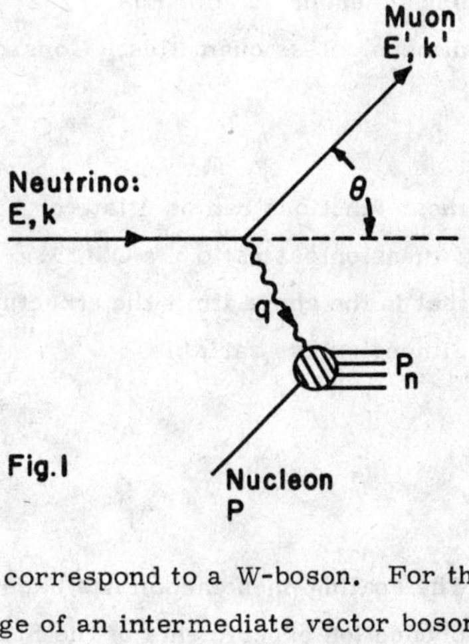

Fig. I

The explicit functional form of the leptonic vertex is known from the effective current-current interaction Lagrangian. The wavy line indicates an exchange force, and may or may not correspond to a W-boson. For the remaining of this article we do not assume the exchange of an intermediate vector boson, unless otherwise stated. All the interesting structure is hidden in the hadronic vertex.

*] Operated by Universities Research Association, Inc. Under contract with the United States Atomic Energy Commission.

The hadronic vertex describes the absorption of a current by a hadron. Since in the experiments the targets are unpolarized, there is no dependence on the spin of the target. The current, however, is a superposition of helicity states. For a space-like current there are three polarization states. The unknown structure functions for the hadronic vertex can be chosen as three total cross sections, corresponding to the absorption of a right-handed, left-handed and scalar current, denoted respectively by

$$\sigma_R(Q^2,\nu),\quad \sigma_L(Q^2,\nu) \text{ and } \sigma_S(Q^2,\nu) \tag{2-1}$$

The double differential cross section[1-3] for incident neutrinos is

$$\frac{d\sigma^\nu}{dQ^2 dE'} = \frac{G^2}{2\pi}\frac{E'}{E} W_2(Q^2,\nu)\{1 + \frac{\nu}{E'}(L) - \frac{\nu}{E}(R)\} \tag{2-2}$$

where

$$\binom{L}{R} = \frac{\binom{\sigma_L}{\sigma_R}}{2\sigma_S + \sigma_L + \sigma_R} \text{ and } F_2(x) = \frac{1}{2\pi} Q^2 \frac{(1-Q^2/2M\nu)}{(1+Q^2/\nu^2)}(2\sigma_S + \sigma_L + \sigma_R) \tag{2-3}$$

The corresponding formula for antineutrinos is

$$\frac{d\sigma^{\bar\nu}}{dQ^2 dE'} = \frac{G^2}{2\pi}\frac{E'}{E}\overline{W_2(Q^2,\nu)}\left\{1 + \frac{\nu}{E'}(\bar R) - \frac{\nu}{E}(\bar L)\right\} \tag{2-4}$$

The bar over the structure functions indicates that in general they are different from those in Eq. (2-2).

A crucial assumption for the remaining of this discussion is Bjorken's scaling phenomenon.[4] From Eqs. (2-2) and (2-3) we observe that $\nu W_2(Q^2,\nu)$, (R) and (L) are dimensionless quantities. Consequently in the limit

$$Q^2 \to \infty \text{ with } Q^2/2M\nu = \text{finite} \tag{2-5}$$

these functions can oscillate or approach zero, infinity, or a non-trivial function of the dimensionless ratio $x = Q^2/2M\nu$. A few years ago Bjorken remarkably predicted[4] that in the above limit the structure functions approach non-trivial functions of a single dimensionless variable

$$\nu W_2(Q^2,\nu) \to F_2(x)$$
$$\binom{R}{L} \to f_{R,L}(x) \tag{2-6}$$

The scaling phenomenon has been observed for limited ranges of Q^2 and ν in the electro-production experiments of the SLAC-MIT group.[5] The pleasant surprise is that the structure functions approach this limit rather fast. It settles in for values of $Q^2 \geq 2(\mathrm{GeV}/c)^2$. Such tests will be extended to larger ranges of Q^2 and ν in the NAL experiments.[6] At this time there is no direct test of scaling in neutrino induced reactions. It has been shown, however, that the conventional theory of weak interactions together with the scaling hypothesis lead to numerous consequences whose verification provide indirect tests of scaling. We discuss these consequences in this and the following section.

Theorem: 1) If all three structure functions scale[4] then

$$\sigma^{\nu} \underset{E\to\infty}{\to} CE_{\nu}$$
$$\sigma^{\bar{\nu}} \underset{E\to\infty}{\to} C'E_{\bar{\nu}} \qquad (2\text{-}7)$$

2) For targets[3] with equal numbers of protons and neutrons (isoscalar) the scaling of all three structure functions implies

$$1/3 \leq \sigma^{\bar{\nu}}/\sigma^{\nu} \leq 3 \qquad (2\text{-}8)$$

Proof: (i) Integrating over Q^2 and appealing to scaling

$$\frac{d\sigma}{dE'} = \frac{G^2}{2\pi}\frac{E'}{E}\, 2M \left\{\int F_2(x)\frac{dQ^2}{2M\nu}\right\}\left\{1 + \frac{\nu}{E'}\langle L\rangle - \frac{\nu}{E}\langle R\rangle\right\} \qquad (2\text{-}9)$$

where

$$0 \leq \langle L, R\rangle \equiv \frac{F_2(x)\,(L,R)\,dx}{F_2(x)\,dx} \leq 1 \qquad (2\text{-}10)$$

Thus scaling decouples the integrations of x and E', so that the dependence in E' is explicitly exhibited. Integrating over E'

$$\sigma^{\nu} = \frac{G^2}{2\pi}\, 2ME\left\{\int F_2(x)\,dx\right\}\left\{\frac{1}{2} + \frac{1}{2}\langle L\rangle - 1/6\,\langle R\rangle\right\} \qquad (2\text{-}11)$$

Similarly for antineutrinos.

(ii) For isoscalar targets

$$F_2(x) = \bar{F}_2(x),\ \langle \bar{L}, \bar{R}\rangle = \langle L, R\rangle \qquad (2\text{-}12)$$

by charge symmetry. Therefore

$$\frac{\sigma^{\bar{\nu}}}{\sigma^{\nu}} = \frac{\frac{1}{2} + \frac{1}{2}\langle R\rangle - 1/6\,\langle L\rangle}{\frac{1}{2} + \frac{1}{2}\langle L\rangle - 1/6\,\langle R\rangle}. \qquad (2\text{-}13)$$

where $0 \leq \langle L\rangle \leq 1$, $0 \leq \langle R\rangle \leq 1$ and $\langle 2S\rangle + \langle L\rangle + \langle R\rangle = 1$. Equation (2-8) now follows with the lower limit corresponding to $\langle L\rangle = 1$ and $\langle R\rangle = 0$.

Figure (2) shows the measurements from the Gargamelle collaboration.[7] The cross sections are consistent with a linear rise. The statistics are too limited to provide undisputed evidence in favor of the linear rise. Consequently tests of other consequences of scaling are desirable. Figure (3) shows the ratio of the cross sections. If we assume that the total cross sections rise linearly with energy starting at 2 GeV, then their ratio is determined with good accuracy:

$$\frac{\sigma^{\bar{\nu}}}{\sigma^{\nu}} \Big/ \exp = 0.38 \pm 0.02 \tag{2-14}$$

It is close to the lowest bound allowed by scaling. Preliminary results at higher energies from NAL[8] are also consistent with the interpretation that the ratio of the total cross sections is in the neighborhood of 1/3. Figure 4 shows the NAL point.

The simplicity of the theorem is not indicative of the stringent constraints that it implies, because semileptonic interactions are not restricted by the bounds which are valid in hadronic interactions. For instance, Froissart's theorem[9] requires that hadronic total cross sections can grow at most like $(\ell n E)^2$. This theorem, however has no implications for semileptonic reactions because two of the basic assumptions required in the proof the theorem do not hold for semileptonic reactions. Namely, Froissart's theorem is based on

(i) absence of zero mass particles

(ii) quadratic unitarity

both of which are absent in semileptonic reactions, because we do have zero mass particles and unitarity is linear.

In addition, the Pomeranchuck theorem[10] refers to the hadronic part of the diagram and the ratio $\sigma^{\bar{\nu}N}/\sigma^{\nu N}$ can be different from unity at very high energies.

III. FLUX INDEPENDENT MOMENTS AND THEIR SIGNIFICANCE

In analyzing neutrino experiments one is faced with several intrinsic problems: (i) difficulties in determining the neutrino flux, (ii) uncertainties in determining the initial energy and (iii) limited statistics for specific regions of phase space. It is of interest, therefore, to ask whether it is possible to reformulate consequences of fundamental principles (locality, charge symmetry, scaling, ...) in terms of quantities which are flux and perhaps energy independent. To achieve this objective we can define mean values of the form

$$\langle f(Q^2, E\mu)\rangle \equiv \frac{1}{\sigma_{tot}} \int f(Q^2, E\mu) \frac{d}{dQ^2 dE\mu} dQ^2 dE\mu \tag{3-1}$$

where $f(Q^2, E\mu)$ can be chosen to be E'/E, $Q^2/2ME = \frac{2E'}{M} \sin^2 \frac{\theta}{2}$, ... Such mean values are useful even with limited statistics since they average over regions of phase space. They are obviously flux independent. In addition, by invoking the scaling hypothesis and utilizing available data we shall show that at high energies they approach constant limits which are accurately determined.

We examine first the mean energy[11] carried by the muon

$$\langle\frac{E'}{E}\rangle = \frac{1}{\sigma_{tot}}\int\left(\frac{E'}{E}\right)\frac{d\sigma}{dQ^2 d\nu}\,dQ^2 d\nu \tag{3-2}$$

The scaling of all three structure functions implies

$$\frac{1}{2} \leq \langle\frac{E'}{E}\rangle_\nu \leq \frac{3}{4} \tag{3-3}$$

where the subscript ν refers to neutrinos. This result is obtained readily by following the same line of reasoning as that of the theorem in the previous sections. After integration over Q^2 and E' one obtains

$$\langle\frac{E'}{E}\rangle_\nu = \frac{\frac{1}{3}+\frac{1}{6}\langle L\rangle - \frac{1}{12}\langle R\rangle}{\frac{1}{2}+\frac{1}{2}\langle L\rangle - \frac{1}{6}\langle R\rangle} \tag{3-4}$$

where the mean values on the right hand side were defined in (2-10). As a result we find the limits of (3-3), with the upper bound corresponding to $\langle R\rangle = 1$, $\langle L\rangle = 0$ and the lower bound to $\langle R\rangle = 0$, $\langle L\rangle = 1$.

In order to study the sensitivity of such bounds on the underlying assumptions, we can proceed in two different directions. First we relax the scaling hypothesis. Consider the case where only $F_2(x) = \nu W_2$ scales and allow for the possibility that (L) and (R) do not scale. Then

$$\langle\frac{E'}{E}\rangle = \frac{\frac{1}{3}+\frac{1}{6}\langle\tilde{L}\rangle - \frac{1}{12}\langle\tilde{R}\rangle}{\frac{1}{2}+\frac{1}{2}\langle L\rangle - \frac{1}{6}\langle R\rangle} \tag{3-5}$$

where $\langle\tilde{L}\rangle$ and $\langle\tilde{R}\rangle$ are again less than unity but independent of $\langle L\rangle$ and $\langle R\rangle$. The corresponding bounds now are

$$\frac{1}{4} \leq \langle E'/E\rangle \leq 1 \quad \text{(Scaling of } \nu W_2 \text{ only)}. \tag{3-6}$$

It is worth emphasizing here the close analogy of the bounds obtained above with the bounds obtained for the ratio of the total cross sections. If all structure functions scale, then the total cross section rises linearly with energy and the ratio of cross sections is a constant. If only νW_2 scales, then the ratio of cross sections is again bounded but it is not required to be a constant.

Proceeding in a different direction we supplement the scaling hypothesis with existing data and try to obtain more restrictive bounds. In view of recent data, it seems reasonable to consider bounds in the case when the ratio of the cross sections is close to 1/3:

$$\frac{\sigma^{\bar{\nu}}}{\sigma^{\nu}} = \frac{1}{3}(1+\epsilon), \quad \epsilon \ll 1. \tag{3-7}$$

Equations (2-13), (3-7) and the trivial identity

$$\langle R\rangle + \langle L\rangle + 2\langle S\rangle = 1 \tag{3-8}$$

imply the constraint equation

$$\frac{\langle R\rangle}{\langle L\rangle} + \frac{3\langle S\rangle}{4\langle L\rangle} = \frac{3}{8}\epsilon + 0(\epsilon^2). \tag{3-9}$$

Maximizing and minimizing (3-4) subject to the constraint equation we obtain [11, 12]

$$\frac{1}{2} + \frac{1}{32}\epsilon \le \langle\frac{E'}{E}\rangle_\nu \le \frac{1}{2} + \frac{1}{12}\epsilon . \tag{3-10}$$

Similar arguments can be carried out for antineutrino induced reactions [11, 12]

$$\frac{3}{4} - \frac{9}{32}\epsilon \le \langle\frac{E'}{E}\rangle_{\bar{\nu}} \le \frac{3}{4} - \frac{\epsilon}{8} . \tag{3-11}$$

Experimentally the mean values have been determined [13] to be

$$\langle\frac{E'}{E}\rangle_\nu = 0.55 \pm 0.10 \text{ and } \langle\frac{E'}{E}\rangle_{\bar{\nu}} = 0.69 \pm 0.09 \tag{3.12}$$

in good agreement with the theoretical expectations.

The same analysis has been extended to other quantities. [11, 12, 14] Of particular interest is the mean value

$$\langle\frac{Q^2}{2ME}\rangle_\nu \equiv \langle\frac{2E'}{M}\sin^2\frac{\theta}{2}\rangle \equiv \langle z\rangle \equiv \langle xy\rangle \tag{3-13}$$

because it is determined by the energy and angle of the outgoing lepton. It has been known for some time, that scaling implies a linear energy dependence of the mean value of Q^2 at high energies. It also implies the bound

$$0 \le \langle\frac{Q^2}{2ME}\rangle_{\nu,\bar{\nu}} \le \frac{3}{5} \tag{3-14}$$

Furthermore, combining scaling with the ratio of the total cross sections we again obtain restrictive bounds [11]

$$2\left(1 + \epsilon - \frac{35}{16}\frac{\epsilon}{\langle x\rangle}\right) \le \frac{\langle\frac{Q^2}{2ME}\rangle_\nu}{\langle\frac{Q^2}{2ME}\rangle_{\bar{\nu}}} \le 2(1 + \epsilon) \text{ provided } \epsilon \ll \langle x\rangle . \tag{3-15}$$

Experimentally the mean values of $\langle Q^2\rangle$ have been determined in the Gargamelle experiment. [7] Figure 5 shows the mean value of $\langle Q^2\rangle$ plotted as a function of the neutrino energy; while Fig. 6 shows the corresponding curve for antineutrinos. The results of linear fits are:

neutrino $\langle Q^2\rangle = 0.12 \pm 0.03 + (0.23 \pm 0.01)\,E$

antineutrino $\langle Q^2\rangle = 0.09 \pm 0.03 + (0.14 + 0.015)\,E$

} $E > 1$ GeV

$$\left.\begin{array}{ll}\text{neutrino} & \langle Q^2\rangle = (0.22 \pm 0.06) + (0.21 \pm 0.02)E \\ \text{antineutrino} & \langle Q^2\rangle = (0.11 \pm 0.08) + (0.14 \pm 0.03)E\end{array}\right\} E > 2\ \text{GeV}$$

The agreement with the theoretical expectation (3-15) is again good.

It has been indicated in this section that the scaling hypothesis combined with charge symmetry leads to several predictions, which can be compared with existing experimental data. So far all such comparisons reveal remarkable agreement between theory and experiment. This situation, however, may change at higher energies. De Rújula and Glashow[12] have emphasized that in addition to the conventional Cabibbo current there may exist another component which is isoscalar and changes charm, a conjectured new quantum number conserved by strong and electromagnetic interactions. By universality these two components may have the same coupling constant, but only the Cabibbo current is effective below the threshold for the production of charmed states. Consequently, one expects violations of both charge symmetry and scaling at the threshold of charmed states. In addition one expects modifications of the sum rules.[15]

IV. FURTHER ANALYSIS OF THE DATA

The constraint equation has two further consequences.

1) $$\langle S\rangle/\langle L\rangle \leq \frac{1}{2}\epsilon \ (\approx 0.06) \tag{4.1}$$

in agreement in the Callan-Gross relation[16] and corresponding ratios in electroproduction. For comparison the electroproduction ratios[17] are

$$\frac{\sigma_S}{\sigma_T} = 0.14 \pm 0.10 \quad \text{proton} \tag{4.2}$$

$$\frac{\sigma_S}{\sigma_T} = 0.15 \pm 0.08 \quad \text{deuteron} \tag{4.3}$$

2) $$\langle R\rangle/\langle L\rangle \leq \delta = (3/8)\epsilon \tag{4.4}$$

This relation may be useful in testing the parton (light cone) relation

$$W_2(V) = W_2(A) \tag{4.5}$$

where V and A indicate the contributions arising from the vector and axial currents respectively. It has been shown[11] from kinematic arguments that

$$1-4\delta^{\frac{1}{2}} + 0(\delta) \leq \frac{\int F_2(V)\,dx}{\int F_2(A)\,dx} \leq 1 + 4\delta^{\frac{1}{2}} + 0(\delta) \tag{4.6}$$

Present data suggest $\delta \simeq 0.05$, so that the above ratio is consistent with the value of one, but Eq. (4.6) is not very restrictive. An accurate determination of such a ratio is rather difficult. The ratio is important, however, because together with the Conserved Vector

Current hypothesis determines the isovector contribution to electroproduction and consequently the isoscalar part.

3) Under slightly different assumptions we can determine from the data two basic integrals. The slopes of the cross sections determine the integral

$$\int F_2^{\nu N}(x)\,dx \simeq 0.47 \pm 0.07 \tag{4.7}$$

where N denotes the average value per nucleon. The slope of $\langle Q^2\rangle$ as a function of the incoming energy determines the next moment

$$\int x\,F_2^{\nu N}(x)\,dx \simeq \left\{\langle Q^2/2ME\rangle_\nu + \frac{\sigma^{\bar\nu}}{\sigma^{\nu}}\langle Q^2/2ME\rangle_{\bar\nu}\right\}\frac{12}{7}\int F_2^{\nu N}(x)\,dx \tag{4.8}$$

$$\approx 0.12$$

To obtain these two results one needs only the scaling of $F_2(x)$ and the conditions $\frac{\sigma_S}{\sigma_R+\sigma_L} \simeq 0$, $\langle R\rangle/\langle L\rangle \approx 0$ suggested by (4.1) and (4.4).

The above integrals can be compared with corresponding integrals in electroproduction. Such comparisons are made by virtue of the following two properties:

(i) $W_2(V) = W_2(A)$

(ii) The parton (light-cone) suggestion that the isoscalar contribution to $F_2^{\gamma p} + F_2^{\gamma n}$ is less than 10%. It then follows

$$4\left[F_2^{\gamma p} + F_2^{\gamma n}\right] \approx F_2^{\nu p} + F_2^{\nu n} \tag{4.9}$$

where the approximate sign indicates the ambiguity associated with the isoscalar contribution. The relation is expected to hold to within 10-20%. Table (1) summarizes the comparisons between electron and neutrino induced reactions. We finally notice that all such comparisons are in agreement with the predictions of the quark-parton model.

Table 1

Feature	Electrons	Neutrinos
Scaling	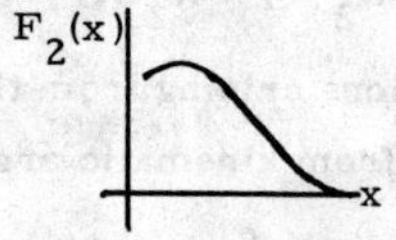	$\sigma^{\nu} \to C \cdot E_\nu$
Spin 1/2	$\sigma_S/\sigma_T = 0.14 \pm 0.10$	$\frac{\langle S\rangle}{\langle L\rangle} \leq \frac{1}{2}\epsilon \approx 0.06$
Momentum carried by Antiparticles	----	$\sigma_R \approx 0 \to \frac{\sigma^{\bar\nu}}{\sigma^{\nu}} \approx \frac{1}{3}$
$\int F_2(x)\,dx$	0.52 ± 0.08	0.47 ± 0.07
$\int xF_2(x)\,dx$	~ 0.12	~ 0.11

V. NEUTRAL CURRENTS

One of the most pleasant aspects of this field is the many implications that it has for other problems of high energy physics. In 1960 Lee and Yang[18] compiled a list of unresolved problems of weak interactions; shown in Table 2.

Table 2

Questions raised by Lee and Yang [Phys. Rev. Letters 4, 307 (1960)]	Experimental answers from ν experiments
1) $\nu_\mu = \nu_e$?	$\nu_\mu \neq \nu_e$
2) Lepton conservations	See Ref. 20
3) Neutral Currents?	See text
4) "Locality"(vector nature of weak interactions)	
5) Universality between ν_μ and ν_e, μ and e?	
6) Charge symmetry?	
7) CVC; isotriplet current?	
8) W?	
9) What happens at high energy ($E_\nu \rightarrow$ "unitarity limit")?	

In the intervening years, a good deal of research has gone in resolving these problems. The question of the two neutrinos has been answered satisfactorily by the discovery of two neutrinos.[19] Lepton number conservation[20] has been tested to some degree of accuracy. We discuss next the progress made in the past few years with regards to the third question.

The revived interest on neutral currents arose from the possibility of constructing a renormalizable gauge theory of weak (and electromagnetic) interactions.[21] Several models have been proposed which can achieve this goal at the expense of introducing neutral currents. Originally, the theories were concerned with leptonic interactions. Subsequently, they were generalized to account, by virtue of universality, for semi-leptonic reactions. We review here the present experimental status together with the corresponding theoretical predictions.

Leptonic Interactions

Models based on the symmetry group SU(2) × U(1) contain a neutral current operator

$$J^{\ell}_{\mu} = \frac{-ig}{\sin\theta_W}\bar{\psi}_L\gamma_\mu(t_3 - Q\sin^2\theta_W)\psi_L \tag{5.1}$$

where $Q = t_3+y$, $e=g\sin\theta_W$, ψ_L a left-handed spinor of a multiplet and $\vec{t}$, y are the weak isospin and hypercharge, respectively. In models where the left-handed leptons belong to a vector,[22, 23] the neutral current decouples completely from the neutrino. When the left-handed leptons belong to a spinor, neutral currents appear[23] in neutrino induced reactions.

A prototype of the latter case is the Weinberg-Salam model, where the effective part of the Lagrangian pertinent to leptonic reactions is

$$\mathscr{L}_L = \frac{G}{\sqrt{2}}\left\{\bar{\nu}\gamma_\mu(1+\gamma_5)\nu\,\bar{e}\gamma^\mu(g_V+g_A\gamma_5)e\right\} \tag{5.2}$$

The effect of the neutral current is to change the values of g_V and g_A from those of the (V-A) theory. For the reactions

$$\nu_\mu e^- \rightarrow \nu_\mu e^- \tag{5.3}$$

$$\bar{\nu}_\mu e^- \rightarrow \bar{\nu}_\mu e^- \tag{5.4}$$

the differential cross section per unit energy of the recoil electron has the form

$$\frac{d\sigma}{dE'} = \frac{G^2 m}{2\pi}\left[(g_V+g_A)^2 + (g_V-g_A)^2\left(1-\frac{E'}{E}\right)^2 + \frac{mE'}{E^2}(g_A^{\,2}-g_V^{\,2})\right] \tag{5.5}$$

where E and E' are the laboratory energies of the incident neutrino and recoiling electron and m is the electron mass.

Searches for the processes (5.3) and (5.4) were made in the Gargamelle experiment. One "good" candidate for reaction (5.4) was found in the antineutrino film satisfying the selection criteria. The probability that this event is due to non-neutral current background is less than 3%. The same experiment also set new upper limits for the cross sections

$$\sigma(\nu_\mu e^- \rightarrow \nu_\mu e^-) \leq 0.26\times 10^{-41}\ E_\nu,\ \text{cm}^2 \tag{5.6}$$

$$\sigma(\bar{\nu}_\mu e^- \rightarrow \bar{\nu}_\mu e^-) \leq 0.88\times 10^{-41}\ E_{\bar{\nu}},\ \text{cm}^2 \tag{5.7}$$

at 90% confidence level. Comparison with the Weinberg-Salam model provides the bounds

$$0.10 \leq \sin^2\theta_W \leq 0.60 \tag{5.8}$$

In a different experiment Gurr, Reines and Sobel[26] use a $\bar{\nu}_e$ beam from a nuclear reactor and search for the reaction $\bar{\nu}_e e^- \rightarrow \bar{\nu}_e e^-$. They observe only a region of the phase space. When their results are translated into total cross sections they imply

$$\frac{\sigma(\bar{\nu}_e+e^- \rightarrow \bar{\nu}_e+e^-)/\ \text{exp}}{\sigma(\nu_e+e^- \rightarrow \bar{\nu}_e+e^-)\ /\ \text{V-A}} \leq 3 \quad \text{(to better than 90\% c.l.)} \tag{5.9}$$

An analysis[27] of this experiment in terms of the Weinberg-Salam model provides the bound

$$\sin^2\theta_W \leq 0.40 \tag{5.10}$$

Semileptonic Interactions:

The term of the effective Lagrangian relevant for semileptonic interactions has the form:

$$\mathscr{L}_{SM} = \frac{G}{\sqrt{2}} \bar{\mu}\gamma^\alpha(1+\gamma_5)\nu(J_\alpha^1+iJ_\alpha^2) + \text{h.c.} + \bar{\nu}\gamma^\alpha(1+\gamma_5)\nu J_\alpha^{(0)} \tag{5.11}$$

with the hadronic neutral current given by

$$\begin{aligned} J_\alpha^{(0)} &= J_\alpha^3 + yJ_\alpha^{em} + zJ_\alpha^s \\ &= A_\alpha^3 + (1+y)\ V_\alpha^3 + J_\alpha^{'s} \end{aligned} \tag{5.12}$$

where J_α^3 is the third component of isospin for the usual weak current

J_α^{em} is the electromagentic current and

J_α^s is an isoscalar current

This form of the hadronic current is representative of several models discussed in the literature. Ignoring strange particles altogether, an extension[28] of the Weinberg-Salam model to hadrons is obtained by identifying $\frac{1}{2}(1+\gamma_5)\binom{p}{n}$ as the left-handed nucleon doublet. Then proceeding as in the leptonic case a neutral current is introduced with $z = 0$ and $y = -2\sin^2\theta_W$. Such a model is obviously unrealistic. The usual quark picture with three quarks (p, n, λ) is also unrealistic, because $\Delta s = 1$ neutral currents are unavoidable. A general solution to this problem is to introduce more quarks[28] and avoid the $\Delta s = 1$ neutral currents by adopting the Glashow-Iliopoulos-Maiani scheme.[29] Several models fall into this category[30] with $y = -2\sin^2\theta_W$ and $z \neq 0$. In order to bound the parameter $\sin^2\theta_W$, we appeal to universality and argue that it is the same parameter which occurs in purely leptonic reactions. The bound (5-10) from the Gurr-Reines-Sobel experiment is, at the moment, the most restrictive and will be used in the subsequent numerical estimates.[31]

We now compare the theoretical predictions with the experimental bounds.

Total Cross Sections

Neutral current candidates in the total cross sections have been observed in the CERN experiment. The events behave as if they arise from neutral current processes induced by neutrinos and antineutrinos. When all the Gargamelle[32] events with only hadrons in the final state and no visible muon are attributed to neutral currents they lead to the ratios

$$R \equiv \frac{\sigma(\nu + \text{freon} \to \nu + x_1)}{\sigma(\nu + \text{freon} \to \mu^- + x_2)} = 0.21 \pm 0.03 \tag{5.13}$$

and

$$R \equiv \frac{\sigma(\bar{\nu} + \text{freon} \rightarrow \bar{\nu} + x_3)}{\sigma(\bar{\nu} + \text{freon} \rightarrow \mu^+ + x_4)} = 0.50 \pm 0.09 \tag{5.14}$$

Within the Weinberg-Salam model the contribution of the neutral currents can be bounded from below. For isoscalar[33, 34] targets

$$R = \frac{[\sigma(\nu+p \rightarrow \nu+X_1) + \sigma(\nu+n \rightarrow \nu+X_2)]}{[\sigma(\nu+p \rightarrow \mu^- + X_3) + \sigma(\nu+n \rightarrow \mu^- + X_4)]} \geq \frac{1}{6}(1+x+x^2) \tag{5.15}$$

and similarly[34]

$$R = \frac{[\sigma(\bar{\nu}+p \rightarrow \bar{\nu}+X_1) + \sigma(\bar{\nu}+n \rightarrow \bar{\nu}+X_2)]}{[\sigma(\bar{\nu}+p \rightarrow \mu^+ + X_3) + \sigma(\bar{\nu}+n \rightarrow \mu^+ + X_4)]} \geq \frac{1}{2}(1-x+x^2) \tag{5.16}$$

where $x = 1 - 2\sin^2\theta_w$. The experiments however are not on isoscalar targets, but since most of the contribution to the cross section comes from large values of Q^2 and comparable values of ν it is safe to assume that the process is incoherent. Then defining $R(Z,A)$ in analogy to the R occurring in (5.15), but on a nucleus with Z protons and (A-Z) neutrons we obtain

$$R(Z, A) \geq \frac{Z}{A-Z} R \geq 0.17 \tag{5.17}$$

The bounds (5.15) and (5.16) should be rather restrictive. When the isoscalar contribution of the neutral current is zero and the V-A interference maximal, (5.15) and (5.16) become equalities. The fact that the V-A interference is almost maximal is confirmed by the ratio of the charged total cross sections. Estimates of the isoscalar contribution improves the bounds, as it is verified by studies[35] of specific models. The bounds from two model calculations are shown in Fig. 7 together with the bounds (5.15) and (5.16). Figure 8 shows the model predictions of $d\sigma/dy$ as a function $y = \frac{\nu}{E_\nu} = \frac{E_{hadrons}}{E_\nu}$.

Single π^o production

There are two experimental upper bounds for the ratio

$$R_1 = \frac{\sigma(\nu + ''p'' \rightarrow \nu p \pi^o) + \sigma(\nu''n'' \rightarrow \nu n \pi^o)}{2\sigma(\nu''n'' \rightarrow p\pi^o\mu^-)} \leq 0.14$$

BNL-Columbia[36]
(W. Lee)

≤ 0.21 (5.18)

Gargamelle[37]

It is important to notice that protons and neutrons in the target are not free but they are bound in nuclei. Consequently charge exchange effects are important[38] and can introduce significant contributions to theoretical estimates of R_1.

There are two types of calculations available at this moment. B. W. Lee[39] calculated a bound in the static model assuming I = 3/2 dominance. The bound obtained is considerably larger than the values allowed by experiment. A more recent calculation by Adler[40] incorporates both I = 3/2 and I = 1/2 contributions. He finds that the I = 1/2 contribution does not substantially modify the static model results. This class of calculations is concerned with free protons and neutrons and should be supplemented by corrections arising from charge exchange effects, before a meaningful comparison can be made with experiment.

In a different approach one considers the scattering from an isospin zero nucleus and describes the final states in terms of the isospin of the resulting nucleus and a pion. In this manner all final state interactions are included, except for electromagnetic effects. The resulting formula is

$$R_1 \geq \frac{1}{4}\left[(r-1)^{\frac{1}{2}} - 2\sin^2\theta_w\left(\frac{V^0_{e.m.}}{\sigma(\nu N \rightarrow \mu^- \pi^0 x_3)}\right)^{\frac{1}{2}}\right]^2 \qquad (5.19)$$

where

$$r = \frac{\sigma(\nu + N \rightarrow \pi^+ \mu + x_1) + \sigma(\nu + N \rightarrow \mu^- + \pi^- + x_2)}{\sigma(\nu N \rightarrow \mu^- \pi^0 x_3)}$$

For numerical estimates one must know the electroproduction of π^0's in nuclei. Data for the electroproduction of π^0's in nuclei is not yeat available and V^0_{em} was estimated making generous allowances for the uncertainties. In addition we need the ratio r. For $2 \leq r \leq 5$ bounds are obtained which vary from 0.07 to 0.44

All other experimental bounds established so far are also consistent with the theoretical expectations of the Weinberg-Salam model. Their status has been reviewed[11] recently and has not changed since that time. Table 3 presents an overview of the present situation. The most striking feature of the table is the proximity of theoretical bounds with the experimental observations. This suggests that new measurements during the next year will supply decisive information.

In summary, high energy neutrino experiments provide an ideal means for probing (i) the structure of hadrons and (ii) the nature of weak interactions. There are numerous experimental measurements in the deep inelastic region, all of which are consistent with the scaling hypothesis and also with the predictions of the quark-parton model. Among the unresolved questions of weak interactions, special attention was paid to that of the existence of neutral currents. A summary of the present situation indicates

Table 3

Comparison between Experimental and Theoretical Bounds for Neutral Current Reactions

Ratio	Experiment (bounds at 90% C. L.)	Theory
$\dfrac{\sigma(\nu N \to \nu x')}{\sigma(\nu N \to \mu^- x'')}$	0.21 ± 0.03	≥ 0.19
$\dfrac{\sigma(\bar{\nu} N \to \bar{\nu} x')}{\sigma(\bar{\nu} N \to \mu^+ x'')}$	0.50 ± 0.09	≥ 0.32
$\dfrac{\sigma(\nu p \to \nu p \pi^0) + \sigma(\nu n \to \nu n \pi^0)}{2\sigma(\nu n \to \mu^- p \pi^0)}$	≤ 0.14 W. Lee ≤ 0.21 Gargamelle	≥ 0.44 to 0.07
$\dfrac{\sigma(\nu p \to n\pi^+) + \sigma(\nu p \to \nu p \pi^0)}{\sigma(\nu p \to \mu^- \Delta^{++})}$	≤ 0.46 Cundy, et al. ≤ 0.31 ANL	≥ 0.10
$\dfrac{\sigma(\nu p \to \nu p)}{\sigma(\nu n \to \mu^- p)}$	≤ 0.24	$0.15 \leq R \leq 0.25$
$\dfrac{\sigma(\nu p \to \nu n \pi^+)}{\sigma(\nu p \to \mu^- \Delta^{++})}$	≤ 0.16	≥ 0.03

that experiments to be performed within a year have the capability of either confirming their existence or eliminating a class of theoretical models. Other questions raised in Table 2 also call for close attention. Progress on several of them has been reported[42] during this summer and considerable improvements are expected soon.

ACKNOWLEDGMENTS

I am very grateful to J. Prentki and B. Zumino for their hospitality at CERN where part of this work was performed.

REFERENCES

1) T.D. Lee and C.N. Yang, Phys. Rev. 126, 2239 (1962).

2) F.J. Gilman, Phys. Rev. 167, 1365 (1968); C.H. Llewellyn-Smith, Phys. Rep. 3C.

3) J.D. Bjorken and E.A. Paschos, Phys. Rev. D1, 3151 (1970).

4) J.D. Bjorken, Phys. Rev. 179, 1547 (1969).

5) G. Miller et al. Phys. Rev. D5, 528 (1972); A. Bodek Ph.D. thesis, M.I.T. Report No. COO-3069-116 (1972).

6) See the NAL experiment E26.

7) T. Eichten et al., CERN preprint (1973).

8) A. Benvenuti, Phys. Rev. Letters 21, 1084 (1973).

9) M. Froissart, Phys. Rev. 123, 1053 (1961). Arguments based on a limited number of partial waves do not apply either because the number of partial waves in the current hadron system is unlimited. See also V.I. Zakharov, in Proceedings of the XVI International Conference on High Energy Physics, (edited by J.D. Jackson), A. Roberts and R. Donaldson, NAL (1973).

10) I. Pomeranchuck, Zh. Eksperim. i Theor. Fiz 34, 725 (58); [English trans. Soviet Phys.-JETP, 7, 499 (58)].

11) E.A. Paschos and V.I. Zakharov, Phys. Rev. D8, 215 (1973). See also E.A. Paschos NAL - preprint NAL-Conf.-73/27 - THY, 1973.

12) The terms $\frac{1}{32}\ \epsilon \approx 0.004$ and $\frac{\epsilon}{8} \approx 0.015$ in (3-10) and (3-11) are improvements on the bounds of Ref. 11. They are due to A. De Rújula and S.L. Glashow, Harvard preprint (1973).

13) P. Musset, APS Meetings, New York (1973).

14) J. D. Bjorken, D. Cline and A.K. Mann, SLAC-PUB-1244 (1973). This article also discusses limits which may be useful in making neutrino-antineutrino comparisons.

15) Reevaluation of the Adler and Gross-Llewellyn-Smith sum rules in the Han-Nambu and other models reveals significant variations.

16) C.G. Gallan and D.J. Gross, Phys. Rev. Letters 22, 156 (1969).

17) E.D. Riordan, Ph. D. Thesis, MIT (1973). See also Ref. 5.

18) T.D. Lee and C.N. Yang, Phys. Rev. Letter 307 (1960).

19) G. Danby et al., Phys. Rev. Letter 9, 36 (1962).

20) For a review see H. Primakoff in Lectures in Particles and Quantum Field Theory, Vol. 2 (MIT press 1970).

21) For a summary see B.W. Lee in Proceedings of the XVI International Conference on High Energy Physics, NAL (1973).

22) Representative models are, B.W. Lee, Phys. Rev. D6, 1188 (1972). J. Prentki and B. Zumino, Nucl. Phys. B47, 99 (1972). Model (3-2) in J.D. Bjorken and C.H. Llewellyn-Smith, Phys. Rev. D7, 887 (1973).

23) This classification and its extension to 0(4)XU(1) has been emphasized by A. Pais, Rockefeller University preprint COO-2232B-32, (1973).

24) Representative models are, A. Salam and J. Ward, Phys. Letters 13, 168 (1964). S. Weinberg, Phys. Rev. Letters 19, 1264 (1969). J. Bjorken and C.H. Llewellyn-Smith, D7, 887 (1973), models (2-2) and (2-3).

25) H. Faissner et al., CERN preprint (1973).

26) The value $\sigma_{exp}/\sigma_{V-A} = 1.1 \pm 0.8$ was present by H.S. Gurr, F. Reines and H.W. Sobel

at the Topical Seminar on Weak Interactions Miramare-Trieste, 1973.

27) H. H. Chen and B. W. Lee, Phys. Rev. D5, 1874 (1972). C. Baltay, in Proceedings of the 1972 Europhysics Neutrino Conference, Balatonfured, Hungary.

28) S. Weinberg, Phys. Rev. D5, 1412 (1972).

29) S. L. Glashow, J. Iliopoulos and L. Maiani, Phys. Rev. D2, 1285 (1972).

30) When the models of Ref. 24 are extended to the hadronic section they all have such a neutral current. In the same category are also included an 8-quark model discussed by Pais, ibid. In the model of M. A. B. Beg and A. Zee the neutral current is proportional to the electromagnetic current, Phys. Rev. Letter 30, 675 (1973) and Rockefeller preprint (1973).

31) This parameter can also be determined by semileptonic reactions alone. Thus bounds from one semileptonic reaction can be the input to predict bounds for other reactions.

32) F. J. Hasert et al., CERN preprint (1973).

33) A. Pais and S. B. Treiman, Phys. Rev. D6, 2700 (1972).

34) E. A. Paschos and L. Wolfenstein, Phys. Rev. D7, 91 (1973).

35) L. M. Sehgal, Aachen preprint (1973), R. Palmer, B. N. L. 18059, NG-258 (1973), C. H. Albright, NAL-PUB-73/ 23-THY (1973).

36) W. Lee, Phys. Letters 40B, 423 (1972).

37) P. Musset, ibid.

38) D. H. Perkins, in Proceedings of the XVI International Conference on High Energy Physics, NAL (1973).

39) B. W. Lee, Phys. Letter 40B, 423 (1972).

40) S. Adler, Institute for Advanced Study preprint, Princeton (1973).

41) C. H. Albright, B. W. Lee, E. A. Paschos and L. Wolfenstein, Phys. Rev. D1, 2220 (1973).

42) B. C. Barish et al., Phys. Rev. Letter 31, 180 (1973) and 31, 410 (1973).

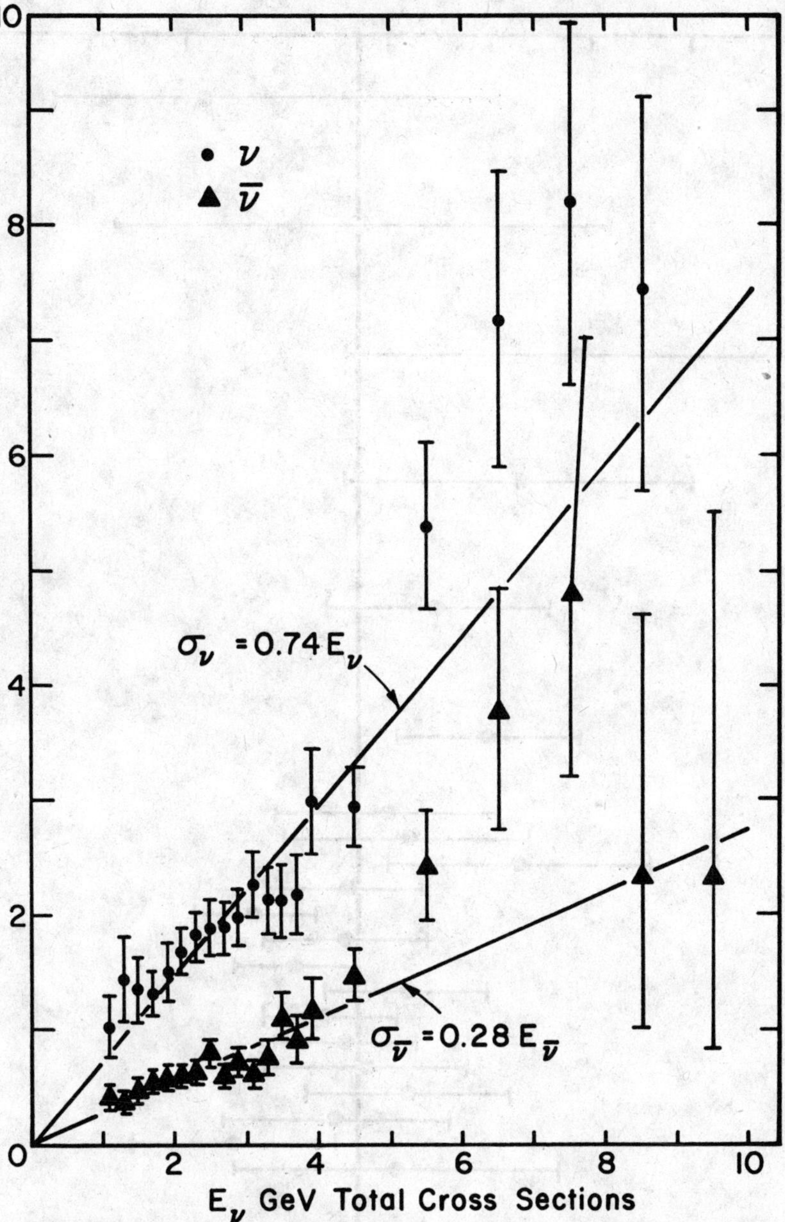

Fig. 2 Neutrino and antineutrino total cross sections as functions of the incident energy (Ref. 7)

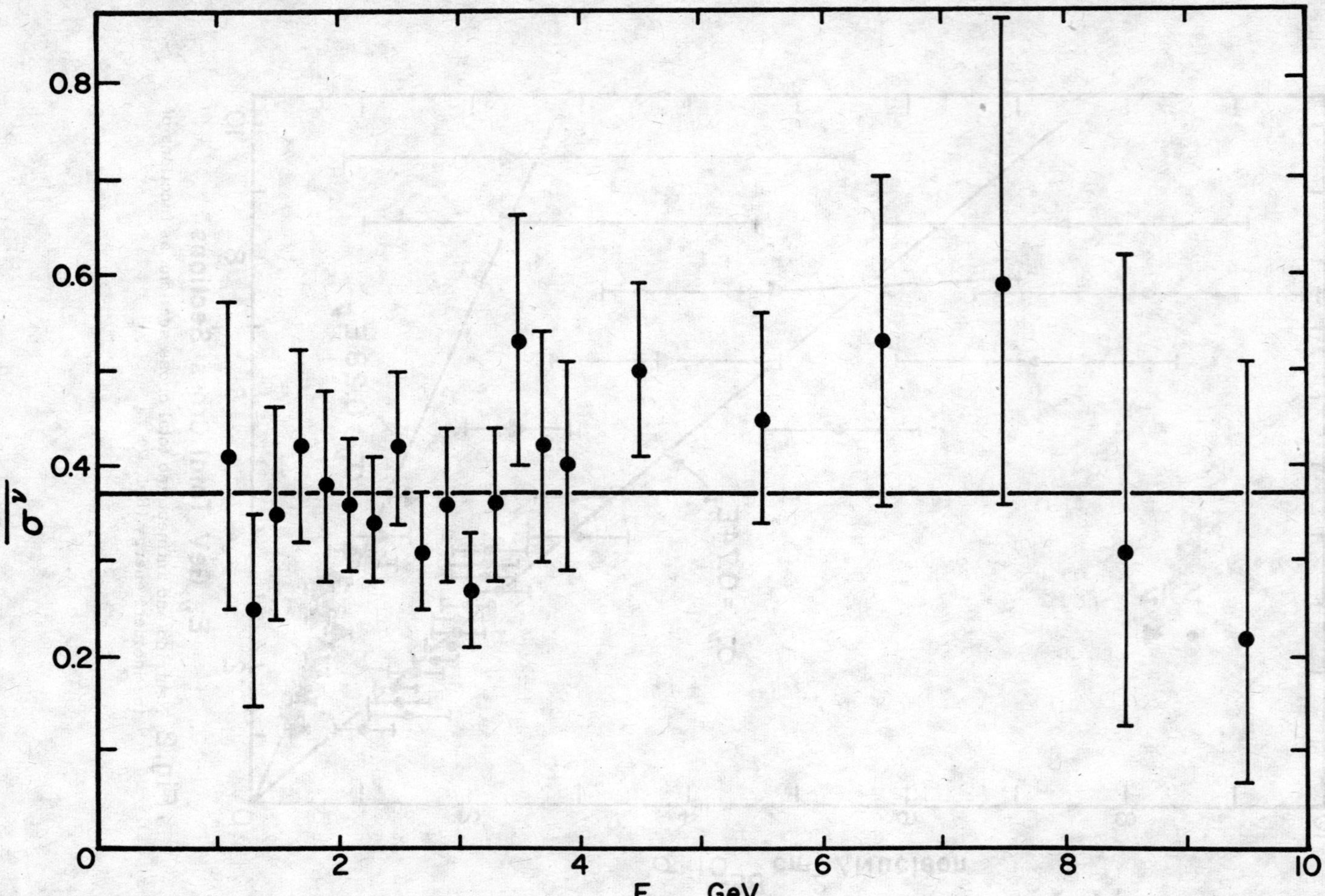

Fig.3 Ratio of the antineutrino to neutrino total cross sections as a function of the incident energy (Ref. 7).

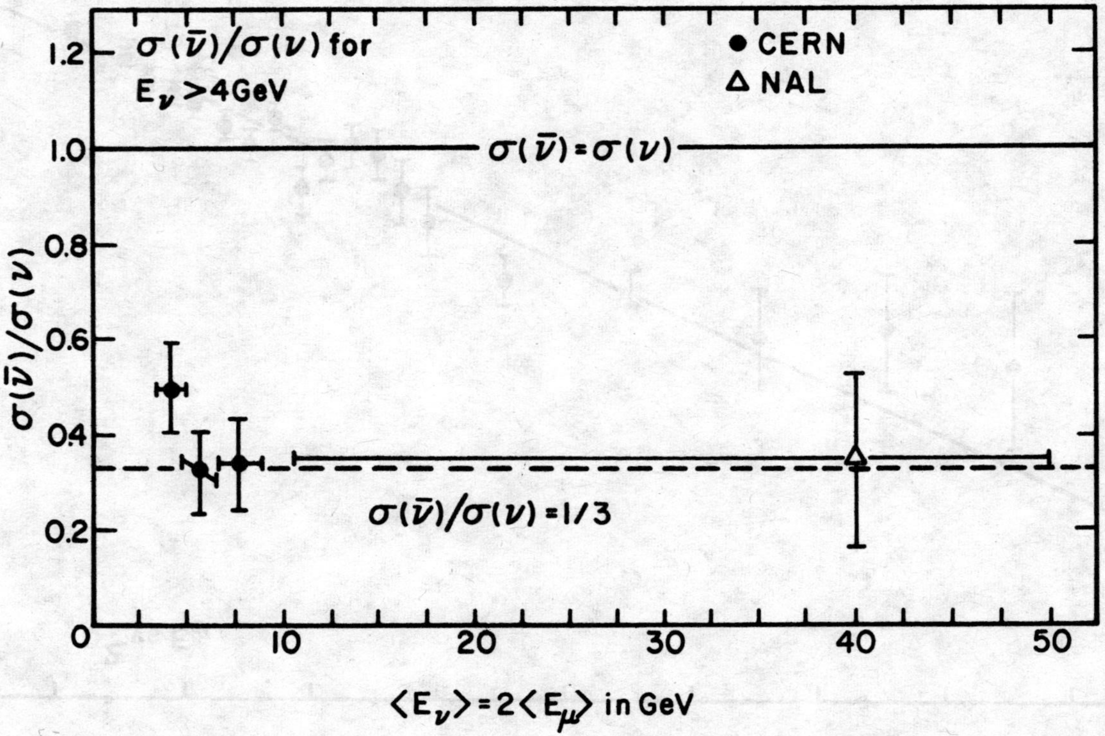

Fig. 4 Ratio of the total cross sections at NAL energies (Ref. 8).

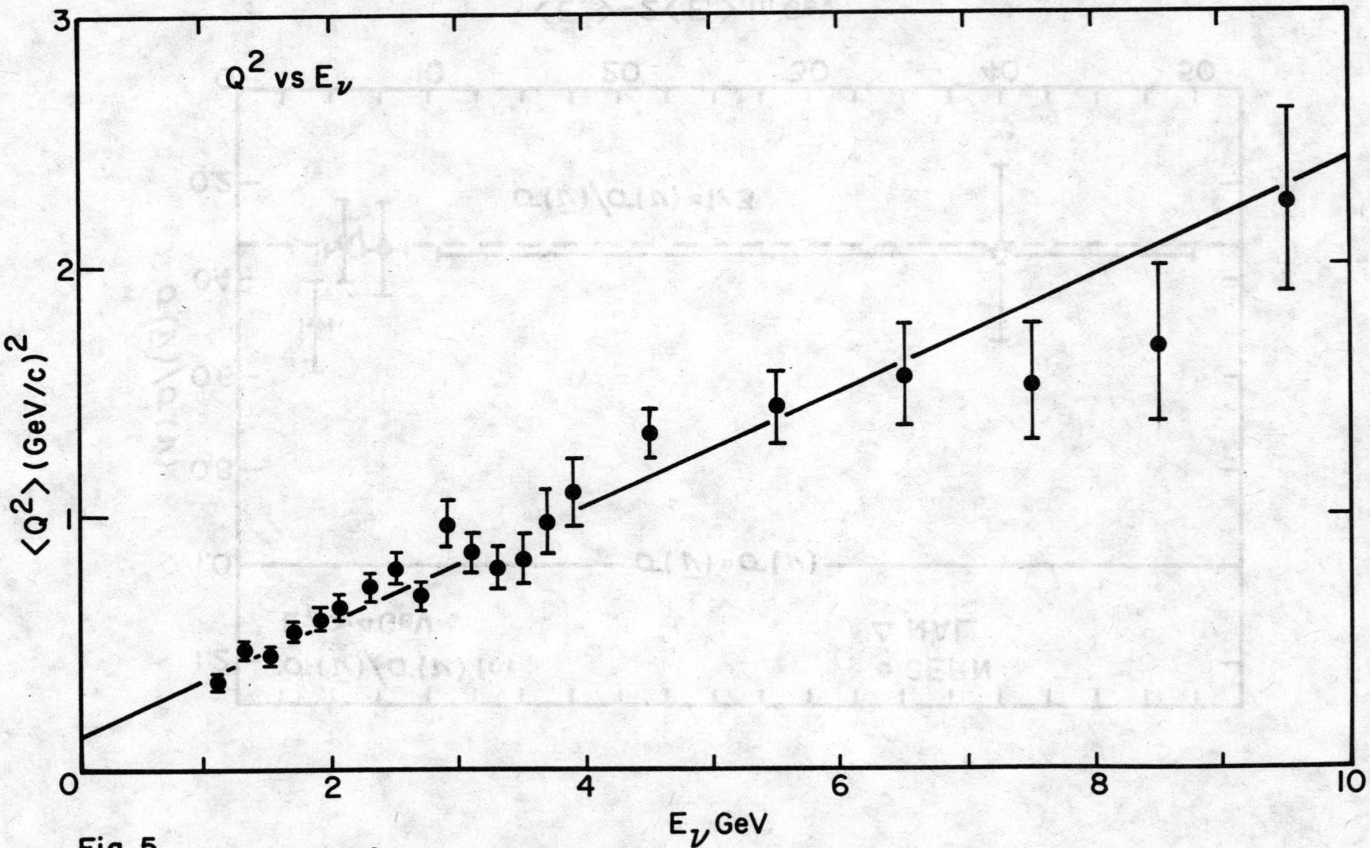

Fig. 5 Mean value of Q^2 as a function of the incident neutrino energy (Ref. 7).

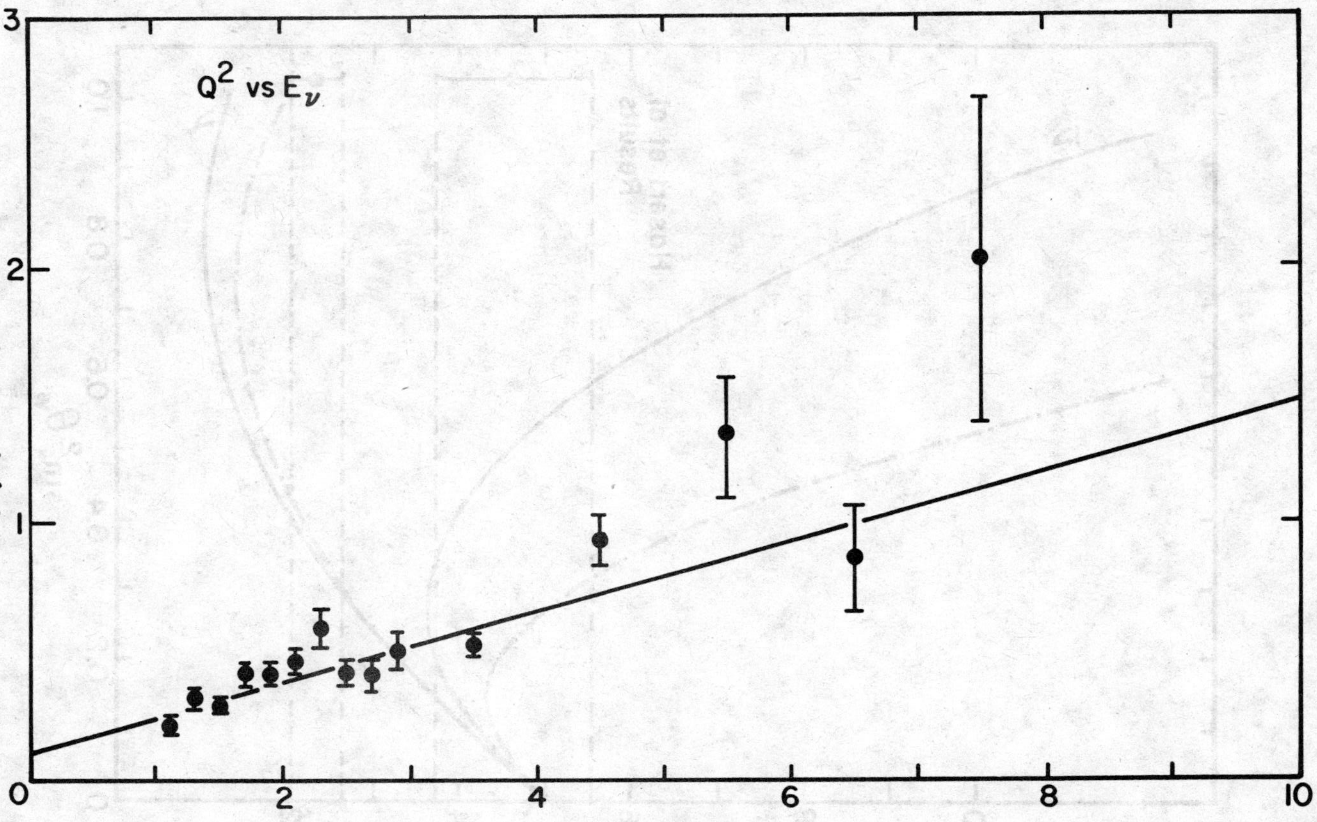

Fig. 6

Mean value of Q^2 as a function of the incident antineutrino energy (Ref. 7).

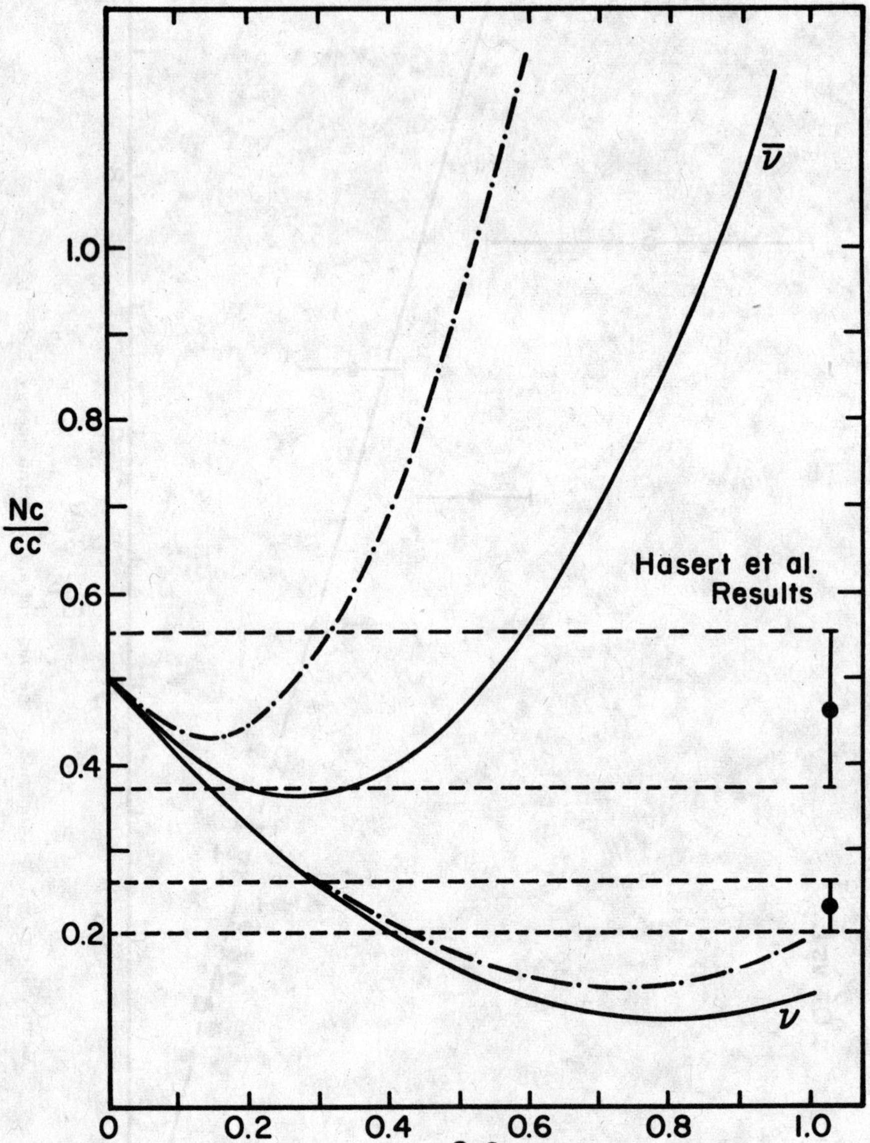

Fig. 7 Lower bounds for the ratio of neutral to charged current contributions. Solid curves are Eqs. (5-15) and (5-16). Dotted curves from Ref. 35.

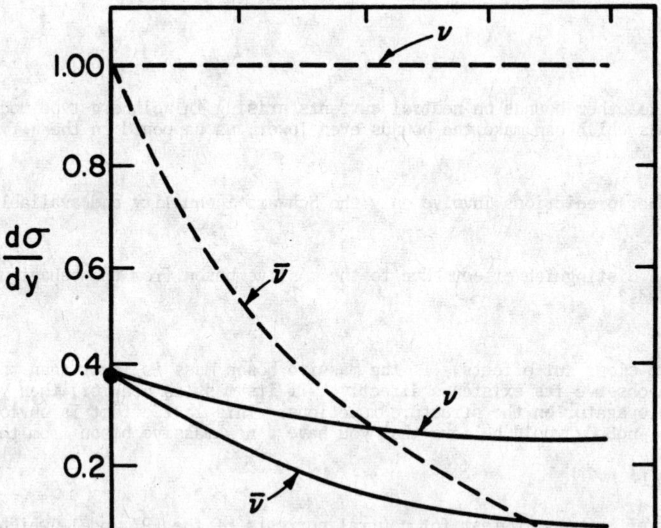

Fig. 8 y-distributions for neutral and charged currents in units of $(G^2ME/\pi)\int F_2(x)dx$. Solid curves correspond to neutral currents and are obtained from Sehgal's paper.

D I S C U S S I O N S

CHAIRMAN: Prof. E. Paschos

Scientific Secretaries: L. Routh, D. Nanopoulos, D.K. Choudhury

DISCUSSION No. 1

- *NANOPOULOS:*

What is the value of Weinberg's angle if the event for neutral currents in $\nu_\mu e \to \nu_\mu e$ is real?

- *PASCHOS:*

$\Theta_W \simeq 45°$.

- *ARONSON:*

Do any of the other bounds on neutral currents arising in Weinberg-type models contain hidden assumptions which can make the bounds even lower, as happened in the W.Y. Lee ratio?

- *PASCHOS:*

No. The other predictions involve only the Schwarz inequality and available data.

- *LOSECCO:*

How does one distinguish effects due to the massive boson from the behaviour of the structure functions?

- *ZICHICHI:*

You have two clear cut effects. If the massive boson mass is lower than the available c.m. energy, you observe its existence directly. If its mass is heavier, then you have the effects of the propagator on the structure functions. This last effect is obviously not very clear; i.e. nobody would believe that you have a new massive boson from this kind of measurement.

- *WILCZEK:*

Do you know of any way to test for neutral currents of the LPZ model, which do not couple to neutrinos, in the near future?

- *PASCHOS:*

By e^+e^- colliding beam experiments, where you measure the polarization or asymmetry of the final muon, or in electron-proton deep inelastic scattering where you look for an interference between the weak and electromagnetic amplitudes. Both tests are, however, very hard.

BEAMS OF MOLECULES, ATOMS, AND NUCLEONS

N.F. Ramsey

Table of Contents

BEAMS OF MOLECULES, ATOMS, AND NUCLEONS

Norman F. Ramsey

Harvard University, Cambridge, Massachusetts, USA
and
Universities Research Association, Washington, D.C., USA

Introduction

My lectures on Beams of Molecules, Atoms and Nucleons cover my recent research activities and those of my associates on five quite different research topics: (1) Molecular Beam Magnetic Resonance Experiments, (2) Molecular Beam Electric Resonance Experiments, (3) Hydrogen Maser Experiments, (4) Search for a Neutron Electric Dipole Moment and (5) The National Accelerator Laboratory.

As a result my scheduled two lectures can best be thought of as five separate lectures on each of the above topics, with the first three being covered in my lecture today on Beams of Molecules and Atoms and the last two in my lecture tomorrow on Beams of Nucleons. My lecture today will be primarily in the field of molecular and atomic physics with only occasional direct bearing on the field of particle physics, the principal concern of this summer school, but my lecture tomorrow will have direct bearing on the field of particle physics.

I. Molecular Beam Magnetic Resonance Experiments

Our molecular beam magnetic resonance experiments at Harvard[1] are done with an apparatus shown schematically in Fig. 1. The molecules being studied are introduced at a pressure of a few torr into the source box (if necessary a heated oven) S and emerge through a slit that is typically 0.02 mm in width. The molecules then pass through an inhomogeneous magnetic field A which deflects them through a uniform magnetic field region C containing a collimator slit, and through a refocussing magnetic field B and into an electron bombardment ionizer. The ions produced are then accelerated, passed through a mass spectrometer and detected with an electron multiplier tube and a phase sensitive detector.

The ability of the second inhomogeneous field B to exactly compensate for the first inhomogeneous field A depends on the assumption that the resultant magnetic moment of the molecule has not changed between the A and B regions. As a result the apparatus can serve as a detector of such transitions. In particular, if an oscillatory magnetic field is introduced in the C region between the A and B fields and if the oscillator frequency ν satisfies the Bohr frequency condition

$$\nu = \frac{W_i - W_J}{h} \quad (1)$$

for allowed transitions between states i and j, some molecules will make the transition and will consequently possess a different magnetic moment when in the second inhomogeneous field than in the first. As a result the deflection in the second inhomogeneous field will no longer exactly compensate for the deflection in the first and the beam intensity at the detector will drop. Consequently, one can study the spectroscopy of the molecule at radiofrequencies by such a molecular beam resonance apparatus.

As in all spectroscopy, the interesting physical results are obtained by interpreting the observed spectrum in terms of a molecular model. In Fig. 2 the different angular momentum vectors of a $^1\Sigma$ diatomic molecule are represented and a list is given of the various interactions within a molecule that can affect the molecular beam resonance spectrum. The rotational magnetic moment of the molecule, for example, can interact with the external magnetic field, a possible electric dipole moment of the molecule can interact with an external electric field, orientation dependent diamagnetic and dielectric interactions can occur, the nuclear magnetic moments can interact with each other and with the magnetic field that results from the rotation of the molecule, etc. From the observed spectrum, the parameters that characterize the various interactions can be obtained.

Some recent observations by Code[2] in our laboratory on the molecule D_2 are given in Fig. 3. These results are the first ones for this molecule in the second rotational state and thereby provide information on the change of the interaction parameters with the centrifugal distortion of the molecules.

One of our more recent studies has been the measurement by Ozier[3] and his associates of the rotational magnetic moments and spinrotational interaction parameter of methane-like molecules. The results of these measurements are given in Fig. 4. With methane a number of other parameters have also been measured as shown in the lower portion of Fig. 4. Particularly interesting measurements have been the coefficient D_T of the fourth order parameter $O^{(4)}$dist that measures the extent to which the molecule is centrifugally distorted by its rotation and the value of the electric dipole moment μ_E that arises from rotational distortion of the CH_4 molecule which by symmetry has no electric dipole moment in the undistorted state.

Recently MacAdam[4] in our laboratory added a static electric field in the resonance region. This has enabled us to observe the dependence of the electric susceptibility of the molecule upon its orientation state. The apparent result is affected by the coupling of the molecule and hence by the strength of the external magnetic field as shown in Fig. 5. Also in Fig. 5 are given for H_2 and D_2 final values for $\gamma = \alpha \; - \alpha_{\perp}$, where α is the electric susceptibility of the molecule for an electric field applied parallel to the molecular axis and $\alpha_{\perp}$ is the susceptibility for a field perpendicular to the axis. The results agree well with theory.

At present our most active experiments on this apparatus are to measure the dependence γ of the electric susceptibility upon orientation for more complicated molecules, including HD and C_2H_2.

II. Molecular Beam Electric Resonance Experiments.

Our newest molecular beam apparatus at Harvard is an electric beam resonance apparatus. The basic principles of this apparatus are essentially the same as those of the magnetic resonance apparatus previously described except that the deflections and refocussings are due to inhomogeneous electric fields interacting with the electric dipole moment of the molecule. One important practical difference results from the much larger deflections that are possible in the electric case. The deflecting and refocussing fields can be quadrupolar in nature (produced by four rods parallel to the beam with adjacent rods being oppositely charged). In such a quadrupolar field, significant space focussing becomes possible and the intensity of the beam in the presence of the fields is significantly enhanced over that in their absence. Since most of our experiments are intensity limited, this increased beam intensity is of real help and enables us to observe individually the spectrum of a single rotational and vibrational state of the molecule.

Gallagher and Hilborn, who first constructed the apparatus, used it to measure the interactions in LiCl, LiBr molecules, with the results shown in Fig. 6. Cecchi has added a uniform magnetic field to the resonance region of the apparatus which makes possible measurements of the rotational magnetic moments of the molecules, the dependence of the molecular magnetic susceptibility upon orientation, and the dependence of the nuclear magnetic shielding upon molecular orientation. Results of this nature obtained by Cecchi and Freeman on LiCl, LiBr, and LiH are shown in Fig. 7. Within the past few weeks such measurements have been made by Freeman on LiD as well. It now appears possible that these measurements may lead to a more accurate value of the deuteron quadrupole moment, since the quadrupole interaction has been accurately measured in this molecule and since a good calculation of the gradient of the electric field in this simple molecule appear to be possible.

III. Hydrogen Maser Experiments

The atomic hydrogen maser operates between the different hyperfine energy levels of atomic hydrogen. These energy levels are shown schematically in Fig. 8. In the $F = 0$ level the electron and proton spins are anti-parallel and in the $F = 1$ level they are parallel and the atom can have a magnetic quantum number m of +1, 0, or -1. If an external magnetic field is increased these energy levels shift with the magnetic field as shown in Fig. 8.

The hydrogen maser, as developed by Kleppner, Goldenberg, and Ramsey,' is shown schematically in Fig. 9. Molecules of hydrogen are dissociated into atoms by an electrical discharge in the source S. The atoms then emerge into an evacuated region and pass through a six-pole focussing magnet whose cross section is shown in Fig. 9. By symmetry the magnetic field is zero along the axis of the focussing magnet and the magnitude of the field consequently increases off the axis. Since the energies of the atoms in the higher energy $F = 1$, $m = +1$ or 0 states by Fig. 8 increase with the magnitude of the magnetic field, atoms in these higher energy states which pass through the magnet off the axis will be deflected back toward the axis, i.e., will be focussed. Inversely off axis atoms in either of the two lower energy states will be defocussed. Consequently, atoms in the higher energy states are focussed onto the teflon coated storage bulb so inside the bulb the population of atoms in these states exceeds that of the lower energy states and the basic requirements for maser oscillation are met since stimulated emission then exceeds absorption. The storage bulb is ordinarily surrounded by a microwave cavity tuned to the 1420 MHZ frequency associated with the $F = 1$, $m = 0$ transition to $F = 0$, $m = 0$. Under these circumstances spontaneous oscillation at this frequency is maintained for a sufficiently intense beam of incident hydrogen atoms. The frequency of the oscillation is highly stable with fractional frequency fluctuations less than 10^{-14} being attainable, though absolute frequency determinations are ordinarily no better than 5×10^{-13} due to uncertainty in the effect of the walls of the storage bulb on the atoms when the atoms collide with the wall. The high stability of the hydrogen maser has led to its frequent use in long base line radioastronomy.

We have used the hydrogen maser in our laboratory for a number of highly accurate measurements of fundamental two-particle system particularly amenable to accurate theoretical calculations, the intercomparisons of theory and experiment are particularly interesting in these cases.

With the hydrogen maser we have measured the hyperfine separations $\Delta\nu$ of atomic hydrogen, deuterium and tritium. Although the hydrogen and tritium measurements were first made some years ago, the accurate measurement of deuterium by Wineland[8] was only recently published. The results are given in Fig. 10 along with the quantum electrodynamical theoretical expression for the hydrogen hyperfine structure. From this expression and from the experimental measurements, the value of the fine structure constant α can be obtained with the value shown in Fig. 10. Although the value of α disagreed by 35 parts per million from the accepted value at the time of the measurement, the result has subsequently been confirmed by measurements of the Josephson effect and the error in the previous value of α has been shown by Appelquist and Brodsky[9] to be due to an error in a quantum electrodynamical correction associated with the earlier measurments.

By means of spin exchange collisions Crampton, Berg and Robinson[10] in our laboratory have used the hydrogen maser to measure the hyperfine structure of atomic nitrogen and Hirsch is currently developing an improved technique to increase the accuracy of this measurement.

Fortson and later Gibbons[11] in our laboratory have measured the effect of an electric field in shifting the hyperfine separation $\Delta\nu$ of atomic hydrogen and have found the results given in Fig. 11. Stuart in our laboratory is currently attempting to improve the accuracy of this measurement by a factor of ten or more to provide a more stringent test of the theory.

Likewise the effect of a magnetic field upon the hyperfine structure has been measured. These measurements have been made in both the strong and weak field limits of Fig. 8. In the strong field limit Myint[12] and later Winkler, Walther, Myint and Kleppner[13] have determined the ratio of the magnetic moment of the proton to that of the electron, with the result shown in Fig. 12. This measurement not only has greater precision than any previous measurement of this ratio, but it is also free from uncertainty from the calculation of the magnetic shielding of the proton in the hydrogen molecule.

At lower magnetic fields, the measurement of the dependence of the transition frequencies upon the field strength provides a value of the magnetic moment of the hydrogen atom. The results of hydrogen maser measurements by Valberg and Larson of the ratios of magnetic moments of Rb, H, D and T are shown in Fig. 13. Agreement between theory and experiment is obtained only if a recent quantum electrodynamic reduced mass correction is included. These quantities have also been recently measured by Robinson in an optical pumping experiment.

As mentioned earlier, the principal disadvantage of the hydrogen maser is the uncertainty in absolute measurements from the influence of the wall coating of the containing vessel on the atom when it is in contact with the wall during a wall collision. Uzgiris[14] has reduced this effect by a factor of ten by using a storage vessel of ten times the diameter and Reinhardt is currently attempting to combine this technique with one suggested by Brenner[15] in which the effect of the wall is measured by changing the storage volume while retaining the same surface as can be done with the apparatus shown schematically in Fig. 14. With this device it is hoped that the precision of the absolute values of hydrogen maser measurements can be further improved.

IV. Search for an Electric Dipole Moment of the Neutron

By time reversal symmetry, it can be shown that the electric dipole moment of the neutron of any other particle or atom with a definite spin should vanish. However, the experiment of Fitch, Cronin, Christianson, and Turlay which showed the existence of a 3 pion decay of the K_L^o demonstrated a violation of CP symmetry and strongly suggested a violation of time reversal symmetry. In fact most of the early theories to account for the CP violating decay of the K_L^o also predicted an electric dipole moment of the neutron whose value when divided by the charge of the electron should be of the order of 10^{-22} cm or larger. For this reason, Miller, Dress and I at Oak Ridge have for sometime been looking for an electric dipole moment of the neutron and have pushed the upper limit of its possible value to successively smaller sizes.

A schematic view of our apparatus is shown in Fig. 15. Slow neutrons of approximately 80 m/sec velocity are totally reflected at glancing angles of less than 5° in coming through a 1 cm x 10 cm pipe. The neutrons are polarized by having them reflected from magnetized iron in which case the neutrons of one direction of polarization are totally reflected while the neutrons of opposite polarization are not. These polarized neutrons are then analyzed by a second magnetized iron mirror and detected. Between the polarizer and analyzer a weak uniform magnetic field is placed and the transition associated with the Larmor precession frequency of the neutrons in that field is observed as in Fig. 16. The intensity of the neutron beam is then carefully measured on the steepest slope of the resonance as an electric field is turned on successively parallel and antiparallel to H. If there were a neutron electric dipole moment the torque on the electric dipole moment would change the resonance frequency and hence the observed neutron intensity when the electric field was reversed.

The principal potential source for an erroneous result is the $(\vec{v}/c) \times \vec{E}$ effect by which the motion of the neutron through the electric field gives rise to an apparent magnetic field which can change the Larmor precession frequency of the neutron due to its known magnetic moment.

This effect can be minimized by making the electric and magnetic fields accurately parallel and it can be measured by changing and reversing the velocity of the neutrons relative to the apparatus.

When we first started measurements of this kind the limit for μ_e/e was 10^{-14} cm and in 1954, as a test for parity violations in an experiment of this general nature, we lowered the upper limit for μ_e/e to 5×10^{-20} cm. Our most recent result has just been published. Although we still find no neutron electric dipole moment we have lowered the experimental upper limit to 10^{-23} cm. This value is well below the 10^{-22} cm predicted by most of the early theories of CP violation and is just at the level predicted by new theories of Lee and Pais. Most superweak theories, however, predict a value of 10^{-26} cm. We have now moved our apparatus from Oak Ridge to the new reactor at Grenoble, France, where we hope to gain at least an additional factor of 10 in sensitivity. At a later stage, with neutrons of about 6 m/s velocity totally reflected in a storage box, we hope to lower the sensitivity of our measurements still further.

V. The National Accelerator Laboratory

The final topic of my talk is one for which I have primarily had administrative responsibilities and have not been personally involved in the individual experiments. Since 1966 I have been part time on leave from my regular position at Harvard University and have been President of Universities Research Association, a group of 52 universities which, under contract with the U. S. Atomic Energy Commission, has constructed and is now operating the National Accelerator Laboratory in Illinois, and its new 300 GeV proton accelerator. The direct management of the laboratory and the design and construction of the accelerator has been the responsibility of the Laboratory Director, Robert Wilson, and his staff.

An overall view of the accelerator is shown in Fig. 17. The accelerator is actually four accelerators in series. The first accelerator is a commercially available 750 keV Cockcroft-Walton electrostatic accelerator which feeds accelerated protons into a 200 MeV linear accelerator which in turn feeds an 8 GeV proton synchrotron booster. The extracted booster beam is then injected into a 1 km radius synchrotron of relatively small aperture where the protons are accelerated to full energy and finally extracted and transferred to the various experimental beam lines.

The original construction schedule was a rapid one calling for a first beam by July 1, 1972. However, in 1970 the construction seemed to be going so well that the scheduled date for the first beam was optimistically advanced to July 1, 1971. This date proved to be excessively optimistic and the first 200 GeV proton beam was achieved in March of 1972.

The original intensity was about 10^9 protons per pulse. This has steadily increased during the past year to the present value of 4×10^{12} protons per pulse. Although this intensity now is about as high as had been anticipated at this stage, it is hoped during the coming year or two that the intensity can be increased to the full value of 5×10^{13} protons per pulse. For many experiments the present intensity is more than adequate even when a number of experiments are operated in parallel, but for some experiments - such as those with neutrino, muon and other secondary beams - the higher the beam intensity the better.

The original design energy of the accelerator was 200 GeV but the design was such that the energy could be later raised to a higher value. The nature of the magnet power supplies has fortunately permitted the transition to higher energy to occur much earlier than anticipated. The accelerator now normally operates at 300 GeV and has had a highly successful two week run at 400 GeV, with a number of research experiments collecting useful data during that run. Before extensive runs can occur at 400 GeV, however, a new agreement will have to be negotiated with the power company and NAL with its present low operating budget can ill afford the increased power bill.

One of the principal NAL efforts at the present time is to increase the operating reliability of the accelerator and beam lines. The accelerator itself now delivers a useful beam about 60% of the time intended but the external beam lines are new and not yet debugged so many users at the end of several kilometers of beam lines and focussing magnets find a much lower reliability for the useful beams actually delivered to their experiments.

The most infamous failures of the NAL accelerator were the electrical shorts in the main ring magnets which plagued operations one year ago. At present, however, magnet failures are no longer a serious problem. Although, main ring magnets are replaced and reworked at an average rate of one per week, most of these failures occur during high voltage tests of the magnet coils during routine maintenance checks.

Magnet failures during operating periods now occur about once a month and cause about four hours per month loss in operating time. The main ring magnet power supplies are now the most frequent sources of failure during operations, followed closely by water leaks, broken insulators, computer control problems and a variety of miscellaneous problems. The principal difficulty arises from the huge number of components which must all operate correctly at the same time and from the inaccessibility of these components because of the large physical distances associated with a main ring 1 km in radius and beam lines several km long.

One attractive feature of the accelerator is its ability with split beams to support a large number of experiments simultaneously. As many as 12 different experimental groups have operated simultaneously and on the average about 8 different groups collect data simultaneously with some of the groups using a short intense spill while others use a slow spill. Initially, the external beams had rather ragged flat tops and large backgrounds and halos but these characteristics have markedly improved in recent weeks and the effective length of the flat top is now about half the actual length. The least satisfactory beam has been with muons where excessive sacrifices in muon intensity were made to permit simultaneous muon and neutrino beams from the same target. This beam is now being improved. A horn to improve the neutrino beam intensity will also soon be installed.

It is a bit early in the history of the Laboratory to give an extensive report on the experimental physics results so far obtained at NAL. Although data taking at that laboratory has been completed for 16 experiments, most of them have not been fully analyzed.

Most of the scientists with counter experiments feel they should complete more calibrations and checks for systematic errors before announcing results, but a few qualitive results have emerged. So far in a quark search experiment no free quarks have been found. Likewise the W-boson has not yet been detected; a neutrino experiment, indicates that M_W 4.4 GeV/c^2 if the branching fraction into leptons is 0.5. The differential cross section measurements with neutrinos are so far consistent with scaling and with SLAC and CERN data. Neutrino events with large momentum transfer occur frequently. One of the neutrino experiments in a relatively short time has collected several thousand, presumably interesting but not yet analyzed, events.

Bare 30 inch bubble chambers photographs have been taken with proton beams at 200, 300 and 400 GeV and one set of bubble chamber runs has been made with π^- beams. These pictures have already been the sources of numerous publications on multiplicity, cross sections etc.. The 15 foot hydrogen bubble chamber has recently been cooled down and filled with hydrogen and the superconducting magnet has been tested. When this device becomes operational it should be a potent tool for both neutrino and hadron physics.

One of the worst problems at NAL is the excessively large and growing number of proposals. So far 230 different experiments have been proposed and of this number 80 have been accepted. The nature and status of these 80 accepted experiments is given in Figs. 19 and 20. Although the brief titles for the experiments are not very informative, I have with me a book summarizing the different proposals which I would be happy to lend to anyone interested. As you can see from the figures, 16 experiments have completed their data runs and 17 are in progress. Most of the latter should finish during the coming year. 15 experiments are in test stage and being installed while the remaining 33 experiments will not be started for another year. Although there will be many complications in running such a heavy experimental program, there should be some exciting scientific results from the new accelerator during the coming year.

REFERENCES

1) N. F. Ramsey, Molecular Beams, Oxford University Press (1971).

2) R. F. Code and N. F. Ramsey, Phys. Rev. A5, 73 (1972).

3) P.N. Yi, I. Ozier and N.F. Ramsey, J. Chem Phys. 55, 5215 (1971) and I. Ozier Phys. Rev. Lett. 27, 1329 (1971) and papers to be published.

4) K. B. MacAdam and N. F. Ramsey, Phys. Rev. A6, 898 (1972).

5) T. F. Gallagher, R. C. Hilborn and N. F. Ramsey, J. Chem. Phys. 56, 5972 (1972).

6) J. Cecchi and N. F. Ramsey, Bul. Am. Phys. Soc. 18, 62 (1973) and J. Chem. Phys. to be published.

7) D. Kleppner, H. M. Goldenberg, and N. F. Ramsey, Phys. Rev. 126, 603 (1962) and Phys. Rev. 138, A 972 (1965).

8) D. Wineland and N. F. Ramsey, Phys. Rev. A5, 821 (1972).

9) T. Appelquist and S. J. Brodsky, Phys. Rev. A2, 2293 (1970) and Phys. Rev. Letters 24, 562 (1970).

10) S. B. Crampton, H. C. Berg, H. Robinson and N. F. Ramsey, Phys. Rev. Letters 24, 195 (1970).

11) P. C. Gibbons and N. F. Ramsey, Phys. Rev. A5, 73 (1972).

12) T. Myint, D. Kleppner, N. F. Ramsey and H. G. Robinson, Phys. Rev. Letters 17, 405 (1966).

13) P. F. Winkler, F. G. Walther, M. T. Myint and D. Kleppner, Bull. Am. Phys. Soc. 15, 44 (1970).

14) E. Uzigiris and N. F. Ramsey, Phys. Rev. A1, 429 (1970).

15) D. Brenner, J. Appl. Phys. 41, 2942 (1970).

FIGURE CAPTIONS

Fig. 1. Schematic diagram of molecular beam resonance apparatus.

Fig. 2. List of nuclear and molecular interactions in molecules. The figure at the left of the table is a schematic representation of a $^1\Sigma$ diatomic molecule showing the different angular momentum vectors that can mutually interact.

Fig. 3. Interaction parameters of D_2 in the first and second rotational states of the molecule.

Fig. 4. Rotational magnetic moments and spin-rotational interactions in methane-like molecules including centrifugal distortion parameter and electric dipole moment of methane.

Fig. 5. Dependence of the apparent γ upon the external magnetic field and final values for $\gamma = \alpha_{||} - \alpha_{\perp}$.

Fig. 6. Interaction parameters in various isotopic combinations of LiCl and LiBr.

Fig. 7. Molecular rotational magnetic moments, dependence of nuclear magnetic shielding upon molecular orientation and dependence of magnetic susceptibility upon orientation.

Fig. 8. Hyperfine energy levels of atomic hydrogen.

Fig. 9. Schematic diagram of atomic hydrogen maser.

Fig. 10. Experimental hyperfine separations $\Delta\nu$ for atomic hydrogen, deuterium and tritium and a comparison of these values with theoretical calculations and the resulting values for the fine structure constant $\alpha = e^2/\hbar c$ and the proton structure and recoil correction $\delta N^{(2)}$.

Fig. 11. Second order stark shift in atomic hydrogen.

Fig. 12. Value of the ratio of the magnetic moment of the proton to that of the electron and the absolute value of the proton magnetic moment in Bohr magnetons.

Fig. 13. Ratios of atomic magnetic moments especially for the hydrogen isotopes.

Fig. 14. Large storage box hydrogen maser with variable volume.

Fig. 15. Schematic view of neutron electric dipole moment apparatus.

Fig. 16. Neutron electric dipole moment resonance.

Fig. 17. Overall view of the accelerator at the National Accelerator Laboratory.

Fig. 18. Experiments at NAL.

Fig. 19. Experimental proposals at NAL.

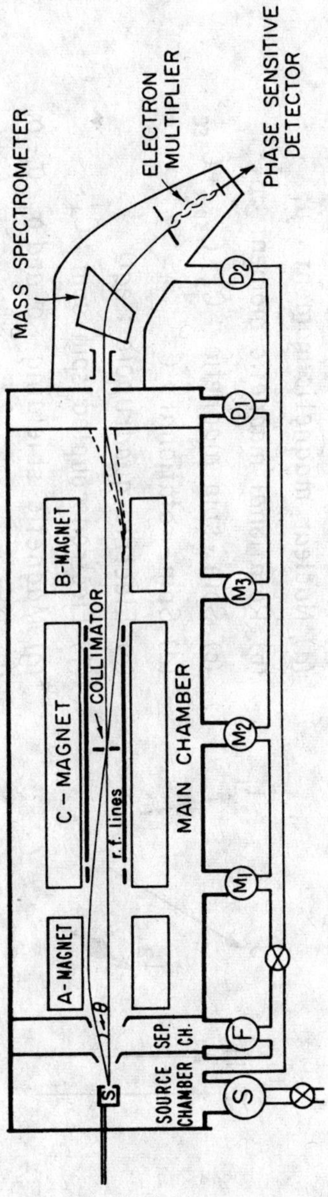
MASS SPECTROMETER
ELECTRON MULTIPLIER
PHASE SENSITIVE DETECTOR
B-MAGNET
COLLIMATOR
C – MAGNET
r.f. lines
MAIN CHAMBER
A-MAGNET
SEP. CH.
SOURCE CHAMBER
S
F
M1
M2
M3
D1
D2

Fig.1

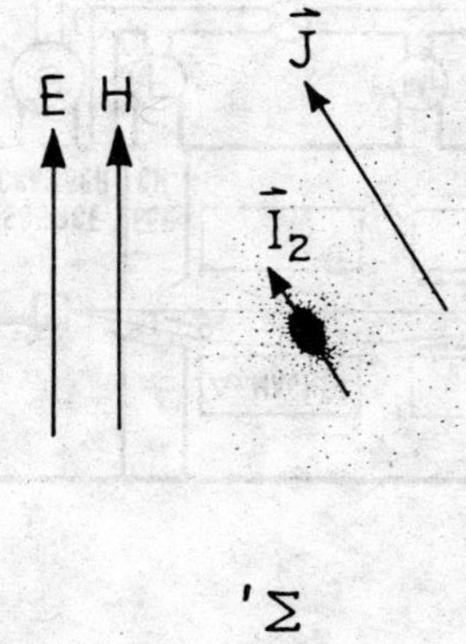

Interactions :

*(a) Nuclear magnetic moment : μ_i

(b) Rotational magnetic moment : μ_J

(c) Spin - spin magnetic : $C_3 = C_{3DIR} + C_{3EC}$

(d) Spin - rotational : C

(e) Electric quadrupole : eqQ

*(f) Electron coupled spin-spin : C_4

*(g) Magnetic shielding : σ and $\sigma_T = \sigma_\perp - \sigma_\parallel$

(h) Diamagnetic susceptibility and its orientation dependence : $\xi_T = \xi_\perp - \xi_\parallel$

(i) Electric susceptibility and its orientation dependence: $\gamma = \alpha_\parallel - \alpha_\perp$

(j) Electric dipole moment : μ_E.

Fig.2

HYPERFINE CONSTANTS OF D_2

$J = 1$ $\quad c = 8.768(2)$ kHz

$eqQ = 225.044(20)$ kHz

$$d'_M = \frac{2}{5}\left(\frac{\mu_d}{I_d}\right)^2 \frac{1}{h} {}_0\!\left\langle \frac{1}{R^3} \right\rangle^{D_2} = 2.737(1) \text{ kHz (calc)}$$

$J = 2$ $\quad c = 8.723(20)$ kHz

$eqQ = 223.380(180)$ kHz

$d'_M = 2.725(14)$ kHz

Fig. 3

MOLECULE	$\frac{\mu_J}{J}$ (nucl. magn.)	C_a (kHz)	C_d (kHz)	d (kHz)
CH_4	$+0.3133 \pm 0.0002$	10.4 ± 0.1	18.5 ± 0.5	21 ± 8
SiH_4	-0.2702 ± 0.0001	3.8 ± 0.23	9.0 ± 3.5	
GeH_4	-0.1082 ± 0.0001	3.62 ± 0.20	5.5 ± 5.0	
CF_4	-0.03126 ± 0.00002	6.85 ± 0.35	$\|C_d\| < 17$	
SiF_4	-0.03193 ± 0.00002	2.42 ± 0.08	$\|C_d\| < 3$	
GeF_4	-0.02659 ± 0.00002	1.88 ± 0.08	$\|C_d\| < 3$	

$$\mathcal{H}/h = B_o\vec{J}^2 - D_\upsilon O^{(3)}_{dist.} - D_T O^{(4)}_{dist.} - (\mu_I/Ih)\, I_z H$$
$$- (\mu_J/Jh)\, J_z H - C_a \vec{I}\cdot\vec{J} - \frac{1}{3} C_d O_{SR} + \frac{3}{16} d\, O_{SS} - \vec{\mu}_\epsilon \cdot \vec{E} - C_{13}\vec{I}_{13}\cdot\vec{J}$$

$D_{T,CH_4} = 132.0$ kHz $\quad \mu_\epsilon = (5.38 \pm 0.10) \times 10^{-6}$ D $\quad C_{13} = (15.94 \pm 0.35)$ kHz

Fig. 4

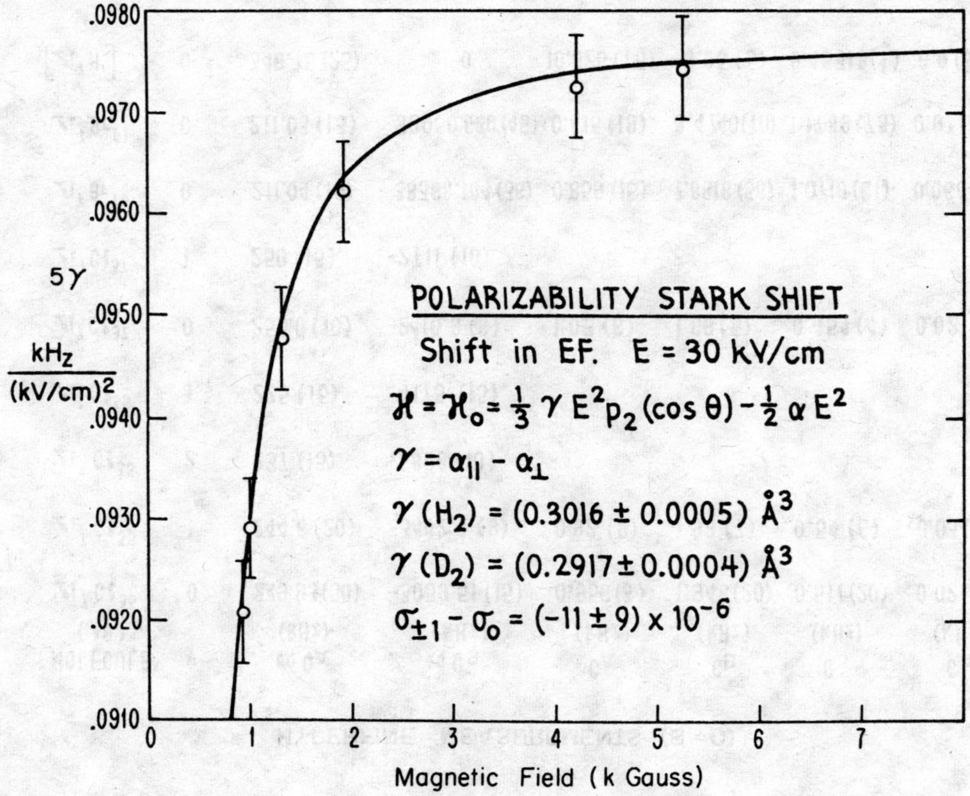
5γ
kHz / (kV/cm)²
.0980
.0970
.0960
.0950
.0940
.0930
.0920
.0910
0
1
2
3
4
5
6
7
Magnetic Field (k Gauss)
POLARIZABILITY STARK SHIFT
Shift in EF. E = 30 kV/cm
$\mathcal{H} = \mathcal{H}_o = \frac{1}{3}\gamma E^2 p_2(\cos\theta) - \frac{1}{2}\alpha E^2$
$\gamma = \alpha_{\parallel} - \alpha_{\perp}$
$\gamma(H_2) = (0.3016 \pm 0.0005)$ Å^3
$\gamma(D_2) = (0.2917 \pm 0.0004)$ Å^3
$\sigma_{\pm 1} - \sigma_0 = (-11 \pm 9) \times 10^{-6}$

Fig.5

HYPERFINE MEASUREMENTS (B = O)

MOLECULE (AB)	v	eq Q_A (kHz)	eq Q_B (kHz)	C_A (kHz)	C_B (kHz)	C_3 (kHz)	C_4 (kHz)
Li^7Cl^{35}	0	249.93(50)	-3059.51 (15)	0.956(3)	1.946(20)	0.511(20)	0.025 (20)
Li^7Cl^{35}	1	244.6 (20)	-3442.1 (9)	0.92 (9)	1.98 (7)	0.54 (6)	0.03 (6)
Li^7Cl^{35}	2	237 (15)	-3808 (10)				
Li^7Cl^{35}	3	225 (15)	-4175 (15)				
Li^7Cl^{37}	0	250.0 (10)	-2410.9 (6)	1.03 (6)	1.66 (5)	0.454 (4)	0.02 (3)
Li^7Cl^{37}	1	250 (15)	-2711 (10)				
Li^7Br^{79}	0	211.04 (10)	38368.104 (36)	0.859 (15)	7.8816 (58)	1.0710 (61)	0.0604 (70)
Li^7Br^{81}	0	211.03 (13)	32050.860 (46)	0.815 (19)	8.4740(70)	1.1789 (78)	0.0711 (89)
$[Li^7H^1]$	0	346.75 (25)	0	10.025 (75)	-9.05 (5)	0.45318(1)	0.0 (3)

Fig. 6

HYPERFINE MEASUREMENTS ($B \neq 0$)

MOLECULE (AB)	v	g_J	σ_{TA}	σ_{TB}	ξ_T (10^{-30} erg gauss^{-1} molecule^{-1})
$\mathcal{L}i^7Cl^{35}$	0	+0.10041 (2)	-0.00004 (1)	-0.00018 (3)	+ 2.22 (30)
$\mathcal{L}i^7Cl^{35}$	1	+0.10063 (4)	-0.00003 (2)	-0.00025 (8)	+ 1.91 (30)
$\mathcal{L}i^7Cl^{37}$	0	+0.10134 (4)	-0.00007 (3)	-0.00025 (10)	+ 2.90 (60)
$\mathcal{L}i^7Br^{79}$	0	+0.112056 (64)	-0.0000480 (42)	-0.000321 (82)	+ 2.35 (50)
$\mathcal{L}i^7Br^{81}$	0	+0.112217 (64)	-0.0000338 (52)	-0.000341 (87)	+ 2.55 (50)
$\mathcal{L}i\,H'$	0	-0.65773 (40)			+0.315 (80)

Fig. 7

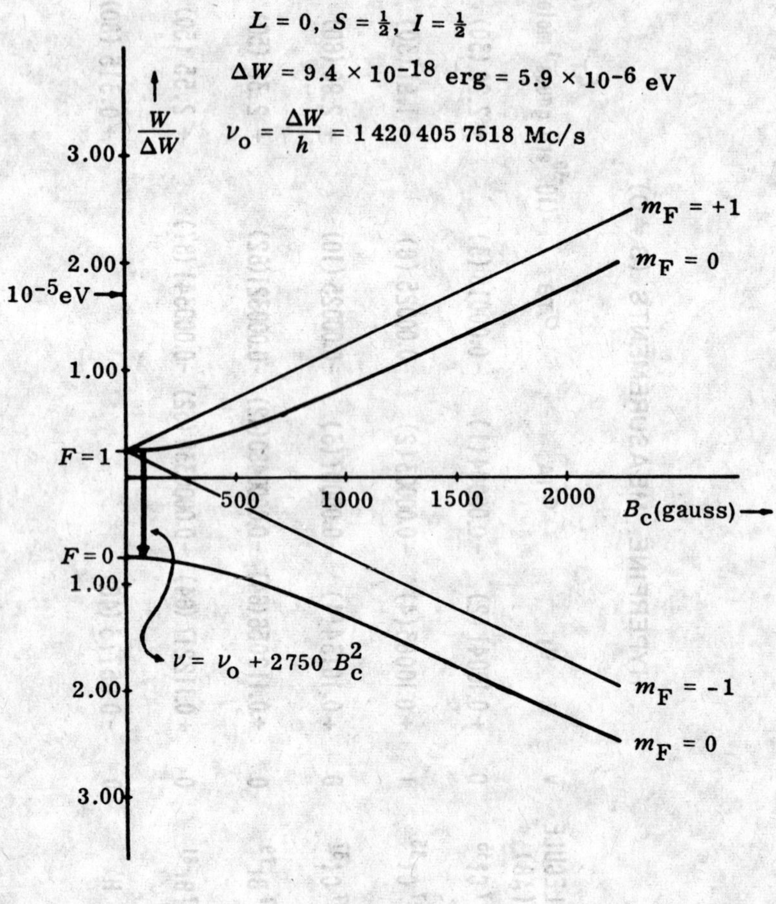

$L = 0,\ S = \frac{1}{2},\ I = \frac{1}{2}$
$\Delta W = 9.4 \times 10^{-18}$ erg $= 5.9 \times 10^{-6}$ eV
$\nu_o = \frac{\Delta W}{h} = 1\,420\,405\,7518$ Mc/s
$\frac{W}{\Delta W}$
3.00
2.00
10^{-5}eV
1.00
$F = 1$
500
1000
1500
2000
B_c(gauss)
$F = 0$
1.00
2.00
3.00
$\nu = \nu_o + 2750\ B_c^2$
$m_F = +1$
$m_F = 0$
$m_F = -1$
$m_F = 0$

Fig. 8

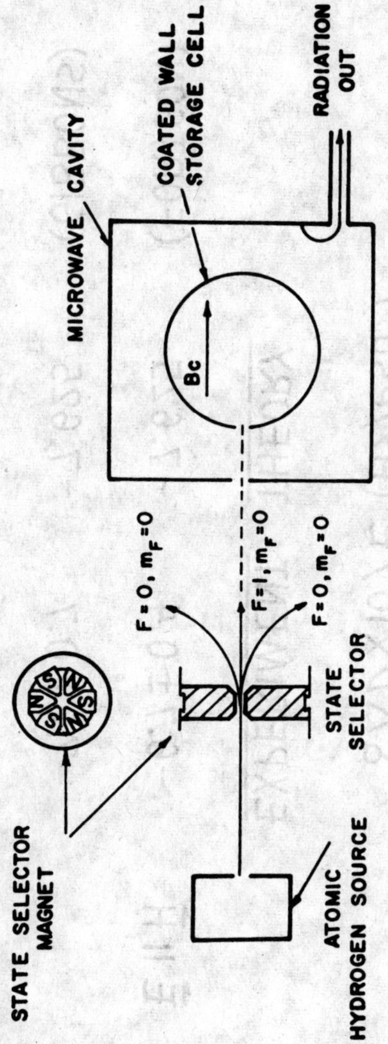
STATE SELECTOR MAGNET
ATOMIC HYDROGEN SOURCE
STATE SELECTOR
F = O, m_F = O
F = I, m_F = O
F = O, m_F = O
MICROWAVE CAVITY
COATED WALL STORAGE CELL
Bc
RADIATION OUT

Fig. 9

STARK SHIFT

$\delta\Delta\nu \times 10^5/E^2 (Hz/esu^2)$

	EXPERIMENT	THEORY	
$\vec{E} \parallel \vec{H}$	-6.7 ± 0.4	-7.625	(FORTSON)
	-7.1 ± 0.7	-7.625	(GIBBONS)
$\vec{E} \perp \vec{H}$	-7.1 ± 0.7	-7.412	(GIBBONS)

Fig.11

$$\Delta\nu_{H\,EXPT} = 1,420,405,751.768 \pm 0.010 \text{ Hz}$$

$$\Delta\nu_{D\,EXPT} = 327,384,352.5222 \pm 0.0017 \text{ Hz}$$

$$\Delta\nu_{T\,EXPT} = 1,516,701,470.7919 \pm 0.0071 \text{ Hz}$$

$$(\Delta\nu_{Cs} = 9,192,631,770 \text{ Hz})$$

$$\Delta\nu_{H\,TH.} = 1,420,403,500 \pm 8000 \text{ Hz}$$

$$\Delta\nu_{D\,TH.} \quad 327,394,000 \pm 33,000 \text{ Hz}$$

$$\Delta\nu_{T\,TH.} \quad 1,516,684,570 \pm 3,000 \text{ Hz}$$

$$\Delta\nu_{H} = \frac{16}{3}\left(1+\frac{m}{M}\right)^{-3} \frac{\mu_p}{\mu_o} \alpha^2 c R_\infty \left\{1+\frac{\alpha}{2\pi} - 0.328\frac{\alpha^2}{\pi^2} - \left(1-\ln 2\right)\alpha^2 + K\right\}$$

$$\alpha^{-1} = \hbar c/e^2 = 137.0361 \pm 0.0008$$

$$\delta N^{(2)} = (2.5 \pm 4.0) \times 10^{-6}$$

Fig.10

$$\mu_e/\mu_p = \frac{(1+1/3\alpha^2)}{(1+1/3\alpha^2)} \frac{(g_J)_H}{(g_p)_H} = 658.210706(6)$$

$$\mu_p = 0.00152103218\ (2)\ \text{Bohr magnetons}$$

Fig. 12

ATOMIC MAGNETIC MOMENT RATIOS

	EXPERIMENT	THEORY
$g_J(Rb^{87})/g_J(H)$	1.00002359 (11) 855 (6)	
$g_J(H)/g_J(D)$	1.0000000089 (14) 722 (10)	1.0000000073
$g_J(H)/g_J(T)$	1.0000000107 (20)	1.0000000097

Fig.13

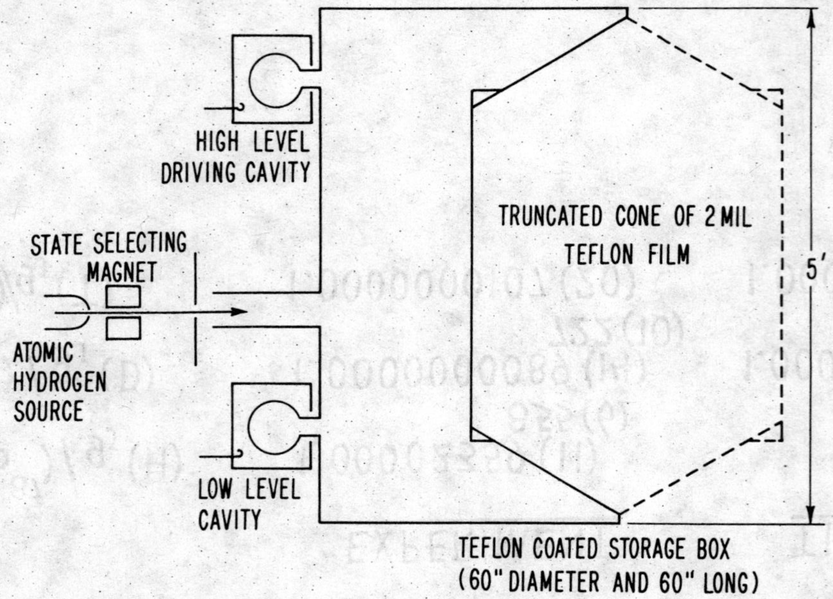

SCHEMATIC DIAGRAM OF THE LARGE STORAGE BOX MASER
WITH FLEXIBLE BOX

Fig.14

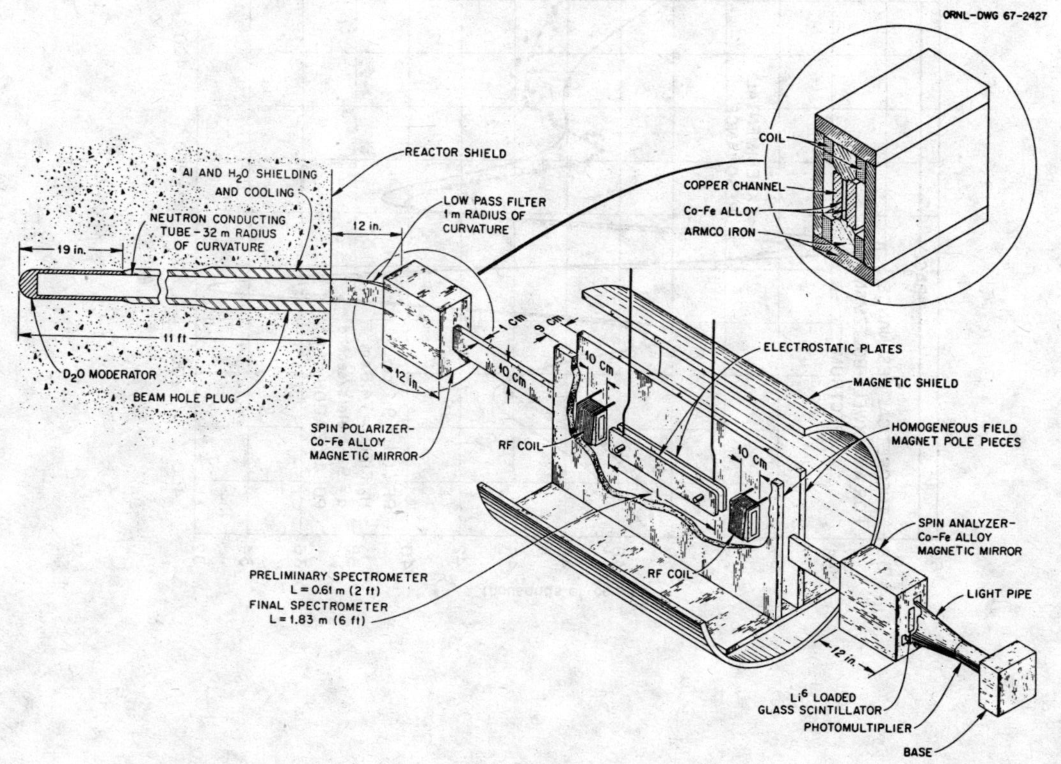
ORNL-DWG 67-2427
REACTOR SHIELD
Al AND H2O SHIELDING
AND COOLING
NEUTRON CONDUCTING
TUBE - 32 m RADIUS
OF CURVATURE
19 in.
11 ft
12 in.
D2O MODERATOR
BEAM HOLE PLUG
LOW PASS FILTER
1 m RADIUS OF
CURVATURE
12 in.
SPIN POLARIZER-
Co-Fe ALLOY
MAGNETIC MIRROR
COIL
COPPER CHANNEL
Co-Fe ALLOY
ARMCO IRON
1 cm
9 cm
10 cm
10 cm
10 cm
RF COIL
RF COIL
ELECTROSTATIC PLATES
MAGNETIC SHIELD
HOMOGENEOUS FIELD
MAGNET POLE PIECES
PRELIMINARY SPECTROMETER
L = 0.61 m (2 ft)
FINAL SPECTROMETER
L = 1.83 m (6 ft)
SPIN ANALYZER-
Co-Fe ALLOY
MAGNETIC MIRROR
LIGHT PIPE
12 in.
Li6 LOADED
GLASS SCINTILLATOR
PHOTOMULTIPLIER
BASE

Fig. 15

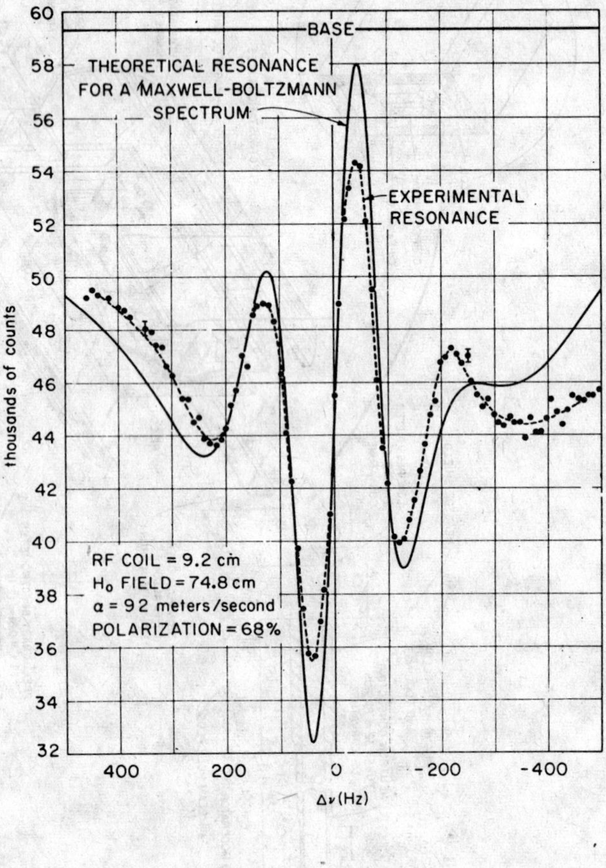

60
58
56
54
52
50
48
46
44
42
40
38
36
34
32
BASE
THEORETICAL RESONANCE
FOR A MAXWELL-BOLTZMANN
SPECTRUM
EXPERIMENTAL
RESONANCE
thousands of counts
RF COIL = 9.2 cm
H₀ FIELD = 74.8 cm
α = 92 meters/second
POLARIZATION = 68%
400
200
0
-200
-400
Δν (Hz)

Fig.16

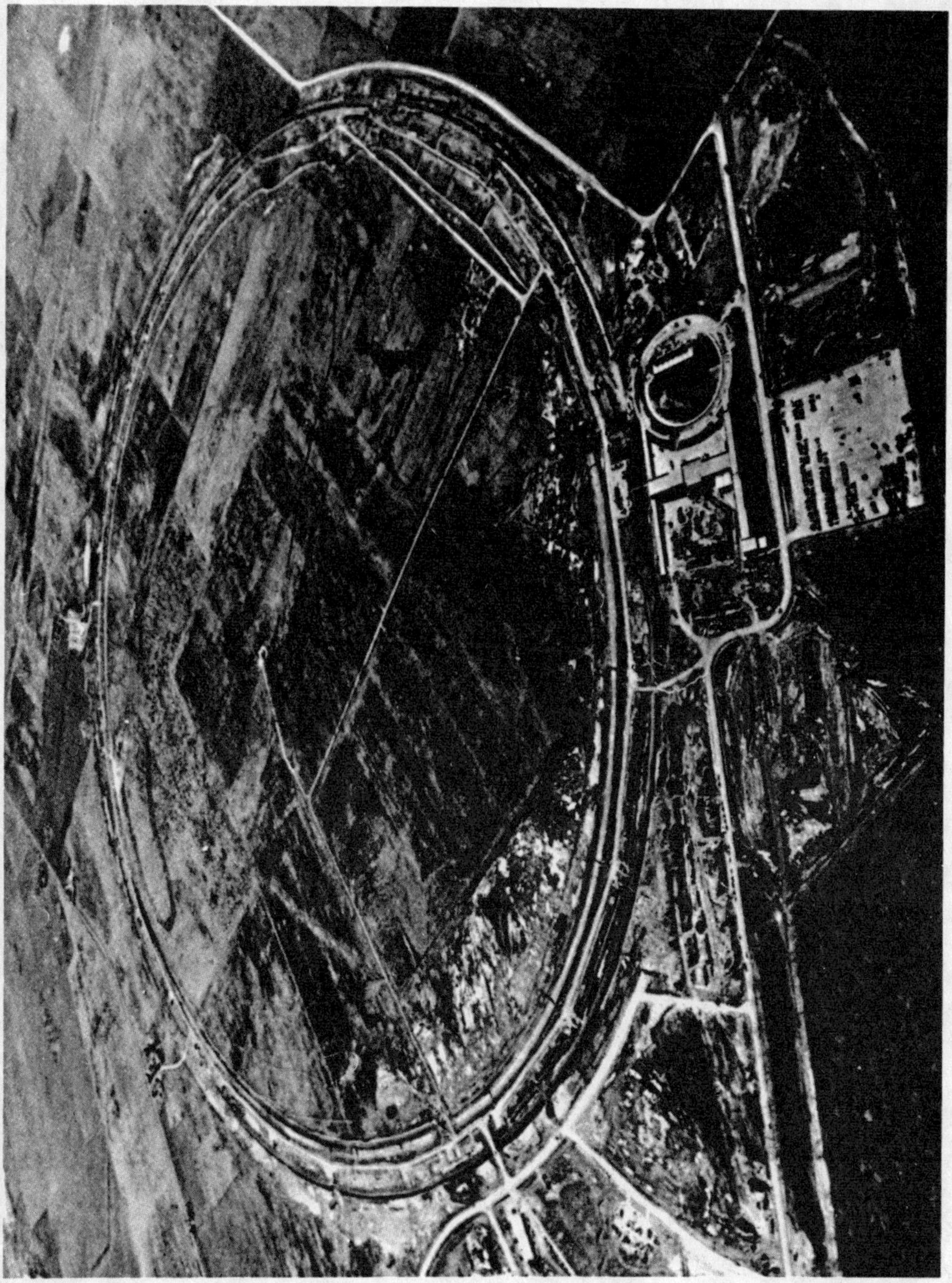

J. R. Sanford
6-18-73

NATIONAL ACCELERATOR LABORATORY
EXPERIMENTAL PROGRAM SITUATION REPORT

The experimental program situation at NAL is summarized below. The experiments are listed under categories that best describe their circumstance as of June 13, 1973. Publications are rendered or talks have been given about the experiments marked *.

A. 16 Experiments that have been Completed:

	Spokesman	Extent of Run to Date	Date Completed
*Proton-proton Inelastic #14A	Franzini	120 hours	May 9, 1973
*30-Inch p-p @ 300 #37A	Malamud	51,000 pix	June 1, 1973
*Quark #72	Leipuner	500 hours	June 11, 1973
Emulsion/protons @ 200 #90	Gierula	4 stacks	Sept. 20, 1972
*Emulsion/protons @ 200 #103	King	1 stack	Sept. 20, 1972
Emulsion/protons @ 200 #105	Malhotra	1 stack	Sept. 20, 1972
*Emulsion/protons @ 200 #114	Jain	1 stack	Sept. 20, 1972
*Emulsion/protons @ 200 #116	Hebert	5 stacks	Sept. 20, 1972
Emulsion/protons @ 200 #117A	Kusumoto	11 stacks	Sept. 20, 1972
*Photon Search #120	Cline	1,200 hours	May 29, 1973
*30-Inch π^--p @ 200 #137	Huson	48,000 pix	March 10, 1973
*30-Inch p-p @ 200 #141A	Fields	67,000 pix	Nov. 27, 1972
Emulsion/protons @ 200 #156	Niu	13 stacks	Sept. 20, 1972
*Emulsion/protons @ 200 #171	Lord	6 stacks	Sept. 20, 1972
Emulsion/protons @ 200 #183	Tretyakova	3 stacks	Sept. 20, 1972
*Proton-nucleon Inclusive #188	Sannes	1,050 hours	May 9, 1973

B. 17 Experiments that are In Progress:

	Spokesman	Extent of Run to Date
*Neutrino #1B	Cline	350 hours
30-Inch Hybrid #2B	Smith	10,000 pix
Neutron Cross Section #4I	Longo	450 hours
*Neutrino #21A	Barish	300 hours
Muon #26	Chen	350 hours
*Proton-proton Elastic #36A	Cool	700 hours
*Particle Production #63A	Walker	1,300 hours
Proton-proton Missing Mass #67A	Sannes	200 hours
Quark #75	Yamanouchi	300 hours
*Monopole #76	Carrigan	3 targets exposed
*Nuclear Chemistry #81A	Weisfield	25 bombardments
Particle Search #100	Piroue	200 hours
*30-Inch π^+-p @ 100 #121A	Lander	Tuning
*30-Inch p-p @ 100 #138I	Ferbel	33,000 pix
30-Inch p-p @ 400 #138II	VanderVelde	12,200 pix
Particle Search #187	Lederman	100 hours
*Massive Particle Search #199	Frankel	2 targets exposed

C. 3 Experiments that are In Test Stage:

	Spokesman	Extent of Run to Date
Lepton #70	Lederman	100 hours
Muon #98	Mo	Starting
30-Inch Hybrid #154 (test)	Pless	(few test pix)

Fig. 18

EXPERIMENTAL PROGRAM SITUATION REPORT (continued) 6-18-73

	Spokesman	Comments
D. 12 Experiments Being Installed:		
Elastic Scattering #7	Meyer	
Neutral Hyperon #8	Pondrom	
Neutron Dissociation #27	Rosen	
Elastic Scattering #69A	Sandweiss	
K^o Regeneration #82	Telegdi	
Diffraction Dissociation #86A	Lubatti	
Photoproduction #87A	Lee	
Elastic Scattering #96	Ritson	
Total Cross Section #104	Kycia	
Beam Dump #108	Awschalom	
Pion Charge Exchange #111	Tollestrup	
30-Inch π^--p @ 100 #125	Morrison	
E. 13 Experiments to be Set Up in the Coming Year:		
15-Foot Anti-Neutrino/H_2 #31A	Derrick	
15-Foot Neutrino/H_2 #45A	Nezrick	
Photon #95A	Cox	
Super Heavy Elements #142	Stoughton	
30-Inch π^--p @ 300 #143	Kalbfleisch	
Super Heavy Elements #147	Hebert	
15-Foot EMI #155	Peterson	
Emulsion/protons @ 100&200 #181	Carey	
Particle Search #184	Mann	
Proton-deuteron Elastic #186	Melissinos	
Emulsion/protons @ <400 #189	Ritson	
Multigamma Test #192	Guiragossian	
Emulsion/protons @ 300 #195	Lim	
F. 20 Other Experiments:		
Monopole #3	Ross	
Neutron Elastic Scattering #4II	Longo	
Proton-proton Elastic #6	Krisch	
Neutron Backward Scattering #12	Reay	
Monopole #22	Collins	
Inclusive Scattering #23A	Rothberg	
Photon Total Cross Section #25A	Caldwell	
15-Foot Neutrino/H_2-Ne #28A	Fry	
TANC #32	Hofstadter	
Detector Development #34	Huggett	
Muon Search #48	Adair	
Multiparticle #51	von Goeler	
15-Foot Neutrino/H_2-Ne #53A	Baltay	
Polarized Scattering #61	Chamberlain	
Monopole #74	Fleischer	
Charged Hyperon #97	Lach	
Multiparticle #110A	Pine	
Long-Lived Particles #115	Stevenson	
15-Foot Anti-Neutrino/H_2-Ne #172	Bingham	
15-Foot Anti-Neutrino/H_2-Ne #180	Mukhin	

Fig. 19

D I S C U S S I O N S

CHAIRMAN: Prof. N. Ramsey

Scientific Secretary: M. Lenecke

DISCUSSION No. 1

- *PAUL:*

Would an induced electric dipole moment of the proton show up in the hyperfine splitting of the atom? If so, what would be the upper limit for the electric polarizability from the error of 10^{-14} in your experiment?

- *RAMSEY:*

In measuring the hyperfine structure of, say, a hydrogen atom, you look for a change in the interaction energy when the spins of the electron and proton change from parallel to antiparallel. Therefore, you would not see an induced proton electric dipole moment in hyperfine structure experiments. The induced electric dipole moment of the proton would show up in Lamb shift experiments.

- *MILLER:*

In your talk this morning you made it clear that one frontier is to measure quantities more and more accurately, but are there other frontiers? Are there any outstanding discrepancies between experimental measurements and theoretical calculations in your field?

- *RAMSEY:*

There are some areas where the situation is rather interesting in the comparison of experiment to theory. In at least one case the experiments have not caught up to the accuracy of the theory. That is the case in some of the measurements of magnetic moments of atoms. In general, however, the problem is that the experimental accuracy exceeds that of the theoretical calculations. An example is the hyperfine interaction in atomic hydrogen. There, the next order QED calculations are needed to reach the current experimental accuracies. In many areas the hope is that we can continue pushing both the experimental and theoretical accuracies until something interesting shows up.

- *MELISSINOS:*

There is an experiment proposed to measure the Lamb shift of the $\pi^+\pi^-$ atom. Could you say what can be learnt from such an experiment?

- *RAMSEY:*

Yes. Dealing with a $\pi^+\pi^-$ atom means that you are dealing with a much smaller atom, and since it is smaller the interactions will be much larger. This will allow you to make higher-order electrodynamic-type corrections to present experiments, such as the hyperfine splittings of hydrogen.

- *MELISSINOS:*

What about applying these resonance-type experiments to particle physics?

- *RAMSEY:*

I think the g-2 experiment on the muon is an example of applying resonance-type experiments to high-energy physics, or more accurately, of moving some particle physics measurements into the realm of atomic physics. There are also other analogues of resonance-type experiments, such as the regeneration of the K^0.

- *MELISSINOS:*

I was thinking of a more direct extension of resonance techniques, for example, using fields to induce transitions between states.

- *RAMSEY:*

There you have a problem with the short lifetime of the states you are dealing with. An example where it has been done successfully is muonium, where one uses an electron to observe changes in the polarization, and even though the lifetime is short you get interesting results.

- *DRECHSLER:*

In Schwinger's dyon model, hadrons are composed of electrically and magnetically charged constituents (dyons). The coupling constant of the magnetic charge to the electromagnetic fields is 137/4, instead of the usual 1/137 for the electric charge interactions. Have there been estimates from molecular or atomic beam experiments as to the interference of magnetic monopoles on the hyperfine spectra?

- *RAMSEY:*

No. However, one might be able to consider the electric dipole moment of the neutrons, if it exists, as due to a circulating magnetic pole. I will speak about the experiments trying to measure the electric dipole moment of the neutron in tomorrow's lecture.

- *MELISSINOS:*

Are your resonance experiments more accurate than zero-crossing experiments.

- *RAMSEY:*

Certainly our accuracy with the hydrogen maser is better than that of the zero-crossing experiments. My own belief is that you usually do not gain accuracy using zero-crossing experiments instead of resonance experiments. The main appeal of zero-crossing experiments is to perform experiments you can not perform otherwise.

- *MELISSINOS:*

But do you not get a width problem by inducing the transition?

- *RAMSEY:*

You get a width whether you use resonance methods or zero crossing. In fact, the technique of separated oscillatory fields gives you the smallest width consistent with the time the molecule is in the apparatus.

DISCUSSION No. 2

- *NAHM:*

To what precision could you determine the electric dipole moment of the neutron with present-day techniques if there were no limitations on costs?

- *RAMSEY:*

The limit on costs determines the quality of the neutron beam. If we get a beam of slow neutrons with a sufficient intensity then the experiment at Grenoble should be able to reach $\mu_e/e < 10^{-26}$ cm.

- *NAHM:*

Did you see any dependence on the linear and the quadratic effect due to the magnetic moment and the velocity of the neutrons?

- *RAMSEY:*

The quadratic effect or any even power effect causes no confusion with the electric dipole moment of the neutron, because one finds no change if one reverses the electric field. So far we have not found any quadratic effect. It is the linear effect which is really quite serious.

The field distortion of the electric field, due to the electric field produced by the moving neutron, really sets a limit to our measurements.

- *NAHM:*

You want the electric and the magnetic field highly parallel. To what degree do you achieve this?

- *RAMSEY:*

We try to reduce the angle between the electric and magnetic field to the order of 10^{-1} degree. This effect is mainly caused by the residual fields of the magnetic material. Possibly you could improve on this effect by using a superconducting shield.

- *RABI:*

What does the theory say about the direction of the electric dipole moment with reference to the angular momentum? Is it parallel, or is it perpendicular to the angular momentum?

- *RAMSEY:*

All time-reversal non-invariant theories say that it is a component along the angular momentum.

- *RABI:*

Which way?

- *RAMSEY:*

The theories give the order of magnitude of the electric dipole moment. However, I think that none of them predict the sign of the dipole moment.

- *RABI:*

Would it not be fair to include that it possibly is directed in both ways, so that you would not see any effect at all?

- *RAMSEY:*

No, because then the theory would not be time reversal non-invariant any more. If a theory predicts both ways, then one way must be dominant.

- *RABI:*

Can you say something about the structure of such a theory, especially where the asymmetry comes from?

- *RAMSEY:*

The asymmetry characteristic is already true in the theories of the CP-violating decay of the K^0. Maybe Prof. Glashow can tell us more about it.

- *GLASHOW:*

First of all a technical remark. A spin-$\frac{1}{2}$ particle can be characterized by a CP-conserving magnetic dipole moment and also be a CP-violating electric dipole moment. There is no question as to the direction of the dipole moment. It is merely a number that can have a sign. It has no meaning to say that the electric dipole moment points in a different direction than the magnetic dipole moment.

The sign is, of course, an interesting question, but it is not predicted by any theory that I am aware of. The magnitude of the electric dipole moment is not reliably predicted by any theory.

The only theories that I am familiar with -- those in the context of the new gauge, renormalizable gauge, unified theories of the weak and electromagnetic interactions -- are the several theories that were introduced by Pais. These are not particularly elegant, but they have lots of beautiful ideas in them and, in particular, they have a more or less natural way of introducing CP-violation. They can calculate the order of magnitude but not the size. Only the order of magnitude of the CP-violation expected in the K^0 system is of the right size. Of course, this is because the theory is cooked up to give the right size, but there is no artificial parameter in such a theory.

Primack and Pais have gone on to compute the expected size of the electric dipole moment of the electron, of the muon, and of the neutron and proton. In the case of the neutron, the various theories predict numbers which vary between 10 and 1000 times smaller than the present experimental limit.

- *RAMSEY:*

Actually between 1 and 100 times smaller.

- *RABI:*

The magnitude of the electric dipole moment might be difficult to get, but it is the sign which must be in the nature of the theory.

- *GLASHOW:*

That may be so. Frankly, I never thought of considering the question of the sign. It is an interesting question worth investigation.

Let me go on a little bit. These theories also predict an electric dipole moment of the electron which is also between a factor of 1 and 100 below the present experimental limits.

One should also consider measuring the electric dipole moment of the muon. It is an interesting historical fact, unexplained to me, that in the original g-2 experiment where they measured the magnetic dipole moment of the muon, they gave their result in the form of a statement about the combined electric and magnetic properties of the muon; that is to say, they found no departure from QED. There could be a departure, but only if the effect due to the electric dipole moment cancelled precisely the effect due to the magnetic dipole moment. They were aware of this ambiguity in their experiment.

The last g-2 measurements were <u>insensitive</u> to the electric dipole moment of the muon, I believe, but the next generation of g-2 experiments will be <u>sensitive</u>. I hope that these future experiments will be used to give a limit on the electric dipole moment of the muon.

- *SALVINI:*

Can you comment on the present limit and the limit of future experiments on the electric dipole moment of the muon?

- *RAMSEY:*

I have not analysed the exact experiments. This you would have to do, knowing to what limit it has been put.

What numbers you infer from the experiments, I do not know.

- *LOSECCO:*

How can you get such great accuracy working with an unstable particle like the neutron?

- *RAMSEY:*

The decay time of the neutron is quite large compared to the time that the particles stay in our apparatus. So this does not influence our beam intensity measurement.

MUON-PROTON AND MUON-NUCLEUS SCATTERING

A.C. Melissinos

Table of Contents

MUON-PROTON AND MUON-NUCLEUS SCATTERING*

A. C. Melissinos
University of Rochester, Rochester, NY, USA.

1. INTRODUCTION

We report on the preliminary results of an extensive program of muon scattering experiments performed at Brookhaven National Laboratory during the past year (1972). The authors of this work are:

A. Entenberg, H. Jöstlein, J. Kostoulas and A. Melissinos
(University of Rochester)

L. Lederman, P. Limon, M. May and P. Rapp
(Columbia University)

H. Gittleson, T. Kirk, M. Murtagh and M. Tannenbaum
(Harvard University) and

J. Sculli, T. White and T. Yamanouchi
(National Accelerator Laboratory)

The processes investigated include the following:

1. Elastic μ-p scattering in the region $0.6 < q^2 < 3.0$ $(GeV/c)^2$
2. Inelastic μ-p scattering up to $q^2 < 3.5$ $(GeV/c)^2$ and in the interval $0.5 < \nu < 6.5$ GeV
3. Study of the hadrons produced in μ-p inelastic collisions
4. Search for excited muons through the missing mass technique
5. Study of the A-dependence of the scattering of muons off heavy nuclei.

The muon beam is produced, as usual, from π-decay and purified by passing it through 20-feet of Beryllium and 4-feet of Carbon, which results in an attenuation of strongly interacting particles by 25 m.f.p. The characteristics of the beam can be summarized as follows.

Momentum	7.2 ± 1 (GeV/c)
Intensity	3×10^6 $\mu^+/2 \times 10^{12}$ protons on target
Beam area	5×5 $(\text{inches})^2$
Halo	1/2 - 1 of useful beam

The incident muons were tagged in momentum and in space by the use of seven scintillation hodoscopes consisting of a total of 240 elements. As a result

$$\Delta p/p \simeq 0.015$$

$$\Delta\theta \simeq 10^{-3} \text{rad.}$$

The apparatus is shown in Fig. 1 and consists of a forward wide aperture spectrometer for detecting the scattered muons. The acceptance is approximately 6% of the total azimuth. It accepts scattering angles from

*This work was supported by the U. S. Atomic Energy Commission.

8° to 24° and has $\Delta p/p = 0.01$, $\Delta\theta \simeq 10^{-3}$ rad. The hadrons emitted in the μ-p collision are detected in another set of chambers placed at a mean angle of 60° from the incident beam. A second spectrometer with an azimuthal acceptance of ~2% is set at 45° to the beam. It is supplemented with time-of-flight counters over a 24-foot path so that protons and pions can be clearly separated up to a momentum of 2 GeV/c.

The total flux collected in this experiment amounted to approximately 10^{12} muons incident on a 120 cm liquid hydrogen target and 0.5×10^{12} muons on heavy nuclei and deuterium. The identification of scattering events is free of background as can be seen in Fig. 2 where we show the distribution of closest approach between the incident and scattered tracks. Furthermore, the empty target rate was less than 1% of the full target rate.

2. μ-e UNIVERSALITY

Elastic scattering events were selected both from a missing mass cut as well as by requiring that the angle between the emitted hadron and the virtual photon be close to zero. These points are illustrated in Fig. 3. In Fig. 3(a) we show the missing hadronic mass for all events where one hadron is detected. The shaded area represents those events which satisfy $\theta(\text{hadron}) = \theta(\gamma_v)$. Events selected on that basis $|\theta_h - \theta_{\gamma_v}| < .050$ and $(MM)^2 < 1.6\ (\text{GeV/c})^2$ were used to form the elastic cross section after applying radiative corrections (typically 3.5%). The differential cross section, $\frac{d\sigma}{dq^2}$, is shown in Fig. 4 where the solid line is the Rosenbluth formula with the dipole form-factor. Note that a cross-section close to 10^{-34} cm^2 is being measured ($q^2 \simeq 3$).

The inelastic data were obtained by measuring $q^2 = 4EE'\sin^2\theta/2$ and $\nu = E - E'$ of the events. The structure function νW_2 was extracted by assuming the ratio of the longitudinal to the transverse cross-section to have the value[1], $R = \sigma_S/\sigma_T = 0.18$. The data plotted against $\omega' = \omega + \frac{m^2}{q^2} = \frac{2m\nu + m^2}{q^2}$ are shown in Fig. 5(a). In this figure the size of each point is proportional to its statistical significance, namely inversely proportional to its error. The solid line is the Breidenbech-Kuti fit[2] to the e-p data, and one notes in general agreement. The range of q^2 is $0.4 < q^2 < 3.6$ and of ν, $0.4 < \nu < 6.2$. The graph contains all the data and one can discern the concentration of elastic events at small values of ω'. The same data are shown in Fig. 5(b) plotted against $x' = 1/\omega'$.

To check μ-e universality we have compared νW_2 over the q^2-ν plane in bins $\Delta q^2 = 0.2$ and $\Delta\nu = 0.4$ forming the ratio

$$R(q^2,\nu) = \frac{\nu W_2(\mu p)}{\nu W_2(ep)}$$

We find no significant ν dependence and since in any case a difference between muon and electron, if it exists, will give rise to a q^2-dependence of the ratio R, we have integrated over ν to obtain $R'(q^2)$ which is shown in

Fig. 6. A linear fit to these data gives

$$R = (0.93 \pm 0.06) - (0.022 \pm 0.025)q^2$$

This implies in terms of a cut-off parameter Λ^2 where

$$R = \frac{1}{(1 + q^2/\Lambda^2)^2} \simeq 1 - 2(\Lambda^{-2})q^2$$

$$\Lambda^2 < 33 \ (\text{GeV/c})^2 \quad 90\% \text{ confidence}$$

The elastic data have been fitted directly to the above relation and give

$$A = 1.02 \pm 0.06 \quad \Lambda^2 < 14 \ (\text{GeV/c})^2 \quad 90\% \text{ confidence}$$

A total of 7500 inelastic and 500 elastic events are included in the data.

3. EXCITED LEPTONS

Next we discuss the study of hadrons passing through the large angle (45°) recoil spectrometer. We used a trigger independent of a muon scatter, so that we are measuring an inclusive spectrum but without knowledge of the virtual γ. Figure 7 shows a typical separation of pions and protons using the time of flight and momentum information. For the protons we can define the 4-momentum transfer

$$t = 2mT_p$$

where T_p is the proton recoil kinetic energy. A plot of the distribution $d\sigma/dt$ is shown in Figure 8 and is typically of the form exp(-6t) but deviates from a pure exponential. One can account for the distribution of these recoil protons, both in shape and magnitude, by $\Delta^+(1238)$ production by low q^2 photons and its subsequent decay.

The recoil protons can also be used to search for excited leptons by the missing mass technique, with the hypothesis

$$\mu + p \to p + \mu^*$$

The missing mass squared of the leptonic system is shown in Fig. 9(a) and 9(b). The elastic peak is useful as a calibration of the mass resolution; the overall shape of the inelastic spectrum is due to the acceptance of the apparatus. We can also make t-cuts in these data in order to enhance any possible signal, as shown in the figures. From a simple search for bumps, we conclude that the production cross section is limited by*

*We assume 30 events in $\Delta t = 0.2$; given an azimuthal acceptance of 2%, a flux of 0.5×10^{12} muons and 8 g/cm^2 of H_2 the above limit is obtained.

$$\left.\frac{d\sigma}{dt}\right|_{t=0.4} \leq 3 \times 10^{-33}\,\mathrm{cm}^2/(\mathrm{GeV/c})^2$$

Using the approximate t-dependence of the recoil protons, exp(-6t), this limit translates into a limit for excited muons up to a mass of 2 GeV

$$\sigma(\mu^*) \leq 5 \times 10^{-33}\ \mathrm{cm}^2$$

To obtain a more quantitative estimate we may assume a particular model for the production of μ^*, as proposed for instance by F. Low[3], through an electromagnetic coupling. This is shown in Fig. 10 where λ is the ratio of the coupling of $\mu\mu^*\gamma$ to the $\mu\mu\gamma$ coupling. Note that the excited lepton cross-section has a slower t-dependence than elastic scattering, so that using high t data places a much better limit on λ^2. The results are shown in Fig. 10(b) and rule out the existence of such a particle with a mass less than 2 GeV/c^2 and a coupling of electromagnetic strength.

4. HEAVY NUCLEI

Another part of the experiment consisted in measuring deep inelastic scattering from nuclear targets in order to study the A dependence of the cross-section. Such measurements are particularly well suited for muon beams in view of the long radiation and interaction lengths which allow the use of thick targets. This can be seen in Fig. 11 which shows the raw data from copper for a 2 day run, and reaches a q^2 of 5(GeV/c)^2.

The data are summarized in Fig. 12 and are fit with an A-dependence of the form A^n. After corrections for neutron-proton difference and multiple scattering, the best fit is

$$n = 1.00 \pm 0.01$$

Note the small error in the exponent n in spite of a generous error of ±4% in the cross-section of the individual elements.

The lack of shadowing of virtual photons can be considered as surprising since real photons do indeed exhibit shadowing with an A-dependence in the vicinity of $A^{0.90}$. This difference is usually attributed to the increased separation of these space-like photons from the vector meson poles. We prefer to explain the difference in terms of a simpler argument based on the fact that the time of interaction of a virtual photon is limited by the uncertainty principle to

$$\Delta t = h/\Delta E = h/\sqrt{q^2}$$

A nucleus of radius R appears in the rest frame[4] of the virtual photon to have a transverse dimension $D = 2R/\gamma = 2R/(\nu/\sqrt{q^2})$ and the time needed in order that the photon probes the entire nucleus is

$$\Delta t = D/c = \frac{2R\sqrt{q^2}}{c\nu}$$

Thus, for shadowing to manifest itself it must hold

$$\Delta\tau > \Delta t \quad \text{or} \quad \frac{h}{\sqrt{q^2}} > \frac{2R\sqrt{q^2}}{c\nu}$$

$$\omega = \frac{2m\nu}{q^2} > R\,\frac{4m}{hc} = 20\ R \quad \text{with R in Fermis}$$

Thus, we should not expect any shadowing for $\omega < 20$ as also obtained by more detailed calculations. Indeed, when we break our data into ω-bins still we observe no shadowing even for values of $\omega \sim 15$. Similar results have been obtained by SLAC using electron beams[5)].

5. DISTRIBUTION OF HADRONS PRODUCED IN μ-p DEEP INELASTIC COLLISIONS

As a final topic we discuss the distribution of hadrons when a high $-q^2$ virtual photon interacts with a proton. In this case the direction, q^2 and ν of the virtual photon are determined, and in addition the direction (but not the momentum) of the hadron is measured. We can therefore plot the angular distribution with respect to the virtual photon and this is shown in Fig. 13. We assume azimuthal symmetry about the q^2-vector and have integrated over 2π. Furthermore, σ_T is the total photoabsorption cross section which is directly related to νW_2 through kinematic transformations*. Thus, an integration of the distributions shown gives directly the mean charged-particle multiplicity.

The data are at relatively large angles from the q^2-vector and we have included a point measured at DESY for the forward direction[6)]. We find that a simple exponential in the recoil angle gives a reasonable parametrization of the data. (This may be related to an exponential dependence in $p_\perp$). We observe no significant difference for the two q^2 intervals shown, even though our multiplicities are lower than those observed for real photons, for all values of W. This is shown in the table below

W or $\sqrt{S}$	$\langle n_{ch}\rangle (q^2 = 0)$	$\langle n_{ch}\rangle (q^2 = 1.3)$ this expt.
1.4 - 2.0	1.7	1.12 ± 0.2
2.0 - 2.5	2.5	1.52 ± 0.2
2.5 - 3.5	3.4	2.34 ± 0.3

The multiplicities obtained in this experiment are in general low[7)] and this may be related to the functional form assumed for the hadronic distribution. It is, however, reasonable to assume that $\langle n_{ch}\rangle$ decreases with q^2.

---

*At $q^2 = 1.16$ and $\omega = 2.63$, $\varepsilon = 0.68$; $\sigma_T = 39.6\ \mu b$.

REFERENCES

1) G. Miller et al., Phys. Rev. D5, 528 (1972).

2) M. Breidenbach and J. Kuti, Physics Letters, 41B, 345 (1972).

3) F. E. Low, Phys. Rev. Letters 14, 238 (1965).

4) Since the virtual photon is space-like no rest-frame exists. What is implied is the frame in which the 3-momentum is zero.

5) E. Bloom, Report at the Nashville Conference on High Energy Interactions (1973).

6) Alder et al., DESY Report 72/31 (1972).

7) J. Ballam et al., SLAC-PUB-1163 (1972).

FIGURE CAPTIONS

Fig. 1. Plan view of the experimental arrangement.

Fig. 2. Distribution of the distance of closest approach in mm between the incident and scattered muons.

Fig. 3. (a) Missing mass spectrum for events where a hadron was detected in addition to the muon. The shaded area represents the events that satisfy coplanarity less than 0.010 and $\theta_{hadron}-\theta_{\gamma_V} \leq 0.025$ rad.
(b) Scatter plot of missing mass squared vs. $\theta_{hadron} - \theta_{\gamma_V}$

Fig. 4. The differential cross-section for muon-proton elastic scattering.

Fig. 5. The structure function νW_2 as obtained from muon-proton inelastic scattering. The size of each point is proportional to its statistical significance. The solid line is the Breidenbach-Kuti fit (reference 2).
(a) Plotted vs. ω' (b) Plotted vs. $x' = 1/\omega'$.

Fig. 6. The ratio of muon/electron inelastic scattering as a function of q^2. The lines represent the best fit to the form $R=A(1+bq^2)$ and the one standard deviation limits.

Fig. 7. Mass spectrum for recoil particles that traverse the hadron spectrometer magnet.

Fig. 8. The distribution of recoil protons as a function of the 4-momentum transferred to them.

Fig. 9. The missing mass (squared) spectrum for the system X in the process $\mu+p \rightarrow p+X$. The presence of an excited lepton is signaled by the presence of a narrow peak in this distribution.
(a) For 5.8 GeV mean incident energy; spectra for different values of t are shown;
(b) For 7.2 GeV mean incident energy.

Fig. 10. (a) Differential cross section for the production of a heavy lepton according to the model of (reference 3). The elastic cross-section is also shown;
(b) Limits on the coupling strength for the electromagnetic production of excited leptons.

Fig. 11. Raw data for μ-nucleus scattering using a 0.50 m copper target.

Fig. 12. A-dependence of muon-nucleus inelastic scattering. For comparison, data on the A-dependence of the absorption of real photons are also shown.

Fig. 13. Angular distributions of hadrons produced in μ-p collisions, with respect to the virtual photon. The angle is in the laboratory system. Data at two different values of q^2 are shown.

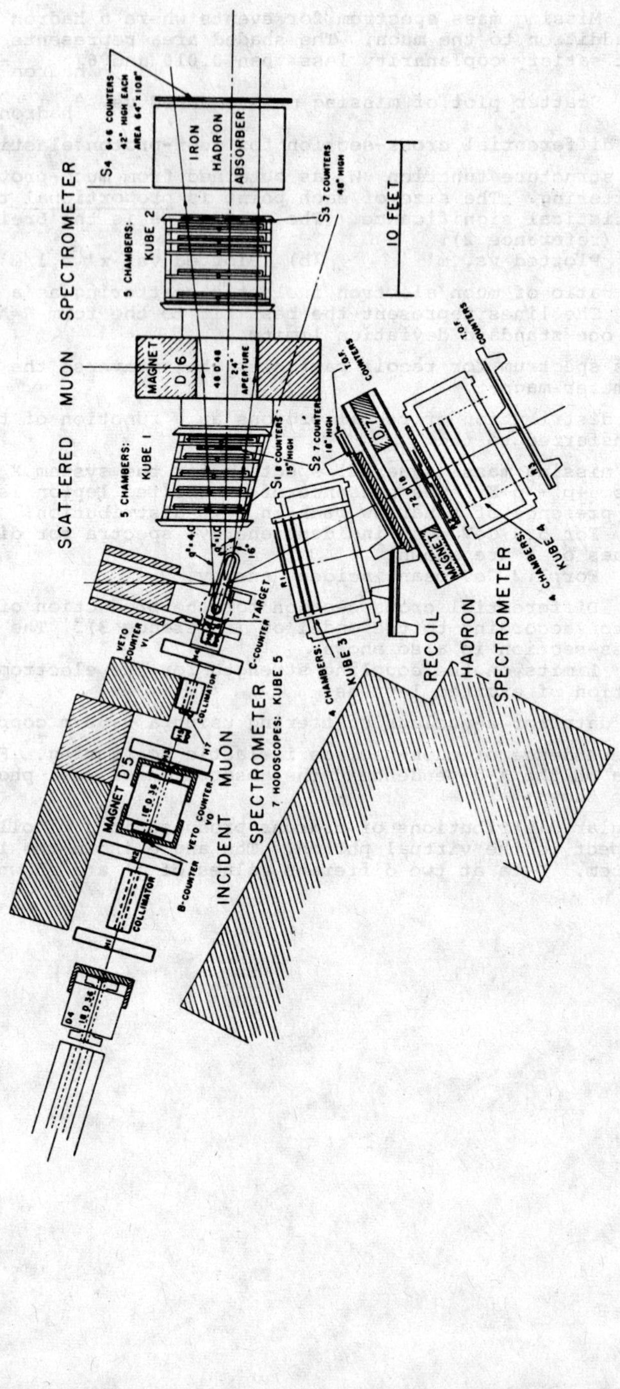
SCATTERED MUON SPECTROMETER
S4 6+6 COUNTERS 42" HIGH EACH AREA 64"x108"
IRON HADRON ABSORBER
S3 7 COUNTERS 48" HIGH
10 FEET
5 CHAMBERS: KUBE 2
MAGNET D6
48 D 48
24" APERTURE
5 CHAMBERS: KUBE 1
S1 8 COUNTERS 15" HIGH
S2 7 COUNTERS 18" HIGH
T.O.F. COUNTERS
D7
72 D 18
MAGNET
4 CHAMBERS: KUBE 4
RECOIL HADRON SPECTROMETER
4 CHAMBERS: KUBE 3
TARGET
C-COUNTER
VETO COUNTER V1
COLLIMATOR
MAGNET D5
18 D 36
VETO COUNTER V0
B-COUNTER
COLLIMATOR
INCIDENT MUON SPECTROMETER
7 HODOSCOPES: KUBE 5
D4
18 D 36

Fig. 1

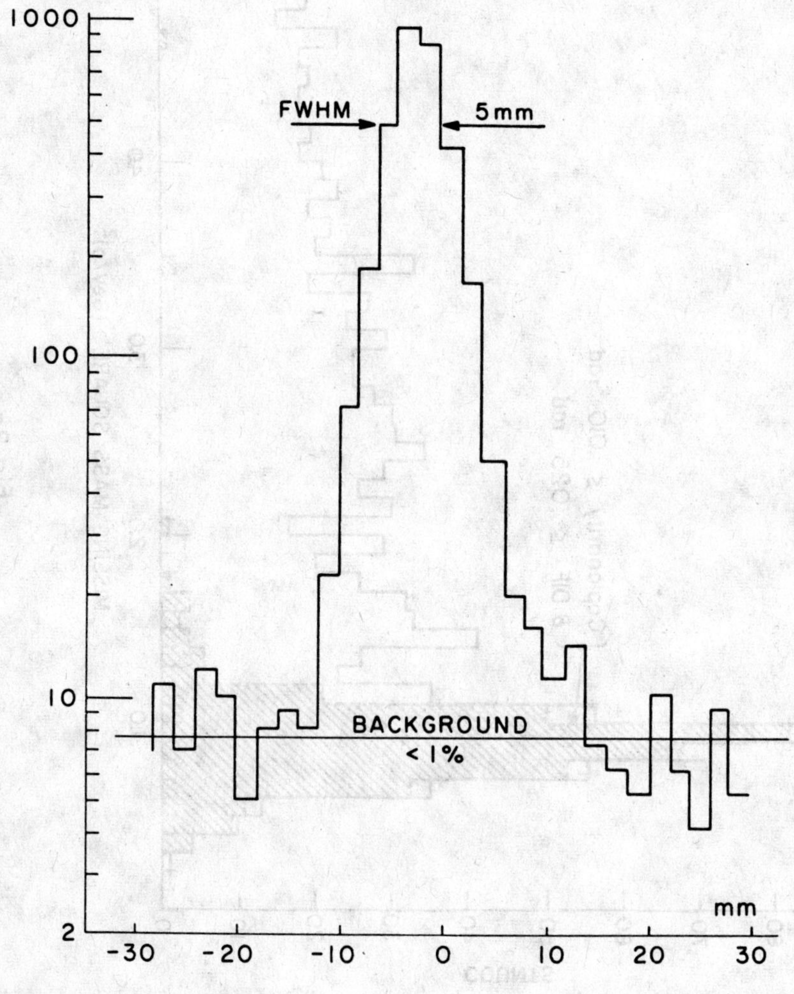

Fig.2

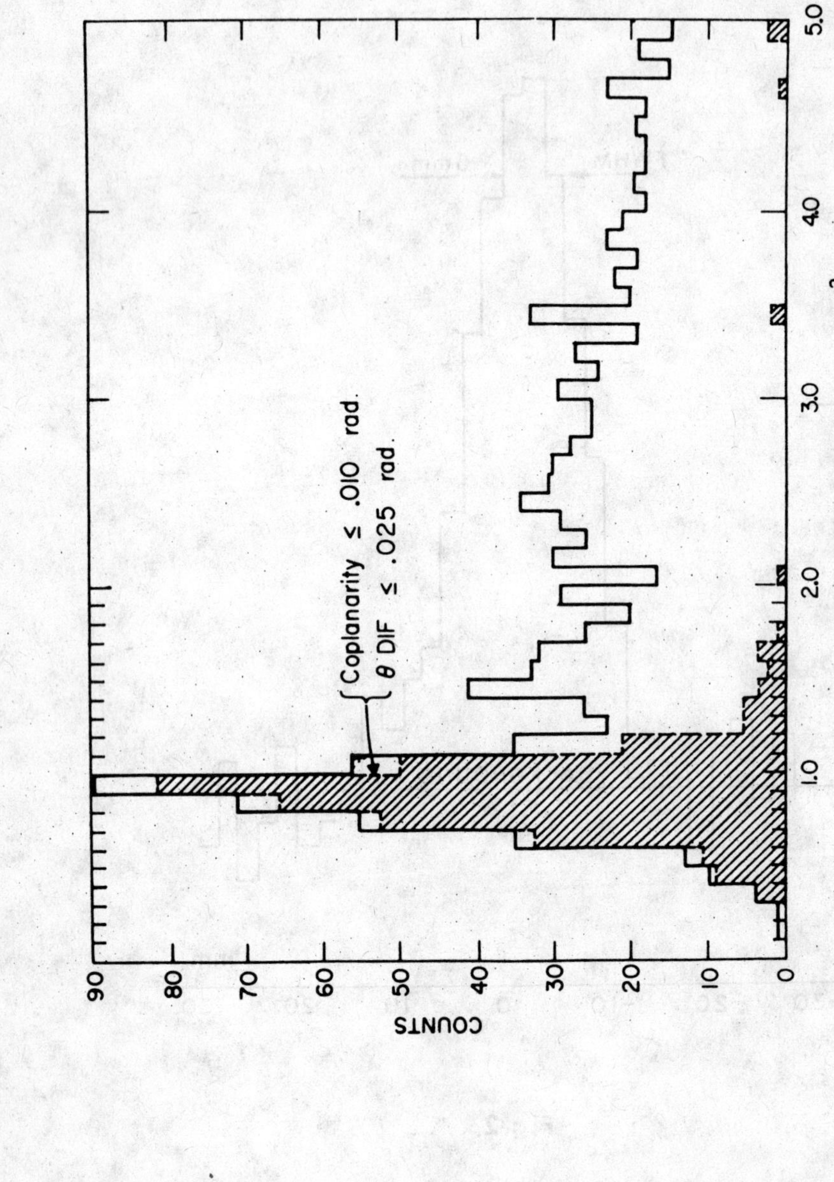
COUNTS
0
10
20
30
40
50
60
70
80
90
1.0
2.0
3.0
4.0
5.0
MISSING MASS SQUARE (GeV/c)^2
Coplanarity ≤ .010 rad.
θ DIF ≤ .025 rad.

Fig. 3a

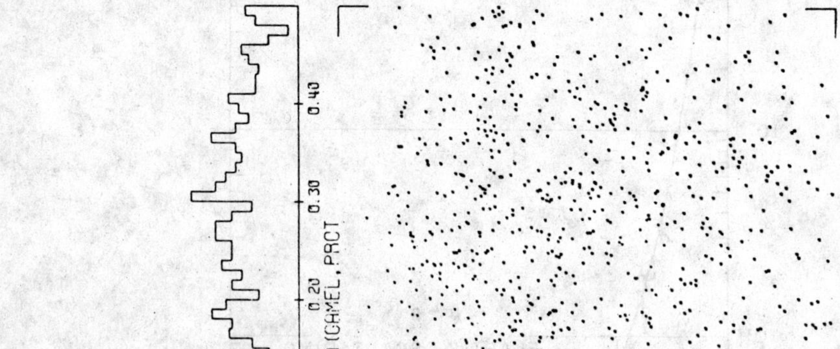
0.40
0.30
0.20
0.10
0.00
THETA (GAMEL, PRCT
60
30
IN BIN
40
20
IN BIN
0.00
2.00
4.00
6.00
8.00
WSQ

Fig.3b

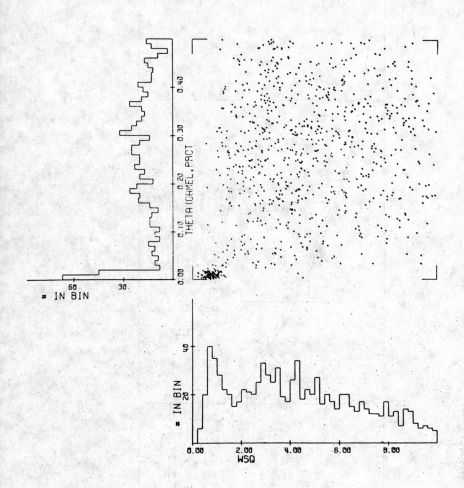

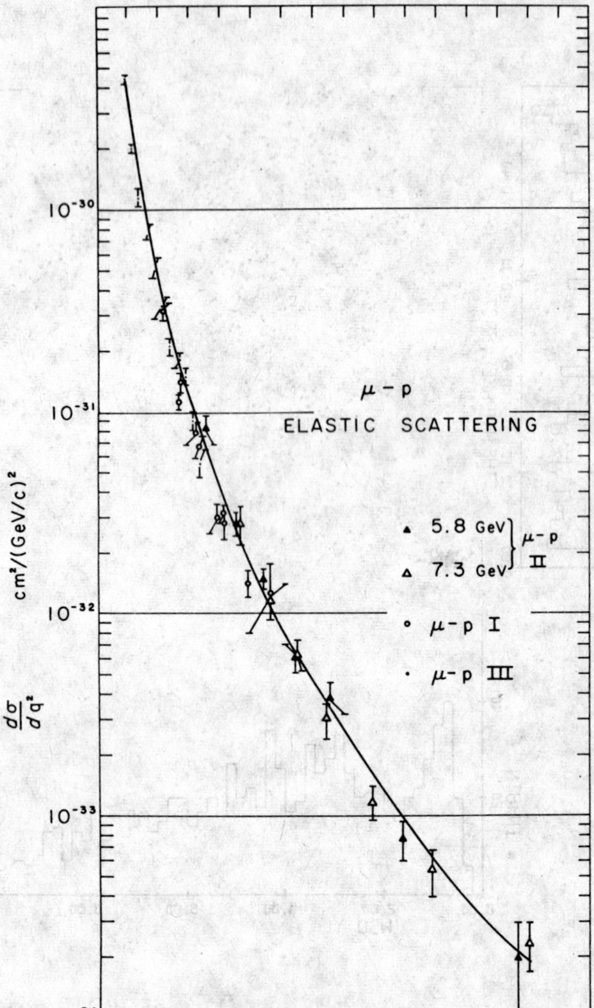

$\mu - p$
ELASTIC SCATTERING
5.8 GeV
7.3 GeV
$\mu - p$ II
$\mu - p$ I
$\mu - p$ III
10^{-30}
10^{-31}
10^{-32}
10^{-33}
10^{-34}
$\frac{d\sigma}{dq^2}$ cm²/(GeV/c)²
1.0
2.0
3.0
q²(GeV)²

Fig.4

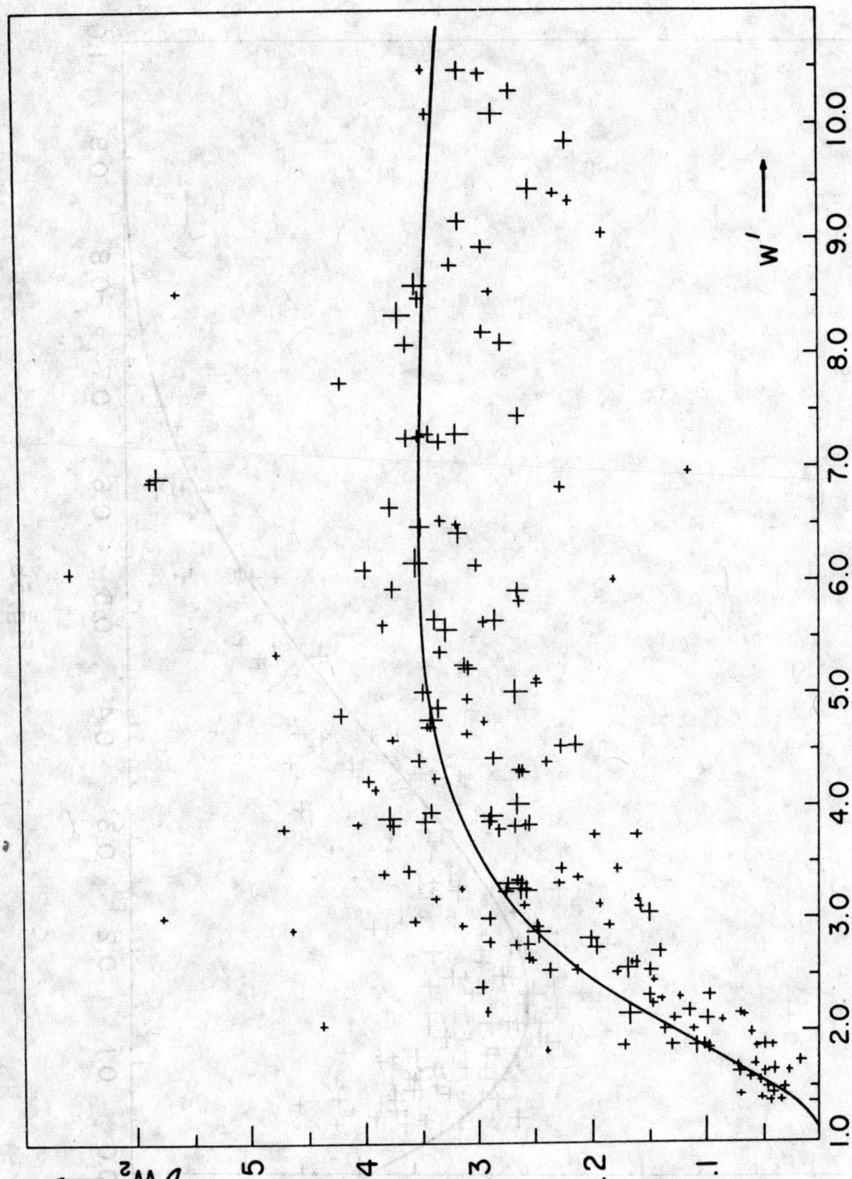

Fig.5a

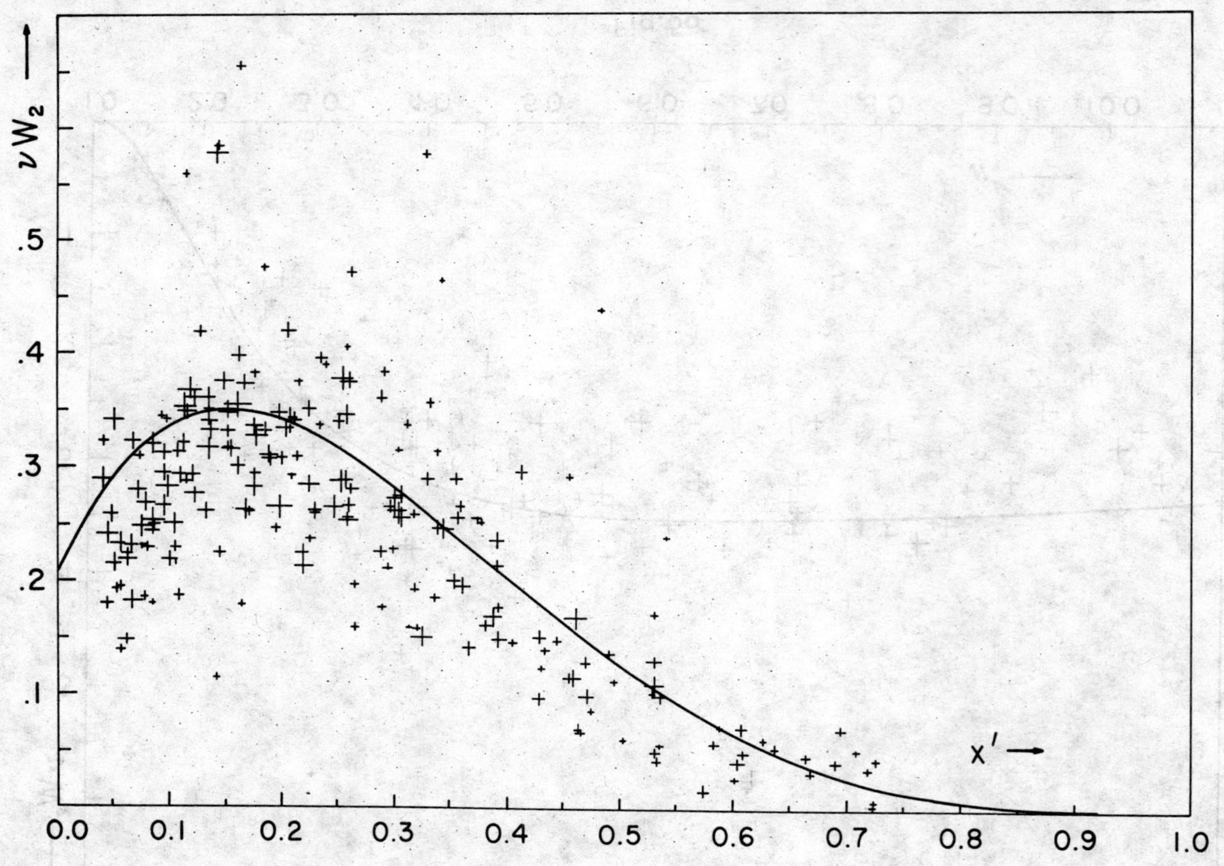

Fig. 5b

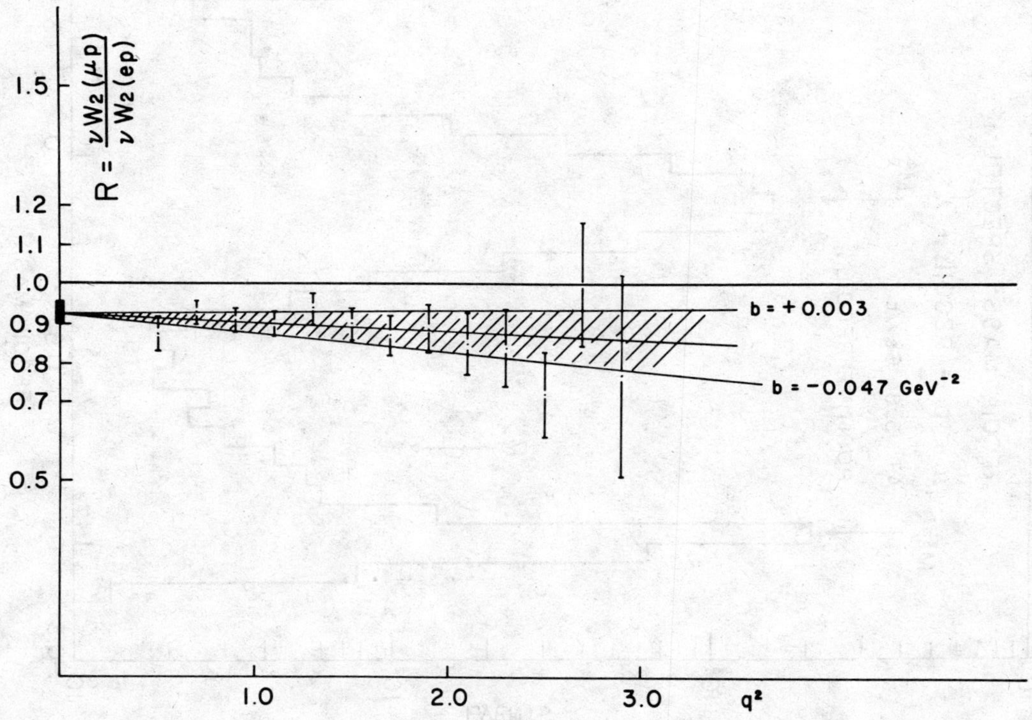
$R = \frac{\nu W_2(\mu p)}{\nu W_2(ep)}$
1.5
1.2
1.1
1.0
0.9
0.8
0.7
0.5
b = +0.003
b = −0.047 GeV^{-2}
1.0
2.0
3.0
q^2

Fig.6

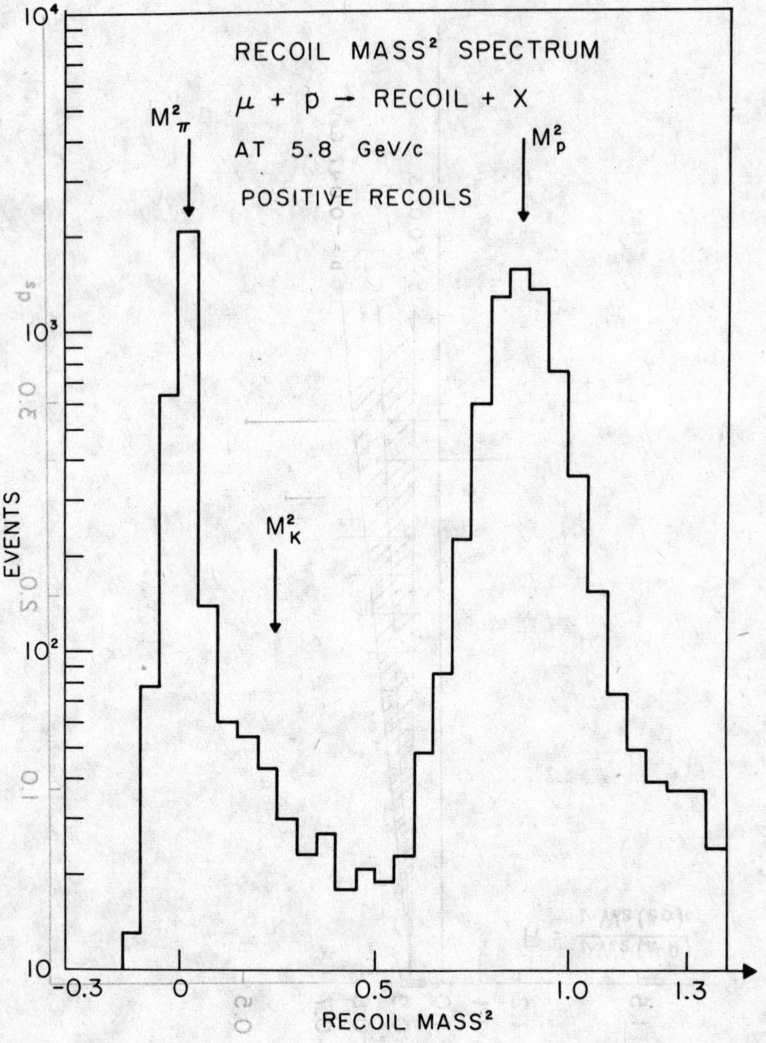
RECOIL MASS² SPECTRUM
$\mu + p \rightarrow$ RECOIL + X
AT 5.8 GeV/c
POSITIVE RECOILS
M^2_π
M^2_p
M^2_K
EVENTS
10^4
10^3
10^2
10
-0.3
0
0.5
1.0
1.3
RECOIL MASS²

Fig. 7

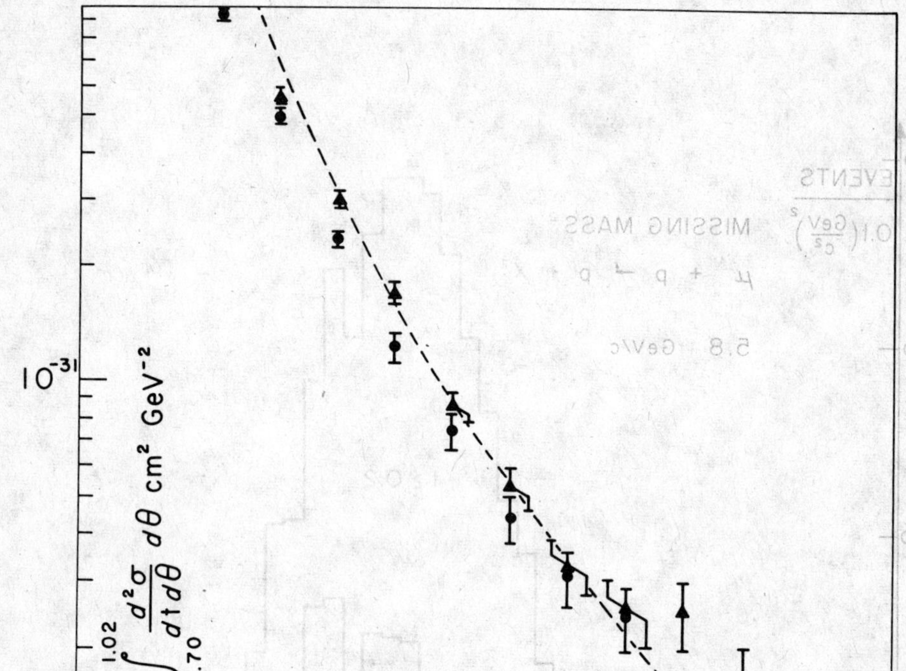
10⁻³¹
10⁻³²
∫ from 0.70 to 1.02 d²σ/dt dθ dθ cm² GeV⁻²
t (GeV/c)²
0
.2
.4
.6
.8
1.0
1.2
1.4

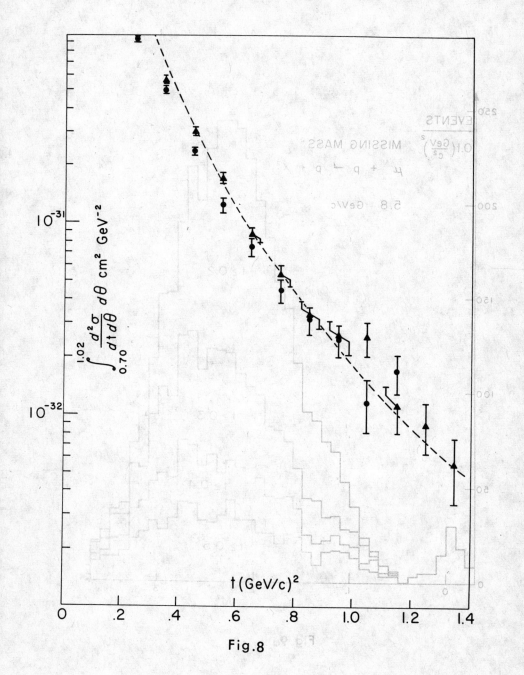

Fig.8

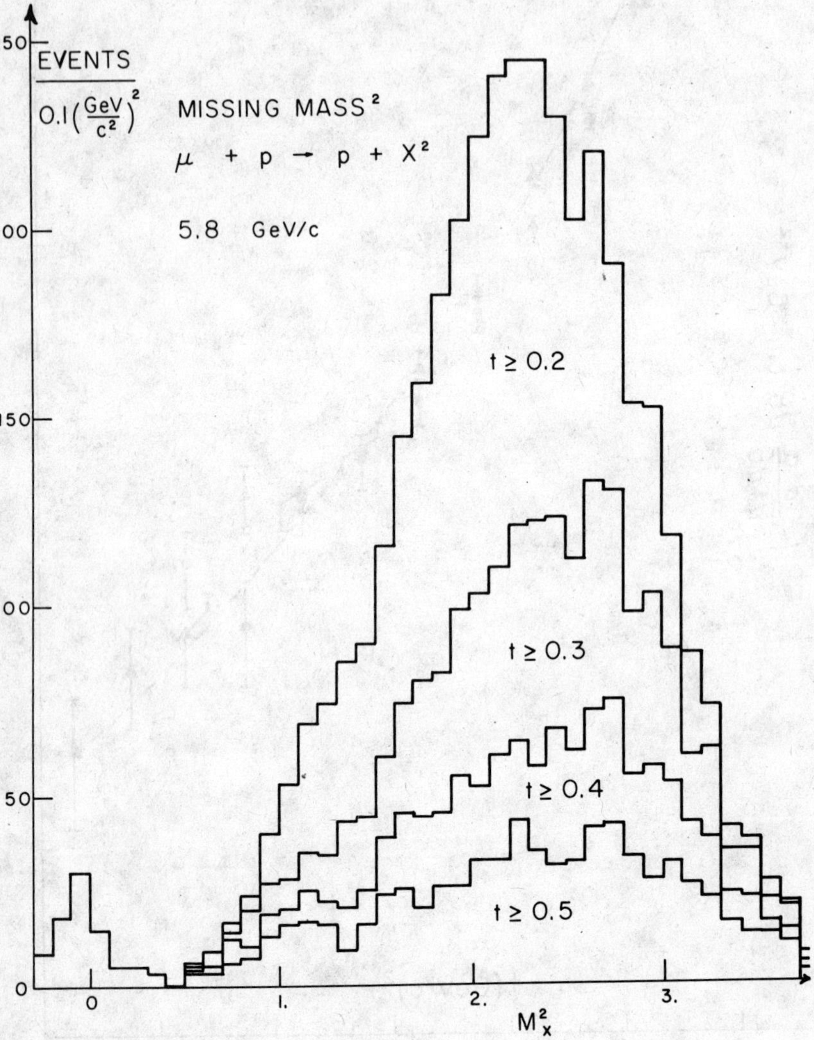
EVENTS
0.1 (GeV/c²)²
MISSING MASS²
μ + p → p + X²
5.8 GeV/c
t ≥ 0.2
t ≥ 0.3
t ≥ 0.4
t ≥ 0.5
250
200
150
100
50
0
0
1.
2.
3.
M²x

Fig. 9a

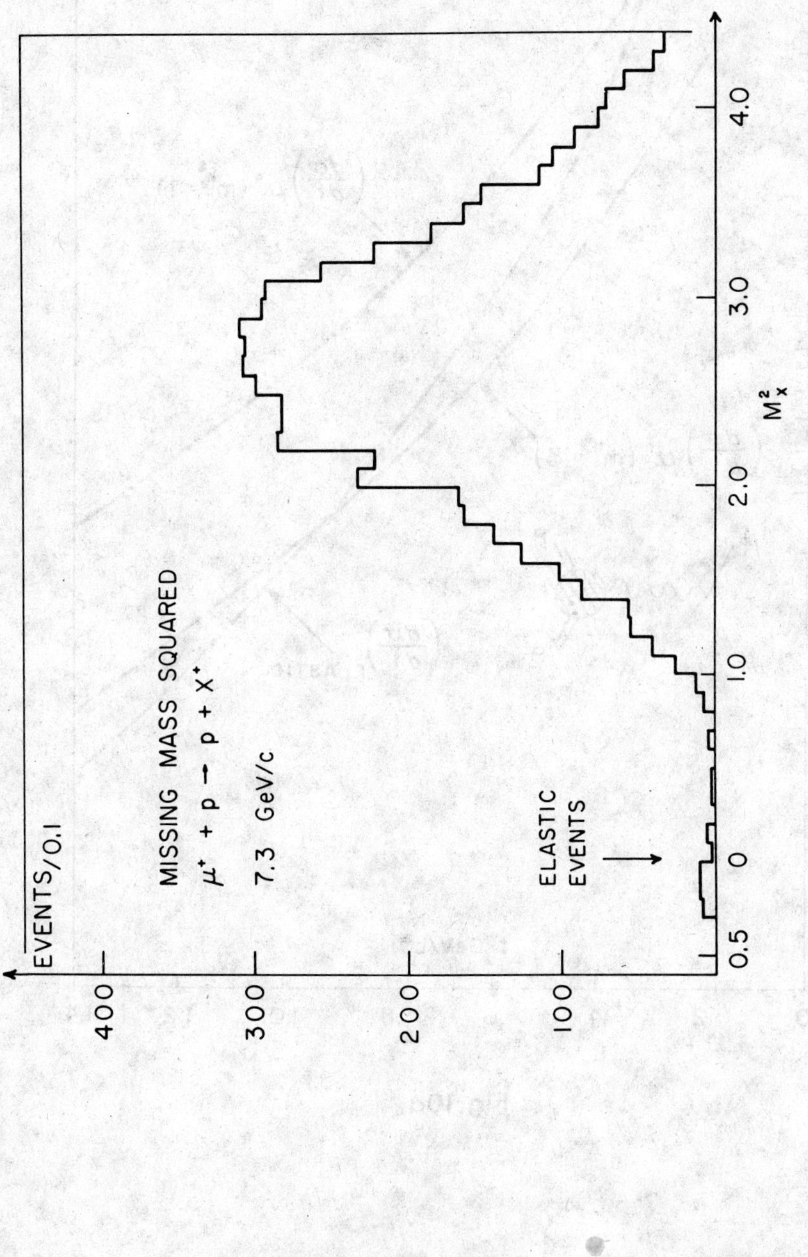
EVENTS/0.1
400
300
200
100
MISSING MASS SQUARED
$\mu^+ + p \rightarrow p + X^+$
7.3 GeV/c
ELASTIC EVENTS
0.5
0
1.0
2.0
3.0
4.0
M_x^2

Fig.9b

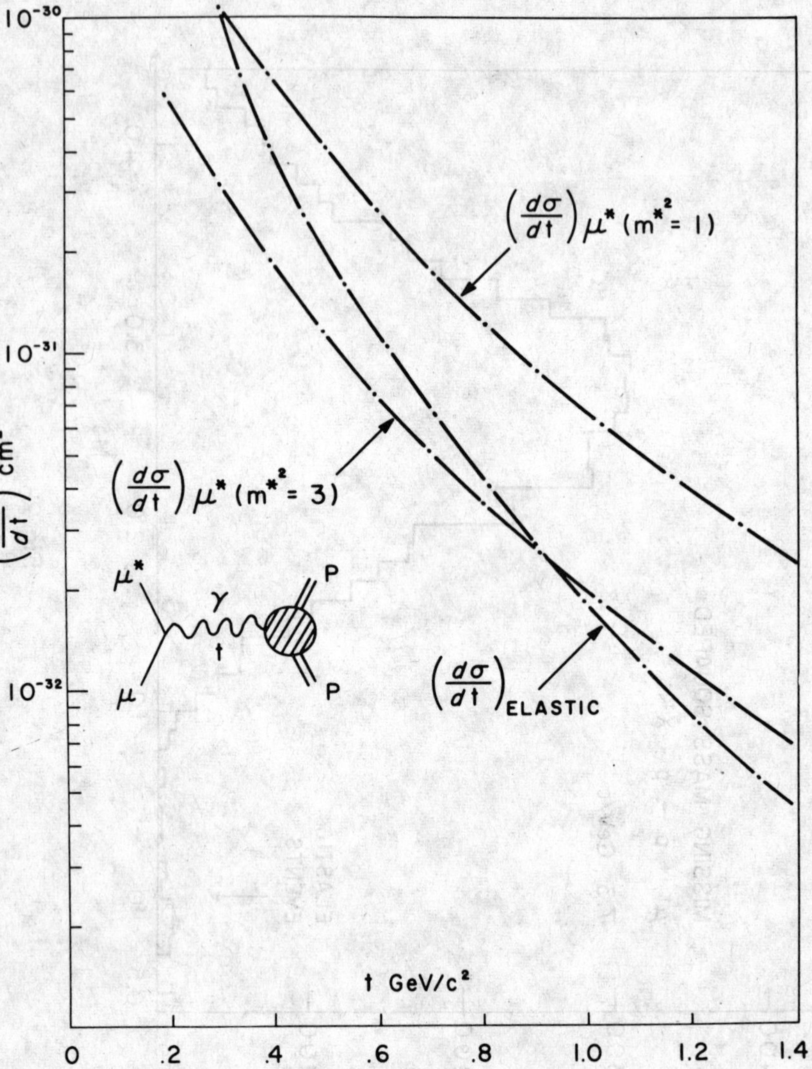

10^{-30}
10^{-31}
10^{-32}
$\left(\frac{d\sigma}{dt}\right)$ cm^2
$\left(\frac{d\sigma}{dt}\right)\mu^*(m^{*2}=1)$
$\left(\frac{d\sigma}{dt}\right)\mu^*(m^{*2}=3)$
$\left(\frac{d\sigma}{dt}\right)_{ELASTIC}$
μ^*
γ
t
μ
P
P
t GeV/c^2
0
.2
.4
.6
.8
1.0
1.2
1.4

Fig.10a

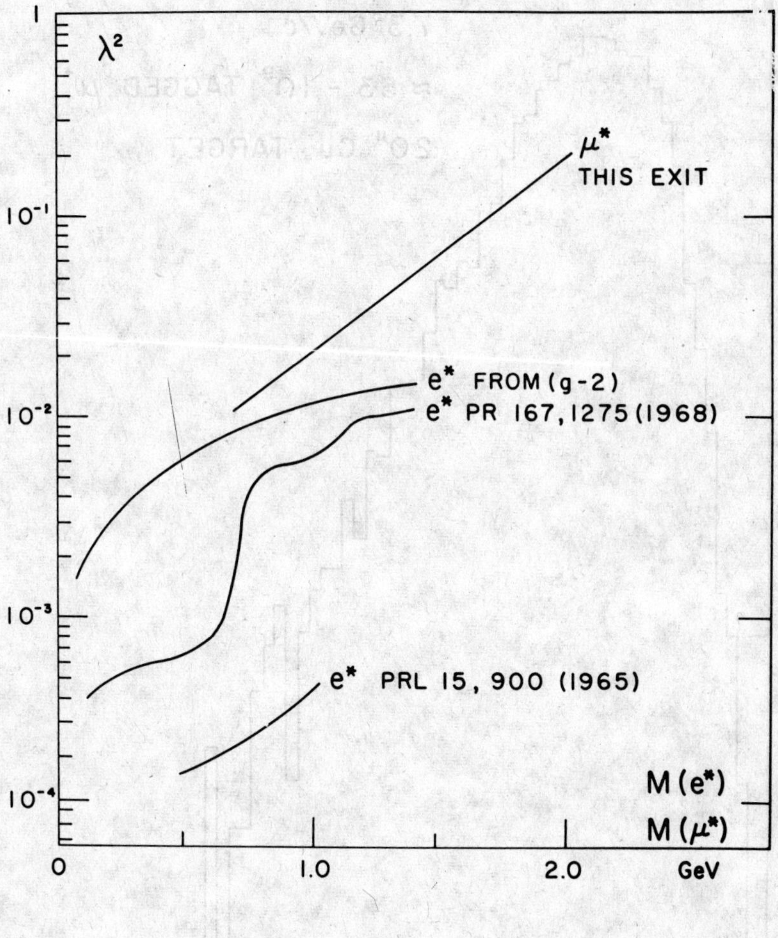
1
λ²
μ*
THIS EXIT
10⁻¹
e* FROM (g-2)
10⁻²
e* PR 167, 1275 (1968)
10⁻³
e* PRL 15, 900 (1965)
M(e*)
10⁻⁴
M(μ*)
0
1.0
2.0
GeV

Fig. 10b

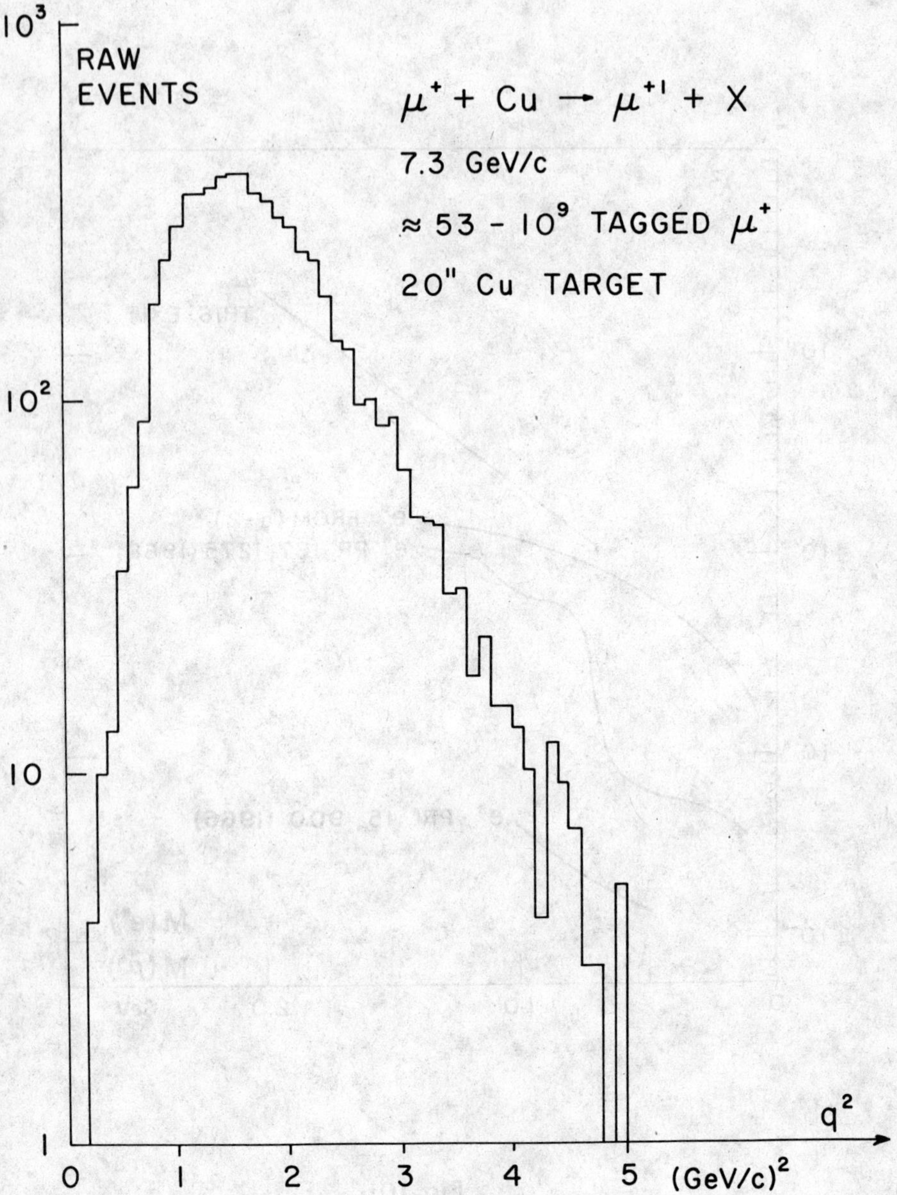

10^3
RAW
EVENTS
$\mu^+ + \mathrm{Cu} \rightarrow \mu^{+\prime} + X$
7.3 GeV/c
$\approx 53 - 10^9$ TAGGED μ^+
20" Cu TARGET
10^2
10
1
q^2
0
1
2
3
4
5
$(\mathrm{GeV/c})^2$

Fig. 11

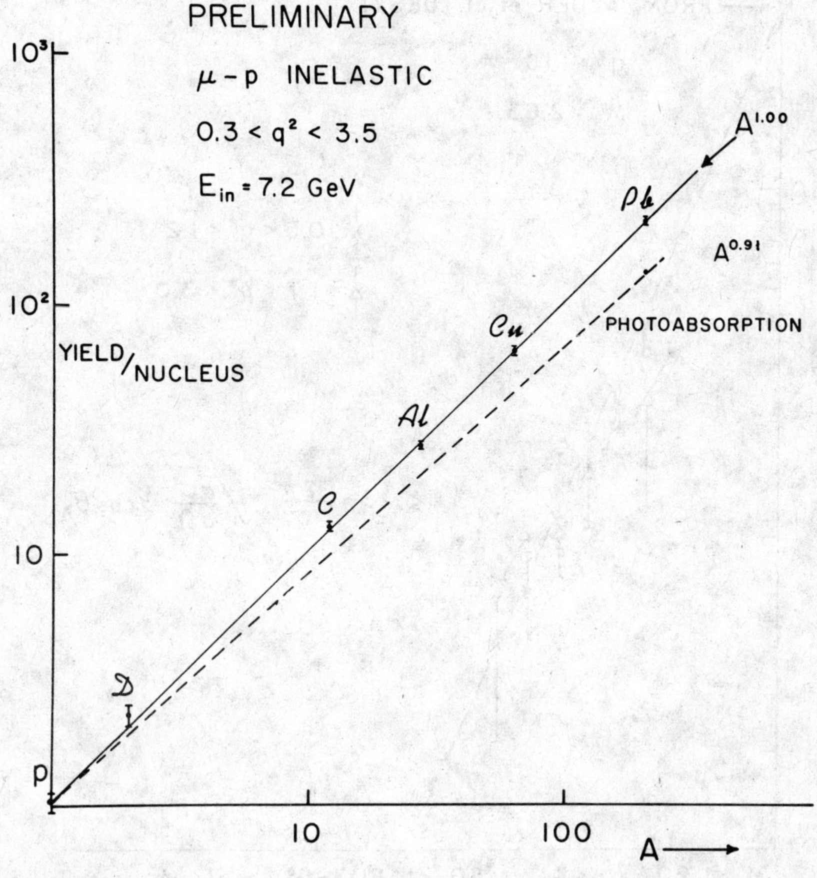
PRELIMINARY
μ - p INELASTIC
0.3 < q² < 3.5
E_in = 7.2 GeV
10³
10²
10
YIELD/NUCLEUS
A^1.00
A^0.91
PHOTOABSORPTION
Pb
Cu
Al
C
D
p
10
100
A

Fig. 12

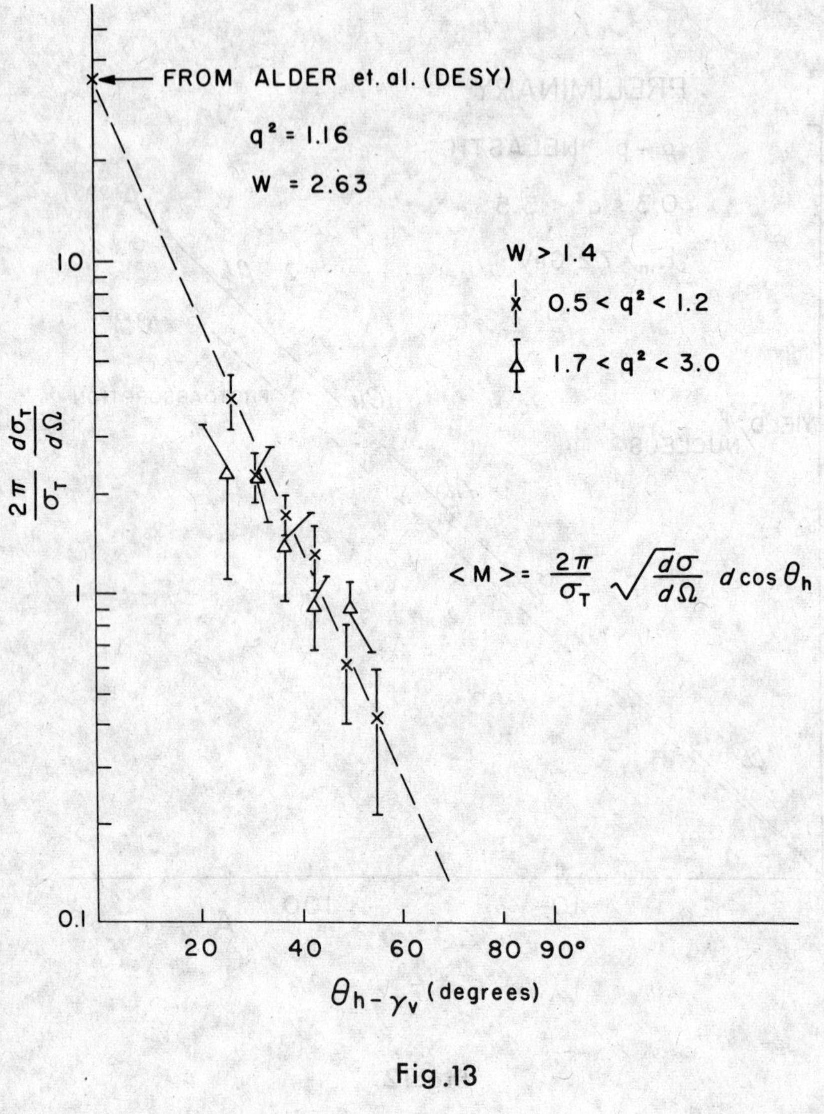
FROM ALDER et. al. (DESY)
q² = 1.16
W = 2.63
W > 1.4
0.5 < q² < 1.2
1.7 < q² < 3.0
<M> = 2π/σ_T √(dσ/dΩ) d cos θ_h
2π/σ_T dσ_T/dΩ
10
1
0.1
20
40
60
80 90°
θ_h - γ_v (degrees)

Fig.13

PHYSICS WITH MIRABELLE

A. Berthelot

Table of Contents

PHYSICS WITH MIRABELLE

André Berthelot
Département de Physique des Particules Elémentaires -Saclay, FRANCE

The large Hydrogen Bubble Chamber Mirabelle has already been described at seminars given at this School in 1969 and 1970 during the construction period. The main characteristics can be found in the Proceedings.Let us simply recall here that the useful volume of hydrogen is a cylinder 4. 6 m long and 1. 6 m in diameter. It is viewed by 8 cameras disposed along three horizontal lines : the higher and the lower ones consist of 3 cameras and the middle one of 2 cameras.

The chamber is installed at Protvino (U. S. S. R.) on the site of the 70 GeV proton synchrotron of the Institute for High Energy Physics and it can be fed with radio-frequency mass separated beams built by CERN. With the present 2 cavities system one can obtain π^- up to 60 GeV , π^+ up to 50 GeV, K^+, K^- and $\bar{p}$ up to 32 GeV. With one more cavity it is expected to push these limits somewhat upward.

The time schedule has been the following :

Summer 1971 : The assembly of Mirabelle has been completed on the site of Protvino.

November 1971 : The first run was made with a proton beam obtained by small angle scattering on an internal target.

Spring 1972 : Photographs have been taken with a π^-beam obtained from an external target and without mass separation.

Summer 1972 : The R. F. mass separated beams were completed making possible to have $\pi^\pm$, $K^\pm$, $\bar{p}$ in Mirabelle.

Although photographs have been obtained with all these particles the study, up to now, has mainly concerned protons(1) (50 and 69 GeV/c), π^-(50 GeV/c)(2) and K^- (32 GeV/c)(2).

With regard to scanning and measuring devices, only tables of the IEP type are in use at present, either specially built for Mirabelle, or adapted to it. Most of them are in Saclay and at the I. H. E. P. but there are also in some other laboratories, in particular at Mons (Belgium) and at the University of Paris.

Automatic devices, both of the H. P.D. and of the C. R. T. type are expected to work in the near future with films from Mirabelle at Saclay, at the Collège de France(Paris), and at the I. H. E. P.

Contribution to the work reported here have come from people belonging to the following laboratories :

For p-p studies: I. H. E. P. (Protvino)- L. P. N. H. E. (University of Paris) - Saclay.

For π^--p and K^- - p studies :Aachen-Berlin Zeuthen-Brussels-CERN - Collège de

France - I. H. E. P. (Protvino) -Imperial College - Mons- Saclay- Vienna .

It is necessary to stress that most results reported at this seminar are preliminary. Work is continuing in order to improve their statistical significance. It is hoped that new results will be presented at the Aix-en-Provence Conference.

STUDY OF TOPOLOGICAL CROSS-SECTIONS

The following values have been obtained

Number of prongs n_c	p-p 50 GeV/c $\sigma(n_c)$ mb	p-p 69 GeV/c $\sigma(n_c)$ mb	K^--p 32 GeV/c $\sigma(n_c)$ mb	π^--p 50 GeV/c $\sigma(n_c)$ mb
0	-	-	0.29 ± 0.03	0.2 ± 0.1
2 inel.	5.97 ± 0.88	4.84 ± 0.42	2.98 ± 0.39	3.1 ± 0.5
4	9.40 ± 0.47	8.63 ± 0.21	5.94 ± 0.14	5.9 ± 0.3
6	7.99 ± 0.43	7.90 ± 0.20	5.03 ± 0.13	5.8 ± 0.2
8	5.02 ± 0.33	5.42 ± 0.16	2.59 ± 0.09	3.5 ± 0.2
10	2.03 ± 0.20	2.75 ± 0.11	0.91 ± 0.06	1.6 ± 0.2
12	0.48 ± 0.10	1.27 ± 0.07	0.22 ± 0.03	0.7 ± 0.1
14	0.20 ± 0.06	0.39 ± 0.04	0.03 ± 0.01	0.2 ± 0.1
16	0.01 ± 0.02	0.11 ± 0.02	0.01 ± 0.00	0.02 ± 0.02
18	-	0.01 ± 0.01	-	-
$\langle n_c \rangle$	5.32 ± 0.13	5.89 ± 0.07	5.16 ± 0.08	7.71 ± 0.13
D	2.58 ± 0.05	2.89 ± 0.03	2.41 ± 0.03	2.72 ± 0.06

The p-p values interpolates between the values at P. S. energies and the recently obtained values from Batavia, while the K^- p and π^- p values are at present the highest energy results available.

The mean value of charged secondaries $\langle n_c \rangle$ seems to be satisfactorily represented as a function of energy by a linear variation with log s for each kind of interaction. This law can be made approximately valid for all kinds of interaction if, instead of s , the variable is the kinetic energy of the initial state in the center of mass : $Q = s^{1/2} - m_A - m_B$

The ratio $\langle n_c \rangle / D$, where $D = \left[\langle n_c^2 \rangle - \langle n_c \rangle^2 \right]^{1/2}$ is the dispersion, is very near to 2 in agreement with the result of Dao et al who have shown that it is approaching the limit 2 from above in case of p-p. The values obtained for K^-(32 GeV/c) and π^- (50 GeV/c) indicates that this is also the case for these particles.

Czyszewski and Rybicki have shown that if instead of plotting $\sigma(n_c)$ as a function of n_c, one plots $\sigma(n_c) \cdot \frac{D}{\sigma \text{ inel.}}$ as a function of $\frac{n_c - \langle n_c \rangle}{D}$, one obtains a universal curve valid for different nature of primaries and different values of s. The points obtained in the experiments reported fit perfectly with the curve established at lower energies per Czyszewski and Rybicki.

It has been shown that neither the Poisson distribution, nor the distribution derived from such models as the Wang ones give sufficient agreement with the experimental data to be considered as satisfactory.

SEARCH FOR CORRELATIONS BETWEEN CHARGED SECONDARY PARTICLES

The f_2 correlation coefficient between any two secondary particles

$$f_2 = D^2 - \langle n_c \rangle$$

which is negative for $Q < 6$ GeV has the following values :

π^- (50 GeV/c) = + 2.10 ± .06 $\quad$ K^- (32GeV/c) = + 2.14 ± .04

p (50 GeV/c) = + 2.08 ± .06 $\quad$ p (69 GeV/c) = + 2.04 ± .03

INCLUSIVE PHOTON PRODUCTION IN p-p COLLISIONS AT 69 GeV/c [3]

Taking account of the fact that the long length of the chamber gives an appreciable value of the probability of conversion of γ's into electron pairs (about 7% after defining a suitable fiducial volume)it is possible to measure the rate of γ emission. Making the hypothesis that all γ's come from π° decay - supported by the fact that when one plots the mass spectrum of γ pairs only the π° peak is found, without any indication of η - the mean value of the number of π° emitted per inelastic collision is found to be $\langle n_{\pi^\circ} \rangle = 2.57 \pm .13$.

This value is in agreement with a linear variation in log s as suggested by a compilation of results including low s points as well as points from Batavia (at 205 GeV/c) and I. S. R.

The mean number of π° as a function of the number of charged prongs n_c is approximately to $n_c/2$ up to $n_c = 8$ and decreases slightly for larger values of n_c .

The correlation parameters between π^- and π° , and between $2\pi^\circ$ have been found :

$$f_2 (\pi^- \pi^\circ) = + 1.0 \pm .2$$
$$f_2 (\pi^\circ \pi^\circ) = - 2 \pm 1$$

It has been verified that photon emission satisfies approximate scaling behaviour in the limited range of values $0.01 < p_t^2 < 0.1\ (\mathrm{GeV}/c)^2$

$$|x| < 0.15$$

but the factorisation of invariant cross section observed at the I. S. R. for $p_t > 0.1$ GeV/c appears to be not valid at 69 GeV/c ($0.01 < p_t < 0.6$ GeV/c).

The 69 GeV/c point on a Ferbel plot interpolates satisfactorily between the 12 GeV/c point and the 205 GeV/c Batavia point plus the I. S. R. points with a linear dependance on $p^{-\frac{1}{4}}$.

INCLUSIVE K° ($\overline{K}°$) and $\Lambda°$ PRODUCTION IN p-p COLLISIONS AT 69 GeV/c

The relevant cross-sections have been found to be :

$p + p \rightarrow K° (\text{or } \overline{K}°) + \ldots.$	8.4 ± 1.1 mb
$p + p \rightarrow \Lambda° + \ldots\ldots$	$3.6 \pm .5$ mb

The ratio K° (or $\overline{K}°$) /$\pi°$ is about 1/10 and approximately independant of n_c . Scaling behaviour is apparently reached for neutral K's and the corresponding point on a Ferbel plot is in agreement with a $p^{-\frac{1}{4}}$ law. The present limited statistics on Λ's does not permit to give a precise statement regarding their behaviour concerning scaling and Ferbel plot.

+

\+ +

REFERENCES

(1) V. V. Amosov et al - Phys. Lett. 42 B (1972) 519

(2) V. V. Amosov et al - To be published by Nuclear Physics

(3) H. Blumenfeld et al - To be published by Physics Letters

(4) H. Blumenfeld et al - To be published by Physics Letters

CLOSING SPEECH

THE ROLE OF THEORETICAL AND EXPERIMENTAL PHYSICS IN THE UNDERSTANDING OF NATURE

A. Zichichi
CERN, Geneva, Switzerland.

Some time ago we had at this School a General Discussion Session, where the problem of theoretical versus experimental students was treated[1]. It was pointed out that perhaps some young theoreticians, busy working with useless models, would do much better if engaged in experimental physics. What is in everybody's mind, but is never clearly stated, is the following: Is theoretical physics harder than experimental physics? i.e. Do you need a higher intellectual level in order to do the former? The answer to this question can play an important role in the choice that a young student makes when deciding in what field to work. We propose to find the answer in the real facts.

For 2,000 years, from Aristotle to Galilei, mankind was doing pure theory (philosophy) and was convinced that no truth can exist which is better than the theoretical truth: 'If I can prove that your theory is wrong using the power of my intellect, I must be right and you wrong'. The revolutionary teaching of Galilei was to put every theory to experimental test, and this produced the great development of modern science. All this is nice but vague.

Let us look at the problem in the following terms: Who contributes most to our understanding of Nature, to the discovery of new laws, experimental or theoretical physics? To answer this question we need only look at the facts.

As is well known, the first discovery of modern science, the universality of motion, is due to Galilei and it is of an experimental nature. He proved that the theory of Aristotle (for 2,000 years believed to be correct), that there are two types of motion, one proper to the sub-lunar world (the linear) and the other proper to the celestial bodies (the circular motion), was not supported by experimental evidence. Again, without the accurate and systematic measurements of Tycho Brahé, it would have been impossible for Kepler to discover his famous three laws, unified later by Newton in his law of universal gravitation. But how could Newton have known that the gravitational attraction Moon-Earth was the same as that of a material body in the Earth (once account is taken of the $1/r^2$ effect) if it was not for the measurements of Galilei? On the other hand, Neptune is a theoretical discovery. Astronomers were asked to point their telescopes at a precise region of the sky, and Neptune was indeed seen. Pluto was discovered in a similar way, and its observation was again based on a theoretical prediction. Somebody could question the validity of referring to such old events. After all, the development of science can introduce drastic changes in the way in which the secret laws of Nature can be mastered.

Let us start with 1947. The Lamb shift was theoretically suggested to be -27 MHz and experimentally measured to be +1,000 MHz. The first-order radiative shift (the key to all self-interaction effects) is therefore an experimental discovery. The pion was theoretically predicted but the muon is an experimental discovery and remains one of the greatest puzzles

1) General discussion in the volume "Theory and Phenomenology in Particle Physics", Proceedings of the 1968 International School of Physics "Ettore Majorana" (Academic Press, New York, 1969), pp. 771-788.

of subnuclear physics. The strange particles were theoretically unwanted, but experimentally discovered; the Ω^- is like Pluto, and with it SU(3) is a theoretical goal.

In the field of weak interactions, theorists are at the forefront with P and C violations, with (V-A), and finally with the famous Cabibbo angle, which allowed the universality principle of weak interactions to be saved. The existence of the K_2^0 was theoretically suggested, even if its lifetime could not (even nowadays) be theoretically predicted. The 'two neutrinos' is a sort of (50-50) case. On the experimental side we had some mysteriously forbidden decay modes of the μ, in particular ($\mu \rightarrow e\gamma$), and on the theoretical side an estimate of the ($\mu \rightarrow e\gamma$) rate which was many orders of magnitude above the experimental limits; thus the suggestion of a new leptonic quantum number. But the 2π decay mode of the K_2^0 is unquestionably an experimental achievement; even if the experiment was motivated by the suspected anomalous regeneration of K_2^0. It was certainly not motivated by a theoretical model of T-violation or of superweak forces. These models came later.

In the field of electromagnetic interactions the hope that radiative effects, in phenomena involving hadrons, would be damped by the hadron structure, was seriously questioned in the celebrated work of Bjorken; and this is the starting point of the deep inelastic physics discovered at SLAC. However, what is probably a strongly connected phenomenon, the discovery of the large (e^+e^-) annihilation cross-section into hadrons, had no theoretical motivation -- more precisely, zero theoretical support -- the expectation being the vanishing tails of the known vector mesons.

The discovery of the antibaryons was theoretically motivated (Dirac). However, it should be emphasized that in so far as we do not understand strong forces we cannot take for granted that "protons" should have antiparticles as electrons do. This is why the existence of the antibaryons lies behind the predictive power of the Dirac theory.

In strong interactions the search for and the discovery of mesonic and baryonic resonances was clearly motivated by theory: the SU(3) of Gell-Mann and Ne'eman. The first observations of shrinkage in (p-p) elastic scattering was due to Reggeology; all subsequent experiments on shrinkage, antishrinkage, etc., including spin-dependent effects, were and still are all motivated by the various versions of the Regge theory. It should, however, be pointed out that the spin-dependence in (π^--p) charge exchange was theoretically unexpected (because only the ρ trajectory was supposed to play a role in the above process), and was immediately cured by *ad hoc* theoretical models. Here we come to a typical feature of strong interaction theoretical physics: the incredible rate of model production by theorists. In spite of this high production rate only two basic features are really understood in strong interactions: i) the existence of a symmetry higher than isospin, SU(3); ii) the fact that the spin of a virtual particle should not be thought of as being untouchable; it changes as function of the four-momentum transfer q^2. If we now go to the very high energy phenomena (those investigated at the CERN ISR), we find two experimental discoveries: the rate of large p_T events; and the rising (p-p) total cross-section; both discussed from the experimental and theoretical points of view at this School. Concerning the 10% rise of the (p-p) total cross-section in the ISR energy range, I would only like to emphasize that the absolute value of the coefficient of the rising term is something like two orders of magnitude below the Froissart bound and that the log-dependence of the rise is not firmly established. In other words, we cannot say today that we understand what is the deep meaning of this rise.

The above examples show that finding keys to the new, is a field which is open to both experimenters and theorists. If you are an experimentalist, in order to find these keys you need to know as much as possible about what is going on in theoretical physics (think of the Ω^- and K_2^0); if you are a theorist you need to know not only those results relevant to your daily model, but all experimental results, because in some corner you can find a new key. For example, in spite of the tremendous theoretical models available on the market, no one is able to predict that the ratio of elastic to total (p-p) cross-section should be $\frac{1}{4}$; or that at energies much higher than their rest masses (and in fact at infinite energies if we extrapolate the latest NAL data) the two SU(3) pseudoscalar brothers, π and K, should show a remarkable difference in their total cross-sections against protons. If we want to have a chance to select something relevant from the bulk of theoretical models or from the bulk of experimental data, we ought to know both. (This is why we need this School.)

As we have seen above, the great development of subnuclear physics in the last quarter of a century is due to the key experiments. Sometimes these have been suggested by theory, but sometimes they had their roots in the experiment itself. So at the highest level of our research, theory and experiment compete well and contribute equally to the development of our knowledge.

However, experimental physics provides formidable challenges to theorists. For example, we do not observe those processes that are predicted by the theorists to have infinite rates, such as all higher-orders weak interaction reactions. Is this because we do not have a renormalizable theory of weak interactions? The negative electron and the positive proton intervene in weak interactions with the same $(1 + \gamma_5)$ coupling. Is this because the leptonic and the baryonic worlds are connected? or are they thoroughly orthogonal? We observe regularities which are beyond SU(3). Is this because unitary spin and Dirac spins are correlated? At relatively low energies (a few GeV's) we observe that the inelastic electromagnetic interaction of the hadrons is point-like. Is this because the hadrons are made of superelementary constituents? and if so, why do they not show up? We observe six types of fundamental interactions. Are they all orthogonal to each other, or have they a common origin?

If you are a theorist and spend your time working on *ad hoc* theoretical models to explain a few experimental results, it is very likely that nothing will remain of your work. But if you are an experimentalist and you measure some quantity, even the most boring one such as total or differential hadronic cross-sections, you can be sure that the measurements will remain, and you even have the chance that something unexpected can come out from your data.

At the key-point of our research, experiment and theory contribute with the same level of intellectual power and pleasure. At the standard level of everyday life, experimental physics is much better off. In other words, if you want to be a theorist you had better be sure that you really are a great genius. Otherwise, you will find yourself cooking-up useless models all your life long.
not sure of being a genius, you can still be of great help to physics not only by doing key experiments but also by performing careful, precise -- even if boring -- measurements.

CLOSING CEREMONY

The Closing Ceremony took place on Tuesday 24th July, 1973. The Director of the School presented the prizes and scholarships to the winners as specified below.

PRIZES AND SCHOLARSHIPS

Prize for *Best Student* - awarded to Dr. F.A. WILCZEK - Princeton University, Princeton, N.J., U.S.A.

The following students received *honorary mentions* for their contributions to the activity of the School:

D. NANOPOULOS -- The University of Sussex, Brighton, Sussex, U.K.

R.F. CAHALAN -- University of Illinois at Urban-Champaign, Urbana, Ill., U.S.A.

A. BOUQUET -- Laboratoire de Physique Théorique et Hautes Energies, Paris, France.

S.A. JACKSON -- Massachusetts Institute of Technology, Cambridge, Mass., U.S.A.

B.R. KIM -- III Physikalisches Institut, Aachen, Germany.

W.Y. YU -- International Centre for Theoretical Physics, Trieste, Italy.

R. VILELA MENDES -- Instituto de Fisica e Matematica, Lisboa, Portugal.

I. MOEN -- University of Cambridge, Cambridge, U.K.

S.J. YELLIN -- DESY, Hamburg, Germany.

The following scholarships were open for competition among the participants. They have been attributed as follows:

Ettore Majorana Scholarship - awarded to Dr. F.A. WILCZEK, Princeton University, Princeton, N.J., U.S.A.

Giulio Racah Scholarship - awarded to Dr. G. CHU, Massachusetts Institute of Technology, Cambridge, Mass., U.S.A.

Amos De-Shalit Scholarship - awarded to Dr. E. HENDRICK, Rockefeller University, New York, U.S.A.

Peter Preiswerk Scholarship - awarded to Dr. P. GENSINI, University of Lecce, Italy.

Enrico Persico Scholarship - awarded to Dr. W. RABINOVICI, The Weizmann Institute of Science, Rehovot, Israel.

Antonio Stanghellini Scholarship - awarded to Dr. Y. EYLON, The Weizmann Institute of Science, Rehovot, Israel.

Gianni Quareni Scholarship - awarded to Dr. M. KUPCZYNSKI, International Centre for Theoretical Physics, Trieste, Italy.

Alberto Tomasini Scholarship - awarded to Dr. A. PATKOS, Roland Eötvös University, Budapest, Hungary.

Giorgio Ghigo Scholarship - awarded to Dr. E. Derman, Columbia University, New York, U.S.A.

The following participants have given their collaboration with the scientific secretarial work:

S.B. Berger
A. Bouquet
R.F. Cahalan
D.K. Choudhury
G. Chu
E. Derman
S. Elitzur
Y. Eylon
P. Gensini
E. Hendrick
S.A. Jackson
S. Kitakado
M. Kupczynski
M. Leneke
J.M. Losecco
R. Madaras
W. Nahm
D. Nanopoulos
J. Nyiri
A. Patkos
E. Rabinovici
J. Rosen
R.T. Ross
L. Routh
F.E. Taylor
E. Tomaselli
J.C. Tompkins
M.N. Tugulea
R. Vilela Mendes
U.M. Wambach
D.M. Webber
F. Wilczeck
D. Wright
W.Y. Yu

LIST OF PARTICIPANTS

LECTURERS

E. AMALDI
Istituto di Fisica dell'Università,
Piazzale delle Scienze, 5,
00185 ROMA, ITALY.

U. AMALDI
Istituto Superiore di Sanità,
Viale Regina Elena, 293,
00161 ROMA, ITALY.

F. AMMAN
Laboratori Nazionali di Frascati,
Casella Postale n. 70,
00044 FRASCATI, ITALY.

E. L. BERGER
Argonne National Laboratory,
9700 Cass Avenue,
ARGONNE, ILL. 60440, U.S.A.

A. BERTHELOT
Centre d'Etudes Nucléaires de Saclay,
Departement de Physique des Particules Elémentaires,
Boîte Postal n. 2,
91 - GIF-SUR-YVETTE, FRANCE.

N. BYERS
University of Oxford,
Department of Theoretical Physics,
12 Parks Road,
OXFORD OX1 3 PG, U.K.

N. CABIBBO
Istituto di Fisica dell'Università,
Piazzale delle Scienze, 5,
00185 ROMA, ITALY.

L. CANESCHI
C.E.R.N.
1211 GENEVE 23, SWITZERLAND

S. COLEMAN
Princeton University,
Department of Physics,
P.O. Box 708,
PRINCETON, N.J. 08540, U.S.A.

R. H. DALITZ
University of Oxford,
Department of Theoretical Physics,
12 Parks Road,
OXFORD OX1 3 PG, U.K.

U. FANO
University of Chicago,
Department of Physics,
1118 East 58th Street,
CHICAGO, Ill. 60637, U.S.A.

P. H. FRAMPTON — University of Bielefeld,
Department of Physics,
Herforder Strasse, 28,
4000 BIELEFELD, GERMANY.

S. L. GLASHOW — Harvard University,
Lyman Laboratory of Physics,
CAMBRIDGE, Mass. 02138, U.S.A.

R. JACKIW — Massachusetts Institute of Technology,
Department of Physics,
CAMBRIDGE, Mass. 02139, U.S.A.

K. JOHNSEN — C.E.R.N.,
1211 GENEVE 23, SWITZERLAND.

H. KLEINERT — Freie Universität Berlin,
Arnimallee, 3,
1000 BERLIN, GERMANY.

G. MANNING — Rutherford High Energy Laboratory,
CHILTON, Didcot, Berks., U.K.

A. C. MELISSINOS — University of Rochester,
Department of Physics,
ROCHESTER, N.Y. 14627, U.S.A.

D. H. MILLER — Purdue University,
Department of Physics,
LAFAYETTE, Indiana 47907, U.S.A.

L. MONTANET — C.E.R.N.,
1211 GENEVE 23, SWITZERLAND.

E. PASCHOS — National Accelerator Laboratory,
P.O. BOX 500,
BATAVIA, Ill. 60510, U.S.A.

W. PAUL — Physikalisches Institut
der Universität Bonn,
2300 BONN, GERMANY.

I. I. RABI — Columbia University,
Department of Physics,
P.O. BOX 133,
NEW YORK, N.Y. 10027, U.S.A.

N. H. RAMSEY — Harvard University,
Department of Physics,
CAMBRIDGE, Mass. 02138, U.S.A.

G. RANDERS — N.A.T.O.,
1110 BRUXELLES, BELGIUM.

C. REBBI — C.E.R.N.,
1211 GENEVE 23, SWITZERLAND.

A. H. ROSENFEL' Lawrence Berkeley Laboratory,
University of California,
East End of Hearts Avenue,
BERKELEY, Calif. 94720, U.S.A.

G. SALVINI Istituto di Fisica dell'Università,
Piazzale delle Scienze, 5,
00185 ROMA, ITALY.

J. J. SAKURAI University of California,
Department of Physics,
405 Hilgard Avenue,
LOS ANGELES, Calif. 90024, U.S.A.

G. H. STAFFORI Rutherford High Energy Laboratory,
CHILTON, Didcot, Berks., U.K.

W. THIRRING Institut für Theoretische Physik
der Universität Wien,
Boltzmanngasse 5,
1090 - WIEN, AUSTRIA.

S. C. C. TING Massachusetts Institute of Technology,
Department of Physics,
CAMBRIDGE, Mass. 02139, U.S.A.

D.E.S.Y.,
Notkestieg 1,
HAMBURG 52, GERMANY.

G. ULLMAN C.E.R.N.,
1211 GENEVE 23, SWITZERLAND.

B. H. WIIK D.E.S.Y.,
Notkestieg 1,
HAMBURG 52, GERMANY.

PARTICIPANTS

S. H. ARONSON — The University of Wisconsin,
Department of Physics,
475 North Charter Street,
MADISON, WI 53706, U.S.A.

S. B. BERGER — Massachusetts Institute of Technology,
Laboratory for Nuclear Science,
CAMBRIDGE, Mass. 02139, U.S.A.

F. BOBISUT — Istituto di Fisica dell'Università,
Via Marzolo, 8,
35100 - PADOVA, ITALY.

A. BOUQUET — Laboratoire de Physique théorique et Hautes En.,
Université Paris VI,
Tour 16 -1er étage,
75230 PARIS CEDEX 05, FRANCE.

G. BRANCO — The City College of the City University of New York,
Department of Physics,
NEW YORK, N.Y. 10031, U.S.A.

A. BRANDT — C.E.R.N. - NP Division,
1211 GENEVE 23, SWITZERLAND.

F. BUCCELLA — Istituto di Fisica dell'Università
Piazzale delle Scienze, 5,
00185 ROMA, ITALY

R. F. CAHALAN — University of Illinois at Urban-Champaign,
Department of Physics,
URBANA, Ill. 61801, U.S.A.

C. CARIMALO — Laboratoire de Physique atomique et moléculaire,
Collège de France,
PARIS 5e, FRANCE.

D. K. CHOUDHURY — University of Oxford,
Department of Theoretical Physics,
12 Parks Road,
OXFORD OX1-3PQ, U.K.

G. CHU — Massachusetts Institute of Technology,
Laboratory for Nuclear Science,
CAMBRIDGE, Mass. 02139, U.S.A.

B. COLUZZI — Istituto di Fisica Sperimentale dell'Università,
Via Antonio Tari, 3,
80138 - NAPOLI, ITALY.

D. CORDS — D.E.S.Y.,
Notkestieg 1,
HAMBURG 52, GERMANY.

C. T. DAY — Cornell University,
Newman Laboratory of Nuclear Studies,
ITHACA, N.Y. 14850, U.S.A.

K. DAMMEIER	University of Tübingen, Institut für theoretische Physik, Morgenstelle, TUBINGEN 7400, GERMANY.
E. DERMAN	Columbia University in the City of New York, Department of Physics, 538 West 120th Street, Box 40, NEW YORK, N.Y. 10027, U.S.A.
P. H. DONDI	University of Cambridge, Department of Applied Mathematics and Theoretical Physics, Silver Street, CAMBRIDGE, CB3 9EW, U.K.
W. DRECHSLER	Max-Planck-Institut für Physik und Astrophisik, Institut für Physik, Föhringer Ring 6, 8 MUNCHEN 40, GERMANY.
S. ELITZUR	Tel-Aviv University, Department of Physics and Astronomy, RAMAT-AVIV, TEL-AVIV, ISRAEL.
R. K. ELLIS	Istituto di Fisica dell'Università, Piazzale delle Scienze, 5, 00185 ROMA, ITALY.
Y. EYLON	The Weizmann Institute of Science, Department of Nuclear Physics, REHOVOT, ISRAEL.
P. FAYET	Laboratoire de Physique théorique et Hautes En., Faculté des Sciences, Batîment 211, 91 ORSAY, FRANCE.
N. FIES	Freie Universität Berlin, Institut für theoretische Physik, Arnimallee 3, 1 BERLIN 33, GERMANY.
M. GARI	Ruhr-Universität Bochum, Institut für theoretische Physik, Lehrstuhl II, Universitätstrasse 150, 463 BOCHUM-QUERENBURG, GERMANY.
P. GAVILLET	C.E.R.N. - TC Division, 1211 GENEVE 23, SWITZERLAND.
P. GENSINI	Istituto di Fisica dell'Università, via Arnesano, 73100 - LECCE, ITALY.
C. GEWENIGER	C.E.R.N. - NP Division, 1211 GENEVE 23, SWITZERLAND.

H. GOTTLIEB	Imperial College, Department of Theoretical Physics, Prince Consort Road, LONDON SW7 2AZ, U.K.
D. GROMES	Institut für theoretische Physik der Universität, Philosophenweg 16, 69 HEIDELBERG, GERMANY.
D. GROSSER	University of Tübingen, Institut für theoretische Physik, Morgenstelle, TUBINGEN 7400, GERMANY.
N.S. HALD	NORDITA Blegdamsvej 17, 2100 KOBENHAVEN, DENMARK
E. HENDRICK	Rockefeller University, Box 14, NEW YORK, N.Y. 10021, U.S.A.
S. A. JACKSON	Massachusetts Institute of Technology, Department of Physics, Room 6-415, CAMBRIDGE, Mass. 02139, U.S.A.
N. O. JOHANNESSON	NORDITA, Blegdamsvej 17, 2100 KOBENHAVN, DENMARK.
W. KAINZ	Institut für theoretische Physik der technischen Hochschule in Wien, Argentinierstrasse 8/4, 1040 WIEN, AUSTRIA.
V. KARIMAKI	C.E.R.N. - TC Division, 1211 GENEVE 23, SWITZERLAND.
K. KELLER	Sektion Physik der Universität München, theoretische Physik, Schellingstrasse 13, 8000 MUNCHEN 13, GERMANY.
R. KEPHART	C.E.R.N. - NP Division, 1211 GENEVE 23, SWITZERLAND.
B. R. KIM	III. Physikalisches Institut, der Rhein-Westf. Technischen Hochschule Aachen, Abteilung Theoretische Elementarteilchenphysik Jägerstrasse, 51 AACHEN, GERMANY.
S. KITAKADO	International Centre for Theoretical Physics, P.O. Box 586 - Miramare, 34100 TRIESTE, ITALY.
K. I. KONISHI	Scuola Normale Superiore, piazza dei Cavalieri, 7, 56100 - PISA, ITALY.

M. KUPCZYNSKI	International Centre for Theoretical Physics, P.O. Box 586 - Miramare, 34100 - TRIESTE, ITALY.
E. LABIE	C.E.R.N. - NP Division, 1211 GENEVE 23, SWITZERLAND.
W. LANG	Universität Karlsruhe, Institut für theoretische Physik, Kaiserstrasse 12, 75 KARLSRUHE, GERMANY.
M. LENEKE	Physikalisches Institut der Universität Bonn, Nussallee 12, 53 BONN, GERMANY.
C. LEROY	Université de Louvain, Institut de Physique théorique, Chemin du Cyclotron, 1348 - LOUVAIN-LA-NEUVE, BELGIUM.
J. M. LOSECCO	Harvard University, Jefferson Physical Laboratory, CAMBRIDGE, Mass. 02138, U.S.A.
R. MADARAS	IN2P3 - Laboratoire de l'Accélérateur Linéaire, Batîment 200 - Centre d'Orsay, 91405 ORSAY, FRANCE.
P. MARCOLUNGO	Istituto di Fisica dell'Università, via Marzolo, 8, 35100 - PADOVA, ITALY.
S. MARCULESCU	Universität Karlsruhe, Institut für theoretische Physik, Kaiserstrasse 12, 75 KARLSRUHE, GERMANY.
H. MATSUMOTO	The University of Wisconsin-Milwaukee, Department of Physics, MILWAUKEE, WI 53201, U.S.A.
T. McCORRISTON	D.E.S.Y., Notkestieg 1, HAMBURG 52, GERMANY.
I. MOEN	University of Cambridge, Department of Physics, Silver Street, CAMBRIDGE, CB3 9EW, U.K.
A. MOLINA	Facultad de Ciencias, Avenida Generalisimo, s/n BARCELONA, SPAIN.
W. NAHM	Physikalisches Institut der Universität Bonn, Endenicher Allee 11-13, 53 BONN, GERMANY.
D. NANOPOULOS	The University of Sussex, School of Mathematical and Physical Sciences Falmer, BRIGHTON BN1 9QH, Sussex, U.K.

J. NYIRI

Hungarian Academy of Sciences,
Central Research Institute for Physics,
P.O. Box 49,
BUDAPEST 114, HUNGARY.

P. OTTERSON

Pontificia Universidade Catolica,
Departamento de Fisica,
Rua Marques de Sao Vicente, 209/263,
ZC-20, RIO DE JANEIRO, BRAZIL.

N. H. PARSONS

University of Glasgow,
Department of Natural Philosophy,
GLASGOW W.2, U.K.

A. PATKOS

Roland Eötvös University,
Department of Atomic Physics,
Puskin utea 5-7,
BUDAPEST VIII., HUNGARY.

R. PETRONZIO

Istituto di Fisica dell'Università,
piazzale delle Scienze, 5,
00185 - ROMA, ITALY.

E. RABINOVICI

The Weizmann Institute of Science,
Department of Nuclear Physics,
REHOVOT, ISRAEL.

J. ROSEN

Tel-Aviv University,
Department of Physics and Astronomy,
RAMAT-AVIV, TEL-AVIV, ISRAEL.

R. T. ROSS

Rutherford High Energy Laboratory,
CHILTON, Didcot, Berks., U.K.

L. ROUTH

The University of Liverpool,
Department of Applied Mathematics,
P.O. Box 147,
LIVERPOOL L69 38X, U.K.

K. SIBOLD

Universität Karlsruhe,
Institut für theoretische Physik,
Kaiserstrasse 12,
75 KARLSRUHE, GERMANY.

N. R. SMITH

University of Southampton,
Department of Physics,
SOUTHAMPTON, SO9 5NH, U.K.

L. SVENSSON

Institute of Physics
LUND, SWEDEN

F. E. TAYLOR

National Accelerator Laboratory,
P.O. Box 500,
BATAVIA, Ill. 60510, U.S.A.

E. TOMASELLI

Universität Wien,
Institut für theoretische Physik,
Boltzmanngasse 5,
1090 WIEN, AUSTRIA.

J. C. TOMPKINS — University of California,
Department of Physics,
405 Hilgard Avenue,
LOS ANGELES, Calif. 90024, U.S.A.

L. TORTORA — Istituto Nazionale di Fisica Nucleare,
via A. Tari, 3,
80138 - NAPOLI, ITALY.

W. TROOST — Universiteit Leuven,
Instituut voor theoretische Fysika,
Department Natuurkunde,
Celestijnenlaan 200 D,
3030 HEVERLEE, BELGIUM.

T. T. W. TSO — Rutherford High Energy Laboratory,
CHILTON, Didcot, Berks., U.K.

M. N. TUGULEA — Bucharest University,
Department of Physics,
14 Academia St.,
BUCHAREST, RUMANIA.

H. C. TZE — Niels Bohr Institute,
Blegdamsvej 17,
2100 KOBENHAVN, DENMARK.

R. VILELA MENDES — Instituto de Fisica e Matematica,
Avenida Do Prof. Gama Pinto,
LISBOA 4, PORTUGAL.

P. VINCIARELLI — University of Maryland,
College of Arts and Sciences,
Department of Physics and Astronomy,
COLLEGE PARK, MD 20742, U.S.A.

U. M. WAMBACH — University of Bielefeld,
Department of Physics,
Herforder Strasse 28,
4800 BIELEFELD, GERMANY.

D. M. WEBBER — University of Durham,
Department of Mathematics,
Science Laboratory,
South Road,
DURHAM, U.K.

F.A. WILCZEK — Princeton University,
Department of Physics,
P.O. Box 708,
PRINCETON, N.J. 08540, U.S.A.

R. M. WILLIAMS — Imperial College of Science and Technology,
Department of Physics,
Prince Consort Road,
LONDON, S.W.7 2BZ, U.K.

D. WRIGHT — University of Toronto,
Department of Physics,
TORONTO 5, Ontario, CANADA.

S. J. YELLIN — D.E.S.Y.,
Notkestieg 1,
HAMBURG 52, GERMANY.

T. P. YIOU — Université Paris VI,
Laboratoire de Physique Nucléaire et des Hautes En.,
11, Quai Saint-Bernard - Tour 32,
PARIS 5e, FRANCE.

W. Y. YU — International Centre for Theoretical Physics,
P.O. Box 586 - Miramare,
34100 - TRIESTE, ITALY.

S. ZERBINI — Istituto di Fisica dell'Università,
via Irnerio, 46,
40126 - BOLOGNA, ITALY.

Finito di stampare nel mese di luglio 1974 dalla Litografia Leschiera
Via Perugino 21 - Cologno Monzese - Milano